Symbol Table

Symbol	Meaning
f	frequency with which a value occurs
Σ	capital sigma; summation
Σx	sum of the values
Σx^2	sum of the squares of the values
$(\Sigma x)^2$	square of the sum of all values
Σxy	sum of the products of each x value multiplied by the corresponding y value
n	number of values in a sample
$n!$	n factorial
N	number of values in a finite population; also used as the size of all samples combined
k	number of samples or populations or categories
$\bar{x}$	mean of the values in a sample
$\bar{R}$	mean of the sample ranges
μ	mu; mean of all values in a population
s	standard deviation of a set of sample values
σ	lowercase sigma; standard deviation of all values in a population
s^2	variance of a set of sample values
σ^2	variance of all values in a population
z	standard score
$z_{\alpha/2}$	critical value of z
t	t distribution
$t_{\alpha/2}$	critical value of t
df	number of degrees of freedom
F	F distribution
χ^2	chi-square distribution
χ^2_R	right-tailed critical value of chi-square
χ^2_L	left-tailed critical value of chi-square
p	probability of an event or the population proportion
q	probability or proportion equal to $1 - p$
$\hat{p}$	sample proportion
$\hat{q}$	sample proportion equal to $1 - \hat{p}$
$\bar{p}$	proportion obtained by pooling two samples
$\bar{q}$	proportion or probability equal to $1 - \bar{p}$
$P(A)$	probability of event A
$P(A\|B)$	probability of event A, assuming event B has occurred
${}_nP_r$	number of permutations of n items selected r at a time
${}_nC_r$	number of combinations of n items selected r at a time

Symbol Table

Symbol	Meaning
$\overline{A}$	complement of event A
H_0	null hypothesis
H_1	alternative hypothesis
α	alpha; probability of a type I error or the area of the critical region
β	beta; probability of a type II error
r	sample linear correlation coefficient
ρ	rho; population linear correlation coefficient
r^2	coefficient of determination
R^2	multiple coefficient of determination
r_s	Spearman's rank correlation coefficient
b_1	point estimate of the slope of the regression line
b_0	point estimate of the y-intercept of the regression line
$\hat{y}$	predicted value of y
d	difference between two matched values
$\overline{d}$	mean of the differences d found from matched sample data
s_d	standard deviation of the differences d found from matched sample data
s_e	standard error of estimate
T	rank sum; used in the Wilcoxon signed-ranks test
H	Kruskal-Wallis test statistic
R	sum of the ranks for a sample; used in the Wilcoxon rank-sum test
μ_R	expected mean rank; used in the Wilcoxon rank-sum test
σ_R	expected standard deviation of ranks; used in the Wilcoxon rank-sum test
G	number of runs in runs test for randomness
μ_G	expected mean number of runs; used in runs test for randomness
σ_G	expected standard deviation for the number of runs; used in runs test for randomness
$\mu_{\overline{x}}$	mean of the population of all possible sample means $\overline{x}$
$\sigma_{\overline{x}}$	standard deviation of the population of all possible sample means $\overline{x}$
E	margin of error of the estimate of a population parameter, or expected value
Q_1, Q_2, Q_3	quartiles
$D_1, D_2, \ldots, D_9$	deciles
$P_1, P_2, \ldots, P_{99}$	percentiles
x	data value

Elementary

STATISTICS

Using the TI-83/84 Plus Calculator

Second Edition

Elementary STATISTICS

Using the TI-83/84 Plus Calculator

Second Edition

Mario F. Triola

Boston San Francisco New York
London Toronto Sydney Tokyo Singapore Madrid
Mexico City Munich Paris Cape Town Hong Kong Montreal

Publisher: Greg Tobin
Editor-in-Chief: Deirdre Lynch
Associate Editor: Sara Oliver Gordus
Senior Managing Editor: Karen Wernholm
Production Supervisor: Sheila Spinney
Senior Designer: Barbara T. Atkinson
Photo Researcher: Beth Anderson
Digital Assets Manager: Marianne Groth
Production Coordinator, Supplements: Kayla Smith-Tarbox
Senior Media Producer: Ceci Fleming
Software Development: Ted Hartman and Janet Wann
Marketing Manager: Phyllis Hubbard
Marketing Assistant: Caroline Celano
Senior Author Support/Technology Specialist: Joe Vetere
Senior Prepress Supervisor: Caroline Fell
Rights and Permissions Advisor: Dana Weightman
Senior Manufacturing Buyer: Evelyn Beaton
Media Buyer: Ginny Michaud
Text and Cover Design: Leslie Haimes
Production Services, Composition, and Illustration: Nesbitt Graphics, Inc.

Cover Photo: Getty Images/Jean Louis Batt

About the cover: Dandelion seeds being scattered in the wind. This cover is symbolic of basic statistical methodology. A variety of random variables affect the dispersion of the seeds, and analysis of those variables can result in predicted locations of next year's flowers. Finding order and predictability in seemingly random events is a hallmark activity of statistics.

Library of Congress Cataloging-in-Publication Data

Triola, Mario F.
Elementary statistics using the TI-83/84 Plus graphing calculator / Mario F. Triola.--2nd ed.
p. cm.
Rev. ed. of: Elementary statistics using the graphing calculator. c2005.
Includes index.
ISBN 0-321-46257-2
1. Mathematical statistics--Textbooks. 2. Statistics--Graphic methods--Textbooks. 3. TI-83 (Calculator)--Textbooks. 4. TI-84 (Calculator)--Textbooks. I. Triola, Mario F. Elementary statistics using the graphing calculator. II. Title.

QA276.12.T76 2008
519.50285--dc22

2006044682

3 4 5 6 7 8 9 10—QWT—10 09 08 07

To Marc and Scott

About the Author

Mario F. Triola is a Professor Emeritus of Mathematics at Dutchess Community College, where he has taught statistics for over 30 years. Marty is the author of *Essentials of Statistics,* Third Edition, *Elementary Statistics Using Excel,* Third Edition, *Elementary Statistics,* Tenth Edition, and he is a co-author of *Biostatistics for the Biological and Health Sciences, Statistical Reasoning for Everyday Life,* and *Business Statistics.* He has written several manuals and workbooks for technology supporting statistics education. Outside of the classroom, Marty has been a speaker at many conferences and colleges. His consulting work includes the design of casino slot machines and fishing rods, and he has worked with attorneys in determining probabilities in paternity lawsuits, identifying salary inequities based on gender, analyzing disputed election results, analyzing medical data, and analyzing medical school surveys. Marty has testified as an expert witness in New York State Supreme Court. The Text and Academic Authors Association has awarded Mario F. Triola a "Texty" for Excellence for his work on *Elementary Statistics.*

Brief Contents

Contents

Preface

Which Calculators Can Be Used with This Book?

This book includes many references to the TI-83/84 Plus calculator. The instructions, screen displays, programs, and the APP (application) for the TI-83/84 Plus calculator apply to the TI-83 Plus Silver Edition, the TI-84 Plus, and the TI-84 Plus Silver Edition calculators. In this book, references to the TI-83/84 Plus calculator apply to any of these calculators.

Other Texas Instruments Calculators

The TI-89, TI-89 Titanium, TI-92 Plus, and Voyager 2000 have the ability to download the free "Statistics with List Editor" application, which is available from the Texas Instruments Web site. After downloading that application, these calculators have at least as much statistics functionality as the TI-83 Plus and TI-84 Plus family of calculators. However, procedures for using the "Statistics with List Editor" application are generally different from the calculator procedures described in this book. Consequently, the TI-89, TI-89 Titanium, TI-92 Plus, and Voyager 2000 calculators could be used for many procedures in this book, but instructions, screen displays, the data application, and programs are not compatible with this book.

Why Use a TI-83/84 Plus Calculator?

The TI-83/84 Plus calculators have dramatically affected the teaching of introductory statistics courses. This family of calculators includes almost all of the important functions required in an introductory statistics course; they are relatively inexpensive and easy to use in class and on tests. There are none of the problems associated with other, more expensive computers. The TI-83/84 Plus calculators simplify otherwise tedious calculations so professors and students can better focus on important concepts of statistics instead of time-consuming arithmetic. These calculators include features that make it possible to conduct classroom experiments that illustrate simulations and other important concepts. Instructors and students enjoy immediate results, instead of waiting to gain access to a laboratory or home computer.

Here are three of many features that have contributed to the notable success of the TI-83/84 Plus calculators:

- In many cases, the calculator is designed to accept either lists of original data, or summary statistics. If the summary statistics are known, there is no need to waste time, energy, and (most of all) patience by manually entering a list of original sample values.

- The calculator has the ability to store APPS (applications). The CD-ROM included with this book contains a TI-83/84 Plus APP that contains all of the data sets found in Appendix B of this textbook. Once this APP has been downloaded to the calculator, instructors and students have easy and ready access to a variety of data sets that can be opened without manual entry of the sample values.
- Students can bring their TI-83/84 Plus calculators to each class, so that everyone has immediate and ready access to the same technology. This makes it easy for instructors to conduct experiments and projects. Also, test problems can be made to be much more realistic with larger sets of real data.

Using the TI-83/84 Plus Calculators

Elementary Statistics Using the TI-83/84 Plus Calculator, Second Edition, is designed to be an introduction to basic statistics. Instead of being a manual of calculator instructions, this book places strong emphasis on understanding concepts of statistics, with the TI-83/84 Plus calculator included throughout as the key supplement. Topics are presented with illustrative examples, identification of required assumptions, and underlying theory. TI-83/84 Plus calculator instructions are provided along with typical displays of results.

This book has key features that are critical to effective use of the calculator as a statistical tool:

- Detailed TI-83/84 Plus calculator **instructions** are included throughout the book.
- Approximately 150 samples of TI-83/84 Plus calculator **displays** are included throughout the book. Also, displays from MINITAB®, Excel®, and STATDISK are occasionally included so students can see different formats for displaying results.
- All **data sets** in Appendix B are available as part of the TI-83/84 Plus APP on the CD-ROM packaged with new copies of this book. Some exercises use those data sets, so students can open TI-83/84 Plus calculator files instead of manually entering long lists of data.
- TI Exercises that are particularly effective for use with the TI-83/84 Plus calculator are identified with a special **icon.**
- In some cases, exercise answers based on table values may be slightly different from answers based on the use of the TI-83/84 Plus calculator. For those exercises, Appendix E generally gives both answers.
- **TI-83/84 Plus calculator Projects** are included among the variety of the end-of-chapter features.
- The CD-ROM included with this book has separate TI-83/84 Plus programs for several procedures that are not included with the calculator. The programs include procedures for hypothesis tests with variances, confidence intervals with variances, multiple regression, two-way analysis of variance, Wilcoxon rank-sum test, Wilcoxon signed-ranks test, and the Kruskal-Wallis test.

For those who choose to further supplement the TI-83/84 Plus calculator, the CD-ROM includes additional **software** that some students may find helpful. STATDISK is a separate computer program designed specifically for the content of this book.

Philosophy

Elementary Statistics Using the TI-83/84 Plus Calculator, Second Edition, is the result of over 30 years of teaching, research, and innovation in statistics education. The goal of this book is to be an engaging and thorough introduction to statistics for students. Although formulas and formal procedures can be found throughout the text, it emphasizes the development of statistical literacy and critical thinking. This book encourages thinking over the blind use of mechanical procedures.

Here are some important features of *Elementary Statistics Using the TI-83/84 Plus Calculator*:

- Emphasis on statistical literacy and critical thinking
- Emphasis on understanding concepts instead of cookbook calculations
- Abundant use of *real* data
- Writing style that is clear, friendly, and occasionally humorous
- Diverse and abundant pedagogical features
- An array of helpful supplements for students and professors
- Addison-Wesley sales, technical support, and editorial professionals who are exceptional in their commitment and expertise

Apart from learning about statistics, another important objective of *Elementary Statistics Using the TI-83/84 Plus Calculator,* Second Edition is to provide a framework that fosters personal growth through the use of technology, work with peers, critical thinking, and the development of communication skills. This book allows students to apply their learned skills beyond the classroom in a real-world context.

This text reflects recommendations from the American Statistical Association and its *Guidelines for Assessment and Instruction in Statistics Education* (GAISE), the Mathematical Association of America, the American Mathematical Association of Two-Year Colleges, and the National Council of Teachers of Mathematics.

Audience/Prerequisites

Elementary Statistics Using the TI-83/84 Plus Calculator is written for students majoring in any subject. Although the use of algebra is minimal, students should have completed at least a high school or college elementary algebra course. In many cases, underlying theory behind topics is included, but this book does not contain the mathematical rigor required of mathematics majors. Because the many examples and exercises cover a wide variety of statistical applications, this book will be interesting and appropriate for students studying disciplines ranging from the social sciences of psychology and sociology to areas such as education, the allied health fields, business, economics, engineering, the humanities, the physical sciences, journalism, communications, and liberal arts.

Changes in This Edition

- The section on Visualizing Data has been divided into two sections, with increased emphasis on statistical graphics:

  Section 2-3: Histograms

  Section 2-4: Statistical Graphics

- The former chapter on Describing, Exploring, and Comparing Data has been divided into two chapters:

  Chapter 2: Summarizing and Graphing Data

  Chapter 3: Statistics for Describing, Exploring, and Comparing Data

- New section: **McNemar's Test for Matched Pairs** (Section 11-4)
- New section on the enclosed CD-ROM: **Bayes' Theorem**
- The text in some sections has been partitioned into Part 1 (Basics) and Part 2 (Beyond the Basics) so that it is easier to focus on core concepts.
- Discussions on certain topics have been expanded: power (Section 8-2); residual plots (Section 10-3); logistic regression (Section 10-5); and interaction plots (Section 12-3).
- **Requirement check:** Where appropriate, solutions begin with a formal check of the requirements that must be verified before a particular method should be used.
- **Statistical Literacy and Critical Thinking:** Each exercise section begins with four exercises that specifically involve statistical literacy and critical thinking. Also, the end of each chapter has another four exercises of this type.
- **Small data sets:** This edition has many more exercises that involve smaller data sets.
- **New exercises and examples:** 68% of the exercises are new, and 53% of the exercises use real data; 66% of the examples are new.
- **Top 20 Topics:** In this edition, we have identified the **Top 20 Topics** that are especially important in any introductory statistics course. These topics are marked with a top 20 in the text. Students using MyStatLab have access to additional resources for learning these topics with definitions, animations, and video lessons.

Flexible Syllabus

The organization of this book reflects the preferences of most statistics instructors, but there are two common variations that can be easily used with this Second Edition:

- **Early coverage of correlation/regression:** Some instructors prefer to cover the basics of correlation and regression early in the course, such as immediately following the topics of Chapter 3. *Sections 10-2 (Correlation) and 10-3 (Regression) can be covered early in the course.* Simply limit coverage to Part 1 (Basic Concepts) in each of those two sections.

- **Minimum probability:** Some instructors feel strongly that coverage of probability should be extensive, while others feel just as strongly that coverage should be kept to a minimum. Instructors preferring minimum coverage can include Section 4-2 while skipping the remaining sections of Chapter 4, as they are not essential for the chapters that follow. Many instructors prefer to cover the fundamentals of probability along with the basics of the addition rule and multiplication rule, and those topics can be covered with Sections 4-1 through 4-4. Section 4-5 includes conditional probability, and the subsequent sections cover simulation methods and counting (including permutations and combinations).

Exercises

There are over 1750 exercises—*68% of them are new!* More exercises use smaller data sets, and many require the *interpretation* of results. Because exercises are of such critical importance to any statistics book, great care has been taken to ensure their usefulness, relevance, and accuracy. Three statisticians have read carefully through the final stages of the book to verify accuracy of the text material and exercise answers. Exercises are arranged in order of increasing difficulty by dividing them into two groups: (1) Basic Skills and Concepts and (2) Beyond the Basics. The Beyond the Basics exercises address more difficult concepts or require a somewhat stronger mathematical background. In a few cases, these exercises also introduce a new concept.

Real data: *53% of the exercises use real data.* (Because this edition has many more exercises in the category of Statistical Literacy and Critical Thinking, the percentage of exercises using real data is less than in the first edition, but the number of exercises using real data is approximately the same.) Because the use of real data is such an important consideration for students, hundreds of hours have been devoted to finding real, meaningful, and interesting data. In addition to the real data included throughout the book, some exercises refer to the 18 large data sets listed in Appendix B.

Hallmark Features

Great care has been taken to ensure that each chapter of *Elementary Statistics Using the TI-83/84 Plus Calculator* will help students understand the concepts presented. The following features are designed to help meet that objective:

- **Chapter-opening features:** A list of chapter sections previews the chapter for the student; a chapter-opening problem, using real data, then motivates the chapter material; and the first section is a chapter overview that provides a statement of the chapter's objectives.
- **End-of-chapter features:** A Chapter Review summarizes the key concepts and topics of the chapter; Statistical Literacy and Critical Thinking exercises address chapter concepts; Review Exercises offer practice on the chapter concepts and procedures—plus new videos show how to work through these exercises.

- **Cumulative Review Exercises** reinforce earlier material.
- **From Data to Decision: Critical Thinking** is a capstone problem that requires critical thinking and writing.

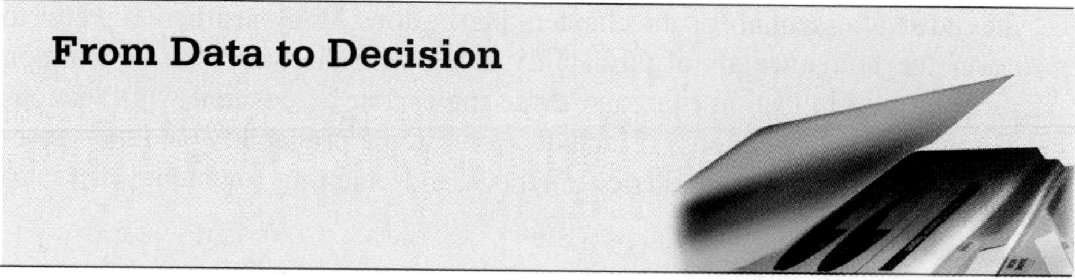

- **Cooperative Group Activities** encourage active learning in groups.
- **TI-83/84 Plus Calculator Projects** are for use with a TI-83/84 Plus calculator.

TI-83/84 Plus Calculator Project

- **Internet Projects** provide students an opportunity to work with Internet data sets and, in some cases, applets.

- **Margin Essays:** The text includes 122 margin essays, which illustrate uses and abuses of statistics in real, practical, and interesting applications. Topics include *Do Boys or Girls Run in the Family, Lefties Die Sooner?, and Picking Lottery Numbers.*
- **Flowcharts:** These appear throughout the text to simplify and clarify more complex concepts and procedures. New for this edition, the flowcharts have been animated and can be accessed in this text's MyStatLab™ (www.mystatlab. com) site.
- **Calculator Instructions:** TI-83/84 Plus instructions and output appear throughout the text.
- **Real Data Sets:** These are used extensively throughout the entire book. Appendix B lists 18 data sets, including 4 that are new and 3 others with new data. These data sets are provided in printed form in Appendix B, and in electronic form on the Web site and the CD-ROM bound in the back of new copies of the book. The data sets include such diverse topics as alcohol and tobacco use in animated children's movies, eruptions of the Old Faithful geyser, and measurements related to secondhand smoke.
- **Interviews:** Every chapter of the text includes interviews with professional men and women in a variety of fields who use statistics in their day-to-day work.

- **Symbol Table:** A symbol table is included at the front of the book for quick and easy reference to key symbols.
- **Detachable Formula/Table Card:** This insert, organized by chapter, gives students a quick reference for studying, or for use when taking tests (if allowed by the instructor).
- **CD-ROM:** The CD-ROM was prepared by Mario F. Triola and is packaged with every new copy of the text. It includes the data sets from Appendix B, which are stored as text files, MINITAB worksheets, SPSS® files, SAS® files, Excel workbooks, and a TI-83/84 Plus application. The CD also includes a section on Bayes' Theorem, programs for the TI-83/84 Plus graphing calculator, STATDISK Statistical Software (Version 10.4), and the Excel add-in DDXL™, which is designed to enhance the capabilities of Excel's statistics programs.

Supplements

The student and instructor supplements packages are intended to be the most complete and helpful learning system available for the introductory statistics course. Instructors should contact their local Addison-Wesley sales representative, or e-mail the company directly at exam@aw.com for examination copies.

For the Instructor

- ***Instructor's Edition,*** by Mario F. Triola, contains answers to all exercises. ISBN: 0-321-51314-2.
- ***Instructor's Solutions Manual,*** by Milton Loyer (Penn State University), contains solutions to all the exercises and sample course syllabi. ISBN: 0-321-46262-9.
- ***Insider's Guide to Teaching with the Triola Statistics Series,*** by Mario F. Triola, contains sample syllabi, and tips for incorporating projects, as well as lesson overviews, extra examples, minimum outcome objectives, and recommended assignments for each chapter. ISBN 0-321-40964-7.
- **MyStatLab** (part of the MyMathLab® and MathXL® product family) is a text-specific, easily customizable online course that integrates interactive multimedia instruction with the textbook content. MyStatLab is powered by CourseCompass™—Pearson Education's online teaching and learning environment—and by MathXL—our online homework, tutorial, and assessment system. MyStatLab gives instructors the tools needed to deliver all or a portion of their course online, whether students are in a lab setting or working from home. MyStatLab provides a rich and flexible set of course materials, featuring free-response tutorial exercises for unlimited practice and mastery. Students can also use online tools, such as video lectures, animations, and a multimedia textbook, to independently improve their understanding and performance. Instructors can use MyStatLab's homework and test managers to select and assign online exercises correlated directly to the textbook, and they can also create and assign their own online exercises and import TestGen tests

for added flexibility. MyStatLab's online gradebook—designed specifically for mathematics and statistics—automatically tracks students' homework and test results and gives the instructor control over how to calculate final grades. Instructors can also add offline (paper-and-pencil) grades to the gradebook. MyStatLab is available to qualified adopters.

For more information, visit www.mystatlab.com or contact your Addison-Wesley sales representative for a demonstration.

- **Testing System:** Great care has been taken to create the most comprehensive testing system possible for the new edition of *Elementary Statistics Using the TI-83/84 Plus Calculator.* Not only is there a printed test bank, there is also a computerized test generator, TestGen, that allows instructors to view and edit testbank questions, transfer them to tests, and print in a variety of formats. The program also offers many options for sorting, organizing, and displaying testbanks and tests. A built-in random number and test generator makes TestGen ideal for creating multiple versions of tests and provides more possible test items than printed testbank questions. Users can export tests to be compatible with a variety of course management systems, or even just to display in a Web browser. Additionally, tests created with TestGen can be used with QuizMaster, which enables students to take exams on a computer network. Printed Testbank ISBN: 0-321-36914-9; TestGen for Mac and Windows ISBN: 0-321-36904-1.
- **PowerPoint® Lecture Presentation CD:** Free to qualified adopters, this classroom lecture presentation software is geared specifically to the sequence and philosophy of *Elementary Statistics Using the TI-83/84 Plus Calculator.* Key graphics from the book are included to help bring the statistical concepts alive in the classroom. These slides are also available to download from the Instructor Resource Center at www.aw-bc.com/irc. Mac and Windows ISBN: 0-321-36905-X.

For the Student

- **TI-83/84 Plus Reference Card:** A six-page summary card of important calculator procedures. ISBN: 0-321-39967-6.
- ***Student's Solutions Manual,*** by Milton Loyer (Penn State University), provides detailed, worked-out solutions to all odd-numbered text exercises. ISBN: 0-321-46258-0.
- **MathXL® for Statistics** is a powerful online homework, tutorial, and assessment system that accompanies Addison-Wesley textbooks in statistics and mathematics. With MathXL for Statistics, instructors can create, edit, and assign online homework created specifically for the Triola textbook and tests using algorithmically generated exercises correlated at the objective level to this book. All student work is tracked in MathXL's online gradebook. Students can take chapter tests in MathXL for Statistics and receive personalized study plans based on their test results. The study plan diagnoses weaknesses and links students directly to tutorial exercises for the objectives they need to study and retest. Students can also access animations and Triola video clips directly from

selected exercises. MathXL for Statistics is available to qualified adopters. For more information, visit our Web site at www.mathxl.com, or contact your Addison-Wesley sales representative.

- **Videos** have been expanded and now supplement most sections in the book, with many topics presented by the author. The videos feature technologies found in the book and the worked-out Chapter Review exercises. This is an excellent resource for students who have missed class or wish to review a topic. It is also an excellent resource for instructors involved with distance learning, individual study, or self-paced learning programs. **Videotape Series** ISBN: 0-321-36913-0. **Digital Video Tutor** (CD-ROM version). ISBN: 0-321-41268-0. The videos on the Digital Video Tutor (CD-ROM version) include captions.
- **Triola *Statistics Series* Web Site:** This Web site may be accessed at http://www.aw-bc.com/triola, and provides Internet projects keyed to every chapter of the text, plus the book's data sets as they appear on the CD.

The following technology manuals include instructions on and examples of the technology's use. Each one has been written to correspond with the text.

- ***Graphing Calculator Manual for the TI-83 Plus, TI-84 Plus, and TI-89,*** by Patricia Humphrey (Georgia Southern University). ISBN: 0-321-36920-3.
- ***Excel Student Laboratory Manual and Workbook,*** written by Johanna Halsey and Ellena Reda (Dutchess Community College), ISBN: 0-321-36909-2.
- ***MINITAB Student Laboratory Manual and Workbook,*** written by Mario F. Triola. ISBN: 0-321-36919-X.
- ***SAS Student Laboratory Manual and Workbook,*** written by Joseph Morgan ISBN 0-321-36910-6.
- ***SPSS Student Laboratory Manual and Workbook,*** ISBN 0-321-36911-4.
- ***STATDISK Student Laboratory Manual and Workbook,*** written by Mario F. Triola. ISBN: 0-321-36912-2.
- ***Graphing Calculator Tutorial for Statistics.*** Organized by topic, this CD-ROM tutorial guides students through the keystrokes needed to most efficiently use their graphing calculator. ISBN: 0-321-41382-2.
- ***ActivStats,*** developed by Paul Velleman and Data Description, Inc., provides complete coverage of introductory statistics topics on CD-ROM, using a full range of multimedia. ***ActivStats*** integrates video, simulation, animation, narration, text, interactive experiments, World Wide Web access, and Data Desk®, a statistical software package. Homework problems and data sets from the Triola text are included on the CD-ROM. ***ActivStats*® for Windows and Macintosh** ISBN: 0-321-30364-4. **Also available in versions for Excel, JMP®, MINITAB, and SPSS.** See your Addison-Wesley sales representative for details or check the Web site at www.aw.com/activstats.
- ***Addison-Wesley Tutor Center:*** Free tutoring is available to students who purchase a new copy of the Second Edition of *Elementary Statistics Using the TI-83/84 Plus Calculator* when bundled with an access code. The Addison-Wesley Tutor Center is staffed by qualified statistics and mathematics instructors who provide students with tutoring on text examples and any exercise with an answer in the back of the book. Tutoring assistance is provided by

toll-free telephone, fax, e-mail, and whiteboard technology—which allows tutors and students to actually see the problems worked while they "talk" in real time over the Internet. This service is available five days a week, seven hours a day. For more information, please contact your Addison-Wesley sales representative.

- ***The Student Edition of MINITAB*** is a condensed version of the professional release of MINITAB Statistical Software. It offers students the full range of MINITAB's statistical methods and graphical capabilities, along with worksheets that can include up to 10,000 data points. It comes with a user's manual that includes case studies and hands-on tutorials, and is perfect for use in any introductory statistics course, including those in the life and social sciences. ***The Student Guide to MINITAB Release 14*** by John D. McKenzie Jr. (Babson College) and Robert Goldman (Simmons College), contains a copy of the most current Student Edition software as well as software instruction. ISBN 0-201-77469-0. MINITAB Student Release 14 statistical software is available for bundling with the Triola textbook. ISBN 0-321-11313-6 (CD only).

Any of these products can be purchased separately, or bundled with Addison-Wesley texts. Instructors can contact local sales representatives for details on purchasing and bundling supplements with the textbook or contact the company at exam@aw.com for examination copies of any of these items.

Acknowledgments

The success of ***Elementary Statistics Using the TI-83/84 Plus Calculator*** is due to the efforts of many dedicated professionals. I wish to express my most sincere thanks to Deirdre Lynch, Sara Oliver Gordus, Greg Tobin, Phyllis Hubbard, Sheila Spinney, Barbara Atkinson, Ceci Fleming, Caroline Celano, Joe Vetere, Beth Anderson, Dana Weightman, and the entire Addison-Wesley staff. I also thank Janet Nuciforo of Nesbitt Graphics for her superb production work.

I am grateful for the many suggestions and comments provided by instructors and students, and I am grateful to the hundreds of researchers who did studies, conducted surveys, and compiled interesting and meaningful data used in examples and exercises throughout this book.

I also take great pride in thanking my wife, Ginny, for her continued support and encouragement. I thank my son Scott for his continued encouragement, and I thank my son Marc Triola, M.D., for reprogramming and supporting STATDISK so that it is now a powerful and quality statistics package.

I would also like to thank the following for their help with the Tenth Edition of *Elementary Statistics,* which is the basis for this book:

Text Accuracy Reviewers

Emily Keaton
David R. Lund, University of Wisconsin-Eau Claire
Tim Mogill
Kimberly Polly, Indiana University Bloomington
Tom Wegleitner

Reviewers of Elementary Statistics

Raid W. Amin, University of West Florida
Keith Carroll, Benedictine University
Monte Cheney, Central Oregon Community College
Christopher Donnelly, Macomb Community College
Theresa DuRapau, Our Lady of Holy Cross
Billy Edwards, University of Tennessee-Chattanooga
Marcos Enriquez, Moorpark College
Angela Everett, Chattanooga State Technical Community College
Joe Franko, Mount San Antonio College
Sanford Geraci, Broward Community College
Laura Heath, Palm Beach Community College
Laura Hillerbrand, Broward Community College
Gary King, Ozarks Technical Community College
Mickey Levendusky, Pima County Community College
Tristan Londre, Blue River Community College
Alma Lopez, South Plains College
Carla Monticelli, Camden County Community College
Julia Norton, California State University—Hayward
Michael Oriolo, Herkimer Community College
Jeanne Osborne, Middlesex Community College
Ali Saadat, University of California-Riverside
Radha Sankaran, Passaic County Community College
Pradipta Seal, Boston University
Sharon Testone, Onondaga Community College
Dave Wallach, University of Findlay
Cheng Wang, Nova Southeastern University
Gail Wiltse, St. John River Community College

Claire Wladis, Borough of Manhattan Community College
Yong Zeng, University of Missouri at Kansas City
Jim Zimmer, Chattanooga State Technical Community College
Cathleen Zucco-Teveloff, Trinity College
Mark Z. Zuiker, Minnesota State University, Mankato

For providing help and suggestions in special areas, I would like to thank the following individuals:

Vincent DiMaso
Rod Elsdon, Chaffey College
David Straayer, Sierra College
Glen Weber, Christopher Newport University

M.F.T.
LaGrange, New York
December, 2006

Index of Applications

In the center column, **CP** = Chapter Problem, **IE** = In-Text Example, **M** = Margin Example, **E** = Exercise, **BB** = Beyond the Basics, **R** = Review Exercise, **CR** = Cumulative Review Exercise, **DD** = Data to Decision, **CGA** = Cooperative Group Activity, **TP** = TI-83/84 Plus Calculator Projector, **SW** = Statistics at Work

Agriculture

Biology

Business and Economics

Education

Engineering

Entertainment

Environment

Finance

Food/Drink

Games

General Interest

Health

Labor

Law

People and Psychology

Politics

Social Issues

Sports

Surveys and Opinion Polls

Technology

Transportation

Elementary

STATISTICS

Using the TI-83/84 Plus Calculator

Second Edition

Introduction to Statistics

1

CHAPTER PROBLEM

Six Degrees of Kevin Bacon: Did the original study use good data?

"Six Degrees of Kevin Bacon" is a recent popular game that involves identifying a movie actor or actress, then linking this person with the actor Kevin Bacon. (As of this writing, the game could be played on the Web site www.cs.virginia.edu/oracle.) Let's consider Richard Gere as an example. Gere was in the movie *Cotton Club* with Laurence Fishburne, who was in the movie *Mystic River* with Kevin Bacon. The linkage of Gere–Fishburne–Bacon has *two* degrees of separation because the target person is not counted. This game, developed by three students (Craig Fass, Brian Turtle, and Mike Ginelli) from Albright College, is a more specialized version of the "Small World Problem," which poses this question: How many intermediaries (friends, relatives, and other acquaintances) are necessary to connect any two randomly selected people on Earth? That is, for any two people on Earth, what is the number of degrees of separation? This problem of connectedness has practical applications to many fields, such as those involving power grids, Internet usage, brain neurons, and the spread of disease.

The concept of "six degrees of separation" grew from a 1967 study conducted by psychologist Stanley Milgram. His original finding was that two random residents in the United States are connected by an average of six intermediaries. In his first experiment, he sent 60 letters to subjects in Wichita, Kansas, and they were asked to forward the letters to a specific woman in Cambridge, Massachusetts. The subjects were instructed to hand deliver the letters to acquaintances who they believed could reach the target person either directly or through other acquaintances. Fifty of the 60 subjects participated, and three of the letters reached the target. Two subsequent experiments had low completion rates, but Milgram eventually reached a 35% completion rate and he found that for completed chains, the mean number of intermediaries was around six. Consequently, Milgram's original data led to the concept referred to as "six degrees of separation."

Here are two key questions: Were Milgram's original data good? Do Milgram's original data justify the concept of "six degrees of separation?" An *extremely* important principle in this chapter, in this book, and in statistics in general is that the method used to collect sample data can make or break the validity of conclusions based on the data.

Today, each of us is bombarded with surveys and results from surveys. Some surveys collect sample data that are helpful in accurately describing important characteristics of populations. Other surveys use sample data that have been collected in ways that condemn the results to the growing garbage heap of misinformation.

In this chapter, we address the question about the quality of the data in Stanley Milgram's experiment, and we discuss and stress the importance of collecting data using sound methods that are likely to result in conclusions that are valid.

1-1 Overview

The Chapter Problem on the previous page involves a study that resulted in sample data. A common goal of such studies is to collect data from a small part of a larger group so that we can learn something about the larger group. This is a common and important goal of the subject of statistics: Learn about a large group by examining data from some of its members. In this context, the terms *sample* and *population* become important. Formal definitions for these and other basic terms are given here.

Definitions

Data are observations (such as measurements, genders, survey responses) that have been collected.

Statistics is a collection of methods for planning studies and experiments, obtaining data, and then organizing, summarizing, presenting, analyzing, interpreting, and drawing conclusions based on the data.

A **population** is the complete collection of all elements (scores, people, measurements, and so on) to be studied. The collection is complete in the sense that it includes all subjects to be studied.

A **census** is the collection of data from *every* member of the population.

A **sample** is a *subcollection* of members selected from a population.

For example, a Gallup poll asked this of 1087 adults: "Do you have occasion to use alcoholic beverages such as liquor, wine, or beer, or are you a total abstainer?" The 1087 survey subjects constitute a *sample*, whereas the *population* consists of the entire collection of all 202,682,345 adult Americans. Every 10 years, the United States Government attempts to obtain a *census* of every citizen, but fails because it is impossible to reach everyone. An ongoing controversy involves the attempt to use sound statistical methods to improve the accuracy of the Census, but political considerations are a key factor causing members of Congress to resist this improvement. Perhaps some readers of this text will one day be members of Congress with the wisdom to bring the Census into the twenty-first century.

An important activity of this book is to demonstrate how we can use sample data to form conclusions about populations. We will see that it is *extremely* critical to obtain sample data that are representative of the population from which the data are drawn. For example, if you survey the alumni who graduated from your college by asking them to write their annual income and mail it back to you, the responses are not likely to be representative of the population of all alumni. Those with low incomes will be less inclined to respond, and those who do respond may be inclined to exaggerate. As we proceed through this chapter, we should focus on these key concepts:

- **Sample data must be collected in an appropriate way, such as through a process of *random* selection.**
- **If sample data are not collected in an appropriate way, the data may be so completely useless that no amount of statistical torturing can salvage them.**

Above all else, we ask that you begin your study of statistics with an open mind. Don't assume that the study of statistics is comparable to a root canal procedure. It has been the author's experience that students are often surprised by the interesting nature of statistics, and they are also surprised by the fact that they can actually master the basic principles without much difficulty, even if they have not excelled in other mathematics courses. We are convinced that by the time you complete this introductory course, you will be firm in your belief that statistics is an interesting and rich subject with applications that are extensive, real, and meaningful. We are also convinced that with regular class attendance and diligence, you will succeed in mastering the basic concepts of statistics presented in this course.

The State of Statistics

The word *statistics* is derived from the Latin word *status* (meaning "state"). Early uses of statistics involved compilations of data and graphs describing various aspects of a state or country. In 1662, John Graunt published statistical information about births and deaths. Graunt's work was followed by studies of mortality and disease rates, population sizes, incomes, and unemployment rates. Households, governments, and businesses rely heavily on statistical data for guidance. For example, unemployment rates, inflation rates, consumer indexes, and birth and death rates are carefully compiled on a regular basis, and the resulting data are used by business leaders to make decisions affecting future hiring, production levels, and expansion into new markets.

1-2 Types of Data

Key Concept The subject of statistics is largely about using sample data to make inferences (or generalizations) about an entire population. We should know and understand the definitions of *population, sample, parameter,* and *statistic* because they are so basic and fundamental. We should also know the difference between *quantitative data* and *qualitative data*. We should know that some numbers, such as zip codes, are not quantities in the sense that they don't really measure or count anything. Zip codes are actually geographic locations, so it makes no sense to perform calculations with them, such as finding an average. This section describes different aspects of the nature of sample data, which can greatly affect the statistical methods that can be used with them.

In Section 1-1 we defined the terms *population* and *sample*. The following two terms are used to distinguish between cases in which we have data for an entire population, and cases in which we have data for a sample only.

Definitions

A **parameter** is a numerical measurement describing some characteristic of a *population.*

A **statistic** is a numerical measurement describing some characteristic of a *sample.*

EXAMPLES

1. **Parameter:** In New York City, there are 3250 walk buttons that pedestrians can press at traffic intersections. It was found that 77% of those buttons do not work (based on data from the article "For Exercise in New York Futility, Push Button," by Michael Luo, *New York Times*). The figure of 77% is a *parameter* because it is based on the entire population of all 3250 pedestrian push buttons.
2. **Statistic:** Based on a sample of 877 surveyed executives, it is found that 45% of them would not hire someone with a typographic error on their job application. That figure of 45% is a *statistic* because it is based on a sample, not the entire population of all executives.

Some data sets consist of numbers (such as heights of 66 inches and 72 inches), while others are nonnumeric (such as eye colors of green and brown). The terms *quantitative data* and *qualitative data* are often used to distinguish between these types.

Definitions

Quantitative data consist of numbers representing counts or measurements.

Qualitative (or **categorical** or **attribute**) **data** can be separated into different categories that are distinguished by some nonnumeric characteristic.

EXAMPLES

1. **Quantitative Data:** The weights of supermodels
2. **Qualitative Data:** The genders (male/female) of professional athletes

When working with quantitative data, it is important to use the appropriate units of measurement, such as dollars, hours, feet, meters, and so on. We should be especially careful to observe such references as "all amounts are in *thousands of dollars*" or "all times are in *hundredths of a second*" or "units are in *kilograms*." To ignore such units of measurement could lead to very wrong conclusions. NASA lost its $125 million Mars Climate Orbiter when it crashed because the controlling software had acceleration data in *English* units, but they were incorrectly assumed to be in *metric* units.

Quantitative data can be further described by distinguishing between *discrete* and *continuous* types.

Definitions

Discrete data result when the number of possible values is either a finite number or a "countable" number. (That is, the number of possible values is 0 or 1 or 2 and so on.)

Continuous (numerical) data result from infinitely many possible values that correspond to some continuous scale that covers a range of values without gaps, interruptions, or jumps.

EXAMPLES

1. **Discrete Data:** The numbers of eggs that hens lay are *discrete* data because they represent counts.
2. **Continuous Data:** The amounts of milk from cows are *continuous* data because they are measurements that can assume any value over a continuous span. During a given time interval, a cow might yield an amount of milk that can be any value between 0 gallons and 5 gallons. It would be possible to get 2.343115 gallons because the cow is not restricted to the discrete amounts of 0, 1, 2, 3, 4, or 5 gallons.

When describing relatively smaller amounts, correct grammar dictates that we use "fewer" for discrete amounts, and "less" for continuous amounts. For example, it is correct to say that we drank *fewer* cans of cola and, in the process, we drank *less* cola. The numbers of cans of cola are discrete data, whereas the actual volume amounts of cola are continuous data.

Another common way of classifying data is to use four levels of measurement: nominal, ordinal, interval, and ratio. In applying statistics to real problems, the level of measurement of the data is an important factor in determining which procedure to use. (See Figure 15-1 on page 784.) There will be some references to these levels of measurement in this book, but the important point here is based on common sense: Don't do computations and don't use statistical methods that are not appropriate for the data. For example, it would not make sense to compute an average of social security numbers, because those numbers are data that are used for identification, and they don't represent measurements or counts of anything. For the same reason, it would make no sense to compute an average of the numbers sewn on the shirts of basketball players.

Measuring Disobedience

How are data collected about something that doesn't seem to be measurable, such as people's level of disobedience? Psychologist Stanley Milgram devised the following experiment: A researcher instructed a volunteer subject to operate a control board that gave increasingly painful "electrical shocks" to a third person. Actually, no real shocks were given, and the third person was an actor. The volunteer began with 15 volts and was instructed to increase the shocks by increments of 15 volts. The disobedience level was the point at which the subject refused to increase the voltage. Surprisingly, two-thirds of the subjects obeyed orders even though the actor screamed and faked a heart attack.

Definition

The **nominal level of measurement** is characterized by data that consist of names, labels, or categories only. The data cannot be arranged in an ordering scheme (such as low to high).

EXAMPLES Here are examples of sample data at the nominal level of measurement.

1. **Yes/no/undecided:** Survey responses of *yes*, *no*, and *undecided*
2. **Colors:** The colors of cars driven by college students (red, black, blue, white, magenta, mauve, and so on)

Because nominal data lack any ordering or numerical significance, they should not be used for calculations. Numbers are sometimes assigned to the different categories (especially when data are coded for computers), but these numbers have no real computational significance and any average calculated with them is meaningless.

Definition

Data are at the **ordinal level of measurement** if they can be arranged in some order, but differences between data values either cannot be determined or are meaningless.

EXAMPLES Here are examples of sample data at the ordinal level of measurement.

1. **Course Grades:** A college professor assigns grades of A, B, C, D, or F. These grades can be arranged in order, but we can't determine differences

continued

between such grades. For example, we know that A is higher than B (so there is an ordering), but we cannot subtract B from A (so the difference cannot be found).

2. **Ranks:** Based on several criteria, a magazine ranks cities according to their "livability." Those ranks (first, second, third, and so on) determine an ordering. However, the differences between ranks are meaningless. For example, a difference of "second minus first" might suggest $2 - 1 = 1$, but this difference of 1 is meaningless because it is not an exact quantity that can be compared to other such differences. The difference between the first city and the second city is not the same as the difference between the second city and the third city. Using the magazine rankings, the *difference* between New York City and Boston cannot be quantitatively compared to the *difference* between St. Louis and Philadelphia.

Ordinal data provide information about relative comparisons, but not the magnitudes of the differences. Usually, ordinal data should not be used for calculations such as an average, but this guideline is sometimes violated (such as when we use letter grades to calculate a grade-point average).

Definition

The **interval level of measurement** is like the ordinal level, with the additional property that the difference between any two data values is meaningful. However, data at this level do not have a *natural* zero starting point (where *none* of the quantity is present).

EXAMPLES The following examples illustrate the interval level of measurement.

1. **Temperatures:** Body temperatures of 98.2°F and 98.6°F are examples of data at this interval level of measurement. Those values are ordered, and we can determine their difference of 0.4°F. However, there is no natural starting point. The value of 0°F might seem like a starting point, but it is arbitrary and does not represent the total absence of heat. Because 0°F is not a natural zero starting point, it is wrong to say that 50°F is *twice* as hot as 25°F.
2. **Years:** The years 1000, 2008, 1776, and 1492. (Time did not begin in the year 0, so the year 0 is arbitrary instead of being a natural zero starting point representing "no time.")

Definition

The **ratio level of measurement** is the interval level with the additional property that there is also a natural zero starting point (where zero indicates that *none* of the quantity is present). For values at this level, differences and ratios are both meaningful.

EXAMPLES The following are examples of data at the ratio level of measurement. Note the presence of the natural zero value, and note the use of meaningful ratios of "twice" and "three times."

1. **Weights:** Weights (in carats) of diamond engagement rings (0 does represent no weight, and 4 carats is twice as heavy as 2 carats.)
2. **Prices:** Prices of college textbooks ($0 does represent no cost, and a $90 book is three times as costly as a $30 book).

This level of measurement is called the ratio level because the zero starting point makes ratios meaningful. Among the four levels of measurement, most difficulty arises with the distinction between the interval and ratio levels. *Hint:* To simplify that distinction, use a simple "ratio test": Consider two quantities where one number is twice the other, and ask whether "twice" can be used to correctly describe the quantities. Because a 200-lb weight is *twice* as heavy as a 100-lb weight, but 50°F is *not twice* as hot as 25°F, weights are at the ratio level while Fahrenheit temperatures are at the interval level. For a concise comparison and review, study Table 1-1 for the differences among the four levels of measurement.

Table 1-1 Levels of Measurement of Data

Level	Summary	Example	
Nominal	Categories only. Data cannot be arranged in an ordering scheme.	Student states: 5 Californians 20 Texans 40 New Yorkers	Categories or names only.
Ordinal	Categories are ordered, but differences can't be found or are meaningless.	Student cars: 5 compact 20 mid-size 40 full-size	An order is determined by "compact, mid-size, full-size."
Interval	Differences are meaningful, but there is no natural zero starting point and ratios are meaningless.	Campus temperatures: 5°F 20°F 40°F	0°F doesn't mean "no heat." 40°F is not twice as hot as 20°F.
Ratio	There is a natural zero starting point and ratios are meaningful.	Student commuting distances: 5 mi 20 mi 40 mi	40 mi is *twice* as far as 20 miles.

1-2 BASIC SKILLS AND CONCEPTS

Statistical Literacy and Critical Thinking

1. Parameter and Statistic What is the difference between a parameter and a statistic?

2. Qualitative/Quantitative Data What is the difference between qualitative data and quantitative data?

3. Discrete/Continuous Data What is the difference between discrete data and continuous data?

4. Continuous/Quantitative Data If an experiment results in data that are continuous in nature, must the data also be quantitative, or can they be qualitative?

In Exercises 5–8, determine whether the given value is a statistic or a parameter.

5. Household Size A sample of households is selected and the average (mean) number of people per household is 2.58 (based on data from the U.S. Census Bureau).

6. Politics Currently, 42% of the governors of the 50 United States are Democrats.

7. Titanic In a study of all 2223 passengers aboard the *Titanic*, it is found that 706 survived when it sank.

8. Television Viewing A sample of Americans is selected and the average (mean) amount of time watching television is 4.6 hours per day.

In Exercises 9–12, determine whether the given values are from a discrete or continuous data set.

9. Mail Experiment In the Chapter Problem, it was noted that when 50 letters were sent as part of an experiment, three of them arrived at the target address.

10. Pedestrian Buttons In New York City, there are 3250 walk buttons that pedestrians can press at traffic intersections, and 2500 of them do not work (based on data from the article "For Exercise in New York Futility, Push Button," by Michael Luo, *New York Times*).

11. Penny Weights The mean weight of pennies currently being minted is 2.5 grams.

12. Gun Ownership In a survey of 1059 adults, it is found that 39% of them have guns in their homes (based on a Gallup poll).

In Exercises 13–20, determine which of the four levels of measurement (nominal, ordinal, interval, ratio) is most appropriate.

13. Marathon Numbers on shirts of marathon runners

14. Consumer Product *Consumer Reports* magazine ratings of "best buy, recommended, not recommended"

15. SSN Social Security Numbers

16. Drinking Survey The number of "yes" responses received when 500 students are asked if they have ever done binge drinking in college

17. Cicadas The years of cicada emergence: 1936, 1953, 1970, 1987, and 2004

18. Women Executives Salaries of women who are chief executive officers of corporations

19. Ratings Movie ratings of one star, two stars, three stars, or four stars

20. Temperatures The current temperatures in the 50 state capitol cities

In Exercises 21–24, identify the (a) sample *and (b)* population. *Also, determine whether the sample is likely to be representative of the population.*

21. Research Project A political scientist randomly selects 25 of the 100 Senators currently serving in Congress, then finds the lengths of time that they have served.

22. Nielsen Rating During the Superbowl game, a survey of 5108 randomly selected households finds that 44% of them have television sets tuned to the Superbowl (based on data from Nielsen Media Research).

23. Gun Ownership In a Gallup poll of 1059 randomly selected adults, 39% answered "yes" when asked "Do you have a gun in your home?"

24. Mail Survey A graduate student at the University of Newport conducts a research project about communication. She mails a survey to all of the 500 adults that she knows. She asks them to mail back a response to this question: "Do you prefer to use e-mail or snail mail (the U.S. Postal Service)?" She gets back 65 responses, with 42 of them indicating a preference for snail mail.

1-2 BEYOND THE BASICS

25. Interpreting Temperature Increase In the "Born Loser" cartoon strip by Art Sansom, Brutus expresses joy over an increase in temperature from 1° to 2°. When asked what is so good about 2°, he answers that "It's twice as warm as this morning." Explain why Brutus is wrong yet again.

26. Interpreting Political Polling A pollster surveys 200 people and asks them their preference of political party. He codes the responses as 0 (for Democrat), 1 (for Republican), 2 (for Independent), or 3 (for any other responses). He then calculates the average (mean) of the numbers and gets 0.95. How can that value be interpreted?

27. Scale for Rating Food A group of students develops a scale for rating the quality of the cafeteria food, with 0 representing "neutral: not good and not bad." Bad meals are given negative numbers and good meals are given positive numbers, with the magnitude of the number corresponding to the severity of badness or goodness. The first three meals are rated as 2, 4, and −5. What is the level of measurement for such ratings? Explain your choice.

1-3 Critical Thinking

Key Concept Success in the introductory statistics course typically requires more *common sense* than mathematical expertise (despite Voltaire's warning that "common sense is not so common"). Because we now have access to calculators and computers, modern applications of statistics no longer require us to master complex algorithms of mathematical manipulations. Instead, we can focus on *interpretation* of data and results. This section is designed to illustrate how common sense is used when we think critically about data and statistics. In this section, instead of memorizing specific methods or procedures, focus on thinking and using common sense in analyzing data. Know that when sample data are collected in an inappropriate way, such as using a *voluntary response sample* (defined later in this section), no statistical methods will be capable of producing valid results.

About a century ago, statesman Benjamin Disraeli famously said, "There are three kinds of lies: lies, damned lies, and statistics." It has also been said that "figures don't lie; liars figure." Historian Andrew Lang said that some people use statistics "as a drunken man uses lampposts—for support rather than illumination." Political cartoonist Don Wright encourages us to "bring back the mystery of life: lie to a pollster." Author Franklin P. Jones wrote that "statistics can be used to support anything—especially statisticians." In *Esar's Comic Dictionary* we find the definition of a statistician to be "a specialist who assembles figures and then leads them astray." These statements refer to instances in which methods of statistics were misused in ways that were ultimately deceptive. There are two main sources of such deception: (1) evil intent on the part of dishonest persons; (2) unintentional errors on the part of people who don't know any better. Regardless of the source, as responsible citizens and as more valuable professional employees, we should have a basic ability to distinguish between statistical conclusions that are likely to be valid and those that are seriously flawed.

To keep this section in proper perspective, know that this is not a book about the misuses of statistics. The remainder of this book will be full of very meaningful uses of valid statistical methods. We will learn general methods for using sample data to make important inferences about populations. We will learn about polls and sample sizes. We will learn about important measures of key characteristics of data. Along with the discussions of these general concepts, we will see many specific real applications, such as the effects of secondhand smoke, the prevalence of alcohol and tobacco in cartoon movies for children, and the quality of consumer products including M&M candies, cereals, Coke, and Pepsi. But even in those meaningful and real applications, we must be careful to correctly interpret the results of valid statistical methods.

We begin our development of critical thinking by considering bad samples. These samples are bad in the sense that the sampling method dooms the sample so that it is likely to be *biased* (not representative of the population from which it has been obtained). In the next section we will discuss in more detail the methods of sampling, and the importance of *randomness* will be described. The first example below describes a sampling procedure that seriously lacks the randomness that is so important. The following definition refers to one of the most common and most serious misuses of statistics.

Definition

A **voluntary response sample** (or **self-selected sample**) is one in which the respondents themselves decide whether to be included.

For example, *Newsweek* magazine ran a survey about the controversial Napster Web site, which had been providing free access to copying music CDs. Readers were asked this question: "Will you still use Napster if you have to pay a fee?" Readers could register their responses on the Web site newsweek.msnbc.com. Among the 1873 responses received, 19% said yes, it is still cheaper than buying CDs. Another 5% said yes, they felt more comfortable using it with a charge. When *Newsweek* or anyone else runs a poll on the Internet, individuals decide themselves whether to participate, so they constitute a voluntary response sample. But people with strong opinions are more likely to participate, so the responses are not representative of the whole population. Here are common examples of voluntary response samples which, by their very nature, are seriously flawed in the sense that we should not make conclusions about a population based on such a biased sample:

- Polls conducted through the Internet, where subjects can decide whether to respond
- Mail-in polls, where subjects can decide whether to reply
- Telephone call-in polls, where newspaper, radio, or television announcements ask that you voluntarily pick up a phone and call a special number to register your opinion

With such voluntary response samples, valid conclusions can be made only about the specific group of people who chose to participate, but a common practice is to incorrectly state or imply conclusions about a larger population. From a statistical viewpoint, such a sample is fundamentally flawed and should not be used for making general statements about a larger population.

Small Samples Conclusions should not be based on samples that are far too small. As one example, the Children's Defense Fund published *Children Out of School in America* in which it was reported that among secondary school students suspended in one region, 67% were suspended at least three times. But that figure is based on a sample of only *three* students! Media reports failed to mention that this sample size was so small. (We will see in Chapters 7 and 8 that we can *sometimes* make some valuable inferences from small samples, but we should be careful to verify that the necessary requirements are satisfied.)

Sometimes a sample might seem relatively large (as in a survey of "2000 randomly selected adult Americans"), but if conclusions are made about subgroups, such as the 21-year-old male Republicans from Pocatello, such conclusions might be based on samples that are too small. Although it is important to have a sample that is sufficiently large, it is just as important to have sample data that have been collected in an appropriate way, such as random selection. Even large samples can be bad samples.

Graphs Graphs—such as bar graphs and pie charts—can be used to exaggerate or understate the true nature of data. (In Chapter 2 we discuss a variety of different graphs.) The two graphs in Figure 1-1 depict the *same data* obtained from the U.S. Bureau of Economic Analysis, but part (b) is designed to exaggerate the difference between the personal income per capita in California and its neighboring state of Nevada. By not starting the horizontal axis at zero, the graph in part (b)

The Literary Digest Poll

In the 1936 presidential race, *Literary Digest* magazine ran a poll and predicted an Alf Landon victory, but Franklin D. Roosevelt won by a landslide. Maurice Bryson notes, "Ten million sample ballots were mailed to prospective voters, but only 2.3 million were returned. As everyone ought to know, such samples are practically always biased." He also states, "Voluntary response to mailed questionnaires is perhaps the most common method of social science data collection encountered by statisticians, and perhaps also the worst." (See Bryson's "The *Literary Digest* Poll: Making of a Statistical Myth," *The American Statistician,* Vol. 30, No. 4.)

Figure 1-1

Comparison of California and Nevada: Personal Income Per Capita

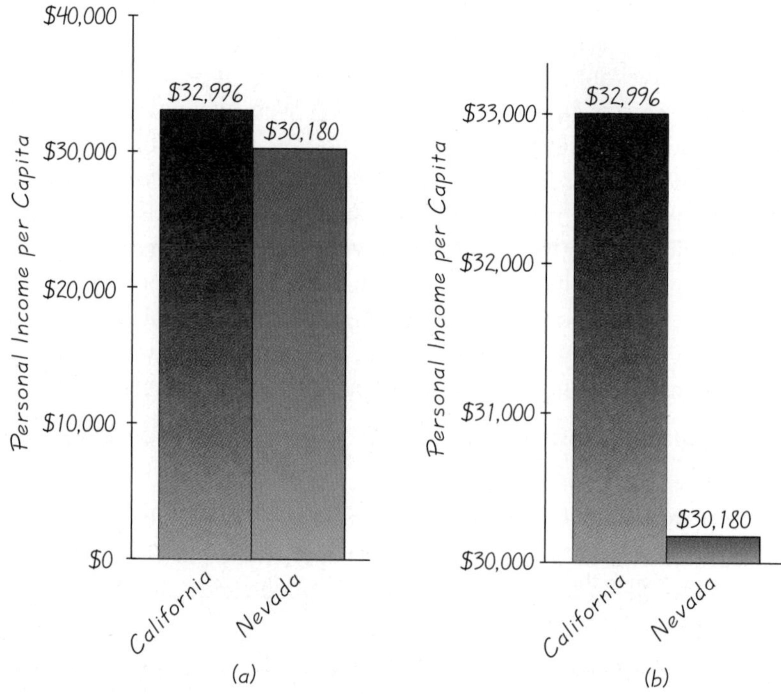

tends to produce a misleading subjective impression, causing readers to incorrectly believe that the difference is much greater than it really is. Figure 1-1 carries this important lesson: To correctly interpret a graph, we should analyze the *numerical* information given in the graph, so that we won't be misled by its general shape.

Pictographs Drawings of objects, called *pictographs,* may also be misleading. Some objects commonly used to depict data include three-dimensional objects, such as moneybags, stacks of coins, army tanks (for military expenditures), barrels (for oil production), and houses (for home construction). When drawing such objects, artists can create false impressions that distort differences. If you double each side of a square, the area doesn't merely double; it increases by a factor of four. If you double each side of a cube, the volume doesn't merely double; it increases by a factor of eight. See Figure 1-2, where part (a) is drawn to correctly depict the relationship between the daily oil consumption of the United States and Japan. In Figure 1-2(a), it appears that the United States consumes roughly four times as much oil as Japan. However, part (b) of Figure 1-2 is drawn using barrels, with *each dimension* drawn in proportion to the actual amounts. See that Figure 1-2(b) grossly exaggerates the difference by creating the false impression that U.S. oil consumption appears to be roughly 50 times that of Japan.

Percentages Misleading or unclear percentages are sometimes used. If you take 100% of some quantity, you are taking it *all.* (It shouldn't require a 110% effort to make sense of the preceding statement.) In referring to lost baggage, Continental Airlines ran ads claiming that this was "an area where we've already improved 100% in the last six months." In an editorial criticizing this statistic, the *New York*

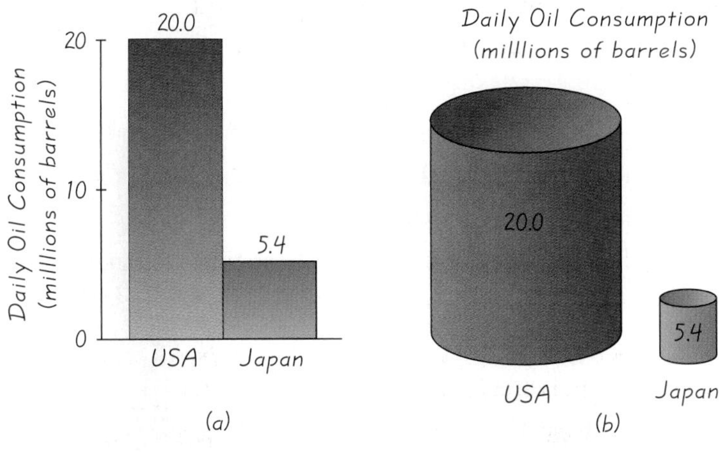

Figure 1-2 Comparison of United States and Japan: Daily Oil Consumption (millions of barrels)

Part (b) is designed to exaggerate the difference by increasing each dimension in proportion to the actual amounts of oil consumption.

Times correctly interpreted the 100% improvement figure to mean that no baggage is now being lost—an accomplishment not yet enjoyed by Continental Airlines.

The following are a few of the key principles to be applied when dealing with percentages. These principles all use the basic notion that % or "percent" really means "divided by 100." The first principle will be used often in this book.

- **Percentage of:** To find some *percentage of* an amount, drop the % symbol and divide the percentage value by 100, then multiply. This example shows that 6% of 1200 is 72:

$$6\% \text{ of } 1200 \text{ responses} = \frac{6}{100} \times 1200 = 72$$

- **Fraction → Percentage:** *To convert from a fraction to a percentage,* divide the denominator into the numerator to get an equivalent decimal number, then multiply by 100 and affix the % symbol. This example shows that the fraction 3/4 is equivalent to 75%:

$$\frac{3}{4} = 0.75 \rightarrow 0.75 \times 100\% = 75\%$$

- **Decimal → Percentage:** To *convert from a decimal to a percentage,* multiply by 100%. This example shows that 0.250 is equivalent to 25.0%:

$$0.250 \rightarrow 0.250 \times 100\% = 25.0\%$$

- **Percentage → Decimal:** *To convert from a percentage to a decimal number,* delete the % symbol and divide by 100. This example shows that 85% is equivalent to 0.85:

$$85\% = \frac{85}{100} = 0.85$$

Misleading Statistics in Journalism

New York Times reporter Daniel Okrant wrote that although every sentence in his newspaper is copyedited for clarity and good writing, "numbers, so alien to so many, don't get nearly this respect. The paper requires no specific training to enhance numeracy, and no specialists whose sole job is to foster it." He cites an example of the *New York Times* reporting about an estimate of more than $23 billion that New Yorkers spend for counterfeit goods each year. Okrant writes that "quick arithmetic would have demonstrated that $23 billion would work out to roughly $8000 per city household, a number ludicrous on its face."

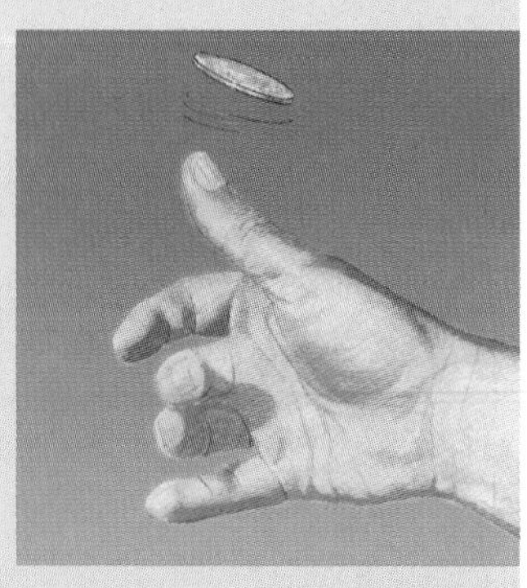

Detecting Phony Data

A class is given the homework assignment of recording the results when a coin is tossed 500 times. One dishonest student decides to save time by just making up the results instead of actually flipping a coin. Because people generally cannot make up results that are really random, we can often identify such phony data. With 500 tosses of an actual coin, it is extremely likely that you will get a run of six heads or six tails, but people almost never include such a run when they make up results.

Another way to detect fabricated data is to establish that the results violate Benford's law: For many collections of data, the leading digits are not uniformly distributed. Instead, the leading digits of 1, 2, . . . , 9 occur with rates of 30%, 18%, 12%, 10%, 8%, 7%, 6%, 5%, and 5%, respectively. (See "The Difficulty of Faking Data", by Theodore Hill, *Chance,* Vol. 12, No. 3.)

Loaded Questions There are many issues affecting survey questions. Survey questions can be "loaded" or intentionally worded to elicit a desired response. See the actual "yes" response rates for the different wordings of a question:

97% yes: "Should the President have the line item veto to eliminate waste?"

57% yes: "Should the President have the line item veto, or not?"

In *The Superpollsters*, David W. Moore describes an experiment in which different subjects were asked if they agree with the following statements:

- Too little money is being spent on welfare.
- Too little money is being spent on assistance to the poor.

Even though it is the poor who receive welfare, only 19% agreed when the word "welfare" was used, but 63% agreed with "assistance to the poor."

Order of Questions Sometimes survey questions are unintentionally loaded by such factors as the order of the items being considered. See these questions from a poll conducted in Germany:

- Would you say that traffic contributes more or less to air pollution than industry?
- Would you say that industry contributes more or less to air pollution than traffic?

When traffic was presented first, 45% blamed traffic and 27% blamed industry; when industry was presented first, 24% blamed traffic and 57% blamed industry.

Nonresponse A *nonresponse* occurs when someone either refuses to respond to a survey question, or the person is unavailable. When people are asked survey questions, some firmly refuse to answer. The refusal rate has been growing in recent years, partly because many persistent telemarketers try to sell goods or services by beginning with a sales pitch that initially sounds like it is part of an opinion poll. (This "selling under the guise" of a poll is often called *sugging.*) In *Lies, Damn Lies, and Statistics,* author Michael Wheeler correctly observes that "people who refuse to talk to pollsters are likely to be different from those who do not. Some may be fearful of strangers and others jealous of their privacy, but their refusal to talk demonstrates that their view of the world around them is markedly different from that of those people who will let poll-takers into their homes."

Missing Data Results can sometimes be dramatically affected by missing data. Sometimes sample data values are missing completely at random, meaning that the chance of being missing is completely unrelated to its values or other values. However, some data are missing because of special factors, such as the tendency of people with low incomes to be less likely to report their incomes. It is well known that the U.S. Census suffers from missing people, and the missing people are often from the homeless or low income groups. In years past, surveys conducted by telephone were often misleading because they suffered from missing people who were not wealthy enough to own telephones.

Correlation and Causality In Chapter 10 of this book we will discuss the statistical association between two variables, such as wealth and IQ. We will use the term *correlation* to indicate that the two variables are related. However, in Chapter 10 we make this important point: *Correlation does not imply causality.* This means that when we find a statistical association between two variables, we cannot conclude that one of the variables is the cause of (or directly affects) the other variable. If we find a correlation between wealth and IQ, we cannot conclude that a person's IQ directly affects his or her wealth, and we cannot conclude that a person's wealth directly affects his or her IQ score. It is quite common for the media to report about a newfound correlation with wording that directly indicates or implies that one of the variables is the cause of the other.

Self-Interest Study Studies are sometimes sponsored by parties with interests to promote. For example, Kiwi Brands, a maker of shoe polish, commissioned a study that resulted in this statement printed in some newspapers: "According to a nationwide survey of 250 hiring professionals, scuffed shoes was the most common reason for a male job seeker's failure to make a good first impression." We should be very wary of such a survey in which the sponsor can enjoy monetary gains from the results. Of growing concern in recent years is the practice of pharmaceutical companies to pay doctors who conduct clinical experiments and report their results in prestigious journals, such as the *Journal of the American Medical Association.*

Precise Numbers "There are now 103,215,027 households in the United States." Because that figure is very precise, many people incorrectly assume that it is also *accurate*. In this case, that number is an estimate and it would be better to state that the number of households is about 103 million.

Partial Pictures "Ninety percent of all our cars sold in this country in the last 10 years are still on the road." Millions of consumers heard that commercial message and didn't realize that 90% of the cars the advertiser sold in this country were sold within the last three years, so most of those cars on the road were quite new. The claim was technically correct, but it was very misleading in not presenting the complete results.

Deliberate Distortions In the book *Tainted Truth,* Cynthia Crossen cites an example of the magazine *Corporate Travel* that published results showing that among car rental companies, Avis was the winner in a survey of people who rent cars. When Hertz requested detailed information about the survey, the actual survey responses disappeared and the magazine's survey coordinator resigned. Hertz sued Avis (for false advertising based on the survey) and the magazine; a settlement was reached.

In addition to the cases cited above, there are many other misuses of statistics. Some of those other cases can be found in books such as Darrell Huff's classic *How to Lie with Statistics,* Robert Reichard's *The Figure Finaglers,* and Cynthia Crossen's *Tainted Truth.* Understanding these practices will be extremely helpful in evaluating the statistical data found in everyday situations.

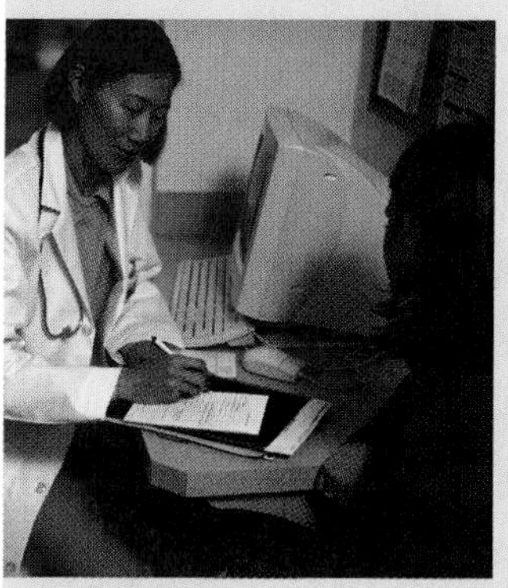

Publication Bias

There is a "publication bias" in professional journals. It is the tendency to publish positive results (such as showing that some treatment is effective) much more often than negative results (such as showing that some treatment has no effect). In the article "Registering Clinical Trials" (*Journal of the American Medical Association*, Vol. 290, No. 4), authors Kay Dickersin and Drummond Rennie state that "the result of not knowing who has performed what (clinical trial) is loss and distortion of the evidence, waste and duplication of trials, inability of funding agencies to plan, and a chaotic system from which only certain sponsors might benefit, and is invariably against the interest of those who offered to participate in trials and of patients in general." They support a process in which *all* clinical trials are registered in one central system.

1-3 BASIC SKILLS AND CONCEPTS

Statistical Literacy and Critical Thinking

1. Voluntary Response Sample What is a voluntary response sample, and why is it generally unsuitable for methods of statistics?

2. Correlation and Causality If we conduct a statistical analysis and find that there is a correlation (or an association) between the lengths of time spent studying and the grades in different courses, can we conclude that more study time causes higher grades? Why or why not?

3. Nonresponses A researcher determines that he needs results from at least 300 subjects to conduct a study. In order to compensate for low return rates, he mails a survey to 10,000 subjects and receives 320 responses. Is his sample of 320 responses a good sample?

4. Loaded Question The phone rings and an automated voice asks whether you are willing to vote for a candidate "with a long history of raising taxes and wasting taxpayer money." Assuming that the calls are made to randomly selected voters, are the results likely to reflect the preference that voters have for this candidate? Why or why not?

In Exercises 5–8, use critical thinking to develop an alternative conclusion. For example, consider a media report that BMW cars cause people to be healthy, because people who drive BMW cars are found to have better health than those who do not. The conclusion that BMW cars cause better health is probably wrong. Here is a better conclusion: BMW drivers tend to be wealthier than other adults, and greater wealth is associated with better health.

5. Height and Exercise Based on a study of heights of men and women who play basketball, a researcher concludes that the exercise from playing basketball causes people to grow taller.

6. College Graduates Live Longer Based on a study showing that college graduates tend to live longer than those who do not graduate from college, a researcher concludes that studying causes people to live longer.

7. Racial Profiling? A study showed that in Orange County, more speeding tickets were issued to minorities than to whites. Conclusion: In Orange County, minorities speed more than whites.

8. Cold Remedy In a study of cold symptoms, every one of the study subjects with a cold was found to be improved two weeks after taking ginger pills. Conclusion: Ginger pills cure colds.

In Exercises 9–20, use critical thinking to address the key issue.

9. Chocolate Health Food The *New York Times* published an article that included these statements: "At long last, chocolate moves toward its rightful place in the food pyramid, somewhere in the high-tone neighborhood of red wine, fruits and vegetables, and green tea. Several studies, reported in the *Journal of Nutrition*, showed that after eating chocolate, test subjects had increased levels of antioxidants in their blood. Chocolate contains flavonoids, antioxidants that have been associated with decreased risk of heart disease and stroke. Mars Inc., the candy company, and the Chocolate Manufacturers Association financed much of the research." What is wrong with this study?

10. Census Data After the last national census was conducted, the *Poughkeepsie Journal* ran this front-page headline: "281,421,906 in America." What is wrong with this headline?

11. Mail Survey When author Shere Hite wrote *Woman and Love: A Cultural Revolution in Progress*, she based conclusions on 4500 replies that she received after mailing 100,000 questionnaires to various women's groups. Are her conclusions likely to be valid in the sense that they can be applied to the general population of all women? Why or why not?

12. "900" Numbers In an ABC "Nightline" poll, 186,000 viewers each paid 50 cents to call a "900" telephone number with their opinion about keeping the United Nations in the United States. The results showed that 67% of those who called were in favor of moving the United Nations out of the United States. Interpret the results by identifying what we can conclude about the way the general population feels about keeping the United Nations in the United States.

13. Conducting Surveys You plan to conduct a survey to find the percentage of people in your state who can name the Lieutenant Governor, who plans to run for the United States Senate. You obtain addresses from telephone directories and you mail a survey to 850 randomly selected people. What is wrong with using the telephone directory as the source of survey subjects?

14. Pictographs Taxes in Newport have doubled over the past 10 years, and a candidate for mayor wants to construct a graph that emphasizes that point. She represents taxes 10 years ago by using a box with a width, length, and height all equal to 1 inch. She then doubles each dimension to show a larger box representing taxes now. What is the volume of the smaller box? What is the volume of the larger box? Does this pictograph correctly depict the relationship between taxes 10 years ago and taxes now?

15. Motorcycle Helmets The Hawaii State Senate held hearings when it was considering a law requiring that motorcyclists wear helmets. Some motorcyclists testified that they had been in crashes in which helmets would not have been helpful. Which important group was not able to testify? (See "A Selection of Selection Anomalies," by Wainer, Palmer, and Bradlow in *Chance*, Vol. 11, No. 2.)

16. Merrill Lynch Client Survey The author received a survey from the investment firm of Merrill Lynch. It was designed to gauge his satisfaction as a client, and it had specific questions for rating the author's personal Financial Consultant. The cover letter included this statement: "Your responses are extremely valuable to your Financial Consultant, Russell R. Smith, and to Merrill Lynch. . . . We will share your name and response with your Financial Consultant." What is wrong with this survey?

17. Average of Averages An economist randomly selects 10 wage earners from each of the 50 states. For each state, he finds the average of the annual incomes, and he then adds those 50 values and divides by 50. Is the result likely to be a good estimate of the average (mean) of all wage earners in the United States? Why or why not?

18. Bad Question A survey includes this item: "Enter your height in inches." It is expected that actual heights of respondents can be obtained and analyzed, but there are two different major problems with this item. Identify them.

19. Number of Household Members You need to conduct a study to determine the average size of a household in your state. You collect data consisting of the number of brothers and sisters from students at your college. What group of households is missed by this approach? Will the results be representative of all households in the state?

20. SIDS In a letter to the editor in the *New York Times*, Moorestown, New Jersey, resident Jean Mercer criticized the statement that "putting infants in the supine position has decreased deaths from SIDS." SIDS refers to sudden infant death syndrome, and the *supine* position is lying on the back with the face upward. She suggested that this statement is better: "Pediatricians advised the supine position during a time when the SIDS rate fell." What is wrong with saying that the supine position *decreased* deaths from SIDS?

Percentages. *In Exercises 21–26, answer the given questions that relate to percentages.*

21. Percentages
- **a.** Convert the fraction 3/20 to an equivalent percentage.
- **b.** Convert 56.7% to an equivalent decimal.
- **c.** What is 34% of 500?
- **d.** Convert 0.789 to an equivalent percentage.

22. Percentages
- **a.** What is 15% of 620?
- **b.** Convert 5% to an equivalent decimal.
- **c.** Convert 0.01 to an equivalent percentage.
- **d.** Convert the fraction 987/1068 to an equivalent percentage. Express the answer to the nearest tenth of a percent.

23. Percentages in a Gallup Poll
- **a.** In a Gallup poll, 52% of 1038 surveyed adults said that secondhand smoke is "very harmful." What is the actual number of adults who said that secondhand smoke is "very harmful"?
- **b.** Among the 1038 surveyed adults, 52 said that secondhand smoke is "not at all harmful." What is the percentage of people who chose "not at all harmful"?

24. Percentages in a Study of Lipitor
- **a.** In a study of the cholesterol drug Lipitor, 270 patients were given a placebo, and 19 of those 270 patients reported headaches. What percentage of this placebo group reported headaches?
- **b.** Among the 270 patients in the placebo group, 3.0% reported back pains. What is the actual number of patients who reported back pains?

25. Percentages in Campus Crime In a study on college campus crimes committed by students high on alcohol or drugs, a mail survey of 1875 students was conducted. A *USA Today* article noted, "Eight percent of the students responding anonymously say they've committed a campus crime. And 62% of that group say they did so under the influence of alcohol or drugs." Assuming that the number of students responding anonymously is 1875, how many actually committed a campus crime while under the influence of alcohol or drugs?

26. Percentages in the Media and Advertising
- **a.** A *New York Times* editorial criticized a chart caption that described a dental rinse as one that "reduces plaque on teeth by over 300%." What is wrong with that statement?
- **b.** In the *New York Times Magazine,* a report about the decline of Western investment in Kenya included this: "After years of daily flights, Lufthansa and Air France had halted passenger service. Foreign investment fell 500 percent during the 1990's." What is wrong with this statement?
- **c.** In an ad for the Club, a device used to discourage car thefts, it was stated that "The Club reduces your odds of car theft by 400%." What is wrong with this statement?

1-3 BEYOND THE BASICS

27. Falsifying Data A researcher at the Sloan-Kettering Cancer Research Center was once criticized for falsifying data. Among his data were figures obtained from 6 groups of mice, with 20 individual mice in each group. These values were given for the percentage of successes in each group: 53%, 58%, 63%, 46%, 48%, 67%. What is the major flaw?

28. What's Wrong with This Picture? Try to identify four major flaws in the following. A daily newspaper ran a survey by asking readers to call in their response to this question: "Do you support the development of atomic weapons that could kill millions of innocent people?" It was reported that 20 readers responded and 87% said "no" while 13% said "yes."

1-4 Design of Experiments

Key Concept This section contains much information and many definitions. Among all of the definitions, the concept of a *simple random sample* is particularly important for its role throughout the remainder of this book and for its role in statistics in general, so be sure to understand that particular definition quite well. Also, clearly understand that the method used to collect data is absolutely and critically important. Recognize that when designing experiments, great thought and care are required to ensure valid results.

If sample data are not collected in an appropriate way, the data may be so completely useless that no amount of statistical torturing can salvage them.

Statistical methods are driven by data. We typically obtain data from two distinct sources: *observational studies* and *experiments.*

Definitions

In an **observational study,** we observe and measure specific characteristics, but we don't attempt to *modify* the subjects being studied.

In an **experiment,** we apply some *treatment* and then proceed to observe its effects on the subjects. (Subjects in experiments are called **experimental units.**)

Types of Observational Studies A Gallup poll is a good example of an observational study, whereas a clinical trial of the drug Lipitor is a good example of an experiment. The Gallup poll is observational in the sense that we merely observe people (often through interviews) without modifying them in any way. But the clinical trial of Lipitor involves treating some people with the drug, so the treated people are modified. There are different types of observational studies, as illustrated in Figure 1-3. These terms, commonly used in many different professional journals, are defined below.

Figure 1-3

Statistical Studies

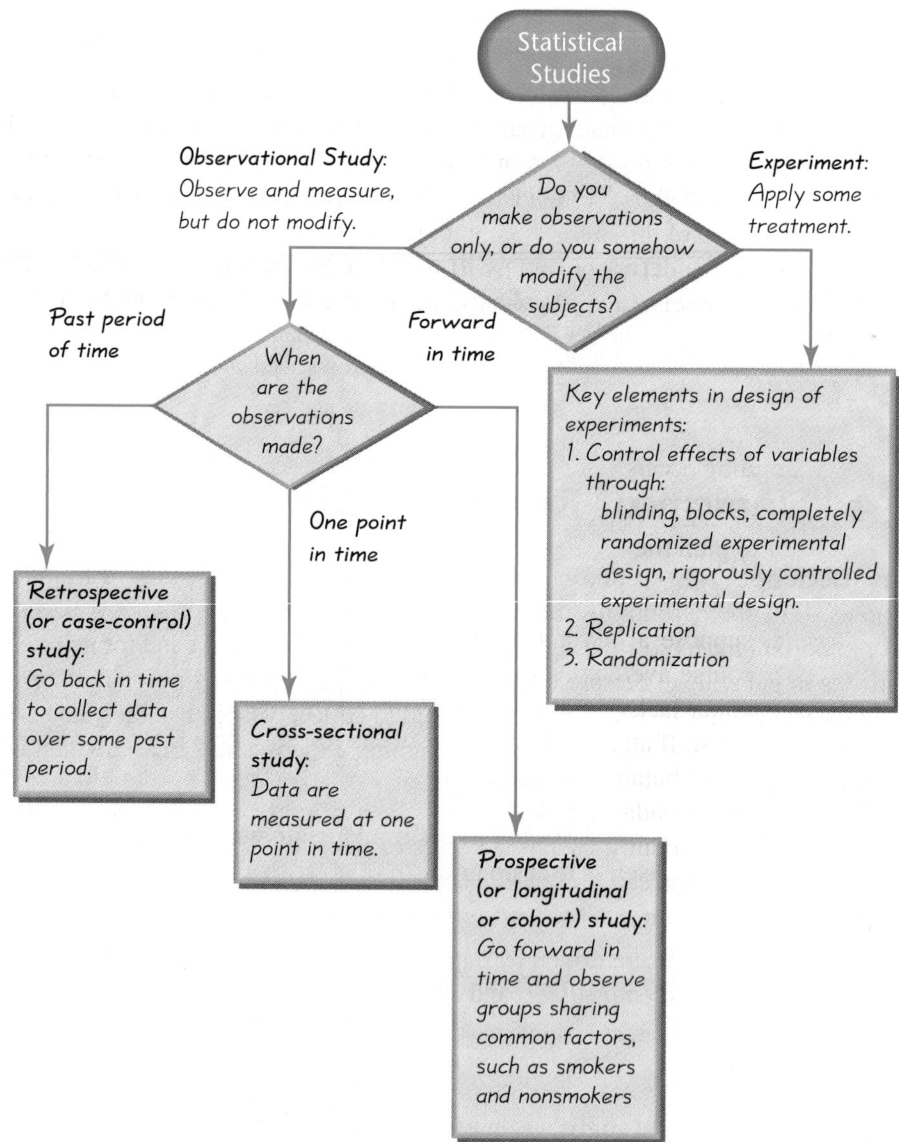

Definitions

In a **cross-sectional study,** data are observed, measured, and collected at one point in time.

In a **retrospective** (or **case-control**) **study,** data are collected from the past by going back in time (through examination of records, interviews, and so on).

In a **prospective** (or **longitudinal** or **cohort**) **study,** data are collected in the future from groups sharing common factors (called *cohorts*).

There is an important distinction between the sampling done in retrospective and prospective studies. In retrospective studies we go back in time to collect data about the resulting characteristic that is of concern, such as a group of drivers who died in car crashes and another group of drivers who did not die in car crashes. In prospective studies we go forward in time by following groups with a potentially causative factor and those without it, such as a group of drivers who use cell phones and a group of drivers who do not use cell phones.

The above three definitions apply to observational studies, but we now shift our focus to experiments. Results of experiments are sometimes ruined because of *confounding*.

Definition

Confounding occurs in an experiment when you are not able to distinguish among the effects of different factors.

Try to plan the experiment so that confounding does not occur.

For example, suppose a Vermont professor experiments with a new attendance policy ("your course average drops one point for each class cut"), but an exceptionally mild winter lacks the snow and cold temperatures that have hindered attendance in the past. If attendance does get better, we can't determine whether the improvement is attributable to the new attendance policy or to the mild winter. The effects of the attendance policy and the weather have been confounded. It is generally very important to control the effects of variables. In addition to confounding, experiments can also be ruined by other factors, such as failure to collect a sample that is representative of the population. In general, great care and extensive planning are required in planning experiments. Figure 1-3 shows that the following are three very important considerations in the design of experiments:

1. Control the effects of variables.
2. Use replication.
3. Use randomization.

We will now focus on the role that these three factors have in the design of experiments.

Clinical Trials vs. Observational Studies

In a *New York Times* article about hormone therapy for women, reporter Denise Grady wrote about a report of treatments tested in randomized controlled trials. She stated that "Such trials, in which patients are assigned at random to either a treatment or a placebo, are considered the gold standard in medical research. By contrast, the observational studies, in which patients themselves decide whether to take a drug, are considered less reliable. . . . Researchers say the observational studies may have painted a falsely rosy picture of hormone replacement because women who opt for the treatments are healthier and have better habits to begin with than women who do not."

Controlling Effects of Variables

Figure 1-3 shows that one of the key elements in the design of experiments is controlling effects of variables. We can gain that control by using such devices as blinding, a randomized block design, a completely randomized experimental design, or a rigorously controlled experimental design, described as follows.

Blinding In 1954, a massive experiment was designed to test the effectiveness of the Salk vaccine in preventing the polio that killed or paralyzed thousands of children. In that experiment, a treatment group was given the actual Salk vaccine, while a second group was given a placebo that contained no drug at all. In experiments

Hawthorne and Experimenter Effects

The well-known placebo effect occurs when an untreated subject incorrectly believes that he or she is receiving a real treatment and reports an improvement in symptoms. The Hawthorne effect occurs when treated subjects somehow respond differently, simply because they are part of an experiment. (This phenomenon was called the "Hawthorne effect" because it was first observed in a study of factory workers at Western Electric's Hawthorne plant.) An experimenter effect (sometimes called a Rosenthall effect) occurs when the researcher or experimenter unintentionally influences subjects through such factors as facial expression, tone of voice, or attitude.

involving placebos, there is often a **placebo effect** that occurs when an untreated subject reports an improvement in symptoms. (The reported improvement in the placebo group may be real or imagined.) This placebo effect can be minimized or accounted for through the use of **blinding,** a technique in which the subject doesn't know whether he or she is receiving a treatment or a placebo. Blinding allows us to determine whether the treatment effect is significantly different from the placebo effect. The polio experiment was **double-blind,** meaning that blinding occurred at two levels: (1) The children being injected didn't know whether they were getting the Salk vaccine or a placebo, and (2) the doctors who gave the injections and evaluated the results did not know either.

Blocks When designing an experiment, we often know in advance that there are some factors that are likely to have a strong effect on the variable being considered. For example, in designing an experiment to test the effectiveness of a new fertilizer on tree growth, we know that the moisture content of the soil (dry or moist) can affect tree growth. In this case, we should use blocks. A **block** is a group of subjects that are similar, but blocks are different in the ways that might affect the outcome of the experiment. Testing the effects of a fertilizer might involve a block of trees in dry soil and a block of trees in moist soil. In designing an experiment to test the effectiveness of a new drug for heart disease, we might form a block of men and a block of women, because it is known that hearts of men and women can behave differently. After identifying the blocks, proceed to *randomly* assign treatments to the subjects in each block.

> ***Randomized block design:* If conducting an experiment of testing one or more different treatments, and there are different groups of similar subjects, but the groups are different in ways that are likely to affect the responses to treatments, use this experimental design:**
>
> 1. **Form blocks (or groups) of subjects with similar characteristics.**
> 2. **Randomly assign treatments to the subjects within each block.**

For example, suppose we want to test the effectiveness of a fertilizer on the weights of 12 poplar trees and we have two plots of land: One plot of land is dry while a second plot of land is moist. A bad experimental design would be to use fertilizer on 6 trees planted in moist soil while the other 6 untreated trees are planted in dry soil. We would not know whether differences are due to the fertilizer treatment or the type of soil or both. See Figure 1-4(a). It would be much better to plant 6 trees in the dry plot and 6 trees in the moist plot. For each group of 6 trees, use randomization to identify 3 trees that are treated with fertilizer while the other 3 trees are not given the fertilizer, as in Figure 1-4(b). The results will allow us to see the effectiveness of the fertilizer. (The choice of 12 for the sample size is totally arbitrary here. More trees would yield better results. Later chapters address issues of determining sample size. Also, the experiment should be carefully designed so that the fertilizer applied to one tree doesn't spread to adjacent trees. Other relevant factors such as watering, temperature, and sunlight must be carefully controlled. The overall experimental design requires much more thought and care than we have described here. A course in the design of experiments would be a great way to learn much more than we can describe here.)

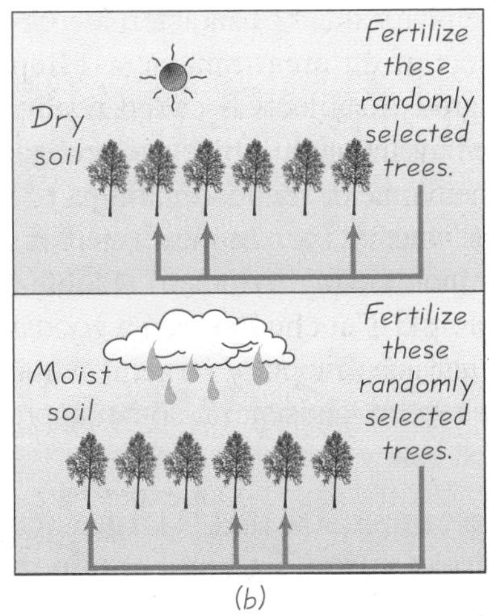

Figure 1-4 Design of Experiment with Different Treatments

(a) Bad experimental design: Use fertilizer to treat all trees in moist soil, and don't use fertilizer for trees in dry soil. There will be no way to know whether the results are attributable to the treatment or the soil type.

(b) Randomized block design: Form a block of trees in moist soil and another block of trees in dry soil. Within each of the blocks, use randomness to determine which trees are treated with fertilizer and which trees are not treated.

(c) Completely randomized experimental design: Use randomness to determine which trees are treated with fertilizer and which trees are not treated.

Completely Randomized Experimental Design With a **completely randomized experimental design,** subjects are assigned to different treatment groups through a process of *random selection.* See Figure 1-4(c) where we show that among 12 trees, 6 are randomly selected for fertilizer treatment. Another example of a completely randomized experimental design is this feature of the polio experiment: Children were assigned to the treatment group or placebo group through a process of random selection (equivalent to flipping a coin).

Rigorously Controlled Design Another approach for assigning subjects to treatments is to use a **rigorously controlled design,** in which subjects are very *carefully chosen* so that those given each treatment are similar in the ways that are important to the experiment. In an experiment testing the effectiveness of a drug designed to lower blood pressure, if the placebo group includes a 30-year-old overweight male smoker who drinks heavily and consumes an abundance of salt and fat, the treatment group should also include a person with similar characteristics (which, in this case, would be easy to find).

Replication and Sample Size

In addition to controlling effects of variables, another key element of experimental design is the size of the samples. Samples should be large enough so that the

Prospective National Children's Study

A good example of a prospective study is the National Children's Study begun in 2005. It is tracking 100,000 children from birth to age 21. The children are from 96 different geographic regions. The objective is to improve the health of children by identifying the effects of environmental factors, such as diet, chemical exposure, vaccinations, movies, and television. It is planned that some preliminary results will be available around 2008. The study will address questions such as these: How do genes and the environment interact to promote or prevent violent behavior in teenagers? Are lack of exercise and poor diet the only reasons why many children are overweight? Do infections impact developmental progress, asthma, obesity, and heart disease? How do city and neighborhood planning and construction encourage or discourage injuries?

erratic behavior that is characteristic of very small samples will not disguise the true effects of different treatments. Repetition of an experiment on sufficiently large groups of subjects is called **replication,** and replication is used effectively when we have enough subjects to recognize differences from different treatments. (In another context, *replication* refers to the repetition or duplication of an experiment so that results can be confirmed or verified.) With replication, the large sample sizes increase the chance of recognizing different treatment effects. However, a large sample is not necessarily a good sample. Although it is important to have a sample that is sufficiently large, it is more important to have a sample in which data have been chosen in some appropriate way, such as random selection (described below).

> **Use a sample size that is large enough so that we can see the true nature of any effects, and obtain the sample using an appropriate method, such as one based on *randomness.***

In the experiment designed to test the Salk vaccine, 200,000 children were given the actual Salk vaccine and 200,000 other children were given a placebo. Because the actual experiment used sufficiently large sample sizes, the effectiveness of the vaccine could be seen. Nevertheless, even though the treatment and placebo groups were very large, the experiment would have been a failure if subjects had not been assigned to the two groups in a way that made both groups similar in the ways that were important to the experiment.

Randomization and Other Sampling Strategies

In statistics as in life, one of the worst mistakes is to collect data in a way that is inappropriate. We cannot overstress this very important point:

> **If sample data are not collected in an appropriate way, the data may be so completely useless that no amount of statistical torturing can salvage them.**

In Section 1-3 we saw that a voluntary response sample is one in which the subjects decide themselves whether to respond. Such samples are very common, but their results are generally useless for making valid inferences about larger populations.

We now define some of the more common methods of sampling.

Definitions

In a **random sample** members from the population are selected in such a way that each *individual member* has an equal chance of being selected.

A **simple random sample** of n subjects is selected in such a way that every possible *sample of the same size n* has the same chance of being chosen.

A **probability sample** involves selecting members from a population in such a way that each member has a known (but not necessarily the same) chance of being selected.

EXAMPLE Random Sample, Simple Random Sample, Probability Sample Picture a classroom with 60 students arranged in six rows of 10 students each. Assume that the professor selects a sample of 10 students by rolling a die and selecting the row corresponding to the outcome. Is the result a random sample? Simple random sample? Probability sample?

SOLUTION The sample is a random sample because each individual student has the same chance (one chance in six) of being selected. The sample is *not* a simple random sample because not all samples of size 10 have the same chance of being chosen. For example, this sampling design of using a die to select a row makes it impossible to select 10 students who are in different rows (but there is one chance in six of selecting the sample consisting of the 10 students in the first row). The sample is a probability sample because each student has a known chance (one chance in six) of being selected.

Important: Throughout this book, we will use a variety of different statistical procedures, and we often have a requirement that we have collected a *simple random sample*, as defined above. With random sampling we expect all components of the population to be (approximately) proportionately represented. Random samples are selected by many different methods, including the use of computers to generate random numbers. (Before computers, tables of random numbers were often used instead. For truly exciting reading, see this book consisting of one million digits that were randomly generated: *A Million Random Digits* published by Free Press. The *Cliff Notes* summary of the plot is not yet available.) Unlike careless or haphazard sampling, random sampling usually requires very careful planning and execution.

In addition to random sampling, there are other sampling techniques in use, and we describe the common ones here. See Figure 1-5 for an illustration depicting the different sampling approaches. Keep in mind that only random sampling and simple random sampling will be used throughout the remainder of the book.

Definitions

In **systematic sampling,** we select some starting point and then select every *k*th (such as every 50th) element in the population.

With **convenience sampling,** we simply use results that are very easy to get.

With **stratified sampling,** we subdivide the population into at least two different subgroups (or strata) so that subjects within the same subgroup share the same characteristics (such as gender or age bracket), then we draw a sample from each subgroup (or stratum).

In **cluster sampling,** we first divide the population area into sections (or clusters), then randomly select some of those clusters, and then choose *all* the members from those selected clusters.

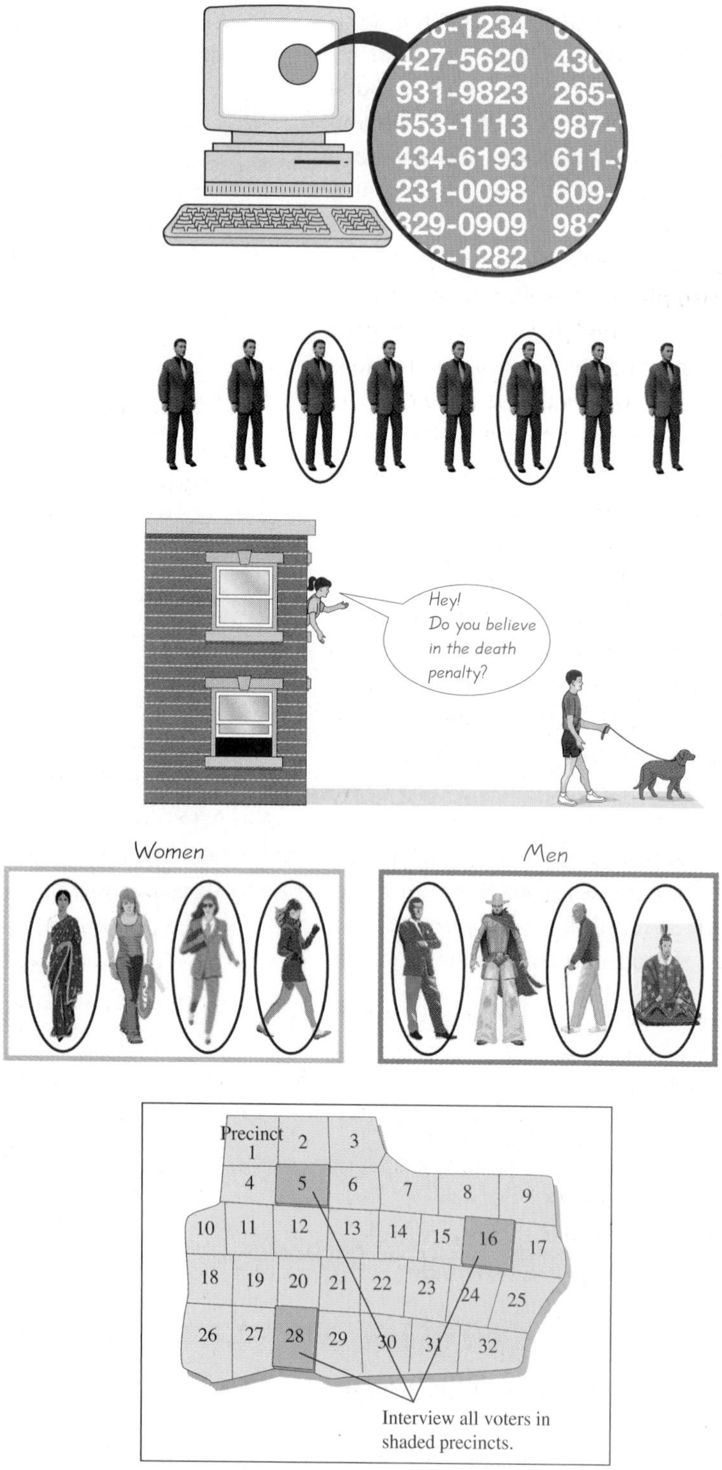

Random Sampling:
Each member of the population has an equal chance of being selected. Computers are often used to generate random telephone numbers.

Simple Random Sampling:
A sample of n subjects is selected in such a way that every possible sample of the same size n has the same chance of being chosen.

Systematic Sampling:
Select some starting point, then select every kth (such as every 50th) element in the population.

Convenience Sampling:
Use results that are easy to get.

Stratified Sampling:
Subdivide the population into at least two different subgroups (or strata) so that subjects within the same subgroup share the same characteristics (such as gender or age bracket), then draw a sample from each subgroup.

Cluster Sampling:
Divide the population into sections (or clusters), then randomly select some of those clusters, and then choose all members from those selected clusters.

Precinct 1, 2, 3, 4, 5, 6, 7, 8, 9, 10, 11, 12, 13, 14, 15, 16, 17, 18, 19, 20, 21, 22, 23, 24, 25, 26, 27, 28, 29, 30, 31, 32

Interview all voters in shaded precincts.

Figure 1-5 Common Sampling Methods

It is easy to confuse stratified sampling and cluster sampling, because they both involve the formation of subgroups. But cluster sampling uses *all* members from a *sample* of clusters, whereas stratified sampling uses a *sample* of members from *all* strata. An example of cluster sampling can be found in a pre-election poll, whereby we randomly select 30 election precincts from a large number of precincts, then survey all the people from each of those precincts. This is much faster and much less expensive than selecting one person from each of the many precincts in the population area. The results of stratified or cluster sampling can be adjusted or weighted to correct for any disproportionate representations of groups.

For a fixed sample size, if you randomly select subjects from different strata, you are likely to get more consistent (and less variable) results than by simply selecting a random sample from the general population. For that reason, stratified sampling is often used to reduce the variation in the results. Many of the methods discussed later in this book have a requirement that sample data be a *simple random sample*, and neither stratified sampling nor cluster sampling satisfies that requirement.

Multistage Sampling Figure 1-5 illustrates five common sampling methods, but professional pollsters and government researchers often collect data by using some combination of the five methods. A **multistage sample design** involves the selection of a sample in different stages that might use different methods of sampling. Let's consider the government's unemployment statistics that result from surveyed households. It is not practical to personally visit each member of a simple random sample, because individual households would be spread all over the country. It would be too expensive and too time consuming to survey such members from a simple random sample. Instead, the U.S. Census Bureau and the Bureau of Labor Statistics combine to conduct a survey called the Current Population Survey. This survey is used to obtain data describing such factors as unemployment rates, college enrollments, and weekly earnings amounts. The survey incorporates a multistage sample design, roughly described as follows:

1. The entire United States is partitioned into 2007 different regions called *primary sampling units* (PSU). The primary sampling units are metropolitan areas, large counties, or groups of smaller counties.
2. In each of the 50 states, a sample of primary sampling units is selected. For the Current Population Survey, 792 of the primary sampling units are used. (All of the 432 primary sampling units with the largest populations are used, and 360 primary sampling units are randomly selected from the other 1575.)
3. Each of the 792 selected primary sampling units is partitioned into blocks, and stratified sampling is used to select a sample of blocks.
4. In each selected block, clusters of households that are close to each other are identified. Clusters are randomly selected, and all households in the selected clusters are interviewed.

Note that the above multistage sample design includes random, stratified, and cluster sampling at different stages. The end result is a complicated sampling

Should you Believe a Statistical Study?

In *Statistical Reasoning for Everyday Life*, 2nd edition, authors Jeff Bennett, William Briggs, and Mario Triola list the following eight guidelines for critically evaluating a statistical study. (1) Identify the goal of the study, the population considered, and the type of study. (2) Consider the source, particularly with regard to a possibility of bias. (3) Analyze the sampling method. (4) Look for problems in defining or measuring the variables of interest. (5) Watch out for confounding variables that could invalidate conclusions. (6) Consider the setting and wording of any survey. (7) Check that graphs represent data fairly, and conclusions are justified. (8) Consider whether the conclusions achieve the goals of the study, whether they make sense, and whether they have practical significance.

design, but it is so much more practical and less expensive than using a simpler design, such as using a simple random sample.

Sampling Errors

No matter how well you plan and execute the sample collection process, there is likely to be some error in the results. For example, randomly select 1000 adults, ask them if they graduated from high school, and record the sample percentage of "yes" responses. If you randomly select another sample of 1000 adults, it is likely that you will obtain a *different* sample percentage.

Definitions

A **sampling error** is the difference between a sample result and the true population result; such an error results from chance sample fluctuations.

A **nonsampling error** occurs when the sample data are incorrectly collected, recorded, or analyzed (such as by selecting a biased sample, using a defective measure instrument, or copying the data incorrectly).

If we carefully collect a sample so that it is representative of the population, we can use methods in this book to analyze the sampling error, but we must exercise extreme care so that nonsampling error is minimized.

In the Chapter Problem, we asked whether original experimental data leading to the "six degrees of separation" concept were *good* data. We noted that Stanley Milgram sent 60 letters to subjects in Wichita, Kansas, and they were asked to forward the letters to a specific woman in Cambridge, Massachusetts. Fifty of the 60 subjects participated, but only three of the letters reached the target. Two subsequent experiments also had low completion rates, but Milgram eventually reached a 35% completion rate and found that the mean number of intermediaries was around 6. Milgram's experiments were seriously flawed in collecting sample data that were not representative of the population of residents in the United States. The original experiment involved people from Wichita. His response and completion rates were too small. Even with a 35% completion rate, it might well be possible that only those interested in the experiment participated, and they might be more likely to know more people and result in smaller numbers of intermediaries. His advertisements were designed to recruit people who were, as the ads stated, "proud of their social skills and confident of their powers to reach someone across class barriers." His recruits were from one specific geographic region, they were likely to be more sociable and wealthier than typical people, and the response rate was too low. Consequently, the result of 6 as the number of degrees of separation was not justified by the data. (That doesn't necessarily mean that the value of 6 is wrong.) The experimental design was seriously flawed and any conclusions based on the results are very suspect. Unfortunately, no well-designed experiment has yet been conducted to verify that 6 is correct or wrong. (In 2001, Duncan Watts from Columbia University began a large-scale e-mail experiment, but subjects using e-mail are not necessarily representative of the world population. However, much can be learned from his results.)

After reading through this section, it is easy to be somewhat overwhelmed by the variety of different definitions. But remember this main point: The method used to collect data is absolutely and critically important, and we should know that *randomness* is particularly important. If sample data are not collected in an appropriate way, the data may be so completely useless that no amount of statistical torturing can salvage them. Also, great thought and care must go into designing an experiment.

1-4 BASIC SKILLS AND CONCEPTS

Statistical Literacy and Critical Thinking

1. What is the difference between a random sample and a simple random sample?

2. What is the difference between an observational study and an experiment?

3. When conducting an experiment to test the effectiveness of a new vaccine, what is blinding, and why is it important?

4. When conducting an experiment to test the effectiveness of a new vaccine, a researcher chooses to use blocking, with men in one block and women in another block. How does this use of blocking help the experiment?

In Exercises 5–8, determine whether the given description corresponds to an observational study *or an* experiment.

5. Touch Therapy Nine-year-old Emily Rosa became an author of an article in the *Journal of the American Medical Association* after she tested professional touch therapists. Using a cardboard partition, she held her hand above the therapist's hand, and the therapist was asked to identify the hand that Emily chose.

6. Treating Syphilis Much controversy arose over a study of patients with syphilis who were *not* given a treatment that could have cured them. Their health was followed for years after they were found to have syphilis.

7. Quality Control The U.S. Food and Drug Administration randomly selects a sample of Bayer aspirin tablets. The amount of aspirin in each tablet is measured for accuracy.

8. Magnetic Bracelets Cruise ship passengers are given magnetic bracelets, which they agree to wear in an attempt to eliminate or diminish the effects of motion sickness.

In Exercises 9–12, identify the type of observational study (cross-sectional, retrospective, prospective).

9. Psychology of Trauma A researcher from Mt. Sinai Hospital in New York City plans to obtain data by following (to the year 2015) siblings of victims who perished in the World Trade Center terrorist attack of September 11, 2001.

10. Drunk Driving Research A researcher from Johns Hopkins University obtains data about the effects of alcohol on driving by examining car crash reports from the past five years.

11. TV Ratings The Nielsen Media Research Company surveys 5000 households to determine the proportion of households tuned to *Saturday Night Live*.

12. Statistics for Success An economist collects income data by selecting and interviewing subjects now, then going back in time to see if they had the wisdom to take a statistics course between the years of 1980 and 2005.

In Exercises 13–24, identify which of these types of sampling is used: random, systematic, convenience, stratified, *or* cluster.

13. Sobriety Checkpoint The author was an observer at a police sobriety checkpoint at which every 5th driver was stopped and interviewed. (He witnessed the arrest of a former student.)

14. Exit Polls On days of presidential elections, the news media organize an exit poll in which specific polling stations are randomly selected and all voters are surveyed as they leave the premises.

15. Education and Sports A researcher for the Spaulding athletic equipment company is studying the relationship between the level of education and participation in any sport. She conducts a survey of 40 randomly selected golfers, 40 randomly selected tennis players, and 40 randomly selected swimmers.

16. Ergonomics An engineering student measures the strength of fingers used to push buttons by testing family members.

17. Cheating An Internal Revenue Service researcher investigates cheating on income tax reports by surveying all waiters and waitresses at 20 randomly selected restaurants.

18. MTV Survey A marketing expert for MTV is planning a survey in which 500 people will be randomly selected from each age group of 10–19, 20–29, and so on.

19. Credit Card Data The author surveyed all of his students to obtain sample data consisting of the number of credit cards students possess.

20. Fund-raising Fund-raisers for the College of Newport test a new telemarketing campaign by obtaining a list of all alumni and selecting every 100th name on that list.

21. Telephone Polls In a Gallup poll of 1059 adults, the interview subjects were selected by using a computer to randomly generate telephone numbers that were then called.

22. Market Research A market researcher has partitioned all California residents into categories of unemployed, employed full time, and employed part time. She is surveying 50 people from each category.

23. Student Drinking Motivated by a student who died from binge drinking, the College of Newport conducts a study of student drinking by randomly selecting 10 different classes and interviewing all of the students in each of those classes.

24. Clinical Trial of Blood Treatment In phase II testing of a new drug designed to increase the red blood cell count, a researcher finds envelopes with the names and addresses of all treated subjects. She wants to increase the dosage in a subsample of 12 subjects, so she thoroughly mixes all of the envelopes in a bin, then pulls 12 of those envelopes to identify the subjects to be given the increased dosage.

Random Samples and Simple Random Samples. *Exercises 25–30 relate to random samples and simple random samples.*

25. Cluster Sample An IRS analyst processes one tax return every 10 minutes, so 240 returns are processed in her first week of work. Her manager checks her work by randomly selecting a day of the week, then reviewing all returns that she processed that day. Does this sampling plan result in a random sample? Simple random sample? Explain.

26. Convenience Sample A statistics professor obtains a sample of students by selecting the first 10 students entering her classroom. Does this sampling plan result in a random sample? Simple random sample? Explain.

27. Systematic Sample A quality control engineer selects every 10,000th M&M plain candy that is produced. Does this sampling plan result in a random sample? Simple random sample? Explain.

28. Stratified Sample A researcher for the Orange County Department of Motor Vehicles plans to test a new online driver registration system by using a sample consisting of 20 randomly selected men and 20 randomly selected women. (Orange County has an equal number of male and female drivers.) Does this sampling plan result in a random sample? Simple random sample? Explain.

29. Sampling Students A classroom consists of 36 students seated in six different rows, with six students in each row. The instructor rolls a die to determine a row, then rolls the die again to select a particular student in the row. This process is repeated until a sample of 6 students is obtained. Does this sampling plan result in a random sample? Simple random sample? Explain.

30. Sampling Vitamin Pills An inspector for the U.S. Food and Drug Administration obtains all vitamin pills produced in an hour at the Health Supply Company. She thoroughly mixes them, then scoops a sample of 10 pills that are to be tested for the exact amount of vitamin content. Does this sampling plan result in a random sample? Simple random sample? Explain.

1-4 BEYOND THE BASICS

31. Sampling Design The Addison-Wesley Publishing Company has commissioned you to survey 100 students who use this book. Describe procedures for obtaining a sample of each type: random, systematic, convenience, stratified, cluster.

32. Confounding Give an example (different from the one in the text) illustrating how confounding occurs.

33. Sample Design In "Cardiovascular Effects of Intravenous Triiodothyronine in Patients Undergoing Coronary Artery Bypass Graft Surgery" (*Journal of the American Medical Association,* Vol. 275, No. 9), the authors explain that patients were assigned to one of three groups: (1) a group treated with triiodothyronine, (2) a group treated with normal saline bolus and dopamine, and (3) a placebo group given normal saline. The authors summarize the sample design as a "prospective, randomized, double-blind, placebo-controlled trial." Describe the meaning of each of those terms in the context of this study.

1-5 Introduction to the TI-83/84 Plus Calculator

This book is designed for TI-83 Plus, TI-83 Plus Silver Edition, TI-84 Plus, and TI/84 Plus Silver Edition calculators. Instead of making repeated reference to each of these calculators throughout this book, we will use "TI-83/84 Plus" as the general reference to the TI-83 Plus, TI-83 Plus Silver Edition, TI-84 Plus, and TI-84 Plus Silver Edition calculators.

Throughout this book, "TI-83/84 Plus" refers to any of the following calculators:

- TI-83 Plus
- TI-83 Plus Silver Edition
- TI-84 Plus
- TI-84 Plus Silver Edition

Instructions, screen displays, programs, and the APP (application) for the "TI-83/84 Plus" calculator also apply to the TI-83 Plus Silver Edition and the TI-84 Plus Silver Edition calculators.

TI-83 Plus versus TI-84 Plus Instructions for TI-83 Plus calculators and TI-84 Plus calculators are essentially identical, but TI-84 Plus calculators have a few functions that are not available on the TI-83 Plus calculators. Users of TI-84 Plus calculators should check to be sure that they have the latest available operating system. (As of this writing, the latest is Version 2.30.)

TI-83 Calculator The older TI-83 calculator (not the TI-83 Plus version) lacks the ability to use APPS (applications), so users of TI-83 calculators will not be able to use the APP included on the CD-ROM accompanying new copies of this textbook. That APP includes the data sets in Appendix B of this textbook, so TI-83 users will sometimes need to manually enter data instead of using downloaded data. However, TI-83 users can download the programs included on the CD-ROM.

TI-89 Calculator The TI-89 calculator has a few statistical functions, but statistics students should download the extensive and free "Statistics with List Editor" application from the Texas Instruments Web site at www.ti.com. (The TI-89 Titanium edition has the statistics application already installed.) There are some fundamental differences between instructions for the TI-89 and the TI-83/84 Plus calculators, but this textbook focuses on procedures for the TI-83/84 Plus calculators. For some more detailed instructions for the TI-89, see this printed supplement for the Triola Statistics Series: *Graphing Calculator Manual for the TI-83 Plus, TI-84 Plus, and TI-89* by Patricia Humphrey.

The "TI-83/84" Plus calculator is an excellent tool that is extremely helpful in an introductory statistics course. It is easy to carry to class, and it has an abundance of statistical tools that can be used throughout the entire course. Using the TI-83/84 Plus calculator makes the statistics course easier by allowing students to focus on the selection of procedures and the interpretation of results, instead of being crushed by arithmetic calculations.

Setting the Calculator Mode

Before starting, it would be helpful to verify that your TI-83/84 Plus calculator has the mode settings suitable for the introductory statistics course. Press the MODE key and compare your display to the one shown in the margin. If your mode settings are not as shown here, change them by using the ⌄ and ⌃ keys to scroll to the desired row, use the 〉 and 〈 keys to highlight the desired row entry, and then press ENTER. When finished, quit by pressing 2ND MODE (which activates the QUIT option, which is depicted above the [MODE] key). This mode setting can be selected once, and it will remain in effect unless it is changed later.

Screen Darkness and Batteries

If the displays on the screen appear to be too light or too dark, adjust the contrast as follows:

Increase darkness of display: Press and release the [2ND] key and then press and hold the [▲] key.

Decrease darkness of display: Press and release the [2ND] key and then press and hold the [▼] key.

The screen display will gradually fade as the batteries become tired, and the darkness of the display will need to be increased. Eventually, a low battery–warning message will be displayed, so replace batteries when necessary. A low-battery message usually indicates that the four AAA batteries need replacement. The lithium battery should last a few years, and it should not be replaced at the same time that the AAA batteries are being replaced.

Special Keys: [2ND], [ALPHA], and [CLEAR]

The typical key on the TI-83/84 Plus can be used for two or three different purposes. For example, the [4] key has 4 on its face, but it also has L4 above it in yellow or blue, as well as the letter T in green. Pressing [4] causes the calculator to use the number 4, pressing [2ND] [4] causes it to refer to the list label L4, and pressing [ALPHA] [4] causes the letter T to be used. The use of the key is easy to follow with the color coding of the yellow or blue [2ND] key and the green [ALPHA] key. The color of the [2ND] key corresponds to the entry above the key being used, and the green [ALPHA] key refers to the green alphabetic character (or other green special symbol) placed above the key being used.

Press [CLEAR] to clear (what else?) the screen display.

Basic Calculations

Many inexpensive calculators use a [=] key to indicate the end of calculation instructions, but the TI-83/84 Plus uses [ENTER] instead. For example, the addition of 2 + 7 is accomplished with this key sequence: [2] [+] [7] [ENTER]. The result should be 9.

Like any good scientific calculator, the TI-83/84 Plus is designed to execute arithmetic operations with a hierarchy that causes multiplications and divisions to be computed before additions and subtractions. This feature can be overcome through the use of parentheses. See the examples below:

2 [+] 3 [×] 4 [ENTER]: *Result:* 14 (because the multiplication is done first)

[(] 2 [+] 3 [)] [×] 4 [ENTER]: *Result:* 20 (the parentheses cause the addition to be done first)

When unsure about the calculator's order of operations, use parentheses to ensure that the calculation is executed exactly as you desire.

Difference Between (−) and ((−))

It is very important to understand the difference between the (−) and ((−)) keys. They are described below:

(−) key: Arithmetic operation of *subtraction,* as in 8 − 5

((−)) key: *Negative sign,* as in −6

For example, the operation of 8 − 5 is done by pressing 8 (−) 5 ENTER, and the result should be 3. If you incorrectly enter 8 ((−)) 5 ENTER, you will get an error message, because the negative sign makes no sense in this context. Be careful to use (−) for the operation of subtraction so that the (−) key is placed between two numbers, whereas the ((−)) key is a negative sign that applies to the number that follows. It is easy to remember that (−) represents the operation of subtraction, because it is located with the other operations of (÷)(×), and (+).

Scientific Notation

Some results in later chapters are displayed by the TI-83/84 Plus calculator as values expressed in *scientific notation.* For example, 3.518437209E13 is a value displayed in scientific notation. This value can be expressed as $3.518437209 \times 10^{13}$ or as 35,184,372,090,000. This might seem complicated, but note that we can express 3.518437209E13 in ordinary notation simply by moving the decimal place to the right 13 places. See the following two examples:

With a *positive* number following E, move the decimal place to the *right,* as in 2.345E15, which can be expressed as 2,345,000,000,000,000.

With a *negative* number following E, move the decimal place to the *left,* as in 2.345E⁻15, which can be expressed as 0.000000000000002345.

Clearing a List

Instead of using single numerical values, the typical statistics course uses collections of data. The TI-83/84 Plus calculator is designed so that we can use *lists* of data. The TI-83/84 Plus calculator already has lists named L1, L2, L3, L4, L5, and L6. Also, you can assign a list of data to a name of your choice. If entering data in L1, or L2, or L3, etc., it is a good practice to first *clear* the list of any data that might be there from some prior procedure. Clear a list using the following sequence of steps:

1. Press STAT.
2. Use (▼) to scroll down to the fourth menu item identified as 4: ClrList.
3. Press ENTER.
4. The screen should now display ClrList, so press 2ND (1) to get the display of ClrList L1, which causes list L1 to be cleared. (Note that 2ND (1) activates the yellow or blue item *above* the (1) key, which is L1 for List 1.)
5. Press ENTER.

Entering a List of Data

Let's assume that we want to work with the following lengths (in seconds) of scenes showing tobacco use in animated movies from Universal Studios.

0 223 0 176 0 548

To enter such a list of values, first clear the list as described above. The following procedure describes the entry of a list labeled as L1, but we could also use L2, L3, L4, L5, or L6, as well as using a name that we can enter. The following procedure applies to the list labeled as L1:

1. Press STAT.
2. We want the first menu item of Edit, and that is the default, so simply press ENTER to select that option.
3. If the cursor is not positioned in column L1, use ◀ and ▶ to get to the column list L1.
4. Enter a data value and then press ENTER. Enter the second data value and press ENTER again. Continue to enter individual data values followed by ENTER. (If you make a mistake in a data value, simply use ▲ and ▼ to highlight the wrong entry and then enter the correct value and press ENTER.) The screen should appear as shown in the accompanying screen display (although lists L2 and L3 might have other data displayed).
5. Press 2ND ENTER to QUIT the list editor.

L1	L2	L3 1
0	------	------
223		
0		
176		
0		
548		

L1(7)=

Even though Step 5 caused us to quit, the list will remain as entered until it is cleared or changed. List L1 will remain in the calculator's memory even after the calculator is turned off.

Naming a List

Sometimes we want to manually enter a long list of data that will be used in future applications. However, the lists L1, L2, L3, L4, L5, and L6 may be used often, and their labels do not suggest the nature of the data. There is a procedure for entering a list and saving it with a name other than L1, L2, L3, L4, L5, or L6. Here is one procedure for saving a list with a desired name:

1. Enter the data enclosed within braces and separated by commas, as in {0, 223, 0, 176, 0, 548}.
2. Press STO▸.
3. Use ALPHA to enter the alphabetic characters that constitute the desired name. The name TIME can be entered with ALPHA T ALPHA I ALPHA M ALPHA E. (*Shortcut hint:* Instead of pressing ALPHA before each letter, you could press 2ND ALPHA to lock the alpha key so that it remains in effect until you unlock it by entering ALPHA a second time.)
4. Press ENTER.

The preceding three steps result in this sequence of keystrokes:

{0, 223, 0, 176, 0, 548} STO› ALPHA T ALPHA I ALPHA M ALPHA E ENTER

and the effect of this sequence of keystrokes is to have the list of values assigned to the name of TIME.

Another way to name a list of data is to first enter the data in list L1 and then enter the expression L1→LTIME, where the symbol L indicates that the name represents a *list* of values (instead of a single value). The symbol L can be obtained by pressing 2ND STAT and selecting OPS and then the last item of L. If you now enter L TIME, you will see the list of values.

APPS

One of the great features of the TI-83/84 Plus is its ability to handle APPs, which are software applications. (APPs cannot be used with the older TI-83 calculators.) One really helpful APP is the collection of data sets found in Appendix B. That data set is an APP on the CD-ROM that comes with this book, and it is also available on the Triola Web site at www.aw.com/triola. It is necessary to install the TriolaXe APP just once. That APP will remain in the calculator until it is removed.

To download the APP from another TI-83/84 Plus calculator to your calculator, you need the cable that came with your calculator. Follow these steps:

1. Use the cable to connect the two calculators.
2. On the receiving calculator, enter 2ND X,T,Θ,n, which calls upon the LINK function. Press ▷ to highlight RECEIVE and then press ENTER to get a message of Waiting.
3. On the sending calculator, enter 2ND X,T,Θ,n, which calls upon the LINK function. Use the ▽ key to select the menu item of C:Apps and press ENTER. Now use ▽ to select the item of TriolaXe and press ENTER. Finally, press ▷ to select TRANSMIT and press ENTER.

To download the APP from a computer to your calculator, first obtain download information from the Texas Instruments Web Site at www.ti.com. The procedure, software, and connection devices will depend on whether you have a TI-83 Plus, TI-83 Plus Silver Edition, TI-84 Plus, or TI-84 Plus Silver Edition calculator. Whereas the TI-83 Plus calculator might require a special TI-GRAPHLINK™ cable, the newer TI-84 Plus and TI-84 Plus Silver Edition have a USB port that simplifies connection to a computer by using a standard cable.

In addition to the TriolaXe application, it is also recommended that you download the CtlgHelp (catalog help) application. This application comes with the TI-83 Plus Silver Edition and the TI-84 Plus Silver Edition. It can be downloaded to the TI-83 Plus calculator and the TI-84 Plus calculator. One nice feature of this catalog help application is that it can be used to identify the format required for many important functions to be described later in this book.

Memory Error It is possible that in the attempt to download an APP, you may get the error message of ERR:MEMORY. If this happens, it is necessary to clear some memory by erasing data stored in lists or by removing other APPs. Because the newer calculators have more memory, they are less likely to create memory errors. (The TI-84 Plus calculator has three times as much memory as the

TI-83 Plus calculator. The TI-83 Plus Silver Edition and the TI-84 Plus Silver Edition have about nine times as much memory as the TI-83 Plus calculator.)

Important: This section contains much information that might seem to be overwhelming. However, it is the author's experience that students quickly and easily learn to make effective use of their TI-83/84 Plus calculators. They find that their statistics course becomes much easier by allowing them to focus on the selection of procedures and the interpretation of results, without being tortured with mind-numbing tedious calculations.

1-5 BASIC SKILLS AND CONCEPTS

Statistical Literacy and Critical Thinking

TI **1.** APP In the terminology of Texas Instruments calculators, what is an APP?

TI **2.** Calculator Result A student estimates that during the first week in her statistics class, the number of absences are around 3, 2, and 2. She then uses a TI-83/84 Plus calculator to find the average (mean) of those values and she gets a result of 2.333333333. Why is it misleading to report that the average number of absences is 2.333333333?

TI **3.** Negative and Subtraction On a TI-83/84 calculator, what is the difference between the keys [−] and [(−)]?

TI **4.** ALPHA Key What does "ALPHA" on the [ALPHA] key represent?

In Exercises 5–12, use a TI-83/84 Plus calculator to find the results.

TI **5.** $4 \times (15 - 20)$

TI **6.** $3 + 4 \times 5$

TI **7.** $5 - (-2)$

TI **8.** $6 - (-3)$

TI **9.** $9 \div (-2)$

TI **10.** $12 \div (-5 - 8)$

TI **11.** $\sqrt{234}$

TI **12.** $\sqrt{37.4}$

In Exercises 13–16, find the TI-83/84 Plus calculator result and explain how the result occurred. For example, the result of 4 + 5 × 6 would be 34, and the calculator used a hierarchy in which the multiplication was executed first so that the result is 4 + 30, or 34.

TI **13.** [2ND] [x^2] 9 [ENTER]

TI **14.** 25 [x^2] [ENTER]

TI **15.** 5 [^] 2 [ENTER]

TI **16.** [2ND] [^] [ENTER]

In Exercises 17–20, enter the values of 0, 223, 0, 176, 0, and 548 in list L1, then press [2ND] [MODE] to quit the LIST editor, and then describe the results of the given expressions. Also, describe how the result relates to the original list.

TI **17.** [2ND] 1 [ENTER]

TI **18.** 5 [+] [2ND] 1 [ENTER]

TI **19.** [2ND] 1 [+] [2ND] 1 [ENTER]

TI **20.** [2ND] 1 [x^2] [ENTER]

Scientific Notation *In Exercises 21–24, use a TI-83/84 Plus calculator to find the result and then express the result in ordinary notation (not scientific notation using* E*).*

TI **21.** 2^{45} (which can be entered as 2 [^] 45 [ENTER])

TI **22.** 0.2^{12} (which can be entered as 0.2 [^] 12 [ENTER])

TI **23.** 0.5^{10}

TI **24.** 5^{16}

1-5 BEYOND THE BASICS

TI **25.** Using an APP Using the CD-ROM included with this book, using the Triola Web site, or using another TI-83/84 Plus calculator, download the TriolaXe application (APP).

a. What does the screen show when the APPS key is pressed?

b. Press the APPS key, use ⌄ to scroll to the entry of TriolaXe, and then press ENTER. Describe the resulting screen display.

c. After doing parts (a) and (b), use ⌄ to scroll to the entry of CIGARET, and press ENTER. Describe the screen display.

d. After completing parts (a)–(c), use 〉 to highlight All and press ENTER. Now use 〉 to highlight Load, use ⌄ to select 2:Load, and press ENTER. Quit and then select List by pressing 2ND STAT. Describe the result.

e. After completing parts (a)–(d), clear the screen and select List by pressing 2ND STAT, use ⌄ to highlight: TAR, and press ENTER twice. Describe the result. (*Hint:* See Data Set 3 in Appendix B.)

TI **26.** Creating a Named List Assign the list of values below to the name ABC and describe the result of entering the following:

a. ABC

b. **L** ABC (where the symbol **L** can be obtained by pressing 2ND STAT, selecting OPS, and selecting the last item of **L**).

c. **L** ABC followed by pressing the x^2.

0 −2 5 8 1

Review

This chapter includes some definitions and concepts that are very basic in the subject of statistics. There were fundamental definitions, such as *sample, population, statistic,* and *parameter*. Section 1-2 discussed different types of data, and the distinction between qualitative data and quantitative data should be well understood. Section 1-3 dealt with the use of critical thinking in analyzing and evaluating statistical results. In particular, we should know that for statistical purposes, some samples (such as voluntary response samples) are very poor. Section 1-4 introduced important elements in the design of experiments. We should understand the definition of a simple random sample. We should also recognize the importance of a carefully planned experimental design. Some sound principles of the design of experiments were briefly discussed, but entire books and courses are devoted to this topic. On completing this chapter, you should be able to do the following:

- Distinguish between a population and a sample and distinguish between a parameter and a statistic
- Understand the importance of good experimental design, including the control of variable effects, replication, and randomization
- Recognize the importance of good sampling methods in general, and recognize the importance of a *simple random sample* in particular. Understand that if sample data are not collected in an appropriate way, the data may be so completely useless that no amount of statistical torturing can salvage them.

Statistical Literacy and Critical Thinking

1. Sample Size Is a large sample necessarily a good sample? Why or why not?

2. Nature of Data A physiologist randomly selects 16 runners who finish the New York Marathon, then she measures the height of each selected person.
   a. Are the data qualitative or quantitative?
   b. Are the data discrete or continuous?
   c. If the researcher uses the sample data to infer something about a population, what is that population?

3. Sampling A graduate student is conducting research in psychology and he needs to obtain the IQ scores of 50 people. He places an ad in the local newspaper and asks for volunteers, each of whom will be paid $50 to take an IQ test. Is his sample a good sample? Why or why not?

4. Sampling A market researcher wants to determine the average value of a car owned by a resident of the United States. She randomly selects and surveys 10 car owners from each state, and obtains a list of 500 sample values. She then adds those 500 values and divides by 500 to get an average. Is the result a good estimate of the average value of a car owned by a resident of the United States? Why or why not?

Review Exercises

1. Sampling Shortly after the World Trade Center towers were destroyed by terrorists, America Online ran a poll of its Internet subscribers and asked this question: "Should the World Trade Center towers be rebuilt?" Among the 1,304,240 responses, 768,731 answered "yes," 286,756 answered "no," and 248,753 said that it was "too soon to decide." Given that this sample is extremely large, can the responses be considered to be representative of the population of the United States? Explain.

2. Sampling Design You have been hired by Verizon to conduct a survey of cell phone usage among the full-time students who attend your college. Describe a procedure for obtaining a sample of each type: random, systematic, convenience, stratified, cluster.

3. Identify the level of measurement (nominal, ordinal, interval, ratio) used in each of the following:
   a. The weights of people being hurled through the air at a lively rock concert
   b. A movie critic's ratings of "must see, recommended, not recommended, don't even think about going"
   c. A movie critic's classification of "drama, comedy, adventure"
   d. Bob, who is different in many ways, measures time in days, with 0 corresponding to his birth date. The day before his birth is -1, the day after his birth is $+1$, and so on. Bob has converted the dates of major historical events to his numbering system. What is the level of measurement of these numbers?

4. Coke The Coca Cola Company has 366,000 stockholders and a poll is conducted by randomly selecting 30 stockholders from each of the 50 states. The number of shares held by each sampled stockholder is recorded.
   a. Are the values obtained discrete or continuous?
   b. Identify the level of measurement (nominal, ordinal, interval, ratio) for the sample data.

c. Which type of sampling (random, systematic, convenience, stratified, cluster) is being used?

d. If the average (mean) number of shares is calculated, is the result a statistic or a parameter?

e. If you are the Chief Executive Officer of the Coca Cola Company, what characteristic of the data set would you consider to be extremely important?

f. What is wrong with gauging stockholder views by mailing a questionnaire that stockholders could complete and mail back?

5. More Coke Identify the type of sampling (random, systematic, convenience, stratified, cluster) used when a sample of the 366,000 Coca Cola shareholders is obtained as described. Then determine whether the sampling scheme is likely to result in a sample that is representative of the population of all 366,000 shareholders.

a. A complete list of all stockholders is compiled and every 500th name is selected.

b. At the annual stockholders' meeting, a survey is conducted of all who attend.

c. Fifty different stockbrokers are randomly selected, and a survey is made of all their clients who own shares of Coca Cola.

d. A computer file of all stockholders is compiled so that they are all numbered consecutively, then random numbers generated by computer are used to select the sample of stockholders.

e. All of the stockholder zip codes are collected, and 5 stockholders are randomly selected from each zip code.

6. Design of Experiment You plan to conduct an experiment to test the effectiveness of Sleepeze, a new drug that is supposed to reduce insomnia. You will use a sample of subjects who are treated with the drug and another sample of subjects who are given a placebo.

a. What is "blinding" and how might it be used in this experiment?

b. Why is it important to use blinding in this experiment?

c. What is a completely randomized design?

d. What is a rigorously controlled design?

e. What is replication, and why is it important?

7. Honest Abe

a. When Abraham Lincoln was first elected to the presidency, he received 39.82% of the 1,865,908 votes cast. The collection of all of those votes is the population being considered. Is the 39.82% a parameter or a statistic?

b. Part (a) gives the total votes cast in the 1860 presidential election. Consider the total numbers of votes cast in all presidential elections. Are those values discrete or continuous?

c. How many votes did Lincoln get when he was first elected to the presidency?

8. Percentages

a. The labels on U-Turn protein energy bars include the statement that these bars contain "125% less fat than the leading chocolate candy brands" (based on data from *Consumer Reports* magazine). What is wrong with that claim?

b. In a study of the presence of radon in homes, 237 homes in LaGrange, New York, were measured, and 19% of them were found to have radon levels above 0.4 pCi/L (picocuries per liter), which is the maximum safe level set by the Environmental Protection Agency. What is the actual number of tested homes that have unsafe levels of radon?

c. When 186 homes in Hyde Park, New York, were tested for radon, 30 were found to have levels above the maximum safe level described in part (b). What is the percentage of the Hyde Park homes above the maximum safe level of radon?

Cumulative Review Exercises

The Cumulative Review Exercises in this book are designed to include topics from preceding chapters. For Chapters 2–15, the Cumulative Review Exercises include topics from preceding chapters. For this chapter, we present calculator warm-up exercises with expressions similar to those found throughout this book. Use your calculator to find the indicated values.

1. Refer to Data Set 14 in Appendix B and consider only the weights of the Indian head pennies. What value is obtained when those weights are added, and the total is then divided by the number of those pennies? (This result, called the *mean*, is discussed in Chapter 3.)

2. $\dfrac{98.20 - 98.60}{0.62}$

3. $\dfrac{98.20 - 98.60}{\dfrac{0.62}{\sqrt{106}}}$

4. $\left[\dfrac{1.96 \cdot 0.25}{0.03}\right]^2$

5. $\dfrac{(50 - 45)^2}{45}$

6. $\dfrac{(2 - 4)^2 + (3 - 4)^2 + (7 - 4)^2}{3 - 1}$

7. $\sqrt{\dfrac{(2 - 4)^2 + (3 - 4)^2 + (7 - 4)^2}{3 - 1}}$

8. $\dfrac{8(151{,}879) - (516.5)(2176)}{\sqrt{8(34{,}525.75) - 516.5^2}\sqrt{8(728{,}520) - 2176^2}}$

In Exercises 9–12, the given expressions are designed to yield results expressed in a form of scientific notation. For example, the calculator displayed result of 1.23E5 can be expressed as 123,000, and the result of 4.56E-4 can be expressed as 0.000456. Perform the indicated operation and express the result as an ordinary number that is not in scientific notation.

9. 0.5^{10} **10.** 2^{40} **11.** 7^{12} **12.** 0.8^{50}

Cooperative Group Activities

1. Out-of-class activity Look through newspapers and magazines to find an example of a graph that is misleading. (See, for example, Figures 1-1 and 1-2.) Describe how the graph is misleading. Redraw the graph so that it depicts the information correctly.

2. In-class activity From the cafeteria, obtain 18 straws. Cut 6 of them in half, cut 6 of them into quarters, and leave the other 6 as they are. There should now be 42 straws of different lengths. Put them in a bag, mix them up, then select one straw, find its length, then replace it. Repeat this until 20 straws have been selected. *Important:* Select the straws without looking into the bag and select the first straw that is touched. Find the average (mean) of the lengths of the sample of 20 straws. Now remove all of the straws and find the mean of the lengths of the population. Did the sample provide an average that was close to the true population average? Why or why not?

3. In-class activity In mid-December of a recent year, the Internet service provider America Online (AOL) ran a survey of its users. This question was asked about Christmas trees: "Which do you prefer?" The response could be "a real tree" or "a fake tree." Among the 7073 responses received by the Internet users, 4650 indicated a real tree, and 2423 indicated a fake tree. We have already noted that because the sample is a voluntary response sample, no conclusions can be made about a population larger than the 7073 people who responded. Identify other problems with this survey question.

4. In-class activity Identify the problems with the following:

- A recent televised report on *CNN Headline News* included a comment that crime in the United States fell in the 1980s because of the growth of abortions in the 1970s, which resulted in fewer unwanted children.
- *Consumer Reports* magazine mailed an Annual Questionnaire about cars and other consumer products. Also included were a request for a voluntary contribution of money and a ballot for the Board of Directors. Responses were to be mailed back in envelopes that required postage stamps.

TI-83/84 Plus Calculator Project

Refer to Data Set 8 in Appendix B for the forecast and actual temperatures. Using a TI-83/84 Plus calculator, download the TriolaXe software application (APP).

a. Open the APP, then open the file named WEATHER, and proceed to load the lists of ACTHI and ACTLO. ACTHI is the list of actual high temperatures and ACTLO is the list of actual low temperatures. (The list names are identified in Data Set 8 in Appendix B.)

b. Use the STO› key to copy the two lists to L1 and L2. Then press STAT ENTER to display the screen showing lists L1 and L2.

c. If a connection to a computer is available, obtain a printout of the screen display from part (c).

From Data to Decision

Critical Thinking

Women currently earn 76 cents for every dollar earned by a man. Construct a graph depicting that information in a way that is fair and unbiased. Construct a second graph that exaggerates the difference in earnings between men and women.

Analyzing the Results

Assuming that you must defend the discrepancy between earnings of men and women, identify at least one factor that could be used to justify that discrepancy. Research and analysis has shown that some of this discrepancy can be explained by legitimate factors, but much of the discrepancy cannot be explained by legitimate factors. Assuming that there is a substantial factor of discrimination based on gender, is it okay to try to fight that discrimination by publishing a graph that exaggerates the difference in earnings between men and women?

Internet Project

Web Site for *Elementary Statistics*

In this section of each chapter, you will be instructed to visit the home page on the Web for this textbook. From there you can reach the pages for all the Internet Projects accompanying *Elementary Statistics, Tenth Edition.* Go to this Web site now and familiarize yourself with all of the available features for the book.

Each Internet Project includes activities, such as exploring data sets, performing simulations, and researching true-to-life examples found at various Web sites. These activities will help you explore and understand the rich nature of statistics and its importance in our world. Visit the book site now and enjoy the explorations!

http://www.aw.com/triola

Statistics @ Work

"We use statistics to determine the degree of isolation between putative stocks."

Sarah Mesnick

Behavioral and Molecular Ecologist

Sarah Mesnick is a National Research Council postdoctoral fellow. In her work as a marine mammal biologist, she conducts research at sea as well as in the Laboratory of Molecular Ecology. Her research focuses on the social organization and population structure of sperm whales. She received her doctorate in evolutionary biology at the University of Arizona.

What do you do?

My research focuses on the relationship between sociality and population structure in sperm whales. We use this information to build better management models for the conservation of this, and other, endangered marine mammal species.

What concepts of statistics do you use?

Currently, I use chi-square and *F*-statistics to examine population structure and regression measures to estimate the degree of relatedness among individuals within whale pods. We use the chi-square and *F*-statistics to determine how many discrete populations of whales are in the Pacific. Discrete populations are managed as independent stocks. The regression analysis of relatedness is used to determine kinship within groups.

Could you cite a specific example illustrating the use of statistics?

I'm currently working with tissue samples obtained from three mass strandings of sperm whales. We use genetic markers to determine the degree of relatedness among individuals within the strandings. This is a striking behavior—entire pods swam up onto the beach following a young female calf, stranded, and subsequently all died. We thought that to do something as dramatic as this, the individuals involved must be very closely related. We're finding, however, that they are not. The statistics enable us to determine the probability that two individuals are related given the number of alleles that they share. Also, sperm whales—and many other marine mammal, bird, and turtle species—are injured or killed incidentally in fishing operations. We need to know the size of the population from which these animals are taken. If the population is small, and the incidental kill large, the marine mammal population may be threatened. We use statistics to determine the degree of isolation between putative stocks. If stocks are found to be isolated, we would use this information to prepare management plans specifically designed to conserve the marine mammals of the region. Human activities may need to protect the health of the marine environment and its inhabitants.

How do you approach your research?

We try not to have preconceived notions about how the animals are dispersed in their environment. In marine mammals in particular, because they are so difficult to study, there are generally accepted notions about what the animals are doing, yet these have not been critically investigated. In the case of relatedness among individuals within sperm whale groups, they were once thought to be matrilineal and accompanied by a "harem master." With the advent of genetic techniques, and dedicated field work, more open minds and more critical analyses—the statistics come in here—we're able to reassess these notions.

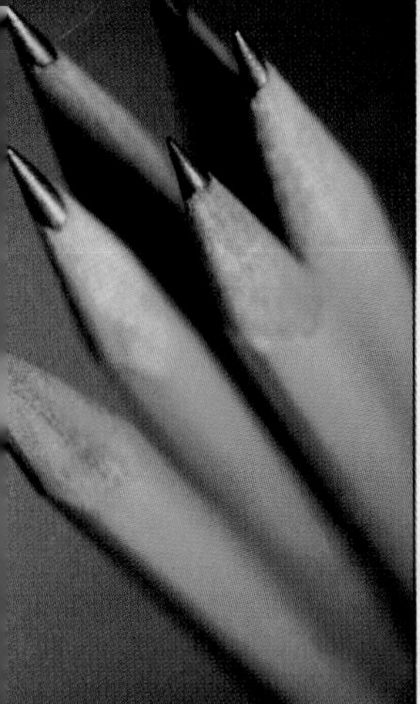

Summarizing and Graphing Data

2

Do the Academy Awards involve discrimination based on age?

Each year, Oscars are awarded to the Best Actress and Best Actor. Table 2-1 lists the ages of those award recipients at the time of the awards ceremony. The ages are listed in order, beginning with the first Academy Awards ceremony in 1928. [*Notes:* In 1968 there was a tie in the Best Actress category, and the average (mean) of the two ages is used; in 1932 there was a tie in the Best Actor category, and the average (mean) of the two ages is used. These data are suggested by the article "Ages of Oscar-winning Best Actors and Actresses," by Richard Brown and Gretchen Davis, *Mathematics Teacher* magazine. In that article, the year of birth of the award winner was subtracted from the year of the awards ceremony, but the ages in Table 2-1 are based on the birth date of the winner and the date of the awards ceremony.]

Here is the key question that we will consider: Are there major and important differences between the ages of the Best Actresses and the ages of the Best Actors? Does it appear that actresses and actors are judged strictly on the basis of their artistic abilities? Or does there appear to be discrimination based on age, with the Best Actresses tending to be younger than the Best Actors? Are there any other notable differences? Apart from being interesting, this issue is important because it potentially gives us some insight into the way that our society perceives women and men in general.

Critical Thinking: A visual comparison of the ages in Table 2-1 might be revealing to those with some special ability to see order in such lists of numbers, but for those of us who are mere mortals, the lists of ages in Table 2-1 probably don't reveal much of anything at all. Fortunately, there are methods for investigating such data sets, and we will soon see that those methods reveal important characteristics that allow us to *understand* the data. We will be able to make intelligent and insightful comparisons. We will learn techniques for summarizing, graphing, describing, exploring, and comparing data sets such as those in Table 2-1.

Table 2-1 Academy Awards: Ages of Best Actresses and Best Actors

The ages (in years) are listed in order, beginning with the first awards ceremony.

Best Actresses

22	37	28	63	32	26	31	27	27	28
30	26	29	24	38	25	29	41	30	35
35	33	29	38	54	24	25	46	41	28
40	39	29	27	31	38	29	25	35	60
43	35	34	34	27	37	42	41	36	32
41	33	31	74	33	50	38	61	21	41
26	80	42	29	33	35	45	49	39	34
26	25	33	35	35	28				

Best Actors

44	41	62	52	41	34	34	52	41	37
38	34	32	40	43	56	41	39	49	57
41	38	42	52	51	35	30	39	41	44
49	35	47	31	47	37	57	42	45	42
44	62	43	42	48	49	56	38	60	30
40	42	36	76	39	53	45	36	62	43
51	32	42	54	52	37	38	32	45	60
46	40	36	47	29	43				

2-1 Overview

In this chapter we present important methods of organizing, summarizing, and graphing sets of data. The ultimate objective is not that of simply obtaining some table or graph. Instead, the ultimate objective is to *understand* the data. When describing, exploring, and comparing data sets, the following characteristics are usually extremely important.

Important Characteristics of Data

1. **Center:** A representative or average value that indicates where the middle of the data set is located.
2. **Variation:** A measure of the amount that the data values vary among themselves.
3. **Distribution:** The nature or shape of the distribution of the data (such as bell-shaped, uniform, or skewed).
4. **Outliers:** Sample values that lie very far away from the vast majority of the other sample values.
5. **Time:** Changing characteristics of the data over time.

Study Hint: Blind memorization is often ineffective for learning or remembering important information. However, the above five characteristics are so important, that they might be better remembered by using a mnemonic for their first letters CVDOT, such as "**C**omputer **V**iruses **D**estroy **O**r **T**erminate." (You might remember the names of the Great Lakes with the mnemonic *homes*, for Huron, Ontario, Michigan, Erie, and Superior.) Such memory devices have been found to be very effective in recalling important keywords that trigger key concepts.

Critical Thinking and Interpretation: Going Beyond Formulas and Manual Calculations

Statistics professors generally believe that it is not so important to memorize formulas or manually perform complex arithmetic calculations and number crunching. Instead, they tend to focus on obtaining results by using some form of technology (calculator or software), then making practical sense of the results through critical thinking. Keep this in mind as you proceed through this chapter, the next chapter, and the remainder of this book. Although this chapter includes detailed steps for important procedures, it is not necessary to master those steps in all cases. However, we recommend that in each case you perform a few manual calculations before using a calculator or computer. Your understanding will be enhanced, and you will acquire a better appreciation for the results obtained from the technology.

2-2 Frequency Distributions

Key Concept When working with large data sets, it is often helpful to organize and summarize the data by constructing a table called a *frequency distribution,* defined below. Because computer software and calculators can automatically generate frequency distributions, the details of constructing them are not as important as understanding what they tell us about data sets. In particular, a frequency distribution helps us understand the nature of the *distribution* of a data set.

Definition

A **frequency distribution** (or **frequency table**) lists data values (either individually or by groups of intervals), along with their corresponding frequencies (or counts).

Table 2-2 is a frequency distribution summarizing the ages of Oscar-winning actresses listed in Table 2-1. The **frequency** for a particular class is the number of original values that fall into that class. For example, the first class in Table 2-2 has a frequency of 28, indicating that 28 of the original ages are between 21 years and 30 years inclusive.

We will first present some standard terms used in discussing frequency distributions, and then we will describe how to construct and interpret them.

Definitions

Lower class limits are the smallest numbers that can belong to the different classes. (Table 2-2 has lower class limits of 21, 31, 41, 51, 61, and 71.)

Upper class limits are the largest numbers that can belong to the different classes. (Table 2-2 has upper class limits of 30, 40, 50, 60, 70, and 80.)

Class boundaries are the numbers used to separate classes, but without the gaps created by class limits. Figure 2-1 shows the gaps created by the class limits from Table 2-2. It is easy to see in Figure 2-1 that the values of 30.5, 40.5, . . . , 70.5 are in the centers of those gaps, and these numbers are referred to as class boundaries. The two unknown class boundaries (indicated in Figure 2-1 by question marks) can be easily identified by simply following the pattern established by the other class boundaries of 30.5, 40.5, . . . , 70.5. The lowest class boundary is 20.5, and the highest class boundary is 80.5. The complete list of class boundaries is therefore 20.5, 30.5, 40.5, 50.5, 60.5, 70.5, and 80.5. Class boundaries will be very useful in the next section when we construct a graph called a histogram.

Class midpoints are the values in the middle of the classes. (Table 2-2 has class midpoints of 25.5, 35.5, 45.5, 55.5, 65.5, and 75.5.) Each class midpoint can be found by adding the lower class limit to the upper class limit and dividing the sum by 2.

Class width is the difference between two consecutive lower class limits or two consecutive lower class boundaries. (Table 2-2 uses a class width of 10.)

Table 2-2 Frequency Distribution: Ages of Best Actresses

Age of Actress	Frequency
21–30	28
31–40	30
41–50	12
51–60	2
61–70	2
71–80	2

The definitions of class width and class boundaries are a bit tricky. Be careful to avoid the easy mistake of making the class width the difference between the lower class limit and the upper class limit. See Table 2-2 and note that the class width is 10, not 9. You can simplify the process of finding class boundaries by

Authors Identified

In 1787–88 Alexander Hamilton, John Jay, and James Madison anonymously published the famous *Federalist Papers* in an attempt to convince New Yorkers that they should ratify the Constitution. The identity of most of the papers' authors became known, but the authorship of 12 of the papers was contested. Through statistical analysis of the frequencies of various words, we can now conclude that James Madison is the *likely* author of these 12 papers. For many of the disputed papers, the evidence in favor of Madison's authorship is overwhelming to the degree that we can be almost certain of being correct.

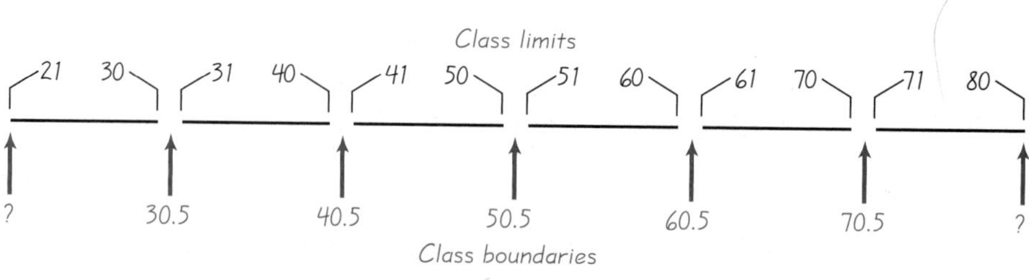

Figure 2-1 Finding Class Boundaries

understanding that they basically split the difference between the end of one class and the beginning of the next class.

Procedure for Constructing a Frequency Distribution

Frequency distributions are constructed for these reasons: (1) Large data sets can be summarized, (2) we can gain some insight into the nature of data, and (3) we have a basis for constructing important graphs (such as *histograms,* introduced in the next section). Many uses of technology allow us to automatically obtain frequency distributions without manually constructing them, but here is the basic procedure:

1. Decide on the number of classes you want. The number of classes should be between 5 and 20, and the number you select might be affected by the convenience of using round numbers.
2. Calculate

$$\text{Class width} \approx \frac{(\text{maximum value}) - (\text{minimum value})}{\text{number of classes}}$$

   Round this result to get a convenient number. (Usually round *up*.) You might need to change the number of classes, but the priority should be to use values that are easy to understand.
3. Starting point: Begin by choosing a number for the lower limit of the first class. Choose either the minimum data value or a convenient value below the minimum data value.
4. Using the lower limit of the first class and the class width, proceed to list the other lower class limits. (Add the class width to the starting point to get the second lower class limit. Add the class width to the second lower class limit to get the third, and so on.)
5. List the lower class limits in a vertical column and proceed to enter the upper class limits, which can be easily identified.
6. Go through the data set putting a tally in the appropriate class for each data value. Use the tally marks to find the total frequency for each class.

When constructing a frequency distribution, be sure that classes do not overlap so that each of the original values must belong to exactly one class. Include all classes, even those with a frequency of zero. Try to use the same width for all classes, although it is sometimes impossible to avoid open-ended intervals, such as "65 years or older."

EXAMPLE Ages of Best Actresses Using the ages of the Best Actresses in Table 2-1, follow the above procedure to construct the frequency distribution shown in Table 2-2. Assume that you want 6 classes.

SOLUTION

Step 1: Begin by selecting 6 as the number of desired classes.

Step 2: Calculate the class width. In the following calculation, 9.833 is rounded up to 10, which is a more convenient number.

$$\text{Class width} \approx \frac{(\text{maximum value}) - (\text{minimum value})}{\text{number of classes}}$$

$$= \frac{80 - 21}{6} = 9.833 \approx 10$$

Step 3: We choose a starting point of 21, which is the minimum value in the list and is also a convenient number, because the first class becomes 21–30.

Step 4: Add the class width of 10 to the starting point of 21 to determine that the second lower class limit is 31. Continue to add the class width of 10 to get the remaining lower class limits of 41, 51, 61, and 71.

Step 5: List the lower class limits vertically as shown in the margin. From this list, we can easily identify the corresponding upper class limits as 30, 40, 50, 60, 70, and 80.

21–
31–
41–
51–
61–
71–

Step 6: After identifying the lower and upper limits of each class, proceed to work through the data set by entering a tally mark for each data value. When the tally marks are completed, add them to find the frequencies shown in Table 2-2.

Relative Frequency Distribution

An important variation of the basic frequency distribution uses **relative frequencies,** which are easily found by dividing each class frequency by the total of all frequencies. A **relative frequency distribution** includes the same class limits as a frequency distribution, but relative frequencies are used instead of actual frequencies. The relative frequencies are often expressed as percents.

$$\text{relative frequency} = \frac{\text{class frequency}}{\text{sum of all frequencies}}$$

In Table 2-3 the actual frequencies from Table 2-2 are replaced by the corresponding relative frequencies expressed as percents. With 28 of the 76 data values falling in the first class, that first class has a relative frequency of $28/76 = 0.368$, or 36.8%, which is often rounded to 37%. The second class has a relative frequency of $30/76 = 0.395$, or 39.5%, and so on. If constructed correctly, the sum of the relative frequencies should total 1 (or 100%), with some small discrepancies allowed for rounding errors. The rounding of results in Table 2-3 causes the sum of the relative frequencies to be 101% instead of 100%.

Table 2-3
Relative Frequency Distribution of Best Actress Ages

Age of Actress	Relative Frequency
21–30	37%
31–40	39%
41–50	16%
51–60	3%
61–70	3%
71–80	3%

Table 2-4
Cumulative Frequency Distribution of Best Actress Ages

Age of Actress	Cumulative Frequency
Less than 31	28
Less than 41	58
Less than 51	70
Less than 61	72
Less than 71	74
Less than 81	76

Because they use simple percentages, relative frequency distributions make it easier for us to understand the distribution of the data and to compare different sets of data.

Cumulative Frequency Distribution

Another variation of the standard frequency distribution is used when cumulative totals are desired. The **cumulative frequency** for a class is the sum of the frequencies for that class and all previous classes. Table 2-4 is the cumulative frequency distribution based on the frequency distribution of Table 2-2. Using the original frequencies of 28, 30, 12, 2, 2, and 2, we add 28 + 30 to get the second cumulative frequency of 58, then we add 28 + 30 + 12 = 70 to get the third, and so on. See Table 2-4 and note that in addition to using cumulative frequencies, the class limits are replaced by "less than" expressions that describe the new ranges of values.

Critical Thinking: Interpreting Frequency Distributions

The transformation of raw data to a frequency distribution is typically a means to some greater end. One important objective is to identify the nature of the distribution, and "normal" distributions are extremely important in the study of statistics.

Normal Distribution In later chapters of this book, there will be frequent reference to data with a *normal distribution*. This use of the word "normal" refers to a special meaning in statistics that is different from the meaning typically used in ordinary language. The concept of a normal distribution will be described later, but for now we can use a frequency distribution to help determine whether the data have a distribution that is approximately normal. One key characteristic of a normal distribution is that when graphed, the result has a "bell" shape, with frequencies that start low, then increase to some maximum, then decrease. For now, we can judge that a frequency distribution is approximately normal by determining whether it has these features:

Normal Distribution

1. The frequencies start low, then increase to some maximum frequency, then decrease to a low frequency.
2. The distribution should be approximately symmetric, with frequencies evenly distributed on both sides of the maximum frequency. (Frequencies of 1, 5, 50, 25, 20, 15, 10, 5, 3, 2, 1 are not symmetric about the maximum of 50 and would not satisfy the requirement of symmetry.)

EXAMPLE **Normal Distribution** One thousand women were randomly selected and their heights were measured. The results are summarized in the frequency distribution of Table 2-5. The frequencies start low, then increase to a maximum frequency, then decrease to low frequencies. Also, the frequencies are roughly symmetric about the maximum frequency of 324. It appears that the distribution is approximately a normal distribution.

Table 2-5 Heights of a Sample of 1000 Women

Normal distribution: The frequencies start low, reach a peak, then become low again.

Height (in.)	Frequency	Normal Distribution:
56.0–57.9	10	← Frequencies start low, . . .
58.0–59.9	64	
60.0–61.9	178	
62.0–63.9	324	← increase to a maximum, . . .
64.0–65.9	251	
66.0–67.9	135	
68.0–69.9	32	
70.0–71.9	6	← decrease to become low again.

Table 2-5 illustrates data with a normal distribution. The following examples illustrate how frequency distributions can be used to describe, explore, and compare data sets. (The following section shows how the construction of a frequency distribution is often the first step in the creation of a graph that visually depicts the nature of the distribution.)

EXAMPLE Describing Data: How Were the Pulse Rates Measured? Refer to Data Set 1 in Appendix B for the pulse rates of 40 randomly selected adult males. Table 2-6 summarizes the *last digits* of those pulse rates. If the pulse rates are measured by counting the number of heartbeats in 1 minute, we expect that those last digits should occur with frequencies that are roughly the same. But note that the frequency distribution shows that the last digits are all *even* numbers; there are *no* odd numbers present. This suggests that the pulse rates were not counted for 1 minute. Perhaps they were counted for 30 seconds and the values were then doubled. (Upon further examination of the *original* pulse rates, we can see that every original value is a multiple of four, suggesting that the number of heartbeats was counted for 15 seconds, then that count was multiplied by 4.) It's fascinating to learn something about the method of data collection by simply describing some characteristics of the data.

EXAMPLE Exploring Data: What Does a Gap Tell Us? Table 2-7 is a frequency table of the weights (grams) of randomly selected pennies. Examination of the frequencies reveals a large gap between the lightest pennies and the heaviest pennies. This suggests that we have two different populations. Upon further investigation, it is found that pennies made before 1983 are 97% copper and 3% zinc, whereas pennies made after 1983 are 3% copper and 97% zinc, which can explain the large gap between the lightest pennies and the heaviest pennies.

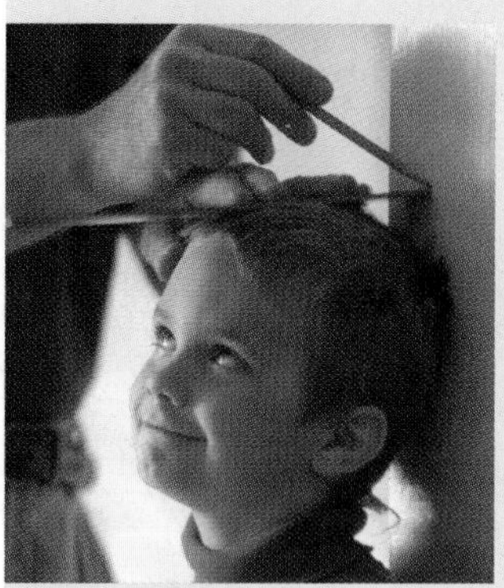

Growth Charts Updated

Pediatricians typically use standardized growth charts to compare their patient's weight and height to a sample of other children. Children are considered to be in the normal range if their weight and height fall between the 5th and 95th percentiles. If they fall outside of that range, they are often given tests to ensure that there are no serious medical problems. Pediatricians became increasingly aware of a major problem with the charts: Because they were based on children living between 1929 and 1975, the growth charts were found to be inaccurate. To rectify this problem, the charts were updated in 2000 to reflect the current measurements of millions of children. The weights and heights of children are good examples of populations that change over time. This is the reason for including changing characteristics of data over time as an important consideration for a population.

Table 2-6
Last Digits of Male Pulse Rates

Last Digit	Frequency
0	7
1	0
2	6
3	0
4	11
5	0
6	9
7	0
8	7
9	0

Table 2-7
Randomly Selected Pennies

Weights (grams) of Pennies	Frequency
2.40–2.49	18
2.50–2.59	19
2.60–2.69	0
2.70–2.79	0
2.80–2.89	0
2.90–2.99	2
3.00–3.09	25
3.10–3.19	8

Table 2-8
Ages of Oscar-Winning Actresses and Actors

Age	Actresses	Actors
21–30	37%	4%
31–40	39%	33%
41–50	16%	39%
51–60	3%	18%
61–70	3%	4%
71–80	3%	1%

Gaps The preceding example suggests that the presence of gaps can reveal the fact that we have data from two or more different populations. However, the converse is not true, because data from different populations do not necessarily result in gaps when histograms are created.

CHAPTER PROBLEM 2

EXAMPLE Comparing Ages of Oscar Winners The Chapter Problem given at the beginning of this chapter includes ages of actresses and actors at the time that they won Academy Award Oscars. Table 2-8 shows the relative frequencies for the two genders. By comparing those relative frequencies, it appears that actresses tend to be somewhat younger than actors. For example, see the first class showing that 37% of the actresses are in the youngest age category, compared to only 4% of the actors.

TI Using the TI-83/84 Plus Calculator to Construct a Frequency Distribution

You can use a TI-83/84 Plus calculator to construct a frequency distribution from a list of sample data. The basic approach is to generate a histogram (a type of graph described in the following section), then use the TRACE and WINDOW keys to identify the class frequencies and boundaries. Use the following procedure.

TI-83/84 Plus

1. Enter or open a list of data. For example, enter the ages of the actresses in Table 2-1 and give that list the name of AGES.
2. Select the STAT PLOT function by pressing 2ND Y=.
3. Press ENTER and use the arrow keys to turn Plot1 to the On state, and also highlight the graph with the bars. The screen display should be as shown in the margin.

4. *Using default settings:* If you want to let the calculator automatically determine the class width and class boundaries, press ZOOM 9.

   Use your own class width and boundaries: If you want to set your own class width and class boundaries, press WINDOW and proceed to enter the values that you desire. The Xscl value will be the class width. The entries in the displayed WINDOW screen include 21 for the minimum x value, 80 for the upper boundary, and a class width of 10 (as in Table 2-2). You might have to estimate (guess) a value for Ymax, which is the highest class frequency, but if the choice doesn't work well, it can be easily changed.

TI-83/84 Plus

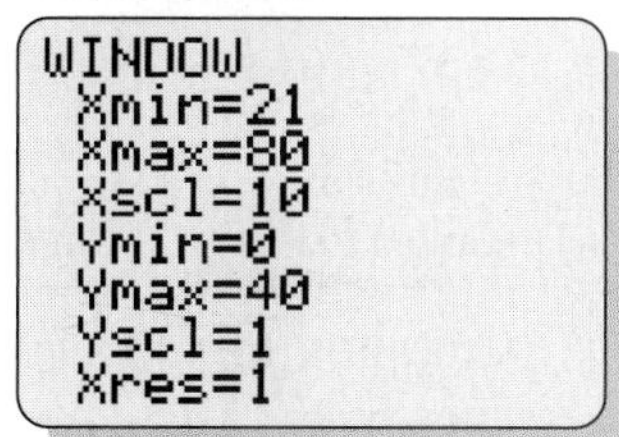

TI-83/84 Plus

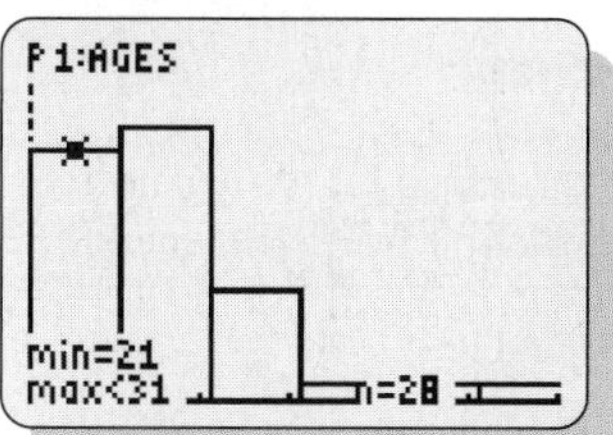

5. Press GRAPH, then press TRACE. If using the settings shown in the display from Step 4, the second display shown above will appear. The blinking cursor is positioned at the top of the first bar, and for this first class we can see from the display that the class frequency is $n = 28$, and the first class includes values greater than or equal to 21 and less than 31 (as indicated by "min = 21" and "max < 31"). Because the data consist of whole numbers only, the first class has limits of 21–30, and it has a frequency of 28.
6. Now press ▷ to move the cursor to the next cell. Using the settings described above, the next cell will display a frequency of $n = 30$, and the next cell will include values greater than or equal to 31 and less than 41. The second cell therefore corresponds to a class with limits of 31–41 and a frequency of 30.
7. Continue to press ▷ to find that the remaining classes have the same class limits and frequencies given in Table 2-2.

2-2 BASIC SKILLS AND CONCEPTS

Statistical Literacy and Critical Thinking

1. Frequency Distribution What is a frequency distribution and why is it useful?
2. Retrieving Original Data Working from a known list of sample values, a researcher constructs a frequency distribution (such as the one shown in Table 2-2). She then discards the original data values. Can she use the frequency distribution to identify all of the original sample values?
3. Overlapping Classes When constructing a frequency distribution, what is the problem created by using these class intervals: 0–10, 10–20, 20–30, . . . , 90–100?
4. Comparing Distributions When comparing two sets of data values, what is the advantage of using relative frequency distributions instead of frequency distributions?

In Exercises 5–8, identify the class width, class midpoints, *and* class boundaries *for the given frequency distribution.*

5.

Daily Low Temperature (°F)	Frequency
35–39	1
40–44	3
45–49	5
50–54	11
55–59	7
60–64	7
65–69	1

6.

Daily Precipitation (inches)	Frequency
0.00–0.49	31
0.50–0.99	1
1.00–1.49	0
1.50–1.99	2
2.00–2.49	0
2.50–2.99	1

7.

Heights (inches) of Men	Frequency
60.0–64.9	4
65.0–69.9	25
70.0–74.9	9
75.0–79.9	1
80.0–84.9	0
85.0–89.9	0
90.0–94.9	0
95.0–99.9	0
100.0–104.9	0
105.0–109.9	1

8.

Weights (lb) of Discarded Plastic	Frequency
0.00–0.99	8
1.00–1.99	12
2.00–2.99	6
3.00–3.99	0
4.00–4.99	0
5.00–5.99	0
6.00–6.99	0
7.00–7.99	5
8.00–8.99	15
9.00–9.99	20

Critical Thinking. *In Exercises 9–12, answer the given questions that relate to Exercises 5–8.*

9. Identifying the Distribution Does the frequency distribution given in Exercise 5 appear to have a normal distribution, as required for several methods of statistics introduced later in this book?

10. Identifying the Distribution Does the frequency distribution given in Exercise 6 appear to have a normal distribution, as required for several methods of statistics introduced later in this book? If we learn that the precipitation amounts were obtained from days randomly selected over the past 200 years, do the results reflect current weather behavior?

11. Outlier Refer to the frequency distribution given in Exercise 7. What is known about the height of the tallest man included in the table? Can the height of the tallest man be a correct value? If the highest value appears to be an error, what can be concluded about the distribution after this error is deleted?

12. Analyzing the Distribution Refer to the frequency distribution given in Exercise 8. There appears to be a large gap between the lowest weights and the highest weights. What does that gap suggest? How might the gap be explained?

In Exercises 13 and 14, construct the relative frequency distribution *that corresponds to the frequency distribution in the exercise indicated.*

13. Exercise 5

14. Exercise 6

In Exercises 15 and 16, construct the cumulative frequency distribution *that corresponds to the frequency distribution in the exercise indicated.*

15. Exercise 5 **16.** Exercise 6

TI **17.** Analysis of Last Digits Heights of statistics students were obtained as part of an experiment conducted for class. The last digits of those heights are listed below. Construct a frequency distribution with 10 classes. Based on the distribution, do the heights appear to be reported or actually measured? What do you know about the accuracy of the results?

0 0 0 0 0 0 0 0 0 0 1 1 2 3 3 3 4 5 5 5 5 5 5 5 5 5 5 5 5 5 5 5 6 6 8 8 8 9

TI **18.** Loaded Die The author drilled a hole in a die and filled it with a lead weight, then proceeded to roll it 180 times. (Yes, the author has too much free time.) The results are given in the frequency distribution in the margin. Construct the frequency distribution for the outcomes that you would expect from a die that is perfectly fair and unbiased. Does the loaded die appear to differ significantly from a fair die that has not been "loaded."

Table for Exercise 18

Outcome	Frequency
1	24
2	28
3	39
4	37
5	25
6	27

TI **19.** Rainfall Amounts Refer to Data Set 10 in Appendix B and use the 52 rainfall amounts for Sunday. Construct a frequency distribution beginning with a lower class limit of 0.00 and use a class width of 0.20. Describe the nature of the distribution. Does the frequency distribution appear to be roughly a normal distribution, as described in this section?

TI **20.** Nicotine in Cigarettes Refer to Data Set 3 in Appendix B and use the 29 measured amounts of nicotine. Construct a frequency distribution with 8 classes beginning with a lower class limit of 0.0, and use a class width of 0.2. Describe the nature of the distribution. Does the frequency distribution appear to be roughly a normal distribution, as described in this section?

TI **21.** BMI Values Refer to Data Set 1 in Appendix B and use the body mass index (BMI) values for the 40 females. Construct a frequency distribution beginning with a lower class limit of 15.0 and use a class width of 6.0. The BMI is calculated by dividing the weight in kilograms by the square of the height in meters. Describe the nature of the distribution. Does the frequency distribution appear to be roughly a normal distribution, as described in this section?

TI **22.** Weather Data Refer to Data Set 8 in Appendix B and use the actual low temperatures to construct a frequency distribution beginning with a lower class limit of 39 and use a class width of 6. The frequency distribution in Exercise 6 represents the precipitation amounts from Data Set 8. Compare the two frequency distributions (for the actual low temperatures and the precipitation amounts). How are they fundamentally different?

TI **23.** Weights of Pennies Refer to Data Set 14 in Appendix B and use the weights of the pre-1983 pennies. Construct a frequency distribution beginning with a lower class limit of 2.9500 and a class width of 0.0500. Do the weights appear to be normally distributed?

TI **24.** Regular Coke and Diet Coke Refer to Data Set 12 in Appendix B. Construct a relative frequency distribution for the weights of regular Coke by starting the first class at 0.7900 lb and use a class width of 0.0050 lb. Then construct another relative frequency distribution for the weights of Diet Coke by starting the first class at 0.7750 lb and use a class width of 0.0050 lb. Then compare the results and determine whether there appears to be a significant difference. If so, provide a possible explanation for the difference.

2-2 BEYOND THE BASICS

TI **25.** Large Data Sets Refer to Data Set 15 in Appendix B. Construct a relative frequency distribution for the 175 axial loads of aluminum cans that are 0.0109 in. thick, then do the same for the 175 axial loads of aluminum cans that are 0.0111 in. thick. Compare the two relative frequency distributions.

TI **26.** Interpreting Effects of Outliers Refer to Data Set 15 in Appendix B for the axial loads of aluminum cans that are 0.0111 in. thick. The load of 504 lb is an *outlier* because it is very far away from all of the other values. Construct a frequency distribution that includes the value of 504 lb, then construct another frequency distribution with the value of 504 lb excluded. In both cases, start the first class at 200 lb and use a class width of 20 lb. Interpret the results by stating a generalization about how much of an effect an outlier might have on a frequency distribution.

27. Number of Classes In constructing a frequency distribution, Sturges' guideline suggests that the ideal number of classes can be approximated by $1 + (\log n)/(\log 2)$, where n is the number of data values. Use this guideline to complete the table in the margin for determining the ideal number of classes.

Table for Exercise 27

Number of Values	Ideal Number of Classes
16–22	5
23–45	6
	7
	8
	9
	10
	11
	12

2-3 Histograms

Key Concept Section 2-2 introduced the frequency distribution as a tool for summarizing and learning the nature of the distribution of a large data set. This section introduces the *histogram* as a very important graph that depicts the nature of the distribution. Because many statistics computer programs and calculators can automatically generate histograms, it is not so important to master the mechanical procedures for constructing them. Instead, we should focus on the *understanding* that can be gained by examining histograms. In particular, we should develop the ability to look at a histogram and understand the nature of the distribution of the data.

Definition

A **histogram** is a bar graph in which the horizontal scale represents classes of data values and the vertical scale represents frequencies. The heights of the bars correspond to the frequency values, and the bars are drawn adjacent to each other (without gaps).

The first step in the construction of a histogram is the construction of a frequency distribution table. The histogram is basically a graphic version of that table. See Figure 2-2, which is the histogram corresponding to the frequency distribution in Table 2-2 given in the preceding section.

On the horizontal scale, each bar of the histogram is marked with its lower class boundary at the left and its upper class boundary at the right, as in Figure 2-2. Instead of using class boundaries along the horizontal scale, it is often more practical

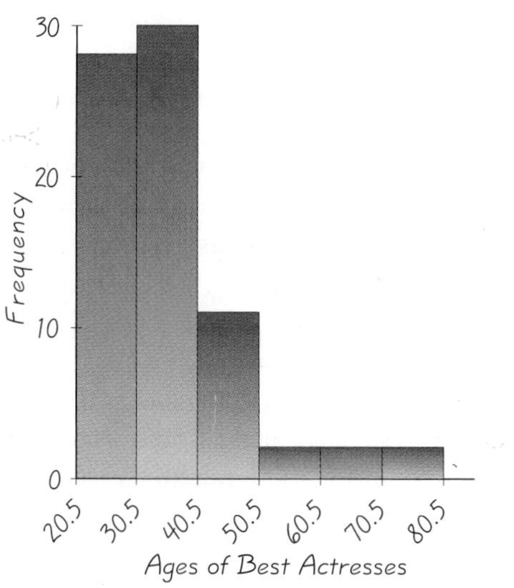

Figure 2-2 Histogram

40%
30%
20%
10%
0%
Relative Frequency
20.5 30.5 40.5 50.5 60.5 70.5 80.5
Ages of Best Actresses

Figure 2-3 Relative Frequency Histogram

to use class midpoint values centered below their corresponding bars. The use of class midpoint values is very common in software packages that automatically generate histograms.

Horizontal Scale: Use class boundaries or class midpoints.

Vertical Scale: Use the class frequencies

Before constructing a histogram from a completed frequency distribution, we must give some thought to the scales used on the vertical and horizontal axes. The maximum frequency (or the next highest convenient number) should suggest a value for the top of the vertical scale; 0 should be at the bottom. In Figure 2-2 we designed the vertical scale to run from 0 to 30. The horizontal scale should be subdivided in a way that allows all the classes to fit well. Ideally, we should try to follow the rule of thumb that the vertical height of the histogram should be about three-fourths of the total width. Both axes should be clearly labeled.

Relative Frequency Histogram

A **relative frequency histogram** has the same shape and horizontal scale as a histogram, but the vertical scale is marked with relative frequencies instead of actual frequencies, as in Figure 2-3.

Critical Thinking: Interpreting Histograms

Remember that the objective is not simply to construct a histogram, but rather to *understand* something about the data. Analyze the histogram to see what can be learned about "CVDOT": the center of the data, the variation (which will be discussed at length in Section 3-3), the shape of the distribution, and whether there are any outliers (values far away from the other values). Examining Figure 2-2, we see that the

Missing Data

Samples are commonly missing some data. Missing data fall into two general categories: (1) Missing values that result from random causes unrelated to the data values, and (2) missing values resulting from causes that are not random. Random causes include factors such as the incorrect entry of sample values or lost survey results. Such missing values can often be ignored because they do not systematically hide some characteristic that might significantly affect results. It's trickier to deal with values missing because of factors that are not random. For example, results of an income analysis might be seriously flawed if people with very high incomes refuse to provide those values because they fear income tax audits. Those missing high incomes should not be ignored, and further research would be needed to identify them.

histogram is centered around 35, the values vary from around 21 to 80, and the shape of the distribution is heavier on the left, which means that actresses who win Oscars tend to be disproportionately younger, with fewer older actresses winning Oscars.

Normal Distribution In Section 2-2 we noted that use of the word "normal" refers to a special meaning in statistics that is different from the meaning typically used in ordinary language. A key characteristic of a normal distribution is that when graphed as a histogram, the result has a "bell" shape, as in the STATDISK-generated histogram and the TI-83/84 Plus histogram shown here. [Key characteristics of the bell shape are (1) the rise in frequencies that reach a maximum, then decrease, and (2) the symmetry with the left half of the graph that is roughly a mirror image of the right half.] These histograms correspond to the frequency distribution of Table 2-5, which was obtained from 1000 randomly selected heights of women. Many statistical methods require that sample data come from a population having a distribution that is not dramatically far from a normal distribution, and we can often use a histogram to judge whether this requirement of a normal distribution is satisfied.

We say that the distribution is *normal* because it is bell-shaped.

STATDISK

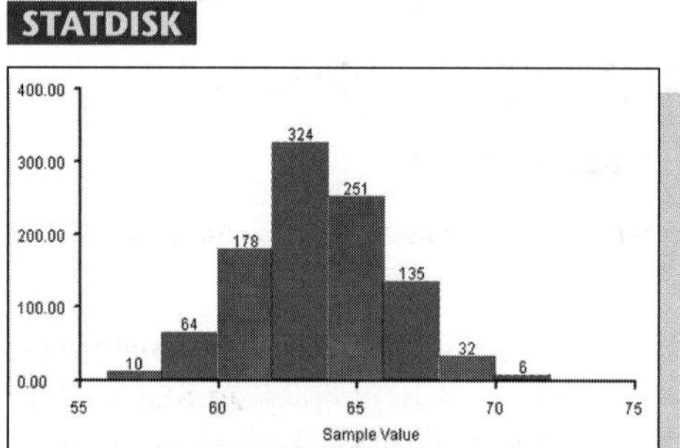

TI-83/84 Plus

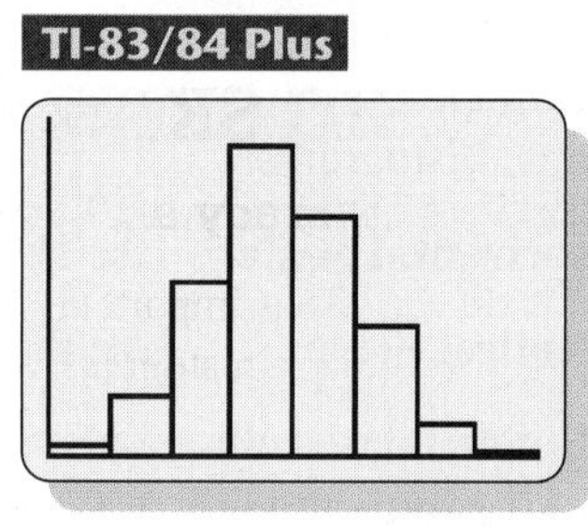

Using the TI-83/84 Plus Calculator to Construct Histograms

A TI-83/84 Plus calculator can be used to generate a histogram, such as the one shown in Figure 2-2. Use the following procedure:

TI-83/84 Plus

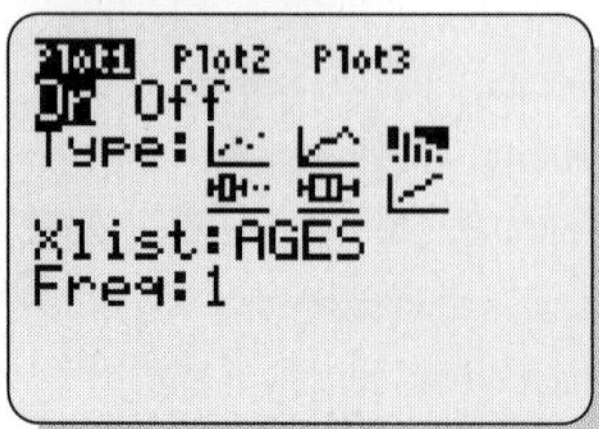

1. Enter or open a list of data. For example, enter the ages of the actresses in Table 2-1 and give that list the name of AGES. We will illustrate the construction of a histogram by using this list of data.
2. Select the "STAT PLOT" function by pressing 2ND Y=.
3. Press ENTER and use the arrow keys to turn Plot1 to the On state, and also highlight the graph with the bars. The screen display should be as shown.
4. *Using default settings:* If you want to let the calculator determine the class width and class boundaries, press ZOOM 9.

   Use your own class width and boundaries: If you want to set your own class width and class boundaries, press WINDOW and proceed to enter the values that you desire. The Xscl value will be the class width. The entries in the displayed WINDOW screen include 21 for the minimum x value, 80 for the upper boundary, and a class width of 10 (as in Table 2-2). You might have to estimate (guess) a value for Ymax, which is the highest class frequency, but if the choice doesn't work well, it can be easily changed.

TI-83/84 Plus

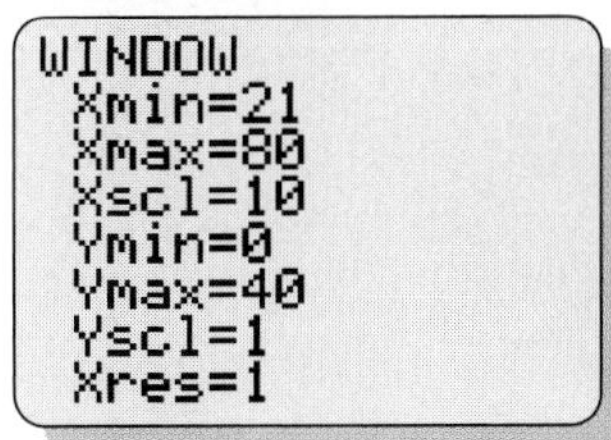

5. Press GRAPH. If using the settings shown in the display from Step 4, the result will be the first display shown below. You can press TRACE to identify the heights of the bars as well as their boundaries. After pressing TRACE, you can move the cursor by pressing ▷. For example, the second display shown below indicates that the third class has a frequency of 12, and it includes values greater than or equal to 41 (indicated by "min = 41") and less than 51 (indicated by "max < 51").

TI-83/84 Plus

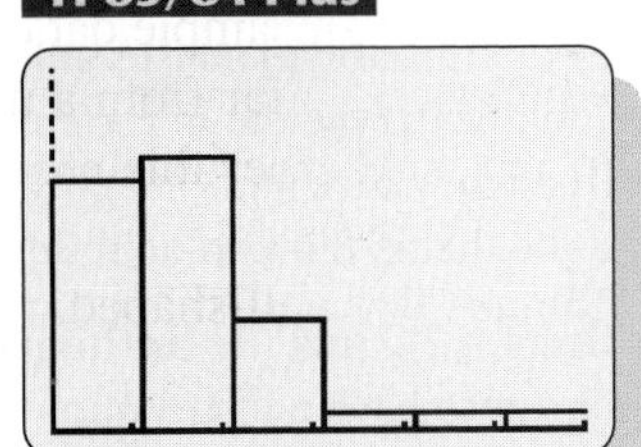

TI-83/84 Plus

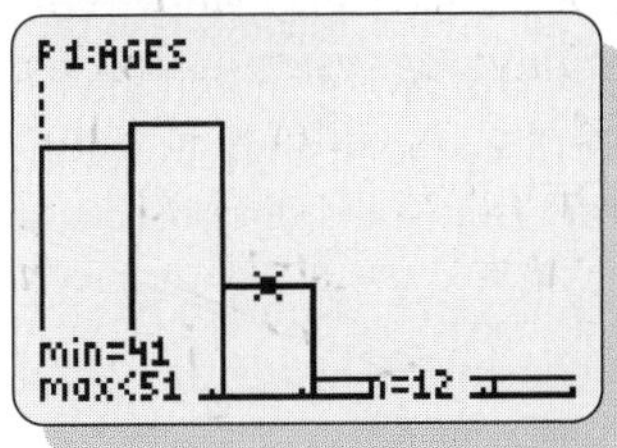

2-3 BASIC SKILLS AND CONCEPTS

Statistical Literacy and Critical Thinking

1. Histogram What important characteristic of data can be better understood through examination of a histogram?

2. Histogram and Frequency Distribution Given that a histogram is essentially a graphic representation of the same data in a frequency distribution, what major advantage does a histogram have over a frequency distribution?

3. Small Data Set If a data set is small, such as one that has only five values, why should we not bother to construct a histogram?

4. Normal Distribution After examining a histogram, what criterion can be used to determine whether the data have a distribution that is approximately normal? Is this criterion totally objective, or does it involve subjective judgment?

In Exercises 5–8, answer the questions by referring to the accompanying TI-83/84 Plus displays. The window settings on the left are used for the histogram on the right. (Hint: *The dots at the left border are one unit apart.) The data represent the weights (in pounds) of coxswains and rowers in a boat race between Oxford and Cambridge. (Based on data from* A Handbook of Small Data Sets, *by D. J. Hand, Chapman & Hall.)*

TI-83/84 Plus

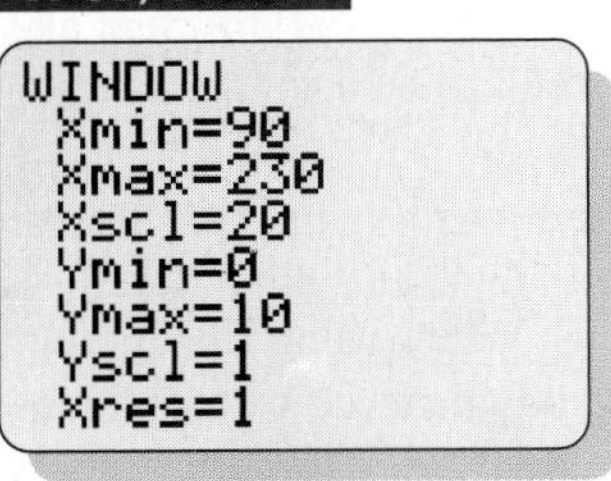

TI-83/84 Plus

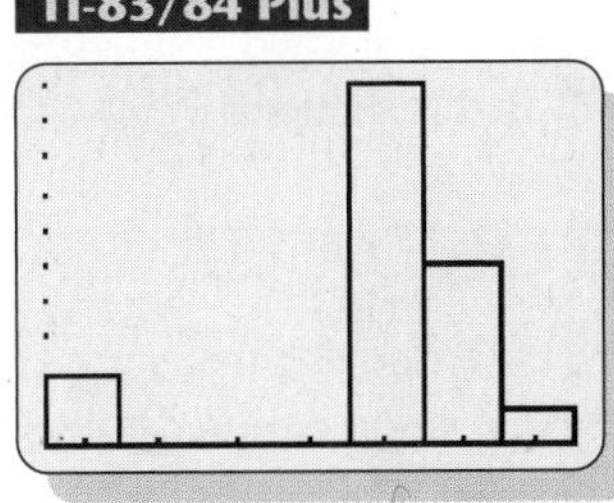

TI **5.** Sample Size How many crew members are included in the histogram?

TI **6.** Variation What is the minimum possible weight? What is the maximum possible weight?

TI **7.** Gap What is a reasonable explanation for the large gap between the leftmost bar and the other bars?

TI **8.** Class Width What is the class width?

TI **9.** Analysis of Last Digits Refer to Exercise 17 from Section 2-2 for the last digits of heights of statistics students that were obtained as part of an experiment conducted for class. Use the frequency distribution from that exercise to construct a histogram. What can be concluded from the distribution of the digits? Specifically, do the heights appear to be reported or actually measured?

TI **10.** Loaded Die Refer to Exercise 18 from Section 2-2 for the results from 180 rolls of a die that the author loaded. Use the frequency distribution to construct the corresponding histogram. What should the histogram look like if the die is perfectly fair and unbiased? Does the histogram for the given frequency distribution appear to differ significantly from a histogram obtained from a die that is fair and unbiased?

TI **11.** Rainfall Amounts Refer to Exercise 19 in Section 2-2 and use the frequency distribution to construct a histogram. Do the data appear to have a distribution that is approximately normal?

TI **12.** Nicotine in Cigarettes Refer to Exercise 20 in Section 2-2 and use the frequency distribution to construct a histogram. Do the data appear to have a distribution that is approximately normal?

TI **13.** BMI Values Refer to Exercise 21 in Section 2-2 and use the frequency distribution to construct a histogram. Do the data appear to have a distribution that is approximately normal?

TI **14.** Weather Data Refer to Exercise 22 in Section 2-2 and use the frequency distribution from the actual low temperatures to construct a histogram. Do the data appear to have a distribution that is approximately normal?

TI **15.** Weights of Pennies Refer to Exercise 23 in Section 2-2 and use the frequency distribution for the weights of the pre-1983 pennies. Construct the corresponding histogram. Do the weights appear to have a normal distribution?

TI **16.** Regular Coke and Diet Coke Refer to Exercise 24 in Section 2-2 and use the two relative frequency distributions to construct the two corresponding relative frequency histograms. Compare the results and determine whether there appears to be a significant difference. If there is a difference, how can it be explained?

TI **17.** Comparing Ages of Actors and Actresses Refer to Table 2-8 and use the relative frequency distribution for the best actors to construct a relative frequency histogram. Compare the result to Figure 2-3, which is the relative frequency histogram for the best actresses. Do the two genders appear to win Oscars at different ages? (See also Exercise 18 in this section.)

2-3 BEYOND THE BASICS

18. Back-to-Back Relative Frequency Histograms When using histograms to compare two data sets, it is sometimes difficult to make comparisons by looking back and forth between the two histograms. A *back-to-back relative frequency histogram* uses a format that makes the comparison much easier. Instead of frequencies, we should use relative frequencies so that the comparisons are not distorted by different sample sizes. Complete the back-to-back relative frequency histograms shown below by using the data from Table 2-8 in Section 2-2. Then use the result to compare the two data sets.

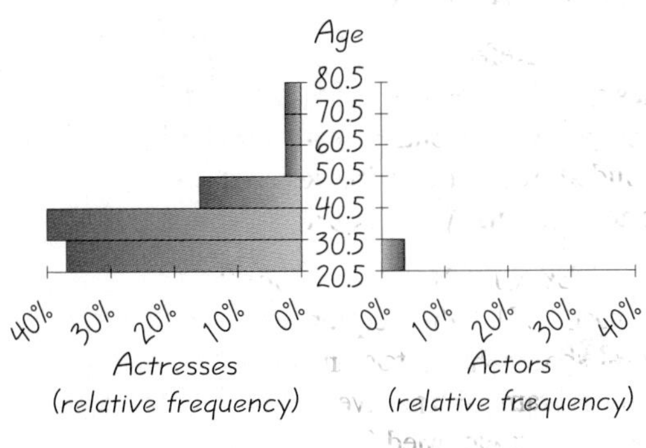

19. Large Data Sets Refer to Exercise 25 in Section 2-2 and construct back-to-back relative frequency histograms for the axial loads of cans that are 0.0109 in. thick and the axial loads of cans that are 0.0111 in. thick. (Back-to-back relative frequency histograms are described in Exercise 18.) Compare the two sets of data. Does the thickness of aluminum cans affect their strength, as measured by the axial loads?

TI **20.** Interpreting Effects of Outliers Refer to Data Set 15 in Appendix B for the axial loads of aluminum cans that are 0.0111 in. thick. The load of 504 lb is an *outlier* because it is very far away from all of the other values. Construct a histogram that includes the value of 504 lb, then construct another histogram with the value of 504 lb excluded. In both cases, start the first class at 200 lb and use a class width of 20 lb. Interpret the results by stating a generalization about how much of an effect an outlier might have on a histogram. (See Exercise 26 in Section 2-2.)

2-4 Statistical Graphics

Key Concept Section 2-3 introduced histograms and relative frequency histograms as graphs that visually display the distributions of data sets. This section presents other graphs commonly used in statistical analyses, as well as some graphs that depict data in ways that are innovative. As in Section 2-3, the main objective is not the generation of a graph. Instead, the main objective is to better *understand* a data set by using a suitable graph that is effective in revealing some important characteristic. Our world needs more people with an ability to construct graphs that clearly and effectively reveal important characteristics of data. Our world also needs more people with an ability to be innovative in creating original graphs that capture key features of data.

This section begins by briefly describing graphs typically included in introductory statistics courses, such as frequency polygons, ogives, dotplots, stemplots,

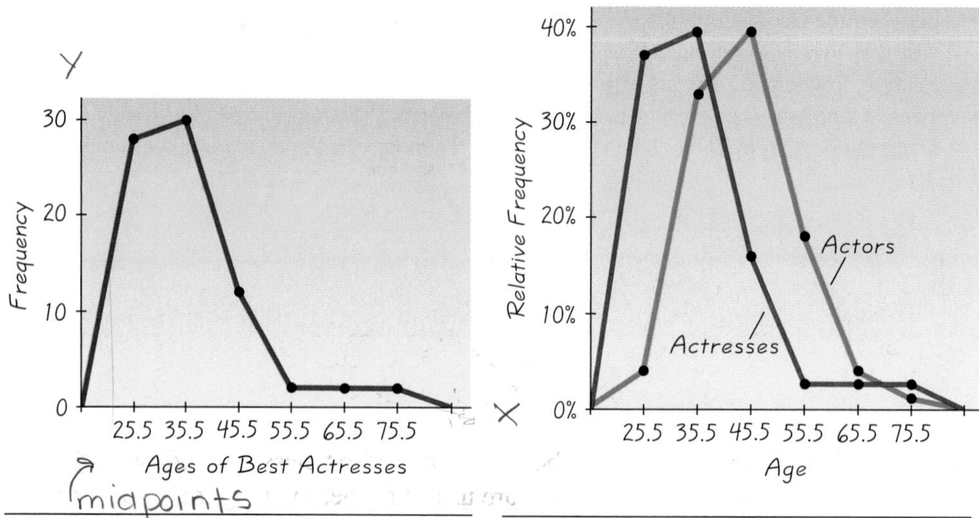

Figure 2-4 Frequency Polygon

Figure 2-5 Relative Frequency Polygons

Pareto charts, pie charts, scatter diagrams, and time-series graphs. We then consider some original and creative graphs. We begin with frequency polygons.

Frequency Polygon

A **frequency polygon** uses line segments connected to points located directly above class midpoint values. See Figure 2-4 for the frequency polygon corresponding to Table 2-2. The heights of the points correspond to the class frequencies, and the line segments are extended to the right and left so that the graph begins and ends on the horizontal axis.

A variation of the basic frequency polygon is the relative frequency polygon, which uses relative frequencies for the vertical scale. When trying to compare two data sets, it is often very helpful to graph two relative frequency polygons on the same axes. See Figure 2-5, which shows the relative frequency polygons for the ages of the Best Actresses and Best Actors as listed in the Chapter Problem. Figure 2-5 makes it visually clear that the actresses tend to be younger than their male counterparts. Figure 2-5 accomplishes something that is truly wonderful: It enables an understanding of data that is not possible with visual examination of the lists of data in Table 2-1. (It's like a good poetry teacher revealing the true meaning of a poem.) For reasons that will not be described here, there does appear to be some type of gender discrimination based on age.

Ogive

An **ogive** (pronounced "oh-jive") is a line graph that depicts *cumulative* frequencies, just as the cumulative frequency distribution (see Table 2-4 in the preceding section) lists cumulative frequencies. Figure 2-6 is an ogive corresponding to Table 2-4. Note that the ogive uses class boundaries along the horizontal scale,

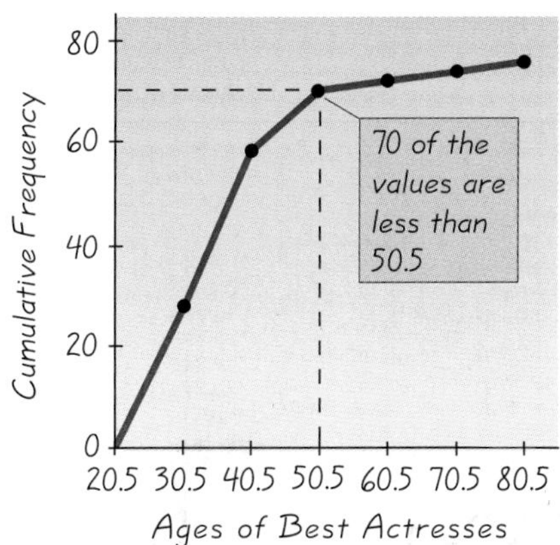

Figure 2-6
Ogive

and the graph begins with the lower boundary of the first class and ends with the upper boundary of the last class. Ogives are useful for determining the number of values below some particular value. For example, see Figure 2-6, where it is shown that 70 of the ages are less than 50.5.

Dotplots

A **dotplot** consists of a graph in which each data value is plotted as a point (or dot) along a scale of values. Dots representing equal values are stacked. See the Minitab-generated dotplot of the ages of the Best Actresses. (The data are from Table 2-1 in the Chapter Problem.) The two dots at the left depict ages of 21 and 22. The next two dots are stacked above 24, indicating that two of the actresses were 24 years of age when they were awarded Oscars. We can see from this dotplot that the ages above 48 are few and far between.

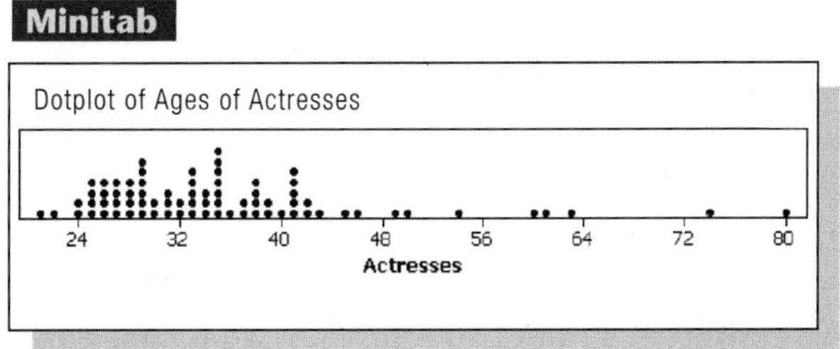

Stemplots

A **stemplot** (or **stem-and-leaf plot**) represents data by separating each value into two parts: the stem (such as the leftmost digit) and the leaf (such as the rightmost digit). The illustration below shows a stem-and-leaf plot for the same ages of the best actresses as listed in Table 2-1 from the Chapter Problem. Those ages sorted according to increasing order are 21, 22, 24, 24, . . . , 80. It is easy to see how the first value of 21 is separated into its stem of 2 and leaf of 1. Each of the remaining values is broken up in a similar way. Note that the leaves are arranged in increasing order, not the order in which they occur in the original list.

By turning the stemplot on its side, we can see a distribution of these data. A great advantage of the stem-and-leaf plot is that we can see the distribution of data

Stemplot

Stem (tens)	Leaves (units)
2	1244555566667777888899999
3	00111122333334445555555677888899
4	011111223569
5	04 ←Values are 50 and 54.
6	013
7	4
8	0 ←Value is 80.

and yet retain all the information in the original list. If necessary, we could reconstruct the original list of values. Another advantage is that construction of a stemplot is a quick and easy way to *sort* data (arrange them in order), and sorting is required for some statistical procedures (such as finding a median, or finding percentiles).

The rows of digits in a stemplot are similar in nature to the bars in a histogram. One of the guidelines for constructing histograms is that the number of classes should be between 5 and 20, and the same guideline applies to stemplots for the same reasons. Better stemplots are often obtained by first rounding the original data values. Also, stemplots can be *expanded* to include more rows and can be *condensed* to include fewer rows. See Exercise 26.

The Power of a Graph

With annual sales approaching $10 billion and with roughly 50 million people using it, Pfizer's prescription drug Lipitor has become the most profitable and most used prescription drug ever. In its early stages of development, Lipitor was compared to other drugs (Zocor, Mevacor, Lescol, and Pravachol) in a process that involved controlled trials. The summary report included a graph showing a Lipitor curve that had a steeper rise than the curves for the other drugs, visually showing that Lipitor was more effective in reducing cholesterol than the other drugs. Pat Kelly, who was then a senior marketing executive for Pfizer, said "I will never forget seeing that chart. . . . It was like 'Aha!' Now I know what this is about. We can communicate this!" The Food and Drug Administration approved Lipitor and allowed Pfizer to include the graph with each prescription. Pfizer sales personnel also distributed the graph to physicians.

top 20 Pareto Charts

The Federal Communications Commission monitors the quality of phone service in the United States. Complaints against phone carriers include *slamming*, which is changing a customer's carrier without the customer's knowledge, and *cramming*, which is the insertion of unauthorized charges. Recently, FCC data showed that complaints against U.S. phone carriers consisted of 4473 for rates and services, 1007 for marketing, 766 for international calling, 614 for access charges, 534 for operator services, 12,478 for slamming, and 1214 for cramming. If you were a print media reporter, how would you present that information? Simply writing the sentence with the numerical data is unlikely to result in understanding. A better approach is to use an effective graph, and a Pareto chart would be suitable here.

A **Pareto chart** is a bar graph for qualitative data, with the bars arranged in order according to frequencies. Vertical scales in Pareto charts can represent frequencies or relative frequencies. The tallest bar is at the left, and the smaller bars are farther to the right. By arranging the bars in order of frequency, the Pareto chart focuses attention on the more important categories. Figure 2-7 is a Pareto chart clearly showing that slamming is by far the most serious issue in customer complaints about phone carriers.

Pie Charts

Pie charts are also used to visually depict qualitative data. Figure 2-8 is an example of a **pie chart**, which is a graph depicting qualitative data as slices of a pie.

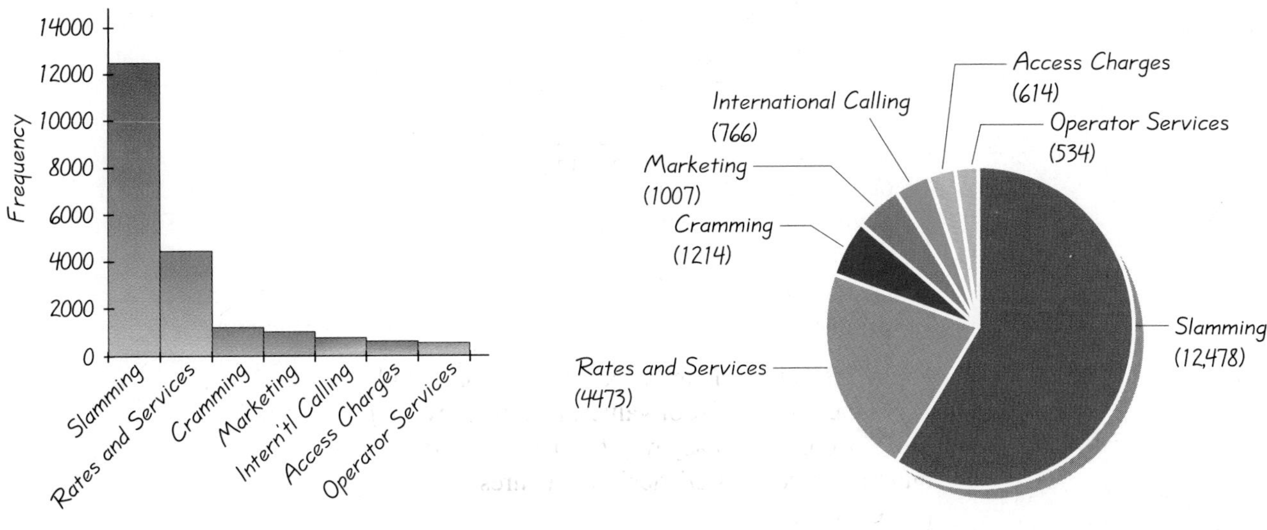

Figure 2-7 Pareto Chart of Phone Company Complaints

Figure 2-8 Pie Chart of Phone Company Complaints

Figure 2-8 represents the same data as Figure 2-7. Construction of a pie chart involves slicing up the pie into the proper proportions. The category of slamming complaints represents 59% of the total, so the wedge representing slamming should be 59% of the total (with a central angle of $0.59 \times 360° = 212°$).

The Pareto chart (Figure 2-7) and the pie chart (Figure 2-8) depict the same data in different ways, but a comparison will probably show that the Pareto chart does a better job of showing the relative sizes of the different components. That helps explain why many companies, such as Boeing Aircraft, make extensive use of Pareto charts.

Scatterplots

A **scatterplot** (or **scatter diagram**) is a plot of paired (x, y) data with a horizontal x-axis and a vertical y-axis. The data are paired in a way that matches each value from one data set with a corresponding value from a second data set. To manually construct a scatterplot, construct a horizontal axis for the values of the first variable, construct a vertical axis for the values of the second variable, then plot the points. The pattern of the plotted points is often helpful in determining whether there is some relationship between the two variables. (This issue is discussed at length when the topic of correlation is considered in Section 10-2.)

One classic use of a scatterplot involves numbers of cricket chirps per minute paired with temperatures (°F). Using data from *The Song of Insects* by George W. Pierce, Harvard University Press, the TI-83/84 Plus scatterplot is shown here. There does appear to be a relationship between chirps and temperature, as shown by the pattern of the points. Crickets can therefore be used as thermometers.

TI-83/84 Plus

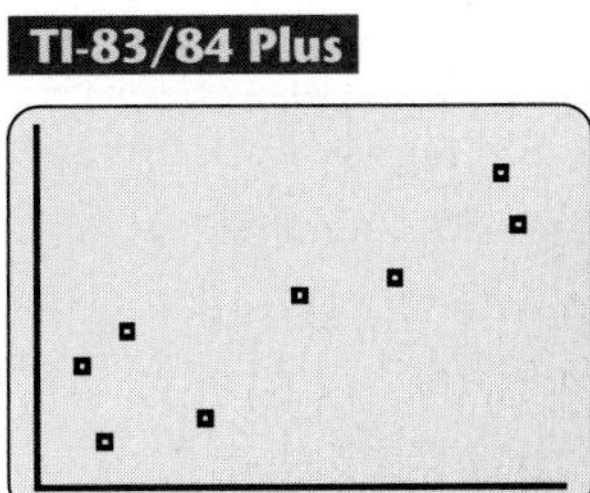

EXAMPLE Clusters Consider the scatterplot of paired data obtained from 16 subjects. For each subject, the weight (in pounds) is measured and the number of times the subject used the television remote control during a period of 1 hour was also recorded. Minitab was used to generate the scatterplot of the paired weight/remote data, and that scatterplot is shown here. This particular scatterplot reveals two very distinct clusters, which can be explained by the inclusion of two different populations: women (with lower weights and less use of the remote control) and men (with higher weights and greater use of the remote control). If we ignored the presence of the clusters, we might think incorrectly that there is a relationship between weight and remote usage. But look at the two groups separately, and it becomes much more obvious that there does *not* appear to be a relationship between weight and usage of the remote control.

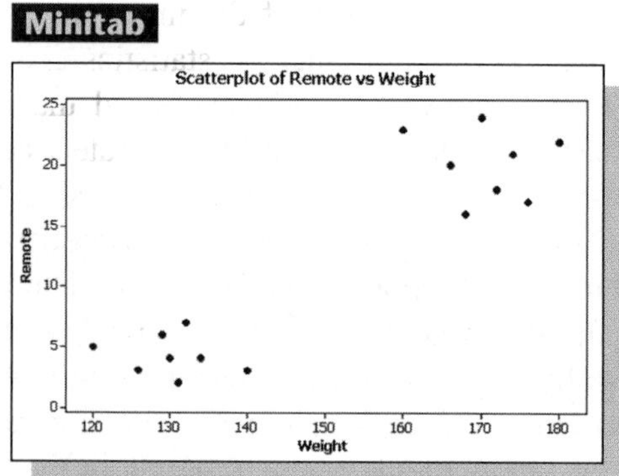

Time-Series Graph

A **time-series graph** is a graph of *time-series data,* which are data that have been collected at different points in time. For example, the accompanying SPSS-generated time-series graph shows the numbers of screens at drive-in movie theaters for a recent period of 17 years (based on data from the National Association of Theater Owners). We can see that for this time period, there is a clear trend of decreasing values. A once significant part of Americana, especially to the author, is undergoing a decline. Fortunately, the rate of decline appears to be less than it was in the late 1980s. It is often critically important to know when population values change over time. Companies have gone bankrupt because they failed to monitor the quality of their goods or services and incorrectly believed that they were dealing with stable data. They did not realize that their products were becoming seriously defective as important population characteristics were changing. Chapter 14 introduces *control charts* as an effective tool for monitoring time-series data.

SPSS Time-Series Graph

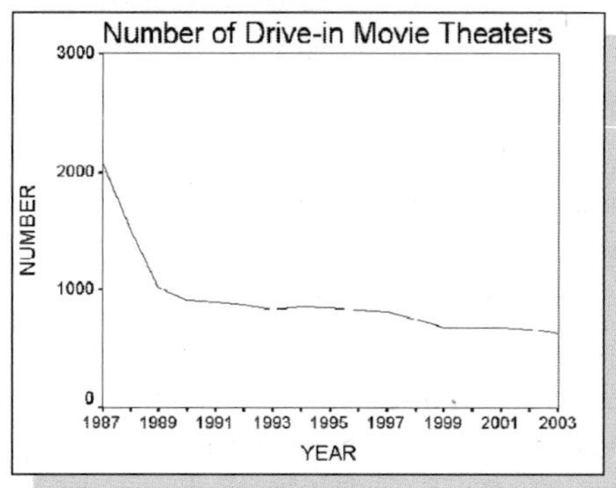

Help Wanted: Statistical Graphics Designer

So far, this section has included some of the important and standard statistical graphs commonly included in introductory statistics courses. There are many other graphs, some of which have not yet been created, that are effective in depicting important and interesting data. The world desperately needs more people with the ability to be creative and original in developing graphs that effectively reveal the nature of data. Currently, graphs found in newspapers, magazines, and television are too often created by reporters with a background in journalism or communications, but with little or no background in formal work with data. It is idealistically but realistically hoped that some readers of this text will recognize that need and, having an interest in this topic, will further study methods of creating statistical graphs. The author strongly recommends careful reading of *The Visual Display of Quantitative Information,* 2nd edition, by Edward Tufte (Graphics Press, PO Box 430, Cheshire, CT 06410). Here are a few of the important principles suggested by Tufte:

- For small data sets of 20 values or fewer, use a table instead of a graph.
- A graph of data should make the viewer focus on the true nature of the data, not on other elements, such as eye-catching but distracting design features.
- Do not distort the data; construct a graph to reveal the true nature of the data.
- Almost all of the ink in a graph should be used for the data, not for other design elements.
- Don't use screening consisting of features such as slanted lines, dots, or cross-hatching, because they create the uncomfortable illusion of movement.
- Don't use areas or volumes for data that are actually one-dimensional in nature. (For example, don't use drawings of dollar bills to represent budget amounts for different years.)
- Never publish pie charts, because they waste ink on non-data components, and they lack an appropriate scale.

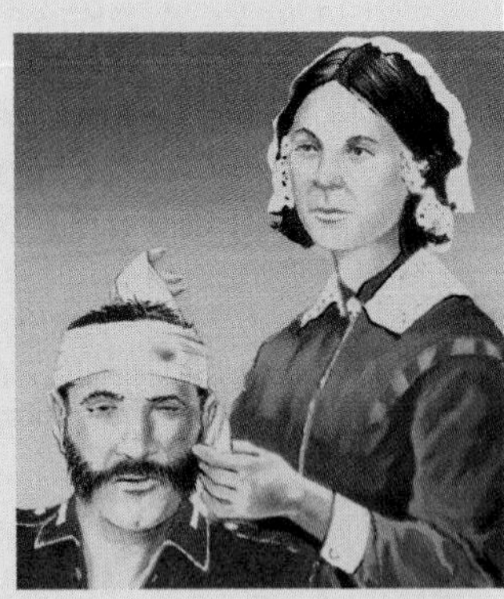

Florence Nightingale

Florence Nightingale (1820–1910) is known to many as the founder of the nursing profession, but she also saved thousands of lives by using statistics. When she encountered an unsanitary and undersupplied hospital, she improved those conditions and then used statistics to convince others of the need for more widespread medical reform. She developed original graphs to illustrate that, during the Crimean War, more soldiers died as a result of unsanitary conditions than were killed in combat. Florence Nightingale pioneered the use of social statistics as well as graphics techniques.

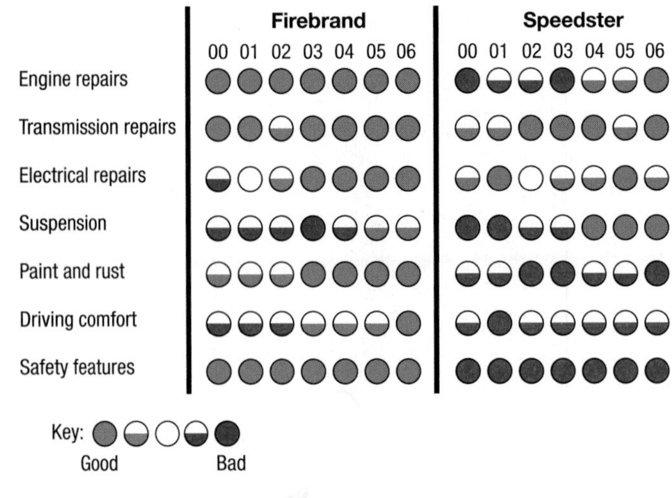

Figure 2-9 Car Reliability Data

Figure 2-9 shows a comparison of two different cars, and it is based on graphs used by *Consumer's Report* magazine. The *Consumer's Report* graphs are based on large numbers of surveys obtained from car owners. Figure 2-9 exemplifies excellence in originality, creativity, and effectiveness in helping the viewer easily see complicated data in a simple format. See the key at the bottom showing that red is used for bad results and green is used for good results, so the color scheme corresponds to the "go" and "stop" used for traffic signals that are so familiar to drivers. (The *Consumer's Report* graphs use red for good results and black for bad results.) We can easily see that over the past several years, the Firebrand car appears to be generally better than the Speedster car. Such information is valuable for consumers considering the purchase of a new or used vehicle.

The figure on the following page has been described as possibly "the best statistical graphic ever drawn." This figure includes six different variables relevant to the march of Napoleon's army to Moscow and back in 1812–1813. The thick band at the left depicts the size of the army when it began its invasion of Russia from Poland. The lower band shows its size during the retreat, along with corresponding temperatures and dates. Although first developed in 1861 by Charles Joseph Minard, this graph is ingenious even by today's standards.

Another notable graph of historical importance is one developed by the world's most famous nurse, Florence Nightingale. This graph, shown in Figure 2-10, is particularly interesting because it actually saved lives when Nightingale used it to convince British officials that military hospitals needed to improve sanitary conditions, treatment, and supplies. It is drawn somewhat like a pie chart, except that the central angles are all the same and different radii are used to show changes in the numbers of deaths each month. The outermost regions of Figure 2-10 represent deaths due to preventable diseases, the innermost regions represent deaths from wounds, and the middle regions represent deaths from other causes.

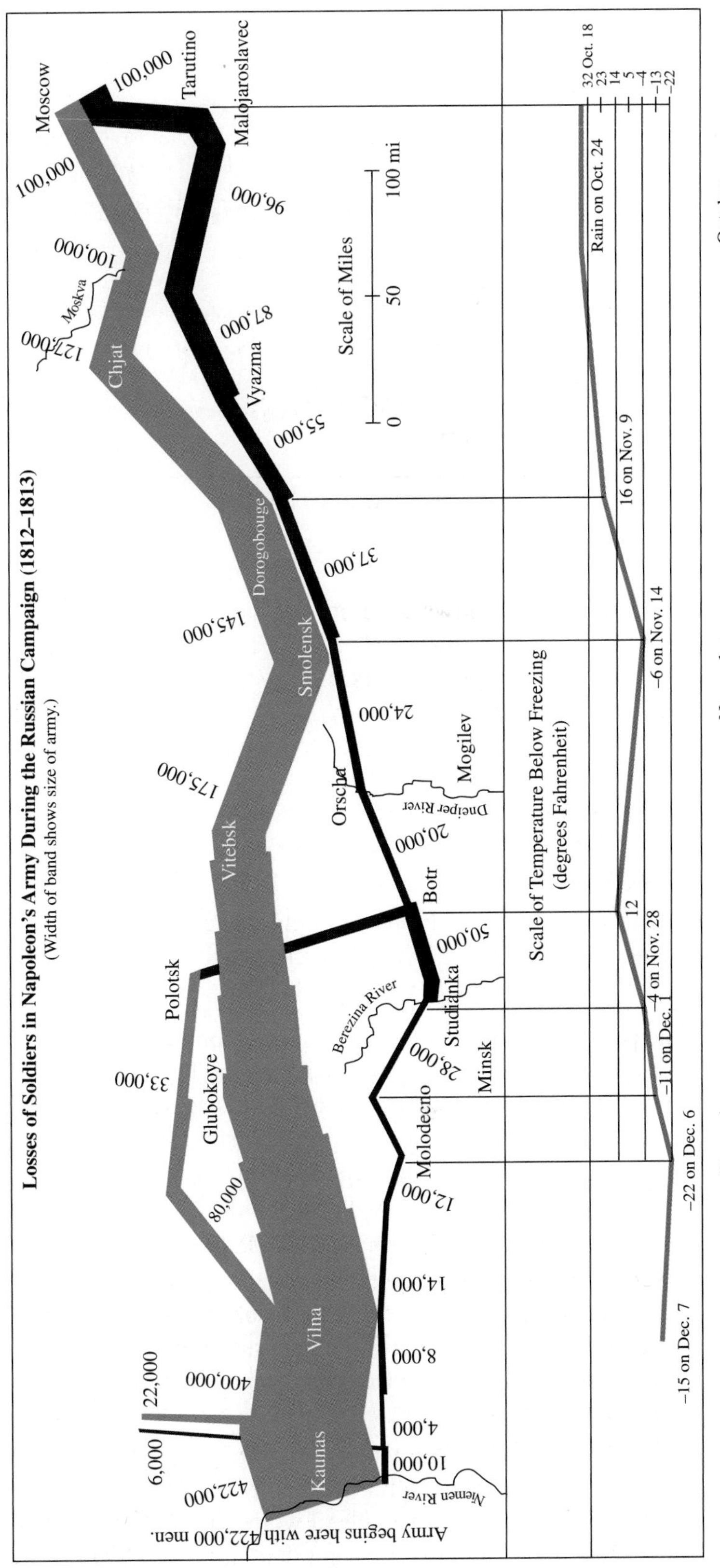

Credit: Edward R. Tufte, *The Visual Display of Quantitative Information* (Cheshire, CT: Graphics Press, 1983). Reprinted with permission.

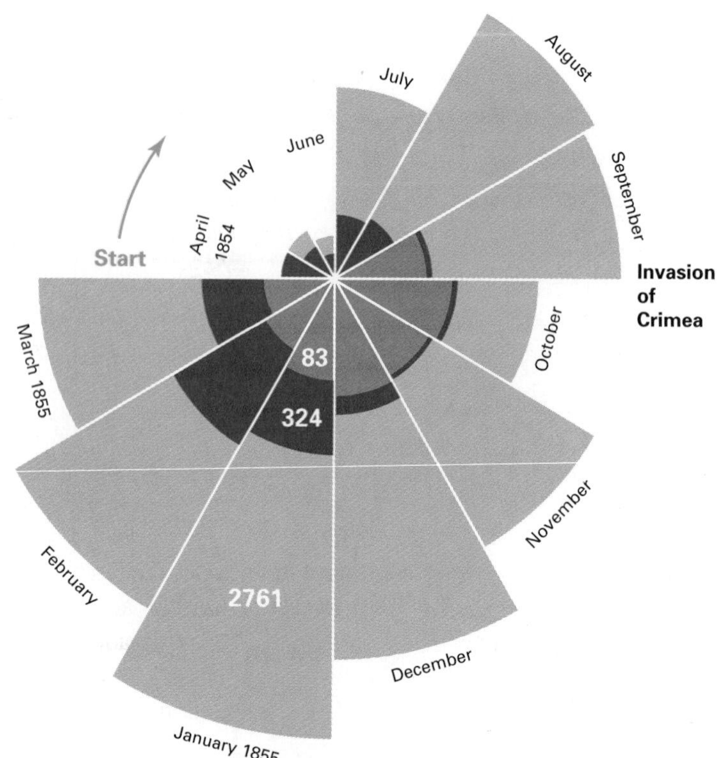

Figure 2-10

Deaths in British Military Hospitals During the Crimean War

Outer region: Deaths due to preventable diseases.
Middle region: Deaths from causes other than wounds or preventable diseases.
Innermost region: Deaths from wounds in battle

Conclusion

The effectiveness of Florence Nightingale's graph illustrates well this important point: A graph is not in itself an end result; it is a tool for describing, exploring, and comparing data, as described below.

Describing data: In a histogram, for example, consider center, variation, distribution, and outliers (CVDOT without the last element of time). What is the approximate value of the center of the distribution, and what is the approximate range of values? Consider the overall shape of the distribution. Are the values evenly distributed? Is the distribution skewed (lopsided) to the right or left? Does the distribution peak in the middle? Is there a large gap, suggesting that the data might come from different populations? Identify any extreme values and any other notable characteristics.

Exploring data: We look for features of the graph that reveal some useful and/or interesting characteristics of the data set. In Figure 2-10, for example, we see that more soldiers were dying from inadequate hospital care than were dying from battle wounds.

Comparing data: Construct similar graphs that make it easy to compare data sets. For example, if you graph a frequency polygon for weights of men and another frequency polygon for weights of women on the same set of axes, the polygon for men should be farther to the right than the polygon for women, showing that men have higher weights.

Using the TI-83/84 Plus Calculator to Construct Frequency Polygons

A TI-83/84 Plus calculator can be used to generate a frequency polygon, such as the one shown in Figure 2-4. Use the following procedure:

1. After constructing a frequency distribution (such as Table 2-2), enter the class midpoint values in list L1 and enter the corresponding frequencies in list L2.
2. Select the "STAT PLOT" function by pressing 2ND Y=.
3. Press ENTER and use the arrow keys to turn Plot1 to the On state, and also highlight the graph type as shown in the display in the margin. Also enter L1 for the Xlist entry and enter L2 for the Ylist entry, as shown in the display.
4. Press ZOOM 9 to obtain the first display shown below. (ZOOM 9 causes the generation of a graph with settings that are automatically determined to fit the screen.) You can use the TRACE and arrow keys to identify frequencies and positions of the points. Note that unlike Figure 2-4, this display does not show a frequency polygon that starts and ends on the horizontal axis. To fix this feature, modify lists L1 and L2 so that they have an additional row at the top and an additional row at the bottom. For these two additional rows, use class midpoint values that follow the same pattern as the other class midpoint values, and use frequencies of 0. The result is the second graph shown below.

TI-83/84 Plus

TI-83/84 Plus

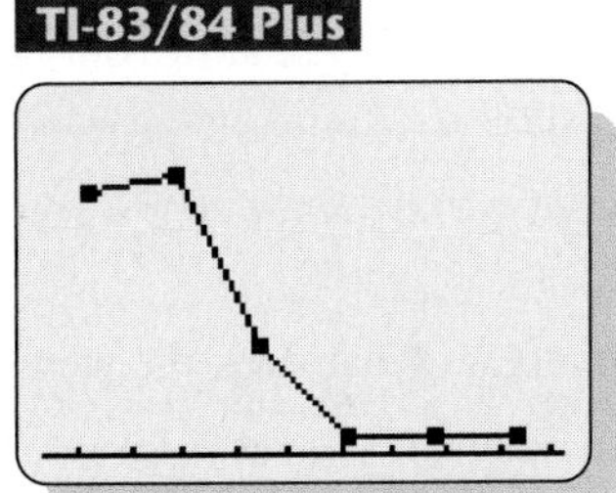

TI-83/84 Plus

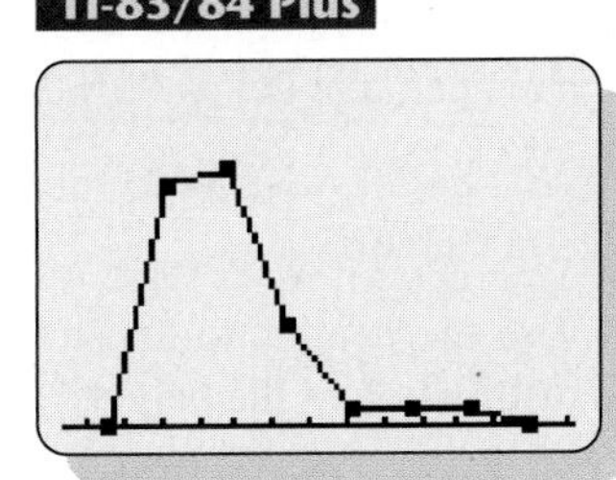

Using the TI-83/84 Plus Calculator to Construct a Scatter Diagram

You can use a TI-83/84 Plus calculator to construct a scatter diagram from a list of paired sample data. Use the following procedure:

1. Enter or open two lists of data with the corresponding entries paired. For example, the TriolaXe software application (APP) includes the lists named MWAST and MWT, which are lists of waist sizes and weights for the 40 males in Data Set 1 in Appendix B. We will illustrate the construction of a scatter diagram by using these two lists.
2. Select the STAT PLOT function by pressing 2ND Y=.
3. Press ENTER and use the arrow keys to turn Plot1 to the On state, and also highlight the graph type as shown in the third line of the display in the margin. Also enter the list names, as shown in the display.
4. Press ZOOM 9 to obtain the scatter diagram. (ZOOM 9 causes the generation of a graph with settings that are automatically determined to fit the screen.) You can press WINDOW to change the settings as you desire.

TI-83/84 Plus

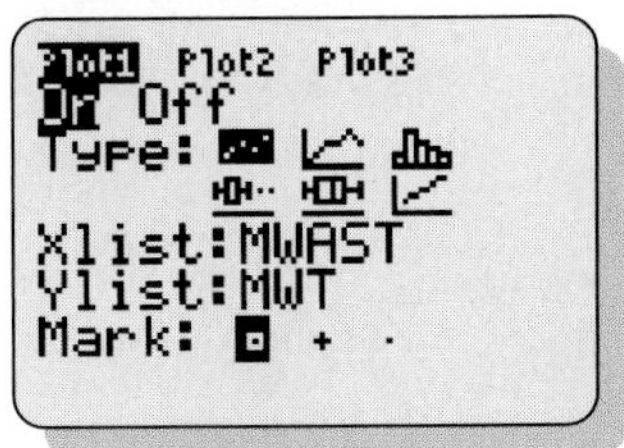

2-4 BASIC SKILLS AND CONCEPTS

Statistical Literacy and Critical Thinking

1. Why Graph? What is the main objective in graphing data?

2. Scatterplot What type of data are required for the construction of a scatterplot, and what does the scatterplot reveal about the data?

3. Time-Series Graph What type of data are required for the construction of a time-series graph, and what does a time-series graph reveal about the data?

4. Pie Chart versus Pareto Chart Why is it generally better to use a Pareto chart instead of a pie chart?

In Exercises 5–8, use the given 35 actual high temperatures listed in Data Set 8 of Appendix B.

5. Dotplot Construct a dotplot of the actual high temperatures. What does the dotplot suggest about the distribution of the high temperatures?

6. Stemplot Use the 35 actual high temperatures to construct a stemplot. What does the stemplot suggest about the distribution of the temperatures?

TI 7. Frequency Polygon Use the 35 actual high temperatures to construct a frequency polygon. For the horizontal axis, use the midpoint values obtained from these class intervals: 50–59, 60–69, 70–79, 80–89.

8. Ogive Use the 35 actual high temperatures to construct an ogive. For the horizontal axis, use these class boundaries: 49.5, 59.5, 69.5, 79.5, 89.5. How many days was the actual high temperature below 80°F?

In Exercises 9–12, use the 40 heights of eruptions of the Old Faithful geyser listed in Data Set 11 of Appendix B.

9. Stemplot Use the heights to construct a stemplot. What does the stemplot suggest about the distribution of the heights?

10. Dotplot Construct a dotplot of the heights. What does the dotplot suggest about the distribution of the heights?

11. Ogive Use the heights to construct an ogive. For the horizontal axis, use these class boundaries: 89.5, 99.5, 109.5, 119.5, 129.5, 139.5, 149.5, 159.5. How many eruptions were below 120 ft?

TI 12. Frequency Polygon Use the heights to construct a frequency polygon. For the horizontal axis, use the midpoint values obtained from these class intervals: 90–99, 100–109, 110–119, 120–129, 130–139, 140–149, 150–159.

13. Jobs A study was conducted to determine how people get jobs. The table below lists data from 400 randomly selected subjects. The data are based on results from the National Center for Career Strategies. Construct a Pareto chart that corresponds to the given data. If someone would like to get a job, what seems to be the most effective approach?

Job Sources of Survey Respondents	Frequency
Help-wanted ads	56
Executive search firms	44
Networking	280
Mass mailing	20

14. Jobs Refer to the data given in Exercise 13, and construct a pie chart. Compare the pie chart to the Pareto chart. Can you determine which graph is more effective in showing the relative importance of job sources?

15. Fatal Occupational Injuries In a recent year, 5524 people were killed while working. Here is a breakdown of causes: transportation (2375); contact with objects or equipment (884); assaults or violent acts (829); falls (718); exposure to harmful substances or a harmful environment (552); fires or explosions (166). (The data are from the Bureau of Labor Statistics.) Construct a pie chart representing the given data.

16. Fatal Occupational Injuries Refer to the data given in Exercise 15 and construct a Pareto chart. Compare the Pareto chart to the pie chart. Which graph is more effective in showing the relative importance of the causes of work-related deaths?

In Exercises 17 and 18, use the given paired data from Appendix B to construct a scatter diagram.

TI **17.** Cigarette Tar/CO In Data Set 3, use tar for the horizontal scale and use carbon monoxide (CO) for the vertical scale. Determine whether there appears to be a relationship between cigarette tar and CO. If so, describe the relationship.

TI **18.** Energy Consumption and Temperature In Data Set 9, use the 10 average daily temperatures and use the corresponding 10 amounts of energy consumption (kWh). (Use the temperatures for the horizontal scale.) Based on the result, is there a relationship between the average daily temperatures and the amounts of energy consumed? Try to identify at least one reason why there is (or is not) a relationship.

In Exercises 19 and 20, use the given data to construct a time-series graph.

19. Runway Near-Hits Given below are the numbers of runway near-hits by aircraft, listed in order for each year beginning with 1990 (based on data from the Federal Aviation Administration). Is there a trend? If so, what is it?

281 242 219 186 200 240 275 292 325 321 421

20. Indoor Movie Theaters Given below are the numbers of indoor movie theaters, listed in order by row for each year beginning with 1987 (based on data from the National Association of Theater Owners). What is the trend? How does this trend compare to the trend for drive-in movie theaters? (A time-series graph for drive-in movie theaters is given in this section.)

20,595 21,632 21,907 22,904 23,740 24,344 24,789 25,830 26,995

28,905 31,050 33,418 36,448 35,567 34,490 35,170 35,361

In Exercises 21–24, refer to the figure in this section that describes Napoleon's 1812 campaign to Moscow and back (see page 71). The thick band at the left depicts the size of the army when it began its invasion of Russia from Poland, and the lower band describes Napoleon's retreat.

21. The number of men who began the campaign is shown as 422,000. Find the number of those men and the percentage of those men who survived the entire campaign.

22. Find the number of men and the percentage of men who died crossing the Berezina River.

23. Of the 320,000 men who marched from Vilna to Moscow, how many of them made it to Moscow? Approximately how far did they travel from Vilna to Moscow?

24. What is the coldest temperature endured by any of the men, and when was that coldest temperature reached?

2-4 BEYOND THE BASICS

25. Back-to-Back Stemplots Refer to the ages of the Best Actresses and Best Actors listed in Table 2-1 in the Chapter Problem. Shown in the margin is a format for *back-to-back stemplots*. The first two ages from each group have been entered. Complete the entries, then compare the results.

Table for Exercise 25

Actresses' Ages (units)	Stem (tens)	Actors' Ages (units)
2	2	
7	3	
	4	14
	5	
	6	
	7	
	8	

26. Expanded and Condensed Stemplots This section includes a stemplot of the ages of the Best Actresses listed in Table 2-1. Refer to that stemplot for the following:

a. The stemplot can be expanded by subdividing rows into those with leaves having digits of 0 through 4 and those with digits 5 through 9. Shown below are the first two rows of the stemplot after it has been expanded. Include the next two rows of the expanded stemplot.

Stem	Leaves	
2	1244	←For leaves of 0 through 4.
2	5555666677778888999999	←For leaves of 5 through 9.

b. The stemplot can be condensed by combining adjacent rows. Shown below is the first row of the condensed stemplot. Note that we insert an asterisk to separate digits in the leaves associated with the numbers in each stem. Every row in the condensed plot must include exactly one asterisk so that the shape of the reduced stemplot is not distorted. Complete the condensed stemplot by identifying the remaining entries.

Stem	Leaves
2-3	1244555566667777888899999 9*0011122333334445555555677888899

Review

In this chapter we considered methods for summarizing and graphing data. When investigating a data set, the characteristics of center, variation, distribution, outliers, and changing pattern over time are generally very important, and this chapter includes a variety of tools for investigating the distribution of the data. After completing this chapter you should be able to do the following:

- Summarize data by constructing a frequency distribution or relative frequency distribution (Section 2-2).
- Visually display the nature of the distribution by constructing a histogram (Section 2-3) or relative frequency histogram.
- Investigate important characteristics of a data set by creating visual displays, such as a frequency polygon, dotplot, stemplot, Pareto chart, pie chart, scatterplot (for paired data), or a time-series graph (Section 2-4).

In addition to creating tables of frequency distributions and graphs, you should be able to *understand* and *interpret* those results. For example, the Chapter Problem includes Table 2-1 with ages of Oscar-winning Best Actresses and Best Actors. Simply examining the two lists of ages probably does not reveal much meaningful information, but frequency distributions and graphs enabled us to see that there does appear to be a significant difference. It appears that the actresses tend to be significantly younger than the actors. This difference can be further explored by considering relevant cultural factors, but methods of statistics give us a great start by pointing us in the right direction.

Statistical Literacy and Critical Thinking

1. Exploring Data When investigating the distribution of a data set, which is more effective: a frequency distribution or a histogram? Why?

2. Comparing Data When comparing two data sets, which is better: frequency distributions or relative frequency distributions? Why?

3. Real Estate A real estate broker is investigating the selling prices of homes in his region over the past 50 years. Which graph would be better: a histogram or a time-series graph? Why?

4. Normal Distribution A histogram is constructed from a set of sample values. What are two key features of the histogram that would suggest that the data have a normal distribution?

Review Exercises

TI **1.** Frequency Distribution of Ages of Best Actors Construct a frequency distribution of the ages of the Oscar-winning actors listed in Table 2-1. Use the same class intervals that were used for the frequency distribution of the Oscar-winning actresses, as shown in Table 2-2. How does the result compare to the frequency distribution for actresses?

TI **2.** Histogram of Ages of Best Actors Construct the histogram that corresponds to the frequency distribution from Exercise 1. How does the result compare to the histogram for actresses (Figure 2-2)?

3. Dotplot of Ages of Best Actors Construct a dotplot of the ages of the Oscar-winning actors listed in Table 2-1. How does the result compare to the dotplot for the ages of the actresses? The dotplot for the ages of the Best Actresses is included in Section 2-4 (see page 65).

4. Stemplot of Ages of Best Actors Construct a stemplot of the ages of the Oscar-winning actors listed in Table 2-1. How does the result compare to the stemplot for the ages of the actresses? The stemplot for the ages of the Best Actresses is included in Section 2-4 (see page 66).

TI **5.** Scatterplot of Ages of Actresses and Actors Refer to Table 2-1 and use only the first 10 ages of actresses and the first 10 ages of actors. Construct a scatterplot. Based on the result, does there appear to be an association between the ages of actresses and the ages of actors?

6. Time-Series Graph Refer to Table 2-1 and use the ages of Oscar-winning actresses. Those ages are listed in order. Construct a time-series graph. Is there a trend? Are the ages systematically changing over time?

Table for Exercises 1–4

Outcome	Frequency
1–5	43
6–10	44
11–15	59
16–20	47
21–25	57
26–30	56
31–35	49
36 or 0 or 00	25

Cumulative Review Exercises

In Exercises 1–4, refer to the frequency distribution in the margin, which summarizes results from 380 spins of a roulette wheel at the Bellagio Hotel and Casino in Las Vegas. American roulette wheels have 38 slots. One slot is labeled 0, another slot is labeled 00, and the remaining slots are numbered 1 through 36.

1. Consider the numbers that result from spins. Do those numbers measure or count anything?

2. What is the level of measurement of the results?

3. Examine the distribution of the results in the table. Given that the last class summarizes results from three slots, is its frequency of 25 approximately consistent with results that would be expected from an unbiased roulette wheel? In general, do the frequencies suggest that the roulette wheel is fair and unbiased?

4. If a gambler learns that the last 500 spins of a particular roulette wheel resulted in numbers that have an average (mean) of 5, can that information be helpful in winning?

5. Consumer Survey The Consumer Advocacy Union mails a survey to 500 randomly selected car owners, and 185 responses are received. One question asks the amount spent for the cars that were purchased. A frequency distribution and histogram are constructed from those amounts. Can those results be used to make valid conclusions about the population of all car owners?

Cooperative Group Activities

1. In-class activity Refer to Figure 2-10 for the graph that Florence Nightingale constructed roughly 150 years ago. That graph illustrates the numbers of soldiers dying from combat wounds, preventable diseases, and other causes. Figure 2-10 is not very easy to understand. Create a new graph that depicts the same data, but create the new graph in a way that greatly simplifies understanding.

2. In-class activity Given below are the ages of motorcyclists at the time they were fatally injured in traffic accidents (based on data from the U.S. Department of Transportation). If your objective is to dramatize the dangers of motorcycles for young people, which would be most effective: histogram, Pareto chart, pie chart, dotplot, stemplot, . . . ? Construct the graph that best meets the objective of dramatizing the dangers of motorcycle driving. Is it okay to deliberately distort data if the objective is one such as saving lives of motorcyclists?

17	38	27	14	18	34	16	42	28
24	40	20	23	31	37	21	30	25
17	28	33	25	23	19	51	18	29

3. Out-of-class activity In each group of three or four students, construct a graph that is effective in addressing this question: Is there a difference between the body mass index (BMI) values for men and for women? (See Data Set 1 in Appendix B.)

TI-83/84 Plus Project

Although manually constructed graphs have a certain primitive charm, they are often considered unsuitable for publications and presentations. Computer-generated graphs or calculator graphs are much better for such purposes. Use a TI-83/84 Plus calculator to generate three histograms: (1) a histogram of the pulse rates of males listed in Data Set 1 in Appendix B; (2) a histogram of the pulse rates of females listed in Data Set 1 in Appendix B; (3) a histogram of the combined list of pulse rates of males and females. If possible, print the three graphs by using a connection to a computer. After obtaining the histograms, compare them. Does it appear that the pulse rates of males and females have similar characteristics? (Later in this book, we will present more formal methods for making such comparisons. See, for example, Section 9-4.)

From Data to Decision

Critical Thinking

Goodness-of-Fit An important issue in statistics is determining whether certain outcomes fit some particular distribution. For example, we could roll a die 60 times to determine whether the outcomes fit the distribution that we would expect with a fair and unbiased die (with all outcomes occurring about the same number of times). Section 11-2 presents a formal method for a *goodness-of-fit* test. This project involves an informal method based on a subjective comparison. We will consider the important issue of car crash fatalities. Car crash fatalities are devastating to the families involved, and they often involve lawsuits and large insurance payments. Listed below are the ages of 100 randomly selected drivers who were killed in car crashes. Also given is a frequency distribution of licensed drivers by age.

Ages (in years) of Drivers Killed in Car Crashes

37	76	18	81	28	29	18	18	27	20
18	17	70	87	45	32	88	20	18	28
17	51	24	37	24	21	18	18	17	40
25	16	45	31	74	38	16	30	17	34
34	27	87	24	45	24	44	73	18	44
16	16	73	17	16	51	24	16	31	38
86	19	52	35	18	18	69	17	28	38
69	65	57	45	23	18	56	16	20	22
77	18	73	26	58	24	21	21	29	51
17	30	16	17	36	42	18	76	53	27

Age	Licensed Drivers (millions)
16–19	9.2
20–29	33.6
30–39	40.8
40–49	37.0
50–59	24.2
60–69	17.5
70–79	12.7
80–89	4.3

Analysis

Convert the given frequency distribution to a relative frequency distribution, then create a relative frequency distribution for the ages of drivers killed in car crashes. Compare the two relative frequency distributions. Which age categories appear to have substantially greater proportions of fatalities than the proportions of licensed drivers? If you were responsible for establishing the rates for auto insurance, which age categories would you select for higher rates? Construct a graph that is effective in identifying age categories that are more prone to fatal car crashes.

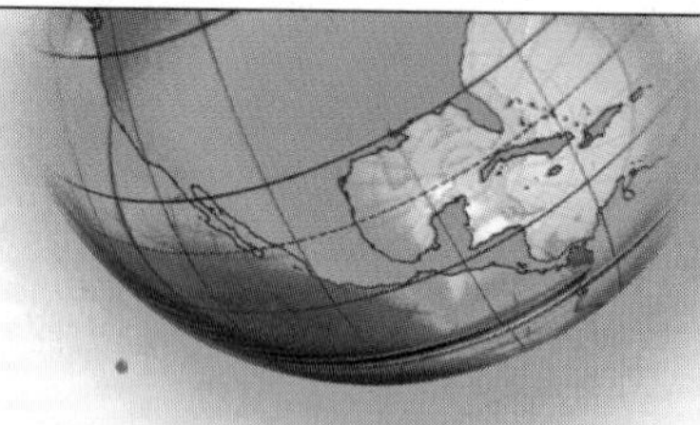

Internet Project

Data on the Internet

The Internet is host to a wealth of information and much of that information comes from raw data that have been collected or observed. Many Web sites summarize such data using the graphical methods discussed in this chapter. For example, we found the following with just a few clicks:

- A bar graph at the site of the U.S. Bureau of Labor Statistics tells us that, at 3%, the unemployment rate is lowest among college graduates versus groups with less education.
- A pie chart provided by the National Collegiate Athletic Association (NCAA) shows that an estimated 89.67% of NCAA revenue in 2004–05 came from television and marketing rights fees.

The Internet Project for this chapter, found at the Elementary Statistics Web site, will further explore graphical representations of data sets found on the Internet. In the process, you will view and collect data sets in the areas of sports, population demographics, and finance, and perform your own graphical analyses.

The Web site for this chapter can be found at

http://www.aw.com/triola

Statistics @ Work

"Statistical applications are tools that can be useful in almost any area of endeavor."

Bob Sehlinger

Publisher, Menasha Ridge Press

Menasha Ridge Press publishes, among many other titles, the *Unofficial Guide* series for John Wiley & Sons (Wiley, Inc.). The Unofficial Guides use statistics extensively to research the experiences that travelers are likely to encounter and to help them make informed decisions that will help them enjoy great vacations.

How do you use statistics in your job and what specific statistical concepts do you use?

We use statistics in every facet of the business: expected value analysis for sales forecasting; regression analysis to determine what books to publish in a series, etc., but we're best known for our research in the areas of queuing and evolutionary computations.

The research methodologies used in the *Unofficial Guide* series are ushering in a truly groundbreaking approach to how travel guides are created. Our research designs and the use of technology from the field of operations research have been cited by academe and reviewed in peer journals for quite some time.

We're using a revolutionary team approach and cutting-edge science to provide readers with extremely valuable information not available in other travel series. Our entire organization is guided by individuals with extensive training and experience in research design as well as data collection and analysis.

From the first edition of the *Unofficial Guide to* our research at Walt Disney World, minimizing our readers' wait in lines has been a top priority. We developed and offered our readers field-tested touring plans that allow them to experience as many attractions as possible with the least amount of waiting in line. We field-tested our approach in the park; the group touring without our plans spent an average of $3\frac{1}{2}$ hours more waiting in line and experienced 37% fewer attractions than did those who used our touring plans.

As we add attractions to our list, the number of possible touring plans grows rapidly. The 44 attractions in the Magic Kingdom One-Day Touring Plan for Adults have a staggering 51,090,942,171,709,440,000 possible touring plans. How good are the new touring plans in the *Unofficial Guide?* Our computer program gets typically within about 2% of the optimal touring plan. To put this in perspective, if the hypothetical "perfect" Adult One-Day touring plan took about 10 hours to complete, the Unofficial touring plan would take about 10 hours and 12 minutes. Since it would take about 30 years for a really powerful computer to find that "perfect" plan, the extra 12 minutes is a reasonable trade-off.

What background in statistics is required to obtain a job like yours?

I work with PhD level statisticians and programmers in developing and executing research designs. I hold an MBA and had a lot of practical experience in operations research before entering publishing, but the main prerequisite in doing the research is knowing enough statistics to see opportunities to use statistics for developing useful information for our readers.

Do you recommend that today's college students study statistics? Why?

Absolutely. In a business context, statistics along with accounting and a good grounding in the mathematics of finance are the quantitative cornerstones. Also, statistics are important in virtually every aspect of life.

Which other skills are important for today's college students?

Good oral and written expression.

Statistics for Describing, Exploring, and Comparing Data

3

CHAPTER PROBLEM

Continued from Chapter 2: Do the Academy Awards involve discrimination based on age?

The opening problem in Chapter 2 included the ages of Oscar-winning Best Actresses and Best Actors. In Chapter 2 we used frequency distributions and graphs to investigate the issue of whether the ages of the actresses are significantly different from the ages of the actors. Based on results found in Chapter 2, the Oscar-winning actresses appear to be somewhat younger than the Oscar-winning actors. In this chapter we continue to investigate the same issue of a discrepancy in ages, but here we introduce new tools that will be helpful in comparing the two sets of ages.

The frequency distributions and graphs in Chapter 2 are not affected by whether the data are a sample or a complete population. However, that distinction does affect some of the tools presented in this chapter. It could be argued that the data constitute a population, because they include the age of *every* Oscar-winning Best Actress and Best Actor from the beginning of the Academy Awards ceremonies in 1928 to the latest results available at the time of this writing. Instead of considering the ages to be population data, we will consider them to be sample data selected from some larger population. Some purists might argue against treating the ages as sample data, but this is a common approach that allows us to address important questions such as this: Is there a significant difference between the average (mean) age of the Best Actresses and the average (mean) age of the Best Actors?

The methods presented in Chapter 2 enabled us to construct frequency distributions and graphs that summarize and visually depict the distribution of data. The methods presented in this chapter enable us to find numerical values of important statistics. (Recall from Chapter 1 that a statistic is a numerical measurement describing some characteristic of a sample, whereas a parameter is a numerical measurement describing some characteristic of a population.) Instead of relying solely on frequency distributions and graphs, we will now proceed to include important statistics as we compare the ages of the Best Actresses and Best Actors. After finding the values of important statistics, we will be better prepared to compare the two sets of ages. We will be better prepared to address this key question: Are there major and important differences between the ages of the Best Actresses and the ages of the Best Actors? We will continue to use the same data from Table 2-1 included with the Chapter Problem from Chapter 2.

3-1 Overview

This chapter is extremely important because it presents basic statistics that describe important characteristics of a set of data. We noted in the Overview (Section 2-1) of Chapter 2 that when describing, exploring, and comparing data sets, these characteristics are usually extremely important: (1) center; (2) variation; (3) distribution; (4) outliers; and (5) changing characteristics of data over time.

Critical Thinking and Interpretation: Going Beyond Formulas

Technology has allowed us to enjoy this principle of modern statistics usage: It is not so important to memorize formulas or manually perform complex arithmetic calculations. Instead, we can focus on obtaining results by using some form of technology (calculator or software), then making practical sense of the results through critical thinking. Keep this in mind as you proceed through this chapter. For example, when studying the extremely important *standard deviation* in Section 3-3, try to see how the key formula serves as a measure of variation, then learn how to find values of standard deviations, but really work on *understanding* and *interpreting* values of standard deviations.

This chapter includes some detailed steps for important procedures, but it is not necessary to master those steps in all cases. However, we recommend that in each case you perform a few manual calculations before using your calculator or computer. Your understanding will be enhanced, and you will acquire a better appreciation of the results obtained from the technology.

The methods of Chapter 2 and this chapter are often called methods of **descriptive statistics,** because the objective is to summarize or *describe* the important characteristics of a set of data. Later in this book we will use methods of **inferential statistics** when we use sample data to make inferences (or generalizations) about a population. With inferential statistics, we are making an inference that goes beyond the known data. Descriptive statistics and inferential statistics are two general divisions of the subject of statistics, and Chapter 2 along with this chapter involve fundamental principles of descriptive statistics.

top 20 3-2 Measures of Center

Key Concept When describing, exploring, and comparing data sets, these characteristics are usually extremely important: center, variation, distribution, outliers, and changes over time. [Remember that the mnemonic of CVDOT (**C**omputer **V**iruses **D**estroy **O**r **T**erminate) is helpful for remembering those characteristics.] The focus of this section is the characteristic of *center*. We want to somehow obtain a number that represents the central value of a data set. The concepts of the *mean* and *median* should be understood extremely well. Specifically, methods for finding the mean and median values should be well known. Also, we should know that the value of the mean can be dramatically affected by the presence of an outlier, but the median is not so sensitive to an outlier. (An outlier is a value that is very far away from almost all of the other values.) Part 1 of this section includes core concepts that should be understood before considering Part 2.

Part 1: Basic Concepts of Measures of Center

Definition

A **measure of center** is a value at the center or middle of a data set.

There are several different ways to determine the center, so we have different definitions of measures of center, including the mean, median, mode, and midrange. We begin with the mean.

Mean

The (arithmetic) mean is generally the most important of all numerical measurements used to describe data, and it is what most people call an *average*.

Definition

The **arithmetic mean** of a set of values is the measure of center found by adding the values and dividing the total by the number of values. This measure of center will be used often throughout the remainder of this text, and it will be referred to simply as the **mean.**

This definition can be expressed as Formula 3-1, which uses the Greek letter Σ (uppercase Greek sigma) to indicate that the data values should be added. That is, Σx represents the sum of all data values. The symbol n denotes the **sample size,** which is the number of values in the data set.

Formula 3-1

$$Mean = \frac{\Sigma x}{n} \quad \begin{array}{l} \leftarrow \text{sum of all sample values} \\ \leftarrow \text{number of sample values} \end{array}$$

The mean is denoted by $\bar{x}$ (pronounced "x-bar") if the data set is a *sample* from a larger population; if all values of the population are used, then we denote the mean by μ (lowercase Greek mu). (Sample statistics are usually represented by English letters, such as $\bar{x}$, and population parameters are usually represented by Greek letters, such as μ.)

Notation

Σ	denotes the *sum* of a set of values.
x	is the *variable* usually used to represent the individual data values.
n	represents the *number of values* in a *sample.*
N	represents the *number of values* in a *population.*
$\bar{x} = \frac{\Sigma x}{n}$	is the mean of a set of *sample* values.
$\mu = \frac{\Sigma x}{N}$	is the mean of all values in a *population.*

STATISTICS IN THE NEWS

Mannequins ≠ Reality

Health magazine compared measurements of mannequins to measurements of women. The following results were reported as "averages," which were presumably means. Height of mannequins: 6 ft; height of women: 5 ft 4 in. Waist of mannequins: 23 in.; waist of women: 29 in. Hip size of mannequins: 34 in.; hip size of women: 40 in. Dress size of mannequins: 6; dress size of women: 11. It becomes apparent that when comparing means, mannequins and real women are very different.

Changing Populations

Included among the five important data set characteristics listed in Chapter 2 is the changing pattern of data over time. Some populations change, and their important statistics change as well. Car seat belt standards haven't changed in 40 years, even though the weights of Americans have increased considerably since then. In 1960, 12.8% of adult Americans were considered obese, compared to 22.6% in 1994.

According to the National Highway Traffic Safety Administration, seat belts must fit a standard crash dummy (designed according to 1960 data) placed in the most forward position, with 4 in. to spare. In theory, 95% of men and 99% of women should fit into seat belts, but those percentages are now lower because of the increases in weight over the last half-century. Some car companies provide seat belt extenders, but some do not.

EXAMPLE Monitoring Lead in Air Lead is known to have some serious adverse affects on health. Listed below are measured amounts of lead (in micrograms per cubic meter, or $\mu g/m^3$) in the air. The Environmental Protection Agency has established an air quality standard for lead: a maximum of 1.5 $\mu g/m^3$. The measurements shown below were recorded at Building 5 of the World Trade Center site on different days immediately following the destruction caused by the terrorist attacks of September 11, 2001. Find the mean for this sample of measured levels of lead in the air.

5.40 1.10 0.42 0.73 0.48 1.10

SOLUTION The mean is computed by using Formula 3-1. First add the values, then divide by the number of values:

$$\bar{x} = \frac{\Sigma x}{n} = \frac{5.40 + 1.10 + 0.42 + 0.73 + 0.48 + 1.10}{6} = \frac{9.23}{6} = 1.538$$

The mean lead level is 1.538 $\mu g/m^3$. Apart from the value of the mean, it is also notable that the data set includes one value (5.40) that is very far away from the others. It would be wise to investigate such an "outlier." In this case, the lead level of 5.40 $\mu g/m^3$ was measured the day after the collapse of the two World Trade Center towers, and there were excessive levels of dust and smoke. Also, some of the lead may have come from the emissions from the large number of vehicles that rushed to the site. These factors provide a reasonable explanation for such an extreme value.

Median

One disadvantage of the mean is that it is sensitive to every value, so one exceptional value can affect the mean dramatically. The median largely overcomes that disadvantage. The median can be thought of as a "middle value" in the sense that about half of the values in a data set are below the median and half are above it. The following definition is more precise.

Definition

The **median** of a data set is the measure of center that is the *middle value* when the original data values are arranged in order of increasing (or decreasing) magnitude. The median is often denoted by $\tilde{x}$ (pronounced "*x*-tilde").

To find the median, first *sort* the values (arrange them in order), then follow one of these two procedures:

1. If the number of values is odd, the median is the number located in the exact middle of the list.
2. If the number of values is even, the median is found by computing the mean of the two middle numbers.

EXAMPLE Monitoring Lead in Air Listed below are measured amounts of lead (in $\mu g/m^3$) in the air. Find the median for this sample.

5.40 1.10 0.42 0.73 0.48 1.10

SOLUTION First sort the values by arranging them in order:

0.42 0.48 0.73 1.10 1.10 5.40

Because the number of values is an even number (6), the median is found by computing the mean of the two middle values of 0.73 and 1.10.

$$\text{Median} = \frac{0.73 + 1.10}{2} = \frac{1.83}{2} = 0.915$$

Because the number of values is an even number (6), the median is the mean of the two middle values, so the median is 0.915 $\mu g/m^3$. Note that the median is very different from the mean of 1.538 $\mu g/m^3$ that was found from the same set of sample data in the preceding example. The reason for this large discrepancy is the effect that 5.40 had on the mean. If this extreme value were reduced to 1.20, the mean would drop from 1.538 $\mu g/m^3$ to 0.838 $\mu g/m^3$, but the median would not change.

EXAMPLE Monitoring Lead in Air Repeat the preceding example after including the measurement of 0.66 $\mu g/m^3$ recorded on another day. That is, find the median of these lead measurements:

5.40 1.10 0.42 0.73 0.48 1.10 0.66

SOLUTION First arrange the values in order:

0.42 0.48 0.66 0.73 1.10 1.10 5.40

Because the number of values is an odd number (7), the median is the value in the exact middle of the sorted list: 0.73 $\mu g/m^3$.

After studying the preceding two examples, the procedure for finding the median should be clear. Also, it should be clear that the mean is dramatically affected by extreme values, whereas the median is not dramatically affected. Because the median is not so sensitive to extreme values, it is often used for data sets with a relatively small number of extreme values. For example, the U.S. Census Bureau recently reported that the *median* household income is \$36,078 annually. The median was used because there is a small number of households with really high incomes.

Class Size Paradox

There are at least two ways to obtain the mean class size, and they can have very different results. At one college, if we take the numbers of students in 737 classes, we get a mean of 40 students. But if we were to compile a list of the class sizes for each student and use this list, we would get a mean class size of 147. This large discrepancy is due to the fact that there are many students in large classes, while there are few students in small classes. Without changing the number of classes or faculty, we could reduce the mean class size experienced by students by making all classes about the same size. This would also improve attendance, which is better in smaller classes.

Mode

Definition

The **mode** of a data set is the value that occurs *most frequently.*

- When two values occur with the same greatest frequency, each one is a mode and the data set is **bimodal.**
- When more than two values occur with the same greatest frequency, each is a mode and the data set is said to be **multimodal.**
- When no value is repeated, we say that there is **no mode.**

EXAMPLE Find the modes of the following data sets:

a. 5.40 1.10 0.42 0.73 0.48 1.10

b. 27 27 27 55 55 55 88 88 99

c. 1 2 3 6 7 8 9 10

SOLUTION

a. The number 1.10 is the mode because it is the value that occurs most often.

b. The numbers 27 and 55 are both modes because they occur with the same greatest frequency. This data set is bimodal because it has two modes.

c. There is no mode because no value is repeated.

In reality, the mode isn't used much with numerical data. But among the different measures of center we are considering, the mode is the only one that can be used with data at the nominal level of measurement. (Recall that the nominal level of measurement applies to data that consist of names, labels, or categories only.)

Midrange

Definition

The **midrange** is the measure of center that is the value midway between the maximum and minimum values in the original data set. It is found by adding the maximum data value to the minimum data value and then dividing the sum by 2, as in the following formula:

$$\text{midrange} = \frac{\text{maximum value} + \text{minimum value}}{2}$$

The midrange is rarely used. Because it uses only the maximum and minimum values, it is too sensitive to those extremes. However, the midrange does have three redeeming features: (1) It is easy to compute; (2) it helps to reinforce the important point that there are several different ways to define the center of a data set; (3) it is sometimes incorrectly used for the median, so confusion can be reduced by clearly defining the midrange along with the median.

Table 3-1 Comparison of Ages of Best Actresses and Best Actors

	Best Actresses	Best Actors
Mean	35.7	43.9
Median	33.5	42.0
Mode	35	41 and 42
Midrange	50.5	52.5

EXAMPLE Monitoring Lead in Air Listed below are measured amounts of lead (in $\mu g/m^3$) in the air from the site of the World Trade Center on different days after September 11, 2001. Find the midrange for this sample.

5.40 1.10 0.42 0.73 0.48 1.10

SOLUTION The midrange is found as follows:

$$\frac{\text{maximum value} + \text{minimum value}}{2} = \frac{5.40 + 0.42}{2} = 2.910$$

The midrange is 2.910 $\mu g/m^3$.

Unfortunately, the term *average* is sometimes used for any measure of center and is sometimes used for the mean. Because of this ambiguity, the term *average* should not be used when referring to a particular measure of center. Instead, we should use the specific term, such as mean, median, mode, or midrange.

In the spirit of describing, exploring, and comparing data, we provide Table 3-1, which summarizes the different measures of center for the ages of Oscar-winning actresses and actors in Table 2-1 in the Chapter Problem for Chapter 2. A comparison of the measures of center suggests that the Best Actresses are younger than the Best Actors. There are methods for determining whether such apparent differences are statistically significant, and we will consider some of those methods later in this book. (See Section 9-3.)

Round-Off Rule

A simple rule for rounding answers is this:

Carry one more decimal place than is present in the original set of values.

When applying this rule, round only the final answer, *not intermediate values that occur during calculations*. Thus the mean of 2, 3, 5, is 3.333333 . . . , which is rounded to 3.3. Because the original values are whole numbers, we rounded to the nearest tenth. As another example, the mean of 80.4 and 80.6 is 80.50 (one more decimal place than was used for the original values).

Critical Thinking

Given a set of sample data, it is relatively easy to calculate measures of center, such as the mean and median. However, we should always think about whether the results make sense. In Section 1-2 we noted that it does not make sense to do numerical calculations with data at the nominal level of measurement. Because data at the nominal level of measurement consist of names, labels, or categories only, it does not make sense to find the value of statistics such as the mean and median. We should also think about the method used to collect the sample data. If the method is not sound, the statistics that we obtain may be misleading.

EXAMPLE Critical Thinking and Measures of Center For each of the following we can find measures of center, such as the mean and median. Identify a major reason why the mean and median are *not* statistics that accurately and effectively serve as measures of center.

a. Zip codes: 12601 90210 02116 76177 19102

b. Ranks of stress levels from different jobs: 2 3 1 7 9

c. Surveyed respondents are coded as 1 (for Democrat), 2 (for Republican), 3 (for Liberal), 4 (for Conservative), or 5 (for any other political party).

SOLUTION

a. The zip codes don't measure or count anything at all. The numbers are actually labels for geographic locations. The mean and median are meaningless statistics.

b. The ranks reflect an ordering, but they don't measure or count anything. The rank of 1 might come from a job that has a stress level substantially greater than the stress level from the job with a rank of 2, so the different numbers don't correspond to the magnitudes of the stress levels. The mean and median are meaningless statistics.

c. The coded results are numbers, but these numbers don't measure or count anything. These numbers are simply different ways of expressing names. Consequently, the mean and median are meaningless statistics.

The preceding example involved data with levels of measurement that do not justify the use of statistics such as the mean or median. The next example involves a more subtle issue.

EXAMPLE Mean Salary of Teachers For each of the 50 states, a researcher obtains the mean salary of secondary school teachers (data from the National Education Association):

$37,200 $49,400 $40,000 . . . $37,800

The mean of the 50 amounts is $42,210, but is this the mean salary of all secondary school teachers in the United States? Why or why not?

SOLUTION No, $42,210 is not necessarily the mean of all secondary school teachers in the United States. The issue here is that some states have many more secondary school teachers than others. The calculation of the mean for the United States should take into account the number of secondary school teachers in each state. (The mean for a frequency distribution is discussed after this example.) The mean for all secondary school teachers in the United States is $45,200, not $42,210.

The preceding example illustrates that the mean of a population is not necessarily equal to the mean of the means found from different subsets of the population.

Instead of mindlessly going through the mechanics of calculating means and medians, the exercises of this section encourage *thinking* about the calculated results.

Part 2: Beyond the Basics of Measures of Center

Mean from a Frequency Distribution

When working with data summarized in a frequency distribution, we don't know the exact values falling in a particular class. To make calculations possible, pretend that in each class, all sample values are equal to the class midpoint. For example, consider the class interval of 21–30 with a frequency of 28 (as in Table 2-2). We pretend that all 28 values are equal to 25.5 (the class midpoint). With the value of 25.5 repeated 28 times, we have a total of $25.5 \cdot 28 = 714$. We can then add those products from each class to find the total of all sample values, which we then divide by the number of sample values. (The number of sample values is the sum of the frequencies Σf.) Formula 3-2 is used to compute the mean when the sample data are summarized in a frequency distribution. Formula 3-2 is not really a new concept; it is simply a variation of Formula 3-1.

First multiply each frequency and class midpoint, then add the products.
↓

Formula 3-2

$$\bar{x} = \frac{\Sigma(f \cdot x)}{\Sigma f} \quad \text{(mean from frequency distribution)}$$

↑
Sum of frequencies

For example, see Table 3-2. The first two columns duplicate the frequency distribution (Table 2-2) for the ages of Oscar-winning actresses. Table 3-2

Table 3-2 Finding the Mean from a Frequency Distribution

Age of Actress	Frequency f	Class Midpoint x	$f \cdot x$
21–30	28	25.5	714
31–40	30	35.5	1065
41–50	12	45.5	546
51–60	2	55.5	111
61–70	2	65.5	131
71–80	2	75.5	151
Totals:	$\Sigma f = 76$		$\Sigma(f \cdot x) = 2718$

$$\bar{x} = \frac{\Sigma(f \cdot x)}{\Sigma f} = \frac{2718}{76} = 35.8$$

illustrates the procedure for using Formula 3-2 when calculating a mean from data summarized in a frequency distribution. Table 3-2 results in $\bar{x} = 35.8$, but we get $\bar{x} = 35.7$ if we use the original list of 76 individual values. Remember, the frequency distribution yields an approximation of $\bar{x}$, because it is not based on the exact original list of sample values.

Weighted Mean

In some cases, the values vary in their degree of importance, so we may want to weight them accordingly. We can then proceed to compute a **weighted mean** of the x values, which is a mean computed with the different values assigned different weights denoted by w in Formula 3-3.

Formula 3-3 $$\text{weighted mean: } \bar{x} = \frac{\Sigma(w \cdot x)}{\Sigma w}$$

Formula 3-3 tells us to multiply each weight w by the corresponding value x, then add the products, then divide that total by the sum of the weights w. For example, suppose we need a mean of 3 test scores (85, 90, 75), but the first test counts for 20%, the second test counts for 30%, and the third test counts for 50% of the final grade. We can assign weights of 20, 30, and 50 to the test scores, then proceed to calculate the mean by using Formula 3-3 with $x = 85, 90, 75$ and the corresponding weights of $w = 20, 30, 50$, as shown here:

$$\begin{aligned}\bar{x} &= \frac{\Sigma(w \cdot x)}{\Sigma w} \\ &= \frac{(20 \times 85) + (30 \times 90) + (50 \times 75)}{20 + 30 + 50} \\ &= \frac{8150}{100} = 81.5\end{aligned}$$

The Best Measure of Center

So far, we have considered the mean, median, mode, and midrange as measures of center. Which one of these is best? Unfortunately, there is no single best answer to that question because there are no objective criteria for determining the most representative measure for all data sets. The different measures of center have different advantages and disadvantages, some of which are shown in Figure 3-1. Throughout the remainder of this book, the mean will be used often, the median will be used occasionally, the mode will not be used, and the midrange will not be used.

The mean is relatively *reliable*, so that when samples are selected from the same population, sample means tend to be more consistent than other measures of center. That is, the means of samples drawn from the same population don't vary as much as the other measures of center. Another important advantage of the mean is that it takes every value into account, but an important disadvantage is that it is sometimes dramatically affected by a few extreme values. This disadvantage can be overcome by using a trimmed mean, as described in Exercise 29.

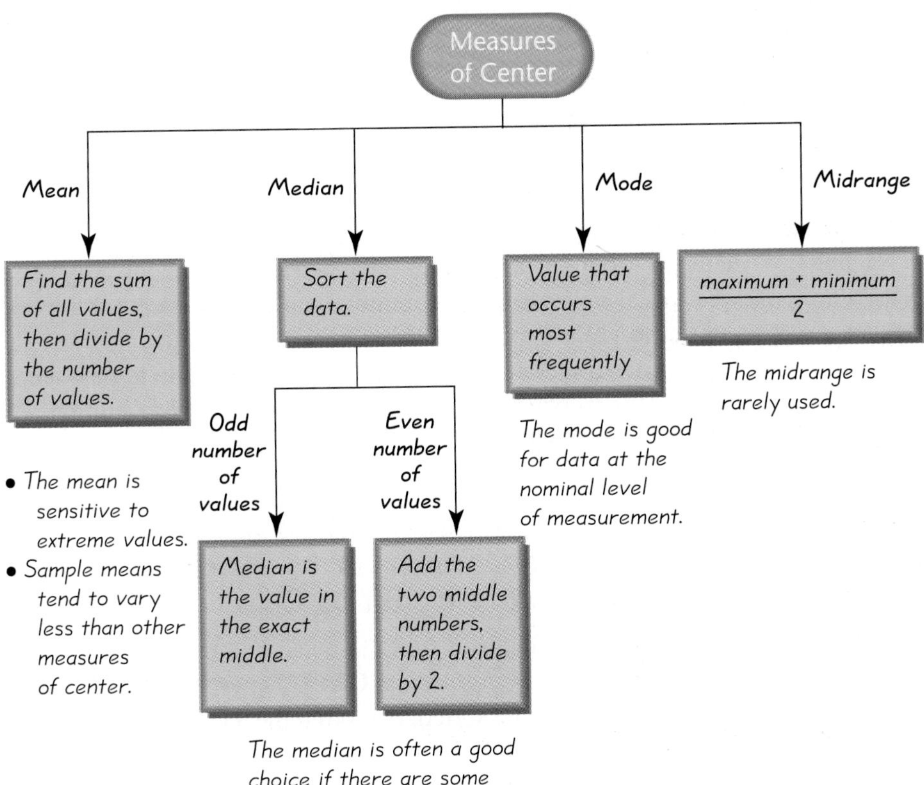

Figure 3-1

Measures of Center

Skewness

A comparison of the mean, median, and mode can reveal information about the characteristic of skewness, defined below and illustrated in Figure 3-2.

Definition

A distribution of data is **skewed** if it is not symmetric and extends more to one side than the other. (A distribution of data is **symmetric** if the left half of its histogram is roughly a mirror image of its right half.)

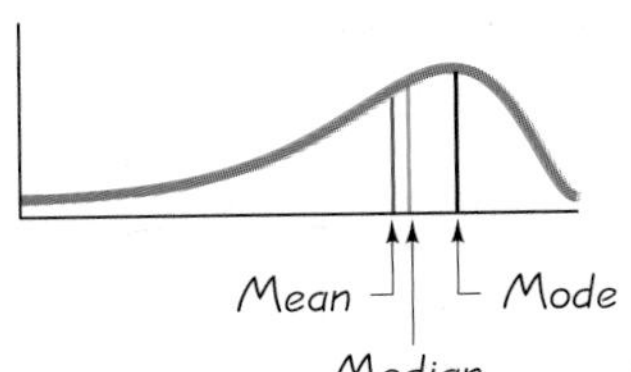

(a) Skewed to the Left (Negatively Skewed): The mean and median are to the *left* of the mode.

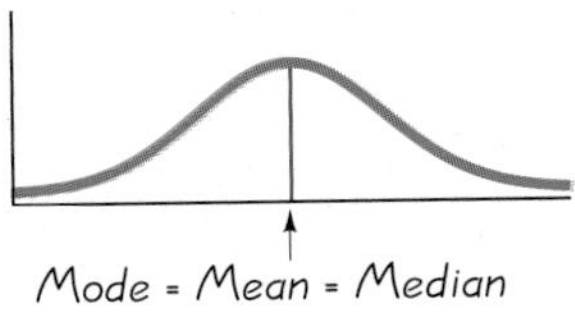

(b) Symmetric (Zero Skewness): The mean, median, and mode are the same.

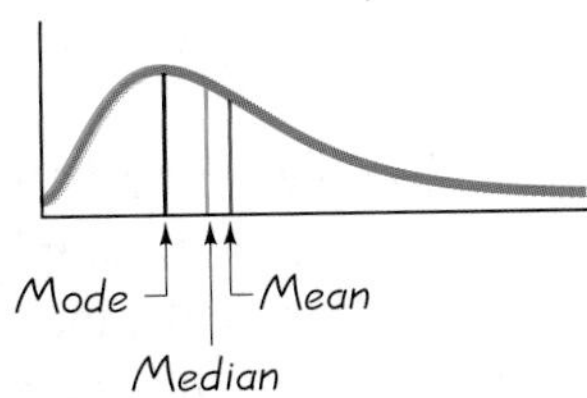

(c) Skewed to the Right (Positively Skewed): The mean and median are to the *right* of the mode.

Figure 3-2

Skewness

Data **skewed to the left** (also called *negatively skewed*) have a longer left tail, and the mean and median are to the left of the mode. [Although not always predictable, data skewed to the left generally have a mean less than the median, as in Figure 3-2(a)]. Data **skewed to the right** (also called *positively skewed*) have a longer right tail, and the mean and median are to the right of the mode. [Again, although not always predictable, data skewed to the right generally have the mean to the right of the median, as in Figure 3-2(c)].

In practice, many distributions of data are symmetric and without skewness. Distributions skewed to the right are more common than those skewed to the left because it's often easier to get exceptionally large values than values that are exceptionally small. With annual incomes, for example, it's impossible to get values below the lower limit of zero, but there are a few people who earn millions of dollars in a year. Annual incomes therefore tend to be skewed to the right, as in Figure 3-2(c).

TI

Using the TI-83/84 Plus Calculator to Find Descriptive Statistics from a List of Data

You can use a TI-83/84 Plus calculator to find the mean and median of a set of data. The basic approach is to press STAT, select **CALC,** select **1-Var Stats,** and enter the name or label of the list to be sorted. Although the TI-83/84 Plus calculator does not find the midrange or mode, it has features that are helpful in finding those values. Use the following procedure:

1. Enter or open a list of data. You could open any of the lists of data in Appendix B by using the APP named TriolaXe, you could use any of the lists L1, L2, . . . , L6, or you could use a list of values assigned to a name. We will proceed by assuming that the 76 ages of actresses in Table 2-1 have been assigned to the name of AGES.
2. Press STAT, then highlight the top menu item of **CALC** by pressing ▷.
3. We want the first menu item of **1-Var Stats,** and it is the default, so press ENTER.
4. The screen should display 1-Var Stats, and we must now identify the list to be used for the calculations. Either enter the name of a list (by pressing 2ND STAT and using the arrow keys to select the desired name) or enter L1 or L2 or L3 or L4 or L5 or L6. If we want to find measures of center for the data listed under the name of AGES, the screen should display 1-Var Stats LAGES.
5. Press ENTER to get results such as those shown in the margin. Note that there are more results than can be shown in one screen, so the arrow on the bottom of the first screen indicates that there are more results available, and you can access them by pressing ▽.
6. *Interpret the results:* The displays shown in the margin indicate that the mean is given by $\bar{x} = 35.68421053$ years, and we should round this result to 35.7 years by using one more decimal place than the original data. The median is 33.5 years. Because the minimum value (21) and the maximum value (80) are displayed, it is easy to find the midrange, which is $(80 + 21)/2 = 50.5$ years. [The midrange is (maximum value + minimum value)/2.]

TI-83/84 Plus

```
1-Var Stats
 x̄=35.68421053
 Σx=2712
 Σx²=105956
 Sx=11.06370707
 σx=10.9906785
↓n=76
```

TI-83/84 Plus

```
1-Var Stats
↑n=76
 minX=21
 Q1=28
 Med=33.5
 Q3=39.5
 maxX=80
```

Using *Sort* to find Mode Although the TI-83/84 Plus calculator does not automatically find the mode from a data set, finding the mode is made easier by using the ability of the calculator to sort data (arrange them in order). To sort a list of data, press STAT. Select the second item of **SortA** to sort the data in *ascending* order, or select the third item of **SortD** to sort the data in *descending* order. After selecting SortA or SortD, press ENTER and enter the name of the list to be sorted, or enter L1, L2, L3 L4, L5, or L6. For example, the command **SortA(LAGES)** causes the values in the list named AGES to be arranged in increasing order. The command **SortA(L1)** causes the values in the list L1 to be arranged in increasing order. After arranging the data in order, view *the data set by scrolling through the* list looking for repeated values. [*Helpful hint:* If the data are in a named list, such as AGES, it is probably easier to store the sorted list in L1 for better viewing in a vertical format. To store the values in the list AGES in the list L1, first select the list name of AGES by pressing 2ND STAT, then scroll to select the name of AGES and press ENTER. You should now see LAGES. Press STO› 2ND 1 so that the screen will display LAGES → L1. The values in the list AGES will now be in list L1. Press STAT ENTER to view the values in list L1.]

TI

Using the TI-83/84 Plus Calculator to Find Descriptive Statistics from a Frequency Distribution

You can use a TI-83/84 Plus calculator to find the mean and median of data represented in the form of a frequency distribution. The basic approach is to enter the class midpoints in list L1, then enter the corresponding class frequencies in list L2, then use STAT and **CALC** and **1-Var Stats** to enter the command **1-Var Stats L1, L2.** Use the following procedure:

1. Given a frequency distribution, enter the class midpoint values in list L1 and enter the corresponding class frequencies in list L2. For example, the frequency distribution of Table 2-2 is reproduced in the margin along with the corresponding TI-83/84 Plus screen display. (Remember, you can enter values in L1 and L2 by pressing STAT and ENTER.)
2. Now press STAT, use the arrow keys to select **CALC,** then select the first menu item of **1-Var Stats,** and press ENTER.
3. With 1-Var Stats displayed on the screen, enter **L1, L2** (including the comma) so that the complete command of 1-Var Stats L1, L2 appears on the screen. (To enter L1, L2, use this sequence of keystrokes: 2ND 1 , 2ND 2.) Press ENTER to get the results, which include the mean and median.

Table 2-2 Frequency Distribution: Ages of Best Actresses

Age of Actress	Frequency
21–30	28
31–40	30
41–50	12
51–60	2
61–70	2
71–80	2

TI-83/84 Plus

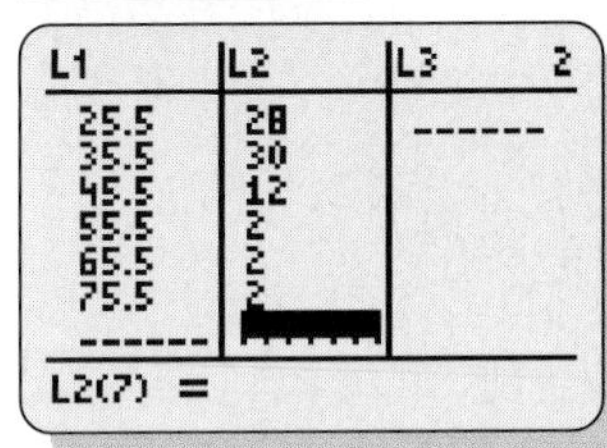

3-2 BASIC SKILLS AND CONCEPTS

Statistical Literacy and Critical Thinking

1. Measures of Center In what sense are the mean, median, mode, and midrange measures of "center"?
2. Mean and Median A statistics class consists of 24 students, all of whom are unemployed or are employed in low-paying part-time jobs. The class also includes a professor who is paid an enormous salary. Which does a better job of describing the

income of a typical person in the class of 25 people (including the professor): the mean or median? Why?

3. Nominal Data In Chapter 1 it was noted that data are at the nominal level of measurement if they consist of names or labels only. A New England Patriots football fan records the number on the jersey of each Patriots player in a Super Bowl game. Does it make sense to calculate the mean of those numbers? Why or why not?

4. Mean Commuting Time A sociologist wants to find the mean commuting time for all working U.S. residents. She knows that it is not practical to survey each of the millions of working people, so she conducts an Internet search and finds the mean commuting time for each of the 50 states. She adds those 50 times and divides by 50. Is the result likely to be a good estimate of the mean commuting time for all workers? Why or why not?

In Exercises 5–12, find the (a) mean, *(b)* median, *(c)* mode, *and (d)* midrange *for the given sample data. Also, answer the given questions.*

TI **5.** Perception of Time Statistics students participated in an experiment to test their ability to determine when 1 minute (or 60 seconds) has passed. The results are given below in seconds. Identify at least one good reason why the mean from this sample might not be a good estimate of the mean for the population of adults.

53 52 75 62 68 58 49 49

TI **6.** Cereal A dietician obtains the amounts of sugar (in centigrams) from 100 centigrams (or 1 gram) in each of 10 different cereals, including Cheerios, Corn Flakes, Fruit Loops, and 7 others. Those values are listed below. Is the mean of those values likely to be a good estimate of the mean amount of sugar in each gram of cereal consumed by the population of all Americans who eat cereal? Why or why not?

3 24 30 47 43 7 47 13 44 39

TI **7.** Phenotypes of Peas An experiment was conducted to determine whether a deficiency of carbon dioxide in the soil affects the phenotypes of peas. Listed below are the phenotype codes, where 1 = smooth-yellow, 2 = smooth-green, 3 = wrinkled-yellow, and 4 = wrinkled-green. Can the measures of center be obtained for these values? Do the results make sense?

2 1 1 1 1 1 1 4 1 2 2 1 2 3 3 2 3 1 3 1 3 1 3 2 2

TI **8.** Old Faithful Geyser Listed below are intervals (in minutes) between eruptions of the Old Faithful geyser in Yellowstone National Park. After each eruption, the National Park Service provides an estimate of the length of time to the next eruption. Based on these values, what appears to be the best single value that could be used as an estimate of the time to the next eruption? Is that estimate likely to be accurate to within 5 minutes?

98 92 95 87 96 90 65 92 95 93 98 94

TI **9.** Blood Pressure Measurements Fourteen different second-year medical students at Bellevue Hospital measured the blood pressure of the same person. The systolic readings (in mmHg) are listed below. What is notable about this data set?

138 130 135 140 120 125 120 130 130 144 143 140 130 150

TI **10.** Body Temperatures Researchers at the University of Maryland collected body temperature readings from a sample of adults, and eight of those temperatures are listed below (in degrees Fahrenheit). Does the mean of this sample equal 98.6, which is commonly believed to be the mean body temperature of adults?

98.6 98.6 98.0 98.0 99.0 98.4 98.4 98.4

TI **11.** Fruit Flies Listed below are the thorax lengths (in millimeters) of a sample of male fruit flies. Fruit flies (*Drosophila*) are a favorite subject of researchers because they have a simple chromosome composition, they reproduce quickly, they have large numbers of offspring, and they are easy to care for. (These are characteristics of the fruit flies, not necessarily the researchers.) If we learn that the listed measurements were obtained from fruit flies hovering over an apple sitting on a kitchen table in Pocatello, does the mean serve as a reasonable estimate of all fruit flies residing in the United States?

0.72 0.90 0.84 0.68 0.84 0.90 0.92 0.84 0.64 0.84 0.76

TI **12.** Personal Income Listed below are the amounts of personal income per capita (in dollars) for the first five states listed in alphabetical order: Alabama, Alaska, Arizona, Arkansas, and California (data from the U.S. Bureau of Economic Analysis). When the 45 amounts from the other states are included, the mean of the 50 state amounts is $29,672.52. Is $29,672.52 the mean amount of personal income per capita for all individuals in the United States? In general, can you find the mean for some characteristic of all residents of the United States by finding the mean for each state, then finding the mean of those 50 results?

$25,128 $32,151 $26,183 $23,512 $32,996

In Exercises 13–20, find the mean *and* median *for each of the two samples, then compare the two sets of results.*

TI **13.** It's Raining Cats Statistics are sometimes used to compare or identify authors of different works. The lengths of the first 20 words in the foreword written by Tennessee Williams in *Cat on a Hot Tin Roof* are listed along with the first 20 words in *The Cat in the Hat* by Dr. Seuss. Does there appear to be a difference?

Cat on a Hot Tin Roof: 2 6 2 2 1 4 4 2 4 2 3 8 4 2 2 7 7 2 3 11

The Cat in the Hat: 3 3 3 3 5 2 3 3 3 2 4 2 2 3 2 3 5 3 4 4

TI **14.** Ages of Stowaways The *Queen Mary* sailed between England and the United States, and stowaways were sometimes found on board. The ages (years) of stowaways from eastbound crossings and westbound crossings are given below (data from the Cunard Steamship Co., Ltd.). Compare the two data sets.

Eastbound: 24 24 34 15 19 22 18 20 20 17

Westbound: 41 24 32 26 39 45 24 21 22 21

TI **15.** Weather Forecast Accuracy In an analysis of the accuracy of weather forecasts, the actual high temperatures are compared to the high temperatures predicted one day earlier and the high temperatures predicted five days earlier. Listed below are the errors between the predicted temperatures and the actual high temperatures for consecutive days in Dutchess County, New York. Do the means and medians indicate that the temperatures predicted one day in advance are more accurate than those predicted five days in advance, as we might expect?

(actual high) − (high predicted one day earlier): 2 2 0 0 −3 −2 1 −2 8 1 0 −1 0 1

(actual high) − (high predicted five days earlier): 0 −3 2 5 −6 −9 4 −1 6 −2 −2 −1 6 −4

TI **16.** Treatment Effect Researchers at Pennsylvania State University conducted experiments with poplar trees. Listed below are weights (kg) of poplar trees given no treatment and poplar trees treated with fertilizer and irrigation. Does there appear to be a difference between the two means? Does the fertilizer and irrigation treatment appear to be effective in increasing the weights of poplar trees?

No treatment:	0.15	0.02	0.16	0.37	0.22
Fertilizer and irrigation:	2.03	0.27	0.92	1.07	2.38

TI **17.** Customer Waiting Times Waiting times (in minutes) of customers at the Jefferson Valley Bank (where all customers enter a single waiting line) and the Bank of Providence (where customers wait in individual lines at three different teller windows) are listed below. Determine whether there is a difference between the two data sets that is not apparent from a comparison of the measures of center. If so, what is it?

Jefferson Valley (single line):	6.5	6.6	6.7	6.8	7.1	7.3	7.4	7.7	7.7	7.7
Providence (individual lines):	4.2	5.4	5.8	6.2	6.7	7.7	7.7	8.5	9.3	10.0

TI **18.** Regular/Diet Coke Weights (in pounds) of samples of the contents in cans of regular Coke and Diet Coke are listed below. Does there appear to be a significant difference between the two data sets? How might such a difference be explained?

Regular:	0.8192	0.8150	0.8163	0.8211	0.8181	0.8247
Diet:	0.7773	0.7758	0.7896	0.7868	0.7844	0.7861

TI **19.** Penny Thoughts U.S. pennies made before 1983 are 97% copper and 3% zinc, whereas pennies made after 1983 are 3% copper and 97% zinc. Listed below are weights (in grams) of pennies from each of those two time periods. (Data were obtained by the author.) Does there appear to be a considerable difference in the means?

Before 1983: 3.1582 3.0406 3.0762 3.0398 3.1043 3.1274 3.0775 3.1038 3.1086 3.0586

After 1983: 2.5113 2.4907 2.5024 2.5298 2.4950 2.5127 2.4998 2.4848 2.4823 2.5163

TI **20.** BMI and Gender It is well known that men tend to weigh more than women, and men tend to be taller than women. The body mass index (BMI) is a measure based on weight and height. Listed below are BMI values from randomly selected men and women. Does there appear to be a notable difference?

Men:	23.8	23.2	24.6	26.2	23.5	24.5	21.5	31.4	26.4	22.7	27.8	28.1
Women:	19.6	23.8	19.6	29.1	25.2	21.4	22.0	27.5	33.5	20.6	29.9	17.7

Large Data Sets. In Exercises 21–24, refer to the data set in Appendix B. Use a TI-83/84 Plus calculator to find the means *and* medians, *then compare the results as indicated.*

TI **21.** Weather Forecast Accuracy Exercise 15 includes temperature forecast data for 14 days. Repeat Exercise 15 after referring to Data Set 8 in Appendix B. Use the data for all 35 days to expand the list of differences shown in Exercise 15. (The data in Exercise 15 are based on the first 14 days in Data Set 8.) Do your conclusions change when the larger data sets are used? Do you have more confidence in the results based on the larger data set?

TI **22.** Regular/Diet Coke Exercise 18 includes weights of regular Coke and Diet Coke listed in Data Set 12 in Appendix B. (The data in Exercise 18 are from the first six cans of regular Coke and the first six cans of Diet Coke.) Repeat Exercise 18 using the complete list of weights of regular Coke and Diet Coke. Do your conclusions change when the larger data sets are used?

TI **23.** Pennies Revisited Exercise 19 includes weights of pennies made before 1983 and after 1983. Repeat Exercise 19 with the complete list of weights of pennies made before 1983 and the complete list of pennies made after 1983 as listed in Data Set 14 of Appendix B. Do your conclusions change when the larger data sets are used?

TI **24.** BMI and Gender Exercise 20 includes BMI values from 12 men and 12 women. Repeat Exercise 20 with the complete list of BMI values found in Data Set 1 in Appendix B. Do your conclusions change when the larger data sets are used?

In Exercises 25–28, find the mean *of the data summarized in the given frequency distribution. Also, compare the* computed means *to the* actual means *obtained by using the original list of data values, which are as follows: (Exercise 25) 53.8; (Exercise 26) 0.230; (Exercise 27) 46.7; (Exercise 28) 98.20.*

TI **25.**

Daily Low Temperature (°F)	Frequency
35–39	1
40–44	3
45–49	5
50–54	11
55–59	7
60–64	7
65–69	1

TI **26.**

Daily Precipitation (inches)	Frequency
0.00–0.49	31
0.50–0.99	1
1.00–1.49	0
1.50–1.99	2
2.00–2.49	0
2.50–2.99	1

Table for Exercise 27

Speed	Frequency
42–45	25
46–49	14
50–53	7
54–57	3
58–61	1

Table for Exercise 28

Temperature	Frequency
96.5–96.8	1
96.9–97.2	8
97.3–97.6	14
97.7–98.0	22
98.1–98.4	19
98.5–98.8	32
98.9–99.2	6
99.3–99.6	4

TI **27.** Speeding Tickets The given frequency distribution describes the speeds of drivers ticketed by the Town of Poughkeepsie police. These drivers were traveling through a 30 mi/h speed zone on Creek Road, which passes the author's college. How does the mean compare to the posted speed limit of 30 mi/h?

TI **28.** Body Temperatures The accompanying frequency distribution summarizes a sample of human body temperatures. (See the temperatures for midnight on the second day, as listed in Data Set 2 in Appendix B.) How does the mean compare to the value of 98.6°F, which is the value assumed to be the mean by most people?

3-2 BEYOND THE BASICS

TI **29.** Trimmed Mean Because the mean is very sensitive to extreme values, we say that it is not a *resistant* measure of center. The **trimmed mean** is more resistant. To find the 10% trimmed mean for a data set, first arrange the data in order, then delete the bottom 10% of the values and the top 10% of the values, then calculate the mean of the remaining values. For the weights of the bears in Data Set 6 from Appendix B, find (a) the mean; (b) the 10% trimmed mean; (c) the 20% trimmed mean. How do the results compare?

30. Degrees of Freedom Ten values have a mean of 75.0. Nine of the values are 62, 78, 90, 87, 56, 92, 70, 70, and 93.

a. Find the 10th value.

b. We need to create a list of n values that have a specific known mean. We are free to select any values we desire for some of the n values. How many of the n values can be freely assigned before the remaining values are determined? (The result is often referred to as the *number of degrees of freedom.*)

31. Censored Data An experiment is conducted to study longevity of trees treated with fertilizer. The experiment is run for a fixed time of five years. (The test is said to be *censored* at five years.) The sample results (in years) are 2.5, 3.4, 1.2, 5+, 5+ (where 5+ indicates that the tree was still alive at the end of the experiment). What can you conclude about the mean tree life?

32. Transformed Data In each of the following, describe how the mean, median, mode, and midrange of a data set are affected.

a. The same constant k is added to each value of the data set.

b. Each value of the data set is multiplied by the same constant k.

33. The **harmonic mean** is often used as a measure of central tendency for data sets consisting of rates of change, such as speeds. It is found by dividing the number of values n by the sum of the *reciprocals* of all values, expressed as

$$\frac{n}{\Sigma \frac{1}{x}}$$

(No value can be zero.) Four students drive from New York to Florida (1200 miles) at a speed of 40 mi/h (yeah, right!). Because they need to make it to statistics class on time, they return at a speed of 60 mi/h. What is their average speed for the round trip? (The harmonic mean is used in averaging speeds.)

34. The **geometric mean** is often used in business and economics for finding average rates of change, average rates of growth, or average ratios. Given n values (all of which are positive), the geometric mean is the nth root of their product. The *average growth factor* for money compounded at annual interest rates of 10%, 8%, 9%, 12%, and 7% can be found by computing the geometric mean of 1.10, 1.08, 1.09, 1.12, and 1.07. Find that average growth factor.

35. The **quadratic mean** (or **root mean square,** or **RMS**) is usually used in physical applications. In power distribution systems, for example, voltages and currents are usually referred to in terms of their RMS values. The quadratic mean of a set of values is obtained by squaring each value, adding the results, dividing by the number of values n, and then taking the square root of that result, as indicated below:

$$\text{Quadratic mean} = \sqrt{\frac{\Sigma x^2}{n}}$$

Find the RMS of these power supplies (in volts) obtained from the author's home generator: 111.2, 108.7, 109.3, 104.8, 112.6.

36. Median When data are summarized in a frequency distribution, the median can be found by first identifying the *median class* (the class that contains the median). We then assume that the values in that class are evenly distributed and we can interpolate. This process can be described by

$$(\text{lower limit of median class}) + (\text{class width})\left(\frac{\left(\frac{n+1}{2}\right) - (m+1)}{\text{frequency of median class}}\right)$$

where n is the sum of all class frequencies and m is the sum of the class frequencies that *precede* the median class. Use this procedure to find the median of the data set summarized in Table 2-2. How does the result compare to the median of the original list of data, which is 33.5? Which value of the median is better: the value computed for the frequency table or the value of 33.5?

top 20 3-3 Measures of Variation

Key Concept Because this section introduces the concept of variation, which is so important in statistics, this is one of the most important sections in the entire book. First read through Part 1 of this section quickly and gain a general understanding of the characteristic of variation. Next, learn how to use a data set for finding the values of the range and standard deviation. Then consider Part 2 and try to understand the range rule of thumb for interpreting values of standard deviations, and also try to understand the reasoning behind the formula for standard deviation, but do not spend much time memorizing formulas or doing arithmetic calculations. Instead, place a high priority on learning how to *interpret* values of standard deviation.

Part 1: Basic Concepts of Variation

For a visual illustration of variation, see Figure 3-3, which shows bar graphs for customer waiting times at three different banks. In the first bank, the manager is compulsive about carefully controlling waiting times by changing the number of tellers as needed. In the second bank, customers all wait in a single line that feeds the available tellers. In the third bank, customers wait in different lines for each of the different tellers. Here are the specific customer waiting times (in minutes) depicted in Figure 3-3:

Bank 1: Variable waiting lines	6	6	6
Bank 2: Single waiting line	4	7	7
Bank 3: Multiple waiting lines	1	3	14

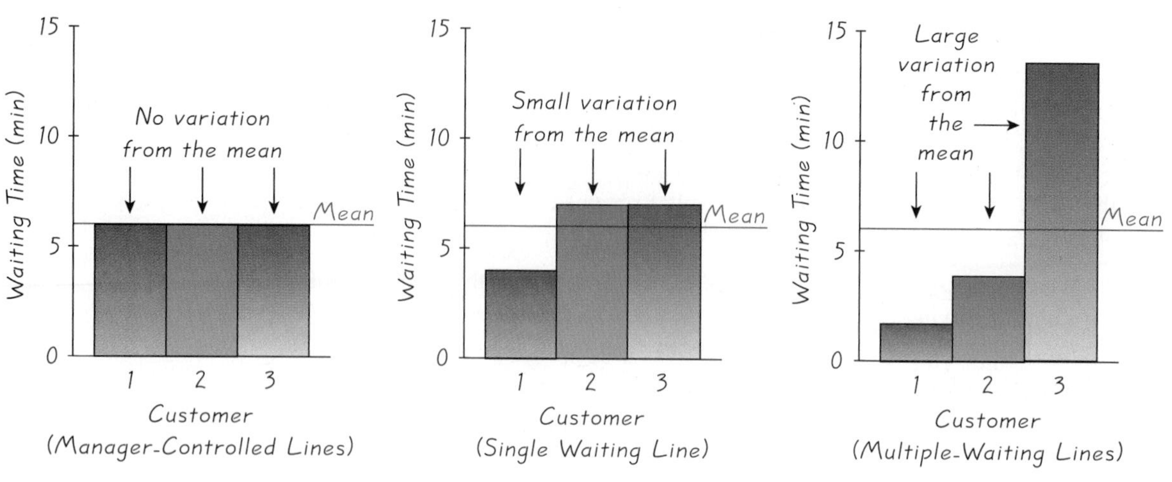

Figure 3-3 Waiting Times (min) of Bank Customers

If we consider only the mean, we would not recognize any differences among the three samples, because they all have a mean of $\bar{x} = 6.0$ min. However, it should be obvious that the samples are very different in the amounts that the waiting times *vary*. In the first bank, all of the waiting times are 6 minutes, so they don't vary at all. The waiting times of the customers in multiple lines vary much more than those in the single line. In this section, we want to develop the ability to measure and understand such variation.

Let's now develop some specific ways of actually measuring variation so that we can use specific numbers instead of subjective judgments. We begin with the range.

Range

Definition

The **range** of a set of data is the difference between the maximum value and the minimum value.

$$\textbf{range} = \textbf{(maximum value)} - \textbf{(minimum value)}$$

To compute the range, simply subtract the lowest value from the highest value. For the customers in the first bank, the range is $6 - 6 = 0$ min. For the single-line waiting times, the range is $7 - 4 = 3$ min. The waiting times with multiple lines have a range of 13 min, and this largest value suggests greatest variation.

The range is very easy to compute, but because it depends on only the maximum and the minimum values, it isn't as useful as the other measures of variation that use every value.

Standard Deviation of a Sample

The standard deviation is the measure of variation that is generally the most important and useful. We define the standard deviation now, but in order to understand it fully you will need to study the subsection of "Interpreting and Understanding Standard Deviation" found later in this section.

Definition

The **standard deviation** of a set of sample values is a measure of variation of values about the mean. It is a type of average deviation of values from the mean that is calculated by using Formulas 3-4 or 3-5. Formula 3-5 is just a different version of Formula 3-4; it is algebraically the same.

Formula 3-4 $$s = \sqrt{\frac{\Sigma(x - \bar{x})^2}{n - 1}}$$ sample standard deviation

Formula 3-5 $$s = \sqrt{\frac{n\Sigma(x^2) - (\Sigma x)^2}{n(n - 1)}}$$ shortcut formula for sample standard deviation

Later in this section we will discuss the rationale for these formulas, but for now we recommend that you use Formula 3-4 for a few examples, then learn how to find standard deviation values using your TI-83/84 Plus calculator. For now, we cite important properties that are consequences of the way in which the standard deviation is defined.

- The standard deviation is a measure of variation of all values from the *mean*.
- The value of the standard deviation s is usually positive. It is zero only when all of the data values are the same number. (It is never negative.) Also, larger values of s indicate greater amounts of variation.
- The value of the standard deviation s can increase dramatically with the inclusion of one or more outliers (data values that are very far away from all of the others).
- The units of the standard deviation s (such as minutes, feet, pounds, and so on) are the same as the units of the original data values.

Procedure for Finding the Standard Deviation with Formula 3-4

Step 1: Compute the mean $\bar{x}$.

Step 2: Subtract the mean from each individual value to get a list of deviations of the form $(x - \bar{x})$.

Step 3: Square each of the differences obtained from Step 2. This produces numbers of the form $(x - \bar{x})^2$.

Step 4: Add all of the squares obtained from Step 3. This is the value of $\Sigma(x - \bar{x})^2$.

continued

Reliability and Validity

The reliability of data refers to the consistency with which results occur, whereas the validity of data refers to how well the data measure what they are supposed to measure. The reliability of an IQ test can be judged by comparing scores for the test given on one date to scores for the same test given at another time. To test the validity of an IQ test, we might compare the test scores to another indicator of intelligence, such as academic performance. Many critics charge that IQ tests are reliable, but not valid; they provide consistent results, but don't really measure intelligence.

Step 5: Divide the total from Step 4 by the number $(n - 1)$, which is 1 less than the total number of values present.

Step 6: Find the square root of the result of Step 5.

EXAMPLE Using Formula 3-4 Use Formula 3-4 to find the standard deviation of the waiting times from the multiple lines. Those times (in minutes) are 1, 3, 14.

SOLUTION We will use the six steps in the procedure just given. Refer to those steps and refer to Table 3-3, which shows the detailed calculations.

Step 1: Obtain the mean of 6.0 by adding the values and then dividing by the number of values:

$$\bar{x} = \frac{\Sigma x}{n} = \frac{18}{3} = 6.0 \text{ min}$$

Step 2: Subtract the mean of 6.0 from each value to get these values of $(x - \bar{x})$: $-5, -3, 8$.

Step 3: Square each value obtained in Step 2 to get these values of $(x - \bar{x})^2$: 25, 9, 64.

Step 4: Sum all of the preceding values to get the value of

$$\Sigma(x - \bar{x})^2 = 98.$$

Step 5: With $n = 3$ values, divide by 1 less than 3:

$$\frac{98}{2} = 49.0$$

Step 6: Find the square root of 49.0. The standard deviation is

$$\sqrt{49.0} = 7.0 \text{ min}$$

Ideally, we would now interpret the meaning of the result, but such interpretations will be discussed later in this section.

Table 3-3 Calculating Standard Deviation

x	$x - \bar{x}$	$(x - \bar{x})^2$
1	−5	25
3	−3	9
14	8	64
Totals: 18		98

$$\bar{x} = \frac{18}{3} = 6.0 \text{ min} \qquad s = \sqrt{\frac{98}{3 - 1}} = \sqrt{49} = 7.0 \text{ min}$$

EXAMPLE Using Formula 3-5 The preceding example used Formula 3-4 for finding the standard deviation of the customer waiting times at the bank with multiple waiting lines. Using the same data set (1, 3, 14), find the standard deviation by using Formula 3-5.

SOLUTION Formula 3-5 requires that we first find values for n, Σx, and Σx^2.

$n = 3$ (because there are 3 values in the sample)

$\Sigma x = 18$ (found by adding the sample values: $1 + 3 + 14 = 18$)

$\Sigma x^2 = 206$ (found by adding the squares of the sample values, as in $1^2 + 3^2 + 14^2 = 206$)

Using Formula 3-5, we get

$$s = \sqrt{\frac{n(\Sigma x^2) - (\Sigma x)^2}{n(n-1)}} = \sqrt{\frac{3(206) - (18)^2}{3(3-1)}} = \sqrt{\frac{294}{6}} = 7.0 \text{ min}$$

A good activity is to stop here and calculate the standard deviation of the waiting times of 4 min, 7 min, and 7 min (for the single line). Follow the same procedures used in the preceding two examples and verify that $s = 1.7$ min. (It will also become important to develop an ability to obtain values of standard deviations by using a calculator and software.) Although the interpretations of these standard deviations will be discussed later, we can now compare them to see that the standard deviation of the times for the single line (1.7 min) is much lower than the standard deviation for multiple lines (7.0 min). This supports our subjective conclusion that the waiting times with the single line have much less variation than the times from multiple lines. The bank with the good fortune of having a compulsive manager who carefully controls waiting times has a standard deviation of 0 min, which is the lowest.

Where Are the 0.400 Hitters?

The last baseball player to hit above 0.400 was Ted Williams, who hit 0.406 in 1941. There were averages above 0.400 in 1876, 1879, 1887, 1894, 1895, 1896, 1897, 1899, 1901, 1911, 1920, 1922, 1924, 1925, and 1930, but none since 1941. Are there no longer great hitters? The late Stephen Jay Gould of Harvard University noted that the mean batting average has been steady at 0.260 for about 100 years, but the standard deviation has been decreasing from 0.049 in the 1870s to 0.031, where it is now. He argued that today's stars are as good as those from the past, but consistently better pitchers now keep averages below 0.400.

Standard Deviation of a Population

In our definition of standard deviation, we referred to the standard deviation of *sample* data. A slightly different formula is used to calculate the standard deviation σ (lowercase Greek sigma) of a population: Instead of dividing by $n - 1$, divide by the population size N, as in the following expression:

$$\sigma = \sqrt{\frac{\Sigma(x-\mu)^2}{N}} \qquad \text{population standard deviation}$$

Because we generally deal with sample data, we will usually use Formula 3-4, in which we divide by $n - 1$. Many calculators give both the sample standard deviation and the population standard deviation, but they use a variety of different notations. Be sure to identify the notation used by your calculator, so that you get the correct result.

Variance of a Sample and Population

We are using the term *variation* as a general description of the amount that values vary among themselves. (The terms *dispersion* and *spread* are sometimes used instead of *variation*.) The term *variance* refers to a specific definition.

Definition

The **variance** of a set of values is a measure of variation equal to the square of the standard deviation.

Sample variance: s^2 square of the standard deviation s.
Population variance: σ^2 square of the population standard deviation σ.

The sample variance s^2 is said to be an **unbiased estimator** of the population variance σ^2, which means that values of s^2 tend to target the value of σ^2 instead of systematically tending to overestimate or underestimate σ^2. For example, consider an IQ test designed so that the variance is 225. If you repeat the process of randomly selecting 100 subjects, giving them IQ tests, and calculating the sample variance s^2 in each case, the sample variances that you obtain will tend to center around 225, which is the population variance.

EXAMPLE Finding Variance In the preceding example, we used the customer waiting times of 1 min, 3 min, and 14 min to find that the standard deviation is given by $s = 7.0$ min. Find the variance of that same sample.

SOLUTION Because the variance is the square of the standard deviation, we get the result shown below. Note that the original data values are in units of minutes, the standard deviation is 7.0 min, but the variance is given in units of min^2.

$$\text{Sample variance} = s^2 = 7.0^2 = 49.0 \text{ min}^2$$

The variance is an important statistic used in some important statistical methods, such as analysis of variance discussed in Chapter 12. For our present purposes, the variance has this serious disadvantage: *The units of variance are different than the units of the original data set.* For example, if the original waiting times are in minutes, the units of the variance are min^2. What is a square minute? Because the variance uses different units, it is extremely difficult to understand variance by relating it to the original data set. Because of this property, we will focus on the standard deviation when we try to develop an understanding of variation later in this section.

We now present the notation and round-off rule we are using.

Notation

s = *sample* standard deviation
s^2 = *sample* variance
σ = *population* standard deviation
σ^2 = *population* variance

Note: Articles in professional journals and reports often use SD for standard deviation and VAR for variance.

Round-Off Rule

We use the same round-off rule given in Section 3-2:

Carry one more decimal place than is present in the original set of data.

Round only the final answer, not in the middle of a calculation. (If it becomes absolutely necessary to round in the middle, carry at least twice as many decimal places as will be used in the final answer.)

Part 2: Beyond the Basics of Variation

Interpreting and Understanding Standard Deviation

This subsection is extremely important, because we will now try to make some intuitive sense of the standard deviation. First, we should clearly understand that the standard deviation measures the *variation* among values. Values close together will yield a small standard deviation, whereas values spread farther apart will yield a larger standard deviation.

One crude but simple tool for understanding standard deviation is the **range rule of thumb,** which is based on the principle that for many data sets, the vast majority (such as 95%) of sample values lie within two standard deviations of the mean. (We could improve the accuracy of this rule by taking into account such factors as the size of the sample and the nature of the distribution, but here we prefer to sacrifice accuracy for the sake of simplicity. Also, we could use three or even four standard deviations instead of two standard deviations, which is a somewhat arbitrary choice. But we want a simple rule that will help us interpret values of standard deviations; later methods will produce more accurate results.)

Range Rule of Thumb

For Estimating a Value of the Standard Deviation s: To roughly estimate the standard deviation from a collection of known sample data, use

$$s \approx \frac{\text{range}}{4}$$

where range = (maximum value) − (minimum value).

For Interpreting a Known Value of the Standard Deviation: If the standard deviation is known, use it to find rough estimates of the minimum and maximum "usual" sample values by using the following:

minimum "usual" value = (mean) − 2 × (standard deviation)
maximum "usual" value = (mean) + 2 × (standard deviation)

When calculating a standard deviation using Formula 3-4 or 3-5, you can use the range rule of thumb as a check on your result, but you must realize that although

More Stocks, Less Risk

In their book *Investments*, authors Zvi Bodie, Alex Kane, and Alan Marcus state that "the average standard deviation for returns of portfolios composed of only one stock was 0.554. The average portfolio risk fell rapidly as the number of stocks included in the portfolio increased." They note that with 32 stocks, the standard deviation is 0.325, indicating much less variation and risk. They make the point that with only a few stocks, a portfolio has a high degree of "firm-specific" risk, meaning that the risk is attributable to the few stocks involved. With more than 30 stocks, there is very little firm-specific risk; instead, almost all of the risk is "market risk," attributable to the stock market as a whole. They note that these principles are "just an application of the well-known law of averages."

the approximation will get you in the general vicinity of the answer, it can be off by a fairly large amount.

EXAMPLE Ages of Best Actresses Use the range rule of thumb to find a rough estimate of the standard deviation of the sample of 76 ages of actresses who won Oscars in the category of Best Actress. Those ages are listed in Table 2-1, which is included with the Chapter Problem for Chapter 2.

SOLUTION In using the range rule of thumb to estimate the standard deviation of sample data, we find the range and divide by 4. By scanning the list of ages of the actresses, we can see that the maximum is 80 and the minimum is 21, so the range rule of thumb can be used to estimate the standard deviation s as follows:

$$s \approx \frac{\text{range}}{4} = \frac{80 - 21}{4} = \frac{59}{4} = 14.75 \approx 15$$

INTERPRETATION This result is in the general neighborhood of the correct value of 11.1 that is found by calculating the exact value of the standard deviation with Formula 3-4 or 3-5, but the result of 15 does miss the actual standard deviation by a considerable amount. This illustrates that the range rule of thumb yields a "crude" estimate that might be off by a fair amount.

The next example is particularly important as an illustration of one way to *interpret* the value of a standard deviation.

EXAMPLE Pulse Rates of Women Past results from the National Health Survey suggest that the pulse rates (beats per minute) have a mean of 76.0 and a standard deviation of 12.5. Use the range rule of thumb to find the minimum and maximum "usual" pulse rates. (These results could be used by a physician who can identify "unusual" pulse rates that might be the result of some disorder.) Then determine whether a pulse rate of 110 would be considered "unusual."

SOLUTION With a mean of 76.0 and a standard deviation of 12.5, we use the range rule of thumb to find the minimum and maximum usual pulse rates as follows:

$$\begin{aligned}\text{minimum "usual" value} &= (\text{mean}) - 2 \times (\text{standard deviation}) \\ &= 76.0 - 2(12.5) = 51 \text{ beats per minute}\end{aligned}$$

$$\begin{aligned}\text{maximum "usual" value} &= (\text{mean}) + 2 \times (\text{standard deviation}) \\ &= 76.0 + 2(12.5) = 101 \text{ beats per minute}\end{aligned}$$

INTERPRETATION Based on these results, we expect that typical women have pulse rates between 51 beats per minute and 101 beats per minute. Because 110 beats per minute does not fall within those limits, it would be considered unusual. With a pulse rate of 110, a physician might try to establish a reason for this unusual reading.

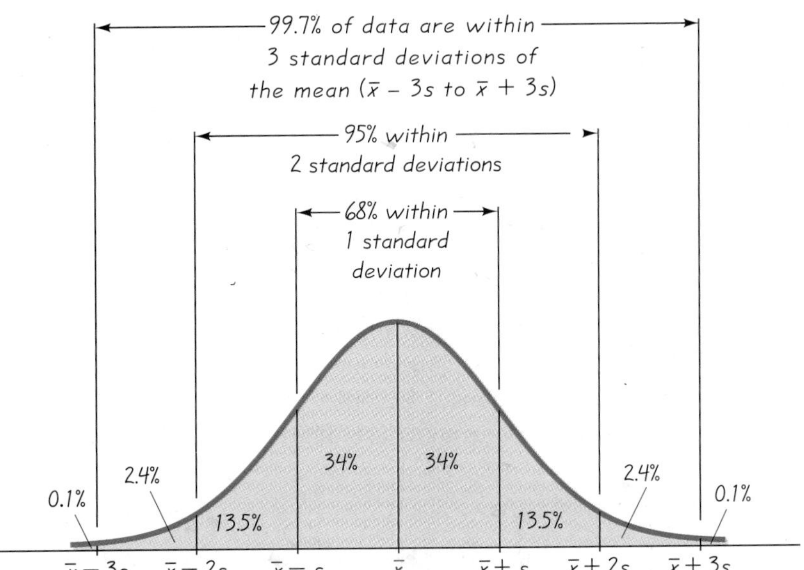

Figure 3-4

The Empirical Rule

Empirical (or 68–95–99.7) Rule for Data with a Bell-Shaped Distribution

Another rule that is helpful in interpreting values for a standard deviation is the **empirical rule.** This rule states that *for data sets having a distribution that is approximately bell-shaped,* the following properties apply. (See Figure 3-4.)

- About 68% of all values fall within 1 standard deviation of the mean.
- About 95% of all values fall within 2 standard deviations of the mean.
- About 99.7% of all values fall within 3 standard deviations of the mean.

EXAMPLE IQ Scores IQ scores have a bell-shaped distribution with a mean of 100 and a standard deviation of 15. What percentage of IQ scores are between 70 and 130?

SOLUTION The key to solving this problem is to recognize that 70 and 130 are each exactly 2 standard deviations away from the mean of 100, as shown below.

$$2 \text{ standard deviations } = 2s = 2(15) = 30$$

Therefore, 2 standard deviations from the mean is

$$100 - 30 = 70$$
$$\text{or } 100 + 30 = 130$$

The empirical rule tells us that about 95% of all values are within 2 standard deviations of the mean, so about 95% of all IQ scores are between 70 and 130.

Hint: Difficulty with applying the empirical rule usually stems from confusion with interpreting phrases such as "within 2 standard deviations of the mean." Stop here and review the preceding example until the meaning of that phrase becomes clear. Also, see the following general interpretations of such phrases.

Phrase	Meaning
Within 1 standard deviation of the mean	Between $(\bar{x} - s)$ and $(\bar{x} + s)$
Within 2 standard deviations of the mean	Between $(\bar{x} - 2s)$ and $(\bar{x} + 2s)$
Within 3 standard deviations of the mean	Between $(\bar{x} - 3s)$ and $(\bar{x} + 3s)$

A third concept that is helpful in understanding or interpreting a value of a standard deviation is **Chebyshev's theorem.** The preceding empirical rule applies only to data sets with a bell-shaped distribution. Instead of being limited to data sets with bell-shaped distributions, Chebyshev's theorem applies to *any* data set, but its results are very approximate. Because the results are lower limits ("at least"), Chebyshev's theorem has limited usefulness.

Chebyshev's Theorem

The proportion (or fraction) of any set of data lying within K standard deviations of the mean is always *at least* $1 - 1/K^2$, where K is any positive number greater than 1. For $K = 2$ and $K = 3$, we get the following statements:

- At least 3/4 (or 75%) of all values lie within 2 standard deviations of the mean.
- At least 8/9 (or 89%) of all values lie within 3 standard deviations of the mean.

EXAMPLE IQ Scores IQ scores have a mean of 100 and a standard deviation of 15. What can we conclude from Chebyshev's theorem?

SOLUTION Applying Chebyshev's theorem with a mean of 100 and a standard deviation of 15, we can reach the following conclusions.

- At least 3/4 (or 75%) of IQ scores are within 2 standard deviations of the mean (between 70 and 130).
- At least 8/9 (or 89%) of all IQ scores are within 3 standard deviations of the mean (between 55 and 145).

When trying to make sense of a value of a standard deviation, we should use one or more of the preceding three concepts. To gain additional insight into the nature of the standard deviation, we now consider the underlying rationale leading to Formula 3-4, which is the basis for its definition. (Formula 3-5 is simply another version of Formula 3-4, derived so that arithmetic calculations can be simplified.)

Rationale for Standard Deviation

The standard deviation of a set of sample data is defined by Formulas 3-4 and 3-5, which are equivalent in the sense that they will always yield the same result. Formula 3-4 has the advantage of reinforcing the concept that the standard deviation is a type of average deviation. Formula 3-5 has the advantage of being easier to use when you must calculate standard deviations on your own. Formula 3-5 also eliminates the intermediate rounding errors introduced in Formula 3-4 when the exact value of the mean is not used. Formula 3-5 is used in calculators and programs because it requires only three memory locations (for n, Σx and Σx^2) instead of a memory location for every value in the data set.

Why define a measure of variation in the way described by Formula 3-4? In measuring variation in a set of sample data, it makes sense to begin with the individual amounts by which values deviate from the mean. For a particular value x, the amount of **deviation** is $x - \bar{x}$, which is the difference between the individual x value and the mean. For the waiting times of 1, 3, 14, the mean is 6.0 so the deviations away from the mean are -5, -3, and 8. It would be good to somehow combine those deviations into a single collective value. Simply adding the deviations doesn't work, because the sum will always be zero. To get a statistic that measures variation (instead of always being zero), we need to avoid the canceling out of negative and positive numbers. One approach is to add absolute values, as in $\Sigma|x - \bar{x}|$. If we find the mean of that sum, we get the **mean absolute deviation** (or **MAD**), which is the mean distance of the data from the mean.

$$\text{Mean absolute deviation} = \frac{\Sigma|x - \bar{x}|}{n}$$

Because the waiting times of 1, 3, 14 have deviations of -5, -3, and 8, the mean absolute deviation is $(5 + 3 + 8)/3 = 16/3 = 5.3$.

Why Not Use the Mean Absolute Deviation? Because the mean absolute deviation requires that we use absolute values, it uses an operation that is not algebraic. (The algebraic operations include addition, multiplication, extracting roots, and raising to powers that are integers or fractions, but absolute value is not included.) The use of absolute values creates algebraic difficulties in inferential methods of statistics. For example, Section 9-3 presents a method for making inferences about the means of two populations, and that method is built around an additive property of variances, but the mean absolute deviation has no such additive property. (Here is a simplified version of the additive property of variances: If you have two independent populations and you randomly select one value from each population and add them, such sums will have a variance equal to the sum of the variances of the two populations.) That same additive property underlies the rationale for regression discussed in Chapter 10 and analysis of variance discussed in Chapter 12. The mean absolute deviation lacks this important additive property. Also, the mean absolute value is *biased*, meaning that when you find mean absolute values of samples, you do not tend to target the mean absolute value of the population. In contrast, the standard deviation uses only algebraic operations. Because it is based on the square root of a sum of squares, the standard deviation closely parallels distance formulas found in algebra. There are many instances where a statistical procedure is based on a similar sum of squares. Therefore, instead of using absolute values, we get a better measure of variation by making all deviations $(x - \bar{x})$ nonnegative by squaring them, and this approach leads to the standard deviation. For these reasons, scientific calculators typically include a standard deviation function, but they almost never include the mean absolute deviation.

Why Divide by $n - 1$? After finding all of the individual values of $(x - \bar{x})^2$, we combine them by finding their sum, then we get an average by dividing by $n - 1$. We divide by $n - 1$ because there are only $n - 1$ independent values. That is, with a given mean, only $n - 1$ values can be freely assigned any number before the last value is determined. See Exercise 38, which provides concrete numbers illustrating that division by $n - 1$ yields a better result than division by n. That exercise shows that if s^2 were defined with division by n, it would systematically underestimate the value of σ^2, so we compensate by increasing its overall value by making its denominator smaller (by using $n - 1$ instead of n). Exercise 38 shows how division by $n - 1$ causes the sample variance s^2 to target the value of the population variance σ^2, whereas division by n causes the sample variance s^2 to underestimate the value of the population variance σ^2.

Step 6 in the above procedure for finding a standard deviation is to find a square root. We take the square root to compensate for the squaring that took place in Step 3. An important consequence of taking the square root is that *the standard deviation has the same units of measurement as the original values*. For example, if customer waiting times are in minutes, the standard deviation of those times will also be in minutes. If we were to stop at Step 5, the result would be in units of "square minutes," which is an abstract concept having no direct link to the real world as we know it.

Comparing Variation in Different Populations

We stated earlier that because the units of the standard deviation are the same as the units of the original data, it is easier to understand the standard deviation than the variance. However, that same property makes it difficult to compare variation for values taken from different populations. By resulting in a value that is free of specific units of measure, the *coefficient of variation* overcomes this disadvantage.

Definition

The **coefficient of variation** (or **CV**) for a set of nonnegative sample or population data, expressed as a percent, describes the standard deviation relative to the mean, and is given by the following:

Sample

$$CV = \frac{s}{\bar{x}} \cdot 100\%$$

Population

$$CV = \frac{\sigma}{\mu} \cdot 100\%$$

EXAMPLE Heights and Weights of Men Using the sample height and weight data for the 40 males included in Data Set 1 in Appendix B, we find the statistics given in the table below. Find the coefficient of variation for heights, then find the coefficient of variation for weights, then compare the two results.

	Mean ($\bar{x}$)	Standard Deviation (s)
Height	68.34 in.	3.02 in.
Weight	172.55 lb	26.33 lb

SOLUTION Because we have sample statistics, we find the two coefficients of variation as follows:

$$\text{Heights:} \quad CV = \frac{s}{\bar{x}} \cdot 100\% = \frac{3.02 \text{ in.}}{68.34 \text{ in.}} \cdot 100\% = 4.42\%$$

$$\text{Weights:} \quad CV = \frac{s}{\bar{x}} \cdot 100\% = \frac{26.33 \text{ lb}}{172.55 \text{ lb}} \cdot 100\% = 15.26\%$$

Although the difference in the units makes it impossible to compare the standard deviation of 3.02 in. to the standard deviation of 26.33 lb, we can compare the coefficients of variation, which have no units. We can see that heights (with $CV = 4.42\%$) have considerably less variation than weights (with $CV = 15.26\%$). This makes intuitive sense, because we routinely see that weights among men vary much more than heights. For example, it is very rare to see two adult men with one of them being twice as tall as the other, but it is much more common to see two men with one of them weighing twice as much as the other.

After studying this section, you should understand that the standard deviation is a measure of variation among values. Given sample data, you should be able to compute the value of the standard deviation. You should be able to interpret the values of standard deviations that you compute. You should know that for typical data sets, it is unusual for a value to differ from the mean by more than two or three standard deviations.

TI

Using the TI-83/84 Plus Calculator to Find Measures of Variation

List of Single Values If using the TI-83/84 Plus calculator to find measures of variation from a list of data values, use the same procedures given at the end of Section 3-2. For a list of single values, enter or open the list, then press STAT, select **CALC,** press ENTER, then enter the list name or label, as in 1-Var Stats L1 or 1-Var Stats LAGES. (If you want to enter the name AGES, access the list of variable names by pressing 2ND STAT, then scroll to the name AGES and Press ENTER.) Press ENTER to obtain the results.

Frequency Distribution If using the TI-83/84 Plus calculator to find measures of variation from a frequency distribution, enter the class midpoints in list L1, enter the frequencies in list L2, press STAT, select **CALC,** press ENTER, then enter **1-Var Stats L1, L2** (including the comma) as described in the procedures at the end of Section 3-2.

TI-83/84 Plus

```
1-Var Stats
x̄=35.68421053
Σx=2712
Σx²=105956
Sx=11.06370707
σx=10.9906785
↓n=76
```

Interpretation and Other Measures The results will include a value indicated by Sx and another value indicated by σx. The Sx value is the *sample* standard deviation, which is the value we usually want. The σx value is the *population* standard deviation, which we rarely want. Also, the range can be easily computed using the minimum and maximum values identified among the results as min$X$ and max$X$. To obtain a value of *variance*, simply square the value of the standard deviation. For example, if the calculator displays $Sx = 11.06370707$, enter 11.06370707 and press **x²** **ENTER** to find that the variance is 122.4056141.

Sx denotes the *sample* standard deviation, so $s = 11.06370707$. (This is the value we usually want.)

To find the sample variance s^2, enter the Sx value and press **x²** **ENTER** to obtain its square.

σx denotes the population standard deviation, so $\sigma = 10.9906785$.

TI-83/84 Plus

```
1-Var Stats
↑n=76
 minX=21
 Q1=28
 Med=33.5
 Q3=39.5
 maxX=80
```

From this part of the display, we find the range by using the minimum and maximum values:

$$\text{range} = \text{maximum} - \text{minimum} = 80 - 21 = 59.$$

3-3 BASIC SKILLS AND CONCEPTS

Statistical Literacy and Critical Thinking

1. **Variation** Why is the standard deviation considered a measure of *variation*? In your own words, describe the characteristic of a data set that is measured by the standard deviation.

2. **Comparing Variation** Which do you think has more variation: the IQ scores of 30 students in a statistics class or the IQ scores of 30 patrons watching a movie? Why?

3. **Unusual Value?** A statistics professor gives a test that has a mean of 50 and a standard deviation of 10. (She is not using a typical test with a maximum score of 100, and she promises to curve the scores.) One student earns a score of 85 on this test. In this context, is the score of 85 "unusual"? Why or why not?

4. **Correct Statement?** In the book *How to Lie with Charts*, Gerald E. Jones writes that "the standard deviation is usually shown as plus or minus the difference between the high and the mean, and the low and the mean. For example, if the mean is 1, the high 3, and the low −1, the standard deviation is ±2." Is that statement correct? Why or why not?

In Exercises 5–12, find the range, variance, *and* standard deviation *for the given sample data. (The same data were used in Section 3-2 where we found measures of center. Here we find measures of variation.) Also, answer the given questions.*

TI 5. **Perception of Time** Statistics students participated in an experiment to test their ability to determine when 1 minute (or 60 seconds) has passed. The results are given below in seconds. Identify at least one good reason why the standard deviation from this sample might not be a good estimate of the standard deviation for the population of adults.

53 52 75 62 68 58 49 49

TI **6.** Cereal A dietician obtains the amounts of sugar (in centigrams) from 100 centigrams (or 1 gram) in each of 10 different cereals, including Cheerios, Corn Flakes, Fruit Loops, and 7 others. Those values are listed below. Is the standard deviation of those values likely to be a good estimate of the standard deviation of the amounts of sugar in each gram of cereal consumed by the population of all Americans who eat cereal? Why or why not?

3 24 30 47 43 7 47 13 44 39

TI **7.** Phenotypes of Peas An experiment was conducted to determine whether a deficiency of carbon dioxide in the soil affects the phenotypes of peas. Listed below are the phenotype codes, where 1 = smooth-yellow, 2 = smooth-green, 3 = wrinkled-yellow, and 4 = wrinkled-green. Can the measures of variation be obtained for these values? Do the results make sense?

2 1 1 1 1 1 1 4 1 2 2 1 2 3 3 2 3 1 3 1 3 1 3 2 2

TI **8.** Old Faithful Geyser Listed below are intervals (in minutes) between eruptions of the Old Faithful geyser in Yellowstone National Park. Based on the results, is an interval of 100 minutes unusual?

98 92 95 87 96 90 65 92 95 93 98 94

TI **9.** Blood Pressure Measurements Fourteen different second-year medical students at Bellevue Hospital measured the blood pressure of the same person. The systolic readings (in mmHg) are listed below. If the subject's blood pressure remains constant and the medical students correctly apply the same measurement technique, what should be the value of the standard deviation?

138 130 135 140 120 125 120 130 130 144 143 140 130 150

TI **10.** Body Temperatures Researchers at the University of Maryland collected body temperature readings from a sample of adults, and eight of those temperatures are listed below (in degrees Fahrenheit). Based on these results, is a body temperature of 104.0°F unusual? Why or why not? What should be done with someone having a body temperature of 104.0°F?

98.6 98.6 98.0 98.0 99.0 98.4 98.4 98.4

TI **11.** Fruit Flies Listed below are the thorax lengths (in millimeters) of a sample of male fruit flies. If we learn that the listed measurements were obtained from fruit flies hovering over an apple sitting on a kitchen table in Pocatello, does the standard deviation serve as a reasonable estimate of the standard deviation for all fruit flies residing in the United States?

0.72 0.90 0.84 0.68 0.84 0.90 0.92 0.84 0.64 0.84 0.76

TI **12.** Personal Income Listed below are the amounts of personal income per capita (in dollars) for the first five states listed in alphabetical order: Alabama, Alaska, Arizona, Arkansas, and California (data from the U.S. Bureau of Economic Analysis). Assume that the 45 amounts from the other states are included and the standard deviation of the 50 state amounts is $4337. Is $4337 necessarily the standard deviation of the amounts of personal income per capita for all individuals in the United States? Why or why not?

$25,128 $32,151 $26,183 $23,512 $32,996

In Exercises 13–20, find the range, variance, *and* standard deviation *for each of the two samples, then compare the two sets of results. (The same data were used in Section 3-2.)*

TI **13.** It's Raining Cats Statistics are sometimes used to compare or identify authors of different works. The lengths of the first 20 words in the foreword written by Tennessee Williams in *Cat on a Hot Tin Roof* are listed along with the first 20 words in *The Cat in the Hat* by Dr. Seuss. Does there appear to be a difference in variation?

Cat on a Hot Tin Roof: 2 6 2 2 1 4 4 2 4 2 3 8 4 2 2 7 7 2 3 11

The Cat in the Hat: 3 3 3 3 5 2 3 3 3 2 4 2 2 3 2 3 5 3 4 4

TI **14.** Ages of Stowaways The *Queen Mary* sailed between England and the United States, and stowaways were sometimes found on board. The ages (in years) of stowaways from eastbound crossings and westbound crossings are given below (data from the Cunard Steamship Co., Ltd.). Compare the variation in the two data sets.

Eastbound: 24 24 34 15 19 22 18 20 20 17

Westbound: 41 24 32 26 39 45 24 21 22 21

TI **15.** Weather Forecast Accuracy In an analysis of the accuracy of weather forecasts, the actual high temperatures are compared to the high temperatures predicted one day earlier and the high temperatures predicted five days earlier. Listed below are the errors between the predicted temperatures and the actual high temperatures for consecutive days in Dutchess County, New York. Do the standard deviations suggest that the temperatures predicted one day in advance are more accurate than those predicted five days in advance, as we might expect?

(actual high) − (high predicted one day earlier): 2 2 0 0 −3 −2 1 −2 8 1 0 −1 0 1

(actual high) − (high predicted five days earlier): 0 −3 2 5 −6 −9 4 −1 6 −2 −2 −1 6 −4

TI **16.** Treatment Effect Researchers at Pennsylvania State University conducted experiments with poplar trees. Listed below are weights (kg) of poplar trees given no treatment and poplar trees treated with fertilizer and irrigation. Does there appear to be a difference between the two standard deviations?

No treatment: 0.15 0.02 0.16 0.37 0.22

Fertilizer and irrigation: 2.03 0.27 0.92 1.07 2.38

TI **17.** Customer Waiting Times Waiting times (in minutes) of customers at the Jefferson Valley Bank (where all customers enter a single waiting line) and the Bank of Providence (where customers wait in individual lines at three different teller windows) are listed below. Compare the variation in the two data sets.

Jefferson Valley (single line): 6.5 6.6 6.7 6.8 7.1 7.3 7.4 7.7 7.7 7.7

Providence (individual lines): 4.2 5.4 5.8 6.2 6.7 7.7 7.7 8.5 9.3 10.0

TI **18.** Regular/Diet Coke Weights (pounds) of samples of the contents in cans of regular Coke and Diet Coke are listed below. Compare the variation in the two data sets.

Regular: 0.8192 0.8150 0.8163 0.8211 0.8181 0.8247

Diet: 0.7773 0.7758 0.7896 0.7868 0.7844 0.7861

TI **19.** Penny Thoughts U.S. pennies made before 1983 are 97% copper and 3% zinc, whereas pennies made after 1983 are 3% copper and 97% zinc. Listed below are weights (in grams) of pennies from each of those two time periods. (Data were obtained by the author.) Does there appear to be a considerable difference in the standard deviations?

Before 1983: 3.1582 3.0406 3.0762 3.0398 3.1043 3.1274 3.0775 3.1038 3.1086 3.0586

After 1983: 2.5113 2.4907 2.5024 2.5298 2.4950 2.5127 2.4998 2.4848 2.4823 2.5163

TI **20.** BMI and Gender It is well known that men tend to weigh more than women, and men tend to be taller than women. The body mass index (BMI) is a measure based on weight and height. Listed below are BMI values from randomly selected men and women. Does there appear to be a difference in variation between the two data sets?

Men: 23.8 23.2 24.6 26.2 23.5 24.5 21.5 31.4 26.4 22.7 27.8 28.1

Women: 19.6 23.8 19.6 29.1 25.2 21.4 22.0 27.5 33.5 20.6 29.9 17.7

Large Data Sets. *In Exercises 21–24, refer to the data set in Appendix B. Use a TI-83/84 Plus calculator to find the range, variance, and standard deviation for each of the two samples, then compare the results.*

TI **21.** Weather Forecast Accuracy Exercise 15 includes temperature forecast data for 14 days. Repeat Exercise 15 after referring to Data Set 8 in Appendix B. Use the data for all 35 days to expand the list of differences shown in Exercise 15. (The data in Exercise 15 are based on the first 14 days in Data Set 8.) Do your conclusions change when the larger data sets are used? Do you have more confidence in the results based on the larger data set?

TI **22.** Regular/Diet Coke Exercise 18 includes weights of regular Coke and Diet Coke listed in Data Set 12 in Appendix B. (The data in Exercise 18 are from the first six cans of regular Coke and the first six cans of Diet Coke.) Repeat Exercise 18 using the complete list of weights of regular Coke and Diet Coke. Do your conclusions change when the larger data sets are used?

TI **23.** Pennies Revisited Exercise 19 includes weights of pennies made before 1983 and after 1983. Repeat Exercise 19 with the complete list of weights of pennies made before 1983 and the complete list of pennies made after 1983 as listed in Data Set 14 of Appendix B. Do your conclusions change when the larger data sets are used?

TI **24.** BMI and Gender Exercise 20 includes BMI values from 12 men and 12 women. Repeat Exercise 20 with the complete list of BMI values found in Data Set 1 in Appendix B. Do your conclusions change when the larger data sets are used?

Finding Standard Deviation from a Frequency Distribution. *In Exercises 25–28, find the standard deviation of sample data summarized in a frequency distribution table by using the formula below, where x represents the class midpoint and f represents the class frequency. Also, compare the computed standard deviations to these standard deviations obtained by using Formula 3-4 with the original list of data values: (Exercise 25) 6.9; (Exercise 26) 0.630; (Exercise 27) 4.0; (Exercise 28) 0.62.*

$$s = \sqrt{\frac{n[\Sigma(f \cdot x^2)] - [\Sigma(f \cdot x)]^2}{n(n-1)}}$$ standard deviation for frequency distribution

TI **25.**

Daily Low Temperature (°F)	Frequency
35–39	1
40–44	3
45–49	5
50–54	11
55–59	7
60–64	7
65–69	1

TI **26.**

Daily Precipitation (inches)	Frequency
0.00–0.49	31
0.50–0.99	1
1.00–1.49	0
1.50–1.99	2
2.00–2.49	0
2.50–2.99	1

TI **27.** Speeding Tickets The given frequency distribution describes the speeds of drivers ticketed by the Town of Poughkeepsie police. These drivers were traveling through a 30 mi/h speed zone on Creek Road, which passes the author's college.

Table for Exercise 27

Speed	Frequency
42–45	25
46–49	14
50–53	7
54–57	3
58–61	1

TI **28.** Body Temperatures The accompanying frequency distribution summarizes a sample of human body temperatures. (See the temperatures for midnight on the second day, as listed in Data Set 2 in Appendix B.)

Table for Exercise 28

Temperature	Frequency
96.5–96.8	1
96.9–97.2	8
97.3–97.6	14
97.7–98.0	22
98.1–98.4	19
98.5–98.8	32
98.9–99.2	6
99.3–99.6	4

29. Range Rule of Thumb Use the range rule of thumb to estimate the standard deviation of ages of all faculty members at your college.

30. Range Rule of Thumb Using the sample data in Data Set 1 from Appendix B, the sample of 40 women have upper leg lengths with a mean of 38.86 cm and a standard deviation of 3.78 cm. Use the range rule of thumb to identify the minimum and maximum "usual" upper leg lengths. Is a length of 47.0 cm considered unusual in this context?

31. Range Rule of Thumb The mean of electrical energy consumption amounts for the author's home during a two-month period is 2838 kWh, and the standard deviation is 504 kWh. (The author actually tracks this stuff.) Use the range rule of thumb to identify minimum and maximum "usual" amounts of electrical energy consumption. For one particular two-month period, the power company recorded consumption of 578 kWh. Is that amount unusual?

32. Range Rule of Thumb Aluminum cans with a thickness of 0.0111 in. have axial loads with a mean of 281.8 lb and a standard deviation of 27.8 lb. The axial load is measured by applying pressure to the top of the can until it collapses. (See Data Set 15 in Appendix B.) Use the range rule of thumb to find the minimum and maximum "usual" axial loads. One particular can had an axial load of 504 lb. Is that unusual?

33. Empirical Rule Heights of men have a bell-shaped distribution with a mean of 176 cm and a standard deviation of 7 cm. Using the empirical rule, what is the approximate percentage of men between

a. 169 cm and 183 cm?

b. 155 cm and 197 cm?

TI **34.** Coefficient of Variation Refer to the data in Exercise 17. Find the coefficient of variation for each of the two samples, then compare the results.

TI **35.** Coefficient of Variation Use the sample data listed below to find the coefficient of variation for each of the two samples, then compare the results.

Heights (in.) of men:	71 66 72 69 68 69
Lengths (mm) of cuckoo eggs:	19.7 21.7 21.9 22.1 22.1 22.3 22.7 22.9 23.9

36. Understanding Units of Measurement If a data set consists of longevity times (in days) of fruit flies, what are the units used for standard deviation? What are the units used for variance?

3-3 BEYOND THE BASICS

37. Interpreting Outliers A data set consists of 20 values that are fairly close together. Another value is included, but this new value is an outlier (very far away from the other values). How is the standard deviation affected by the outlier? No effect? A small effect? A large effect?

TI **38.** Why Divide by $n - 1$? Let a population consist of the values 3, 6, 9. Assume that samples of 2 values are randomly selected *with replacement.*

a. Find the variance σ^2 of the population {3, 6, 9}.

b. List the nine different possible samples of 2 values selected with replacement, then find the sample variance s^2 (which includes division by $n - 1$) for each of them. If you repeatedly select 2 sample values, what is the mean value of the sample variances s^2?

c. For each of the nine samples, find the variance by treating each sample as if it is a population. (Be sure to use the formula for population variance, which includes division by n.) If you repeatedly select 2 sample values, what is the mean value of the population variances?

d. Which approach results in values that are better estimates of σ^2: part (b) or part (c)? Why? When computing variances of samples, should you use division by n or $n - 1$?

e. The preceding parts show that s^2 is an unbiased estimator of σ^2. Is s an unbiased estimator of σ?

3-4 Measures of Relative Standing

Key Concept This section introduces measures that can be used to compare values from different data sets, or to compare values within the same data set. The most important concept in this section is the z score, so we should understand the role of z scores (for comparing values from different data sets) and we should develop the ability to convert data values to z scores. We also discuss quartiles and percentiles that are used for comparing values within the same data set.

z Scores

A z score (or standardized value) is found by converting a value to a standardized scale, as given in the following definition. We will use z scores extensively in Chapter 6 and later chapters, so they are extremely important.

Cost of Laughing Index

There really is a Cost of Laughing Index (CLI), which tracks costs of such items as rubber chickens, Groucho Marx glasses, admissions to comedy clubs, and 13 other leading humor indicators. This is the same basic approach used in developing the Consumer Price Index (CPI), which is based on a weighted average of goods and services purchased by typical consumers. While standard scores and percentiles allow us to compare different values, they ignore any element of time. Index numbers, such as the CLI and CPI, allow us to compare the value of some variable to its value at some base time period. The value of an index number is the current value, divided by the base value, multiplied by 100.

Definition

A z **score** (or **standardized value**), is the number of standard deviations that a given value x is above or below the mean. It is found using the following expressions:

Sample $\quad z = \dfrac{x - \bar{x}}{s}$ or *Population* $\quad z = \dfrac{x - \mu}{\sigma}$

(Round z to two decimal places.)

The following example illustrates how z scores can be used to compare values, even though they might come from different populations.

EXAMPLE Comparing Heights With a height of 75 in., Lyndon Johnson was the tallest president of the past century. With a height of 85 in., Shaquille O'Neal is the tallest player on the Miami Heat basketball team. Who is *relatively* taller: Lyndon Johnson among the presidents of the past century, or Shaquille O'Neal among the players on his Miami Heat team? Presidents of the past century have heights with a mean of 71.5 in. and a standard deviation of 2.1 in. Basketball players for the Miami Heat have heights with a mean of 80.0 in. and a standard deviation of 3.3 in.

SOLUTION The heights of presidents and basketball players are from very different populations, so a comparison requires that we standardize heights by converting them to z scores.

$$\text{Lyndon Johnson: } z = \frac{x - \mu}{\sigma} = \frac{75 - 71.5}{2.1} = 1.67$$

$$\text{Shaquille O'Neal: } z = \frac{x - \mu}{\sigma} = \frac{85 - 80.0}{3.3} = 1.52$$

INTERPRETATION Lyndon Johnson's height is 1.67 standard deviations above the mean, and Shaquille O'Neal's height is 1.52 standard deviations above the mean. Lyndon Johnson's height among presidents of the past century is relatively greater than Shaquille O'Neal's height among the Miami Heat basketball players. Shaquille O'Neal is much taller than Lyndon Johnson, but Johnson is relatively taller when compared to colleagues.

z Scores and Unusual Values

In Section 3-3 we used the range rule of thumb to conclude that a value is "unusual" if it is more than 2 standard deviations away from the mean. It follows that unusual values have z scores less than -2 or greater than $+2$. (See Figure 3-5.) Using this criterion, Lyndon Johnson is not unusually tall when compared to

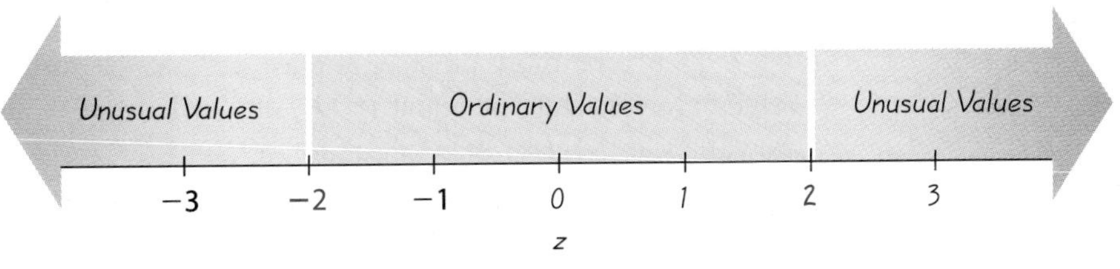

Figure 3-5 Interpreting z Scores
Unusual values are those with z scores less than −2.00 or greater than 2.00.

presidents of the past century, and Shaquille O'Neal is not unusually tall when compared to his teammates, because neither of them has a height with a z score greater than 2.

Ordinary values: $-2 \leq z$ score ≤ 2
Unusual values: z score < -2 *or* z score > 2

While considering Miami Heat basketball players, the shortest player is Damon Jones with a height of 75 in. His z score is -1.52, as shown in the calculation below. (We again use $\mu = 80.0$ in. and $\sigma = 3.3$ in. for the Miami Heat.)

$$\text{Damon Jones: } z = \frac{x - \mu}{\sigma} = \frac{75 - 80.0}{3.3} = -1.52$$

Damon Jones' height illustrates this principle about values that are below the mean:

Whenever a value is less than the mean, its corresponding z score is negative.

z scores are measures of position in the sense that they describe the location of a value (in terms of standard deviations) relative to the mean. A z score of 2 indicates that a value is two standard deviations *above* the mean, and a z score of -3 indicates that a value is three standard deviations *below* the mean. Quartiles and percentiles are also measures of position, but they are defined differently than z scores and they are useful for comparing values within the same data set or between different sets of data.

Quartiles and Percentiles

From Section 3-2 we know that the median of a data set is the middle value, so that 50% of the values are equal to or less than the median and 50% of the values are greater than or equal to the median. Just as the median divides the data into two equal parts, the three **quartiles,** denoted by Q_1, Q_2, and Q_3, divide the sorted values into four equal parts. (Values are *sorted* when they are arranged in order.) Here are descriptions of the three quartiles:

Q_1 (First quartile): Separates the bottom 25% of the sorted values from the top 75%. (To be more precise, at least 25% of the sorted values are less than or equal to Q_1, and at least 75% of the values are greater than or equal to Q_1.)

Q_2 (Second quartile): Same as the median; separates the bottom 50% of the sorted values from the top 50%.

Q_3 (Third quartile): Separates the bottom 75% of the sorted values from the top 25%. (To be more precise, at least 75% of the sorted values are less than or equal to Q_3, and at least 25% of the values are greater than or equal to Q_3.)

We will describe a procedure for finding quartiles after we discuss percentiles. There is not universal agreement on a single procedure for calculating quartiles, and different computer programs often yield different results. For example, if you use the data set of 1, 3, 6, 10, 15, 21, 28, 36, you will get these results:

	Q_1	Q_2	Q_3
STATDISK	4.5	12.5	24.5
Minitab	3.75	12.5	26.25
SPSS	3.75	12.5	26.25
Excel	5.25	12.5	22.75
SAS	4.5	12.5	24.5
TI-83/84 Plus	4.5	12.5	24.5

If you use a calculator or computer software for exercises involving quartiles, you may get results that differ slightly from the answers given in the back of the book.

Just as there are three quartiles separating a data set into four parts, there are also 99 **percentiles,** denoted $P_1, P_2, \ldots, P_{99}$, which partition the data into 100 groups with about 1% of the values in each group. (Quartiles and percentiles are examples of *quantiles*—or *fractiles*—which partition data into groups with roughly the same number of values.)

The process of finding the percentile that corresponds to a particular value x is fairly simple, as indicated in the following expression:

$$\text{percentile of value } x = \frac{\text{number of values less than } x}{\text{total number of values}} \cdot 100$$

(round the result to the nearest whole number)

CHAPTER 3 PROBLEM

EXAMPLE Ages of Best Actresses Table 3-4 lists the 76 sorted ages of Oscar-winning Best Actresses included in Table 2-1. (Table 2-1 is included with the Chapter Problem for Chapter 2.) Find the percentile corresponding to the age of 30 years.

Table 3-4 *Sorted* Ages of 76 Best Actresses

21	22	24	24	25	25	25	25	26	26
26	26	27	27	27	27	28	28	28	28
29	29	29	29	29	29	30	30	31	31
31	32	32	33	33	33	33	33	34	34
34	35	35	35	35	35	35	35	36	37
37	38	38	38	38	39	39	40	41	41
41	41	41	42	42	43	45	46	49	50
54	60	61	63	74	80				

SOLUTION From Table 3-4 we see that there are 26 ages less than 30, so

$$\text{percentile of } 30 = \frac{26}{76} \cdot 100 = 34$$

INTERPRETATION The age of 30 years is the 34th percentile.

The preceding example shows how to convert from a given sample value to the corresponding percentile. There are several different methods for the reverse procedure of converting a given percentile to the corresponding value in the data set. The procedure we will use is summarized in Figure 3-6, which uses the following notation.

Notation

- n total number of values in the data set
- k percentile being used (Example: For the 25th percentile, $k = 25$.)
- L locator that gives the *position* of a value (Example: For the 12th value in the sorted list, $L = 12$.)
- P_k kth percentile (Example: P_{25} is the 25th percentile.)

EXAMPLE Ages of Best Actresses Refer to the sorted ages of the Best Actresses in Table 3-4 and use Figure 3-6 to find the value of the 20th percentile, P_{20}.

SOLUTION Referring to Figure 3-6, we see that the sample data are already sorted, so we can proceed to find the value of the locator L. In this computation we use $k = 20$ because we are trying to find the value of the 20th percentile. We use $n = 76$ because there are 76 data values.

$$L = \frac{k}{100} \cdot n = \frac{20}{100} \cdot 76 = 15.2$$

continued

STATISTICS IN THE NEWS

The Growth of Statistics

Reporter Richard Rothstein wrote in the *New York Times* that the study of algebra, trigonometry, and geometry in high school "leaves too little room for the study of statistics and probability. Yet students need grounding in data analysis." He observed that calculus plays a prominent role in college studies, even though "only a few jobs, mostly in technical fields actually use it." Rothstein cited a study conducted by University of Massachusetts professor Clifford Konold, who counted data displays in the *New York Times.* In 1972 issues of the *Times*, Dr. Konold found four graphs or tables in each of 10 weekday editions (not including the sports and business sections), but in 1982 there were 8, and in 1992 there were 44, and "next year, he (Dr. Konold) could find more than100." The growth of statistics as a discipline is fostered in part by the increasing use of such data displays in the media.

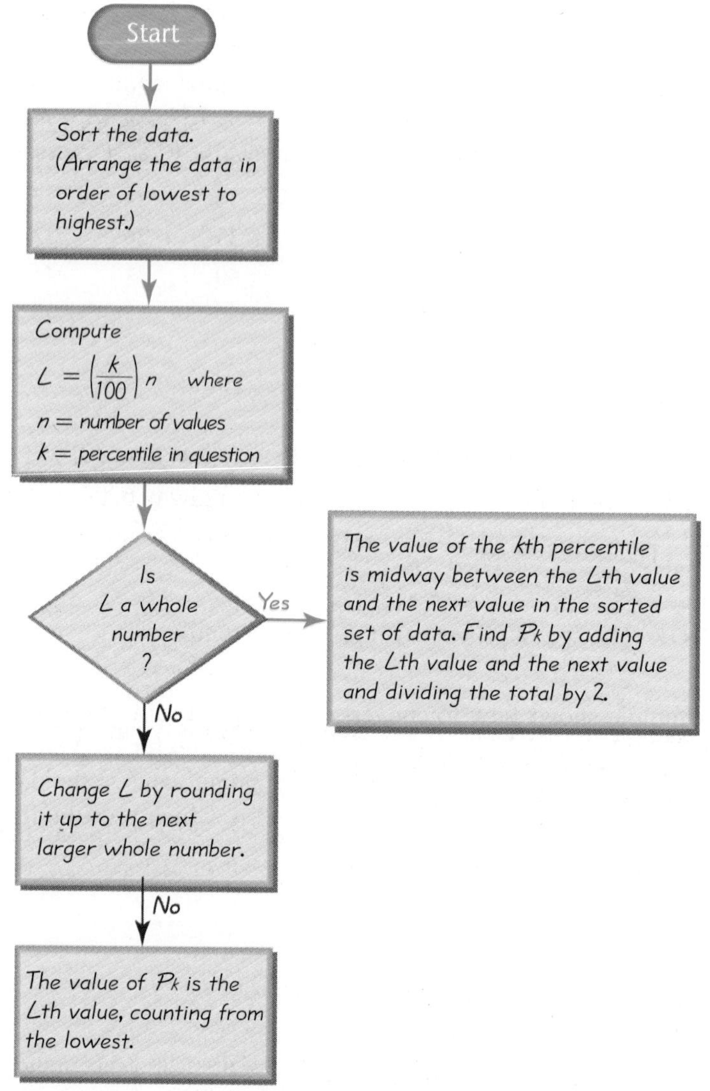

Figure 3-6 Converting from the kth Percentile to the Corresponding Data Value

Next, we are asked if L is a whole number and we answer no, so we proceed to the next lower box where we change L by rounding it up from 15.2 to 16. (In this book we typically round off the usual way, but this is one of two cases where we round *up* instead of rounding *off*.) Finally, the bottom box shows that the value of P_{20} is the 16th value, counting from the lowest. In Table 3-4, the 16th value is 27. That is, $P_{20} = 27$ years of age.

EXAMPLE Ages of Best Actresses Refer to the ages of the Best Actresses given in Table 3-4. Use Figure 3-6 to find the value of Q_3, which is the third quartile.

SOLUTION First we note that Q_3 is the same as P_{75}, so we can proceed with the objective of finding the value of the 75th percentile. Referring to Figure 3-6, we see that the sample data are already sorted, so we can proceed to compute the value of the locator L. In this computation, we use $k = 75$ because we are attempting to find the value of the 75th percentile, and we use $n = 76$ because there are 76 data values.

$$L = \frac{k}{100} \cdot n = \frac{75}{100} \cdot 76 = 57$$

Next, we are asked if L is a whole number and we answer yes, so we proceed to the box located at the right. We now see that the value of the kth (75th) percentile is midway between the Lth (57th) value and the next value in the original set of data. That is, the value of the 75th percentile is midway between the 57th value and the 58th value. The 57th value is 39 years and the 58th value is 40 years, so the value midway between them is 39.5 years. We conclude that the 75th percentile is $P_{75} = 39.5$ years. The value of the third quartile Q_3 is also 39.5 years.

The preceding example showed that when finding a quartile value (such as Q_3), we can use the equivalent percentile value (such as P_{75}) instead. See the margin for relationships relating quartiles to equivalent percentiles.

$Q_1 = P_{25}$
$Q_2 = P_{50}$
$Q_3 = P_{75}$

In earlier sections of this chapter we described several statistics, including the mean, median, mode, range, and standard deviation. Some other statistics are defined using quartiles and percentiles, as in the following:

$$\text{interquartile range (or IQR)} = Q_3 - Q_1$$

$$\text{semi-interquartile range} = \frac{Q_3 - Q_1}{2}$$

$$\text{midquartile} = \frac{Q_3 + Q_1}{2}$$

$$\text{10–90 percentile range} = P_{90} - P_{10}$$

After completing this section, you should be able to convert a value into its corresponding z score (or standard score) so that you can compare it to other values, which may be from different data sets. You should be able to convert a value into its corresponding percentile value so that you can compare it to other values in some data set. You should be able to convert a percentile to the corresponding data value. And finally, you should understand the meanings of quartiles and be able to relate them to their corresponding percentile values (as in $Q_3 = P_{75}$).

Using the TI-83/84 Plus Calculator to Find *z* Scores, Percentiles, and Quartiles

z Scores Calculation of a z score is easily done with the TI-83/84 Plus basic operations. To find $z = (x - \bar{x})/s$, simply use the − and ÷ keys, but *be sure to use parentheses to group the numerator*. (Without parentheses, the calculator will do the division before the subtraction, and the result will be wrong.) Given a mean of 100 and a standard deviation of 15, conversion of 130 to a z score can be accomplished by entering (130 − 100) ÷ 15 ENTER. The result will be a z score of 2.

Percentiles The TI-83/84 Plus calculator does not have a function for finding percentiles, but determination of percentiles is made easier using the **Sort** function, which arranges the values in order. To sort a list of values, press STAT, then select **SortA** (for ascending order) or **SortD** (for descending order). The command SortA(L1) rearranges the entries in list L1 to be in ascending (increasing) order. The command Sort(LAGES) rearranges the entries in the list named AGES to be in ascending (increasing) order. The percentile procedures described in this section are much easier with a sorted list.

Quartiles Quartiles are included among the descriptive statistics obtained by using STAT, **CALC, 1-Var Stats** as described at the end of Section 3-2.

From this display we know that the quartiles are $Q_1 = 28$, $Q_2 = 33.5$, and $Q_3 = 39.5$. These are the same results that would be obtained by following the procedures of this section. However, the procedures of this section are simplified, so they will sometimes yield values that differ somewhat from those obtained by using a TI-83/84 Plus calculator.

TI-83/84 Plus

```
1-Var Stats
↑n=76
 minX=21
 Q1=28
 Med=33.5
 Q3=39.5
 maxX=80
```

3-4 BASIC SKILLS AND CONCEPTS

Statistical Literacy and Critical Thinking

1. z Scores A value from a large data set is found to have a z score of −2. Is the value above the mean or below the mean? How many standard deviations away from the mean is this value?

2. z Scores A sample consists of lengths of American bald eagles, measured in centimeters. If the length of one particular eagle is converted to a z score, what units are used for the z score? Centimeters?

3. Quartiles For a large data set, the first quartile Q_1 is found to be 15. What does it mean when we say that 15 is the first quartile?

4. ZIP Codes ZIP codes are arranged from east to west. Eastern states, such as Maine, have ZIP codes beginning with 0, but western states, such as Hawaii, Alaska, and California have ZIP codes beginning with 9. If the first quartile Q_1 is computed from

the list of all ZIP codes, does the result correspond to the location that is 25% of the distance from the location that is farthest east to the location that is farthest west?

In Exercises 5–8, express all z scores with two decimal places.

TI **5.** Darwin's Height Men have heights with a mean of 176 cm and a standard deviation of 7 cm. Charles Darwin had a height of 182 cm.

a. What is the difference between Darwin's height and the mean?
b. How many standard deviations is that [the difference found in part (a)]?
c. Convert Darwin's height to a z score.
d. If we consider "usual" heights to be those that convert to z scores between -2 and 2, is Darwin's height usual or unusual?

TI **6.** Einstein's IQ Stanford Binet IQ scores have a mean of 100 and a standard deviation of 16. Albert Einstein reportedly had an IQ of 160.

a. What is the difference between Einstein's IQ and the mean?
b. How many standard deviations is that [the difference found in part (a)]?
c. Convert Einstein's IQ score to a z score.
d. If we consider "usual" IQ scores to be those that convert to z scores between -2 and 2, is Einstein's IQ usual or unusual?

TI **7.** Heights of Presidents With a height of 67 in., William McKinley was the shortest president of the past century. The presidents of the past century have a mean height of 71.5 in. and a standard deviation of 2.1 in.

a. What is the difference between McKinley's height and the mean height of presidents from the past century?
b. How many standard deviations is that [the difference found in part (a)]?
c. Convert McKinley's height to a z score.
d. If we consider "usual" heights to be those that convert to z scores between –2 and 2, is McKinley's height usual or unusual?

TI **8.** World's Tallest Woman Sandy Allen is the world's tallest woman with a height of 91.25 in. (or 7 ft, 7.25 in.). Women have heights with a mean of 63.6 in. and a standard deviation of 2.5 in.

a. What is the difference between Sandy Allen's height and the mean height of women?
b. How many standard deviations is that [the difference found in part (a)]?
c. Convert Sandy Allen's height to a z score.
d. Does Sandy Allen's height meet the criterion of being unusual by corresponding to a z score that does not fall between -2 and 2?

TI **9.** Body Temperatures Human body temperatures have a mean of 98.20°F and a standard deviation of 0.62°F. Convert the given temperatures to z scores.

a. 97.5°F
b. 98.60°F
c. 98.20°F

TI **10.** Heights of Women The Beanstalk Club is limited to women and men who are very tall. The minimum height requirement for women is 70 in. Women's heights have a mean of 63.6 in. and a standard deviation of 2.5 in. Find the z score corresponding to a woman with a height of 70 in. and determine whether that height is unusual.

TI **11.** Length of Pregnancy A woman wrote to *Dear Abby* and claimed that she gave birth 308 days after a visit from her husband, who was in the Navy. Lengths of pregnancies have a mean of 268 days and a standard deviation of 15 days. Find the z score for 308 days. Is such a length unusual? What do you conclude?

TI **12.** Cholesterol Levels For men aged between 18 and 24 years, serum cholesterol levels (in mg/100 mL) have a mean of 178.1 and a standard deviation of 40.7 (based on data from the National Health Survey). Find the z score corresponding to a male, aged 18–24 years, who has a serum cholesterol level of 259.0 mg/100 mL. Is this level unusually high?

TI **13.** Comparing Test Scores Which is relatively better: a score of 85 on a psychology test or a score of 45 on an economics test? Scores on the psychology test have a mean of 90 and a standard deviation of 10. Scores on the economics test have a mean of 55 and a standard deviation of 5.

TI **14.** Comparing Scores Three students take equivalent stress tests. Which is the highest relative score?

a. A score of 144 on a test with a mean of 128 and a standard deviation of 34.
b. A score of 90 on a test with a mean of 86 and a standard deviation of 18.
c. A score of 18 on a test with a mean of 15 and a standard deviation of 5.

In Exercises 15–18, use the 76 sorted ages of Best Actresses listed in Table 3-4. Find the percentile corresponding to the given age.

TI **15.** 25 TI **16.** 35 TI **17.** 40 TI **18.** 50

In Exercises 19–26, use the 76 sorted ages of Best Actresses listed in Table 3-4. Find the indicated percentile or quartile.

TI **19.** P_{10} TI **20.** Q_1 TI **21.** P_{25} TI **22.** Q_2

TI **23.** P_{33} TI **24.** P_{66} TI **25.** P_1 TI **26.** P_{85}

3-4 BEYOND THE BASICS

27. Ages of Best Actresses Use the 76 sorted ages of Best Actresses listed in Table 3-4.

a. Find the interquartile range.
b. Find the midquartile.
c. Find the 10–90 percentile range.
d. Does $P_{50} = Q_2$? If so, does P_{50} *always* equal Q_2?
e. Does $Q_2 = (Q_1 + Q_3)/2$? If so, does Q_2 *always* equal $(Q_1 + Q_3)/2$?

28. Interpolation When finding percentiles using Figure 3-6, if the locator L is not a whole number, we round it up to the next larger whole number. An alternative to this procedure is to *interpolate*. For example, using interpolation with a locator of $L = 23.75$ leads to a value that is 0.75 (or 3/4) of the way between the 23rd and 24th values. Use this method of interpolation to find P_{35} for the ages of the Best Actresses in Table 3-4. How does the result compare to the value that would be found by using Figure 3-6 without interpolation?

29. Deciles and Quintiles For a given data set, there are nine **deciles,** denoted by $D_1, D_2, \ldots, D_9$ which separate the sorted data into 10 groups, with about 10% of the values in each group. There are also four **quintiles,** which divide the sorted data into 5 groups, with about 20% of the values in each group. (Note the difference between quintiles and quantiles, which were described earlier in this section.)

a. Which percentile is equivalent to D_1? D_5? D_8?
b. Using the sorted ages of the Best Actresses in Table 3-4, find the nine deciles.
c. Using the sorted ages of the Best Actresses in Table 3-4, find the four quintiles.

3-5 Exploratory Data Analysis (EDA)

Key Concept This section discusses outliers, then introduces a new statistical graph called a boxplot, which is helpful for visualizing the distribution of data. (Boxplots were not included in Section 2-4 because they use quartiles, which were not introduced until Section 3-4.) We should know how to construct a simple boxplot. This section also focuses on the important principle that we should *explore* data before jumping to specific methods of statistics.

We begin this section by first defining exploratory data analysis, then we introduce outliers, 5-number summaries, and boxplots. Modified boxplots, which are introduced near the end of this section, are somewhat more complicated, but they provide more specific information about outliers.

Data Mining

The term *data mining* is commonly used to describe the now popular practice of analyzing an existing large set of data for the purpose of finding relationships, patterns, or any interesting results that were not found in the original studies of the data set. Some statisticians express concern about ad hoc inference—a practice in which a researcher goes on a fishing expedition through old data, finds something significant, and then identifies an important question that has already been answered. Robert Gentleman, a column editor for *Chance* magazine, writes that "there are some interesting and fundamental statistical issues that data mining can address. We simply hope that its current success and hype don't do our discipline (statistics) too much damage before its limitations are discussed."

Definition

Exploratory data analysis is the process of using statistical tools (such as graphs, measures of center, measures of variation) to investigate data sets in order to understand their important characteristics.

Recall that in Section 2-1 we listed five important characteristics of data: center, variation, distribution, outliers, and changing pattern over time. We can investigate center with measures such as the mean and median. We can investigate variation with measures such as the standard deviation and range. We can investigate the distribution of data by using tools such as frequency distributions and histograms. We have seen that some important statistics (such as the mean and standard deviation) can be strongly affected by the presence of an outlier. It is generally important to further investigate the data set to identify any notable features, especially those that could strongly affect results and conclusions. One such feature is the presence of outliers. We now consider outliers in more detail.

Outliers

Throughout this book we will consider an **outlier** to be a value that is located very far away from almost all of the other values. (There is a more specific alternative definition in the subsection "Modified Boxplots" that is near the end of this section.) Relative to the other data, an outlier is an *extreme* value that falls well outside the general pattern of almost all of the data. When exploring a data set, outliers should be considered because they may reveal important information, and they may strongly affect the value of the mean and standard deviation, as well as seriously distorting a histogram. The following example uses an incorrect entry as an example of an outlier, but not all outliers are errors; some outliers are correct values.

EXAMPLE Ages of Best Actresses When using computer software or a calculator, it is often easy to make keystroke errors. Refer to the ages of the Best Actresses listed in Table 2-1. (Table 2-1 is included with the Chapter Problem for Chapter 2.) Assume that when entering the ages, the first entry of 22 is incorrectly entered as 2222 because you held

continued

the 2 key down too long when you were distracted by a meteorite landing on your porch. The incorrect entry of 2222 is an outlier because it is located very far away from the other values. How does that outlier affect the mean, standard deviation, and histogram?

SOLUTION When the entry of 22 is replaced by the outlier value of 2222, the mean changes from 35.7 years to 64.6 years, so the effect of the outlier is very substantial. The incorrect entry of 2222 causes the standard deviation to change from 11.1 to 251.0, so the effect of the outlier here is also substantial. Figure 2-2 in Section 2-3 depicts the histogram for the correct ages of the actresses in Table 2-1, but the STATDISK and TI-83/84 Plus calculator displays below show the histogram that results from using the same data with the value of 22 replaced by the incorrect value of 2222. Compare this histogram to Figure 2-2 and you can easily see that the presence of the outlier dramatically affects the shape of the distribution.

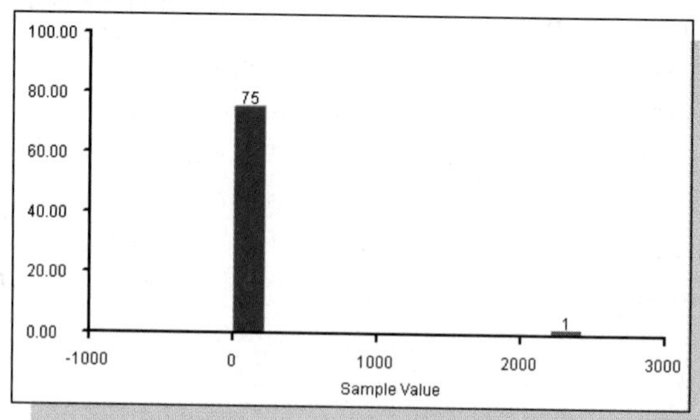

TI-83/84 Plus

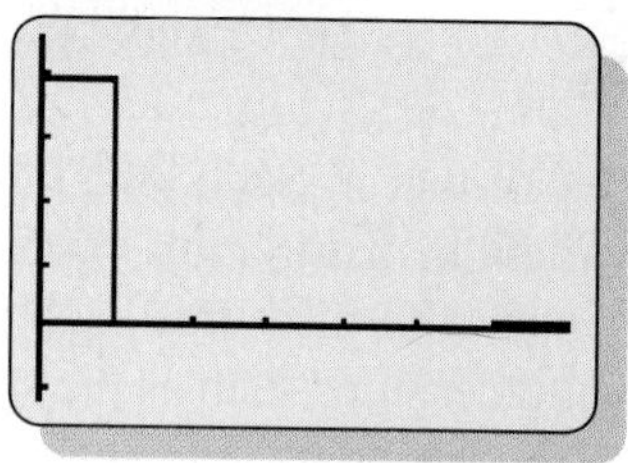

The preceding example illustrates these important principles:

1. **An outlier can have a dramatic effect on the mean.**
2. **An outlier can have a dramatic effect on the standard deviation.**
3. **An outlier can have a dramatic effect on the scale of the histogram so that the true nature of the distribution is totally obscured.**

An easy procedure for finding outliers is to examine a *sorted* list of the data. In particular, look at the minimum and maximum sample values and determine whether they are very far away from the other typical values. Some outliers are correct values and some are errors, as in the preceding example. If we are sure that an outlier is an error, we should correct it or delete it. If we include an outlier because we know that it is correct, we might study its effects by constructing graphs and calculating statistics with and without the outlier included.

Boxplots

In addition to the graphs presented in Sections 2-3 and 2-4, a boxplot is another graph that is used often. Boxplots are useful for revealing the center of the data, the spread of the data, the distribution of the data, and the presence of outliers.

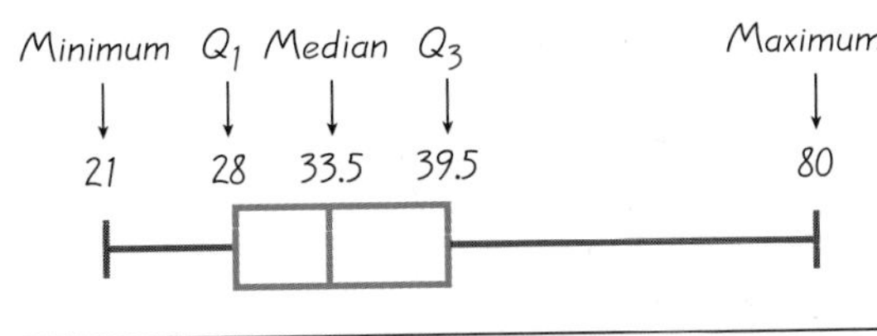

Figure 3-7 Boxplot of Ages of Best Actresses

The construction of a boxplot requires that we first obtain the minimum value, the maximum value, and quartiles, as defined in the *5-number summary*.

Definition

For a set of data, the **5-number summary** consists of the minimum value, the first quartile Q_1, the median (or second quartile Q_2), the third quartile Q_3, and the maximum value.

A **boxplot** (or **box-and-whisker diagram**) is a graph of a data set that consists of a line extending from the minimum value to the maximum value, and a box with lines drawn at the first quartile Q_1, the median, and the third quartile Q_3. (See Figure 3-7.)

Procedure for Constructing a Boxplot

1. Find the 5-number summary consisting of the minimum value, Q_1, the median, Q_3, and the maximum value.
2. Construct a scale with values that include the minimum and maximum data values.
3. Construct a box (rectangle) extending from Q_1 to Q_3, and draw a line in the box at the median value.
4. Draw lines extending outward from the box to the minimum and maximum data values.

Boxplots don't show as much detailed information as histograms or stem-and-leaf plots, so they might not be the best choice when dealing with a single data set. They are often great for comparing two or more data sets. When using two or more boxplots for comparing different data sets, it is important to use the same scale so that correct comparisons can be made.

EXAMPLE Ages of Actresses Refer to the 76 ages of Best Actresses in Table 2-1 (without the error of 2222 used in place of 22, as in the preceding example).

a. Find the values constituting the 5-number summary.

b. Construct a boxplot.

continued

An Outlier Tip

Outliers are important to consider because, in many cases, one extreme value can have a dramatic effect on statistics and conclusions derived from them. In some cases an outlier is a mistake that should be corrected or deleted. In other cases, an outlier is a valid data value that should be investigated for any important information. Students of the author collected data consisting of restaurant bills and tips, and no notable outliers were found among their sample data. However, one such outlier is the tip of $16,000 that was left for a restaurant bill of $8,899.78. The tip was left by an unidentified London executive to waiter Lenny Lorando at Nello's restaurant in New York City. Lorando said that he had waited on the customer before and "He's always generous, but never anything like that before. I have to tell my sister about him."

SOLUTION

a. The 5-number summary consists of the minimum, Q_1, median, Q_3, and maximum. To find those values, first sort the data (by arranging them in order from lowest to highest). The minimum of 21 and the maximum of 80 are easy to identify from the sorted list. Now proceed to find the quartiles. For the first quartile we have $Q_1 = P_{25} = 28$. [Using the flowchart of Figure 3-6, the locator is $L = (25/100)76 = 19$, which is a whole number, so Figure 3-6 indicates that Q_1 is the value midway between the 19th value and the 20th value in the sorted list.] The median is 33.5, which is the value midway between the 38th and 39th values. We also find that $Q_3 = 39.5$ by using Figure 3-6 to find the 75th percentile. The 5-number summary is therefore 21, 28, 33.5, 39.5, 80.

b. In Figure 3-7 we graph the boxplot for the data. We use the minimum (21) and the maximum (80) to determine a scale of values, then we plot the values from the 5-number summary as shown.

Boxplots are particularly helpful for comparing data sets, especially when they are drawn on the same scale. Shown below are the STATDISK and TI-83/84 Plus boxplots for the ages of the Best Actresses (top boxplot) and the Best Actors displayed on the same scale. We can see from the two boxplots that the ages of the actresses tend to vary more, and the ages of the actresses tend to be lower than the actors.

STATDISK

21 33.5 80
28 39.5
29 42 76
38 49
20 40 60 80

TI-83/84 Plus

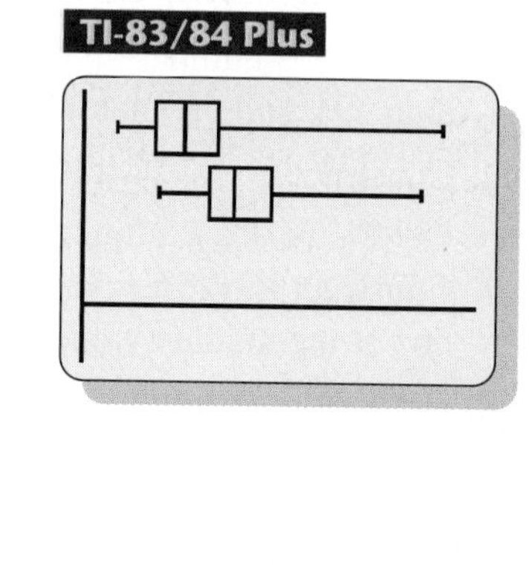

Modified Boxplots

The preceding description of boxplots describes **skeletal** (or **regular**) **boxplots,** but some statistical software packages provide modified boxplots, which represent outliers as special points. For example, the TI-83/84 Plus calculator uses asterisks to depict outliers that are identified using a specific criterion that uses the *interquartile range* (IQR). In Section 3-4 it was noted that the interquartile range is this: Interquartile range (IQR) $= Q_3 - Q_1$. The value of the IQR can then be used to identify outliers as follows:

A data value is an outlier if it is . . .

above Q_3 by an amount greater than 1.5 × IQR

or

below Q_1 by an amount greater than 1.5 × IQR

A **modified boxplot** is a boxplot constructed with these modifications: (1) A special symbol (such as an asterisk or dot) is used to identify outliers as defined here, and (2) the solid horizontal line extends only as far as the minimum data value that is not an outlier and the maximum data value that is not an outlier. (*Note: Exercises involving modified boxplots are found in the* "Beyond the Basics" *Exercises only.*)

EXAMPLE Do Males and Females Have the Same Pulse Rates? It has been well documented that there are important physiological differences between males and females. Males tend to weigh more and they tend to be taller than females. But is there a difference in pulse rates of males and females? Data Set 1 in Appendix B lists pulse rates for a sample of 40 males and a sample of 40 females. Later in this book we will describe important statistical methods that can be used to formally test for differences, but for now, let's explore the data to see what can be learned. (Even if we already know how to apply those formal statistical methods, it would be wise to first explore the data before proceeding with the formal analysis.)

SOLUTION Let's begin with an investigation into the key elements of center, variation, distribution, outliers, and characteristics over time (the same "CVDOT" list introduced in Section 2-1). Listed below are measures of center (mean), measures of variation (standard deviation), and the 5-number summary for the pulse rates listed in Data Set 1. The accompanying displays show modified boxplots for each of the two samples (with + denoting values that are outliers), a histogram of the male pulse rates, and a histogram of the female pulse rates.

	Mean	Standard Deviation	Minimum	Q_1	Median	Q_3	Maximum
Males	69.4	11.3	56	60.0	66.0	76.0	96
Females	76.3	12.5	60	68.0	74.0	80.0	124

TI-83/84 Plus Modified Boxplots

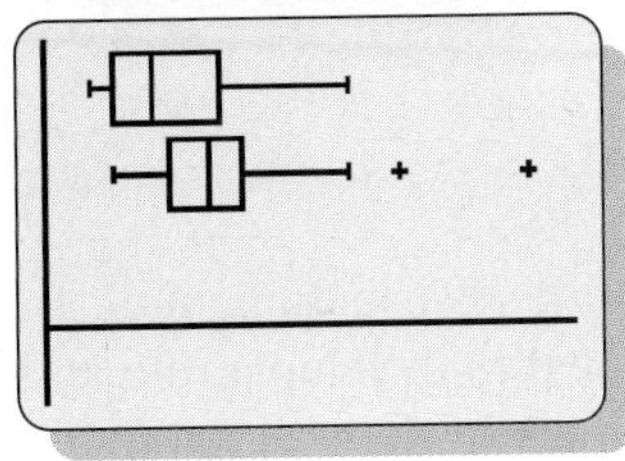

TI-83/84 Plus Histogram of Male Pulse Rates

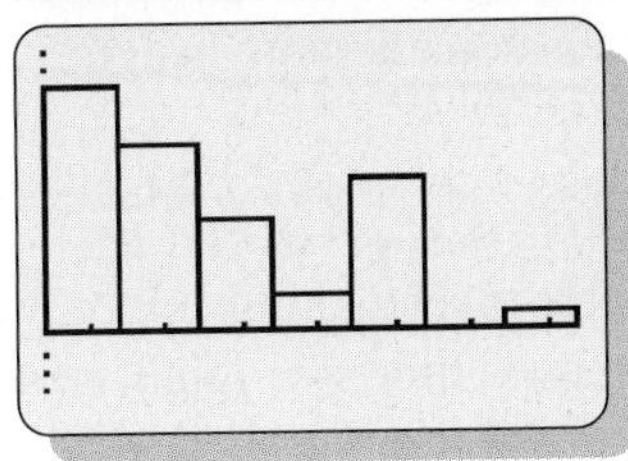

TI-83/84 Plus Histogram of Female Pulse Rates

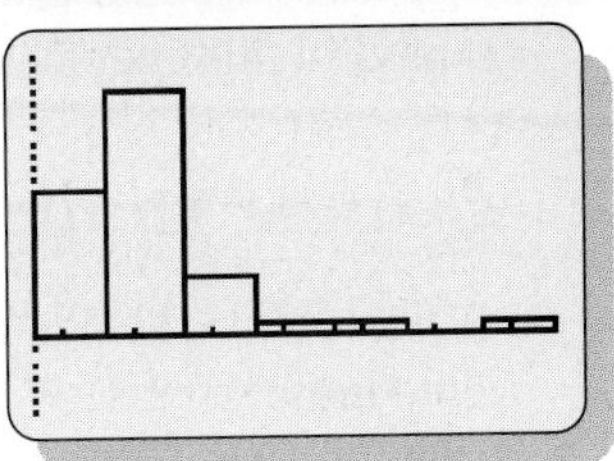

Human Lie Detectors

Researchers tested 13,000 people for their ability to determine when someone is lying. They found 31 people with exceptional skills at identifying lies. These human lie detectors had accuracy rates around 90%. They also found that federal officers and sheriffs were quite good at detecting lies, with accuracy rates around 80%. Psychology Professor Maureen O'-Sullivan questioned those who were adept at identifying lies, and she said that "all of them pay attention to nonverbal cues and the nuances of word usages and apply them differently to different people. They could tell you eight things about someone after watching a two-second tape. It's scary, the things these people notice." Methods of statistics can be used to distinguish between people unable to detect lying and those with that ability.

INTERPRETATION Examining and comparing the statistics and graphs, we make the following important observations.

- ***Center:*** The mean pulse rates of 69.4 for males and 76.3 for females appear to be very different. The boxplots do not depict the values of the two means, but they do suggest that the pulse rates of males are somewhat lower than those of females, so it follows that the mean pulse rate for males does appear to be lower than the mean for females.
- ***Variation:*** The standard deviations 11.3 and 12.5 do not appear to be dramatically different. Also, the boxplots depict the spread of the data. The widths of the boxplots do not appear to be very different, further supporting the observation that the standard deviations do not appear to be dramatically different.
- The values listed for the minimums, first quartiles, medians, third quartiles, and maximums suggest that the values for males are lower in each case, so that male pulse rates appear to be lower than females. There is a very dramatic difference between the maximum values of 96 (for males) and 124 (for females), but the maximum of 124 is an outlier because it is very different from almost all of the other pulse rates. We should now question whether that pulse rate of 124 is correct or is an error, and we should also investigate the effect of that outlier on our overall results. For example, if the outlier of 124 is removed, do the male pulse rates continue to appear less than the female pulse rates? See the following comments about outliers.
- ***Outliers:*** The modified boxplots include two + symbols corresponding to two female pulse rates considered to be outliers. The highest female pulse rate of 124 is an outlier that does not dramatically affect the results. If we delete the value of 124, the mean female pulse rate changes from 76.3 to 75.1, so the mean pulse rate for males continues to be considerably lower. Even if we delete both outliers of 104 and 124, the mean female pulse rate changes from 76.3 to 74.3, so the mean value of 69.4 for males continues to appear lower.
- ***Distributions:*** The histogram for the female pulse rates appears to be roughly bell-shaped, as we might have expected, but the distribution of male pulse rates appears to be somewhat different.

We now have considerable insight into the nature of the pulse rates for males and females. Based on our exploration, we can conclude that males appear to have pulse rates with a mean that is less than that of females. There are more advanced methods we could use later (such as the methods of Section 9-3), but the tools presented in this chapter give us considerable insight.

Critical Thinking

Armed with a list of tools for investigating center, variation, distribution, outliers, and characteristics of data over time, we might be tempted to develop a rote and mindless procedure, but critical thinking is critically important. In addition to using the tools presented in this chapter, we should consider any other relevant factors that might be crucial to the conclusions we form. We might pose questions such as these: Is the sample likely to be representative of the population, or is the sample somehow biased? What is the source of the data, and might the source be someone with an interest that could affect the quality of the data? Suppose, for example, that we want to estimate the mean income of statistics professors. Also suppose that we mail questionnaires to 200 statistics professors and receive 20 responses. We could calculate the mean, standard deviation, construct graphs, identify outliers, and so on, but the results will be what statisticians refer to as hogwash. The sample is a voluntary response sample, and it is not likely to be representative of the population of all statistics professors. In addition to the specific statistical tools presented in this chapter, we should also *think!*

TI

Using the TI-83/84 Plus Calculator for Five-Number Summaries and Boxplots

Five-Number Summaries To obtain a five-number summary from a list of data, use STAT, **CALC,** and **1-Var Stats** to obtain the descriptive statistics. (Detailed instructions are included at the end of Section 3-2.) Scrolling down through the displayed statistics, we find that the last five entries constitute the five-number summary consisting of the minimum, first quartile Q_1, median (or second quartile Q_2), third quartile Q_3, and the maximum.

Boxplot To obtain a boxplot for a single list of values, first enter or open the list. Then follow this procedure:

1. Press 2ND Y= to select the STAT PLOT feature.
2. Press ENTER and turn on Plot1. (Plot2 should be off, Plot3 should be off, Plot4 should be off, Plot5 should be off, and Plot6 should be off.)
3. Select the graph type that is in the middle of the second row, as shown in the accompanying screen display. (Selection of the boxplot icon in the beginning of the second row would result in a modified boxplot.)
4. For Xlist, enter the label of the list storing the data, or enter the name of the list, such as AGES.
5. Press ZOOM 9 to select the ZoomStat graphing option, which configures the window so that all of the data are included.

TI-83/84 Plus

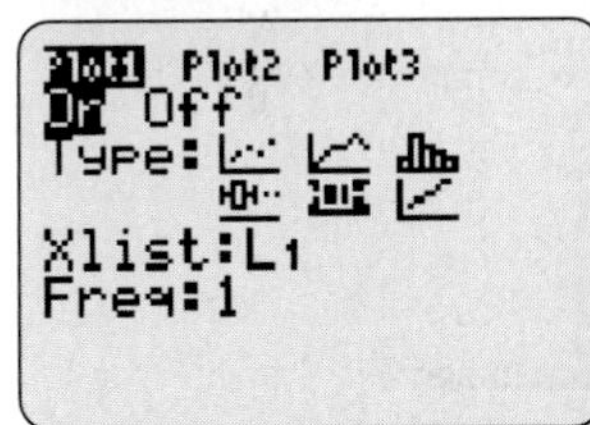

To obtain boxplots for two or three lists of data, follow the above steps, but turn on Plot2 and/or Plot3.

Whether displaying one, two, or three boxplots in the window, pressing the TRACE key and using the arrow keys will allow you to see the actual values of the minimum, maximum, and quartiles.

3-5 BASIC SKILLS AND CONCEPTS

Statistical Literacy and Critical Thinking

1. Boxplot Refer to the STATDISK-generated boxplot on the following page. What do the values of 2, 5, 10, 12, 20 tell us about the data set from which the boxplot was constructed?

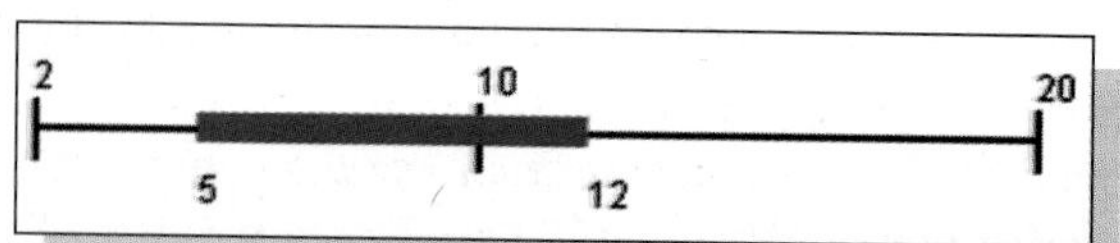

2. Boxplot Comparisons Refer to the two STATDISK-generated boxplots shown below that are drawn on the same scale. One boxplot represents heights of randomly selected men and the other represents heights of randomly selected women. Which boxplot represents women? How do you know?

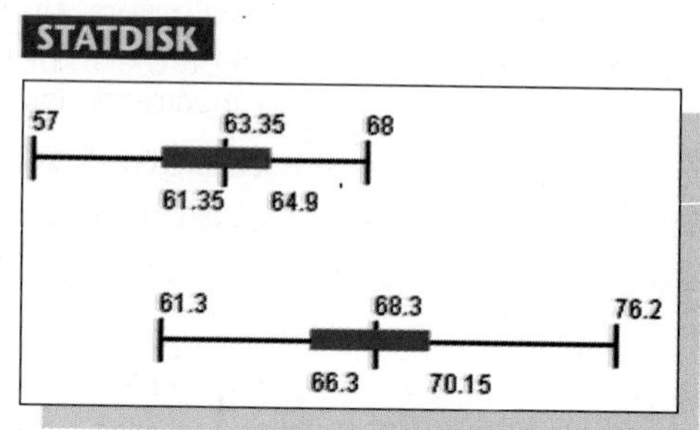

3. Variation The two boxplots shown below correspond to the service times from two different companies that repair air conditioning units. They are drawn on the same scale. The top boxplot corresponds to the Sigma Air Conditioning Company, and the bottom boxplot corresponds to the Newport Repair Company. Which company has less variation in repair times? Which company should have more predictable costs? Why?

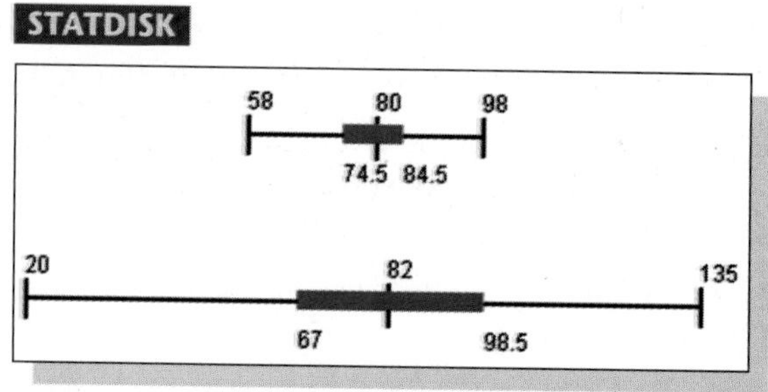

4. Outliers A set of 20 sample values includes one outlier that is very far away from the other 19 values. How much of an effect does that outlier have on each of these statistics: mean, median, standard deviation, and range?

TI **5.** Testing Corn Seeds In 1908, William Gosset published the article "The Probable Error of a Mean" under the pseudonym of "Student" (*Biometrika*, Vol. 6, No. 1). He included the data listed below for two different types of corn seed (regular and kiln dried) that were used on adjacent plots of land. The listed values are the yields of head corn in pounds per acre. Using the yields from *regular seed*, find the 5-number summary and construct a boxplot.

Regular	1903	1935	1910	2496	2108	1961	2060	1444	1612	1316	1511
Kiln Dried	2009	1915	2011	2463	2180	1925	2122	1482	1542	1443	1535

TI **6.** Testing Corn Seeds Using the yields from the *kiln-dried seed* listed in Exercise 5, find the 5-number summary and construct a boxplot. Do the results appear to be substantially different from those obtained in Exercise 5?

TI **7.** Weights of Quarters Refer to Data Set 14 in Appendix B and use the weights of the silver quarters (pre-1964) to find the 5-number summary and construct a boxplot.

TI **8.** Weights of Quarters Refer to Data Set 14 in Appendix B and use the weights of the post-1964 quarters to find the 5-number summary and construct a boxplot. Do the results appear to be substantially different from those obtained in Exercise 7?

TI **9.** Bear Lengths Refer to Data Set 6 for the lengths (in inches) of the 54 bears that were anesthetized and measured. Find the 5-number summary and construct a boxplot. Does the distribution of the lengths appear to be symmetric or does it appear to be skewed?

TI **10.** Body Temperatures Refer to Data Set 2 in Appendix B for the 106 body temperatures for 12 A.M. on day 2. Find the 5-number summary and construct a boxplot, then determine whether the sample values support the common belief that the mean body temperature is 98.6°F.

TI **11.** BMI Values Refer to Data Set 1 in Appendix B and use the body mass index (BMI) values of males to find the 5-number summary and construct a boxplot.

TI **12.** BMI Values Refer to Data Set 1 in Appendix B and use the body mass index (BMI) values of females to find the 5-number summary and construct a boxplot. Do the results appear to be substantially different from those obtained in Exercise 11?

3-5 BEYOND THE BASICS

13. Refer to the accompanying STATDISK display of three boxplots that represent the measure of longevity (in months) of samples of three different car batteries. If you are the manager of a fleet of cars and you must select one of the three brands, which boxplot represents the brand you should choose? Why?

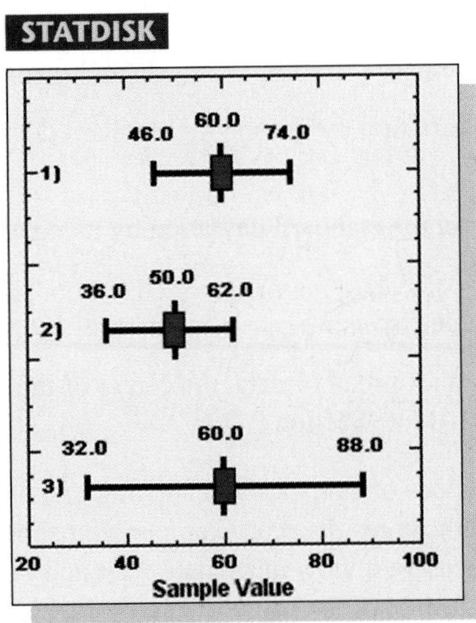

TI **14.** Outliers Instead of considering a data value to be an outlier if it is "very far away from almost all of the other data values," consider an outlier to be a value that is above Q_3 by an amount greater than 1.5 × IQR or below Q_1 by an amount greater than 1.5 × IQR. Use the data set given below and find the following:

a. 5-number summary

b. Interquartile range (IQR)

c. Outliers

2 3 4 5 7 9 14 15 15 16 19 32 50

15. Mild and Extreme Outliers Some statistics software packages construct boxplots that identify individual mild outliers (often plotted as solid dots) and extreme outliers (often plotted as hollow circles). **Mild outliers** are below Q_1 or above Q_3 by an amount that is greater than 1.5 × IQR but not greater than 3 × IQR. Extreme outliers are either below Q_1 by more than 3 × IQR, or above Q_3 by more than 3 × IQR. Refer to Data Set 10 in Appendix B and use the Monday rainfall amounts to identify any mild outliers and extreme outliers.

Review

In this chapter we discussed measures of center, measures of variation, measures of relative standing, and general methods of describing, exploring, and comparing data sets. When investigating a data set, these characteristics are generally very important:

1. *Center:* A representative or average value.
2. *Variation:* A measure of the amount that the values vary.
3. *Distribution:* The nature or shape of the distribution of the data (such as bell-shaped, uniform, or skewed).
4. *Outliers:* Sample values that lie very far away from the vast majority of the other sample values.
5. *Time:* Changing characteristics of the data over time.

The following are particularly important skills that should be learned or concepts that should be understood:

- Calculate measures of center by finding the mean and median (Section 3-2).
- Calculate measures of variation by finding the standard deviation, variance, and range (Section 3-3).
- *Understand* and *interpret* the standard deviation by using tools such as the range rule of thumb (Section 3-3).
- Compare individual values by using z scores, quartiles, or percentiles (Section 3-4).
- Investigate and explore the spread of data, the center of the data, and the range of values by constructing a boxplot (Section 3-5).

In addition to finding values of statistics and creating graphs, it is extremely important to *understand* and *interpret* those results. For example, you should clearly understand that the standard deviation is a measure of how much data vary, and you should be able to use the standard deviation to distinguish between values that are usual and those that are unusual.

Statistical Literacy and Critical Thinking

1. Center and Variation A quality control engineer redesigns repair procedures so that the standard deviation of the repair times is reduced. Does this imply that the repairs are being done in less time? Why or why not?

2. Production Specifications When designing the production procedure for batteries used in heart pacemakers, an engineer specifies that "the batteries must have a mean life greater than 10 years, and the standard deviation of the battery lives can be ignored." If the mean battery life is greater than 10 years, can the standard deviation be ignored? Why or why not?

3. Outlier After 50 credit card holders are randomly selected, the amounts that they currently owe are found. The values of the mean, median, and standard deviation are then determined. An additional amount of $1,000,000 is then included. How much of an effect will this additional amount have on the mean, median, and standard deviation?

4. Internet Survey An Internet service provider (ISP) conducts an anonymous online survey of its subscribers and 2500 of them respond by reporting the values of the cars that they currently own. Given that the sample size is so large, are the results likely to result in a mean that is fairly close to the mean value of all cars owned by Americans? Why or why not?

Review Exercises

TI 1. Tree Heights In a study of the relationship between heights and trunk diameters of trees, botany students collected sample data. Listed below are the tree circumferences (in feet). The data are based on results in "Tree Measurements," by Stanley Rice, *American Biology Teacher*, Vol. 61, No. 9. Using the circumferences listed below, find the (a) mean; (b) median; (c) mode; (d) midrange; (e) range; (f) standard deviation; (g) variance; (h) Q_1; (i) Q_3; (j) P_{10}.

1.8 1.9 1.8 2.4 5.1 3.1 5.5 5.1 8.3 13.7 5.3 4.9 3.7 3.8 4.0 3.4 5.2 4.1 3.7 3.9

TI 2. **a.** Using the results from Exercise 1, convert the circumference of 13.7 ft to a z score.
b. In the context of these sample data, is the circumference of 13.7 ft "unusual"? Why or why not?
c. Using the range rule of thumb, identify any other listed circumferences that are unusual.

TI 3. Frequency Distribution Using the same tree circumferences listed in Exercise 1, construct a frequency distribution. Use seven classes with 1.0 as the lower limit of the first class, and use a class width of 2.0.

TI 4. Histogram Using the frequency distribution from Exercise 3, construct a histogram and identify the general nature of the distribution (such as uniform, bell-shaped, skewed).

TI 5. Boxplot Using the same circumferences listed in Exercise 1, construct a boxplot and identify the values constituting the 5-number summary.

TI 6. Comparing Scores An industrial psychologist for the Citation Corporation develops two different tests to measure job satisfaction. Which score is better: A score of 72 on the management test, which has a mean of 80 and a standard deviation of 12, or a score of 19 on the test for production employees, which has a mean of 20 and a standard deviation of 5? Explain.

7. Estimating Mean and Standard Deviation

a. Estimate the mean age of cars driven by students at your college.

b. Use the range rule of thumb to make a rough estimate of the standard deviation of the ages of cars driven by students at your college.

8. Interpreting Standard Deviation The mean height of men is 69.0 in. and the standard deviation is 2.5 in. Use the range rule of thumb to identify the minimum "usual" height and the maximum "usual" height. In this context, is a height of 72 in. (or 6 ft) unusual? Why or why not?

Cumulative Review Exercises

1. Tree Measurements Refer to the sample of tree circumferences (in feet) listed in Review Exercise 1.

a. Are the given values from a population that is discrete or continuous?

b. What is the level of measurement of these values? (Nominal, ordinal, interval, ratio)

2. a. A set of data is at the nominal level of measurement and you want to obtain a representative data value. Which of the following is most appropriate: mean, median, mode, or midrange? Why?

b. A botanist wants to obtain data about the plants being grown in homes. A sample is obtained by telephoning the first 250 people listed in the local telephone directory; what type of sampling is being used? (Random, stratified, systematic, cluster, convenience)

c. An exit poll is conducted by surveying everyone who leaves the polling booth at 50 randomly selected election precincts. What type of sampling is being used? (Random, stratified, systematic, cluster, convenience)

d. A manufacturer makes fertilizer sticks to be used for growing plants. A manager finds that the amounts of fertilizer placed in the sticks are not very consistent, so that some fertilization lasts longer than claimed, but others don't last long enough. She wants to improve quality by making the amounts of fertilizer more consistent. When analyzing the amounts of fertilizer, which of the following statistics is most relevant: mean, median, mode, midrange, standard deviation, first quartile, third quartile? Should the value of that statistic be raised, lowered, or left unchanged?

Cooperative Group Activities

1. Out-of-class activity Are estimates influenced by anchoring numbers? In the article "Weighing Anchors" in *Omni* magazine, author John Rubin observed that when people estimate a value, their estimate is often "anchored" to (or influenced by) a preceding number, even if that preceding number is totally unrelated to the quantity being estimated. To demonstrate this, he asked people to give a quick estimate of the value of $8 \times 7 \times 6 \times 5 \times 4 \times 3 \times 2 \times 1$. The average answer given was 2250, but when the order of the numbers was reversed, the average became 512. Rubin explained that when we begin calculations with larger numbers (as in $8 \times 7 \times 6$), our estimates tend to be larger. He noted that both 2250 and 512 are far below the correct product, 40,320. The article suggests that irrelevant numbers can play a role in influencing real estate appraisals, estimates of car values, and estimates of the likelihood of nuclear war.

   Conduct an experiment to test this theory. Select some subjects and ask them to quickly estimate the value of

   $$8 \times 7 \times 6 \times 5 \times 4 \times 3 \times 2 \times 1$$

   Then select other subjects and ask them to quickly estimate the value of

   $$1 \times 2 \times 3 \times 4 \times 5 \times 6 \times 7 \times 8$$

   Record the estimates along with the particular order used. Carefully design the experiment so that conditions are uniform and the two sample groups are selected in a way that minimizes any bias. Don't describe the theory to subjects until after they have provided their estimates. Compare the two sets of sample results by using the methods of this chapter. Provide a printed report that includes the data collected, the detailed methods used, the method of analysis, any relevant graphs and/or statistics, and a statement of conclusions. Include a critique of reasons why the results might not be correct and describe ways in which the experiment could be improved.

2. Out-of-class activity In each group of three or four students, collect an original data set of values at the interval or ratio level of measurement. Provide the following: (1) a list of sample values, (2) printed computer results of descriptive statistics and graphs, and (3) a written description of the nature of the data, the method of collection, and important characteristics.

TI-83/84 Plus Project

When dealing with larger data sets, manual entry of data can become quite tedious and time consuming. There are better things to do with your time, such as mastering the aerodynamic principles of a Frisbee. Refer to Data Set 17 in Appendix B, which includes home-run distances of three exceptional baseball players: Barry Bonds (2001 season), Mark McGwire (1998 season), and Sammy Sosa (1998 season). Instead of manually entering the 209 distances in the three sets of data, use a TI-83/84 Plus calculator to load the data sets, which are available on the CD included with this book. Proceed to generate histograms and find appropriate statistics that allow you to compare the three sets of data. Are there any significant differences? Are there any outliers? Does it appear that more home runs are hit by players who hit farther? Why or why not? Analyze the last digits of the distances and determine whether the values appear to be estimates or measurements. Write a brief report including your conclusions and supporting graphs.

From Data to Decision

Critical Thinking

Is there a keyboard configuration that is more efficient than the one that most of us now use? The traditional keyboard configuration is called a *QWERTY* keyboard because of the positioning of the letters QWERTY on the top row of letters. Developed in 1872, the QWERTY configuration was designed to force people to type slower so that the early typewriters would not jam. Developed in 1936, the Dvorak keyboard supposedly provides a more efficient arrangement by positioning the most used keys on the middle row (or "home" row) where they are more accessible.

A *Discover* magazine article suggested that you can measure the ease of typing by using this point rating system: Count each letter on the middle row as 0, count each letter on the top row as 1, and count each letter on the bottom row as 2. (See "Typecasting," by Scott Kim, *Discover*.) Using this rating system with each of the 52 words in the Preamble to the Constitution, we get the rating values shown below.

QWERTY Keyboard Word Ratings for the Preamble to the Constitution

2	2	5	1	2	6	3	3	4	2
4	0	5	7	7	5	6	6	8	10
7	2	2	10	5	8	2	5	4	2
6	2	6	1	7	2	7	2	3	8
1	5	2	5	2	14	2	2	6	3
1	7								

Dvorak Keyboard Word Ratings for the Preamble to the Constitution

2	0	3	1	0	0	0	0	2	0
4	0	3	4	0	3	3	1	3	5
4	2	0	5	1	4	0	3	5	0
2	0	4	1	5	0	4	0	1	3
0	1	0	3	0	1	2	0	0	0
1	4								

Interpreting Results

Visual comparison of the two data sets might not reveal very much, so use the methods of this chapter and Chapter 2 to describe, explore, and compare them. Address at least one of the following questions:

a. Does there appear to be a significant difference between the ease of typing with the two different keyboard configurations? Does the Dvorak configuration appear to be more efficient?

b. If there is a significant difference and the Dvorak configuration is more efficient, why has it not been adopted?

c. Are the words in the Preamble representative? Would the same conclusion be reached with a sample of words from this textbook, which was written much more recently?

d. Each sample value is a total rating for the letters in a *word*, but should we be using words, or should we simply use the individual letters?

e. Is the rating system appropriate? (Letters on the top row are assigned 1, letters on the middle row are assigned 0, and letters on the bottom row are assigned 2.) Is there a different rating system that would better reflect the difficulty of typing?

Internet Project

Using Statistics to Summarize Data

The importance of statistics as a tool to summarize data cannot be underestimated. For example, consider data sets such as the ages of all the students at your school or the annual incomes of every person in the United States. On paper, these data sets would be lengthy lists of numbers, too lengthy to be absorbed and interpreted on their own. In the previous chapter, you learned a variety of graphical tools used to represent such data sets. This chapter focused on the use of numbers or statistics to summarize various aspects of data.

Just as important as being able to summarize data with statistics is the ability to *interpret* such statistics when presented. Given a number such as the arithmetic mean, you need not only to understand what it is telling you about the underlying data, but also what additional statistics you need to put the value of the mean in context.

The Internet Project for this chapter will help you develop these skills using data from such diverse fields as meteorology, entertainment, and health. You will also discover uses for such statistics as the geometric mean that you might not have expected.

The Web site for this chapter can be found at

http://www.aw.com/triola

Statistics @ Work

"A knowledge of basic probability, data summarization, and the principles of inferential statistics is essential to our understanding of the scientific method and our evaluation of reports of scientific studies."

Robert S. Holzman, MD

Professor of Medicine and Environmental Medicine, NYU School of Medicine; Hospital Epidemiologist, Bellevue Hospital Center, New York City

Dr. Holzman is an internist specializing in Infectious Diseases. He is responsible for the Infection Control Program at Bellevue Hospital. He also teaches medical students and postdoctoral trainees in Clinical Infectious Diseases and Epidemiology.

How do you use statistics in your job and what specific statistical concepts do you use?

Much of what I do on a day-to-day basis is applied statistical analysis, including determination of sample size and analysis of clinical trials and laboratory experiments, and the development of regression models for retrospective studies, primarily using logistic regression. I also track hospital infection rates using control charts.

Please describe a specific example of how the use of statistics was helpful in improving a practice or service.

A surveillance nurse detected an increase in the isolation of a certain type of bacteria among patients in an intensive care unit 2 months ago. We took action to remedy that increase, and control charts were used to show us that the bacteria levels were returning to their baseline levels.

What background in statistics is required to obtain a job like yours? What other educational requirements are there?

To be an academic physician requires college and medical school, followed by at least 5 years of postgraduate training, often more. In many cases students today are combining their medical education with a research-oriented PhD training program. I acquired my own statistical and epidemiologic knowledge through a combination of on-the-job association with statistical professionals, reading statistical texts, and some graduate course work. Today such knowledge is introduced as part of standard training programs, but additional work is still important for mastery.

At your place of work, do you feel job applicants are viewed more favorably if they have studied some statistics?

Ability to apply statistical and epidemiologic knowledge to evaluate the medical literature is definitely considered a plus for physicians, even those who work as clinical caregivers. To work as an epidemiologist requires additional statistical study.

Do you recommend that today's college students study statistics? Why?

A knowledge of basic probability, data summarization, and the principles of inferential statistics is essential to our understanding of the scientific method and our evaluation of reports of scientific studies. Such reports are found daily in newspapers, and knowledge of what is left out of the report helps temper uncritical acceptance of new "facts."

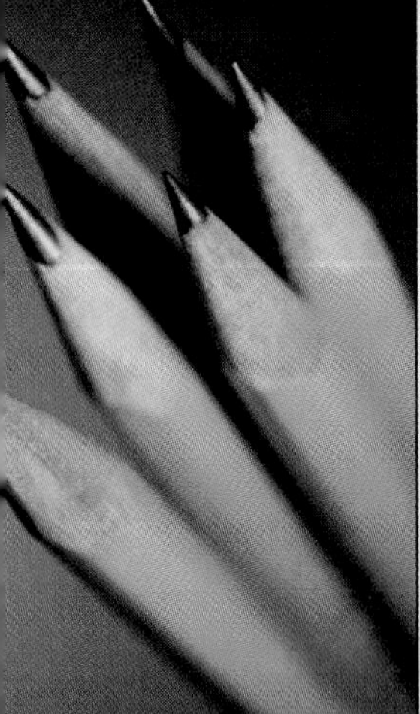

Probability

4

4 CHAPTER PROBLEM

When applying for a job, should you be concerned about drug testing?

According to the American Management Association, about 70% of U.S. companies now test at least some employees and job applicants for drug use. The U.S. National Institute on Drug Abuse claims that about 15% of people in the 18–25 age bracket use illegal drugs. Quest Diagnostics estimates that 3% of the general workforce in the United States uses marijuana. Let's assume that you applied for a job with excellent qualifications (including successful completion of a statistics course), you took a test for marijuana usage, and you were not offered a job. You might suspect that you failed the marijuana test, even though you do not use marijuana.

Analyzing the Results

Table 4-1 shows data from a test of the "1-Panel-THC" test for the screening of marijuana usage. This test device costs $5.95 and is provided by the company Drug Test Success. The test results were confirmed using gas chromatography and mass spectrometry, which the company describes as "the preferred confirmation method." (These results are based on using 50 ng/mL as the cutoff level for determining the presence of marijuana.) Based on the results given in Table 4-1, how likely is it that the test indicates that you use marijuana, even though you do not? When a test shows the presence of some condition, such as a disease or the presence of some drug, the test result is called *positive*. When the test shows a positive result, but the condition is not actually present, the result is called a *false positive*. That is, a false positive is a mistake whereby the test indicates the presence of a condition when that condition is not actually present. In this case, the job applicant might be concerned about the likelihood of a false positive, because that would be an error that would unfairly result in job denial. (The employer might be concerned about another type of error, a *false negative*, which consists of a test showing that the applicant does not use marijuana when he or she does use it. Such a false negative might result in hiring someone who uses marijuana, and that mistake might be critical for some jobs, such as those involving pilots, surgeons, or train engineers.)

In this chapter, we will address important questions, such as these: Given the sample results in Table 4-1, what is the likelihood of a false positive result? What is the likelihood of a false negative result? Are those probabilities low enough so that job applicants and employers need not be concerned about wrong decisions caused by incorrect test results?

Table 4-1 Results from Tests for Marijuana Use

	Did the Subject Actually Use Marijuana? Yes	No
Positive test result (Test indicated that marijuana is *present*.)	119 (true positive)	24 (false positive)
Negative test result (Test indicated that marijuana is *absent*.)	3 (false negative)	154 (true negative)

4-1 Overview

Probability is the underlying foundation on which the important methods of inferential statistics are built. As a simple example, suppose that you have developed a gender-selection procedure and you claim that it greatly increases the likelihood of a baby being a girl. Suppose that independent test results from 100 couples show that your procedure results in 98 girls and only 2 boys. Even though there is a chance of getting 98 girls in 100 births with no special treatment, that chance is so incredibly low that it would be rejected as a reasonable explanation. Instead, it would be generally recognized that the results provide strong support for the claim that the gender-selection technique is effective. This is exactly how statisticians think: They reject explanations based on very low probabilities. Statisticians use the *rare event rule for inferential statistics.*

Rare Event Rule for Inferential Statistics

If, under a given assumption, the probability of a particular observed event is extremely small, we conclude that the assumption is probably not correct.

The main objective in this chapter is to develop a sound understanding of probability values, which will be used in later chapters of this book. A secondary objective is to develop the basic skills necessary to determine probability values in a variety of important circumstances.

4-2 Fundamentals

Key Concept This section introduces the basic concept of the *probability* of an event. Three different methods for finding probability values will be presented. We will see that probability values are expressed as numbers between 0 and 1 inclusive. However, the most important objective of this section is to learn how to *interpret* probability values. For example, we should understand that a small probability, such as 0.001, corresponds to an event that is *unusual* in the sense that it rarely occurs. Later chapters will refer to specific values called "*P*-values," and we will see that they play an extremely important role in various methods of inferential statistics. However, those *P*-values are just probability values, as described in this section. Focus on developing an intuitive sense for interpreting probability values, especially those that are relatively small.

In considering probability, we deal with procedures (such as rolling a die, answering a multiple-choice test question, or undergoing a test for drug use) that produce outcomes.

Definitions

An **event** is any collection of results or outcomes of a procedure.

A **simple event** is an outcome or an event that cannot be further broken down into simpler components.

The **sample space** for a procedure consists of all possible *simple* events. That is, the sample space consists of all outcomes that cannot be broken down any further.

EXAMPLES In the following display, we use f to denote a female baby and we use m to denote a male baby.

Procedure	Example of Event	Complete Sample Space
Single birth	female (simple event)	{f, m}
3 births	2 females and a male (ffm, fmf, mff are all simple events resulting in 2 females and a male)	{fff, ffm, fmf, fmm, mff, mfm, mmf, mmm}

With one birth, the result of a female is a *simple event* because it cannot be broken down any further. With three births, the event of "2 females and a male" is *not a simple event* because it can be broken down into simpler events, such as ffm, fmf, or mff. With three births, the *sample space* consists of the 8 simple events listed above. With three births, the outcome of ffm is considered a simple event, because it is an outcome that cannot be broken down any further. We might incorrectly think that ffm can be further broken down into the individual results of f, f, and m, but f, f, and m are not individual outcomes from three births. With three births, there are exactly 8 outcomes that are simple events: fff, ffm, fmf, fmm, mff, mfm, mmf, and mmm.

There are different ways to define the probability of an event, and we will present three approaches. First, however, we list some basic notation.

Notation for Probabilities

P denotes a probability.
A, B, and *C* denote specific events.
P(*A*) denotes the probability of event *A* occurring.

Rule 1: Relative Frequency Approximation of Probability

Conduct (or observe) a procedure, and count the number of times that event *A* actually occurs. Based on these actual results, *P*(*A*) is *estimated* as follows:

$$P(A) = \frac{\text{number of times } A \text{ occurred}}{\text{number of times the trial was repeated}}$$

Probabilities That Challenge Intuition

In certain cases, our subjective estimates of probability values are dramatically different from the actual probabilities. Here is a classical example: If you take a deep breath, there is better than a 99% chance that you will inhale a molecule that was exhaled in dying Caesar's last breath. In that same morbid and unintuitive spirit, if Socrates' fatal cup of hemlock was mostly water, then the next glass of water you drink will likely contain one of those same molecules. Here's another less morbid example that can be verified: In classes of 25 students, there is better than a 50% chance that at least two students will share the same birthday (day and month).

You Bet

In the typical state lottery, the "house" has a 65% to 70% advantage, since only 30% to 35% of the money bet is returned as prizes. The house advantage at racetracks is usually around 15%. In casinos, the house advantage is 5.26% for roulette, 5.9% for blackjack, 1.4% for craps, and 3% to 22% for slot machines. Some professional gamblers can systematically win at blackjack by using complicated card-counting techniques. They know when a deck has disproportionately more high cards, and this is when they place large bets. Many casinos react by ejecting card counters or by shuffling the decks more frequently.

Rule 2: Classical Approach to Probability (Requires Equally Likely Outcomes)

Assume that a given procedure has n different simple events and that *each of those simple events has an equal chance of occurring*. If event A can occur in s of these n ways, then

$$P(A) = \frac{\text{number of ways } A \text{ can occur}}{\text{number of different simple events}} = \frac{s}{n}$$

Rule 3: Subjective Probabilities

$P(A)$, the probability of event A, is *estimated* by using knowledge of the relevant circumstances.

It is very important to note that the classical approach (*Rule 2*) requires *equally likely outcomes.* If the outcomes are not equally likely, we must use the relative frequency estimate or we must rely on our knowledge of the circumstances to make an *educated guess*. Figure 4-1 illustrates the three approaches.

When finding probabilities with the relative frequency approach (Rule 1), we obtain an *approximation* instead of an exact value. As the total number of observations increases, the corresponding approximations tend to get closer to the

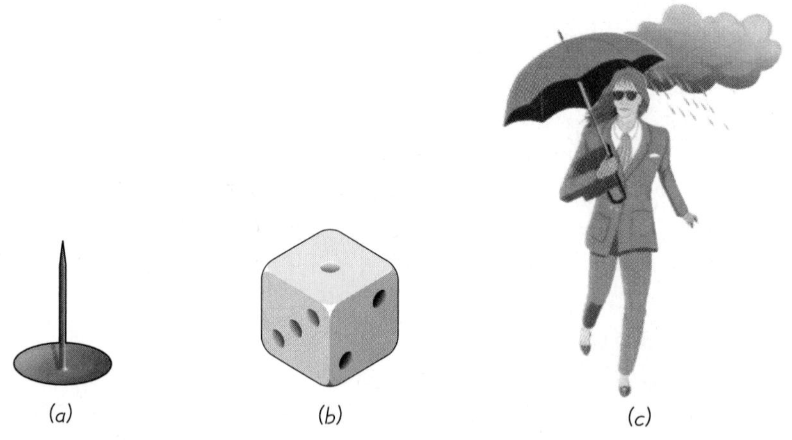

Figure 4-1 Three Approaches to Finding Probability

(a) Relative Frequency Approach (Rule 1): When trying to determine: P(tack lands point up), we must repeat the procedure of tossing the tack many times and then find the ratio of the number of times the tack lands with the point up to the number of tosses.
(b) Classical Approach (Rule 2): When trying to determine $P(2)$ with a balanced and fair die, each of the six faces has an equal chance of occurring.

$$P(2) = \frac{\text{number of ways 2 can occur}}{\text{number of simple events}} = \frac{1}{6}$$

(c) Subjective Probability (Rule 3): When trying to estimate the probability of rain tomorrow, meteorologists use their expert knowledge of weather conditions to develop an estimate of the probability.

actual probability. This property is stated as a theorem commonly referred to as the *law of large numbers.*

Law of Large Numbers

As a procedure is repeated again and again, the relative frequency probability (from Rule 1) of an event tends to approach the actual probability.

The law of large numbers tells us that the relative frequency approximations from Rule 1 tend to get better with more observations. This law reflects a simple notion supported by common sense: A probability estimate based on only a few trials can be off by substantial amounts, but with a very large number of trials, the estimate tends to be much more accurate. For example, suppose that we want to survey people to estimate the probability that someone can simultaneously pat their head while rubbing their stomach. If we survey only five people, the estimate could easily be in error by a large amount. But if we survey thousands of randomly selected people, the estimate is much more likely to be fairly close to the true population value.

Probability and Outcomes That Are Not Equally Likely One common mistake is to incorrectly assume that outcomes are equally likely just because we know nothing about the likelihood of each outcome. For example, the probability that a Republican will win the next presidential election is *not* $1/2$. The value of $1/2$ is often given as the probability, based on the incorrect reasoning that either a Republican will win or a Republican will not win, and those two outcomes are equally likely. They aren't equally likely. The probability of a Republican winning the next presidential election depends on factors such as the ability of the candidates, the amounts of money raised, the state of the economy, the status of national security, and the weather on election day. Similarly, it is incorrect to conclude that there is a $1/2$ probability of passing your next statistics test. With adequate preparation, the probability should be much higher than $1/2$. When you know nothing about the likelihood of different possible outcomes, do not assume that they are equally likely.

How Probable?

How do we interpret such terms as *probable, improbable,* or *extremely improbable*? The FAA interprets these terms as follows. *Probable:* A probability on the order of 0.00001 or greater for each hour of flight. Such events are expected to occur several times during the operational life of each airplane. *Improbable:* A probability on the order of 0.00001 or less. Such events are not expected to occur during the total operational life of a single airplane of a particular type, but may occur during the total operational life of all airplanes of a particular type. *Extremely improbable;* A probability on the order of 0.000000001 or less. Such events are so unlikely that they need not be considered to ever occur.

EXAMPLE Sinking a Free Throw Find the probability that NBA basketball player Reggie Miller makes a free throw after being fouled. At one point in his career, he made 5915 free throws in 6679 attempts (based on data from the NBA).

SOLUTION The sample space consists of two simple events: Miller makes the free throw or he does not. Because the sample space consists of events that are not equally likely, we can't use the classical approach (Rule 2). We can use the relative frequency approach (Rule 1) with his past results. We get the following result:

$$P(\text{Miller makes free throw}) = \frac{5915}{6679} = 0.886$$

Subjective Probabilities at the Racetrack

Researchers studied the ability of racetrack bettors to develop realistic subjective probabilities. (See "Racetrack Betting: Do Bettors Understand the Odds?" by Brown, D' Amato, and Gertner, *Chance* magazine, Vol. 7, No. 3.) After analyzing results for 4400 races, they concluded that although bettors slightly overestimate the winning probabilities of "long shots" and slightly underestimate the winning probabilities of "favorites," their general performance is quite good. The subjective probabilities were calculated from the payoffs, which are based on the amounts bet, and the actual probabilities were calculated from the actual race results.

EXAMPLE Genotypes As part of a study of the genotypes AA, Aa, aA, and aa, you write each individual genotype on an index card, then you shuffle the four cards and randomly select one of them. What is the probability that you select the genotype Aa?

SOLUTION Because the sample space (AA, Aa, aA, aa) in this case includes equally likely outcomes, we use the classical approach (Rule 2) to get

$$P(\text{Aa}) = \frac{1}{4}$$

EXAMPLE Crashing Meteorites What is the probability that your car will be hit by a meteorite this year?

SOLUTION In the absence of historical data on meteorites hitting cars, we cannot use the relative frequency approach of Rule 1. There are two possible outcomes (hit or no hit), but they are not equally likely, so we cannot use the classical approach of Rule 2. That leaves us with Rule 3, whereby we make a subjective estimate. In this case, we all know that the probability in question is very, very small. Let's estimate it to be, say, 0.000000000001 (equivalent to 1 in a trillion). That subjective estimate, based on our general knowledge, is likely to be in the general ballpark of the true probability.

In basic probability problems of the type we are now considering, it is very important to examine the available information carefully and to identify the total number of possible outcomes correctly. In some cases, the total number of possible outcomes is given, but in other cases it must be calculated, as in the following example, which requires us to find the total number of possible outcomes.

EXAMPLE Cloning of Humans Adults are randomly selected for a Gallup poll, and they are asked if they think that cloning of humans should or should not be allowed. Among the randomly selected adults surveyed, 91 said that cloning of humans should be allowed, 901 said that it should not be allowed, and 20 had no opinion. Based on these results, estimate the probability that a randomly selected person believes that cloning of humans should be allowed.

SOLUTION *Hint:* Instead of trying to formulate an answer directly from the written statement, summarize the given information in a format that allows you to better understand it. For example, use this format:

91	cloning of humans should be allowed
901	cloning of humans should not be allowed
20	no opinion
1012	total responses

We can now use the relative frequency approach (Rule 1) as follows:

$$P(\text{cloning of humans should be allowed}) = \frac{\text{number believing that cloning of humans should be allowed}}{\text{total number of people surveyed}} = \frac{91}{1012} = 0.0899$$

We estimate that there is a 0.0899 probability that when an adult is randomly selected, he or she believes that cloning of humans should be allowed. As with all surveys, the accuracy of this result depends on the quality of the sampling method and the survey procedure. Because the poll was conducted by the Gallup organization, the results are likely to be reasonably accurate. Chapter 7 will include more advanced procedures for analyzing such survey results.

EXAMPLE Gender of Children Find the probability that when a couple has 3 children, they will have exactly 2 boys. Assume that boys and girls are equally likely and that the gender of any child is not influenced by the gender of any other child.

SOLUTION The biggest obstacle here is correctly identifying the sample space. It involves more than working only with the numbers 2 and 3 that were given in the statement of the problem. The sample space consists of 8 different ways that 3 children can occur, and we list them in the margin. Those 8 outcomes are equally likely, so we use Rule 2. Of those 8 different possible outcomes, 3 correspond to exactly 2 boys, so

$$P(2 \text{ boys in } 3 \text{ births}) = \frac{3}{8} = 0.375$$

INTERPRETATION There is a 0.375 probability that if a couple has 3 children, exactly 2 will be boys.

	1st 2nd 3rd
	boy-boy-boy
exactly 2 boys →	boy-boy-girl
exactly 2 boys →	boy-girl-boy
	boy-girl-girl
exactly 2 boys →	girl-boy-boy
	girl-boy-girl
	girl-girl-boy
	girl-girl-girl

The statements of the three rules for finding probabilities and the preceding examples might seem to suggest that we should always use Rule 2 when a procedure has equally likely outcomes. In reality, many procedures are so complicated that the classical approach (Rule 2) is impractical to use. In the game of solitaire, for example, the outcomes (hands dealt) are all equally likely, but it is extremely frustrating to try to use Rule 2 to find the probability of winning. In such cases we can more easily get good estimates by using the relative frequency approach (Rule 1). Simulations are often helpful when using this approach. (A **simulation** of a procedure is a process that behaves in the same ways as the procedure itself, so that similar results are produced. See Section 4-6 and the TI-83/84 Plus Project near the end of this chapter.) For example, it's much easier to use Rule 1 for estimating the probability of winning at solitaire—that is, to play the game many times (or to run a computer simulation)—than to perform the extremely complex calculations required with Rule 2.

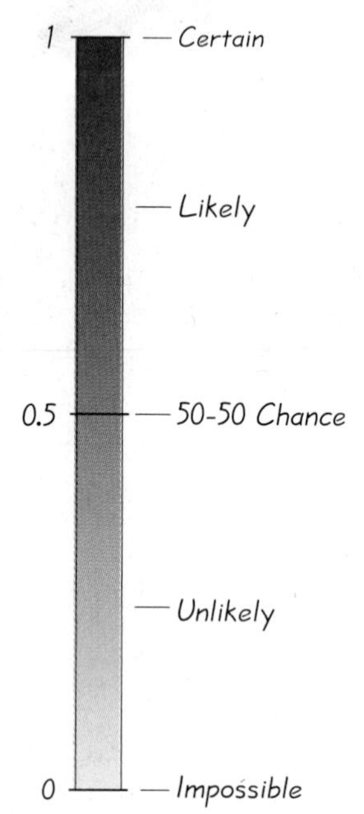

Figure 4-2 Possible Values for Probabilities

EXAMPLE Thanksgiving Day If a year is selected at random, find the probability that Thanksgiving Day will be on a (a) Wednesday or (b) Thursday.

SOLUTION

a. Thanksgiving Day always falls on the fourth Thursday in November. It is therefore impossible for Thanksgiving to be on a Wednesday. When an event is impossible, we say that its probability is 0.

b. It is certain that Thanksgiving will be on a Thursday. When an event is certain to occur, we say that its probability is 1.

Because any event imaginable is impossible, certain, or somewhere in between, it follows that the mathematical probability of any event is 0, 1, or a number between 0 and 1 (see Figure 4-2).

- **The probability of an impossible event is 0.**
- **The probability of an event that is certain to occur is 1.**
- **For any event A, the probability of A is between 0 and 1 inclusive. That is, $0 \leq P(A) \leq 1$.**

In Figure 4-2, the scale of 0 through 1 is shown, and the more familiar and common expressions of likelihood are included.

Complementary Events

Sometimes we need to find the probability that an event A does *not* occur.

Definition

The **complement** of event A, denoted by $\overline{A}$, consists of all outcomes in which event A does *not* occur.

EXAMPLE Birth Genders In reality, more boys are born than girls. In one typical group, there are 205 newborn babies, 105 of whom are boys. If one baby is randomly selected from the group, what is the probability that the baby is *not* a boy?

SOLUTION Because 105 of the 205 babies are boys, it follows that 100 of them are girls, so

$$P(\text{not selecting a boy}) = P(\overline{\text{boy}}) = P(\text{girl}) = \frac{100}{205} = 0.488$$

Although it is difficult to develop a universal rule for rounding off probabilities, the following guide will apply to most problems in this text.

Rounding Off Probabilities

When expressing the value of a probability, either give the *exact* fraction or decimal or round off final decimal results to three significant digits. (*Suggestion:* When a probability is not a simple fraction such as 2/3 or 5/9, express it as a decimal so that the number can be better understood.)

All digits in a number are significant except for the zeros that are included for proper placement of the decimal point.

EXAMPLES

- The probability of 0.021491 has five significant digits (21491), and it can be rounded to three significant digits as 0.0215.
- The probability of $1/3$ can be left as a fraction, or rounded to 0.333. (Do *not* round to 0.3.)
- The probability of heads in a coin toss can be expressed as $1/2$ or 0.5; because 0.5 is exact, there's no need to express it as 0.500.
- The fraction $432/7842$ is exact, but its value isn't obvious, so express it as the decimal 0.0551.

An important concept in this section is the mathematical expression of probability as a number between 0 and 1. This type of expression is fundamental and common in statistical procedures, and we will use it throughout the remainder of this text. A typical computer output, for example, may include a "*P*-value" expression such as "significance less than 0.001." We will discuss the meaning of *P*-values later, but they are essentially probabilities of the type discussed in this section. For now, you should recognize that a probability of 0.001 (equivalent to $1/1000$) corresponds to an event so rare that it occurs an average of only once in a thousand trials.

Most Common Birthday: October 5

A Web site stated that "a recent in depth database query conducted by Anybirthday.com suggests that October 5 is the United States' most popular birth date." It was noted that a New Year's Eve conception would likely result in an October 5 birth date. The least common birth date was identified as May 22. Apparently, August 18 does not have the same charm as New Year's Eve.

Odds

Expressions of likelihood are often given as *odds,* such as 50:1 (or "50 to 1"). A serious disadvantage of odds is that they make many calculations extremely difficult. As a result, statisticians, mathematicians, and scientists prefer to use probabilities. The advantage of odds is that they make it easier to deal with money transfers associated with gambling, so they tend to be used in casinos, lotteries, and racetracks. Note that in the three definitions that follow, the actual odds against and the actual odds in favor describe the actual likelihood of some event, but the payoff odds describe the relationship between the bet and the amount of the payoff. The actual odds correspond to actual outcomes, but the payoff odds are set by racetrack and casino operators. Racetracks and casinos are in business to make a profit, so the payoff odds will not be the same as the actual odds.

Definition

The **actual odds against** event *A* occurring are the ratio $P(\overline{A})/P(A)$, usually expressed in the form of *a*:*b* (or "*a* to *b*"), where *a* and *b* are integers having no common factors.

The **actual odds in favor** of event *A* are the reciprocal of the actual odds against that event. If the odds against *A* are *a*:*b,* then the odds in favor of *A* are *b*:*a*.

The **payoff odds** against event *A* represent the ratio of net profit (if you win) to the amount bet.

$$\text{payoff odds against event } A = (\text{net profit}) : (\text{amount bet})$$

EXAMPLE If you bet \$5 on the number 13 in roulette, your probability of winning is $1/38$ and the payoff odds are given by the casino as 35:1.

a. Find the actual odds against the outcome of 13.

b. How much net profit would you make if you win by betting on 13?

c. If the casino were operating just for the fun of it, and the payoff odds were changed to match the actual odds against 13, how much would you win if the outcome were 13?

SOLUTION

a. With $P(13) = 1/38$ and $P(\text{not } 13) = 37/38$, we get

$$\text{actual odds against 13} = \frac{P(\text{not } 13)}{P(13)} = \frac{37/38}{1/38} = \frac{37}{1} \text{ or } 37{:}1$$

b. Because the payoff odds against 13 are 35:1, we have

$$35{:}1 = (\text{net profit}){:}(\text{amount bet})$$

so that there is a \$35 profit for each \$1 bet. For a \$5 bet, the net profit is \$175. The winning bettor would collect \$175 plus the original \$5 bet. That is, the winning bettor of \$5 would receive the \$5 bet plus another \$175. The total amount returned would be \$180, for a net profit of \$175.

c. If the casino were operating for fun and not for profit, the payoff odds would be equal to the actual odds against the outcome of 13. If the payoff odds were changed from 35:1 to 37:1, you would obtain a net profit of \$37 for each \$1 bet. If you bet \$5, your net profit would be \$185. (The casino makes its profit by paying only \$175 instead of the \$185 that would be paid with a roulette game that is fair instead of favoring the casino.)

4-2 BASIC SKILLS AND CONCEPTS

Statistical Literacy and Critical Thinking

1. Interpreting Probability What does it mean when we say that "the probability of winning the grand prize in the Illinois lottery is 1/20,358,520"? Is such a win *unusual*?

2. Probability of Rain When writing about the probability that it will rain in Boston on July 4 of next year, a newspaper reporter states that the probability is 1/2, because either it will rain or it will not. Is this reasoning correct? Why or why not?

3. Probability and Unusual Events A news reporter states that a particular event is *unusual* because its probability is only 0.001. Is that a correct statement? Why or why not?

4. Subjective Probability Use subjective judgment to estimate the probability that the next time you ride an elevator, it gets stuck between floors.

In Exercises 5 and 6, express the indicated degree of likelihood as a probability value between 0 and 1.

5. Identifying Probability Values

a. "Because you have studied diligently and understand the concepts, you will surely pass the statistics test."

b. "The forecast for tomorrow includes a 10% chance of rain."

c. "You have a snowball's chance in hell of marrying my daughter."

6. Identifying Probability Values

a. "When flipping a quarter, there is a 50–50 chance that the outcome will be heads."

b. "You have one chance in five of guessing the correct answer."

c. "You have a 1% chance of getting a date with the person who just entered the room."

7. Identifying Probability Values Which of the following values *cannot* be probabilities?

$$0, \quad 1, \quad -1, \quad 2, \quad 0.0123, \quad 3/5, \quad 5/3, \quad \sqrt{2}$$

8. Identifying Probability Values

a. What is the probability of an event that is certain to occur?

b. What is the probability of an impossible event?

c. A sample space consists of 10 separate events that are equally likely. What is the probability of each?

d. On a true/false test, what is the probability of answering a question correctly if you make a random guess?

e. On a multiple-choice test with five possible answers for each question, what is the probability of answering a question correctly if you make a random guess?

9. Gender of Children In this section we gave an example that included a list of the eight outcomes that are possible when a couple has three children. Refer to that list, and find the probability of each event.

a. Among three children, there is exactly one girl.

b. Among three children, there are exactly two girls.

c. Among three children, all are girls.

10. Cell Phones and Brain Cancer In a study of 420,095 cell phone users in Denmark, it was found that 135 developed cancer of the brain or nervous system. Estimate the probability that a randomly selected cell phone user will develop such a cancer. Is the result very different from the probability of 0.000340 that was found for the general population? What does the result suggest about cell phones as a cause of such cancers, as has been claimed?

11. Mendelian Genetics When Mendel conducted his famous genetics experiments with peas, one sample of offspring consisted of 428 green peas and 152 yellow peas. Based on those results, estimate the probability of getting an offspring pea that is green. Is the result reasonably close to the value of 3/4 that was expected?

12. Struck by Lightning In a recent year, 389 of the 281,421,906 people in the United States were struck by lightning. Estimate the probability that a randomly selected person in the United States will be struck by lightning this year.

13. Gender Selection In a test of the MicroSort gender-selection technique, results consisted of 295 baby girls and 30 baby boys (based on data from the Genetics & IVF Institute). Based on this result, what is the probability of a girl born to a couple using the MicroSort method? Does it appear that the technique is effective in increasing the likelihood that a baby will be a girl?

14. Brand Recognition

a. In a study of brand recognition, 831 consumers knew of Campbell's Soup, and 18 did not (based on data from Total Research Corporation). Use these results to estimate the probability that a randomly selected consumer will recognize Campbell's Soup.

b. *Estimate* the subjective probability that a randomly selected adult American consumer will recognize the brand name of McDonald's, most notable as a fast-food restaurant chain.

c. *Estimate* the subjective probability that a randomly selected adult American consumer will recognize the brand name of Veeco Instruments, a manufacturer of microelectronic products.

15. Blue M&M Plain Candies

a. Refer to the 100 M&Ms listed in Data Set 13 in Appendix B and estimate the probability that when a plain M&M candy is randomly selected, it is one that is blue.

b. The Mars Company claims that 24% of its plain M&M candies are blue. Does the estimate from part (a) agree roughly with this claim, or does there appear to be substantial disagreement?

16. Pedestrian Walk Buttons New York City has 750 pedestrian walk buttons that work, and another 2500 that do not work (based on data from "For Exercise in New York Futility, Push Button," by Michael Luo, *New York Times*). If a pedestrian walk button is randomly selected in New York City, what is the probability that it works? Is the same probability likely to be a good estimate for a different city, such as Chicago?

Using Probability to Identify Unusual Events. *In Exercises 17–24, consider an event to be "unusual" if its probability is less than or equal to 0.05. (This is equivalent to the same criterion commonly used in inferential statistics, but the value of 0.05 is not absolutely rigid, and other values such as 0.01 are sometimes used instead.)*

17. Guessing Birthdays On their first date, Kelly asks Mike to guess the date of her birth, not including the year.

a. What is the probability that Mike will guess correctly? (Ignore leap years.)

b. Would it be unusual for him to guess correctly on his first try?

c. If you were Kelly, and Mike did guess correctly on his first try, would you believe his claim that he made a lucky guess, or would you be convinced that he already knew when you were born?

d. If Kelly asks Mike to guess her age, and Mike's guess is too high by 15 years, what is the probability that Mike and Kelly will have a second date?

18. IRS Accuracy The U.S. General Accounting Office tested the Internal Revenue Service for correctness of answers to taxpayers' questions. For 1733 trials, the IRS was correct 1107 times and wrong 626 times.

a. Estimate the probability that a randomly selected taxpayer's question will be answered incorrectly.

b. Is it unusual for the IRS to provide a wrong answer to a taxpayer's question? Should it be unusual?

19. Probability of a Car Crash Among 400 randomly selected drivers in the 20–24 age bracket, 136 were in a car crash during the last year (based on data from the National Safety Council). If a driver in that age bracket is randomly selected, what is the approximate probability that he or she will be in a car accident during the next year? Is it unusual for a driver in that age bracket to be involved in a car crash during a year? Is the resulting value high enough to be of concern to those in the 20–24 age bracket?

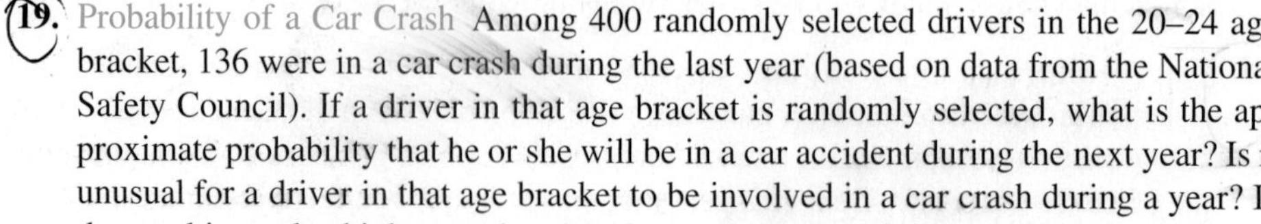

20. Adverse Effect of Lipitor In a clinical trial of Lipitor (atorvastatin), a common drug used to lower cholesterol, one group of patients was given a treatment of 10-mg atorvastatin tablets. That group consists of 19 patients who experienced flu symptoms and 844 patients who did not (based on data from Pfizer, Inc.).
a. Estimate the probability that a patient taking the drug will experience flu symptoms.
b. Is it "unusual" for a patient taking the drug to experience flu symptoms?

21. Adverse Effect of Viagra When the drug Viagra was clinically tested, 117 patients reported headaches and 617 did not (based on data from Pfizer, Inc.). Use this sample to estimate the probability that a Viagra user will experience a headache. Is it unusual for a Viagra user to experience headaches? Is the probability high enough to be of concern to Viagra users?

22. Interpreting Effectiveness of a Treatment A double-blind experiment is designed to test the effectiveness of the drug Statisticzene as a treatment for number blindness. When treated with Statisticzene, subjects seem to show improvement. Researchers calculate that there is a 0.04 probability that the treatment group would show improvement if the drug has no effect. Is it unusual for someone treated with an ineffective drug to show improvement? What should you conclude about the effectiveness of Statisticzene?

23. Probability of a Wrong Result Table 4-1 shows that among 178 subjects who did *not* use marijuana, the test result for marijuana usage was wrong 24 times.
a. Based on the available results, find the probability of a wrong test result for a person who does not use marijuana.
b. Is it "unusual" for the test result to be wrong for those not using marijuana?

24. Probability of a Wrong Result Table 4-1 shows that among 122 subjects who did use marijuana, the test result for marijuana usage was wrong 3 times.
a. Based on the available results, find the probability of a wrong test result for a person who does use marijuana.
b. Is it "unusual" for the test result to be wrong for those using marijuana?

Constructing Sample Space. *In Exercises 25–28, construct the indicated sample space and answer the given questions.*

25. Gender of Children: Constructing Sample Space This section included a table summarizing the gender outcomes for a couple planning to have three children.
a. Construct a similar table for a couple planning to have *two* children.
b. Assuming that the outcomes listed in part (a) are equally likely, find the probability of getting two girls.
c. Find the probability of getting exactly one child of each gender.

26. Genetics: Constructing Sample Space Both parents have the brown/blue pair of eye color genes, and each parent contributes one gene to a child. Assume that if the child has at least one brown gene, that color will dominate and the eyes will be brown. (The actual determination of eye color is somewhat more complicated.)
a. List the different possible outcomes. Assume that these outcomes are equally likely.
b. What is the probability that a child of these parents will have the blue/blue pair of genes?
c. What is the probability that the child will have brown eyes?

27. Genetics: Constructing Sample Space Repeat Exercise 26 assuming that one parent has a brown/brown pair of eye color genes while the other parent has a brown/blue pair of eye color genes.

28. Genetics: Constructing Sample Space Repeat Exercise 26 assuming that one parent has a brown/brown pair of eye color genes while the other parent has a blue/blue pair of eye color genes.

Odds. *In Exercises 29–32, answer the given questions that involve odds.*

29. Solitaire Odds Because the calculations involved with solitaire are so complex, the game was played 500 times so that the probability of winning could be estimated. (The results are from the Microsoft solitaire game, and the Vegas rules of "draw 3" with \$52 bet and a return of \$5 per card are used.) Among the 500 trials, the game was won 77 times. Based on these results, find the odds against winning.

30. Kentucky Derby Odds The probability of the horse Outta Here winning the 129th Kentucky Derby was 1/50. What were the actual odds against Outta Here winning that race?

31. Finding Odds in Roulette A roulette wheel has 38 slots. One slot is 0, another is 00, and the others are numbered 1 through 36, respectively. You are placing a bet that the outcome is an odd number.

a. What is your probability of winning?

b. What are the actual odds against winning?

c. When you bet that the outcome is an odd number, the payoff odds are 1:1. How much profit do you make if you bet \$18 and win?

d. How much profit would you make on the \$18 bet if you could somehow convince the casino to change its payoff odds so that they are the same as the actual odds against winning? (*Recommendation:* Don't actually try to convince any casino of this; their sense of humor is remarkably absent when it comes to things of this sort.)

32. Kentucky Derby Odds When the horse Funny Cide won the 129th Kentucky Derby, a \$2 bet that Funny Cide would win resulted in a return of \$27.60.

a. How much net profit was made from a \$2 win bet on Funny Cide?

b. What were the payoff odds against a Funny Cide win?

c. Based on preliminary wagering before the race, bettors collectively believed that Funny Cide had a 2/33 probability of winning. Assuming that 2/33 was the true probability of a Funny Cide victory, what were the actual odds against his winning?

d. If the payoff odds were the actual odds found in Part (c), how much would a \$2 ticket be worth after the Funny Cide win?

4-2 BEYOND THE BASICS

33. Finding Probability from Odds If the actual odds against event A are $a{:}b$, then $P(A) = b/(a + b)$. Find the probability of the horse Buddy Gil winning the 129th Kentucky Derby, given that the actual odds against his winning that race were 9:1.

34. Relative Risk and Odds Ratio In a clinical trial of 734 subjects treated with Viagra, 117 reported headaches. In a control group of 725 subjects not treated with Viagra, 29 reported headaches. Denoting the proportion of headaches in the treatment group by p_t and denoting the proportion of headaches in the control group by p_c, the **relative risk** is p_t/p_c. The relative risk is a measure of the strength of the effect of the Viagra treatment. Another such measure is the **odds ratio,** which is the ratio of the odds in favor of a headache for the treatment group to the odds in favor of a headache for the control group, found by evaluating the following:

$$\frac{p_t/(1 - p_t)}{p_c/(1 - p_c)}$$

The relative risk and odds ratios are commonly used in medicine and epidemiological studies. Find the relative risk and odds ratio for the headache data.

35. Flies on an Orange If two flies land on an orange, find the probability that they are on points that are within the same hemisphere.

36. Points on a Stick Two points along a straight stick are randomly selected. The stick is then broken at those two points. Find the probability that the three resulting pieces can be arranged to form a triangle. (This is possibly the most difficult exercise in this book.)

Boys and Girls Are Not Equally Likely

In many probability calculations, good results are obtained by assuming that boys and girls are equally likely to be born. In reality, a boy is more likely to be born (with probability 0.512) than a girl (with probability 0.488). These results are based on recent data from the National Center for Health Statistics, which showed that the 4,058,814 births in one year included 2,076,969 boys and 1,981,845 girls. Researchers monitor these probabilities for changes that might suggest such factors as changes in the environment and exposure to chemicals.

4-3 Addition Rule

Key Concept The main objective of this section is to present the *addition rule* as a device for finding probabilities that can be expressed as $P(A \text{ or } B)$, the probability that either event A occurs or event B occurs (or they both occur) as the single outcome of a procedure. To find the probability of event A occurring or event B occurring, we begin by finding the total number of ways that A can occur and the number of ways that B can occur, but we find that total without counting any outcomes more than once.

The key word in this section is "or." Throughout this text we use the *inclusive or,* which means either one or the other or both. (Except for Exercise 26, we will not consider the *exclusive or,* which means either one or the other but not both.)

In the previous section we presented the fundamentals of probability and considered events categorized as *simple.* In this and the following section we consider *compound events.*

Definition

A **compound event** is any event combining two or more simple events.

Notation for Addition Rule

$P(A \text{ or } B) = P(\text{in a single trial, event } A \text{ occurs or event } B \text{ occurs or they both occur})$

Understanding the Notation In this section, $P(A \text{ and } B)$ denotes the probability that A and B both occur in the same trial, but in the following section we will use $P(A \text{ and } B)$ to denote the probability that event A occurs on one trial followed by event B on another trial. The true meaning of $P(A \text{ and } B)$ can therefore be determined only by knowing whether we are referring to one trial that can have outcomes of A and B, or two trials with event A occurring on the first trial and event B occurring on the second trial. The meaning denoted by $P(A \text{ and } B)$ therefore depends upon the context.

Shakespeare's Vocabulary

According to Bradley Efron and Ronald Thisted, Shakespeare's writings included 31,534 different words. They used probability theory to conclude that Shakespeare probably knew at least another 35,000 words that he didn't use in his writings. The problem of estimating the size of a population is an important problem often encountered in ecology studies, but the result given here is another interesting application. (See "Estimating the Number of Unseen Species: How Many Words Did Shakespeare Know?", in *Biometrika,* Vol. 63, No. 3.)

Table 4-1 Results from Tests for Marijuana Use

	Did the Subject Actually Use Marijuana? Yes	No
Positive test result (Test indicated that marijuana is *present*.)	119 (true positive)	24 (false positive)
Negative test result (Test indicated that marijuana is *absent*.)	3 (false negative)	154 (true negative)

Refer to Table 4-1, reproduced here for your convenience. In the sample of 300 subjects represented in Table 4-1, how many of them tested positive *or* used marijuana? (Remember, "tested positive or used marijuana" really means "tested positive, or used marijuana, or both.") Examination of Table 4-1 should show that a total of 146 subjects tested positive or used marijuana. (*Important note*: It is *wrong* to add the 143 subjects who tested positive to the 122 subjects who used marijuana, because this total of 265 would have counted 119 of the subjects twice, but they are individuals that should be counted only once each.) See the role that the correct total of 146 plays in the following example.

EXAMPLE Drug Testing Refer to Table 4-1, reproduced here for your convenience. Assuming that 1 person is randomly selected from the 300 people that were tested, find the probability of selecting a subject who had a positive test result or used marijuana.

SOLUTION From Table 4-1 we see that there are 146 subjects who had a positive test result or used marijuana. We obtain that total of 146 by adding the subjects who tested positive to the subjects who used marijuana, being careful to count everyone only once. Dividing the total of 146 by the overall total of 300, we get this result: $P(\text{positive test result or used marijuana}) = 146/300$ or 0.487.

In the preceding example, there are several strategies you could use for counting the subjects who tested positive or used marijuana. Any of the following would work:

- Color the cells representing subjects who tested positive or used marijuana, then add the numbers in those colored cells, being careful to add each number only once. This approach yields

$$119 + 24 + 3 = 146$$

- Add the 143 subjects who tested positive to the 122 subjects who used marijuana, but the total of 265 involves double-counting of 119 subjects, so compensate for the double-counting by subtracting the overlap consisting of the 119 subjects who tested positive and used marijuana. This approach yields a result of

$$143 + 122 - 119 = 146$$

- Start with the total of 143 subjects who tested positive, then add those subjects who used marijuana and were not yet included in that total, to get a result of

$$143 + 3 = 146$$

Carefully study the preceding example to understand this essential feature of finding the probability of an event *A* or event *B:* use of the word "or" suggests addition, and the addition must be done without double-counting.

The preceding example suggests a general rule whereby we add the number of outcomes corresponding to each of the events in question:

When finding the probability that event *A* occurs or event *B* occurs, find the total of the number of ways *A* can occur and the number of ways *B* can occur, but *find that total in such a way that no outcome is counted more than once.*

One way to formalize the rule is to combine the number of ways event *A* can occur with the number of ways event *B* can occur and, if there is any overlap, compensate by subtracting the number of outcomes that are counted twice, as in the following rule.

Formal Addition Rule

$$P(A \text{ or } B) = P(A) + P(B) - P(A \text{ and } B)$$

where $P(A \text{ and } B)$ denotes the probability that *A* and *B* both occur at the same time as an outcome in a trial of a procedure.

The formal addition rule is presented as a formula, but the blind use of formulas is not recommended. It is generally better to *understand* the spirit of the rule and use that understanding, as follows.

Intuitive Addition Rule

To find $P(A \text{ or } B)$, find the sum of the number of ways event *A* can occur and the number of ways event *B* can occur, *adding in such a way that every outcome is counted only once.* $P(A \text{ or } B)$ is equal to that sum, divided by the total number of outcomes in the sample space.

Because the overlapping of events is such a critical consideration in the addition rule, there is a special term that describes it:

Definition

Events *A* and *B* are **disjoint** (or **mutually exclusive**) if they cannot occur at the same time. (That is, disjoint events do not overlap.)

EXAMPLE Drug Testing Again refer to Table 4-1.

a. Consider the procedure of randomly selecting 1 of the 300 subjects included in Table 4-1. Determine whether the following events are disjoint: *A*: Getting a subject with a negative test result; *B*: getting a subject who did not use marijuana.

b. Assuming that 1 person is randomly selected from the 300 people that were tested, find the probability of selecting a subject who had a negative test result or did not use marijuana.

SOLUTION

a. In Table 4-1 we see that there are 157 subjects with negative test results and there are 178 subjects who did not use marijuana. The event of getting a subject with a negative test result and getting a subject who did not use marijuana can occur at the same time (because there are 154 subjects who had negative test results and did not use marijuana). Because those events overlap, they can occur at the same time and we say that the events are *not* disjoint.

b. In Table 4-1 we must find the total number of subjects who had negative test results or did not use marijuana, but we must find that total without double-counting anyone. We get a total of 181.

Because 181 subjects had negative test results or did not use marijuana, and because there are 300 total subjects included, we get

$$P(\text{negative test result or did not use marijuana}) = \frac{181}{300} = 0.603$$

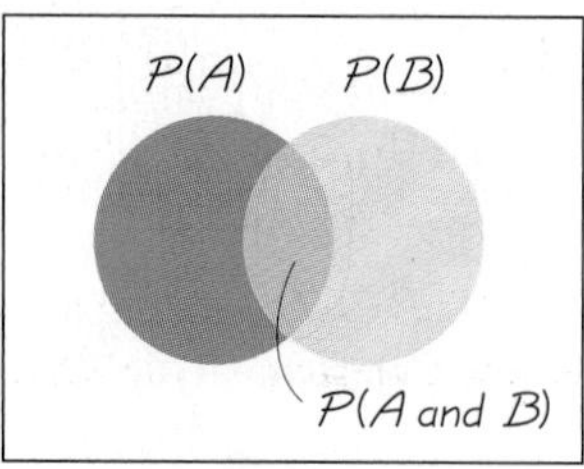

Figure 4-3 Venn Diagram for Events That Are Not Disjoint

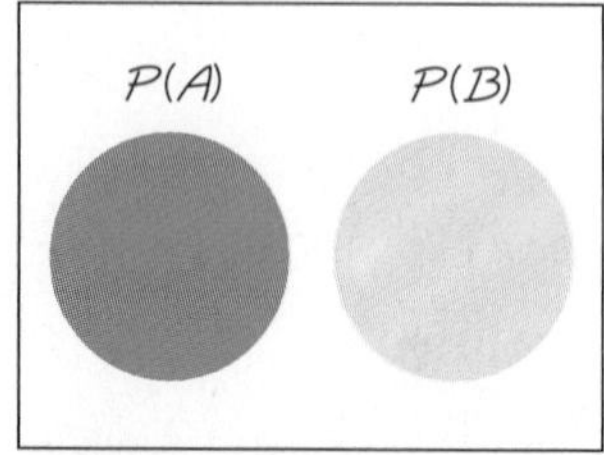

Figure 4-4 Venn Diagram for Disjoint Events

Figure 4-3 shows a Venn diagram that provides a visual illustration of the formal addition rule. In this figure we can see that the probability of *A* or *B* equals the probability of *A* (left circle) plus the probability of *B* (right circle) minus the probability of *A* and *B* (football-shaped middle region). This figure shows that the addition of the areas of the two circles will cause double-counting of the football-shaped middle region. This is the basic concept that underlies the addition rule. Because of the relationship between the addition rule and the Venn diagram shown in Figure 4-3, the notation $P(A \cup B)$ is sometimes used in place of *P*(*A* or *B*). Similarly, the notation $P(A \cap B)$ is sometimes used in place of *P*(*A* and *B*) so the formal addition rule can be expressed as

$$P(A \cup B) = P(A) + P(B) - P(A \cap B)$$

The addition rule is simplified whenever *A* and *B* are disjoint (cannot occur simultaneously), so *P*(*A* and *B*) becomes zero. Figure 4-4 illustrates that when *A* and *B* are disjoint, we have $P(A \text{ or } B) = P(A) + P(B)$.

We can summarize the key points of this section as follows:

1. To find *P*(*A* or *B*), begin by associating use of the word "or" with addition.
2. Consider whether events *A* and *B* are disjoint; that is, can they happen at the same time? If they are not disjoint (that is, they can happen at the same time),

be sure to avoid (or at least compensate for) double-counting when adding the relevant probabilities. If you understand the importance of not double-counting when you find $P(A \text{ or } B)$, you don't necessarily have to calculate the value of $P(A) + P(B) - P(A \text{ and } B)$.

Errors made when applying the addition rule often involve double-counting; that is, events that are not disjoint are treated as if they were. One indication of such an error is a total probability that exceeds 1; however, errors involving the addition rule do not always cause the total probability to exceed 1.

Complementary Events

In Section 4-2 we defined the complement of event A and denoted it by $\overline{A}$. We said that $\overline{A}$ consists of all the outcomes in which event A does *not* occur. Events A and $\overline{A}$ must be disjoint, because it is impossible for an event and its complement to occur at the same time. Also, we can be absolutely certain that A either does or does not occur, which implies that either A or $\overline{A}$ must occur. These observations let us apply the addition rule for disjoint events as follows:

$$P(A \text{ or } \overline{A}) = P(A) + P(\overline{A}) = 1$$

We justify $P(A \text{ or } \overline{A}) = P(A) + P(\overline{A})$ by noting that A and $\overline{A}$ are disjoint; we justify the total of 1 by our certainty that A either does or does not occur. This result of the addition rule leads to the following three equivalent expressions.

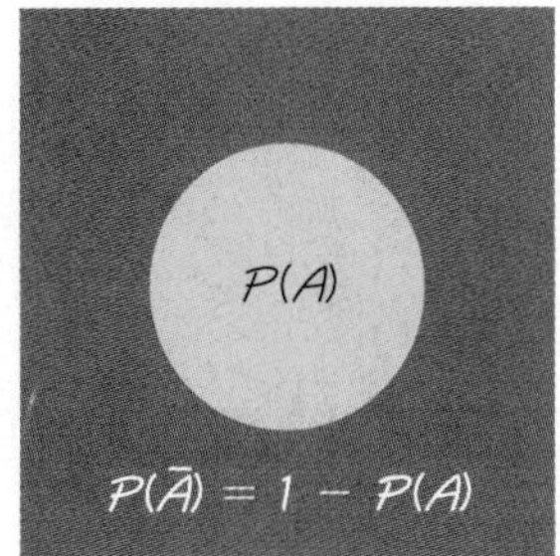

Figure 4-5 Venn Diagram for the Complement of Event *A*

Rule of Complementary Events

$$P(A) + P(\overline{A}) = 1$$
$$P(\overline{A}) = 1 - P(A)$$
$$P(A) = 1 - P(\overline{A})$$

Figure 4-5 visually displays the relationship between $P(A)$ and $P(\overline{A})$.

EXAMPLE In reality, when a baby is born, $P(\text{boy}) = 0.512$. Find $P(\overline{\text{boy}})$.

SOLUTION Using the rule of complementary events, we get

$$P(\overline{\text{boy}}) = 1 - P(\text{boy}) = 1 - 0.512 = 0.488$$

That is, the probability of not getting a boy, which is the probability of a girl, is 0.488.

A major advantage of the *rule of complementary events* is that its use can greatly simplify certain problems. We will illustrate this particular advantage in Section 4-5.

4-3 BASIC SKILLS AND CONCEPTS

Statistical Literacy and Critical Thinking

1. Disjoint Events In your own words, describe what it means for two events to be *disjoint*.

2. Addition Rule In your own words, describe how the addition rule is applied to finding the probability that event *A* occurs or event *B* occurs.

3. Survey For a research project, you need to find the probability that someone is left-handed or drives a car. What is wrong with surveying 500 of your closest friends and relatives?

4. Disjoint Events and Complements If an event is the complement of another event, must those two events be disjoint? Why or why not?

Determining Whether Events Are Disjoint. *For each part of Exercises 5 and 6, are the two events disjoint for a single trial? (Hint: Consider "disjoint" to be equivalent to "separate" or "not overlapping.")*

5. **a.** Electing a president of the United States
   Electing a female candidate
   b. Randomly selecting someone who smokes cigars
   Randomly selecting a male
   c. Randomly selecting someone treated with the cholesterol-reducing drug Lipitor
   Randomly selecting someone in a control group given no medication

6. **a.** Randomly selecting a fruit fly with red eyes
   Randomly selecting a fruit fly with sepian (dark brown) eyes
   b. Receiving a phone call from a volunteer survey subject who opposes all cloning
   Receiving a phone call from a volunteer survey subject who approves of cloning of sheep
   c. Randomly selecting a nurse
   Randomly selecting a male

7. Finding Complements
   a. If $P(A) = 0.05$, find $P(\overline{A})$.
   b. Women have a 0.25% rate of red/green color blindness. If a woman is randomly selected, what is the *probability* that she does *not* have red/green color blindness? (*Hint:* The decimal equivalent of 0.25% is 0.0025, not 0.25.)

8. Finding Complements
   a. Find $P(\overline{A})$ given that $P(A) = 0.01$.
   b. A Reuters/Zogby poll showed that 61% of Americans say they believe that life exists elsewhere in the galaxy. What is the probability of randomly selecting someone not having that belief?

In Exercises 9–12, use the data in the following table, which summarizes results from 985 pedestrian deaths that were caused by accidents (based on data from the National Highway Traffic Safety Administration).

		Pedestrian Intoxicated?	
		Yes	No
Driver intoxicated?	Yes	59	79
	No	266	581

9. Pedestrian Deaths If one of the pedestrian deaths is randomly selected, find the probability that the pedestrian was intoxicated or the driver was intoxicated.

10. Pedestrian Deaths If one of the pedestrian deaths is randomly selected, find the probability that the pedestrian was not intoxicated or the driver was not intoxicated.

11. Pedestrian Deaths If one of the pedestrian deaths is randomly selected, find the probability that the pedestrian was intoxicated or the driver was not intoxicated.

12. Pedestrian Deaths If one of the pedestrian deaths is randomly selected, find the probability that the driver was intoxicated or the pedestrian was not intoxicated.

In Exercises 13–20, use the data in the following table, which summarizes blood groups and Rh types for 100 typical people. These values may vary in different regions according to the ethnicity of the population.

		Group			
		O	A	B	AB
Type	Rh^+	39	35	8	4
	Rh^-	6	5	2	1

13. Blood Groups and Types If one person is randomly selected, find the probability of getting someone who is not group A.

14. Blood Groups and Types If one person is randomly selected, find the probability of getting someone who is type Rh^-.

15. Blood Groups and Types If one person is randomly selected, find the probability of getting someone who is group A or type Rh^-.

16. Blood Groups and Types If one person is randomly selected, find the probability of getting someone who is group A or group B.

17. Blood Groups and Types If one person is randomly selected, find $P(\text{not type Rh}^+)$.

18. Blood Groups and Types If one person is randomly selected, find $P(\text{group B or type Rh}^+)$.

19. Blood Groups and Types If one person is randomly selected, find $P(\text{group AB or type Rh}^+)$.

20. Blood Groups and Types If one person is randomly selected, find $P(\text{group A or O or type Rh}^+)$.

In Exercises 21 and 22, refer to the figure (on the top of the next page) depicting peas used in a genetics study. (Probabilities play a prominent role in genetics, and Mendel conducted famous hybridization experiments with peas, such as those depicted in the figure.)

21. Constructing Table Use the figure to identify the frequencies in the accompanying table. (The flowers are the top portions of the peas, and the pods are the bottom portions.)

Table for Exercise 21

		Flower	
		Purple	White
Pod	Green	?	?
	Yellow	?	?

22. Hybridization Experiment Assume that one of the peas is randomly selected.

a. Refer to the figure and find $P(\text{green pod or purple flower})$.

b. Refer to the table completed in Exercise 21 and find $P(\text{green pod or purple flower})$.

c. Which format is easier to use: the figure or the table?

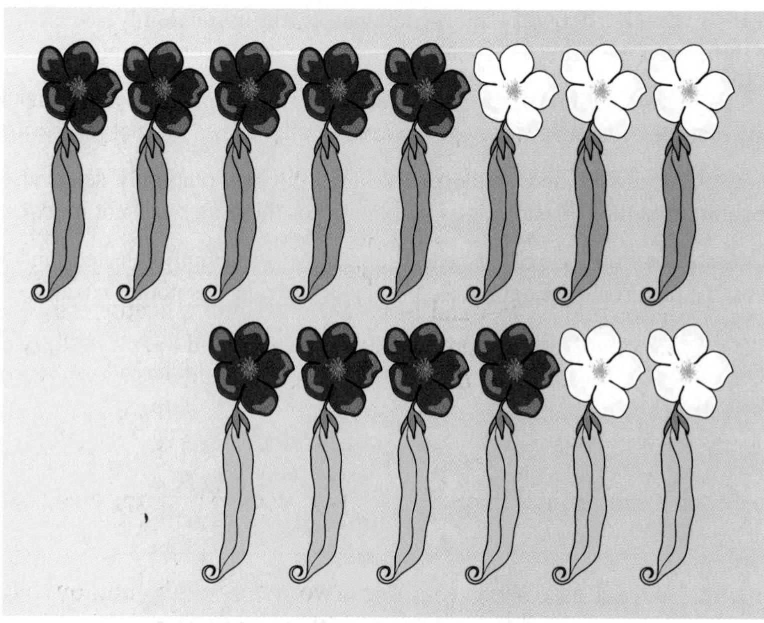

Peas Used in a Hybridization Experiment

23. Poll Resistance Pollsters are concerned about declining levels of cooperation among persons contacted in surveys. A pollster contacts 84 people in the 18–21 age bracket and finds that 73 of them respond and 11 refuse to respond. When 275 people in the 22–29 age bracket are contacted, 255 respond and 20 refuse to respond (based on data from "I Hear You Knocking but You Can't Come In," by Fitzgerald and Fuller, *Sociological Methods and Research,* Vol. 11, No. 1). Assume that 1 of the 359 people is randomly selected. Find the probability of getting someone in the 18–21 age bracket or someone who refused to respond.

24. Poll Resistance Refer to the same data set as in Exercise 23. Assume that 1 of the 359 people is randomly selected, and find the probability of getting someone who is in the 18–21 age bracket or someone who responded.

4-3 BEYOND THE BASICS

25. Disjoint Events If events A and B are disjoint and events B and C are disjoint, must events A and C be disjoint? Give an example supporting your answer.

26. Exclusive *Or* How is the addition rule changed if the *exclusive or* is used instead of the *inclusive or?* In this section it was noted that the *exclusive or* means either one or the other but not both.

27. Extending the Addition Rule The formal addition rule included in this section expressed the probability of A or B as follows: $P(A \text{ or } B) = P(A) + P(B) - P(A \text{ and } B)$. Extend that formal rule to develop an expression for $P(A \text{ or } B \text{ or } C)$. (*Hint:* Draw a Venn diagram.)

4-4 Multiplication Rule: Basics

Key Concept In Section 4-3 we presented the addition rule for finding $P(A \text{ or } B)$, the probability that a single trial has an outcome of A or B or both. This section presents the basic multiplication rule, which is used for finding $P(A \text{ and } B)$, the probability that event A occurs in a first trial and event B occurs in a second trial. If the outcome of the first event A somehow affects the probability of the second event B, it is important to adjust the probability of B to reflect the occurrence of event A. The rule for finding $P(A \text{ and } B)$ is called the multiplication rule because it involves the multiplication of the probability of event A and the probability of event B (where the probability of event B is adjusted because of the outcome of event A).

Notation

$P(A \text{ and } B) = P(\text{event } A \text{ occurs in a first trial and event } B \text{ occurs in a second trial})$

In Section 4-3 we associated use of the word "or" with addition, but in this section we associate use of the word "and" with multiplication.

Probability theory is used extensively in the analysis and design of standardized tests, such as the SAT, ACT, MCAT (for medicine), and the LSAT (for law). For ease of grading, such tests typically use true/false or multiple-choice questions. Let's assume that the first question on a test is a true/false type, while the second question is a multiple-choice type with five possible answers (a, b, c, d, e). We will use the following two questions. Try them!

1. True or false: A pound of feathers is heavier than a pound of gold.
2. Which one of the following has had the most influence on our understanding of genetics?
   - **a.** Gene Hackman
   - **b.** Gene Simmons
   - **c.** Gregor Mendel
   - **d.** jeans
   - **e.** Jean-Jacques Rousseau

The answers to the two questions are T (for "true") and c. (The first answer is true. Weights of feathers are expressed in Avoirdupois pounds, but weights of gold are expressed in Troy pounds.) Let's find the probability that if someone makes random guesses for both answers, the first answer will be correct *and* the second answer will be correct. One way to find that probability is to list the sample space as follows:

T,a	T,b	T,c	T,d	T,e
F,a	F,b	F,c	F,d	F,e

If the answers are random guesses, then the 10 possible outcomes are equally likely, so

$$P(\text{both correct}) = P(T \text{ and } c) = \frac{1}{10} = 0.1$$

Redundancy

Reliability of systems can be greatly improved with redundancy of critical components. Race cars in the NASCAR Winston Cup series have two ignition systems so that if one fails, the other can be used. Airplanes have two independent electrical systems, and aircraft used for instrument flight typically have two separate radios. The following is from a *Popular Science* article about stealth aircraft: "One plane built largely of carbon fiber was the Lear Fan 2100 which had to carry two radar transponders. That's because if a single transponder failed, the plane was nearly invisible to radar." Such redundancy is an application of the multiplication rule in probability theory. If one component has a 0.001 probability of failure, the probability of two independent components both failing is only 0.000001.

Independent Jet Engines

Soon after departing from Miami, Eastern Airlines Flight 855 had one engine shut down because of a low oil pressure warning light. As the L-1011 jet turned to Miami for landing, the low pressure warning lights for the other two engines also flashed. Then an engine failed, followed by the failure of the last working engine. The jet descended without power from 13,000 ft to 4000 ft when the crew was able to restart one engine, and the 172 people on board landed safely. With independent jet engines, the probability of all three failing is only 0.0001^3, or about one chance in a trillion. The FAA found that the same mechanic who replaced the oil in all three engines failed to replace the oil plug sealing rings. The use of a single mechanic caused the operation of the engines to become dependent, a situation corrected by requiring that the engines be serviced by different mechanics.

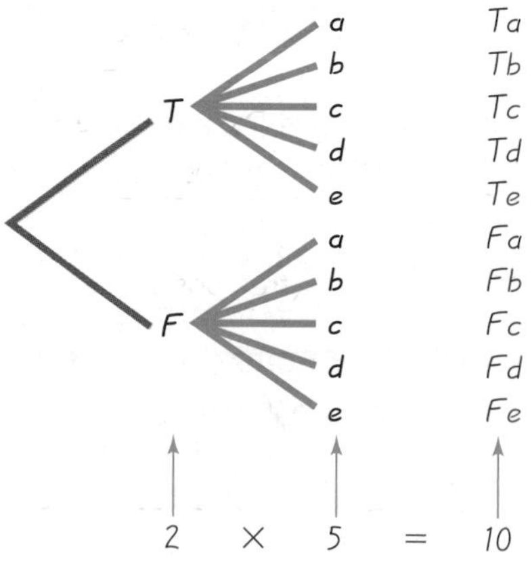

Figure 4-6 Tree Diagram of Test Answers

Now note that $P(T \text{ and } c) = 1/10$, $P(T) = 1/2$, and $P(c) = 1/5$, from which we see that

$$\frac{1}{10} = \frac{1}{2} \cdot \frac{1}{5}$$

so that

$$P(T \text{ and } c) = P(T) \times P(c)$$

This suggests that, in general, $P(A \text{ and } B) = P(A) \cdot P(B)$, but let's consider another example before making that generalization.

First, note that tree diagrams are sometimes helpful in determining the number of possible outcomes in a sample space. A **tree diagram** is a picture of the possible outcomes of a procedure, shown as line segments emanating from one starting point. These diagrams are sometimes helpful if the number of possibilities is not too large. The tree diagram shown in Figure 4-6 summarizes the outcomes of the true/false and multiple-choice questions. From Figure 4-6 we see that if both answers are random guesses, all 10 branches are equally likely and the probability of getting the correct pair (T,c) is 1/10. For each response to the first question, there are 5 responses to the second. *The total number of outcomes is 5 taken 2 times, or 10.* The tree diagram in Figure 4-6 therefore provides a visual illustration of the reason for the use of multiplication.

Our first example of the true/false and multiple-choice questions suggests that $P(A \text{ and } B) = P(A) \cdot P(B)$, but the next example will introduce another important element.

EXAMPLE Drug Testing The Chapter Problem includes Table 4-1, which is reproduced here. If two of the subjects included in the table are randomly selected *without replacement*, find the probability that the first selected person had a positive test result and the second selected person had a negative test result.

Table 4-1 Results from Tests for Marijuana Use

	Did the Subject Actually Use Marijuana? Yes	No
Positive test result (Test indicated that marijuana is *present.*)	119 (true positive)	24 (false positive)
Negative test result (Test indicated that marijuana is *absent.*)	3 (false negative)	154 (true negative)

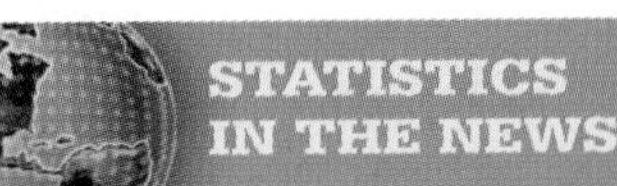

Lottery Advice

New York Daily News columnist Stephen Allensworth recently provided tips for selecting numbers in New York State's Daily Numbers game. In describing a winning system, he wrote that "it involves double numbers matched with cold digits. (A cold digit is one that hits once or not at all in a seven-day period.)" Allensworth proceeded to identify some specific numbers that "have an excellent chance of being drawn this week."

Allensworth assumes that some numbers are "overdue," but the selection of lottery numbers is independent of past results. The system he describes has no basis in reality and will not work. Readers who follow such poor advice are being misled and they might lose more money because they incorrectly believe that their chances of winning are better.

SOLUTION

First selection:
P(positive test result) = 143/300 (because there are 143 subjects who tested positive, and the total number of subjects is 300)

Second selection:
P(negative test result) = 157/299 (after the first selection of a subject with a positive test result, there are 299 subjects remaining, 157 of whom had negative test results)

With P(first subject has positive test result) = 143/300 and P(second subject has negative test result) = 157/299, we have

$$P\left(\begin{array}{l}\text{1st subject has positive test result}\\ \text{and 2nd subject has negative result}\end{array}\right) = \frac{143}{300} \cdot \frac{157}{299} = 0.250$$

The key point is that *we must adjust the probability of the second event to reflect the outcome of the first event.* Because selection of the second subject is made without replacement of the first subject, the second probability must take into account the fact that the first selection removed a subject who tested positive, so only 299 subjects are available for the second selection, and 157 of them had a negative test result.

This example illustrates the important principle that *the probability for the second event B should take into account the fact that the first event A has already occurred.* This principle is often expressed using the following notation.

Notation for Conditional Probability

$P(B|A)$ represents the probability of event B occurring after it is assumed that event A has already occurred. (We can read $B|A$ as "B given A" or as "event B occurring after event A has already occurred.")

Definitions

Two events A and B are **independent** if the occurrence of one does not affect the probability of the occurrence of the other. (Several events are similarly independent if the occurrence of any does not affect the probabilities of the occurrence of the others.) If A and B are not independent, they are said to be **dependent.**

For example, playing the California lottery and then playing the New York lottery are *independent* events because the result of the California lottery has absolutely no effect on the probabilities of the outcomes of the New York lottery. In contrast, the event of having your car start and the event of getting to your statistics class on time are *dependent* events, because the outcome of trying to start your car does affect the probability of getting to the statistics class on time.

Using the preceding notation and definitions, along with the principles illustrated in the preceding examples, we can summarize the key concept of this section as the following *formal multiplication rule*, but it is recommended that you work with the *intuitive multiplication rule*, which is more likely to reflect *understanding* instead of blind use of a formula.

Formal Multiplication Rule

$$P(A \text{ and } B) = P(A) \cdot P(B|A)$$

If A and B are independent events, $P(B|A)$ is really the same as $P(B)$. See the following *intuitive multiplication rule.* (Also see Figure 4-7.)

Intuitive Multiplication Rule

When finding the probability that event A occurs in one trial and event B occurs in the next trial, multiply the probability of event A by the probability of event B, but be sure that the probability of event B takes into account the previous occurrence of event A.

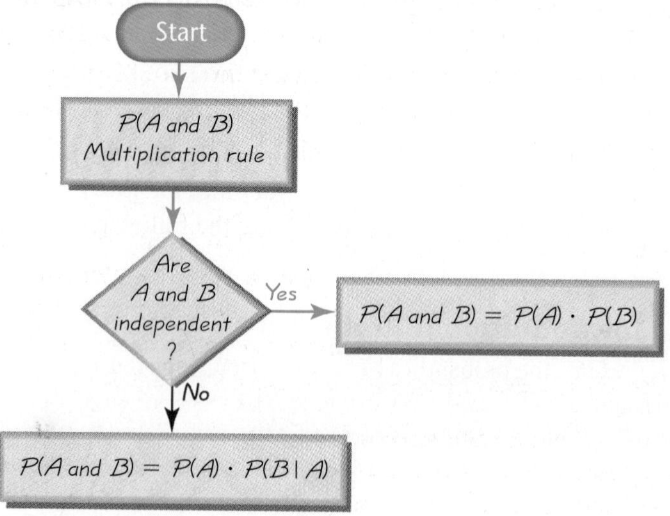

Figure 4-7

Applying the Multiplication Rule

EXAMPLE Plants A biologist experiments with a sample of two vascular plants (denoted here by V) and four nonvascular plants (denoted here by N). Listed below are the codes for the six plants being studied. She wants to randomly select two of the plants for further experimentation. Find the probability that the first selected plant is nonvascular (N) and the second plant is also nonvascular (N). Assume that the selections are made (a) with replacement; (b) without replacement.

V V N N N N

SOLUTION

a. If the two plants are selected *with replacement*, the two selections are independent because the second event is not affected by the first outcome. In each of the two selections there are four nonvascular (N) plants among the six plants, so we get

$$P(\text{first plant is N and second plant is N}) = \frac{4}{6} \cdot \frac{4}{6} = \frac{16}{36} \text{ or } 0.444$$

b. If the two plants are selected *without replacement*, the two selections are dependent because the probability of the second event is affected by the first outcome. In the first selection, four of the six plants are nonvascular (N). After selecting a nonvascular plant on the first selection, we are left with five plants including three that are nonvascular. We therefore get

$$P(\text{first plant is N and second plant is N}) = \frac{4}{6} \cdot \frac{3}{5} = \frac{12}{30} = \frac{2}{5} \text{ or } 0.4$$

Note that in this case, we adjust the second probability to take into account the selection of a nonvascular plant (N) in the first outcome. After selecting N the first time, there would be three Ns among the five plants that remain.

So far we have discussed two events, but the multiplication rule can be easily extended to several events. In general, the probability of any sequence of independent events is simply the product of their corresponding probabilities. For example, the probability of tossing a coin three times and getting all heads is $0.5 \cdot 0.5 \cdot 0.5 = 0.125$. We can also extend the multiplication rule so that it applies to several dependent events; simply adjust the probabilities as you go along.

Treating Dependent Events as Independent Part (b) of the last example involved selecting items without replacement, and we therefore treated the events as being dependent. However, it is a common practice to treat events as independent when *small samples* are drawn from *large populations*. In such cases, it is rare to select the same item twice. Here is a common guideline:

If a sample size is no more than 5% of the size of the population, treat the selections as being *independent* (even if the selections are made without replacement, so they are technically dependent).

Convicted by Probability

A witness described a Los Angeles robber as a Caucasian woman with blond hair in a ponytail who escaped in a yellow car driven by an African-American male with a mustache and beard. Janet and Malcolm Collins fit this description, and they were convicted based on testimony that there is only about 1 chance in 12 million that any couple would have these characteristics. It was estimated that the probability of a yellow car is 1/10, and the other probabilities were estimated to be 1/4, 1/10, 1/3, 1/10, and 1/1000. The convictions were later overturned when it was noted that no evidence was presented to support the estimated probabilities or the independence of the events. However, because the couple was not randomly selected, a serious error was made in not considering the probability of *other* couples being in the same region with the same characteristics.

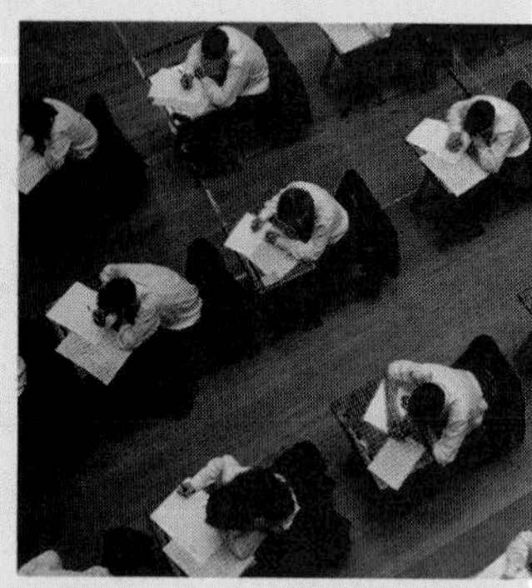

Perfect SAT Score

If an SAT subject is randomly selected, what is the probability of getting someone with a perfect score? What is the probability of getting a perfect SAT score by guessing? These are two very different questions.

The SAT test changed from two sections to three in 2005, and among the 300,000 students who took the first test in March of 2005, 107 achieved perfect scores of 2400 by getting 800 on each of the three sections of writing, critical reading, and math. Based on these results, the probability of getting a perfect score from a randomly selected test subject is 107/300,000 or about 0.000357. In one year with the old SAT test, 1.3 million took the test, and 587 of them received perfect scores of 1600, for a probability of about 0.000452. Just one portion of the SAT consists of 35 multiple-choice questions, and the probability of answering all of them correct by guessing is $(1/5)^{35}$, which is so small that when written as a decimal, 24 zeros follow the decimal point before a nonzero digit appears.

Pollsters use this guideline when they survey roughly 1000 adults from a population of millions. They assume independence, even though they sample without replacement.

The following example gives us some insight into the important procedure of *hypothesis testing* that is introduced in Chapter 8.

EXAMPLE Effectiveness of Gender Selection A geneticist develops a procedure for increasing the likelihood of female babies. In an initial test, 20 couples use the method and the results consist of 20 females among 20 babies. Assuming that the gender-selection procedure has no effect, find the probability of getting 20 females among 20 babies by chance. Based on the result, is there strong evidence to support the geneticist's claim that the procedure is effective in increasing the likelihood that babies will be females?

SOLUTION We want to find P(all 20 babies are female) with the assumption that the procedure has no effect, so that the probability of any individual offspring being a female is 0.5. Because separate pairs of parents were used, we will treat the events as if they are independent. We get this result:

$$
\begin{aligned}
&P(\text{all 20 offspring are female})\\
&= P(\text{1st is female and 2nd is female and 3rd is female} \ldots\\
&\quad \text{and 20th is female})\\
&= P(\text{female}) \cdot P(\text{female}) \cdot \cdots \cdot P(\text{female})\\
&= 0.5 \cdot 0.5 \cdot \cdots \cdot 0.5\\
&= 0.5^{20} = 0.000000954
\end{aligned}
$$

The low probability of 0.000000954 indicates that instead of getting 20 females by chance, a more reasonable explanation is that females appear to be more likely with the gender-selection procedure. Because there is such a small chance (0.000000954) of getting 20 females in 20 births, we do have sufficient evidence to conclude that the gender-selection procedure appears to be effective in increasing the likelihood that an offspring is female. That is, the procedure does appear to be effective.

Calculator Hint: The preceding example required evaluation of 0.5 multiplied by itself 20 times, which is 0.5^{20}. Many calculators have a "power" key labeled x^y, but the TI-83/84 Plus calculator uses ^ for the power key. To evaluate 0.5^{20}, enter 0.5 ^ 20 ENTER to get a result of 9.536743164E⁻7, which is 0.0000009536743164 when expressed in ordinary notation. In the preceding example, this result was rounded to three significant digits as 0.000000954.

We can summarize the fundamentals of the addition and multiplication rules as follows:

- In the addition rule, the word "or" in $P(A \text{ or } B)$ suggests addition. Add $P(A)$ and $P(B)$, being careful to add in such a way that every outcome is counted only once.
- In the multiplication rule, the word "and" in $P(A \text{ and } B)$ suggests multiplication. Multiply $P(A)$ and $P(B)$, but be sure that the probability of event B takes into account the previous occurrence of event A.

4-4 BASIC SKILLS AND CONCEPTS

Statistical Literacy and Critical Thinking

1. Independent Events In your own words, state what it means for two events to be independent.

2. Sampling With Replacement The professor in a class of 25 students randomly selects a student, then randomly selects a second student. If all 25 students are available for the second selection, is this sampling with replacement or sampling without replacement? Is the second outcome independent of the first?

3. Sampling Without Replacement The professor in a class of 25 students randomly selects a student, then randomly selects a second student. If 24 students are available for the second selection, is this sampling with replacement or sampling without replacement? Is the second outcome independent of the first?

4. Notation What does the notation $P(B|A)$ represent?

Identifying Events as Independent or Dependent. *In Exercises 5 and 6, for each given pair of events, classify the two events as* independent *or* dependent. *(If two events are technically dependent but can be treated as if they are independent, consider them to be independent.)*

5. **a.** Randomly selecting a quarter made before 2001
Randomly selecting a second quarter made before 2001
b. Randomly selecting a TV viewer who is watching *The Barry Manilow Biography*
Randomly selecting a second TV viewer who is watching *The Barry Manilow Biography*
c. Wearing plaid shorts with black sox and sandals
Asking someone on a date and getting a positive response

6. **a.** Finding that your calculator is working
Finding that your cell phone is working
b. Finding that your kitchen toaster is not working
Finding that your refrigerator is not working
c. Drinking or using drugs until your driving ability is impaired
Being involved in a car crash

7. Guessing A quick quiz consists of a true/false question followed by a multiple-choice question with four possible answers (a, b, c, d). If both questions are answered with random guesses, find the probability that both responses are correct. Does guessing appear to be a good strategy on this quiz?

8. Letter and Digit A new computer owner creates a password consisting of two characters. She randomly selects a letter of the alphabet for the first character and a digit (0, 1, 2, 3, 4, 5, 6, 7, 8, 9) for the second character. What is the probability that her password is "K9"? Would this password be effective as a deterrent against someone trying to gain access to her computer?

9. Wearing Hunter Orange A study of hunting injuries and the wearing of "hunter" orange clothing showed that among 123 hunters injured when mistaken for game, 6 were wearing orange (based on data from the Centers for Disease Control). If a follow-up study begins with the random selection of hunters from this sample of 123, find the probability that the first two selected hunters were both wearing orange.

continued

a. Assume that the first hunter is replaced before the next one is selected.
b. Assume that the first hunter is not replaced before the second one is selected.
c. Given a choice between selecting with replacement and selecting without replacement, which choice makes more sense in this situation? Why?

10. Selecting U.S. Senators In the 108th Congress, the Senate consists of 51 Republicans, 48 Democrats, and 1 Independent. If a lobbyist for the tobacco industry randomly selects three different Senators, what is the probability that they are all Republicans? Would a lobbyist be likely to use random selection in this situation?

11. Acceptance Sampling With one method of a procedure called *acceptance sampling,* a sample of items is randomly selected without replacement and the entire batch is accepted if every item in the sample is okay. The Niko Electronics Company has just manufactured 5000 CDs, and 100 are defective. If 4 of these CDs are randomly selected for testing, what is the probability that the entire batch will be accepted?

12. Poll Confidence Level It is common for public opinion polls to have a "confidence level" of 95%, meaning that there is a 0.95 probability that the poll results are accurate within the claimed margins of error. If six different organizations conduct independent polls, what is the probability that all six of them are accurate within the claimed margins of error? Does the result suggest that with a confidence level of 95%, we can expect that almost all polls will be within the claimed margin of error?

13. Testing Effectiveness of Gender-Selection Method Recent developments appear to make it possible for couples to dramatically increase the likelihood that they will conceive a child with the gender of their choice. In a test of a gender-selection method, 12 couples try to have baby girls. If this gender-selection method has no effect, what is the probability that the 12 babies will be all girls? If there are actually 12 girls among 12 children, does this gender-selection method appear to be effective? Why?

14. Voice Identification of Criminal In a Riverhead, New York, case, nine different crime victims listened to voice recordings of five different men. All nine victims identified the same voice as that of the criminal. If the voice identifications were made by random guesses, find the probability that all nine victims would select the same person. Does this constitute reasonable doubt?

15. Redundancy The principle of redundancy is used when system reliability is improved through redundant or backup components. Assume that your alarm clock has a 0.975 probability of working on any given morning.
a. What is the probability that your alarm clock will *not* work on the morning of an important final exam?
b. If you have two such alarm clocks, what is the probability that they both fail on the morning of an important final exam?
c. With one alarm clock, you have a 0.975 probability of being awakened. What is the probability of being awakened if you use two alarm clocks?
d. Does a second alarm clock result in greatly improved reliability?

16. Social Skills Bob reasons that when he asks a woman for a date, she can accept or reject his request, so he assumes that he has a 0.5 probability of getting a date. If his assumption is correct, what is the probability of getting five rejections when Bob asks five different women for dates? Is that result the correct probability that Bob will get five rejections when he asks five different women for dates? Why or why not?

In Exercises 17–20, use the data in the following table, which summarizes results from 985 pedestrian deaths that were caused by accidents (based on data from the National Highway Traffic Safety Administration).

		Pedestrian Intoxicated?	
		Yes	No
Driver intoxicated?	Yes	59	79
	No	266	581

17. Intoxicated Drivers If two *different* pedestrian deaths are randomly selected, find the probability that they both involved intoxicated drivers.

18. Intoxicated Pedestrians If two *different* pedestrian deaths are randomly selected, find the probability that they both involved intoxicated pedestrians.

19. Pedestrian Deaths

a. If one of the pedestrian deaths is randomly selected, what is the probability that it involves a case in which neither the pedestrian nor the driver was intoxicated?

b. If two *different* pedestrian deaths are randomly selected, what is the probability that in both cases, neither the pedestrian nor the driver was intoxicated?

c. If two pedestrian deaths are randomly selected with replacement, what is the probability that in both cases, neither the pedestrian nor the driver was intoxicated?

d. Compare the results from parts (b) and (c).

20. Pedestrian Deaths

a. If one of the pedestrian deaths is randomly selected, what is the probability that it involves an intoxicated pedestrian and an intoxicated driver?

b. If two *different* pedestrian deaths are randomly selected, what is the probability that in both cases, both the pedestrian and the driver were intoxicated?

c. If two pedestrian deaths are randomly selected with replacement, what is the probability that in both cases, both the pedestrian and the driver were intoxicated?

d. Compare the results from parts (b) and (c).

4-4 BEYOND THE BASICS

21. Same Birthdays Find the probability that no two people have the same birthday when the number of randomly selected people is

a. 3

b. 5

c. 25

22. Gender of Children

a. If a couple plans to have eight children, find the probability that they are all of the same gender.

b. Assuming that boys and girls are equally likely, find the probability of getting all girls when 1000 babies are born. Does the result indicate that such an event is impossible?

23. Drawing Cards Two cards are to be randomly selected without replacement from a shuffled deck. Find the probability of getting an ace on the first card and a spade on the second card.

24. Complements and the Addition Rule

a. Develop a formula for the probability of not getting either A or B on a single trial. That is, find an expression for $P(\overline{A \text{ or } B})$.

b. Develop a formula for the probability of not getting A or not getting B on a single trial. That is, find an expression for $P(\overline{A} \text{ or } \overline{B})$.

c. Compare the results from parts (a) and (b). Does $P(\overline{A \text{ or } B}) = P(\overline{A} \text{ or } \overline{B})$?

4-5 Multiplication Rule: Complements and Conditional Probability

Key Concept Section 4-4 introduced the basic concept of the multiplication rule, but in this section we extend our use of that rule to two other special applications. First, when we want to find the probability that among several trials, we get *at least one* of some specified event, one easy approach is to find the probability that *none* of the events occur, then find the complement of that event. Second, we consider conditional probability, which is the probability of an event given the additional information that some other event has already occurred.

We begin with situations in which we want to find the probability that among several trials, *at least one* will result in some specified outcome.

Complements: The Probability of "At Least One"

The multiplication rule and the rule of complements can be used together to greatly simplify the solution to this type of problem: Find the probability that among several trials, *at least one* will result in some specified outcome. In such cases, it is critical that the meaning of the language be clearly understood:

- "At least one" is equivalent to "one or more."
- The complement of getting at least one item of a particular type is that you get *no* items of that type.

Suppose a couple plans to have three children and they want to know the probability of getting at least one girl. See the following interpretations:

At least 1 girl among 3 children = 1 or more girls.

The complement of "at least 1 girl" = no girls = all 3 children are boys.

We could easily find the probability from a list of the entire sample space of eight outcomes, but we want to illustrate the use of complements, which can be used in many other problems that cannot be solved so easily.

EXAMPLE Gender of Children Find the probability of a couple having at least 1 girl among 3 children. Assume that boys and girls are equally likely and that the gender of a child is independent of the gender of any brothers or sisters.

SOLUTION

Step 1: Use a symbol to represent the event desired. In this case, let A = at least 1 of the 3 children is a girl.

Step 2: Identify the event that is the complement of A.

$$\begin{aligned}\overline{A} &= \textit{not} \text{ getting at least 1 girl among 3 children} \\ &= \text{all 3 children are boys} \\ &= \text{boy and boy and boy}\end{aligned}$$

Step 3: Find the probability of the complement.

$$\begin{aligned}P(\overline{A}) &= P(\text{boy and boy and boy}) \\ &= \frac{1}{2} \cdot \frac{1}{2} \cdot \frac{1}{2} = \frac{1}{8}\end{aligned}$$

Step 4: Find $P(A)$ by evaluating $1 - P(\overline{A})$.

$$P(A) = 1 - P(\overline{A}) = 1 - \frac{1}{8} = \frac{7}{8}$$

INTERPRETATION There is a $7/8$ probability that if a couple has 3 children, at least 1 of them is a girl.

The principle used in this example can be summarized as follows:

To find the probability of *at least one* of something, calculate the probability of *none*, then subtract that result from 1. That is,

$$\mathbf{P(\text{at least one}) = 1 - P(\text{none}).}$$

Conditional Probability

Next we consider the second major point of this section, which is based on the principle that the probability of an event is often affected by knowledge of circumstances. For example, if you randomly select someone from the general population, the probability of getting a male is 0.5, but if you then learn that the selected person smokes cigars, there is a dramatic increase in the probability that the selected person is a male (because 85% of cigar smokers are males). A *conditional probability* of an event is used when the probability is affected by the knowledge of other circumstances. The conditional probability of event B occurring, given that event A has already occurred, can be found by using the multiplication rule $[P(A \text{ and } B) = P(A) \cdot P(B|A)]$ and solving for $P(B|A)$ by dividing both sides of the equation by $P(A)$.

Definition

A **conditional probability** of an event is a probability obtained with the additional information that some other event has already occurred. $P(B|A)$ denotes the conditional probability of event B occurring, given that event A has already occurred, and it can be found by dividing the probability of events A and B both occurring by the probability of event A:

$$P(B|A) = \frac{P(A \text{ and } B)}{P(A)}$$

Coincidences?

John Adams and Thomas Jefferson (the second and third presidents) both died on July 4, 1826. President Lincoln was assassinated in Ford's Theater; President Kennedy was assassinated in a Lincoln car made by the Ford Motor Company. Lincoln and Kennedy were both succeeded by vice presidents named Johnson. Fourteen years *before* the sinking of the *Titanic*, a novel described the sinking of the *Titan*, a ship that hit an iceberg; see Martin Gardner's *The Wreck of the Titanic Foretold?* Gardner states, "In most cases of startling coincidences, it is impossible to make even a rough estimate of their probability."

The preceding formula is a formal expression of conditional probability, but blind use of formulas is not recommended. We recommend the following intuitive approach:

Intuitive Approach to Conditional Probability

The conditional probability of B given A can be found by assuming that event A has occurred and, working under that assumption, calculating the probability that event B will occur.

EXAMPLE Drug Test Refer to Table 4-1, reproduced here for your convenience. Find the following:

a. If 1 of the 300 test subjects is randomly selected, find the probability that the person tested positive, given that he or she actually used marijuana.

b. If 1 of the 300 test subjects is randomly selected, find the probability that the person actually used marijuana, given that he or she tested positive.

SOLUTION

a. We want $P(\text{positive} \mid \text{marijuana use})$, the probability of getting someone who tested positive, *given that the selected person used marijuana.* Here is the key point: If we assume that the selected person used marijuana, we are dealing only with the 122 subjects in the first column of Table 4-1. Among those 122 subjects, 119 tested positive, so

$$P(\text{positive} \mid \text{marijuana use}) = \frac{119}{122} = 0.975$$

The same result can be found by using the formula given with the definition of conditional probability. In the following calculation, we use the fact that 119 of the 300 subjects were both marijuana users and tested positive. Also, 122 of

Table 4-1 Results from Tests for Marijuana Use

	Did the Subject Actually Use Marijuana? Yes	No
Positive test result (Test indicated that marijuana is *present*.)	119 (true positive)	24 (false positive)
Negative test result (Test indicated that marijuana is *absent*.)	3 (false negative)	154 (true negative)

the 300 subjects used marijuana. We get

$$P(\text{positive} \mid \text{marijuana use}) = \frac{P(\text{marijuana use and positive})}{P(\text{marijuana use})}$$

$$= \frac{119/300}{122/300} = 0.975$$

b. Here we want $P(\text{marijuana use} \mid \text{positive})$. If we assume that the person selected tested positive, we are dealing with the 143 subjects in the first row of Table 4-1. Among those 143 subjects, 119 used marijuana, so

$$P(\text{marijuana use} \mid \text{positive}) = \frac{119}{143} = 0.832$$

Again, the same result can be found by applying the formula for conditional probability:

$$P(\text{marijuana use} \mid \text{positive}) = \frac{P(\text{positive and marijuana use})}{P(\text{positive})}$$

$$= \frac{119/300}{143/300} = 0.832$$

By comparing the results from parts (a) and (b), we see that $P(\text{positive} \mid \text{marijuana use})$ is not the same as $P(\text{marijuana use} \mid \text{positive})$.

INTERPRETATION The first result of $P(\text{positive} \mid \text{marijuana use}) = 0.975$ indicates that a marijuana user has a 0.975 probability of testing positive. The second result of $P(\text{marijuana use} \mid \text{positive}) = 0.832$ indicates that for someone who tests positive, there is an 0.832 probability that this person actually used marijuana.

Confusion of the Inverse

Note that in the preceding example, $P(\text{positive} \mid \text{marijuana use}) \neq P(\text{marijuana use} \mid \text{positive})$. To incorrectly believe that $P(B \mid A)$ and $P(A \mid B)$ are the same, or to incorrectly use one value for the other is often called *confusion of the inverse*. Studies have shown that physicians often give very misleading information when they confuse the inverse. Based on real studies, they tended to confuse $P(\text{cancer} \mid \text{positive test})$ with $P(\text{positive test} \mid \text{cancer})$. About 95% of physicians estimated $P(\text{cancer} \mid \text{positive test})$ to be about 10 times too high, with the result that patients were given diagnoses that were very misleading, and patients were unnecessarily distressed by the incorrect information.

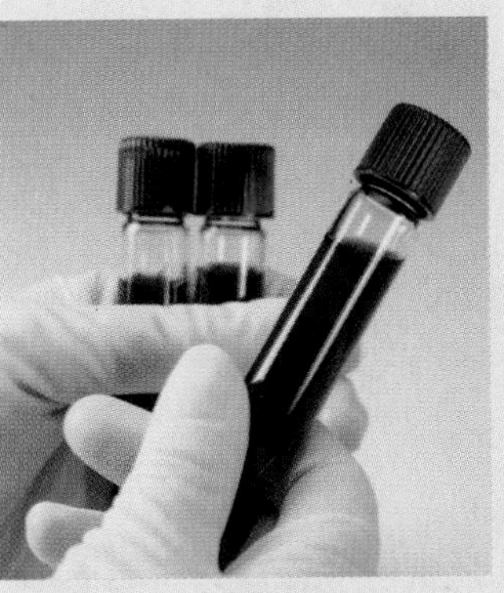

Composite Sampling

The U.S. Army once tested for syphilis by giving each inductee an individual blood test that was analyzed separately. One researcher suggested mixing pairs of blood samples. After the mixed pairs were tested, syphilitic inductees could be identified by retesting the few blood samples that were in the pairs that tested positive. The total number of analyses was reduced by pairing blood specimens, so why not put them in groups of three or four or more? Probability theory was used to find the most efficient group size, and a general theory was developed for detecting the defects in any population. This technique is known as *composite sampling.*

4-5 BASIC SKILLS AND CONCEPTS

Statistical Literacy and Critical Thinking

1. Probability of at Least One You want to find the probability of getting at least 1 defect when 10 heart pacemakers are randomly selected and tested. What do you know about the exact number of defects if "at least one" of the 10 items is defective?

2. Conditional Probability In your own words, describe conditional probability and give an example.

3. Finding Probability A market researcher needs to find the probability that a shopper is male, given that a credit card was used for a purchase. He reasons that there are two outcomes (male, female), so the probability is 1/2. Is he correct? What important information is not included in his reasoning process?

4. Confusion of the Inverse What is confusion of the inverse?

Describing Complements. *In Exercises 5–8, provide a written description of the complement of the given event.*

5. Blood Testing When six job applicants are tested for use of marijuana, at least one of them tests positive.

6. Quality Control When 50 electrocardiograph units are shipped, all of them are free of defects.

7. X-Linked Disorder When 12 males are tested for a particular X-linked recessive gene, none of them are found to have the gene.

8. A Hit with the Misses When Brutus asks 12 different women for a date, at least one of them accepts.

9. Subjective Conditional Probability Use subjective probability to estimate the probability that a credit card is being used fraudulently, given that today's charges were made in several different countries.

10. Subjective Conditional Probability Use subjective probability to estimate the probability of randomly selecting an adult and getting a male, given that the selected person owns a motorcycle. If a criminal investigator finds that a motorcycle is registered to Pat Ryan, is it reasonable to believe that Pat is a male?

11. Probability of At Least One Girl If a couple plans to have four children, what is the probability that they will have at least one girl? Is that probability high enough for the couple to be very confident that they will get at least one girl in four children?

12. Probability of At Least One Girl If a couple plans to have 10 children (it could happen), what is the probability that there will be at least one girl? If the couple eventually has 10 children and they are all boys, what can the couple conclude?

13. Probability of a Girl Find the probability of a couple having a baby girl when their third child is born, given that the first two children were both girls. Is the result the same as the probability of getting three girls among three children?

14. Drug Testing Refer to Table 4-1 and assume that 1 of the 300 test subjects is randomly selected. Find the probability of getting someone who tests positive, given that he or she did not use marijuana. Why is this particular case problematic for test subjects?

15. Drug Testing Refer to Table 4-1 and assume that 1 of the 300 test subjects is randomly selected. Find the probability of getting someone who tests negative, given that he or she did not use marijuana.

16. Drug Testing Refer to Table 4-1 and assume that 1 of the 300 test subjects is randomly selected. Find the probability of getting someone who did not use marijuana, given that he or she tested negative. Compare this result and the result found in Exercise 15.

17. Redundancy in Alarm Clocks A statistics professor wants to ensure that she is not late for an early class because of a malfunctioning alarm clock. Instead of using one alarm clock, she decides to use three. What is the probability that at least one of her alarm clocks works correctly if each individual alarm clock has a 95% chance of working correctly? Does the professor really gain much by using three alarm clocks instead of only one?

18. Acceptance Sampling With one method of the procedure called *acceptance sampling,* a sample of items is randomly selected without replacement, and the entire batch is rejected if there is at least one defect. The Medtyme Company has just manufactured 5000 blood pressure monitors, and 4% are defective. If 3 of them are selected and tested, what is the probability that the entire batch will be rejected?

19. Using Composite Blood Samples When doing blood testing for HIV infections, the procedure can be made more efficient and less expensive by combining samples of blood specimens. If samples from three people are combined and the mixture tests negative, we know that all three individual samples are negative. Find the probability of a positive result for three samples combined into one mixture, assuming the probability of an individual blood sample testing positive is 0.1 (the probability for the "at-risk" population, based on data from the New York State Health Department).

20. Using Composite Water Samples The Orange County Department of Public Health tests water for contamination due to the presence of *E. coli* (*Escherichia coli*) bacteria. To reduce laboratory costs, water samples from six public swimming areas are combined for one test, and further testing is done only if the combined sample fails. Based on past results, there is a 2% chance of finding *E. coli* bacteria in a public swimming area. Find the probability that a combined sample from six public swimming areas will reveal the presence of *E. coli* bacteria.

Conditional Probabilities. *In Exercises 21–24, use the following data from the 100 Senators from the 108th Congress of the United States.*

	Republican	Democrat	Independent
Male	46	39	1
Female	5	9	0

21. If we randomly select one Senator, what is the probability of getting a Republican, given that a male was selected?

22. If we randomly select one Senator, what is the probability of getting a male, given that a Republican was selected? Is this the same result found in Exercise 21?

23. If we randomly select one Senator, what is the probability of getting a female, given that an Independent was selected?

24. If we randomly select one Senator, what is the probability of getting a Democrat or Independent, given that a male was selected?

4-5 BEYOND THE BASICS

25. Shared Birthdays Find the probability that of 25 randomly selected people,

a. no 2 share the same birthday.

b. at least 2 share the same birthday.

26. Whodunnit? The Atlanta plant of the Medassist Pharmaceutical Company manufactures 400 heart pacemakers, of which 3 are defective. The Baltimore plant of the same company manufactures 800 pacemakers, of which 2 are defective. If 1 of the 1200 pacemakers is randomly selected and is found to be defective, what is the probability that it was manufactured in Atlanta?

27. Roller Coaster The Rock 'n' Roller Coaster at Disney–MGM Studios in Orlando has two seats in each of 12 rows. Riders are assigned to seats in the order that they arrive. If you ride this roller coaster once, what is the probability of getting the coveted first row? How many times must you ride in order to have at least a 95% chance of getting a first-row seat at least once?

28. Unseen Coins A statistics professor tosses two coins that cannot be seen by any students. One student asks this question: "Did one of the coins turn up heads?" Given that the professor's response is "yes," find the probability that both coins turned up heads.

Probability of an Event That Has Never Occurred

Some events are possible, but are so unlikely that they have never occurred. Here is one such problem of great interest to political scientists: Estimate the probability that your single vote will determine the winner in a U.S. Presidential election. Andrew Gelman, Gary King, and John Boscardin write in the *Journal of the American Statistical Association* (Vol. 93, No. 441) that "the exact value of this probability is of only minor interest, but the number has important implications for understanding the optimal allocation of campaign resources, whether states and voter groups receive their fair share of attention from prospective presidents, and how formal 'rational choice' models of voter behavior might be able to explain why people vote at all." The authors show how the probability value of 1 in 10 million is obtained for close elections.

4-6 Probabilities Through Simulations

Key Concept So far in this chapter we have identified several basic and important rules commonly used for finding probabilities, but in this section we introduce a very different approach that can overcome much of the difficulty encountered with the formal methods in the preceding sections of this chapter. Instead of using formal rules for finding probabilities, this alternative approach consists of developing a simulation, whereby we use some different procedure that behaves the same way as the procedure we are considering.

Definition

A **simulation** of a procedure is a process that behaves the same way as the procedure, so that similar results are produced.

Consider the following examples to better understand how simulations can be used.

EXAMPLE Gender Selection When testing techniques of gender selection, medical researchers need to know probability values of different outcomes, such as the probability of getting at least 60 girls among 100 children. Assuming that male and female births are equally likely, describe a simulation that results in the genders of 100 newborn babies.

SOLUTION One approach is simply to flip a coin 100 times, with heads representing females and tails representing males. Another approach is to use a calculator or computer to randomly generate 0s and 1s, with 0 representing a male and 1 representing a female. The numbers must be generated in such a way that they are equally likely.

EXAMPLE Same Birthdays Exercise 25 in Section 4-5 refers to the classical birthday problem, in which we find the probability that in a randomly selected group of 25 people, at least 2 share the same birthday. The theoretical solution is somewhat difficult. It isn't practical to survey many different groups of 25 people, so a simulation is a helpful alternative. Describe a simulation that could be used to find the probability that among 25 randomly selected people, at least 2 share the same birthday.

SOLUTION Begin by representing birthdays by integers from 1 through 365, where 1 = January 1, 2 = January 2, . . . , 365 = December 31. Then use a calculator or computer program to generate 25 random numbers, each between 1 and 365. Those numbers can then be sorted, so it becomes easy to survey the list to determine whether any 2 of the simulated birth dates are the same. We can repeat the process as many times as we like, until we are satisfied that we have a good estimate of the probability. Our estimate of the probability is the number of times we did get at least 2 birth dates that are the same, divided by the total number of groups of 25 that were generated.

To Win, Bet Boldly

The *New York Times* published an article by Andrew Pollack in which he reported lower than expected earnings for the Mirage casino in Las Vegas. He wrote that "winnings for Mirage can be particularly volatile, because it caters to high rollers, gamblers who might bet $100,000 or more on a hand of cards. The law of averages does not work as consistently for a few large bets as it does for thousands of smaller ones . . ." This reflects the most fundamental principle of gambling: To win, place one big bet instead of many small bets! With the right game, such as craps, you have just under a 50% chance of doubling your money if you place one big bet. With many small bets, your chance of doubling your money drops substantially.

There are several ways of obtaining randomly generated numbers from 1 through 365, including a table of random digits or a TI-83/84 Plus calculator, as follows:

- **A Table of Random Digits:** Refer, for example, to the *CRC Standard Probability and Statistics Tables and Formulae,* which contains a table of 14,000 digits. (In such a table there are many ways to extract numbers from 1 through 365. One way is by referring to the digits in the first three columns and ignoring 000 as well as anything above 365.)

TI TI-83/84 Plus Calculator for Simulation of Birth Dates

Use the following procedure to generate 25 simulated birthdays, where each birthday is represented by a whole number between 1 and 365.

1. Press MATH, then use the ▶ key to select **PRB,** then use the ▼ key to select the menu item of randInt (for random integers).
2. The format of the **randInt** command requires a minimum number, a maximum number, and the number of values to be generated. To simulate 25 birthdays, enter 1 for the minimum, 365 for the maximum, and 25 for the number of generated values, so that the command is entered as **randInt(1, 365, 25).** Then press ENTER.
3. For ease of viewing, store the list of generated values in L1 by entering this sequence of key strokes: STO› 2ND 1 ENTER.

4. It is much easier to inspect the list of values for matches if the list has been arranged in order, so sort the values in list L1 by entering STAT, then selecting the menu item of **SortA** (to arrange the values in ascending order). Press ENTER. Now identify L1 as the list to be sorted by pressing 2ND 1 ENTER, so that the screen displays the command of **SortA(L1).** Press ENTER, and the screen should provide the message of Done.
5. Inspect list L1 for matches by pressing STAT ENTER, so that the entries in L1 are displayed in a single column of values arranged in order. Use the ▽ key to scroll through the list of values searching for matches. Matches are easy to identify because they will be adjacent equal numbers.

The above steps will result in displays similar to those shown here, although the generated numbers will be different, because they are randomly generated.

TI-83/84 Plus

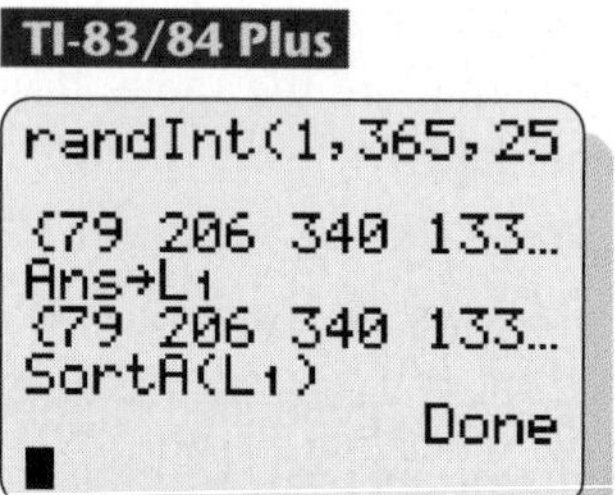

TI-83/84 Plus

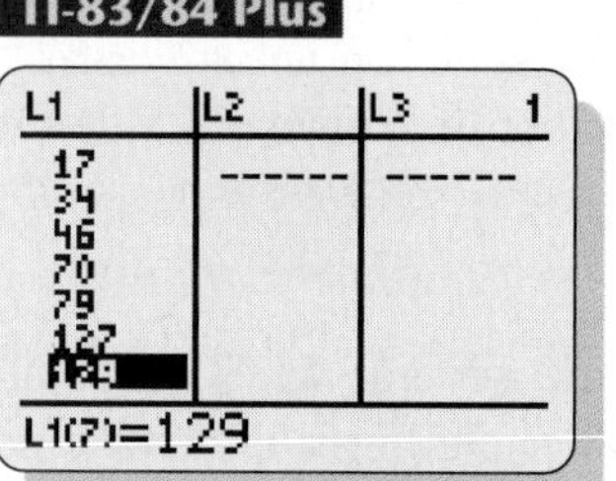

It is extremely important to construct a simulation so that it behaves just like the real procedure. In the next example we demonstrate the right way and a wrong way to construct a simulation.

EXAMPLE Simulating Dice Describe a procedure for simulating the rolling of a pair of dice.

SOLUTION In the procedure of rolling a pair of dice, each of the two dice yields a number between 1 and 6 (inclusive), and those two numbers are then added. Any simulation should do exactly the same thing. There is a right way and a wrong way to simulate rolling two dice.

The right way: Randomly generate one number between 1 and 6, randomly generate another number between 1 and 6, and then add the two results.

The wrong way: Randomly generate a number between 2 and 12. This procedure is similar to rolling dice in the sense that the results are always between 2 and 12, but these outcomes between 2 and 12 are equally likely. With real dice, the values between 2 and 12 are *not* equally likely. This simulation would produce very misleading results.

Some probability problems can be solved only by estimating the probability from actual observations or constructing a simulation. The widespread availability of calculators and computers has made it very easy to use simulation methods, so that simulations are now used often for determining probability values.

Monkey Typists

A classical claim is that a monkey randomly hitting a keyboard would eventually produce the complete works of Shakespeare, assuming that it continues to type century after century. The multiplication rule for probability has been used to find such estimates. One result of 1,000,000,000,000,000,000,000,000,000,000,000,000 years is considered by some to be too short. In the same spirit, Sir Arthur Eddington wrote this poem: "There once was a brainy baboon, who always breathed down a bassoon. For he said, 'It appears that in billions of years, I shall certainly hit on a tune.'"

4-6 BASIC SKILLS AND CONCEPTS

Statistical Literacy and Critical Thinking

1. Simulations What is a simulation? If a simulation method is used for a probability problem, is the result the exact correct answer?

2. Simulation When three babies are born, there can be 0 girls, 1 girl, 2 girls, or 3 girls. A researcher simulates three births as follows: The number 0 is written on one index card, the number 1 is written on another index card, 2 is written on a third card, and 3 is written on a fourth card. The four index cards are then mixed in a bowl and one card is randomly selected. Considering only the outcome of the number of girls in three births, does this process simulate the three births in such a way that it behaves the same way as actual births? Why or why not?

3. Simulation A student wants to simulate 25 birthdays as described in this section, but she does not have a calculator or software program available, so she makes up 25 numbers between 1 and 365. Is it okay to conduct the simulation this way? Why or why not?

4. Simulation A student wants to simulate three births, so she writes "male" on one index card and "female" on another. She shuffles the cards, then selects one and records the gender. She shuffles the cards a second time, selects one and records the gender. She shuffles the cards a third time, selects one and records the gender. Is this process okay for simulating three births?

In Exercises 5–8, describe the simulation procedure. (For example, to simulate 10 births, use a random number generator to generate 10 integers between 0 and 1 inclusive, and consider 0 to be a male and 1 to be a female.)

5. Simulating Motorcycle Safety Study In a study of fatalities caused by motorcycle crashes, it was found that 95% of motorcycle drivers are men (based on data from "Motorcycle Rider Conspicuity and Crash Related Injury," by Wells et al., *BJM USA*). Describe a procedure for using a TI-83/84 Plus calculator to simulate the random selection of 20 motorcycle drivers. Each individual outcome should consist of an indication of whether the motorcycle driver is a man or woman.

6. Simulating Hybridization When Mendel conducted his famous hybridization experiments, he used peas with green pods and yellow pods. One experiment involved crossing peas in such a way that 25% of the offspring peas were expected to have yellow pods, and 75% of the offspring peas were expected to have green pods. Describe a procedure for using a TI-83/84 Plus calculator to simulate 12 peas in such a hybridization experiment.

7. Simulating Manufacturing Describe a procedure for using a TI-83/84 Plus calculator to simulate 500 manufactured cell phones. For each cell phone, the result should consist of an indication of whether the cell phone is good or defective. The manufacturing process has a defect rate of 2%.

8. Simulating Left-Handedness Fifteen percent of U.S. men are left-handed (based on data from a Scripps Survey Research Center poll). Describe a procedure for using a TI-83/84 Plus calculator to simulate the random selection of 200 men. The outcomes should consist of an indication of whether each man is left-handed or is not.

In Exercises 9–12, develop a simulation using a TI-83/84 Plus calculator.

TI **9.** Simulating Motorcycle Safety Study Refer to Exercise 5, which required a description of a simulation.

a. Conduct the simulation and record the number of male motorcycle drivers. If possible, obtain a printed copy of the results. Is the percentage of males from the simulation reasonably close to the value of 95%?

b. Repeat the simulation until it has been conducted a total of 10 times. Record the numbers of males in each case. Based on the results, do the numbers of males appear to be very consistent? Based on the results, would it be *unusual* to randomly select 20 motorcycle drivers and find that half of them are females?

TI **10.** Simulating Hybridization Refer to Exercise 6, which required a description of a hybridization simulation.

a. Conduct the simulation and record the number of yellow peas. If possible, obtain a printed copy of the results. Is the percentage of yellow peas from the simulation reasonably close to the value of 25%?

b. Repeat the simulation until it has been conducted a total of 10 times. Record the numbers of peas with yellow pods in each case. Based on the results, do the numbers of peas with yellow pods appear to be very consistent? Based on the results, would it be *unusual* to randomly select 12 such offspring peas and find that none of them have yellow pods?

TI **11.** Simulating Manufacturing Refer to Exercise 7, which required a description of a manufacturing simulation.

a. Conduct the simulation and record the number of defective cell phones. Is the percentage of defective cell phones from the simulation reasonably close to the value of 2%?

b. Repeat the simulation until it has been conducted a total of 4 times. Record the numbers of defective cell phones in each case. Based on the results, would it be *unusual* to randomly select 500 such cell phones and find that none of them are defective?

TI **12.** Simulating Left-Handedness Refer to Exercise 8, which required a description of a simulation.

a. Conduct the simulation and record the number of left-handed men. Is the percentage of left-handed men from the simulation reasonably close to the value of 15%?

b. Repeat the simulation until it has been conducted a total of 5 times. Record the numbers of left-handed men in each case. Based on the results, would it be *unusual* to randomly select 200 men and find that none of them are left-handed?

TI **13.** Analyzing the Effectiveness of a Drug It has been found that when someone tries to stop smoking under certain circumstances, the success rate is 20%. A new nicotine substitute drug has been designed to help those who wish to stop smoking. In a trial of 50 smokers who use the drug while trying to stop, it was found that 12 successfully stopped. The drug manufacturer argues that the 12 successes are better than the 10 that would be expected without the drug, so the drug is effective. Conduct a simulation of 50 smokers trying to stop, and assume that the drug has no effect, so the success rate continues to be 20%. Repeat the simulation several times and determine whether 12 successes could easily occur with an ineffective drug. What do you conclude about the effectiveness of the drug?

TI **14.** Analyzing the Effectiveness of a Gender-Selection Method When testing the effectiveness of a gender-selection technique, a trial was conducted with 20 couples trying to have a baby girl. Among the 20 babies that were born, there were 18 girls. Conduct a simulation of 20 births assuming that the gender-selection method has no effect. Repeat the simulation several times and determine whether 18 girls could easily occur with an ineffective gender-selection method. What do you conclude about the effectiveness of the method?

4-6 BEYOND THE BASICS

15. Simulating the Monty Hall Problem A problem that has attracted much attention in recent years is the *Monty Hall problem,* based on the old television game show "Let's Make a Deal," hosted by Monty Hall. Suppose you are a contestant who has selected one of three doors after being told that two of them conceal nothing, but that a new red Corvette is behind one of the three. Next, the host opens one of the doors you didn't select and shows that there is nothing behind it. He then offers you the choice of sticking with your first selection or switching to the other unopened door. Should you stick with your first choice or should you switch? Develop a simulation of this game and determine whether you should stick or switch. (According to *Chance* magazine, business schools at such institutions as Harvard and Stanford use this problem to help students deal with decision making.)

16. Simulating Birthdays

a. Develop a simulation for finding the probability that when 50 people are randomly selected, at least 2 of them have the same birth date. Describe the simulation and estimate the probability.

b. Develop a simulation for finding the probability that when 50 people are randomly selected, at least 3 of them have the same birth date. Describe the simulation and estimate the probability.

17. Genetics: Simulating Population Control A classical probability problem involves a king who wanted to increase the proportion of women by decreeing that after a mother gives birth to a son, she is prohibited from having any more children. The king reasons that some families will have just one boy, whereas other families will have a few girls and one boy, so the proportion of girls will be increased. Is his reasoning correct? Will the proportion of girls increase?

4-7 Counting

Key Concept In many probability problems, the big obstacle is finding the total number of outcomes, and this section presents several different methods for finding such numbers. For example, California's Fantasy 5 lottery involves the selection of five different (whole) numbers between 1 and 39 inclusive. Because winning the jackpot requires that you select the same five numbers that are drawn when the lottery is run, the probability of winning the jackpot is 1 divided by the number of different possible ways to select five numbers out of 39. This section presents methods for finding numbers of outcomes, such as the number of different possible ways to select five numbers between 1 and 39.

This section introduces different methods for finding numbers of different possible outcomes without directly listing and counting the possibilities. We begin with the *fundamental counting rule.*

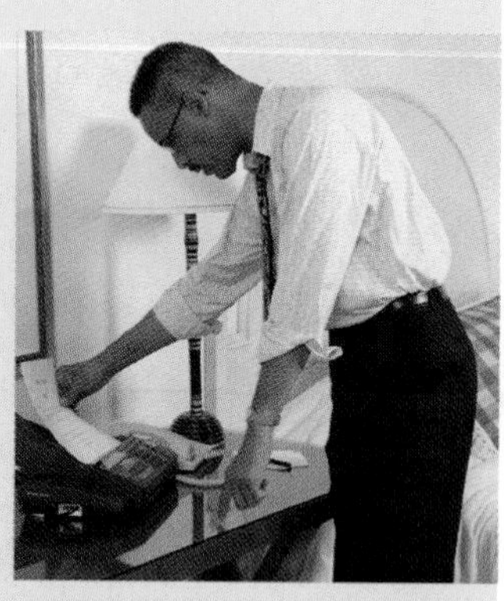

The Phone Number Crunch

Telephone companies often split regions with one area code into regions with two or more area codes because new fax and Internet lines have nearly exhausted the possible numbers that can be listed under a single code. Because a seven-digit telephone number cannot begin with a 0 or 1, there are $8 \cdot 10 \cdot 10 \cdot 10 \cdot 10 \cdot 10 \cdot 10 = 8{,}000{,}000$ different possible telephone numbers.

Before cell phones, fax machines, and the Internet, all toll-free numbers had a prefix of 800. Those 800 numbers lasted for 29 years before they were all assigned. The 888 prefix was introduced to help meet the demand for toll-free numbers, but it was estimated that it would take only 2.5 years for the 888 numbers to be exhausted. Next up: toll-free numbers with a prefix of 877. The counting techniques of this section are used to determine the number of different possible toll-free numbers with a given prefix, so that future needs can be met.

Fundamental Counting Rule

For a sequence of two events in which the first event can occur m ways and the second event can occur n ways, the events together can occur a total of $m \cdot n$ ways.

The fundamental counting rule easily extends to situations involving more than two events, as illustrated in the following examples.

EXAMPLE Identity Theft It is a good practice to not reveal social security numbers, because they are often used by criminals attempting identity theft that allows them to use other people's money. Assume that a criminal is found using your social security number and claims that all of the digits were randomly generated. What is the probability of getting your social security number when randomly generating nine digits? Is the criminal's claim that your number was randomly generated likely to be true?

SOLUTION Each of the 9 digits has 10 possible outcomes: 0, 1, 2, . . . , 9. By applying the fundamental counting rule, we get

$$10 \cdot 10 \cdot 10 \cdot 10 \cdot 10 \cdot 10 \cdot 10 \cdot 10 \cdot 10 = 1{,}000{,}000{,}000$$

Only one of those 1,000,000,000 possibilities corresponds to your social security number, so the probability of randomly generating a social security number and getting yours is $1/1{,}000{,}000{,}000$. It is extremely unlikely that a criminal would generate your social security by chance, assuming that only one social security number is generated. (Even if the criminal could generate thousands of social security numbers and try to use them, it is highly unlikely that your number would be generated.) If someone is found using your social security number, it was probably accessed through some other means, such as spying on Internet transactions or searching through your mail or garbage.

EXAMPLE Cotinine in Smokers Data Set 4 in Appendix B lists measured cotinine levels for a sample of people from each of three groups: smokers (denoted here by S), nonsmokers who were exposed to tobacco smoke (denoted by E), and nonsmokers not exposed to tobacco smoke (denoted by N). When nicotine is absorbed by the body, cotinine is produced. If we calculate the mean cotinine level for each of the three groups, then arrange those means in order from low to high, we get the sequence NES. An antismoking lobbyist claims that this is evidence that tobacco smoke is unhealthy, because the presence of cotinine increases as exposure to and use of tobacco increase. How many ways can the three groups denoted by N, E, and S be arranged? If an arrangement is selected at random, what is the probability of getting the sequence of NES? Is the probability low enough to conclude that the sequence of NES indicates that the presence of cotinine increases as exposure to and use of tobacco increase?

SOLUTION In arranging sequences of the groups N, E, and S, there are 3 possible choices for the first group, 2 remaining choices for the second group, and only 1 choice for the third group. The total number of possible arrangements is therefore

$$3 \cdot 2 \cdot 1 = 6$$

There are six different ways to arrange the N, E, and S groups. (They can be listed as NES, NSE, ESN, ENS, SNE, and SEN.) If we randomly select one of the six possible sequences, there is a probability of 1/6 that the sequence NES is obtained. Because that probability of 1/6 (or 0.167) is relatively high, we know that the sequence of NES could easily occur by chance. The probability is not low enough to conclude that the sequence of NES indicates that the presence of cotinine increases as exposure to and use of tobacco increase. We would need a smaller probability, such as 0.01.

In the preceding example, we found that 3 groups can be arranged $3 \cdot 2 \cdot 1 = 6$ different ways. This particular solution can be generalized by using the following notation for the symbol ! and the following *factorial rule*.

Notation

The **factorial symbol !** denotes the product of decreasing positive whole numbers. For example, $4! = 4 \cdot 3 \cdot 2 \cdot 1 = 24$. By special definition, $0! = 1$. (Many calculators have a factorial key. On the TI-83/84 Plus calculator, first enter the number, press MATH and select **PRB,** and then select menu item 4 and press ENTER.)

TI-83/84 Plus

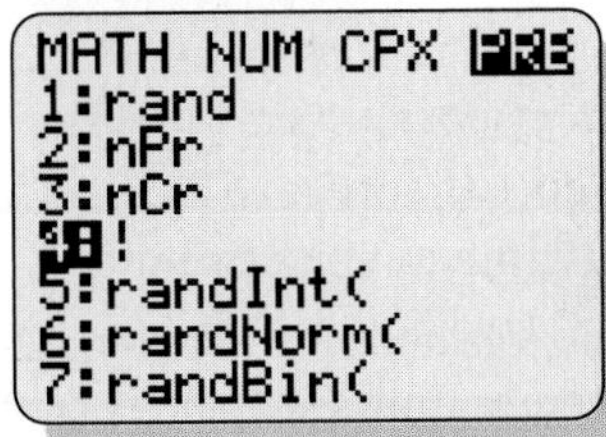

Factorial Rule

A collection of n different items can be arranged in order $n!$ different ways. (This **factorial rule** reflects the fact that the first item may be selected n different ways, the second item may be selected $n - 1$ ways, and so on.)

Making Cents of the Lottery

Many people spend large sums of money buying lottery tickets, even though they don't have a realistic sense for their chances of winning. Brother Donald Kelly of Marist College suggests this analogy: Winning the lottery is equivalent to correctly picking the "winning" dime from a stack of dimes that is 21 miles tall! Commercial aircraft typically fly at altitudes of 6 miles, so try to image a stack of dimes more than three times higher than those high-flying jets, then try to imagine selecting the one dime in that stack that represents a winning lottery ticket. Using the methods of this section, find the probability of winning your state's lottery, then determine the height of the corresponding stack of dimes.

Choosing Personal Security Codes

All of us use personal security codes for ATM machines, computer Internet accounts, and home security systems. The safety of such codes depends on the large number of different possibilities, but hackers now have sophisticated tools that can largely overcome that obstacle. Researchers found that by using variations of the user's first and last names along with 1800 other first names, they could identify 10% to 20% of the passwords on typical computer systems. When choosing a password, *do not* use a variation of any name, a word found in a dictionary, a password shorter than seven characters, telephone numbers, or social security numbers. Do include nonalphabetic characters, such as digits or punctuation marks.

Routing problems often involve application of the factorial rule. Verizon wants to route telephone calls through the shortest networks. Federal Express wants to find the shortest routes for its deliveries. American Airlines wants to find the shortest route for returning crew members to their homes. See the following example.

EXAMPLE Routes to Rides You are planning a trip to Disney World and you want to get through these five rides the first day: Space Mountain, Tower of Terror, Rock 'n' Roller Coaster, Mission Space, and Dinosaur. The rides can sometimes have long waiting times that vary throughout the day, so planning an efficient route could help maximize the pleasure of the day. How many different routes are possible?

SOLUTION By applying the factorial rule, we know that 5 different rides can be arranged in order 5! different ways. The number of different routes is $5! = 5 \cdot 4 \cdot 3 \cdot 2 \cdot 1 = 120$.

The preceding example is a variation of a classical problem called the *traveling salesman problem.* Because routing problems are so important to so many different companies and because the number of different routes can be very large, there is a continuing effort to simplify the method of finding the most efficient routes.

According to the factorial rule, n different items can be arranged $n!$ different ways. Sometimes we have n different items, but we need to select *some* of them instead of all of them. For example, if we must conduct surveys in state capitals, but we have time to visit only four capitals, the number of different possible routes is $50 \cdot 49 \cdot 48 \cdot 47 = 5{,}527{,}200$. Another way to obtain this same result is to evaluate

$$\frac{50!}{46!} = 50 \cdot 49 \cdot 48 \cdot 47 = 5{,}527{,}200$$

In this calculation, note that the factors in the numerator divide out with the factors in the denominator, except for the factors of 50, 49, 48, and 47 that remain. We can generalize this result by noting that if we have n different items available and we want to select r of them, the number of different arrangements possible is $n!/(n-r)!$ as in $50!/46!$. This generalization is commonly called the *permutations rule.*

Permutations Rule (When Items Are All Different)

Requirements

1. There are n *different* items available. (This rule does not apply if some of the items are identical to others.)
2. We select r of the n items (without replacement).
3. We consider rearrangements of the same items to be different sequences. (The permutation of *ABC* is different from *CBA* and is counted separately)

If the preceding requirements are satisfied, the number of **permutations** (or sequences) of r items selected from n different available items (without replacement) is

$${}_nP_r = \frac{n!}{(n-r)!}$$

When we use the term *permutations, arrangements,* or *sequences,* we imply that *order is taken into account* in the sense that different orderings of the same items are counted separately. The letters *ABC* can be arranged six different ways: *ABC, ACB, BAC, BCA, CAB, CBA*. (Later, we will refer to *combinations,* which do not count such arrangements separately.) In the following example, we are asked to find the total number of different sequences that are possible. That suggests use of the permutations rule.

How Many Shuffles?

After conducting extensive research, Harvard mathematician Persi Diaconis found that it takes seven shuffles of a deck of cards to get a complete mixture. The mixture is complete in the sense that all possible arrangements are equally likely. More than seven shuffles will not have a significant effect, and fewer than seven are not enough. Casino dealers rarely shuffle as often as seven times, so the decks are not completely mixed. Some expert card players have been able to take advantage of the incomplete mixtures that result from fewer than seven shuffles.

EXAMPLE Clinical Trial of New Drug When testing a new drug, Phase I involves only 8 volunteers, and the objective is to assess the drug's safety. To be very cautious, you plan to treat the 8 subjects in sequence, so that any particularly adverse effect can allow for stopping the treatments before any other subjects are treated. If 10 volunteers are available and 8 of them are to be selected, how many different sequences of 8 subjects are possible?

SOLUTION We have $n = 10$ different subjects available, and we plan to select $r = 8$ of them without replacement. The number of different sequences of arrangements is found as shown:

$$_nP_r = \frac{n!}{(n-r)!} = \frac{10!}{(10-8)!} = 1,814,400$$

There are 1,814,400 different possible arrangements of 8 subjects selected from the 10 that are available. The size of that result indicates that it is not practical to list the sequences or somehow consider each one of them individually.

We sometimes need to find the number of permutations, but some of the items are identical to others. The following variation of the permutations rule applies to such cases.

Permutations Rule (When Some Items Are Identical to Others)

Requirements

1. There are n items available, and some items are identical to others.
2. We select all of the n items (without replacement).
3. We consider rearrangements of distinct items to be different sequences.

If the preceding requirements are satisfied, and if there are n_1 alike, n_2 alike, . . . , n_k alike, the number of **permutations** (or sequences) of all items selected without replacement is

$$\frac{n!}{n_1!n_2! \cdots n_k!}$$

The Random Secretary

One classical problem of probability goes like this: A secretary addresses 50 different letters and envelopes to 50 different people, but the letters are randomly mixed before being put into envelopes. What is the probability that at least one letter gets into the correct envelope? Although the probability might seem like it should be small, it's actually 0.632. Even with a million letters and a million envelopes, the probability is 0.632. The solution is beyond the scope of this text—way beyond.

EXAMPLE Gender Selection The classical examples of the permutations rule are those showing that the letters of the word *Mississippi* can be arranged 34,650 different ways and that the letters of the word *statistics* can be arranged 50,400 ways. We will consider a different application.

In designing a test of a gender-selection method with 10 couples, a researcher knows that there are 1024 different possible sequences of genders when 10 babies are born. (Using the fundamental counting rule, the number of possibilities is $2 \cdot 2 \cdot 2 \cdot 2 \cdot 2 \cdot 2 \cdot 2 \cdot 2 \cdot 2 \cdot 2 = 1024$.) Ten couples use a method of gender selection with the result that their 10 babies consist of 8 girls and 2 boys.

a. How many ways can 8 girls and 2 boys be arranged in sequence?

b. What is the probability of getting 8 girls and 2 boys among 10 births?

c. Is that probability from part (b) useful for assessing the effectiveness of the method of gender selection?

SOLUTION

a. We have $n = 10$ births, with $n_1 = 8$ alike (girls) and $n_2 = 2$ others (boys) that are alike. The number of permutations is computed as follows:

$$\frac{n!}{n_1!n_2!} = \frac{10!}{8!2!} = \frac{3{,}628{,}800}{80{,}640} = 45$$

b. Because there are 45 different ways that 8 girls and 2 boys can be arranged, and there is a total of 1024 different possible arrangements, the probability of 8 females and 2 males is given by $P(\text{8 females and 2 males}) = 45/1024 = 0.0439$.

c. The probability of 0.0439 is *not* the probability that should be used in assessing the effectiveness of the gender-selection method. Instead of the probability of 8 females in 10 births, we should consider the probability of 8 *or more* females in 10 births, which is 0.0547. (In Section 5-2 we will clarify the reason for using the probability of *8 or more* females instead of the probability of *exactly 8* females in 10 births.)

The preceding example involved n items, each belonging to one of two categories. When there are only two categories, we can stipulate that x of the items are alike and the other $n - x$ items are alike, so the permutations formula simplifies to

$$\frac{n!}{(n-x)!x!}$$

This particular result will be used for the discussion of binomial probabilities, which are introduced in Section 5-3.

When we intend to select r items from n different items but *do not take order into account,* we are really concerned with possible combinations rather than permutations. That is, **when different orderings of the same items are counted separately, we have a permutation problem, but when different orderings of the same items are not counted separately, we have a combination problem** and may apply the following rule.

Combinations Rule

Requirements

1. There are *n different* items available.
2. We select *r* of the *n* items (without replacement).
3. We consider rearrangements of the same items to be the same. (The combination *ABC* is the same as *CBA*.)

If the preceding requirements are satisfied, the number of **combinations** of *r* items selected from *n* different items is

$$_nC_r = \frac{n!}{(n-r)!r!}$$

Because choosing between the permutations rule and the combinations rule can be confusing, we provide the following example, which is intended to emphasize the difference between them.

EXAMPLE Phase I of a Clinical Trial When testing a new drug on humans, a clinical test is normally done in three phases. Phase I is conducted with a relatively small number of healthy volunteers. Let's assume that we want to treat 8 healthy humans with a new drug, and we have 10 suitable volunteers available.

a. If the subjects are selected and treated *in sequence*, so that the trial is discontinued if anyone presents with a particularly adverse reaction, how many different sequential arrangements are possible if 8 people are selected from the 10 that are available?

b. If 8 subjects are selected from the 10 that are available, and the 8 selected subjects are all treated at the same time, how many different treatment groups are possible?

SOLUTION Note that in part (a), order is relevant because the subjects are treated sequentially and the trial is discontinued if anyone exhibits a particularly adverse reaction. However, in part (b) the order of selection is irrelevant because all of the subjects are treated at the same time.

a. Because order does count, we want the number of *permutations* of $r = 8$ people selected from the $n = 10$ available people. In a preceding example in this section, we found that the number of permutations is 1,814,400.

b. Because order does *not* count, we want the number of *combinations* of $r = 8$ people selected from the $n = 10$ available people. We get

$$_nC_r = \frac{n!}{(n-r)!r!} = \frac{10!}{(10-8)!8!} = 45$$

With order taken into account, there are 1,814,400 permutations, but without order taken into account, there are 45 combinations.

Too Few Bar Codes

In 1974, a pack of gum was the first item to be scanned in a supermarket. That scanning required that the gum be identified with a bar code. Bar codes or Universal Product Codes are used to identify individual items to be purchased. Bar codes used 12 digits that allowed scanners to automatically list and record the price of each item purchased. The use of 12 digits became insufficient as the number of different products increased, so the codes were recently modified to include 13 digits.

Similar problems are encountered when telephone area codes are split because there are too many different telephones for one area code in a region. Methods of counting are used to design systems to accommodate future numbers of units that must be processed or served.

This section presented these five different tools for finding total numbers of outcomes: fundamental counting rule, factorial rule, permutations rule, permutations rule when some items are identical, and the combinations rule. Not all counting problems can be solved with one of these five rules, but they do provide a strong foundation for many real and important applications.

TI

Using a TI-83/84 Plus Calculator for Factorials, Permutations, and Combinations

Factorials First enter the number, then press MATH, select **PRB,** then select menu item 4 (which is the ! symbol) and press ENTER. For example, 5! is evaluated with MATH **PRB** 4 ENTER, and the result is 120.

Permutations When evaluating an expression of the form ${}_nP_r$, follow these steps:

1. First enter the value of n.
2. Press MATH, then select **PRB,** then select the second item in the menu, which is ${}_nP_r$. Press ENTER.
3. Now enter the value of r and press ENTER.

For example, to evaluate ${}_9P_3$, first enter 9, then execute Step 2 above, then enter 3 and press ENTER to obtain a result of 504. That is, ${}_9P_3 = 504$.

Combinations When evaluating an expression of the form ${}_nC_r$, follow these steps:

1. First enter the value of n.
2. Press MATH, then select **PRB,** then select the third item in the menu, which is ${}_nC_r$ Press ENTER.
3. Now enter the value of r and press ENTER.

For example, to evaluate ${}_9C_3$, first enter 9, then execute Step 2 above, then enter 3 and press ENTER to obtain a result of 84. That is, ${}_9C_3 = 84$.

4-7 BASIC SKILLS AND CONCEPTS

Statistical Literacy and Critical Thinking

1. Permutations and Combinations What is the basic difference between a situation requiring application of the permutations rule and one that requires the combinations rule?

2. Counting When trying to find the probability of winning the California Fantasy 5 lottery, it becomes necessary to find the number of different outcomes that can occur when 5 numbers between 1 and 39 are selected. Why can't that number be found by simply listing all of the possibilities?

3. Relative Frequency A researcher is analyzing a large sample of text in order to find the relative frequency of the word "zip" among three-letter words. That is, she wants to estimate the probability of getting the word "zip" when a three-letter word is

randomly selected from typical English text. Can that probability be found by using the methods of this section?

4. Probability Someone reasons that when a coin is tossed, there are three possible outcomes: It comes up heads or tails or it lands on its edge. With three outcomes on each toss, the fundamental counting rule suggests that there are nine possibilities (from $3 \cdot 3 = 9$) for two tosses of a coin. It therefore follows that the probability of two heads in two tosses is $1/9$. Is this reasoning correct? If not, what is wrong?

Calculating Factorials, Combinations, Permutations. *In Exercises 5–12, evaluate the given expressions and express all results using the usual format for writing numbers (instead of scientific notation).*

TI **5.** $5!$ TI **6.** $8!$ TI **7.** ${}_{24}C_4$ TI **8.** ${}_{24}P_4$

TI **9.** ${}_{52}P_2$ TI **10.** ${}_{52}C_2$ TI **11.** ${}_{30}C_3$ TI **12.** ${}_{10}P_3$

Probability of Winning the Lottery *Because the California Fantasy 5 lottery is won by selecting the correct five numbers (in any order) between 1 and 39, there are 575,757 different 5-number combinations that could be played, and the probability of winning this lottery is 1/575,757. In Exercises 13–16, find the probability of winning the indicated lottery.*

TI **13.** Massachusetts Mass Cash Lottery Select the five winning numbers from 1, 2, . . . , 35.

TI **14.** New York Lotto Select the six winning numbers from 1, 2, . . . , 59.

TI **15.** Pennsylvania Lucky for Life Lotto Select the six winning numbers from 1, 2, . . . , 38.

TI **16.** Texas Cash Five Select the five winning numbers from 1, 2, . . . , 37.

TI **17.** California Fantasy 5 The California Fantasy 5 lotto is won by selecting the correct five numbers from 1, 2, . . . , 39. The probability of winning that game is 1/575,757. What is the probability of winning if the rules are changed so that in addition to selecting the correct five numbers, you must now select them in the same order as they are drawn?

TI **18.** DNA Nucleotides DNA (deoxyribonucleic acid) is made of nucleotides, and each nucleotide can contain any one of these nitrogenous bases: A (adenine), G (guanine), C (cytosine), T (thymine). If one of those four bases (A, G, C, T) must be selected three times to form a linear triplet, how many different triplets are possible? Note that all four bases can be selected for each of the three components of the triplet.

TI **19.** Age Discrimination The Cytertonics Communications Company reduced its management staff from 15 managers to 10. The company claimed that five managers were randomly selected for job termination. However, the five managers chosen are the five oldest managers among the 15 that were employed. Find the probability that when five managers are randomly selected from a group of 15, the five oldest are selected. Is that probability low enough to charge that instead of using random selection, the company actually fired the oldest employees?

TI **20.** Computer Design In designing a computer, if a *byte* is defined to be a sequence of 8 bits and each bit must be a 0 or 1, how many different bytes are possible? (A byte is often used to represent an individual character, such as a letter, digit, or punctuation symbol. For example, one coding system represents the letter *A* as 01000001.) Are there enough different bytes for the characters that we typically use, including lowercase letters, capital letters, digits, punctuation symbols, dollar sign, and so on?

TI **21.** Tree Growth Experiment When designing an experiment to study tree growth, the following four treatments are used: none, irrigation only, fertilization only, irrigation and fertilization. A row of 10 trees extends from a moist creek bed to a dry land area. If one of the four treatments is randomly assigned to each of the 10 trees, how many different treatment arrangements are possible?

TI **22.** Design of Experiments In designing an experiment involving a treatment applied to 12 test subjects, researchers plan to use a simple random sample of 12 subjects selected from a pool of 20 available subjects. (Recall that with a simple random sample, all samples of the same size have the same chance of being selected.) How many different simple random samples are possible? What is the probability of each simple random sample in this case?

TI **23.** Probability of Defective Pills A batch of pills consists of 7 that are good and 3 that are defective (because they contain the wrong amount of the drug).

a. How many different permutations are possible when all 10 pills are randomly selected (without replacement)?

b. If 3 pills are randomly selected without replacement, find the probability that all three of the defective pills are selected.

TI **24.** Air Routes You have just started your own airline company named Air Me (motto: "To us, you are not just another statistic"). So far, you have one plane for a route connecting Austin, Boise, and Chicago. One route is Austin–Boise–Chicago and a second route is Chicago–Boise–Austin. How many total routes are possible if service is expanded to include a total of eight cities?

TI **25.** Testing a Claim Mike claims that he has developed the ability to roll a 6 almost every time that he rolls a die. You test his claim by having Mike roll a die five times, and he gets a 6 each time. If Mike has no ability to affect the outcomes, find the probability that he will roll five consecutive 6s when a die is rolled five times. Is that probability low enough to rule out chance as an explanation for Mike's results?

TI **26.** Gender Selection In a test of a gender-selection method, 14 babies are born and 10 of them are girls.

a. Find the number of different possible sequences of genders that are possible when 14 babies are born.

b. How many ways can 10 girls and 4 boys be arranged in a sequence?

c. If 14 babies are randomly selected, what is the probability that they consist of 10 girls and 4 boys?

d. Does the gender-selection method appear to yield a result that is significantly different from a result that might be expected by random chance?

TI **27.** Elected Board of Directors There are 12 members on the board of directors for the Newport General Hospital.

a. If they must elect a chairperson, first vice chairperson, second vice chairperson, and secretary, how many different slates of candidates are possible?

b. If they must form an ethics subcommittee of four members, how many different subcommittees are possible?

TI **28.** Jumble Puzzle Many newspapers carry "Jumble," a puzzle in which the reader must unscramble letters to form words. For example, the letters TAISER were included in newspapers on the day this exercise was written. How many ways can the letters of TAISER be arranged? Identify the correct unscrambling, then determine the probability of getting that result by randomly selecting an arrangement of the given letters.

TI **29.** Finding the Number of Possible Melodies In Denys Parsons' *Directory of Tunes and Musical Themes,* melodies for more than 14,000 songs are listed according to the following scheme: The first note of every song is represented by an asterisk *, and successive notes are represented by *R* (for repeat the previous note), *U* (for a note that goes up), or *D* (for a note that goes down). Beethoven's Fifth Symphony begins as **RRD*. Classical melodies are represented through the first 16 notes. With this scheme, how many different classical melodies are possible?

TI **30.** Combination Locks A typical "combination" lock is opened with the correct sequence of three numbers between 0 and 49 inclusive. (A number can be used more than once.) What is the probability of guessing those three numbers and opening the lock with the first try?

TI **31.** Finding the Number of Area Codes *USA Today* reporter Paul Wiseman described the old rules for the three-digit telephone area codes by writing about "possible area codes with 1 or 0 in the second digit. (Excluded: codes ending in 00 or 11, for toll-free calls, emergency services, and other special uses.)" Codes beginning with 0 or 1 should also be excluded. How many different area codes were possible under these old rules?

TI **32.** Cracked Eggs A carton contains 12 eggs, 3 of which are cracked. If we randomly select 5 of the eggs for hard boiling, what is the probability of the following events?

a. All of the cracked eggs are selected.

b. None of the cracked eggs are selected.

c. Two of the cracked eggs are selected.

TI **33.** NCAA Basketball Tournament Each year, 64 college basketball teams compete in the NCAA tournament. Sandbox.com recently offered a prize of $10 million to anyone who could correctly pick the winner in each of the tournament games. The president of that company also promised that, in addition to the cash prize, he would eat a bucket of worms. Yuck.

a. How many games are required to get one championship team from the field of 64 teams?

b. If someone makes random guesses for each game of the tournament, find the probability of picking the winner in each game.

c. In an article about the $10 million prize, *The New York Times* wrote that "Even a college basketball expert who can pick games at a 70 percent clip has a 1 in __________ chance of getting all the games right." Fill in the blank.

TI **34.** ATM Machine You want to obtain cash by using an ATM machine, but it's dark and you can't see your card when you insert it. The card must be inserted with the front side up and the printing configured so that the beginning of your name enters first.

a. What is the probability of selecting a random position and inserting the card, with the result that the card is inserted correctly?

b. What is the probability of randomly selecting the card's position and finding that it is incorrectly inserted on the first attempt, but it is correctly inserted on the second attempt?

c. How many random selections are required to be absolutely sure that the card works because it is inserted correctly?

TI **35.** California Lottery In California's Super Lotto Plus lottery game, winning the jackpot requires that you select the correct five numbers between 1 and 47 inclusive and, in a separate drawing, you must also select the correct single number between 1 and 27 inclusive. Find the probability of winning the jackpot.

TI **36.** Power Ball Lottery The Power Ball lottery is run in 27 states. Winning a Power Ball lottery jackpot requires that you select the correct five numbers between 1 and 53 inclusive and, in a separate drawing, you must also select the correct single number between 1 and 42 inclusive. Find the probability of winning the jackpot.

4-7 BEYOND THE BASICS

TI **37.** Finding the Number of Computer Variable Names A common computer programming rule is that names of variables must be between 1 and 8 characters long. The first character can be any of the 26 letters, while successive characters can be any of the 26 letters or any of the 10 digits. For example, allowable variable names are A, BBB, and M3477K. How many different variable names are possible?

38. Handshakes and Round Tables

a. Five managers gather for a meeting. If each manager shakes hands with each other manager exactly once, what is the total number of handshakes?

b. If n managers shake hands with each other exactly once, what is the total number of handshakes?

c. How many different ways can five managers be seated at a round table? (Assume that if everyone moves to the right, the seating arrangement is the same.)

d. How many different ways can n managers be seated at a round table?

TI **39.** Evaluating Large Factorials Many calculators or computers cannot directly calculate 70! or higher. When n is large, $n!$ can be approximated by $n = 10^k$, where $K = (n + 0.5) \log n + 0.39908993 - 0.43429448n$.

a. You have been hired to visit the capitol of each of the 50 states. How many different routes are possible? Evaluate the answer using the factorial key on a calculator and also by using the approximation given here.

b. The Bureau of Fisheries once asked Bell Laboratories for help finding the shortest route for getting samples from 300 locations in the Gulf of Mexico. If you compute the number of different possible routes, how many digits are used to write that number?

TI **40.** Computer Intelligence Can computers "think"? According to the *Turing test,* a computer can be considered to think if, when a person communicates with it, the person believes he or she is communicating with another person instead of a computer. In an experiment at Boston's Computer Museum, each of 10 judges communicated with four computers and four other people and was asked to distinguish between them.

a. Assume that the first judge cannot distinguish between the four computers and the four people. If this judge makes random guesses, what is the probability of correctly identifying the four computers and the four people?

b. Assume that all 10 judges cannot distinguish between computers and people, so they make random guesses. Based on the result from part (a), what is the probability that all 10 judges make all correct guesses? (That event would lead us to conclude that computers cannot "think" when, according to the Turing test, they can.)

41. Change for a Dollar How many different ways can you make change for a dollar?

4-8 Bayes' Theorem (on CD-ROM)

The CD-ROM enclosed in this book includes another section dealing with conditional probability. This additional section discusses applications of *Bayes' theorem* (or *Bayes' rule*), which we use for revising a probability value based on additional information that is later obtained. See the CD-ROM for the discussion, examples, and exercises describing applications of Bayes' theorem.

Review

We began this chapter with the basic concept of probability, which is so important for methods of inferential statistics introduced later in this book. The single most important concept to learn from this chapter is the rare event rule for inferential statistics: If, under a given assumption, the probability of a particular event is extremely small, we conclude that the assumption is probably not correct. As an example of the basic approach used, consider a test of a method of gender selection. If we conduct a trial of a gender-selection technique and get 20 girls in 20 births, we can make one of two inferences from these sample results:

1. The technique of gender selection is not effective, and the string of 20 consecutive girls is an event that could easily occur by chance.
2. The technique of gender selection is effective (or there is some other explanation for why boys and girls are not occurring with the same frequencies).

Statisticians use the rare event rule when deciding which inference is correct: In this case, the probability of getting 20 consecutive girls is so small (1/1,048,576) that the inference of an effective technique of gender selection is the better choice. Here we can see the important role played by probability in the standard methods of statistical inference.

In Section 4-2 we presented the basic definitions and notation, including the representation of events by letters such as A. We should know that a probability value, which is expressed as a number between 0 and 1, reflects the likelihood of some event. We defined probabilities of simple events as

$$P(A) = \frac{\text{number of times that A occurred}}{\text{number of times trial was repeated}} \qquad \text{(relative frequency)}$$

$$P(A) = \frac{\text{number of ways A can occur}}{\text{number of different simple events}} = \frac{s}{n} \qquad \text{(for equally likely outcomes)}$$

We noted that the probability of any impossible event is 0, the probability of any certain event is 1, and for any event A, $0 \leq P(A) \leq 1$. Also, $\overline{A}$ denotes the complement of event A. That is, $\overline{A}$ indicates that event A does *not* occur.

In Sections 4-3, 4-4, and 4-5 we considered compound events, which are events combining two or more simple events. We associate use of the word "or" with addition and associate use of the word "and" with multiplication. Always keep in mind the following key considerations:

- When conducting one trial, do we want the probability of event A *or* B? If so, use the addition rule, but be careful to avoid counting any outcomes more than once.

- When finding the probability that event A occurs on one trial *and* event B occurs on a second trial, use the multiplication rule. Multiply the probability of event A by the probability of event B. *Caution:* When calculating the probability of event B, be sure to take into account the fact that event A has already occurred.

Section 4-6 described simulation techniques that are often helpful in determining probability values, especially in situations where formulas or theoretical calculations are extremely difficult.

In some probability problems, the biggest obstacle is finding the total number of possible outcomes. Section 4-7 was devoted to the following counting techniques:

- Fundamental counting rule
- Factorial rule
- Permutations rule (when items are all different)
- Permutations rule (when some items are identical to others)
- Combinations rule

Statistical Literacy and Critical Thinking

1. Probability Value A statistics student reports that when tossing a fair coin, the probability that the coin turns up heads is 50–50. What is wrong with that statement? What is the correct statement?

2. Interpreting Probability Value Medical researchers conduct a clinical trial of a new drug designed to lower cholesterol. They determine that there is a 0.27 probability that their results could occur by chance. Based on that probability value, can chance be ruled out as a reasonable explanation? Why or why not?

3. Probability of Life on Alfa Romeo Astronomers identify a new planet in a solar system far, far away. An astronomer reasons that life either exists on this new planet or does not. Because there are two outcomes (life exists, life does not exist), he concludes that the probability of life on this planet is $1/2$ or 0.5. Is this reasoning correct? Why or why not?

4. Disjoint Events and Independent Events What does it mean when we say that two events are disjoint? What does it mean when we say that two events are independent?

Review Exercises

Clinical Test of Lipitor. In Exercises 1–8, use the data in the accompanying table (based on data from Parke-Davis). The cholesterol-reducing drug Lipitor consists of atorvastatin calcium.

	Treatment	
	10-mg Atorvastatin	Placebo
Headache	15	65
No headache	17	3

1. If 1 of the 100 subjects is randomly selected, find the probability of getting someone who had a headache.

2. If 1 of the 100 subjects is randomly selected, find the probability of getting someone who was treated with 10 mg of atorvastatin.

3. If 1 of the 100 subjects is randomly selected, find the probability of getting someone who had a headache or was treated with 10 mg of atorvastatin.

4. If 1 of the 100 subjects is randomly selected, find the probability of getting someone who was given a placebo or did not have a headache.

5. If two different subjects are randomly selected, find the probability that they both used placebos.

6. If two different subjects are randomly selected, find the probability that they both had a headache.

7. If one subject is randomly selected, find the probability that he or she had a headache, given that the subject was treated with 10 mg of atorvastatin.

8. If one subject is randomly selected, find the probability that he or she was treated with 10 mg of atorvastatin, given that the subject had a headache.

9. National Statistics Day
   a. If a person is randomly selected, find the probability that his or her birthday is October 18, which is National Statistics Day in Japan. Ignore leap years.
   b. If a person is randomly selected, find the probability that his or her birthday is in October. Ignore leap years.
   c. Estimate a subjective probability for the event of randomly selecting an adult American and getting someone who knows that October 18 is National Statistics Day in Japan.
   d. Is it unusual to randomly select an adult American and get someone who knows that October 18 is National Statistics Day in Japan?

10. Fruitcake Survey In a Bruskin-Goldring Research poll, respondents were asked how a fruitcake should be used. The respondents consist of 132 people indicating that it should be used for a doorstop, and 880 other people who gave other uses, including birdfeed, landfill, and a gift. If one of these respondents is randomly selected, what is the probability of getting someone who would use the fruitcake as a doorstop?

11. Testing a Claim The Biogene Research Company claims that it has developed a technique for ensuring that a baby will be a girl. In a test of that technique, 12 couples all have baby girls. Find the probability of getting 12 baby girls by chance, assuming that boys and girls are equally likely and that the gender of any child is independent of the others. Does that result appear to support the company's claim?

12. Life Insurance The New England Life Insurance Company issues one-year policies to 12 men who are all 27 years of age. Based on data from the Department of Health and Human Services, each of these men has a 99.82% chance of living through the year. What is the probability that they all survive the year?

TI 13. Electrifying When testing for electrical current in a cable with five color-coded wires, the author used a meter to test two wires at a time. What is the probability that the two live wires are located with the first random selection of two wires?

14. Acceptance Sampling With one method of acceptance sampling, a sample of items is randomly selected without replacement, and the entire batch is rejected if there is at least one defect. The Medtyme Pharmaceutical Company has just manufactured 2500 aspirin tablets, and 2% are defective because they contain too much or too little

aspirin. If 4 of the tablets are selected and tested, what is the probability that the entire batch will be rejected?

15. Chlamydia Rate For a recent year, the rate of chlamydia was reported as 278.32 per 100,000 population.

a. Find the probability that a randomly selected person has chlamydia.

b. If two people are randomly selected, find the probability that they both have chlamydia, and express the result using three significant digits.

c. If two people are randomly selected, find the probability that neither of them have chlamydia, and express the result using seven decimal places.

TI **16.** Bar Codes On January 1, 2005, the bar codes put on retail products were changed so that they now represent 13 digits instead of 12. How many different products can now be identified with the new bar codes?

Cumulative Review Exercises

TI **1.** Treating Chronic Fatigue Syndrome Patients suffering from chronic fatigue syndrome were treated with medication, then their change in fatigue was measured on a scale from -7 to $+7$, with positive values representing improvement and 0 representing no change. The results are listed below (based on data from "The Relationship Between Neurally Mediated Hypotension and the Chronic Fatigue Syndrome," by Bou-Holaigah, Rowe, Kan, and Calkins, *Journal of the American Medical Association*, Vol. 274, No. 12.)

6 5 0 5 6 7 3 3 2 4 4 0 7 3 4 3 6 0 5 5 6

a. Find the mean.

b. Find the median.

c. Find the standard deviation.

d. Find the variance.

e. Based on the results, does it appear that the treatment was effective?

f. If one value is randomly selected from this sample, find the probability that it is positive.

g. If two different values are randomly selected from this sample, find the probability that they are both positive.

h. Ignore the three values of 0 and assume that only positive or negative values are possible. Assuming that the treatment is ineffective and that positive and negative values are equally likely, find the probability that 18 subjects all have positive values (as in this sample group). Is that probability low enough to justify rejection of the assumption that the treatment is ineffective? Does the treatment appear to be effective?

TI **2.** High Temperatures The actual high temperatures (in degrees Fahrenheit) for September are described with this 5-number summary: 62, 72, 76, 80, 85. (The values are based on Data Set 8 in Appendix B.) Use these values from the 5-number summary to answer the following:

a. What is the median?

b. If a high temperature is found for some day randomly selected in some September, find the probability that it is between 72°F and 76°F.

c. If a high temperature is obtained for some day randomly selected in some September, find the probability that it is below 72°F or above 76°F.

d. If two different days are randomly selected from September, find the probability that they are both days with high temperatures between 72°F and 76°F.

e. If two *consecutive* days in September are randomly selected, are the events of getting both high temperatures above 80°F independent? Why or why not?

Cooperative Group Activities

1. In-class activity Divide into groups of three or four and use coin tossing to develop a simulation that emulates the kingdom that abides by this decree: After a mother gives birth to a son, she will not have any other children. If this decree is followed, does the proportion of girls increase?

2. In-class activity Divide into groups of three or four and use actual thumbtacks to estimate the probability that when dropped, a thumbtack will land with the point up. How many trials are necessary to get a result that appears to be reasonably accurate when rounded to the first decimal place?

3. Out-of-class activity Marine biologists often use the *capture-recapture method* as a way to estimate the size of a population, such as the number of fish in a lake. This method involves capturing a sample from the population, tagging each member in the sample, then returning them to the population. A second sample is later captured, and the tagged members are counted along with the total size of this second sample. The results can be used to estimate the size of the population.

Instead of capturing real fish, simulate the procedure using some uniform collection of items such as BB's, colored beads, M&Ms, Fruit Loop cereal pieces, or index cards. Start with a large collection of such items. Collect a sample of 50 and use a magic marker to "tag" each one. Replace the tagged items, mix the whole population, then select a second sample and proceed to estimate the population size. Compare the result to the actual population size obtained by counting all of the items.

4. In-class activity Divide into groups of two. Refer to Exercise 15 in Section 4-6 for a description of the "Monty Hall problem." Simulate the contest and record the results for sticking and switching, then determine which of those two strategies is better.

5. Out-of-class activity Divide into groups of two for the purpose of doing an experiment designed to show one approach to dealing with sensitive survey questions, such as those related to drug use, sexual activity (or inactivity), stealing, or cheating. Instead of actually using a controversial question that would reap wrath upon the author, we will use this innocuous question: "Were you born in a month that has the letter *r* in it?" About $2/3$ of all responses should be "yes," but let's pretend that the question is very sensitive and that survey subjects are reluctant to answer honestly. Survey people by asking them to flip a coin and respond as follows:

- Answer "yes" if the coin turns up tails *or* you were born in a month containing the letter *r*.
- Answer "no" if the coin turns up heads *and* you were born in a month not containing the letter *r*.

Supposedly, respondents tend to be more honest because the coin flip protects their privacy. Survey people and analyze the results to determine the proportion of people born in a month containing the letter *r*. The accuracy of the results could be checked against their actual birth dates, which can be obtained from a second question. The experiment could be repeated with a question that is more sensitive, but such a question is not given here because the author already receives enough mail.

TI-83/84 Plus Project
Using Simulations for Probabilities and Variation in Manufacturing

Students typically find that the topic of probability is the single most difficult topic in an introductory statistics course. Some probability problems might sound simple while their solutions are incredibly complex. In this chapter we have identified several basic and important rules commonly used for finding probabilities, but in this project we use a very different approach that can overcome much of the difficulty encountered with the application of formal rules. This alternative approach consists of developing a simulation, which is a process that behaves the same way as the procedure, so that similar results are produced. (See Section 4-6.)

In Exercise 11 from Section 4-6, we referred to a process of manufacturing cell phones. We assumed that a batch consists of 500 cell phones and the overall rate of defective cell phones is 2%. We can conduct a simulation by generating 500 numbers, with each number between 1 and 100 inclusive. Because the defect rate is 2%, we can consider any outcome of 1 or 2 to be a defective cell phone, while outcomes of 3, 4, 5, . . . , 100 represent good cell phones. The mean number of defects in batches of 500 should be 10. However, some batches will have exactly 10 defects, but some batches will have fewer than 10 defects, and other batches will have more than 10 defects.

a. Use a TI-83/84 Plus calculator to simulate the manufacture of 500 cell phones. Record the number of defects in this simulated batch. [*Hint:* It would be helpful to sort the results, so that the defects (represented by outcomes of 1 or 2) can be easily identified.]

b. Repeat part (a) 19 more times, so that a total of 20 simulated batches have been generated. List the number of defects in each of the 20 batches.

c. Using the results from part (b), estimate the probability that the number of defects in a batch is exactly 10. Do you think that this estimate is somewhat accurate? Why or why not?

d. Using the results from part (b), estimate the probability that the number of defects in a batch is exactly 9.

e. After examining the results from part (b), how much do the numbers of defects vary? Are the numbers of defects in batches somewhat predictable, or do they vary by large amounts?

f. A quality control engineer claims that a new manufacturing process reduces the numbers of defects, and a test of the new process results in a batch having no defects. Based on the results from part (b), does the absence of defects in a batch appear to suggest that the new method is better, or could random chance be a reasonable explanation for the absence of defective cell phones? Explain.

From Data to Decision

Critical Thinking: As a physician, what should you tell a woman after she has taken a test for pregnancy?

It is important for a woman to know if she becomes pregnant so that she can discontinue any activities, medications, exposure to toxicants at work, smoking, or alcohol consumption that could be potentially harmful to the baby. Pregnancy tests, like almost all health tests, do not yield results that are 100% accurate. In clinical trials of a blood test for pregnancy, the results shown in the accompanying table were obtained for the Abbot blood test (based on data from "Specificity and Detection Limit of Ten Pregnancy Tests," by Tiitinen and Stenman, *Scandinavion Journal of Clinical Laboratory Investigation*, Vol. 53, Supplement 216). Other tests are more reliable than the test with results given in this table.

Analyzing the Results

1. Based on the results in the table, what is the probability of a woman being pregnant if the test indicates a negative result? If you are a physician and you have a patient who tested negative, what advice would you give?
2. Based on the results in the table, what is the probability of a false positive? That is, what is the probability of getting a positive result if the woman is not actually pregnant? If you are a physician and you have a patient who tested positive, what advice would you give?

Pregnancy Test Results

	Positive Test Result (Pregnancy is indicated)	Negative Test Result (Pregnancy is not indicated)
Subject is pregnant	80	5
Subject is not pregnant	3	11

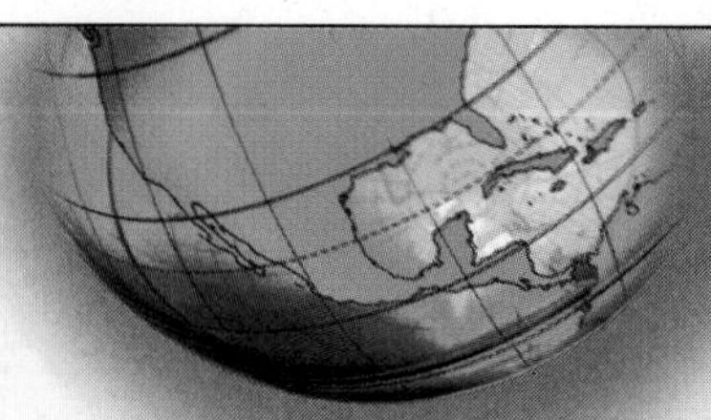

Internet Project

Computing Probabilities

Finding probabilities when rolling dice is easy. With one die, there are six possible outcomes, so each outcome, such as a roll of 2, has probability 1/6. For a card game the calculations are more involved, but they are still manageable. But what about a more complicated game, such as the board game Monopoly? What is the probability of landing on a particular space on the board? The probability depends on the space your piece currently occupies, the roll of the dice, the drawing of cards, as well as other factors. Now consider a more true-to-life example, such as the probability of having an auto accident. The number of factors involved is too large to even consider, yet such probabilities are nonetheless quoted, for example, by insurance companies.

The Internet Project for this chapter considers methods for computing probabilities in complicated situations. Go to the Internet Project for this chapter which can be found at this site:

http://www.aw.com/triola

You will be guided in the research of probabilities for a board game. Then you compute such probabilities yourself. Finally, you will estimate a health-related probability using empirical data.

Statistics @ Work

"We must have sound knowledge in statistical theory, good understanding of experimental designs, . . . , and how statistical thinking can be applied to the various stages of a drug development process."

Christy Chuang-Stein

Senior Director at Pfizer Inc.

Christy works as a statistical consultant within a group called the Statistical Research and Consulting Center (SRCC). All members of the SRCC work to provide strategic and tactical advice on issues related to statistical policy and applications within Pfizer. In addition, members of the SRCC conduct independent and collaborative research on problems that address the company's business needs.

How do you use statistics in your job and what specific statistical concepts do you use?

Working for a pharmaceutical company, we use statistics extensively to help support the discovery and development of new medical products. This includes identifying promising compounds, testing the compounds, investigating the safety and efficacy of product candidates in clinical trials, and manufacturing the products according to predetermined specifications. The statistical concepts we use include sampling, variability, efficiency, controlling for bias, reducing sources of variability, estimation of parameters, and hypothesis testing.

Please describe a specific example of how the use of statistics was useful in improving a product or service.

The attrition rate of compounds in the pharmaceutical industry is extremely high. Less than 12% of compounds entering the human testing phase will eventually make it to the marketplace. Because of the high attrition rate and the extraordinarily high cost of developing a pharmaceutical product, one important success factor is to make sound go/no go decisions on product candidates, and to do so as soon as possible. We have successfully used group sequential designs to help terminate trials early, because the trials are not likely to meet their objectives even if they were to continue. Stopping the trials early has allowed us to save resources that we can then apply to the development of other promising pharmaceutical products.

What do you find exciting, interesting, or rewarding about your work?

The most rewarding aspect of my work is the knowledge that by introducing new and innovative medicines, we are helping millions of people live a longer life with a higher quality. During the past 50 years, the value of medicine has been clearly demonstrated in applications such as diabetes, heart diseases, osteoporosis, cancer, HIV infection, schizophrenia, epilepsy, and childhood vaccines, just to name a few.

At your company, do you feel job applicants are viewed more favorably if they studied some statistics?

This depends on the job an individual is interviewing for. Even so, since statistical principles are applicable to so many non-statistical functions in a pharmaceutical company (such as portfolio evaluation, project management and improvement, study and data management, tracking of metrics), people with some training in statistics often excel in jobs that require quantitative skills and deductive reasoning. As a result, I would think applicants with some basic understanding of statistics will be viewed favorably in many areas within my company.

Do you recommend that today's college students study statistics? Why?

I would definitely recommend the study of statistics for students hoping to work in an environment that includes research and development activities. It is amazing how very basic concepts such as population, sample, variability, bias, and estimation are used with such a high frequency even in an average working environment.

Discrete Probability Distributions

5

Can statistical methods show that a jury selection process is discriminatory?

After a defendant has been convicted of some crime, appeals are sometimes filed on the grounds that the defendant was not convicted by a jury of his or her peers. One criterion is that the jury selection process should result in jurors that represent the population of the region. In one notable case, Dr. Benjamin Spock, who wrote the popular *Baby and Child Care* book, was convicted of conspiracy to encourage resistance to the draft during the Vietnam War. His defense argued that Dr. Spock was handicapped by the fact that all 12 jurors were men. Women would have been more sympathetic, because opposition to the war was greater among women and Dr. Spock was so well known as a baby doctor. A statistician testified that the presiding judge had a consistently lower proportion of women jurors than the other six judges in the same district. Dr. Spock's conviction was overturned for other reasons, but federal court jurors are now supposed to be randomly selected.

In 1972, Rodrigo Partida, a Mexican-American, was convicted of burglary with intent to commit rape. His conviction took place in Hidalgo County, which is in Texas on the border with Mexico. Hidalgo County had 181,535 people eligible for jury duty, and 80% of them were Mexican-American. (Because the author recently renewed his poetic license, he will use 80% throughout this chapter instead of the more accurate value of 79.1%.) Among 870 people selected for grand jury duty, 39% (339) were Mexican-American. Partida's conviction was later appealed (*Castaneda v. Partida*) on the basis of the large discrepancy between the 80% of the Mexican-Americans eligible for grand jury duty and the fact that only 39% of Mexican-Americans were actually selected.

We will consider the *Castaneda v. Partida* issue in this chapter. Here are key questions that will be addressed:

1. Given that Mexican-Americans constitute 80% of the population, and given that Partida was convicted by a jury of 12 people with only 58% of them (7 jurors) that were Mexican-American, can we conclude that his jury was selected in a process that discriminates against Mexican-Americans?
2. Given that Mexican-Americans constitute 80% of the population of 181,535 and, over a period of 11 years, only 39% of those selected for grand jury duty were Mexican-Americans, can we conclude that the process of selecting grand jurors discriminated against Mexican-Americans? (We know that because of random chance, samples naturally vary somewhat from what we might theoretically expect. But is the discrepancy between the 80% rate of Mexican-Americans in the population and the 39% rate of Mexican-Americans selected for grand jury duty a discrepancy that is just too large to be explained by chance?)

This example illustrates well the importance of a basic understanding of statistical methods in the field of law. Attorneys with no statistical background might not be able to serve some of their clients well. The author once testified in New York State Supreme Court and observed from his cross-examination that a lack of understanding of basic statistical concepts can be very detrimental to an attorney's client.

5-1 Overview

In this chapter we combine the methods of *descriptive statistics* presented in Chapters 2 and 3 and those of *probability* presented in Chapter 4. Figure 5-1 presents a visual summary of what we will accomplish in this chapter. As the figure shows, using the methods of Chapters 2 and 3, we would repeatedly roll the die to collect sample data, which then can be described with graphs (such as a histogram or boxplot), measures of center (such as the mean), and measures of variation (such as the standard deviation). Using the methods of Chapter 4, we could find the probability of each possible outcome. In this chapter we will combine those concepts as we develop probability distributions that describe what will *probably* happen instead of what actually *did* happen. In Chapter 2 we constructed frequency tables and histograms using *observed* sample values that were actually collected, but in this chapter we will construct probability distributions by presenting possible outcomes along with the relative frequencies we *expect*. In this chapter we consider *discrete* probability distributions, but Chapter 6 includes *continuous* probability distributions.

The table at the extreme right in Figure 5-1 represents a probability distribution that serves as a model of a theoretically perfect population frequency distribution. In essence, we can describe the relative frequency table for a die rolled an infinite number of times. With this knowledge of the population of outcomes, we are able to find its important characteristics, such as the mean and standard deviation. The remainder of this book and the very core of inferential statistics are based on some knowledge of probability distributions. We begin by examining the concept of a random variable, and then we consider important distributions that have many real applications.

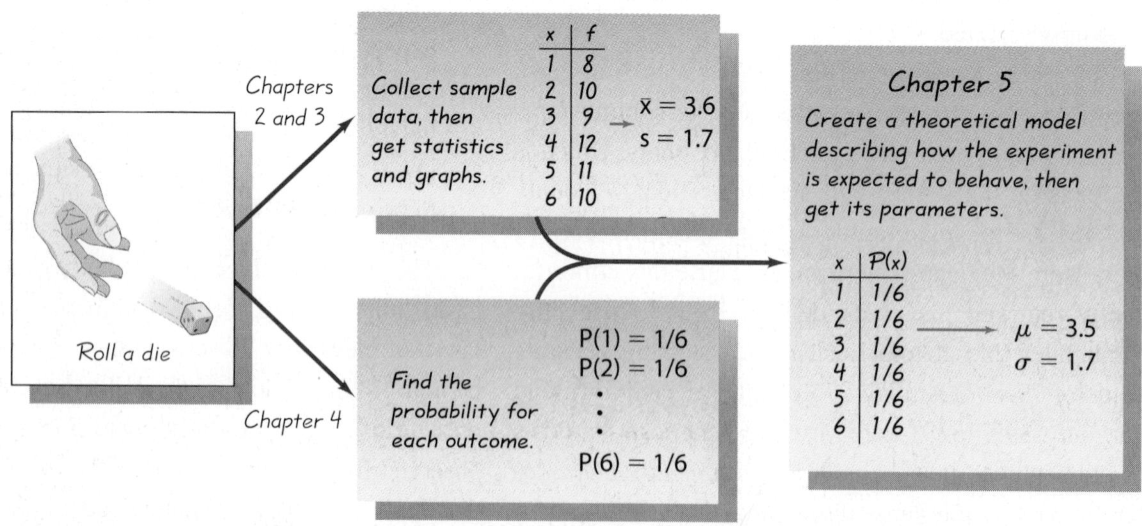

Figure 5-1 Combining Descriptive Methods and Probabilities to Form a Theoretical Model of Behavior

5-2 Random Variables

Key Concept This section introduces the important concept of a probability distribution, which gives the probability for each value of a variable that is determined by chance. This section also includes procedures for finding the mean and standard deviation for a probability distribution. In addition to the concept of a probability distribution, particular attention should be given to methods for distinguishing between outcomes that are likely to occur by chance and outcomes that are "unusual" in the sense that they are not likely to occur by chance.

We begin with the related concepts of *random variable* and *probability distribution.*

Definitions

A **random variable** is a variable (typically represented by x) that has a single numerical value, determined by chance, for each outcome of a procedure.

A **probability distribution** is a description that gives the probability for each value of the random variable. It is often expressed in the format of a graph, table, or formula.

EXAMPLE Jury Selection Twelve jurors are to be randomly selected from a population in which 80% of the jurors are Mexican-American. If we assume that jurors are randomly selected without bias, and if we let

$$x = \text{number of Mexican-American jurors among 12 jurors}$$

then x is a random variable because its value depends on chance. The possible values of x are 0, 1, 2, . . . , 12. Table 5-1 lists the values of x along with the corresponding probabilities. Probability values that are very small, such as 0.000000123 are represented by 0+. (In Section 5-3 we will see how to find the probability values, such as those listed in Table 5-1.) Because Table 5-1 gives the probability for each value of the random variable x, that table describes a probability distribution.

Table 5-1 Probability Distribution: Probabilities of Numbers of Mexican-Americans on a Jury of 12, Assuming That Jurors Are Randomly Selected from a Population in Which 80% of the Eligible People Are Mexican-Americans

x (Mexican-Americans)	$P(x)$
0	0+
1	0+
2	0+
3	0+
4	0.001
5	0.003
6	0.016
7	0.053
8	0.133
9	0.236
10	0.283
11	0.206
12	0.069

In Section 1-2 we made a distinction between discrete and continuous data. Random variables may also be discrete or continuous, and the following two definitions are consistent with those given in Section 1-2.

Definitions

A **discrete random variable** has either a finite number of values or a countable number of values, where "countable" refers to the fact that there might be infinitely many values, but they can be associated with a counting process.

A **continuous random variable** has infinitely many values, and those values can be associated with measurements on a continuous scale without gaps or interruptions.

Picking Lottery Numbers

In a typical state lottery, you select six different numbers. After a random drawing, any entries with the correct combination share in the prize. Since the winning numbers are randomly selected, any choice of six numbers will have the same chance as any other choice, but some combinations are better than others. The combination of 1, 2, 3, 4, 5, 6 is a poor choice because many people tend to select it. In a Florida lottery with a $105 million prize, 52,000 tickets had 1, 2, 3, 4, 5, 6; if that combination had won, the prize would have been only $1000. It's wise to pick combinations not selected by many others. Avoid combinations that form a pattern on the entry card.

(a) Discrete Random Variable: Count of the number of movie patrons.

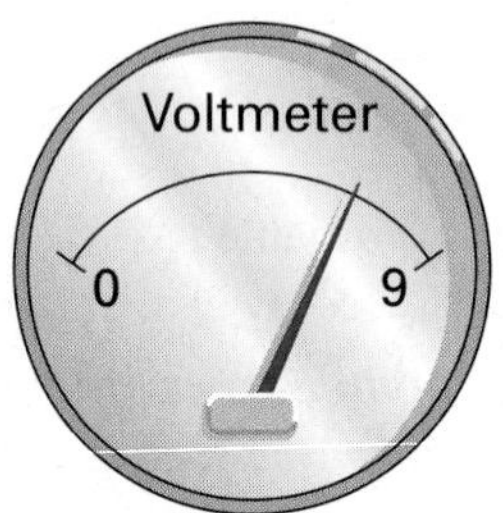

(b) Continuous Random Variable: The measured voltage of a smoke detector battery.

Figure 5-2 Devices Used to Count and Measure Discrete and Continuous Random Variables

This chapter deals exclusively with discrete random variables, but the following chapters will deal with continuous random variables.

EXAMPLES The following are examples of discrete and continuous random variables.

1. Let $x =$ the number of eggs that a hen lays in a day. This is a *discrete* random variable because its only possible values are 0, or 1, or 2, and so on. No hen can lay 2.343115 eggs, which would have been possible if the data had come from a continuous scale.
2. The count of the number of statistics students present in class on a given day is a whole number and is therefore a discrete random variable. The counting device shown in Figure 5-2(a) is capable of indicating only a finite number of values, so it is used to obtain values for a *discrete* random variable.
3. Let $x =$ the amount of milk a cow produces in one day. This is a *continuous* random variable because it can have any value over a continuous span. During a single day, a cow might yield an amount of milk that can be any value between 0 gallons and 5 gallons. It would be possible to get 4.123456 gallons, because the cow is not restricted to the discrete amounts of 0, 1, 2, 3, 4, or 5 gallons.

4. The measure of voltage for a particular smoke detector battery can be any value between 0 volts and 9 volts. It is therefore a continuous random variable. The voltmeter shown in Figure 5-2(b) is capable of indicating values on a continuous scale, so it can be used to obtain values for a *continuous* random variable.

Graphs

There are various ways to graph a probability distribution, but we will consider only the **probability histogram.** Figure 5-3 is a probability histogram that is very similar to the relative frequency histogram discussed in Chapter 2, but the vertical scale shows *probabilities* instead of relative frequencies based on actual sample results.

In Figure 5-3, note that along the horizontal axis, the values of 0, 1, 2, . . . , 12 are located at the centers of the rectangles. This implies that the rectangles are each 1 unit wide, so the areas of the rectangles are 0+, 0+, 0+, 0+, 0.001, 0.003, . . . , 0.069. The areas of these rectangles are the same as the *probabilities* in Table 5-1. We will see in Chapter 6 and future chapters that such a correspondence between area and probability is very useful in statistics.

Every probability distribution must satisfy each of the following two requirements.

Requirements for a Probability Distribution

1. $\Sigma P(x) = 1$ where x assumes all possible values. (That is, the sum of all probabilities must be 1.)
2. $0 \le P(x) \le 1$ for every individual value of x. (That is, each probability value must be between 0 and 1 inclusive.)

The first requirement comes from the simple fact that the random variable x represents all possible events in the entire sample space, so we are certain (with probability 1) that one of the events will occur. (In Table 5-1, the sum of the

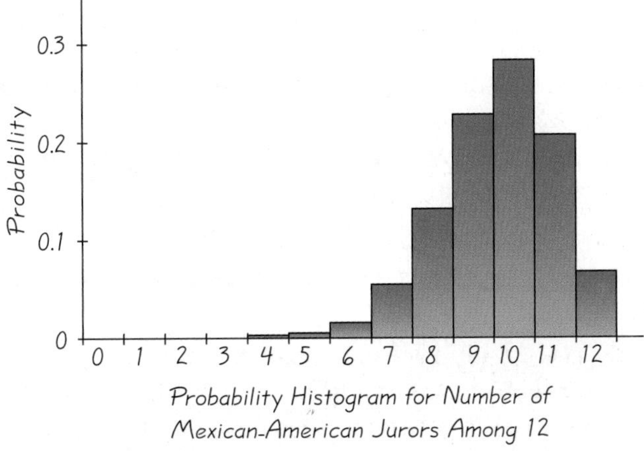

Figure 5-3 Probability Histogram for Number of Mexican-American Jurors Among 12 Jurors

probabilities is 1, but in other cases values such as 0.999 or 1.001 are acceptable because they result from rounding errors.) Also, the probability rule stating $0 \leq P(x) \leq 1$ for any event A implies that $P(x)$ must be between 0 and 1 for any value of x. Because Table 5-1 does satisfy both of the requirements, it is an example of a probability distribution. A probability distribution may be described by a table, such as Table 5-1, or a graph, such as Figure 5-3, or a formula.

Table 5-2 Probabilities for a Random Variable

x	$P(x)$
0	0.2
1	0.5
2	0.4
3	0.3

EXAMPLE Does Table 5-2 describe a probability distribution?

SOLUTION To be a probability distribution, $P(x)$ must satisfy the preceding two requirements. But

$$\begin{aligned}\Sigma P(x) &= P(0) + P(1) + P(2) + P(3) \\ &= 0.2 + 0.5 + 0.4 + 0.3 \\ &= 1.4 \quad [\text{showing that } \Sigma P(x) \neq 1]\end{aligned}$$

Because the first requirement is not satisfied, we conclude that Table 5-2 does not describe a probability distribution.

EXAMPLE Does $P(x) = x/3$ (where x can be 0, 1, or 2) determine a probability distribution?

SOLUTION For the given function we find that $P(0) = 0/3$, $P(1) = 1/3$ and $P(2) = 2/3$, so that

1. $\Sigma P(x) = \frac{0}{3} + \frac{1}{3} + \frac{2}{3} = \frac{3}{3} = 1$
2. Each of the $P(x)$ values is between 0 and 1.

Because both requirements are satisfied, the $P(x)$ function given in this example is a probability distribution.

Mean, Variance, and Standard Deviation

In Chapter 2 we described the following important characteristics of data (which can be remembered with the mnemonic of CVDOT for "**C**omputer **V**iruses **D**estroy **O**r **T**erminate"): (1) center; (2) variation; (3) distribution; (4) outliers; and (5) time (changing characteristics of data over time). The probability histogram can give us insight into the nature or shape of the distribution. Also, we can often find the mean, variance, and standard deviation of data, which provide insight into the other characteristics. The mean, variance, and standard deviation for a probability distribution can be found by applying Formulas 5-1, 5-2, 5-3, and 5-4.

Formula 5-1 $\mu = \Sigma[x \cdot P(x)]$ Mean for a probability distribution

Formula 5-2 $\sigma^2 = \Sigma[(x - \mu)^2 \cdot P(x)]$ Variance for a probability distribution

Formula 5-3 $\sigma^2 = \Sigma[x^2 \cdot P(x)] - \mu^2$ Variance for a probability distribution

Formula 5-4 $\sigma = \sqrt{\Sigma[x^2 \cdot P(x)] - \mu^2}$ Standard deviation for a probability distribution

Caution: Evaluate $\Sigma[x^2 \cdot P(x)]$ by first squaring each value of x, then multiplying each square by the corresponding probability $P(x)$, then adding.

The TI-83/84 Plus calculator can be used to find the mean and standard deviation corresponding to Formulas 5-1 and 5-4. Shown here is the screen display that results from the probability distribution summarized in Table 5-1. In this TI-83/84 Plus display, the value shown as $\bar{x}$ is actually the value of the mean μ, and the value shown as σx is the value of the standard deviation σ. Based on the displayed results, we get $\mu = 9.6$ and $\sigma = 1.4$ after we use the round-off rule that follows shortly.

TI-83/84 Plus

```
1-Var Stats
 x̄=9.598
 Σx=9.598
 Σx²=94.054
 Sx=
 σx=1.390106471
↓n=1
```

Rationale for Formulas 5-1 through 5-4

Instead of blindly accepting and using formulas, it is much better to have some understanding of why they work. Formula 5-1 accomplishes the same task as the formula for the mean of a frequency table. (Recall that f represents class frequency and N represents population size.) Rewriting the formula for the mean of a frequency table so that it applies to a population and then changing its form, we get

$$\mu = \frac{\Sigma(f \cdot x)}{N} = \Sigma\left[\frac{f \cdot x}{N}\right] = \Sigma\left[x \cdot \frac{f}{N}\right] = \Sigma[x \cdot P(x)]$$

In the fraction f/N, the value of f is the frequency with which the value x occurs and N is the population size, so f/N is the probability for the value of x.

Similar reasoning enables us to take the variance formula from Chapter 3 and apply it to a random variable for a probability distribution; the result is Formula 5-2. Formula 5-3 is a shortcut version that will always produce the same result as Formula 5-2. Although Formula 5-3 is usually easier to work with, Formula 5-2 is easier to understand directly. Based on Formula 5-2, we can express the standard deviation as

$$\sigma = \sqrt{\Sigma[(x - \mu)^2 \cdot P(x)]}$$

or as the equivalent form given in Formula 5-4.

When applying Formulas 5-1 through 5-4, use this rule for rounding results.

Round-off Rule for μ, σ, and σ^2

Round results by carrying one more decimal place than the number of decimal places used for the random variable x. If the values of x are integers, round μ, σ, and σ^2 to one decimal place.

It is sometimes necessary to use a different rounding rule because of special circumstances, such as results that require more decimal places to be meaningful. For example, with four-engine jets the mean number of jet engines working successfully throughout a flight is 3.999714286, which becomes 4.0 when rounded to one more decimal place than the original data. Here, 4.0 would be misleading because it suggests that all jet engines always work successfully. We need more precision to correctly reflect the true mean, such as the precision in the number 3.999714.

Identifying *Unusual* Results with the Range Rule of Thumb

The range rule of thumb (discussed in Section 3-3) may also be helpful in interpreting the value of a standard deviation. According to the range rule of thumb, most values should lie within 2 standard deviations of the mean; it is unusual for a value to differ from the mean by more than 2 standard deviations. (The use of 2 standard deviations is not an absolutely rigid value, and other values such as 3 could be used instead.) We can therefore identify "unusual" values by determining that they lie outside of these limits:

Range Rule of Thumb
maximum usual value = $\mu + 2\sigma$
minimum usual value = $\mu - 2\sigma$

EXAMPLE Table 5-1 describes the probability distribution for the number of Mexican-Americans among 12 randomly selected jurors in Hidalgo County, Texas. Assuming that we repeat the process of randomly selecting 12 jurors and counting the number of Mexican-Americans each time, find the mean number of Mexican-Americans (among 12), the variance, and the standard deviation. Use those results and the range rule of thumb to find the maximum and minimum usual values. Based on the results, determine whether a jury consisting of 7 Mexican-Americans among 12 jurors is usual or unusual.

SOLUTION In Table 5-3, the two columns at the left describe the probability distribution given earlier in Table 5-1, and we create the three columns at the right for the purposes of the calculations required.

Using Formulas 5-1 and 5-3 and the table results, we get

$$\mu = \Sigma[x \cdot P(x)] = 9.598 = 9.6 \quad \text{(rounded)}$$

$$\sigma^2 = \Sigma[x^2 \cdot P(x)] - \mu^2$$

$$= 94.054 - 9.598^2 = 1.932396 = 1.9 \quad \text{(rounded)}$$

The standard deviation is the square root of the variance, so

$$\sigma = \sqrt{1.932396} = 1.4 \quad \text{(rounded)}$$

We now know that when randomly selecting 12 jurors, the mean number of Mexican-Americans is 9.6, the variance is 1.9 "Mexican-Americans squared," and the standard deviation is 1.4 Mexican-Americans. Using the range rule of thumb, we can now find the maximum and minimum usual values as follows:

$$\text{maximum usual value: } \mu + 2\sigma = 9.6 + 2(1.4) = 12.4$$

$$\text{minimum usual value: } \mu - 2\sigma = 9.6 - 2(1.4) = 6.8$$

INTERPRETATION Based on these results, we conclude that for groups of 12 jurors randomly selected in Hidalgo County, the number of Mexican-Americans should usually fall between 6.8 and 12.4. If a jury consists of 7 Mexican-Americans, it would not be unusual and would not be a basis for a charge that the jury was selected in a way that it discriminates against Mexican-Americans. (The jury that convicted Roger Partida included 7 Mexican-Americans, but the charge of an unfair selection process was based on the process for selecting grand juries, not the specific jury that convicted him.)

Table 5-3 Calculating μ, σ, and σ^2 for a Probability Distribution

x	$P(x)$	$x \cdot P(x)$	x^2	$x^2 \cdot P(x)$
0	0+	0.000	0	0.000
1	0+	0.000	1	0.000
2	0+	0.000	4	0.000
3	0+	0.000	9	0.000
4	0.001	0.004	16	0.016
5	0.003	0.015	25	0.075
6	0.016	0.096	36	0.576
7	0.053	0.371	49	2.597
8	0.133	1.064	64	8.512
9	0.236	2.124	81	19.116
10	0.283	2.830	100	28.300
11	0.206	2.266	121	24.926
12	0.069	0.828	144	9.936
Total		9.598 ↑ $\Sigma[x \cdot P(x)]$		94.054 ↑ $\Sigma[x^2 \cdot P(x)]$

Identifying *Unusual* Results with Probabilities

Strong recommendation: Take time to carefully read and understand the rare event rule and the paragraph that follows it. This brief discussion presents an extremely important approach used often in statistics.

Rare Event Rule

If, under a given assumption (such as the assumption that a coin is fair), the probability of a particular observed event (such as 992 heads in 1000 tosses of a coin) is extremely small, we conclude that the assumption is probably not correct.

Probabilities can be used to apply the rare event rule as follows:

Using Probabilities to Determine When Results Are Unusual

- **Unusually high number of successes:** x successes among n trials is an *unusually high* number of successes if $P(x \text{ or more}) \leq 0.05$.*
- **Unusually low number of successes:** x successes among n trials is an *unusually low* number of successes if $P(x \text{ or fewer}) \leq 0.05$.*

Suppose you were flipping a coin to determine whether it favors heads, and suppose 1000 tosses resulted in 501 heads. This is not evidence that the coin

*The value of 0.05 is commonly used, but is not absolutely rigid. Other values, such as 0.01, could be used to distinguish between events that can easily occur by chance and events that are very unlikely to occur by chance.

favors heads, because it is very easy to get a result like 501 heads in 1000 tosses just by chance. Yet, the probability of getting *exactly* 501 heads in 1000 tosses is actually quite small: 0.0252. This low probability reflects the fact that with 1000 tosses, any *specific* number of heads will have a very low probability. However, we do not consider 501 heads among 1000 tosses to be *unusual*, because the probability of getting *at least* 501 heads is high: 0.487.

CHAPTER 5 PROBLEM

EXAMPLE Jury Selection If 80% of those eligible for jury duty in Hidalgo County are Mexican-American, then a jury of 12 randomly selected people should have around 9 or 10 who are Mexican-American. (The mean number of Mexican-Americans on juries should be 9.6.) Is 7 Mexican-American jurors among 12 an unusually low number? Does the selection of only 7 Mexican-Americans among 12 jurors suggest that there is discrimination in the selection process?

SOLUTION We will use the criterion that 7 Mexican-Americans among 12 jurors is unusually low if $P(7 \text{ or fewer Mexican-Americans}) \leq 0.05$. If we refer to Table 5-1, we get this result:

$$\begin{aligned}
&P(7 \text{ or fewer Mexican-Americans among 12 jurors}) \\
&\quad = P(7 \text{ or } 6 \text{ or } 5 \text{ or } 4 \text{ or } 3 \text{ or } 2 \text{ or } 1 \text{ or } 0) \\
&\quad = P(7) + P(6) + P(5) + P(4) + P(3) + P(2) + P(1) + P(0) \\
&\quad = 0.053 + 0.016 + 0.003 + 0.001 + 0 + 0 + 0 + 0 \\
&\quad = 0.073
\end{aligned}$$

INTERPRETATION Because the probability 0.073 is greater than 0.05, we conclude that the result of 7 Mexican-Americans is *not unusual*. There is a high likelihood (0.073) of getting 7 Mexican-Americans by random chance. (Only a probability of 0.05 or less would indicate that the event is unusual.) No court of law would rule that under these circumstances, the selection of only 7 Mexican-American jurors is discriminatory.

Expected Value

The mean of a discrete random variable is the theoretical mean outcome for infinitely many trials. We can think of that mean as the *expected value* in the sense that it is the average value that we would expect to get if the trials could continue indefinitely. The uses of expected value (also called *expectation*, or *mathematical expectation*) are extensive and varied, and they play a very important role in an area of application called *decision theory*.

Definition

The **expected value** of a discrete random variable is denoted by E, and it represents the average value of the outcomes. It is obtained by finding the value of $\Sigma[x \cdot P(x)]$.

$$E = \Sigma[x \cdot P(x)]$$

From Formula 5-1 we see that $E = \mu$. That is, the mean of a discrete random variable is the same as its expected value. See Table 5-3 and note that when selecting 12 jurors from a population in which 80% of the people are Mexican-Americans, the *mean* number of Mexican-Americans is 9.6, so it follows that the *expected value* of the number of Mexican-Americans is also 9.6.

EXAMPLE Kentucky Pick 4 Lottery If you bet $1 in Kentucky's Pick 4 lottery game, you either lose $1 or gain $4999. (The winning prize is $5000, but your $1 bet is not returned, so the net gain is $4999.) The game is played by selecting a four-digit number between 0000 and 9999. If you bet $1 on 1234, what is your expected value of gain or loss?

SOLUTION For this bet, there are two outcomes: You either lose $1 or you gain $4999. Because there are 10,000 four-digit numbers and only one of them is the winning number, the probability of losing is 9,999/10,000 and the probability of winning is 1/10,000. Table 5-4 summarizes the probability distribution, and we can see that the expected value is $E = -50¢$.

Table 5-4 Kentucky Pick 4 Lottery

Event	x	$P(x)$	$x \cdot P(x)$
Lose	−$1	0.9999	−$0.9999
Gain (net)	$4999	0.0001	$0.4999
Total			−$0.50 (or −50¢)

INTERPRETATION In any individual game, you either lose $1 or have a net gain of $4999, but the expected value shows that in the long run, you can expect to lose an average of 50¢ for each $1 bet. This lottery might have some limited entertainment value, but it is definitely an extremely poor financial investment.

In this section we learned that a random variable has a numerical value associated with each outcome of some random procedure, and a probability distribution has a probability associated with each value of a random variable. We examined methods for finding the mean, variance, and standard deviation for a probability distribution. We saw that the expected value of a random variable is really the same as the mean. Finally, an extremely important concept of this section is the use of probabilities for determining when outcomes are *unusual*.

TI Using a TI-83/84 Plus Calculator for Probability Distributions

Entering a Probability Distribution Let's assume that we have a probability distribution in the format of a table, such as Table 5-1. We can enter that table in a TI-83/84 Plus calculator by using the following procedure:

Entering the Probability Distribution of Table 5-1

L1	L2	L3 2
0	0	------
1	0	
2	0	
3	0	
4	.001	
5	.003	
6	.016	

L2(7) =.016

1. First clear lists L1 and L2 as follows: Press STAT, select **4:CrlList,** then enter **L1, 2** (with the comma included), and press ENTER.
2. Press STAT and ENTER to enter the edit mode, and proceed to enter the values of the random variable x in list L1, and enter the corresponding probabilities in list L2. See the first display in the margin for the screen showing the top portion of Table 5-1.

Finding the Mean and Standard Deviation First enter the probability distribution by using the procedure described in Steps 1 and 2 above. Then find the mean and standard deviation by using the following procedure.

1. After completing the entry of the probability distribution as described above, quit the edit mode by pressing 2ND MODE (for "QUIT").
2. Now enter the command of **1-Var Stats L1, L2** by using this sequence of keystrokes: Press STAT, then select **CALC** by pressing the ▷ key, and press ENTER. The screen should now show 1-Var Stats.
3. Enter **L1, L2** (including the comma), so that the screen shows 1-Var Stats L1, L2. (Enter L1 by pressing 2ND 1, press ,, then enter L2 by pressing 2ND 2.)
4. With the command of **1-Var Stats L1, L2** displayed, press ENTER to get the results shown in the second screen display in the margin.

   Interpreting results: Note that the value shown as $\bar{x}$ is actually the population mean μ, so $\mu = 9.598$, which we round to 9.6. Also, there is no value shown for the sample standard deviation indicated by Sx, but the population standard deviation is indicated by σx, so we know that $\sigma = 1.390106471$, which we round to 1.4.

Finding the Mean and Standard Deviation

```
1-Var Stats
 x̄=9.598
 Σx=9.598
 Σx²=94.054
 Sx=
 σx=1.390106471
↓n=1
```

Obtaining a Probability Histogram Enter the probability distribution as described above, then use the following procedure:

1. After completing the entry of the probability distribution as described above, quit the edit mode by pressing 2ND MODE (for "QUIT").
2. Get to the STAT PLOT menu by pressing 2ND Y=.
3. Press ENTER, then press ENTER again to turn on the first plot (with the other plots turned off). Select the graph type that is the third entry of the first row.
4. For Xlist, enter **L1.**
5. For Freq (frequency), enter **L2.**
6. Press WINDOW and adjust the settings. For the probability distribution of Table 5-1, use the settings shown in the screen display, then press GRAPH to get the probability histogram shown on the following page. You can also press TRACE and use the arrow keys to find values corresponding to individual elements of the graph.

Entries for Graph

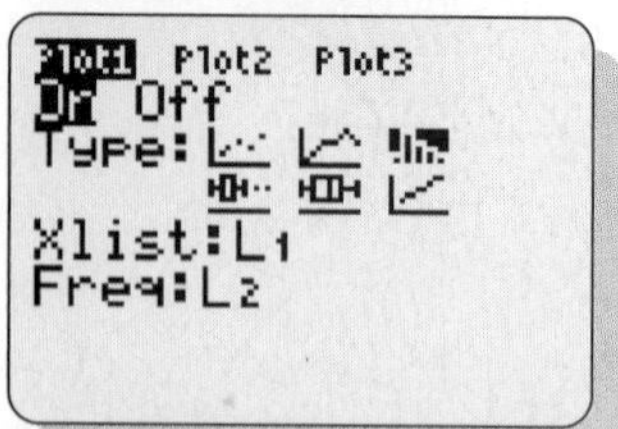

Window Settings for Table 5-1

```
WINDOW
 Xmin=0
 Xmax=13
 Xscl=1
 Ymin=0
 Ymax=.3
 Yscl=0
 Xres=1
```

Shown in the margin is the probability histogram that corresponds to Table 5-1. This screen display shows the probability histogram that is essentially the same as Figure 5-3. The graph shown here does not include values along the horizontal scale, nor does it show values along the vertical scale. However, the TRACE key could be used to identify those values. Press the TRACE key, then press the ▷ key to scroll to the right. For example, press the TRACE key, then press ▷ ten times, and the screen will indicate that the tallest bar represents a probability of 0.283 for the value of $x = 10$.

Probability Histogram for Table 5-1

5-2 BASIC SKILLS AND CONCEPTS

Statistical Literacy and Critical Thinking

1. Probability Distribution Consider the trial of rolling a single die, with outcomes of 1, 2, 3, 4, 5, 6. Construct a table representing the probability distribution.

2. Probability Distribution One of the requirements of a probability distribution is that the sum of the probabilities must be 1 (with a small amount of leeway allowed for rounding errors). What is the justification for this requirement?

3. Probability Distribution A professional gambler claims that he has loaded a die so that the outcomes of 1, 2, 3, 4, 5, 6 have corresponding probabilities of 0.1, 0.2, 0.3, 0.4, 0.5, and 0.6. Can he actually do what he has claimed? Is a probability distribution described by listing the outcomes along with their corresponding probabilities?

4. Expected Value A researcher calculates the expected value for the number of girls in five births. He gets a result of 2.5. He then rounds the result to 3, saying that it is not possible to get 2.5 girls when five babies are born. Is this reasoning correct?

Identifying Discrete and Continuous Random Variables. *In Exercises 5 and 6, identify the given random variable as being* discrete *or* continuous.

5. **a.** The height of a randomly selected giraffe living in Kenya
b. The number of bald eagles located in New York State
c. The exact time it takes to evaluate 27 + 72.
d. The number of textbook authors now sitting at a computer
e. The number of statistics students now reading a book

6. **a.** The cost of conducting a genetics experiment
b. The number of supermodels who ate pizza yesterday
c. The exact life span of a kitten
d. The number of statistics professors who read a newspaper each day
e. The weight of a feather

Identifying Probability Distributions. *In Exercises 7–12, determine whether a probability distribution is given. In those cases where a probability distribution is not described, identify the requirements that are not satisfied. In those cases where a probability distribution is described, find its mean and standard deviation.*

7. Genetic Disorder Three males with an X-linked genetic disorder have one child each. The random variable x is the number of children among the three who inherit the X-linked genetic disorder.

x	$P(x)$
0	0.4219
1	0.4219
2	0.1406
3	0.0156

8. Numbers of Girls A researcher reports that when groups of four children are randomly selected from a population of couples meeting certain criteria, the probability distribution for the number of girls is as given in the accompanying table.

x	$P(x)$
0	0.502
1	0.365
2	0.098
3	0.011
4	0.001

9. Genetics Experiment A genetics experiment involves offspring peas in groups of four. A researcher reports that for one group, the number of peas with white flowers has a probability distribution as given in the accompanying table.

x	$P(x)$
0	0.04
1	0.16
2	0.80
3	0.16
4	0.04

10. Mortality Study For a group of four men, the probability distribution for the number x who live through the next year is as given in the accompanying table.

x	$P(x)$
0	0.0000
1	0.0001
2	0.0006
3	0.0387
4	0.9606

11. Number of Games in a Baseball World Series Based on past results found in the *Information Please Almanac*, there is a 0.1818 probability that a baseball World Series contest will last four games, a 0.2121 probability that it will last five games, a 0.2323 probability that it will last six games, and a 0.3737 probability that it will last seven games. Is it unusual for a team to "sweep" by winning in four games?

12. Brand Recognition In a study of brand recognition of Sony, groups of four consumers are interviewed. If x is the number of people in the group who recognize the Sony brand name, then x can be 0, 1, 2, 3, or 4, and the corresponding probabilities are 0.0016, 0.0250, 0.1432, 0.3892, and 0.4096. Is it unusual to randomly select four consumers and find that none of them recognizes the brand name of Sony?

13. Determining Whether a Jury Selection Process Discriminates Assume that 12 jurors are randomly selected from a population in which 80% of the people are Mexican-Americans. Refer to Table 5-1 and find the indicated probabilities.

binompdf **a.** Find the probability of exactly 5 Mexican-Americans among 12 jurors.

binomcdf **b.** Find the probability of 5 or fewer Mexican-Americans among 12 jurors.

c. Which probability is relevant for determining whether 5 jurors among 12 is unusually low: the result from part (a) or part (b)?

d. Does 5 Mexican-Americans among 12 jurors suggest that the selection process discriminates against Mexican-Americans? Why or why not?

14. Determining Whether a Jury Selection Process Discriminates Assume that 12 jurors are randomly selected from a population in which 80% of the people are Mexican-Americans. Refer to Table 5-1 and find the indicated probabilities.

a. Find the probability of exactly 6 Mexican-Americans among 12 jurors.

b. Find the probability of 6 or fewer Mexican-Americans among 12 jurors.

c. Which probability is relevant for determining whether 6 jurors among 12 is unusually low: the result from part (a) or part (b)?

d. Does 6 Mexican-Americans among 12 jurors suggest that the selection process discriminates against Mexican-Americans? Why or why not?

15. Determining Whether a Jury Selection Process Discriminates Assume that 12 jurors are randomly selected from a population in which 80% of the people are Mexican-Americans. Refer to Table 5-1 and find the indicated probability.

a. Using the probability values in Table 5-1, find the probability value that should be used for determining whether the result of 8 Mexican-Americans among 12 jurors is unusually low.

b. Does the result of 8 Mexican-American jurors suggest that the selection process discriminates against Mexican-Americans? Why or why not?

16. Determining Whether a Jury Selection Process Is Biased Assume that 12 jurors are randomly selected from a population in which 80% of the people are Mexican-Americans. Refer to Table 5-1 and find the indicated probability.

a. Using the probability values in Table 5-1, find the probability value that should be used for determining whether the result of 11 Mexican-Americans among 12 jurors is unusually high.

b. Does the selection of 11 Mexican-American jurors suggest that the selection process favors Mexican-Americans? Why or why not?

TI **17.** Expected Value in Roulette When you give the Venetian casino in Las Vegas $5 for a bet on the number 7 in roulette, you have a 37/38 probability of losing $5 and you have a 1/38 probability of making a net gain of $175. (The prize is $180, including your $5 bet, so the net gain is $175.) If you bet $5 that the outcome is an odd number, the probability of losing $5 is 20/38 and the probability of making a net gain of $5 is 18/38. (If you bet $5 on an odd number and win, you are given $10 that includes your bet, so the net gain is $5.)

a. If you bet $5 on the number 7, what is your expected value?

b. If you bet $5 that the outcome is an odd number, what is your expected value?

c. Which of these options is best: bet on 7, bet on an odd number, or don't bet? Why?

TI **18.** Expected Value in Casino Dice When you give a casino $5 for a bet on the "pass line" in a casino game of dice, there is a 251/495 probability that you will lose $5 and there is a 244/495 probability that you will make a net gain of $5. (If you win, the casino gives you $5 and you get to keep your $5 bet, so the net gain is $5.) What is your expected value? In the long run, how much do you lose for each dollar bet?

TI **19.** Expected Value for a Life Insurance Policy The CNA Insurance Company charges a 21-year-old male a premium of $250 for a one-year $100,000 life insurance policy. A 21-year-old male has a 0.9985 probability of living for a year (based on data from the National Center for Health Statistics).

a. From the perspective of a 21-year-old male (or his estate), what are the values of the two different outcomes?

b. What is the expected value for a 21-year-old male who buys the insurance?

c. What would be the cost of the insurance policy if the company just breaks even (in the long run with many such policies), instead of making a profit?

d. Given that the expected value is negative (so the insurance company can make a profit), why should a 21-year-old male or anyone else purchase life insurance?

TI **20.** Expected Value for a Magazine Sweepstakes *Reader's Digest* ran a sweepstakes in which prizes were listed along with the chances of winning: $1,000,000 (1 chance in

90,000,000), $100,000 (1 chance in 110,000,000), $25,000 (1 chance in 110,000,000), $5,000 (1 chance in 36,667,000), and $2,500 (1 chance in 27,500,000).

a. Assuming that there is no cost of entering the sweepstakes, find the expected value of the amount won for one entry.

b. Find the expected value if the cost of entering this sweepstakes is the cost of a postage stamp. Is it worth entering this contest?

TI **21.** Finding Mean and Standard Deviation Let the random variable x represent the number of girls in a family of three children. Construct a table describing the probability distribution, then find the mean and standard deviation. (*Hint:* List the different possible outcomes.) Is it unusual for a family of three children to consist of three girls?

TI **22.** Finding Mean and Standard Deviation Let the random variable x represent the number of girls in a family of four children. Construct a table describing the probability distribution, then find the mean and standard deviation. (*Hint:* List the different possible outcomes.) Is it unusual for a family of four children to consist of four girls?

TI **23.** Telephone Surveys Computers are often used to randomly generate digits of telephone numbers to be called for surveys. Each digit has the same chance of being selected. Construct a table representing the probability distribution for the digits selected, find its mean, find its standard deviation, and describe the shape of the probability histogram.

TI **24.** Home Sales Refer to the numbers of bedrooms in homes sold, as listed in Data Set 18 in Appendix B. Use the frequency distribution to construct a table representing the probability distribution, then find the mean and standard deviation. Also, describe the shape of the probability histogram.

5-2 BEYOND THE BASICS

25. Frequency Distribution and Probability Distribution What is the fundamental difference between a frequency distribution (as defined in Section 2-2) and a probability distribution (as defined in this section)?

TI **26.** Junk Bonds Kim Hunter has $1000 to invest, and her financial analyst recommends two types of junk bonds. The A bonds have a 6% annual yield with a default rate of 1%. The B bonds have an 8% annual yield with a default rate of 5%. (If the bond defaults, the $1000 is lost.) Which of the two bonds is better? Why? Should she select either bond? Why or why not?

TI **27.** Defective Parts: Finding Mean and Standard Deviation The Sky Ranch is a supplier of aircraft parts. Included in stock are eight altimeters that are correctly calibrated and two that are not. Three altimeters are randomly selected without replacement. Let the random variable x represent the number that are not correctly calibrated. Find the mean and standard deviation for the random variable x.

28. Labeling Dice to Get a Uniform Distribution Assume that you have two blank dice, so that you can label the 12 faces with any numbers. Describe how the dice can be labeled so that, when the two dice are rolled, the totals of the two dice are uniformly distributed so that the outcomes of 1, 2, 3, . . . , 12 each have probability 1/12. (See "Can One Load a Set of Dice So That the Sum Is Uniformly Distributed?" by Chen, Rao, and Shreve, *Mathematics Magazine*, Vol. 70, No. 3.)

5-3 Binomial Probability Distributions

Key Concept Section 5-2 discussed discrete probability distributions in general, but in this section we focus on one specific type: binomial probability distributions. Because binomial probability distributions involve proportions used with methods of inferential statistics discussed later in this book, it becomes important to understand fundamental properties of this particular class of probability distributions. This section presents a basic definition of a binomial probability distribution along with notation, and it presents methods for finding probability values.

Binomial probability distributions allow us to deal with circumstances in which the outcomes belong to *two* relevant categories, such as acceptable/defective or survived/died. Other requirements are given in the following definition.

Definition

A **binomial probability distribution** results from a procedure that meets all the following requirements:

1. The procedure has a *fixed number of trials.*
2. The trials must be *independent.* (The outcome of any individual trial doesn't affect the probabilities in the other trials.)
3. Each trial must have all outcomes classified into *two categories* (commonly referred to as *success* and *failure*).
4. The probability of a success remains the same in all trials.

If a procedure satisfies these four requirements, the distribution of the random variable x (number of successes) is called a *binomial probability distribution* (or *binomial distribution*). The following notation is commonly used.

Notation for Binomial Probability Distributions

S and F (success and failure) denote the two possible categories of all outcomes; p and q will denote the probabilities of S and F, respectively, so

$P(S) = p$	(p = probability of a success)
$P(F) = 1 - p = q$	(q = probability of a failure)
n	denotes the fixed number of trials.
x	denotes a specific number of successes in n trials, so x can be any whole number between 0 and n, inclusive.
p	denotes the probability of *success* in *one* of the n trials.
q	denotes the probability of *failure* in *one* of the n trials.
$P(x)$	denotes the probability of getting exactly x successes among the n trials.

Not At Home

Pollsters cannot simply ignore those who were not at home when they were called the first time. One solution is to make repeated callback attempts until the person can be reached. Alfred Politz and Willard Simmons describe a way to compensate for those missing results without making repeated callbacks. They suggest weighting results based on how often people are not at home. For example, a person at home only two days out of six will have a 2/6 or 1/3 probability of being at home when called the first time. When such a person is reached the first time, his or her results are weighted to count three times as much as someone who is always home. This weighting is a compensation for the other similar people who are home two days out of six and were not at home when called the first time. This clever solution was first presented in 1949.

The word *success* as used here is arbitrary and does not necessarily represent something good. Either of the two possible categories may be called the success S as long as its probability is identified as p. Once a category has been designated as the success S, be sure that p is the probability of a success and x is the number of successes. That is, *be sure that the values of p and x refer to the* same *category designated as a success.* (The value of q can always be found by subtracting p from 1; if $p = 0.95$, then $q = 1 - 0.95 = 0.05$.) Here is an important hint for working with binomial probability problems:

Be sure that x and p both refer to the *same* category being called a success.

When selecting a sample (such as a survey) for some statistical analysis, we usually sample without replacement, and sampling without replacement involves dependent events, which violates the second requirement in the above definition. However, the following rule of thumb is commonly used (because errors are negligible): When sampling without replacement, the events can be treated as if they are independent if the sample size is no more than 5% of the population size.

When sampling without replacement, consider events to be independent if $n \le 0.05N$.

EXAMPLE Jury Selection In the case of *Castaneda v. Partida* it was noted that although 80% of the population in a Texas county is Mexican-American, only 39% of those summoned for grand juries were Mexican-American. Let's assume that we need to select 12 jurors from a population that is 80% Mexican-American, and we want to find the probability that among 12 randomly selected jurors, exactly 7 are Mexican-Americans.

a. Does this procedure result in a binomial distribution?

b. If this procedure does result in a binomial distribution, identify the values of n, x, p, and q.

SOLUTION

a. This procedure does satisfy the requirements for a binomial distribution, as shown below.

1. The number of trials (12) is fixed.
2. The 12 trials are independent. (Technically, the 12 trials involve selection without replacement and are not independent, but we can assume independence because we are randomly selecting only 12 members from a very large population.)
3. Each of the 12 trials has two categories of outcomes: The juror selected is either Mexican-American or is not.
4. For each juror selected, the probability that he or she is Mexican-American is 0.8 (because 80% of this population is Mexican-American). That probability of 0.8 remains the same for each of the 12 jurors.

b. Having concluded that the given procedure does result in a binomial distribution, we now proceed to identify the values of n, x, p, and q.

1. With 12 jurors selected, we have $n = 12$.
2. We want the probability of exactly 7 Mexican-Americans, so $x = 7$.
3. The probability of success (getting a Mexican-American) for one selection is 0.8, so $p = 0.8$.
4. The probability of failure (not getting a Mexican-American) is 0.2, so $q = 0.2$.

Again, it is very important to be sure that x and p both refer to the same concept of "success." In this example, we use x to count the number of Mexican-Americans, so p must be the probability of a Mexican-American. Therefore, x and p do use the same concept of success (Mexican-American) here.

We will now discuss three methods for finding the probabilities corresponding to the random variable x in a binomial distribution. The first method involves calculations using the *binomial probability formula* and is the basis for the other two methods. The second method involves the use of Table A-1, and the third method involves the use of statistical software or a calculator. If you are using software or a calculator that automatically produces binomial probabilities, we recommend that you solve one or two exercises using Method 1 to ensure that you understand the basis for the calculations. Understanding is always infinitely better than blind application of formulas.

Method 1: Using the Binomial Probability Formula In a binomial probability distribution, probabilities can be calculated by using the binomial probability formula.

Formula 5-5 $$P(x) = \frac{n!}{(n-x)!x!} \cdot p^x \cdot q^{n-x} \quad \text{for } x = 0, 1, 2, \ldots, n$$

where
n = number of trials
x = number of successes among n trials
p = probability of success in any one trial
q = probability of failure in any one trial ($q = 1 - p$)

The factorial symbol !, introduced in Section 4-7, denotes the product of decreasing factors. Two examples of factorials are $3! = 3 \cdot 2 \cdot 1 = 6$ and $0! = 1$ (by definition).

EXAMPLE Jury Selection Use the binomial probability formula to find the probability of getting exactly 7 Mexican-Americans when 12 jurors are randomly selected from a population that is 80% Mexican-American. That is, find $P(7)$ given that $n = 12$, $x = 7$, $p = 0.8$, and $q = 0.2$.

continued

From Table A-1:

n	x	p 0.80
12	0	0+
	1	0+
	2	0+
	3	0+
	4	0.001
	5	0.003
	6	0.016
	7	0.053
	8	0.133
	9	0.236
	10	0.283
	11	0.206
	12	0.069

↓

Binomial probability distribution for $n = 12$ and $p = 0.8$

x	p
0	0+
1	0+
2	0+
3	0+
4	0.001
5	0.003
6	0.016
7	0.053
8	0.133
9	0.236
10	0.283
11	0.206
12	0.069

SOLUTION Using the given values of n, x, p, and q in the binomial probability formula (Formula 5-5), we get

$$P(7) = \frac{12!}{(12-7)!7!} \cdot 0.8^7 \cdot 0.2^{12-7}$$
$$= \frac{12!}{5!7!} \cdot 0.2097152 \cdot 0.00032$$
$$= (792)(0.2097152)(0.00032) = 0.0531502203$$

The probability of getting exactly 7 Mexican-American jurors among 12 randomly selected jurors is 0.0532 (rounded to three significant digits).

Calculation hint: When computing a probability with the binomial probability formula, it's helpful to get a single number for $n!/[(n-x)!x!]$, a single number for p^x and a single number for q^{n-x}, then simply multiply the three factors together as shown at the end of the calculation for the preceding example. Don't round too much when you find those three factors; round only at the end.

Method 2: Using Table A-1 in Appendix A In some cases, we can easily find binomial probabilities by simply referring to Table A-1 in Appendix A. (Part of Table A-1 is shown in the margin.) First locate n and the corresponding value of x that is desired. At this stage, one row of numbers should be isolated. Now align that row with the proper probability of p by using the column across the top. The isolated number represents the desired probability. A very small probability, such as 0.000064, is indicated by 0+.

EXAMPLE Use the portion of Table A-1 (for $n = 12$ and $p = 0.8$) shown in the margin to find the following:

a. The probability of exactly 7 successes

b. The probability of 7 *or fewer* successes.

SOLUTION

a. The display in the margin from Table A-1 shows that when $n = 12$ and $p = 0.8$, the probability of $x = 7$ is given by $P(7) = 0.053$, which is the same value (except for rounding) computed with the binomial probability formula in the preceding example.

b. "7 or fewer" successes means that the number of successes is 7 or 6 or 5 or 4 or 3 or 2 or 1 or 0.

$$\begin{aligned} P(7 \text{ or fewer}) &= P(7 \text{ or } 6 \text{ or } 5 \text{ or } 4 \text{ or } 3 \text{ or } 2 \text{ or } 1 \text{ or } 0) \\ &= P(7) + P(6) + P(5) + P(4) + P(3) + P(2) + P(1) + P(0) \\ &= 0.053 + 0.016 + 0.003 + 0.001 + 0 + 0 + 0 + 0 \\ &= 0.073 \end{aligned}$$

Because the probability of 0.073 is not small (it is not 0.05 or less), it suggests that if 12 jurors are randomly selected, the result of 7 Mexican-Americans is not unusually low and could easily occur by random chance.

In part (b) of the preceding solution, if we wanted to find P(7 or fewer) by using the binomial probability formula, we would need to apply that formula eight times to compute eight different probabilities, which would then be added. Given this choice between the formula and the table, it makes sense to use the table. Unfortunately, Table A-1 includes only limited values of n as well as limited values of p, so the table doesn't always work, and we must then find the probabilities by using the binomial probability formula, software, or a calculator, as in the following method.

Method 3: Using Technology The TI-83/84 Plus calculator can be used to find binomial probabilities. We will present details for its use at the end of this section. For now, see the typical screen display that lists binomial probabilities for $n = 4$ and $p = 0.2$.

TI-83/84 Plus

Given that we now have three different methods for finding binomial probabilities, here is an effective and efficient strategy:

1. Use computer software or a TI-83/84 Plus calculator, if available.
2. If neither software nor the TI-83/84 Plus calculator is available, use Table A-1, if possible.
3. If neither software nor the TI-83/84 Plus calculator is available and the probabilities can't be found using Table A-1, use the binomial probability formula.

Rationale for the Binomial Probability Formula

The binomial probability formula is the basis for all three methods presented in this section. Instead of accepting and using that formula blindly, let's see why it works.

Earlier in this section, we used the binomial probability formula for finding the probability of getting exactly 7 Mexican-Americans when 12 jurors are randomly selected from a population that is 80% Mexican-American. For each selection, the probability of getting a Mexican-American is 0.8. If we use the multiplication rule from Section 4-4, we get the following result:

$$\begin{aligned} &P(\text{selecting 7 Mexican-Americans followed by 5 people}\\ &\quad \text{who are not Mexican-American})\\ &\qquad = 0.8 \cdot 0.8 \cdot 0.8 \cdot 0.8 \cdot 0.8 \cdot 0.8 \cdot 0.8 \cdot 0.2 \cdot 0.2 \cdot 0.2 \cdot 0.2 \cdot 0.2\\ &\qquad = 0.8^7 \cdot 0.2^5\\ &\qquad = 0.0000671 \end{aligned}$$

This result isn't correct because it assumes that the *first* seven jurors are Mexican-Americans and the *last* five are not, but there are other arrangements possible for seven Mexican-Americans and five people who are not Mexican-American.

In Section 4-7 we saw that with seven subjects identical to each other (such as Mexican-Americans) and five other subjects identical to each other (such as non-Mexican-Americans), the total number of arrangements (permutations) is 12!/

$[(7 - 5)!7!]$ or 792. Each of those 792 different arrangements has a probability of $0.8^7 \cdot 0.2^5$, so the total probability is as follows:

$$P(7 \text{ Mexican-Americans among 12 jurors}) = \frac{12!}{(12-7)!7!} \cdot 0.8^7 \cdot 0.2^5$$

Generalize this result as follows: Replace 12 with n, replace 7 with x, replace 0.8 with p, replace 0.2 with q, and express the exponent of 5 as $12 - 7$, which can be replaced with $n - x$. The result is the binomial probability formula. That is, the binomial probability formula is a combination of the multiplication rule of probability and the counting rule for the number of arrangements of n items when x of them are identical to each other and the other $n - x$ are identical to each other. (See Exercises 13 and 14.)

$$P(x) = \overbrace{\frac{n!}{(n-x)!x!}}^{\text{The number of outcomes with exactly } x \text{ successes among } n \text{ trials}} \cdot \overbrace{p^x \cdot q^{n-x}}^{\text{The probability of } x \text{ successes among } n \text{ trials for any one particular order}}$$

TI Using a TI-83/84 Plus Calculator for the Binomial Probability Distribution

Finding a Single Probability Value In the first example of this section, we want to find the probability of exactly 7 Mexican-American jurors among 12 jurors randomly selected from a population that is 80% Mexican-American. Using the notation introduced in this section, we have $n = 12$, $x = 7$, and $p = 0.8$. That is, given $n = 12$ trials with probability of success $p = 0.8$, we want to find the probability of $x = 7$ successes. This is accomplished with the command binompdf(12, 0.8, 7), which is entered with the following procedure:

1. Gain access to the DISTR (distributions) menu by pressing **2ND** **VARS**.
2. The DISTR menu includes the following two items that relate to this section, but we will focus on the first of these two items:
   - **binompdf** Gives binomial probabilities for *individual* values of x.
   - **binomcdf** Gives *cumulative* probabilities for values from 0 up to, and including, a desired value of x.

To find the probability of $x = 7$ successes in $n = 12$ trials with probability $p = 0.8$, use the arrow keys to select **binompdf(**, then enter 12, 0.8, 7 and press **ENTER** to get a result of 0.0531502203.

The general format of the binompdf (binomial probability density function) command is **binompdf(*n*, *p*, *x*)** where n, p, and x are as described in this section. (n is the total number of trials, p is the probability of success on a *single* trial, and x is the number of successes among the n trials.)

Finding a Probability Distribution You can also obtain all of the probabilities for all values of x by omitting a specific entry for x, as in binompdf(n, p). For example, the command of binompdf(4, 0.2) will yield a list of these values: 0.4096, 0.4096, 0.1536, 0.0256, and 0.0016.

Calculator Hint: Store the list of probabilities in list L2 (by pressing STO▸ L2, then manually enter the x values of 0, 1, 2, . . ., n in L1, so that lists L1 and L2 combine to form a probability distribution, as in the sample screen display shown here. You could then use the TI-83/84 Plus calculator procedures described in Section 5-2 to find the mean and standard deviation or to generate a probability histogram.

Calculator Hint: If you know that you want to use your TI-83/84 Plus calculator to find a binomial probability, but you cannot remember the correct format of n, p, x, use the calculator's HELP feature. It is included with the TI-83/84 Plus Silver Edition, and it is an APP that can be downloaded to a TI-83 Plus calculator. To use HELP, scroll to the binompdf menu item, but press + *before* pressing ENTER. The screen will indicate that the correct format is **(numtrials, p, [,x]).** This is a reminder that the format of the required entry is n, p, and x, where the entry of x is optional.

Binomial Probability Distribution

5-3 BASIC SKILLS AND CONCEPTS

Statistical Literacy and Critical Thinking

1. Notation When using the binomial probability distribution for analyzing guesses on a multiple-choice quiz, what is wrong with letting p denote the probability of getting a correct answer while x counts the number of wrong answers?

2. Independence Assume that we want to use the binomial probability distribution for analyzing the genders when 12 jurors are randomly selected from a large population of potential jurors. If selection is made without replacement, are the selections independent? Can the selections be treated as being independent so that the binomial probability distribution can be used?

3. Table A-1 Because the binomial probabilities in Table A-1 are so easy to find, why don't we use that table every time that we need to find a binomial probability?

4. Binomial Probabilities When trying to find the probability of getting exactly two 6s when a die is rolled five times, why can't the answer be found as follows: Use the multiplication rule to find the probability of getting two 6s followed by three outcomes that are not 6, which is $(1/6)(1/6)(5/6)(5/6)(5/6)$?

Identifying Binomial Distributions. In Exercises 5–12, determine whether the given procedure results in a binomial distribution. For those that are not binomial, identify at least one requirement that is not satisfied.

5. Randomly selecting 12 jurors and recording their nationalities

6. Surveying 12 jurors and recording whether there is a "no" response when they are asked if they have ever been convicted of a felony

7. Treating 50 smokers with Nicorette and asking them how their mouth and throat feel

8. Treating 50 smokers with Nicorette and recording whether there is a "yes" response when they are asked if they experience any mouth or throat soreness

9. Recording the genders of 250 newborn babies

10. Recording the number of children in 250 families

11. Surveying 250 married couples and recording whether there is a "yes" response when they are asked if they have any children

12. Determining whether each of 500 defibrillators is acceptable or defective

13. Finding Probabilities When Guessing Answers Multiple-choice questions each have five possible answers (a, b, c, d, e), one of which is correct. Assume that you guess the answers to three such questions.

a. Use the multiplication rule to find the probability that the first two guesses are wrong and the third is correct. That is, find *P*(WWC), where C denotes a correct answer and W denotes a wrong answer.

b. Beginning with WWC, make a complete list of the different possible arrangements of two wrong answers and one correct answer, then find the probability for each entry in the list.

c. Based on the preceding results, what is the probability of getting exactly one correct answer when three guesses are made?

14. Finding Probabilities When Guessing Answers A test consists of multiple-choice questions, each having four possible answers (a, b, c, d), one of which is correct. Assume that you guess the answers to six such questions.

a. Use the multiplication rule to find the probability that the first two guesses are wrong and the last four guesses are correct. That is, find *P*(WWCCCC), where C denotes a correct answer and W denotes a wrong answer.

b. Beginning with WWCCCC, make a complete list of the different possible arrangements of two wrong answers and four correct answers, then find the probability for each entry in the list.

c. Based on the preceding results, what is the probability of getting exactly four correct answers when six guesses are made?

Using Table A-1. *In Exercises 15–20, assume that a procedure yields a binomial distribution with a trial repeated n times. Use Table A-1 to find the probability of x successes given the probability p of success on a given trial.*

15. $n = 3, x = 0, p = 0.05$ **16.** $n = 4, x = 3, p = 0.30$

17. $n = 8, x = 4, p = 0.05$ **18.** $n = 8, x = 7, p = 0.20$

19. $n = 14, x = 2, p = 0.30$ **20.** $n = 15, x = 12, p = 0.90$

Using the Binomial Probability Formula. *In Exercises 21–24, assume that a procedure yields a binomial distribution with a trial repeated n times. Use the binomial probability formula to find the probability of x successes given the probability p of success on a single trial.*

21. $n = 5, x = 2, p = 0.25$ **22.** $n = 6, x = 4, p = 0.75$

23. $n = 9, x = 3, p = 1/4$ **24.** $n = 10, x = 2, p = 2/3$

TI-83/84 Plus

L1	L2	L3
0	.3341	
1	.40188	
2	.20142	
3	.05384	
4	.0081	
5	6.5E-4	
6	2.2E-5	

L3(1)=

Using Calculator Results. *In Exercises 25–28, refer to the TI-83/84 Plus calculator display in the margin. The probabilities were obtained by entering the values of n = 6 and p = 0.167. In a clinical test of the drug Lipitor, 16.7% of the subjects treated with 10 mg of atorvastatin experienced headaches (based on data from Parke-Davis). In each case, assume that 6 subjects are randomly selected and treated with 10 mg of atorvastatin, then find the indicated probability.*

TI **25.** Find the probability that at least five of the subjects experience headaches. Is it unusual to have at least five of six subjects experience headaches?

TI **26.** Find the probability that at most two subjects experience headaches. Is it unusual to have at most two of six subjects experience headaches?

TI **27.** Find the probability that more than one subject experiences headaches. Is it unusual to have more than one of six subjects experience headaches?

TI **28.** Find the probability that at least one subject experiences headaches. Is it unusual to have at least one of six subjects experience headaches?

TI **29.** TV Viewer Surveys The CBS television show *60 Minutes* has been successful for many years. That show recently had a share of 20, meaning that among the TV sets in use, 20% were tuned to *60 Minutes* (based on data from Nielsen Media Research). Assume that an advertiser wants to verify that 20% share value by conducting its own survey, and a pilot survey begins with 10 households having TV sets in use at the time of a *60 Minutes* broadcast.

a. Find the probability that none of the households are tuned to *60 Minutes.*

b. Find the probability that at least one household is tuned to *60 Minutes.*

c. Find the probability that at most one household is tuned to *60 Minutes.*

d. If at most one household is tuned to *60 Minutes,* does it appear that the 20% share value is wrong? Why or why not?

TI **30.** IRS Audits The Hemingway Financial Company prepares tax returns for individuals. (Motto: "We also write great fiction.") According to the Internal Revenue Service, individuals making \$25,000–\$50,000 are audited at a rate of 1%. The Hemingway Company prepares five tax returns for individuals in that tax bracket, and three of them are audited.

a. Find the probability that when 5 people making \$25,000–\$50,000 are randomly selected, exactly 3 of them are audited.

b. Find the probability that at least 3 people are audited.

c. Based on the preceding results, what can you conclude about the Hemingway customers? Are they just unlucky, or are they being targeted for audits?

TI **31.** Acceptance Sampling The Medassist Pharmaceutical Company receives large shipments of aspirin tablets and uses this acceptance sampling plan: Randomly select and test 24 tablets, then accept the whole batch if there is only one or none that doesn't meet the required specifications. If a particular shipment of thousands of aspirin tablets actually has a 4% rate of defects, what is the probability that this whole shipment will be accepted?

TI **32.** Affirmative Action Programs A study was conducted to determine whether there were significant differences between medical students admitted through special programs (such as affirmative action) and medical students admitted through the regular admissions criteria. It was found that the graduation rate was 94% for the medical students admitted through special programs (based on data from the *Journal of the American Medical Association*).

a. If 10 of the students from the special programs are randomly selected, find the probability that at least 9 of them graduated.

b. Would it be unusual to randomly select 10 students from the special programs and get only 7 that graduate? Why or why not?

TI **33.** Overbooking Flights Air America has a policy of booking as many as 15 persons on an airplane that can seat only 14. (Past studies have revealed that only 85% of the booked passengers actually arrive for the flight.) Find the probability that if Air America books 15 persons, not enough seats will be available. Is this probability low enough so that overbooking is not a real concern for passengers?

TI **34.** Author's Slot Machine The author purchased a slot machine that is configured so that there is a $1/2000$ probability of winning the jackpot on any individual trial. Although no one would seriously consider tricking the author, suppose that a guest claims that she played the slot machine 5 times and hit the jackpot twice.

a. Find the probability of exactly two jackpots in 5 trials.
b. Find the probability of at least two jackpots in 5 trials.
c. Does the guest's claim of two jackpots in 5 trials seem valid? Explain.

TI **35.** Identifying Gender Discrimination After being rejected for employment, Kim Kelly learns that the Bellevue Credit Company has hired only two women among the last 20 new employees. She also learns that the pool of applicants is very large, with an approximately equal number of qualified men and women. Help her address the charge of gender discrimination by finding the probability of getting two or fewer women when 20 people are hired, assuming that there is no discrimination based on gender. Does the resulting probability really support such a charge?

TI **36.** Improving Quality The Write Right Company manufactures ballpoint pens and has been experiencing a 5% rate of defective pens. Modifications are made to the manufacturing process in an attempt to improve quality, and the manager claims that the modified procedure is better, because a test of 50 pens shows that only one is defective.

a. Assuming that the 5% rate of defects has not changed, find the probability that among 50 pens, exactly one is defective.
b. Assuming that the 5% rate of defects has not changed, find the probability that among 50 pens, none are defective.
c. What probability value should be used for determining whether the modified process results in a defect rate that is less than 5%?
d. What do you conclude about the effectiveness of the modified production process?

5-3 BEYOND THE BASICS

37. Geometric Distribution If a procedure meets all the conditions of a binomial distribution except that the number of trials is not fixed, then the **geometric distribution** can be used. The probability of getting the first success on the xth trial is given by $P(x) = p(1 - p)^{x-1}$ where p is the probability of success on any one trial. Assume that the probability of a defective computer component is 0.2. Find the probability that the first defect is found in the seventh component tested.

38. Hypergeometric Distribution If we sample from a small finite population without replacement, the binomial distribution should not be used because the events are not independent. If sampling is done without replacement and the outcomes belong to one of two types, we can use the **hypergeometric distribution.** If a population has A objects of one type, while the remaining B objects are of the other type, and if n objects are sampled without replacement, then the probability of getting x objects of type A and $n - x$ objects of type B is

$$P(x) = \frac{A!}{(A - x)!x!} \cdot \frac{B!}{(B - n + x)!(n - x)!} \div \frac{(A + B)!}{(A + B - n)!n!}$$

In Lotto 54, a bettor selects six numbers from 1 to 54 (without repetition), and a winning six-number combination is later randomly selected. Find the probability of getting

a. all six winning numbers.
b. exactly five of the winning numbers.
c. exactly three of the winning numbers.
d. no winning numbers.

39. Multinomial Distribution The binomial distribution applies only to cases involving two types of outcomes, whereas the **multinomial distribution** involves more than two categories. Suppose we have three types of mutually exclusive outcomes denoted by A, B, and C. Let $P(A) = p_1$, $P(B) = p_2$, and $P(C) = p_3$. In n independent trials, the probability of x_1 outcomes of type A, x_2 outcomes of type B, and x_3 outcomes of type C is given by

$$\frac{n!}{(x_1!)(x_2!)(x_3!)} \cdot p_1^{x_1} \cdot p_2^{x_2} \cdot p_3^{x_3}$$

A genetics experiment involves six mutually exclusive genotypes identified as A, B, C, D, E, and F, and they are all equally likely. If 20 offspring are tested, find the probability of getting exactly five A's, four B's, three C's, two D's, three E's, and three F's by expanding the above expression so that it applies to six types of outcomes instead of only three.

5-4 Mean, Variance, and Standard Deviation for the Binomial Distribution

Key Concept Section 5-3 introduced the binomial probability distribution, and in this section we consider important characteristics of a binomial distribution, including center, variation, and distribution. That is, given a particular binomial probability distribution, we will present methods for finding its mean, variance, and standard deviation. As in earlier sections, the objective is not to simply find those values, but to *interpret* them and *understand* them.

Section 5-2 included methods for finding the mean, variance, and standard deviation from a discrete probability distribution. Because a binomial distribution is a special type of probability distribution, we could use Formulas 5-1, 5-3, and 5-4 (from Section 5-2) for finding the mean, variance, and standard deviation, but those formulas can be greatly simplified for binomial distributions, as shown below.

For Any Discrete Probability Distribution		**For Binomial Distributions**	
Formula 5-1	$\mu = \Sigma[x \cdot P(x)]$	Formula 5-6	$\mu = np$
Formula 5-3	$\sigma^2 = \Sigma[x^2 \cdot P(x)] - \mu^2$	Formula 5-7	$\sigma^2 = npq$
Formula 5-4	$\sigma = \sqrt{\Sigma[x^2 \cdot P(x)] - \mu^2}$	Formula 5-8	$\sigma = \sqrt{npq}$

As in earlier sections, finding values for μ and σ is fine, but it is especially important to *interpret* and *understand* those values, so the range rule of thumb can be very helpful. Using the range rule of thumb, we can consider values to be unusual if they fall outside of the limits obtained from the following:

maximum usual value: $\mu + 2\sigma$
minimum usual value: $\mu - 2\sigma$

CHAPTER 5 PROBLEM

EXAMPLE Selecting Jurors In Section 5-2 we included an example illustrating calculations for μ and σ. We used the example of the random variable x representing the number of Mexican-Americans on a jury of 12 people. (We are assuming that the jurors are

continued

randomly selected from a population that is 80% Mexican-American. See Table 5-3 on page 217 for the calculations that illustrate Formulas 5-1 and 5-4.) Use Formulas 5-6 and 5-8 to find the mean and standard deviation for the numbers of Mexican-Americans on juries selected from this population that is 80% Mexican-American.

SOLUTION Using the values $n = 12$, $p = 0.8$, and $q = 0.2$, Formulas 5-6 and 5-8 can be applied as follows:

$$\mu = np = (12)(0.8) = 9.6$$

$$\sigma = \sqrt{npq} = \sqrt{(12)(0.8)(0.2)} = 1.4 \quad \text{(rounded)}$$

If you compare these calculations to the calculations in Table 5-3, it should be obvious that Formulas 5-6 and 5-8 are substantially easier to use.

Formula 5-6 for the mean makes sense intuitively. If 80% of a population is Mexican-American and 12 people are randomly selected, we expect to get around $12 \cdot 0.8 = 9.6$ Mexican-Americans, and this result can be easily generalized as $\mu = np$. The variance and standard deviation are not so easily justified, and we will omit the complicated algebraic manipulations that lead to Formulas 5-7 and 5-8. Instead, refer again to the preceding example and Table 5-3 to verify that for a binomial distribution, Formulas 5-6, 5-7, and 5-8 will produce the same results as Formulas 5-1, 5-3, and 5-4.

EXAMPLE Grand Jury Selection The Chapter Problem notes that in Hidalgo County, Texas, 80% of those eligible for jury duty were Mexican-Americans. It was also noted that during a period of 11 years, 870 people were selected for duty on a grand jury.

a. Assuming that groups of 870 grand jurors are randomly selected, find the mean and standard deviation for the numbers of Mexican-Americans.

b. Use the range rule of thumb to find the minimum usual number and the maximum usual number of Mexican-Americans. Based on those numbers, can we conclude that the actual result of 339 Mexican-Americans is *unusual*? Does this suggest that the selection process discriminated against Mexican-Americans?

SOLUTION

a. Assuming that jurors were randomly selected, we have $n = 870$ people selected with $p = 0.80$, and $q = 0.20$. We can find the mean and standard deviation for the number of Mexican-Americans by using Formulas 5-6 and 5-8 as follows:

$$\mu = np = (870)(0.80) = 696.0$$

$$\sigma = \sqrt{npq} = \sqrt{(870)(0.80)(0.20)} = 11.8$$

For groups of 870 randomly selected jurors, the mean number of Mexican-Americans is 696.0 and the standard deviation is 11.8.

b. We must now interpret the results to determine whether 339 Mexican-Americans is a result that could easily occur by chance, or whether that result is so unlikely that the selection process appears to be discriminatory. We will use the range rule of thumb as follows:

$$\text{maximum usual value: } \mu + 2\sigma = 696.0 + 2(11.8) = 719.6$$

$$\text{minimum usual value: } \mu - 2\sigma = 696.0 - 2(11.8) = 672.4$$

INTERPRETATION According to the range rule of thumb, values are considered to be usual if they are between 672.4 and 719.6, so 339 Mexican-Americans is an unusual result because it is not between those two values. It is very unlikely that we would get as few as 339 Mexican-Americans just by chance. In fact, the Supreme Court ruled that the result of only 339 Mexican-Americans was significant evidence of a jury selection process that is biased. The *Castaneda v. Partida* decision became an important judicial ruling, and it was actually based on application of the binomial probability distribution.

Remember that finding values for the mean μ and standard deviation σ is important, but it is particularly important to be able to *interpret* those values by using such devices as the range rule of thumb for identifying a range of usual values.

5-4 BASIC SKILLS AND CONCEPTS

Statistical Literacy and Critical Thinking

1. Identifying Unusual Values If we consider an experiment of generating 100 births and recording the genders of the babies, the mean number of girls is 50 and the standard deviation is 5 girls. Would it be *unusual* to get 70 girls in 100 births? Why or why not?

2. Identifying Unusual Values A manufacturing process has a defect rate of 10%, meaning that 10% of the items produced are defective. If batches of 80 items are produced, the mean number of defects per batch is 8.0 and the standard deviation is 2.7. Would it be *unusual* to get only five defects in a batch? Why or why not?

3. Variance A researcher plans an experimental design in such a way that when randomly selecting treatment groups of people, the mean number of females is 3.0 and the standard deviation is 1.2 females. What is the variance? (Express the answer including the appropriate units.)

4. Mean and Standard Deviation A researcher conducts an observational study, then uses the methods of this section to find that the mean is 5.0 while the standard deviation is −2.0. What is wrong with these results?

Finding μ, σ, and Unusual Values. *In Exercises 5–8, assume that a procedure yields a binomial distribution with n trials and the probability of success for one trial is p. Use the given values of n and p to find the mean μ and standard deviation σ. Also, use the range rule of thumb to find the minimum usual value $\mu - 2\sigma$ and the maximum usual value $\mu + 2\sigma$.*

5. $n = 200, p = 0.4$

6. $n = 60, p = 0.25$

continued

7. $n = 1492, p = 1/4$

8. $n = 1068, p = 2/3$

9. Guessing Answers Several psychology students are unprepared for a surprise true/false test with 16 questions, and all of their answers are guesses.

a. Find the mean and standard deviation for the number of correct answers for such students.

b. Would it be unusual for a student to pass by guessing and getting at least 10 correct answers? Why or why not?

10. Guessing Answers Several economics students are unprepared for a multiple-choice quiz with 25 questions, and all of their answers are guesses. Each question has five possible answers, and only one of them is correct.

a. Find the mean and standard deviation for the number of correct answers for such students.

b. Would it be unusual for a student to pass by guessing and getting at least 15 correct answers? Why or why not?

11. Are 20% of M&M Candies Orange? Mars, Inc. claims that 20% of its M&M plain candies are orange, and a sample of 100 such candies is randomly selected.

a. Find the mean and standard deviation for the number of orange candies in such groups of 100.

b. Data Set 13 in Appendix B consists of a random sample of 100 M&Ms in which 25 are orange. Is this result unusual? Does it seem that the claimed rate of 20% is wrong?

12. Are 14% of M&M Candies Yellow? Mars, Inc. claims that 14% of its M&M plain candies are yellow, and a sample of 100 such candies is randomly selected.

a. Find the mean and standard deviation for the number of yellow candies in such groups of 100.

b. Data Set 13 in Appendix B consists of a random sample of 100 M&Ms in which 8 are yellow. Is this result unusual? Does it seem that the claimed rate of 14% is wrong?

13. Gender Selection In a test of the MicroSort method of gender selection, 325 babies are born to couples trying to have baby girls, and 295 of those babies are girls (based on data from the Genetics & IVF Institute).

a. If the gender-selection method has no effect and boys and girls are equally likely, find the mean and standard deviation for the numbers of girls born in groups of 325.

b. Is the result of 295 girls unusual? Does it suggest that the gender-selection method appears to be effective?

14. Gender Selection In a test of the MicroSort method of gender selection, 51 babies are born to couples trying to have baby boys, and 39 of those babies are boys (based on data from the Genetics & IVF Institute).

a. If the gender-selection method has no effect and boys and girls are equally likely, find the mean and standard deviation for the numbers of boys born in groups of 51.

b. Is the result of 39 boys unusual? Does it suggest that the gender-selection method appears to be effective?

15. Deciphering Messages The Central Intelligence Agency has specialists who analyze the frequencies of letters of the alphabet in an attempt to decipher intercepted messages. In standard English text, the letter *r* is used at a rate of 7.7%.

a. Find the mean and standard deviation for the number of times the letter *r* will be found on a typical page of 2600 characters.

b. In an intercepted message sent to Iraq, a page of 2600 characters is found to have the letter *r* occurring 175 times. Is this unusual?

16. Mendelian Genetics When Mendel conducted his famous genetics experiments with peas, one sample of offspring consisted of 580 peas, and Mendel theorized that 25% of them would be yellow peas.

a. If Mendel's theory is correct, find the mean and standard deviation for the numbers of yellow peas in such groups of 580 offspring peas.

b. The actual results consisted of 152 yellow peas. Is that result unusual? What does this result suggest about Mendel's theory?

17. Voting In a past presidential election, the actual voter turnout was 61%. In a survey, 1002 subjects were asked if they voted in the presidential election.

a. Find the mean and standard deviation for the numbers of actual voters in groups of 1002.

b. In the survey of 1002 people, 701 *said* that they voted in the last presidential election (based on data from ICR Research Group). Is this result consistent with the actual voter turnout, or is this result unlikely to occur with an actual voter turnout of 61%? Why or why not?

c. Based on these results, does it appear that accurate voting results can be obtained by asking voters how they acted?

18. Cell Phones and Brain Cancer In a study of 420,095 cell phone users in Denmark, it was found that 135 developed cancer of the brain or nervous system. If we assume that such cancer is not affected by cell phones, the probability of a person having such a cancer is 0.000340.

a. Assuming that cell phones have no effect on cancer, find the mean and standard deviation for the numbers of people in groups of 420,095 that can be expected to have cancer of the brain or nervous system.

b. Based on the results from part (a), is it unusual to find that among 420,095 people, there are 135 cases of cancer of the brain or nervous system? Why or why not?

c. What do these results suggest about the publicized concern that cell phones are a health danger because they increase the risk of cancer of the brain or nervous system?

19. Cholesterol Drug In a clinical trial of Lipitor (atorvastatin), a common drug used to lower cholesterol, 863 patients were given a treatment of 10-mg atorvastatin tablets. That group consists of 19 patients who experienced flu symptoms (based on data from Pfizer, Inc.). The probability of flu symptoms for a person not receiving any treatment is 0.019.

a. Assuming that Lipitor has no effect on flu symptoms, find the mean and standard deviation for the numbers of people in groups of 863 that can be expected to have flu symptoms.

b. Based on the result from part (a), is it unusual to find that among 863 people, there are 19 who experience flu symptoms? Why or why not?

c. Based on the preceding results, do flu symptoms appear to be an adverse reaction that should be of concern to those who use Lipitor?

20. Test of Touch Therapy Nine-year-old Emily Rosa conducted this test: A professional touch therapist put both hands through a cardboard partition and Emily would use a coin toss to randomly select one of the hands. Emily would place her hand just above the hand of the therapist, who was then asked to identify the hand that Emily had selected. The touch therapists believed that they could sense the energy field and identify the hand that Emily had selected. The trial was repeated 280 times. (Based on data from "A Close Look at Therapeutic Touch," by Rosa et al., *Journal of the American Medical Association,* Vol. 279, No. 13.)

a. Assuming that the touch therapists have no special powers and made random guesses, find the mean and standard deviation for the numbers of correct responses in groups of 280 trials.

b. The professional touch therapists identified the correct hand 123 times in the 280 trials. Is that result unusual? What does the result suggest about the ability of touch therapists to select the correct hand by sensing an energy field?

5-4 BEYOND THE BASICS

21. Using the Empirical Rule An experiment is designed to test the effectiveness of the MicroSort method of gender selection, and 100 couples try to have baby girls using the MicroSort method. Assume that boys and girls are equally likely and also assume that the method of gender selection has no effect.

a. Using the methods of this section, what are the minimum and maximum usual numbers of girls in groups of 100 randomly selected babies?

b. The empirical rule (see Section 3-3) applies to distributions that are bell-shaped. Is the binomial probability distribution for this experiment (approximately) bell-shaped? How do you know?

c. Assuming that the distribution is bell-shaped, how likely is it that the number of girls will fall between 40 and 60 (according to the empirical rule)?

22. Acceptable/Defective Products Mario's Pizza Parlor has just opened. Due to a lack of employee training, there is only a 0.8 probability that a pizza will be edible. An order for five pizzas has just been placed. What is the minimum number of pizzas that must be made in order to be at least 99% sure that there will be five that are edible?

Skip

5-5 Poisson Probability Distributions

Key Concept This section introduces the *Poisson distribution*, which is an important discrete probability distribution. It is important because it is often used for describing the behavior of rare events (with small probabilities). We should know the requirements for using the Poisson distribution, and we should know how to calculate probabilities using Formula 5-9. We should also know that when the Poisson distribution applies to a variable with mean μ, the standard deviation is $\sqrt{\mu}$.

The Poisson distribution is used for describing behavior such as radioactive decay, arrivals of people in a line, eagles nesting in a region, patients arriving at an emergency room, and Internet users logging onto a Web site. For example, suppose your local hospital experiences a mean of 2.3 patients arriving at the emergency room on Fridays between 10:00 P.M. and 11:00 P.M. We can find the probability that for a randomly selected Friday between 10:00 P.M. and 11:00 P.M., exactly four patients arrive. We use the Poisson distribution, defined as follows.

Definition

The **Poisson distribution** is a discrete probability distribution that applies to occurrences of some event *over a specified interval.* The random variable x is the number of occurrences of the event in an interval. The interval can be time, distance, area, volume, or some similar unit. The probability of the event occurring x times over an interval is given by Formula 5-9.

Formula 5-9 $$P(x) = \frac{\mu^x \cdot e^{-\mu}}{x!} \quad \text{where } e \approx 2.71828$$

Requirements for the Poisson Distribution

- The random variable x is the number of occurrences of an event *over some interval.*
- The occurrences must be *random.*
- The occurrences must be *independent* of each other.
- The occurrences must be *uniformly distributed* over the interval being used.

Parameters of the Poisson Distribution

- The mean is μ.
- The standard deviation is $\sigma = \sqrt{\mu}$.

A Poisson distribution differs from a binomial distribution in these fundamental ways:

1. The binomial distribution is affected by the sample size n and the probability p, whereas the Poisson distribution is affected only by the mean μ.
2. In a binomial distribution, the possible values of the random variable x are 0, 1, . . . , n, but a Poisson distribution has possible x values of 0, 1, 2, . . . , with no upper limit.

EXAMPLE World War II Bombs In analyzing hits by V-1 buzz bombs in World War II, South London was subdivided into 576 regions, each with an area of 0.25 km^2. A total of 535 bombs hit the combined area of 576 regions.

a. If a region is randomly selected, find the probability that it was hit exactly twice.

b. Based on the probability found in part (a), how many of the 576 regions are expected to be hit exactly twice?

continued

Queues

Queuing theory is a branch of mathematics that uses probability and statistics. The study of queues, or waiting lines, is important to businesses such as supermarkets, banks, fast-food restaurants, airlines, and amusement parks. Grand Union supermarkets try to keep checkout lines no longer than three shoppers. Wendy's introduced the "Express Pak" to expedite servicing its numerous drive-through customers. Disney conducts extensive studies of lines at its amusement parks so that it can keep patrons happy and plan for expansion. Bell Laboratories uses queuing theory to optimize telephone network usage, and factories use it to design efficient production lines.

SOLUTION

a. The Poisson distribution applies because we are dealing with the occurrences of an event (bomb hits) over some interval (a region with area of $0.25\ \text{km}^2$). The mean number of hits per region is

$$\mu = \frac{\text{number of bomb hits}}{\text{number of regions}} = \frac{535}{576} = 0.929$$

Because we want the probability of exactly two hits in a region, we let $x = 2$ and use Formula 5-9 as follows:

$$P(x) = \frac{\mu^x \cdot e^{-\mu}}{x!} = \frac{0.929^2 \cdot 2.71828^{-0.929}}{2!} = \frac{0.863 \cdot 0.395}{2} = 0.170$$

The probability of a particular region being hit exactly twice is $P(2) = 0.170$.

b. Because there is a probability of 0.170 that a region is hit exactly twice, we expect that among the 576 regions, the number that is hit exactly twice is $576 \cdot 0.170 = 97.9$.

In the preceding example, we can also calculate the probabilities and expected values for 0, 1, 3, 4, and 5 hits. (We stop at $x = 5$ because no region was hit more than five times, and the probabilities for $x > 5$ are 0.000 when rounded to three decimal places.) Those probabilities and expected values are listed in Table 5-5. The fourth column of Table 5-5 describes the results that actually occurred during World War II. There were 229 regions that had no hits, 211 regions that were hit once, and so on. We can now compare the frequencies *predicted* with the Poisson distribution (third column) to the *actual* frequencies (fourth column) to conclude that there is very good agreement. In this case, the Poisson distribution does a good job of predicting the results that actually occurred. (Section 11-2 describes a statistical procedure for determining whether such expected frequencies constitute a good "fit" to the actual frequencies. That procedure does suggest that there is a good fit in this case.)

Poisson as Approximation to Binomial

The Poisson distribution is sometimes used to approximate the binomial distribution when n is large and p is small. One rule of thumb is to use such an approximation when the following two conditions are both satisfied.

Requirements for Using the Poisson Distribution as an Approximation to the Binomial

1. $n \geq 100$
2. $np \leq 10$

Table 5-5	V-1 Buzz Bomb Hits for 576 Regions in South London		
Number of Bomb Hits	Probability	Expected Number of Regions	Actual Number of Regions
0	0.395	227.5	229
1	0.367	211.4	211
2	0.170	97.9	93
3	0.053	30.5	35
4	0.012	6.9	7
5	0.002	1.2	1

If both of these conditions are satisfied and we want to use the Poisson distribution as an approximation to the binomial distribution, we need a value for μ, and that value can be calculated by using Formula 5-6 (first presented in Section 5-4):

Formula 5-6 $$\mu = np$$

EXAMPLE Kentucky Pick 4 In Kentucky's Pick 4 game, you pay $1 to select a sequence of four digits, such as 2283. If you play this game once every day, find the probability of winning exactly once in 365 days.

SOLUTION Because the time interval is 365 days, $n = 365$. Because there is one winning set of numbers among the 10,000 that are possible (from 0000 to 9999), $p = 1/10{,}000$. The conditions $n \geq 100$ and $np \leq 10$ are both satisfied, so we can use the Poisson distribution as an approximation to the binomial distribution. We first need the value of μ, which is found as follows:

$$\mu = np = 365 \cdot \frac{1}{10{,}000} = 0.0365$$

Having found the value of μ, we can now find $P(1)$:

$$P(1) = \frac{\mu^x \cdot e^{-\mu}}{x!} = \frac{0.0365^1 \cdot 2.71828^{-0.0365}}{1!} = \frac{0.0352}{1} = 0.0352$$

Using the Poisson distribution as an approximation to the binomial distribution, we find that there is a 0.0352 probability of winning exactly once in 365 days. If we use the binomial distribution, we again get 0.0352, so we can see that the Poisson approximation is quite good here. (Carrying more decimal places would show that the Poisson approximation yields 0.03519177 and the more accurate binomial result is 0.03519523.)

TI Using the TI-83/84 Plus Calculator for the Poisson Distribution

To find an individual probability value from the Poisson distribution, use the following procedure:

1. Gain access to the DISTR (distributions) menu by pressing 2ND VARS.
2. Use the arrow keys to select **poissonpdf.**
3. The general format of the **poissonpdf** (Poisson probability density function) command is **poissonpdf(μ, x),** where μ is the mean and x is the desired number of successes. For the last example in this section, $\mu = 0.0365$ and $x = 1$, so enter poissonpdf(0.0365, 1) to get a result of 0.0351917704, which can be rounded to 0.0352.

Calculator Hint: If you know that you want to use your TI-83/84 Plus calculator to find a probability using the Poisson distribution, but you cannot remember the correct format of μ, x, use the calculator's HELP feature. It is included with the TI-83/84 Plus Silver Edition, and it is an APP that can be downloaded to a TI-83 Plus or TI-84 Plus calculator. To use HELP, scroll to the poissonpdf menu item, but press + before pressing ENTER. The screen will indicate that the correct format is (μ, x).

5-5 BASIC SKILLS AND CONCEPTS

Statistical Literacy and Critical Thinking

1. Poisson Distribution What are the conditions for using the Poisson distribution?

2. Poisson Distribution The random variable x represents the number of phone calls received in an hour, and it has a Poisson distribution with a mean of 9. What is its standard deviation? What is its variance?

3. Parameters When attempting to apply the Poisson distribution, which of the following must be known: mean, standard deviation, variance, shape of the distribution?

4. Poisson/Binomial An experiment involves rolling a die 6 times and counting the number of 2s that occur. If we calculate the probability of $x = 0$ occurrences of 2 using the Poisson distribution, we get 0.368, but we get 0.335 if we use the binomial distribution. Which is the correct probability of getting no 2s when a die is rolled 6 times? Why is the other probability wrong?

Using a Poisson Distribution to Find Probability. *In Exercises 5–8, assume that the Poisson distribution applies and proceed to use the given mean to find the indicated probability.*

TI 5. If $\mu = 5$, find $P(4)$.

TI 6. If $\mu = 3/4$, find $P(2)$.

TI 7. If $\mu = 0.5$, find $P(3)$.

TI 8. If $\mu = 3.25$, find $P(5)$.

In Exercises 9–14, use the Poisson distribution to find the indicated probabilities.

TI 9. Dandelions Dandelions are studied for their effects on crop production and lawn growth. In one region, the mean number of dandelions per square meter was found to be 7.0 (based on data from Manitoba Agriculture and Food).
a. Find the probability of no dandelions in an area of 1 m^2.
b. Find the probability of at least one dandelion in an area of 1 m^2.
c. Find the probability of at most two dandelions in an area of 1 m^2.

TI **10.** Phone Calls The author found that in one month (30 days), he made 47 cell phone calls, which were distributed as follows: No calls were made on 17 days, 1 call was made on each of 7 days, 3 calls were made on each of two days, 4 calls were made on each of two days, 12 calls were made on one day, and 14 calls were made on one day.

a. Find the mean number of calls per day.

b. Use the Poisson distribution to find the probability of no calls on a day, and compare the result to the actual relative frequency for the number of days with no calls.

c. Use the Poisson distribution to find the probability of one call on a day, and compare the result to the actual relative frequency for the number of days with one call.

d. Based on the preceding results, does it appear that the author's cell phone calls made in a day fit the Poisson distribution reasonably well? Why or why not?

TI **11.** Radioactive Decay Radioactive atoms are unstable because they have too much energy. When they release their extra energy, they are said to decay. When studying cesium-137, it is found that during the course of decay over 365 days, 1,000,000 radioactive atoms are reduced to 977,287 radioactive atoms.

a. Find the mean number of radioactive atoms lost through decay in a day.

b. Find the probability that on a given day, 50 radioactive atoms decayed.

TI **12.** Deaths from Horse Kicks A classical example of the Poisson distribution involves the number of deaths caused by horse kicks to men in the Prussian Army between 1875 and 1894. Data for 14 corps were combined for the 20-year period, and the 280 corps-years included a total of 196 deaths. After finding the mean number of deaths per corps-year, find the probability that a randomly selected corps-year has the following numbers of deaths.

a. 0 **b.** 1 **c.** 2 **d.** 3 **e.** 4

The actual results consisted of these frequencies: 0 deaths (in 144 corps-years); 1 death (in 91 corps-years); 2 deaths (in 32 corps-years); 3 deaths (in 11 corps-years); 4 deaths (in 2 corps-years). Compare the actual results to those expected from the Poisson probabilities. Does the Poisson distribution serve as a good device for predicting the actual results?

TI **13.** Homicide Deaths In one year, there were 116 homicide deaths in Richmond, Virginia (based on "A Classroom Note on the Poisson Distribution: A Model for Homicidal Deaths in Richmond, Va for 1991," by Winston A. Richards in *Mathematics and Computer Education*). For a randomly selected day, find the probability that the number of homicide deaths is

a. 0 **b.** 1 **c.** 2 **d.** 3 **e.** 4

Compare the calculated probabilities to these actual results: 268 days (no homicides); 79 days (1 homicide); 17 days (2 homicides); 1 day (3 homicides); no days with more than 3 homicides.

TI **14.** Earthquakes For a recent period of 100 years, there were 93 major earthquakes (at least 6.0 on the Richter scale) in the world (based on data from the *World Almanac and Book of Facts*). Assuming that the Poisson distribution is a suitable model, find the mean number of major earthquakes per year, then find the probability that the number of earthquakes in a randomly selected year is

a. 0 **b.** 1 **c.** 2 **d.** 3 **e.** 4
f. 5 **g.** 6 **h.** 7

continued

Here are the actual results: 47 years (0 major earthquakes); 31 years (1 major earthquake); 13 years (2 major earthquakes); 5 years (3 major earthquakes); 2 years (4 major earthquakes); 0 years (5 major earthquakes); 1 year (6 major earthquakes); 1 year (7 major earthquakes). After comparing the calculated probabilities to the actual results, is the Poisson distribution a good model?

5-5 BEYOND THE BASICS

TI **15.** Poisson Approximation to Binomial The Poisson distribution can be used to approximate a binomial distribution if $n \geq 100$ and $np \leq 10$. Assume that we have a binomial distribution with $n = 100$ and $p = 0.1$. It is impossible to get 101 successes in such a binomial distribution, but we *can* compute the probability that $x = 101$ with the Poisson approximation. Find that value. How does the result agree with the impossibility of having $x = 101$ with a binomial distribution?

TI **16.** Poisson Approximation to Binomial For a binomial distribution with $n = 10$ and $p = 0.5$, we should not use the Poisson approximation because the conditions $n \geq 100$ and $np \leq 10$ are not both satisfied. Suppose we go way out on a limb and use the Poisson approximation anyway. Are the resulting probabilities unacceptable approximations? Why or why not?

Review

The concept of a probability distribution is a key element of statistics. A probability distribution describes the probability for each value of a random variable. This chapter includes only discrete probability distributions, but the following chapters will include continuous probability distributions. The following key points were discussed:

- A *random variable* has values that are determined by chance.
- A *probability distribution* consists of all values of a random variable, along with their corresponding probabilities. A probability distribution must satisfy two requirements: $\Sigma P(x) = 1$ and, for each value of x, $0 \leq P(x) \leq 1$.
- Important characteristics of a *probability distribution* can be explored by constructing a probability histogram and by computing its mean and standard deviation using these formulas:

$$\mu = \Sigma[x \cdot P(x)]$$

$$\sigma = \sqrt{\Sigma[x^2 \cdot P(x)] - \mu^2}$$

- In a *binomial distribution,* there are two categories of outcomes and a fixed number of independent trials with a constant probability. The probability of x successes among n trials can be found by using the binomial probability formula, or Table A-1, or software (such as STATDISK, Minitab, or Excel), or a TI-83/84 Plus calculator.
- In a binomial distribution, the mean and standard deviation can be easily found by calculating the values of $\mu = np$ and $\sigma = \sqrt{npq}$.
- A *Poisson probability distribution* applies to occurrences of some event over a specific interval, and its probabilities can be computed with Formula 5-9.

- *Unusual outcomes:* This chapter stressed the importance of interpreting results by distinguishing between outcomes that are usual and those that are unusual. We used two different criteria: the range rule of thumb and the use of probabilities.

Using the range rule of thumb to identify unusual values:

$$\textbf{maximum usual value} = \mu + 2\sigma$$

$$\textbf{minimum usual value} = \mu - 2\sigma$$

Using probabilities to identify unusual values:

- **Unusually high number of successes:** x successes among n trials is an *unusually high* number of successes if $P(x \text{ or more}) \leq 0.05$.*
- **Unusually low number of successes:** x successes among n trials is an *unusually low* number of successes if $P(x \text{ or fewer}) \leq 0.05$.*

Statistical Literacy and Critical Thinking

1. Probability Distribution What is a probability distribution?

2. Probability Distribution What are the requirements of a probability distribution?

3. Discrete Probability Distribution The probability distributions described in this chapter are *discrete*. What makes them *discrete?* What other type of probability distribution is there?

4. Probability Distributions This chapter described the concept of a discrete probability distribution, and then described the binomial and Poisson probability distributions. Are all discrete probability distributions either binomial or Poisson? Why or why not?

Review Exercises

TI **1.** Multiple-Choice Test Because they are so easy to correct, multiple-choice questions are commonly used for class tests, SAT tests, MCAT tests for medical schools, and many other circumstances. The table in the margin describes the probability distribution for the number of correct responses when someone makes random guesses for 10 multiple-choice questions on an SAT test. Each question has 5 possible answers (a, b, c, d, e), one of which is correct. Assume that random guesses are made for each of the 10 questions.

a. Verify that the table satisfies the requirements necessary for a probability distribution.

b. Find the mean number of correct responses.

c. Find the standard deviation for the numbers of correct responses when random guesses are made for the 10 questions by many different subjects.

d. What is the probability that someone gets at least half of the questions correct?

continued

x	$P(x)$
0	0.107
1	0.268
2	0.302
3	0.201
4	0.088
5	0.026
6	0.006
7	0.001
8	0+
9	0+
10	0+

*The value of 0.05 is commonly used, but is not absolutely rigid. Other values, such as 0.01, could be used to distinguish between events that can easily occur by chance and events that are very unlikely to occur by chance.

e. When someone makes guesses for all 10 answers, what is the expected number of correct answers?

f. What is the probability of getting at least 1 answer correct?

g. If someone gets at least 1 answer correct, does that mean that this person knows something about the subject matter being tested?

TI **2.** TV Ratings The television show *Cold Case* has a 15 share, meaning that while it is being broadcast, 15% of the TV sets in use are tuned to *Cold Case* (based on data from Nielsen Media Research). A special focus group consists of 12 randomly selected households (each with one TV set in use during the time of a *Cold Case* broadcast).

a. What is the expected number of sets tuned to *Cold Case*?

b. In such groups of 12, what is the mean number of sets tuned to *Cold Case*?

c. In such groups of 12, what is the standard deviation for the number of sets tuned to *Cold Case*?

d. For such a group of 12, find the probability that exactly 3 TV sets are tuned to *Cold Case.*

e. For such a group of 12, would it be unusual to find that no sets are tuned to *Cold Case*? Why or why not?

TI **3.** Reasons for Being Fired "Inability to get along with others" is the reason cited in 17% of worker firings (based on data from Robert Half International, Inc.). Concerned about her company's working conditions, the personnel manager at the Boston Finance Company plans to investigate the five employee firings that occurred over the past year.

a. Assuming that the 17% rate applies, find the probability that among those five employees, the number fired because of an inability to get along with others is at least four.

b. If the personnel manager actually does find that at least four of the firings are due to an inability to get along with others, does this company appear to be very different from other typical companies? Why or why not?

TI **4.** Deaths Currently, an average of 7 residents of the village of Westport (population 760) die each year (based on data from the National Center for Health Statistics).

a. Find the mean number of deaths per day.

b. Find the probability that on a given day, there are no deaths.

c. Find the probability that on a given day, there is one death.

d. Find the probability that on a given day, there is more than one death.

e. Based on the preceding results, should Westport have a contingency plan to handle more than one death per day? Why or why not?

Cumulative Review Exercises

TI **1.** Weights: Analysis of Last Digits The accompanying table lists the last digits of weights of the subjects listed in Data Set 1 in Appendix B. The last digits of a data set can sometimes be used to determine whether the data have been measured or simply reported. The presence of disproportionately more 0s and 5s is often a sure indicator that the data have been reported instead of measured.

x	f
0	7
1	14
2	5
3	11
4	8
5	4
6	5
7	6
8	12
9	8

a. Find the mean and standard deviation of those last digits.

b. Construct the relative frequency table that corresponds to the given frequency table.

c. Construct a table for the probability distribution of randomly selected digits that are all equally likely. List the values of the random variable x (0, 1, 2, . . . , 9) along with their corresponding probabilities (0.1, 0.1, 0.1, . . . , 0.1), then find the mean and standard deviation of this probability distribution.

d. Recognizing that sample data naturally deviate from the results we theoretically expect, does it seem that the given last digits roughly agree with the distribution we expect with random selection? Or does it seem that there is something about the sample data (such as disproportionately more 0s and 5s) suggesting that the given last digits are not random? (In Chapter 11, we will present a method for answering such questions much more objectively.)

2. Determining the Effectiveness of an HIV Training Program The New York State Health Department reports a 10% rate of the HIV virus for the "at-risk" population. In one region, an intensive education program is used in an attempt to lower that 10% rate. After running the program, a follow-up study of 150 at-risk individuals is conducted.

a. Assuming that the program has no effect, find the mean and standard deviation for the number of HIV cases in groups of 150 at-risk people.

b. Among the 150 people in the follow-up study, 8% (or 12 people) tested positive for the HIV virus. If the program has no effect, is that rate unusually low? Does this result suggest that the program is effective?

TI **3.** Credit Card Usage A student of the author conducted a survey of credit card usage by 25 of her friends. Each subject was asked how many times he or she used a credit card within the past seven days, and the results are listed as relative frequencies in the accompanying table.

x	*Relative Frequency*
0	0.16
1	0.24
2	0.40
3	0.16
4	0.04

a. Does the table constitute a probability distribution? Why or why not?

b. Assuming that the table does describe a probability distribution, what is the population that it represents? Is it the population of all credit card holders in the United States?

c. Does the type of sampling limit the usefulness of the data?

d. Find the mean number of credit card uses in the past seven days.

e. Find the standard deviation by assuming that the table is a relative frequency table summarizing results from a sample of 25 subjects.

Cooperative Group Activities

1. In-class activity Win $1,000,000! The James Randi Educational Foundation offers a $1,000,000 prize to anyone who can show, "under proper observing conditions, evidence of any paranormal, supernatural, or occult power or event." Divide into groups of three. Select one person who will be tested for extrasensory perception (ESP) by trying to correctly identify a digit randomly selected by another member of the group. Another group member should record the randomly selected digit, the digit guessed by the subject, and whether the guess was correct or wrong. Construct the table for the probability distribution of randomly generated digits, construct the relative frequency table for the random digits that were actually obtained, and construct a relative frequency table for the guesses that were made. After comparing the three tables, what do you conclude? What proportion of guesses are correct? Does it seem that the subject has the ability to select the correct digit significantly more often than would be expected by chance?

2. In-class activity See the preceding activity and *design an experiment* that would be effective in testing someone's claim that they have the ability to identify the color of a card selected from a standard deck of playing cards. Describe the experiment with great detail. Because the prize of $1,000,000 is at stake, we want to be careful to avoid the serious mistake of concluding that the person has the paranormal power when that power is not actually present. There will likely be some chance that the subject could make random guesses and be correct every time, so identify a probability that is reasonable for the event of the subject passing the test with random guesses. Be sure that the test is designed so that this probability is equal to or less than the probability value selected.

3. In-class activity Suppose we want to identify the probability distribution for the number of children born to randomly selected couples. For each student in the class, find the number of brothers and sisters and record the total number of children (including the student) in each family. Construct the relative frequency table for the result obtained. (The values of the random variable x will be 1, 2, 3, . . .). What is wrong with using this relative frequency table as an estimate of the probability distribution for the number of children born to randomly selected couples?

4. Out-of-class activity See Cumulative Review Exercise 1, which suggests that an analysis of the last digits of data can sometimes reveal whether the data have been collected through actual measurements or reported by the subjects. Refer to an almanac or the Internet and find a collection of data (such as lengths of rivers in the world), then analyze the distribution of last digits to determine whether the values were obtained through actual measurements.

TI-83/84 Plus Calculator Project

American Airlines Flight 179 from New York to San Francisco uses a Boeing 767-300 with 213 seats. Because some people with reservations don't show up, American Airlines can overbook by accepting more than 213 reservations. If the flight is not overbooked, the airline will lose revenue due to empty seats, but if too many seats are sold and some passengers are denied seats, the airline loses money from the compensation that must be given to the bumped passengers. Assume that there is a 0.0995 probability that a passenger with a reservation will not show up for the flight (based on data from the IBM research paper "Passenger-Based Predictive Modeling of Airline No-Show Rates," by Lawrence, Hong, and Cherrier). Also assume that the airline accepts 236 reservations for the 213 seats that are available.

Find the probability that when 236 reservations are accepted for Flight 179, there are more passengers showing up than there are seats available. That is, find the probability of more than 213 people showing up with reservations, assuming that 236 reservations were accepted. Because of the values involved, Table A-1 cannot be used, and calculations with the binomial probability formula would be extremely time-consuming and painfully tedious. The best approach is to use statistics software or a TI-83/84 Plus calculator. See Section 5-3 for instructions describing the use of a TI-83/84 Plus calculator. Is the probability of overbooking small enough so that it does not happen very often, or does it seem too high so that changes must be made to make it lower? Now use trial and error to find the maximum number of reservations that could be accepted so that the probability of having more passengers than seats is 0.05 or less.

From Data to Decision

Critical Thinking: Determining criteria for concluding that a gender-selection method is effective

You are responsible for analyzing results from a clinical trial of the effectiveness of a new method of gender selection. Assume that the sample size of $n = 50$ couples has already been established, and each couple will have one child. Further assume that each of the couples will be subjected to a treatment that supposedly increases the likelihood that the child will be a girl.

There is a danger in obtaining results first, then making conclusions about the results. If the results are close to showing the effectiveness of a treatment, it might be tempting to conclude that there is an effect when, in reality, there is no effect. It is better to establish criteria *before* obtaining results. Using the methods of this chapter, identify the criteria that should be used for concluding that the treatment is effective in increasing the likelihood of a girl. Among the 50 births, how many girls would you require in order to conclude that the gender-selection procedure is effective? Explain how you arrived at this result.

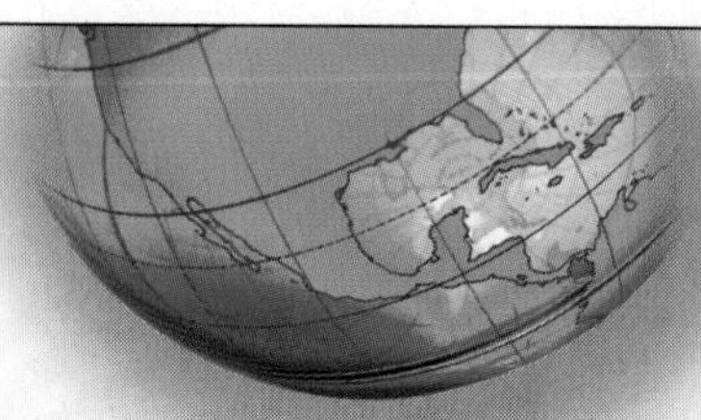

Internet Project

Probability Distributions and Simulation

Probability distributions are used to predict the outcome of the events they model. For example, if we toss a fair coin, the distribution for the outcome is a probability of 0.5 for heads and 0.5 for tails. If we toss the coin ten consecutive times, we expect five heads and five tails. We might not get this exact result, but in the long run, over hundreds or thousands of tosses, we expect the split between heads and tails to be very close to "50–50". Go to the Web site for this textbook:

http://www.aw.com/triola

Proceed to the Internet Project for Chapter 5 where you will find two explorations. In the first exploration you are asked to develop a probability distribution for a simple experiment, and use that distribution to predict the outcome of repeated trial runs of the experiment. In the second exploration, we will analyze a more complicated situation: the paths of rolling marbles as they move in pinball-like fashion through a set of obstacles. In each case, a dynamic visual simulation will allow you to compare the predicted results with a set of experimental outcomes.

Statistics @ Work

"Our program is really an education program, but it has wide recognition because the results are released publicly."

Barbara Carvalho

Director of the Marist College Poll

Lee Miringoff

Director of the Marist College Institute for Public Opinion

Barbara Carvalho and Lee Miringoff report on their poll results in many interviews for print and electronic media, including news programs for NBC, CBS, ABC, FOX, and public television. Lee Miringoff appears regularly on NBC's *Today* show.

What do you do?

We do public polling. We survey public issues, approval ratings of public officials in New York City, New York State, and nationwide. We don't do partisan polling for political parties, political candidates, or lobby groups. We are independently funded by Marist College and we have no outside funding that in any way might suggest that we are doing research for any particular group on any one issue.

How do you select survey respondents?

For a statewide survey we select respondents in proportion to county voter registrations. Different counties have different refusal rates and if we were to select people at random throughout the state, we would get an uneven model of what the state looks like. We stratify by county and use random digit dialing so that we get listed and unlisted numbers.

You mentioned refusal rates. Are they a real problem?

One of the issues that we deal with extensively is the issue of people who don't respond to surveys. That has been increasing over time and there has been much attention from the survey research community. As a research center we do quite well when compared to others. But when you do face-to-face interviews and have refusal rates of 25% to 50%, there's a real concern to find out who is refusing and why they are not responding, and the impact that has on the representativeness of the studies that we're doing.

Would you recommend a statistics course for students?

Absolutely. All numbers are not created equally. Regardless of your field of study or career interests, an ability to critically evaluate research information that is presented to you, to use data to improve services, or to interpret results to develop strategies is a very valuable asset. Surveys, in particular, are everywhere. It is vital that as workers, managers, and citizens we are able to evaluate their accuracy and worth. Statistics cuts across disciplines. Students will inevitably find it in their careers at some point.

Do you have any other recommendations for students?

It is important for students to take every opportunity to develop their communication and presentation skills. Sharpen not only your ability to speak and write, but also raise your comfort level with new technologies.

Normal Probability Distributions

6

6 CHAPTER PROBLEM

How do we identify safe limits for passengers?

"We have an emergency for Air Midwest fifty-four eighty," said pilot Katie Leslie, just before her plane crashed in Charlotte, North Carolina. The crash of the Beech 1900 plane killed all of the 21 people aboard. In the subsequent investigation, the weight of the passengers was suspected as a factor that contributed to the crash. This prompted the Federal Aviation Administration to order airlines to collect weight information from randomly selected flights, so that the old assumptions about passenger weights could be updated.

A water taxi recently sank in Baltimore's Inner Harbor. Among the 25 people on board, 5 died and 16 were injured. An investigation revealed that the safe passenger load for the water taxi was 3500 pounds. Assuming a mean passenger weight of 140 pounds, the boat was allowed to carry 25 passengers, but the mean of 140 pounds was determined 44 years ago when people were not as heavy as they are today. (The mean weight of the 25 passengers aboard the boat that sank was found to be 168 pounds.) The National Transportation and Safety Board suggested that the old estimated mean of 140 pounds be updated to 174 pounds, so the safe load of 3500 pounds would now allow only 20 passengers instead of 25. In this chapter, we will investigate weights of passengers and the role that those weights play in setting safe passenger load limits.

The examples of the aircraft and water taxi crashes illustrate critically important issues that affect all of us. One issue is the changing weight of people over time. In Chapter 2 we noted that in addition to center, variation, distribution, and outliers, an important characteristic of a population is the change that might be occurring over time. Results from the National Health and Nutrition Examination Survey show that adult Americans weigh about 25 pounds more than in 1960. Continued use of weight assumptions from many years ago can therefore result in incorrect calculations and unsafe circumstances.

The problems of determining safe loads in aircraft and boats are examples of the type of problem studied in a relatively new discipline called *ergonomics,* which is the study of people fitting into their environments. Good ergonomic design results in an environment that is safe, functional, efficient, and comfortable. Ergonomics is used in a wide variety of applications including the design of car dashboards, caskets, kayaks, cycle helmets, screw bottle tops, door handles, manhole covers, keyboards, air traffic control centers, and computer assembly lines. Working with the issue of safe passenger loads will provide some real experience with the role of statistics in ergonomics.

6-1 Overview

In Chapter 2 we considered the distribution of data, and in Chapter 3 we considered some important measures of data sets, including measures of center and variation. In Chapter 4 we discussed basic principles of probability, and in Chapter 5 we presented the following concepts:

- A *random variable* is a variable having a single numerical value, determined by chance, for each outcome of some procedure.
- A *probability distribution* describes the probability for each value of the random variable.
- A *discrete* random variable has either a finite number of values or a countable number of values. That is, the *number* of possible values that x can assume is 0, or 1, or 2, and so on.
- A *continuous* random variable has infinitely many values, and those values are often associated with measurements on a continuous scale with no gaps or interruptions.

In Chapter 5 we considered only *discrete* probability distributions, but in this chapter we present *continuous* probability distributions. Although we begin with a uniform distribution, most of the chapter focuses on *normal distributions.* Normal distributions are extremely important because they occur so often in real applications and they play such an important role in methods of inferential statistics. Normal distributions will be used often throughout the remainder of this text.

Definition

If a continuous random variable has a distribution with a graph that is symmetric and bell-shaped, as in Figure 6-1, and it can be described by the equation given as Formula 6-1, we say that it has a **normal distribution.**

Formula 6-1

$$y = \frac{e^{-\frac{1}{2}\left(\frac{x-\mu}{\sigma}\right)^2}}{\sigma\sqrt{2\pi}}$$

Yes, Formula 6-1 is intimidating and complex, but here is great news: It isn't really necessary for us to actually use Formula 6-1. We show Formula 6-1 to illustrate that any particular normal distribution is determined by two parameters: the mean, μ, and standard deviation, σ. Once specific values are selected for μ and σ, we can graph Formula 6-1 as we would graph any equation relating x and y; the result is a continuous probability distribution with the same bell shape shown in Figure 6-1.

Figure 6-1

The Normal Distribution

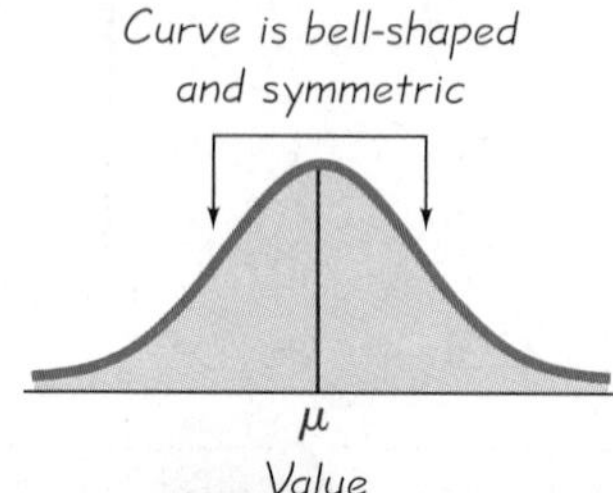

6-2 The Standard Normal Distribution

Key Concept This section presents the standard normal distribution, which has these three properties: (1) It is bell-shaped (as in Figure 6-1); (2) it has a mean equal to 0; (3) it has a standard deviation equal to 1. In this section, it is extremely important to develop the skill to find areas (or probabilities or relative frequencies) corresponding to various regions under the graph of the standard normal distribution. It is also important to find values of the variable z that correspond to areas under the graph. Given the nature of Formula 6-1, these skills might seem unreachable but, in fact, it will be relatively *easy* to master them.

The focus of this chapter is the concept of a normal probability distribution, but we begin with a *uniform distribution.* The uniform distribution makes it easier for us to see these two very important properties: (1) The area under the graph of a probability distribution is equal to 1; (2) there is a correspondence between area and probability (or relative frequency), so some probabilities can be found by identifying the corresponding areas.

Uniform Distributions

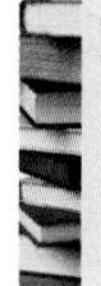

Definition

A continuous random variable has a **uniform distribution** if its values spread *evenly* over the range of possibilities. The graph of a uniform distribution results in a rectangular shape.

EXAMPLE Class Length A statistics professor plans classes so carefully that the lengths of her classes are uniformly distributed between 50.0 min and 52.0 min. (Because statistics classes are so interesting, they usually seem much shorter.) That is, any time between 50.0 min and 52.0 min is possible, and all of the possible values are equally likely. If we randomly select one of her classes and let x be the random variable representing the length of that class, then x has a distribution that can be graphed as in Figure 6-2.

When we discussed *discrete* probability distributions in Section 5-2, we identified two requirements: (1) $\Sigma P(x) = 1$ and (2) $0 \leq P(x) \leq 1$ for all values of x. Also in Section 5-2, we stated that the graph of a discrete probability *continued*

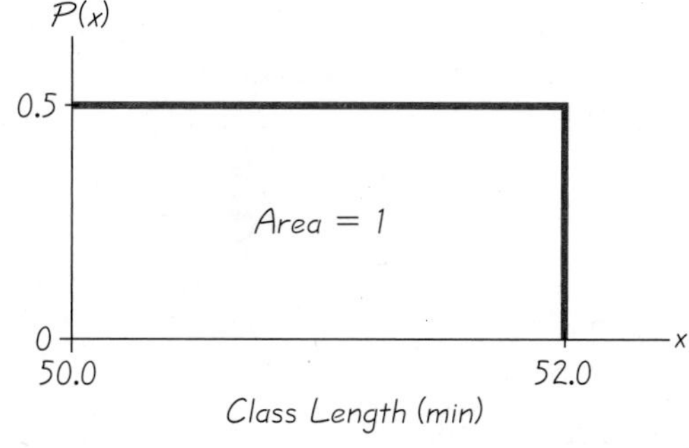

Figure 6-2 Uniform Distribution of Class Times

STATISTICS IN THE NEWS

Good Advice for Journalists

Columnist Max Frankel wrote in the *New York Times* that "most schools of journalism give statistics short shrift and some let students graduate without any numbers training at all. How can such reporters write sensibly about trade and welfare and crime, or air fares, health care and nutrition? The media's sloppy use of numbers about the incidence of accidents or disease frightens people and leaves them vulnerable to journalistic hype, political demagoguery, and commercial fraud." He cites several cases, including an example of a full-page article about New York City's deficit with a promise by the mayor to close a budget gap of \$2.7 billion; the entire article never once mentioned the *total* size of the budget, so the \$2.7 billion figure had no context.

distribution is called a *probability histogram.* The graph of a continuous probability distribution, such as that of Figure 6-2, is called a *density curve,* and it must satisfy two properties similar to the requirements for discrete probability distributions, as listed in the following definition.

Definition

A **density curve** is a graph of a continuous probability distribution. It must satisfy the following properties:

1. The total area under the curve must equal 1.
2. Every point on the curve must have a vertical height that is 0 or greater. (That is, the curve cannot fall below the x-axis.)

By setting the height of the rectangle in Figure 6-2 to be 0.5, we force the enclosed area to be $2 \times 0.5 = 1$, as required. (In general, the area of the rectangle becomes 1 when we make its height equal to the value of 1/range.) This property (area = 1) makes it very easy to solve probability problems, so the following statement is important:

Because the total area under the density curve is equal to 1, there is a correspondence between *area* and *probability*.

EXAMPLE Class Length Kim, who has developed a habit of living on the edge, has scheduled a job interview immediately following her statistics class. If the class runs longer than 51.5 minutes, she will be late for the job interview. Given the uniform distribution illustrated in Figure 6-2, find the probability that a randomly selected class will last longer than 51.5 minutes.

SOLUTION See Figure 6-3, where we shade the region representing times that are longer than 51.5 minutes. Because the total area under the density curve is equal to 1, there is a correspondence between area and probability. We

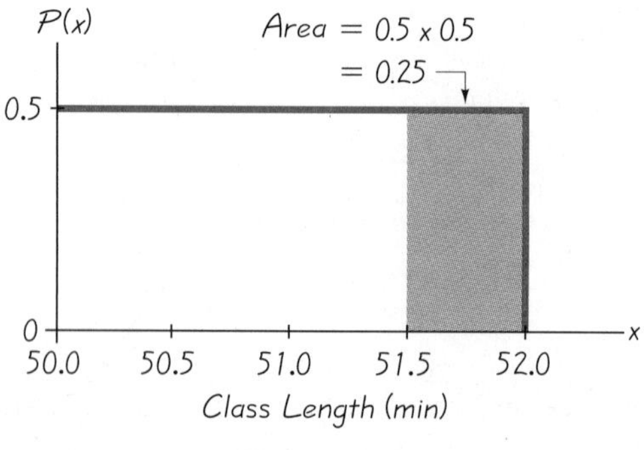

Figure 6-3 Using Area to Find Probability

can therefore find the desired probability by using areas as follows:

$$\begin{aligned} P(\text{class longer than 51.5 minutes}) &= \text{area of shaded region in Figure 6-3} \\ &= 0.5 \times 0.5 \\ &= 0.25 \end{aligned}$$

INTERPRETATION The probability of randomly selecting a class that runs longer than 51.5 minutes is 0.25. Because that probability is so high, Kim should consider making contingency plans that will allow her to get to her job interview on time. Nobody should ever be late for a job interview.

Standard Normal Distribution

The density curve of a uniform distribution is a horizontal line, so it's easy to find the area of any rectangular region: multiply width and height. The density curve of a normal distribution has the more complicated bell shape shown in Figure 6-1, so it's more difficult to find areas, but the basic principle is the same: *There is a correspondence between area and probability.*

Just as there are many different uniform distributions (with different ranges of values), there are also many different normal distributions, with each one depending on two parameters: the population mean, μ, and the population standard deviation, σ. (Recall from Chapter 1 that a *parameter* is a numerical measurement describing some characteristic of a *population.*) Figure 6-4 shows density curves for heights of adult women and men. Because men have a larger mean height, the peak of the density curve for men is farther to the right. Because men's heights have a slightly larger standard deviation, the density curve for men is slightly wider. Figure 6-4 shows two different possible normal distributions. There are infinitely many other possibilities, but one is of special interest.

Definition

The **standard normal distribution** is a normal probability distribution with $\mu = 0$ and $\sigma = 1$, and the total area under its density curve is equal to 1. (See Figure 6-5.)

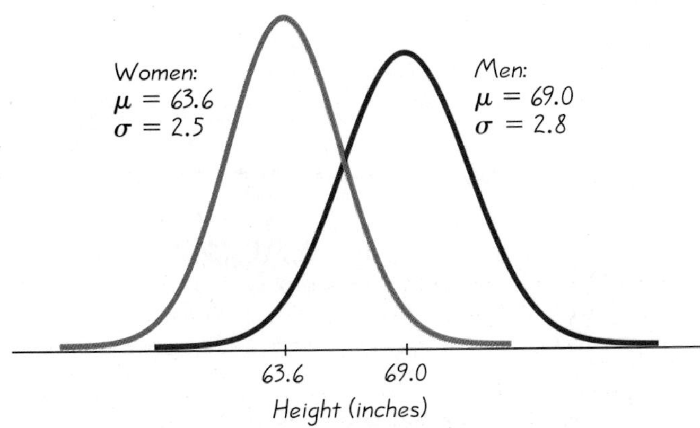

Figure 6-4

Heights of Adult Women and Men

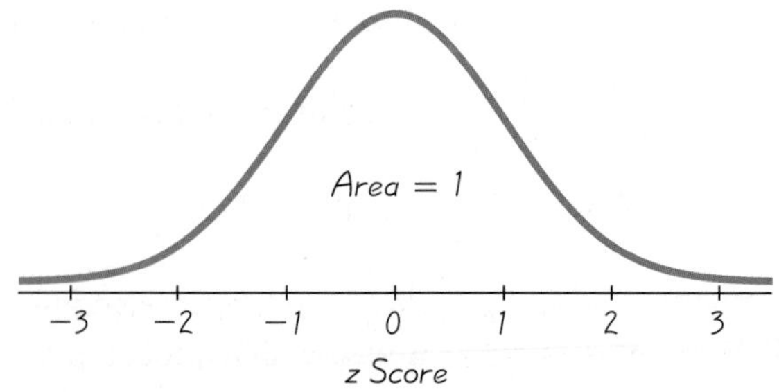

Figure 6-5

Standard Normal Distribution: $\mu = 0$ and $\sigma = 1$

Suppose that we were hired to perform calculations using Formula 6-1. We would quickly see that the easiest values for μ and σ are $\mu = 0$ and $\sigma = 1$. By letting $\mu = 0$ and $\sigma = 1$, mathematicians have calculated many different areas under the curve. As shown in Figure 6-5, the area under the curve is 1, and this allows us to make the correspondence between area and probability, as we did in the preceding example with the uniform distribution.

Finding Probabilities When Given *z* Scores

Using Table A-2 (in Appendix A and the *Formulas and Tables* insert card), we can find areas (or probabilities) for many different regions. Such areas can be found by using Table A-2, a TI-83/84 Plus calculator, or software. The key features of the different methods are summarized in Table 6-1. It is not necessary to know both methods; you only need to know the method you will be using for class and tests. Because calculators or software often give more accurate results than Table A-2, the use of technology is recommended. (When there are discrepancies, answers in Appendix E will generally include results based on Table A-2 as well as answers based on the TI-83/84 Plus calculator.)

Table 6-1 Methods for Finding Normal Distribution Areas

Method	Graph	Procedure
Table A-2 Gives the cumulative area from the left up to a vertical line above a specific value of *z*.	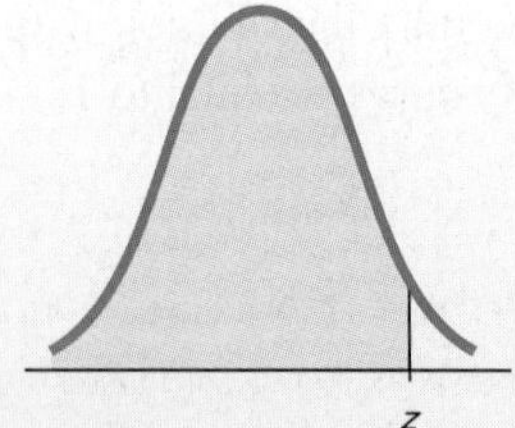	Table A-2: The procedure for using Table A-2 is described in the text.
TI-83/84 Plus Calculator Gives area bounded on the left and bounded on the right by vertical lines above any specific values.	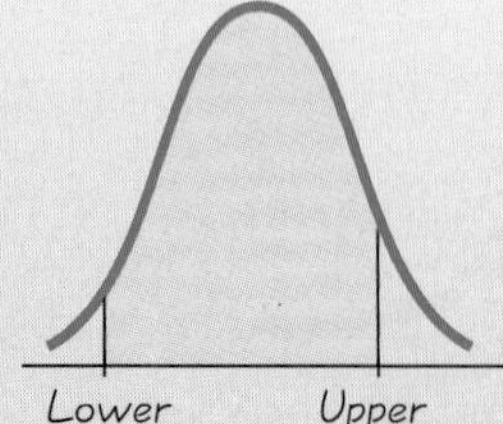	TI-83/84: Press 2ND VARS **[2: normal cdf(]**, then enter the two *z* scores separated by a comma, as in (left *z* score, right *z* score).

If using Table A-2, it is essential to understand these points:

1. Table A-2 is designed only for the *standard* normal distribution, which has a mean of 0 and a standard deviation of 1.
2. Table A-2 is on two pages, with one page for *negative* z scores and the other page for *positive* z scores.
3. Each value in the body of the table is a *cumulative area from the left* up to a vertical boundary above a specific z score.
4. When working with a graph, avoid confusion between z scores and areas.

   z score: ***Distance*** **along the horizontal scale of the standard normal distribution; refer to the leftmost column and top row of Table A-2.**

   Area: ***Region*** **under the curve; refer to the values in the body of Table A-2.**
5. The part of the z score denoting hundredths is found across the top row of Table A-2.

The following example requires that we find the probability associated with a value less than 1.58. Begin with the z score of 1.58 by locating 1.5 in the left column; next find the value in the adjoining row of probabilities that is directly below 0.08, as shown in the margin excerpt from Table A-2.

z	. . . . 0.08
.	.
.	.
.	.
1.5	. . . . 0.9429

The area (or probability) value of 0.9429 indicates that there is a probability of 0.9429 of randomly selecting a z score less than 1.58. (The following sections will consider cases in which the mean is not 0 or the standard deviation is not 1.)

EXAMPLE Scientific Thermometers The Precision Scientific Instrument Company manufactures thermometers that are supposed to give readings of 0°C at the freezing point of water. Tests on a large sample of these instruments reveal that at the freezing point of water, some thermometers give readings below 0° (denoted by negative numbers) and some give readings above 0° (denoted by positive numbers). Assume that the mean reading is 0°C and the standard deviation of the readings is 1.00°C. Also assume that the readings are normally distributed. If one thermometer is randomly selected, find the probability that, at the freezing point of water, the reading is less than 1.58°.

SOLUTION The probability distribution of readings is a standard normal distribution, because the readings are normally distributed with $\mu = 0$ and $\sigma = 1$. We need to find the area in Figure 6-6 below $z = 1.58$. The *area*

continued

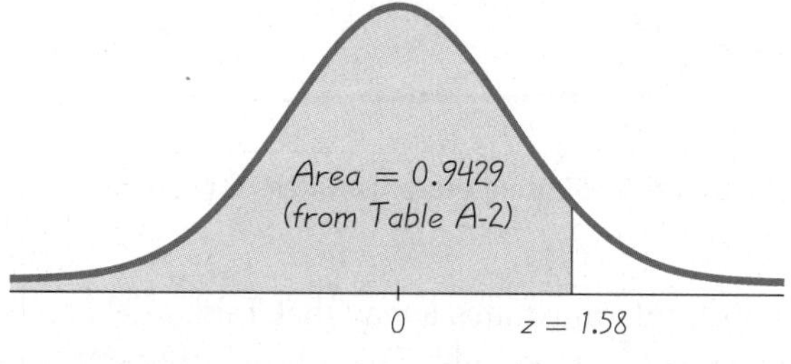

Figure 6-6 Finding the Area Below $z = 1.58$

below $z = 1.58$ is equal to the *probability* of randomly selecting a thermometer with a reading less than 1.58°. From Table A-2 we find that this area is 0.9429. The TI-83/84 Plus calculator gives an area of 0.942946563.

INTERPRETATION The *probability* of randomly selecting a thermometer with a reading less than 1.58° (at the freezing point of water) is equal to the area of 0.9429 shown as the shaded region in Figure 6-6. Another way to interpret this result is to conclude that 94.29% of the thermometers will have readings below 1.58°.

EXAMPLE Scientific Thermometers Using the thermometers from the preceding example, find the probability of randomly selecting one thermometer that reads (at the freezing point of water) above −1.23°.

SOLUTION We again find the desired *probability* by finding a corresponding *area*. We are looking for the area of the region that is shaded in Figure 6-7, but Table A-2 is designed to apply only to cumulative areas from the *left*. Referring to Table A-2 for the page with *negative* z scores, we find that the cumulative area from the left up to $z = -1.23$ is 0.1093 as shown. Knowing that the total area under the curve is 1, we can find the shaded area by subtracting 0.1093 from 1. The result is 0.8907. Even though Table A-2 is designed only for cumulative areas from the left, we can use it to find cumulative areas from the right, as shown in Figure 6-7.

If we use a TI-83/84 Plus calculator as described at the end of this section, the shaded area in Figure 6-7 is 0.8906513833.

INTERPRETATION Because of the correspondence between probability and area, we conclude that the *probability* of randomly selecting a thermometer with a reading above −1.23° at the freezing point of water is 0.8907 (which is the *area* to the right of $z = -1.23$). In other words, 89.07% of the thermometers have readings above −1.23°.

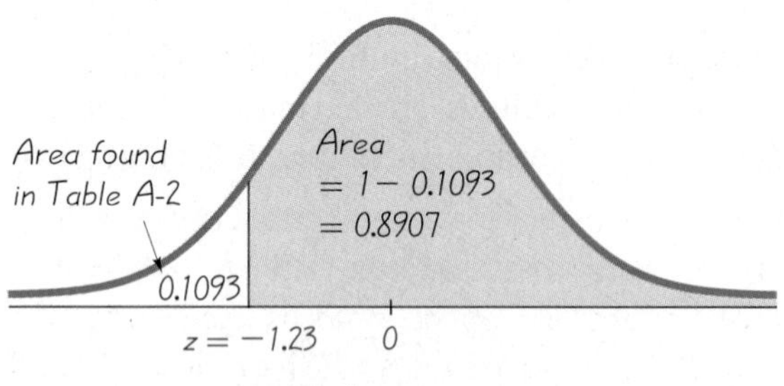

Figure 6-7 Finding the Area Above $z = -1.23$

The preceding example illustrates a way that Table A-2 can be used indirectly to find a cumulative area from the right. The following example illustrates another way that we can find an area by using Table A-2.

EXAMPLE Scientific Thermometers Once again, make a random selection from the same sample of thermometers. Find the probability that the chosen thermometer reads (at the freezing point of water) between $-2.00°$ and $1.50°$.

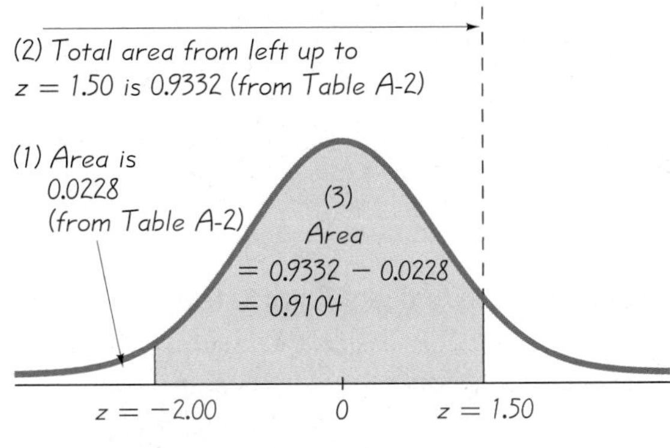

Figure 6-8 Finding the Area Between Two Values

SOLUTION We are again dealing with normally distributed values having a mean of 0° and a standard deviation of 1°. The probability of selecting a thermometer that reads between $-2.00°$ and $1.50°$ corresponds to the shaded area in Figure 6-8. Table A-2 cannot be used to find that area directly, but we can use the table to find that $z = -2.00$ corresponds to the area of 0.0228, and $z = 1.50$ corresponds to the area of 0.9332, as shown in the figure. Refer to Figure 6-8 to see that the shaded area is the difference between 0.9332 and 0.0228. The shaded area is therefore $0.9332 - 0.0228 = 0.9104$.

If using a TI-83/84 Plus calculator, the shaded area in Figure 6-8 is found to be 0.9104427093.

INTERPRETATION Using the correspondence between probability and area, we conclude that there is a probability of 0.9104 of randomly selecting one of the thermometers with a reading between $-2.00°$ and $1.50°$ at the freezing point of water. Another way to interpret this result is to state that if many thermometers are selected and tested at the freezing point of water, then 0.9104 (or 91.04%) of them will read between $-2.00°$ and $1.50°$.

The preceding example can be generalized as a rule stating that the area corresponding to the region between two specific z scores can be found by finding the difference between the two areas found in Table A-2. See Figure 6-9, which shows that the shaded region B can be found by finding the *difference* between two areas found from Table A-2: area A and B combined (found in Table A-2 as the area corresponding to z_{Right}) and area A (found in Table A-2 as the area corresponding to z_{Left}). *Recommendation:* Don't try to memorize a rule or formula for this case, because it is infinitely better to *understand* the procedure. Understand how Table A-2 works, then draw a graph, shade the desired area, and think of a way to find that area given the condition that Table A-2 provides only cumulative areas from the left.

Figure 6-9

Finding the Area Between Two z Scores

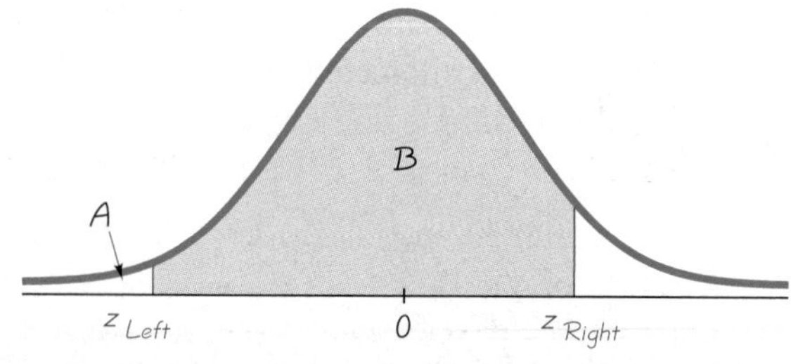

Shaded area B = (areas A and B combined) – (area A)
= (area from Table A-2 using z_{Right}) – (area from Table A-2 using z_{Left})

The preceding example concluded with the statement that the probability of a reading between −2.00° and 1.50° is 0.9104. Such probabilities can also be expressed with the following notation.

Notation

$P(a < z < b)$	denotes the probability that the z score is between a and b.
$P(z > a)$	denotes the probability that the z score is greater than a.
$P(z < a)$	denotes the probability that the z score is less than a.

Using this notation, we can express the result of the last example as: $P(-2.00 < z < 1.50) = 0.9104$, which states in symbols that the probability of a z score falling between −2.00 and 1.50 is 0.9104. With a continuous probability distribution such as the normal distribution, the probability of getting any single *exact* value is 0. That is, $P(z = a) = 0$. For example, there is a 0 probability of randomly selecting someone and getting a person whose height is exactly 68.12345678 in. In the normal distribution, any single point on the horizontal scale is represented not by a region under the curve, but by a vertical line above the point. For $P(z = 1.50)$ we have a vertical line above $z = 1.50$, but that vertical line by itself contains no area, so $P(z = 1.50) = 0$. With any continuous random variable, the probability of any one exact value is 0, and it follows that $P(a \leq z \leq b) = P(a < z < b)$. It also follows that the probability of getting a z score of *at most b* is equal to the probability of getting a z score *less than b*. It is important to correctly interpret key phrases such as *at most, at least, more than, no more than,* and so on.

Finding *z* Scores from Known Areas

So far, the examples of this section involving the standard normal distribution have all followed the same format: Given z scores, we found areas under the curve. These areas correspond to probabilities. In many other cases, we want a reverse process because we already know the area (or probability), but we need to find the corresponding z score. In such cases, it is very important to avoid confusion between z scores and areas. Remember, z scores are *distances* along the horizontal scale, and they are represented by the numbers in Table A-2 that are in the extreme left column and across the top row. Areas (or probabilities) are regions under the curve, and they are represented by the values in the body of Table A-2. Also, z scores positioned in the left half of the curve are always negative. If we already know a probability and want to determine the corresponding z score, we find it as follows.

Procedure for Finding a *z* Score from a Known Area

1. Draw a bell-shaped curve and identify the region under the curve that corresponds to the given probability. If that region is not a cumulative region from the left, work instead with a known region that is a cumulative region from the left.
2. ***Using Table A-2:*** Using the cumulative area from the left, locate the closest probability in the body of Table A-2 and identify the corresponding z score.

   Using a TI-83/84 Plus Calculator: Press 2ND VARS to get the DISTR (distribution) menu, then select **3:invNorm,** and then proceed to enter the total area to the *left* of the desired z score.

EXAMPLE Scientific Thermometers Use the same thermometers as earlier, with temperature readings at the freezing point of water that are normally distributed with a mean of 0°C and a standard deviation of 1°C. Find the temperature corresponding to P_{95}, the 95th percentile. That is, find the temperature separating the bottom 95% from the top 5%. See Figure 6-10.

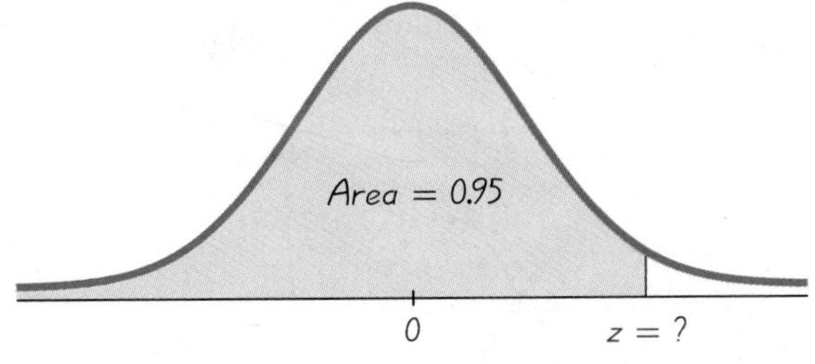

Figure 6-10 Finding the 95th Percentile

SOLUTION Figure 6-10 shows the z score that is the 95th percentile, with 95% of the area (or 0.95) below it. *Important:* Remember that when referring to Table A-2, the body of the table includes *cumulative areas from the left.* Referring to Table A-2, we search for the area of 0.95 *in the body* of the table and then find the corresponding z score. In Table A-2 we find the areas of 0.9495 and 0.9505, but there's an asterisk with a special note indicating that 0.9500 corresponds to a z score of 1.645. We can now conclude that the z score in Figure 6-10 is 1.645, so the 95th percentile is the temperature reading of 1.645°C.

If using a TI-83/84 Plus calculator, the z score is found to be 1.644853626.

INTERPRETATION When tested at freezing, 95% of the readings will be less than or equal to 1.645°C, and 5% of them will be greater than or equal to 1.645°C.

Note that in the preceding solution, Table A-2 led to a z score of 1.645, which is midway between 1.64 and 1.65. When using Table A-2, we can usually

z Score	Cumulative Area from the Left
1.645	0.9500
−1.645	0.0500
2.575	0.9950
−2.575	0.0050

avoid interpolation by simply selecting the closest value. There are special cases listed in the accompanying table that are important because they are used so often in a wide variety of applications. (The value of $z = 2.576$ gives an area slightly closer to the area of 0.9950, but $z = 2.575$ has the advantage of being the value midway between $z = 2.57$ and $z = 2.58$.) Except in these special cases, we can select the closest value in the table. (If a desired value is midway between two table values, select the larger value.) Also, for z scores above 3.49, we can use 0.9999 as an approximation of the cumulative area from the left; for z scores below −3.49, we can use 0.0001 as an approximation of the cumulative area from the left.

EXAMPLE Scientific Thermometers Using the same thermometers, find the temperatures separating the bottom 2.5% and the top 2.5%.

SOLUTION Refer to Figure 6-11 where the required z scores are shown. To find the z score located to the left, refer to Table A-2 and search the *body of the table* for an area of 0.025. The result is $z = -1.96$. To find the z score located to the right, refer to Table A-2 and search the *body of the table* for an area of 0.975. (Remember that Table A-2 always gives cumulative areas from the *left*.) The result is $z = 1.96$. The values of $z = -1.96$ and $z = 1.96$ separate the bottom 2.5% and the top 2.5% as shown in Figure 6-11.

If using a TI-83/84 Plus calculator, the z scores are found to be −1.959963986 and 1.959963986.

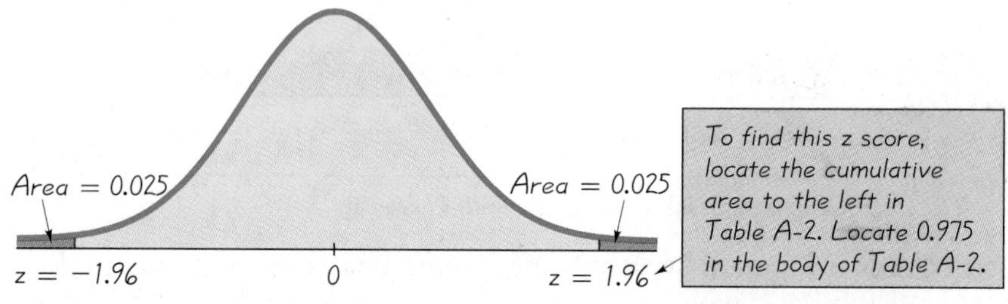

Figure 6-11 Finding z Scores

INTERPRETATION When tested at freezing, 2.5% of the thermometer readings will be equal to or less than −1.96°, and 2.5% of the readings will be equal to or greater than 1.96°. Another interpretation is that at the freezing level of water, 95% of all thermometer readings will fall between −1.96° and 1.96°.

The examples in this section were contrived so that the mean of 0 and the standard deviation of 1 coincided exactly with the parameters of the standard normal distribution. In reality, it is unusual to find such convenient parameters, because typical normal distributions involve means different from 0 and standard deviations different from 1. In the next section we introduce methods for working with such normal distributions, which are much more realistic.

TI

Using the TI-83/84 Plus Calculator to Find an Area Between Two *z* Scores

Important: The following procedure can be used with *z* scores, but the procedure given at the end of Section 6-3 is more general, so it is recommended that you use the procedure at the end of that section. To find the area between two *z* scores, press 2ND VARS to get to the DISTR (distribution) menu, then select **2:normalcdf.** You must now enter the lower boundary and the upper, separated by a comma. *Hint:* If there is no actual lower boundary or no upper boundary, enter a very large value, as shown below.

- Cumulative area from the left (with no actual lower boundary): Enter −999999 for the lower boundary.
- Cumulative area from the right (with no actual upper boundary): Enter 999999 for the upper boundary.

TI

Using the TI-83/84 Plus Calculator to Find a *z* Score Corresponding to a Known Probability

Important: The following procedure can be used with *z* scores, but the procedure given at the end of Section 6-3 is more general, so it is recommended that you use the procedure at the end of that section. To find a *z* score corresponding to a known probability, press 2ND VARS to get to the DISTR (distribution) menu, then select **3:invNorm.** Proceed to enter the *cumulative area to the left* of the desired *z* score.

6-2 BASIC SKILLS AND CONCEPTS

Statistical Literacy and Critical Thinking

1. Normal Distribution When we refer to a "normal" distribution, does the word "normal" have the same meaning as in ordinary language, or does it have a special meaning in statistics? What exactly is a normal distribution?

2. Normal Distribution A normal distribution is informally and loosely described as a probability distribution that is "bell-shaped" when graphed. What is the "bell shape"?

3. Standard Normal Distribution What requirements are necessary for a normal probability distribution to be a *standard* normal probability distribution?

4. Areas If you determine that for the graph of a standard normal distribution the cumulative area to the left of a *z* score is 0.4, what is the cumulative area to the right of that *z* score?

Continuous Uniform Distribution. *In Exercises 5–8, refer to the continuous uniform distribution depicted in Figure 6-2, assume that a class length between 50.0 min and 52.0 min is randomly selected, and find the probability that the given time is selected.*

5. Less than 51.5 min

6. Greater than 50.5 min

7. Between 50.5 min and 51.5 min

8. Between 51.5 min and 51.6 min

Standard Normal Distribution. *In Exercises 9–28, assume that the readings on the thermometers are normally distributed with a mean of 0° and a standard deviation of 1.00°C.*

A thermometer is randomly selected and tested. In each case, draw a sketch, and find the probability of each reading. (The given values are in Celsius degrees.)

TI **9.** Less than −1.00

TI **10.** Less than −2.50

TI **11.** Less than 1.00

TI **12.** Less than 2.50

TI **13.** Greater than 1.25

TI **14.** Greater than 1.96

TI **15.** Greater than −1.75

TI **16.** Greater than −1.96

TI **17.** Between 1.00 and 2.00

TI **18.** Between 0.50 and 1.50

TI **19.** Between −2.45 and −2.00

TI **20.** Between 1.05 and 2.05

TI **21.** Between −2.11 and 1.55

TI **22.** Between −1.80 and 2.08

TI **23.** Between −1.00 and 4.00

TI **24.** Between −3.90 and 1.50

TI **25.** Greater than 3.52

TI **26.** Less than −3.75

TI **27.** Greater than 0

TI **28.** Less than 0

Basis for Empirical Rule. *In Exercises 29–32, find the indicated area under the curve of the standard normal distribution, then convert it to a percentage and fill in the blank. The results form the basis for the empirical rule introduced in Section 3-3.*

TI **29.** About _____% of the area is between $z = -1$ and $z = 1$ (or within 1 standard deviation of the mean).

TI **30.** About _____% of the area is between $z = -2$ and $z = 2$ (or within 2 standard deviations of the mean).

TI **31.** About _____% of the area is between $z = -3$ and $z = 3$ (or within 3 standard deviations of the mean).

TI **32.** About _____% of the area is between $z = -3.5$ and $z = 3.5$ (or within 3.5 standard deviations of the mean).

Finding Probability. *In Exercises 33–36, assume that the readings on the thermometers are normally distributed with a mean of 0° and a standard deviation of 1.00°. Find the indicated probability, where z is the reading in degrees.*

TI **33.** $P(-1.96 < z < 1.96)$

TI **34.** $P(z > 1.645)$

TI **35.** $P(z < -2.575 \text{ or } z > 2.575)$

TI **36.** $P(z < -1.96 \text{ or } z > 1.96)$

Finding Temperature Values. *In Exercises 37–40, assume that the readings on the thermometers are normally distributed with a mean of 0° and a standard deviation of 1.00°C. A thermometer is randomly selected and tested. In each case, draw a sketch, and find the temperature reading corresponding to the given information.*

TI **37.** Find P_{90}, the 90th percentile. This is the temperature reading separating the bottom 90% from the top 10%.

TI **38.** Find P_{25}, the 25th percentile.

TI **39.** If 2.5% of the thermometers are rejected because they have readings that are too high and another 2.5% are rejected because they have readings that are too low, find the two readings that are cutoff values separating the rejected thermometers from the others.

TI **40.** If 1.0% of the thermometers are rejected because they have readings that are too high and another 1.0% are rejected because they have readings that are too low, find the two readings that are cutoff values separating the rejected thermometers from the others.

6-2 BEYOND THE BASICS

TI **41.** For a standard normal distribution, find the percentage of data that are
a. within 1 standard deviation of the mean.
b. within 1.96 standard deviations of the mean.
c. between $\mu - 3\sigma$ and $\mu + 3\sigma$.
d. between 1 standard deviation below the mean and 2 standard deviations above the mean.
e. more than 2 standard deviations away from the mean.

TI **42.** If a continuous uniform distribution has parameters of $\mu = 0$ and $\sigma = 1$, then the minimum is $-\sqrt{3}$ and the maximum is $\sqrt{3}$.
a. For this distribution, find $P(-1 < x < 1)$.
b. Find $P(-1 < x < 1)$ if you incorrectly assume that the distribution is normal instead of uniform.
c. Compare the results from parts (a) and (b). Does the distribution affect the results very much?

TI **43.** Assume that z scores are normally distributed with a mean of 0 and a standard deviation of 1.
a. If $P(0 < z < a) = 0.3907$, find a.
b. If $P(-b < z < b) = 0.8664$, find b.
c. If $P(z > c) = 0.0643$, find c.
d. If $P(z > d) = 0.9922$, find d.
e. If $P(z < e) = 0.4500$, find e.

44. In a continuous uniform distribution,

$$\mu = \frac{\text{minimum} + \text{maximum}}{2} \quad \text{and} \quad \sigma = \frac{\text{range}}{\sqrt{12}}$$

Find the mean and standard deviation for the uniform distribution represented in Figure 6-2.

45. Sketch a graph representing a cumulative distribution for (a) a uniform distribution and (b) a normal distribution.

46. Refer to the graph of the triangular probability distribution of the continuous random variable x. (See the margin graph.)
a. Find the value of the constant c.
b. Find the probability that x is between 0 and 3.
c. Find the probability that x is between 2 and 9.

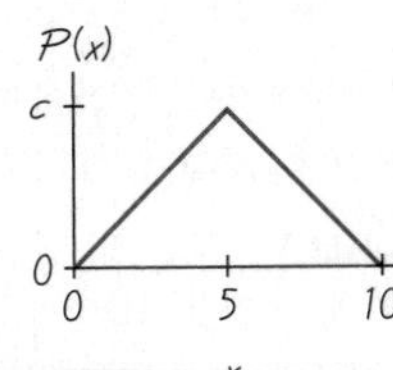

6-3 Applications of Normal Distributions

Key Concept The preceding section involved applications of the normal distribution, but all of the examples and exercises were based on the *standard* normal distribution with $\mu = 0$ and $\sigma = 1$. This section presents methods for working with normal distributions that are not standard. That is, either the mean is not 0 or

Multiple Lottery Winners

Evelyn Marie Adams won the New Jersey Lottery twice in four months. This happy event was reported in the media as an incredible coincidence with a likelihood of only 1 chance in 17 trillion. But Harvard mathematicians Persi Diaconis and Frederick Mosteller note that there is 1 chance in 17 trillion that a particular person with one ticket in each of two New Jersey lotteries will win both times. However, there is about 1 chance in 30 that someone in the United States will win a lottery twice in a four-month period. Diaconis and Mosteller analyzed coincidences and conclude that "with a large enough sample, any outrageous thing is apt to happen." More recently, according to the *Detroit News,* Joe and Dolly Hornick won the Pennsylvania lottery four times in 12 years for prizes of \$2.5 million, \$68,000, \$206,217, and \$71,037.

the standard deviation is not 1, or both. The key concept is that we can use a simple conversion (Formula 6-2) that allows us to standardize any normal distribution so that the same methods of the preceding section can be used. This standardization process therefore allows us to tackle much more realistic and meaningful applications.

To work with some nonstandard normal distribution, simply standardize values so that we can continue to use the same procedures from Section 6-2.

If we convert values to standard scores using Formula 6-2, then procedures for working with all normal distributions are the same as those for the standard normal distribution.

Formula 6-2 $z = \dfrac{x - \mu}{\sigma}$ (round z scores to 2 decimal places)

Continued use of Table A-2 requires understanding and application of the above principle. (If you use certain calculators or software programs, the conversion to z scores is not necessary because probabilities can be found directly.) Regardless of the method used, you need to clearly understand the above basic principle, because it is an important foundation for concepts introduced in the following chapters.

Figure 6-12 illustrates the conversion from a nonstandard to a standard normal distribution. The area in any normal distribution bounded by some score x [as in Figure 6-12(a)] is the same as the area bounded by the equivalent z score in the standard normal distribution [as in Figure 6-12(b)]. This means that when working with a nonstandard normal distribution, you can use Table A-2 the same way it was used in Section 6-2 as long as you first convert the values to z scores. When finding areas with a nonstandard normal distribution use either of the following two procedures.

Using Table A-2 to Find Areas

1. Sketch a normal curve, label the mean, label the x value at the lower boundary of the desired area, and label the x value at the upper boundary of the desired area. Also, *shade* the region representing the desired probability.
2. For each of the two values x that is a boundary for the shaded region, use Formula 6-2 to convert that value to the equivalent z score.
3. Refer to Table A-2 to find the area of the shaded region. This area is the desired probability.

Figure 6-12

Converting from a Nonstandard to a Standard Normal Distribution

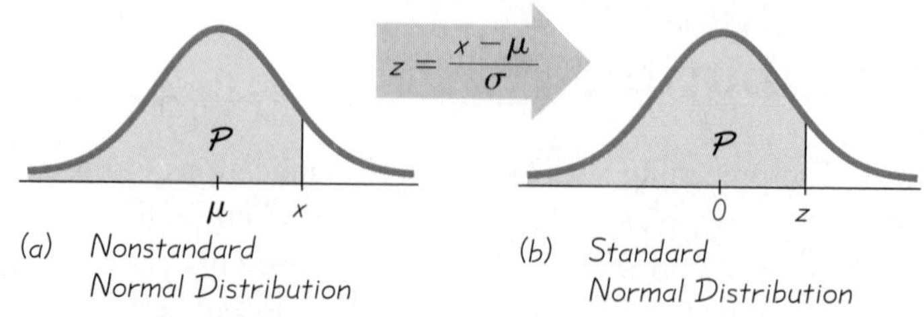

Using the TI-83/84 Plus Calculator to Find Areas

1. Sketch a normal curve, label the mean, label the x value at the lower boundary of the desired area, and label the x value at the upper boundary of the desired area. Also, *shade* the region representing the desired probability.
2. Press 2ND VARS to get the DISTR (distribution) menu, then select **2:normalcdf** and proceed to enter the two x values, the mean, and the standard deviation, all separated by commas. That is, use this format: **normalcdf(lower bound, upper bound, μ, σ).**
3. Press ENTER and the area will be displayed.

Hint: If there is no lower bound or there is no upper bound, enter a very large value, as indicated below.

- Cumulative area from the left (with no actual lower bound): Enter −999999 for the lower bound.
- Cumulative area from the right (with no actual upper bound): Enter 999999 for the upper bound.

The following example illustrates the relationship between a typical nonnormal distribution and the standard normal distribution.

EXAMPLE Weights of Water Taxi Passengers In the Chapter Problem, we noted that the safe load for a water taxi was found to be 3500 pounds. We also noted that the mean weight of a passenger was assumed to be 140 pounds. Let's assume a "worst case" scenario in which all of the passengers are adult men. (This could easily occur in a city that hosts conventions in which people of the same gender often travel in groups.) Based on data from the National Health and Nutrition Examination Survey, assume that weights of men are normally distributed with a mean of 172 pounds and a standard deviation of 29 pounds. If one man is randomly selected, find the probability that he weighs less than 174 pounds (the value suggested by the National Transportation and Safety Board).

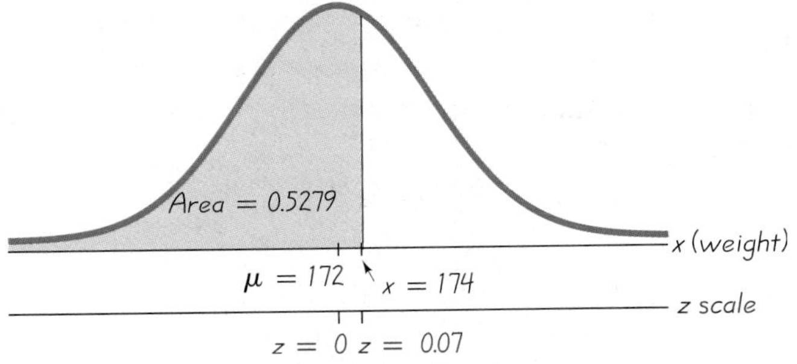

Figure 6-13 Weights (in pounds) of Men

SOLUTION

Step 1: See Figure 6-13, which incorporates this information: Men have weights that are normally distributed with a mean of 172 pounds and a standard deviation of 29 pounds, and the shaded region represents the men with weights less than 174 pounds.

continued

Clinical Trial Cut Short

What do you do when you're testing a new treatment and, before your study ends, you find that it is clearly effective? You should cut the study short and inform all participants of the treatment's effectiveness. This happened when hydroxyurea was tested as a treatment for sickle cell anemia. The study was scheduled to last about 40 months, but the effectiveness of the treatment became obvious and the study was stopped after 36 months. (See "Trial Halted as Sickle Cell Treatment Proves Itself," by Charles Marwick, *Journal of the American Medical Association,* Vol. 273, No. 8.)

Step 2: To use Table A-2, we first must use Formula 6-2 to convert from the nonstandard normal distribution to the standard normal distribution. The weight of 174 pounds is converted to a z score as follows:

$$z = \frac{x - \mu}{\sigma} = \frac{174 - 172}{29} = 0.07$$

Step 3: Referring to Table A-2 and using $z = 0.07$, we find that the cumulative area to the left of $z = 0.07$ is an area of 0.5279.

If using a TI-83/84 Plus calculator, press 2ND VARS and proceed to enter **normalcdf(−999999, 174, 172, 29)** to get an area of 0.527491537.

INTERPRETATION There is a probability of 0.5279 of randomly selecting a man with a weight less than 174 pounds. It follows that 52.79% of men have weights less than 174 pounds. It also follows that 47.21% of men have weights greater than 174 pounds.

EXAMPLE IQ Scores A psychologist is designing an experiment to test the effectiveness of a new training program for airport security screeners. She wants to begin with a homogeneous group of subjects having IQ scores between 85 and 125. Given that IQ scores are normally distributed with a mean of 100 and a standard deviation of 15, what percentage of people have IQ scores between 85 and 125?

SOLUTION See Figure 6-14, which shows the shaded region representing IQ scores between 85 and 125. We can't find that shaded area directly from Table A-2, but we can find it indirectly by using the same basic procedures presented in Section 6-2. Here is the strategy for finding the shaded area: First find the cumulative area from the left up to 85, then find the cumulative area from the left up to 125, then find the difference between those two areas.

Finding the cumulative area up to 85:

$$z = \frac{x - \mu}{\sigma} = \frac{85 - 100}{15} = -1.00$$

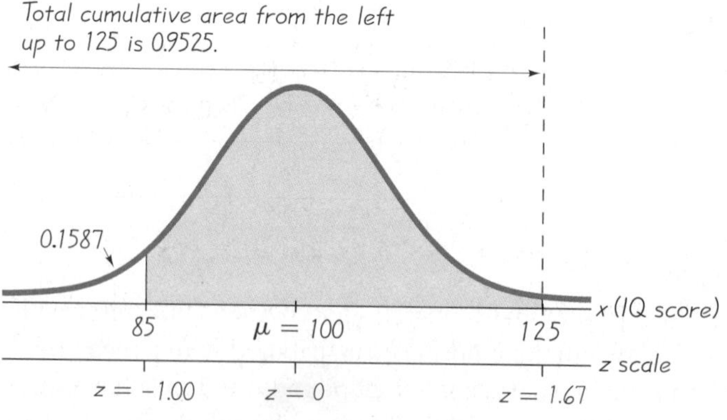

Figure 6-14 IQ Scores

Using Table A-2, we find that $z = -1.00$ corresponds to an area of 0.1587, as shown in Figure 6-14.

Finding the cumulative area up to 125:

$$z = \frac{x - \mu}{\sigma} = \frac{125 - 100}{15} = 1.67$$

Using Table A-2, we find that $z = 1.67$ corresponds to an area of 0.9525, as shown in Figure 6-14.

Finding the shaded area between 85 and 125:

$$\text{Shaded area} = 0.9525 - 0.1587 = 0.7938$$

If using a TI-83/84 Plus calculator, press 2ND VARS and proceed to enter **normalcdf(85, 125, 100, 15)** to get an area of 0.7935544101.

INTERPRETATION Expressing the result as a percentage, we conclude that 79.38% of people have IQ scores between 85 and 125.

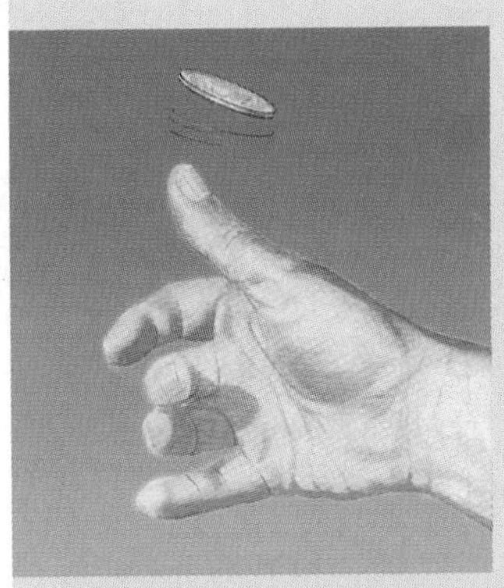

Sensitive Surveys

Survey respondents are sometimes reluctant to honestly answer questions on a sensitive topic, such as employee theft or sex. Stanley Warner (York University, Ontario) devised a scheme that leads to more accurate results in such cases. As an example, ask employees if they stole within the past year and also ask them to flip a coin. The employees are instructed to answer no if they didn't steal and the coin turns up heads. Otherwise, they should answer yes. The employees are more likely to be honest because the coin flip helps protect their privacy. Probability theory can then be used to analyze responses so that more accurate results can be obtained.

Finding Values from Known Areas

Here are helpful hints for those cases in which the area (or probability or percentage) is known and we must find the relevant value(s):

1. *Don't confuse z scores and areas.* Remember, z scores are *distances* along the horizontal scale, but areas are *regions* under the normal curve. Table A-2 lists z scores in the left columns and across the top row, but areas are found in the body of the table.
2. *Choose the correct (right/left) side of the graph.* A value separating the top 10% from the others will be located on the right side of the graph, but a value separating the bottom 10% will be located on the left side of the graph.
3. A z score must be *negative* whenever it is located in the *left* half of the normal distribution.
4. Areas (or probabilities) are positive or zero values, but they are never negative.

Graphs are extremely helpful in visualizing, understanding, and successfully working with normal probability distributions, so they should be used whenever possible.

Using Table A-2 and Formula 6-2 to Find Values

1. Sketch a normal distribution curve, enter the given probability or percentage in the appropriate region of the graph, and identify the x value(s) being sought.
2. Use Table A-2 to find the z score corresponding to the cumulative left area bounded by x. Refer to the *body* of Table A-2 to find the closest area, then identify the corresponding z score.

3. Using Formula 6-2, enter the values for μ, σ, and the z score found in Step 2, then solve for x. Based on the format of Formula 6-2, we can solve for x as follows:

$$x = \mu + (z \cdot \sigma) \quad \text{(another form of Formula 6-2)}$$

(If z is located to the left of the mean, be sure that it is a negative number.)

4. Refer to the sketch of the curve to verify that the solution makes sense in the context of the graph and in the context of the problem.

Using a TI-83/84 Plus Calculator to Find Values

1. Sketch a normal distribution curve, enter the given probability or percentage in the appropriate region of the graph, and identify the x value being sought.
2. Press 2ND VARS to get the DISTR (distribution) menu, then select **3:invNorm** and proceed to enter the *cumulative area to the left* of the desired x value, the mean, and the standard deviation, all separated by commas. That is, use this format: **invNorm(area to the left, μ, σ).**
3. Press ENTER and the value of x will be displayed.

The following example uses the procedures just outlined.

EXAMPLE Weights of Water Taxi Passengers A previous example in this section showed that 52.79% of men have weights less than the value of 174 pounds set by the National Transportation and Safety Board. What weight separates the lightest 99.5% of men from the heaviest 0.5%? Again assume that weights of men are normally distributed with a mean of 172 pounds and a standard deviation of 29 pounds.

SOLUTION

Step 1: We begin with the graph shown in Figure 6-15. We have entered the mean of 172 pounds, shaded the area representing the lightest 99.5% of men, and identified the desired value as x.

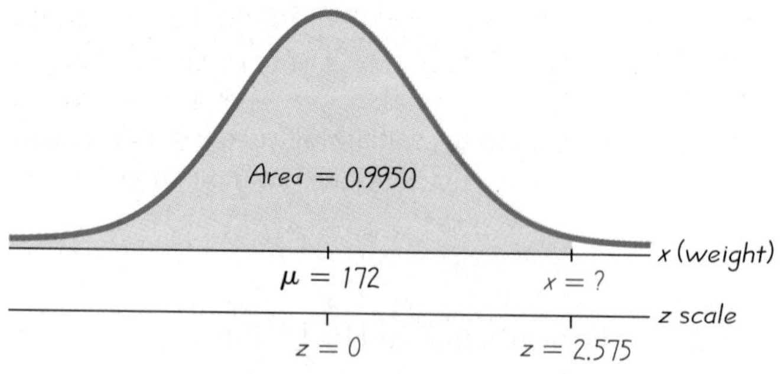

Figure 6-15 Finding Weight

Step 2: In Table A-2 we search for an area of 0.9950 in the *body* of the table. (The area of 0.9950 shown in Figure 6-15 is a cumulative area from the left, and that is exactly the type of area listed in Table A-2.) The area of 0.9950 is between the Table A-2 areas of 0.9949 and 0.9951, but there is an asterisk and footnote indicating that an area of 0.9950 corresponds to a z score of 2.575.

Step 3: With $z = 2.575$, $\mu = 172$, and $\sigma = 29$, we can solve for x by using Formula 6-2:

$$z = \frac{x - \mu}{\sigma} \quad \text{becomes} \quad 2.575 = \frac{x - 172}{29}$$

The result of $x = 246.675$ pounds can be found directly or by using the following version of Formula 6-2:

$$x = \mu + (z \cdot \sigma) = 172 + (2.575 \cdot 29) = 246.675$$

Step 4: The solution of $x = 247$ pounds (rounded) in Figure 6-15 is reasonable because it is greater than the mean of 172 pounds.

If using a TI-83/84 Plus calculator, the command **invNorm(0.9950, 172, 29)** yields an x value of 246.6990498.

INTERPRETATION The weight of 247 pounds separates the lightest 99.5% of men from the heaviest 0.5%.

EXAMPLE Designing Car Dashboards When designing the placement of a CD player in a new model car, engineers must consider the forward grip reach of the driver. If the CD player is placed beyond the forward grip reach, the driver must move his or her body in a way that could be distracting and dangerous. (We wouldn't want anyone injured while trying to hear the best of Barry Manilow.) Design engineers decide that the CD should be placed so that it is within the forward grip reach of 95% of women. Women have forward grip reaches that are normally distributed with a mean of 27.0 in. and a standard deviation of 1.3 in. (based on anthropometric survey data from Gordon, Churchill, et al.). Find the forward grip reach of women that separates the longest 95% from the others.

SOLUTION

Step 1: We begin with the graph shown in Figure 6-16. We have entered the mean of 27.0 in., and we have identified the area representing the longest 95% of forward grip reaches. Even though the problem refers to the top 95%, Table A-2 requires that we work with a cumulative *left* area, so we subtract 0.95 from 1 to get 0.05, which is shown as the shaded region.

continued

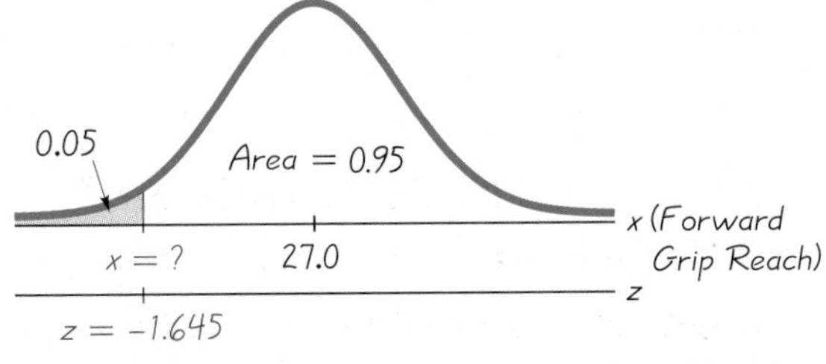

Figure 6-16 Finding the Value Separating the Top 95%

Step 2: In Table A-2 we search for an area of 0.05 in the *body* of the table. The areas closest to 0.05 are 0.0505 and 0.0495, but there is an asterisk indicating that an area of 0.05 corresponds to a z score of -1.645.

Step 3: With $z = -1.645$, $\mu = 27.0$, and $\sigma = 1.3$, we solve for x by using Formula 6-2 directly or by using the following version of Formula 6-2:

$$x = \mu + (z \cdot \sigma) = 27.0 + (-1.645 \cdot 1.3) = 24.8615$$

Step 4: If we let $x = 24.8615$ in Figure 6-16, we see that this solution is reasonable because the forward grip reach separating the top 95% from the bottom 5% should be less than the mean of 27.0 in.

If using a TI-83/84 Plus calculator, the command **invNorm(0.05, 27.0, 1.3)** yields an x value of 24.86169029.

INTERPRETATION The forward grip reach of 24.9 in. (rounded) separates the top 95% from the others, since 95% of women have forward grip reaches that are longer than 24.9 in. and 5% of women have forward grip reaches shorter than 24.9 in.

TI Using a TI-83/84 Plus Calculator to Find Areas

normalcdf on the DISTR menu

```
DISTR DRAW
1:normalpdf(
2:normalcdf(
3:invNorm(
4:tpdf(
5:tcdf(
6:X²pdf(
7↓X²cdf(
```

Entries for First Example of This Section

```
normalcdf(-99999
9,174,172,29)
        .527491537
```

1. Sketch a normal curve, label the mean, label the x value at the lower boundary of the desired area, and label the x value at the upper boundary of the desired area. Also, *shade* the region representing the desired probability.
2. Press 2ND VARS to get the DISTR (distribution) menu, then select **2:normalcdf** and proceed to enter the two x values, the mean, and the standard deviation, all separated by commas. That is, use this format: **normalcdf(lower bound, upper bound, μ, σ).**
3. Press ENTER and the area will be displayed.

Hint: If there is no lower bound or there is no upper bound, enter a very large value, as indicated below.

- Cumulative area from the left (with no actual lower bound): Enter -999999 for the lower bound.
- Cumulative area from the right (with no actual upper bound): Enter 999999 for the upper bound.

Example: The first screen in the margin shows the DISTR (distribution) menu with 2:normalcdf selected. The second screen shows the entries used for the first example of this section. Pressing ENTER yields 0.527491537, which is the area under the normal distribution from the left up to 174, assuming that the mean is 172 and the standard deviation is 29.

TI Using a TI-83/84 Plus Calculator to Find Values

1. Sketch a normal distribution curve, enter the given probability or percentage in the appropriate region of the graph, and identify the x value being sought.
2. Press 2ND VARS to get the DISTR (distribution) menu, then select **3:invNorm** and proceed to enter the *cumulative area to the left* of the desired x value, the

mean, and the standard deviation, all separated by commas. That is, use this format: **invNorm(area to the left, μ, σ).**

3. Press ENTER and the value of x will be displayed.

Example: The first screen in the margin shows the DISTR (distribution) menu with invNorm selected, and the second screen shows the entries for the last example in this section.

TI-83/84 Plus

```
DISTR DRAW
1:normalpdf(
2:normalcdf(
3:invNorm(
4:tpdf(
5:tcdf(
6:X²pdf(
7↓X²cdf(
```

TI-83/84 Plus

```
invNorm(.05,27.0
,1.3)
         24.86169029
```

6-3 BASIC SKILLS AND CONCEPTS

Statistical Literacy and Critical Thinking

1. Normal Distributions What is the difference between a standard normal distribution and a nonstandard normal distribution?

2. Normal Distributions The distribution of IQ scores is a nonstandard normal distribution with a mean of 100 and a standard deviation of 15. What are the values of the mean and standard deviation after all IQ scores have been standardized using $z = (x - \mu)/\sigma$?

3. Random Digits Computers are often used to randomly generate digits of telephone numbers to be called when conducting a survey. Can the methods of this section be used to find the probability that when one digit is randomly generated, it is less than 5? Why or why not? What is the probability of getting a digit less than 5?

4. z Scores and Areas What is the difference between a z score and an area under the graph of a normal probability distribution? Can a z score be negative? Can an area be negative?

IQ Scores. In Exercises 5–12, assume that adults have IQ scores that are normally distributed with a mean of 100 and a standard deviation of 15 (as on the Wechsler test). (Hint: Draw a graph in each case.)

TI 5. Find the probability that a randomly selected adult has an IQ that is less than 130.

TI 6. Find the probability that a randomly selected adult has an IQ greater than 131.5 (the requirement for membership in the Mensa organization).

TI 7. Find the probability that a randomly selected adult has an IQ between 90 and 110 (referred to as the *normal* range).

TI 8. Find the probability that a randomly selected adult has an IQ between 110 and 120 (referred to as *bright normal*).

TI 9. Find P_{10}, which is the IQ score separating the bottom 10% from the top 90%.

TI 10. Find P_{60}, which is the IQ score separating the bottom 60% from the top 40%.

TI 11. Find the IQ score separating the top 35% from the others.

TI 12. Find the IQ score separating the top 85% from the others.

In Exercises 13–16, use this information (based on data from the National Health Survey):

- ***Men's** heights are normally distributed with mean 69.0 in. and standard deviation 2.8 in.*
- ***Women's** heights are normally distributed with mean 63.6 in. and standard deviation 2.5 in.*

TI **13.** Beanstalk Club Height Requirement The Beanstalk Club, a social organization for tall people, has a requirement that women must be at least 70 in. (or 5 ft 10 in.) tall. What percentage of women meet that requirement?

TI **14.** Height Requirement for Women Soldiers The U.S. Army requires women's heights to be between 58 in. and 80 in. Find the percentage of women meeting that height requirement. Are many women being denied the opportunity to join the Army because they are too short or too tall?

TI **15.** Designing Doorways The standard doorway height is 80 in.

a. What percentage of men are too tall to fit through a standard doorway without bending, and what percentage of women are too tall to fit through a standard doorway without bending? Based on those results, does it appear that the current doorway design is adequate?

b. If a statistician designs a house so that all of the doorways have heights that are sufficient for all men except the tallest 5%, what doorway height would be used?

TI **16.** Designing Caskets The standard casket has an inside length of 78 in.

a. What percentage of men are too tall to fit in a standard casket, and what percentage of women are too tall to fit in a standard casket? Based on those results, does it appear that the standard casket size is adequate?

b. A manufacturer of caskets wants to reduce production costs by making smaller caskets. What inside length would fit all men except the tallest 1%?

TI **17.** Birth Weights Birth weights in the United States are normally distributed with a mean of 3420 g and a standard deviation of 495 g. If a hospital plans to set up special observation conditions for the lightest 2% of babies, what weight is used for the cutoff separating the lightest 2% from the others?

TI **18.** Birth Weights Birth weights in Norway are normally distributed with a mean of 3570 g and a standard deviation of 500 g. Repeat Exercise 17 for babies born in Norway. Is the result very different from the result found in Exercise 17?

TI **19.** Eye Contact In a study of facial behavior, people in a control group are timed for eye contact in a 5-minute period. Their times are normally distributed with a mean of 184.0 s and a standard deviation of 55.0 s (based on data from "Ethological Study of Facial Behavior in Nonparanoid and Paranoid Schizophrenic Patients," by Pittman, Olk, Orr, and Singh, *Psychiatry*, Vol. 144, No. 1). For a randomly selected person from the control group, find the probability that the eye contact time is greater than 230.0 s, which is the mean for paranoid schizophrenics.

TI **20.** Body Temperatures Based on the sample results in Data Set 2 of Appendix B, assume that human body temperatures are normally distributed with a mean of 98.20°F and a standard deviation of 0.62°F.

a. Bellevue Hospital in New York City uses 100.6°F as the lowest temperature considered to be a fever. What percentage of normal and healthy persons would be considered to have a fever? Does this percentage suggest that a cutoff of 100.6°F is appropriate?

b. Physicians want to select a minimum temperature for requiring further medical tests. What should that temperature be, if we want only 5.0% of healthy people to exceed it? (Such a result is a *false positive,* meaning that the test result is positive, but the subject is not really sick.)

TI **21.** Lengths of Pregnancies The lengths of pregnancies are normally distributed with a mean of 268 days and a standard deviation of 15 days.

a. One classical use of the normal distribution is inspired by a letter to "Dear Abby" in which a wife claimed to have given birth 308 days after a brief visit from her husband, who was serving in the Navy. Given this information, find the probability of a pregnancy lasting 308 days or longer. What does the result suggest?

b. If we stipulate that a baby is *premature* if the length of pregnancy is in the lowest 4%, find the length that separates premature babies from those who are not premature. Premature babies often require special care, and this result could be helpful to hospital administrators in planning for that care.

TI **22.** Hip Breadths and Aircraft Seats Engineers want to design seats in commercial aircraft so that they are wide enough to fit 98% of all males. (Accommodating 100% of males would require very wide seats that would be much too expensive.) Men have hip breadths that are normally distributed with a mean of 14.4 in. and a standard deviation of 1.0 in. (based on anthropometric survey data from Gordon, Clauser, et al.). Find P_{98}. That is, find the hip breadth for men that separates the smallest 98% from the largest 2%.

TI **23.** Designing Helmets Engineers must consider the breadths of male heads when designing motorcycle helmets. Men have head breadths that are normally distributed with a mean of 6.0 in. and a standard deviation of 1.0 in. (based on anthropometric survey data from Gordon, Churchill, et al.). Due to financial constraints, the helmets will be designed to fit all men except those with head breadths that are in the smallest 2.5% or largest 2.5%. Find the minimum and maximum head breadths that the helmets will fit.

TI **24.** Sitting Distance A common design requirement is that an item (such as an aircraft or theater seat) must fit the range of people who fall between the 5th percentile for women and the 95th percentile for men. If this requirement is adopted, what is the minimum sitting distance and what is the maximum sitting distance? For the sitting distance, use the buttock-to-knee length. Men have buttock-to-knee lengths that are normally distributed with a mean of 23.5 in. and a standard deviation of 1.1 in. Women have buttock-to-knee lengths that are normally distributed with a mean of 22.7 in. and a standard deviation of 1.0 in.

TI **25.** Appendix B Data Set: Systolic Blood Pressure Refer to Data Set 1 in Appendix B and use the systolic blood pressure levels for males.

a. Using the systolic blood pressure levels for males, find the mean, standard deviation, and verify that the data have a distribution that is roughly normal.

b. Assuming that systolic blood pressure levels of males are normally distributed, find the 5th percentile and the 95th percentile. [Treat the statistics from part (a) as if they were population parameters.] Such percentiles could be helpful when physicians try to determine whether blood pressure levels are too low or too high.

TI **26.** Appendix B Data Set: Systolic Blood Pressure Repeat Exercise 25 for females.

6-3 BEYOND THE BASICS

27. Units of Measurement Heights of women are normally distributed.

a. If heights of individual women are expressed in units of inches, what are the units used for the z scores that correspond to individual heights?

b. If heights of all women are converted to z scores, what are the mean, standard deviation, and distribution of these z scores?

TI **28.** Using Continuity Correction There are many situations in which a normal distribution can be used as a good approximation to a random variable that has only *discrete* values. In such cases, we can use this *continuity correction:* Represent each whole number by the interval extending from 0.5 below the number to 0.5 above it. Assume that IQ scores are all whole numbers having a distribution that is approximately normal with a mean of 100 and a standard deviation of 15.

a. Without using any correction for continuity, find the probability of randomly selecting someone with an IQ score greater than 105.

b. Using the correction for continuity, find the probability of randomly selecting someone with an IQ score greater than 105.

c. Compare the results from parts (a) and (b).

TI **29.** Curving Test Scores A statistics professor gives a test and finds that the scores are normally distributed with a mean of 25 and a standard deviation of 5. She plans to curve the scores.

a. If she curves by adding 50 to each grade, what is the new mean? What is the new standard deviation?

b. Is it fair to curve by adding 50 to each grade? Why or why not?

c. If the grades are curved according to the following scheme (instead of adding 50), find the numerical limits for each letter grade.

A: Top 10%
B: Scores above the bottom 70% and below the top 10%
C: Scores above the bottom 30% and below the top 30%
D: Scores above the bottom 10% and below the top 70%
F: Bottom 10%

d. Which method of curving the grades is fairer: Adding 50 to each grade or using the scheme given in part (c)? Explain.

TI **30.** SAT and ACT Tests Scores by women on the SAT test are normally distributed with a mean of 998 and a standard deviation of 202. Scores by women on the ACT test are normally distributed with a mean of 20.9 and a standard deviation of 4.6. Assume that the two tests use different scales to measure the same aptitude.

a. If a woman gets a SAT score that is the 67th percentile, find her actual SAT score and her equivalent ACT score.

b. If a woman gets a SAT score of 1220, find her equivalent ACT score.

top 20 6-4 Sampling Distributions and Estimators

Key Concept The main objective of this section is to understand the concept of a *sampling distribution of a statistic,* which is the distribution of all values of that statistic when all possible samples of the same size are taken from the same population. Specifically, we discuss the sampling distribution of the proportion and the sampling distribution of the mean. We will also see that some statistics (such as the proportion and mean) are good for estimating values of population parameters, whereas other statistics (such as the median) don't make good estimators of population parameters.

In a survey conducted by the Education and Resources Institute, 750 college students were randomly selected, and 64% (or 0.64) of them said that they had at least one credit card. This survey involves only 750 respondents from a population of roughly 15 million college students. We know that sample statistics naturally vary from sample to sample, so another sample of 750 college students will probably yield a sample proportion different from the 64% found in the first survey. In this context, we might think of the sample proportion as a random variable.

Given that the sample of 750 college students is such a tiny percentage (0.005%) of the population, can we really expect the sample proportion to be a reasonable estimate of the actual proportion of all college students who have credit cards? Yes! Statisticians, being the clever creatures that they are, have devised methods for using sample results to estimate population parameters with fairly good accuracy. How do they do it? Their approach is based on an understanding of how statistics behave. By understanding the *distribution* of a sample proportion, statisticians can determine how accurate an individual sample proportion is likely to be. They also understand the distribution of sample means. In general, they understand the concept of a *sampling distribution of a statistic*.

Do Boys or Girls Run in the Family?

The author of this book, his siblings, and his siblings' children consist of 11 males and only one female. Is this an example of a phenomenon whereby one particular gender runs in a family? This issue was studied by examining a random sample of 8770 households in the United States. The results were reported in the *Chance* magazine article "Does Having Boys or Girls Run in the Family?" by Joseph Rodgers and Debby Doughty. Part of their analysis involves use of the binomial probability. Their conclusion is that "We found no compelling evidence that sex bias runs in the family."

Definition

The **sampling distribution of a statistic** (such as a sample proportion or sample mean) is the distribution of all values of the statistic when all possible samples of the same size n are taken from the same population. (The sampling distribution of a statistic is typically represented as a probability distribution in the format of a table, probability histogram, or formula.)

Sampling Distribution of Proportion

The preceding definition can be applied to the specific statistic of a sample proportion.

Definition

The **sampling distribution of the proportion** is the distribution of sample proportions, with all samples having the same sample size n taken from the same population.

We will better understand the important concept of a sampling distribution of the proportion if we consider some specific examples.

EXAMPLE Sampling Distribution of the Proportion of Girls from Two Births When two births are randomly selected, the sample space is bb, bg, gb, gg. Those four equally likely outcomes suggest that the probability of 0 girls is 0.25, the probability of 1 girl is 0.50, and the probability of 2 girls is 0.25. The accompanying display shows the probability distribution for the *number* of girls, followed by two different formats (table and graph) describing the sampling distribution for the *proportion* of girls.

continued

Prophets for Profits

Many books and computer programs claim to be helpful in predicting winning lottery numbers. Some use the theory that particular numbers are "due" (and should be selected) because they haven't been coming up often; others use the theory that some numbers are "cold" (and should be avoided) because they haven't been coming up often; and still others use astrology, numerology, or dreams. Because selections of winning lottery number combinations are independent events, such theories are worthless. A valid approach is to choose numbers that are "rare" in the sense that they are not selected by other people, so that if you win, you will not need to share your jackpot with many others. For this reason, the combination of 1, 2, 3, 4, 5, and 6 is a bad choice because many people use it, whereas 12, 17, 18, 33, 40, 46 is a much better choice, at least until it was published in this book.

In addition to a table or graph, a sampling distribution can also be expressed as a formula (see Exercise 15), or it might be described some other way, such as this: "The sampling distribution of the sample mean is a normal distribution with $\mu = 100$ and $\sigma = 15$." In this section, we usually describe a sampling distribution using a table that lists values of the sample statistic along with their corresponding probabilities. In later chapters we will use some of the other descriptions.

Although typical surveys involve sample sizes around 1000 to 2000 and population sizes often in the millions, the next example involves a population with only three values so that we can easily list every possible sample.

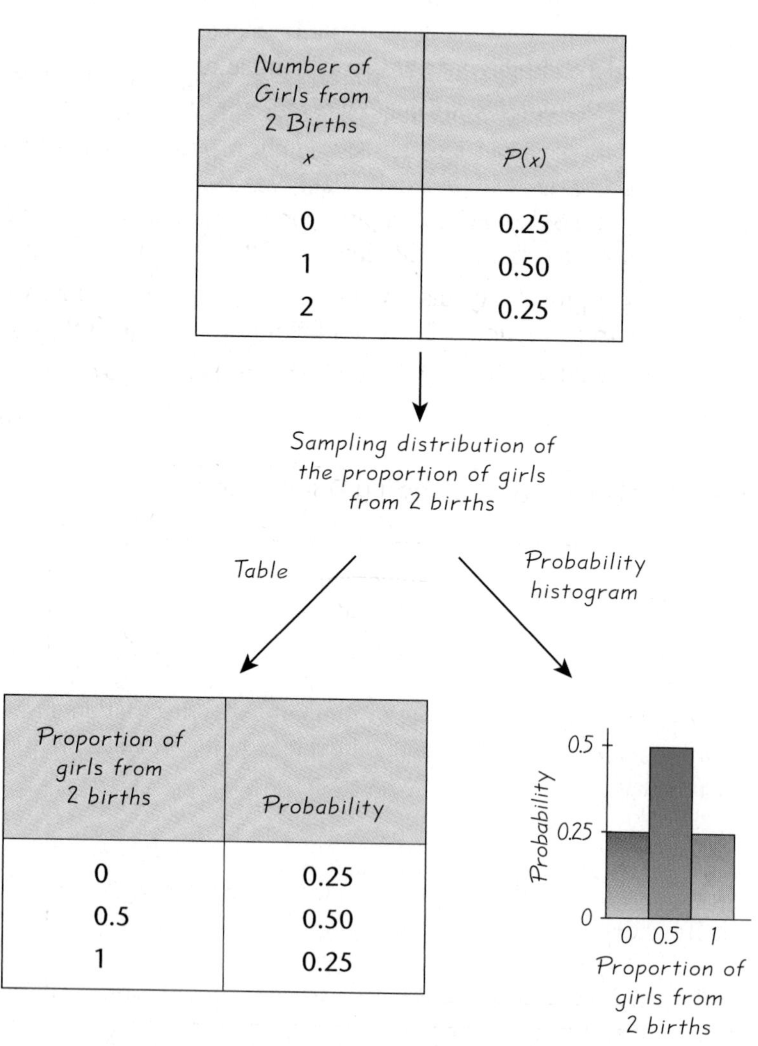

EXAMPLE Sampling Distribution of Proportions A quarterback threw 1 interception in his first game, 2 interceptions in his second game, 5 interceptions in his third game, and he then retired. Consider the *population* consisting of the values 1, 2, 5. Note that two of the values (1 and 5) are odd, so the proportion of odd numbers in the population is 2/3.

a. List all of the different possible samples of size $n = 2$ selected with replacement. (Later, we will explain why sampling *with replacement* is so important.) For each sample, find the proportion of numbers that are *odd*. Use a table to represent the sampling distribution for the proportion of odd numbers.

b. Find the mean of sampling distribution for the proportion of odd numbers.

c. For the population of 1, 2, 5, the proportion of odd numbers is 2/3. Is the mean of the sampling distribution for the proportion of odd numbers also equal to 2/3? Do sample proportions target the value of the population proportion? That is, do the sample proportions have a mean that is equal to the population proportion?

SOLUTION

a. In Table 6-2 we list the nine different possible samples of size $n = 2$ taken with replacement from the population of 1, 2, 5. Table 6-2 also shows the numbers of sample values that are odd numbers, and it includes their probabilities. (Because there are 9 equally likely samples, each sample has probability 1/9.) Table 6-3, which is simply a condensed version of Table 6-2, concisely represents the sampling distribution of the proportion of odd numbers.

b. Table 6-3 is a probability distribution, so we can find its mean by using Formula 5-1 from Section 5-2. We get the mean of 2/3 as follows:

$$\mu = \Sigma[x \cdot P(x)] = (0 \cdot 1/9) + (0.5 \cdot 4/9) + (1 \cdot 4/9) = 6/9 = 2/3$$

continued

Table 6-2 Sampling Distribution of Proportions of Odd Numbers

Sample	Proportion of Odd Numbers	Probability
1, 1	1	1/9
1, 2	0.5	1/9
1, 5	1	1/9
2, 1	0.5	1/9
2, 2	0	1/9
2, 5	0.5	1/9
5, 1	1	1/9
5, 2	0.5	1/9
5, 5	1	1/9

Table 6-3 Condensed Version of Table 6-2

Proportion of Odd Numbers	Probability
0	1/9
0.5	4/9
1	4/9

c. The mean of the sampling distribution of proportions is 2/3, and 2/3 of the numbers in the population are odd. This is no coincidence. In general, the sampling distribution of proportions will have a mean that is equal to the population proportion. It is in this sense that we say that sample proportions "target" the population proportion.

INTERPRETATION For the case of selecting two values (with replacement) from the population of 1, 2, 5, we have identified the sampling distribution (Table 6-3). We also found that the mean of the sampling distribution is 2/3, which is equal to the proportion of odd numbers in the population. Sample proportions therefore tend to target the population proportion, instead of systematically tending to underestimate or overestimate that value.

The preceding example involves a fairly small population, so let's now consider the genders of the Senators in the 107th Congress. Because there are only 100 members [13 females (F) and 87 males (M)], we can list the entire population:

M **F** M M **F** M M M M M M M **F** M M M M M M M
M M M M M M M M M M M M **F** **F** M M M M M M
M M M **F** M M M M M **F** M M M M M M M M M M
F M M M M M M M M M M M M M M M M **F** **F** **F**
M M M **F** M **F** M M M M M M M M M M M M M M

The population proportion of female Senators is $p = 13/100 = 0.13$. Usually, we don't know all of the members of the population, so we must estimate it from a sample. For the purpose of studying the behavior of sample proportions, we list a few samples of size $n = 10$, and we show the corresponding proportion of females.

Sample 1: M **F** M M **F** M M M M M → sample proportion is 0.2
Sample 2: M **F** M M M M M M M M → sample proportion is 0.1
Sample 3: M M M M M M **F** M M M → sample proportion is 0.1
Sample 4: M M M M M M M M M M → sample proportion is 0
Sample 5: M M M M M M M M **F** M → sample proportion is 0.1

Table 6-4
Results from 100 Samples

Proportion of Female Senators	Frequency
0.0	26
0.1	41
0.2	24
0.3	7
0.4	1
0.5	1
Mean:	0.119
Standard deviation:	0.100

We prefer not to list all of the 100,000,000,000,000,000,000 different possible samples. Instead, the author randomly selected just 95 additional samples before stopping to rotate his car tires. Combining these additional 95 samples with the 5 listed here, we get 100 samples summarized in Table 6-4.

We can see from Table 6-4 that the mean of the 100 sample proportions is 0.119, but if we were to include all other possible samples of size 10, the mean of the sample proportions would equal 0.13, which is the value of the population proportion. Figure 6-17 shows the distribution of the 100 sample proportions summarized in Table 6-4. The shape of that distribution is reasonably close to the shape that would have been obtained with all possible samples of size 10. We can see that the distribution depicted in Figure 6-17 is somewhat skewed to the right, but with a bit of a stretch, it might be approximated very roughly by a normal distribution. In Figure 6-18 we show the results obtained from 10,000 samples of size 50 randomly selected with replacement from the above list of 100 genders. Figure 6-18 very strongly suggests that the distribution is approaching the characteristic bell shape of a normal distribution. The results from Table 6-4 and Figure 6-18 therefore suggest the following:

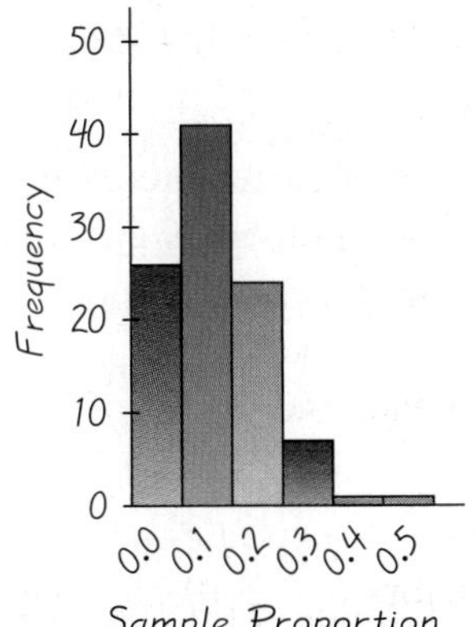

Figure 6-17

100 Sample Proportions with $n = 10$ in Each Sample

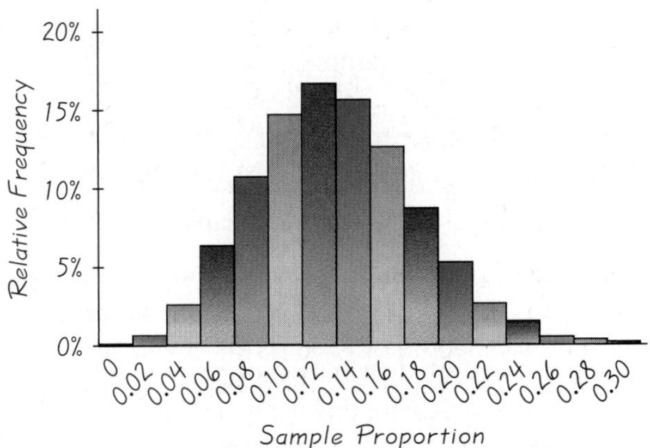

Figure 6-18

10,000 Sample Proportions with $n = 50$ in Each Sample

Properties of the Sampling Distribution of the Sample Proportion

- **Sample proportions tend to target the value of the population proportion. (That is, all possible sample proportions have a mean equal to the population proportion.)**
- **Under certain conditions, the distribution of the sample proportion can be approximated by a normal distribution.**

Sampling Distribution of the Mean

Now let's consider the sampling distribution of the mean.

Definition

The **sampling distribution of the mean** is the distribution of sample means, with all samples having the same sample size n taken from the same population. (The sampling distribution of the mean is typically represented as a probability distribution in the format of a table, probability histogram, or formula.)

Table 6-5
Sampling Distribution of $\bar{x}$

Sample	Mean $\bar{x}$	Probability
1, 1	1.0	1/9
1, 2	1.5	1/9
1, 5	3.0	1/9
2, 1	1.5	1/9
2, 2	2.0	1/9
2, 5	3.5	1/9
5, 1	3.0	1/9
5, 2	3.5	1/9
5, 5	5.0	1/9

↓

Table 6-6
Condensed Version of Table 6-5

$\bar{x}$	Probability
1.0	1/9
1.5	2/9
2.0	1/9
3.0	2/9
3.5	2/9
5.0	1/9

Again, instead of getting too abstract, we use a small population to illustrate the important properties of this distribution.

EXAMPLE Sampling Distribution of the Mean A *population* consists of the values 1, 2, 5. Note that the mean of this population is $\mu = 8/3$.

a. List all of the possible samples (with replacement) of size $n = 2$ along with the sample means and their individual probabilities.

b. Find the mean of the sampling distribution.

c. The population mean is 8/3. Do the sample means target the value of the population mean?

SOLUTION

a. In Table 6-5 we list the nine different possible samples of size $n = 2$ taken with replacement from the population of 1, 2, 5. Table 6-5 also shows the sample means, and it includes their probabilities. (Because there are 9 equally likely samples, each sample has probability 1/9.) Table 6-6, which is simply a condensed version of Table 6-5, concisely represents the sampling distribution of the sample means.

b. Table 6-6 is a probability distribution, so we can find its mean by using Formula 5-1 from Section 5-2. We get the mean of 8/3 as follows:

$$\mu = \Sigma[x \cdot P(x)] = (1.0 \cdot 1/9) + (1.5 \cdot 2/9) + \cdots + (5.0 \cdot 1/9)$$
$$= 24/9 = 8/3$$

c. The mean of the sampling distribution of proportions is 8/3, and the population mean is also 8/3. Again, this is not a coincidence. In general, the distribution of sample means will have a mean equal to the population mean. The sample means therefore tend to target the population mean instead of systematically being too low or too high.

INTERPRETATION For the case of selecting two values (with replacement) from the population of 1, 2, 5, we have identified the sampling distribution (Table 6-6). We also found that the mean of the sampling distribution is 8/3, which is equal to the population mean. Sample means therefore tend to target the population mean.

From the preceding example we see that the mean of all of the different possible sample means is equal to the mean of the original population, which is $\mu = 8/3$. We can generalize this as a property of sample means: For a fixed sample size, the mean of all possible sample means is equal to the mean of the population. We will revisit this important property in the next section.

Let's now make an obvious but important observation: *Sample means vary.* See Table 6-5 and observe how the sample means are different. The first sample mean is 1.0, the second sample mean is 1.5, and so on. This leads to the following definition.

Definition

The value of a statistic, such as the sample mean $\bar{x}$, depends on the particular values included in the sample, and it generally varies from sample to sample. This variability of a statistic is called **sampling variability.**

In Chapter 2 we introduced the important characteristics of a data set: center, variation, distribution, outliers, and pattern over time. In examining the samples in Table 6-5, we have already identified a property describing the behavior of sample means: The mean of sample means is equal to the mean of the population. This property addresses the characteristic of center, and we will investigate other characteristics in the next section. We will see that as the sample size increases, the sampling distribution of sample means tends to become a *normal distribution.* Consequently, the normal distribution assumes an importance that goes far beyond the applications illustrated in Section 6-3. The normal distribution will be used for many cases in which we want to use a sample mean $\bar{x}$ for the purpose of making some inference about a population mean μ.

Which Statistics Make Good Estimators of Parameters?

Chapter 7 will introduce formal methods for using sample statistics to make estimates of the values of population parameters. Some statistics work much better than others, and we can judge their value by examining their sampling distributions, as in the following example.

EXAMPLE Sampling Distributions A *population* consists of the values 1, 2, 5. If we randomly select samples of size 2 with replacement, there are nine different possible samples, and they are listed in Table 6-7. Because the nine different samples are equally likely, each sample has probability 1/9.

a. For each sample, find the mean, median, range, variance, standard deviation, and the proportion of sample values that are odd. (For each statistic, this will generate nine values which, when associated with nine probabilities of 1/9 each, will combine to form a *sampling distribution* for the statistic.)

b. For each statistic, find the mean of the results from part (a).

c. Compare the means from part (b) to the corresponding population parameters, then determine whether each statistic targets the value of the population parameter. For example, the sample means tend to center about the value of the population mean, which is 8/3, so the sample means target the value of the population mean.

SOLUTION

a. See Table 6-7. The individual statistics are listed for each sample.

b. The means of the sample statistics are shown near the bottom of Table 6-7. The mean of the sample means is 8/3, the mean of the sample medians is 8/3, and so on.

continued

Table 6-7 Sampling Distributions of Statistics (for Samples of Size 2 Drawn with Replacement from the Population 1, 2, 5)

Sample	Mean $\bar{x}$	Median	Range	Variance s^2	Standard Deviation s	Proportion of Odd Numbers	Probability
1, 1	1.0	1.0	0	0.0	0.000	1	1/9
1, 2	1.5	1.5	1	0.5	0.707	0.5	1/9
1, 5	3.0	3.0	4	8.0	2.828	1	1/9
2, 1	1.5	1.5	1	0.5	0.707	0.5	1/9
2, 2	2.0	2.0	0	0.0	0.000	0	1/9
2, 5	3.5	3.5	3	4.5	2.121	0.5	1/9
5, 1	3.0	3.0	4	8.0	2.828	1	1/9
5, 2	3.5	3.5	3	4.5	2.121	0.5	1/9
5, 5	5.0	5.0	0	0.0	0.000	1	1/9
Mean of Statistic Values	8/3	8/3	16/9	26/9	1.3	2/3	
Population Parameter	8/3	2	4	26/9	1.7	2/3	
Does the sample statistic target the population parameter?	Yes	No	No	Yes	No	Yes	

c. The bottom row of Table 6-7 is based on a comparison of the population parameter and results from the sample statistics. For example, the population mean of 1, 2, 5 is $\mu = 8/3$, and the sample means "target" that value of 8/3 in the sense that the mean of the sample means is also 8/3.

INTERPRETATION Based on the results in Table 6-7, we can see that when using a sample statistic to estimate a population parameter, some statistics are good in the sense that they target the population parameter and are therefore likely to yield good results. Such statistics are called *unbiased estimators.* Other statistics are not so good (because they are *biased estimators*). Here is a summary.

- **Statistics that target population parameters:** Mean, Variance, Proportion
- **Statistics that do not target population parameters:** Median, Range, Standard Deviation

Although the sample standard deviation does not target the population standard deviation, the bias is relatively small in large samples, so s is often used to estimate σ. Consequently, means, proportions, variances, and standard deviations will all be considered as major topics in following chapters, but the median and range will rarely be used.

Why sample *with* replacement? For small samples of the type that we have considered so far in this section, sampling *without replacement* would have the very practical advantage of avoiding wasteful duplication whenever the same item is selected more than once. However, we are particularly interested in sampling *with replacement* for these reasons: (1) When selecting a relatively small sample from a large population, it makes no significant difference whether we sample with replacement or without replacement. (2) Sampling with replacement results in independent events that are unaffected by previous outcomes, and independent events are easier to analyze and they result in simpler formulas. We therefore focus on the behavior of samples that are randomly selected *with replacement*. Many of the statistical procedures discussed in the following chapters are based on the assumption that sampling is conducted with replacement.

The key point of this section is to introduce the concept of a sampling distribution of a statistic. Consider the goal of trying to find the mean body temperature of all adults. Because that population is so large, it is not practical to measure the temperature of every adult. Instead, we obtain a sample of body temperatures and use it to estimate the population mean. Data Set 2 in Appendix B includes a sample of 106 such body temperatures, and the mean for that sample is $\bar{x} = 98.20°\text{F}$. Conclusions that we make about the population mean temperature of all adults require that we understand the behavior of the sampling distribution of all such sample means. Even though it is not practical to obtain every possible sample and we are stuck with just one sample, we can form some very meaningful and important conclusions about the population of all body temperatures. A major goal of the following sections and chapters is to learn how we can effectively use a sample to form conclusions about a population. In Section 6-5 we consider more details about the sampling distribution of sample means, and in Section 6-6 we consider more details about the sampling distribution of sample proportions.

6-4 BASIC SKILLS AND CONCEPTS

Statistical Literacy and Critical Thinking

1. Sampling Distribution Use your own words to answer this question: "What is a sampling distribution?"

2. Unbiased Estimator What does it mean when we say that sample means "target" the population mean, or that the sample mean is an *unbiased estimator* of the population mean?

3. Unbiased Estimators Which of the following statistics are unbiased estimators?

a. Sample mean used to estimate a population mean
b. Sample median used to estimate a population median
c. Sample proportion used to estimate a population proportion
d. Sample variance used to estimate a population variance
e. Sample standard deviation used to estimate a population standard deviation
f. Sample range used to estimate a population range

4. Sampling with Replacement Give at least one reason why statistical methods tend to be based on the assumption that sampling is conducted *with* replacement, instead of without replacement.

5. Survey of Voters Based on a random sample of $n = 400$ voters, the NBC news division predicts that the Democratic candidate for the presidency will get 49% of the vote, but she actually gets 51%. Should we conclude that the survey was done incorrectly? Why or why not?

6. Sampling Distribution of Body Temperatures Data Set 2 in Appendix B includes a sample of 106 body temperatures of adults. If we were to construct a histogram to depict the shape of the distribution of that sample, would that histogram show the shape of a *sampling distribution* of sample means? Why or why not?

In Exercises 7–14, represent sampling distributions in the format of a table that lists the different values of the sample statistic along with their corresponding probabilities.

7. Phone Center The Nome Ice Company was in business for only three days (guess why). Here are the numbers of phone calls received on each of those days: 10, 6, 5. Assume that samples of size 2 are randomly selected *with replacement* from this population of three values.

a. List the 9 different possible samples and find the mean of each of them.

b. Identify the probability of each sample and describe the sampling distribution of sample means. (*Hint:* See Table 6-6.)

c. Find the mean of the sampling distribution.

d. Is the mean of the sampling distribution [from part (c)] equal to the mean of the population of the three listed values? Are those means *always* equal?

8. Telemarketing Here are the numbers of sales per day that were made by Kim Ryan, a courteous telemarketer who worked four days before being fired: 1, 11, 9, 3. Assume that samples of size 2 are randomly selected *with replacement* from this population of four values.

a. List the 16 different possible samples and find the mean of each of them.

b. Identify the probability of each sample, then describe the sampling distribution of sample means. (*Hint:* See Table 6-6.)

c. Find the mean of the sampling distribution.

d. Is the mean of the sampling distribution (from part [c]) equal to the mean of the population of the four listed values? Are those means *always* equal?

9. Wealthiest People The assets (in billions of dollars) of the five wealthiest people in the United States are 47 (Bill Gates), 43 (Warren Buffet), 21 (Paul Allen), 20 (Alice Walton), and 20 (Helen Walton). Assume that samples of size 2 are randomly selected *with replacement* from this population of five values.

a. After listing the 25 different possible samples and finding the mean of each sample, use a table to describe the sampling distribution of the sample means. (*Hint:* See Table 6-6.)

b. Find the mean of the sampling distribution.

c. Is the mean of the sampling distribution [from part (b)] equal to the mean of the population of the five listed values? Are those means *always* equal?

10. Military Presidents Here is the population of all five U.S. presidents who had professions in the military, along with their ages at inauguration: Eisenhower (62), Grant (46), Harrison (68), Taylor (64), and Washington (57). Assume that samples of size 2 are randomly selected *with replacement* from the population of five ages.

a. After listing the 25 different possible samples and finding the mean of each sample, use a table to describe the sampling distribution of the sample means. (*Hint:* See Table 6-6.)

b. Find the mean of the sampling distribution.

c. Is the mean of the sampling distribution [from part (b)] equal to the mean of the population of the five listed values? Are those means *always* equal?

11. Genetics A genetics experiment involves a population of fruit flies consisting of 1 male named Mike and 3 females named Anna, Barbara, and Chris. Assume that two fruit flies are randomly selected *with replacement.*
    a. After listing the 16 different possible samples, find the proportion of females in each sample, then use a table to describe the sampling distribution of the proportions of females. (*Hint:* See Table 6-3.)
    b. Find the mean of the sampling distribution.
    c. Is the mean of the sampling distribution [from part (b)] equal to the population proportion of females? Does the mean of the sampling distribution of proportions *always* equal the population proportion?

12. Quality Control After constructing a new manufacturing machine, 5 prototype car headlights are produced and it is found that 2 are defective (D) and 3 are acceptable (A). Assume that two headlights are randomly selected *with replacement* from this population.
    a. After identifying the 25 different possible samples, find the proportion of defects in each of them, then use a table to describe the sampling distribution of the proportions of defects. (*Hint:* See Table 6-3.)
    b. Find the mean of the sampling distribution.
    c. Is the mean of the sampling distribution [from part (b)] equal to the population proportion of defects? Does the mean of the sampling distribution of proportions *always* equal the population proportion?

13. Ranks of Olympic Triathlon Competitors U.S. women competed in the triathlon in the Olympic games held in Athens, and their final rankings were 3, 9, and 23. Assume that samples of size 2 are randomly selected with replacement.
    a. Use a table to describe the sampling distribution of the sample means.
    b. Given that the data consist of ranks, does it really make sense to identify the sampling distribution of the sample means?

14. Median and Moons of Jupiter Jupiter has 4 large moons and 12 small moons. The 4 large moons have these orbit times (in days): 1.8 (Io), 3.6 (Europa), 7.2 (Ganymede), and 16.7 (Callisto). Assume that two of these values are randomly selected with replacement.
    a. After identifying the 16 different possible samples, find the median in each of them, then use a table to describe the sampling distribution of the medians.
    b. Find the mean of the sampling distribution.
    c. Is the mean of the sampling distribution [from part (b)] equal to the population median? Is the median an unbiased estimator of the population median?

6-4 BEYOND THE BASICS

15. Using a Formula to Describe a Sampling Distribution See the first example in this section, which includes a table and graph to describe the sampling distribution of the proportions of girls from two births. Consider the formula shown below, and evaluate that formula using sample proportions x of 0, 0.5, and 1. Based on the results, does the formula describe the sampling distribution? Why or why not?

$$P(x) = \frac{1}{2(2 - 2x)!(2x)!} \quad \text{where } x = 0, 0.5, 1$$

16. Mean Absolute Deviation The population of 1, 2, 5 was used to develop Table 6-7. Identify the sampling distribution of the mean absolute deviation (defined in Section 3-3), then determine whether the mean absolute deviation of a sample is a good statistic for estimating the mean absolute deviation of the population.

6-5 The Central Limit Theorem

Key Concept Section 6-4 included some discussion of the sampling distribution of $\bar{x}$, and this section describes procedures for using that sampling distribution in very real and practical applications. The procedures of this section form the foundation for estimating population parameters and hypothesis testing—topics discussed at length in the following chapters. When selecting a simple random sample from a population with mean μ and standard deviation σ, it is essential to know these principles:

1. If $n > 30$, then the sample means have a distribution that can be approximated by a normal distribution with mean μ and standard deviation $\sigma/\sqrt{n}$. (This guideline is commonly used, regardless of the distribution of the original population.)
2. If $n \leq 30$ and the original population has a normal distribution, then the sample means have a normal distribution with mean μ and standard deviation $\sigma/\sqrt{n}$.
3. If $n \leq 30$ but the original population does not have a normal distribution, then the methods of this section do not apply.

Try to keep this big picture in mind: As we sample from a population, we want to know the behavior of the sample means. The *central limit theorem* tells us that if the sample size is large enough, the distribution of sample means can be approximated by a *normal distribution*, even if the original population is not normally distributed. Although we discuss a "theorem," we do not include rigorous proofs. Instead, we focus on the *concepts* and how to apply them. Here are the key points that form an important foundation for the following chapters.

The Central Limit Theorem and the Sampling Distribution of $\bar{x}$

Given:

1. The random variable x has a distribution (which may or may not be normal) with mean μ and standard deviation σ.
2. Simple random samples all of the same size n are selected from the population. (The samples are selected so that all possible samples of size n have the same chance of being selected.)

Conclusions:

1. The distribution of sample means $\bar{x}$ will, as the sample size increases, approach a *normal* distribution.
2. The mean of all sample means is the population mean μ. (That is, the normal distribution from Conclusion 1 has mean μ.)
3. The standard deviation of all sample means is $\sigma/\sqrt{n}$. (That is, the normal distribution from Conclusion 1 has standard deviation $\sigma/\sqrt{n}$.)

Practical Rules Commonly Used

1. If the original population is not itself normally distributed, here is a common guideline: For samples of size n greater than 30, the distribution of the sample means can be approximated reasonably well by a normal distribution. (There are exceptions, such as populations with very nonnormal distributions requiring sample sizes larger than 30, but such exceptions are relatively rare.) The approximation gets better as the sample size n becomes larger.
2. If the original population is itself normally distributed, then the sample means will be normally distributed for *any* sample size n (not just the values of n larger than 30).

The central limit theorem involves two different distributions: the distribution of the original population and the distribution of the sample means. As in previous chapters, we use the symbols μ and σ to denote the mean and standard deviation of the original population, but we use the following new notation for the mean and standard deviation of the distribution of sample means.

Notation for the Sampling Distribution of $\bar{x}$

If all possible random samples of size n are selected from a population with mean μ and standard deviation σ, the mean of the sample means is denoted by $\mu_{\bar{x}}$, so

$$\mu_{\bar{x}} = \mu$$

Also, the standard deviation of the sample means is denoted by $\sigma_{\bar{x}}$, so

$$\sigma_{\bar{x}} = \frac{\sigma}{\sqrt{n}}$$

$\sigma_{\bar{x}}$ is often called the **standard error of the mean.**

EXAMPLE Simulation with Random Digits Computers are often used to randomly generate digits of telephone numbers to be called for polling purposes. (For example, the Pew Research Center randomly generates the last two digits of telephone numbers so that a "listing bias" can be avoided.) The digits 0, 1, 2, 3, 4, 5, 6, 7, 8, 9 are generated in such a way that they are all equally likely. The following Minitab display shows the histogram of 500,000 generated digits. Observe that the distribution appears to be a uniform distribution, as we expect.

Now group the 500,000 digits into 5000 samples, with each sample having $n = 100$ values. We find the mean for each sample and show the histogram of the 5000 sample means. See this absolutely astounding effect: **Even though the original 500,000 digits have a *uniform* distribution, the distribution of the 5000 sample means is approximately a *normal* distribution!** It's a truly fascinating and intriguing phenomenon in statistics that by sampling from any distribution, we can create a distribution of sample means that is normal or at least approximately normal.

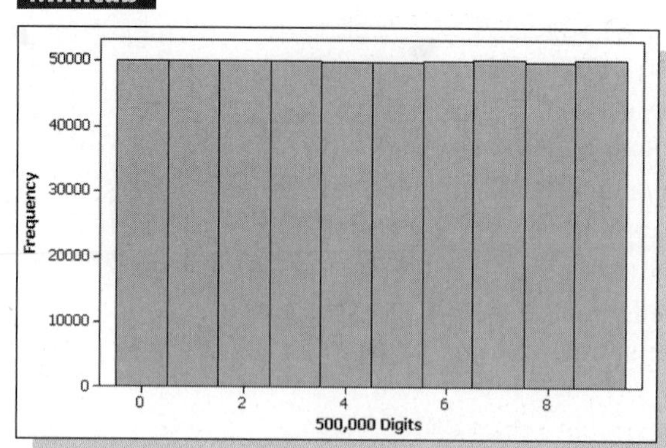

Minitab

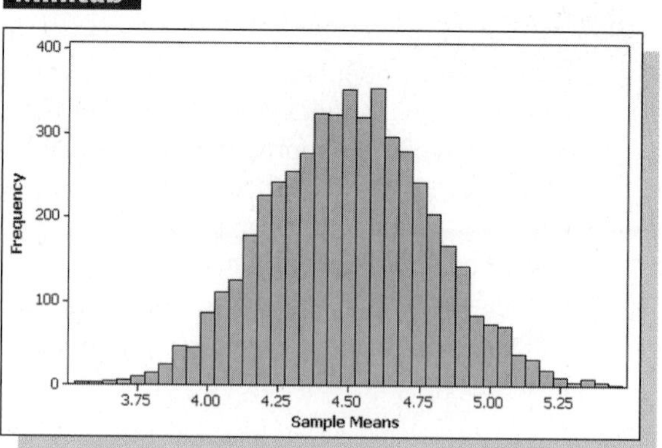

Applying the Central Limit Theorem

Many important and practical problems can be solved with the central limit theorem. When working with such problems, remember that if the sample size is greater than 30, or if the original population is normally distributed, treat the distribution of sample means as if it were a normal distribution with mean μ and standard deviation $\sigma/\sqrt{n}$.

In the following example, part (a) involves an *individual* value, but part (b) involves the mean for a *sample* of 20 men, so we must use the central limit theorem in working with the random variable $\bar{x}$. Study this example carefully to understand the significant difference between the procedures used in parts (a) and (b). See how this example illustrates the following working procedure:

- **When working with an *individual* value from a normally distributed population, use the methods of Section 6-3. Use $z = \dfrac{x - \mu}{\sigma}$.**
- **When working with a mean for some *sample* (or group), be sure to use the value of $\sigma/\sqrt{n}$ for the standard deviation of the sample means. Use $z = \dfrac{\bar{x} - \mu}{\dfrac{\sigma}{\sqrt{n}}}$.**

EXAMPLE Water Taxi Safety In the Chapter Problem we noted that some passengers died when a water taxi sank in Baltimore's Inner Harbor. Men are typically heaver than women and children, so when loading a water taxi, let's assume a worst-case scenario in which all passengers are men. Based on data from the National Health and Nutrition Examination Survey, assume that weights of men are normally distributed with a mean of 172 lb and a standard deviation of 29 lb.

a. Find the probability that if an individual man is randomly selected, his weight will be greater than 175 lb.

b. Find the probability that 20 randomly selected men will have a mean that is greater than 175 lb (so that their total weight exceeds the safe capacity of 3500 lb).

SOLUTION

a. *Approach: Use the methods presented in Section 6-3* (because we are dealing with an individual value from a normally distributed population). We seek the area of the green-shaded region in Figure 6-19(a). If using Table A-2, we convert the weight of 175 to the corresponding z score:

$$z = \frac{x - \mu}{\sigma} = \frac{175 - 172}{29} = 0.10$$

Refer to Table A-2 using $z = 0.10$ and find that the cumulative area to the left of 175 lb is 0.5398. The green-shaded region is therefore $1 - 0.5398 = 0.4602$. The probability of a randomly selected man weighing more than 175 lb is 0.4602. (If using a TI-84/84 Plus calculator, use the DISTR menu to enter **normalcdf(175, 999999, 172, 29)** to get the more accurate result of 0.4588.

b. *Approach: Use the central limit theorem* (because we are dealing with the *mean for a sample* of 20 men, not an individual man). Even though the sample size is not greater than 30, we use a normal distribution for this reason: The original population of men has a normal distribution, so samples of *any* size will yield means that are normally distributed. Because we are now dealing with a distribution of sample means, we must use the parameters $\mu_{\bar{x}}$ and $\sigma_{\bar{x}}$, which are evaluated as follows:

$$\mu_{\bar{x}} = \mu = 172$$

$$\sigma_{\bar{x}} = \frac{\sigma}{\sqrt{n}} = \frac{29}{\sqrt{20}} = 6.4845971$$

Here is a really important point: We must use the computed standard deviation of 6.4845971, not the original standard deviation of 29 (because we are working with the distribution of sample means for which the standard deviation is 6.4845971, not the distribution of individual weights for which the standard deviation is 29). We want to find the green-shaded area shown in Figure 6-19(b). If using Table A-2, we find the relevant z score, which is calculated as follows:

$$z = \frac{\bar{x} - \mu_{\bar{x}}}{\sigma_{\bar{x}}} = \frac{175 - 172}{\frac{29}{\sqrt{20}}} = \frac{3}{6.4845971} = 0.46$$

Referring to Table A-2, we find that $z = 0.46$ corresponds to a cumulative left area of 0.6772, so the green-shaded region is $1 - 0.6772 = 0.3228$. The probability that the 20 men have a mean weight greater than 175 lb is 0.3228.

If using the TI-83/84 Plus calculator, enter **normalcdf(175, 999999, 172, 6.4845971)** to get the more accurate result of 0.3218.

INTERPRETATION There is a 0.4602 probability that an individual man will weigh more than 175 lb, and there is a 0.3228 probability that 20 men will have a mean weight of more than 175 lb. Given that the safe capacity of the water taxi is 3500 lb, there is a fairly good chance (with probability 0.3228)

continued

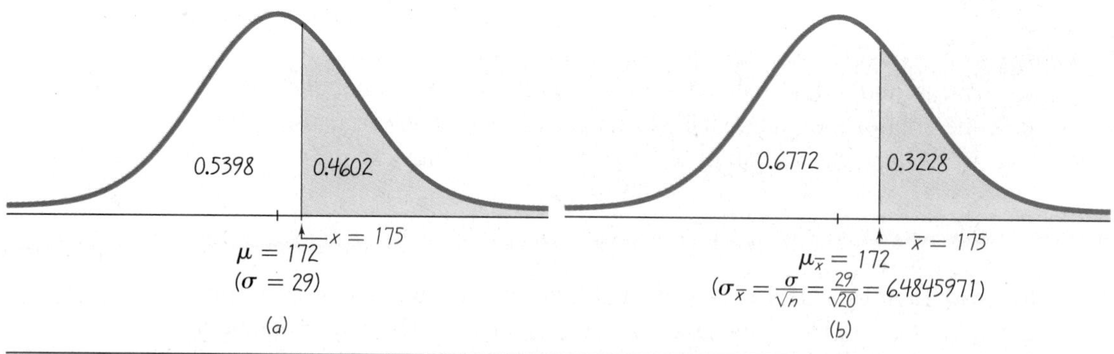

Figure 6-19 Men's Weights

(a) Distribution of Individual Men's Weights; (b) Distribution of Sample Means

that it will be overweight if is filled with 20 randomly selected men. Given that 21 people have already died, and given the high chance of overloading, it would be wise to limit the number of passengers to some level below 20. The capacity of 20 passengers is just not safe enough.

The calculations used here are exactly the type of calculations used by engineers when they design ski lifts, elevators, escalators, airplanes, and other devices that carry people.

Interpreting Results

The next example illustrates another application of the central limit theorem, but carefully examine the conclusion that is reached. This example shows the type of thinking that is the basis for the important procedure of hypothesis testing (discussed in Chapter 8). This example illustrates the rare event rule for inferential statistics, first presented in Section 4-1.

Rare Event Rule

If, under a given assumption, the probability of a particular observed event is exceptionally small, we conclude that the assumption is probably not correct.

EXAMPLE Body Temperatures Assume that the population of human body temperatures has a mean of 98.6°F, as is commonly believed. Also assume that the population standard deviation is 0.62°F (based on data from University of Maryland researchers). If a sample of size $n = 106$ is randomly selected, find the probability of getting a mean of 98.2°F or lower. (The value of 98.2°F was actually obtained; see the midnight temperatures for Day 2 in Data Set 2 of Appendix B.)

SOLUTION

We weren't given the distribution of the population, but because the sample size $n = 106$ exceeds 30, we use the central limit theorem and conclude that the distribution of sample means is a normal distribution with these parameters:

$$\mu_{\bar{x}} = \mu = 98.6 \qquad \text{(by assumption)}$$

$$\sigma_{\bar{x}} = \frac{\sigma}{\sqrt{n}} = \frac{0.62}{\sqrt{106}} = 0.0602197$$

Figure 6-20 shows the shaded area (see the tiny left tail of the graph) corresponding to the probability we seek. Having already found the parameters that apply to the distribution shown in Figure 6-20, we can now find the shaded area by using the same procedures developed in Section 6-3. Using Table A-2, we first find the z score:

$$z = \frac{\bar{x} - \mu_{\bar{x}}}{\sigma_{\bar{x}}} = \frac{98.20 - 98.6}{0.0602197} = -6.64$$

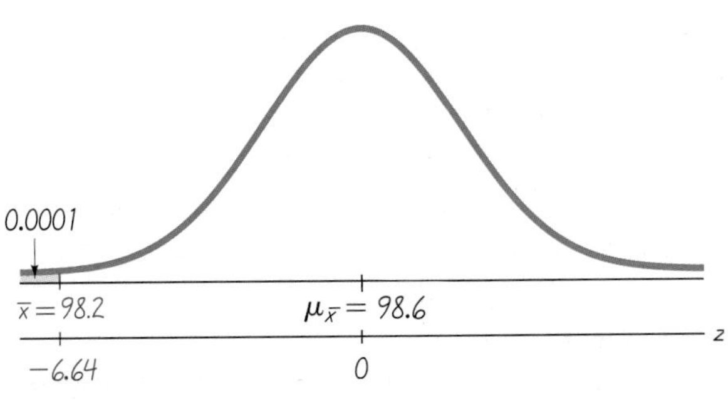

Figure 6-20

Distribution of Mean Body Temperatures for Samples of Size n = 106

Referring to Table A-2, we find that $z = -6.64$ is off the chart, but for values of z below -3.49, we use an area of 0.0001 for the cumulative left area up to $z = -3.49$. We therefore conclude that the shaded region in Figure 6-20 is 0.0001. (If using a TI-83/84 Plus calculator or software, the area of the shaded region is closer to 0.00000000002, but even those results are only approximations. We can safely report that the probability is quite small, such as less than 0.001.)

INTERPRETATION The result shows that if the mean of our body temperatures is really 98.6°F, then there is an extremely small probability of getting a sample mean of 98.2°F or lower when 106 subjects are randomly selected. University of Maryland researchers did obtain such a sample mean, and there are two possible explanations: Either the population mean really is 98.6°F and their sample represents a chance event that is extremely rare, or the population mean is actually lower than 98.6°F and so their sample is typical. Because the probability is so low, it seems more reasonable to conclude that the population mean is lower than 98.6°F. This is the type of reasoning used in *hypothesis testing,* to be introduced in Chapter 8. For now, we should focus on the use of the central limit theorem for finding the probability of 0.0001, but we should also observe that this theorem will be used later in developing some very important concepts in statistics.

The Placebo Effect

It has long been believed that placebos actually help some patients. In fact, some formal studies have shown that when given a placebo (a treatment with no medicinal value), many test subjects show some improvement. Estimates of improvement rates have typically ranged between one-third and two-thirds of the patients. However, a more recent study suggests that placebos have no real effect. An article in the *New England Journal of Medicine* (Vol. 334, No. 21) was based on research of 114 medical studies over 50 years. The authors of the article concluded that placebos appear to have some effect only for relieving pain, but not for other physical conditions. They concluded that apart from clinical trials, the use of placebos "cannot be recommended."

Correction for a Finite Population

In applying the central limit theorem, our use of $\sigma_{\bar{x}} = \sigma/\sqrt{n}$ assumes that the population has infinitely many members. When we sample with replacement (that is, put back each selected item before making the next selection), the population is effectively infinite. Yet many realistic applications involve sampling without replacement, so successive samples depend on previous outcomes. In manufacturing, quality-control inspectors typically sample items from a finite production run without replacing them. For such a finite population, we may need to adjust $\sigma_{\bar{x}}$. Here is a common rule of thumb:

> **When sampling without replacement and the sample size n is greater than 5% of the finite population size N (that is, $n > 0.05N$), adjust the standard deviation of sample means $\sigma_{\bar{x}}$ by multiplying it by the *finite population correction factor:***

$$\sqrt{\frac{N - n}{N - 1}}$$

Except for Exercises 22 and 23, the examples and exercises in this section assume that the finite population correction factor does *not* apply, because we are sampling with replacement or the population is infinite or the sample size doesn't exceed 5% of the population size.

The central limit theorem is so important because it allows us to use the basic normal distribution methods in a wide variety of different circumstances. In Chapter 7, for example, we will apply the theorem when we use sample data to estimate means of populations. In Chapter 8 we will apply it when we use sample data to test claims made about population means. Such applications of estimating population parameters and testing claims are extremely important uses of statistics, and the central limit theorem makes them possible.

The Fuzzy Central Limit Theorem

In *The Cartoon Guide to Statistics,* by Gonick and Smith, the authors describe the Fuzzy Central Limit Theorem as follows: "Data that are influenced by many small and unrelated random effects are approximately normally distributed. This explains why the normal is everywhere: stock market fluctuations, student weights, yearly temperature averages, SAT scores: All are the result of many different effects." People's heights, for example, are the results of hereditary factors, environmental factors, nutrition, health care, geographic region, and other influences which, when combined, produce normally distributed values.

6-5 BASIC SKILLS AND CONCEPTS

Statistical Literacy and Critical Thinking

1. Standard Error of the Mean What is the standard error of the mean?

2. Small Sample If selecting samples of size $n = 2$ from a population with a known mean and standard deviation, what requirement must be satisfied in order to assume that the distribution of the sample means is a normal distribution?

3. Notation Large ($n > 30$) samples are randomly selected from a population with mean μ and standard deviation σ. What notation is used for the mean of the population consisting of all sample means? What notation is used for the standard deviation of the population consisting of all sample means?

4. Convenience Sample Because a statistics student waited until the last minute to do a project, she has only enough time to collect heights from female friends and female relatives. She then calculates the mean height of the females in her sample. Assuming that females have heights that are normally distributed with a mean of 63.6 in. and a standard deviation of 2.5 in., can she use the central limit theorem when analyzing the mean height of her sample?

Using the Central Limit Theorem. *In Exercises 5–8, assume that women's heights are normally distributed with a mean given by $\mu = 63.6$ in. and a standard deviation given by $\sigma = 2.5$ in. (based on data from the National Health Survey).*

TI **5. a.** If 1 woman is randomly selected, find the probability that her height is less than 64 in.
b. If 36 women are randomly selected, find the probability that they have a mean height less than 64 in.

TI **6. a.** If 1 woman is randomly selected, find the probability that her height is greater than 63 in.
b. If 100 women are randomly selected, find the probability that they have a mean height greater than 63 in.

TI **7. a.** If 1 woman is randomly selected, find the probability that her height is between 63.5 in. and 64.5 in.
b. If 9 women are randomly selected, find the probability that they have a mean height between 63.5 in. and 64.5 in.
c. Why can the central limit theorem be used in part (b), even though the sample size does not exceed 30?

TI **8. a.** If 1 woman is randomly selected, find the probability that her height is between 60 in. and 65 in.
b. If 16 women are randomly selected, find the probability that they have a mean height between 60 in. and 65 in.
c. Why can the central limit theorem be used in part (b), even though the sample size does not exceed 30?

TI **9.** Gondola Safety A ski gondola in Vail, Colorado, carries skiers to the top of a mountain. It bears a plaque stating that the maximum capacity is 12 people or 2004 pounds. That capacity will be exceeded if 12 people have weights with a mean greater than $2004/12 = 167$ pounds. Because men tend to weigh more than women, a "worst case" scenario involves 12 passengers who are all men. Men have weights that are normally distributed with a mean of 172 lb and a standard deviation of 29 lb (based on data from the National Health Survey).
a. Find the probability that if an individual man is randomly selected, his weight will be greater than 167 pounds.
b. Find the probability that 12 randomly selected men will have a mean that is greater than 167 pounds (so that their total weight is greater than the gondola maximum capacity of 2004 lb).
c. Does the gondola appear to have the correct weight limit? Why or why not?

TI **10.** Casino Buses The new Lucky Lady Casino wants to increase revenue by providing buses that can transport gamblers from other cities. Research shows that these gamblers tend to be older, they tend to play slot machines only, and they have losses with a mean of \$182 and a standard deviation of \$105. The buses carry 35 gamblers per trip. The casino gives each bus passenger \$50 worth of vouchers that can be converted to cash, so the casino needs to recover that cost in order to make a profit. Find the probability that if a bus is filled with 35 passengers, the mean amount lost by a passenger will exceed \$50. Based on the result, does the casino gamble when it provides such buses?

 **11.** Amounts of Coke Assume that cans of Coke are filled so that the actual amounts have a mean of 12.00 oz and a standard deviation of 0.11 oz.

a. Find the probability that a sample of 36 cans will have a mean amount of at least 12.19 oz, as in Data Set 12 in Appendix B.

b. Based on the result from part (a), is it reasonable to believe that the cans are actually filled with a mean of 12.00 oz? If the mean is not 12.00 oz, are consumers being cheated?

TI **12.** Coaching for the SAT Scores for men on the verbal portion of the SAT-I test are normally distributed with a mean of 509 and a standard deviation of 112 (based on data from the College Board). Randomly selected men are given the Columbia Review Course before taking the SAT test. Assume that the course has no effect.

a. If 1 of the men is randomly selected, find the probability that his score is at least 590.

b. If 16 of the men are randomly selected, find the probability that their mean score is at least 590.

c. In finding the probability for part (b), why can the central limit theorem be used even though the sample size does not exceed 30?

d. If the random sample of 16 men does result in a mean score of 590, is there strong evidence to support the claim that the course is actually effective? Why or why not?

TI **13.** Designing Strobe Lights An aircraft strobe light is designed so that the times between flashes are normally distributed with a mean of 3.00 s and a standard deviation of 0.40 s.

a. Find the probability that an individual time is greater than 4.00 s.

b. Find the probability that the mean for 60 randomly selected times is greater than 4.00 s.

c. Given that the strobe light is intended to help other pilots see an aircraft, which result is more relevant for assessing the safety of the strobe light: The result from part (a) or part (b)? Why?

TI **14.** Designing Motorcycle Helmets Engineers must consider the breadths of male heads when designing motorcycle helmets. Men have head breadths that are normally distributed with a mean of 6.0 in. and a standard deviation of 1.0 in. (based on anthropometric survey data from Gordon, Churchill, et al.).

a. If one male is randomly selected, find the probability that his head breadth is less than 6.2 in.

b. The Safeguard Helmet company plans an initial production run of 100 helmets. Find the probability that 100 randomly selected men have a mean head breadth less than 6.2 in.

c. The production manager sees the result from part (b) and reasons that all helmets should be made for men with head breadths less than 6.2 in., because they would fit all but a few men. What is wrong with that reasoning?

TI **15.** Blood Pressure For women aged 18–24, systolic blood pressures (in mm Hg) are normally distributed with a mean of 114.8 and a standard deviation of 13.1 (based on data from the National Health Survey). Hypertension is commonly defined as a systolic blood pressure above 140.

a. If a woman between the ages of 18 and 24 is randomly selected, find the probability that her systolic blood pressure is greater than 140.

b. If 4 women in that age bracket are randomly selected, find the probability that their mean systolic blood pressure is greater than 140.

c. Given that part (b) involves a sample size that is not larger than 30, why can the central limit theorem be used?

d. If a physician is given a report stating that 4 women have a mean systolic blood pressure below 140, can she conclude that none of the women have hypertension (with a blood pressure greater than 140)?

TI **16.** Staying Out of Hot Water In planning for hot water requirements, the manager of the Luxurion Hotel finds that guests spend a mean of 11.4 min each day in the shower (based on data from the Opinion Research Corporation). Assume that the shower times are normally distributed with a standard deviation of 2.6 min.

a. Find the percentage of guests who shower more than 12 min.

b. The hotel has installed a system that can provide enough hot water provided that the mean shower time for 84 guests is less than 12 min. If the hotel currently has 84 guests, find the probability that there will not be enough hot water. Does the current system appear to be effective?

TI **17.** Redesign of Ejection Seats When women were allowed to become pilots of fighter jets, engineers needed to redesign the ejection seats because they had been designed for men only. The ACES-II ejection seats were designed for men weighing between 140 lb and 211 lb. The population of women has normally distributed weights with a mean of 143 lb and a standard deviation of 29 lb (based on data from the National Health Survey).

a. If 1 woman is randomly selected, find the probability that her weight is between 140 lb and 211 lb.

b. If 36 different women are randomly selected, find the probability that their mean weight is between 140 lb and 211 lb.

c. When redesigning the fighter jet ejection seats to better accommodate women, which probability is more relevant: The result from part (a) or the result from part (b)? Why?

TI **18.** Labeling of M&M Packages M&M plain candies have a mean weight of 0.8565 g and a standard deviation of 0.0518 g (based on Data Set 13 in Appendix B). The M&M candies used in Data Set 13 came from a package containing 465 candies, and the package label stated that the net weight is 396.9 g. (If every package has 465 candies, the mean weight of the candies must exceed $396.9/465 = 0.8535$ for the net contents to weigh at least 396.9 g.)

a. If 1 M&M plain candy is randomly selected, find the probability that it weighs more than 0.8535 g.

b. If 465 M&M plain candies are randomly selected, find the probability that their mean weight is at least 0.8535 g.

c. Given these results, does it seem that the Mars Company is providing M&M consumers with the amount claimed on the label?

TI **19.** Vending Machines Currently, quarters have weights that are normally distributed with a mean of 5.670 g and a standard deviation of 0.062 g. A vending machine is configured to accept only those quarters with weights between 5.550 g and 5.790 g.

a. If 280 different quarters are inserted into the vending machine, what is the expected number of rejected quarters?

b. If 280 different quarters are inserted into the vending machine, what is the probability that the mean falls between the limits of 5.550 g and 5.790 g?

c. If you own the vending machine, which result would concern you more? The result from part (a) or the result from part (b)? Why?

TI **20.** Aircraft Safety Standards Under older Federal Aviation Administration rules, airlines had to estimate the weight of a passenger as 185 pounds. (That amount is for an adult traveling in winter, and it includes 20 pounds of carry-on baggage.) Current rules require an estimate of 195 pounds. Men have weights that are normally distributed with a mean of 172 pounds and a standard deviation of 29 pounds.

a. If 1 adult male is randomly selected and is assumed to have 20 pounds of carry-on baggage, find the probability that his total is greater than 195 pounds.

b. If a Boeing 767-300 aircraft is full of 213 adult male passengers and each is assumed to have 20 pounds of carry-on baggage, find the probability that the mean passenger weight (including carry-on baggage) is greater than 195 pounds. Does a pilot have to be concerned about exceeding this weight limit?

6-5 BEYOND THE BASICS

TI **21.** Seating Design You need to build a bench that will seat 18 male college football players, and you must first determine the length of the bench. Men have hip breadths that are normally distributed with a mean of 14.4 in. and a standard deviation of 1.0 in.

a. What is the minimum length of the bench if you want a 0.975 probability that it will fit the combined hip breadths of 18 randomly selected men?

b. What would be wrong with actually using the result from part (a) as the bench length?

TI **22.** Correcting for a Finite Population The Boston Women's Club needs an elevator limited to 8 passengers. The club has 120 women members with weights that approximate a normal distribution with a mean of 143 lb and a standard deviation of 29 lb. (*Hint:* See the discussion of the finite population correction factor.)

a. If 8 different members are randomly selected, find the probability that their total weight will not exceed the maximum capacity of 1300 lb.

b. If we want a 0.99 probability that the elevator will not be overloaded whenever 8 members are randomly selected as passengers, what should be the maximum allowable weight?

TI **23.** Population Parameters A *population* consists of these values: 2, 3, 6, 8, 11, 18.

a. Find μ and σ.

b. List all samples of size $n = 2$ that can be obtained without replacement.

c. Find the population of all values of $\bar{x}$ by finding the mean of each sample from part (b).

d. Find the mean $\mu_{\bar{x}}$ and standard deviation $\sigma_{\bar{x}}$ for the population of sample means found in part (c).

e. Verify that

$$\mu_{\bar{x}} = \mu \quad \text{and} \quad \sigma_{\bar{x}} = \frac{\sigma}{\sqrt{n}}\sqrt{\frac{N-n}{N-1}}$$

6-6 Normal as Approximation to Binomial

Key Concept This section presents a method for using a normal distribution as an approximation to a binomial probability distribution. If the conditions $np \geq 5$ and $nq \geq 5$ are both satisfied, then probabilities from a binomial probability distribution can be approximated reasonably well by using a normal distribution with mean $\mu = np$ and standard deviation $\sigma = \sqrt{npq}$. Because a binomial probability distribution typically uses only whole numbers for the random variable x, while the normal approximation is continuous, we must use a "continuity correction" with a whole number x represented by the interval from $x - 0.5$ to $x + 0.5$. *Important note:* Instead of using a normal distribution as an approximation to a binomial probability distribution, most practical applications of the binomial distribution can be handled with software or a calculator, but this section introduces the important principle that a binomial distribution can be approximated by a normal distribution, and that principle will be used in later chapters.

Consider the loading of an American Airlines Boeing 767-300, which carries 213 passengers. In analyzing the safe load that can be carried, we must consider the weight of the passengers. We know that a typical man weighs about 30 lb more than a typical woman, so the number of male passengers becomes an important issue. We can use the binomial probability distribution with $n = 213$, $p = 0.5$, and $q = 0.5$ (assuming that men and women are equally likely). See the accompanying Minitab display showing a graph of the probability for each number of male passengers from 0 through 213, and notice how the graph appears to be a normal distribution, even though the plotted points are from a binomial distribution. This graph suggests that we can use a normal distribution to approximate the binomial distribution.

Minitab

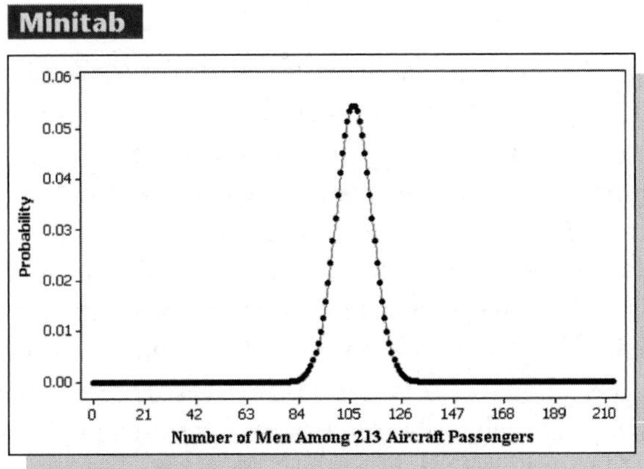

Normal Distribution as Approximation to Binomial Distribution

If a binomial probability distribution satisfies the requirements that $np \geq 5$ and $nq \geq 5$, then that binomial probability distribution can be approximated by a normal distribution with mean $\mu = np$ and standard deviation $\sigma = \sqrt{npq}$, and with the discrete whole number x adjusted with a *continuity correction*, so that x is represented by the interval from $x - 0.5$ to $x + 0.5$.

When using a normal approximation to a binomial distribution, follow this procedure.

Procedure for Using a Normal Distribution to Approximate a Binomial Distribution

1. Establish that the normal distribution is a suitable approximation to the binomial distribution by checking the requirement that $np \geq 5$ and $nq \geq 5$. (If these conditions are not both satisfied, then you must use software, or a calculator, or Table A-1, or calculations using the binomial probability formula.)

Voltaire Beats Lottery

In 1729, the philosopher Voltaire became rich by devising a scheme to beat the Paris lottery. The government ran a lottery to repay municipal bonds that had lost some value. The city added large amounts of money with the net effect that the prize values totaled more than the cost of all tickets. Voltaire formed a group that bought all the tickets in the monthly lottery and won for more than a year. A bettor in the New York State Lottery tried to win a share of an exceptionally large prize that grew from a lack of previous winners. He wanted to write a $6,135,756 check that would cover all combinations, but the state declined and said that the nature of the lottery would have been changed.

2. Find the values of the parameters μ and σ by calculating $\mu = np$ and $\sigma = \sqrt{npq}$.
3. Identify the discrete value x (the number of successes). Change the *discrete* value x by replacing it with the *interval* from $x - 0.5$ to $x + 0.5$. (For further clarification, see the discussion under the subheading "Continuity Corrections" found later in this section.) Draw a normal curve and enter the values of μ, σ, and either $x - 0.5$ or $x + 0.5$, as appropriate.
4. Change x by replacing it with $x - 0.5$ or $x + 0.5$, as appropriate.
5. Using $x - 0.5$ or $x + 0.5$ (as appropriate) in place of x, find the area corresponding to the desired probability by first finding the z score: $z = (x - \mu)/\sigma$. Now use that z score to find the area to the left of the adjusted value of x. That area can now be used to identify the area corresponding to the desired probability.

We will illustrate this normal approximation procedure with the following example.

EXAMPLE Passenger Load on a Boeing 767-300 An American Airlines Boeing 767-300 aircraft has 213 seats. When fully loaded with passengers, baggage, cargo, and fuel, the pilot must verify that the gross weight is below the maximum allowable limit, and the weight must be properly distributed so that the balance of the aircraft is within safe acceptable limits. Instead of weighing each passenger, their weights are estimated according to Federal Aviation Administration rules. In reality, we know that men have a mean weight of 172 pounds, whereas women have a mean weight of 143 pounds, so disproportionately more male passengers might result in an unsafe overweight situation. Assume that if there are at least 122 men in a roster of 213 passengers, the load must be somehow adjusted. Assuming that passengers are booked randomly, male passengers and female passengers are equally likely, and the aircraft is full of adults, find the probability that a Boeing 767-300 with 213 passengers has at least 122 men.

SOLUTION The given problem does involve a binomial distribution with a fixed number of trials ($n = 213$), which are presumably independent, two categories (man, woman) of outcome for each trial, and a probability of a male ($p = 0.5$) that presumably remains constant from trial to trial. Calculations with the binomial probability formula are not practical, because we would have to apply it 92 times (once for each value of x from 122 to 213 inclusive). Instead, we proceed with the five-step approach of using a normal distribution to approximate the binomial distribution.

Step 1: Requirement check: We must first verify that it is reasonable to approximate the binomial distribution by the normal distribution because $np \geq 5$ and $nq \geq 5$. With $n = 213$, $p = 0.5$, and $q = 1 - p = 0.5$, we verify the required conditions as follows:

$$np = 213 \cdot 0.5 = 106.5 \quad \text{(Therefore } np \geq 5.\text{)}$$
$$nq = 213 \cdot 0.5 = 106.5 \quad \text{(Therefore } nq \geq 5.\text{)}$$

Step 2: We now proceed to find the values for μ and σ that are needed for the normal distribution. We get the following:

$$\mu = np = 213 \cdot 0.5 = 106.5$$
$$\sigma = \sqrt{npq} = \sqrt{213 \cdot 0.5 \cdot 0.5} = 7.2972598$$

Step 3: We want the probability of "at least 122 males," and the discrete value of 122 is adjusted by using the continuity correction as follows: Represent $x = 122$ by the vertical strip bounded by 121.5 and 122.5.

Step 4: Because we want the probability of *at least* 122 men, we want the area representing the discrete whole number of 122 (the region bounded by 121.5 and 122.5), as well as the area to the right, as shown in Figure 6-21.

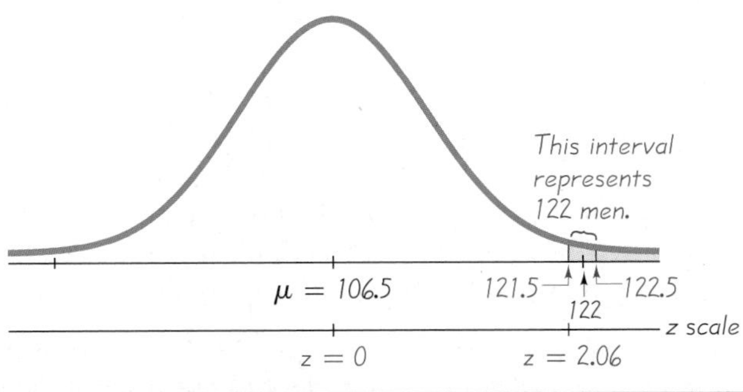

Figure 6-21 Finding the Probability of "At Least 122 Men" Among 213 Passengers

Step 5: We can now proceed to find the shaded area of Figure 6-21 by using the same methods used in Section 6-3. If we use Table A-2 for the standard normal distribution, we must first convert 121.5 to a z score, then use the table to find the area to the left of 121.5, which is then subtracted from 1. The z score is found as follows:

$$z = \frac{x - \mu}{\sigma} = \frac{121.5 - 106.5}{7.2972598} = 2.06$$

Using Table A-2, we find that $z = 2.06$ corresponds to an area of 0.9803, so the shaded region is $1 - 0.9803 = 0.0197$.

If using a TI-83/84 Plus calculator, the command **normalcdf(121.5, 999999, 106.5, 7.2972598)** yields 0.019912111.

INTERPRETATION There is a 0.0197 probability of getting at least 122 men among 213 passengers. Because that probability is so small, we know that a roster of 213 passengers will rarely include at least 122 men, so the adjustments to the aircraft loading will not have to be made very often.

Is Parachuting Safe?

About 30 people die each year as more than 100,000 people make about 2.25 million parachute jumps. In comparison, a typical year includes about 200 scuba diving fatalities, 7000 drownings, 900 bicycle deaths, 800 lightning deaths, and 1150 deaths from bee stings. Of course, these figures don't necessarily mean that parachuting is safer than bike riding or swimming. A fair comparison should involve fatality rates, not just the total number of deaths.

The author, with much trepidation, made two parachute jumps but quit after missing the spacious drop zone both times. He has also flown in a hang glider, hot air balloon, ultralight, sailplane, and Goodyear blimp.

Continuity Corrections

The procedure for using a normal distribution to approximate a binomial distribution includes an adjustment in which we change a discrete whole number to an interval that is 0.5 below and 0.5 above. This particular step, called a *continuity correction,* is usually difficult to understand, so we now consider it in more detail.

Definition

When we use the normal distribution (which is a *continuous* probability distribution) as an approximation to the binomial distribution (which is *discrete*), a **continuity correction** is made to a discrete whole number x in the binomial distribution by representing the single value x by the *interval* from $x - 0.5$ to $x + 0.5$ (that is, adding and subtracting 0.5).

The following practical suggestions should help you use continuity corrections properly.

Procedure for Continuity Corrections

1. When using the normal distribution as an approximation to the binomial distribution, *always* use the continuity correction. (It is required because we are using the *continuous* normal distribution to approximate the *discrete* binomial distribution.)
2. In using the continuity correction, first identify the discrete whole number x that is relevant to the binomial probability problem. For example, if you're trying to find the probability of getting at least 122 men among 213 randomly selected people, the discrete whole number of concern is $x = 122$. First focus on the x value itself, and temporarily ignore whether you want at least x, more than x, fewer than x, or whatever.
3. Draw a normal distribution centered about μ, then draw a *vertical strip area* centered over x. Mark the left side of the strip with the number equal to $x - 0.5$, and mark the right side with the number equal to $x + 0.5$. With $x = 122$, for example, draw a strip from 121.5 to 122.5. *Consider the entire area of the entire strip to represent the probability of the discrete whole number x itself.*
4. Now determine whether the value of x itself should be included in the probability you want. (For example, "at least x" does include x itself, but "more than x" does not include x itself.) Next, determine whether you want the probability of at least x, more than x, at most x, fewer than x, or exactly x. Shade the area to the right or left of the strip, as appropriate; also shade the interior of the strip itself *if and only if x itself* is to be included. This total shaded region corresponds to the probability being sought.

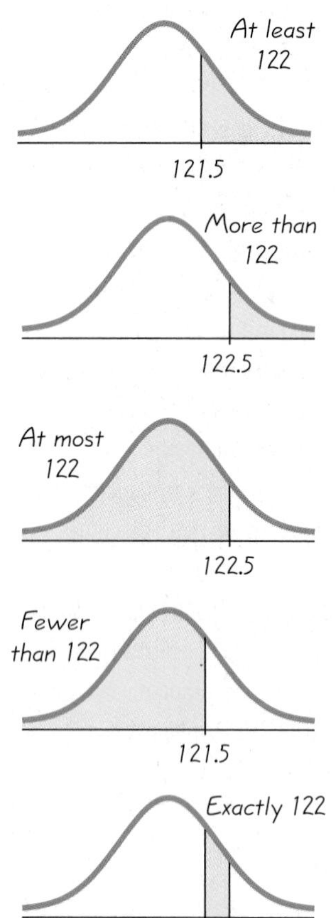

Figure 6-22

Using Continuity Corrections

To see how this procedure results in continuity corrections, see the common cases illustrated in Figure 6-22. Those cases correspond to the statements in the following list.

Statement	Area
At least 122 (includes 122 and above)	To the *right* of 121.5
More than 122 (doesn't include 122)	To the *right* of 122.5
At most 122 (includes 122 and below)	To the *left* of 122.5
Fewer than 122 (doesn't include 122)	To the *left* of 121.5
Exactly 122	Between 121.5 and 122.5

EXAMPLE Internet Use A recent survey showed that among 2013 randomly selected adults, 1358 (or 67.5%) stated that they are Internet users (based on data from Pew Research Center). If the proportion of all adults using the Internet is actually 2/3, find the probability that a random sample of 2013 adults will result in exactly 1358 Internet users.

SOLUTION We have $n = 2013$ independent survey subjects, $x = 1358$ of them are Internet users, and we assume that the population proportion is $p = 2/3$, so it follows that $q = 1/3$. We will use a normal distribution to approximate the binomial distribution.

Step 1: *Requirement check:* First we determine whether the normal approximation is suitable:

$$np = 2013 \cdot 2/3 = 1342 \quad \text{(Therefore } np \geq 5.\text{)}$$
$$nq = 2013 \cdot 1/3 = 671 \quad \text{(Therefore } nq \geq 5.\text{)}$$

Step 2: We now proceed to find the values for μ and σ that are needed for the normal distribution. We get the following:

$$\mu = np = 2013 \cdot 2/3 = 1342$$
$$\sigma = \sqrt{npq} = \sqrt{2013 \cdot (2/3) \cdot (1/3)} = 21.150256$$

Step 3: We draw the normal curve shown in Figure 6-23. The shaded region of the figure represents the probability of getting exactly 1358 Internet users. Using the continuity correction, we represent $x = 1358$ by the region between 1357.5 and 1358.5.

continued

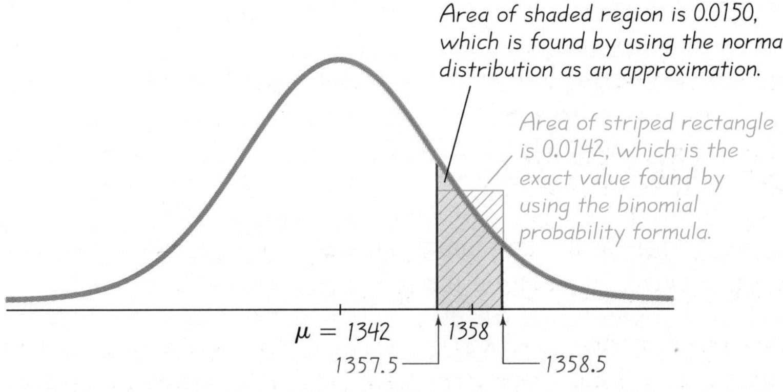

Figure 6-23 Using the Continuity Correction

Step 4: Here is the approach used to find the shaded region in Figure 6-23: First find the total area to the left of 1358.5, then find the total area to the left of 1357.5, then find the *difference* between those two areas. Let's begin with the total area to the left of 1358.5. If we want to use Table A-2 we must first find the z score corresponding to 1358.5. We get

$$z = \frac{1358.5 - 1342}{21.150256} = 0.78$$

We use Table A-2 to find that $z = 0.78$ corresponds to a probability of 0.7823, which is the total area to the left of 1358.5. Now we proceed to find the area to the left of 1357.5 by first finding the z score corresponding to 1357.5:

$$z = \frac{1357.5 - 1342}{21.150256} = 0.73$$

We use Table A-2 to find that $z = 0.73$ corresponds to a probability of 0.7673, which is the total area to the left of 1357.5. The shaded area is $0.7823 - 0.7673 = 0.0150$.

If using the TI-83/84 Plus calculator, the command of **normalcdf (1357.5, 1358.5, 1342, 21.150256)** yields 0.0141679792.

INTERPRETATION If we assume that 2/3 of all adults use the Internet, the probability of getting exactly 1358 Internet users among 2013 randomly selected people is 0.0150. This probability tells us that if the proportion of Internet users in the population is 2/3, then it is highly unlikely that we will get exactly 1358 Internet users when we survey 2013 people. Actually, when surveying 2013 people, the probability of *any* single number of Internet users will be very small.

If we solve the preceding example using the TI-83/84 Plus calculator, we get a result of 0.0142, but the normal approximation method resulted in a value of 0.0150. The discrepancy of 0.0008 results from two factors: (1) The use of the normal distribution results in an *approximate* value that is the area of the shaded region in Figure 6-23, whereas the exact correct area is a rectangle centered above 1358 (Figure 6-23 illustrates this discrepancy); (2) the use of Table A-2 forced us to find one of a limited number of table values based on a rounded z score. The area of the rectangle is 0.0142, but the area of the approximating shaded region is 0.0150.

Interpreting Results

In reality, when we use a normal distribution as an approximation to a binomial distribution, our ultimate goal is not simply to find a probability number. We often need to make some *judgment* based on the probability value. For example, suppose a newspaper reporter sees the sample data in the preceding example and, after observing that 67.5% of the surveyed adults used the Internet, she writes the headline that "More Than 2/3 of Adults Use the Internet." That headline is not justified by the sample data, for the following reason: If the true population proportion *equals* 2/3 (instead of being greater than 2/3), there is a high likelihood (0.2327) of getting at least 1358 Internet users among the 2013 adults surveyed. (The area to the left of 1357.5 is 0.7673, so the probability of getting at least 1358 Internet users is 1 − 0.7673 = 0.2327.) That is, with a population proportion equal to 2/3, the results of 1358 Internet users is *not unusually high.* Such conclusions might seem a bit confusing at this point, but following chapters will introduce systematic methods that will make them much easier. For now, we should understand that low probabilities correspond to events that are very unlikely, whereas large probabilities correspond to likely events. The probability value of 0.05 is often used as a cutoff to distinguish between unlikely events and likely events. The following criterion (from Section 5-2) describes the use of probabilities for distinguishing between results that could easily occur by chance and those results that are highly unusual.

Using Probabilities to Determine When Results Are Unusual

- **Unusually high:** x successes among n trials is an *unusually high* number of successes if $P(x$ or more) is very small (such as 0.05 or less).
- **Unusually low:** x successes among n trials is an *unusually low* number of successes if $P(x$ or fewer) is very small (such as 0.05 or less).

The Role of the Normal Approximation

In reality, almost all practical applications of the binomial probability distribution can now be handled well with software or a TI-83/84 Plus calculator. This section presents methods for dealing with cases in which software cannot be used, but, more importantly, it also illustrates the principle that under appropriate circumstances, the binomial probability distribution can be approximated by a normal distribution. Later chapters will include procedures based on the use of a normal distribution as an approximation to a binomial distribution, so this section forms a foundation for those important procedures.

TI

Using a TI-83/84 Plus Calculator for the Binomial Distribution

The main focus of this section is the use of the normal distribution as an *approximation* to the binomial distribution, and the TI-83/84 Plus calculator can be used for such approximations, as illustrated in examples from this section. However, the TI-83/84 Plus calculator can often be used to find *exact* values, so that approximations are not necessary. The following procedure can be used to find such exact values:

1. Gain access to the "DISTR" (distributions) menu by pressing 2ND VARS.
2. The DISTR menu includes the options of binompdf and binomcdf for a single probability and a cumulative probability.
   - **Single Value** To find the binomial probability for a *single* value of x:

     binompdf(n, p, x) gives the binomial probability of exactly x successes among n trials, where each trial has a probability of success p.

- **Cumulative Values** To find a *cumulative* probability for values from 0 up to (and including) the desired value of x:

  binomcdf(n, p, x) gives a *cumulative* probability from 0 up to (and including) the desired value of x, where n is the number of trials and p is the probability of success for each trial.

6-6 BASIC SKILLS AND CONCEPTS

Statistical Literacy and Critical Thinking

1. Distribution of Sample Proportions Consider a study in which we obtain records from the next 50 babies that are born, then compute the proportion of girls in this sample. Assume that this study is repeated many times, and the sample proportions are used to construct a histogram. What is the shape of the histogram?

2. Continuity Correction The Wechsler test is used to measure IQ scores. It is designed so that the mean is 100 and the standard deviation is 16. It is known that IQ scores have a normal distribution. Assume that we want to find the probability that a randomly selected person has an IQ equal to 107. What is the continuity correction, and how would it be applied in finding that probability?

3. Distribution of Sample Proportions The Newport Bottling plant produces bottles of cola that are packaged in six-packs. The probability of a defective bottle is 0.001. Can we approximate the distribution of defects in six-packs as a normal distribution? Why or why not?

4. Interpreting Binomial Probability In a test of a method of gender selection, 80 couples are given a treatment designed to increase the likelihood that a baby will be a girl. There were 47 girls among the 80 babies born. If the gender-selection method has no effect, the probability of getting exactly 47 girls is 0.0264, and the probability of getting 47 or more girls is 0.0728. Which probability should be used to assess the effectiveness of the gender-selection method? Does it appear that the method is effective?

Applying Continuity Correction. *In Exercises 5–12, the given values are discrete. Use the continuity correction and describe the region of the normal distribution that corresponds to the indicated probability. For example, the probability of "more than 20 defective items" corresponds to the area of the normal curve described with this answer: "the area to the right of 20.5."*

5. Probability of more than 15 people in prison for removing caution labels from pillows

6. Probability of at least 12 adult males on an elevator in the Empire State Building

7. Probability of fewer than 12 crying babies on an American Airlines Flight

8. Probability that the number of working vending machines in the United States is exactly 27

9. Probability of no more than 4 absent students in a statistics class

10. Probability that the number of incorrect statistical procedures in Excel is between 15 and 20 inclusive

11. Probability that the number of truly honest politicians is between 8 and 10 inclusive

12. Probability that exactly 3 employees were fired for inappropriate use of the Internet

Using Normal Approximation. *In Exercises 13–16, do the following: (a) Find the indicated binomial probability by using Table A-1 in Appendix A. (b) If $np \geq 5$ and $nq \geq 5$, also estimate the indicated probability by using the normal distribution as an approximation to the binomial distribution; if $np < 5$ or $nq < 5$, then state that the normal approximation is not suitable.*

13. With $n = 12$ and $p = 0.6$, find $P(7)$.

14. With $n = 14$ and $p = 0.4$, find $P(6)$.

15. With $n = 11$ and $p = 0.5$, find $P(\text{at least } 4)$.

16. With $n = 13$ and $p = 0.3$, find $P(\text{fewer than } 5)$.

TI 17. Probability of More Than 36 Girls Assuming that boys and girls are equally likely, estimate the probability of getting more than 36 girls in 64 births. Is it unusual to get more than 36 girls in 64 births?

TI 18. Probability of at Least 42 Girls Assuming that boys and girls are equally likely, estimate the probability of getting at least 42 girls in 64 births. Is it unusual to get at least 42 girls in 64 births?

TI 19. Voters Lying? In a survey of 1002 people, 701 said that they voted in a recent presidential election (based on data from ICR Research Group). Voting records show that 61% of eligible voters actually did vote. Given that 61% of eligible voters actually did vote, find the probability that among 1002 randomly selected eligible voters, at least 701 actually did vote. What does the result suggest?

TI 20. TV Advertising Charges for advertising on a TV show are based on the number of viewers, which is measured by the rating. The rating is a percentage of the population of 110 million TV households. The CBS television show *60 Minutes* recently had a rating of 7.8, indicating that 7.8% of the households were tuned to that show. An advertiser conducts an independent survey of 100 households and finds that only 4 were tuned to *60 Minutes*. Assuming that the 7.8 rating is correct, find the probability of surveying 100 randomly selected households and getting 4 or fewer tuned to *60 Minutes*. Does the result suggest that the rating of 7.8 is too high? Does the advertiser have grounds for claiming a refund on the basis that the size of the audience was exaggerated?

TI 21. Mendel's Hybridization Experiment When Mendel conducted his famous hybridization experiments, he used peas with green pods and yellow pods. One experiment involved crossing peas in such a way that 25% (or 145) of the 580 offspring peas were expected to have yellow pods. Instead of getting 145 peas with yellow pods, he obtained 152. Assuming that Mendel's 25% rate is correct, estimate the probability of getting at least 152 peas with yellow pods among the 580 offspring peas. Is there strong evidence to suggest that Mendel's rate of 25% is wrong?

TI 22. Cholesterol Reducing Drug The probability of flu symptoms for a person not receiving any treatment is 0.019. In a clinical trial of Lipitor, a common drug used to lower cholesterol, 863 patients were given a treatment of 10-mg atorvastatin tablets, and 19 of those patients experienced flu symptoms (based on data from Pfizer, Inc.). Assuming that these tablets have no effect on flu symptoms, estimate the probability that at least 19 of 863 people experience flu symptoms. What do these results suggest about flu symptoms as an adverse reaction to the drug?

TI **23.** Cell Phones and Brain Cancer In a study of 420,095 cell phone users in Denmark, it was found that 135 developed cancer of the brain or nervous system. Assuming that cell phones have no effect, there is a 0.000340 probability of a person developing cancer of the brain or nervous system. We therefore expect about 143 cases of such cancer in a group of 420,095 randomly selected people. Estimate the probability of 135 or fewer cases of such cancer in a group of 420,095 people. What do these results suggest about media reports that cell phones cause cancer of the brain or nervous system?

TI **24.** Overbooking Flights Air America is considering a new policy of booking as many as 400 persons on an airplane that can seat only 350. (Past studies have revealed that only 85% of the booked passengers actually arrive for the flight.) Estimate the probability that if Air America books 400 persons, not enough seats will be available. Is that probability low enough to be workable, or should the policy be changed?

TI **25.** Identifying Gender Discrimination After being rejected for employment, Kim Kelly learns that the Bellevue Advertising Company has hired only 21 women among its last 62 new employees. She also learns that the pool of applicants is very large, with an equal number of qualified men and women. Help her in her charge of gender discrimination by estimating the probability of getting 21 or fewer women when 62 people are hired, assuming no discrimination based on gender. Does the resulting probability really support such a charge?

TI **26.** M&M Candies: Are 20% Orange? According to Mars (the candy company, not the planet), 20% of M&M plain candies are orange. Data Set 13 in Appendix B shows that among 100 M&Ms chosen, 25 are orange. Assuming that the claimed orange M&Ms rate of 20% is correct, estimate the probability of randomly selecting 100 M&Ms and getting 25 or more that are orange. Based on the result, is it unusual to get 25 or more orange M&Ms when 100 are randomly selected?

TI **27.** Blood Group Forty-five percent of us have Group O blood, according to data provided by the Greater New York Blood Program. Providence Memorial Hospital is conducting a blood drive because its supply of Group O blood is low, and it needs 177 donors of Group O blood. If 400 volunteers donate blood, estimate the probability that the number with Group O blood is at least 177. Is the pool of 400 volunteers likely to be sufficient?

TI **28.** Acceptance Sampling Some companies monitor quality by using a method of acceptance sampling, whereby an entire batch of items is rejected if a random sample of a particular size includes more than some specified number of defects. The Dayton Machine Company buys machine bolts in batches of 5000 and rejects a batch if, when 50 of them are sampled, at least 2 defects are found. Estimate the probability of rejecting a batch if the supplier is manufacturing the bolts with a defect rate of 10%. Is this monitoring plan likely to identify the unacceptable rate of defects?

TI **29.** Cloning Survey A recent Gallup poll consisted of 1012 randomly selected adults who were asked whether "cloning of humans should or should not be allowed." Results showed that 89% of those surveyed indicated that cloning should not be allowed. A news reporter wants to determine whether these survey results constitute strong evidence that the majority (more than 50%) of people are opposed to such cloning. Assuming that 50% of all people are opposed, estimate the probability of getting at least 89%

opposed in a survey of 1012 randomly selected people. Based on that result, is there strong evidence supporting the claim that the majority is opposed to such cloning?

TI **30.** Bias in Jury Selection In Orange County, 12% of those eligible for jury duty are left-handed. Among 250 people selected for jury duty, 25 (or 10%) are lefties. Find the probability of getting at most 25 lefties asssuming that they are chosen with a process designed to yield a 12% rate of lefties. Can we conclude that this process of selecting jurors discriminates against lefties?

TI **31.** Detecting Credit Card Fraud The Dynamic Credit company issues credit cards and uses software to detect fraud. After tracking the spending habits of one consumer, it is found that charges over $100 constitute 35.8% of the credit transactions. Among 30 charges made this month, 18 involve totals that exceed $100. Does this constitute an unusual spending pattern that should be verified? Explain.

TI **32.** Detecting Fraud When working for the Brooklyn District Attorney, investigator Robert Burton analyzed the leading digits of amounts on checks from companies that were suspected of fraud. Among 784 checks, 479 had amounts with leading digits of 5, but checks issued in the normal course of honest transactions were expected to have 7.9% of the checks with amounts having leading digits of 5. Is there strong evidence to indicate that the check amounts are significantly different from amounts that are normally expected? Explain?

6-6 BEYOND THE BASICS

TI **33.** Winning at Roulette Marc Taylor plans to place 200 bets of $1 each on the number 7 at roulette. A win pays off with odds of 35:1 and, on any one spin, there is a probability of $1/38$ that 7 will be the winning number. Among the 200 bets, what is the minimum number of wins needed for Marc to make a profit? Estimate the probability that Marc will make a profit.

TI **34.** Replacement of TVs Replacement times for TV sets are normally distributed with a mean of 8.2 years and a standard deviation of 1.1 year (based on data from "Getting Things Fixed," *Consumer Reports*). Estimate the probability that for 250 randomly selected TV sets, at least 15 of them have replacement times greater than 10.0 years.

TI **35.** Joltin' Joe Assume that a baseball player hits .350, so his probability of a hit is 0.350. (Ignore the complications caused by walks.) Also assume that his hitting attempts are independent of each other.

a. Find the probability of at least 1 hit in 4 tries in 1 game.

b. Assuming that this batter gets up to bat 4 times each game, estimate the probability of getting a total of at least 56 hits in 56 games.

c. Assuming that this batter gets up to bat 4 times each game, find the probability of at least 1 hit in each of 56 consecutive games (Joe DiMaggio's 1941 record).

d. What minimum batting average would be required for the probability in part (c) to be greater than 0.1?

TI **36.** Overbooking Flights Vertigo Airlines works only with advance reservations and experiences a 7% rate of no-shows. How many reservations could be accepted for an airliner with a capacity of 250 if there is at least a 0.95 probability that all reservation holders who show will be accommodated?

TI **37.** Normal Approximation Required This section included the statement that "In reality, almost all practical applications of the binomial probability distribution can now be handled well with software or a TI-83/84 Plus calculator." Using a TI-83/84 Plus calculator, identify a case in which the calculator fails so that a normal approximation to a binomial distribution is required.

6-7 Assessing Normality

Key Concept The following chapters include some very important statistical methods requiring sample data randomly selected from a population with a *normal* distribution. This section provides criteria for determining whether the requirement of a normal distribution is satisfied. The criteria involve visual inspection of a histogram to see if it is roughly bell-shaped, identifying any outliers, and constructing a new graph called a *normal quantile plot.*

Definition

A **normal quantile plot** (or **normal probability plot**) is a graph of points (x, y) where each x value is from the original set of sample data, and each y value is the corresponding z score that is a quantile value expected from the standard normal distribution. (See Step 3 in the following procedure for details on finding these z scores.)

Procedure for Determining Whether Data Have a Normal Distribution

1. *Histogram:* Construct a histogram. Reject normality if the histogram departs dramatically from a bell shape.
2. *Outliers:* Identify outliers. Reject normality if there is more than one outlier present. (Just one outlier could be an error or the result of chance variation, but be careful, because even a single outlier can have a dramatic effect on results.)
3. *Normal quantile plot:* If the histogram is basically symmetric and there is at most one outlier, construct a *normal quantile plot.* The following steps describe one relatively simple procedure for constructing a normal quantile plot, but different statistical packages use various other approaches. (STATDISK and the TI-83/84 Plus calculator use the procedure given here.) This procedure is messy enough so that we usually use software or a calculator to generate the graph, and the end of this section includes instructions for using a TI-83/84 Plus calculator.
   a. First sort the data by arranging the values in order from lowest to highest.
   b. With a sample of size n, each value represents a proportion of $1/n$ of the sample. Using the known sample size n, identify the areas of $1/2n$, $3/2n$, $5/2n$, $7/2n$, and so on. These are the cumulative areas to the left of the corresponding sample values.

c. Use the standard normal distribution (Table A-2 or software or a calculator) to find the z scores corresponding to the cumulative left areas found in Step (b). (These are the z scores that are expected from a normally distributed sample.)
d. Match the original sorted data values with their corresponding z scores found in Step (c), then plot the points (x, y), where each x is an original sample value and y is the corresponding z score.
e. Examine the normal quantile plot using these criteria: If the points do not lie close to a straight line, or if the points exhibit some systematic pattern that is not a straight-line pattern, then the data appear to come from a population that does *not* have a normal distribution. If the pattern of the points is reasonably close to a straight line, then the data appear to come from a population that has a normal distribution. (These criteria can be used loosely for small samples, but they should be used more strictly for large samples.)

Steps 1 and 2 are straightforward, but we illustrate the construction of a normal quantile plot (Step 3) in the following example.

EXAMPLE Heights of Men Data Set 1 in Appendix B includes heights (in inches) of randomly selected men. Let's consider only the first five heights listed for men: 70.8, 66.2, 71.7, 68.7, 67.6. With only five values, a histogram will not be very helpful in revealing the distribution of the data. Instead, construct a normal quantile plot for these five values and determine whether they appear to come from a population that is normally distributed.

SOLUTION The following steps correspond to those listed in the above procedure for constructing a normal quantile plot.

a. First, sort the data by arranging them in order. We get 66.2, 67.6, 68.7, 70.8, 71.7.
b. With a sample of size $n = 5$, each value represents a proportion of $1/5$ of the sample, so we proceed to identify the cumulative areas to the left of the corresponding sample values. Those cumulative left areas, which are expressed in general as $1/2n$, $3/2n$, $5/2n$, $7/2n$, and so on, become these specific areas for this example with $n = 5$: $1/10$, $3/10$, $5/10$, $7/10$, and $9/10$. Those same cumulative left areas expressed in decimal form are 0.1, 0.3, 0.5, 0.7, and 0.9.
c. We now search the body of Table A-2 for the cumulative left areas of 0.1000, 0.3000, 0.5000, 0.7000, and 0.9000. We find these corresponding z scores: -1.28, -0.52, 0, 0.52, and 1.28.
d. We now pair the original sorted heights with their corresponding z scores, and we get these (x, y) coordinates which are plotted as in the accompanying TI-83/84 Plus and STATDISK displays: (66.2, -1.28), (67.6, -0.52), (68.7, 0), (70.8, 0.52), and (71.7, 1.28).

continued

States Rig Lottery Selections

Many states run a lottery in which players select four digits, such as 1127 (the author's birthday). If a player pays \$1 and selects the winning sequence in the correct order, a prize of \$5000 is won. States monitor the number selections and, if one particular sequence is selected too often, players are prohibited from placing any more bets on it. The lottery machines are rigged so that once a popular sequence reaches a certain level of sales, the machine will no longer accept that particular sequence. This prevents states from paying out more than they take in. Critics say that this practice is unfair. According to William Thompson, a gambling expert at the University of Nevada in Las Vegas, "They're saying that they (the states) want to be in the gambling business, but they don't want to be gamblers. It just makes a sham out of the whole numbers game."

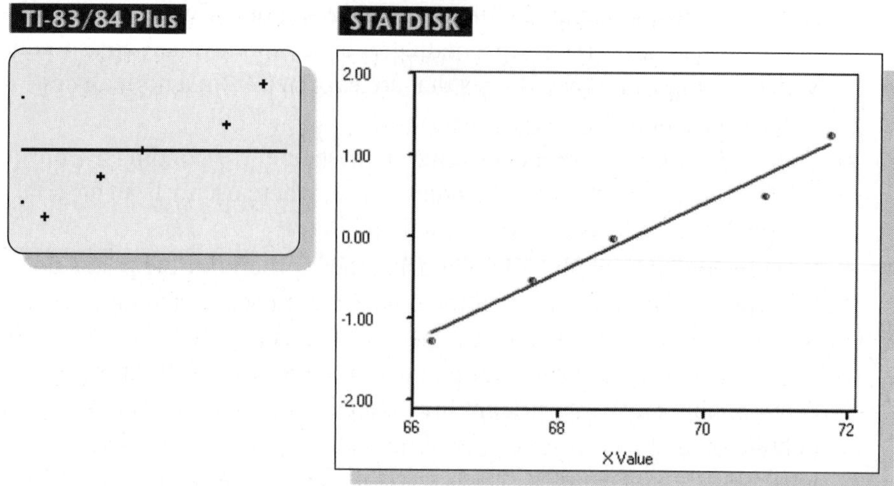

INTERPRETATION We examine the normal quantile plot in the TI-83/84 Plus and STATDISK displays. Because the points appear to lie reasonably close to a straight line and there does not appear to be a systematic pattern that is not a straight-line pattern, we conclude that the sample of five heights appears to come from a normally distributed population.

The following TI-83/84 Plus and STATDISK displays show the normal quantile plot for the same data from the preceding example, with one change: The largest value of 71.7 is changed to 717, so it becomes an outlier. Note how the change in that one value affects the graph. This normal quantile plot does *not* result in points that reasonably approximate a straight-line pattern. The following displays suggests that the values of 66.2, 67.6, 68.7, 70.8, 717 are from a population with a distribution that is *not* a normal distribution.

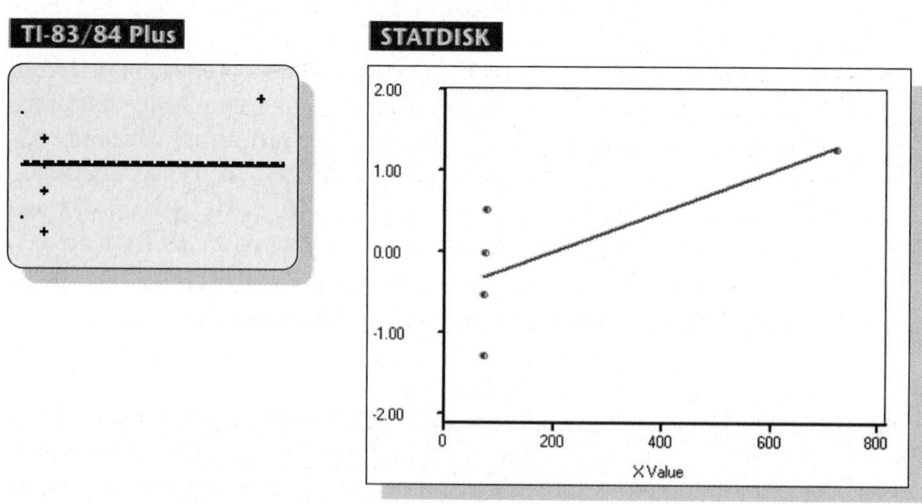

The following example illustrates the use of a histogram and quantile plot with a larger data set. Such larger data sets typically require the use of software.

EXAMPLE Heights of Men The preceding normal quantile plots referred to heights of men, but they involved only five sample values. Consider the following 100 heights of men supplied by a researcher who was instructed to randomly select 100 men and measure each of their heights.

63.3	63.4	63.5	63.6	63.7	63.8	63.9	64.0	64.1	64.2
64.3	64.4	64.5	64.6	64.7	64.8	64.9	65.0	65.1	65.2
65.3	65.4	65.5	65.6	65.7	65.8	65.9	66.0	66.1	66.2
66.3	66.4	66.5	66.6	66.7	66.8	66.9	67.0	67.1	67.2
67.3	67.4	67.5	67.6	67.7	67.8	67.9	68.0	68.1	68.2
68.3	68.4	68.5	68.6	68.7	68.8	68.9	69.0	69.1	69.2
69.3	69.4	69.5	69.6	69.7	69.8	69.9	70.0	70.1	70.2
70.3	70.4	70.5	70.6	70.7	70.8	70.9	71.0	71.1	71.2
71.3	71.4	71.5	71.6	71.7	71.8	71.9	72.0	72.1	72.2
72.3	72.4	72.5	72.6	72.7	72.8	72.9	73.0	73.1	73.2

SOLUTION

Step 1: Construct a histogram. The accompanying TI-83/84 Plus display shows the histogram of the 100 heights, and that histogram suggests that those heights are *not* normally distributed.

Step 2: Identify outliers. Examining the list of 100 heights, we find that no values appear to be outliers.

Step 3: Construct a normal quantile plot. The accompanying TI-83/84 Plus display shows the normal quantile plot. Examination of the normal quantile plot reveals a systematic pattern that is not a straight-line pattern, suggesting that the data are not from a normally distributed population.

Histogram

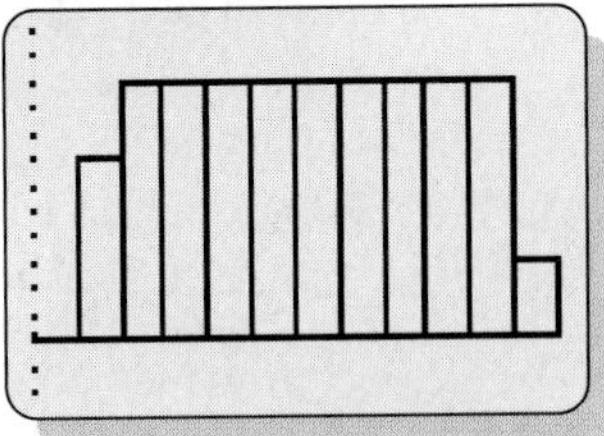

Normal Quantile Plot

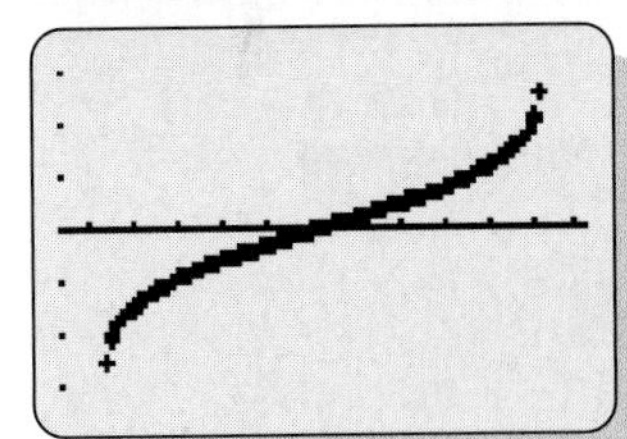

INTERPRETATION Because the histogram does not appear to be bell-shaped, and because the normal quantile plot reveals a pattern of points that is not a straight-line pattern, we conclude that the heights do not appear to be normally distributed. Some of the statistical procedures in later chapters require that sample data be normally distributed, but that requirement is not satisfied for this data set. We expect that 100 randomly selected heights of men should have a distribution that is approximately normal, so the researcher should be investigated. We could also examine the data more closely. Note that the values, when arranged in order, increase consistently by 0.1. This is further evidence that the heights are not the result of measurements obtained from randomly selected men.

Data Transformations Many data sets have a distribution that is not normal, but we can *transform* the data so that the modified values have a normal distribution. One common transformation is to replace each value of x with $\log(x + 1)$. [Instead of using $\log(x + 1)$, we could use the more direct transformation of replacing each value x with $\log x$, but the use of $\log(x + 1)$ has some advantages, including the property that if $x = 0$, then $\log(x + 1)$ can be evaluated, whereas $\log x$ is undefined.]

If the distribution of the $\log(x + 1)$ values is a normal distribution, the distribution of the x values is referred to as a **lognormal distribution.** (See Exercise 22.) In addition to replacing each x value with $\log(x + 1)$, there are other transformations, such as replacing each x value with $\sqrt{x}$, or $1/x$, or x^2. In addition to getting a required normal distribution when the original data values are not normally distributed, such transformations can be used to correct other deficiencies, such as a requirement (found in later chapters) that different data sets have the same variance.

Here are a few final comments about procedures for determining whether data are from a normally distributed population:

- If the requirement of a normal distribution is not too strict, examination of a histogram and consideration of outliers may be all that you need to assess normality.
- Normal quantile plots can be difficult to construct on your own, but they can be generated with a TI-83/84 Plus calculator or suitable software, such as STATDISK, SPSS, SAS, Minitab, and Excel.
- In addition to the procedures discussed in this section, there are other more advanced procedures for assessing normality, such as the chi-square goodness-of-fit test, the Kolmogorov-Smirnov test, and the Lilliefors test.

TI Using a TI-83/84 Plus Calculator to Generate a Normal Quantile Plot

To generate a normal quantile plot using a TI-83/84 Plus calculator, use the following procedure:

1. First enter the data in list L1, or assign the data to a list name. (Remember that the data sets in Appendix B are available in the APPS that is on the CD-ROM included with this book.)
2. Get to the **STAT PLOT** menu by pressing 2ND Y=.
3. Press ENTER ENTER to turn on Plot1, and be sure that Plot2 and Plot3 are in the off state.
4. For the "type" item, select the last item in the second row of displayed options, then enter L1 or a list name for the "Data List." Select X for the Data Axis, and select the icon to be used to depict the plotted points. See the accompanying screen display showing the screen for generating a normal quantile plot using the sample values in list L1.
5. Press ZOOM 9 to generate the normal quantile plot.

TI-83/84 Plus

6-7 BASIC SKILLS AND CONCEPTS

Statistical Literacy and Critical Thinking

1. Normal Quantile Plot What is the purpose of constructing a normal quantile plot?
2. Rejecting Normality Identify two different characteristics of a normal quantile plot, where each characteristic would lead to the conclusion that the data are not from a normally distributed population.

3. Normal Quantile Plot If you randomly select 100 women aged 21–30 and then construct a normal quantile plot of their heights, describe the normal quantile plot that you would expect.

4. Outlier If you have 49 data values randomly selected from a normally distributed population, and there is also a 50th value that is an outlier, will that outlier somehow stand out in the graph of the normal quantile plot, or will it seem to fit in because it is only one value among 50?

Interpreting Normal Quantile Plots. *In Exercises 5–8, examine the normal quantile plot and determine whether it depicts sample data from a population with a normal distribution.*

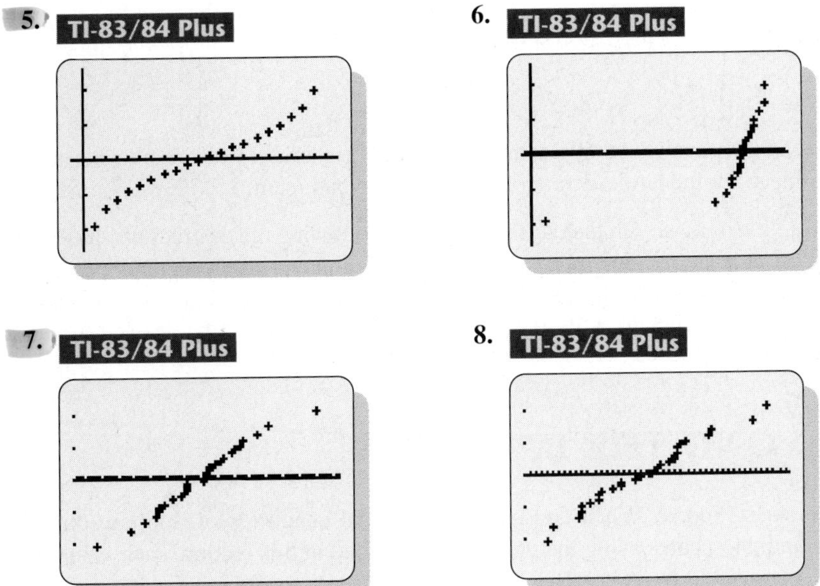

Determining Normality. *In Exercises 9–12, refer to the indicated data set and determine whether the requirement of a normal distribution is satisfied. Assume that this requirement is loose in the sense that the population distribution need not be exactly normal, but it must be a distribution that is basically symmetric with only one mode.*

TI **9.** BMI The measured BMI (body mass index) values of a sample of men, as listed in Data Set 1 in Appendix B.

TI **10.** Weights of Pennies The weights of the wheat pennies, as listed in Data Set 14 in Appendix B.

TI **11.** Precipitation The precipitation amounts, as listed in Data Set 8 in Appendix B.

TI **12.** Temperatures The average daily temperatures, as listed in Data Set 9 in Appendix B.

Using Technology to Generate Normal Quantile Plots. *In Exercises 13–16, use the data from the indicated exercise in this section. Use a TI-83/84 Plus calculator to generate normal quantile plots, then determine whether the data come from a normally distributed population.*

TI **13.** Exercise 9

TI **14.** Exercise 10

TI **15.** Exercise 11

TI **16.** Exercise 12

TI **17.** Comparing Data Sets Using the heights of women and the cholesterol levels of women, as listed in Data Set 1 in Appendix B, analyze each of the two data sets and determine whether each appears to come from a normally distributed population. Compare the results and give a possible explanation for any notable differences between the two distributions.

TI **18.** Comparing Data Sets Using the systolic blood pressure levels and the elbow breadths of women, as listed in Data Set 1 in Appendix B, analyze each of the two data sets and determine whether each appears to come from a normally distributed population. Compare the results and give a possible explanation for any notable differences between the two distributions.

Constructing Normal Quantile Plots. *In Exercises 19 and 20, use the given data values and identify the corresponding z scores that are used for a normal quantile plot, then construct the normal quantile plot and determine whether the data appear to be from a population with a normal distribution.*

TI **19.** Heights of LA Lakers Use this sample of heights (in inches) of the players in the starting lineup for the LA Lakers professional basketball team: 85, 79, 82, 73, 78.

TI **20.** Monitoring Lead in Air On the days immediately following the destruction caused by the terrorist attacks on September 11, 2001, lead amounts (in micrograms per cubic meter) in the air were recorded at Building 5 of the World Trade Center site, and these values were obtained: 5.40, 1.10, 0.42, 0.73, 0.48, 1.10.

6-7 BEYOND THE BASICS

21. Using Standard Scores When constructing a normal quantile plot, suppose that instead of finding z scores using the procedure described in this section, each value in a sample is converted to its corresponding standard score using $z = (x - \bar{x})/s$. If the (x, z) points are plotted in a graph, can this graph be used to determine whether the sample comes from a normally distributed population? Explain.

TI **22.** Lognormal Distribution Test the following phone call times (in seconds) for normality, then replace each x value with $\log(x + 1)$ and test the transformed values for normality. What do you conclude?

31.5	75.9	31.8	87.4	54.1	72.2	138.1	47.9	210.6	127.7
160.8	51.9	57.4	130.3	21.3	403.4	75.9	93.7	454.9	55.1

Review

We introduced the concept of probability distributions in Chapter 5, but included only *discrete* distributions. In this chapter we introduced *continuous* probability distributions and focused on the most important category: normal distributions. Normal distributions will be used extensively in the following chapters.

In Section 6-2 we observed that normal distributions are approximately bell-shaped when graphed. The total area under the density curve of a normal distribution is 1, so there is a convenient correspondence between areas and probabilities. Specific areas can be found using Table A-2 or a TI-83/84 Plus calculator or software. (We do not use Formula 6-1, the equation that is used to define the normal distribution.)

In Section 6-3 we presented important methods for working with normal distributions, including those that use the standard score $z = (x - \mu)/\sigma$ for solving problems such as these:

- Given that IQ scores are normally distributed with $\mu = 100$ and $\sigma = 15$, find the probability of randomly selecting someone with an IQ above 90.
- Given that IQ scores are normally distributed with $\mu = 100$ and $\sigma = 15$, find the IQ score separating the bottom 85% from the top 15%.

In Section 6-4 we introduced the concept of a sampling distribution. The sampling distribution of the mean is the probability distribution of sample means, with all samples having the same sample size n. The sampling distribution of the proportion is the probability distribution of sample proportions, with all samples having the same sample size n. In general, the sampling distribution of any statistic is the probability distribution of that statistic.

In Section 6-5 we presented the following important points associated with the central limit theorem:

1. The distribution of sample means will, as the sample size n increases, approach a normal distribution.
2. The mean of the sample means is the population mean μ.
3. The standard deviation of the sample means is $\sigma/\sqrt{n}$.

In Section 6-6 we noted that we can sometimes approximate a binomial probability distribution by a normal distribution. If both $np \geq 5$ and $nq \geq 5$, the binomial random variable x is approximately normally distributed with the mean and standard deviation given as $\mu = np$ and $\sigma = \sqrt{npq}$. Because the binomial probability distribution deals with discrete data and the normal distribution deals with continuous data, we apply the continuity correction, which should be used in normal approximations to binomial distributions.

Finally, in Section 6-7 we presented a procedure for determining whether sample data appear to come from a population that has a normal distribution. Some of the statistical methods covered later in this book have a loose requirement of a normally distributed population. In such cases, examination of a histogram and outliers might be all that is needed. In other cases, normal quantile plots might be necessary because of a very strict requirement that the population must have a normal distribution.

Statistical Literacy and Critical Thinking

1. Normal Distribution What is a normal distribution? What is a standard normal distribution?

2. Distribution of Sample Means A process consists of randomly selecting 250 adults, measuring the grip strength (right hand only), then finding the sample mean. Assuming that this process is repeated many times, what important fact do we know about the distribution of the resulting sample means?

3. Simple Random Sample A researcher has collected a large sample of IQ scores from friends and relatives. He claims that because his sample is large and the distribution of his sample scores is very close to the bell shape of a normal distribution, his sample is representative of the population. Is this reasoning correct?

4. Central Limit Theorem What does the central limit theorem tell us?

Review Exercises

TI **1.** Weighing Errors A scale is designed so that when items are weighed, the errors in the indicated weights are normally distributed with a mean of 0 g and a standard deviation of 1 g. (If the scale reading is too low, the error is negative. If the scale reading is too high, the error is positive.)

a. If an item is randomly selected and weighed, what is the probability that it has an error between -0.5 g and 0.5 g?

b. If 16 items are randomly selected and weighed, what is the probability that the mean of the errors is between -0.5 g and 0.5 g?

c. What is the 90th percentile for the errors?

TI **2.** Boston Beanstalk Club The Boston Beanstalk Club has a minimum height requirement of 74 in. for men. Heights of men are normally distributed with a mean of 69.0 in. and a standard deviation of 2.8 in. (based on data from the National Health Survey).

a. What percentage of men meet the minimum height requirement?

b. If four men are randomly selected, what is the probability that their mean height is at least 74 in.?

c. If the minimum height requirement for men is to be changed so that only the tallest 10% of men are eligible, what is the new height requirement?

TI **3.** High Cholesterol Levels The serum cholesterol levels in men aged 18–24 are normally distributed with a mean of 178.1 and a standard deviation of 40.7. Units are in mg/100 mL, and the data are based on the National Health Survey.

a. If one man aged 18–24 is randomly selected, find the probability that his serum cholesterol level is greater than 260, a value considered to be "moderately high."

b. If one man aged 18–24 is randomly selected, find the probability that his serum cholesterol level is between 170 and 200.

c. If 9 men aged 18–24 are randomly selected, find the probability that their mean serum cholesterol level is between 170 and 200.

d. The Providence Health Maintenance Organization wants to establish a criterion for recommending dietary changes if cholesterol levels are in the top 3%. What is the cutoff for men aged 18–24?

TI **4.** Detecting Gender Bias in a Test Question When analyzing one particular question on an IQ test, it is found that among the 20 people who answered incorrectly, 18 are women. Given that the test was given to the same number of men and women, all carefully chosen so that they have the same intellectual ability, find the probability that among 20 wrong answers, at least 18 are made by women. Does the result provide strong evidence that the question is biased in favor of men?

TI **5.** Gender Discrimination When several women are not hired at the Telektronics Company, they do some research and find that among the many people who applied, 30% were women. However, the 20 people who were hired consist of only 3 women and 17 men. Find the probability of randomly selecting 20 people from a large pool of applicants (30% of whom are women) and getting 3 or fewer women. Based on the result, does it appear that the company is discriminating based on gender?

TI **6.** Testing for Normality Listed below are the lengths of time (in days) that the New York State budget has been late for each of 19 consecutive and recent years. Do those lengths of time appear to come from a population that has a normal distribution? Why or why not?

4 4 10 19 18 48 64 1 4 68 67 103 125 13 125 34 124 45 44

Cumulative Review Exercises

TI **1.** Carbohydrates in Food Some standard food items are randomly selected and the carbohydrate contents (in grams) are measured with the results listed below (based on data from the U.S. Department of Agriculture). (The items are 12 oz of regular coffee, one cup of whole milk with 3.3% fat, one egg, one banana, one plain doughnut, one tablespoon of peanut butter, one carrot, and 10 potato chips.)

0 12 1 27 24 3 7 10

a. Find the mean $\bar{x}$.
b. Find the median.
c. Find the standard deviation s.
d. Find the variance s^2.
e. Convert the value of 3 g to a z score.
f. Find the actual percentage of these sample values that exceeds 3 g.
g. Assuming a normal distribution, find the percentage of *population* amounts that exceeds 3 g. Use the sample values of $\bar{x}$ and s as estimates of μ and σ.
h. What level of measurement (nominal, ordinal, interval, ratio) describes this data set?
i. The listed measurements appear to be rounded to the nearest gram, but are the exact unrounded amounts discrete data or continuous data?
j. Does it make sense to use the sample mean as an estimate of the carbohydrate content of the food items consumed by the average adult American? Why or why not?

TI **2.** Left-Handedness According to data from the American Medical Association, 10% of us are left-handed.
a. If three people are randomly selected, find the probability that they are all left-handed.
b. If three people are randomly selected, find the probability that at least one of them is left-handed.
c. Why can't we solve the problem in part (b) by using the normal approximation to the binomial distribution?
d. If groups of 50 people are randomly selected, what is the mean number of left-handed people in such groups?
e. If groups of 50 people are randomly selected, what is the standard deviation for the numbers of left-handed people in such groups?
f. Would it be unusual to get 8 left-handed people in a randomly selected group of 50 people? Why or why not?

Cooperative Group Activities

1. Out-of-class activity Divide into groups of three or four students. In each group, develop an original procedure to illustrate the central limit theorem. The main objective is to show that when you randomly select samples from a population, the means of those samples tend to be *normally* distributed, regardless of the nature of the population distribution.

2. In-class activity Divide into groups of three or four students. Using a coin to simulate births, each individual group member should simulate 25 births and record the number of simulated girls. Combine all the results in the group and record n = total number of births and x = number of girls. Given batches of n births, compute the mean and standard deviation for the number of girls. Is the simulated result usual or unusual? Why?

3. In-class activity Divide into groups of three or four students. Select a set of data from Appendix B (excluding Data Sets 1, 8, 9, and 14, which were used in examples or exercises in Section 6-7). Use the methods of Section 6-7 and construct a histogram and normal quantile plot, then determine whether the data set appears to come from a normally distributed population.

TI-83/84 Plus Calculator Project

The first example of Section 6–3 involved weights of men. The solution showed that 52.79% of men have weights less than 174 lb. In that example it was assumed that the weights of men are normally distributed with $\mu = 172$ lb and $\sigma = 29$ lb. This project describes a different method of solution that is based on a simulation technique. We will use the TI-83/84 Plus calculator to randomly generate 500 weights of men from a normally distributed population with $\mu = 172$ lb and $\sigma = 29$ lb. Then we will find the percentage of those simulated weights that are less than 174 lb. The TI-83/84 Plus procedure is described as follows:

1. Generate 500 values from a normally distributed population with $\mu = 172$ lb and $\sigma = 29$ lb. This can be accomplished by pressing MATH, selecting **PRB,** then selecting the option of **6:randNorm.**

2. The format of the randNorm command requires the mean, followed by the standard deviation, followed by the number of desired values, all separated by commas. To generate 500 values from a normally distributed population with $\mu = 172$ lb and $\sigma = 29$ lb, enter the command as **randNorm(172, 29, 500).**

3. Enter **STO→L1** to store the data in list L1. (**STO→L1** can be entered by pressing STO› 2ND 1.)

4. Arrange the data in order by using this procedure: Press STAT, select **2:Sort(A** and enter L1 to obtain the command of **Sort(L1).** (Remember that L1 is entered by pressing 2ND 1.)

5. The sorted entries in list L1 can now be examined by pressing STAT ENTER. Visually examine the entries in list L1 and find the number of simulated weights that are less than 174 lb, and divide that number by 500 to find the percentage of simulated values that are less than 174 lb. Compare the result to the theoretical value of 52.79% that was found in Section 6-3.

From Data to Decision

Critical Thinking: Designing aircraft seating

In this project we consider the issue of determining the "sitting distance" shown in Figure 6-24(a). We define the sitting distance to be the length between the back of the seat cushion and the seat in front. Determining the sitting distance must take into account human body measurements. Specifically, we must consider the "buttock-to-knee length," as shown in Figure 6-24(b). Determining the sitting distance for an aircraft is extremely important. If the sitting distance is unnecessarily large, rows of seats might be eliminated. It has been estimated that removing a single row of six seats can cost around $8 million over the life of the aircraft. If the sitting distance is too small, passengers might be uncomfortable and might prefer to fly other aircraft, or their safety might be compromised because of their limited mobility.

In determining the sitting distance in our aircraft, we will use previously collected data from measurements of large numbers of people. Results from those measurements are summarized in the given table. We can use the data in the table to determine the required sitting distance, but we must make some hard choices. If we are to accommodate *everyone* in the population, we will have a sitting distance that is so costly in terms of reduced seating that it might not be economically feasible. One of the hard decisions that we must make is this: What percentage of the population are we willing to exclude? Another necessary decision is this: How much extra room do we want to provide for passenger comfort and safety? Use the available information to determine the sitting distance. Identify the choices and decisions that were made in that determination.

Buttock-to-Knee Length (inches)

	Mean	Standard Deviation	Distribution
Males	23.5 in.	1.1 in.	Normal
Females	22.7 in.	1.0 in.	Normal

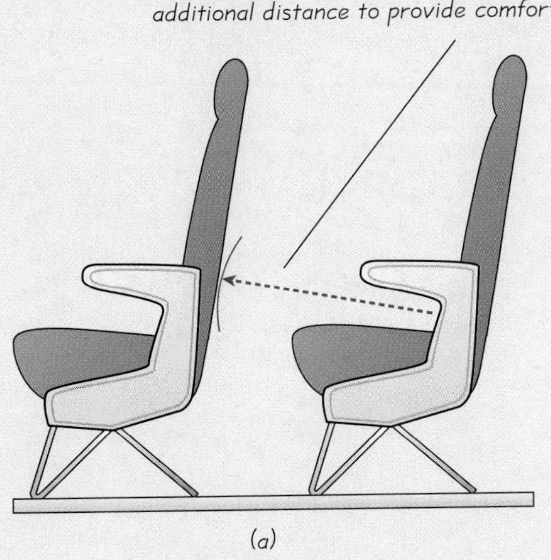

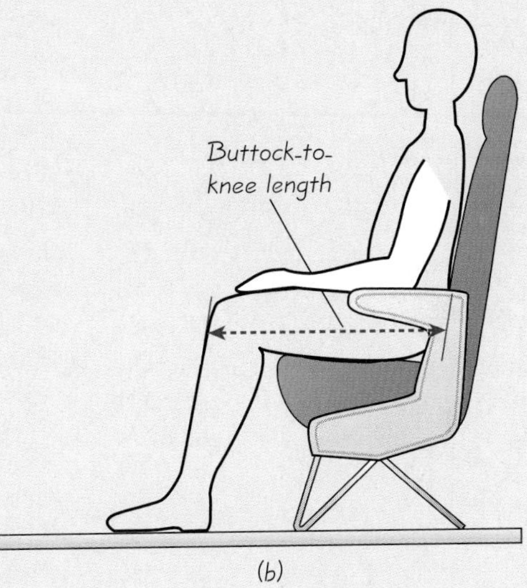

Figure 6-24 Sitting Distance and Buttock-to-Knee Length

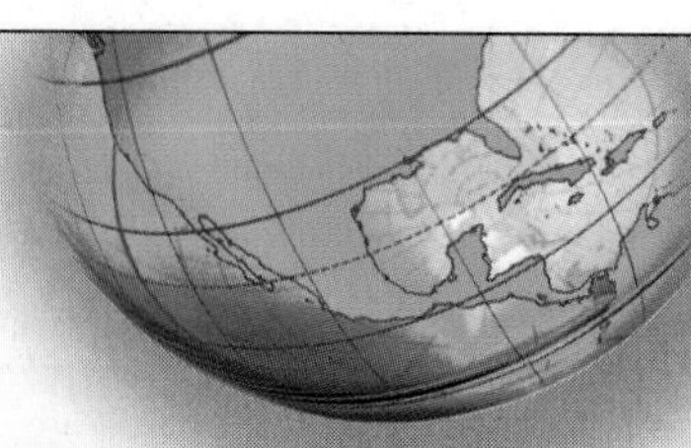

Internet Project

Exploring the Central Limit Theorem

The central limit theorem is one of the most important results in statistics. It also may be one of the most surprising. Informally, the central limit theorem says that the normal distribution is everywhere. No matter what probability distribution underlies an experiment, there is a corresponding distribution of means that will be approximately normal in shape.

The best way to both understand and appreciate the central limit theorem is to see it in action.

The Internet Project for this chapter, found at the *Elementary Statistics* Web site

http://www.aw.com/triola

will allow you to do just that. You will be asked to view, interpret, and discuss a demonstration of the central limit theorem as part of a dice rolling experiment. In addition, you will be guided in a search through the Internet for other such demonstrations.

Statistics @ Work

"It is possible to be a journalist and not be comfortable with statistics, but you're definitely limited in what you can do."

Joel B. Obermayer

Newspaper reporter for The News & Observer

Joel B. Obermayer writes about medical issues and health affairs for *The News & Observer*, a newspaper that covers the eastern half of North Carolina. He reports on managed care, public health, and research at academic medical centers including Duke University and the University of North Carolina at Chapel Hill.

What concepts of statistics do you use?

I use ideas like statistical significance, error rates, and probability. I don't need to do anything incredibly sophisticated, but I need to be very comfortable with the math and with asking questions about it.

I use statistics to look at medical research to decide whether different studies are significant and to decide how to write about them. Mostly, I need to be able to read statistics and understand them, rather than develop them myself. I use statistics to develop good questions and to bolster the arguments I make in print. I also use statistics to decide if someone is trying to give me a positive spin on something that might be questionable. For example, someone at a local university once sent me a press release about miracle creams that supposedly help slim you down by dissolving fat cells. Well, I doubt that those creams work. The study wasn't too hot either. They were trying to make claims based on a study of only 11 people. The researcher argued that 11 people were enough to make good empirical health claims. Not too impressive. People try to manipulate the media all the time. Good verifiable studies with good verifiable statistical bases make it easier to avoid being manipulated.

Is your use of probability and statistics increasing, decreasing, or remaining stable?

Increasing. People's interest in new therapies that may only be in the clinical trial stage is increasing, partly because of the emphasis on AIDS research and on getting new drugs approved and delivered to patients sooner. It's more important than ever for a medical writer to use statistics to make sure that the studies really prove what the public relations people say they prove.

Should prospective employees have studied some statistics?

It is possible to be a journalist and not be comfortable with statistics, but you're definitely limited in what you can do. If you write about government-sponsored education programs and whether they're effective, or if you write about the dangers of particular contaminants in the environment, you're going to need to use statistics.

In my field, editors often don't think about statistics in the interview process. They worry more about writing skills. A knowledge of statistics is more important for what you can do once you are on the job.

Estimates and Sample Sizes

7 CHAPTER PROBLEM

Does touch therapy work?

Many patients pay $25 to $50 for a session of touch therapy in which the touch therapist moves his or her hands within a few inches of the patient's body without actually making physical contact. The objective is to cure a wide variety of medical conditions, including cancer, AIDS, asthma, heart disease, headaches, burns, and bone fractures. The basic theory is that a professionally trained touch therapist can detect poor alignments in the patient's energy field, and can then reposition energy fields to create an energy balance that fosters the healing process.

When she was in the fourth grade, nine-year-old Emily Rosa chose the topic of touch therapy for a science fair project. She convinced 21 experienced touch therapists to participate in a simple test of their ability to detect a human energy field. Emily constructed a cardboard partition with two holes for hands. Each touch therapist would put both hands through the two holes, and Emily would place her hand just above one of the therapist's hands; then the therapist was asked to identify the hand that Emily had selected. Emily used a coin toss to randomly select the hand to be used. This test was repeated 280 times. If the touch therapists really did have the ability to sense a human energy field, they should have identified the correct hand much more than 50% of the time. If they did not have the ability to detect the energy field and they just guessed, they should have been correct about 50% of the time. Here are the results that Emily obtained: Among the 280 trials, the touch therapists identified the correct hand 123 times, for a success rate of 44%. Emily, with the help of her mother, a statistician, and a physician, submitted her findings for publication in the prestigious *Journal of the American Medical Association.* After a careful and thorough review of the experimental design and results, the article "A Close Look at Therapeutic Touch" was published (*Journal of the American Medical Association*, Vol. 279, No. 13). Emily became the youngest researcher to be published in that magazine. And she also won a blue ribbon for her science fair project.

Let's consider the key results from Emily's project. Among the 280 trials, the touch therapists were correct 123 times. We have a sample proportion with $n = 280$ and $x = 123$. Arguments against the validity of the study might include the claim that the number of trials is too small to be meaningful, or that the touch therapists just had a bad day and, because of chance, they were not as successful as the population of all touch therapists. We will consider such issues in this chapter.

It should also be noted that Emily Rosa's project was relatively simple. Remember, she did the project while she was a fourth-grade student. Emily's project is the type of activity that could be conducted by any student in an introductory statistics course. After understanding concepts taught in the typical introductory statistics course, students have the ability to accomplish significant and meaningful work.

7-1 Overview

In this chapter we begin working with the true core of inferential statistics as we use sample data to make inferences about populations. Specifically, we will use sample data to make estimates of population parameters. For example, the Chapter Problem refers to touch therapists who correctly identified the human energy field in only 44% of 280 trials. Based on the sample statistic of 44%, we will estimate the percentage of correct identifications for the entire population of all touch therapists.

The two major applications of inferential statistics involve the use of sample data to (1) estimate the value of a population parameter, and (2) test some claim (or hypothesis) about a population. In this chapter we introduce methods for estimating values of these important population parameters: proportions, means, and variances. We also present methods for determining the sample sizes necessary to estimate those parameters. In Chapter 8 we will introduce the basic methods for testing claims (or hypotheses) that have been made about a population parameter.

7-2 Estimating a Population Proportion

Key Concept In this section we present important methods for using a sample proportion to estimate the value of a population proportion with a *confidence interval*. We also present methods for finding the size of the sample that is needed to estimate a population proportion. This section introduces general concepts that are used in the following sections and following chapters, so it is important to understand this section well.

A Study Strategy: The time devoted to this section will be well spent because we introduce the concept of a confidence interval, and that same general concept will be applied to the following sections of this chapter. We suggest that you first read this section with the limited objective of simply trying to understand what confidence intervals are, what they accomplish, and why they are needed. Second, try to develop the ability to construct confidence interval estimates of population proportions. Third, learn how to interpret a confidence interval correctly. Fourth, read the section once again and try to understand the underlying theory. You will always enjoy much greater success if you understand what you are doing, instead of blindly applying mechanical steps in order to obtain an answer that may or may not make any sense.

This section will consider only cases in which the normal distribution can be used to approximate the sampling distribution of sample proportions. The following requirements apply to the methods of this section.

Requirements

1. The sample is a simple random sample.
2. The conditions for the binomial distribution are satisfied. That is, there is a fixed number of trials, the trials are independent, there are two categories of outcomes, and the probabilities remain constant for each trial. (See Section 5-3.)
3. There are at least 5 successes and at least 5 failures. (With p and q unknown, we estimate their values using the sample proportion, so this requirement is a

way of verifying that $np \geq 5$ and $nq \geq 5$ are both satisfied, so the normal distribution is a suitable approximation to the binomial distribution. Also, there are procedures for dealing with situations in which the normal distribution is not a suitable approximation. See Exercise 51.)

Notation for Proportions

$p =$ *population* proportion

$\hat{p} = \frac{x}{n} =$ sample proportion of x *successes* in a sample of size n

$\hat{q} = 1 - \hat{p} =$ sample proportion of *failures* in a sample of size n

Proportion, Probability, and Percent This section focuses on the population proportion p, but we can also work with probabilities or percentages. When working with a percentage, express it in decimal form. (For example, express 44% as 0.44, so that $\hat{p} = 0.44$.) If we want to estimate a population proportion with a single value, the best estimate is $\hat{p}$. Because $\hat{p}$ consists of a single value, it is called a point estimate.

Definition

A **point estimate** is a single value (or point) used to approximate a population parameter.

The sample proportion $\hat{p}$ is the best point estimate of the population proportion p.

We use $\hat{p}$ as the point estimate of p because it is unbiased and is the most consistent of the estimators that could be used. It is unbiased in the sense that the distribution of sample proportions tends to center about the value of p; that is, sample proportions $\hat{p}$ do not systematically tend to underestimate p, nor do they systematically tend to overestimate p. (See Section 6-4.) The sample proportion $\hat{p}$ is the most consistent estimator in the sense that the standard deviation of sample proportions tends to be smaller than the standard deviation of any other unbiased estimators.

EXAMPLE Touch Therapy Success Rate In the Chapter Problem we noted that in 280 trials involving touch therapists, the correct hand was selected in 123 trials, so the success rate is $\hat{p} = 123/280 = 0.44$. Using these test results, find the best point estimate of the proportion of all correct selections that would be made if all touch therapists were tested.

SOLUTION Because the sample proportion is the best point estimate of the population proportion, we conclude that the best point estimate of p is 0.44. (Using common sense, we might examine the design of the experiment and we might conclude that the true population proportion is actually 0.5, but using only the sample results leads to 0.44 as the best estimate of the population proportion p.)

Small Sample

The Children's Defense Fund was organized to promote the welfare of children. The group published *Children Out of School in America,* which reported that in one area, 37.5% of the 16- and 17-year-old children were out of school. This statistic received much press coverage, but it was based on a sample of only 16 children. Another statistic was based on a sample size of only 3 students. (See "Firsthand Report: How Flawed Statistics Can Make an Ugly Picture Look Even Worse," *American School Board Journal*, Vol. 162.)

Why Do We Need Confidence Intervals?

In the preceding example we saw that 0.44 was our *best* point estimate of the population proportion p, but we have no indication of just how *good* our best estimate is. Because a point estimate has the serious flaw of not revealing anything about how good it is, statisticians have cleverly developed another type of estimate. This estimate, called a *confidence interval* or *interval estimate,* consists of a range (or an interval) of values instead of just a single value.

Definition

A **confidence interval** (or **interval estimate**) is a range (or an interval) of values used to estimate the true value of a population parameter. A confidence interval is sometimes abbreviated as CI.

A confidence interval is associated with a confidence level, such as 0.95 (or 95%). The confidence level gives us the success rate of the procedure used to construct the confidence interval. The confidence level is often expressed as the probability or area $1 - \alpha$ (lowercase Greek alpha), where α is the complement of the *confidence level*. For a 0.95 (or 95%) confidence level, $\alpha = 0.05$. For a 0.99 (or 99%) confidence level, $\alpha = 0.01$.

Definition

The **confidence level** is the probability $1 - \alpha$ (often expressed as the equivalent percentage value) that is the proportion of times that the confidence interval actually does contain the population parameter, assuming that the estimation process is repeated a large number of times. (The confidence level is also called the **degree of confidence,** or the **confidence coefficient.**)

The most common choices for the confidence level are 90% (with $\alpha = 0.10$), 95% (with $\alpha = 0.05$), and 99% (with $\alpha = 0.01$). The choice of 95% is most common because it provides a good balance between precision (as reflected in the width of the confidence interval) and reliability (as expressed by the confidence level).

Here's an example of a confidence interval based on the sample data of 280 trials of touch therapists, with 44% of the trials resulting in correct identification of the hand that was selected:

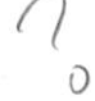

The 0.95 (or 95%) confidence interval estimate of the population proportion p is $0.381 < p < 0.497$.

Interpreting a Confidence Interval

We must be careful to interpret confidence intervals correctly. There is a correct interpretation and many different and creative wrong interpretations of the confidence interval $0.381 < p < 0.497$.

Correct: "We are 95% confident that the interval from 0.381 to 0.497 actually does contain the true value of p." This means that if we were to select many different samples of size 280 and construct the corresponding confidence intervals, 95% of them would actually contain the value of the population proportion p. (Note that in this correct interpretation, the level of 95% refers to the success rate of the *process* being used to estimate the proportion, and it does not refer to the population proportion itself.)

Wrong: "There is a 95% chance that the true value of p will fall between 0.381 and 0.497."

At any specific point in time, a population has a fixed and constant value p, and a confidence interval constructed from a sample either includes p or does not. Similarly, if a baby has just been born and the doctor is about to announce its gender, it's wrong to say that there is probability of 0.5 that the baby is a girl; the baby is a girl or is not, and there's no probability involved. A population proportion p is like the baby that has been born—the value of p is fixed, so the confidence interval limits either contain p or do not, and that is why it's wrong to say that there is a 95% chance that p will fall between values such as 0.381 and 0.497.

A confidence level of 95% tells us that the *process* we are using will, in the long run, result in confidence interval limits that contain the true population proportion 95% of the time. Suppose that the true proportion of all correct hand identifications made by touch therapists is $p = 0.5$. Then the confidence interval obtained from the given sample data would not contain the population proportion, because the true population proportion 0.5 is not between 0.381 and 0.497. This is illustrated in Figure 7-1. Figure 7-1 shows typical confidence intervals resulting from 20 different samples. With 95% confidence, we expect that 19 out of 20 samples should result in confidence intervals that do contain the true value of p, and Figure 7-1 illustrates this with 19 of the confidence intervals containing p, while one confidence interval does not contain p.

Using Confidence Intervals for Comparisons *Caution: Confidence intervals can be used informally to compare different data sets, but the overlapping of confidence intervals should not be used for making formal and final conclusions about equality of proportions.* (See "On Judging the Significance of Differences by Examining the Overlap Between Confidence Intervals," by Schenker and Gentleman, *American Statistician,* Vol. 55, No. 3.)

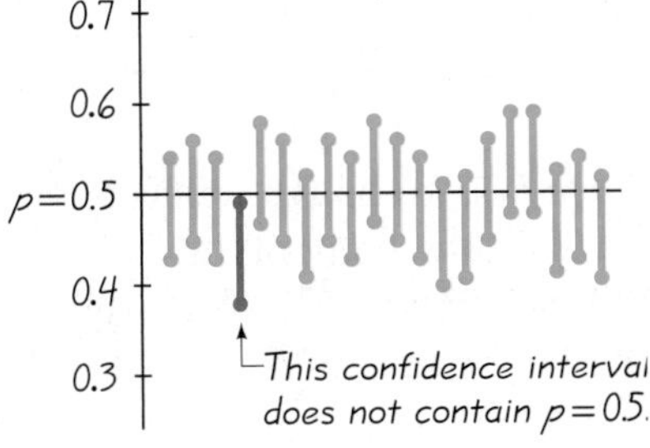

Figure 7-1

Confidence Intervals from 20 Different Samples

Critical Values

The methods of this section and many of the other statistical methods found in the following chapters include the use of a standard z score that can be used to distinguish between sample statistics that are likely to occur and those that are unlikely. Such a z score is called a *critical value* (defined below). Critical values are based on the following observations:

1. We know from Section 6-6 that under certain conditions, the sampling distribution of sample proportions can be approximated by a normal distribution, as in Figure 7-2.
2. Sample proportions have a relatively small chance (with probability denoted by α) of falling in one of the red tails of Figure 7-2.
3. Denoting the area of each shaded tail by $\alpha/2$, there is a total probability of α that a sample proportion will fall in either of the two red tails.
4. By the rule of complements (from Chapter 4), there is a probability of $1 - \alpha$ that a sample proportion will fall within the inner green-shaded region of Figure 7-2.
5. The z score separating the right-tail region is commonly denoted by $z_{\alpha/2}$, and is referred to as a *critical value* because it is on the borderline separating sample proportions that are likely to occur from those that are unlikely to occur.

These observations can be formalized with the following notation and definition.

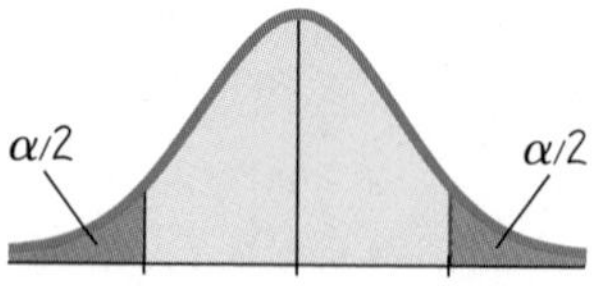

$\alpha/2$ $\alpha/2$

$z = 0$ $z_{\alpha/2}$

Found from Table A-2 (corresponds to area of $1 - \alpha/2$)

Figure 7-2

Critical Value $z_{\alpha/2}$ in the Standard Normal Distribution

Notation for Critical Value

The critical value $z_{\alpha/2}$ is the positive z value that is at the vertical boundary separating an area of $\alpha/2$ in the right tail of the standard normal distribution. (The value of $-z_{\alpha/2}$ is at the vertical boundary for the area of $\alpha/2$ in the left tail.) The subscript $\alpha/2$ is simply a reminder that the z score separates an area of $\alpha/2$ in the right tail of the standard normal distribution.

Definition

A **critical value** is the number on the borderline separating sample statistics that are likely to occur from those that are unlikely to occur. The number $z_{\alpha/2}$ is a critical value that is a z score with the property that it separates an area of $\alpha/2$ in the right tail of the standard normal distribution. (See Figure 7-2.)

EXAMPLE Finding a Critical Value Find the critical value $z_{\alpha/2}$ corresponding to a 95% confidence level.

SOLUTION *Using Table A-2:* To find the critical z value for a 95% confidence level, do *not* look up 0.95 in the body of Table A-2. A 95% confidence level corresponds to $\alpha = 0.05$. See Figure 7-3, where we show that the area in each of the red-shaded tails is $\alpha/2 = 0.025$. We find $z_{\alpha/2} = 1.96$ by noting that all of the

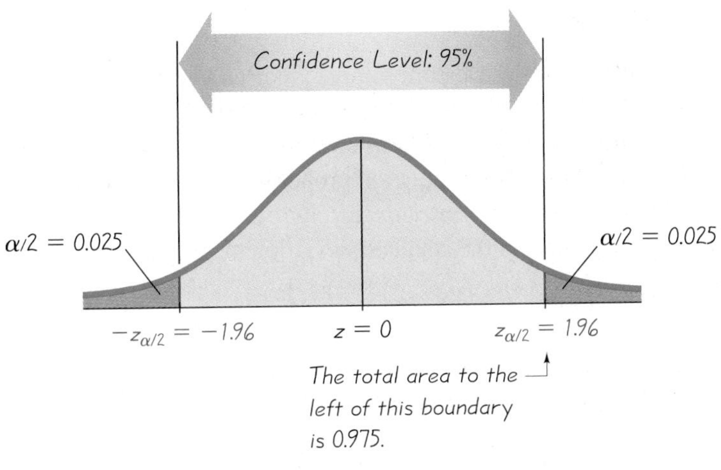

Figure 7-3 Finding $z_{\alpha/2}$ for a 95% Confidence Level

area to its left must be 1 − 0.025, or 0.975. We can refer to Table A-2 and find that the area of 0.9750 (found in the *body* of the table) corresponds exactly to a z score of 1.96. For a 95% confidence level, the critical value is therefore $z_{\alpha/2} = 1.96$. To find the critical z score for a 95% confidence level, look up 0.9750 in the body of Table A-2, not 0.95.

Using the TI-83/84 Plus calculator: To find a critical z value for a 95% confidence level, use the command **invNorm(0.975)** to get the result of 1.959963986. The invNorm command is in the DISTR (distribution) menu, which is accessed by pressing 2ND VARS.

The preceding example showed that a 95% confidence level results in a critical value of $z_{\alpha/2} = 1.96$. This is the most common critical value, and it is listed with two other common values in the table that follows.

Confidence Level	α	Critical Value, $z_{\alpha/2}$
90%	0.10	1.645
95%	0.05	1.96
99%	0.01	2.575

Margin of Error

When we collect a set of sample data, such as Emily Rosa's touch therapy data given in the Chapter Problem (with 44% of 280 trials resulting in correct identifications), we can calculate the sample proportion $\hat{p}$ and that sample proportion is typically different from the population proportion p. The difference between the sample proportion and the population proportion can be thought of as an error. We now define the *margin of error E* as follows.

Curbstoning

The glossary for the 2000 Census defines *curbstoning* as "the practice by which a census enumerator fabricates a questionnaire for a residence without actually visiting it." Curbstoning occurs when a census enumerator sits on a curbstone (or anywhere else) and fills out survey forms by making up responses. Because data from curbstoning are not real, they can affect the validity of the Census. The extent of curbstoning has been investigated in several studies, and one study showed that about 4% of Census enumerators practiced curbstoning at least some of the time.

The methods of Section 7-2 assume that the sample data have been collected in an appropriate way, so if much of the sample data have been obtained through curbstoning, then the resulting confidence interval estimates might be very flawed.

Definition

When data from a simple random sample are used to estimate a population proportion p, the **margin of error,** denoted by E**,** is the maximum likely (with probability $1 - \alpha$) difference between the observed sample proportion $\hat{p}$ and the true value of the population proportion p. The margin of error E is also called the *maximum error of the estimate* and can be found by multiplying the critical value and the standard deviation of sample proportions, as shown in Formula 7-1.

Formula 7-1 $$E = z_{\alpha/2}\sqrt{\frac{\hat{p}\hat{q}}{n}}$$ margin of error for proportions

Given the way that the margin of error E is defined, there is a probability of α that the sample proportion will be in error by more than E.

Confidence Interval (or Interval Estimate) for the Population Proportion p

$$\hat{p} - E < p < \hat{p} + E \quad \text{where} \quad E = z_{\alpha/2}\sqrt{\frac{\hat{p}\hat{q}}{n}}$$

The confidence interval is often expressed in the following equivalent formats:

$$\hat{p} \pm E$$

or

$$(\hat{p} - E, \hat{p} + E)$$ This is the format used by the TI-83/84 Plus calculator.

In Chapter 4, when probabilities were given in decimal form, we rounded to three significant digits. We use that same rounding rule here.

Round-Off Rule for Confidence Interval Estimates of p

Round the confidence interval limits for p to three significant digits.

Based on the preceding results, we can summarize the procedure for constructing a confidence interval estimate of a population proportion p as follows.

Procedure for Constructing a Confidence Interval for p

1. Verify that the requirements are satisfied. (The sample is a simple random sample, the conditions for the binomial distribution are satisfied, and there are at least 5 successes and at least 5 failures.)
2. Refer to Table A-2 or use a TI-83/84 Plus calculator to find the critical value $z_{\alpha/2}$ that corresponds to the desired confidence level. (For example, if the confidence level is 95%, the critical value is $z_{\alpha/2} = 1.96$.)

3. Evaluate the margin of error $E = z_{\alpha/2}\sqrt{\hat{p}\hat{q}/n}$.
4. Using the value of the calculated margin of error E and the value of the sample proportion $\hat{p}$, find the values of $\hat{p} - E$ and $\hat{p} + E$. Substitute those values in the general format for the confidence interval:

$$\hat{p} - E < p < \hat{p} + E$$

or

$$\hat{p} \pm E$$

or

$$(\hat{p} - E, \hat{p} + E)$$

5. Round the resulting confidence interval limits to three significant digits.

The TI-83/84 Plus calculator can automatically generate confidence interval estimates for the population proportion p. See the procedure at the end of this section.

EXAMPLE Touch Therapy Success Rate In the Chapter Problem we noted that touch therapists participated in 280 trials of their ability to sense a human energy field. In each trial, a touch therapist was asked to identify which hand was just below the hand of Emily Rosa. Among the 280 trials, there were 123 correct identifications. The sample results are $n = 280$, and $\hat{p} = 123/280 = 0.439286$. (Instead of using 0.44 for the sample proportion, we carry extra decimal places so that subsequent calculations will not be affected by a rounding error.)

a. Find the margin of error E that corresponds to a 95% confidence level.

b. Find the 95% confidence interval estimate of the population proportion p.

c. Based on the results, what can we conclude about the effectiveness of touch therapy?

SOLUTION

REQUIREMENT ✔ We should first verify that the necessary requirements are satisfied. (Earlier in this section we listed the requirements for using a normal distribution as an approximation to a binomial distribution.) Given the design of the experiment, it is reasonable to assume that the sample is a simple random sample. The conditions for a binomial experiment are satisfied, because there is a fixed number of trials (280), the trials are independent (because the result from one trial doesn't affect the probability of the result of another trial), there are two categories of outcome (correct, wrong), and the probability of being correct remains constant. Also, with 123 correct identifications in 280 trials, there are 157 wrong identifications, so the number of successes (123) and the number of failures (157) are both at least 5. The check of requirements has been successfully completed. ✔

continued

a. The margin of error is found by using Formula 7-1 with $z_{\alpha/2} = 1.96$ (as found in the preceding example), $\hat{p} = 0.439286$, $\hat{q} = 1 - 0.439286 = 0.560714$, and $n = 280$.

$$E = z_{\alpha/2}\sqrt{\frac{\hat{p}\hat{q}}{n}} = 1.96\sqrt{\frac{(0.439286)(0.560714)}{280}} = 0.058133$$

b. Constructing the confidence interval is quite easy now that we have the values of $\hat{p}$ and E. We simply substitute those values to obtain this result:

$$\hat{p} - E < p < \hat{p} + E$$

$$0.439286 - 0.058133 < p < 0.439286 + 0.058133$$

$$0.381 < p < 0.497 \quad \text{(rounded to three significant digits)}$$

This same result could be expressed in the format of 0.439 ± 0.058 or $(0.381, 0.497)$. If we want the 95% confidence interval for the true population *percentage,* we could express the result as $38.1\% < p < 49.7\%$. This confidence interval is often reported with a statement such as this: "It is estimated that the success rate is 44%, with a margin of error of plus or minus 6 percentage points." That statement is a verbal expression of this format for the confidence interval: $44\% \pm 6\%$. The level of confidence should also be reported, but it rarely is in the media. The media typically use a 95% confidence level but omit any reference to it.

c. To interpret the results, note that pure guessing would result in about 50% of the identifications being correct. If the touch therapists actually had an ability to sense a human energy field, their success rate should have been greater than 50% by a significant amount. However, the touch therapists did slightly worse than what we might expect from coin tossing. Here is a statement from the article published in the *Journal of the American Medical Association:* "Their (touch therapists) failure to substantiate TT's (touch therapy's) most fundamental claim is unrefuted evidence that the claims of TT (touch therapy) are groundless and that further professional use is unjustified." Based on the results from Emily Rosa's science fair project, it appears that touch therapy is ineffective.

Rationale for the Margin of Error Because the sampling distribution of proportions is approximately normal (because the conditions $np \geq 5$ and $nq \geq 5$ are both satisfied), we can use results from Section 6-6 to conclude that μ and σ are given by $\mu = np$ and $\sigma = \sqrt{npq}$. Both of these parameters pertain to n trials, but we convert them to a per-trial basis by dividing by n as follows:

$$\text{Mean of sample proportions: } \mu = \frac{np}{n} = p$$

$$\text{Standard deviation of sample proportions: } \sigma = \frac{\sqrt{npq}}{n} = \sqrt{\frac{npq}{n^2}} = \sqrt{\frac{pq}{n}}$$

The first result may seem trivial, because we already stipulated that the true population proportion is p. The second result is nontrivial and is useful in describing the margin of error E, but we replace the product pq by $\hat{p}\hat{q}$ because we don't know the value of p (it is the value we are trying to estimate). Formula 7-1 for the margin of error reflects the fact that $\hat{p}$ has a probability of $1 - \alpha$ of being within $z_{\alpha/2}\sqrt{pq/n}$ of p. The confidence interval for p, as given previously, reflects the fact that there is a probability of $1 - \alpha$ that $\hat{p}$ differs from p by less than the margin of error $E = z_{\alpha/2}\sqrt{pq/n}$.

Determining Sample Size

Suppose we want to collect sample data with the objective of estimating some population proportion. How do we know *how many* sample items must be obtained? If we take the expression for the margin of error E (Formula 7-1), then solve for n, we get Formula 7-2. Formula 7-2 requires $\hat{p}$ as an estimate of the population proportion p, but if no such estimate is known (as is often the case), we replace $\hat{p}$ by 0.5 and replace $\hat{q}$ by 0.5, with the result given in Formula 7-3.

Sample Size for Estimating Proportion *p*

When an estimate $\hat{p}$ is known: **Formula 7-2** $n = \dfrac{[z_{\alpha/2}]^2\hat{p}\hat{q}}{E^2}$

When no estimate $\hat{p}$ is known: **Formula 7-3** $n = \dfrac{[z_{\alpha/2}]^2 \cdot 0.25}{E^2}$

Round-Off Rule for Determining Sample Size

In order to ensure that the required sample size is at least as large as it should be, if the computed sample size is not a whole number, round it up to the next *higher* whole number.

Use Formula 7-2 when reasonable estimates of $\hat{p}$ can be made by using previous samples, a pilot study, or someone's expert knowledge. Otherwise, use Formula 7-3. Note that Formulas 7-2 and 7-3 do not include the population size N, so the size of the population is irrelevant. (*Exception:* When sampling is without replacement from a relatively small finite population. See Exercise 49.)

EXAMPLE Sample Size for E-Mail Survey The ways that we communicate have been dramatically affected by the use of answering machines, fax machines, voice mail, and e-mail. Suppose a sociologist wants to determine the current percentage of U.S. households using e-mail. How many households must be surveyed in order to be 95% confident that the sample percentage is in error by no more than four percentage points?

a. Use this result from an earlier study: In 1997, 16.9% of U.S. households used e-mail (based on data from the *World Almanac and Book of Facts*).

b. Assume that we have no prior information suggesting a possible value of $\hat{p}$.

continued

SOLUTION

a. The prior study suggests that $\hat{p} = 0.169$, so $\hat{q} = 0.831$ (found from $\hat{q} = 1 - 0.169$). With a 95% level of confidence, we have $\alpha = 0.05$, so $z_{\alpha/2} = 1.96$. Also, the margin of error is $E = 0.04$ (the decimal equivalent of "four percentage points"). Because we have an estimated value of $\hat{p}$, we use Formula 7-2 as follows:

$$n = \frac{[z_{\alpha/2}]^2 \hat{p}\hat{q}}{E^2} = \frac{[1.96]^2(0.169)(0.831)}{0.04^2}$$

$$= 337.194 = 338 \quad \text{(rounded up)}$$

We must survey at least 338 randomly selected households.

b. As in part (a), we again use $z_{\alpha/2} = 1.96$ and $E = 0.04$, but with no prior knowledge of $\hat{p}$ (or $\hat{q}$), we use Formula 7-3 as follows:

$$n = \frac{[z_{\alpha/2}]^2 \cdot 0.25}{E^2} = \frac{[1.96]^2 \cdot 0.25}{0.04^2}$$

$$= 600.25 = 601 \quad \text{(rounded up)}$$

INTERPRETATION To be 95% confident that our sample percentage is within four percentage points of the true percentage for all households, we should randomly select and survey 601 households. By comparing this result to the sample size of 338 found in part (a), we can see that if we have no knowledge of a prior study, a larger sample is required to achieve the same results as when the value of $\hat{p}$ can be estimated. But now let's use some common sense: We know that the use of e-mail is growing so rapidly that the 1997 estimate is too old to be of much use. Today, substantially more than 16.9% of households use e-mail. Realistically, we need a sample larger than 338 households. Assuming that we don't really know the current rate of e-mail usage, we should randomly select 601 households. With 601 households, we will be 95% confident that we are within four percentage points of the true percentage of households using e-mail.

Common Errors Try to avoid these two common errors when calculating sample size: (1) Don't make the mistake of using $E = 4$ as the margin of error corresponding to "four percentage points." (2) Be sure to substitute the critical z score for $z_{\alpha/2}$. For example, if you are working with 95% confidence, be sure to replace $z_{\alpha/2}$ with 1.96. Don't make the mistake of replacing $z_{\alpha/2}$ with 0.95 or 0.05.

Population Size Many people incorrectly believe that the sample size should be some percentage of the population, but Formula 7-3 shows that the population size is irrelevant. (In reality, the population size is sometimes used, but only in cases in which we sample without replacement from a relatively small population. See Exercise 49.) Polls commonly use sample sizes in the range of 1000 to 2000 and, even though such polls may involve a very small percentage of the total population, they can provide results that are quite good.

Finding the Point Estimate and E from a Confidence Interval Sometimes we want to better understand a confidence interval that might have been obtained from a journal article, or it might have been generated using software or a calculator. If we already know the confidence interval limits, the sample proportion $\hat{p}$ and the margin of error E can be found as follows:

Point estimate of p:

$$\hat{p} = \frac{(\text{upper confidence limit}) + (\text{lower confidence limit})}{2}$$

Margin of error:

$$E = \frac{(\text{upper confidence limit}) - (\text{lower confidence limit})}{2}$$

EXAMPLE The article "High-Dose Nicotine Patch Therapy," by Dale, Hurt, et al. (*Journal of the American Medical Association,* Vol. 274, No. 17) includes this statement: "Of the 71 subjects, 70% were abstinent from smoking at 8 weeks (95% confidence interval [CI], 58% to 81%)." Use that statement to find the point estimate $\hat{p}$ and the margin of error E.

SOLUTION From the given statement, we see that the 95% confidence interval is $0.58 < p < 0.81$. The point estimate $\hat{p}$ is the value midway between the upper and lower confidence interval limits, so we get

$$\hat{p} = \frac{(\text{upper confidence limit}) + (\text{lower confidence limit})}{2}$$

$$= \frac{0.81 + 0.58}{2} = 0.695$$

The margin of error can be found as follows:

$$E = \frac{(\text{upper confidence limit}) - (\text{lower confidence limit})}{2}$$

$$= \frac{0.81 - 0.58}{2} = 0.115$$

Better-Performing Confidence Intervals

Important note: The exercises for this section are based on the confidence interval described above, not the confidence intervals described in the following discussion.

The confidence interval described in this section has the format typically presented in introductory statistics courses, but it does not perform as well as some other confidence intervals. The *adjusted Wald confidence interval* performs better in the sense that its probability of containing the true population proportion p is closer to the confidence level that is used. The adjusted Wald confidence interval uses this simple procedure: Add 2 to the number of successes x, add 2 to the number of failures (so that the number of trials n is increased by 4), then find the confidence interval as described in this section.

For example, if we use the methods of this section with $x = 10$ and $n = 20$, we get this 95% confidence interval: $0.281 < p < 0.719$. With $x = 10$ and $n = 20$ we use the adjusted Wald confidence interval by letting $x = 12$ and $n = 24$ to get this confidence interval: $0.300 < p < 0.700$. The chance that the confidence interval $0.300 < p < 0.700$ contains p is closer to 95% than the chance that $0.281 < p < 0.719$ contains p.

Another confidence interval that performs better than the one described in this section and the adjusted Wald confidence interval is the *Wilson score confidence interval*. It has the lower confidence interval limit shown below, and the upper confidence interval limit is expressed by changing the minus sign to a plus sign. (It is easy to see why this approach is not used much in introductory courses.) Using $x = 10$ and $n = 20$, the 95% Wilson score confidence interval is $0.299 < p < 0.701$.

$$\frac{\hat{p} + \dfrac{z_{\alpha/2}^2}{2n} - z_{\alpha/2}\sqrt{\dfrac{\hat{p}\hat{q} + \dfrac{z_{\alpha/2}^2}{4n}}{n}}}{1 + \dfrac{z_{\alpha/2}^2}{n}}$$

For a discussion of these and other confidence intervals for p, see "Approximation Is Better than 'Exact' for Interval Estimation of Binomial Proportions," by Agresti and Coull, *American Statistician*, Vol. 52, No. 2.

TI

Using the TI-83/84 Plus Calculator for Confidence Interval Estimates of *p*

Here is the procedure for using the TI-83/84 Plus calculator for the construction of a confidence interval estimate of a population proportion p:

1. You must first identify the number of successes x and the number of trials n. (If you know only the sample size n and the sample proportion $\hat{p}$, the number of successes x can be calculated by using this expression: $x = n\hat{p}$. If the resulting value of x is not a whole number, round it to the nearest whole number.)
2. Press STAT, then use the ▷ key to select the option of **TESTS.**
3. The menu option of **A: 1-PropZInt** identifies the confidence interval for the case involving one proportion (as described in this section), which is constructed using a critical z score. Select this option by pressing 3 or by using the ▽ key to scroll down to **1-PropZInt.**
4. The screen requires an entry for x, an entry for n, and an entry for the C-Level (confidence level, such as 0.95). Enter those values, then scroll down to **Calculate** and press ENTER. A typical screen display of these entries is shown here. After pressing ENTER, the confidence interval will be expressed in a format with the confidence interval limits enclosed within parentheses, such as

TI-83/84 Plus

```
1-PropZInt
 (.38115,.49742)
 p̂=.4392857143
 n=280
```

(.38115, .49742). See the accompanying screen that corresponds to the example in this section. The TI-83/84 Plus confidence interval of (.38115, .49742) can also be expressed as $0.381 < p < 0.497$.

Using Technology for Sample Size

Although the TI-83/84 Plus calculator can be used to calculate the sample size required to estimate a population proportion p, it does not yet include a special function designed specifically for that purpose. STATDISK, a computer software program designed to accompany this textbook, does have a module for determining sample size. STATDISK is on the CD-ROM included with this book or it can be downloaded from www.aw.com/triola. After starting STATDISK, click on **Analysis** on the main menu at the top, then click on the menu item of **Sample Size Determination.** That item has three options, so select **Estimate Proportion,** then proceed to enter the items in the dialog box. Click on **Evaluate** to obtain the required sample size.

7-2 BASIC SKILLS AND CONCEPTS

Statistical Literacy and Critical Thinking

1. Critical Value What is a critical value for a normal distribution?

2. Margin of Error What is a margin of error?

3. Confidence Interval When surveying 500 people, we get 200 yes responses to a particular question, so the proportion of yes responses from the whole population is estimated to be 0.4. Given that we have the estimated value of 0.4, why would we need a confidence interval? That is, what additional information does the confidence interval provide?

4. Sampling A student surveys 100 classmates by asking each if they have outstanding loans. After finding the sample proportion for this sample of $n = 100$ subjects, can the methods of this section be used to estimate the proportion of all adults who have outstanding loans? Why or why not?

Finding Critical Values. *In Exercises 5–8, find the critical value $z_{\alpha/2}$ that corresponds to the given confidence level.*

TI **5.** 99% TI **6.** 90%

TI **7.** 98% TI **8.** 99.5%

9. Express the confidence interval $0.222 < p < 0.444$ in the form of $\hat{p} \pm E$.

10. Express the confidence interval $0.600 < p < 0.800$ in the form of $\hat{p} \pm E$.

11. Express the confidence interval (0.206, 0.286) in the form of $\hat{p} \pm E$.

12. Express the confidence interval 0.337 ± 0.050 in the form of $\hat{p} - E < p < \hat{p} + E$.

Interpreting Confidence Interval Limits. *In Exercises 13–16, use the given confidence interval limits to find the point estimate $\hat{p}$ and the margin of error E.*

13. (0.868, 0.890) **14.** $0.325 < p < 0.375$

15. $0.607 < p < 0.713$ **16.** $0.0144 < p < 0.0882$

Finding Margin of Error. *In Exercises 17–20, assume that a sample is used to estimate a population proportion p. Find the margin of error E that corresponds to the given statistics and confidence level.*

17. $n = 500$, $x = 200$, 95% confidence

18. $n = 1200$, $x = 800$, 99% confidence

19. 98% confidence; the sample size is 1068, of which 25% are successes.

20. 90% confidence; the sample size is 2107, of which 65% are successes.

Constructing Confidence Intervals. *In Exercises 21–24, use the sample data and confidence level to construct the confidence interval estimate of the population proportion p.*

TI **21.** $n = 500$, $x = 200$, 95% confidence

TI **22.** $n = 1200$, $x = 800$, 99% confidence

TI **23.** $n = 1068$, $x = 267$, 98% confidence

TI **24.** $n = 4500$, $x = 2925$, 90% confidence

Determining Sample Size. *In Exercises 25–28, use the given data to find the minimum sample size required to estimate a population proportion or percentage.*

25. Margin of error: 0.020; confidence level: 95%; $\hat{p}$ and $\hat{q}$ unknown

26. Margin of error: 0.050; confidence level: 99%; $\hat{p}$ and $\hat{q}$ unknown

27. Margin of error: three percentage points; confidence level: 95%; from a prior study, $\hat{p}$ is estimated by the decimal equivalent of 27%.

28. Margin of error: five percentage points; confidence level: 90%; from a prior study, $\hat{p}$ is estimated by the decimal equivalent of 65%.

TI **29.** Gender Selection The Genetics and IVF Institute conducted a clinical trial of the XSORT method designed to increase the probability of conceiving a girl. As this book was being written, 325 babies were born to parents using the XSORT method, and 295 of them were girls. Use the sample data to construct a 99% confidence interval estimate of the percentage of girls born to parents using the XSORT method. Based on the result, does the XSORT method appear to be effective? Why or why not?

TI **30.** Gender Selection The Genetics and IVF Institute conducted a clinical trial of the YSORT method designed to increase the probability of conceiving a boy. As this book was being written, 51 babies were born to parents using the YSORT method, and 39 of them were boys. Use the sample data to construct a 99% confidence interval estimate of the percentage of boys born to parents using the YSORT method. Based on the result, does the YSORT method appear to be effective? Why or why not?

TI **31.** Postponing Death An interesting and popular hypothesis is that individuals can temporarily postpone their death to survive a major holiday or important event such as a birthday. In a study of this phenomenon, it was found that in the week before and the week after Thanksgiving, there were 12,000 total deaths, and 6062 of them occurred in the week before Thanksgiving (based on data from "Holidays, Birthdays, and Postponement of Cancer Death," by Young and Hade, *Journal of the American Medical*

Association, Vol. 292, No. 24.) Construct a 95% confidence interval estimate of the proportion of deaths in the week before Thanksgiving to the total deaths in the week before and the week after Thanksgiving. Based on the result, does there appear to be any indication that people can temporarily postpone their death to survive the Thanksgiving holiday? Why or why not?

TI **32.** Medical Malpractice An important issue facing Americans is the large number of medical malpractice lawsuits and the expenses that they generate. In a study of 1228 randomly selected medical malpractice lawsuits, it is found that 856 of them were later dropped or dismissed (based on data from the Physician Insurers Association of America). Construct a 99% confidence interval estimate of the proportion of medical malpractice lawsuits that are dropped or dismissed. Does it appear that the majority of such suits are dropped or dismissed?

TI **33.** Mendelian Genetics When Mendel conducted his famous genetics experiments with peas, one sample of offspring consisted of 428 green peas and 152 yellow peas.

a. Find a 95% confidence interval estimate of the percentage of yellow peas.

b. Based on his theory of genetics, Mendel expected that 25% of the offspring peas would be yellow. Given that the percentage of offspring yellow peas is not 25%, do the results contradict Mendel's theory? Why or why not?

TI **34.** Misleading Survey Responses In a survey of 1002 people, 701 said that they voted in a recent presidential election (based on data from ICR Research Group). Voting records show that 61% of eligible voters actually did vote.

a. Find a 99% confidence interval estimate of the proportion of people who say that they voted.

b. Are the survey results consistent with the actual voter turnout of 61%? Why or why not?

TI **35.** Cell Phones and Cancer A study of 420,095 Danish cell phone users found that 135 of them developed cancer of the brain or nervous system. Prior to this study of cell phone use, the rate of such cancer was found to be 0.0340% for those not using cell phones. The data are from the *Journal of the National Cancer Institute*.

a. Use the sample data to construct a 95% confidence interval estimate of the percentage of cell phone users who develop cancer of the brain or nervous system.

b. Do cell phone users appear to have a rate of cancer of the brain or nervous system that is different from the rate of such cancer among those not using cell phones? Why or why not?

TI **36.** Cloning Survey A recent Gallup poll consisted of 1012 randomly selected adults who were asked whether "cloning of humans should or should not be allowed." Results showed that 901 of those surveyed indicated that cloning should not be allowed. A news reporter wants to determine whether these survey results constitute strong evidence that the majority (more than 50%) of people are opposed to such cloning. Construct a 99% confidence interval estimate of the proportion of adults believing that cloning of humans should not be allowed. Based on that result, is there strong evidence supporting the claim that the majority is opposed to such cloning?

TI **37.** Bias in Jury Selection In the case of *Casteneda v. Partida,* it was found that during a period of 11 years in Hidalgo County, Texas, 870 people were selected for grand jury duty, and 39% of them were Mexican-Americans. Use the sample data to construct a 99% confidence interval estimate of the proportion of grand jury members who were

Mexican-Americans. Given that among the people eligible for jury duty, 79.1% of them were Mexican-Americans, does it appear that the jury selection process was somehow biased against Mexian-Americans? Why or why not?

TI **38.** Detecting Fraud When working for the Brooklyn District Attorney, investigator Robert Burton analyzed the leading digits of amounts on checks from companies that were suspected of fraud. Among 784 checks, 61% had amounts with leading digits of 5. Construct a 99% confidence interval estimate of the proportion of checks having amounts with leading digits of 5. When checks are issued in the normal course of honest transactions, it is expected that 7.9% of the checks will have amounts with leading digits of 5. What does the confidence interval suggest?

TI **39.** Telephone Households In 1920 only 35% of U.S. households had telephones, but that rate is now much higher. A recent survey of 4276 randomly selected households showed that 94% of them had telephones (based on data from the U.S. Census Bureau). Using those survey results, construct a 99% confidence interval estimate of the proportion of households with telephones. Given that the survey involves only 4276 households out of 115 million households, do we really have enough evidence to say that the percentage of households with telephones is now more than the 35% rate in 1920?

TI **40.** Internet Shopping In a Gallup poll, 1025 randomly selected adults were surveyed and 29% of them said that they used the Internet for shopping at least a few times a year.

a. Find the point estimate of the *percentage* of adults who use the Internet for shopping.

b. Find a 99% confidence interval estimate of the *percentage* of adults who use the Internet for shopping.

c. If a traditional retail store wants to estimate the percentage of adult Internet shoppers in order to determine the maximum impact of Internet shoppers on its sales, what percentage of Internet shoppers should be used?

Determining Sample Size. *In Exercises 41–44, find the minimum sample size required to estimate a population proportion or percentage.*

41. Sample Size for Internet Purchases Many states are carefully considering steps that would help them collect sales taxes on items purchased through the Internet. How many randomly selected sales transactions must be surveyed to determine the percentage that transpired over the Internet? Assume that we want to be 99% confident that the sample percentage is within two percentage points of the true population percentage for all sales transactions.

42. Sample Size for Downloaded Songs The music industry must adjust to the growing practice of consumers downloading songs instead of buying CDs. It therefore becomes important to estimate the proportion of songs that are currently downloaded. How many randomly selected song purchases must be surveyed to determine the percentage that were obtained by downloading? Assume that we want to be 95% confident that the sample percentage is within one percentage point of the true population percentage of songs that are downloaded.

43. Nitrogen in Tires A recent campaign was designed to convince car owners that they should fill their tires with nitrogen instead of air. At a cost of about $5 per tire, nitrogen supposedly has the advantage of leaking at a much slower rate than air, so that the ideal tire pressure can be maintained more consistently. Before spending huge sums to advertise the nitrogen, it would be wise to conduct a survey to determine the percentage of car owners who would pay for the nitrogen. How many randomly selected car owners should be surveyed? Assume that we want to be 98% confident that the sample percentage is within three percentage points of the true percentage of all car owners who would be willing to pay for the nitrogen.

44. Sunroof and Side Air Bags Toyota provides an option of a sunroof and side air bag package for its Corolla model. This package costs $1400 ($1159 invoice price). Assume that prior to offering this option package, Toyota wants to determine the percentage of Corolla buyers who would pay $1400 extra for the sunroof and side air bags. How many Corolla buyers must be surveyed if we want to be 95% confident that the sample percentage is within four percentage points of the true percentage for all Corolla buyers?

Using Appendix B Data Sets. *In Exercises 45–48, use the indicated data set from Appendix B.*

TI **45.** Blue M&M Candies Refer to Data Set 13 in Appendix B and find the sample proportion of M&Ms that are blue. Use that result to construct a 95% confidence interval estimate of the population percentage of M&Ms that are blue. Is the result consistent with the 24% rate that is reported by the candy maker Mars?

TI **46.** Alcohol and Tobacco Use in Children's Movies Refer to Data Set 5 in Appendix B.

a. Construct a 95% confidence interval estimate of the percentage of animated children's movies showing any tobacco use.

b. Construct a 95% confidence interval estimate of the percentage of animated children's movies showing any alcohol use.

c. Compare the preceding results. Does either tobacco or alcohol appear in a greater percentage of animated children's movies?

d. In using the results from parts (a) and (b) as measures of the depiction of unhealthy habits, what important characteristic of the data is not included?

TI **47.** Precipitation in Boston Refer to Data Set 10 in Appendix B, and consider days with precipitation values different from 0 to be days with precipitation. Construct a 95% confidence interval estimate of the proportion of Wednesdays with precipitation, and also construct a 95% confidence interval estimate of the proportion of Sundays with precipitation. Compare the results. Does precipitation appear to occur more on either day?

TI **48.** Accuracy of Forecast Temperatures Refer to Data Set 8 in Appendix B. Construct a 95% confidence interval estimate of the proportion of days with an actual high temperature that is more than 2° different from the high temperature that was forecast one day before. Then construct a 95% confidence interval estimate of the proportion of days with an actual high temperature that is more than 2° different from the high temperature that was forecast five days before. Compare the results.

7-2 BEYOND THE BASICS

49. Using Finite Population Correction Factor This section presented Formulas 7-2 and 7-3, which are used for determining sample size. In both cases we assumed that the population is infinite or very large and that we are sampling with replacement. When we have a relatively small population with size N and sample without replacement, we modify E to include the *finite population correction factor* shown here, and we can solve for n to obtain the result given here. Use this result to repeat Exercise 44, assuming that we limit our population to 1250 Toyota Corolla buyers in one region.

$$E = z_{\alpha/2}\sqrt{\frac{\hat{p}\hat{q}}{n}}\sqrt{\frac{N-n}{N-1}} \qquad n = \frac{N\hat{p}\hat{q}[z_{\alpha/2}]^2}{\hat{p}\hat{q}[z_{\alpha/2}]^2 + (N-1)E^2}$$

50. One-Sided Confidence Interval A *one-sided confidence interval* for p can be expressed as $p < \hat{p} + E$ or $p > \hat{p} - E$, where the margin of error E is modified by replacing $z_{\alpha/2}$ with z_α. If Air America wants to report an on-time performance of at least x percent with 95% confidence, construct the appropriate one-sided confidence interval and then find the percent in question. Assume that a simple random sample of 750 flights results in 630 that are on time.

TI **51.** Confidence Interval from Small Sample Special tables are available for finding confidence intervals for proportions involving small numbers of cases, where the normal distribution approximation cannot be used. For example, given $x = 3$ successes among $n = 8$ trials, the 95% confidence interval found in *Standard Probability and Statistics Tables and Formulae* (CRC Press) is $0.085 < p < 0.755$. Find the confidence interval that would result if you were to use the normal distribution incorrectly as an approximation to the binomial distribution. Are the results reasonably close?

TI **52.** Interpreting Confidence Interval Limits Assume that a coin is modified so that it favors heads, and 100 tosses result in 95 heads. Find the 99% confidence interval estimate of the proportion of heads that will occur with this coin. What is unusual about the results obtained by the methods of this section? Does common sense suggest a modification of the resulting confidence interval?

53. Rule of Three Suppose n trials of a binomial experiment result in no successes. According to the *Rule of Three,* we have 95% confidence that the true population proportion has an upper bound of $3/n$. (See "A Look at the Rule of Three," by Jovanovic and Levy, *American Statistician,* Vol. 51, No. 2.)

a. If n independent trials result in no successes, why can't we find confidence interval limits by using the methods described in this section?

b. If 20 patients are treated with a drug and there are no adverse reactions, what is the 95% upper bound for p, the proportion of all patients who experience adverse reactions to this drug?

54. Poll Accuracy A *New York Times* article about poll results states, "In theory, in 19 cases out of 20, the results from such a poll should differ by no more than one percentage point in either direction from what would have been obtained by interviewing all voters in the United States." Find the sample size suggested by this statement.

7-3 Estimating a Population Mean: σ Known

Key Concept Section 7-2 introduced the point estimate and confidence interval as tools for using a sample proportion to estimate a population proportion, and this section presents methods for using sample data to find a point estimate and confidence interval estimate of a population mean. A key requirement in this section is that in addition to having sample data, we also know σ, the standard deviation of the population. This section also presents a method for finding the sample size that would be required to estimate a population mean.

Requirements

1. The sample is a simple random sample. (All samples of the same size have an equal chance of being selected.)
2. The value of the population standard deviation σ is known.
3. Either or both of these conditions is satisfied: The population is normally distributed or $n > 30$.

Knowledge of σ The above requirements include knowledge of the population standard deviation σ, but the following section presents methods for estimating a population mean without knowledge of the value of σ.

Normality Requirement The requirements include the property that either the population is normally distributed or $n > 30$. If $n \leq 30$, the population need not have a distribution that is exactly normal, but it should be approximately normal. We can consider the normality requirement to be satisfied if there are no outliers and a histogram of the sample data is not too far from being bell-shaped. (The methods of this section are said to be *robust,* which means that these methods are not strongly affected by departures from normality, provided that those departures are not too extreme.)

Sample Size Requirement This section uses the normal distribution as the distribution of sample means. If the original population is not itself normally distributed, then we say that means of samples with size $n > 30$ have a distribution that can be approximated by a normal distribution. The condition that the sample size is $n > 30$ is commonly used as a guideline, but it is not possible to identify a specific minimum sample size that is sufficient for all cases. The minimum sample size actually depends on how much the population distribution departs from a normal distribution. Sample sizes of 15 to 30 are adequate if the population appears to have a distribution that is not far from being normal, but some other populations have distributions that are extremely far from normal and sample sizes of 50 or even 100 or higher might be necessary. We will use the simplified criterion of $n >$ 30 as justification for treating the distribution of sample means as a normal distribution.

In Section 7-2 we saw that the sample proportion $\hat{p}$ is the best point estimate of the population proportion p. For similar reasons, the sample mean $\bar{x}$ is the best point estimate of the population mean μ.

The sample mean $\bar{x}$ is the best point estimate of the population mean.

The sample mean $\bar{x}$ usually provides the best estimate, for the following two reasons:

1. For all populations, the sample mean $\bar{x}$ is an **unbiased estimator** of the population mean μ, meaning that the distribution of sample means tends to center

about the value of the population mean μ. [That is, sample means do not systematically tend to overestimate the value of μ, nor do they systematically tend to underestimate μ. Instead, they tend to target the value of μ itself (as illustrated in Section 6-4).]

2. For many populations, the distribution of sample means $\bar{x}$ tends to be more consistent (with *less variation*) than the distributions of other sample statistics.

Estimating Wildlife Population Sizes

The National Forest Management Act protects endangered species, including the northern spotted owl, with the result that the forestry industry was not allowed to cut vast regions of trees in the Pacific Northwest. Biologists and statisticians were asked to analyze the problem, and they concluded that survival rates and population sizes were decreasing for the female owls, known to play an important role in species survival. Biologists and statisticians also studied salmon in the Snake and Columbia Rivers in Washington State, and penguins in New Zealand. In the article "Sampling Wildlife Populations" (*Chance*, Vol. 9, No. 2), authors Bryan Manly and Lyman McDonald comment that in such studies, "biologists gain through the use of modeling skills that are the hallmark of good statistics. Statisticians gain by being introduced to the reality of problems by biologists who know what the crucial issues are."

EXAMPLE Pulse Rates of Females Pulse rates of people are quite important. Without them, where would we be? Data Set 1 in Appendix B includes pulse rates (in beats per minute) of randomly selected women, and here are the statistics: $n = 40$, $\bar{x} = 76.3$, and $s = 12.5$. Use this sample to find the best point estimate of the population mean μ of pulse rates for all women.

SOLUTION For the sample data, $\bar{x} = 76.3$. Because the sample mean $\bar{x}$ is the best point estimate of the population mean μ, we conclude that the best point estimate of the pulse rate for all women is 76.3.

Confidence Intervals

Although a point estimate is the *best* single value for estimating a population parameter, it does not give us any indication of just how *good* the best estimate is. However, a confidence interval gives us information that enables us to better understand the accuracy of the estimate. The confidence interval is associated with a confidence level, such as 0.95 (or 95%). The confidence level gives us the success rate of the procedure used to construct the confidence interval. As in Section 7-2, α is the complement of the confidence level. For a 0.95 (or 95%) confidence level, $\alpha = 0.05$. For a 0.99 (or 99%) confidence level, $\alpha = 0.01$.

Margin of Error When we collect a set of sample data, such as the set of 40 pulse rates of women listed in Data Set 1 from Appendix B, we can calculate the sample mean $\bar{x}$ and that sample mean is typically different from the population mean μ. The difference between the sample mean and the population mean is an error. In Section 6-5 we saw that $\sigma/\sqrt{n}$ is the standard deviation of sample means. Using $\sigma/\sqrt{n}$ and the $z_{\alpha/2}$ notation introduced in Section 7-2, we now use the *margin of error E* expressed as follows:

Formula 7-4 $$E = z_{\alpha/2} \cdot \frac{\sigma}{\sqrt{n}}$$ margin of error for mean (based on known σ)

Formula 7-4 reflects the fact that the sampling distribution of sample means $\bar{x}$ is *exactly* a normal distribution with mean μ and standard deviation $\sigma/\sqrt{n}$ whenever the population has a normal distribution with mean μ and standard deviation σ. If the population is not normally distributed, large samples yield sample means with a distribution that is *approximately* normal. (Formula 7-4 requires that you know the population standard deviation σ, but Section 7-4 will present a method for calculating the margin of error E when σ is not known.)

Using the margin of error E, we can now identify the confidence interval for the population mean μ (if the requirements for this section are satisfied). The three commonly used formats for expressing the confidence interval are shown in the following box.

Confidence Interval Estimate of the Population Mean μ (With σ Known)

$$\bar{x} - E < \mu < \bar{x} + E \qquad \text{where} \qquad E = z_{\alpha/2} \cdot \frac{\sigma}{\sqrt{n}}$$

or

$$\bar{x} \pm E$$

or

$$(\bar{x} - E, \bar{x} + E)$$

This is the format used by the TI-83/84 Plus calculator.

Definition

The two values $\bar{x} - E$ and $\bar{x} + E$ are called **confidence interval limits.**

Procedure for Constructing a Confidence Interval for μ (with Known σ)

1. Verify that the requirements are satisfied. (We have a simple random sample, σ is known, and either the population appears to be normally distributed or $n > 30$.)
2. Refer to Table A-2 or use the TI-83/84 Plus calculator to find the critical value $z_{\alpha/2}$ that corresponds to the desired confidence level. (For example, if the confidence level is 95%, the critical value is $z_{\alpha/2} = 1.96$.)
3. Evaluate the margin of error $E = z_{\alpha/2} \cdot \sigma/\sqrt{n}$.
4. Using the value of the calculated margin of error E and the value of the sample mean $\bar{x}$, find the values of $\bar{x} - E$ and $\bar{x} + E$. Substitute those values in the general format for the confidence interval:

$$\bar{x} - E < \mu < \bar{x} + E$$

or

$$\bar{x} \pm E$$

or

$$(\bar{x} - E, \bar{x} + E)$$

5. Round the resulting values by using the following round-off rule.

The TI-83/84 Plus calculator can automatically generate confidence interval estimates for the population mean μ. See the procedure at the end of this section.

Captured Tank Serial Numbers Reveal Population Size

During World War II, Allied intelligence specialists wanted to determine the number of tanks Germany was producing. Traditional spy techniques provided unreliable results, but statisticians obtained accurate estimates by analyzing serial numbers on captured tanks. As one example, records show that Germany actually produced 271 tanks in June 1941. The estimate based on serial numbers was 244, but traditional intelligence methods resulted in the extreme estimate of 1550. (See "An Empirical Approach to Economic Intelligence in World War II," by Ruggles and Brodie, *Journal of the American Statistical Association,* Vol. 42.)

Round-Off Rule for Confidence Intervals Used to Estimate μ

1. When using the *original set of data* to construct a confidence interval, round the confidence interval limits to one more decimal place than is used for the original set of data.
2. When the original set of data is unknown and only the *summary statistics* (n, $\bar{x}$, s) are used, round the confidence interval limits to the same number of decimal places used for the sample mean.

Interpreting a Confidence Interval As in Section 7-2, be careful to interpret confidence intervals correctly. After obtaining a confidence interval estimate of the population mean μ, such as a 95% confidence interval of $72.4 < \mu < 80.2$, there is a correct interpretation and many wrong interpretations.

Correct: "We are 95% confident that the interval from 72.4 to 80.2 actually does contain the true value of μ." This means that if we were to select many different samples of the same size and construct the corresponding confidence intervals, in the long run 95% of them would actually contain the value of μ. (As in Section 7-2, this correct interpretation refers to the success rate of the *process* being used to estimate the population mean.)

Wrong: Because μ is a fixed constant, it would be wrong to say "there is a 95% chance that μ will fall between 72.4 and 80.2." It would also be wrong to say that "95% of all data values are between 72.4 and 80.2." It would also be wrong to say that "95% of sample means fall between 72.4 and 80.2."

EXAMPLE Pulse Rates of Females For the sample of pulse rates of women in Data Set 1 in Appendix B, we have $n = 40$ and $\bar{x} = 76.3$, and the sample is a simple random sample. Assume that σ is known to be 12.5. Using a 0.95 confidence level, find both of the following:

a. The margin of error E

b. The confidence interval for μ.

SOLUTION

REQUIREMENT ✔ We must first verify that the requirements are satisfied. The sample is a simple random sample. The value of σ is assumed to be known (12.5). With $n > 30$, we satisfy the requirement that "the population is normally distributed or $n > 30$." The requirements are therefore satisfied and we can proceed with the methods of this section. ✔

a. The 0.95 confidence level implies that $\alpha = 0.05$, so $z_{\alpha/2} = 1.96$ (as was shown in an example in Section 7-2). The margin of error E is calculated by using Formula 7-4 as follows. Extra decimal places are used to minimize rounding errors in the confidence interval found in part (b).

$$E = z_{\alpha/2} \cdot \frac{\sigma}{\sqrt{n}} = 1.96 \cdot \frac{12.5}{\sqrt{40}} = 3.8737901$$

b. With $\bar{x} = 76.3$ and $E = 3.8737901$ we construct the confidence interval as follows:

$$\bar{x} - E < \mu < \bar{x} + E$$

$$76.3 - 3.8737901 < \mu < 76.3 + 3.8737901$$

$$72.4 < \mu < 80.2 \quad \text{(rounded to one decimal place as in } \bar{x}\text{)}$$

INTERPRETATION This result could also be expressed as 76.3 ± 3.9 or as (72.4, 80.2). Based on the sample with $n = 40$, $\bar{x} = 76.3$ and σ assumed to be 12.5, the confidence interval for the population mean μ is $72.4 < \mu < 80.2$ and this interval has a 0.95 confidence level. This means that if we were to select many different random samples of 40 women and construct the confidence intervals as we did here, 95% of them would actually contain the value of the population mean μ.

Rationale for the Confidence Interval The basic idea underlying the construction of confidence intervals relates to the central limit theorem, which indicates that if we collect simple random samples of the same size from a normally distributed population, sample means are normally distributed with mean μ and standard deviation $\sigma/\sqrt{n}$. If we collect simple random samples all of size $n > 30$ from any population, the distribution of sample means is approximately normal with mean μ and standard deviation $\sigma/\sqrt{n}$. The confidence interval format is really a variation of the equation that was already used with the central limit theorem. In the expression $z = (\bar{x} - \mu_{\bar{x}})/\sigma_{\bar{x}}$, replace $\sigma_{\bar{x}}$ with $\sigma/\sqrt{n}$, replace $\mu_{\bar{x}}$ with μ, then solve for μ to get

$$\mu = \bar{x} - z\frac{\sigma}{\sqrt{n}}$$

Using the positive and negative values for z results in the confidence interval limits we are using.

Let's consider the specific case of a 95% confidence level, so $\alpha = 0.05$ and $z_{\alpha/2} = 1.96$. For this case, there is a probability of 0.05 that a sample mean will be more than 1.96 standard deviations (or $z_{\alpha/2} \cdot \sigma/\sqrt{n}$, which we denote by E) away from the population mean μ. Conversely, there is a 0.95 probability that a sample mean will be within 1.96 standard deviations (or $z_{\alpha/2} \cdot \sigma/\sqrt{n}$) of μ. (See Figure 7-4.) If the sample mean $\bar{x}$ is within $z_{\alpha/2} \cdot \sigma/\sqrt{n}$ of the population mean μ, then μ must be between $\bar{x} - z_{\alpha/2} \cdot \sigma/\sqrt{n}$ and $\bar{x} + z_{\alpha/2} \cdot \sigma/\sqrt{n}$; this is expressed in the general format of our confidence interval (with $z_{\alpha/2} \cdot \sigma/\sqrt{n}$ denoted as E): $\bar{x} - E < \mu < \bar{x} + E$.

Determining Sample Size Required to Estimate μ

We now address this important question: When we plan to collect a simple random sample of data that will be used to estimate a population mean μ, *how many* sample values must be obtained? For example, suppose we want to estimate the mean weight of airline passengers (an important value for reasons of safety). How many passengers must be randomly selected and weighed? Determining the size of a simple random sample is a very important issue, because samples that are

Figure 7-4

Distribution of Sample Means with Known σ

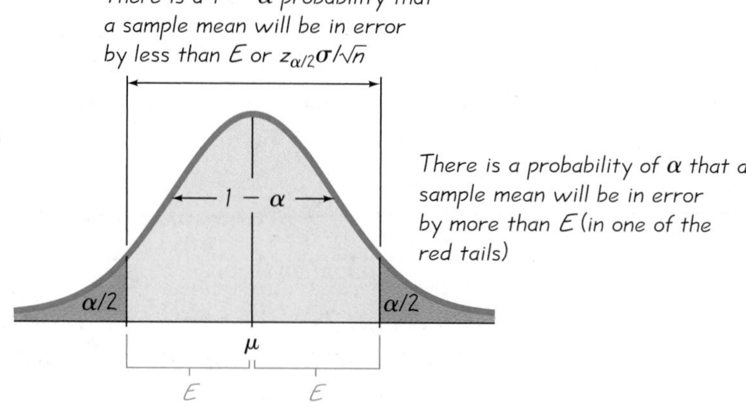

needlessly large waste time and money, and samples that are too small may lead to poor results.

If we begin with the expression for the margin of error E (Formula 7-4) and solve for the sample size n, we get the following.

Sample Size for Estimating Mean μ

Formula 7-5

$$n = \left[\frac{z_{\alpha/2}\sigma}{E}\right]^2$$

where $z_{\alpha/2}$ = critical z score based on the desired confidence level

E = desired margin of error

σ = population standard deviation

Formula 7-5 is remarkable because it shows that the sample size does not depend on the size (N) of the population; the sample size depends on the desired confidence level, the desired margin of error, and the value of the standard deviation σ. (See Exercise 40 for dealing with cases in which a relatively large sample is selected without replacement from a finite population.)

The sample size must be a whole number, because it represents the number of sample values that must be found. However, Formula 7-5 usually gives a result that is not a whole number, so we use the following round-off rule. (It is based on the principle that when rounding is necessary, the required sample size should be rounded *upward* so that it is at least adequately large as opposed to slightly too small.)

Round-Off Rule for Sample Size n

When finding the sample size n, if the use of Formula 7-5 does not result in a whole number, always *increase* the value of n to the next *larger* whole number.

Dealing with Unknown σ When Finding Sample Size When applying Formula 7-5, there is a practical dilemma: The formula requires that we substitute some value for the population standard deviation σ, but in reality, it is usually unknown. When determining a required sample size (not constructing a confidence interval), here are some ways that we can work around this problem:

1. Use the range rule of thumb (see Section 3-3) to estimate the standard deviation as follows: $\sigma \approx \text{range}/4$. (With a sample of 87 or more values randomly selected from a normally distributed population, range/4 will yield a value that is greater than or equal to σ at least 95% of the time. See "Using the Sample Range as a Basis for Calculating Sample Size in Power Calculations," by Richard Browne, *American Statistician,* Vol. 55, No. 4.)
2. Conduct a pilot study by starting the sampling process. Start the sample collection process and, using the first several values, calculate the sample standard deviation s and use it in place of σ. The estimated value of σ can then be improved as more sample data are obtained, and the sample size can be refined accordingly.
3. Estimate the value of σ by using the results of some other study that was done earlier.

In addition, we can sometimes be creative in our use of other known results. For example, IQ tests are typically designed so that the mean is 100 and the standard deviation is 15. Statistics professors have IQ scores with a mean greater than 100 and a standard deviation less than 15 (because they are a more homogeneous group than people randomly selected from the general population). We do not know the specific value of σ for statistics professors, but we can play it safe by using $\sigma = 15$. Using a value for σ that is larger than the true value will make the sample size larger than necessary, but using a value for σ that is too small would result in a sample size that is inadequate. *When calculating the sample size n, any errors should always be conservative in the sense that they make n too large instead of too small.*

EXAMPLE IQ Scores of Statistics Professors Assume that we want to estimate the mean IQ score for the population of statistics professors. How many statistics professors must be randomly selected for IQ tests if we want 95% confidence that the sample mean is within 2 IQ points of the population mean?

SOLUTION The values required for Formula 7-5 are found as follows:

$z_{\alpha/2} = 1.96$ (This is found by converting the 95% confidence level to $\alpha = 0.05$, then finding the critical z score as described in Section 7-2.)

$E = 2$ (Because we want the sample mean to be within 2 IQ points of μ, the desired margin of error is 2.)

$\sigma = 15$ (See the discussion in the paragraph that immediately precedes this example.)

continued

With $z_{\alpha/2} = 1.96$, $E = 2$, and $\sigma = 15$, we use Formula 7-5 as follows:

$$n = \left[\frac{z_{\alpha/2}\sigma}{E}\right]^2 = \left[\frac{1.96 \cdot 15}{2}\right]^2 = 216.09 = 217 \quad \text{(rounded } up\text{)}$$

INTERPRETATION Among the thousands of statistics professors, we need to obtain a simple random sample of at least 217 of them, then we need to get their IQ scores. With a simple random sample of only 217 statistics professors, we will be 95% confident that the sample mean $\bar{x}$ is within 2 IQ points of the true population mean μ.

If we are willing to settle for less accurate results by using a larger margin of error, such as 4, the sample size drops to 54.0225, which is rounded *up* to 55. Doubling the margin of error causes the required sample size to decrease to one-fourth its original value. Conversely, halving the margin of error quadruples the sample size. Consequently, if you want more accurate results, the sample size must be substantially increased. Because large samples generally require more time and money, there is often a need for a tradeoff between the sample size and the margin of error E.

TI Using the TI-83/84 Plus Calculator for Confidence Interval Estimates of μ (With σ Known)

Here is the procedure for using a TI-83/84 Plus calculator for the construction of a confidence interval estimate of a population mean μ, *assuming that the population standard deviation σ is known:*

1. Press STAT, then use the ▷ key to select the option of **TESTS.**
2. If the population standard deviation σ is known, select the menu option of **ZInterval.** Select this option by pressing ▽ to scroll down. (If σ is not known, use the procedure described in Section 7-4.)
3. The first line, identified as Inpt (for input), gives you the option of using Data or Stats. Use the arrow keys ◁ and ▷ to select either Data or Stats.
   - **Data:** Use this option if you have the sample data values stored as a list.
   - **Stats:** Use this option if you know the values of σ, $\bar{x}$, and n.
4. What happens next depends on the choice made in Step 3. If you choose to use Stats, you must enter the values of σ, $\bar{x}$, and n.

   If you choose to use Data, you must enter the known value of σ, the list where the data are stored, and a value for Freq (frequency). Enter 1 for the frequency. (However, if the sample data are stored in two columns as a frequency distribution, the Freq entry should be the list containing frequencies).
5. Enter a confidence level, such as 0.95, then scroll down to Calculate and press ENTER. A typical screen display of these entries is shown here. After pressing ENTER, the confidence interval will be expressed in a format with the

TI-83/84 Plus

```
ZInterval
 (72.426,80.174)
 x̄=76.3
 n=40
```

confidence interval limits enclosed within parentheses, such as (72.426, 80.174). See the accompanying screen display, which corresponds to the example in this section. The TI-83/84 Plus confidence interval of (72.426, 80.174) can also be expressed as $72.4 < \mu < 80.2$.

Using Technology for Sample Size

Although the TI-83/84 Plus calculator can be used to calculate the sample size required to estimate a population mean μ, it does not yet include a special function designed specifically for that purpose. STATDISK, a computer software program designed to accompany this textbook, does have a module for determining sample size. STATDISK is on the CD-ROM included with this book or it can be downloaded from www.aw.com/triola. After starting STATDISK, click on **Analysis** on the main menu at the top, then click on the menu item of **Sample Size Determination.** That item has three options, so select **Estimate Mean,** then proceed to enter the items in the dialog box. Click on **Evaluate** to obtain the required sample size.

7-3 BASIC SKILLS AND CONCEPTS

Statistical Literacy and Critical Thinking

1. Confidence Interval Based on sample data, the following 95% confidence interval is obtained: $2.5 < \mu < 6.0$. Write a statement that correctly interprets that confidence interval.

2. Unbiased Estimator One of the features of the sample mean that makes it a good estimator of a population mean μ is that the sample mean is an unbiased estimator. What does it mean for a statistic to be an unbiased estimator of a population parameter?

3. Confidence Interval A manufacturer of amusement park rides needs a confidence interval estimate of the force that can be exerted when riders push on a leg safety restraint. Unable to find data, a sample is obtained by measuring the force from 100 high school students participating in a science fair. Will the resulting confidence interval be a good estimate of the mean force for the population of all potential riders? Why or why not?

4. Sample Size A researcher calculates the sample size needed to estimate the force that can be exerted by legs of people on amusement park rides, and the sample size of 120 is obtained. If the researcher cannot obtain a random sample and must rely instead on a convenience sample consisting of friends and relatives, can he or she compensate and get good results by using a much larger sample size?

Finding Critical Values. *In Exercises 5–8, find the critical value $z_{\alpha/2}$ that corresponds to the given confidence level.*

TI **5.** 95% TI **6.** 96%

TI **7.** 92% TI **8.** 99%

Verifying Requirements and Finding the Margin of Error. *In Exercises 9–12, calculate the margin of error $E = z_{\alpha/2}\sigma/\sqrt{n}$ if the necessary requirements are satisfied. If the requirements are not all satisfied, state that the margin of error cannot be calculated by using the methods of this section.*

9. The confidence level is 95%, the sample size is $n = 100$, and $\sigma = 15$.

10. The confidence level is 95%, the sample size is $n = 9$, and σ is not known.

11. The confidence level is 99%, the sample size is $n = 9$, $\sigma = 15$, and the original population is normally distributed.

12. The confidence level is 99%, the sample size is $n = 12$, σ is not known, and the original population is normally distributed.

Finding a Confidence Interval. *In Exercises 13–16, use the given confidence level and sample data to find a confidence interval for estimating the population mean μ.*

TI **13.** Salaries of college graduates who took a statistics course in college: 95% confidence; $n = 41$, $\bar{x} = \$67{,}200$, and σ is known to be \$18,277.

TI **14.** Speeds of drivers ticketed in a 55 mi/h zone: 95% confidence; $n = 90$, $\bar{x} = 66.2$ mi/h, and σ is known to be 3.4 mi/h.

TI **15.** FICO (Fair, Isaac, and Company) credit rating scores of applicants for credit cards: 99% confidence; $n = 70$, $\bar{x} = 688$, and σ is known to be 68.

TI **16.** Amounts lost by gamblers who took a bus to an Atlantic City casino: 99% confidence; $n = 40$, $\bar{x} = \$189$, and σ is known to be \$87.

Finding Sample Size. *In Exercises 17–20, use the given margin of error, confidence level, and population standard deviation σ to find the minimum sample size required to estimate an unknown population mean μ.*

17. Margin of error: 0.5 in., confidence level: 95%, $\sigma = 2.5$ in.

18. Margin of error: 0.25 sec, confidence level: 99%, $\sigma = 5.40$ sec.

19. Margin of error: \$1, confidence level: 90%, $\sigma = \$12$.

20. Margin of error: 1.5 mm, confidence level: 95%, $\sigma = 8.7$ mm.

Interpreting Results. *In Exercises 21–24, refer to the accompanying Minitab display of a 95% confidence interval generated using the methods of this section. The sample display results from using a random sample of speeds of drivers ticketed on a section of Interstate 95 in Connecticut.*

MINITAB

```
Variable  N    Mean    StDev  SE Mean          95% CI

Speed     81  67.3849  3.3498  0.3722  (66.6554, 68.1144)
```

21. Identify the value of the point estimate of the population mean μ.

22. Express the confidence interval in the format of $\bar{x} - E < \mu < \bar{x} + E$.

23. Express the confidence interval in the format of $\bar{x} \pm E$.

24. Write a statement that interprets the 95% confidence interval.

TI **25.** Length of Time to Earn Bachelor's Degree In a study of the length of time that students require to earn bachelor's degrees, 80 students are randomly selected and they are found to have a mean of 4.8 years (based on data from the National Center for Education Statistics). Assuming that $\sigma = 2.2$ years, construct a 95% confidence interval estimate of the population mean. Does the resulting confidence interval contradict the fact that 39% of students earn their bachelor's degrees in four years?

TI **26.** Ages of Motorcyclists Killed in Crashes A study of the ages of motorcyclists killed in crashes involves the random selection of 150 drivers with a mean of 37.1 years (based on data from the Insurance Institute for Highway Safety). Assuming that $\sigma = 12.0$ years, construct a 99% confidence interval estimate of the mean age of all motorcyclists killed in crashes. If the confidence interval limits do not include ages below 20 years, does it mean that motorcyclists under the age of 20 rarely die in crashes?

TI **27.** Perception of Time Randomly selected statistics students of the author participated in an experiment to test their ability to determine when 1 min (or 60 seconds) has passed. Forty students yielded a sample mean of 58.3 sec. Assuming that $\sigma = 9.5$ sec, construct a 95% confidence interval estimate of the population mean of all statistics students. Based on the result, is it likely that their estimates have a mean that is reasonably close to 60 sec?

TI **28.** Cotinine Levels of Smokers When people smoke, the nicotine they absorb is converted to cotinine, which can be measured. A sample of 40 smokers has a mean cotinine level of 172.5. Assuming that σ is known to be 119.5, find a 90% confidence interval estimate of the mean cotinine level of all smokers. What aspect of this problem is not realistic?

TI **29.** Blood Pressure Levels When 14 different second-year medical students at Bellevue Hospital measured the blood pressure of the same person, they obtained the results listed below. Assuming that the population standard deviation is known to be 10 mmHg, construct a 95% confidence interval estimate of the population mean. Ideally, what should the confidence interval be in this situation?

138 130 135 140 120 125 120 130 130 144 143 140 130 150

TI **30.** World's Smallest Mammal The world's smallest mammal is the bumblebee bat, also known as the Kitti's hog-nosed bat (or *Craseonycteris thonglongyai*). Such bats are roughly the size of a large bumblebee. Listed below are weights (in grams) from a sample of these bats. Assuming that the weights of all such bats have a standard deviation of 0.30 g, construct a 95% confidence interval estimate of their mean weight. Use the confidence interval to determine whether this sample of bats is from the same population with a known mean of 1.8 g.

1.7 1.6 1.5 2.0 2.3 1.6 1.6 1.8 1.5 1.7 2.2 1.4 1.6 1.6 1.6

TI **31.** Weights of Quarters from Appendix B Use the weights of post-1964 quarters listed in Data Set 14 from Appendix B. Assuming that quarters are minted to produce weights with a population standard deviation of 0.068 g, use the sample of weights to construct a 99% confidence interval estimate of the mean weight. U.S. mint specifications require that quarters have weights between 5.443 g and 5.897 g. What does the confidence interval suggest about the production process?

TI **32.** Forecast Errors from Appendix B Refer to Data Set 8 in Appendix B and subtract each actual high temperature from the high temperature that was forecast one day before. The result is a list of errors. Assuming that all such errors have a standard deviation of 2.5°, construct a 95% confidence interval estimate of the mean of all such errors. What does the result suggest about the accuracy of the forecast temperatures?

Finding Sample Size. *In Exercises 33–38, find the indicated sample size.*

33. Sample Size for Mean IQ of Statistics Students The Wechsler IQ test is designed so that the mean is 100 and the standard deviation is 15 for the population of normal adults. Find the sample size necessary to estimate the mean IQ score of statistics students. We want to be 95% confident that our sample mean is within 2 IQ points of the true mean. The mean for this population is clearly greater than 100. The standard deviation for this population is probably less than 15 because it is a group with less variation than a group randomly selected from the general population; therefore, if we use $\sigma = 15$, we are being conservative by using a value that will make the sample size at least as large as necessary. Assume then that $\sigma = 15$ and determine the required sample size.

34. Sample Size for Weights of Quarters The Tyco Video Game Corporation finds that it is losing income because of slugs used in its video games. The machines must be adjusted to accept coins only if they fall within set limits. In order to set those limits, the mean weight of quarters in circulation must be estimated. A sample of quarters will be weighed in order to determine the mean. How many quarters must we randomly select and weigh if we want to be 99% confident that the sample mean is within 0.025 g of the true population mean for all quarters? Based on results from a sample of quarters, we can estimate the population standard deviation as 0.068 g.

35. Sample Size for Atkins Diet You want to estimate the mean weight loss of people one year after using the Atkins diet. How many dieters must be surveyed if we want to be 95% confident that the sample mean weight loss is within 0.25 lb of the true population mean? Assume that the population standard deviation is known to be 10.6 lb (based on data from "Comparison of the Atkins, Ornish, Weight Watchers, and Zone Diets for Weight Loss and Heart Disease Risk Reduction," by Dansinger et al., *Journal of the American Medical Association,* Vol. 293, No. 1).

36. Sample Size for Television Viewing Nielsen Media Research wants to estimate the mean amount of time (in minutes) that full-time college students spend watching television each weekday. Find the sample size necessary to estimate that mean with a 15-min margin of error. Assume that a 96% confidence level is desired. Also assume that a pilot study showed that the standard deviation is estimated to be 112.2 min.

37. Sample Size Using Range Rule of Thumb You have just been hired by the marketing division of General Motors to estimate the mean amount of money now being spent on the purchase of new cars in the United States. First use the range rule of thumb to make a rough estimate of the standard deviation of the amounts spent. It is reasonable to assume that typical amounts range from $12,000 to about $70,000. Then use that estimated standard deviation to determine the sample size corresponding to 95% confidence and a $100 margin of error. Is the sample size practical? If not, what should be changed to get a practical sample size?

38. Sample Size Using Sample Data You want to estimate the mean pulse rate of adult males. Refer to Data Set 1 in Appendix B and find the maximum and minimum pulse rates for males, then use those values with the range rule of thumb to estimate σ. How many adult males must you randomly select and test if you want to be 95% confident that the sample mean pulse rate is within 2 beats (per minute) of the true population mean μ? If, instead of using the range rule of thumb, the standard deviation of the male pulse rates in Data Set 1 is used as an estimate of σ, is the required sample size very different? Which sample size is likely to be closer to the correct sample size?

7-3 BEYOND THE BASICS

39. Confidence Interval with Finite Population Correction Factor The standard error of the mean is $\sigma/\sqrt{n}$ provided that the population size is infinite. If the population size is finite and is denoted by N, then the correction factor $\sqrt{(N-n)/(N-1)}$ should be used whenever $n > 0.05N$. This correction factor multiplies the margin of error E given in Formula 7-4, so that the margin of error is as shown below. Find the 95% confidence interval for the mean of 250 IQ scores if a sample of 35 of those scores produces a mean of 110. Assume that $\sigma = 15$.

$$E = z_{\alpha/2}\frac{\sigma}{\sqrt{n}}\sqrt{\frac{N-n}{N-1}}$$

40. Sample Size with Finite Population Correction Factor In Formula 7-4 for the margin of error E, we assume that the population is infinite, that we are sampling with replacement, or that the population is very large. If we have a relatively small population and sample without replacement, we should modify E to include a *finite population correction factor*, so that the margin of error is as shown in Exercise 39, where N is the population size. That expression for the margin of error can be solved for n to yield

$$n = \frac{N\sigma^2(z_{\alpha/2})^2}{(N-1)E^2 + \sigma^2(z_{\alpha/2})^2}$$

Repeat Exercise 33, assuming that the statistics students are randomly selected without replacement from a population of $N = 200$ statistics students.

7-4 Estimating a Population Mean: σ Not Known

Key Concept This section presents methods for finding a confidence interval estimate of a population mean when the population standard deviation is *not known.* (Section 7-3 presented methods for estimating μ when σ is known.) With σ unknown, we will use the *Student t distribution* (instead of the normal distribution), assuming that certain requirements (given below) are satisfied. Because σ is typically unknown in real circumstances, the methods of this section are very realistic, practical, and they are used often.

Requirements

1. The sample is a simple random sample.
2. Either the sample is from a normally distributed population or $n > 30$.

As in Section 7-3, the requirement of a normally distributed population is not a strict requirement, so we can usually consider the population to be normally distributed after using the sample data to confirm that there are no outliers and the histogram has a shape that is not very far from a normal distribution. Also, as in Section 7-3, the requirement that the sample size is $n > 30$ is commonly used as a guideline, but the minimum sample size actually depends on how much the population distribution departs from a normal distribution. [If a population is known to be normally distributed, distribution of sample means $\bar{x}$ is *exactly* a normal distribution with mean μ and standard deviation $\sigma/\sqrt{n}$; if the population is not normally distributed, large ($n > 30$) samples yield sample means with a distribution that is *approximately* normal with mean μ and standard deviation $\sigma/\sqrt{n}$.]

As in Section 7-3, the sample mean $\bar{x}$ is the best point estimate (or single-valued estimate) of the population mean μ.

The sample mean $\bar{x}$ is the best point estimate of the population mean μ.

Here is a major point of this section: If σ is not known, but the above requirements are satisfied, we use a *Student t distribution* (instead of a normal distribution) developed by William Gosset (1876–1937). Gosset was a Guinness Brewery employee who needed a distribution that could be used with small samples. The Irish brewery where he worked did not allow the publication of research results, so Gosset published under the pseudonym *Student*. (In the interest of research and better serving his readers, the author visited the Guinness Brewery and sampled some of the product. Such commitment.)

Because we do not know the value of σ, we estimate it with the value of the sample standard deviation s, but this introduces another source of unreliability, especially with small samples. In order to keep a confidence interval at some desired level, such as 95%, we compensate for this additional unreliability by making the confidence interval wider: We use critical values $t_{\alpha/2}$ (from a Student t distribution) that are larger than the critical values of $z_{\alpha/2}$ from the normal distribution.

Student *t* Distribution

If a population has a normal distribution, then the distribution of

$$t = \frac{\bar{x} - \mu}{\frac{s}{\sqrt{n}}}$$

is a **Student *t* distribution** for all samples of size n. A Student t distribution, often referred to as a ***t* distribution,** is used to find critical values denoted by $t_{\alpha/2}$.

We will soon discuss some of the important properties of the t distribution, but we first present the components needed for the construction of confidence intervals. Let's start with the critical value denoted by $t_{\alpha/2}$. A value of $t_{\alpha/2}$ can be found in Table A-3 by locating the appropriate number of *degrees of freedom* in the left column and proceeding across the corresponding row until reaching the number directly below the appropriate area at the top.

Definition

The number of **degrees of freedom** for a collection of sample data is the number of sample values that can vary after certain restrictions have been imposed on all data values.

For example, if 10 students have quiz scores with a mean of 80, we can freely assign values to the first 9 scores, but the 10th score is then determined. The sum of the 10 scores must be 800, so the 10th score must equal 800 minus the sum of the first 9 scores. Because those first 9 scores can be freely selected to be any values, we say that there are 9 degrees of freedom available. For the applications of this section, the number of degrees of freedom is simply the sample size minus 1.

$$\textbf{degrees of freedom} = n - 1$$

EXAMPLE Finding a Critical Value A sample of size $n = 23$ is a simple random sample selected from a normally distributed population. Find the critical value $t_{\alpha/2}$ corresponding to a 95% confidence level.

SOLUTION Because $n = 23$, the number of degrees of freedom is given by $n - 1 = 22$. Using Table A-3, we locate the 22nd row by referring to the column at the extreme left. As in Section 7-2, a 95% confidence level corresponds to $\alpha = 0.05$, so we find the column listing values for an *area of 0.05 in two tails*. The value corresponding to the row for 22 degrees of freedom and the column for an area of 0.05 in two tails is 2.074, so $t_{\alpha/2} = 2.074$.

Now that we know how to find critical values denoted by $t_{\alpha/2}$, we can describe the margin of error E and the confidence interval.

Margin of Error E for the Estimate of μ (With σ Not Known)

Formula 7-6 $$E = t_{\alpha/2}\frac{s}{\sqrt{n}}$$

where $t_{\alpha/2}$ has $n - 1$ degrees of freedom. Table A-3 lists values of $t_{\alpha/2}$.

Confidence Interval for the Estimate of μ (With σ Not Known)

$$\bar{x} - E < \mu < \bar{x} + E$$

where

$$E = t_{\alpha/2}\frac{s}{\sqrt{n}}$$

The following procedure uses the above margin of error in the construction of confidence interval estimates of μ.

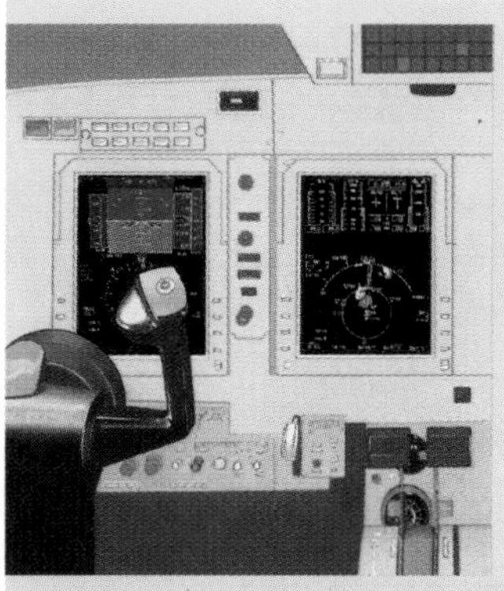

Excerpts from a Department of Transportation Circular

The following excerpts from a Department of Transportation circular concern some of the accuracy requirements for navigation equipment used in aircraft. Note the use of the confidence interval. "The total of the error contributions of the airborne equipment, when combined with the appropriate flight technical errors listed, should not exceed the following with a 95% confidence (2-sigma) over a period of time equal to the update cycle." "The system of airways and routes in the United States has widths of route protection used on a VOR system with accuracy of ±4.5 degrees on a 95% probability basis."

Procedure for Constructing a Confidence Interval for μ (With σ Unknown)

1. Verify that the requirements are satisfied. (We have a simple random sample, and either the population appears to be normally distributed or $n > 30$.)
2. Using $n - 1$ degrees of freedom, refer to Table A-3 and find the critical value $t_{\alpha/2}$ that corresponds to the desired confidence level. (For the confidence level, refer to the "Area in Two Tails.")
3. Evaluate the margin of error $E = t_{\alpha/2} \cdot s/\sqrt{n}$.
4. Using the value of the calculated margin of error E and the value of the sample mean $\bar{x}$, find the values of $\bar{x} - E$ and $\bar{x} + E$. Substitute those values in the general format for the confidence interval:

$$\bar{x} - E < \mu < \bar{x} + E$$

or

$$\bar{x} \pm E$$

or

$$(\bar{x} - E, \bar{x} + E)$$

5. Round the resulting confidence interval limits. If using the original set of data, round to one more decimal place than is used for the original set of data. If using summary statistics (n, $\bar{x}$, s), round the confidence interval limits to the same number of decimal places used for the sample mean.

The TI-83/84 Plus calculator can automatically generate confidence interval estimates for the population mean μ. See the procedure at the end of this section.

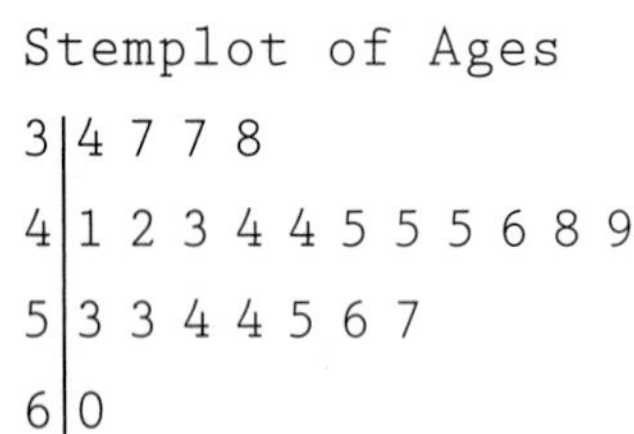

EXAMPLE Constructing a Confidence Interval Listed in the accompanying stemplot are the ages of applicants who were unsuccessful in winning promotion (based on data from "Debating the Use of Statistical Evidence in Allegations of Age Discrimination," by Barry and Boland, *American Statistician,* Vol. 58, No. 2). There is an important larger issue of whether these applicants suffered age discrimination, but for now we will focus on the simple issue of using those values as a sample for the purpose of estimating the mean of a larger population. Assume that the sample is a simple random sample and use the sample data with a 95% confidence level to find both of the following:

a. The margin of error E
b. The confidence interval for μ

SOLUTION

REQUIREMENT ✔ We must first verify that the requirements are satisfied. We are assuming that the sample is a simple random sample. We now address the requirement that "the population is normally distributed or $n > 30$." Because $n = 23$, we must check that the distribution is approximately normal. The shape of the stemplot does suggest a normal distribution. Also, a normal quantile plot confirms that the sample data are from a population with a distribution that is approximately normal. The requirements are therefore satisfied and we can proceed with the methods of this section. ✔

a. The 0.95 confidence level implies that $\alpha = 0.05$, so $t_{\alpha/2} = 2.074$ (use Table A-3 with df $= n - 1 = 22$, as was shown in the preceding example). After finding that the sample statistics are $n = 23$, $\bar{x} = 47.0$, and $s = 7.2$, the margin of error E is calculated by using Formula 7-6 as follows. Extra decimal places are used to minimize rounding errors in the confidence interval found in part (b).

$$E = t_{\alpha/2}\frac{s}{\sqrt{n}} = 2.074 \cdot \frac{7.2}{\sqrt{23}} = 3.11370404$$

b. With $\bar{x} = 47.0$ and $E = 3.11370404$, we construct the confidence interval as follows:

$$\bar{x} - E < \mu < \bar{x} + E$$

$$47.0 - 3.11370404 < \mu < 47.0 + 3.11370404$$

$$43.9 < \mu < 50.1$$ (rounded to one more decimal place than the original data)

INTERPRETATION This result could also be expressed in the format of 47.0 ± 3.1 or (43.9, 50.1). On the basis of the given sample results, we are 95% confident that the limits of 43.9 years and 50.1 years actually do contain the value of the population mean μ.

We now list the important properties of the t distribution that we are using in this section.

Important Properties of the Student *t* Distribution

1. The Student t distribution is different for different sample sizes. (See Figure 7-5 for the cases $n = 3$ and $n = 12$.)
2. The Student t distribution has the same general symmetric bell shape as the standard normal distribution, but it reflects the greater variability (with wider distributions) that is expected with small samples.

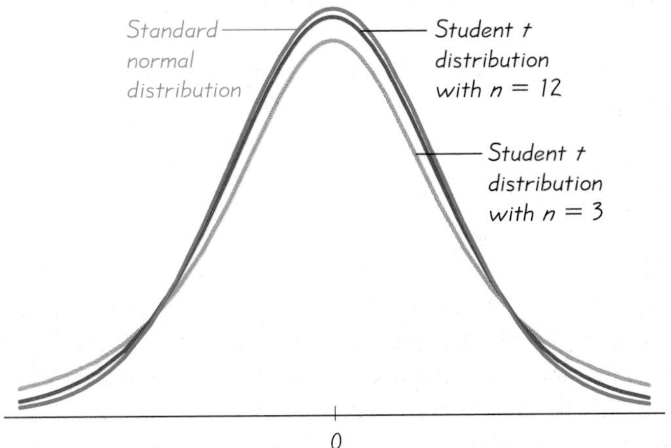

Figure 7-5

Student *t* Distributions for $n = 3$ and $n = 12$

The Student t distribution has the same general shape and symmetry as the standard normal distribution, but it reflects the greater variability that is expected with small samples.

3. The Student t distribution has a mean of $t = 0$ (just as the standard normal distribution has a mean of $z = 0$).
4. The standard deviation of the Student t distribution varies with the sample size, but it is greater than 1 (unlike the standard normal distribution, which has $\sigma = 1$).
5. As the sample size n gets larger, the Student t distribution gets closer to the standard normal distribution.

Choosing the Appropriate Distribution

It is sometimes difficult to decide whether to use the standard normal z distribution or the Student t distribution. The flowchart in Figure 7-6 and the accompanying Table 7-1 both summarize the key points to be considered when constructing confidence intervals for estimating μ, the population mean. In Figure 7-6 or Table 7-1, note that if we have a small ($n \leq 30$) sample drawn from a distribution that differs dramatically from a normal distribution, we can't use the methods described in this chapter. One alternative is to use nonparametric methods (see Chapter 13), and another alternative is to use the computer bootstrap method. In both of those approaches, no assumptions are made about the original population. The bootstrap method is described in the TI-83/84 Plus Calculator Project at the end of this chapter.

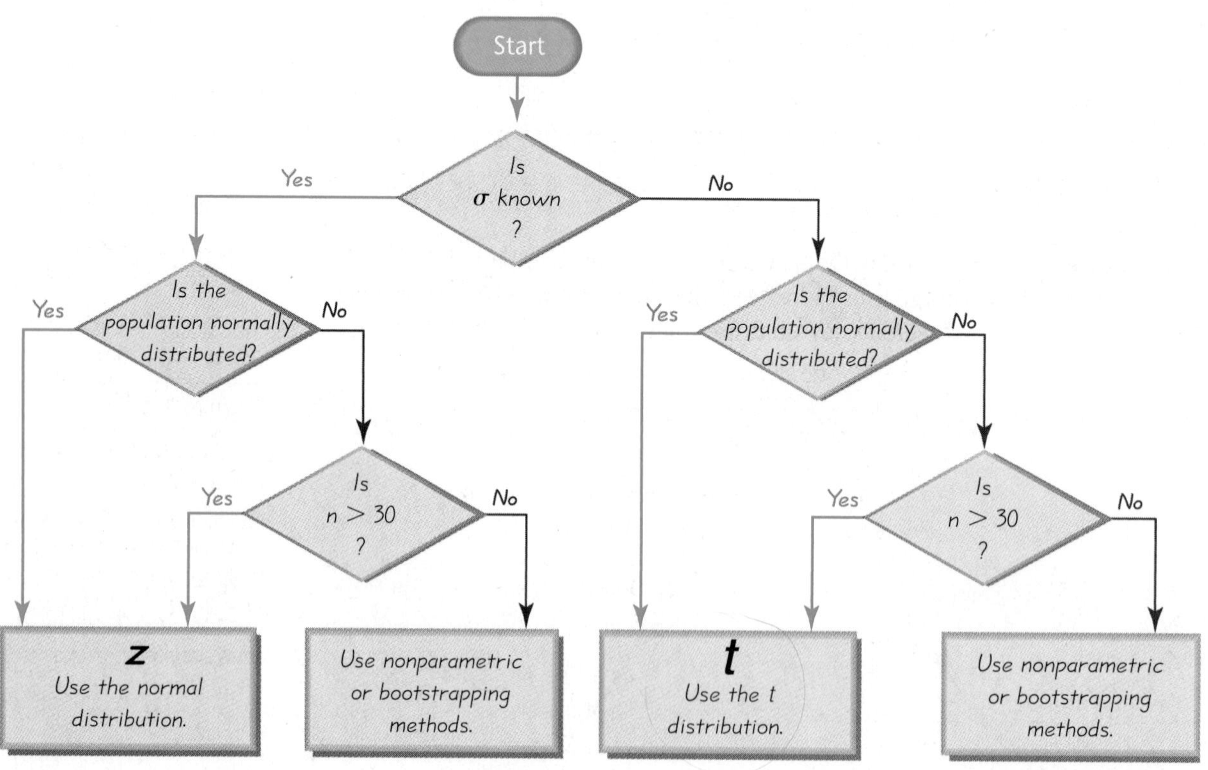

Figure 7-6 Choosing Between z and t

Table 7-1 Choosing between z and t

Method	Conditions
Use normal (z) distribution.	σ known and normally distributed population *or* σ known and $n > 30$
Use t distribution.	σ not known and normally distributed population *or* σ not known and $n > 30$
Use a nonparametric method or bootstrapping.	Population is not normally distributed and $n \leq 30$.

Notes: 1. **Criteria for deciding whether the population is normally distributed:** Population need not be exactly normal, but it should appear to be somewhat symmetric with one mode and no outliers.

2. **Sample size $n > 30$:** This is a commonly used guideline, but sample sizes of 15 to 30 are adequate if the population appears to have a distribution that is not far from being normal and there are no outliers. For some population distributions that are extremely far from normal, the sample size might need to be larger than 50 or even 100.

The following example focuses on choosing the correct approach by using the methods of this section and Section 7-3.

EXAMPLE Choosing Distributions Assuming that you plan to construct a confidence interval for the population mean μ, use the given data to determine whether the margin of error E should be calculated using a critical value of $z_{\alpha/2}$ (from the normal distribution), a critical value of $t_{\alpha/2}$ (from a t distribution), or neither (so that the methods of Sections 7-3 and this section cannot be used).

a. $n = 150, \bar{x} = 100, s = 15$, and the population has a skewed distribution.

b. $n = 8, \bar{x} = 100, s = 15$, and the population has a normal distribution.

c. $n = 8, \bar{x} = 100, s = 15$, and the population has a very skewed distribution.

d. $n = 150, \bar{x} = 100, \sigma = 15$, and the distribution is skewed. (This situation almost never occurs.)

e. $n = 8, \bar{x} = 100, \sigma = 15$, and the distribution is extremely skewed. (This situation almost never occurs.)

SOLUTION Refer to Figure 7-6 or Table 7-1 to determine the following:

a. Because the population standard deviation σ is not known and the sample is large ($n > 30$), the margin of error is calculated using $t_{\alpha/2}$ in Formula 7-6.

b. Because the population standard deviation σ is not known and the population is normally distributed, the margin of error is calculated using $t_{\alpha/2}$ in Formula 7-6.

continued

Estimating Sugar in Oranges

In Florida, members of the citrus industry make extensive use of statistical methods. One particular application involves the way in which growers are paid for oranges used to make orange juice. An arriving truckload of oranges is first weighed at the receiving plant, then a sample of about a dozen oranges is randomly selected. The sample is weighed and then squeezed, and the amount of sugar in the juice is measured. Based on the sample results, an estimate is made of the total amount of sugar in the entire truckload. Payment for the load of oranges is based on the estimate of the amount of sugar because sweeter oranges are more valuable than those less sweet, even though the amounts of juice may be the same.

Estimates to Improve the Census

In the decennial Census, not everyone is counted, and some people are counted more than once. Methods of statistics can be used to improve population counts with adjustments in each county of each state. Some argue that the Constitution specifies that the Census be an "actual enumeration" which does not allow for adjustment. A Supreme Court ruling prohibits use of adjusted population counts for reapportionment of congressional seats, but a recent ruling by a federal appeals court ordered that the adjusted counts be released, even if they can't be used for that purpose. According to the Associated Press, "The Census Bureau has left open the possibility of using adjusted data for federal funding in the future," so the use of powerful statistics methods might eventually result in better allocation of federal and state funds.

c. Because the sample is small and the population does not have a normal distribution, the margin of error E should not be calculated using a critical value of $z_{\alpha/2}$ or $t_{\alpha/2}$. The methods of Section 7-3 and this section do not apply.

d. Because the population standard deviation σ is known and the sample is large ($n > 30$), the margin of error is calculated using $z_{\alpha/2}$ in Formula 7-4.

e. Because the population is not normally distributed and the sample is small ($n \leq 30$), the margin of error E should not be calculated using a critical value of $z_{\alpha/2}$ or $t_{\alpha/2}$. The methods of Section 7-3 and this section do not apply.

EXAMPLE Confidence Interval for Birth Weights In a study of the effects of prenatal cocaine use on infants, the following sample data were obtained for weights at birth: $n = 190$, $\bar{x} = 2700$ g, $s = 645$ g (based on data from "Cognitive Outcomes of Preschool Children with Prenatal Cocaine Exposure," by Singer et al., *Journal of the American Medical Association,* Vol. 291, No. 20). The design of the study justifies the assumption that the sample can be treated as a simple random sample. Use the sample data to construct a 95% confidence interval estimate of μ, the mean birth weight of all infants born to mothers who used cocaine.

SOLUTION

REQUIREMENT ✔ We must first verify that the requirements are satisfied. The sample is a simple random sample. Because $n = 190$, we satisfy the requirement that "the population is normally distributed or $n > 30$." The requirements are therefore satisfied. (This is Step 1 in the five-step procedure listed earlier, and we can now proceed with the remaining steps.) ✔

Step 2: The critical value is $t_{\alpha/2} = 1.972$. It is found in Table A-3 as the critical value corresponding to $n - 1 = 189$ degrees of freedom (left column of Table A-3) and an area in two tails of 0.05. (Because Table A-3 does not include df $= 189$, we use the closest critical value of 1.972. We can use software to find that a more accurate critical value is 1.973, so the approximation is quite good here.)

Step 3: *Find the margin of error E:* The margin of error $E = 2.97355$ is computed using Formula 7-6 as shown below, with extra decimal places used to minimize rounding error in the confidence interval found in Step 4.

$$E = t_{\alpha/2}\frac{s}{\sqrt{n}} = 1.972 \cdot \frac{645}{\sqrt{190}} = 92.276226$$

Step 4: *Find the confidence interval:* The confidence interval can now be found by using $\bar{x} = 2700$ and $E = 92.276226$ as shown below:

$$\bar{x} - E < \mu < \bar{x} + E$$
$$2700 - 92.276226 < \mu < 2700 + 92.276226$$
$$2607.7238 < \mu < 2792.2762$$

Step 5: Round the confidence interval limits. Because the sample mean is rounded as a whole number, round the confidence interval limits to get this result: $2608 < \mu < 2792$.

INTERPRETATION On the basis of the sample data, we are 95% confident that the limits of 2608 g and 2792 g actually do contain the value of the mean birth weight. We can now compare this result to a confidence interval constructed for birth weights of children whose mothers did not use cocaine. (See Exercise 17.)

Finding Point Estimate and *E* from a Confidence Interval

Later in this section we will describe how the TI-83/84 Plus calculator can be used to find a confidence interval. A typical usage requires that you enter a confidence level and sample statistics, and the display shows the confidence interval limits. The sample mean $\bar{x}$ is the value midway between those limits, and the margin of error E is one-half the difference between those limits (because the upper limit is $\bar{x} + E$ and the lower limit is $\bar{x} - E$, the distance separating them is $2E$).

Point estimate of μ:

$$\bar{x} = \frac{(\text{upper confidence limit}) + (\text{lower confidence limit})}{2}$$

Margin of error:

$$E = \frac{(\text{upper confidence limit}) - (\text{lower confidence limit})}{2}$$

EXAMPLE Ages of Stowaways In analyzing the ages of all *Queen Mary* stowaways (based on data from the Cunard Line), the Minitab display shown below is obtained. Use the given confidence interval to find the point estimate $\bar{x}$ and the margin of error E. Treat the values as sample data randomly selected from a large population.

Minitab

```
95.0% CI
(24.065, 27.218)
```

SOLUTION In the following calculations, results are rounded to one decimal place, which is one additional decimal place beyond the rounding used for the original list of ages.

$$\bar{x} = \frac{(\text{upper confidence limit}) + (\text{lower confidence limit})}{2}$$

$$= \frac{27.218 + 24.065}{2} = 25.6 \text{ years}$$

$$E = \frac{(\text{upper confidence limit}) - (\text{lower confidence limit})}{2}$$

$$= \frac{27.218 - 24.065}{2} = 1.6 \text{ years}$$

Estimating Crowd Size

There are sophisticated methods of analyzing the size of a crowd. Aerial photographs and measures of people density can be used with reasonably good accuracy. However, reported crowd size estimates are often simple guesses. After the Boston Red Sox won the World Series for the first time in 86 years, Boston city officials estimated that the celebration parade was attended by 3.2 million fans. Boston police provided an estimate of around 1 million, but it was admittedly based on guesses by police commanders. A photo analysis led to an estimate of around 150,000. Boston University Professor Farouk El-Baz used images from the U.S. Geological Survey to develop an estimate of at most 400,000. MIT physicist Bill Donnelly said that "it's a serious thing if people are just putting out any number. It means other things aren't being vetted that carefully."

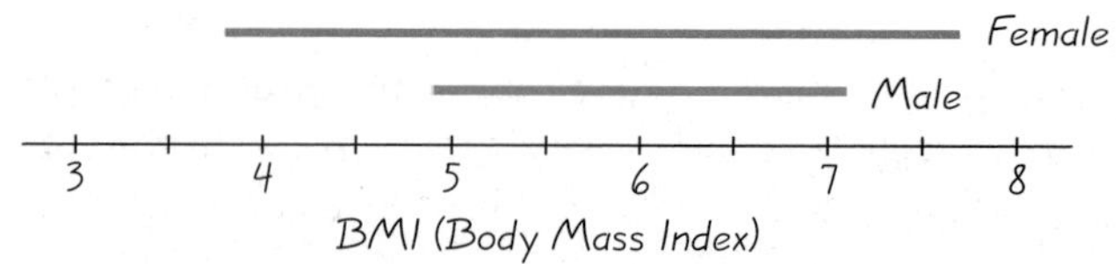

Figure 7-7

BMI Indexes of Females and Males

Using Confidence Intervals to Describe, Explore, or Compare Data

In some cases, we might use a confidence interval to achieve an ultimate goal of estimating the value of a population parameter. In other cases, a confidence interval might be one of several different tools used to describe, explore, or compare data sets. Figure 7-7 shows graphs of confidence intervals for the BMI indexes of a sample of females and males (see Data Set 1 in Appendix B). Because the confidence intervals overlap, there does not appear to be a significant difference between the mean BMI index of females and males.

Caution: As in Sections 7-2 and 7-3, confidence intervals can be used informally to compare different data sets, but the overlapping of confidence intervals should not be used for making formal and final conclusions about equality of means. Later chapters will include procedures for deciding whether two populations have equal means, and those methods will not have the pitfalls associated with comparisons based on the overlap of confidence intervals.

Do not use the overlapping of confidence intervals as the basis for making formal conclusions about the equality of means.

TI

Using a TI-83/84 Plus Calculator for Confidence Interval Estimates of μ (With σ Not Known)

Here is the procedure for using a TI-83/84 Plus calculator for the construction of a confidence interval estimate of a population mean μ, assuming that the population standard deviation σ is not known:

1. Press ENTER, then use the ▷ key to select the option of **TESTS.**
2. If the population standard deviation σ is not known, select the menu option of **TInterval.** Select this option by pressing ▽ to scroll down to **TInterval.** (If σ is known, use the procedure described in Section 7-3.)
3. The first line, identified as Inpt (for input), gives you the option of using Data or Stats. Use the arrow keys ◁ and ▷ to select either Data or Stats.
   - **Data:** Use this option if you have the sample data values stored as a list.
   - **Stats:** Use this option if you know the values of $\bar{x}$, s, and n.
4. What happens next depends on the choice made in Step 3. If you choose to use Stats, you must enter the values of $\bar{x}$, s, and n.

   If you choose to use Data, you must enter the list where the data are stored, and a value for Freq (frequency). Enter 1 for the frequency. (However, if the sample data are stored in two columns as a frequency distribution, the Freq entry should be the list containing frequencies).

5. Enter a confidence level, such as 0.95, then scroll down to **Calculate** and press ENTER. After pressing ENTER, the confidence interval will be expressed in a format with the confidence interval limits enclosed within parentheses, such as (2607.7, 2792.3). See the accompanying screen display which corresponds to the fourth example in this section. The TI-83/84 Plus confidence interval of (2607.7, 2792.3) can also be expressed as $2608 < \mu < 2792$.

TI-83/84 Plus

```
TInterval
 (2607.7,2792.3)
 x̄=2700
 Sx=645
 n=190
```

7-4 BASIC SKILLS AND CONCEPTS

Statistical Literacy and Critical Thinking

1. What's Wrong? A "snapshot" in *USA Today* noted that "Consumers will spend an estimated average of \$483 on merchandise" for back-to-school spending. It was reported that the value is based on a survey of 8453 consumers, and the margin of error is "±1 percentage point." What's wrong with this information?

2. Confidence Interval The *Newport Chronicle* issued a report stating that based on a sample of homes, the mean tax bill is \$4626 with a margin of error of \$591. Express the confidence interval in the format of $\bar{x} - E < \mu < \bar{x} + E$.

3. Interpreting a Confidence Interval Using the systolic blood pressure levels of the 40 men listed in Data Set 1 in Appendix B, we get this 99% confidence interval: $114.4 < \mu < 123.4$. Write a statement that correctly interprets that confidence interval.

4. Checking Requirements Suppose that we want to construct a confidence interval estimate of the amounts of precipitation on Mondays in Boston, and we plan to use the amounts listed in Data Set 10 from Appendix B. We can examine those amounts to see that among the 52 Mondays, there are 33 that have amounts of 0. Based on that observation, do the amounts of precipitation on Mondays appear to be normally distributed? Assuming that the sample can be treated as a simple random sample, can we use the methods of this section to construct a confidence interval estimate of the population mean? Why or why not?

Using Correct Distribution. *In Exercises 5–12, do one of the following, as appropriate: (a) Find the critical value $z_{\alpha/2}$, (b) find the critical value $t_{\alpha/2}$, (c) state that neither the normal nor the t distribution applies.*

5. 95%; $n = 12$; σ is unknown; population appears to be normally distributed.

6. 99%; $n = 15$; σ is unknown; population appears to be normally distributed.

7. 99%; $n = 4$; σ is known; population appears to be very skewed.

8. 95%; $n = 50$; σ is known; population appears to be very skewed.

9. 90%; $n = 200$; σ is unknown; population appears to be normally distributed.

10. 98%; $n = 16$; $\sigma = 5.0$; population appears to be very skewed.

11. 98%; $n = 18$; $\sigma = 21.5$; population appears to be normally distributed.

12. 90%; $n = 33$; σ is unknown; population appears to be normally distributed.

Finding Confidence Intervals. In Exercises 13 and 14, use the given confidence level and sample data to find (a) the margin of error and (b) the confidence interval for the population mean μ. Assume that the population has a normal distribution.

TI **13.** Weight lost on Weight Watchers diet: 95% confidence; $n = 40, \bar{x} = 3.0$ kg, $s = 4.9$ kg.

TI **14.** Life span of desktop PC: 99% confidence; $n = 21, \bar{x} = 6.8$ years, $s = 2.4$ years.

Interpreting Display. In Exercises 15 and 16, use the given data and the corresponding display to express the confidence interval in the format of $\bar{x} - E < \mu < \bar{x} + E$. Also write a statement that interprets the confidence interval.

15. IQ scores of statistics students: 95% confidence; $n = 25, \bar{x} = 118.0, s = 10.7$.

Minitab

N	Mean	StDev	SE Mean	95% CI
25	118.000	10.700	2.140	(113.583, 122.417)

TI-83/84 Plus

```
TInterval
 (3.5839,5.6161)
 x̄=4.6
 Sx=1.9
 n=27
```

TI **16.** Life span of cell phone: 99% confidence; $n = 27, \bar{x} = 4.6$ years, $s = 1.9$ years. (See the TI-83/84 Plus calculator display in the margin.)

Constructing Confidence Intervals. In Exercises 17–26, construct the confidence interval.

TI **17.** Birth Weights A random sample of the birth weights of 186 babies has a mean of 3103 g and a standard deviation of 696 g (based on data from "Cognitive Outcomes of Preschool Children with Prenatal Cocaine Exposure," by Singer et al., *Journal of the American Medical Association,* Vol. 291, No. 20). These babies were born to mothers who did not use cocaine during their pregnancies. Construct a 95% confidence interval estimate of the mean birth weight for all such babies. Compare the result to the confidence interval obtained in the example in this section that involved birth weights of babies born to mothers who used cocaine during pregnancy. Does cocaine use appear to affect the birth weight of a baby?

TI **18.** Mean Body Temperature Data Set 2 in Appendix B includes 106 body temperatures for which $\bar{x} = 98.20$°F and $s = 0.62$°F. Using the sample statistics, construct a 99% confidence interval estimate of the mean body temperature of all healthy humans. Do the confidence interval limits contain 98.6°F? What does the sample suggest about the use of 98.6°F as the mean body temperature?

TI **19.** Forecast and Actual Temperatures Data Set 8 in Appendix B includes a list of actual high temperatures and the corresponding list of three-day-forecast high temperatures. If the difference for each day is found by subtracting the three-day-forecast high temperature from the actual high temperature, the result is a list of 35 values with a mean of −1.3° and a standard deviation of 4.7°.

a. Construct a 99% confidence interval estimate of the mean difference between all actual high temperatures and three-day-forecast high temperatures.

b. Does the confidence interval include 0°? If a meteorologist claims that three-day-forecast high temperatures tend to be too high because the mean difference of the sample is −1.3°, does that claim appear to be valid? Why or why not?

TI **20.** Shoveling Heart Rates Because cardiac deaths appear to increase after heavy snowfalls, an experiment was designed to compare cardiac demands of snow shoveling to those of using an electric snow thrower. Ten subjects cleared tracts of snow using both methods, and their maximum heart rates (beats per minute) were recorded during both activities. The following results were obtained (based on data from "Cardiac Demands of Heavy Snow Shoveling," by Franklin et al., *Journal of the American Medical Association,* Vol. 273, No. 11):

Manual snow shoveling maximum heart rates: $n = 10, \bar{x} = 175, s = 15.$

Electric snow thrower maximum heart rates: $n = 10, \bar{x} = 124, s = 18.$

a. Find the 95% confidence interval estimate of the population mean for those people who shovel snow manually.

b. Find the 95% confidence interval estimate of the population mean for those people who use the electric snow thrower.

c. If you are a physician with concerns about cardiac deaths fostered by manual snow shoveling, what single value in the confidence interval from part (a) would be of greatest concern?

d. Compare the confidence intervals from parts (a) and (b) and interpret your findings.

TI **21.** Monitoring Lead in Air Listed below are measured amounts of lead (in micrograms per cubic meter, or $\mu g/m^3$) in the air. The Environmental Protection Agency has established an air quality standard for lead of 1.5 $\mu g/m^3$. The measurements shown below were recorded at Building 5 of the World Trade Center site on different days immediately following the destruction caused by the terrorist attacks of September 11, 2001. After the collapse of the two World Trade Center buildings, there was considerable concern about the quality of the air. Use the given values to construct a 95% confidence interval estimate of the mean amount of lead in the air. Is there anything about this data set suggesting that the confidence interval might not be very good? Explain.

5.40 1.10 0.42 0.73 0.48 1.10

TI **22.** Constructing a Confidence Interval The stemplot below lists the ages of applicants who were successful in winning promotion (based on data from "Debating the Use of Statistical Evidence in Allegations of Age Discrimination," by Barry and Boland, *American Statistician,* Vol. 58, No. 2). Assume that the sample is a simple random sample and construct a 95% confidence interval estimate of the mean age of all such successful people. Compare the result to the confidence interval for the ages of those who were not successful (see the example in this section).

```
3 | 3 6 7 8 8 9
4 | 2 2 3 3 4 4 4 5 5 5 5 6 6 7 7 8 8 9 9
5 | 1 1 2 4
```

TI **23.** Credit Rating When consumers apply for credit, their credit is rated using FICO (Fair, Isaac, and Company) scores. Credit ratings are given below for a sample of applicants for car loans. Use the sample data to construct a 99% confidence interval for the mean FICO score of all applicants for credit. If one bank requires a credit rating of at least 620 for a car loan, does it appear that almost all applicants will have suitable credit ratings?

661 595 548 730 791 678 672 491 492 583 762 624 769 729 734 706

TI **24.** World's Smallest Mammal The world's smallest mammal is the bumblebee bat, also known as the Kitti's hog-nosed bat (or *Craseonycteris thonglongyai*). Such bats are roughly the size of a large bumblebee. Listed below are weights (in grams) from a sample of these bats. Construct a 95% confidence interval estimate of their mean weight. Are the confidence interval limits very different from the limits of 1.56 and 1.87 that are found when assuming that σ is known to be 0.30 g?

1.7 1.6 1.5 2.0 2.3 1.6 1.6 1.8 1.5 1.7 2.2 1.4 1.6 1.6 1.6

TI **25.** Estimating Car Pollution In a sample of seven cars, each car was tested for nitrogen-oxide emissions (in grams per mile) and the following results were obtained: 0.06, 0.11, 0.16, 0.15, 0.14, 0.08, 0.15 (based on data from the Environmental Protection Agency). Assuming that this sample is representative of the cars in use, construct a 98% confidence interval estimate of the mean amount of nitrogen-oxide emissions for all cars. If the Environmental Protection Agency requires that nitrogen-oxide emissions be less than 0.165 g/mi, can we safely conclude that this requirement is being met?

TI **26.** Skull Breadths Maximum breadths of samples of male Egyptian skulls from 4000 B.C. and 150 A.D. (based on data from *Ancient Races of the Thebaid* by Thomson and Randall-Maciver):

4000 B.C.:	131	119	138	125	129	126	131	132	126	128	128	131
150 A.D.:	136	130	126	126	139	141	137	138	133	131	134	129

Changes in head sizes over time suggest interbreeding with people from other regions. Use confidence intervals to determine whether the head sizes appear to have changed from 4000 B.C. to 150 A.D. Explain your result.

Appendix B Data Sets. In Exercises 27 and 28, use the data sets from Appendix B.

TI **27.** Pulse Rates A physician wants to develop criteria for determining whether a patient's pulse rate is atypical, and she wants to determine whether there are significant differences between males and females. Use the sample pulse rates in Data Set 1 from Appendix B.

a. Construct a 95% confidence interval estimate of the mean pulse rate for males.

b. Construct a 95% confidence interval estimate of the mean pulse rate for females.

c. Compare the preceding results. Can we conclude that the population means for males and females are different? Why or why not?

TI **28.** Comparing Regular and Diet Pepsi Refer to Data Set 12 in Appendix B and use the sample data.

a. Construct a 95% confidence interval estimate of the mean weight of cola in cans of regular Pepsi.

b. Construct a 95% confidence interval estimate of the mean weight of cola in cans of Diet Pepsi.

c. Compare the results from parts (a) and (b) and interpret them. Does there appear to be a difference? If so, identify a reason for the difference.

7-4 BEYOND THE BASICS

TI **29.** Effect of an Outlier Test the effect of an outlier as follows: Use the sample data from Exercise 22 to find a 95% confidence interval estimate of the population mean, after changing the last age from 54 years to 540 years. This value is not realistic, but such an error can easily occur during a data entry process. Does the confidence interval change much when 54 years is changed to 540 years? Are confidence interval limits sensitive to outliers? How should you handle outliers when they are found in sample data sets that will be used for the construction of confidence intervals?

TI **30.** Alternative Method Figure 7-6 and Table 7-1 summarize the decisions made when choosing between the normal and t distributions. An alternative method included in some textbooks (but almost never included in professional journals) is based on this criterion: Substitute the sample standard deviation s for σ whenever $n > 30$, then proceed as if σ is known. Assume that for a simple random sample, $n = 35$, $\bar{x} = 50.0$, and $s = 10.0$, then construct 95% confidence interval estimates of μ using the method of this section and using the alternative method. Compare the results.

TI **31.** Finite Population Correction Factor If a simple random sample of size n is selected without replacement from a finite population of size N, and the sample size is more than 5% of the population size ($n > 0.05N$), better results can be obtained by using the finite population correction factor, which involves multiplying the margin of error E by $\sqrt{(N - n)/(N - 1)}$. For the sample of 100 weights of M&M candies in Data Set 13 from Appendix B, we get $\bar{x} = 0.8565$ g and $s = 0.0518$ g. First construct a 95% confidence interval estimate of μ assuming that the population is large, then construct a 95% confidence interval estimate of the mean weight of M&Ms in the full bag from which the sample was taken. The full bag has 465 M&Ms. Compare the results.

32. Using the Wrong Distribution Assume that a small simple random sample is selected from a normally distributed population for which σ is unknown. Construction of a confidence interval should use the t distribution, but how are the confidence interval limits affected if the normal distribution is incorrectly used instead?

33. Confidence Interval for Sample of Size $n = 1$ When a manned NASA spacecraft lands on Mars, the astronauts encounter a single adult Martian, who is found to be 12.0 ft tall. It is reasonable to assume that the heights of all Martians are normally distributed.

a. The methods of this chapter require information about the variation of a variable. If only one sample value is available, can it give us any information about the variation of the variable?

b. When using the methods of this section, what happens when you try to use the single height in constructing a 95% confidence interval?

c. Based on the article "An Effective Confidence Interval for the Mean with Samples of Size One and Two," by Wall, Boen, and Tweedie (*American Statistician,* Vol. 55, No. 2), a 95% confidence interval for μ can be found (using methods not discussed in this book) for a sample of size $n = 1$ randomly selected from a normally distributed population, and it can be expressed as $x \pm 9.68|x|$. Use this result to construct a 95% confidence interval using the single sample value of 12.0 ft, and express it in the format of $\bar{x} - E < \mu < \bar{x} + E$. Based on the result, is it likely that some other randomly selected Martian might be 50 ft tall?

7-5 Estimating a Population Variance

Key Concept This section presents methods for (1) finding a confidence interval estimate of a population standard deviation or variance and (2) determining the sample size required to estimate a population standard deviation or variance. In this section we introduce the chi-square distribution, which is used for finding a confidence interval estimate of σ or σ^2.

Requirements

1. The sample is a simple random sample.
2. The population must have normally distributed values (even if the sample is large).

The assumption of a normally distributed population was made in earlier sections, but that requirement is much more critical here. For the methods of this section, departures from normal distributions can lead to gross errors. Consequently, the requirement of having a normal distribution is much stricter, and we should check the distribution of data by constructing histograms and normal quantile plots, as described in Section 6-7.

When we considered estimates of proportions and means, we used the normal and Student t distributions. When developing estimates of variances or standard deviations, we use another distribution, referred to as the chi-square distribution. We will examine important features of that distribution before proceeding with the development of confidence intervals.

Chi-Square Distribution

In a normally distributed population with variance σ^2, assume that we randomly select independent samples of size n and, for each sample, compute the sample variance s^2 (which is the square of the sample standard deviation s). The sample statistic $\chi^2 = (n - 1)s^2/\sigma^2$ has a sampling distribution called the **chi-square distribution.**

Chi-Square Distribution

Formula 7-7 $$\chi^2 = \frac{(n - 1)s^2}{\sigma^2}$$

where n = sample size
s^2 = sample variance
σ^2 = population variance

We denote chi-square by χ^2, pronounced "kigh square." To find critical values of the chi-square distribution, refer to Table A-4. The chi-square distribution is determined by the number of degrees of freedom, and in this chapter we use $n - 1$ degrees of freedom.

degrees of freedom = $n - 1$

In later chapters we will encounter situations in which the degrees of freedom are not $n - 1$, so we should not make the incorrect generalization that the number of degrees of freedom is always $n - 1$.

Properties of the Distribution of the Chi-Square Statistic

1. The chi-square distribution is not symmetric, unlike the normal and Student t distributions (see Figure 7-8). (As the number of degrees of freedom

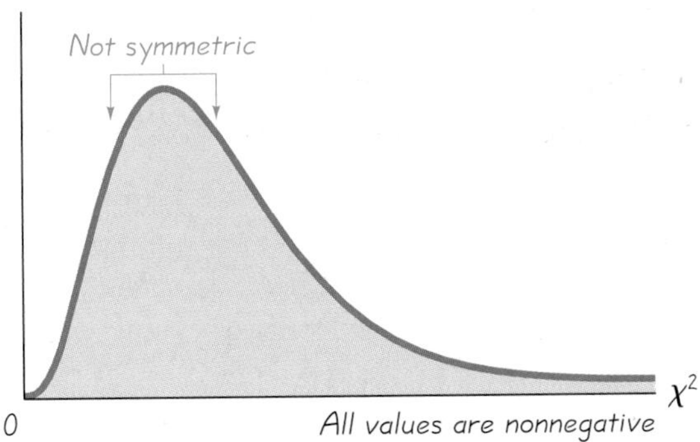

Figure 7-8

Chi-Square Distribution

increases, the distribution becomes more symmetric, as Figure 7-9 illustrates.)

2. The values of chi-square can be zero or positive, but they cannot be negative (see Figure 7-8).
3. The chi-square distribution is different for each number of degrees of freedom (see Figure 7-9), and the number of degrees of freedom is given by df $= n - 1$ in this section. As the number of degrees of freedom increases, the chi-square distribution approaches a normal distribution.

Because the chi-square distribution is skewed instead of symmetric, the confidence interval does not fit a format of $s^2 \pm E$ and we must do separate calculations for the upper and lower confidence interval limits. If using Table A-4 for finding critical values, note the following feature of that table:

In Table A-4, each critical value of χ^2 corresponds to an area given in the top row of the table, and that area represents the *cumulative area located to the right* of the critical value.

Table A-2 for the standard normal distribution provides cumulative areas from the *left*, but Table A-4 for the chi-square distribution provides cumulative areas from the *right*.

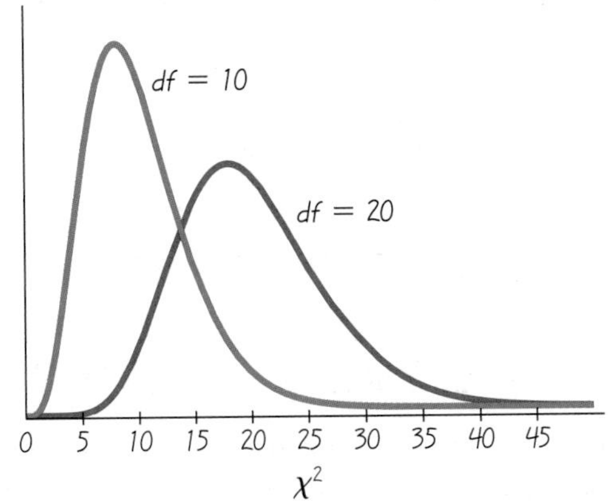

Figure 7-9

Chi-Square Distribution for df = 10 and df = 20

EXAMPLE Critical Values Find the critical values of χ^2 that determine critical regions containing an area of 0.025 in each tail. Assume that the relevant sample size is 10 so that the number of degrees of freedom is 10 − 1, or 9.

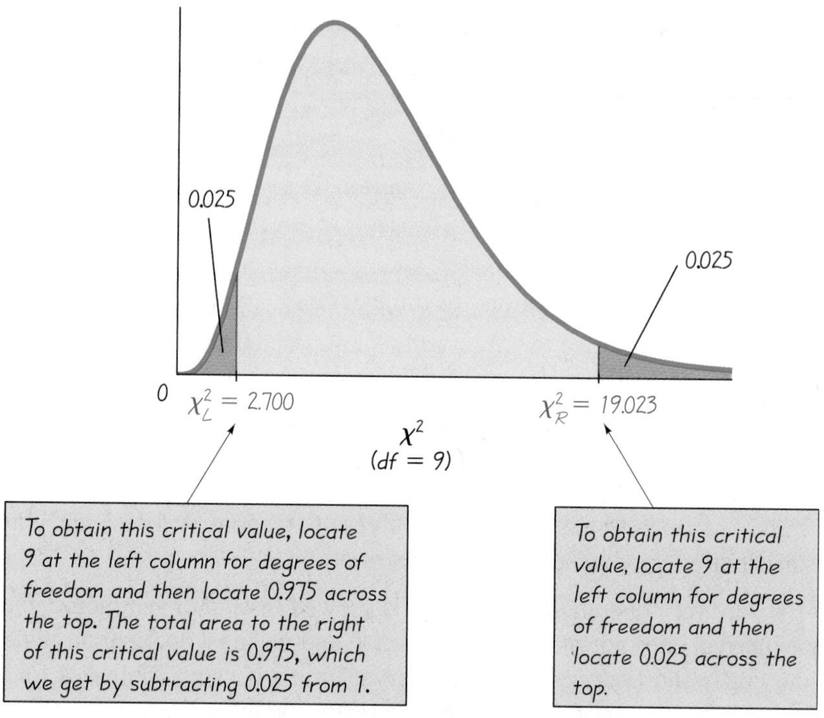

Figure 7-10 Critical Values of the Chi-Square Distribution

SOLUTION See Figure 7-10 and refer to Table A-4. The critical value to the right ($\chi^2 = 19.023$) is obtained in a straightforward manner by locating 9 in the degrees-of-freedom column at the left and 0.025 across the top. The critical value of $\chi^2 = 2.700$ to the left once again corresponds to 9 in the degrees-of-freedom column, but we must locate 0.975 (found by subtracting 0.025 from 1) across the top because the values in the top row are always *areas to the right* of the critical value. Refer to Figure 7-10 and see that the total area to the right of $\chi^2 = 2.700$ is 0.975. Figure 7-10 shows that, for a sample of 10 values taken from a normally distributed population, the chi-square statistic $(n - 1)s^2/\sigma^2$ has a 0.95 probability of falling between the chi-square critical values of 2.700 and 19.023.

When obtaining critical values of χ^2 from Table A-4, note that the numbers of degrees of freedom are consecutive integers from 1 to 30, followed by 40, 50, 60, 70, 80, 90, and 100. When a number of degrees of freedom (such as 52) is not found in the table, you can usually use the closest critical value. For example, if the number of degrees of freedom is 52, refer to Table A-4 and use 50 degrees of freedom. (If the number of degrees of freedom is exactly midway between table values, such as 55, simply find the mean of the two χ^2 values.) For numbers of degrees of freedom greater than 100, use the equation given in Exercise 27, or a more detailed table, or a statistical software package.

Estimators of σ^2

In Section 6-4 we showed that sample variances s^2 tend to target (or center on) the value of the population variance σ^2, so we say that s^2 is an *unbiased estimator* of σ^2. That is, sample variances s^2 do not systematically tend to overestimate the value of σ^2, nor do they systematically tend to underestimate σ^2. Instead, they tend to target the value of σ^2 itself. Also, the values of s^2 tend to produce smaller errors by being closer to σ^2 than do other unbiased measures of variation. For these reasons, s^2 is generally used to estimate σ^2. [However, there are other estimators of σ^2 that could be considered better than s^2. For example, even though $(n-1)s^2/(n+1)$ is a biased estimator of σ^2, it has the very desirable property of minimizing the mean of the squares of the errors and therefore has a better chance of being closer to σ^2. See Exercise 28.]

The sample variance s^2 is the best point estimate of the population variance σ^2.

Because s^2 is an unbiased estimator of σ^2, we might expect that s would be an unbiased estimator of σ, but this is not the case. (See Section 6-4.) If the sample size is large, however, the bias is small so that we can use s as a reasonably good estimate of σ. Even though it is a biased estimate, s is often used as a point estimate of σ.

The sample standard deviation s is commonly used as a point estimate of σ (even though it is a biased estimate).

Although s^2 is the best point estimate of σ^2, there is no indication of how good it actually is. To compensate for that deficiency, we develop an interval estimate (or confidence interval) that is more informative.

Confidence Interval (or Interval Estimate) for the Population Variance σ^2

$$\frac{(n-1)s^2}{\chi_R^2} < \sigma^2 < \frac{(n-1)s^2}{\chi_L^2}$$

This expression is used to find a confidence interval for the variance σ^2, but the confidence interval (or interval estimate) for the standard deviation σ is found by taking the square root of each component, as shown below.

$$\sqrt{\frac{(n-1)s^2}{\chi_R^2}} < \sigma < \sqrt{\frac{(n-1)s^2}{\chi_L^2}}$$

The notations χ_R^2 and χ_L^2 in the preceding expressions are described as follows. (Note that some other texts use $\chi_{\alpha/2}^2$ in place of χ_R^2 and they use $\chi_{1-\alpha/2}^2$ in place of χ_L^2.)

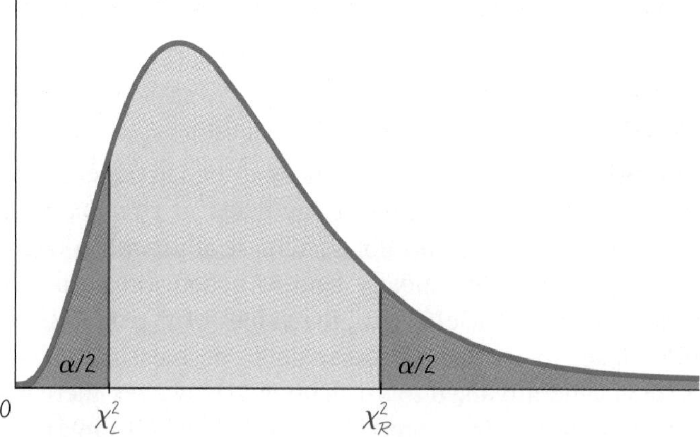

Figure 7-11

Chi-Square Distribution with Critical Values χ^2_L and χ^2_R

The critical values χ^2_L and χ^2_R separate the extreme areas corresponding to sample variances that are unlikely (with probability α).

Notation

With a total area of α divided equally between the two tails of a chi-square distribution, χ^2_L denotes the left-tailed critical value and χ^2_R denotes the right-tailed critical value (as illustrated in Figure 7-11).

Based on the preceding results, we can summarize the procedure for constructing a confidence interval estimate of σ or σ^2 as follows.

Procedure for Constructing a Confidence Interval for σ or σ^2

1. Verify that the requirements are satisfied. (The sample is a simple random sample and a histogram or normal quantile plot suggests that the population has a distribution that is very close to a normal distribution.)
2. Using $n - 1$ degrees of freedom, refer to Table A-4 and find the critical values χ^2_R and χ^2_L that correspond to the desired confidence level.
3. Evaluate the upper and lower confidence interval limits using this format of the confidence interval:

$$\frac{(n-1)s^2}{\chi^2_R} < \sigma^2 < \frac{(n-1)s^2}{\chi^2_L}$$

4. If a confidence interval estimate of σ is desired, take the square root of the upper and lower confidence interval limits and change σ^2 to σ.
5. Round the resulting confidence interval limits. If using the original set of data, round to one more decimal place than is used for the original set of data. If using the sample standard deviation or variance, round the confidence interval limits to the same number of decimal places.

Caution: Confidence intervals can be used informally to compare different data sets, but the overlapping of confidence intervals should not be used for making formal and final conclusions about equality of variances or standard deviations. Later chapters will include procedures for deciding whether two populations have equal variances or standard deviations, and those methods will not have the pitfalls associated with comparisons based on the overlap of confidence intervals.

Do not use the overlapping of confidence intervals as the basis for making definitive conclusions about the equality of variances or standard deviations.

EXAMPLE Confidence Interval for Penny Weights Pennies are currently being minted with a standard deviation of 0.0165 g (based on data from Data Set 14 in Appendix B). New equipment is being tested in an attempt to improve quality by reducing variation. A simple random sample of 10 pennies is obtained from pennies manufactured with the new equipment. A normal quantile plot and histogram show that the weights are from a normally distributed population, and the sample has a standard deviation of 0.0125 g. Use the sample results to construct a 95% confidence interval estimate of σ, the standard deviation of the weights of pennies made with the new equipment. Based on the results, does the new equipment appear to be effective in reducing the variation of the weights?

SOLUTION

REQUIREMENT ✔ We first verify that the requirements are satisfied. It was noted that the sample is a simple random sample. Based on the descriptions of the histogram and normal quantile plot, the requirement of a normal distribution is also satisfied. Both requirements are therefore satisfied. (This check of requirements is Step 1 in the process of finding a confidence interval of σ, so we proceed next with Step 2.) ✔

Step 2: Using $n - 1$ degrees of freedom, we now find the critical values of χ^2. The sample size is $n = 10$, so there are 9 degrees of freedom. We refer to Table A-4 for the row corresponding to 9 degrees of freedom, and we refer to the columns with areas of 0.975 and 0.025. (For a 95% confidence level, we divide $\alpha = 0.05$ equally between the two tails of the chi-square distribution, and we refer to the values of 0.975 and 0.025 across the top row of Table A-4.) The critical values of χ^2 are $\chi^2_L = 2.700$ and $\chi^2_R = 19.023$. (See Figure 7-10.)

Step 3: Using the critical values of 2.700 and 19.023, the sample standard deviation of $s = 0.0125$, and the sample size of 10, we construct the 95% confidence interval by evaluating the following:

$$\frac{(10 - 1)(0.0125)^2}{19.023} < \sigma^2 < \frac{(10 - 1)(0.0125)^2}{2.700}$$

Step 4: Evaluating the above expression results in $0.0000739237 < \sigma^2 < 0.000520833$. Finding the square root of each part (before rounding), then rounding to four decimal places yields $0.0086 \text{ g} < \sigma < 0.0228 \text{ g}$.

INTERPRETATION Based on this result, we have 95% confidence that the limits of 0.0086 g and 0.0228 g contain the true value of σ. Note that this interval includes the standard deviation of 0.0165 g for weights of pennies currently

continued

being manufactured. It does not appear that the new equipment significantly reduces the variation. Even though the standard deviation of the sample (0.0125 g) is less than the current standard deviation of 0.0165 g, it is not low enough to be significant. Based on the available data, the new equipment does not appear to be effective.

The confidence interval $0.0086 < \sigma < 0.0228$ can also be expressed as (0.0086, 0.0228), but the format of $s \pm E$ *cannot* be used because the confidence interval does not have s at its center.

Rationale We now explain why the confidence intervals for σ and σ^2 have the forms just given. If we obtain samples of size n from a population with variance σ^2, the distribution of the $(n-1)s^2/\sigma^2$ values will be as shown in Figure 7-11. For a simple random sample, there is a probability of $1-\alpha$ that the statistic $(n-1)s^2/\sigma^2$ will fall between the critical values of χ_L^2 and χ_R^2. In other words (and symbols), there is a $1-\alpha$ probability that both of the following are true:

$$\frac{(n-1)s^2}{\sigma^2} < \chi_R^2 \quad \text{and} \quad \frac{(n-1)s^2}{\sigma^2} > \chi_L^2$$

If we multiply both of the preceding inequalities by σ^2 and divide each inequality by the appropriate critical value of χ^2, we see that the two inequalities can be expressed in the equivalent forms:

$$\frac{(n-1)s^2}{\chi_R^2} < \sigma^2 \quad \text{and} \quad \frac{(n-1)s^2}{\chi_L^2} > \sigma^2$$

These last two inequalities can be combined into one inequality:

$$\frac{(n-1)s^2}{\chi_R^2} < \sigma^2 < \frac{(n-1)s^2}{\chi_L^2}$$

Determining Sample Size

The procedures for finding the sample size necessary to estimate σ^2 are much more complex than the procedures given earlier for means and proportions. Instead of using very complicated procedures, we will use Table 7-2. STATDISK also provides sample sizes.

EXAMPLE We want to estimate σ, the standard deviation of all body temperatures. We want to be 95% confident that our estimate is within 10% of the true value of σ. How large should the sample be? Assume that the population is normally distributed.

SOLUTION From Table 7-2, we can see that 95% confidence and an error of 10% for σ correspond to a sample of size 191. We should randomly select 191 values from the population of body temperatures.

Table 7-2

Sample Size for σ^2		Sample Size for σ	
To be 95% confident that s^2 is within	of the value of σ^2, the sample size n should be at least	To be 95% confident that s is within	of the value of σ, the sample size n should be at least
1%	77,207	1%	19,204
5%	3,148	5%	767
10%	805	10%	191
20%	210	20%	47
30%	97	30%	20
40%	56	40%	11
50%	37	50%	7
To be 99% confident that s^2 is within	of the value of σ^2, the sample size n should be at least	To be 99% confident that s is within	of the value of σ, the sample size n should be at least
1%	133,448	1%	33,218
5%	5,457	5%	1,335
10%	1,401	10%	335
20%	368	20%	84
30%	171	30%	37
40%	100	40%	21
50%	67	50%	13

TI

Using the TI-83/84 Plus Calculator for Confidence Interval Estimates for σ or σ^2

The TI-83/84 Plus calculator is not yet designed to automatically generate confidence intervals for the population standard deviation σ or the population variance σ^2, but the program **S2INT** can be used. That program was written by Michael Lloyd of Henderson State University, and it is on the CD-ROM included with this book or it can be downloaded from www.aw.com/triola. The program S2INT uses the program ZZINEWT, so that program must also be installed. Use the following procedure:

1. Use the enclosed CD-ROM or the Web site www.aw.com/Triola to download the programs S2INT and ZZINEWT. (Download instructions can be found at the site www.ti.com, and they typically involve the use of software such as TI GRAPHLINK or TI CONNECT.)
2. Press PRGM, select the program **S2INT,** and proceed to enter the sample variance s^2, the sample size n, and the confidence level (such as 0.95). Press ENTER, and wait a while (such as 60 seconds) for the display of the confidence interval limits for σ^2 and σ.

Entering Data

Confidence Interval Results

The accompanying screens result from a sample with $n = 25$ and $s = 3.0$. Note that the program requires entry of the sample variance s^2. The first screen shows the entry of the sample statistics ($s^2 = 3.0^2$ and $n = 25$) and the confidence level of 0.95. The second screen shows the resulting confidence intervals. *Note:* You may have to press ENTER and/or press ▷ to see all of the results. In the second screen, for example, you must press ▷ to scroll to the right in order to see the value of the upper confidence interval limits. Also, the program initially displayed the confidence interval for σ^2, but the confidence interval for σ can be seen by pressing ENTER. Based on the calculator results, the following 95% confidence intervals are obtained:

$$5.5 < \sigma^2 < 17.4$$
$$2.3 < \sigma < 4.2$$

Using Technology for Sample Sizes for Estimating σ or σ^2

The TI-83/84 Plus calculator is not yet designed to automatically generate sample sizes required to estimate the population standard deviation σ or the population variance σ^2, but STATDISK can be used instead. With STATDISK, select **Analysis** from the main menu at the top, then select **Sample Size Determination** and select the option of **Estimate St Dev.** Proceed to enter the confidence level and the percent margin of error (as in Table 7-2), then click **Evaluate.**

7-5 BASIC SKILLS AND CONCEPTS

Statistical Literacy and Critical Thinking

1. Interpreting Confidence Interval Using the heights of females listed in Data Set 1 from Appendix B, we use the standard deviation of the sample ($s = 2.741$ in.) to obtain the following 95% confidence interval estimate of the standard deviation of the heights of all females: 2.25 in. $< \sigma <$ 3.52 in. Write a statement that correctly interprets that confidence interval.

2. Expressing Confidence Intervals The confidence interval given in Exercise 1 can also be expressed as (2.25, 3.52), but it cannot be expressed as 2.885 ± 0.635. Given that 2.885 ± 0.635 results in values of 2.25 and 3.52, why can't we express the confidence interval as 2.885 ± 0.635?

3. Interpreting Confidence Intervals For each of the 50 states, a researcher obtains a random sample of credit card debt and calculates the mean to obtain 50 representative values. She then uses the 50 sample means to construct a confidence interval estimate of σ. Is the result an estimate of the standard deviation of credit card debts for all statistics students in the United States? Why or why not?

4. Unbiased Estimators What is an unbiased estimator? Is the sample variance an unbiased estimator of the population variance? Is the sample standard deviation an unbiased estimator of the population standard deviation?

Finding Critical Values. *In Exercises 5–8, find the critical values* χ^2_L *and* χ^2_R *that correspond to the given confidence level and sample size.*

5. 95%; $n = 27$

6. 95%; $n = 7$

7. 99%; $n = 41$

8. 90%; $n = 91$

Finding Confidence Intervals. *In Exercises 9–12, use the given confidence level and sample data to find a confidence interval for the population standard deviation σ. In each case, assume that a simple random sample has been selected from a population that has a normal distribution.*

TI **9.** Salaries of college graduates who took a statistics course in college: 95% confidence; $n = 41, \bar{x} = \$67{,}200, s = \$18{,}277$.

TI **10.** Speeds of drivers ticketed in a 55 mi/h zone: 95% confidence; $n = 90$, $\bar{x} = 66.2$ mi/h, $s = 3.4$ mi/h.

TI **11.** FICO (Fair, Isaac, and Company) credit rating scores of applicants for credit cards: 99% confidence; $n = 70, \bar{x} = 688, s = 68$.

TI **12.** Amounts lost by gamblers who took a bus to an Atlantic City casino: 99% confidence; $n = 40, \bar{x} = \$189, s = \$87$.

Determining Sample Size. *In Exercises 13–16, assume that each sample is a simple random sample obtained from a normally distributed population.*

13. Find the minimum sample size needed to be 95% confident that the sample standard deviation s is within 5% of σ.

14. Find the minimum sample size needed to be 95% confident that the sample standard deviation s is within 20% of σ.

15. Find the minimum sample size needed to be 99% confident that the sample variance is within 10% of the population variance. Is such a sample size practical in most cases?

16. Find the minimum sample size needed to be 95% confident that the sample variance is within 30% of the population variance.

Finding Confidence Intervals. *In Exercises 17–24, assume that each sample is a simple random sample obtained from a population with a normal distribution.*

TI **17.** Birth Weights In a study of the effects of prenatal cocaine use on infants, the following sample data were obtained for weights at birth: $n = 190, \bar{x} = 2700$ g, $s = 645$ g (based on data from "Cognitive Outcomes of Preschool Children with Prenatal Cocaine Exposure," by Singer et al., *Journal of the American Medical Association,* Vol. 291, No. 20). Use the sample data to construct a 95% confidence interval estimate of the standard deviation of all birth weights of infants born to mothers who used cocaine during pregnancy. (Because Table A-4 has a maximum of 100 degrees of freedom while we require 189 degrees of freedom, use these critical values obtained from STATDISK: $\chi^2_L = 152.8222$ and $\chi^2_R = 228.9638$.) Based on the result, does the standard deviation appear to be different from the standard deviation of 696 g for birth weights of babies born to mothers who did not use cocaine during pregnancy?

TI **18.** Minting Quarters Quarters are currently minted with weights having a mean of 5.670 g and a standard deviation of 0.062 g. New equipment is being tested in an attempt to improve quality by reducing variation. A simple random sample of 24 quarters is obtained from those manufactured with the new equipment, and this sample has a standard deviation of 0.049 g. Use the sample results to construct a 95% confidence interval estimate of σ, the standard deviation of the weights of quarters made with the new equipment. Based on the confidence interval, does the new equipment appear to produce a standard deviation that is clearly lower than the standard deviation of 0.062 g from the old equipment? Based on the results, does the new equipment appear to be effective in reducing the variation of the weights?

TI **19.** Body Temperature Data Set 2 in Appendix B includes 106 body temperatures for which $\bar{x} = 98.20°F$ and $s = 0.62°F$. Using the sample statistics, construct a 99% confidence interval estimate of the standard deviation of body temperature of all healthy humans. Based on the result, can we safely conclude that the population standard deviation is less than 2.10°F? (If the population standard deviation is 2.10 or greater, the variation is large enough so that the sample mean of 98.20°F does not differ from 98.60°F by a significant amount.)

TI **20.** Shoveling Heart Rates Because cardiac deaths appear to increase after heavy snowfalls, an experiment was designed to compare cardiac demands of snow shoveling to those of using an electric snow thrower. Ten subjects cleared tracts of snow using both methods, and their maximum heart rates (beats per minute) were recorded during both activities. The results shown below were obtained (based on data from "Cardiac Demands of Heavy Snow Shoveling," by Franklin et al., *Journal of the American Medical Association,* Vol. 273, No. 11).

Manual snow shoveling maximum heart rates: $n = 10, \bar{x} = 175, s = 15.$

Electric snow thrower maximum heart rates: $n = 10, \bar{x} = 124, s = 18.$

a. Construct a 95% confidence interval estimate of the population standard deviation σ for those who did manual snow shoveling.

b. Construct a 95% confidence interval estimate of the population standard deviation σ for those who used the automated electric snow thrower.

c. Compare the results. Does the variation appear to be different for the two groups?

TI **21.** Credit Rating When consumers apply for credit, their credit is rated using FICO (Fair, Isaac, and Company) scores. Credit ratings are given below for a sample of applicants for car loans. Use the sample data to construct a 99% confidence interval for the standard deviation of FICO scores for all applicants for credit.

661 595 548 730 791 678 672 491 492 583 762 624 769 729 734 706

TI **22.** World's Smallest Mammal The world's smallest mammal is the bumblebee bat, also known as the Kitti's hog-nosed bat (or *Craseonycteris thonglongyai*). Such bats are roughly the size of a large bumblebee. Listed below are weights (in grams) from a sample of these bats. Construct a 95% confidence interval estimate of the standard deviation of weights for all such bats.

1.7 1.6 1.5 2.0 2.3 1.6 1.6 1.8 1.5 1.7 2.2 1.4 1.6 1.6 1.6

TI **23.** Monitoring Lead in Air Listed below are measured amounts of lead (in micrograms per cubic meter, or $\mu g/m^3$) in the air. The Environmental Protection Agency has established an air quality standard for lead of 1.5 $\mu g/m^3$. The measurements shown below were recorded at Building 5 of the World Trade Center site on different days immediately following the destruction caused by the terrorist attacks of September 11, 2001. After the collapse of the two World Trade Center buildings, there was considerable concern about the quality of the air. Use the given values to construct a 95% confidence interval estimate of the standard deviation of the amounts of lead in the air. Is there anything about this data set suggesting that the confidence interval might not be very good? Explain.

5.40 1.10 0.42 0.73 0.48 1.10

TI **24. a.** Comparing Waiting Lines The listed values are waiting times (in minutes) of customers at the Jefferson Valley Bank, where customers enter a single waiting line that feeds three teller windows. Construct a 95% confidence interval for the population standard deviation σ.

6.5 6.6 6.7 6.8 7.1 7.3 7.4 7.7 7.7 7.7

b. The listed values are waiting times (in minutes) of customers at the Bank of Providence, where customers may enter any one of three different lines that have formed at three teller windows. Construct a 95% confidence interval for the population standard deviation σ.

4.2 5.4 5.8 6.2 6.7 7.7 7.7 8.5 9.3 10.0

c. Interpret the results found in parts (a) and (b). Do the confidence intervals suggest a difference in the variation among waiting times? Which arrangement seems better: the single-line system or the multiple-line system?

TI **25.** Appendix B Data for Body Mass Index (BMI) Refer to Data Set 1 in Appendix B and use the sample data.

a. Construct a 99% confidence interval estimate of the standard deviation of BMIs for men.

b. Construct a 99% confidence interval estimate of the standard deviation of BMIs for women.

c. Compare and interpret the results.

TI **26.** Appendix B Data for Weights of Quarters Refer to Data Set 14 in Appendix B and use the sample data.

a. Construct a 99% confidence interval estimate of the standard deviation of weights of quarters made after 1964.

b. Construct a 99% confidence interval estimate of the standard deviation of weights of silver quarters made before 1964.

c. Compare and interpret the results.

7-5 BEYOND THE BASICS

27. Finding Critical Values In constructing confidence intervals for σ or σ^2, we use Table A-4 to find the critical values χ_L^2 and χ_R^2, but that table applies only to cases in which $n \leq 101$, so the number of degrees of freedom is 100 or smaller. For larger numbers of degrees of freedom, we can approximate χ_L^2 and χ_R^2 by using

$$\chi^2 = \frac{1}{2}\left[\pm z_{\alpha/2} + \sqrt{2k - 1}\right]^2$$

where k is the number of degrees of freedom and $z_{\alpha/2}$ is the critical z score first described in Section 7-2. STATDISK was used to find critical values for 189 degrees of freedom with a confidence level of 95%, and those critical values are given in Exercise 17. Use the approximation shown here to find the critical values and compare the results to those found from STATDISK.

TI **28.** Finding the Best Estimator We noted that values of s^2 tend to produce smaller errors by being closer to σ^2 than do other unbiased measures of variation. Let's now consider the biased estimator of $(n - 1)s^2/(n + 1)$. Given the population of values {2, 3, 7}, use the value of σ^2, and use the nine different possible samples of size $n = 2$ (with sampling done with replacement) for the following.

a. Find s^2 for each of the nine samples, then find the error $s^2 - \sigma^2$ for each sample, then square those errors, then find the mean of those squares. The result is the value of the mean square error.

b. Find $(n - 1)s^2/(n + 1)$ for each of the nine samples, then find the error $(n - 1)s^2/(n + 1) - \sigma^2$ for each sample, then square those errors, then find the mean of those squares. The result is the mean square error.

c. The mean square error can be used to measure how close an estimator comes to the population parameter. Which estimator does a better job by producing the smaller mean square error? Is that estimator biased or unbiased?

Review

The two main activities of inferential statistics are estimating population parameters and testing claims made about population parameters. In this chapter we introduced basic methods for finding *estimates* of population proportions, means, and variances and developed procedures for finding each of the following:

- point estimate
- confidence interval
- required sample size

We discussed point estimate (or single-valued estimate) and formed these conclusions:

- Proportion: The best point estimate of p is $\hat{p}$
- Mean: The best point of estimate of μ is $\bar{x}$.
- Variation: The value of s is commonly used as a point estimate of σ, even though it is a biased estimate. Also, s^2 is the best point estimate of σ^2.

Because the above point estimates consist of single values, they have the serious disadvantage of not revealing how good they are, so confidence intervals (or interval estimates) are commonly used as more revealing and useful estimates. We also considered ways of determining the sample sizes necessary to estimate parameters to within given margins of error. This chapter also introduced the Student t and chi-square distributions. We must be careful to use the correct probability distribution for each set of circumstances. This chapter used the following criteria for selecting the appropriate distribution:

Confidence interval for proportion p:	Use the *normal* distribution (assuming that the required conditions are satisfied and there are at least 5 successes and at least 5 failures so that the normal distribution can be used to approximate the binomial distribution).
Confidence interval for μ:	See Figure 7-6 or Table 7-1 to choose between the *normal* or t distributions (or conclude that neither applies).

Confidence interval for σ or σ^2: Use the *chi-square* distribution (assuming that the required conditions are satisfied).

For the confidence interval and sample size procedures in this chapter, it is very important to verify that the requirements are satisfied. If they are not, then we cannot use the methods of this chapter and we may need to use other methods, such as the bootstrap method described in the TI-83/84 Plus Calculator Project at the end of this chapter or nonparametric methods, such as those discussed in Chapter 13.

This chapter also introduced the concept of critical values. It is important to learn how to obtain critical values for the normal, *t*, and chi-square distributions, because they will be used in the following chapters.

Statistical Literacy and Critical Thinking

1. Critical Values When working with a normal distribution, the critical value of $z = 1.96$ is obtained for a 95% confidence level. What is the relationship between $z = 1.96$ and the confidence level of 95%?

2. Confidence Interval When trying to estimate the mean weight of garbage discarded by households in one week, we obtain these sample results: $n = 62$, $\bar{x} = 27.44$ lb, $s = 12.46$ lb. Knowing that the sample mean is an unbiased estimator of the population mean, we correctly conclude that our best estimate of μ is 27.44 lb. Why do we then need a confidence interval? What does the confidence interval tell us that is missing from the estimate of 27.44 lb?

3. Margin of Error A newspaper reports survey results by stating that "65% of those surveyed favored the proposition, with a margin of error of ± 3 percentage points." What confidence interval is suggested by that statement?

4. Interpreting Confidence Interval Use the confidence interval found in Exercise 3. The media often fail to report the confidence level, but assuming that the confidence interval was found using a 95% confidence level, write a statement that correctly interprets the confidence interval.

Review Exercises

TI 1. Alcohol Service Policy In a Gallup poll of 1004 adults, 93% indicated that restaurants and bars should refuse service to patrons who have had too much to drink. Construct a 95% confidence interval estimate of the percentage of all adults who believe that restaurants and bars should refuse service for those who have had too much to drink. Write a statement interpreting the confidence interval.

2. Determining Sample Size for a Survey See the survey described in Exercise 1. If you plan to conduct a new poll to confirm that the percentage continues to be correct, how many randomly selected adults must you survey if you want 95% confidence that the margin of error is four percentage points?

TI 3. Vending Machine Specifications for vending machines made by the Newton Machine Company require that they dispense amounts of coffee having a mean of 12 oz. Listed below are amounts of coffee (in ounces) randomly selected from different machines. Use these sample results to construct a 95% confidence interval for the mean amount of coffee in all dispensed cups. Does this confidence interval suggest that the machines are working properly? Is there anything else about the data suggesting that there is a problem?

11.5	10.8	9.7	13.0	11.5	11.1	$n = 16$
13.2	11.1	11.1	12.6	6.8	11.4	$\bar{x} = 10.89$
9.0	10.6	10.1	10.8			$s = 1.56$

TI **4.** Confidence Interval for σ Use the same sample data from Exercise 3 to construct a 95% confidence interval estimate of σ. New specifications are being considered to control the variation of the amounts of coffee that are dispensed. We want almost all of the dispensed amounts to be within 0.5 oz of 12 oz and, using the range rule of thumb, this suggests that the standard deviation should be 0.25 oz. Based on the confidence interval, is 0.25 oz a feasible value of the population standard deviation? Does the machine require modifications to reduce the variation?

5. Sample Size You have been hired by a consortium of dairy farmers to conduct a survey about the consumption of milk.

a. If you want to estimate the percentage of adults who drink milk daily, how many adults must you survey if you want 95% confidence that your sample percentage is in error by no more than two percentage points?

b. If you want to estimate the mean amount of milk consumed daily by adults, how many adults must you survey if you want 95% confidence that your sample mean is in error by no more than 0.5 oz? (Based on results from a pilot study, assume that $\sigma = 8.7$ oz.)

c. If you plan to obtain the estimates described in parts (a) and (b) with a single survey having several questions, how many people must be surveyed?

TI **6.** Estimating Length of Car Ownership A NAPA Auto Parts supplier wants information about how long car owners plan to keep their cars. A simple random sample of 25 car owners results in $\bar{x} = 7.01$ years and $s = 3.74$ years, respectively (based on data from a Roper poll). Assume that the sample is drawn from a normally distributed population.

a. Find a 95% confidence interval estimate of the population mean.

b. Find a 95% confidence interval estimate of the population standard deviation.

c. If several years pass and you want to conduct a new survey to estimate the mean length of time that car owners plan to keep their cars, how many randomly selected car owners must you survey? Assume that you want 99% confidence that the sample mean is within 0.25 year (or 3 months) of the population mean, and also assume that $\sigma = 3.74$ years (based on the latest result).

d. When conducting the survey described in part (c), you find that the survey process can be simplified with a substantially reduced cost if you use an available database consisting of people who purchased a General Motors car within the past 10 years. Would good results be obtained from this population?

TI **7.** Smoking and College Education The tobacco industry closely monitors all surveys that involve smoking. One survey showed that among 785 randomly selected subjects who completed four years of college, 18.3% smoke (based on data from the American Medical Association).

a. Construct the 98% confidence interval for the true percentage of smokers among all people who completed four years of college.

b. Based on the result from part (a), does the smoking rate for those with four years of college appear to be substantially different than the 27% rate for the general population?

TI **8.** Crash Hospital Costs A study was conducted to estimate hospital costs for accident victims who wore seat belts. Twenty randomly selected cases have a distribution that appears to be bell-shaped with a mean of $9004 and a standard deviation of $5629 (based on data from the U.S. Department of Transportation).

a. Construct the 99% confidence interval for the mean of all such costs.

b. If you are a manager for an insurance company that provides lower rates for drivers who wear seat belts, and you want a conservative estimate for a worst case scenario, what amount should you use as the possible hospital cost for an accident victim who wears seat belts?

Cumulative Review Exercises

TI **1.** Analyzing Weights of Supermodels Supermodels are sometimes criticized on the grounds that their low weights encourage unhealthy eating habits among young women. Listed below are the weights (in pounds) of nine randomly selected supermodels.

125 (Taylor)	119 (Auermann)	128 (Schiffer)	128 (MacPherson)
119 (Turlington)	127 (Hall)	105 (Moss)	123 (Mazza)
115 (Hume)			

Find each of the following:

a. mean
b. median
c. mode
d. midrange
e. range
f. variance
g. standard deviation
h. Q_1
i. Q_2
j. Q_3
k. What is the level of measurement of these data (nominal, ordinal, interval, ratio)?
l. Construct a boxplot for the data.
m. Construct a 99% confidence interval for the population mean.
n. Construct a 99% confidence interval for the standard deviation σ.
o. Find the sample size necessary to estimate the mean weight of all supermodels so that there is 99% confidence that the sample mean is in error by no more than 2 lb. Use the sample standard deviation s from part (g) as an estimate of the population standard deviation σ.
p. When women are randomly selected from the general population, their weights are normally distributed with a mean of 143 lb and a standard deviation of 29 lb (based on data from the National Health and Examination Survey). Based on the given sample values, do the weights of supermodels appear to be substantially less than the weights of randomly selected women? Explain.

TI **2.** X-Linked Recessive Disorders A genetics expert has determined that for certain couples, there is a 0.25 probability that any child will have an X-linked recessive disorder.

a. Find the probability that among 200 such children, at least 65 have the X-linked recessive disorder.

b. A subsequent study of 200 actual births reveals that 65 of the children have the X-linked recessive disorder. Based on these sample results, construct a 95% confidence interval for the proportion of all such children having the disorder.

continued

c. Based on parts (a) and (b), does it appear that the expert's determination of a 0.25 probability is correct? Explain.

TI **3.** Estimating Theme Park Attendance Each year, billions of dollars are spent at theme parks owned by Disney, Universal Studios, Sea World, Busch Gardens, and others. A survey includes 111 people who took trips that included visits to theme parks, and there were 1122 other respondents who took trips without visits to a theme park (based on data from the Travel Industry Association of America).
a. Find the point estimate of the percentage of people who visit a theme park when they take a trip.
b. Find a 95% confidence interval estimate of the percentage of all people who visit a theme park when they take a trip.
c. The survey was conducted among people who took trips, but no information was given about the percentage of people who take trips for pleasure. If you want to estimate the percentage of adults who take a pleasure trip in a year, how many people must you survey if you want to be 99% confident that your sample percentage is within 2.5 percentage points of the correct population percentage?

Cooperative Group Activities

1. Out-of-class activity Collect sample data, and use the methods of this chapter to construct confidence interval estimates of population parameters. Here are some suggestions for parameters:

- Proportion of students at your college who can raise one eyebrow without raising the other eyebrow. [These sample results are easy to obtain because survey subjects tend to raise one eyebrow (if they can) when they are approached by someone asking questions.]
- Mean age of cars driven by statistics students and/or the mean age of cars driven by faculty.
- Mean length of words in *New York Times* editorials and mean length of words in editorials found in your local newspaper.
- Mean lengths of words in *Time* magazine, *Newsweek* magazine, and *People* magazine.
- Proportion of students at your college who can correctly identify the president, vice president, and secretary of state.
- Proportion of students at your college who are over the age of 18 and are registered to vote.
- Mean age of full-time students at your college.
- Proportion of motor vehicles in your region that are cars.

2. In-class activity Without using any measuring device, each student should draw a line believed to be 3 in. long and another line that is 3 cm long. Then use rulers to measure and record the lengths of the lines drawn. Find the means and standard deviations of the two sets of lengths. Use the sample data to construct a confidence interval for the length of the line estimated to be 3 in., then do the same for the length of the line estimated to be 3 cm. Do the confidence interval limits actually contain the correct length? Compare the results. Do the estimates of the 3-in. line appear to be more accurate than those for the 3-cm line?

3. In-class activity Assume that a method of gender selection can affect the probability of a baby being a girl, so that the probability becomes 1/4. Each student should simulate 20 births by drawing 20 cards from a shuffled deck. Replace each card after it has been drawn, then reshuffle. Consider the hearts to be girls and consider all other cards to be boys. After making 20 selections and recording the "genders" of the babies, construct a confidence interval estimate of the proportion of girls. Does the result appear to be effective in identifying the true value of the population proportion? (If decks of cards are not available, use some other way to simulate the births, such as using the random number generator on a calculator or using digits from phone numbers or social security numbers.)

4. Out-of-class activity Groups of three or four students should go to the library and collect a sample consisting of the ages of books (based on copyright dates). Plan and describe the sampling plan, execute the sampling procedure, then use the results to construct a confidence interval estimate of the mean age of all books in the library.

5. In-class activity Each student should write an estimate of the age of the current President of the United States. All estimates should be collected and the sample mean and standard deviation should be calculated. Then use the sample results to construct a confidence interval. Do the confidence interval limits contain the correct age of the President?

6. In-class activity A class project should be designed to conduct a test in which each student is given a taste of Coke and a taste of Pepsi. The student is then asked to identify which sample is Coke. After all of the results are collected, analyze the claim that the success rate is better than the rate that would be expected with random guesses.

7. In-class activity Each student should estimate the length of the classroom. The values should be based on visual estimates, with no actual measurements being taken. After the estimates have been collected, construct a confidence interval, then measure the length of the room. Does the confidence interval contain the actual length of the classroom? Is there a "collective wisdom," whereby the class mean is approximately equal to the actual room length?

8. In-class activity Divide into groups of three or four. Examine a current magazine such as *Time* or *Newsweek*, and find the proportion of pages that include advertising. Based on the results, construct a 95% confidence interval estimate of the percentage of all such pages that have advertising. Compare results with other groups.

9. In-class activity Divide into groups of two. First find the sample size required to estimate the proportion of times that a coin turns up heads when tossed, assuming that you want 80% confidence that the sample proportion is within 0.08 of the true population proportion. Then toss a coin the required number of times and record your results. What percentage of such confidence intervals should actually contain the true value of the population proportion, which we know is $p = 0.5$? Verify this last result by comparing your confidence interval with the confidence intervals found in other groups.

10. Out-of-class activity Identify a topic of general interest and coordinate with all members of the class to conduct a survey. Instead of conducting a "scientific" survey using sound principles of random selection, use a convenience sample consisting of respondents that are readily available, such as friends, relatives, and other students. Analyze and interpret the results. Identify the population. Identify the shortcomings of using a convenience sample, and try to identify how the sample might be different for a sample of subjects randomly selected from the population.

11. Out-of-class activity Each student should find an article in a professional journal that includes a confidence interval of the type discussed in this chapter. Write a brief report describing the confidence interval and its role in the context of the article.

TI-83/84 Plus Calculator Project

Bootstrap Resampling The *bootstrap method* can be used to construct confidence intervals for situations in which traditional methods cannot (or should not) be used. For example, the following sample of 10 values was randomly selected from a population with a distribution that is very far from normal, so any methods requiring a normal distribution should not be used.

2.9 564.2 1.4 4.7 67.6 4.8 51.3 3.6 18.0 3.6

If we want to use the above sample data for the construction of a confidence interval estimate of the population mean μ, we note that the sample is small, and there is an outlier. Also, see the TI-83/84 Plus–generated normal quantile plot and note that the pattern of the 10 points is dramatically different from a straight-line pattern, which suggests that the sample is *not* from a population with a normal distribution. The bootstrap method, which makes no assumptions about the original population, requires that we build a bootstrap population by replicating (duplicating) a sample many times. We can draw from the sample with replacement, thereby creating an approximation of the original population. In this way, we pull the sample up "by its own bootstraps" to simulate the original population. Using the sample data given above, we will proceed to construct a 95% confidence interval estimate of the population mean μ by using the bootstrap method as described in the following steps.

continued

TI-83/84 Plus **Normal Quantile Plot**

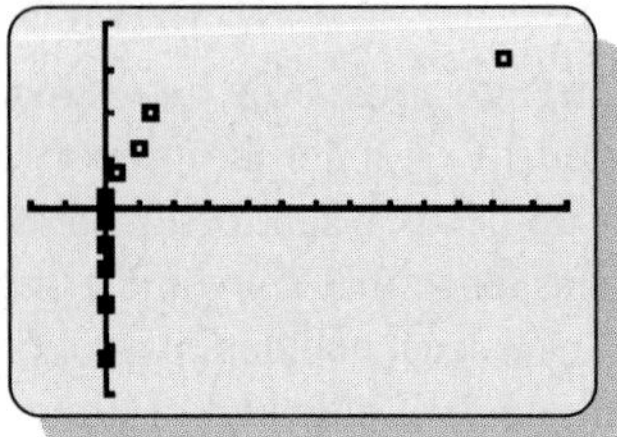

a. Create a new sample of size 10 by selecting 10 values with replacement from the 10 sample values given above. Accomplish this as follows:

i. Press MATH, select **PRB,** and then select the menu item of **randInt.**

ii. Press ENTER and proceed to enter (1, 10, 10) so that the command of **randInt(1, 10, 10)** is displayed. This command tells the calculator to randomly select 10 integers between 1 and 10 inclusive, with repetition allowed. A typical result might be this: {3 7 3 2 2 8 9 4 2 5}.

iii. Consider the 10 randomly generated integers to be *positions* of data values from the original list given above. For example, the randomly generated list of {3 7 3 2 2 8 9 4 2 5} indicates the 3rd value in the original list, the 7th value, 3rd value, and so on. The list of {3 7 3 2 2 8 9 4 2 5} therefore corresponds to this list of values from the original data set: {1.4, 51.3, 1.4, 564.2, 564.2, 3.6, 18, 4.7, 564.2, 67.6}. This constitutes a *resampling* of the original data set.

iv. Now find the mean of the resampled data set. For example, if the resampled data set is {1.4, 51.3, 1.4, 564.2, 564.2, 3.6, 18, 4.7, 564.2, 67.6}, its mean is 184.06.

b. Repeat the procedure in Step (a) until you have a list of 5 sample means. Ideally, we would repeat the process in Step (a) thousands of times. However, in the interest of preserving the sanity of readers, the author has compiled 95 such means, with the *sorted* list of 95 means given below.

c. Combine your 5 sample means with the 95 sample means in the list below. Be sure to enter them where they belong in the sorted list.

d. Now find the percentiles $P_{2.5}$ and $P_{97.5}$ for the sorted list of 100 means. ($P_{2.5}$ is the 3rd value in the sorted list; $P_{97.5}$ is the 98th value in the sorted list.) Identify the resulting confidence interval by substituting the values for $P_{2.5}$ and $P_{97.5}$ in $P_{2.5} < \mu < P_{97.5}$. Does this confidence interval contain the true value of μ, which is 148?

2.77	4.45	8.53	8.81	9.75	9.76	9.91	11.23	11.33	11.38
11.43	14.53	14.75	14.75	14.89	15.85	16.07	16.10	16.18	16.34
17.11	19.08	19.24	20.73	20.76	20.89	21.11	22.47	24.11	27.50
28.93	29.10	29.25	30.90	31.83	59.54	59.75	59.79	61.27	65.70
65.86	67.14	70.14	70.61	71.76	72.14	72.33	72.76	73.01	73.31
73.32	73.61	73.81	73.94	75.05	75.13	75.53	75.72	76.71	77.06
77.38	78.28	80.48	81.26	82.17	87.99	89.74	120.40	120.50	121.82
123.50	123.90	124.68	124.87	125.16	125.33	126.65	127.89	128.35	128.56
129.75	129.85	129.96	131.13	131.55	131.91	132.91	137.87	171.90	173.38
178.11	179.59	184.51	184.74	236.94					

The same procedure described above can be used to find a 95% confidence interval for the population standard deviation σ. (Instead of resampling to find means, resample to find standard deviations.)

An alternative to using the somewhat (?) laborious procedure described above is to use special software designed specifically for bootstrap-resampling methods. The author recommends Resampling Stats, available from Resampling Stats, Inc., 612 N. Jackson St., Arlington, VA, 22201; telephone number: (703) 522-2713.

From Data to Decision

Critical Thinking: What do the "photo-cop" survey results tell us?

Surveys have become an important component of American life. They directly affect the television shows we watch, the products we buy, the political leaders we elect, and the clothes that we wear. Because surveys are now such an integral part of all of our lives, it is important that every citizen has the ability to interpret survey results. Surveys are the focus of this exercise.

The *Star Tribune,* a Minneapolis–St. Paul newspaper, sponsored a poll designed to reveal opinions about the "photo-cop," which consists of cameras positioned to catch drivers who run red lights. The cameras photograph the license plates of cars passing through red lights, and those car owners are later mailed traffic violations. The newspaper sponsored the poll because of pending Minnesota legislation that would approve the use of cameras for issuing traffic violations. (Thanks to Beth Hentges who provided the newspaper information.) Pollsters surveyed 829 adult Minnesotans and found that 51% were opposed to the photo-cop legislation.

Analyzing the Data

1. Use the survey results to construct a 95% confidence interval estimate of the percentage of all Minnesotans opposed to the photo-cop legislation.
2. Given that 51% of the 829 Minnesotans surveyed were opposed to the photo-cop legislation, explain why it would or would not be okay for a newspaper to make this statement: "Based on results from a recent survey, the majority of Minnesotans are opposed to the photo-cop legislation."
3. A common criticism of surveys is that they poll only a very small percentage of the population and therefore cannot be accurate. Is a sample of only 829 people taken from a population of 3.4 million adult Minnesotans a sample size that is too small? Write an explanation of why the sample size of 829 is or is not too small.
4. In reference to another survey, the president of a company wrote to the Associated Press about a nationwide survey of 1223 subjects. Here is what he wrote:

> When you or anyone else attempts to tell me and my associates that 1223 persons account for our opinions and tastes here in America, I get mad as hell! How dare you! When you or anyone else tells me that 1223 people represent America, it is astounding and unfair and should be outlawed.

The writer of that letter then proceeds to claim that because the sample size of 1223 people represents 120 million people, his single letter represents 98,000 (120 million divided by 1223) who share the same views. Do you agree or disagree with this claim? Write a response that either supports or refutes this claim.

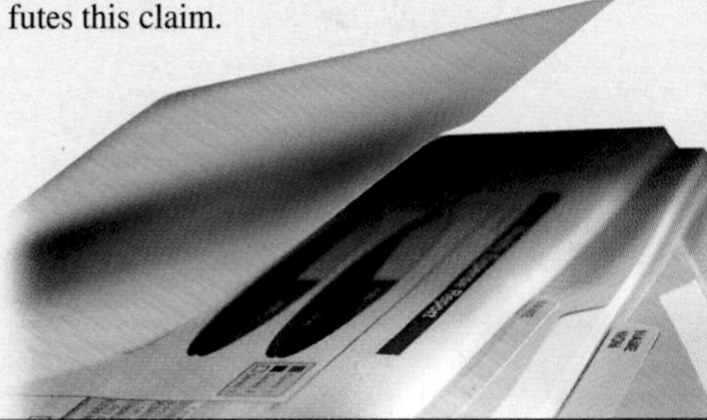

Internet Project

Confidence Intervals

The confidence intervals in this chapter illustrate an important point in the science of statistical estimation. Namely, estimations based on sample data are made with certain degrees of confidence. In the Internet Project for this chapter, you will use confidence intervals to make a statement about the temperature where you live. Go to the Web site for this textbook:

http://www.aw.com/triola

Locate the project for this chapter. There you will find instructions on how to use the Internet to find temperature data collected by the weather station nearest your home. With this data in hand, you will construct confidence intervals for temperatures during different time periods and attempt to draw conclusions about temperature change in your area. In addition, you will learn more about the relationship between confidence and probability.

Statistics @ Work

"For research and teaching in the field of ecology, animal behavior, and ecotoxicology, knowledge of statistics is essential to obtain a good job, and hold it."

Joanna Burger

Distinguished professor of Biology at Rutgers University, and member of the Environmental and Occupational Health Sciences Institute.

Joanna Burger teaches, does research, and serves on many national and international environmental committees dealing with endangered species, contaminants in wildlife, the effects of chemicals on animal behavior, and the effects of people on ecosystems.

What concepts of statistics do you use in your work?

I use a variety of statistical approaches including both parametric and nonparametric methods. Without a firm understanding of statistics I would not be able to test whether environmental factors affect reproductive success. I use statistics to test hypotheses that I generate by observing animals within their natural environments. While observation leads to hypotheses, only with the use of well-designed experiments and statistical tests is it possible to answer the questions. For research and teaching in the field of ecology, animal behavior, and ecotoxicology, knowledge of statistics is essential to obtain a good job, and hold it.

Could you give a specific example of how you have used statistics in the past?

Statistics is very helpful in identifying which factors are important in influencing animal behavior. Birds nest in particular habitats, but we questioned whether they nest randomly or select very specific nest sites. This is important because conservation requires knowing what animals need in order to create, protect, and/or manage that habitat. I tested the hypothesis that common terns were selecting particular salt marsh islands. By comparing statistically a wide range of environmental factors (such as island height, island size, and vegetation type and density) on all islands, with the same set of factors on islands used for nesting by terns, we could show that terns actually selected a very specific set of characteristics. Although there were over 250 islands in the bay where this study was conducted, only 36 met the criteria used by the terns. The characteristics they were selecting actually resulted in their choosing islands that were high enough to avoid summer storm tides, but low enough so that predators could not survive over the winter. Islands that are high enough to avoid winter storm tides can have viable predator populations, such as fox and raccoons, which will eat the eggs and chicks of the terns.

Is a knowledge of statistics critical for your work?

A firm understanding of statistics is absolutely essential to conducting research with humans and animals. Using hypothesis testing, and multiple regression analysis, it is possible to begin to identify and evaluate the factors that affect behavior, such as fishing and consumption behavior of people, foraging behavior of shorebirds, and nesting behavior of seabirds.

In terms of statistics, what would you recommend for prospective employees?

Anyone who wants to be in the field of conservation biology, ecotoxicology, animal behavior, or ecology needs a wide range of statistical skills. Two or three courses would be best, including general statistics, regression, and nonparametrics. The nature of each problem and the characteristics of the data will determine the statistics that are required, and one should not be limited by a lack of statistical knowledge.

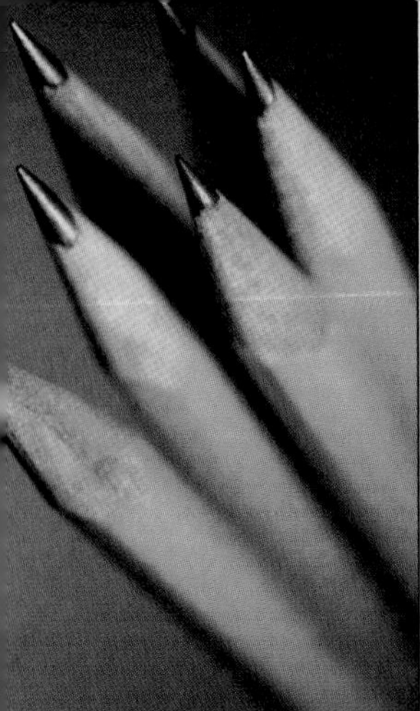

8 Hypothesis Testing

8 CHAPTER PROBLEM

What is the best way to go about finding a job?

After completing a college degree, graduates are faced with the important project of finding a job. Options include networking through friends and relatives, pursuing newspaper ads, applying in person, conducting an Internet job search, using a professional employment agency or recruiter, and using a college placement office. Some of these avenues are likely to be much more productive than others. Specifically, networking has been found to be one of the most productive approaches. With networking, a job seeker develops contacts and exchanges information through an informal network of people.

One recent survey involved 703 randomly selected subjects who were all working. Among those subjects, 61% said that they found their job through networking (based on data from Taylor Nelson Sofres Intersearch). Another effective approach appeared to be newspaper ads, with 16% of the respondents finding jobs through such ads. Based on these survey results, can a newspaper article publish the headline with the claim that "Most Workers Find Jobs Through Networking"? Some might argue that although 61% is greater than 50%, this survey involves only 703 people from the millions of workers, so the survey doesn't provide sufficient justification for the claim that most workers find jobs through networking. But do the survey results provide sufficient justification for the claim? In this chapter, we introduce very standard methods used to formally and objectively test such claims.

8-1 Overview

The two main activities of inferential statistics are using sample data to (1) *estimate* a population parameter (as in Chapter 7), and (2) test a hypothesis or claim about a population parameter (as in this chapter).

Definition

In statistics, a **hypothesis** is a claim or statement about a property of a population.

A **hypothesis test** (or **test of significance**) is a standard procedure for testing a claim about a property of a population.

Here are examples of hypotheses that can be tested by the procedures we develop in this chapter.

- **Business** A newspaper headline makes the claim that most workers get their jobs through networking.
- **Medicine** Medical researchers claim that the mean body temperature of healthy adults is not equal to 98.6°F.
- **Aircraft Safety** The Federal Aviation Administration claims that the mean weight of an airline passenger (with carry-on baggage) is greater than the 185 lb that it was 20 years ago.
- **Quality Control** When new equipment is used to manufacture aircraft altimeters, the new altimeters are better because the variation in the errors is reduced so that the readings are more consistent. (In many industries, the quality of goods and services can often be improved by reducing variation.)

The methods presented in this chapter are based on the rare event rule (Section 4-1) for inferential statistics, so let's review it before proceeding.

Rare Event Rule for Inferential Statistics

If, under a given assumption, the probability of a particular observed event is exceptionally small, we conclude that the assumption is probably not correct.

Following this rule, we test a claim by analyzing sample data in an attempt to distinguish between results that can *easily occur by chance* and results that are *highly unlikely to occur by chance*. We can explain the occurrence of highly unlikely results by saying that either a rare event has indeed occurred or that the underlying assumption is not true. Let's apply this reasoning in the following example.

EXAMPLE **Gender Selection** ProCare Industries, Ltd. once provided a product called "Gender Choice," which, according to advertising claims, allowed couples to "increase your chances of having a girl up to 80%." Suppose we conduct an experiment with 100 couples who want to have baby girls, and they all follow the Gender Choice "easy-to-use in-home system" described in the pink package designed for girls. Assuming that Gender Choice has no effect and using only common sense and no formal statistical methods, what should we conclude about the assumption of "no effect" from Gender Choice if 100 couples using Gender Choice have 100 babies consisting of

a. 52 girls?

b. 97 girls?

SOLUTION

a. We normally expect around 50 girls in 100 births. The result of 52 girls is close to 50, so we should not conclude that the Gender Choice product is effective. The result of 52 girls could easily occur by chance, so there isn't sufficient evidence to say that Gender Choice is effective.

b. The result of 97 girls in 100 births is extremely unlikely to occur by chance. We could explain the occurrence of 97 girls in one of two ways: Either an *extremely* rare event has occurred by chance, or Gender Choice is effective. The extremely low probability of getting 97 girls suggests that Gender Choice is effective.

The key point of the preceding example is that we should conclude that the product is effective only if we get *significantly* more girls than we would normally expect. Although the outcomes of 52 girls and 97 girls are both "above average," the result of 52 girls is not significant, whereas 97 girls is a significant result.

This brief example illustrates the kind of thinking used in testing hypotheses. The formal method involves a variety of standard terms and conditions incorporated into an organized procedure. We suggest that you begin by first reading Sections 8-2 and 8-3 casually to obtain a general idea of their concepts and then rereading Section 8-2 more carefully to become familiar with the terminology.

8-2 Basics of Hypothesis Testing

Key Concept This section presents individual components of a hypothesis test, and the following sections use those components in comprehensive procedures. The role of the following components should be understood: null hypothesis, alternative hypothesis, test statistic, critical region, significance level, critical value, P-value, type I error, and type II error. These basic concepts are presented in Part 1 of this section, and they should be understood before considering the concept of power of a test, which is discussed in Part 2.

Part 1: Basic Concepts of Hypothesis Testing

Here are objectives for this section that should be met before considering the discussion of power in Part 2.

Objectives for this Section

- Given a claim, identify the null hypothesis and the alternative hypothesis, and express them both in symbolic form.

- Given a claim and sample data, calculate the value of the test statistic.
- Given a significance level, identify the critical value(s).
- Given a value of the test statistic, identify the P-value.
- State the conclusion of a hypothesis test in simple, nontechnical terms.

You should study the following example until you thoroughly understand it. Once you do, you will have captured a major concept of statistics.

Lie Detectors

Why not require all criminal suspects to take lie detector tests and dispense with trials by jury? The Council of Scientific Affairs of the American Medical Association states, "It is established that classification of guilty can be made with 75% to 97% accuracy, but the rate of false positives is often sufficiently high to preclude use of this (polygraph) test as the sole arbiter of guilt or innocence." A "false positive" is an indication of guilt when the subject is actually innocent. Even with accuracy as high as 97%, the percentage of false positive results can be 50%, so half of the innocent subjects incorrectly appear to be guilty.

EXAMPLE Gender Selection and Probability Let's again refer to the pink packages of Gender Choice. ProCare Industries claimed that couples using the pink packages of Gender Choice would have girls at a rate that is greater than 50% or 0.5. Let's again consider an experiment whereby 100 couples use Gender Choice in an attempt to have a baby girl, and let's assume that the 100 babies include exactly 52 girls. We will proceed to formalize some of the analysis, but there are two points that can be confusing:

1. **Assume $p = 0.5$:** In attempting to determine whether 52 girls in 100 births is evidence of Gender Choice's effectiveness, we assume that $p = 0.5$ so that we can determine whether the result of 52 girls can easily occur by chance (with no treatment effect) or whether the result of 52 girls is unlikely to occur by chance (so that the treatment appears to be effective).
2. **Use P(52 or more girls):** When determining whether "52 girls" is likely to occur by chance, use the probability of *52 or more* girls. [Review the subsection of "Using Probabilities to Determine When Results Are Unusual" in Section 5-2, where we noted that "x successes among n trials is an *unusually high* number of successes if $P(x \text{ or more}) \leq 0.05$."]

Under normal circumstances the proportion of girls is $p = 0.5$, so a claim that Gender Choice is effective can be expressed as $p > 0.5$. We support the claim of $p > 0.5$ only if a result such as 52 girls is unlikely (with a small probability, such as less than or equal to 0.05). Using a normal distribution as an approximation to the binomial distribution (see Section 6-6), we find $P(52 \text{ or more girls in 100 births}) = 0.3821$. Figure 8-1 shows that with a probability of 0.5, the outcome of 52 girls in 100 births is not unusual, so we do *not* reject random chance as a reasonable explanation. We conclude that the proportion of girls born to couples using Gender Choice is *not* significantly greater than the number that we would expect by random chance. Here are the key points of this example:

- Claim: For couples using Gender Choice, the proportion of girls is $p > 0.5$.
- Working assumption: The proportion of girls is $p = 0.5$ (with no effect from Gender Choice).
- The sample resulted in 52 girls among 100 births, so the sample proportion is $\hat{p} = 52/100 = 0.52$.
- Assuming that $p = 0.5$, we use a normal distribution as an approximation to the binomial distribution to find that $P(\text{at least 52 girls in 100 births}) = 0.3821$.

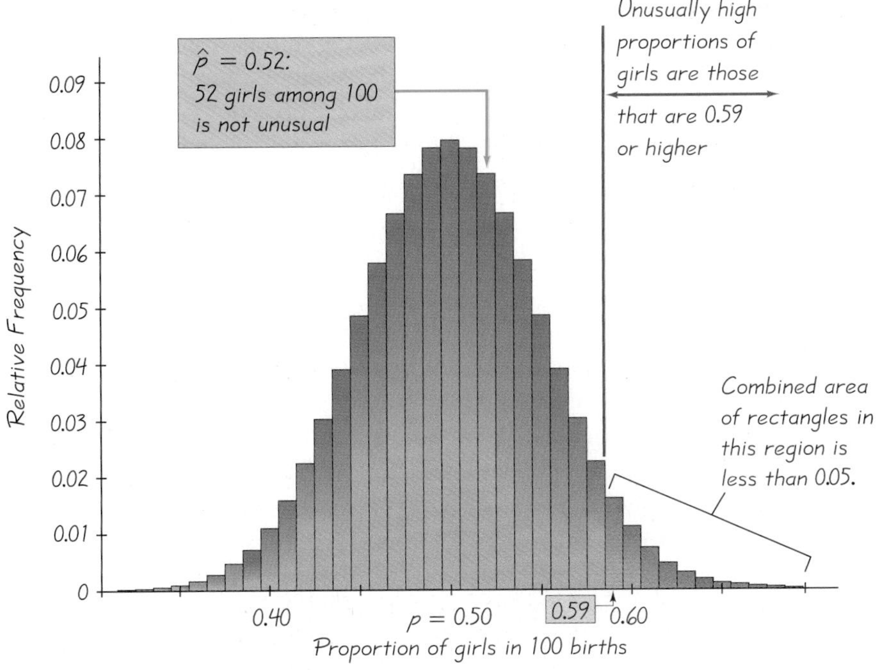

Figure 8-1 Sampling Distribution of Proportions of Girls in 100 Births

- There are two possible explanations for the result of 52 girls in 100 births: Either a random chance event (with probability 0.3821) has occurred, or the proportion of girls born to couples using Gender Choice is greater than 0.5. Because the probability of getting at least 52 girls by chance is so high (0.3821), we go with random chance as a reasonable explanation. There isn't sufficient evidence to support a claim that Gender Choice is effective in producing more girls than expected by chance. (It was actually this type of analysis that led to the removal of Gender Choice from the market.)

In Section 8-3 we will describe the specific steps used in hypothesis testing, but let's first describe the components of a formal **hypothesis test,** or **test of significance.** These terms are often used in a wide variety of disciplines when statistical methods are required.

Components of a Formal Hypothesis Test

Null and Alternative Hypotheses

- The **null hypothesis** (denoted by H_0) is a statement that the value of a population parameter (such as proportion, mean, or standard deviation) is *equal to* some claimed value. Here are some typical null hypotheses of the type considered in this chapter:

$$H_0\text{: } p = 0.5 \qquad H_0\text{: } \mu = 98.6 \qquad H_0\text{: } \sigma = 15$$

We test the null hypothesis directly in the sense that we assume it is true and reach a conclusion to either reject H_0 or fail to reject H_0.

- The **alternative hypothesis** (denoted by H_1 or H_a or H_A) is the statement that the parameter has a value that somehow differs from the null hypothesis. For the methods of this chapter, the symbolic form of the alternative hypothesis must use one of these symbols: $<$ or $>$ or $\neq$. Here are nine different examples of alternative hypotheses involving proportions, means, and standard deviations:

Proportions:	$H_1: p > 0.5$	$H_1: p < 0.5$	$H_1: p \neq 0.5$
Means:	$H_1: \mu > 98.6$	$H_1: \mu < 98.6$	$H_1: \mu \neq 98.6$
Standard Deviations:	$H_1: \sigma > 15$	$H_1: \sigma < 15$	$H_1: \sigma \neq 15$

Note About Always Using the Equal Symbol in H_0: A few textbooks use the symbols $\leq$ and $\geq$ in the null hypothesis H_0, but most professional journals use only the equal symbol for equality. We conduct the hypothesis test by assuming that the proportion, mean, or standard deviation is *equal to* some specified value so that we can work with a single distribution having a specific value.

Note About Forming Your Own Claims (Hypotheses): If you are conducting a study and want to use a hypothesis test to *support* your claim, the claim must be worded so that it becomes the alternative hypothesis (and can be expressed using only the symbols $<$ or $>$ or $\neq$). You can never support a claim that some parameter is *equal to* some specified value.

For example, if you have developed a gender-selection method that increases the likelihood of a girl, state your claim as $p > 0.5$ so that your claim can be supported. (In this context of trying to support the goal of the research, the alternative hypothesis is sometimes referred to as the *research hypothesis*.) You will assume for the purpose of the test that $p = 0.5$, but you hope that $p = 0.5$ gets rejected so that $p > 0.5$ is supported.

Note About Identifying H_0 and H_1: Figure 8-2 summarizes the procedures for identifying the null and alternative hypotheses. Note that the original state-

Figure 8-2

Identifying H_0 and H_1

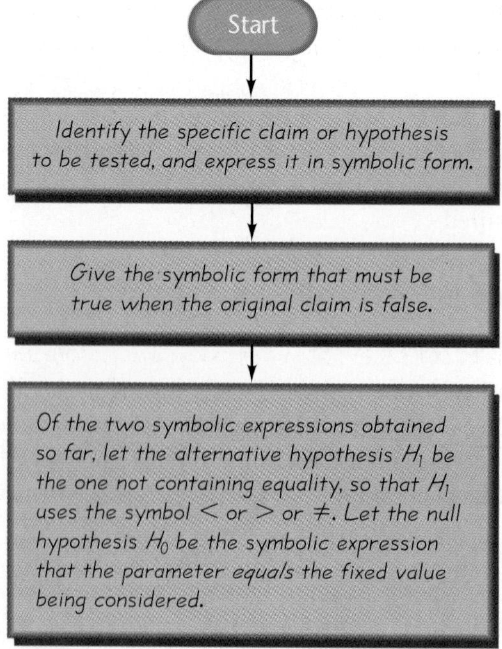

ment could become the null hypothesis, it could become the alternative hypothesis, or it might not correspond exactly to either the null hypothesis or the alternative hypothesis.

For example, we sometimes test the validity of someone else's claim, such as the claim of the Coca-Cola Bottling Company that "the mean amount of Coke in cans is at least 12 oz." That claim can be expressed in symbols as $\mu \geq 12$. In Figure 8-2 we see that if that original claim is false, then $\mu < 12$. The alternative hypothesis becomes $\mu < 12$, but the null hypothesis is $\mu = 12$. We will be able to address the original claim after determining whether there is sufficient evidence to reject the null hypothesis of $\mu = 12$.

EXAMPLE Identifying the Null and Alternative Hypotheses
Refer to Figure 8-2 and use the given claims to express the corresponding null and alternative hypotheses in symbolic form.

a. The proportion of workers who get jobs through networking is greater than 0.5.

b. The mean weight of airline passengers with carry-on baggage is at most 195 lb (the current figure used by the Federal Aviation Administration).

c. The standard deviation of IQ scores of actors is equal to 15.

SOLUTION See Figure 8-2, which shows the three-step procedure.

a. In Step 1 of Figure 8-2, we express the given claim as $p > 0.5$. In Step 2 we see that if $p > 0.5$ is false, then $p \leq 0.5$ must be true. In Step 3, we see that the expression $p > 0.5$ does not contain equality, so we let the alternative hypothesis H_1 be $p > 0.5$, and we let H_0 be $p = 0.5$.

b. In Step 1 of Figure 8-2, we express "a mean of at most 195 lb" in symbols as $\mu \leq 195$. In Step 2 we see that if $\mu \leq 195$ is false, then $\mu > 195$ must be true. In Step 3, we see that the expression $\mu > 195$ does not contain equality, so we let the alternative hypothesis H_1 be $\mu > 195$, and we let H_0 be $\mu = 195$.

c. In Step 1 of Figure 8-2, we express the given claim as $\sigma = 15$. In Step 2 we see that if $\sigma = 15$ is false, then $\sigma \neq 15$ must be true. In Step 3, we let the alternative hypothesis H_1 be $\sigma \neq 15$, and we let H_0 be $\sigma = 15$.

Test Statistic

- The **test statistic** is a value used in making a decision about the null hypothesis, and it is found by converting the sample statistic (such as the sample proportion $\hat{p}$, or the sample mean $\bar{x}$, or the sample standard deviation s) to a score (such as z, t, or χ^2) with the assumption that the null hypothesis is true. In this chapter we use the following test statistics:

Test statistic for proportion $$z = \frac{\hat{p} - p}{\sqrt{\frac{pq}{n}}}$$

Large Sample Size Isn't Good Enough

Biased sample data should not be used for inferences, no matter how large the sample is. For example, in *Women and Love: A Cultural Revolution in Progress*, Shere Hite bases her conclusions on 4500 replies that she received after mailing 100,000 questionnaires to various women's groups. A *random* sample of 4500 subjects would usually provide good results, but Hite's sample is biased. It is criticized for overrepresenting women who join groups and women who feel strongly about the issues addressed. Because Hite's sample is biased, her inferences are not valid, even though the sample size of 4500 might seem to be sufficiently large.

Test statistic for mean $$z = \frac{\bar{x} - \mu}{\frac{\sigma}{\sqrt{n}}} \quad \text{or} \quad t = \frac{\bar{x} - \mu}{\frac{s}{\sqrt{n}}}$$

Test statistic for standard deviation $$\chi^2 = \frac{(n-1)s^2}{\sigma^2}$$

The test statistic for a mean uses the normal or Student t distribution, depending on the conditions that are satisfied. This chapter will use the same criteria described in Section 7-4. (See Figure 7-6 and Table 7-1.)

EXAMPLE Finding the Test Statistic A survey of $n = 703$ randomly selected workers showed that 61% (or $\hat{p} = 0.61$) of those respondents found their job through networking. Find the value of the test statistic for the claim that most (more than 50%) workers get their jobs through networking. (In Section 8-3 we will see that there are requirements that must be verified. For this example, assume that the requirements are satisfied and focus on finding the indicated test statistic.)

SOLUTION The preceding example showed that the given claim results in the following null and alternative hypotheses: H_0: $p = 0.5$ and H_1: $p > 0.5$. Because we work under the assumption that the null hypothesis is true with $p = 0.5$, we get the following test statistic:

$$z = \frac{\hat{p} - p}{\sqrt{\frac{pq}{n}}} = \frac{0.61 - 0.5}{\sqrt{\frac{(0.5)(0.5)}{703}}} = 5.83$$

INTERPRETATION We know from previous chapters that a z score of 5.83 is "unusual" (because it is greater than 2). It appears that in addition to being greater than 50%, the sample result of 61% is *significantly* greater than 50%. See Figure 8-3 where we show that the sample proportion of 0.61 (from 61%) does fall within the range of values considered to be significant because they are so far above 0.5 that they are not likely to occur by chance (assuming that the population proportion is $p = 0.5$).

Critical Region, Significance Level, Critical Value, and *P*-Value

- The **critical region** (or **rejection region**) is the set of all values of the test statistic that cause us to reject the null hypothesis. For example, see the red-shaded region in Figure 8-3.
- The **significance level** (denoted by α) is the probability that the test statistic will fall in the critical region when the null hypothesis is actually true. If the test statistic falls in the critical region, we reject the null hypothesis, so α is the probability of making the mistake of rejecting the null hypothesis when it is true. This is the same α introduced in Section 7-2, where

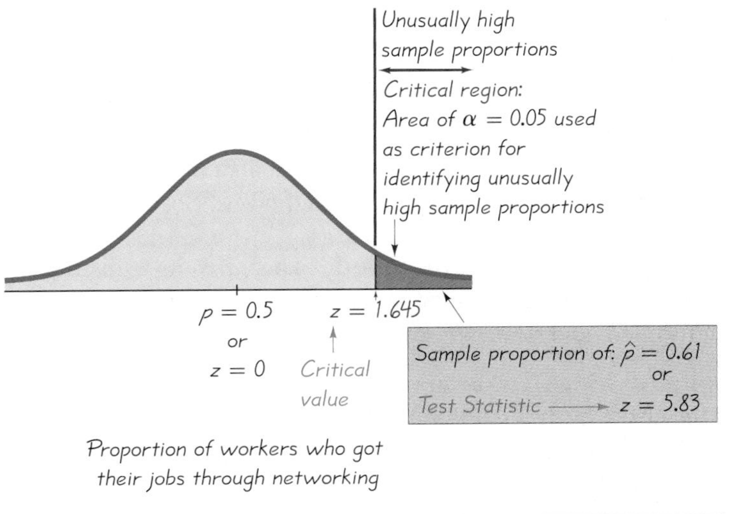

Figure 8-3 Critical Region, Critical Value, Test Statistic

Win $1,000,000 for ESP

Magician James Randi instituted an educational foundation that offers a prize of $1 million to anyone who can demonstrate paranormal, supernatural, or occult powers. Anyone possessing power such as fortune telling, ESP (extrasensory perception), or the ability to contact the dead, can win the prize by passing testing procedures. A preliminary test is followed by a formal test, but so far, no one has passed the preliminary test. The formal test would be designed with sound statistical methods, and it would likely involve analysis with a formal hypothesis test. According to the foundation, "We consult competent statisticians when an evaluation of the results, or experiment design, is required." Information about the application can be found at the foundation's Web site, randi.org.

we defined the confidence level for a confidence interval to be the probability $1 - \alpha$. Common choices for α are 0.05, 0.01, and 0.10, with 0.05 being most common.

- A **critical value** is any value that separates the critical region (where we reject the null hypothesis) from the values of the test statistic that do not lead to rejection of the null hypothesis. The critical values depend on the nature of the null hypothesis, the sampling distribution that applies, and the significance level of α. See Figure 8-3 where the critical value of $z = 1.645$ corresponds to a significance level of $\alpha = 0.05$. (Critical values were first discussed in Chapter 7.)

EXAMPLE Finding Critical Values Using a significance level of $\alpha = 0.05$, find the critical z values for each of the following alternative hypotheses (assuming that the normal distribution can be used to approximate the binomial distribution):

a. $p \neq 0.5$ (so the critical region is in *both* tails of the normal distribution)

b. $p < 0.5$ (so the critical region is in the *left* tail of the normal distribution)

c. $p > 0.5$ (so the critical region is in the *right* tail of the normal distribution)

SOLUTION

a. See Figure 8-4(a). The shaded tails contain a total area of $\alpha = 0.05$, so each tail contains an area of 0.025. Using the methods of Section 6-2, the values of $z = 1.96$ and $z = -1.96$ separate the right and left tail regions. The critical values are therefore $z = 1.96$ and $z = -1.96$.

continued

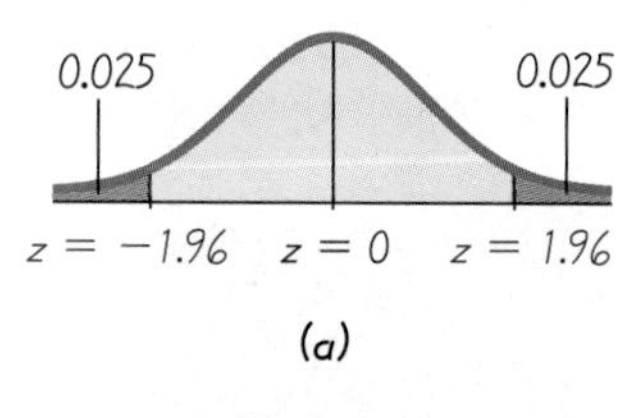

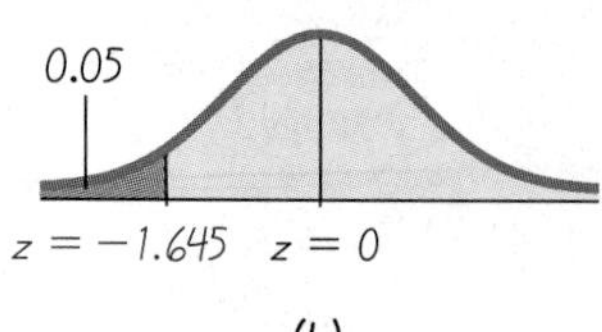

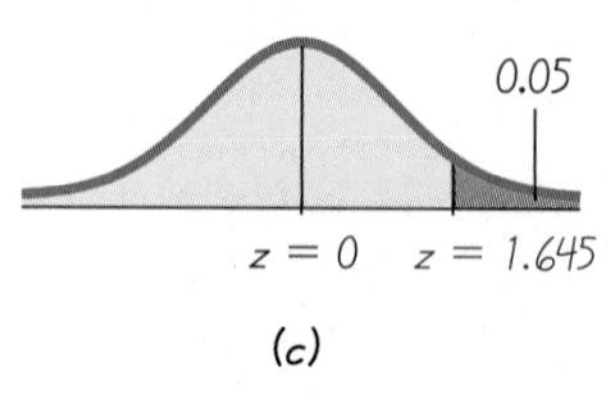

Figure 8-4
Finding Critical Values

b. See Figure 8-4(b). With an alternative hypothesis of $p < 0.5$, the critical region is in the left tail. With a left-tailed area of 0.05, the critical value is found to be $z = -1.645$ (by using the methods of Section 6-2).

c. See Figure 8-4(c). With an alternative hypothesis of $p > 0.5$, the critical region is in the right tail. With a right-tailed area of 0.05, the critical value is found to be $z = 1.645$ (by using the methods of Section 6-2).

Two-Tailed, Left-Tailed, Right-Tailed The *tails* in a distribution are the extreme regions bounded by critical values. Some hypothesis tests are two-tailed, some are left-tailed, and some are right-tailed.

- **Two-tailed test:** The critical region is in the two extreme regions (tails) under the curve [as in Figure 8-4(a)].
- **Left-tailed test:** The critical region is in the extreme left region (tail) under the curve [as in Figure 8-4(b)].
- **Right-tailed test:** The critical region is in the extreme right region (tail) under the curve [as in Figure 8-4(c)].

In two-tailed tests, the significance level α is divided equally between the two tails that constitute the critical region. For example, in a two-tailed test with a significance level of $\alpha = 0.05$, there is an area of 0.025 in each of the two tails. In tests that are right- or left-tailed, the area of the critical region in one tail is α. (See Figure 8-4.)

By examining the alternative hypothesis, we can determine whether a test is right-tailed, left-tailed, or two-tailed. The tail will correspond to the critical region containing the values that would conflict significantly with the null hypothesis. A useful check is summarized in the margin figures (see Figure 8-5), which show that the inequality sign in H_1 points in the direction of the critical region. The symbol $\neq$ is often expressed in programming languages as $< >$, and this reminds us that an alternative hypothesis such as $p \neq 0.5$ corresponds to a two-tailed test.

- The ***P*-value** (or ***p*-value** or **probability value**) is the probability of getting a value of the test statistic that is *at least as extreme* as the one representing the sample data, assuming that the null hypothesis is true. The null hypothesis is rejected if the *P*-value is very small, such as 0.05 or less. *P*-values can be found by using the procedure summarized in Figure 8-6.

Decisions and Conclusions

The standard procedure of hypothesis testing requires that we always test the null hypothesis, so our initial conclusion will always be one of the following:

1. Reject the null hypothesis.
2. Fail to reject the null hypothesis.

Decision Criterion The decision to reject or fail to reject the null hypothesis is usually made using either the traditional method (or classical method) of testing hypotheses, the *P*-value method, or the decision is sometimes based on confidence intervals. In recent years, use of the *P*-value method has been increasing along with the inclusion of *P*-values in results from software packages.

Traditional method: *Reject* H_0 if the test statistic falls within the critical region.
Fail to reject H_0 if the test statistic does not fall within the critical region.

***P*-value method:** *Reject* H_0 if the *P*-value $\leq \alpha$ (where α is the significance level, such as 0.05).
Fail to reject H_0 if the *P*-value $> \alpha$.

Another option: Instead of using a significance level such as $\alpha = 0.05$, simply identify the *P*-value and leave the decision to the reader.

Confidence intervals: Because a confidence interval estimate of a population parameter contains the likely values of that parameter, reject a claim that the population parameter has a value that is not included in the confidence interval.

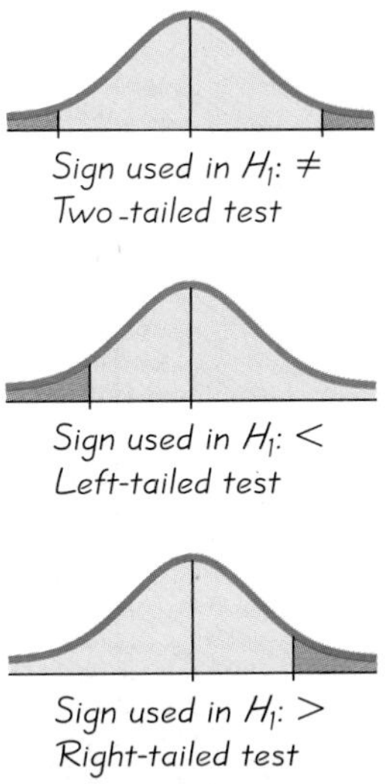

Sign used in H_1: $\neq$
Two-tailed test

Sign used in H_1: $<$
Left-tailed test

Sign used in H_1: $>$
Right-tailed test

Figure 8-5
Two-Tailed, Left-Tailed, Right-Tailed Tests

EXAMPLE Finding *P*-Values First determine whether the given conditions result in a right-tailed test, a left-tailed test, or a two-tailed test, then use Figure 8-6 to find the *P*-value, then state a conclusion about the null hypothesis.

a. A significance level of $\alpha = 0.05$ is used in testing the claim that $p > 0.25$, and the sample data result in a test statistic of $z = 1.18$.

b. A significance level of $\alpha = 0.05$ is used in testing the claim that $p \neq 0.25$, and the sample data result in a test statistic of $z = 2.34$.

SOLUTION

a. With a claim of $p > 0.25$, the test is right-tailed. Using Figure 8-6 for a right-tailed test, we see that the *P*-value is the area to the right of the test statistic $z = 1.18$. We refer to Table A-2 and find that the area to the *right* of $z = 1.18$ is 0.1190. The *P*-value of 0.1190 is greater than the significance level $\alpha = 0.05$, so we fail to reject the null hypothesis. The *P*-value of 0.1190 is relatively large, indicating that the sample results could easily occur by chance.

b. With a claim of $p \neq 0.25$, the test is two-tailed. Using Figure 8-6 for a two-tailed test, we see that the *P*-value is *twice* the area to the right of $z = 2.34$. We refer to Table A-2 and find that the area to the right of $z = 2.34$ is 0.0096, so the *P*-value $= 2 \times 0.0096 = 0.0192$. The *P*-value of 0.0192 is less than or equal to the significance level, so we reject the null hypothesis. The small *P*-value of 0.0192 shows that the sample results are not likely to occur by chance.

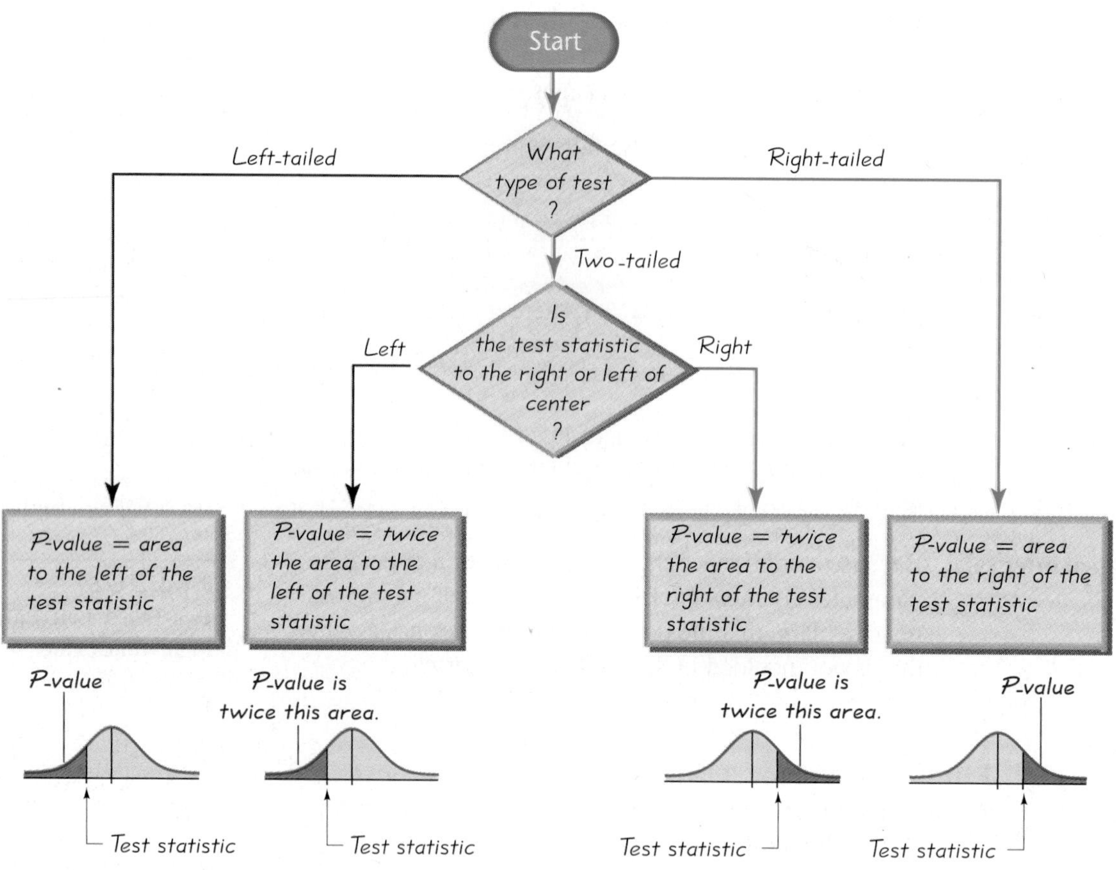

Figure 8-6 Procedure for Finding *P*-Values

Wording the Final Conclusion The conclusion of rejecting the null hypothesis or failing to reject it is fine for those of us with the wisdom to take a statistics course, but we should use simple, nontechnical terms in stating what the conclusion really means. Figure 8-7 summarizes a procedure for wording of the final conclusion. Note that only one case leads to wording indicating that the sample data actually *support* the conclusion. If you want to support some claim, state it in such a way that it becomes the alternative hypothesis, and then hope that the null hypothesis gets rejected.

Accept/Fail to Reject Some texts say "accept the null hypothesis" instead of "fail to reject the null hypothesis." Whether we use the term *accept* or *fail to reject,* we should recognize that *we are not proving the null hypothesis;* we are merely saying that the sample evidence is not strong enough to warrant rejection of the null hypothesis. (When a jury does not find enough evidence to convict a suspect, it returns a verdict of not guilty; it does not return a verdict of innocent.) The term *accept* is somewhat misleading, because it seems to imply incorrectly that the null

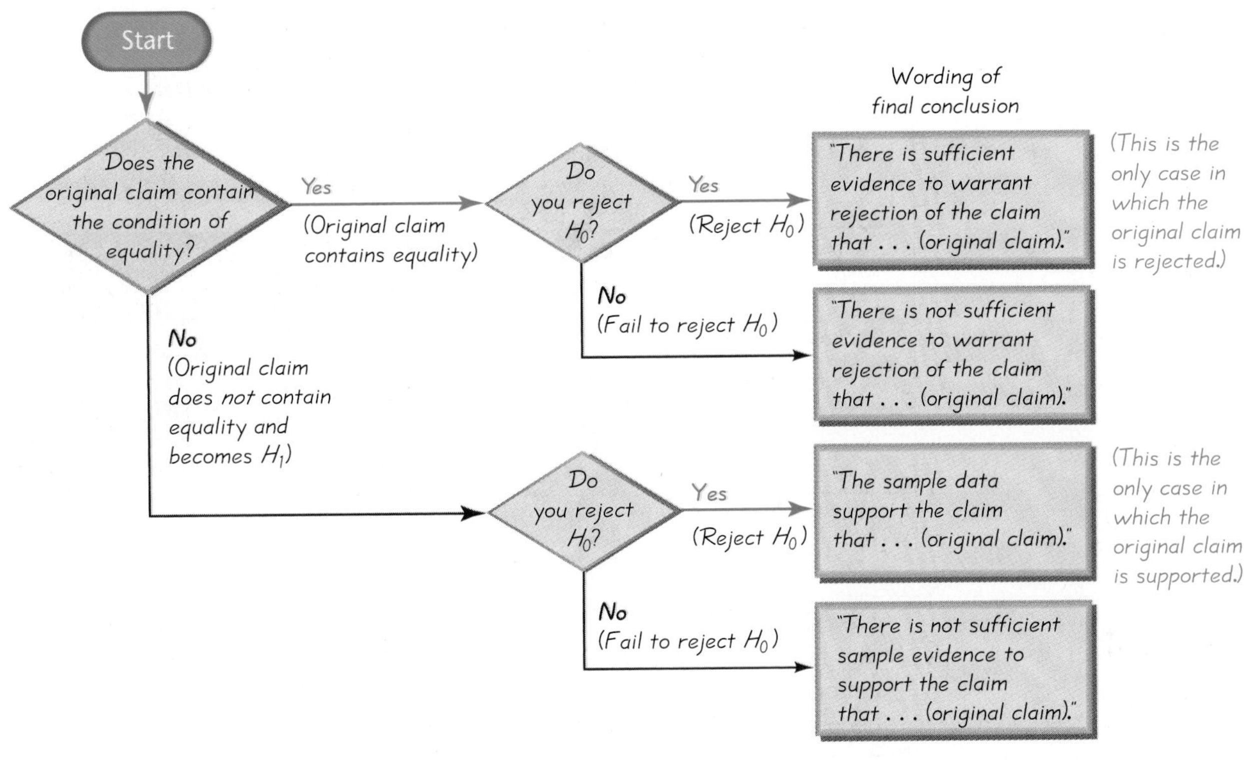

Figure 8-7 Wording of Final Conclusion

hypothesis has been proved. (It is misleading to state that "there is sufficient evidence to accept the null hypothesis.") The phrase *fail to reject* says more correctly that the available evidence isn't strong enough to warrant rejection of the null hypothesis. In this text we will use the terminology *fail to reject the null hypothesis,* instead of *accept the null hypothesis.*

Multiple Negatives When stating the final conclusion in nontechnical terms, it is possible to get correct statements with up to three negative terms. (Example: "There is *not* sufficient evidence to warrant *rejection* of the claim of *no* difference between 0.5 and the population proportion.") Such conclusions with so many negative terms can be confusing, so it would be good to restate them in a way that makes them understandable, but care must be taken to not change the meaning. For example, instead of saying that "there is not sufficient evidence to warrant rejection of the claim of no difference between 0.5 and the population proportion," better statements would be these:

- Fail to reject the claim that the population proportion is equal to 0.5.
- Until stronger evidence is obtained, continue to assume that the population proportion is equal to 0.5.

Table 8-1 Type I and Type II Errors			
		True State of Nature	
		The null hypothesis is true	The null hypothesis is false
Decision	We decide to reject the null hypothesis	**Type I error** (rejecting a true null hypothesis) α	Correct decision
	We fail to reject the null hypothesis	Correct decision	**Type II error** (failing to reject a false null hypothesis) β

EXAMPLE Stating the Final Conclusion Suppose a reporter claims that most (more than 50%) workers find their jobs through networking. This claim of $p > 0.5$ becomes the alternative hypothesis, while the null hypothesis becomes $p = 0.5$. Further suppose that the sample evidence causes us to reject the null hypothesis of $p = 0.5$. State the conclusion in simple, nontechnical terms.

SOLUTION Refer to Figure 8-7. The original claim does not contain the condition of equality, and we do reject the null hypothesis. The wording of the final conclusion should therefore be as follows: "The sample data support the claim that most workers find their jobs through networking."

Type I and Type II Errors When testing a null hypothesis, we arrive at a conclusion of rejecting it or failing to reject it. Such conclusions are sometimes correct and sometimes wrong (even if we do everything correctly). Table 8-1 summarizes the two different types of errors that can be made, along with the two different types of correct decisions. We distinguish between the two types of errors by calling them type I and type II errors.

- **Type I error:** The mistake of rejecting the null hypothesis when it is actually true. The symbol α (alpha) is used to represent the probability of a type I error.
- **Type II error:** The mistake of failing to reject the null hypothesis when it is actually false. The symbol β (beta) is used to represent the probability of a type II error.

Because it can be difficult to remember which error is type I and which is type II, we recommend a mnemonic device, such as "ROUTINE FOR FUN." Using only the consonants from those words (**R**ou**T**i**N**e **F**o**R** **F**u**N**), we can easily remember that a type I error is RTN: Reject True Null (hypothesis), whereas a type II error is FRFN: Failure to Reject a False Null (hypothesis).

Notation

α (alpha) = probability of a type I error (the probability of rejecting the null hypothesis when it is true)

β (beta) = probability of a type II error (the probability of failing to reject a null hypothesis when it is false)

EXAMPLE Identifying Type I and Type II Errors Assume that we are conducting a hypothesis test of the claim that $p < 0.5$. Here are the null and alternative hypotheses:

$$H_0: p = 0.5$$
$$H_1: p < 0.5$$

Give statements identifying

a. a type I error.

b. a type II error.

SOLUTION

a. A type I error is the mistake of rejecting a true null hypothesis, so this is a type I error: Conclude that there is sufficient evidence to support $p < 0.5$, when in reality $p = 0.5$.

b. A type II error is the mistake of failing to reject the null hypothesis when it is false, so this is a type II error: Fail to reject $p = 0.5$ (and therefore fail to support $p < 0.5$) when in reality $p < 0.5$.

Controlling Type I and Type II Errors: One step in our standard procedure for testing hypotheses involves the selection of the significance level α, which is the probability of a type I error. However, we don't select β [P(type II error)]. It would be great if we could always have $\alpha = 0$ and $\beta = 0$, but in reality that is not possible, so we must attempt to manage the α and β error probabilities. Mathematically, it can be shown that α, β, and the sample size n are all related, so when you choose or determine any two of them, the third is automatically determined. The usual practice in research and industry is to select the values of α and n, so the value of β is determined. Depending on the seriousness of a type I error, try to use the largest α that you can tolerate. For type I errors with more serious consequences, select smaller values of α. Then choose a sample size n as large as is reasonable, based on considerations of time, cost, and other relevant factors. (Sample size determinations were discussed in Chapter 7.) The following practical considerations may be relevant:

1. For any fixed α, an increase in the sample size n will cause a decrease in β. That is, a larger sample will lessen the chance that you make the error of not rejecting the null hypothesis when it's actually false.
2. For any fixed sample size n, a decrease in α will cause an increase in β. Conversely, an increase in α will cause a decrease in β.
3. To decrease both α and β, increase the sample size.

To make sense of these abstract ideas, let's consider M&Ms (produced by Mars, Inc.) and Bufferin brand aspirin tablets (produced by Bristol-Myers Products).

- The mean weight of the M&M candies is supposed to be at least 0.8535 g (in order to conform to the weight printed on the package label).
- The Bufferin tablets are supposed to have a mean weight of 325 mg of aspirin.

Because M&Ms are candies used for enjoyment, whereas Bufferin tablets are drugs used for treatment of health problems, we are dealing with two very different levels of seriousness. In testing the claim that $\mu = 0.8535$ g for M&Ms, we might choose $\alpha = 0.05$ and a sample size of $n = 100$; in testing the claim that $\mu = 325$ mg for Bufferin tablets, we might choose $\alpha = 0.01$ and a larger sample size of $n = 500$. The smaller significance level α and larger sample size n are chosen because of the more serious consequences associated with testing a commercial drug.

Comprehensive Hypothesis Test In this section we describe the individual components used in a hypothesis test, but the following sections will combine those components in comprehensive procedures. We can test claims about population parameters by using the *P*-value method summarized in Figure 8-8, the traditional method summarized in Figure 8-9, or we can use a confidence interval. For two-tailed hypothesis tests construct a confidence interval with a confidence level of $1 - \alpha$; but for a one-tailed hypothesis test with significance level α, construct a confidence interval with a confidence level of $1 - 2\alpha$. (See Table 8-2 for common cases.) After constructing the confidence interval, use this criterion:

A confidence interval estimate of a population parameter contains the likely values of that parameter. We should therefore reject a claim that the population parameter has a value that is not included in the confidence interval. ***Caution:*** **In some cases, a conclusion based on a confidence interval may be different from a conclusion based on a hypothesis test. See the comments in the individual sections.**

The exercises for this section involve isolated components of hypothesis tests, but the following sections will involve complete and comprehensive hypothesis tests.

Part 2: Beyond the Basics of Hypothesis Testing: The *Power* of a Test

Power of a Test: We use β to denote the probability of failing to reject a false null hypothesis (type II error). It follows that $1 - \beta$ is the probability of rejecting a false null hypothesis. Statisticians refer to this probability as the *power* of a test, and they often use it to gauge the test's effectiveness in recognizing that a null hypothesis is false.

Definition

The **power** of a hypothesis test is the probability $(1 - \beta)$ of rejecting a false null hypothesis, which is computed by using a particular significance level α and a particular value of the population parameter that is an alternative to the value assumed true in the null hypothesis. That is, the power of a hypothesis test is the probability of supporting an alternative hypothesis that is true.

Figure 8-8 ***P*-Value Method**

Figure 8-9 **Traditional Method**

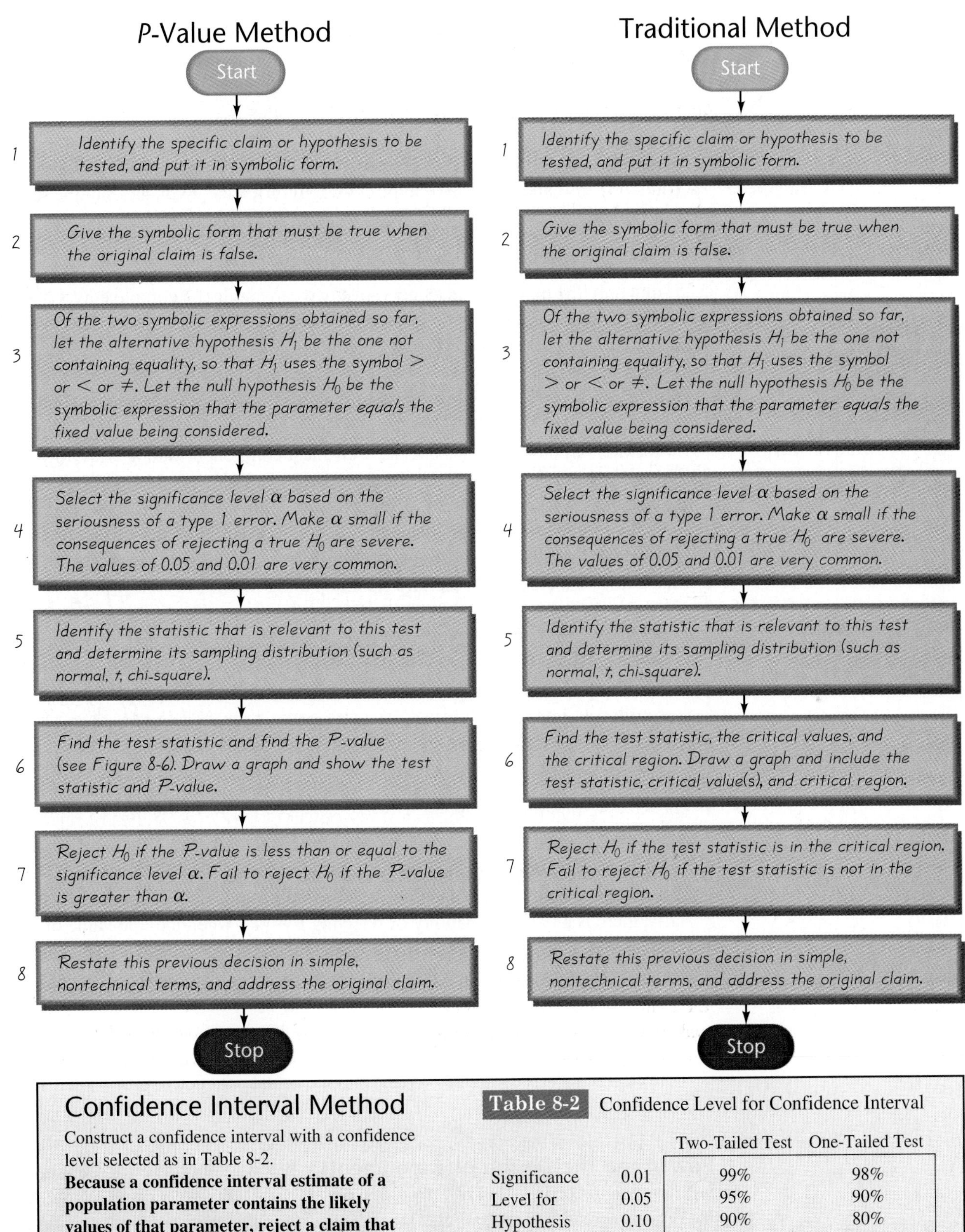

Confidence Interval Method

Construct a confidence interval with a confidence level selected as in Table 8-2.

Because a confidence interval estimate of a population parameter contains the likely values of that parameter, reject a claim that the population parameter has a value that is not included in the confidence interval.

Table 8-2 Confidence Level for Confidence Interval

		Two-Tailed Test	One-Tailed Test
Significance Level for Hypothesis Test	0.01	99%	98%
	0.05	95%	90%
	0.10	90%	80%

Note that in the above definition, determination of power requires a specific value that is an alternative to the value assumed in the null hypothesis. Consequently, a hypothesis test can have many different values of power, depending on the particular values chosen as alternatives to the null hypothesis.

EXAMPLE Power of a Hypothesis Test Assume that we have the following null and alternative hypotheses, significance level, and sample data.

H_0: $p = 0.5$ H_1: $p \neq 0.5$ Significance level: $\alpha = 0.05$
Sample size: $n = 100$ Sample proportion: $\hat{p} = 0.57$

Using only the components listed above, we can conduct a complete hypothesis test. (The test statistic is $z = 1.4$, the critical values are $z = \pm 1.96$, the *P*-value is 0.1616, we fail to reject the null hypothesis, and we conclude that there is not sufficient evidence to warrant rejection of the claim that the population proportion is equal to 0.5.) However, determination of power requires another item: a *specific* value of p to be used as an alternative to the value of $p = 0.5$ assumed in the null hypothesis. If we use the above test components along with different alternative values of p, we get the following examples of power values. (The values of power can be found using some software packages, such as Minitab, or values of power can be manually calculated. Because the calculations of power are complicated, in this section only Exercise 47 deals with power, and that exercise includes a procedure for calculating power.)

Specific Alternative Value of p	β	Power of Test $(1 - \beta)$
0.3	0.013	0.987
0.4	0.484	0.516
0.6	0.484	0.516
0.7	0.013	0.987

INTERPRETATION OF POWER Based on the above list of power values, we see that this hypothesis test has power of 0.987 (or 98.7%) of rejecting H_0: $p = 0.5$ when the population proportion p is actually 0.3. That is, if the true population proportion is actually equal to 0.3, there is a 98.7% chance of making the correct conclusion of rejecting the false null hypothesis that $p = 0.5$. Similarly, there is a 0.516 probability of rejecting $p = 0.5$ when the true value of p is actually 0.4. It makes sense that this test is more effective in rejecting the claim of $p = 0.5$ when the population proportion is actually 0.3 than when the population proportion is actually 0.4. [When identifying animals assumed to be horses, there's a better chance of rejecting an elephant as a horse (because of the greater difference) than rejecting a mule as a horse.] In general, increasing the difference between the assumed parameter value and the actual parameter value results in an increase in power.

Power and the Design of Experiments If a hypothesis test is conducted with sample data consisting of only a few observations, the power will be low, but the power increases as the sample size increases (and the other components remain the same). In addition to increasing the sample size, there are other ways to increase the power, such as increasing the significance level, using a more extreme value for the population parameter, or decreasing the standard deviation. Just as 0.05 is a common choice for a significance level, a power of at least 0.80 is a common requirement for determining that a hypothesis test is effective. (Some statisticians argue that the power should be higher, such as 0.85 or 0.90.)

When designing an experiment, we might consider how much of a difference between the claimed value of a parameter and its true value is an important amount of difference. If testing the effectiveness of a gender-selection method, a change in the proportion of girls from 0.5 to 0.501 is likely to be unimportant. A change in the proportion of girls from 0.5 to 0.6 might be important. Such magnitudes of differences affect power. When designing an experiment, a goal of having a power value of at least 0.80 can often be used to determine the minimum required sample size. For example, here is a statement similar to one taken from an article in the *Journal of the American Medical Association:* "The trial design assumed that with a 0.05 significance level, 153 randomly selected subjects would be needed to achieve 80% power to detect a reduction in the coronary heart disease rate from 0.5 to 0.4." Before conducting the experiment, the researchers determined that in order to achieve power of at least 0.80, they needed at least 153 randomly selected subjects. Due to factors such as dropout rates, the researchers are likely to need somewhat more than 153 subjects.

8-2 BASIC SKILLS AND CONCEPTS

Statistical Literacy and Critical Thinking

1. Proving That $p = 0.5$ A newspaper article states that "based on a recent survey, it has been proved that 50% of all truck drivers smoke." What is wrong with that statement?

2. Null Hypothesis The quality control manager of a cola bottling company claims that "the mean amount of cola in our cans is at least 12 ounces." In testing that claim, express the null hypothesis and alternative hypothesis in symbolic form. Does the original claim correspond to the null hypothesis, alternative hypothesis, or neither?

3. Test Statistic and Critical Value What is the difference between a test statistic and a critical value?

4. *P*-Value You have developed a new drug and you claim that it lowers cholesterol, so you express the claim as $\mu < 100$ (where 100 is a standardized index of cholesterol). Which *P*-value would you prefer to get: 0.04 or 0.01? Why?

Stating Conclusions About Claims. *In Exercises 5–8, make a decision about the given claim. (Don't use formal procedures and exact calculations. Use only the rare event rule described in Section 8-1, and make subjective estimates to determine whether events are likely.)*

5. Claim: A coin favors heads when tossed, and there are 11 heads in 20 tosses.

6. Claim: The majority of people in the 18–29 age bracket voted in the last presidential election, and polls showed that among 1000 voters in that age bracket, 170 voted.

7. Claim: The mean pulse rate (in beats per minute) of males is greater than 60, and a random sample of 400 males has a mean pulse rate of 69.4.

8. Claim: College students have IQ scores that vary less than the general population for which $\sigma = 15$, and a random sample of 500 college students results in IQ scores with $s = 10.2$.

Identifying H_0 and H_1. *In Exercises 9–16, examine the given statement, then express the null hypothesis H_0 and alternative hypothesis H_1 in symbolic form. Be sure to use the correct symbol (μ, p, σ) for the indicated parameter.*

9. More than 25% of Internet users pay bills online.

10. Most households have telephones.

11. The mean weight of women who won Miss America titles is equal to 121 lb.

12. The mean top of knee height of a sitting male is 20.7 in.

13. IQ scores of college professors have a standard deviation less than 15, which is the standard deviation for the general population.

14. High school teachers have incomes with a standard deviation that is less than $20,000.

15. Plain M&M candies have a mean weight that is at least 0.8535 g.

16. The percentage of workers who got a job through their college is no more than 2%.

Finding Critical Values. *In Exercises 17–24, find the critical z values. In each case, assume that the normal distribution applies.*

17. Two-tailed test; $\alpha = 0.05$.

18. Two-tailed test; $\alpha = 0.01$.

19. Right-tailed test; $\alpha = 0.01$.

20. Left-tailed test; $\alpha = 0.05$.

21. $\alpha = 0.10$; H_1 is $p \neq 0.17$.

22. $\alpha = 0.10$; H_1 is $p > 0.18$.

23. $\alpha = 0.02$; H_1 is $p < 0.19$.

24. $\alpha = 0.005$; H_1 is $p \neq 0.20$.

Finding Test Statistics. *In Exercises 25–28, find the value of the test statistic z using*

$$z = \frac{\hat{p} - p}{\sqrt{\dfrac{pq}{n}}}$$

25. Gallup Poll on Smoking The claim is that the proportion of adults who smoked a cigarette in the past week is less than 0.25, and the sample statistics include $n = 1018$ subjects with 224 saying that they smoked a cigarette in the past week.

26. Genetics Experiment The claim is that the proportion of peas with yellow pods is equal to 0.25 (or 25%), and the sample statistics include $n = 580$ peas with 152 of them having yellow pods.

27. Gallup Poll on Job Satisfaction The claim is that more than 75% of workers are satisfied with their job, and sample statistics include 580 employed adults, with 516 of them saying that they are satisfied with their job.

28. Medical Malpractice Lawsuits The claim is that the majority of medical malpractice lawsuits are dropped or dismissed, and a random sample of 500 lawsuits includes 349 that were dropped or dismissed.

Finding P-values. *In Exercises 29–36, use the given information to find the P-value. (Hint: See Figure 8-6.)*

29. The test statistic in a right-tailed test is $z = 1.00$.

30. The test statistic in a left-tailed test is $z = -2.00$.

31. The test statistic in a two-tailed test is $z = 1.96$.

32. The test statistic in a two-tailed test is $z = -0.50$.

33. With H_1: $p > 0.333$, the test statistic is $z = 1.50$.

34. With H_1: $p \neq 0.667$, the test statistic is $z = 2.05$.

35. With H_1: $p \neq 1/4$, the test statistic is $z = -1.75$.

36. With H_1: $p < 2/3$, the test statistic is $z = -0.45$.

Stating Conclusions. *In Exercises 37–40, state the final conclusion in simple nontechnical terms. Be sure to address the original claim. (Hint: See Figure 8-7.)*

37. Original claim: The proportion of male golfers is less than 0.5.
Initial conclusion: Reject the null hypothesis.

38. Original claim: The percentage of female coaches is greater than 50%.
Initial conclusion: Reject the null hypothesis.

39. Original claim: The proportion of red M&Ms is different from 0.13.
Initial conclusion: Fail to reject the null hypothesis.

40. Original claim: The proportion of smokers who experience sleeping difficulty is equal to 0.34.
Initial conclusion: Reject the null hypothesis.

Identifying Type I and Type II Errors. *In Exercises 41–44, identify the type I error and the type II error that correspond to the given hypothesis.*

41. The proportion of settled medical malpractice suits is 0.25.

42. The proportion of unlisted telephones in Nevada is 0.524.

43. The proportion of murders cleared by arrests is 0.62.

44. The proportion of car crashes occurring less than a mile from home is 0.23.

8-2 BEYOND THE BASICS

45. Unnecessary Test When testing a claim that the majority of adult Americans are against the death penalty for a person convicted of murder, a random sample of 491 adults is obtained, and 27% of them are against the death penalty (based on data from a Gallup poll). Find the *P*-value. Why is it not necessary to go through the steps of conducting a formal hypothesis test?

46. Significance Level If a null hypothesis is rejected with a significance level of 0.05, is it also rejected with a significance level of 0.01? Why or why not?

47. Power of a Test Assume that you are using a significance level of $\alpha = 0.05$ to test the claim that $p > 0.5$ and that your sample is a simple random sample of size $n = 64$.

a. Assuming that the true population proportion is 0.65, find the power of the test, which is the probability of rejecting the null hypothesis when it is false. (In the procedure below, we refer to $p = 0.5$ as the "assumed" value, because it is assumed in the null hypothesis; we refer to $p = 0.65$ as the "alternative" value, because it is the value of the population proportion used as an alternative to 0.5.) Use the following procedure and see the figure below.

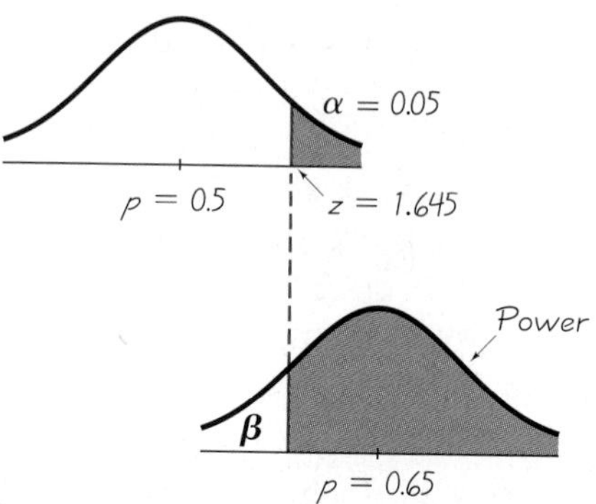

Step 1: Using the significance level, find the critical z value(s). (For a right-tailed test, there is a single critical z value that is positive; for a left-tailed test, there is a single critical value of z that is negative; and a two-tailed test will have a critical z value that is negative along with another critical z value that is positive.)

Step 2: In the expression for the test statistic below, substitute the assumed value of p (used in the null hypothesis). Evaluate $1 - p$ and substitute that result for the entry of q. Also substitute the critical value(s) for z. Then solve for the sample statistic $\hat{p}$. (If the test is two-tailed, substitute the critical value of z that is positive, then solve for the sample statistic $\hat{p}$. Next, substitute the critical value of z that is negative, and solve for the sample statistic $\hat{p}$. A two-tailed test should therefore result in two different values of $\hat{p}$.) The resulting value(s) of $\hat{p}$ separate the region(s) where the null hypothesis is rejected from the region where we fail to reject the null hypothesis.

$$z = \frac{\hat{p} - p}{\sqrt{\frac{pq}{n}}}$$

Step 3: The calculation of power requires a specific value of p that is to be used as an alternative to the value assumed in the null hypothesis. Identify this alternative value of p (not the value used in the null hypothesis), draw a normal curve with this alternative value at the center, and plot the value(s) of $\hat{p}$ found in Step 2.

Step 4: Refer to the graph in Step 3, and find the area of the new critical region bounded by the value(s) of $\hat{p}$ found in Step 2. (*Caution*: When evaluating $\sqrt{pq/n}$, be sure to use the alternative value of p, not the value of p used for the null hypothesis.) This is the probability of rejecting the null hypothesis, given that the alternative value of p is the true value of the population proportion. Because this is the probability of rejecting the false null hypothesis, it is the *power* of the test.

b. Find β, which is the probability of *failing* to reject the false null hypothesis. The value of β is easily determined by finding the complement of the power.

48. Finding Sample Size A researcher plans to conduct a hypothesis test using the alternative hypothesis of H_1: $p < 0.4$, and she plans to use a significance level of $\alpha = 0.05$. Find the sample size required to achieve at least 80% power in detecting a reduction in p from 0.4 to 0.3. (*This is a very difficult exercise. Hint: See Exercise 47.*)

8-3 Testing a Claim About a Proportion

Key Concept This section presents complete procedures for testing a hypothesis (or claim) made about a population proportion. This section uses the components introduced in Section 8-2 for the *P*-value method, the traditional method, and the use of confidence intervals. In addition to testing claims about population proportions, we can use the same procedures for testing claims about probabilities or the decimal equivalents of percents. The methods of this section use the normal distribution as an approximation to the binomial probability distribution.

The following are examples of the types of claims we will be able to test:

- More than 50% of workers get their jobs through networking.
- Subjects taking the cholesterol-reducing drug Lipitor experience headaches at a rate that is greater than the 7% rate for people who do not take Lipitor.
- The percentage of late-night television viewers who watch *The Late Show with David Letterman* is equal to 20%.

Claims about a population proportion are usually tested by using a normal distribution as an approximation to the binomial distribution (see Section 6-6). Instead of using the same exact methods of Section 6-6, we use a different but equivalent form of the test statistic shown below, and we don't include the correction for continuity (because its effect tends to be very small with large samples). If the requirements are not all satisfied, we must use other methods not described in this section. In this section, all examples and exercises involve cases in which the requirements are satisfied, so the sampling distribution of sample proportions can be approximated by the normal distribution.

Testing Claims About a Population Proportion *p*

Requirements

1. The sample observations are a simple random sample.
2. The conditions for a *binomial distribution* are satisfied. (There are a fixed number of independent trials having constant probabilities, and each trial has two outcome categories of "success" and "failure.")
3. The conditions $np \geq 5$ and $nq \geq 5$ are both satisfied, so **the binomial distribution of sample proportions can be approximated by a normal distribution with $\mu = np$ and $\sigma = \sqrt{npq}$** (as described in Section 6-6). Note that p is the *assumed* proportion used in the claim, not the sample proportion.

continued

Process of Drug Approval

Gaining FDA approval for a new drug is expensive and time consuming. It begins with a **Phase I study** in which the safety of the drug is tested with a small (20–100) group of volunteers. In **Phase II,** the drug is tested for effectiveness in randomized trials involving a larger (100–300) group of subjects. This phase often has subjects randomly assigned to either a treatment group or a placebo group. In **Phase III,** the goal is to better understand the effectiveness of the drug as well as its adverse reactions. Phase III typically involves 1000–3000 subjects, and this phase alone might require several years of testing. Lisa Gibbs wrote in *Money* magazine that "the (drug) industry points out that for every 5,000 treatments tested, only five make it to clinical trials and only one ends up in drugstores." Total cost estimates vary from a low of $40 million to as much as $1.5 billion.

Notation

n = sample size or number of trials

$\hat{p} = \frac{x}{n}$ (*sample* proportion)

p = population proportion (used in the null hypothesis)

$q = 1 - p$

Test Statistic for Testing a Claim About a Proportion

$$z = \frac{\hat{p} - p}{\sqrt{\frac{pq}{n}}}$$

***P*-values:** Use the standard normal distribution (Table A-2) and refer to Figure 8-6.

Critical values: Use the standard normal distribution (Table A-2).

TI-83/84 Plus Calculator: The TI-83/84 Plus calculator can be used to automatically generate the test statistic and *P*-value. See the procedure at the end of this section.

EXAMPLE Finding a Job Through Networking The Chapter Problem included these survey results: Among 703 randomly selected workers, 61% got their jobs through networking. Use the sample data with a 0.05 significance level to test the claim that most (more than 50%) workers get their jobs through networking. Here is a summary of the claim and the sample data:

Claim: Most workers get their jobs through networking. That is, $p > 0.5$.

Sample data: $n = 703$ and $\hat{p} = 0.61$

We will provide solutions using the *P*-value method, the traditional method, and confidence intervals. Before proceeding, however, we should verify that the necessary requirements are satisfied.

REQUIREMENT ✔ See the three requirements listed above.

1. The survey appears to have been conducted using sound sampling methods, so we can assume that the sample is a simple random sample.
2. There is a fixed number (703) of independent trials with two categories (either the worker got the job through networking or did not). (Technically, the trials are not independent, but they can be treated as independent by using this guideline presented in Section 5-3: "When sampling without replacement, the events can be treated as if they are independent if the sample size is no more than 5% of the population size. That is, $n \le 0.05N$.")
3. The requirements $np \ge 5$ and $nq \ge 5$ are both satisfied with $n = 703$, $p = 0.5$, and $q = 0.5$. [We get $np = (703)(0.5) = 351.5 \ge 5$ and $nq = (703)(0.5) = 351.5 \ge 5$.]

With the three requirements all satisfied, we can now proceed to conduct formal hypothesis tests. The *P*-value method, the traditional method, and the use of confidence intervals are illustrated in the following discussion. ✔

The *P*-Value Method

When testing the claim $p > 0.5$ given in the preceding example, the following steps correspond to the *P*-value method of testing hypotheses that is summarized in Figure 8-8.

Step 1: The original claim in symbolic form is $p > 0.5$.

Step 2: The opposite of the original claim is $p \leq 0.5$.

Step 3: Of the preceding two symbolic expressions, the expression $p > 0.5$ does not contain equality, so it becomes the alternative hypothesis. The null hypothesis is the statement that p equals the fixed value of 0.5. We can therefore express H_0 and H_1 as follows:

$$H_0: p = 0.5$$
$$H_1: p > 0.5$$

Step 4: In the absence of any special circumstances, we will select $\alpha = 0.05$ for the significance level.

Step 5: The normal distribution is used for this test, because we are testing a claim about a population proportion p, the sample statistic $\hat{p}$ is relevant to this test, and the sampling distribution of sample proportions $\hat{p}$ is approximated by a normal distribution.

Step 6: The test statistic is $z = 5.83$, which is calculated as shown here:

$$z = \frac{\hat{p} - p}{\sqrt{\frac{pq}{n}}} = \frac{0.61 - 0.5}{\sqrt{\frac{(0.5)(0.5)}{703}}} = 5.83$$

We now find the *P*-value by using the following procedure, which is shown in Figure 8-6:

Right-tailed test: *P*-value = area to right of test statistic z

Left-tailed test: *P*-value = area to left of test statistic z

Two-tailed test: *P*-value = *twice* the area of the extreme region bounded by the test statistic z

Because the hypothesis test we are considering is right-tailed with a test statistic of $z = 5.83$, the *P*-value is the area to the right of $z = 5.83$. Referring to Table A-2, we see that for values of $z = 3.50$ and higher, we use 0.0001 for the cumulative area to the *right* of the test statistic. The *P*-value is therefore 0.0001. Figure 8-10 shows the test statistic and *P*-value for this example. (If the TI-83/84 Plus calculator is used instead of Table A-2, a more accurate *P*-value of 0.000000003 is obtained.)

Figure 8-10

***P*-Value Method**

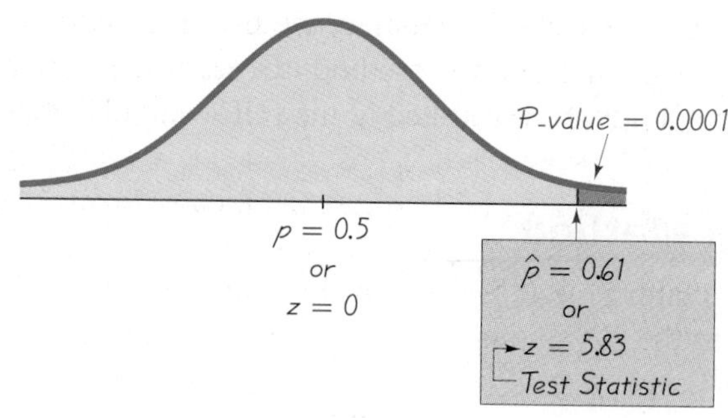

Step 7: Because the *P*-value of 0.0001 is less than or equal to the significance level of $\alpha = 0.05$, we reject the null hypothesis.

Step 8: We conclude that there is sufficient sample evidence to support the claim that most workers got their jobs through networking. (See Figure 8-7 for help with wording this final conclusion.)

The Traditional Method

The traditional method of testing hypotheses is summarized in Figure 8-9, and Steps 1 through 5 are the same as in Steps 1 through 5 for the *P*-value method, as shown above. We therefore continue with Step 6 of the traditional method.

Step 6: The test statistic is $z = 5.83$ as shown for the preceding *P*-value method. We now find the critical value (instead of the *P*-value). This is a right-tailed test, so the area of the critical region is an area of $\alpha = 0.05$ in the right tail. Referring to Table A-2 and applying the methods of Section 6-2, we find that the critical value of $z = 1.645$ is at the boundary of the critical region. See Figure 8-3 on page 407, which shows the critical region, critical value, and test statistic.

Step 7: Because the test statistic falls within the critical region, we reject the null hypothesis.

Step 8: We conclude that there is sufficient sample evidence to support the claim that most workers get their jobs through networking. (See Figure 8-7 for help with wording this final conclusion.)

Confidence Interval Method

The claim of $p > 0.5$ can be tested with a 0.05 significance level by constructing a 90% confidence interval. (See Table 8-2. In general, for two-tailed hypothesis tests construct a confidence interval with a confidence level of $1 - \alpha$; but one-tailed hypothesis tests with significance level α require a confidence interval with a confidence level of $1 - 2\alpha$.)

Let's now use the confidence interval method to test the claim of $p > 0.5$, with sample data consisting of $n = 703$ and $\hat{p} = 0.61$ (as in the Chapter Problem). If we want a significance level of $\alpha = 0.05$ in a right-tailed test, we use a 90% confidence level with the methods of Section 7-2 to get this result: $0.580 < p < 0.640$. Because we are 90% confident that the true value of p is contained within the limits of 0.580 and 0.640, we have sufficient evidence to support the claim that $p > 0.5$.

Caution: When testing claims about a population proportion, the traditional method and the *P*-value method are equivalent in the sense that they always yield the same results, but the confidence interval method is somewhat different. Both the traditional method and *P*-value method use the same standard deviation based on the *claimed proportion* p, but the confidence interval uses an estimated standard deviation based on the *sample proportion* $\hat{p}$. Consequently, it is possible that in some cases, the traditional and *P*-value methods of testing a claim about a proportion might yield a different conclusion than the confidence interval method. (See Exercise 29.) A good strategy is to use a confidence interval to estimate a population proportion, but use the *P*-value method or traditional method for testing a hypothesis.

When testing a claim about a population proportion p, be careful to identify correctly the sample proportion $\hat{p}$. The sample proportion $\hat{p}$ is sometimes given directly, but in other cases it must be calculated. See the following examples:

Given Statement	Finding $\hat{p}$
10% of the observed sports cars are red	$\hat{p}$ is given directly: $\hat{p} = 0.10$
96 surveyed households have cable TV and 54 do not	$\hat{p}$ must be calculated using $\hat{p} = x/n$. $\hat{p} = \frac{x}{n} = \frac{96}{96 + 54} = 0.64$

EXAMPLE Mendel's Genetics Experiments When Gregor Mendel conducted his famous hybridization experiments with peas, one such experiment resulted in offspring consisting of 428 peas with green pods and 152 peas with yellow pods. According to Mendel's theory, 1/4 of the offspring peas should have yellow pods. Use a 0.05 significance level with the *P*-value method to test the claim that the proportion of peas with yellow pods is equal to 1/4.

SOLUTION

REQUIREMENT ✔ (1) Based on the design of the experiment, it is reasonable to assume that the sample is a simple random sample. (2) There is a fixed number ($428 + 152 = 580$) of independent trials with two categories (green pods or yellow pods). (3) The requirements $np \geq 5$ and $nq \geq 5$ are both satisfied with $n = 580$, $p = 1/4$, and $q = 3/4$. [We get $np = (580)(1/4) = 145 \geq 5$ and $nq = (580)(3/4) = 435 \geq 5$.] With the three requirements all satisfied, we can now proceed to conduct a formal hypothesis test. ✔

We use the *P*-value method summarized in Figure 8-8 found in Section 8-2. Note that $n = 428 + 152 = 580$, $\hat{p} = 152/580 = 0.262$, and, for the purposes of the test, we assume that $p = 1/4 = 0.25$.

Step 1: The original claim is that the proportion of peas with yellow pods is equal to 1/4. We express this in symbolic form as $p = 0.25$.

Step 2: The opposite of the original claim is $p \neq 0.25$

Step 3: Because $p \neq 0.25$ does not contain equality, it becomes H_1. We get

$H_0: p = 0.25$ (null hypothesis and original claim)

$H_1: p \neq 0.25$ (alternative hypothesis)

Step 4: The significance level is $\alpha = 0.05$.

continued

Step 5: Because the claim involves the proportion p, the statistic relevant to this test is the sample proportion $\hat{p}$, and the sampling distribution of sample proportions is approximated by the normal distribution.

Step 6: The test statistic of $z = 0.67$ is found as follows:

$$z = \frac{\hat{p} - p}{\sqrt{\frac{pq}{n}}} = \frac{0.262 - 0.25}{\sqrt{\frac{(0.25)(0.75)}{580}}} = 0.67$$

Refer to Figure 8-6 for the procedure for finding the P-value. Figure 8-6 shows that for this two-tailed test with the test statistic located to the right of the center (because $z = 0.67$ is positive), the P-value is *twice* the area to the right of the test statistic. Using Table A-2, $z = 0.67$ has an area of 0.7486 to its left, so the area to the right of $z = 0.67$ is $1 - 0.7486 = 0.2514$, which we double to get 0.5028. (If using the TI-83/84 Plus calculator, we get the more accurate P-value of 0.5021.)

Step 7: Because the P-value of 0.5028 is greater than the significance level of 0.05, we fail to reject the null hypothesis.

INTERPRETATION The methods of hypothesis testing never allow us to support a claim of equality, so we cannot conclude that the proportion of peas with yellow pods is equal to 1/4. Here is the correct conclusion: There is not sufficient evidence to warrant rejection of the claim that 1/4 of the offspring peas have yellow pods.

Traditional Method: If we were to repeat the preceding example using the traditional method of testing hypotheses, we would see that in Step 6, the critical values are found to be $z = -1.96$ and $z = 1.96$. In Step 7, we would fail to reject the null hypothesis because the test statistic of $z = 0.67$ would not fall within the critical region. See the accompanying STATDISK display. We would reach the same conclusion from the P-value method: There is not sufficient evidence to warrant rejection of the claim that 1/4 of the offspring peas have yellow pods.

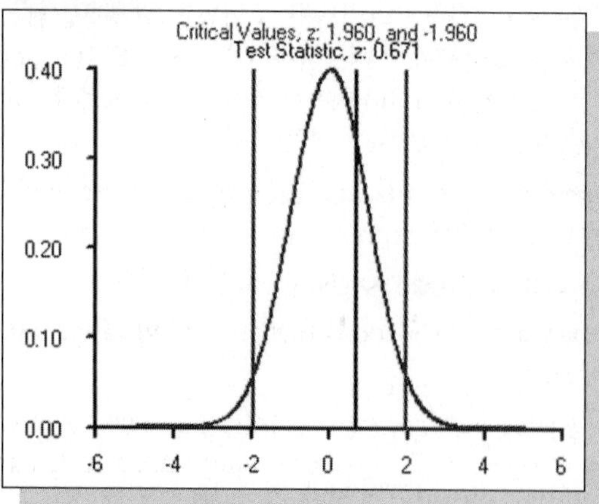

Confidence Interval Method: If we were to repeat the preceding example using the confidence interval method, we would obtain this 95% confidence interval: $0.226 < p < 0.298$. Because the confidence interval limits do contain the claimed value of 0.25, we conclude that there is not sufficient evidence to warrant rejection of the claim that 1/4 of the offspring peas have yellow pods. In this case, the *P*-value method, traditional method, and confidence interval method all lead to the same conclusion. In some other relatively rare cases, the *P*-value method and the traditional method might lead to a conclusion that is different from the conclusion reached through the confidence interval method.

Rationale for the Test Statistic: The test statistic used in this section is justified by noting that when using the normal distribution to approximate a binomial distribution, we use $\mu = np$ and $\sigma = \sqrt{npq}$ to get

$$z = \frac{x - \mu}{\sigma} = \frac{x - np}{\sqrt{npq}}$$

We used the above expression in Section 6-6 along with a correction for continuity, but when testing claims about a population proportion, we make two modifications. First, we don't use the correction for continuity because its effect is usually very small for the large samples we are considering. Also, instead of using the above expression to find the test statistic, we use an equivalent expression obtained by dividing the numerator and denominator by n, and we replace x/n by the symbol $\hat{p}$ to get the test statistic we are using. The end result is that the test statistic is simply the same standard score (from Section 3-4) of $z = (x - \mu)/\sigma$, but modified for the binomial notation.

TI Using a TI-83/84 Plus Calculator for Testing a Claim About a Proportion

Requirements: When using a TI-83/84 Plus calculator for testing a claim about a population proportion p, you must know the sample size n and the number of successes x. If you do not know the value of x, but you know the value of the sample size n and the sample proportion $\hat{p}$, calculate the value of x as follows: $x = n \cdot \hat{p}$, then round the result to the nearest whole number. For example, the Chapter Problem gives a sample size of $n = 703$ subjects and a sample proportion of $\hat{p} = 0.61$ (the decimal equivalent of 61%). We find 61% of 703 by calculating $x = n \cdot \hat{p} = 703 \cdot 0.61 = 428.83$, which we round to the nearest whole number of 429. When we must enter a value for x, we enter 429.

Here is the procedure for using a TI-83/84 Plus calculator for testing a claim about a population proportion:

1. Press STAT, then press ▷ twice to select **TESTS.**
2. Select the menu item of **1-PropZTest** (for a hypothesis test of a claim involving one proportion).
3. For the first required entry indicated by $p0$, enter the claimed value of the population proportion. It is the same value used in the statement of the null and alternative hypotheses.

4. For the second required entry of x, enter the number of successes. (See the above comment about calculating x when only the sample size n and sample proportion $\hat{p}$ are known.)
5. For the third required entry of n, enter the sample size.
6. For the next line, select the type of test (two-tailed, left-tailed, right-tailed). Select the option that uses the same symbol as the alternative hypothesis. For a two-tailed case select the option including $\neq$, for a left-tailed test select the option of $<p0$, and for a right-tailed test select the option of $>p0$.
7. Use the arrow keys to scroll down to the bottom line, select **Calculate,** then press ENTER.

TI-83/84 Plus

```
1-PropZTest
 p0:.5
 x:429
 n:703
 prop≠p0 <p0 >p0
 Calculate Draw
```

See the Chapter Problem and the examples in this section that relate to it. That problem requires that we test the claim of $p > 0.5$, and we use the sample data consisting of $n = 703$ surveyed subjects with 61% of them getting jobs through networking. The TI-83/84 Plus calculator requires that we find the value of x, which is 429. (See the discussion preceding the above list of steps.) Using $n =$ 703 and $x = 429$, we proceed to use the calculator as described above. Shown here are the screen displays for the required entries and the generated results.

TI-83/84 Plus

```
1-PropZTest
 prop>.5
 z=5.84593572
 p=2.526737E-9
 p̂=.6102418208
 n=703
```

Interpreting the results: See the displayed results and note that the TI-83/84 Plus calculator provides the test statistic of $z = 5.84593572$ and the P-value of $2.526737\text{E}^{-}9$, which can be expressed in standard format as 0.000000003 (rounded). Because the P-value is less than the significance level of 0.05, we reject the null hypothesis and conclude that there is sufficient sample evidence to conclude that most workers got their jobs through networking.

Because the TI-83/84 Plus calculator provides the test statistic and P-value (but not the critical value), we use the P-value method of testing hypotheses.

8-3 BASIC SKILLS AND CONCEPTS

Statistical Literacy and Critical Thinking

1. Distribution Assuming that the listed requirements of this section are satisfied, what distribution is used to test a claim about a population proportion? Why?

2. Sample Proportion When respondents are asked a question on a survey, 40 of them answer *yes*, 60 of them answer *no,* and there are no other responses. What is the sample proportion of *yes* responses, and what notation is used to represent it?

3. Sampling America Online conducts a survey in which Internet users are asked to respond to a question. Among the 96,772 responses, there are 76,885 responses of "yes." Is it valid to use these sample results for testing the claim that the majority of the general population answers "yes"?

4. *P*-Value Method A P-value of 0.00001 is obtained when using sample data to test the claim that the majority of car crashes occur within 5 miles of home. Interpret this P-value in the context of this hypothesis test. That is, what does the P-value tell us?

In Exercises 5–8, identify the indicated values or interpret the given display.

TI 5. Mendel's Hybridization Experiments In one of Mendel's famous hybridization experiments, 8023 offspring peas were obtained, and 24.94% of them had green flowers.

The others had white flowers. Consider a hypothesis test that uses a 0.05 significance level to test the claim that green-flowered peas occur at a rate of 25%.

a. What is the test statistic?

b. What are the critical values?

c. What is the P-value?

d. What is the conclusion?

e. Can a hypothesis test be used to "prove" that the rate of green-flowered peas is 25%, as claimed?

TI **6.** Survey of Workers In a survey of 703 randomly selected workers, 15.93% got their jobs through newspaper ads (based on data from Taylor Nelson Sofres Intereach). Consider a hypothesis test that uses a 0.05 significance level to test the claim that less than 20% of workers get their jobs through newspaper ads.

a. What is the test statistic?

b. What is the critical value?

c. What is the P-value?

d. What is the conclusion?

e. Based on the preceding results, can we conclude that 15.93% is significantly less than 20% for all such hypothesis tests? Why or why not?

TI **7.** Interpreting Display When 109,857 arrests for federal offenses were randomly selected, it was found that 31,969 of them were drug offenses. When testing the claim that more than 29% of federal crimes were for drug offenses, the accompanying TI-83/84 Plus calculator display was obtained. Use the results from the display to test the given claim.

TI-83/84 Plus

```
1-PropZTest
 prop>.29
 z=.7345177804
 p=.2313165397
 p̂=.29100558
 n=109857
```

8. Percentage of Telephone Users A survey of 4276 randomly selected households showed that 4019 of them had telephones (based on data from the U.S. Census Bureau). Minitab was used to test the claim that the percentage of households is now greater than the 35% rate that was found in 1920. The Minitab display is given below. The current rate of 4019/4276 (or 94%) appears to be significantly greater than the 1920 rate of 35%, but is there sufficient evidence to support that claim? Use a 0.01 significance level.

Minitab Display for Exercise 8

Test of p = 0.35 vs p > 0.35

Sample	X	N	Sample p	95% Lower Bound	Z-Value	P-Value
1	4019	4276	0.939897	0.933919	80.87	0.000

Testing Claims About Proportions. *In Exercises 9–24, test the given claim. Identify the null hypothesis, alternative hypothesis, test statistic, P-value or critical value(s), conclusion about the null hypothesis, and final conclusion that addresses the original claim. Use the P-value method unless your instructor specifies otherwise.*

TI **9.** Gender Selection for Girls The Genetics and IVF Institute conducted a clinical trial of the XSORT method designed to increase the probability of conceiving a girl. As this book was being written, 325 babies were born to parents using the XSORT method, and 295 of them were girls. Use the sample data with a 0.01 significance level to test the claim that with this method, the probability of a baby being a girl is greater than 0.5. Does the method appear to work?

TI **10.** Gender Selection for Boys The Genetics and IVF Institute conducted a clinical trial of the YSORT method designed to increase the probability of conceiving a boy. As this book was being written, 51 babies were born to parents using the YSORT method, and 39 of them were boys. Use the sample data with a 0.01 significance level to test the claim that with this method, the probability of a baby being a boy is greater than 0.5. Does the method appear to work?

TI **11.** Car Crashes In a study of 11,000 car crashes, it was found that 5720 of them occurred within 5 miles of home (based on data from Progressive Insurance). Use a 0.01 significance level to test the claim that more than 50% of car crashes occur within 5 miles of home. Are the results questionable because they are based on a survey sponsored by an insurance company?

TI **12.** Travel Through the Internet Among 734 randomly selected Internet users, it was found that 360 of them use the Internet for making travel plans (based on data from a Gallup poll). Use a 0.01 significance level to test the claim that among Internet users, less than 50% use it for making travel plans. Are the results important for travel agents?

TI **13.** Percentage of E-Mail Users Technology is dramatically changing the way we communicate. In 1997, a survey of 880 U.S. households showed that 149 of them use e-mail (based on data from *The World Almanac and Book of Facts*). Use those sample results to test the claim that more than 15% of U.S. households use e-mail. Use a 0.05 significance level. Is the conclusion valid today? Why or why not?

TI **14.** Drug Testing of Job Applicants In 1990, 5.8% of job applicants who were tested for drugs failed the test. At the 0.01 significance level, test the claim that the failure rate is now lower if a simple random sample of 1520 current job applicants results in 58 failures (based on data from the American Management Association). Does the result suggest that fewer job applicants now use drugs?

TI **15.** Cell Phones and Cancer In a study of 420,095 Danish cell phone users, 135 subjects developed cancer of the brain or nervous system (based on data from the *Journal of the National Cancer Institute* as reported in *USA Today*). Test the claim of a once popular belief that such cancers are affected by cell phone use. That is, test the claim that cell phone users develop cancer of the brain or nervous system at a rate that is different from the rate of 0.0340% for people who do not use cell phones. Because this issue has such great importance, use a 0.005 significance level. Should cell phone users be concerned about cancer of the brain or nervous system?

TI **16.** Testing Effectiveness of Nicotine Patches In one study of smokers who tried to quit smoking with nicotine patch therapy, 39 were smoking one year after the treatment, and 32 were not smoking one year after the treatment (based on data from "High-Dose Nicotine Patch Therapy," by Dale et al., *Journal of the American Medical Association,* Vol. 274, No. 17). Use a 0.10 significance level to test the claim that among smokers who try to quit with nicotine patch therapy, the majority are smoking a year after the treatment. Do these results suggest that the nicotine patch therapy is ineffective?

TI **17.** Store Checkout Scanner Accuracy In a study of store checkout scanners, 1234 items were checked and 20 checked items were found to be overcharges, and 1214 checked items were not overcharges (based on data from "UPC Scanner Pricing Systems: Are They Accurate?" by Goodstein, *Journal of Marketing,* Vol. 58). Use a 0.05 significance level to test the claim that with scanners, 1% of sales are overcharges. (Before scanners were used, the overcharge rate was estimated to be about 1%.) Based on these results, do scanners appear to help consumers avoid overcharges?

TI **18.** Postponing Death An interesting and popular hypothesis is that individuals can temporarily postpone their death to survive a major holiday or important event such as a birthday. In a study of this phenomenon, it was found that there were 6062 deaths in the week before Thanksgiving, and there were 5938 deaths the week after Thanksgiving (based on data from "Holidays, Birthdays, and Postponement of Cancer Death," by Young and Hade, *Journal of the American Medical Association,* Vol. 292, No. 24). If people can postpone their deaths until after Thanksgiving, then the proportion of deaths in the week before should be less than 0.5. Use a 0.05 significance level to test the claim that the proportion of deaths in the week before Thanksgiving is less than 0.5. Based on the result, does there appear to be any indication that people can temporarily postpone their death to survive the Thanksgiving holiday?

TI **19.** Drinking Survey A recent Gallup poll of 976 randomly selected adults showed that 312 of them never drink. Use those survey results to test the claim that less than 1/3 of all adults never drink. Use a 0.05 significance level. Also, examine the following wording of the actual question and determine whether it is likely to elicit honest responses: "How often, if ever, do you drink alcoholic beverages such as liquor, wine or beer—every day, a few times a week, about once a week, less than once a week, only on special occasions such as New Year's and holidays, or never?"

TI **20.** Smoking In a Gallup poll of 1018 adults, it was found that 22% smoked cigarettes in the past week. Use a 0.05 significance level to test the claim that less than 25% of adults have smoked within the past week. Would the conclusion change if, instead of a Gallup poll, the results were obtained from an Internet survey in which Internet users were asked to respond?

TI **21.** Flying In a Gallup poll of 1125 adults, it was found that 47% fly never or rarely. Use a 0.05 significance level to test the claim that the percentage of adults who fly never or rarely is equal to 50%. Given that the survey subjects volunteered an answer to a question about flying, is it possible that the sample results are in error by a considerable amount?

TI **22.** Bias in Jury Selection In the case of *Casteneda v. Partida,* it was found that during a period of 11 years in Hidalgo County, Texas, 870 people were selected for grand jury duty, and 39% of them were Mexican-Americans. Among the people eligible for grand jury duty, 79.1% were Mexican-Americans. Use a 0.01 significance level to test the claim that the selection process is biased against Mexican-Americans.

TI **23.** Testing Clarinex for Adverse Reaction Clarinex is a drug used to treat asthma. In clinical tests of this drug, 1655 patients were treated with 5-mg doses of Clarinex, and 2.1% of them experienced fatigue (based on data from the Schering Corporation). Use a 0.01 significance level to test the claim that the percentage of Clarinex users experiencing fatigue is greater than the 1.2% rate for those not using Clarinex. Does it appear that fatigue is an adverse reaction of Clarinex?

TI **24.** Smoking and College Education One survey showed that among 785 randomly selected subjects who completed four years of college, 18.3% smoke and 81.7% do not smoke (based on data from the American Medical Association). Use a 0.01 significance level to test the claim that the rate of smoking among those with four years of college is less than the 27% rate for the general population. Why would college graduates smoke at a lower rate than others?

Using Appendix B Data Sets. *In Exercises 25–28, use the Data Set from Appendix B to test the given claim.*

TI **25.** Using M&M Data Refer to Data Set 13 in Appendix B and find the sample proportion of M&Ms that are blue. Use that result to test the claim of Mars, Inc. that 24% of its plain M&M candies are blue.

TI **26.** Precipitation in Boston Refer to Data Set 10 in Appendix B, and note that days with any precipitation have values different from 0. Use a 0.05 significance level to test the claim that on Sundays in Boston, there is precipitation more than 25% of the days.

TI **27.** Accuracy of Forecast Temperatures Refer to Data Set 8 in Appendix B. Find the proportion of days with an actual high temperature that is more than 2° different from the high temperature that was forecast one day before. Let p represent the proportion of days with an actual high temperature that is more than 2° different from the high temperature that was forecast one day before. Use a 0.05 significance level to test the claim that $p < 0.5$. What does the result indicate about the accuracy of the forecast?

TI **28.** Alcohol and Tobacco Use in Animated Children's Movies Using results listed in Data Set 5 in Appendix B, test the claim that the majority of animated children's movies show the use of alcohol or tobacco (or both). Use a 0.05 significance level.

8-3 BEYOND THE BASICS

TI **29.** Using Confidence Intervals to Test Hypotheses When analyzing the last digits of telephone numbers in Port Jefferson, it is found that among 1000 randomly selected digits, 119 are zeros. If the digits are randomly selected, the proportion of zeros should be 0.1.

a. Use the traditional method with a 0.05 significance level to test the claim that the proportion of zeros equals 0.1.

b. Use the *P*-value method with a 0.05 significance level to test the claim that the proportion of zeros equals 0.1.

c. Use the sample data to construct a 95% confidence interval estimate of the proportion of zeros. What does the confidence interval suggest about the claim that the proportion of zeros equals 0.1?

d. Compare the results from the traditional method, the *P*-value method, and the confidence interval method. Do they all lead to the same conclusion?

TI **30.** Using the Continuity Correction Repeat Exercise 28, but include the correction for continuity that was described in Section 6-6. How are the results affected by including the continuity correction?

31. Alternative Method of Testing a Claim About p In a study of perception, 80 men are tested and 7 are found to have red/green color blindness (based on data from *USA Today*). We want to use a 0.01 significance level to test the claim that men have a red/green color-blindness rate that is greater than the 0.25% rate for women.

a. Why can't we use the methods of this section?

b. Assuming that the red/green color-blindness rate for men is equal to the 0.25% rate for women, find the probability that among 80 randomly selected men, at least 7 will have that type of color blindness. Describe the method used to find that probability.

c. Based on the result from part (b), what do you conclude?

32. Coping with No Successes In a simple random sample of 50 plain M&M candies, it is found that none of them are blue. We want to use a 0.01 significance level to test the claim of Mars, Inc., that the proportion of M&M candies that are blue is equal to 0.10. Can the methods of this section be used? If so, test the claim. If not, explain why not.

33. Power For a hypothesis test with a specified significance level α, the probability of a type I error is α, whereas the probability β of a type II error depends on the particular value of p that is used as an alternative to the null hypothesis.

a. Using an alternative hypothesis of $p < 0.4$, a sample size of $n = 50$, and assuming that the true value of p is 0.25, find the power of the test. Use the procedure given in part (a) of Exercise 47 in Section 8-2. [*Hint:* In Step 3, use the values $p = 0.25$ and $pq/n = (0.25)(0.75)/50$.]

b. Find the value of β, the probability of making a type II error.

c. Given the conditions cited in part (a), what do the results indicate about the effectiveness of the hypothesis test?

8-4 Testing a Claim About a Mean: σ Known

Key Concept This section presents methods for testing a claim about a population mean, given that the population standard deviation is a known value. The following section presents methods for testing a claim about a mean when σ is not known. This section uses the normal distribution with the same components of hypothesis tests that were introduced in Section 8-2.

The requirements, test statistic, critical values, and *P*-value are summarized as follows:

Testing Claims About a Population Mean (with σ Known)

Requirements

1. The sample is a simple random sample.
2. The value of the population standard deviation σ is known.
3. Either or both of these conditions is satisfied: The population is normally distributed or $n > 30$.

Test Statistic for Testing a Claim About a Mean (with σ Known)

$$z = \frac{\bar{x} - \mu_{\bar{x}}}{\frac{\sigma}{\sqrt{n}}}$$

***P*-values:** Use the standard normal distribution (Table A-2) and refer to Figure 8-6.

Critical values: Use the standard normal distribution (Table A-2).

TI-83/84 Plus Calculator: The TI-83/84 Plus calculator can be used to automatically generate the test statistic and *P*-value. See the procedure at the end of this section.

Commercials

Television networks have their own clearance departments for screening commercials and verifying claims. The National Advertising Division, a branch of the Council of Better Business Bureaus, investigates advertising claims. The Federal Trade Commission and local district attorneys also become involved. In the past, Firestone had to drop a claim that its tires resulted in 25% faster stops, and Warner Lambert had to spend $10 million informing customers that Listerine doesn't prevent or cure colds. Many deceptive ads are voluntarily dropped, and many others escape scrutiny simply because the regulatory mechanisms can't keep up with the flood of commercials.

Before beginning any hypothesis testing procedure, we should first *explore* the data set and we should verify that the requirements are satisfied for the particular test being used. Using the methods introduced in Chapters 2 and 3, investigate center, variation, and distribution by drawing a graph; find the mean, standard deviation, and 5-number summary; and identify any outliers.

EXAMPLE ***P*-Value Method** Data Set 13 in Appendix B includes weights of 13 red M&M candies randomly selected from a bag containing 465 M&Ms. The standard deviation of the weights of all of the M&Ms in the bag is $\sigma = 0.0565$ g. The sample weights (in grams) are listed below, and they have a mean of $\bar{x} = 0.8635$. The bag states that the net weight of the contents is 396.9 g, so the M&Ms must have a mean weight that is at least $396.9/465 = 0.8535$ g in order to provide the amount claimed. Use the sample data with a 0.05 significance level to test the claim of a production manager that the M&Ms have a mean that is actually greater than 0.8535 g, so consumers are being given more than the amount indicated on the label. Use the *P*-value method by following the procedure outlined in Figure 8-8.

0.751	0.841	0.856	0.799	0.966	0.859	0.857
0.942	0.873	0.809	0.890	0.878	0.905	

SOLUTION

REQUIREMENTS  See the three requirements listed above. The sample is a simple random sample. The value of σ is known (0.0565 g). The sample size is $n = 13$, which is not greater than 30, but there are no outliers and the accompanying STATDISK and TI-83/84 Plus calculator displays of the normal quantile plot suggest that the weights are normally distributed (because the points are close to the straight line and they show no systematic pattern). A histogram would also suggest that the weights are normally distributed. The requirements are satisfied, and we can proceed with the hypothesis test. ✔

STATDISK

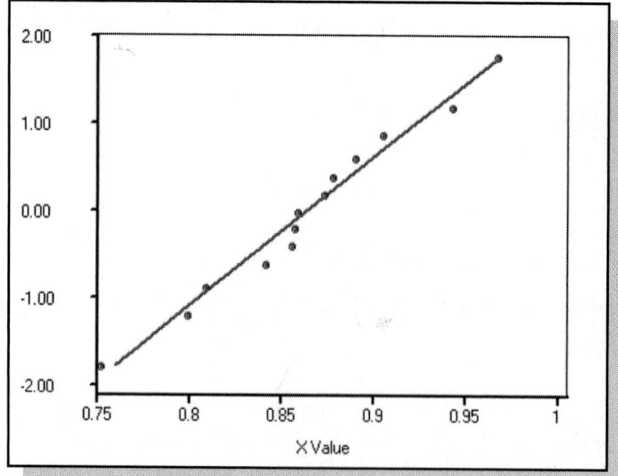

TI-83/84 Plus

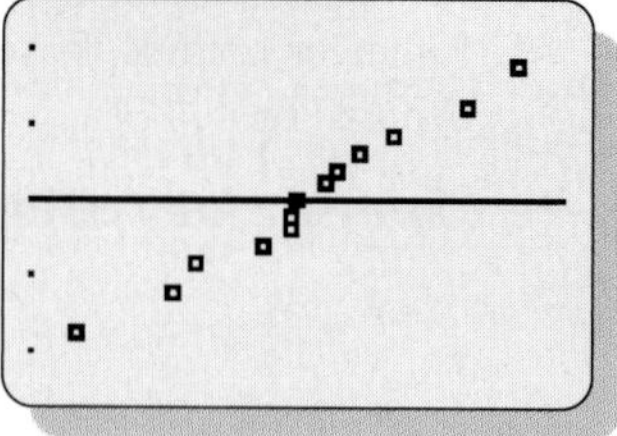

We follow the P-value procedure summarized in Figure 8-8.

Step 1: The claim that the mean is more than 0.8535 g is expressed in symbolic form as $\mu > 0.8535$.

Step 2: The alternative (in symbolic form) to the original claim is $\mu \leq 0.8535$.

Step 3: Because the statement $\mu > 0.8535$ does not contain the condition of equality, it becomes the alternative hypothesis. The null hypothesis is the statement that $\mu = 0.8535$.

$$H_0: \mu = 0.8535$$

$$H_1: \mu > 0.8535 \quad \text{(original claim)}$$

Step 4: As specified in the statement of the problem, the significance level is $\alpha = 0.05$.

Step 5: Because the claim is made about the *population mean* μ, the sample statistic most relevant to this test is the *sample mean* $\bar{x} = 0.8635$. Because σ is assumed to be known (0.0565) and the population appears to be normally distributed, the central limit theorem indicates that the distribution of sample means can be approximated by a *normal* distribution.

Step 6: The test statistic is calculated as follows:

$$z = \frac{\bar{x} - \mu_{\bar{x}}}{\frac{\sigma}{\sqrt{n}}} = \frac{0.8635 - 0.8535}{\frac{0.0565}{\sqrt{13}}} = 0.64$$

Using the test statistic of $z = 0.64$, we now proceed to find the P-value. See Figure 8-6 for the flowchart summarizing the procedure for finding P-values. This is a right-tailed test, so the P-value is the area to the right of $z = 0.64$. We now refer to Table A-2 to find that the area to the left of $z = 0.64$ is 0.7389, so the P-value is $1 - 0.7389 = 0.2611$. (Using the TI-83/84 Plus calculator, a more accurate P-value is 0.2609.) (See Figure 8-11.)

Step 7: Because the P-value of 0.2611 is greater than the significance level of $\alpha = 0.05$, we fail to reject the null hypothesis.

continued

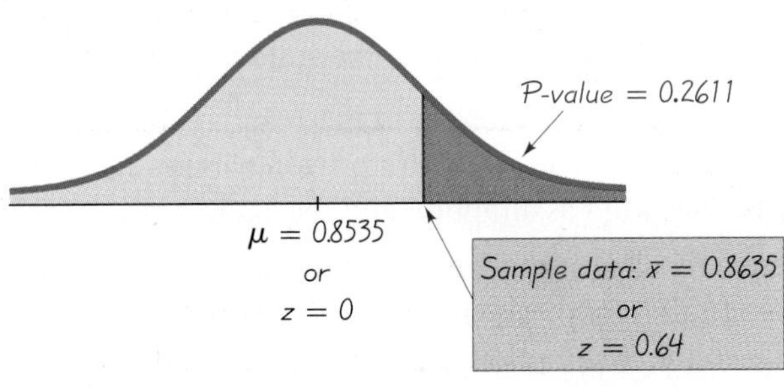

Figure 8-11 Testing the Claim that $\mu > 0.8535$

INTERPRETATION The P-value of 0.2611 tells us that if $\mu = 0.8535$ g, there is a good chance (0.2611) of getting a sample mean such as the one found from the sample. That is, a sample mean such as 0.8635 could easily occur by chance with a population mean of 0.8535. There is not sufficient evidence to support a conclusion that the population mean is greater than 0.8535 as claimed by the production manager.

Traditional Method If the traditional method of testing hypotheses is used for the preceding example, the first five steps would be the same. In Step 6 we would find the critical value of $z = 1.645$ instead of finding the P-value. We would again fail to reject the null hypothesis because the test statistic of $z = 0.64$ would not fall in the critical region. The final conclusion will be the same.

Confidence Interval Method We can use a confidence interval for testing a claim about μ when σ is known. For a one-tailed hypothesis test with a 0.05 significance level, we construct a 90% confidence interval. If we use the sample data in the preceding example ($n = 13$ and $\bar{x} = 0.8635$) and assume that $\sigma = 0.0565$ is known, we can test the claim that $\mu > 0.8535$ by using the methods of Section 7-3 to construct this 90% confidence interval: $0.8377 < \mu < 0.8893$. Because that confidence interval contains 0.8535, we cannot support a claim that μ is greater than 0.8535. Based on the confidence interval, μ might equal 0.8535. As in the hypothesis test, we conclude that there is not sufficient evidence to support the claim that the mean is greater than 0.8535 g.

In Section 8-3 we saw that when testing a claim about a population proportion, the traditional method and P-value method are equivalent, but the confidence interval method is somewhat different. When testing a claim about a population mean, there is no such difference, and all three methods are equivalent.

In the remainder of the text, we will apply methods of hypothesis testing to other circumstances. It is easy to become entangled in a complex web of steps without ever understanding the underlying rationale of hypothesis testing. The key to that understanding lies in the rare event rule for inferential statistics: **If, under a given assumption, there is an exceptionally small probability of getting sample results at least as extreme as the results that were obtained, we conclude that the assumption is probably not correct.** When testing a claim, we make an assumption (null hypothesis) of equality. We then compare the assumption and the sample results to form one of the following conclusions:

- If the sample results (or more extreme results) can easily occur when the assumption (null hypothesis) is true, we attribute the relatively small discrepancy between the assumption and the sample results to chance.
- If the sample results (or more extreme results) cannot easily occur when the assumption (null hypothesis) is true, we explain the relatively large discrepancy between the assumption and the sample results by concluding that the assumption is not true, so we reject the assumption.

TI

Using a TI-83/84 Plus Calculator for Testing a Claim About a Mean (With σ Known)

When using a TI-83/84 Plus calculator for testing a claim about a population mean μ *and the value of the population standard deviation* σ *is known,* you have the option of using either the summary statistics of n and $\bar{x}$, or you can use a list of original sample data values. Here is the procedure:

1. Press STAT, then press ▷ twice to select **TESTS.**
2. Select the first menu item of **Z-Test.**
3. The first line, identified as Inpt (for input), gives you the option of using Data or Stats. Use the arrow keys ◁ and ▷ to select either Data or Stats.
   - **Data:** Use this option if you have the sample data values stored as a list.
   - **Stats:** Use this option if you know the values of $\bar{x}$ and n.
4. For the value indicated by $\mu 0$, enter the claimed value of the population mean. This is the same value used in the null and alternative hypotheses.
5. For the value indicated by σ, enter the known value of the *population* standard deviation. (If only the sample standard deviation s is known, see the procedure in the following section.)
6. If you selected Stats in Step 3, proceed to enter the values of the sample mean $\bar{x}$ and the sample size n.

   If you selected Data in Step 3, proceed to enter the name of the list where the data are stored (such as L1 or a named list such as AGES), and a value for Freq (frequency). Enter 1 for the frequency. (However, if the sample data are stored in two columns as a frequency distribution, the Freq entry should be the list containing frequencies).
7. For the next line, select the type of test (two-tailed, left-tailed, right-tailed). Select the option that uses the same symbol as the alternative hypothesis. For a two-tailed case select the option including $\neq$, for a left-tailed test select the option of $<\mu 0$, and for a right-tailed test select the option of $>\mu 0$.
8. Use the arrow keys to scroll down to the bottom line, select **Calculate,** then press ENTER.

Examples in this section involved the claim of $\mu > 0.8535$, the 13 sample values are listed, and it is assumed that $\sigma = 0.0565$. Shown here are the screen displays for the required entries and the generated results. (The sample values are in list L1.)

Interpreting the results: See the displayed results and note that the TI-83/84 Plus calculator provides the test statistic of $z = 0.6406050939$ and the P-value of 0.2608895792, which can be rounded to 0.2609. Because the P-value is greater than the significance level of 0.05, we fail to reject the null hypothesis and conclude that there is not sufficient sample evidence to support the claim that the population mean is greater than 0.8535.

Because the TI-83/84 Plus calculator provides the test statistic and P-value (but not the critical value), we use the P-value method of testing hypotheses.

TI-83/84 Plus

```
Z-Test
 Inpt:Data Stats
 μ0:.8535
 σ:.0565
 List:L1
 Freq:1
 μ:≠μ0 <μ0 >μ0
 Calculate Draw
```

TI-83/84 Plus

```
Z-Test
 μ>.8535
 z=.6406050939
 p=.2608895792
 x̄=.8635384615
 Sx=.0575942928
 n=13
```

8-4 BASIC SKILLS AND CONCEPTS

Statistical Literacy and Critical Thinking

1. Requirements Must you have a sample size of $n > 30$ in order to use the methods of hypothesis testing presented in this section? If a simple random sample has fewer than 31 values, what requirement must be satisfied to justify using the methods of this section?

2. Verifying Requirements A simple random sample consists of $n = 12$ values, so the requirement of normality must be checked. How can you check that requirement of normality?

3. Confidence Interval You want to test the claim that $\mu < 100$ by constructing a confidence interval. If the hypothesis test is to be conducted with a 0.01 significance level, what confidence level should be used for the confidence interval?

4. Systematic Sampling A dean obtains a sample at a college by selecting every 50th name in the list of all 6000 current full-time students. She then proceeds to test the claim that the mean grade-point average is greater than 2.50. Is the sample a simple random sample? Is the sample likely to be representative, or is it somehow biased?

Verifying Assumptions. *In Exercises 5–8, determine whether the given conditions justify using the methods of this section when testing a claim about the population mean μ.*

5. The sample size is $n = 27$, $\sigma = 6.44$, and a histogram of the sample data is approximately bell-shaped.

6. The sample size is $n = 9$, $\sigma = 2.5$, and a histogram of the sample data shows a distribution that is far from being bell-shaped.

7. The sample size is $n = 24$, σ is not known, and a histogram of the sample data reveals a distribution that is approximately bell-shaped.

8. The sample size is $n = 121$, $\sigma = 0.25$, and a histogram of the sample data reveals a distribution that is uniform instead of being bell-shaped.

In Exercises 9–12, identify the indicated values or interpret the given display.

TI 9. M&Ms Data Set 13 in Appendix B includes a sample of 27 blue M&Ms with a mean weight of 0.8560 g. Assume that σ is known to be 0.0565 g. Consider a hypothesis test that uses a 0.05 significance level to test the claim that the mean weight of all M&Ms is equal to 0.8535 g (the weight necessary so that bags of M&Ms have the weight printed on the package).
   a. What is the test statistic?
   b. What are the critical values?
   c. What is the P-value?
   d. What is the conclusion about the null hypothesis (reject, fail to reject)?
   e. What is the final conclusion in simple nontechnical terms?

TI 10. Human Body Temperature Data Set 2 in Appendix B includes a sample of 106 body temperatures with a mean of 98.20°F. Assume that σ is known to be 0.62°F. Consider a hypothesis test that uses a 0.05 significance level to test the claim that the mean body temperature of the population is less than 98.6°F.
   a. What is the test statistic?
   b. What is the critical value?
   c. What is the P-value?

d. What is the conclusion about the null hypothesis (reject, fail to reject)?
e. What is the final conclusion in simple nontechnical terms?

11. Everglades Temperatures In order to monitor the ecological health of the Florida Everglades, various measurements are recorded at different times. The bottom temperatures are recorded at the Garfield Bight station and the mean of 30.377°C is obtained for 61 temperatures recorded on 61 different days. Assuming that σ = 1.7°C, test the claim that the population mean is greater than 30.0°C. Use a 0.05 significance level, and use the displayed results from the DDXL add-in for Excel.

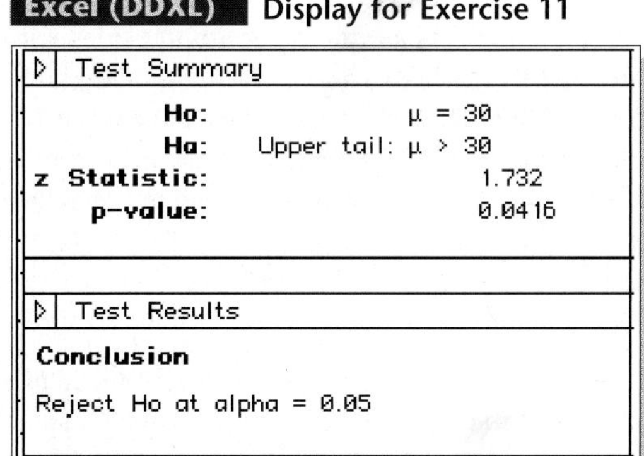

TI **12.** Cotinine Levels of Smokers When people smoke, the nicotine they absorb is converted to cotinine, which can be measured. A sample of 40 smokers has a mean cotinine level of 172.5 ng/ml. Assuming that σ is known to be 119.5 ng/ml, test the claim that the mean cotinine level of all smokers is equal to 200.0 ng/ml. Use a 0.05 significance level, and use the displayed results from a TI-83/84 Plus calculator.

TI-83/84 Plus

```
Z-Test
 μ≠200
 z=-1.455441601
 p=.1455471254
 x̄=172.5
 n=40
```

Testing Hypotheses. In Exercises 13–20, test the given claim. Identify the null hypothesis, alternative hypothesis, test statistic, P-value or critical value(s), conclusion about the null hypothesis, and final conclusion that addresses the original claim. Use the P-value method unless your instructor specifies otherwise.

TI **13.** Perception of Time Randomly selected statistics students of the author participated in an experiment to test their ability to determine when 1 min (or 60 sec) has passed. Forty students yielded a sample mean of 58.3 sec. Assuming that σ = 9.5 sec, use a 0.05 significance level to test the claim that the population mean is equal to 60 sec. Based on the result, does there appear to be an overall perception of 1 min that is reasonably accurate?

TI **14.** Analysis of Last Digits Analysis of the last digits of sample data values sometimes reveals whether the data have been accurately measured and reported. When single digits 0 through 9 are randomly selected with replacement, the mean should be 4.50 and the standard deviation should be 2.87. Reported data (such as weights or heights) are often rounded so that the last digits include disproportionately more 0s and 5s. Use the last digits in the reported lengths (in feet) of the 73 home runs hit by Barry Bonds in 2001 to test the claim that they come from a population with a mean of 4.50. Use a 0.05 significance level. The 73 home-run distances (listed in Data Set 17 in Appendix B) have last digits with a mean of 1.753. Based on the results, does it appear that the distances were accurately measured?

TI **15.** Does the Diet Work? When 40 people used the Atkins diet for one year, their mean weight change was −2.1 lb (based on data from "Comparison of the Atkins, Ornish, Weight Watchers, and Zone Diets for Weight Loss and Heart Disease Reduction," by Dansinger, et al., *Journal of the American Medical Association,* Vol. 293, No. 1). Assume that the standard deviation of all such weight changes is $\sigma = 4.8$ lb and use a 0.05 significance level to test the claim that the mean weight change is less than 0. Based on these results, does the diet appear to be effective? Does the mean weight change appear to be substantial enough to justify the special diet?

TI **16.** Are Thinner Aluminum Cans Weaker? An axial load of an aluminum can is the maximum weight that the sides can support before collapsing. The axial load is an important measure, because the top lids are pressed onto the sides with pressures that vary between 158 lb and 165 lb. Pepsi experimented with thinner aluminum cans, and a random sample of 175 of the thinner cans has a mean axial load of 267.11 lb. The standard cans have a mean axial load of 281.81 lb and a standard deviation of 27.77 lb. Use a 0.01 significance level to test the claim that the thinner cans have a mean axial load that is less than 281.81 lb. Assume that $\sigma = 27.77$ lb. Do the thinner cans appear to have a mean axial load less than 281.81 lb? Do the thinner cans appear to be strong enough so that they are not crushed when the top lids are pressed onto the sides?

TI **17.** Blood Pressure Levels When 14 different second-year medical students at Bellevue Hospital measured the systolic blood pressure of the same person, they obtained the results listed below (in mmHg). Assuming that the population standard deviation is known to be 10 mmHg, use a 0.05 significance level to test the claim that the mean blood pressure level is less than 140 mmHg. Hypertension is defined to be a blood pressure level that is too high because it is 140 mmHg or greater. Based on the hypothesis test results, can it be safely concluded that the person does not have hypertension?

138 130 135 140 120 125 120 130 130 144 143 140 130 150

TI **18.** World's Smallest Mammal The world's smallest mammal is the bumblebee bat, also known as the Kitti's hog-nosed bat (or *Craseonycteris thonglongyai*). Such bats are roughly the size of a large bumblebee. Listed below are weights (in grams) from a sample of these bats. Assuming that the weights of all such bats have a standard deviation of 0.30 g, use a 0.05 significance level to test the claim that these bats are from the same population with a known mean of 1.8 g. Do the bats appear to come from the same population?

1.7 1.6 1.5 2.0 2.3 1.6 1.6 1.8 1.5 1.7 2.2 1.4 1.6 1.6 1.6

TI **19.** Appendix B Data Set: Weights of Quarters Use the weights of the post-1964 quarters listed in Data Set 14 from Appendix B. Assuming that quarters are minted to produce weights with a population standard deviation of 0.068 g, use the sample of weights with a 0.01 significance level to test the claim that the quarters are from a population with a mean of 5.670 g. Do the quarters appear to be manufactured according to the U.S. mint specification that the mean is equal to 5.670 g?

TI **20.** Appendix B Data Set: Forecast Errors Refer to Data Set 8 in Appendix B and subtract each actual high temperature from the high temperature that was forecast one day before. The result is a list of errors. Assuming that all such errors have a standard deviation of 2.5°, use a 0.05 significance level to test the claim that all such errors have a mean equal to 0. What does the result suggest about the accuracy of the forecast temperatures?

8-4 BEYOND THE BASICS

21. Power of a Test The procedure for finding the power in a hypothesis test involving a proportion is given in Exercise 47 from Section 8-2. Use the same procedure to find the power of a hypothesis test of the claim that $\mu > 100$, given a sample of size 40, a known population standard deviation of 15, and given that the population mean is actually 108. (*Hint:* In Step 2 of the procedure, use the test statistic given in this section instead of the test statistic used for a proportion and, in Step 4 of the procedure, use $\mu = 108$.) What does the value of the power indicate about the effectiveness of the test to recognize that the mean is greater than 100 when the mean is actually 108? Also find β, the probability of a type II error.

8-5 Testing a Claim About a Mean: σ Not Known

Key Concept The previous section presented methods for testing a claim about a population mean, given that the value of the population standard deviation σ is known, and this section presents methods for testing a claim about a population mean when we do not know the value of σ. The methods of this section are very practical and realistic because σ is not typically known. The methods of this section use the Student t distribution that was introduced in Section 7-4. The requirements, test statistic, P-value, and critical values are summarized as follows.

Testing Claims About a Population Mean (with σ Not Known)

Requirements

1. The sample is a simple random sample.
2. The value of the population standard deviation σ is *not* known.
3. Either or both of these conditions is satisfied: The population is normally distributed or $n > 30$.

Test Statistic for Testing a Claim About a Mean (with σ Not Known)

$$t = \frac{\bar{x} - \mu_{\bar{x}}}{\dfrac{s}{\sqrt{n}}}$$

***P*-values and Critical values:** Use Table A-3 and use df $= n - 1$ for the number of degrees of freedom. (See Figure 8-6 for P-value procedures.)

TI-83/84 Plus Calculator: The TI-83/84 Plus calculator can be used to automatically generate the test statistic and P-value. See the procedure at the end of this section.

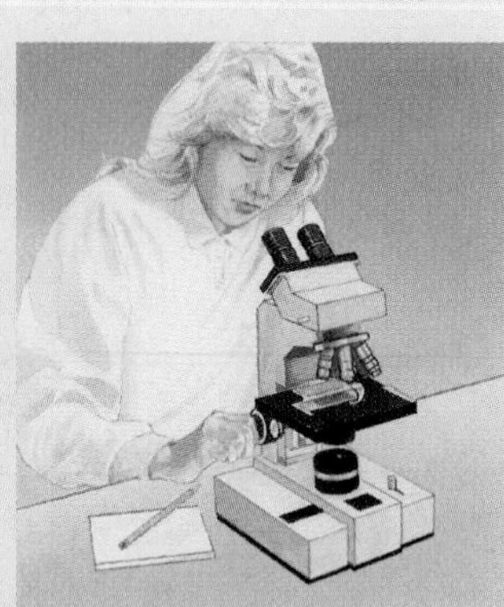

Ethics in Experiments

Sample data can often be obtained by simply observing or surveying members selected from the population. Many other situations require that we somehow manipulate circumstances to obtain sample data. In both cases ethical questions may arise. Researchers in Tuskegee, Alabama, withheld the effective penicillin treatment to syphilis victims so that the disease could be studied. This experiment continued for a period of 27 years!

Normality Requirement The requirement of a normally distributed population is not a strict requirement, and we can usually consider the population to be normally distributed after using the sample data to confirm that there are no outliers and the histogram has a shape that is not very far from a normal distribution. This t test is said to be *robust* against a departure from normality, meaning that the test works reasonably well if the departure from normality is not too extreme.

Sample Size We use the simplified criterion of $n > 30$ as justification for treating the distribution of sample means as a normal distribution, but the minimum sample size actually depends on how much the population distribution departs from a normal distribution. Because we do not know the value of σ, we estimate it with the value of the sample standard deviation s, but this introduces another source of unreliability, especially with small samples. We compensate for this added unreliability by finding P-values and critical values using the t distribution instead of the normal distribution that was used in Section 8-4 where σ was known. Here are the important properties of the Student t distribution:

Important Properties of the Student t Distribution

1. The Student t distribution is different for different sample sizes (see Figure 7-5 in Section 7-4).
2. The Student t distribution has the same general bell shape as the standard normal distribution; its wider shape reflects the greater variability that is expected when s is used to estimate σ.
3. The Student t distribution has a mean of $t = 0$ (just as the standard normal distribution has a mean of $z = 0$).
4. The standard deviation of the Student t distribution varies with the sample size and is greater than 1 (unlike the standard normal distribution, which has $\sigma = 1$).
5. As the sample size n gets larger, the Student t distribution gets closer to the standard normal distribution.

Choosing the Appropriate Distribution

When testing claims made about population means, sometimes the normal distribution applies, sometimes the Student t distribution applies, and sometimes neither applies, so we must use nonparametric methods or bootstrap resampling techniques. (Nonparametric methods, which do not require a particular distribution, are discussed in Chapter 13; the bootstrap resampling technique is described in the TI-83/84 Plus Calculator Project at the end of Chapter 7.) See pages 366–367 where Figure 7-6 and Table 7-1 both summarize the decisions to be made in choosing between the normal and Student t distributions. They show that when testing claims about population means, the Student t distribution is used under these conditions:

Use the Student t distribution when σ is not known and either or both of these conditions is satisfied:

The population is normally distributed or $n > 30$.

EXAMPLE Quality Control of M&Ms Data Set 13 in Appendix B includes weights of 13 red M&M candies randomly selected from a bag containing 465 M&Ms. Those weights (in grams) are listed below, and they have a mean of $\bar{x} = 0.8635$ and a standard deviation of $s = 0.0576$ g. The bag states that the net weight of the contents is 396.9 g, so the M&Ms must have a mean weight that is at least $396.9/465 = 0.8535$ g in order to provide the amount claimed. Use the sample data with a 0.05 significance level to test the claim of a production manager that the M&Ms have a mean that is actually greater than 0.8535 g, so consumers are being given more than the amount indicated on the label. Use the traditional method by following the procedure outlined in Figure 8-9.

0.751	0.841	0.856	0.799	0.966	0.859	0.857
0.942	0.873	0.809	0.890	0.878	0.905	

SOLUTION

REQUIREMENTS ✔ See the three requirements listed above. The sample is a simple random sample. We are not using a known value of σ. The sample size is $n = 13$, which is not greater than 30, but there are no outliers and a normal quantile plot suggests that the weights are normally distributed because the points are close to the straight line and they show no systematic pattern. (The normal quantile plot is shown in the preceding section.) A histogram would also suggest that the weights are normally distributed. The requirements are satisfied, and we can proceed with the hypothesis test. ✔

We follow the traditional procedure summarized in Figure 8-9.

Step 1: The claim that the mean is more than 0.8535 g is expressed in symbolic form as $\mu > 0.8535$.

Step 2: The alternative (in symbolic form) to the original claim is $\mu \leq 0.8535$.

Step 3: Because the statement $\mu > 0.8535$ does not contain the condition of equality, it becomes the alternative hypothesis. The null hypothesis is the statement that $\mu = 0.8535$.

$$H_0: \mu = 0.8535$$
$$H_1: \mu > 0.8535 \quad \text{(original claim)}$$

Step 4: As specified in the statement of the problem, the significance level is $\alpha = 0.05$.

Step 5: Because the claim is made about the *population mean* μ, and because the requirements for using the t test statistic are satisfied, we use the t distribution. (Refer to Figure 7-6 or Table 7-1 for the criteria for choosing between the normal and t distributions.)

Step 6: The test statistic is calculated as follows:

$$t = \frac{\bar{x} - \mu_{\bar{x}}}{\frac{s}{\sqrt{n}}} = \frac{0.8635 - 0.8535}{\frac{0.0576}{\sqrt{13}}} = 0.626$$

continued

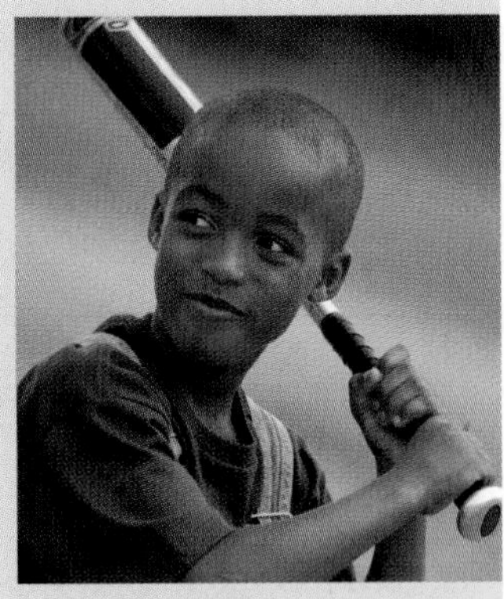

Lefties Die Sooner?

A study by psychologists Diane Halpern and Stanley Coren received considerable media attention and generated considerable interest when it concluded that left-handed people don't live as long as right-handed people. Based on their study, it appeared that left-handed people live an average of nine years less than righties. The Halpern/Coren study has been criticized for using flawed data. They used secondhand data by surveying relatives about people who had recently died. The myth of lefties dying younger became folklore that has survived many years. However, more recent studies show that left-handed people do *not* have shorter lives than those who are right-handed.

The critical value of $t = 1.782$ is found by referring to Table A-3. First locate the correct number of degrees of freedom in the column at the left. We use df $= n - 1 = 12$ in this example. Because this test is right-tailed with $\alpha = 0.05$, refer to the column indicating an area of 0.05 in one tail. The test statistic and critical value are shown in the accompanying STATDISK display.

Step 7: Because the test statistic of $t = 0.626$ does not fall in the critical region, we fail to reject the null hypothesis.

INTERPRETATION (Refer to Figure 8-7 for help in wording the final conclusion.) We fail to reject the null hypothesis. Based on the available sample data, we do not have sufficient evidence to support the claim that the mean weight of M&Ms is greater than 0.8535 g. Although the sample mean of 0.8635 g does exceed 0.8535 g, it does not exceed it by a *significant* amount.

The critical value in the preceding example was $t = 1.782$, but if the normal distribution were being used, the critical value would have been $z = 1.645$. The Student t critical value is larger (farther to the right), showing that with the Student t distribution, the sample evidence must be *more extreme* before we consider it to be significant.

STATDISK

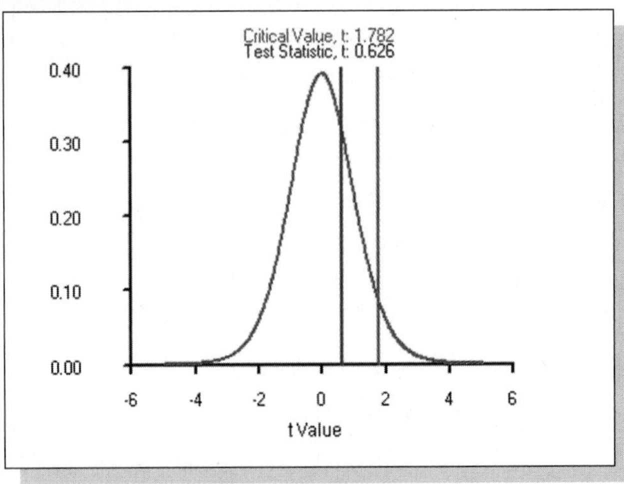

Finding *P*-Values with the Student *t* Distribution

The preceding example followed the traditional approach to hypothesis testing, but the TI-83/84 Plus calculator and many articles in professional journals will display *P*-values. For the preceding example, the TI-83/84 Plus calculator displays a *P*-value of 0.2715 if the given summary statistics are used. With a significance level of 0.05 and a *P*-value greater than 0.05, we fail to reject the null hypothesis, as we did using the traditional method in the preceding example. If a TI-83/84 Plus calculator is not available, we can use Table A-3 to identify a *range of values* containing the *P*-value. We recommend this strategy for finding *P*-values using the t distribution:

1. Use software or a TI-83/84 Plus calculator. (STATDISK, Minitab, the DDXL add-in for Excel, and the TI-83/84 Plus calculator, SPSS, and SAS all provide P-values for t tests.)
2. If the technology is not available, use Table A-3 to identify a range of P-values as follows: Use the number of degrees of freedom to locate the relevant row of Table A-3, then determine where the test statistic lies relative to the t values in that row. Based on a comparison of the t test statistic and the t values in the row of Table A-3, identify a range of values by referring to the area values given at the top of Table A-3.

EXAMPLE Finding *P*-Value Assuming that neither software nor a TI-83/84 Plus calculator is available, use Table A-3 to find a range of values for the P-value corresponding to the test statistic of $t = 0.626$ from the preceding example. Note that the test is right-tailed, and the number of degrees of freedom is 12.

SOLUTION This is a right-tailed test, so the P-value is the area to the right of the test statistic. Refer to Table A-3 and locate the row corresponding to 12 degrees of freedom. The test statistic of $t = 0.626$ is smaller than every value in that row of Table A-3, so the "area in one tail" (to the right of the test statistic) is greater than 0.10. See Figure 8-12, which shows the location of the test statistic $t = 0.626$ relative to the next lowest t value in the 12th row of Table A-3. From Figure 8-12, we can see clearly that the area to the right of $t = 0.626$ is greater than 0.10. Although we can't find the exact P-value, we can conclude that the P-value > 0.10.

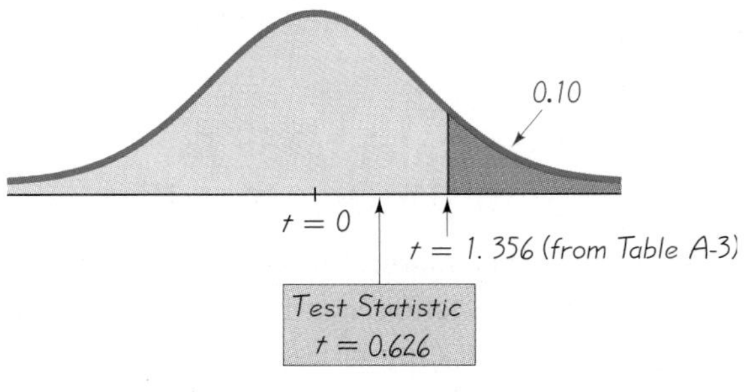

Figure 8-12 Finding Range for *P*-Value

EXAMPLE Finding *P*-Value Assuming that neither software nor a TI-83/84 Plus calculator is available, use Table A-3 to find a range of values for the P-value corresponding to a t test with these components: Test statistic is $t = 2.777$, sample size is $n = 20$, significance level is $\alpha = 0.05$, alternative hypothesis is H_1: $\mu \neq 5$.

SOLUTION Because the sample size is 20, refer to Table A-3 and locate the row corresponding to 19 degrees of freedom. Because the test statistic of $t = 2.777$ lies between the table values of 2.861 and 2.539, the P-value is between

continued

Meta-Analysis

The term *meta-analysis* refers to a technique of doing a study that essentially combines results of other studies. It has the advantage that separate smaller samples can be combined into one big sample, making the collective results more meaningful. It also has the advantage of using work that has already been done. Meta-analysis has the disadvantage of being only as good as the studies that are used. If the previous studies are flawed, the "garbage in, garbage out" phenomenon can occur. The use of meta-analysis is currently popular in medical research and psychological research. As an example, a study of migraine headache treatments was based on data from 46 other studies. (See "Meta-Analysis of Migraine Headache Treatments: Combining Information from Heterogeneous Designs," by Dominici et al., *Journal of the American Statistical Association,* Vol. 94, No. 445.)

the corresponding "areas in two tails" of 0.01 and 0.02. (Be sure to use the "area in two tails" if the test is two-tailed.) Although we can't find the exact P-value, we can conclude that $0.01 < P\text{-value} < 0.02$. (Software or the TI-83/84 Plus calculator provide the exact P-value of 0.0120.)

Remember, P-values can be easily found by using software or a TI-83/84 Plus calculator. Also, the traditional method of testing hypotheses can be used instead of the P-value method.

Confidence Interval Method We can use a confidence interval for testing a claim about μ when σ is not known. For a two-tailed hypothesis test with a 0.05 significance level, we construct a 95% confidence interval, but for a one-tailed hypothesis test with a 0.05 significance level we construct a 90% confidence interval. (See Table 8-2.) Using the sample data from the example in this section ($n = 13$, $\bar{x} = 0.8635$ g, $s = 0.0576$ g) with σ not known, and using a 0.05 significance level, we can test the claim that $\mu > 0.8535$ by using the confidence interval method. Construct this 90% confidence interval: $0.8350 < \mu < 0.8920$ (see Section 7-4). Because the assumed value of $\mu = 0.8535$ is contained within the confidence interval, we cannot reject that assumption. Based on the 13 sample values given in the example, we do not have sufficient evidence to support the claim that the mean weight is greater than 0.8535 g. Based on the confidence interval, the true value of μ is likely to be any value between 0.8350 g and 0.8920 g, including 0.8535 g.

In Section 8-3 we saw that when testing a claim about a population proportion, the traditional method and P-value method are equivalent, but the confidence interval method is somewhat different. When testing a claim about a population mean, there is no such difference, and all three methods are equivalent.

TI Using a TI-83/84 Plus Calculator for Testing a Claim About a Mean (With σ Not Known)

When using a TI-83/84 Plus calculator for testing a claim about a population mean μ and the value of the population standard deviation σ is *not* known, you have the option of using either the summary statistics of $\bar{x}$, s, n, or you can use a list of original sample data values. Here is the procedure:

1. Press STAT, then press ▷ twice to select **TESTS.**
2. Select the menu item of **T-Test.**
3. The first line, identified as Inpt (for input), gives you the option of using Data or Stats. Use the arrow keys ◁ and ▷ to select either Data or Stats.
   - **Data:** Use this option if you have the sample data values stored as a list.
   - **Stats:** Use this option if you know the values of $\bar{x}$, s, n.
4. For the value indicated by $\mu 0$, enter the claimed value of the population mean. This is the same value used in the null and alternative hypotheses.
5. If you selected Stats in Step 3, proceed to enter the values of the sample mean $\bar{x}$, the sample standard deviation s, and the sample size n.

Using a List of Sample Data

If you selected Data in Step 3, proceed to enter the list where the data are stored (such as L1 or a named list such as AGES), and a value for Freq (frequency). Enter 1 for the frequency. (However, if the sample data are stored in two columns as a frequency distribution, the Freq entry should be the list containing frequencies).

6. For the next line, select the type of test (two-tailed, left-tailed, right-tailed). Select the option that uses the same symbol as the alternative hypothesis. For a two-tailed case select the option including $\neq$, for a left-tailed test select the option of $<\mu 0$, and for a right-tailed test select the option of $>\mu 0$.
7. Use the arrow keys to scroll down to the bottom line, select **Calculate,** then press ENTER.

Using Summary Statistics

The first example in this section involved the claim of $\mu > 0.8535$, and the list of sample data was given as 0.751, 0.841, . . . , 0.905. If the sample data are first stored in list L1, we can then proceed to enter the data as shown in the first screen display. If the summary statistics of $\bar{x} = 0.8635$, $s = 0.0576$, and $n = 13$ are used, the entry screen should appear as shown in the second screen display. The results from using the list of data values are shown in the third screen display.

Hypothesis Test Results

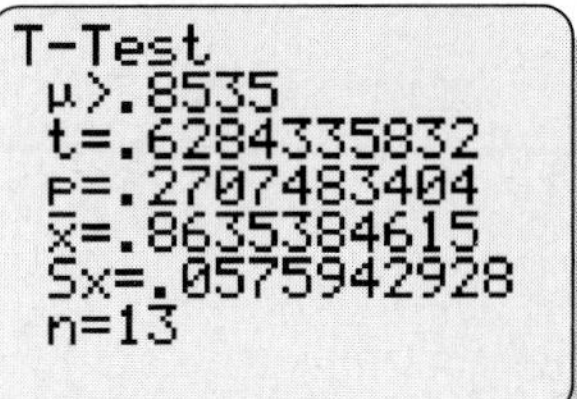

Interpreting the results: See the displayed results and note that the TI-83/84 Plus calculator provides the test statistic of $t = 0.628$ (rounded) and the P-value of 0.2707 (rounded). Because the P-value is greater than the significance level of 0.05, we fail to reject the null hypothesis and conclude that there is not sufficient sample evidence to support the claim that the mean is greater than 0.8535.

Because the TI-83/84 Plus calculator provides the test statistic and P-value (but not the critical value), we use the P-value method of testing hypotheses.

8-5 BASIC SKILLS AND CONCEPTS

Statistical Literacy and Critical Thinking

1. Degrees of Freedom When using Table A-3 to find critical values, we must use the appropriate number of degrees of freedom. If a sample consists of five values, what is the appropriate number of degrees of freedom? If you don't know any of the five sample values, but you know that their mean is exactly 20.0, how many values can you make up before the remaining values are determined by the restriction that the mean is 20.0?

2. Normal and t Distributions Identify two characteristics that the standard normal and t distributions have in common, and identify two characteristics that are different for the normal and t distributions.

3. Unnecessary Test A quality control manager claims that cans of cola are being filled with amounts having a mean that is less than 12 oz. If the sample data consist of 24 cans with a mean of 12.13 oz and a standard deviaton of 0.12 oz, why is it not necessary to conduct a formal hypothesis test in order to conclude that the sample data do not support the manager's claim?

4. Reality Check Unlike the preceding section, this section does not include a requirement that the value of the population standard deviation must be known. Which section is more likely to apply in realistic situations: this section or the preceding section? Why?

Using Correct Distribution. In Exercises 5–8, determine whether the hypothesis test involves a sampling distribution of means that is a normal distribution, Student t distribution, or neither. (Hint: See Figure 7-6 and Table 7-1.)

5. Claim: $\mu = 2.55$. Sample data: $n = 7$, $\bar{x} = 2.41$, $s = 0.66$. The sample data appear to come from a normally distributed population with unknown μ and σ.

6. Claim: $\mu = 1002$. Sample data: $n = 200$, $\bar{x} = 1045$, $s = 85$. The sample data appear to come from a population with a distribution that is not normal, and σ is unknown.

7. Claim: $\mu = 0.0105$. Sample data: $n = 17$, $\bar{x} = 0.0134$, $s = 0.0022$. The sample data appear to come from a population with a distribution that is very far from normal, and σ is unknown.

8. Claim: $\mu = 75$. Sample data: $n = 15$, $\bar{x} = 66$, $s = 12$. The sample data appear to come from a normally distributed population with $\sigma = 14$.

Finding P-values. In Exercises 9–12, either use the TI-83/84 Plus calculator to find the P-value or use Table A-3 to find a range of values for the P-value.

TI **9.** Right-tailed test with $n = 7$ and test statistic $t = 3.500$.

TI **10.** Left-tailed test with $n = 27$ and test statistic $t = -1.500$.

TI **11.** Two-tailed test with $n = 21$ and test statistic $t = 9.883$.

TI **12.** Two-tailed test with $n = 11$ and test statistic $t = -2.000$.

Finding Test Components. In Exercises 13 and 14, assume that a simple random sample has been selected from a normally distributed population. Find the null and alternative hypotheses, test statistic, P-value (or range of P-values), critical value(s), and state the final conclusion.

TI **13.** Claim: The mean IQ score of statistics professors is greater than 120.
Sample data: $n = 12$, $\bar{x} = 132$, $s = 12$. The significance level is $\alpha = 0.05$.

TI **14.** Claim: The mean life span of desktop PCs is less than 7 years.
Sample data: $n = 21$, $\bar{x} = 6.8$ years, $s = 2.4$ years. The significance level is $\alpha = 0.05$.

Interpreting Display. In Exercises 15 and 16, test the given claim by interpreting the accompanying display of hypothesis test results.

15. Use a 0.05 significance level to test the claim that statistics students have a mean IQ greater than 110. The sample data are summarized by the statistics $n = 25$, $\bar{x} = 118.0$, and $s = 10.7$. See the accompanying Minitab display.

Minitab

```
Test of mu = 110 vs > 110

                                       95%
                                     Lower
 N      Mean    StDev  SE Mean       Bound      T      P
25   118.000   10.700    2.140     114.339   3.74  0.001
```

16. Use a 0.05 significance level to test the claim that the mean life span of cell phones is equal to 5 years. The sample data are summarized by the statistics $n = 27$, $\bar{x} = 4.6$ years, $s = 1.9$ years. See the accompanying SPSS display.

SPSS

	Test Value = 5					
					95% Confidence Interval of the Difference	
	t	df	Sig. (2-tailed)	Mean Difference	Lower	Upper
AGE	-1.094	26	.284	-.4000007	********	********

Testing Hypotheses. *In Exercises 17–32, assume that a simple random sample has been selected from a normally distributed population and test the given claim. Unless specified by your instructor, use either the traditional method or P-value method for testing hypotheses.*

TI **17.** Body Temperatures Data Set 2 in Appendix B includes 106 body temperatures with a mean of 98.20°F and a standard deviation of 0.62°F. Use a 0.05 significance level to test the claim that the mean body temperature is less than 98.6°F. Based on these results, does it appear that the commonly used mean of 98.6°F is wrong?

TI **18.** Baseballs In previous tests, baseballs were dropped 24 ft onto a concrete surface, and they bounced an average of 92.84 in. In a test of a sample of 40 new balls, the bounce heights had a mean of 92.67 in. and a standard deviation of 1.79 in. (based on data from Brookhaven National Laboratory and *USA Today*). Use a 0.05 significance level to determine whether there is sufficient evidence to support the claim that the new balls have bounce heights with a mean different from 92.84 in. Does it appear that the new baseballs are different?

TI **19.** Birth Weights In a study of the effects of prenatal cocaine use on infants, the following sample data were obtained for weights at birth: $n = 190$, $\bar{x} = 2700$ g, $s = 645$ g (based on data from "Cognitive Outcomes of Preschool Children with Prenatal Cocaine Exposure," by Singer, et al., *Journal of the American Medical Association,* Vol. 291, No. 20). Use a 0.01 significance level to test the claim that weights of babies born to cocaine users have a mean that is less than the mean of 3103 g for babies born to mothers who do not use cocaine. Based on the results, does it appear that birth weights are affected by cocaine use?

TI **20.** Credit Rating When consumers apply for credit, their credit is rated using FICO (Fair, Isaac, and Company) scores. A random sample of credit ratings is obtained, and the FICO scores are summarized with these statistics: $n = 18$, $\bar{x} = 660.3$, $s = 95.9$. Use a 0.05 significance level to test the claim that these credit ratings are from a population with a mean that is equal to 700. If the Bank of Newport requires a credit rating of 700 or higher for a car loan, do the results indicate that everyone will be eligible for a car loan? Why or why not?

TI **21.** Treating Chronic Fatigue Syndrome Patients with chronic fatigue syndrome were tested, then retested after being treated with fludrocortisone. A standard scale from -7 to $+7$ is used to measure fatigue before and after the treatment. The changes are summarized with these statistics: $n = 21$, $\bar{x} = 4.00$, $s = 2.17$ (based on data from "The Relationship Between Neurally Mediated Hypotension and the Chronic Fatigue Syndrome," by Bou-Holaigah, Rowe, Kan, and Calkins, *Journal of the American Medical Association,* Vol. 274, No. 12). The changes were computed in a way that makes positive values represent improvements. Use a 0.01 significance level to test the claim that the mean change is positive. Does the treatment appear to be effective?

TI **22.** Effectiveness of Diet Forty subjects followed the Weight Watchers diet for a year. Their weight changes are summarized by these statistics: $\bar{x} = -6.6$ lb, $s = 10.8$ lb (based on data from "Comparison of the Atkins, Ornish, Weight Watchers, and Zone

Diets for Weight Loss and Heart Disease Risk Reduction," by Dansinger et al., *Journal of the American Medical Association,* Vol. 293, No. 1). Use a 0.01 significance level to test the claim that the diet has no effect. Based on the results, does the diet appear to be effective?

TI **23.** Heights of Supermodels The heights are measured for supermodels Niki Taylor, Nadia Avermann, Claudia Schiffer, Elle MacPherson, Christy Turlington, Bridget Hall, Kate Moss, Valeria Mazza, and Kristy Hume. They have a mean of 70.2 in. and a standard deviation of 1.5 in. Use a 0.01 significance level to test the claim that supermodels have heights with a mean that is greater than the mean of 63.6 in. for women from the general population. Given that there are only nine heights represented, can we really conclude that supermodels are taller than the typical woman?

TI **24.** Conductor Life Span A *New York Times* article noted that the mean life span for 35 male symphony conductors was 73.4 years, in contrast to the mean of 69.5 years for males in the general population. Assuming that the 35 males have life spans with a standard deviation of 8.7 years, use a 0.05 significance level to test the claim that male symphony conductors have a mean life span that is greater than 69.5 years. Does it appear that male symphony conductors live longer than males from the general population? Why doesn't the experience of being a male symphony conductor cause men to live longer? (*Hint*: Are male symphony conductors born, or do they become conductors at a much later age?)

TI **25.** Monitoring Lead in Air Listed below are measured amounts of lead (in micrograms per cubic meter, or $\mu g/m^3$) in the air. The Environmental Protection Agency has established an air quality standard for lead of 1.5 $\mu g/m^3$. The measurements shown below were recorded at Building 5 of the World Trade Center site on different days immediately following the destruction caused by the terrorist attacks of September 11, 2001. After the collapse of the two World Trade Center buildings, there was considerable concern about the quality of the air. Use a 0.05 significance level to test the claim that the sample is from a population with a mean greater than the EPA standard of 1.5 $\mu g/m^3$. Is there anything about this data set suggesting that the assumption of a normally distributed population might not be valid?

5.40 1.10 0.42 0.73 0.48 1.10

TI **26.** Sugar in Cereal Different cereals are randomly selected, and the sugar content (grams of sugar per gram of cereal) is obtained for each cereal, with the results given below for Cheerios, Harmony, Smart Start, Cocoa Puffs, Lucky Charms, Corn Flakes, Fruit Loops, Wheaties, Cap'n Crunch, Frosted Flakes, Apple Jacks, Bran Flakes, Special K, Rice Krispies, Corn Pops, and Trix. Use a 0.05 significance level to test the claim of a cereal lobbyist that the mean for all cereals is less than 0.3 g.

0.03 0.24 0.30 0.47 0.43 0.07 0.47 0.13
0.44 0.39 0.48 0.17 0.13 0.09 0.45 0.43

TI **27.** World's Smallest Mammal The world's smallest mammal is the bumblebee bat, also known as the Kitti's hog-nosed bat (or *Craseonycteris thonglongyai*). Such bats are roughly the size of a large bumblebee. Listed below are weights (in grams) from a sample of these bats. Test the claim that these bats come from the same population having a mean weight equal to 1.8 g.

1.7 1.6 1.5 2.0 2.3 1.6 1.6 1.8 1.5 1.7 2.2 1.4 1.6 1.6 1.6

TI **28.** Olympic Winners Listed below are the winning times (in seconds) of men in the 100-meter dash for consecutive summer Olympic games, listed in order by row. Assuming that these results are sample data randomly selected from the population of all past and future Olympic games, test the claim that the mean time is less than 10.5 sec. What do you observe about the precision of the numbers? What extremely important characteristic of the data set is not considered in this hypothesis test? Do the results from the hypothesis test suggest that future winning times should be around 10.5 sec, and is such a conclusion valid?

12.0	11.0	11.0	11.2	10.8	10.8	10.8	10.6	10.8	10.3	10.3	10.3
10.4	10.5	10.2	10.0	9.95	10.14	10.06	10.25	9.99	9.92	9.96	

Using Appendix B Data Sets. *In Exercises 29–32, use the Data Set from Appendix B to test the given claim.*

TI **29.** Appendix B Data Set: Weights of Quarters Use the weights of the post-1964 quarters listed in Data Set 14 from Appendix B. Test the claim that the quarters are manufactured according to the U.S. mint specification that the mean is equal to 5.670 g.

TI **30.** Appendix B Data Set: Forecast Errors Refer to Data Set 8 in Appendix B and subtract each actual high temperature from the high temperature that was forecast one day before. The result is a list of errors. Test the claim that all such errors have a mean equal to 0. What does the result suggest about the accuracy of the forecast temperatures?

TI **31.** Appendix B Data Set: Pulse Rates The author, at the peak of an exercise program, claimed that his pulse rate was lower than the mean pulse rate of a typical male. The author's pulse rate was measured to be 60 beats per minute. Use the pulse rates of the males listed in Data Set 1 in Appendix B and test the claim that those pulse rates are from a population with a mean greater than 60.

TI **32.** Appendix B Data Set: Tobacco Use in Children's Movies Refer to Data Set 5 in Appendix B and use only those movies that show some use of tobacco. Test the claim of a movie critic that "among those movies that show the use of tobacco, the mean exposure time is 2 minutes." Given the sample data, is that claim deceptive?

8-5 BEYOND THE BASICS

TI **33.** Alternative Method When testing a claim about the population mean μ using a simple random sample from a normally distributed population with unknown σ, an alternative method (not used in this book) is to use the methods of this section if the sample is small ($n \leq 30$), but if the sample is large ($n > 30$) substitute s for σ and proceed as if σ is known (as in Section 8-4). A sample of size $n = 32$ has $\bar{x} = 105.3$ and $s = 15.0$. Use a 0.05 significance level to test the claim that the sample is from a population with a mean equal to 100. Use the alternative method and compare the results to those obtained by using the method of this section.

34. Using the Wrong Distribution When testing a claim about a population mean with a simple random sample selected from a normally distributed population with unknown σ, the Student t distribution should be used for finding critical values and/or a P-value. If the standard normal distribution is incorrectly used instead, does that mistake make you more or less likely to reject the null hypothesis, or does it not make a difference? Explain.

TI **35.** Effect of an Outlier Repeat Exercise 25 after changing the first value from 5.40 to 540. Based on the results, describe the effect of an outlier on a t test.

36. Finding Critical t Values When finding critical values, we sometimes need significance levels other than those available in Table A-3. Some computer programs approximate critical t values by calculating

$$t = \sqrt{\text{df} \cdot (e^{A^2/\text{df}} - 1)}$$

where $\text{df} = n - 1$, $e = 2.718$, $A = z(8 \cdot \text{df} + 3)/(8 \cdot \text{df} + 1)$, and z is the critical z score. Use this approximation to find the critical t score corresponding to $n = 10$ and a significance level of 0.05 in a right-tailed case. Compare the results to the critical t value found in Table A-3.

37. Power of a Test Refer to the sample data in Exercise 27 and assume that you're using a 0.05 significance level for testing the claim that $\mu < 1.8$ g. Minitab is used to find that $\beta = 0.5873$, given that the actual mean is 1.7 g. Find the power of the test and the probability of a type II error. Does the power appear to be high enough to make the test very effective in rejecting $\mu = 1.8$ g when, in reality, $\mu = 1.7$ g?

8-6 Testing a Claim About a Standard Deviation or Variance

Key Concept This section introduces methods for testing a claim made about a population standard deviation σ or population variance σ^2. The methods of this section use the chi-square distribution that was first introduced in Section 7-5. The assumptions, test statistic, P-value, and critical values are summarized as follows.

Testing Claims About σ or σ^2

Requirements

1. The sample is a simple random sample.
2. The population has a normal distribution. (This is a much stricter requirement than the requirement of a normal distribution when testing claims about means, as in Sections 8-4 and 8-5.)

Test Statistic for Testing a Claim About σ or σ^2

$$\chi^2 = \frac{(n-1)s^2}{\sigma^2}$$

***P*-values and Critical values:** Use Table A-4 with $\text{df} = n - 1$ for the number of degrees of freedom. (Table A-4 is based on *cumulative areas from the right.*)

Do not use the methods of this section with a population having a distribution that is far from normal. In Sections 8-4 and 8-5 we saw that the methods of testing claims about means require a normally distributed population, and those methods work reasonably well as long as the population distribution is not very far from

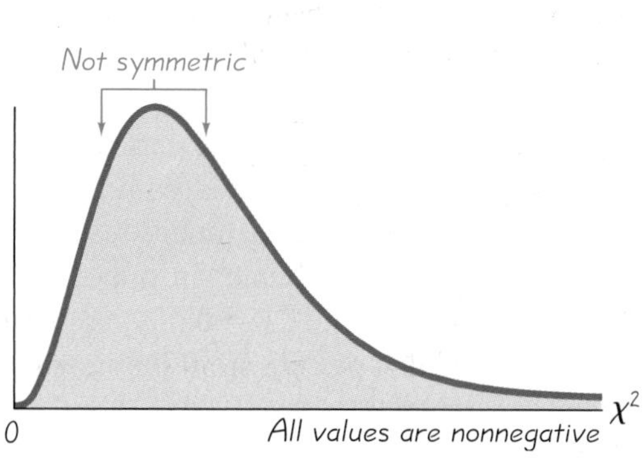

Figure 8-13 Properties of the Chi-Square Distribution

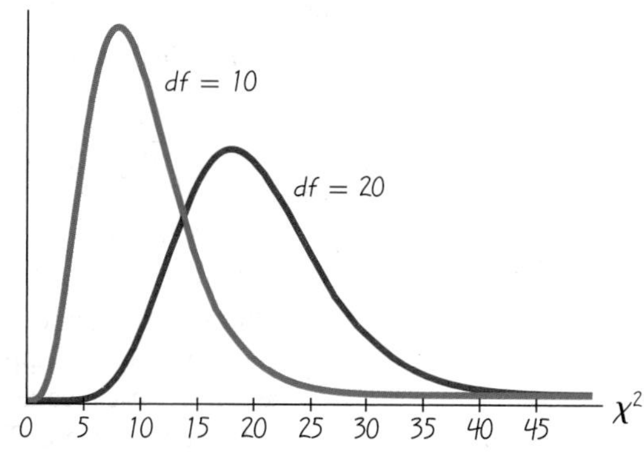

Figure 8-14 Chi-Square Distribution for df = 10 and df = 20

being normal. However, tests of claims about standard deviations or variances are not as *robust* against departures from normality, meaning that they don't work too well with distributions that are far from normal. The condition of a normally distributed population is therefore a much stricter requirement in this section.

The chi-square distribution was introduced in Section 7-5, where we noted the following important properties.

Properties of the Chi-Square Distribution

1. All values of χ^2 are nonnegative, and the distribution is not symmetric (see Figure 8-13).
2. There is a different χ^2 distribution for each number of degrees of freedom (see Figure 8-14).
3. The critical values are found in Table A-4 using

$$\textbf{degrees of freedom} = n - 1$$

Table A-4 is based on cumulative areas from the *right* (unlike the entries in Table A-2, which are cumulative areas from the left). Critical values are found in Table A-4 by first locating the row corresponding to the appropriate number of degrees of freedom (where df $= n - 1$). Next, the significance level α is used to determine the correct column. The following examples are based on a significance level of $\alpha = 0.05$, but any other significance level can be used in a similar manner.

Right-tailed test: Because the area to the *right* of the critical value is 0.05, locate 0.05 at the top of Table A-4.

Left-tailed test: With a left-tailed area of 0.05, the area to the *right* of the critical value is 0.95, so locate 0.95 at the top of Table A-4.

Two-tailed test: Divide the significance level of 0.05 between the left and right tails, so the areas to the *right* of the two critical values are 0.975 and 0.025, respectively. Locate 0.975 and 0.025 at the top of Table A-4. (See Figure 7-10 and the example on page 378.)

EXAMPLE Quality Control The industrial world shares this common goal: Improve quality by reducing variation. Quality control engineers want to ensure that a product has an acceptable mean, but they also want to produce items of *consistent* quality so that there will be few defects. The Newport Bottling Company had been manufacturing cans of cola with amounts having a standard deviation of 0.051 oz. A new bottling machine is tested, and a simple random sample of 24 cans results in the amounts (in ounces) listed below. (Those 24 amounts have a standard deviation of $s = 0.039$ oz.) Use a 0.05 significance level to test the claim that cans of cola from the new machine have amounts with a standard deviation that is less than 0.051 oz.

11.98	11.98	11.99	11.98	11.90	12.02	11.99	11.93
12.02	12.02	12.02	11.98	12.01	12.00	11.99	11.95
11.95	11.96	11.96	12.02	11.99	12.07	11.93	12.05

SOLUTION

REQUIREMENTS ✔ See the two requirements listed above.

1. The sample is a simple random sample.
2. Based on the accompanying STATDISK-generated histogram and normal quantile plot, the sample does appear to come from a population having a normal distribution. The histogram appears to be bell-shaped. The points on the normal quantile plot are quite close to a straight-line pattern, and there is no other pattern. There are no outliers. The departure from an exact normal distribution is relatively minor. With both requirements satisfied, we can now proceed to use the methods of this section to test the claim that the amounts of cola are from a population with a standard deviation less than 0.051 oz. ✔

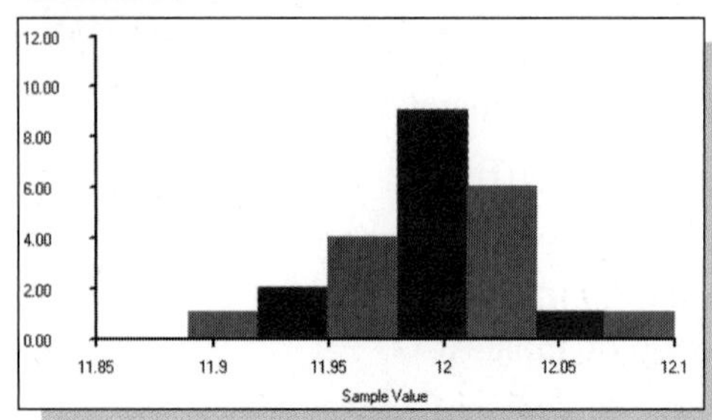

STATDISK

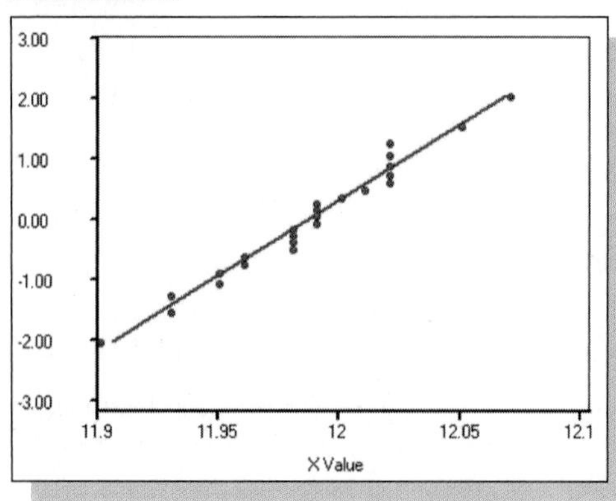

We will use the traditional method of testing hypotheses as outlined in Figure 8-9.

Step 1: The claim is expressed in symbolic form as $\sigma < 0.051$.

Step 2: If the original claim is false, then $\sigma \geq 0.051$.

Step 3: The expression $\sigma < 0.051$ does not contain equality, so it becomes the alternative hypothesis. The null hypothesis is the statement that $\sigma = 0.051$.

$$H_0: \sigma = 0.051$$
$$H_1: \sigma < 0.051 \quad \text{(original claim)}$$

Step 4: The significance level is $\alpha = 0.05$.

Step 5: Because the claim is made about σ we use the chi-square distribution.

Step 6: The test statistic is

$$\chi^2 = \frac{(n-1)s^2}{\sigma^2} = \frac{(24-1)(0.039)^2}{0.051^2} = 13.450$$

The critical value of 13.091 is found in Table A-4, in the 23rd row (degrees of freedom $= n - 1 = 23$) in the column corresponding to 0.95. See the test statistic and critical values shown in Figure 8-15.

Step 7: Because the test statistic is not in the critical region, we fail to reject the null hypothesis.

INTERPRETATION There is not sufficient evidence to support the claim that the standard deviation of amounts with the new machine is less than 0.051 oz. Perhaps the new machine does produce amounts of cola that are more consistent with a lower standard deviation than 0.051 oz, but we do not yet have enough evidence to support that claim.

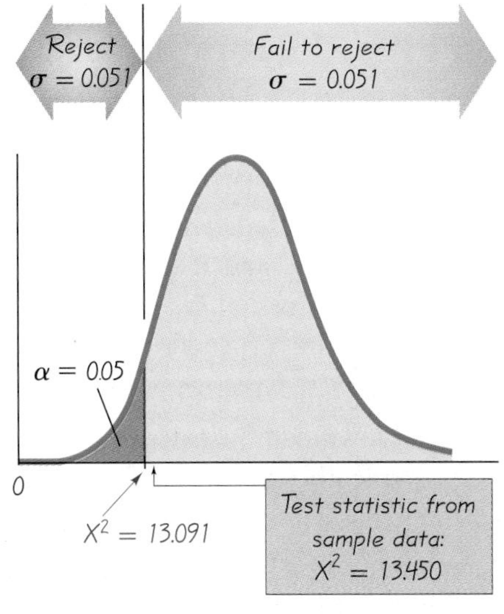

Figure 8-15 Testing the Claim that $\sigma < 0.051$

Push Polling

"Push polling" is the practice of political campaigning under the guise of a poll. Its name is derived from its objective of pushing voters away from opposition candidates by asking loaded questions designed to discredit them. Here's an example of one such question that was used: "Please tell me if you would be more likely or less likely to vote for Roy Romer if you knew that Gov. Romer appoints a parole board which has granted early release to an average of four convicted felons per day every day since Romer took office." The National Council on Public Polls characterizes push polls as unethical, but some professional pollsters do not condemn the practice as long as the questions do not include outright lies.

P-Value Method

Instead of using the traditional approach to hypothesis testing for the preceding example, we can also use the *P*-value approach summarized in Figures 8-6 and 8-8. If STATDISK or a TI-83/84 Plus calculator is used for the preceding example, the *P*-value of 0.0584 will be found. Because the *P*-value is greater than the significance level of 0.05, we fail to reject the null hypothesis and arrive at the same conclusion. When using Table A-4, we usually cannot find *exact* *P*-values because that chi-square distribution table includes only selected values of α. (Because of this limitation, testing claims about σ or σ^2 with Table A-4 is easier with the traditional method than with the *P*-value method.) If using Table A-4, we can identify limits that contain the *P*-value. The test statistic from the last example is $\chi^2 = 13.450$ and we know that the test is left-tailed with 23 degrees of freedom. Refer to the 23rd row of Table A-4 and see that the test statistic of 13.450 is between the row entries of 13.091 and 14.848, which means that the area to the right of the test statistic is between 0.95 and 0.90. It follows that the area to the left of the test statistic is between 0.05 and 0.10, so we conclude that $0.05 < P\text{-value} < 0.10$. Because the *P*-value is greater than the significance level of $\alpha = 0.05$, we fail to reject the null hypothesis. Again, the traditional method and the *P*-value method are equivalent in the sense that they always lead to the same conclusion.

Confidence Interval Method

The preceding example can also be solved with the confidence interval method of testing hypotheses. Using the methods described in Section 7-5, we can use the sample data ($n = 24$, $s = 0.039$) to construct this 90% confidence interval: $0.032 < \sigma < 0.052$. (Remember, a one-tailed hypothesis test with significance level 0.05 is tested using a 90% confidence interval. See Table 8-2.) Based on this confidence interval, we cannot conclude that σ is less than 0.051. We reach the same conclusion from the traditional and *P*-value methods.

Using a TI-83/84 Plus Calculator for Testing a Claim About a Standard Deviation or Variance

The TI-83/84 Plus calculator is not designed to automatically generate results for a hypothesis test about a population standard deviation σ or a population variance σ^2, but the program **S2TEST** can be used. That program was written by Michael Lloyd of Henderson State University, and it can be downloaded from the CD-ROM included with this book, or it can be downloaded from the Web site www.aw.com/triola. Use the following procedure.

1. Use the enclosed CD-ROM or the Web site www.aw.com/triola to download the program S2TEST. (Download instructions can be found at the site www.ti.com, and they typically involve the use of software such as TI-GRAPHLINK or TI CONNECT.)
2. Press PRGM, select the program **S2TEST,** then press ENTER.
3. Proceed to enter the value of the population variance σ^2. (The value of the population standard deviation σ is used in the null and alternative hypotheses.) Be sure to enter the value of σ^2 when prompted, then press ENTER.

4. Next, there will be a prompt for the entry of the sample variance s^2. Enter the square of the sample standard deviation, then press ENTER.
5. There will be a prompt for the entry of the sample size n, so enter that value and press ENTER.
6. Now select the type of test as follows:
   - For a two-tailed hypothesis test, select the first option that includes the symbol $\neq$.
   - For a left-tailed test, select the option that includes the symbol $<$.
   - For a right-tailed test, select the option that includes the symbol $>$.
7. Press ENTER and the test statistic and P-value will be provided.

If using the above procedure with the example in this section, the test statistic of 13.450 and the P-value of 0.0584 will be provided.

Another method is to manually calculate the value of the test statistic and number of degrees of freedom $(n - 1)$. After finding the test statistic and number of degrees of freedom, the P-value can be found from the TI-83/84 calculator as follows:

1. Press 2ND VARS to get to the DISTR (distribution) menu, then select **χ^2 cdf.**
2. The χ^2 cdf command requires three items separated by commas in this format: **(lower bound, upper bound, degrees of freedom).** Find the P-value by using the number of degrees of freedom and the test statistic as shown below.

   Left-tailed test: The P-value is the result of χ^2 cdf $(-999999$, test statistic, degrees of freedom).

   Right-tailed test: The P-value is the result of χ^2 cdf (test statistic, 999999, degrees of freedom).

   Two-tailed test: The P-value is the smaller of the following two results:
   - Twice the value obtained by entering χ^2 cdf $(-999999$, test statistic, degrees of freedom)
   - Twice the value obtained by entering χ^2 cdf (test statistic, 999999, degrees of freedom)

The example in this section involved a left-tailed test with the test statistic of $\chi^2 = 13.450$ and degrees of freedom of $n - 1 = 23$. Using the above procedure, the P-value is the result of χ^2 cdf $(-999999, 13.450, 23)$, which is 0.0584.

STATDISK does have the ability to conduct hypothesis tests involving claims about a population standard deviation σ or a population variance σ^2. Select **Analysis** from the main menu at the top, then select **Hypothesis Testing** and proceed to the case identified as **St Dev One Sample.**

8-6 BASIC SKILLS AND CONCEPTS

Statistical Literacy and Critical Thinking

1. Robust What does it mean when we say that the chi-square test of this section is not *robust* against departures from normality? How does that affect the conditions that must be satisfied for the chi-square test of this section?

2. Using Confidence Interval Assume that you must use a 0.01 significance level to test the claim that $\sigma > 5.00$. If you plan to test that claim by constructing a confidence interval, what level of confidence should be used for the confidence interval? Will the conclusion based on the confidence interval be the same as the conclusion based on a hypothesis test that uses the traditional method or the *P*-value method?

3. Requirements When rolling a fair die, the results have mean $\mu = 3.5$ and standard deviation $\sigma = \sqrt{35/12} \approx 1.7$. A die is rolled 100 times in an attempt to verify that it behaves like a fair die. If you calculate the standard deviation of the 100 outcomes, can you use that value with the methods of this section to test the claim that $\sigma = 1.7$ for this die? Why or why not?

4. Standard Deviation and Variance Is a test of the claim that $\sigma = 2.00$ equivalent to a test of the claim that $\sigma^2 = 4.00$? When testing the claim that $\sigma = 2.00$, do you use a different test statistic than the one used for testing the claim that $\sigma^2 = 4.00$?

Finding Test Components. *In Exercises 5–8, calculate the test statistic, then use Table A-4 to find critical value(s) of χ^2, then use Table A-4 to find limits containing the P-value, then determine whether there is sufficient evidence to support the given alternative hypothesis.*

5. $H_1: \sigma \neq 2.00$, $\alpha = 0.05$, $n = 10$, $s = 3.00$.

6. $H_1: \sigma > 15$, $\alpha = 0.05$, $n = 5$, $s = 30$.

7. $H_1: \sigma < 15$, $\alpha = 0.01$, $n = 21$, $s = 10$.

8. $H_1: \sigma \neq 75$, $\alpha = 0.01$, $n = 51$, $s = 70$.

Testing Claims About Variation. *In Exercises 9–20, test the given claim. Assume that a simple random sample is selected from a normally distributed population.*

TI 9. Birth Weights A study was conducted of babies born to mothers who use cocaine during pregnancy, and the following sample data were obtained for weights at birth: $n = 190$, $\bar{x} = 2700$ g, $s = 645$ g (based on data from "Cognitive Outcomes of Preschool Children with Prenatal Cocaine Exposure," by Singer et al., *Journal of the American Medical Association,* Vol. 291, No. 20). Use a 0.05 significance level to test the claim that the standard deviation of birth weights of infants born to mothers who use cocaine is different from the standard deviation of 696 g for birth weights of infants born to mothers who do not use cocaine during pregnancy. (Because Table A-4 has a maximum of 100 degrees of freedom while we require 189 degrees of freedom, use these critical values obtained from STATDISK: $\chi^2_L = 152.8222$ and $\chi^2_R = 228.9638$.) Based on the result, does cocaine use by mothers appear to affect the variation of the weights of their babies?

TI 10. Minting Quarters Quarters are currently minted with weights having a mean of 5.670 g and a standard deviation of 0.062 g. New equipment is being tested in an attempt to improve quality by reducing variation. A simple random sample of 24 quarters is obtained from those manufactured with the new equipment, and this sample has a standard deviation of 0.049 g. Use a 0.05 significance level to test the claim that quarters manufactured with the new equipment have weights with a standard deviation less than 0.062 g. Does the new equipment appear to be effective in reducing the variation of weights? What would be an adverse consequence of having quarters with weights that vary too much?

TI 11. Variation in Peanut M&Ms Use a 0.01 significance level to test the claim that peanut M&M candies have weights that vary more than the weights of plain M&M candies.

The standard deviation for the weights of plain M&M candies is 0.056 g. A sample of 41 peanut M&Ms has weights with a standard deviation of 0.31 g. Why should peanut M&M candies have weights that vary more than the weights of plain M&M candies?

TI **12.** Manufacturing Aircraft Altimeters The Stewart Aviation Products Company uses a new production method to manufacture aircraft altimeters. A simple random sample of 81 altimeters is tested in a pressure chamber, and the errors in altitude are recorded as positive values (for readings that are too high) or negative values (for readings that are too low). The sample has a standard deviation of $s = 52.3$ ft. At the 0.05 significance level, test the claim that the new production line has errors with a standard deviation different from 43.7 ft, which was the standard deviation for the old production method. If it appears that the standard deviation has changed, does the new production method appear to be better or worse than the old method?

TI **13.** Credit Rating When consumers apply for credit, their credit is rated using FICO (Fair, Isaac, and Company) scores. Credit ratings are given below for a sample of applicants for car loans, and these applicants are all from a new branch of the Bank of Newport. Use the sample data to test the claim that these credit ratings are from a population with a standard deviation different from 83, which is the standard deviation for applicants from the main bank. Use a 0.05 significance level. Based on the results, do applicants from the branch appear to have credit ratings that vary more than those at the main bank?

661 595 548 730 791 678 672 491 492 583 762 624 769 729 734 706

TI **14.** World's Smallest Mammal The world's smallest mammal is the bumblebee bat, also known as the Kitti's hog-nosed bat (or *Craseonycteris thonglongyai*). Such bats are roughly the size of a large bumblebee. Listed below are weights (in grams) from a sample of these bats. Using a 0.05 significance level, test the claim that these weights come from a population with a standard deviation equal to 0.30 g, which is the standard deviation of weights of the bumblebee bats from one region in Thailand. Do these bats appear to have weights with the same variation as the bats from that region in Thailand?

1.7 1.6 1.5 2.0 2.3 1.6 1.6 1.8 1.5 1.7 2.2 1.4 1.6 1.6 1.6

TI **15.** Supermodel Weights Use a 0.01 significance level to test the claim that weights of female supermodels vary less than the weights of women in general. The standard deviation of weights of the population of women is 29 lb. Listed below are the weights (in pounds) of nine randomly selected supermodels.

125 (Taylor)	119 (Auermann)	128 (Schiffer)	128 (MacPherson)
119 (Turlington)	127 (Hall)	105 (Moss)	123 (Mazza)
115 (Hume)			

TI **16.** Supermodel Heights Use a 0.05 significance level to test the claim that heights of female supermodels vary less than the heights of women in general. The standard deviation of heights of the population of women is 2.5 in. Listed below are the heights (in inches) of randomly selected supermodels (Taylor, Harlow, Mulder, Goff, Evangelista, Auermann, Schiffer, MacPherson, Turlington, Hall, Crawford, Campbell, Herzigova, Seymour, Banks, Moss, Mazza, Hume).

71	71	70	69	69.5	70.5	71	72	70
70	69	69.5	69	70	70	66.5	70	71

TI **17.** Monitoring Lead in Air Listed below are measured amounts of lead (in micrograms per cubic meter, or $\mu g/m^3$) in the air. The Environmental Protection Agency has established an air quality standard for lead of 1.5 $\mu g/m^3$. The measurements shown below were recorded at Building 5 of the World Trade Center site on different days immediately following the destruction caused by the terrorist attacks of September 11, 2001. Test the claim that these amounts are from a population with a standard deviation greater than 0.4 $\mu g/m^3$. Use a significance level of 0.05. Is there anything about the sample values that might violate a requirement for using the chi-square test?

5.40 1.10 0.42 0.73 0.48 1.10

TI **18.** Bank Waiting Lines The listed values are waiting times (in minutes) of customers at the Jefferson Valley Bank, where customers enter a single waiting line that feeds three teller windows. Test the claim that the standard deviation of waiting times is less than 1.9 min, which is the standard deviation of waiting times at the same bank when separate waiting lines are used at each teller window. Use a significance level of 0.05. Does the use of the single line appear to reduce the variation among waiting times? What is an advantage of reducing the variation among waiting times?

6.5 6.6 6.7 6.8 7.1 7.3 7.4 7.7 7.7 7.7

TI **19.** Appendix B Data for Weights of Quarters Refer to Data Set 14 in Appendix B and use the sample data consisting of weights of quarters made after 1964. Test the claim that these quarters come from a population with a standard deviation equal to the specified value of 0.068 g. Use a significance level of 0.05. Does the variation of the weights appear to be as desired?

TI **20.** Appendix B Data for Precipitation Amounts Refer to Data Set 8 in Appendix B and use the sample data consisting of the precipitation amounts. Test the claim that these amounts are from a population with a standard deviation less than 1.00 in. Use a significance level of 0.05. Does the requirement of a normal distribution appear to be satisfied, and how does that affect the results?

8-6 BEYOND THE BASICS

21. Finding Critical Values of χ^2 For large numbers of degrees of freedom, we can approximate critical values of χ^2 as follows:

$$\chi^2 = \frac{1}{2}\left(z + \sqrt{2k - 1}\right)^2$$

Here k is the number of degrees of freedom and z is the critical value, found in Table A-2. For example, if we want to approximate the two critical values of χ^2 in a two-tailed hypothesis test with $\alpha = 0.05$ and a sample size of 190, we let $k = 189$ with $z = -1.96$ followed by $k = 189$ and $z = 1.96$. Use this approximation to estimate the critical values of χ^2 in a two-tailed hypothesis test with $n = 190$ and $\alpha = 0.05$. Compare the results to the critical values given in Exercise 9.

22. Finding Critical Values of χ^2 Repeat Exercise 21 using this approximation (with k and z as described in Exercise 21):

$$\chi^2 = k\left(1 - \frac{2}{9k} + z\sqrt{\frac{2}{9k}}\right)^3$$

23. Effect of Outlier When using the hypothesis testing procedure of this section, will the result be dramatically affected by the presence of an outlier? Describe how you arrived at your response.

Review

This chapter presented basic methods for testing claims about a population proportion, population mean, or population standard deviation (or variance). The methods of this chapter are used by professionals in a wide variety of disciplines, as illustrated in their many professional journals.

In Section 8-2 we presented the fundamental concepts of a hypothesis test: null hypothesis, alternative hypothesis, test statistic, critical region, significance level, critical value, *P*-value, type I error, and type II error. We also discussed two-tailed tests, left-tailed tests, right-tailed tests, and the statement of conclusions. We used those components in identifying three different methods for testing hypotheses:

1. The *P*-value method (summarized in Figure 8-8)
2. The traditional method (summarized in Figure 8-9)
3. Confidence intervals (discussed in Chapter 7)

In Sections 8-3 through 8-6 we discussed specific methods for dealing with different parameters. Because it is so important to be correct in selecting the distribution and test statistic, we provide Table 8-3, which summarizes the hypothesis testing procedures of this chapter.

Table 8-3 Hypothesis Tests

Parameter	Requirements: Simple Random Sample and . . .	Distribution and Test Statistic	Critical and *P*-Values
Proportion	$np \geq 5$ and $nq \geq 5$	Normal: $z = \dfrac{\hat{p} - p}{\sqrt{\dfrac{pq}{n}}}$	Table A-2
Mean	σ known and normally distributed population *or* σ known and $n > 30$	Normal: $z = \dfrac{\bar{x} - \mu_{\bar{x}}}{\dfrac{\sigma}{\sqrt{n}}}$	Table A-2
	σ not known and normally distributed population *or* σ not known and $n > 30$	Student *t*: $t = \dfrac{\bar{x} - \mu_{\bar{x}}}{\dfrac{s}{\sqrt{n}}}$	Table A-3
	Population not normally distributed and $n \leq 30$	Use a nonparametric method or bootstrapping	
Standard Deviation or Variance	Population normally distributed	Chi-square: $\chi^2 = \dfrac{(n-1)s^2}{\sigma^2}$	Table A-4

Statistical Literacy and Critical Thinking

1. Practical Significance When testing the claim that the mean amount lost on a diet is greater than 0 lb, we use sample results showing that 10,000 people lost amounts with a mean of 0.1 lb. Also, the sample standard deviation is 0.8 lb. The hypothesis test results include a test statistic of $t = 12.500$ and a P-value of 0.0000 (rounded to four decimal places). Is the mean weight loss statistically significant? Does the mean weight loss suggest that this diet is practical? Should it be recommended?

2. Sample You have just collected a very large ($n = 2575$) sample of responses obtained from adult Americans who mailed responses to a questionnaire printed in *Fortune* magazine. If this sample is used to make inferences about a population, what is the population? A hypothesis test conducted at the 0.01 significance level leads to the conclusion that most (more than 50%) adults are opposed to estate taxes. Can we conclude that most adult Americans are opposed to estate taxes? Why or why not?

3. *P*-Value You have just developed a new cure for the common cold and you plan to conduct a formal test to justify its effectiveness. Which P-value would you most prefer: 0.99, 0.05, 0.5, 0.01, or 0.001? Why?

4. *P*-Value In testing the claim that the mean amount of cola in cans is greater than 12 oz, you use sample data to obtain a P-value of 0.250. What should you conclude? What does the probability of 0.250 represent?

Review Exercises

1. Identifying Hypotheses, Distributions, and Verifying Requirements Based on the given conditions, identify the alternative hypothesis. Also, either identify the sampling distribution (normal, t, chi-square) of the test statistic, or state that the methods of this chapter should not be used.
   a. Claim: The mean annual income of full-time college students is below $10,000. Sample data: For 750 randomly selected college students, the mean is $3662 and the standard deviation is $2996. The sample data suggest that the population has a distribution that is not normal.
   b. Claim: The majority of college students have at least one credit card. Sample data: Of 500 randomly selected college students, 82% have at least one credit card.
   c. Claim: The mean IQ of adults is equal to 100. Sample data: $n = 150$ and $\bar{x} = 98.8$. It is reasonable to assume that $\sigma = 15$.
   d. Claim: With manual assembly of telephone parts, the assembly times vary more than the times for automated assembly, which are known to have a mean of 27.6 sec and a standard deviation of 1.8 sec. Sample data: For 24 randomly selected times, the mean is 25.4 sec, the standard deviation is 2.5 sec, and the sample data suggest that the population has a distribution that is far from normal.
   e. Claim: Statistics professors have IQ scores that vary less than IQ scores from the adult population, which has a standard deviation of 15. Sample data: For 24 randomly selected statistics professors, the standard deviation is 10.2, and the sample data suggest that the population has a distribution that is very close to a normal distribution.

TI 2. Interviewing Mistakes An Accountemps survey of 150 executives showed that 44% of them say that "little or no knowledge of the company" is the most common mistake made by candidates during job interviews (based on data from *USA Today*). Use a

0.05 significance level to test the claim that less than half of all executives identify that error as being the most common job interviewing error.

TI **3.** *Glamour* Magazine Survey *Glamour* magazine sponsored a survey of 2500 prospective brides and found that 60% of them spent less than \$750 on their wedding gown. Use a 0.01 significance level to test the claim that less than 62% of brides spend less than \$750 on their wedding gown. How are the results affected if it is learned that the responses were obtained from magazine readers who decided to respond to the survey through an Internet Web site?

TI **4.** Weights of Sugar Packets A student of the author randomly selected 70 packets of sugar and weighed the contents of each packet, getting a mean of 3.586 g and a standard deviation of 0.074 g. Test the claim that the weights of the sugar packets have a mean equal to 3.5 g, as indicated on the label. If the mean does not appear to equal 3.5 g, does the difference appear to be large enough to create health problems for consumers?

TI **5.** Are Consumers Being Cheated? The Orange County Bureau of Weights and Measures received complaints that the Windsor Bottling Company was cheating consumers by putting less than 12 oz of root beer in its cans. When 24 cans are randomly selected and measured, the amounts are found to have a mean of 11.4 oz and a standard deviation of 0.62 oz. The company president, Harry Windsor, claims that the sample is too small to be meaningful. Use the sample data to test the claim that consumers are being cheated. Does Harry Windsor's argument have any validity?

TI **6.** Testing Lipitor for Adverse Effects In clinical tests of the drug Lipitor (generic name, atorvastatin), 863 patients were treated with 10-mg doses of atorvastatin, and 19 of those patients experienced flu symptoms (based on data from Parke-Davis). Use a 0.01 significance level to test the claim that the percentage of treated patients with flu symptoms is greater than the 1.9% rate for patients not given the treatments. Does it appear that flu symptoms are an adverse reaction of the treatment?

TI **7.** IQ Scores For a simple random sample of adults, IQ scores are normally distributed with a mean of 100 and a standard deviation of 15. A simple random sample of 13 statistics professors yields a standard deviation of $s = 7.2$. A psychologist is quite sure that statistics professors have IQ scores that have a mean greater than 100. He doesn't understand the concept of standard deviation very well and does not realize that the standard deviation should be lower than 15 (because statistics professors have less variation than the general population). Instead, he claims that statistics professors have IQ scores with a standard deviation equal to 15, the same standard deviation for the general population. Assume that IQ scores of statistics professors are normally distributed and use a 0.05 significance level to test the claim that $\sigma = 15$. Based on the result, what do you conclude about the standard deviation of IQ scores for statistics professors?

TI **8.** Random Generation of Data The TI-83/84 Plus calculator can be used to generate random data from a normally distributed population. The command **randNorm(100,15,50)** generates 50 values from a normally distributed population with $\mu = 100$ and $\sigma = 15$. One such generated sample of 50 values has a mean of 98.4 and a standard deviation of 16.3. Use a 0.10 significance level to test the claim that the sample actually does come from a population with a mean equal to 100. Assume that σ is known to be 15. Based on the results, does it appear that the calculator's random number generator is working correctly?

TI **9.** Body Temperature A premed student in a statistics class is required to do a class project. Intrigued by the body temperatures in Data Set 2 of Appendix B, she plans to collect her own sample data to test the claim that the mean body temperature is less than 98.6°F, as is commonly believed. Because of time constraints imposed by other courses and the desire to maintain a social life that goes beyond talking in her sleep, she finds that she has time to collect data from only 12 people. After carefully planning a procedure for obtaining a simple random sample of 12 healthy adults, she measures their body temperatures and obtains the results listed below. Use a 0.05 significance level to test the claim that these body temperatures come from a population with a mean that is less than 98.6°F.

98.0 97.5 98.6 98.8 98.0 98.5 98.6 99.4 98.4 98.7 98.6 97.6

TI **10.** Heights of Women Anthropometric survey data are used to publish values that can be used in designing products that are suitable for use by adults. According to Gordon, Churchill, et al., women have heights with a mean of 64.1 in. and a standard deviation of 2.52 in. The sample of 40 heights of women in Data Set 1 in Appendix B has a mean of 63.20 in. and a standard deviation of 2.74 in. Use a 0.05 significance level to test the claim that this sample is from a population with a standard deviation equal to 2.52 in. When designing car seats for women, what would be a consequence of believing that heights of women vary less than they really vary?

TI **11.** Umpire Strike Rate In a recent year, some professional baseball players complained that umpires were calling more strikes than the average rate of 61.0% called the previous year. At one point in the season, umpire Dan Morrison called strikes in 2231 of 3581 pitches (based on data from *USA Today*). Use a 0.05 significance level to test the claim that his strike rate is greater than 61.0%.

TI **12.** Nicotine in Cigarettes The Carolina Tobacco Company advertised that its best-selling nonfiltered cigarettes contain at most 40 mg of nicotine, but *Consumer Advocate* magazine ran tests of 10 randomly selected cigarettes and found the amounts (in mg) shown in the accompanying list. It's a serious matter to charge that the company advertising is wrong, so the magazine editor chooses a significance level of $\alpha = 0.01$ in testing her belief that the mean nicotine content is greater than 40 mg. Using a 0.01 significance level, test the editor's belief that the mean is greater than 40 mg.

47.3 39.3 40.3 38.3 46.3 43.3 42.3 49.3 40.3 46.3

Cumulative Review Exercises

TI **1.** Heights of Presidents Listed below are the heights (in inches) of the Presidents who served in the 20th century. Assume that these values are sample data from some larger population.

67	70	72	71	72	70	71	74	69
70.5	72	75	71.5	69.5	73	74	74.5	

a. Find the mean.
b. Find the median.
c. Find the standard deviation.
d. Find the variance.
e. Find the range.
f. Assuming that the values are sample data, construct a 95% confidence interval estimate of the population mean.

g. The mean height of men is 69.0 in. Use a 0.05 significance level to test the claim that this sample comes from a population with a mean greater than 69.0 in. Do Presidents appear to be taller than the typical man?

TI **2.** Normal Quantile Plot Refer to the data listed in Exercise 1 and construct a normal quantile plot. Based on the result, do the heights appear to come from a population that is normally distributed?

TI **3.** SAT Math Scores of Women The math SAT scores for women are normally distributed with a mean of 496 and a standard deviation of 108.

a. If a woman who takes the math portion of the SAT is randomly selected, find the probability that her score is above 500.

b. If five math SAT scores are randomly selected from the population of women who take the test, find the probability that all five of the scores are above 500.

c. If five women who take the math portion of the SAT are randomly selected, find the probability that their mean is above 500.

d. Find P_{90}, the score separating the bottom 90% from the top 10%.

Cooperative Group Activities

1. In-class activity Without using any measuring device, each student should draw a line believed to be 3 in. long and another line that is 3 cm long. Then use rulers to measure and record the lengths of the lines drawn. Find the means and standard deviations of the two sets of lengths. Test the claim that the lines estimated to be 3 in. have a mean length that is equal to 3 in. Test the claim that the lines estimated to be 3 cm have a mean length that is equal to 3 cm. Compare the results. Do the estimates of the 3-in. line appear to be more accurate than those for the 3-cm line?

2. In-class activity Assume that a method of gender selection can affect the probability of a baby being a girl, so that the probability becomes 1/4. Each student should simulate 20 births by drawing 20 cards from a shuffled deck. Replace each card after it has been drawn, then reshuffle. Consider the hearts to be girls and consider all other cards to be boys. After making 20 selections and recording the "genders" of the babies, use a 0.10 significance level to test the claim that the proportion of girls is equal to 1/4. How many students are expected to get results leading to the wrong conclusion that the proportion is not 1/4? How does that relate to the probability of a type I error? Does this procedure appear to be effective in identifying the effectiveness of the gender-selection method? (If decks of cards are not available, use some other way to simulate the births, such as using the random number generator on a calculator or using digits from phone numbers or social security numbers.)

3. Out-of-class activity Groups of three or four students should go to the library and collect a sample consisting of the ages of books (based on copyright dates). Plan and describe the sampling plan, execute the sampling procedure, then use the results to test the claim that the mean of the ages of all books in the library is greater than 20 years.

4. In-class activity Each student should write an estimate of the age of the current President of the United States. All estimates should be collected and the sample mean and standard deviation should be calculated. Then test the hypothesis that the mean of all such estimates is equal to the actual current age of the President.

5. In-class activity A class project should be designed to conduct a test in which each student is given a taste of Coke and a taste of Pepsi. The student is then asked to identify which sample is Coke. After all of the results are collected, test the claim that the success rate is better than the rate that would be expected with random guesses.

6. In-class activity Each student should estimate the length of the classroom. The values should be based on visual estimates, with no actual measurements being taken. After the estimates have been collected, measure the length of the room, then test the claim that the sample mean is equal to the actual length of the classroom. Is there a "collective wisdom," whereby the class mean is approximately equal to the actual room length?

7. Out-of-class activity Using a wristwatch that is reasonably accurate, set the time to be exact. Use a radio station or telephone time report which states that "at the tone, the time is . . . ". If you cannot set the time to the nearest second, record the error for the watch you are using. Now compare the time on your watch to the time on others. Record the errors with positive signs for watches that are ahead of the actual time and negative signs for those watches that are behind the actual time. Use the data to test the claim that the mean error of all wristwatches is equal to 0. Do we collectively run on time, or are we early or late? Also test the claim that the standard deviation of errors is less than 1 min. What are the practical implications of a standard deviation that is excessively large?

8. In-class activity In a group of three or four people, conduct an ESP experiment by selecting one of the group members as the subject. Draw a circle on one small piece of paper and draw a square on another sheet of the same size. Repeat this experiment 20 times: Randomly select the circle or the square and place it in the subject's hand behind his or her back so that it cannot be seen, then ask the subject to identify the shape (without looking at it); record whether the response is correct. Test the claim that the subject has ESP because the proportion of correct responses is greater than 0.5.

9. In-class activity After dividing into groups with sizes between 10 and 20 people, each group member should record the number of heartbeats in a minute. After calculating $\bar{x}$ and s, each group should proceed to test the claim that the mean is greater than 60, which is the author's result. (When people exercise, they tend to have lower pulse rates, and the author runs five miles a few times each week. What a guy.)

10. Out-of-class activity As part of a Gallup poll, subjects were asked "Are you in favor of the death penalty for persons convicted of murder?" Sixty-five percent of the respondents said that they were in favor, while 27% were against and 8% had no opinion. Use the methods of Section 7-2 to determine the sample size necessary to estimate the proportion of students at your college who are in favor. The class should agree on a confidence level and margin of error. Then divide the sample size by the number of students in the class, and conduct the survey by having each class member ask the appropriate number of students at the college. Analyze the results to determine whether the students differ significantly from the results found in the Gallup poll.

11. Out-of-class activity Each student should find an article in a professional journal that includes a hypothesis test of the type discussed in this chapter. Write a brief report describing the hypothesis test and its role in the context of the article.

TI-83/84 Plus Calculator Project

This project introduces simulations as another way to test hypotheses. Suppose that we want to test the claim that residents of Maine have a mean IQ greater than 100, and we obtain a random sample of 50 IQ scores from residents of Maine, with these results: $n = 50$, $\bar{x} = 106.0$, $s = 15.0$. Using a 0.05 significance level, we can test the claim that $\mu > 100$ by using the *P*-value method, the traditional method, and by constructing a 90% confidence interval.

The basic idea underlying a hypothesis test is the rare event rule for inferential statistics first introduced in Chapter 4. Using that rule, we need to determine whether the IQ score of 106.0 is "unusual" or if it can easily occur by chance. Make that determination by using simulations. Repeatedly generate 50 IQ scores from a normally distributed population having the assumed mean of 100 (as in the null hypothesis) and a standard deviation of 15. Then, based on the sample means that are found, determine whether a mean such as 106.0 or greater is "unusual" or can easily occur by chance. If a sample mean of 106.0 (or greater) is found in fewer than 5% of the samples, conclude that a value such as 106.0 cannot easily occur by chance, so there is sufficient evidence to support the claim that the mean is greater than 100.

Use the TI-83/84 Plus calculator to repeat the process of generating a random sample of 50 values from a normal distribution with a mean of 100 and a standard deviation of 15. Find the mean of each sample. Generate enough samples to be reasonably confident in determining that a sample mean such as 106.0 or greater is unusual or can easily occur by chance. Record all of your results. What do these results suggest about the claim that the mean IQ is greater than 100?

Here are TI-83/84 Plus calculator instructions for generating random values from a normally distributed population.

1. First clear list L1 by pressing STAT, then select **ClrList** and enter L1 and press ENTER.
2. Now press MATH, then select **PRB** and select the menu item of **randNorm(** and press ENTER.
3. Proceed to enter the mean, standard deviation, and sample size. For example, to generate 50 IQ scores from a normally distributed population with a mean of 100 and a standard deviation of 15, the entry should be **randNorm(100, 15, 50).** Press ENTER.
4. Store the 50 simulated IQ scores in list L1. (This can be accomplished by pressing STO› 2ND 1 ENTER.)
5. Now find the mean of the sample values in L1 as follows: Press STAT, select CALC, select the first menu item of **1-VAR Stats,** press ENTER, enter L1, then press ENTER.

From Data to Decision

Critical Thinking: Can dogs be used to detect diseases?

A study published in the *British Medical Journal* described how dogs were used in an attempt to identify patients having bladder cancer. A trial involved six different samples of urine from healthy people plus another sample of urine from a person known to have bladder cancer. The dogs were trained to identify the sample from the patient with bladder cancer. The trial was repeated 54 times with 22 correct identifications and 32 wrong identifications (based on data from the *New York Times*).

Analyzing the Results

a. Given that each trial involved six healthy samples and one sample from a patient with bladder cancer, what is the probability that a dog would select the cancer sample if it made a random guess?

b. Given that the trial was conducted 54 times, what is the expected number of correct identifications, assuming that random guesses were made?

c. Among the 54 trials, there were 22 correct identifications. Test the hypothesis that the dogs did significantly better than what would be expected with random guessing. Does it appear that the dogs were guessing, or do they appear to have some ability to identify the cancer sample?

d. Assuming that the dogs did better than what would be expected with random guessing, did they do well enough to be used for actual medical diagnoses? Why or why not?

Internet Project

Hypothesis Testing

This chapter introduced the methodology behind hypothesis testing, an essential technique of inferential statistics. This Internet Project will have you conduct tests using a variety of data sets in different areas of study. For each subject, you will be asked to

- collect data available on the Internet.
- formulate a null and alternative hypothesis based on a given question.
- conduct a hypothesis test at a specified level of significance.
- summarize your conclusion.

Go to the *Elementary Statistics* Web site at

http://www.aw.com/triola

and locate the Internet Project for this chapter. There you will find guided investigations in the fields of education, economics, and sports, and a classical example from the physical sciences.

Statistics @ Work

"It is extremely important for everyone to have an understanding of statistics in order to effectively process the huge amounts of information that we are presented with each day in our professional and personal lives."

Michael Saccucci

Director of Statistics and Quality Management for Consumers Union, which tests products and services and provides ratings and recommendations to consumers in Consumer Reports *magazine*

What statistical concepts and procedures do you use at Consumers Union?

On any given day, the statisticians may have to use any number of statistical procedures, many of which are discussed in this text. For example, in a recent study to evaluate the quality and safety of chicken, we developed a complex sampling scheme so that the different manufacturers were fairly represented. In a recent study of sunscreens, we used the normal distribution to help determine the appropriate number of replicates necessary to fairly evaluate the products. Depending on the type of test, the statistician might have to construct a completely randomized design, a randomized block design, or some other type of experimental design to ensure that our results are accurate and unbiased. During the analysis phase, the statistician might use any number of techniques, such as analysis of variance, regression analysis, time series analysis, categorical analysis, and/or nonparametric analysis.

What do the statisticians do at Consumers Union?

Statisticians perform a variety of tasks. Early in a project, the statistician works with the project team to develop the test protocol and help select which products to test. Next, the statistician helps develop an appropriate experimental design for use during the testing. Once the test data have been obtained, the statistician analyzes the results and presents findings in a statistical report. The statistician also gets involved with a variety of special projects, depending on the needs of the organization. Consumers rely on the information we provide, so it's important that we use proper statistical techniques to make sure that our ratings are correct.

What steps do you take to ensure objectivity in your testing procedures?

It is Consumers Union's policy that all tests must be performed in an objective, scientific manner, and with due regard for the safety of test personnel. We go to great lengths to adhere to this policy. For example, we don't accept any type of outside advertising in our publications. We employ anonymous shoppers located throughout the United States to purchase our test samples in ways normally available to consumers. We don't accept free samples from anyone, including vendors. And we don't test unsolicited samples sent from a manufacturer. In addition, technicians use randomized experimental designs to ensure that our testing is done with scientific integrity and objectivity. When practical, tested items are blind coded so that the testers do not know which brands they are evaluating.

Are the ratings and recommendations in *Consumer Reports* magazine based on statistical significance alone?

No. The information we provide must be useful to consumers. Our technicians perform a variety of tests to evaluate a product's performance. These tests are designed to simulate conditions of predictable consumer usage. If it turns out that there is a statistically significant, but not meaningful difference in test results, we would not rate one brand over another. In testing water sealants, for example, we might find that there is a

Author's Note: The author met with Mike Saccucci and the other statisticians at Consumers Union: Keith Newsom-Stewart, Martin Romm, and Eric Rosenberg. The author toured the product-testing facilities and observed several different experiments in progress. He was extremely impressed with the involvement of statisticians at the various stages of product testing, the extreme and detailed care taken in the design of experiments, and the careful and effective use of statistical analyses of test results.

statistically significant difference between the amounts of water that seep through two different brands of sealant. However, if the difference amounts to a few drops of water, we would rate the products similarly for that attribute.

Do you feel that job applicants are viewed more favorably if they have studied some statistics?

Given the level of innumeracy that exists today, I believe that a basic knowledge of statistics would be viewed favorably for just about any field of study. This would be especially true of quantitative fields, such as the sciences, engineering, and business. It is extremely important for everyone to have an understanding of statistics in order to effectively process the huge amounts of information that we are presented with each day in our professional and personal lives. A focus on statistical thinking would be especially useful.

How critical do you find your background for performing your responsibilities with excellence?

Consumers Union's mission is to advance the interests of consumers by providing information and advice about products and services and about issues affecting their welfare, and by advocating a consumer point of view. To stay competitive, we've had to look for more efficient ways to provide more information to consumers in less time. My background in both statistics and quality management have been extremely valuable in helping Consumers Union achieve this mission.

While a college student, did you expect to be using statistics on the job?

I started out as a math major and didn't really get interested in statistics until my senior year. It wasn't until graduate school, while working under the direction of Professor Hoerl at the University of Delaware, that I realized how interesting a career in statistics would be. And despite the negative feelings a lot of students have about statistics, I think I have one of the most interesting jobs. I never know what to expect on a given day. One day I may be sitting in on the training session for wine tasting to learn about the testing procedures. On another day, I may be discussing various ways to test paints. On most days, though, I spend a significant amount of time using a computer to help design an upcoming study or to sift through large amounts of data that will ultimately be used as the basis for product ratings.

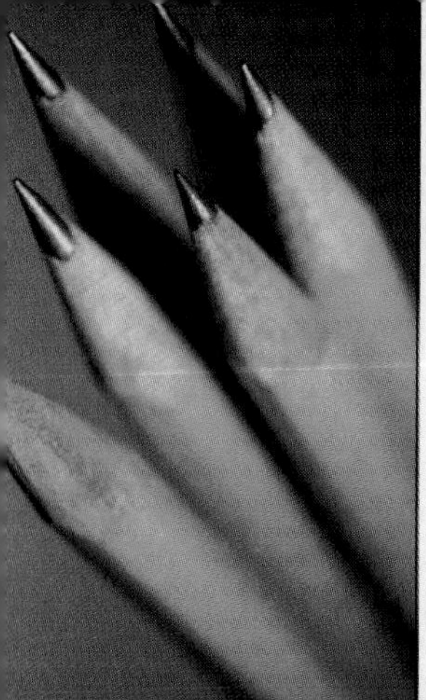

Inferences from Two Samples

9

CHAPTER PROBLEM

Which is better for treating carpal tunnel syndrome: splinting or surgery?

Carpal tunnel syndrome is a common wrist complaint resulting from a compressed nerve. It is often the result of extended use of repetitive wrist movements, such as those associated with the use of a keyboard. Among the various treatments available, two are common: apply a splint or perform surgery. The splint treatment has the advantages of being noninvasive, simpler, quicker, and much less expensive. But do those advantages justify the splint treatment instead of the surgery treatment? A critical factor is the success of the treatment. In one randomized controlled trial, 156 patients were identified as having carpal tunnel syndrome, they were treated with either splinting or surgery, then they were evaluated one year later. Success was defined to be "completely recovered" or "much improved," and it was determined by using patient scores and other measured outcomes, such as the numbers of nights that patients awoke from symptoms. Among 73 patients treated with surgery and evaluated one year later, 67 were found to have successful treatments. Among 83 patients treated with splints and evaluated one year later, 60 were found to have successful treatments. These results are summarized in Table 9-1 (based on data from "Splinting vs Surgery in the Treatment of Carpal Tunnel Syndrome," by Gerritsen et al., *Journal of the American Medical Association,* Vol. 288, No. 10).

Examining the results of the trials in Table 9-1, it appears that surgery is a better treatment because its success rate is 92%, compared to only 72% for the splint treatment. However, we should not form a conclusion based on those success rates alone. We should also take into account the sample sizes and the magnitude of the difference between the two rates. We should also consider the sampling distribution that applies. We need a procedure that takes all of these relevant factors into account.

Table 9-1 includes two sample proportions: 67/73 (for the surgery treatment group) and 60/83 (for the splint treatment group). In a journal article about the trial, authors claimed that "treatment with open carpal tunnel release surgery resulted in better outcomes than treatment with wrist splinting for patients with CTS (carpal tunnel syndrome)." Do the sample results really support the claim that surgery is better? This chapter will introduce methods for testing such claims. We will then be able to determine whether the surgery treatment is *significantly* better than the splint treatment.

This Chapter Problem involves two population proportions, but the methods of this chapter will also allow us to compare two independent means, differences from paired data, and two independent standard deviations or variances.

Table 9-1 Treatments of Carpal Tunnel Syndrome

	Treatment	
	Surgery	Splint
Success one year after treatment	67	60
Total number treated	73	83
Success Rate	**92%**	**72%**

9-1 Overview

Chapters 7 and 8 introduced important and fundamental concepts of inferential statistics: methods for *estimating* values of population parameters, and methods for *testing hypotheses* (or claims) made about population parameters. Chapters 7 and 8 both involve methods that apply to a single sample used for making an inference about a single population parameter. In reality, however, there are many important and meaningful situations in which it becomes necessary to compare *two* sets of sample data. The following are examples typical of those found in this chapter, which presents methods for using sample data from two populations so that inferences can be made about those populations.

- Test the claim that when treating carpal tunnel syndrome, surgery is more successful than applying a splint.
- When testing the effectiveness of the Salk vaccine in preventing paralytic polio, determine whether the treatment group had a lower incidence of polio than the group given a placebo.
- When testing the effectiveness of Lipitor, determine whether subjects have lower levels of cholesterol after taking the drug.
- Given two similar groups of subjects with bipolar depression, determine whether the group treated with paroxetine has lower scores on the Hamilton depression scale than the group given a placebo.

Chapters 7 and 8 included methods that were applied to proportions, means, and measures of variation (standard deviation and variance), and this chapter will address those same parameters. This chapter extends the same methods introduced in Chapters 7 and 8 to situations involving comparisons of two samples instead of only one.

9-2 Inferences About Two Proportions

Key Concept This section presents methods for using two sample proportions for constructing a confidence interval estimate of the difference between the corresponding population proportions, or testing a claim made about the two population proportions. This section is based on proportions, but we can use the same methods for dealing with probabilities or the decimal equivalents of percentages. For example, we might want to determine whether there is a difference between the percentage of adverse reactions in a placebo group and the percentage of adverse reactions in a drug treatment group. We can convert the percentages to their corresponding decimal values and proceed to use the methods of this section.

When testing a hypothesis made about two population proportions or when constructing a confidence interval estimate of the difference between two population proportions, the requirements and notation are as follows. Note that when testing the null hypothesis of $p_1 = p_2$, there is no need to estimate the individual parameters p_1 and p_2, but we estimate their common value with the pooled sample proportion described below in the summary box.

Requirements

1. We have proportions from two simple random samples that are *independent.* (Samples are independent if the sample values selected from one population are not related to or somehow paired or matched with the sample values selected from the other population.)
2. For each of the two samples, the number of successes is at least 5 and the number of failures is at least 5.

Notation for Two Proportions

For population 1 we let

p_1 = *population* proportion

n_1 = size of the sample

x_1 = number of successes in the sample

$\hat{p} = \dfrac{x_1}{n_1}$ (the *sample* proportion)

$\hat{q}_1 = 1 - \hat{p}_1$

The corresponding meanings are attached to p_2, n_2, x_2, $\hat{p}_2$, and $\hat{q}_2$, which come from population 2.

Pooled Sample Proportion

The **pooled sample proportion** is denoted by $\bar{p}$ and is given by:

$$\bar{p} = \frac{x_1 + x_2}{n_1 + n_2}$$

We denote the complement of $\bar{p}$ by $\bar{q}$, so $\bar{q} = 1 - \bar{p}$.

Test Statistic for Two Proportions (with H_0: $p_1 = p_2$)

$$z = \frac{(\hat{p}_1 - \hat{p}_2) - (p_1 - p_2)}{\sqrt{\dfrac{\bar{p}\,\bar{q}}{n_1} + \dfrac{\bar{p}\,\bar{q}}{n_2}}}$$

where $p_1 - p_2 = 0$ (assumed in the null hypothesis)

$$\hat{p}_1 = \frac{x_1}{n_1} \quad \text{and} \quad \hat{p}_2 = \frac{x_2}{n_2}$$

$$\bar{p} = \frac{x_1 + x_2}{n_1 + n_2}$$

$$\bar{q} = 1 - \bar{p}$$

P-value: Use Table A-2. (Use the computed value of the test statistic z and find the P-value by following the procedure summarized in Figure 8-6.)

Critical values: Use Table A-2. (Based on the significance level α, find critical values by using the procedures introduced in Section 8-2.)

continued

The Lead Margin of Error

Authors Stephen Ansolabehere and Thomas Belin wrote in their article "Poll Faulting" (*Chance* magazine) that "our greatest criticism of the reporting of poll results is with the margin of error of a single proportion (usually ±3%) when media attention is clearly drawn to the *lead* of one candidate." They point out that the lead is really the *difference* between two proportions $(p_1 - p_2)$ and go on to explain how they developed the following rule of thumb: The lead is approximately $\sqrt{3}$ times larger than the margin of error for any one proportion. For a typical preelection poll, a reported ±3% margin of error translates to about ±5% for the lead of one candidate over the other. They write that the margin of error for the lead should be reported.

Confidence Interval Estimate of $p_1 - p_2$

The confidence interval estimate of the difference $p_1 - p_2$ is:

$$(\hat{p}_1 - \hat{p}_2) - E < (p_1 - p_2) < (\hat{p}_1 - \hat{p}_2) + E$$

where the margin of error E is given by $E = z_{\alpha/2}\sqrt{\dfrac{\hat{p}_1\hat{q}_1}{n_1} + \dfrac{\hat{p}_2\hat{q}_2}{n_2}}$

TI-83/84 Plus Calculator: The TI-83/84 Plus calculator can be used to automatically generate the test statistic and P-value for a hypothesis test, and it can also generate a confidence interval. See the procedures at the end of this section.

Finding the Numbers of Successes x_1 and x_2 The calculations for hypothesis tests and confidence intervals require that we have specific values for x_1, n_1, x_2, and n_2. Sometimes the available sample data include those specific numbers, but sometimes it is necessary to calculate the values of x_1 and x_2. For example, consider the statement that "when 1125 people are surveyed, 47% of them said that they fly never or rarely." From that statement we can see that $n_1 =$ 1125 and $\hat{p}_1 = 0.47$, but the actual number of successes x_1 is not given. However, from $\hat{p}_1 = x_1/n_1$, we know that

$$x_1 = n_1 \cdot \hat{p}_1$$

so that $x_1 = 1125 \cdot 0.47 = 528.75$. But you cannot have 528.75 people who have flown never or rarely, so the number of successes x_1 must be the whole number 529. We can now use $x_1 = 529$ in the calculations that require its value. It's really quite simple: 47% of 1125 means $0.47 \cdot 1125$, which results in 528.75, which we round to 529.

Hypothesis Tests

We will now consider tests of hypotheses made about two population proportions, but we will consider only tests having a null hypothesis of $p_1 = p_2$. (For claims that the difference between p_1 and p_2 is equal to a nonzero constant, see Exercise 35 in this section.) The following example will help clarify the roles of x_1, n_1, $\hat{p}_1$, $\bar{p}$, and so on. In particular, you should recognize that under the assumption of equal proportions, the best estimate of the common proportion is obtained by pooling both samples into one big sample, so that $\bar{p}$ is the estimator of the common population proportion.

EXAMPLE Is Surgery Better Than Splinting? The Chapter Problem includes results from a clinical trial in which patients were treated for carpal tunnel syndrome, and Table 9-1 summarizes results. Use the sample data from Table 9-1 and use a 0.05 significance level to test the claim that the success rate with surgery is better than the success rate with splinting.

SOLUTION

REQUIREMENTS ✔ We should first verify that the necessary requirements are satisfied. Given the design of the experiment, it is reasonable to assume that the sample is a simple random sample. Also, the surgery treatment group is independent of the splint treatment group. For the second requirement, note that the surgery treatment group has 67 successes among 73 patients, so there are 6 failures. The surgery treatment group therefore has at least 5 successes and at least 5 failures. Also, the splint treatment group has 60 successes among 83 patients, so the number of failures is 23. The splint treatment group therefore has at least 5 successes and at least 5 failures. For each of the two samples, we have verified that the number of successes is at least 5 and the number of failures is at least 5. The check of requirements has been successfully completed and we can proceed with the hypothesis test. ✔

We will now use the *P*-value method of hypothesis testing, as summarized in Figure 8-8. In the following steps we stipulate that the surgery patients are denoted as Sample 1, and the splint patients are denoted as Sample 2.

Step 1: The claim of a greater success rate for the surgery treatment group can be expressed as $p_1 > p_2$.

Step 2: If $p_1 > p_2$ is false, then $p_1 \leq p_2$.

Step 3: Because our claim of $p_1 > p_2$ does not contain equality, it becomes the alternative hypothesis. The null hypothesis is the statement of equality, so we have

$$H_0: p_1 = p_2 \qquad H_1: p_1 > p_2 \quad \text{(original claim)}$$

Step 4: The significance level is $\alpha = 0.05$.

Step 5: We will use the normal distribution (with the test statistic given above) as an approximation to the binomial distribution. We estimate the common value of p_1 and p_2 with the pooled sample estimate $\bar{p}$ calculated as shown below, with extra decimal places used to minimize rounding errors in later calculations.

$$\bar{p} = \frac{x_1 + x_2}{n_1 + n_2} = \frac{67 + 60}{73 + 83} = 0.81410256$$

With $\bar{p} = 0.81410256$, it follows that $\bar{q} = 1 - 0.81410256 = 0.18589744$.

Step 6: We can now find the value of the test statistic:

$$z = \frac{(\hat{p}_1 - \hat{p}_2) - (p_1 - p_2)}{\sqrt{\dfrac{\bar{p}\,\bar{q}}{n_1} + \dfrac{\bar{p}\,\bar{q}}{n_2}}}$$

$$= \frac{\left(\dfrac{67}{73} - \dfrac{60}{83}\right) - 0}{\sqrt{\dfrac{(0.81410256)(0.18589744)}{73} + \dfrac{(0.81410256)(0.18589744)}{83}}}$$

$$= 3.12$$

continued

Author as a Witness

The author was asked to testify in New York State Supreme Court by a former student who was contesting a lost reelection to the office of Dutchess County Clerk. The author testified by using statistics to show that the voting behavior in one contested district was significantly different from the behavior in all other districts. When the opposing attorney asked about results of a confidence interval, he asked if the 5% error (from a 95% confidence level) could be added to the three percentage point margin of error to get a total error of 8%, thereby indicating that he did not understand the basic concept of a confidence interval. The judge cited the author's testimony, upheld the claim of the former student, and ordered a new election in the contested district. That judgment was later overturned by the appellate court on the grounds that the ballot irregularities should have been contested before the election, not after.

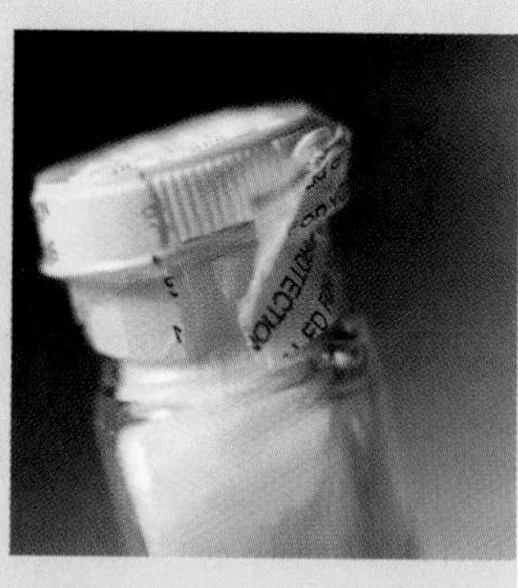

Does Aspirin Help Prevent Heart Attacks?

In a recent study of 22,000 male physicians, half were given regular doses of aspirin while the other half were given placebos. The study ran for six years at a cost of $4.4 million. Among those who took the aspirin, 104 suffered heart attacks. Among those who took the placebos, 189 suffered heart attacks. (The figures are based on data from *Time* and the *New England Journal of Medicine,* Vol. 318, No. 4.) This is a classic experiment involving a treatment group (those who took the aspirin) and a placebo group (those who took pills that looked and tasted like the aspirin pills, but no aspirin was present). We can use methods presented in this chapter to address the issue of whether the results show a statistically significant lower rate of heart attacks among the sample group who took aspirin.

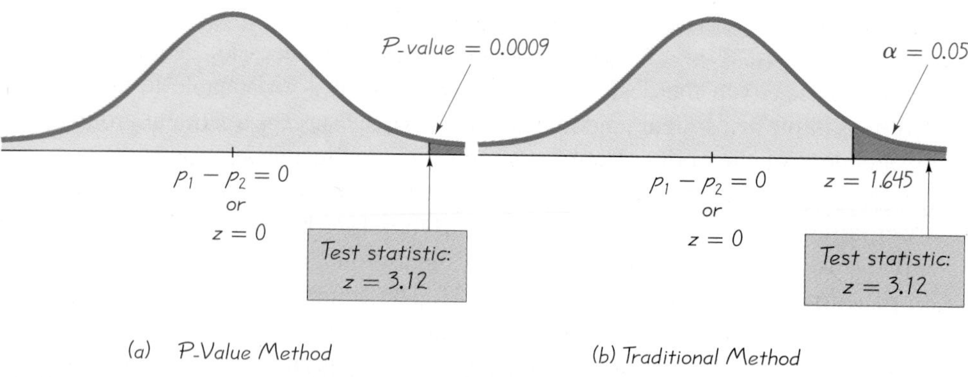

Figure 9-1 Testing Claim That Surgery Is Better Than Splinting

The *P*-value of 0.0009 is found as follows: This is a right-tailed test, so the *P*-value is the area to the right of the test statistic $z = 3.12$ (as indicated by Figure 8-6). Refer to Table A-2 and find that the area to the left of the test statistic $z = 3.12$ is 0.9991, so the *P*-value is $1 - 0.9991 = 0.0009$. The test statistic and *P*-value are shown in Figure 9-1(a).

Step 7: Because the *P*-value of 0.0009 is less than the significance level of $\alpha = 0.05$, we reject the null hypothesis of $p_1 = p_2$.

INTERPRETATION We must address the original claim that the success rate with surgery is greater than the success rate with splinting. Because we reject the null hypothesis, we conclude that there is sufficient evidence to support the claim that the proportion of successes with surgery is greater than that for splinting. (See Figure 8-7 for help in wording the final conclusion.) The researchers were justified when they wrote in a journal article that "treatment with open carpal tunnel release surgery resulted in better outcomes than treatment with wrist splinting for patients with CTS (carpal tunnel syndrome)." Based on the clinical trial and the subsequent statistical analysis of the results, physicians have much better guidance when recommending a treatment for carpal tunnel syndrome that can be so painful.

Traditional Method of Testing Hypotheses

The preceding example illustrates the *P*-value approach to hypothesis testing, but it would be quite easy to use the traditional approach instead. In Step 6, instead of finding the *P*-value, find the critical value. With a significance level of $\alpha = 0.05$ in a right-tailed test based on the normal distribution, refer to Table A-2 to find that an area of $\alpha = 0.05$ in the right tail corresponds to the critical value of $z = 1.645$. See Figure 9-1(b) where we can see that the test statistic does fall in the critical region bounded by the critical value of $z = 1.645$. We again reject the null hypothesis. Again, we conclude that there is sufficient evidence to support the claim that surgery has a better success rate than splinting.

Confidence Intervals

We can construct a confidence interval estimate of the difference between population proportions $(p_1 - p_2)$ by using the format given above. If a confidence interval estimate of $p_1 - p_2$ does not include 0, we have evidence suggesting that p_1 and

p_2 have different values. However, the standard deviation used for a confidence interval is different from the standard deviation used for a hypothesis test. When testing claims about the difference between two population proportions, the traditional method and the *P*-value method are equivalent in the sense that they always yield the same results, but the confidence interval estimate of the difference might suggest a different conclusion. (See Exercise 34.) If different conclusions are obtained, know that the traditional and *P*-value methods use an *exact* standard deviation based on the assumption that there is no difference between the population proportions (as stated in the null hypothesis). However, the confidence interval is constructed using a standard deviation based on *estimated* values of the two population proportions. Consequently, if you want to estimate the difference between two population proportions, do so by constructing a confidence interval, but if you want to test some claim about two population proportions, use the *P*-value method or the traditional method.

Also, *don't test for equality of two population proportions by determining whether there is an overlap between two individual confidence interval estimates of the two individual population proportions.* When compared to the confidence interval estimate of $p_1 - p_2$, the analysis of overlap between two individual confidence intervals is more conservative (by rejecting equality less often), and it has less power (because it is less likely to reject $p_1 = p_2$ when in reality $p_1 \neq p_2$). (See "On Judging the Significance of Differences by Examining the Overlap Between Confidence Intervals," by Schenker and Gentleman, *American Statistician,* Vol. 55, No. 3.) See Exercise 33.

Polio Experiment

In 1954 an experiment was conducted to test the effectiveness of the Salk vaccine as protection against the devastating effects of polio. Approximately 200,000 children were injected with an ineffective salt solution, and 200,000 other children were injected with the vaccine. The experiment was "double blind" because the children being injected didn't know whether they were given the real vaccine or the placebo, and the doctors giving the injections and evaluating the results didn't know either. Only 33 of the 200,000 vaccinated children later developed paralytic polio, whereas 115 of the 200,000 injected with the salt solution later developed paralytic polio. Statistical analysis of these and other results led to the conclusion that the Salk vaccine was indeed effective against paralytic polio.

EXAMPLE Is Surgery Better than Splinting? Use the sample data given in Table 9-1 to construct a 90% confidence interval estimate of the difference between the two population proportions. (The confidence level of 90% is comparable to the significance level of $\alpha = 0.05$ used in the preceding right-tailed hypothesis test. See Table 8-2 in Section 8-2.)

SOLUTION

REQUIREMENT ✔ The solution to the preceding example begins with the same requirement check needed here. That check of requirements applies here, so we can proceed with the construction of the confidence interval. ✔

With a 90% confidence level, $z_{\alpha/2} = 1.645$ (from Table A-2). We first calculate the value of the margin of error E as shown.

$$E = z_{\alpha/2}\sqrt{\frac{\hat{p}_1\hat{q}_1}{n_1} + \frac{\hat{p}_2\hat{q}_2}{n_2}} = 1.645\sqrt{\frac{\left(\frac{67}{73}\right)\left(\frac{6}{73}\right)}{73} + \frac{\left(\frac{60}{83}\right)\left(\frac{23}{83}\right)}{83}} = 0.0966$$

With $\hat{p}_1 = 67/73 = 0.9178$, $\hat{p}_2 = 60/83 = 0.7299$, and $E = 0.0966$, the confidence interval is evaluated as follows:

$$(\hat{p}_1 - \hat{p}_2) - E < (p_1 - p_2) < (\hat{p}_1 - \hat{p}_2) + E$$

$$(0.9178 - 0.7229) - 0.0966 < (p_1 - p_2) < (0.9178 - 0.7229) + 0.0966$$

$$0.0983 < (p_1 - p_2) < 0.292$$

continued

Death Penalty as Deterrent

A common argument supporting the death penalty is that it discourages others from committing murder. Jeffrey Grogger of the University of California analyzed daily homicide data in California for a four-year period during which executions were frequent. Among his conclusions published in the *Journal of the American Statistical Association* (Vol. 85, No. 410): "The analyses conducted consistently indicate that these data provide no support for the hypothesis that executions deter murder in the short term." This is a major social policy issue, and the efforts of people such as Professor Grogger help to dispel misconceptions so that we have accurate information with which to address such issues.

INTERPRETATION The confidence interval limits do not contain 0, suggesting that there is a significant difference between the two proportions. Also, based on these results, we have 90% confidence that the percentage of successes with surgery is greater than the percentage of successes with splinting by an amount that is between 9.83% and 29.2%.

Rationale: Why Do the Procedures of This Section Work? The test statistic given for hypothesis tests is justified by the following:

1. With $n_1p_1 \geq 5$ and $n_1q_1 \geq 5$, the distribution of $\hat{p}_1$ can be approximated by a normal distribution with mean p_1 and standard deviation $\sqrt{p_1q_1/n_1}$ and variance p_1q_1/n_1. These conclusions are based on Sections 6-6 and 7-2, and they also apply to the second sample.
2. Because $\hat{p}_1$ and $\hat{p}_2$ are each approximated by a normal distribution, $\hat{p}_1 - \hat{p}_2$ will also be approximated by a normal distribution with mean $p_1 - p_2$ and variance

$$\sigma^2_{(\hat{p}_1-\hat{p}_2)} = \sigma^2_{\hat{p}_1} + \sigma^2_{\hat{p}_2} = \frac{p_1q_1}{n_1} + \frac{p_2q_2}{n_2}$$

(The above result is based on this property: The variance of the *differences* between two independent random variables is the *sum* of their individual variances.)

3. Because the values of p_1, q_1, p_2, and q_2 are typically unknown and from the null hypothesis, we assume that $p_1 = p_2$ and we can pool (or combine) the sample data. The pooled estimate of the common value of p_1 and p_2 is $\bar{p} = (x_1 + x_2)/(n_1 + n_2)$. If we replace p_1 and p_2 by $\bar{p}$ and replace q_1 and q_2 by $\bar{q} = 1 - \bar{p}$, the variance from Step 2 leads to the following standard deviation:

$$\sigma_{(\hat{p}_1-\hat{p}_2)} = \sqrt{\frac{\bar{p}\,\bar{q}}{n_1} + \frac{\bar{p}\,\bar{q}}{n_2}}$$

4. We now know that the distribution of $p_1 - p_2$ is approximately normal, with mean $p_1 - p_2$ and standard deviation as shown in Step 3, so that the z test statistic has the form given earlier.

The form of the confidence interval requires an expression for the variance different from the one given in Step 3. In Step 3 we are assuming that $p_1 = p_2$, but if we don't make that assumption (as in the construction of a confidence interval), we estimate the variance of $p_1 - p_2$ as

$$\sigma^2_{(\hat{p}_1-\hat{p}_2)} = \sigma^2_{\hat{p}_1} + \sigma^2_{\hat{p}_2} = \frac{\hat{p}_1\hat{q}_1}{n_1} + \frac{\hat{p}_2\hat{q}_2}{n_2}$$

and the standard deviation becomes

$$\sqrt{\frac{\hat{p}_1\hat{q}_1}{n_1} + \frac{\hat{p}_2\hat{q}_2}{n_2}}$$

In the test statistic

$$z = \frac{(\hat{p}_1 - \hat{p}_2) - (p_1 - p_2)}{\sqrt{\frac{\hat{p}_1\hat{q}_1}{n_1} + \frac{\hat{p}_2\hat{q}_2}{n_2}}}$$

use the positive and negative values of z (for two tails) and solve for $p_1 - p_2$. The results are the limits of the confidence interval given earlier.

TI Using a TI-83/84 Plus Calculator for Inferences About Two Proportions

Requirements: When using a TI-83/84 Plus calculator for inferences about two population proportions, you must know the sample size n and the number of successes x for the samples from each of the two populations. That is, you must know the values of x_1, n_1, x_2, and n_2. For either sample, if you do not know the value of x, but you know the value of the sample size n and the sample proportion $\hat{p}$, calculate the value of x as follows: $x = n \cdot \hat{p}$ then round the result to the nearest whole number. For example, if we know that $n = 73$ and the success rate is 92%, then x is 92% of 73, which is $0.92 \times 73 = 67.16$, which we round to 67.

Hypothesis Testing Given below is the procedure for using a TI-83/84 Plus calculator for testing a claim about two population proportions. *Because the TI-83/84 Plus calculator provides the test statistic and P-value (but not the critical value), we use the P-value method of testing hypotheses:*

1. Press STAT, then press ▷ twice to select **TESTS.**
2. Select the menu item of **2-PropZTest** (for a hypothesis test of a claim involving two proportions).
3. Enter the values of x_1, n_1, x_2, and n_2. All of those values must be whole numbers.
4. For the next line, select the type of test (two-tailed, left-tailed, right-tailed) by selecting the expression of the alternative hypothesis:

$$p_1 \neq p_2 \qquad p_1 < p_2 \qquad p_1 > p_2$$

5. Use the arrow keys to scroll down to the bottom line, select **Calculate,** then press ENTER.

Hypothesis Test

```
2-PropZTest
 p1>p2
 z=3.122560496
 p=8.9649363E-4
 p̂1=.9178082192
 p̂2=.7228915663
↓p̂=.8141025641
```

The first example in this section uses the sample data summarized in Table 9-1, and the accompanying display shows the results of the hypothesis test of the claim that the success rate for surgery is greater than the success rate with splinting. The displayed *P*-value of 8.9649363E⁻4 can be expressed in standard form as 0.0009 (rounded). The small *P*-value suggests that we should reject the null hypothesis and support the stated claim.

Confidence Intervals Here is the procedure for using a TI-83/84 Plus calculator for finding a confidence interval estimate of the difference $(p_1 - p_2)$ between two population proportions:

1. Press STAT, then press ▷ twice to select **TESTS.**
2. Select the menu item of **2-PropZInt** (for a confidence interval involving two proportions).

3. Enter the values of x_1, n_1, x_2, and n_2. All of those values must be whole numbers.
4. For the next line labeled C-Level, enter the confidence level. Enter 0.95 for a 95% confidence level.
5. Use the arrow keys to scroll down to the bottom line, select **Calculate,** then press ENTER. The result will consist of confidence interval limits enclosed within parentheses.

Confidence Interval

```
2-PropZInt
 (.09835,.29149)
 p̂1=.9178082192
 p̂2=.7228915663
 n1=73
 n2=83
```

The second example in this section illustrates the construction of a 90% confidence interval estimate of the difference $p_1 - p_2$ based on the sample data in Table 9-1. The accompanying display shows the resulting confidence interval. Note that the TI-83/84 Plus calculator expresses the confidence interval in the format of (.09835, .29149), and it is easy to express the confidence interval as $0.0984 < (p_1 - p_2) < 0.291$. There is a very small discrepancy between this result and the one shown in the example, and this discrepancy is due to rounding in the example. The TI-83/84 Plus calculator result is better.

9-2 BASIC SKILLS AND CONCEPTS

Statistical Literacy and Critical Thinking

1. Pooled Sample Proportion What is a pooled sample proportion and what is the symbol that represents it? Is it used for hypothesis tests and confidence intervals?

2. $\hat{p}, \bar{p}, p$ What do the symbols $\hat{p}, \bar{p}, p$ represent?

3. Interpreting Confidence Interval Sample data from two different populations are used to construct this 95% confidence interval: $0.200 < p_1 - p_2 < 0.300$. Write a statement interpreting that confidence interval.

4. Equivalence of Hypothesis Tests and Confidence Intervals Given sample proportions from two different populations, we want to use a 0.05 significance level to test the claim that $p_1 = p_2$. One approach is to use the *P*-value method of hypothesis testing, a second approach is to use the traditional method of hypothesis testing, and a third approach is to base the conclusion on the 95% confidence interval estimate of $p_1 - p_2$. Will all three approaches always result in the same conclusion? Explain.

Finding Number of Successes. *In Exercises 5–8, find the number of successes x suggested by the given statement.*

5. From a *New York Times* article: Among 3250 pedestrian walk buttons in New York City, 23% of them work.

6. From a Gallup poll: Among 1018 survey subjects, 22% smoked cigarettes in the past week.

7. From a Gallup poll: Among 976 survey subjects, 7% have an alcoholic drink every day.

8. From a CNN/USA Today/Gallup poll: Among 1003 survey subjects, 11% say that the opinion of a celebrity would influence their own opinion.

Calculations for Testing Claims. *In Exercises 9 and 10, assume that you plan to use a significance level of $\alpha = 0.05$ to test the claim that $p_1 = p_2$. Use the given sample sizes and numbers of successes to find (a) the pooled estimate $\overline{p}$, (b) the z test statistic, (c) the critical z values, and (d) the P-value.*

TI **9.**

Treatment Group	Placebo Group
$n_1 = 500$	$n_2 = 400$
$x_1 = 100$	$x_2 = 50$

TI **10.**

Males	Females
$n_1 = 1068$	$n_2 = 1220$
$x_1 = 332$	$x_2 = 420$

TI **11.** Home Field Advantage When games were sampled from throughout a season, it was found that the home team won 127 of 198 professional *basketball* games, and the home team won 57 of 99 professional *football* games (based on data from "Predicting Professional Sports Game Outcomes from Intermediate Game Scores," by Cooper et al., *Chance,* Vol. 5, No. 3–4). See the accompanying TI-83/84 Plus display that results from testing the claim of equal proportions. Based on the results, does there appear to be a significant difference between the proportions of home wins? What do you conclude about the home field advantage?

TI-83/84 Plus

```
2-PropZTest
 p1≠p2
 z=1.098647687
 p=.2719218487
 p̂1=.6414141414
 p̂2=.5757575758
↓p̂=.6195286195
```

12. Testing Laboratory Gloves The *New York Times* ran an article about a study in which Professor Denise Korniewicz and other Johns Hopkins researchers subjected laboratory gloves to stress. Among 240 vinyl gloves, 63% leaked viruses. Among 240 latex gloves, 7% leaked viruses. See the accompanying display of the Minitab results. Using a 0.005 significance level, test the claim that vinyl gloves have a larger virus leak rate than latex gloves.

Minitab

```
Difference = p (1) - p (2)
Estimate for difference:  0.558333
95% lower bound for difference:  0.500263
Test for difference = 0 (vs > 0):  Z = 12.82  P-Value = 0.000
```

TI **13.** Gender Selection The Genetics and IVF Institute conducted a clinical trial of its methods for gender selection. As this book was written, results included 325 babies born to parents using the XSORT method to increase the probability of conceiving a girl, and 295 of those babies were girls. Also, 51 babies were born to parents using the YSORT method to increase the probability of conceiving a boy, and 39 of those babies were boys.

a. Construct a 95% confidence interval estimate of the difference between the proportion of girls and the proportion of boys.

b. Does there appear to be a difference? Does the XSORT method appear to be effective? Does the YSORT method appear to be effective?

TI **14.** Do Bednets Reduce Malaria? In a randomized controlled trial in Kenya, insecticide-treated bednets were tested as a way to reduce malaria. Among 343 infants who used the bednets, 15 developed malaria. Among 294 infants not using bednets, 27 developed malaria (based on data from "Sustainability of Reductions in Malaria Transmission and Infant Mortality in Western Kenya with Use of Insecticide-Treated Bednets," by Lindblade et al., *Journal of the American Medical Association,* Vol. 291, No. 21). Use a 0.01 significance level to test the claim that the incidence of malaria is lower for infants who use the bednets. Do the bednets appear to be effective?

TI **15.** Telephone Survey Methods In a study of the accuracy of telephone surveys, 720 people refused to respond when they were among the 1720 people included in a "standard" 5-day survey. In the same study, 429 people refused to respond when they were among the 1640 people included in a "rigorous" 8-week survey. (The data are based on results from "Consequences of Reducing Nonresponse in a National Telephone Survey," by Keeter et al., *Public Opinion Quarterly,* Vol. 64, No. 2.) Use a 0.01 significance level to test the claim that the refusal rate is lower with the rigorous survey. Does the rigorous survey appear to be more likely to produce accurate results?

TI **16.** Telephone Survey Methods The preceding exercise involves a one-sided hypothesis test with a 0.01 significance level. If you plan to test the claim using a confidence interval, what confidence level should be used? (*Hint:* See Table 8-2 in Section 8-2.) Using the appropriate confidence level, construct a confidence interval of the difference between the two refusal rates. Based on the confidence interval, does it appear that the refusal rate is lower with the rigorous survey? Why or why not?

TI **17.** E-mail and Privacy A survey of 436 workers showed that 192 of them said that it was seriously unethical to monitor employee e-mail. When 121 senior-level bosses were surveyed, 40 said that it was seriously unethical to monitor employee e-mail (based on data from a Gallup poll). Use a 0.05 significance level to test the claim that for those saying that monitoring e-mail is seriously unethical, the proportion of employees is greater than the proportion of bosses.

TI **18.** E-mail and Privacy Refer to the sample data given in Exercise 17 and construct a 90% confidence interval estimate of the difference between the two population proportions. Is there a substantial gap between the employees and bosses?

TI **19.** Effectiveness of Smoking Bans The Joint Commission on Accreditation of Healthcare Organizations mandated that hospitals ban smoking by 1994. In a study of the effects of this ban, subjects who smoke were randomly selected from two different populations. Among 843 smoking employees of hospitals with the smoking ban, 56 quit smoking one year after the ban. Among 703 smoking employees from workplaces without a smoking ban, 27 quit smoking a year after the ban (based on data from "Hospital Smoking Bans and Employee Smoking Behavior," by Longo, Brownson, et al., *Journal of the American Medical Association,* Vol. 275, No. 16). Is there a significant difference between the two proportions at a 0.05 significance level? Is there a significant difference between the two proportions at a 0.01 significance level? Does it appear that the ban had an effect on the smoking quit rate?

TI **20.** Testing Effectiveness of Vaccine In a *USA Today* article about an experimental nasal spray vaccine for children, the following statement was presented: "In a trial involving 1602 children only 14 (1%) of the 1070 who received the vaccine developed the flu, compared with 95 (18%) of the 532 who got a placebo." The article also referred to a study claiming that the experimental nasal spray "cuts children's chances of getting the flu." Is there sufficient sample evidence to support the stated claim?

TI **21.** Adverse Effects of Clarinex The drug Clarinex is used to treat symptoms from allergies. In a clinical trial of this drug, 2.1% of the 1655 treated subjects experienced fatigue. Among the 1652 subjects given placebos, 1.2% experienced fatigue (based on data from Schering Corporation). Use a 0.05 significance level to test the claim that the incidence of fatigue is greater among those who use Clarinex. Does fatigue appear to be a major concern for those who use Clarinex?

TI **22.** Adverse Effects of Clarinex Use the same sample data as in the preceding exercise. If you plan to test the given claim by constructing a confidence interval, what confidence level should be used? (*Hint:* See Table 8-2 in Section 8-2.) Using the appropriate confidence level, construct a confidence interval estimate of the difference between the fatigue rates for Clarinex users and those given placebos. Does fatigue appear to be a major concern for those who use Clarinex?

TI **23.** Driving to Work In a survey of commuting habits, it was found that among 1068 homeowners, 82.4% drive themselves to work. Among 1064 renters, 68.1% drive themselves to work (based on data from The U.S. Census American Housing Survey). Construct a 95% confidence interval estimate of the difference between the proportion of homeowners who drive themselves to work and the proportion of renters who drive themselves to work. Based on the result, does there appear to be a significant difference between those two proportions? Identify at least one major factor that might explain any such difference.

TI **24.** Lost Baggage Among 5000 items of randomly selected baggage handled by American Airlines, 22 were lost. Among 4000 items of randomly selected baggage handled by Delta Airlines, 15 were lost (based on data from the U.S. Department of Transportation). Use the sample data to construct a 95% confidence interval estimate of the difference between the two rates of lost baggage. Based on the result, does there appear to be a difference between the two rates of lost baggage? Why or why not?

TI **25.** Gender Gap for Seat Belt Use? Among 2200 randomly selected male car occupants over the age of 8, 72% wear seat belts. Among 2380 randomly selected female car occupants over the age of 8, 84% wear seat belts (based on data from the U.S. Department of Transportation). Use a 0.05 significance level to test the claim that both genders have the same rate of seat belt use. Based on the result, does there appear to be a gender gap?

TI **26.** Incidence of Radon Radon is a gas produced when radium decays, and it can enter homes where it can become a health threat. Among 186 homes in Hyde Park, New York (home of Franklin D. Roosevelt), 16% were found to have unsafe radon levels (above 4 picocuries per liter). Among 237 homes in LaGrange, New York (home of the author), 19% were found to have unsafe radon levels (based on data from the New York State Department of Health). Use a 0.05 significance level to test the claim that the two Dutchess County regions have different rates of unsafe radon levels. Does the presence of unsafe radon levels appear to vary by region?

TI **27.** Attitudes Toward Marriage In a Time/CNN survey, 24% of 205 single women said that they "definitely want to get married." In the same survey, 27% of 260 single men gave that same response. Construct a 99% confidence interval estimate of the difference between the proportions of single women and single men who definitely want to get married. Is there a gender gap on this issue?

TI **28.** Attitudes Toward Marriage Refer to the same sample data as in the preceding exercise and use a 0.01 significance level to test the claim that there is a difference between the proportion of men and the proportion of women who definitely want to get married. Does there appear to be a difference?

TI **29.** Appendix B Data Set: Precipitation Data Set 10 in Appendix B lists precipitation amounts for Boston. Consider a day to have precipitation if the amount is any positive value. Test the claim that the percentage of weekdays (Monday through Friday) with precipitation is the same as the percentage of weekend days (Saturday and Sunday) with precipitation. Several newspaper articles reported that it rains more on weekends. Do the Boston data support that claim?

TI **30.** Appendix B Data Set: Precipitation Data Set 8 in Appendix B lists precipitation amounts for the author's home in Dutchess County, New York. Data Set 10 in Appendix B lists precipitation amounts for Boston. Consider a day to have precipitation if the amount is any positive value. Test the claim that the percentage of days with precipitation in Dutchess County is the same as the percentage of days with precipitation in Boston.

TI **31.** Appendix B Data Set: Alcohol and Tobacco in Children's Movies Test the claim that the proportion of 25 of 50 randomly selected children's movies showing some use of alcohol is significantly less than the sample proportion of 28 of 50 other such movies showing some use of tobacco. Are the results valid if they are taken from Data Set 5 in Appendix B?

TI **32.** Appendix B Data Set: Health Survey Refer to Data Set 1 in Appendix B and use the sample data to test the claim that the proportion of men over the age of 30 is equal to the proportion of women over the age of 30.

9-2 BEYOND THE BASICS

TI **33.** Interpreting Overlap of Confidence Intervals In the article "On Judging the Significance of Differences by Examining the Overlap Between Confidence Intervals," by Schenker and Gentleman (*American Statistician,* Vol. 55, No. 3), the authors consider sample data in this statement: "Independent simple random samples, each of size 200, have been drawn, and 112 people in the first sample have the attribute, whereas 88 people in the second sample have the attribute."

a. Use the methods of this section to construct a 95% confidence interval estimate of the difference $p_1 - p_2$. What does the result suggest about the equality of p_1 and p_2?

b. Use the methods of Section 7-2 to construct individual 95% confidence interval estimates for each of the two population proportions. After comparing the overlap between the two confidence intervals, what do you conclude about the equality of p_1 and p_2?

c. Use a 0.05 significance level to test the claim that the two population proportions are equal. What do you conclude?

d. Based on the preceding results, what should you conclude about equality of p_1 and p_2? Which of the three preceding methods is least effective in testing for equality of p_1 and p_2?

TI **34.** Equivalence of Hypothesis Test and Confidence Interval Two different simple random samples are drawn from two different populations. The first sample consists of 20 people with 10 having a common attribute. The second sample consists of 2000 people with 1404 of them having the same common attribute. Compare the results from a hypothesis test of $p_1 = p_2$ (with a 0.05 significance level) and a 95% confidence interval estimate of $p_1 - p_2$.

35. Testing for Constant Difference To test the null hypothesis that the difference between two population proportions is equal to a nonzero constant c, use the test statistic

$$z = \frac{(\hat{p}_1 - \hat{p}_2) - c}{\sqrt{\dfrac{\hat{p}_1(1 - \hat{p}_1)}{n_1} + \dfrac{\hat{p}_2(1 - \hat{p}_2)}{n_2}}}$$

As long as n_1 and n_2 are both large, the sampling distribution of the test statistic z will be approximately the standard normal distribution. Refer to Exercise 12 and use a 0.05 significance level to test the claim that the virus leak rate for vinyl gloves is 50 percentage points greater than the virus leak rate for latex gloves.

36. Determining Sample Size The sample size needed to estimate the difference between two population proportions to within a margin of error E with a confidence level of $1 - \alpha$ can be found as follows. In the expression

$$E = z_{\alpha/2}\sqrt{\frac{p_1 q_1}{n_1} + \frac{p_2 q_2}{n_2}}$$

replace n_1 and n_2 by n (assuming that both samples have the same size) and replace each of p_1, q_1, p_2, and q_2 by 0.5 (because their values are not known). Then solve for n.

Use this approach to find the size of each sample if you want to estimate the difference between the proportions of men and women who own cars. Assume that you want 95% confidence that your error is no more than 0.03.

37. Verifying Property of Variances When discussing the rationale for the methods of this section, it was stated that because $\hat{p}_1$ and $\hat{p}_2$ are each approximated by a normal distribution, $\hat{p}_1 - \hat{p}_2$ will also be approximated by a normal distribution with mean $p_1 - p_2$ and variance $\sigma^2_{(\hat{p}_1 - \hat{p}_2)} = \sigma^2_{\hat{p}_1} + \sigma^2_{\hat{p}_2}$. Do the following to verify that the variance of the *difference* between two independent random variables is the *sum* of their individual variances.

a. Assuming that two dimes are tossed, list the sample space of four simple events, then find the proportion of heads in each of the four cases. Use the formula $\sigma^2 = \Sigma(x - \mu)^2/N$ to find the variance for the population of the four proportions.

b. Assuming that two quarters are tossed, the sample space and variance will be the same as in part (a). List the 16 differences in proportions $(\hat{p}_D - \hat{p}_Q)$ that are possible when every outcome of the two dimes is matched with every possible outcome of the two quarters. Find the variance of σ^2 of the population of 16 differences in proportions.

c. Use the preceding results to verify that the variance of the *difference* between two independent random variables is the *sum* of their individual variances.

9-3 Inferences About Two Means: Independent Samples

Key Concept This section presents methods for using sample data from two independent samples to test hypotheses made about two population means or to construct confidence interval estimates of the difference between two population means. This section includes Part 1 for situations in which the standard deviations of the two populations are unknown and are not assumed to be equal. Part 2 involves two other situations: (1) The two population standard deviations are both known; (2) the two population standard deviations are unknown but are assumed to be equal. For reasons to be explained later, most attention should be given to the methods described in Part 1 because they are most realistic and perform best overall.

Part 1: Independent Samples with σ_1 and σ_2 Unknown and Not Assumed Equal

We begin with formal definitions that distinguish between *independent* and *dependent* samples.

Definitions

Two samples are **independent** if the sample values selected from one population are not related to or somehow paired or matched with the sample values selected from the other population.

Two samples are **dependent** (or consist of **matched pairs)** if the members of one sample can be used to determine the members of the other sample. [Samples consisting of matched pairs (such as husband/wife data) are dependent. In addition to matched pairs of sample data, dependence could also occur with samples related through associations such as family members.] (In this book, we will use *matched pairs,* which best describes the nature of the data.)

EXAMPLE Drug Testing

Independent samples: One group of subjects is treated with the cholesterol-reducing drug Lipitor, while a second and separate group of subjects is given a placebo. These two sample groups are independent because the individuals in the treatment group are in no way paired or matched with corresponding members in the placebo group.

Matched pairs (or dependent samples): The effectiveness of a diet is tested using weights of subjects measured before and after the diet treatment. Each "before" value is matched with the "after" value because each before/after pair of measurements comes from the same person.

This section considers two independent samples, and the following section addresses matched pairs. When using two independent samples to test a claim about the difference $\mu_1 - \mu_2$, or to construct a confidence interval estimate of $\mu_1 - \mu_2$, use the following summary box.

Requirements

1. σ_1 and σ_2 are unknown and no assumption is made about the equality of σ_1 and σ_2.
2. The two samples are *independent.*
3. Both samples are *simple random samples.*
4. Either or both of these conditions is satisfied: The two sample sizes are both *large* (with $n_1 > 30$ and $n_2 > 30$) or both samples come from populations having normal distributions. (For small samples, the normality requirement is loose in the sense that the procedures perform well as long as there are no outliers and departures from normality are not too extreme.)

Hypothesis Test Statistic for Two Means: Independent Samples

$$t = \frac{(\bar{x}_1 - \bar{x}_2) - (\mu_1 - \mu_2)}{\sqrt{\dfrac{s_1^2}{n_1} + \dfrac{s_2^2}{n_2}}}$$

Degrees of Freedom: When finding critical values or *P*-values, use the following for determining the number of degrees of freedom, denoted by df. (Although these two methods typically result in different numbers of degrees of freedom, the conclusion of a hypothesis test is rarely affected by the choice.)

1. In this book we use this simple and conservative estimate: df = smaller of $n_1 - 1$ and $n_2 - 1$.
2. Statistical software packages typically use the more accurate but more difficult estimate given in Formula 9-1. (We will not use Formula 9-1 for the examples and exercises in this book.)

Formula 9-1

$$\text{df} = \frac{(A + B)^2}{\dfrac{A^2}{n_1 - 1} + \dfrac{B^2}{n_2 - 1}}$$

where $A = \dfrac{s_1^2}{n_1}$ and $B = \dfrac{s_2^2}{n_2}$

***P*-values:** Refer to Table A-3. Use the procedure summarized in Figure 8-6.

Critical values: Refer to Table A-3.

Confidence Interval Estimate of $\mu_1 - \mu_2$: Independent Samples

The confidence interval estimate of the difference $\mu_1 - \mu_2$ is

$$(\bar{x}_1 - \bar{x}_2) - E < (\mu_1 - \mu_2) < (\bar{x}_1 - \bar{x}_2) + E$$

where $E = t_{\alpha/2}\sqrt{\dfrac{s_1^2}{n_1} + \dfrac{s_2^2}{n_2}}$

and the number of degrees of freedom df is as described above for hypothesis tests. (In this book, we use df = smaller of $n_1 - 1$ and $n_2 - 1$.)

TI-83/84 Plus Calculator: The TI-83/84 Plus calculator can be used for hypothesis tests and confidence intervals. See the procedures at the end of this section.

Because the hypothesis test and confidence interval use the same distribution and standard error, they are equivalent in the sense that they result in the same conclusions. Consequently, the null hypothesis of $\mu_1 = \mu_2$ (or $\mu_1 - \mu_2 = 0$) can be tested by determining whether the confidence interval includes 0. For a two-tailed hypothesis test with a 0.05 significance level, use a 95% confidence interval. For a one-tailed test with a 0.05 significance level, use a 90% confidence interval. (See Table 8-2 for common cases.)

Exploring the Data Sets

We should verify the requirements when using two independent samples to make inferences about two population means. Instead of immediately conducting a hypothesis test or constructing a confidence interval, we should first *explore* the two samples using the methods described in Chapters 2 and 3. For each of the two samples, we should investigate center, variation, distribution, outliers, and whether the population appears to be changing over time (CVDOT). It could be very helpful to do the following:

- Find descriptive statistics for both data sets, including n, $\bar{x}$, and s.
- Create boxplots of both data sets, drawn on the same scale so that they can be compared.
- Create histograms of both data sets, so that their distributions can be seen and compared.
- Identify any outliers.

EXAMPLE **Discrimination Based on Age** The Revenue Commissioners in Ireland conducted a contest for promotion. The ages of the unsuccessful and successful applicants are given below (based on data from "Debating the Use of Statistical Evidence in Allegations of Age Discrimination," by Barry and Boland, *American Statistician,* Vol. 58, No. 2). Some of the applicants who were unsuccessful in getting the promotion charged that the competition involved discrimination based on age. Treat the data as samples from larger populations and use a 0.05 significance level to test the claim that the unsuccessful applicants are from a population with a greater mean age than the mean age of successful applicants. Based on the result, does there appear to be discrimination based on age?

Ages of Unsuccessful Applicants	Ages of Successful Applicants
34 37 37 38 41 42 43 44 44 45	27 33 36 37 38 38 39 42 42 43
45 45 46 48 49 53 53 54 54 55	43 44 44 44 45 45 45 45 46 46
56 57 60	47 47 48 48 49 49 51 51 52 54

SOLUTION

REQUIREMENT ✔ The values of the two population standard deviations are not known and we are not making an assumption that they are equal. The two samples are independent because values from one sample are in no way matched or paired with values from the other sample. We can assume that the samples are simple random samples. Both samples are small, so we must verify that each sample appears to come from a population with a normal distribution. There do not appear to be any outliers. The back-to-back stemplots shown below suggest that the data are normally distributed. Also, the normal quantile plots show patterns of points that are reasonably close to straight-line patterns, without having any other systematic pattern. Based on these displays, we can safely conclude that the distributions are not far from normal distributions. The requirements are satisfied and we can proceed with the hypothesis test. ✔

Unsuccessful	Successful
$n = 23$	$n = 30$
$\bar{x} = 47.0$	$\bar{x} = 43.9$
$s = 7.2$	$s = 5.9$

Unsuccessful		Successful
	2	7
4	3	3
877	3	67889
44321	4	2233444
986555	4	555566778899
4433	5	1124
765	5	
0	6	

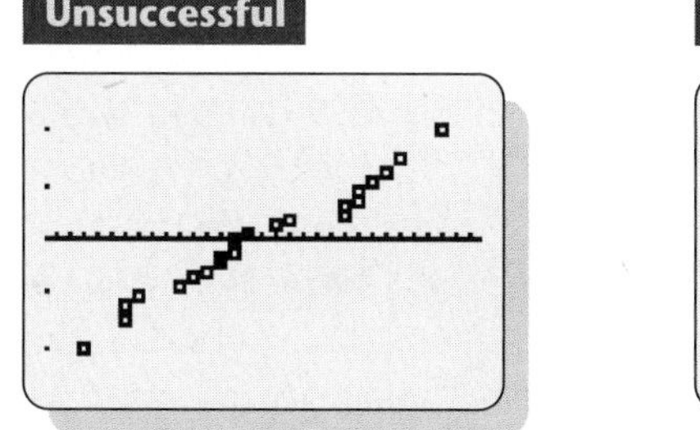

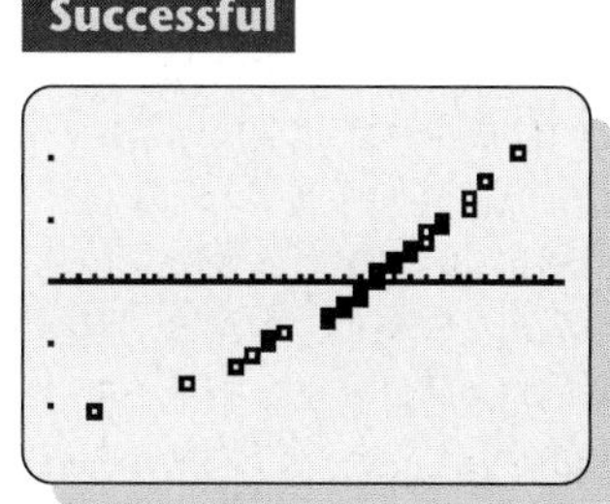

Having verified that the requirements are satisfied, we proceed with the hypothesis test. We use the traditional method summarized in Figure 8-9.

Step 1: The claim that unsuccessful applicants have a mean age greater than the mean age of successful applicants can be expressed symbolically as $\mu_1 > \mu_2$.

Step 2: If the original claim is false, then $\mu_1 \leq \mu_2$.

continued

Using Statistics to Identify Thieves

Methods of statistics can be used to determine that an employee is stealing, and they can also be used to estimate the amount stolen. The following are some of the indicators that have been used. For comparable time periods, samples of sales have means that are significantly different. The mean sale amount decreases significantly. There is a significant increase in the proportion of "no sale" register openings. There is a significant decrease in the ratio of cash receipts to checks. Methods of hypothesis testing can be used to identify such indicators. (See "How To Catch a Thief," by Manly and Thomson, *Chance,* Vol. 11, No. 4.)

Step 3: The alternative hypothesis is the expression not containing equality, and the null hypothesis is an expression of equality, so we have

$$H_0: \mu_1 = \mu_2 \qquad H_1: \mu_1 > \mu_2 \qquad \text{(original claim)}$$

We now proceed with the assumption that $\mu_1 = \mu_2$, or $\mu_1 - \mu_2 = 0$.

Step 4: The significance level is $\alpha = 0.05$.

Step 5: Because we have two independent samples and we are testing a claim about the two population means, we use a t distribution with the test statistic given earlier in this section.

Step 6: The test statistic is calculated as follows:

$$t = \frac{(\bar{x}_1 - \bar{x}_2) - (\mu_1 - \mu_2)}{\sqrt{\frac{s_1^2}{n_1} + \frac{s_2^2}{n_2}}} = \frac{(47.0 - 43.9) - 0}{\sqrt{\frac{7.2^2}{23} + \frac{5.9^2}{30}}} = 1.678$$

Because we are using a t distribution, the critical value of $t = 1.717$ is found from Table A-3. (With an area of 0.05 in the right tail, we want the t value corresponding to 22 degrees of freedom, which is the smaller of $n_1 - 1$ and $n_2 - 1$ [or the smaller of 22 and 29].) The test statistic, critical value, and critical region are shown in Figure 9-2.

Using a TI-83/84 Plus calculator, we can also find that the P-value is 0.0548 (based on df = 41.868).

Step 7: Because the test statistic does not fall within the critical region, fail to reject the null hypothesis $\mu_1 = \mu_2$ (or $\mu_1 - \mu_2 = 0$).

INTERPRETATION There is not sufficient evidence to support the claim that the mean age of unsuccessful applicants is greater than the mean age of successful applicants. Based on this hypothesis test, there does not appear to be discrimination based on age. However, the hypothesis test considers one facet of the overall picture and there are other factors that we might consider. For

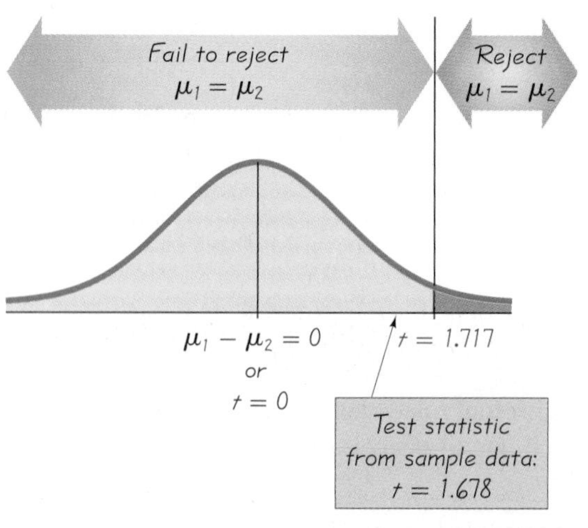

Figure 9-2 Testing the Claim of Age Discrimination

example, we might compare the proportion of applicants over age 50 who were successful to the proportion of applicants under age 50 who were successful. Two different statisticians presented evidence in the court case, and they reached different conclusions. The court ruled that there was a "link between the age of candidates and their success or failure in the competition."

EXAMPLE Confidence Interval for Ages of Applicants for Promotion Using the sample data given in the preceding example, construct a 90% confidence interval estimate of the difference between the mean age of unsuccessful applicants and the mean age of successful applicants. (Remember, a one-sided hypothesis test with significance level 0.05 can be tested by using a 90% confidence interval.)

SOLUTION

REQUIREMENT ✔ Because we are using the same data from the preceding example, the same requirement check applies here. The requirements are satisfied, and we can proceed to construct the confidence interval. ✔

We first find the value of the margin of error E. We use $t_{\alpha/2} = 1.717$, which is found in Table A-3 as the t score corresponding to an area of 0.10 in two tails and df = 22. (As in the preceding example, we want the t score corresponding to 22 degrees of freedom, which is the smaller of $n_1 - 1$ and $n_2 - 1$ [or the smaller of 22 and 29].)

$$E = t_{\alpha/2}\sqrt{\frac{s_1^2}{n_1} + \frac{s_2^2}{n_2}} = 1.717\sqrt{\frac{7.2^2}{23} + \frac{5.9^2}{30}} = 3.2$$

We now find the desired confidence interval as follows:

$$(\bar{x}_1 - \bar{x}_2) - E < (\mu_1 - \mu_2) < (\bar{x}_1 - \bar{x}_2) + E$$

$$(47.0 - 43.9) - 3.2 < (\mu_1 - \mu_2) < (47.0 - 43.9) + 3.2$$

$$-0.1 < (\mu_1 - \mu_2) < 6.3$$

If we use the TI-83/84 Plus calculator to obtain more accurate results, we get the confidence interval of $-0.1 < (\mu_1 - \mu_2) < 6.1$, so we can see that the above confidence interval is quite good.

INTERPRETATION We are 90% confident that the limits of −0.1 year and 6.3 years actually do contain the difference between the two population means. Because those limits do contain 0, this confidence interval suggests that it is very possible that the two population means are equal. There is not a significant difference between the two means.

Rationale: Why Do the Test Statistic and Confidence Interval Have the Particular Forms We Have Presented? If the given assumptions are satisfied, the sampling distribution of $\bar{x}_1 - \bar{x}_2$ can be approximated by a t distribution with mean equal to $\mu_1 - \mu_2$ and standard deviation equal to $\sqrt{s_1^2/n_1 + s_2^2/n_2}$. This last expression for the standard deviation is based on the property that the variance of the *differences* between two independent random variables equals the variance of the first random variable *plus* the variance of the second random variable. That is,

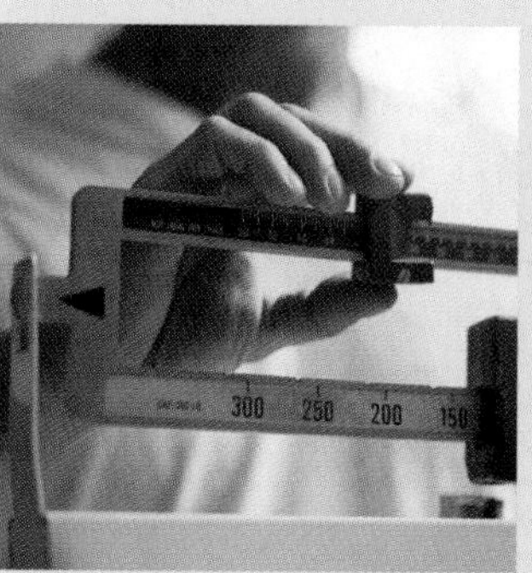

Expensive Diet Pill

There are many past examples in which ineffective treatments were marketed for substantial profits. Capsules of "Fat Trapper" and "Exercise in a Bottle," manufactured by the Enforma Natural Products company, were advertised as being effective treatments for weight reduction. Advertisements claimed that after taking the capsules, fat would be blocked and calories would be burned, even without exercise. Because the Federal Trade Commission identified claims that appeared to be unsubstantiated, the company was fined $10 million for deceptive advertising.

The effectiveness of such treatments can be determined with experiments in which one group of randomly selected subjects is given the treatment, while another group of randomly selected subjects is given a placebo. The resulting weight losses can be compared using statistical methods, such as those described in this section.

the variance of sample values $\bar{x}_1 - \bar{x}_2$ will tend to equal $s_1^2/n_1 + s_2^2/n_2$ provided that $\bar{x}_1$ and $\bar{x}_2$ are independent. (See Exercise 35.)

Part 2: Alternative Methods

Part 1 of this section dealt with situations in which the two population standard deviations are unknown and are not assumed to be equal. In Part 2 we address two other situations: (1) The two population standard deviations are both known; (2) the two population standard deviations are unknown but are assumed to be equal. We now describe the procedures for these alternative cases.

Alternative Method: σ_1 and σ_2 Are Known

In reality, the population standard deviations σ_1 and σ_2 are almost never known, but if they are known, the test statistic and confidence interval are based on the normal distribution instead of the *t* distribution. See the summary box below.

Requirements

1. The two population standard deviations are both known.
2. The two samples are *independent.*
3. Both samples are *simple random samples.*
4. Either or both of these conditions is satisfied: The two sample sizes are both *large* (with $n_1 > 30$ and $n_2 > 30$) or both samples come from populations having normal distributions. (For small samples, the normality requirement is loose in the sense that the procedures perform well as long as there are no outliers and departures from normality are not too extreme.)

Hypothesis Test for Two Means: Independent Samples with σ_1 and σ_2 Both Known

Test statistic: $$z = \frac{(\bar{x}_1 - \bar{x}_2) - (\mu_1 - \mu_2)}{\sqrt{\dfrac{\sigma_1^2}{n_1} + \dfrac{\sigma_2^2}{n_2}}}$$

***P*-values** and **Critical values:** Refer to Table A-2.

Confidence Interval Estimate of $\mu_1 - \mu_2$: Independent Samples with σ_1 and σ_2 Both Known

Confidence interval: $(\bar{x}_1 - \bar{x}_2) - E < (\mu_1 - \mu_2) < (\bar{x}_1 - \bar{x}_2) + E$

where $$E = z_{\alpha/2}\sqrt{\frac{\sigma_1^2}{n_1} + \frac{\sigma_2^2}{n_2}}$$

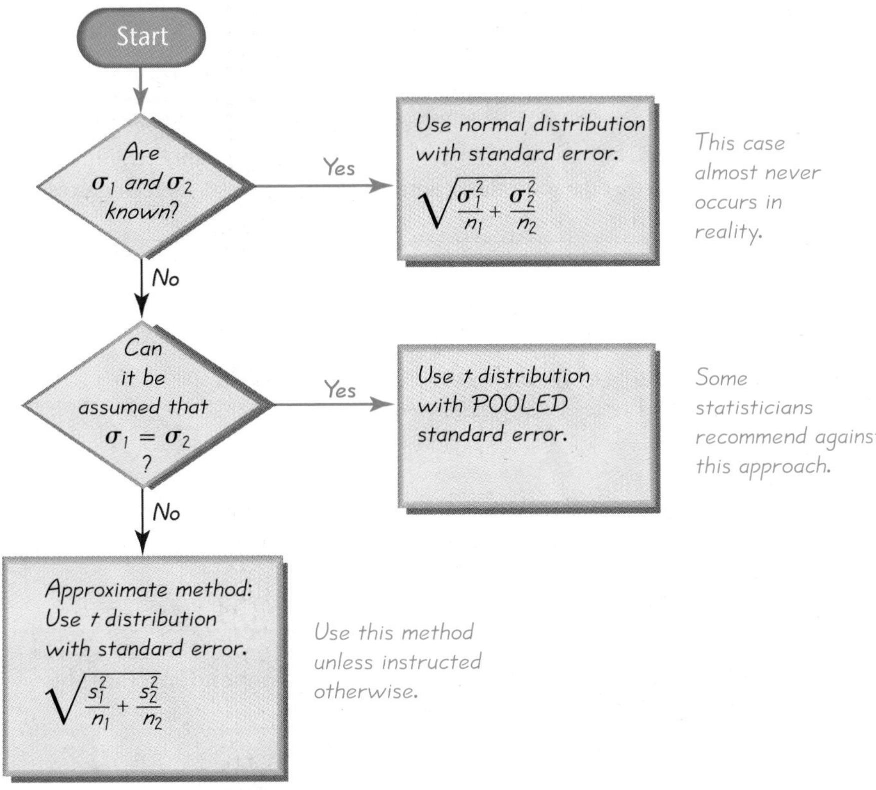

Figure 9-3 Methods for Inferences About Two Independent Means

One alternative method (not used in this book) is to use the above expressions if σ_1 and σ_2 are not known but both samples are large (with $n_1 > 30$ and $n_2 > 30$). This alternative method is used with σ_1 replaced by s_1 and σ_2 replaced by s_2. Because σ_1 and σ_2 are rarely known in reality, this book will not use this alternative method. See Figure 9-3 for the procedure used in this book.

Alternative Method: Assume That $\sigma_1 = \sigma_2$ and *Pool* the Sample Variances

Even when the specific values of σ_1 and σ_2 are not known, if it can be assumed that they have the *same* value, the sample variances s_1^2 and s_2^2 can be *pooled* to obtain an estimate of the common population variance σ^2. The **pooled estimate of σ^2** is denoted by s_p^2 and is a weighted average of s_1^2 and s_2^2, which is included in the following box.

Super Bowls

Students were invited to a Super Bowl game and half of them were given large 4-liter snack bowls while the other half were given smaller 2-liter bowls. Those using the large bowls consumed 56% more than those using the smaller bowls. (See "Super Bowls: Serving Bowl Size and Food Consumption," by Wansink and Cheney, *Journal of the American Medical Association,* Vol. 293, No. 14.)

A separate study showed that there is "a significant increase in fatal motor vehicle crashes during the hours following the Super Bowl telecast in the United States." Researchers analyzed 20,377 deaths on 27 Super Bowl Sundays and 54 other Sundays used as controls. They found a 41% increase in fatalities after Super Bowl games. (See "Do Fatal Crashes Increase Following a Super Bowl Telecast?" by Redelmeier and Stewart, *Chance,* Vol. 18, No. 1.)

Better Results with Smaller Class Size

An experiment at the State University of New York at Stony Brook found that students did significantly better in classes limited to 35 students than in large classes with 150 to 200 students. For a calculus course, failure rates were 19% for the small classes compared to 50% for the large classes. The percentages of A's were 24% for the small classes and 3% for the large classes. These results suggest that students benefit from smaller classes, which allow for more direct interaction between students and teachers.

Requirements

1. The two population standard deviations are not known, but they are assumed to be equal. That is, $\sigma_1 = \sigma_2$.
2. The two samples are *independent.*
3. Both samples are *simple random samples.*
4. Either or both of these conditions is satisfied: The two sample sizes are both *large* (with $n_1 > 30$ and $n_2 > 30$) or both samples come from populations having normal distributions. (For small samples, the normality requirement is loose in the sense that the procedures perform well as long as there are no outliers and departures from normality are not too extreme.)

Hypothesis Test Statistic for Two Means: Independent Samples and $\boldsymbol{\sigma_1 = \sigma_2}$

Test statistic: $t = \dfrac{(\bar{x}_1 - \bar{x}_2) - (\mu_1 - \mu_2)}{\sqrt{\dfrac{s_p^2}{n_1} + \dfrac{s_p^2}{n_2}}}$

where $\quad s_p^2 = \dfrac{(n_1 - 1)s_1^2 + (n_2 - 1)s_2^2}{(n_1 - 1) + (n_2 - 1)}$ (*Pooled* variance)

and the number of degrees of freedom is given by df $= n_1 + n_2 - 2$.

Confidence Interval Estimate of $\boldsymbol{\mu_1 - \mu_2}$: Independent Samples and $\boldsymbol{\sigma_1 = \sigma_2}$

Confidence interval: $(\bar{x}_1 - \bar{x}_2) - E < (\mu_1 - \mu_2) < (\bar{x}_1 - \bar{x}_2) + E$

where $\quad E = t_{\alpha/2}\sqrt{\dfrac{s_p^2}{n_1} + \dfrac{s_p^2}{n_2}}$ and s_p^2 is as given in the above test statistic and the number of degrees of freedom is df $= n_1 + n_2 - 2$.

If we want to use this method, how do we determine that $\sigma_1 = \sigma_2$? One approach is to use a hypothesis test of the null hypothesis $\sigma_1 = \sigma_2$, as given in Section 9-5, but that approach is not recommended and, in this book, we will not use the preliminary test of $\sigma_1 = \sigma_2$. In the article "Homogeneity of Variance in the Two-Sample Means Test" (by Moser and Stevens, *American Statistician,* Vol. 46, No. 1), the authors note that we rarely know that $\sigma_1 = \sigma_2$. They analyze the performance of the different tests by considering sample sizes and powers of the tests. They conclude that more effort should be spent learning the method given in Part 1, and less emphasis should be placed on the method based on the assumption of $\sigma_1 = \sigma_2$. Unless instructed otherwise, we use the following strategy, which is consistent with the recommendations in the article by Moser and Stevens:

> **Assume that $\boldsymbol{\sigma_1}$ and $\boldsymbol{\sigma_2}$ are unknown, do *not* assume that $\boldsymbol{\sigma_1 = \sigma_2}$, and use the test statistic and confidence interval given in Part 1 of this section. (See Figure 9-3.)**

TI

Using the TI-83/84 Plus Calculator for Inferences About Two Means (With Independent Samples)

1. Press STAT, then press ▷ twice to select **TESTS.**
2. Select one of the following as appropriate:

2-SampTTest	Hypothesis test (with σ_1 and σ_2 not known)
2-SampTInt	Confidence interval (with σ_1 and σ_2 not known)
2-SampZTest	Hypothesis test (with σ_1 and σ_2 known)
2-SampZInt	Confidence interval (with σ_1 and σ_2 known)

3. Select **Data** if the two sets of sample data are already stored in the calculator as lists, or select **Stats** if you know the summary statistics of $\bar{x}_1$, s_1, n_1, $\bar{x}_2$, s_2, and n_2.
4. If you selected **Data** in Step 3, proceed to enter the list names or labels, such as L1 and L2. If you selected **Stats** in Step 3, proceed to enter the values of the statistics for each sample.
5. The next step depends on whether you are conducting a hypothesis test or constructing a confidence interval.
   - *Hypothesis test:* If conducting a hypothesis test, select the type of test (two-tailed, left-tailed, or right-tailed). Select the format corresponding to the alternative hypothesis ($\mu_1 \neq \mu_2$, $\mu_1 < \mu_2$, or $\mu_1 > \mu_2$).
   - *Confidence interval:* If creating a confidence interval, you must enter a confidence level, such as 0.95.
6. If using the t distribution (by selecting either 2-SampTTest or 2-SampTInt in Step 2), you will see a line displaying **Pooled: No Yes.** It is recommended that you select **No** so that the sample variances are not pooled. (You would pool the sample variances if you were assuming that the two population variances are equal.)
7. Use the arrow keys to scroll down to the bottom line, select **Calculate,** then press ENTER.

Hypothesis Test

```
2-SampTTest
 μ1>μ2
 t=1.634613932
 p=.0548138719
 df=41.86847492
 x̄1=46.95652174
↓x̄2=43.93333333
```

Confidence Interval

```
2-SampTInt
 (-.0878,6.1341)
 df=41.86847492
 x̄1=46.95652174
 x̄2=43.93333333
 Sx1=7.22041462
↓Sx2=5.88354417
```

The accompanying screen displays correspond to the hypothesis test and confidence interval examples in this section. Note the P-value of 0.0548 in the hypothesis test. Also note that the confidence interval limits can be rounded and expressed in this format:

$$-0.1 < (\mu_1 - \mu_2) < 6.1$$

9-3 BASIC SKILLS AND CONCEPTS

Statistical Literacy and Critical Thinking

1. Requirements See the age data used in the hypothesis testing and confidence interval examples of this section. If we include one more age of 80 years for an unsuccessful applicant, are the requirements given in Part 1 satisfied? Why or why not?

2. Confidence Interval for Hypothesis Testing You plan to construct a confidence interval to be used for testing the claim that one population has a mean greater than the other. If the test is to have a 0.01 significance level, what confidence level should be used for the confidence interval? What confidence level should be used to test the claim that two populations have different means (again using a 0.01 significance level)?

3. Degrees of Freedom In the hypothesis test example in this section, the critical value of $t = 1.717$ was obtained by using df = smaller of $n_1 - 1$ and $n_2 - 1$. With sample sizes of 23 and 30, we used df = 22. If we calculate df using Formula 9-1, we get df = 41.868, and the corresponding critical value is 1.682. How is using a critical value of $t = 1.717$ "more conservative" than using a critical value of 1.682?

4. Before–After Data A random sample of subjects is treated with a drug intended to lower their cholesterol levels. For each subject, the cholesterol is measured once before the treatment and once after the treatment. Can we use the methods of this section to test the claim that the "before" cholesterol levels have a mean that is greater than the "after" levels? Why or why not?

Independent Samples and Matched Pairs. *In Exercises 5–8, determine whether the samples are independent or consist of matched pairs.*

5. The effectiveness of Prilosec for treating heartburn is tested by measuring gastric acid secretion in a group of patients treated with Prilosec and another group of patients given a placebo.

6. The effectiveness of Prilosec for treating heartburn is tested by measuring gastric acid secretion in patients before and after the drug treatment. The data consist of the before/after measurements for each patient.

7. The effectiveness of the Weight Watchers diet is tested in an experiment, and for each subject, the weight before the diet and the weight after the diet are recorded.

8. The effectiveness of a flu vaccine is tested by treating one group of subjects with the vaccine while another group of subjects is given placebos.

In Exercises 9–28, assume that the two samples are independent simple random samples selected from normally distributed populations. Do not assume that the population standard deviations are equal, unless your instructor stipulates otherwise.

TI 9. Hypothesis Test of Effectiveness of Echinacea In a randomized, double-blind, placebo-controlled trial of children, echinacea was tested as a treatment for upper respiratory infections in children. "Days of fever" was one criterion used to measure effects. Among 337 children treated with echinacea, the mean number of days with fever was 0.81, with a standard deviation of 1.50 days. Among 370 children given a placebo, the mean number of days with fever was 0.64 with a standard deviation of 1.16 days (based on data from "Efficacy and Safety of Echinacea in Treating Upper Respiratory Tract Infections in Children," by Taylor et al., *Journal of the American Medical Association,* Vol. 290, No. 21). Use a 0.05 significance level to test the claim that echinacea affects the number of days with fever. Based on these results, does echinacea appear to be effective?

TI 10. Hypothesis Test of Effects of Cocaine on Children A study was conducted to assess the effects that occur when children are exposed to cocaine before birth. Children were tested at age 4 for object assembly skill, which was described as "a task requiring visual-spatial skills related to mathematical competence." The 190 children born to cocaine users had a mean of 7.3 and a standard deviation of 3.0. The 186 children not exposed to cocaine had a mean score of 8.2 with a standard deviation of 3.0. (The data are based on "Cognitive Outcomes of Preschool Children with Prenatal Cocaine Exposure," by Singer et al., *Journal of the American Medical Association,* Vol. 291, No. 20.) Use a 0.05 significance level to test the claim that prenatal cocaine exposure is associated with lower scores of 4-year-old children on the test of object assembly.

TI **11.** Confidence Interval for Effect of Birth Weight on IQ Score When investigating a relationship between birth weight and IQ, researchers found that 258 subjects with extremely low birth weights (less than 1000 g) had Wechsler IQ scores at age 8 with a mean of 95.5 and a standard deviation of 16.0. For 220 subjects with normal birth weights, the mean at age 8 is 104.9 and the standard deviation is 14.1. (Based on data from "Neurobehavioral Outcomes of School-age Children Born Extremely Low Birth Weight or Very Preterm in the 1990s," by Anderson et al., *Journal of the American Medical Association,* Vol. 289, No. 24.) Construct a 95% confidence interval estimate of the difference between the mean IQ score of 8-year-old children born with low birth weight and the mean score of 8-year-old children born with normal birth weight. Does IQ score appear to be affected by birth weight?

TI **12.** Confidence Interval for Comparing Diets A randomized trial tested the effectiveness of diets on adults. Among 40 subjects using the Weight Watchers diet, the mean weight loss after one year was 3.0 lb with a standard deviation of 4.9 lb. Among 40 subjects using the Atkins diet, the mean weight loss after one year was 2.1 lb with a standard deviation of 4.8 lb. Construct a 95% confidence interval estimate of the difference between the population means. Does there appear to be a difference in the effectiveness of the two diets? Does the amount of weight loss appear to justify either diet?

TI **13.** Hypothesis Test for Effect of Marijuana Use on College Students Many studies have been conducted to test the effects of marijuana use on mental abilities. In one such study, groups of light and heavy users of marijuana in college were tested for memory recall, with the results given below (based on data from "The Residual Cognitive Effects of Heavy Marijuana Use in College Students," by Pope and Yurgelun-Todd, *Journal of the American Medical Association*, Vol. 275, No. 7). Use a 0.01 significance level to test the claim that the population of heavy marijuana users has a lower mean than the light users. Should marijuana use be of concern to college students?

Items sorted correctly by light marijuana users: $n = 64, \bar{x} = 53.3, s = 3.6$
Items sorted correctly by heavy marijuana users: $n = 65, \bar{x} = 51.3, s = 4.5$

TI **14.** Confidence Interval for Effects of Marijuana Use on College Students Refer to the sample data used in Exercise 13 and construct a 98% confidence interval for the difference between the two population means. Does the confidence interval include zero? What does the confidence interval suggest about the equality of the two population means?

TI **15.** Confidence Interval for Bipolar Depression Treatment In clinical experiments involving different groups of independent samples, it is important that the groups be similar in the important ways that affect the experiment. In an experiment designed to test the effectiveness of paroxetine for treating bipolar depression, subjects were measured using the Hamilton depression scale with the results given below (based on data from "Double-Blind, Placebo-Controlled Comparison of Imipramine and Paroxetine in the Treatment of Bipolar Depression," by Nemeroff et al., *American Journal of Psychiatry,* Vol. 158, No. 6). Construct a 95% confidence interval for the difference between the two population means. Based on the results, does it appear that the two populations have different means? Should paroxetine be recommended as a treatment for bipolar depression?

Placebo group:	$n = 43, \bar{x} = 21.57, s = 3.87$
Paroxetine treatment group:	$n = 33, \bar{x} = 20.38, s = 3.91$

TI **16.** Hypothesis Test for Bipolar Depression Treatment Refer to the sample data in Exercise 15 and use a 0.05 significance level to test the claim that the treatment group and placebo group come from populations with the same mean. What does the result of the hypothesis test suggest about paroxetine as a treatment for bipolar depression?

TI **17.** Hypothesis Test for Magnet Treatment of Pain People spend huge sums of money (currently around $5 billion annually) for the purchase of magnets used to treat a wide variety of pains. Researchers conducted a study to determine whether magnets are effective in treating back pain. Pain was measured using the visual analog scale, and the results given below are among the results obtained in the study (based on data from "Bipolar Permanent Magnets for the Treatment of Chronic Lower Back Pain: A Pilot Study," by Collacott, Zimmerman, White, and Rindone, *Journal of the American Medical Association,* Vol. 283, No. 10). Use a 0.05 significance level to test the claim that those treated with magnets have a greater reduction in pain than those given a sham treatment (similar to a placebo). Does it appear that magnets are effective in treating back pain? Is it valid to argue that magnets might appear to be effective if the sample sizes are larger?

Reduction in pain level after magnet treatment:	$n = 20, \bar{x} = 0.49, s = 0.96$
Reduction in pain level after sham treatment:	$n = 20, \bar{x} = 0.44, s = 1.4$

TI **18.** Confidence Interval for Magnet Treatment of Pain Refer to the sample data from Exercise 17 and construct a 90% confidence interval estimate of the difference between the mean reduction in pain for those treated with magnets and the mean reduction in pain for those given a sham treatment. Based on the result, does it appear that the magnets are effective in reducing pain?

TI **19.** Confidence Interval for Identifying Psychiatric Disorders Are severe psychiatric disorders related to biological factors that can be physically observed? One study used x-ray computed tomography (CT) to collect data on brain volumes for a group of patients with obsessive-compulsive disorders and a control group of healthy persons. Sample results for volumes (in mL) follow for the right cordate (based on data from "Neuroanatomical Abnormalities in Obsessive-Compulsive Disorder Detected with Quantitative X-Ray Computed Tomography," by Luxenberg et al., *American Journal of Psychiatry*, Vol. 145, No. 9). Construct a 99% confidence interval estimate of the difference between the mean brain volume for the healthy control group and the mean brain volume for the obsessive-compulsive group. What does the confidence interval suggest about the difference between the two population means? Based on this result, does it seem that obsessive-compulsive disorders have a biological basis?

Control group:	$n = 10, \bar{x} = 0.45, s = 0.08$
Obsessive-compulsive patients:	$n = 10, \bar{x} = 0.34, s = 0.08$

TI **20.** Hypothesis Test for Identifying Psychiatric Disorders Refer to the sample data in Exercise 19 and use a 0.01 significance level to test the claim that there is a difference between the two population means. Based on the result, does it seem that obsessive-compulsive disorders have a biological basis?

TI **21.** Confidence Interval for Effects of Alcohol An experiment was conducted to test the effects of alcohol. The *errors* were recorded in a test of visual and motor skills for a treatment group of people who drank ethanol and another group given a placebo. The results are shown in the accompanying table (based on data from "Effects of Alcohol Intoxication on Risk Taking, Strategy, and Error Rate in Visuomotor Performance," by Streufert et al., *Journal of Applied Psychology,* Vol. 77, No. 4). Construct a 95% confidence interval estimate of the difference between the two population means. Do the results support the common belief that drinking is hazardous for drivers, pilots, ship captains, and so on? Why or why not?

Treatment Group	Placebo Group
$n_1 = 22$	$n_2 = 22$
$\bar{x}_1 = 4.20$	$\bar{x}_2 = 1.71$
$s_1 = 2.20$	$s_2 = 0.72$

TI **22.** Hypothesis Test for Effects of Alcohol Refer to the sample data in Exercise 21 and use a 0.05 significance level to test the claim that there is a difference between the treatment group and control group. If there is a significant difference, can we conclude that the treatment causes a decrease in visual and motor skills?

TI **23.** Hypothesis Test for Difference in Home Values Listed below are fair market values (in thousands of dollars) of randomly selected homes on Long Beach Island in New Jersey. Use a 0.05 significance level to test a realtor's claim that oceanfront homes (directly on the beach) have greater value than oceanside homes, which are not directly on the beach. Given that there are only five values in each sample, can we really conclude that oceanfront homes have a greater mean value?

Oceanfront:	2199	3750	1725	2398	2799
Oceanside:	700	1355	795	1575	759

TI **24.** Hypothesis Test for Difference in Car/Taxi Ages When the author visited Dublin in Ireland (home of Guinness Brewery employee William Gosset, who first developed the *t* distribution), he recorded the ages of randomly selected passenger cars and randomly selected taxis. (There is no end to the fun of traveling with the author.) The ages (in years) are listed below. Use a 0.05 significance level to test the claim that there is a difference between the mean age of a Dublin car and the mean age of a Dublin taxi. We might expect that taxis would be newer, but what do the results suggest?

Cars	Taxis
4 0 8 11 14 3 4 4 3 5	8 8 0 3 8 4 3 3 6 11
8 3 3 7 4 6 6 1 8 2	7 7 6 9 5 10 8 4 3 4
15 11 4 1 6 1 8	

TI **25.** Tar and Cigarettes Refer to the sample data listed below and use a 0.05 significance level to test the claim that the mean amount of tar in filtered king-size cigarettes is *less than* the mean amount of tar in nonfiltered king-size cigarettes. All measurements are in milligrams, and the data are from the Federal Trade Commission.

Filtered	16	15	16	14	16	1	16	18	10	14	12
	11	14	13	13	13	16	16	8	16	11	
Nonfiltered	23	23	24	26	25	26	21	24			

TI **26.** Blanking Out on Tests Many students have had the unpleasant experience of panicking on a test because the first question was exceptionally difficult. The arrangement of test items was studied for its effect on anxiety. The following scores are measures of "debilitating test anxiety," which most of us call panic or blanking out (based on data from "Item Arrangement, Cognitive Entry Characteristics, Sex and Test Anxiety as

Predictors of Achievement in Examination Performance," by Klimko, *Journal of Experimental Education,* Vol. 52, No. 4). Is there sufficient evidence to support the claim that the two populations of scores have the same mean? Is there sufficient evidence to support the claim that the arrangement of the test items has an effect on the score?

Questions Arranged from Easy to Difficult					Questions Arranged from Difficult to Easy			
24.64	39.29	16.32	32.83	28.02	33.62	34.02	26.63	30.26
33.31	20.60	21.13	26.69	28.90	35.91	26.68	29.49	35.32
26.43	24.23	7.10	32.86	21.06	27.24	32.34	29.34	33.53
28.89	28.71	31.73	30.02	21.96	27.62	42.91	30.20	32.54
25.49	38.81	27.85	30.29	30.72				

TI **27.** Appendix B Data Set: Weights of Quarters Weights of quarters are used by vending machines as one way to detect counterfeit coins. Refer to Data Set 14 in Appendix B and test the claim that the mean weight of pre-1964 silver quarters is equal to the mean weight of post-1964 quarters. Given the relatively small sample sizes from the large populations of millions of quarters, can we really conclude that the mean weights are different?

TI **28.** Appendix B Data Set: Weights of Coke Refer to Data Set 12 in Appendix B and test the claim that because they contain the same amount of cola, the weights of cans of regular Coke have the same mean as the weights of Diet Coke. If there is a difference in the mean weights, identify the most likely explanation for that difference.

Pooling. *In Exercises 29–32, assume that the two samples are independent simple random samples selected from normally distributed populations. Also assume that the population standard deviations are equal ($\sigma_1 = \sigma_2$) so that the standard error of the differences between means is obtained by pooling the sample variances.*

TI **29.** Confidence Interval with Pooling Do Exercise 12 with the additional assumption that $\sigma_1 = \sigma_2$. How are the results affected by this additional assumption?

TI **30.** Hypothesis Test with Pooling Do Exercise 13 with the additional assumption that $\sigma_1 = \sigma_2$. How are the results affected by this additional assumption?

TI **31.** Hypothesis Test with Pooling Do Exercise 17 with the additional assumption that $\sigma_1 = \sigma_2$. How are the results affected by this additional assumption?

TI **32.** Confidence Interval with Pooling Do Exercise 18 with the additional assumption that $\sigma_1 = \sigma_2$. How are the results affected by this additional assumption?

9-3 BEYOND THE BASICS

TI **33.** Effects of an Outlier Refer to Exercise 24 and include an outlier consisting of a car that is 50 years old. Is the hypothesis test dramatically affected by the presence of the outlier?

TI **34.** Effects of Units of Measurement How are the results of Exercise 24 affected if all of the ages are converted from years to months? In general, does the choice of the scale affect the conclusions about equality of the two population means, and does the choice of scale affect the confidence interval?

TI **35.** Verifying a Property of Variances

a. Find the variance for this *population* of x values: 5, 10, 15. (See Section 3-3 for the variance σ^2 of a population.)

b. Find the variance for this *population* of y values: 1, 2, 3.

c. List the *population* of all possible differences $x - y$, and find the variance of this population.

d. Use the results from parts (a), (b), and (c) to verify that the variance of the *differences* between two independent random variables is the *sum* of their individual variances ($\sigma^2_{x-y} = \sigma^2_x + \sigma^2_y$). (This principle is used to derive the test statistic and confidence interval given in this section.)

e. How is the *range* of the differences $x - y$ related to the range of the x values and the range of the y values?

36. Effect of No Variation in Sample An experiment was conducted to test the effects of alcohol. The breath alcohol levels were measured for a treatment group of people who drank ethanol and another group given a placebo. The results are given in the accompanying table. Use a 0.05 significance level to test the claim that the two sample groups come from populations with the same mean. The given results are based on data from "Effects of Alcohol Intoxication on Risk Taking, Strategy, and Error Rate in Visuomotor Performance," by Streufert et al., *Journal of Applied Psychology,* Vol. 77, No. 4.

Treatment Group	Placebo Group
$n_1 = 22$	$n_2 = 22$
$\bar{x}_1 = 0.049$	$\bar{x}_2 = 0.000$
$s_1 = 0.015$	$s_2 = 0.000$

37. Calculating Degrees of Freedom How is the number of degrees of freedom for Exercises 19 and 20 affected if Formula 9-1 is used instead of selecting the smaller of $n_1 - 1$ and $n_2 - 1$? If Formula 9-1 is used for the number of degrees of freedom instead of the smaller of $n_1 - 1$ and $n_2 - 1$, how are the *P*-value and the width of the confidence interval affected? In what sense is "df = smaller of $n_1 - 1$ and $n_2 - 1$" a more conservative estimate of the number of degrees of freedom than the estimate obtained with Formula 9-1?

9-4 Inferences from Matched Pairs

Key Concept Two samples are **dependent** (or consist of **matched pairs**) if the members of one sample can be used to determine the members of the other sample. In this section we develop methods for testing claims about the mean difference from matched pairs. For each matched pair of sample values, we find the difference between the two values, then we use those sample differences to test claims about the population difference or to construct confidence interval estimates of the population difference. Chapter 10 will also deal with paired data, but in Chapter 10 we are interested in the *association* between two variables, not the *mean* of the differences.

With matched pairs, there is some relationship so that each value in one sample is paired with a corresponding value in the other sample. Here are some typical examples of matched pairs:

- The sample data are matched pairs of low-density lipoprotein (LDL) cholesterol measurements taken before and after Lipitor treatments. Example: LDL before Lipitor = 182; LDL after Lipitor = 155.
- The sample data are matched pairs of husband/wife body mass indexes (BMIs). Example: BMI of husband = 25.1; BMI of wife = 19.7.
- The sample data are heights of winning presidential candidates matched with the heights of the candidates who received the second highest number of votes. Example: Height of Truman = 69 in.; height of Dewey = 68 in.

When dealing with inferences about the means of matched pairs, the requirements, notation, hypothesis test statistic, and confidence interval are given in the summary box below. Because the hypothesis test and confidence interval use the same distribution and standard error, they are equivalent in the sense that they result in the same conclusions. Consequently, the null hypothesis that the mean difference equals 0 can be tested by determining whether the confidence interval includes 0. A two-tailed hypothesis test with a 0.05 significance level can be tested with a 95% confidence, but a one-tailed hypothesis test with a 0.05 significance level can be tested with a 90% confidence interval. (See Table 8-2 for common cases.)

Requirements

1. The sample data consist of matched pairs.
2. The samples are simple random samples.
3. Either or both of these conditions is satisfied: The number of matched pairs of sample data is large ($n > 30$) or the pairs of values have differences that are from a population having a distribution that is approximately normal. (If there is a radical departure from a normal distribution, we should not use the methods given in this section, but we may be able to use nonparametric methods discussed in Chapter 13.)

Notation for Matched Pairs

d = individual difference between the two values in a single matched pair

μ_d = mean value of the differences d for the *population* of all matched pairs

$\bar{d}$ = mean value of the differences d for the paired *sample* data (equal to the mean of the $x - y$ values)

s_d = standard deviation of the differences d for the paired *sample* data

n = number of *pairs* of data

Hypothesis Test Statistic for Matched Pairs

$$t = \frac{\bar{d} - \mu_d}{\dfrac{s_d}{\sqrt{n}}}$$

where degrees of freedom = $n - 1$.

***P*-values** and **Critical values:** Table A-3 (t distribution)

Confidence Intervals for Matched Pairs

$$\bar{d} - E < \mu_d < \bar{d} + E$$

where $$E = t_{\alpha/2}\frac{s_d}{\sqrt{n}}$$

Critical values of $t_{\alpha/2}$: Use Table A-3 with $n - 1$ degrees of freedom.

TI-83/84 Plus Calculator: The TI-83/84 Plus calculator can be used for hypothesis tests and confidence intervals. See the procedure at the end of this section.

EXAMPLE Hypothesis Test with Actual and Forecast Temperatures Table 9-2 consists of five actual low temperatures and the corresponding low temperatures that were predicted five days earlier (based on data recorded near the author's home). The data consist of matched pairs, because each pair of values represents the same day. The forecast temperatures appear to be very different from the actual temperatures, but is there sufficient evidence to conclude that the mean difference is not zero? Use a 0.05 significance level to test the claim that there is a difference between the actual low temperatures and the low temperatures that were forecast five days earlier.

Table 9-2 Actual and Forecast Temperature

Actual low	54	54	55	60	64
Low forecast five days earlier	56	57	59	56	64
Difference d = actual − forecast	−2	−3	−4	4	0

SOLUTION

REQUIREMENTS ✔ We consider each of the three requirements. (1) The sample data consist of matched pairs referring to the same day. The actual low is the low temperature reached on the given day, and the other value is the low temperature for the same day that was forecast five days earlier. (2) Instead of being a simple random sample, we have results from the first five consecutive days listed in Data Set 8 in Appendix B. This could be a problem arising from such factors as an exceptionally bad (or good) forecaster who happened to make the forecast for these five days only. We will assume that there is a more general forecasting system, and the five days are typical of those that would result from a simple random sample. (3) The number of matched pairs is not large, so we should check for normality of the differences. The accompanying displays show the normal quantile plot of the differences, and we conclude that they are from a normally distributed population because the points are reasonably close to a straight-line pattern without showing some other systematic pattern. The requirements are satisfied and we can proceed with the hypothesis test. ✔

STATDISK

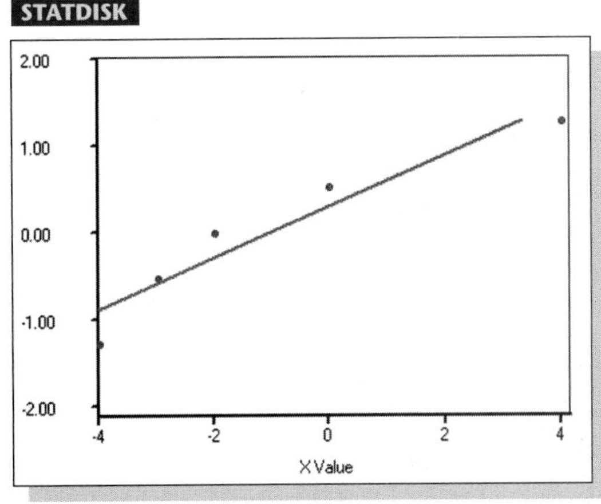

TI-83/84 Plus

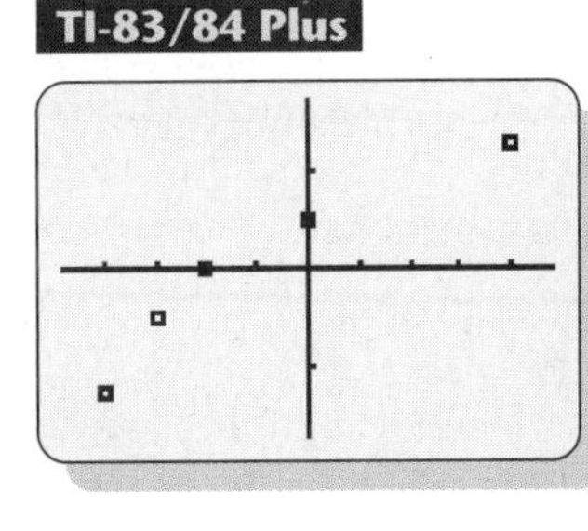

continued

Crest and Dependent Samples

In the late 1950s, Procter & Gamble introduced Crest toothpaste as the first such product with fluoride. To test the effectiveness of Crest in reducing cavities, researchers conducted experiments with several sets of twins. One of the twins in each set was given Crest with fluoride, while the other twin continued to use ordinary toothpaste without fluoride. It was believed that each pair of twins would have similar eating, brushing, and genetic characteristics. Results showed that the twins who used Crest had significantly fewer cavities than those who did not. This use of twins as dependent samples allowed the researchers to control many of the different variables affecting cavities.

Do Air Bags Save Lives?

The National Highway Transportation Safety Administration reported that for a recent year, 3,448 lives were saved because of air bags. It was reported that for car drivers involved in frontal crashes, the fatality rate was reduced 31%; for passengers, there was a 27% reduction. It was noted that "calculating lives saved is done with a mathematical analysis of the real-world fatality experience of vehicles with air bags compared with vehicles without air bags. These are called double-pair comparison studies, and are widely accepted methods of statistical analysis."

We will follow the same basic method of hypothesis testing that was introduced in Chapter 8, but we will use the test statistic for matched pairs that was given earlier in this section.

Step 1: The claim that there is a difference between the actual low temperatures and the five-day-predicted low temperatures can be expressed as $\mu_d \neq 0$.

Step 2: If the original claim is not true, we have $\mu_d = 0$.

Step 3: The null hypothesis must express equality and the alternative hypothesis cannot include equality, so we have

$$H_0: \mu_d = 0 \qquad H_1: \mu_d \neq 0 \qquad \text{(original claim)}$$

Step 4: The significance level is $\alpha = 0.05$.

Step 5: We use the Student t distribution.

Step 6: Before finding the value of the test statistic, we must first find the values of $\bar{d}$, and s_d. Refer to Table 9-2 and use the differences of -2, -3, -4, 4, 0 to find these sample statistics: $\bar{d} = -1.0$ and $s_d = 3.2$. Using these sample statistics and the assumption of the hypothesis test that $\mu_d = 0$, we can now find the value of the test statistic.

$$t = \frac{\bar{d} - \mu_d}{\frac{s_d}{\sqrt{n}}} = \frac{-1.0 - 0}{\frac{3.2}{\sqrt{5}}} = -0.699$$

The critical values of $t = \pm 2.776$ are found from Table A-3 as follows: Use the column for 0.05 (Area in Two Tails), and use the row with degrees of freedom of $n - 1 = 4$. Figure 9-4 shows the test statistic, critical values, and critical region.

Step 7: Because the test statistic does not fall in the critical region, we fail to reject the null hypothesis.

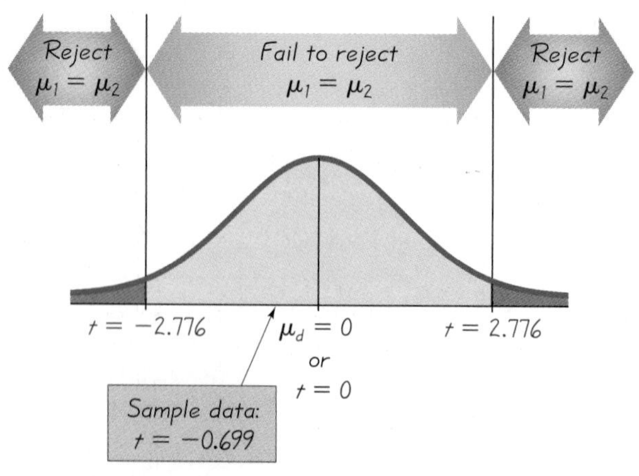

Figure 9-4 Distribution of Differences *d* Between Values in Matched Pairs

INTERPRETATION The sample data in Table 9-2 do not provide sufficient evidence to support the claim that actual and five-day-forecast low temperatures are different. This does *not* establish that the actual and forecast temperatures are equal. Perhaps additional sample data might provide the necessary evidence to conclude that the actual and forecast low temperatures are different. (See Exercise 21 where results for 35 days are used.)

P-Value Method The preceding example used the traditional method, but the *P*-value approach could be used. Using a TI-83/84 Plus calculator, we can find the *P*-value of 0.5185. (Using Table A-3 with the test statistic of $t = -0.699$, we can determine that the *P*-value is greater than 0.20.) We again fail to reject the null hypothesis, because the *P*-value is greater than the significance level of $\alpha = 0.05$.

EXAMPLE Confidence Interval with Actual and Forecast Temperatures Using the same sample matched pairs in Table 9-2, construct a 95% confidence interval estimate of μ_d, which is the mean of the differences between actual low temperatures and five-day-forecast low temperatures. Interpret the result.

SOLUTION

REQUIREMENT ✔ The solution for the preceding example includes verification that the requirements are satisfied. We can proceed with the construction of the confidence interval. ✔

We use the values of $\bar{d} = -1.0$, $s_d = 3.2$, $n = 5$, and $t_{\alpha/2} = 2.776$ (found from Table A-3 with $n - 1 = 4$ degrees of freedom and an area of 0.05 in two tails). We first find the value of the margin of error E.

$$E = t_{\alpha/2} \frac{s_d}{\sqrt{n}} = 2.776 \cdot \frac{3.2}{\sqrt{5}} = 4.0$$

The confidence interval can now be found.

$$\bar{d} - E < \mu_d < \bar{d} + E$$
$$-1.0 - 4.0 < \mu_d < -1.0 + 4.0$$
$$-5.0 < \mu_d < 3.0$$

INTERPRETATION The result is sometimes expressed as -1.0 ± 4.0 or as $(-5.0, 3.0)$. (If we use a TI-83/84 Plus calculator, we get a result of $-4.9 < \mu_d < 2.9$. The discrepancy is due to use of the rounded standard deviation of 3.2 instead of the unrounded value of 3.16227766.) In the long run, 95% of such samples will lead to confidence interval limits that actually do contain the true population mean of the differences. Note that the confidence interval limits do contain 0, indicating that the true value of μ_d is not significantly different from 0. We cannot conclude that there is a significant difference between the actual and forecast low temperatures.

Experimental Design Suppose that we want to conduct an experiment with two different fertilizers (new and old) and we have 20 plots of land having the same area. Instead of using 10 plots with the new fertilizer and 10 plots with the old fertilizer, it would be better to divide each of the 20 plots in half and treat each half with the new fertilizer while the other half is treated with the old fertilizer. The yields can then be matched by the plots that they share. The big advantage of using matched pairs is that extraneous variation is reduced. There is less of a chance of being misled by yields that are higher because of plots with richer soil. If one particular plot happens to have extremely rich soil, that plot would be shared by both fertilizers instead of having one fertilizer that appears to be very effective, when it is the soil responsible for the high yield, not the fertilizer. When designing an experiment or planning an observational study, using matched pairs is generally better than using two independent samples.

TI

Using a TI-83/84 Plus Calculator for Inferences from Matched Pairs

Caution: For inferences from matched pairs, do *not* use the menu items of **2-SampTTest** or **2-SampTInt,** because they are designed for *independent samples.* Instead, use the following procedure, which begins by creating a list of the differences between the values in each pair of sample data.

1. Clear lists L1, L2, and L3 by pressing STAT, selecting the menu item of **ClrList,** and entering L1, L2, L3 (with the commas included).
2. Enter the data for the first variable in list L1 and also enter the data for the second variable in list L2.
3. Create a list of the differences and store the list in L3 by clearing the screen and entering **L1 − L2 → L3.** This expression can be entered with the following sequence of keystrokes: 2ND 1 − 2ND 2 STO› 2ND 3 ENTER.
4. Press STAT, then press ▷ twice to select **TESTS.**
5. Select one of the following:

   T-Test for a hypothesis test

   TInterval for a confidence interval
6. For the Inpt option, select **Data.**
7. Proceed to enter the required items.
   - For the List entry, enter **L3** (by pressing 2ND 3).
   - If conducting a hypothesis test, enter 0 for the value of μ_0, which corresponds to a null hypothesis of $\mu_d = 0$. That is, the test is conducted by assuming that the mean difference is equal to 0.
   - For the Freq (frequency) value, enter 1.
   - For a confidence interval, enter the confidence level, such as 0.95. For a hypothesis test, select the type of test (two-tailed, left-tailed, right-tailed).
8. Use the arrow keys to scroll down to the bottom line, select **Calculate,** then press ENTER.

Hypothesis Test

```
T-Test
 μ≠0
 t=-.7071067812
 p=.5185185185
 x̄=-1
 Sx=3.16227766
 n=5
```

Confidence Interval

```
TInterval
 (-4.926,2.9265)
 x̄=-1
 Sx=3.16227766
 n=5
```

The accompanying screen displays correspond to the hypothesis test example and the confidence interval example in this section. Note the *P*-value of 0.5185 for the hypothesis test. The test statistic of $t = -0.707$ differs slightly from the test statistic of $t = -0.699$ because of rounding used in the example. Also note that the displayed confidence interval limits of (−4.926, 2.9265) can be rounded and expressed in this format: $-4.9 < \mu_d < 2.9$.

9-4 BASIC SKILLS AND CONCEPTS

Statistical Literacy and Critical Thinking

1. Bear Heights and Weights Sample data consist of heights and weights of 12 randomly selected bears. When analyzing the relationship between height and weight, do the methods of this section apply? Why or why not?

2. Matching with Coin Tosses A researcher is investigating measures of intelligence and obtains IQ scores of 24 subjects configured as 12 matched pairs, with the matching based on coin tosses. Can the methods of this section be used?

3. Notation What do the symbols $\bar{d}$ and s_d denote?

4. Boxplots The examples in this section are based on matched actual/forecast temperature data. For those examples, would it make sense to construct a boxplot of the actual temperatures and another boxplot of the forecast temperatures?

Calculations for Matched Pairs. *In Exercises 5 and 6, assume that you want to use a 0.05 significance level to test the claim that the paired sample data come from a population for which the mean difference is $\mu_d = 0$. Find (a) $\bar{d}$, (b) s_d, (c) the t test statistic, and (d) the critical values.*

TI 5.

x	2	5	7	1	1
y	2	3	9	4	0

TI 6.

x	8	8	5	2	7	3
y	9	6	5	5	3	8

TI 7. Confidence Interval Using the sample paired data in Exercise 5, construct a 95% confidence interval for the population mean of all differences $x - y$.

TI 8. Confidence Interval Using the sample paired data in Exercise 6, construct a 99% confidence interval for the population mean of all differences $x - y$.

9. Interpreting Display of Test for Treating Motion Sickness The following Minitab display resulted from an experiment in which 10 subjects were tested for motion sickness before and after taking the drug astemizole. The Minitab results are based on differences in the number of head movements that the subjects could endure without becoming nauseous. (The differences were obtained by subtracting the "after" values from the "before" values.)
   a. Use a 0.05 significance level to test the claim that astemizole has an effect (for better or worse) on vulnerability to motion sickness. Based on the result, would you use astemizole if you were concerned about motion sickness while on a cruise ship?
   b. Instead of testing for some effect (for better or worse), suppose we want to test the claim that astemizole is effective in *preventing* motion sickness? What is the P-value, and what do you conclude?

```
95% CI for mean difference: (-48.8, 33.8)
T-Test of mean difference = 0 (vs not = 0):
T-value = -0.41 P-value = 0.691
```

10. Interpreting Display for Test with Before–After Treatment Results Captopril is a drug designed to lower systolic blood pressure. When subjects were tested with this drug, their systolic blood pressure readings (in mmHg) were measured before and after the drug was taken. Excel was used to conduct a t test using the matched pairs of before–after readings, and the results are shown below (based on data from "Essential Hypertension: Effect of an Oral Inhibitor of Angiotensin-Converting Enzyme," by MacGregor et al., *British Medical Journal,* Vol. 2). Use the Excel results to test the claim that captopril is effective in lowering systolic blood pressure. Is it effective?

Excel

t-Test: Paired Two Sample for Means		
	Variable 1	*Variable 2*
Mean	185.3333333	166.75
Variance	291.3333333	220.9318182
Observations	12	12
Pearson Correlation	0.808392337	
Hypothesized Mean Difference	0	
df	11	
t Stat	6.371428571	
P(T<=t) one-tail	2.64274E-05	
t Critical one-tail	1.795884814	
P(T<=t) two-tail	5.28547E-05	
t Critical two-tail	2.200985159	

TI **11.** Forecast Temperatures Examples in this section included actual low temperatures and low temperatures that were forecast five days earlier. Listed below are actual high temperatures and the high temperatures that were forecast one day earlier (based on data recorded near the author's home).

a. Use a 0.05 significance level to test the claim of a zero mean difference between the actual high temperatures and the high temperatures that were forecast one day earlier. What do the results suggest about the accuracy of the forecast temperatures?

b. Construct a 95% confidence interval estimate of the mean difference between actual high temperatures and high temperatures forecast one day earlier. Interpret the resulting confidence interval, and comment on the implications of whether the confidence interval limits contain 0.

Actual high	80	77	81	85	73
High forecast one day earlier	78	75	81	85	76

TI **12.** Is Friday the 13th Unlucky? Researchers collected data on the numbers of hospital admissions resulting from motor vehicle crashes, and results are given below for Fridays on the 6th of a month and Fridays on the following 13th of the same month (based on data from "Is Friday the 13th Bad for Your Health?" by Scanlon et al., *British Medical Journal,* Vol. 307, as listed in the *Data and Story Line* online resource of data sets). Use a 0.05 significance level to test the claim that when the 13th day of a month falls on a Friday, the numbers of hospital admissions from motor vehicle crashes are not affected.

Friday the 6th:	9	6	11	11	3	5
Friday the 13th:	13	12	14	10	4	12

TI **13.** Measuring Blood Pressure Fourteen different medical students measured the blood pressure of the same patient, then repeated the measurement the following day. Listed below are the systolic readings in mmHg (based on data from Bellevue Hospital in New York City). Construct a 95% confidence interval estimate of the mean difference between systolic readings taken one day and those taken on the following day. What does the result suggest?

Day 1:	138	130	135	140	120	125	120	130	130	144	143	140	130	150
Day 2:	116	120	125	110	120	135	124	118	120	130	140	140	130	138

TI **14.** Heights of Winners and Runners-Up Listed below are the heights of candidates who won presidential elections and the heights of the candidates with the next highest number of popular votes. The data are in chronological order, so the corresponding heights from the two lists are matched. For candidates who won more than once, only the heights from the first election are included, and no elections before 1900 are included.

a. A well-known theory is that winning candidates tend to be taller than the corresponding losing candidates. Use a 0.05 significance level to test that theory. Does height appear to be an important factor in winning the presidency?

b. If you plan to test the claim in part (a) by using a confidence interval, what confidence level should be used? Construct a confidence interval using that confidence level, then interpret the result.

Won Presidency	**Runner-Up**
71 74.5 74 73 69.5 71.5 75 72	73 74 68 69.5 72 71 72 71.5
70.5 69 74 70 71 72 70 67	70 68 71 72 70 72 72 72

TI **15.** SAT Training Course The article "An SAT Coaching Program That Works," by Kaplan (*Chance*, Vol. 15, No. 1) included a graph depicting SAT scores for 50 subjects in a control group. Nine of the 50 points were randomly selected, with each point representing the score on the SAT test taken the first time and the score on the SAT test taken a second time, with no preparatory course taken between the two tests. The graph was used to identify the scores listed below. Test the claim that the differences have a mean of 0. What do the results suggest?

Student	A	B	C	D	E	F	G	H	I
First SAT score	480	510	530	540	550	560	600	620	660
Second SAT score	460	500	530	520	580	580	560	640	690

TI **16.** Self-Reported and Measured Male Heights As part of the National Health and Nutrition Examination Survey conducted by the Department of Health and Human Services, self-reported heights and measured heights were obtained for males aged 12–16. Listed below are sample results.

a. Is there sufficient evidence to support the claim that there is a difference between self-reported heights and measured heights of males aged 12–16? Use a 0.05 significance level.

b. Construct a 95% confidence interval estimate of the mean difference between reported heights and measured heights. Interpret the resulting confidence interval, and comment on the implications of whether the confidence interval limits contain 0.

Reported height	68	71	63	70	71	60	65	64	54	63	66	72
Measured height	67.9	69.9	64.9	68.3	70.3	60.6	64.5	67.0	55.6	74.2	65.0	70.8

TI **17.** Effectiveness of Hypnotism in Reducing Pain A study was conducted to investigate the effectiveness of hypnotism in reducing pain. Results for randomly selected subjects are given in the accompanying table (based on "An Analysis of Factors That Contribute to the Efficacy of Hypnotic Analgesia," by Price and Barber, *Journal of Abnormal Psychology,* Vol. 96, No. 1). The values are before and after hypnosis; the measurements are in centimeters on a pain scale.

a. Construct a 95% confidence interval for the mean of the "before–after" differences.

b. Use a 0.05 significance level to test the claim that the sensory measurements are lower after hypnotism.

c. Does hypnotism appear to be effective in reducing pain?

Subject	A	B	C	D	E	F	G	H
Before	6.6	6.5	9.0	10.3	11.3	8.1	6.3	11.6
After	6.8	2.4	7.4	8.5	8.1	6.1	3.4	2.0

TI **18.** Measuring Intelligence in Children Mental measurements of young children are often made by giving them blocks and telling them to build a tower as tall as possible. One experiment of block building was repeated a month later, with the times (in seconds) listed in the accompanying table (based on data from "Tower Building," by Johnson and Courtney, *Child Development,* Vol. 3).

a. Is there sufficient evidence to support the claim that there is a difference between the two times? Use a 0.01 significance level.

b. Construct a 99% confidence interval for the mean of the differences. Do the confidence interval limits contain 0, indicating that there is not a significant difference between the times of the first and second trials?

Child	A	B	C	D	E	F	G	H	I	J	K	L	M	N	O
First trial	30	19	19	23	29	178	42	20	12	39	14	81	17	31	52
Second trial	30	6	14	8	14	52	14	22	17	8	11	30	14	17	15

TI **19.** Testing Corn Seeds In 1908, William Gosset published the article "The Probable Error of a Mean" under the pseudonym of "Student" (*Biometrika,* Vol. 6, No. 1). He included the data listed below for two different types of corn seed (regular and kiln dried) that were used on adjacent plots of land. The listed values are the yields of head corn in pounds per acre.

a. Using a 0.05 significance level, test the claim that there is no difference between the yields from the two types of seed.

b. Construct a 95% confidence interval estimate of the mean difference between the yields from the two types of seed.

c. Does it appear that either type of seed is better?

Regular	1903	1935	1910	2496	2108	1961	2060	1444	1612	1316	1511
Kiln dried	2009	1915	2011	2463	2180	1925	2122	1482	1542	1443	1535

TI **20.** Confidence Interval for Comparing Keyboards The traditional keyboard configuration is referred to as a QWERTY configuration because of the positions of the letters QWERTY on the top row of keys. Developed in 1936, the Dvorak keyboard supposedly provides a more efficient arrangement by positioning the most used keys on the middle row where they are more accessible. A *Discover* magazine article suggested that you can measure the ease of typing by using this point-rating system: Count each letter on the home row as 0, count each letter on the top row as 1, and count each letter on the bottom row as 2. Using this rating system with each of the 52 words of the Preamble to the Constitution, we get the values listed in order below. Construct a 95% confidence interval estimate of the difference between the means of the ratings for words using both systems. Does the result appear to support the claim that the Dvorak keyboard configuration is easier?

QWERTY Keyboard	Dvorak Keyboard
2 2 5 1 2 6 3 3 4 2 4 0 5 7 7 5 6 6 8 10	2 0 3 1 0 0 0 0 2 0 4 0 3 4 0 3 3 1 3 5
7 2 2 10 5 8 2 4 4 2 6 2 6 1 7 2 7 2 3 8	2 0 4 1 5 0 4 0 1 3 0 1 0 3 0 1 2 0 0 0
1 5 2 5 2 14 2 2 6 3 1 7	0 1 0 3 0 1 2 0 0 0 1 4

TI **21.** Appendix B Data Set: Forecast Temperatures The examples in this section used only five pairs of sample data so that the calculations would be easy. Refer to Data Set 8 in Appendix B and use all of the actual low temperatures and the low temperatures that were forecast five days earlier.

a. Using a 0.05 significance level, test the claim that there is no difference between the actual low temperatures and the low temperatures that were forecast five days earlier.

b. Construct a 95% confidence interval estimate of the mean difference between the actual low temperatures and the low temperatures that were forecast five days earlier.

c. Compare the results to those obtained in the examples of this section. Does it appear that the forecast low temperatures are accurate?

TI **22.** Appendix B Data Set: Tobacco and Alcohol in Children's Movies Refer to Data Set 5 in Appendix B. Use the paired data consisting of times that the movies showed tobacco use and the times that they showed alcohol use.

a. Is there sufficient evidence to conclude that the times are different?

b. Construct a 99% confidence interval estimate of the mean of the differences between the times of tobacco use and alcohol use. Based on the result, is there a significant difference in the times that children are exposed to tobacco use and the times that they are exposed to alcohol use?

TI **23.** Appendix B Data Set: Home Prices Refer to Data Set 18 in Appendix B and use the selling prices and list prices of homes sold.

a. Using a 0.05 significance level, test the claim of a realtor that the housing market is so hot that there is no difference between selling prices and list prices.

b. Construct a 95% confidence interval estimate of the mean difference between the selling prices and list prices.

TI **24.** Appendix B Data Set: Old Faithful Refer to Data Set 11 in Appendix B and use the time intervals before and after each eruption.

a. Using a 0.05 significance level, test the claim that the before–after differences have a mean equal to 0. What does the result suggest about the timing of eruptions?

b. Construct a 95% confidence interval estimate of the mean difference between the times before eruptions and the times after eruptions.

9-4 BEYOND THE BASICS

TI **25.** SAT Training Course The article "An SAT Coaching Program That Works" (by Kaplan, *Chance,* Vol. 15, No. 1) included a graph similar to the one shown below. The points represent the SAT scores of 9 subjects before and after taking an SAT preparation course. For each subject, identify the before and after SAT scores, then test the claim that the preparatory course is effective in raising SAT scores.

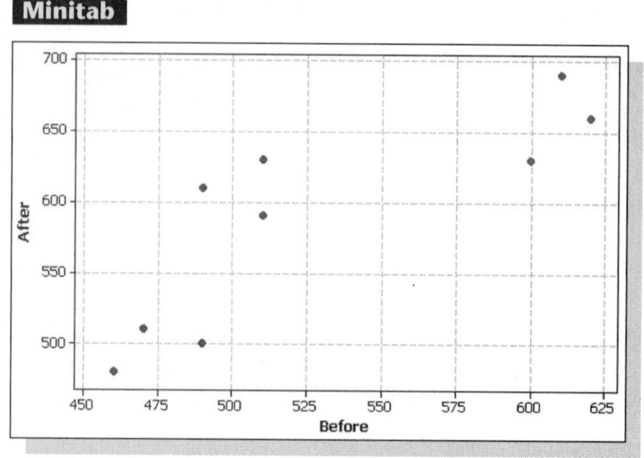

26. Effects of an Outlier and Units of Measurement

a. When using the methods of this section, can an outlier have a dramatic effect on the hypothesis test and confidence interval?

b. The examples in this section used temperatures measured in degrees Fahrenheit. If we convert all sample temperatures from Fahrenheit degrees to Celsius degrees, is the hypothesis test affected by such a change in units? Is the confidence interval affected by such a change in units? How?

TI **27.** Using the Correct Procedure

a. Consider the sample data given below to be matched pairs and use a 0.05 significance level to test the claim that $\mu_d > 0$.

b. Consider the sample data given below to be two independent samples. Use a 0.05 significance level to test the claim that $\mu_1 > \mu_2$.

c. Compare the results from parts (a) and (b). Is it critical that the correct method be used? Why or why not?

x	1	3	2	2	1	2	3	3	2	1
y	1	2	1	2	1	2	1	2	1	2

9-5 Comparing Variation in Two Samples

Key Concept Because the characteristic of variation among data is extremely important, this section presents the F test for using two samples to compare two population variances (or standard deviations). In this section we introduce the F distribution that is used for the F test. It is extremely important to be aware of a serious shortcoming of this procedure: The F test for comparing two population variances (or standard deviations) is *very* sensitive to departures from normal distributions. At the end of this section we briefly discuss alternatives to the F test that are not so sensitive to departures from normality.

F Test for Comparing Variances

In this section we use sample variances (or standard deviations) to compare two population variances (or standard deviations). We should know that the variance is the square of the standard deviation, and we should know the following notation.

Measures of Variation

s = standard deviation of *sample* s^2 = variance of *sample* (sample standard deviation squared)

σ = standard deviation of *population* σ^2 = variance of *population* (population standard deviation squared)

The computations of this section will be greatly simplified if we designate the two samples so that s_1^2 represents the *larger* of the two sample variances. Mathematically, it doesn't really matter which sample is designated as Sample 1, so life will be better if we let s_1^2 represent the larger of the two sample variances, as in the test statistic included in the summary box below.

Requirements

1. The two populations are *independent* of each other. (Recall from Section 9-2 that two samples are independent if the sample selected from one population is not related to the sample selected from the other population. The samples are not matched or paired.)
2. The two populations are each *normally distributed.* (This assumption is important because the methods of this section are not robust, meaning that they are extremely sensitive to departures from normality.)

Notation for Hypothesis Tests with Two Variances or Standard Deviations

s_1^2 = larger of the two sample variances

n_1 = size of the sample with the *larger* variance

σ_1^2 = variance of the population from which the sample with the *larger* variance was drawn

The symbols s_2^2, n_2, and σ_2^2 are used for the other sample and population.

Test Statistic for Hypothesis Tests with Two Variances

$$F = \frac{s_1^2}{s_2^2} \quad \text{(where } s_1^2 \text{ is the } \textit{larger} \text{ of the two sample variances)}$$

Critical values: Use Table A-5 to find critical F values that are determined by the following:

1. The significance level α (Table A-5 has four pages of critical values for α = 0.025 and 0.05.)
2. **Numerator degrees of freedom = $n_1 - 1$**
3. **Denominator degrees of freedom = $n_2 - 1$**

Lower Variation, Higher Quality

Ford and Mazda were producing similar transmissions that were supposed to be made with the same specifications. But the American-made transmissions required more warranty repairs than the Japanese-made transmissions. When investigators inspected samples of the Japanese transmission gearboxes, they first thought that their measuring instruments were defective because they weren't detecting any variability among the Mazda transmission gearboxes. They realized that although the American transmissions were within the specifications, the Mazda transmissions were not only within the specifications, but consistently close to the desired value. By reducing variability among transmission gearboxes, Mazda reduced the costs of inspection, scrap, rework, and warranty repair.

TI-83/84 Plus Calculator: The TI-83/84 Plus calculator can be used for hypothesis tests. See the procedure at the end of this section.

For two normally distributed populations with equal variances (that is, $\sigma_1^2 = \sigma_2^2$), the sampling distribution of the test statistic $F = s_1^2/s_2^2$ is the ***F* distribution** shown in Figure 9-5 with critical values listed in Table A-5. If you continue to repeat an experiment of randomly selecting samples from two normally distributed populations with equal variances, the distribution of the ratio s_1^2/s_2^2 of the sample variances is the F distribution.

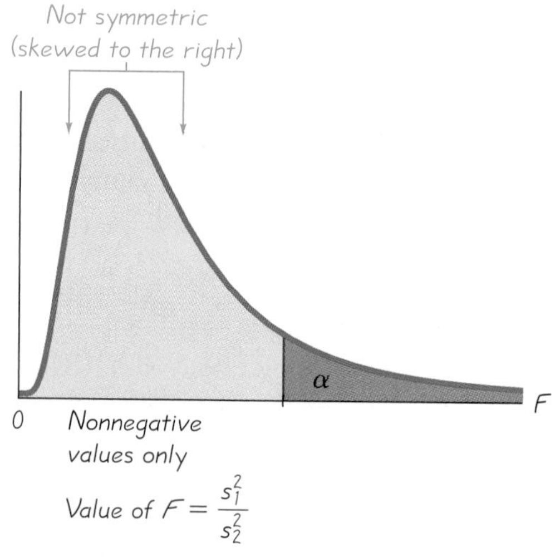

Figure 9-5 *F* Distribution

There is a different F distribution for each different pair of degrees of freedom for the numerator and denominator.

In Figure 9-5, note these properties of the F distribution:

- The F distribution is not symmetric.
- Values of the F distribution cannot be negative.
- The exact shape of the F distribution depends on two different degrees of freedom.

Critical Values To find a critical value, first refer to the part of Table A-5 corresponding to α (for a one-tailed test) or $\alpha/2$ (for a two-tailed test), then intersect the column representing the degrees of freedom for s_1^2 with the row representing the degrees of freedom for s_2^2. Because we are stipulating that the larger sample variance is s_1^2, all one-tailed tests will be right-tailed and all two-tailed tests will require that we find only the critical value located to the right. Good news: We have no need to find a critical value separating a left-tailed critical region. (Because the F distribution is not symmetric and has only nonnegative values, a left-tailed critical value cannot be found by using the negative of the right-tailed critical value; instead, a left-tailed critical value is found by using the reciprocal of the right-tailed value with the numbers of degrees of freedom reversed. See Exercise 23.)

We often have numbers of degrees of freedom that are not included in Table A-5. We could use linear interpolation to approximate the missing values, but in most cases that's not necessary because the F test statistic is either less than the lowest possible critical value or greater than the largest possible critical value. For example, Table A-5 shows that for $\alpha = 0.025$ in the right tail, 20 degrees of freedom for the numerator, and 34 degrees of freedom for the denominator, the critical F value is between 2.0677 and 2.1952. Any F test statistic below 2.0677 will result in failure to reject the null hypothesis, any F test statistic above 2.1952 will result in rejection of the null hypothesis, and interpolation is necessary only if the F test statistic happens to fall between 2.0677 and 2.1952. The use of an appropriate technology completely eliminates this problem by providing P-values or critical values for any numbers of degrees of freedom.

Interpreting the F Test Statistic If the two populations really do have equal variances, then the ratio s_1^2/s_2^2 tends to be close to 1 because s_1^2 and s_2^2 tend to be close in value. But if the two populations have radically different variances, s_1^2 and s_2^2 tend to be very different numbers. Denoting the larger of the sample variances by s_1^2, we see that the ratio s_1^2/s_2^2 will be a large number whenever s_1^2 and s_2^2 are far apart in value. Consequently, a value of F near 1 will be evidence in favor of the conclusion that $\sigma_1^2 = \sigma_2^2$, but a large value of F will be evidence against the conclusion of equality of the population variances.

***Large* values of F are evidence against $\sigma_1^2 = \sigma_2^2$.**

Claims About Standard Deviations The F test statistic applies to a claim made about two variances, but we can also use it for claims about two population standard deviations. Any claim about two population standard deviations can be restated in terms of the corresponding variances.

Exploring the Data

Because the requirement of normal distributions is so important and so strict, we should begin by comparing the two sets of sample data by using tools such as histograms, boxplots, and normal quantile plots (see Section 6-7), and we should search for outliers. We should find the values of the sample statistics, especially the standard deviations. For example, consider the 36 weights of regular Coke in 36 different cans. (The weights are listed in Data Set 12 in Appendix B.) Shown here are a histogram and a normal quantile plot from a TI-83/84 Plus calculator. The histogram shows that the data have a distribution that is approximately normal and that there is one value that is a potential outlier. The normal quantile plot shows that the points are reasonably close to a straight line, but they don't fit the straight line perfectly. This data set clearly satisfies a requirement of a distribution that is *approximately* normal, but it isn't absolutely certain that this data set satisfies the stricter requirements of normality that apply to the methods of this section.

Twins in Twinsburg

During the first weekend in August of each year, Twinsburg, Ohio celebrates its annual "Twins Days in Twinsburg" festival. Thousands of twins from around the world have attended this festival in the past. Scientists saw the festival as an opportunity to study identical twins. Because they have the same basic genetic structure, identical twins are ideal for studying the different effects of heredity and environment on a variety of traits, such as male baldness, heart disease, and deafness—traits that were recently studied at one Twinsburg festival. A study of twins showed that myopia (nearsightedness) is strongly affected by hereditary factors, not by environmental factors such as watching television, surfing the Internet, or playing computer or video games.

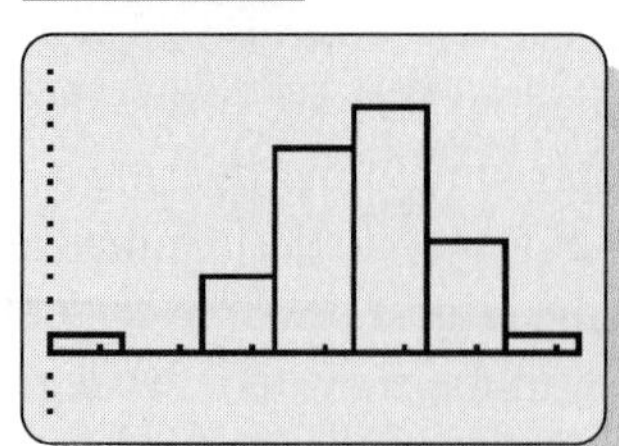

Normal Quantile Plot

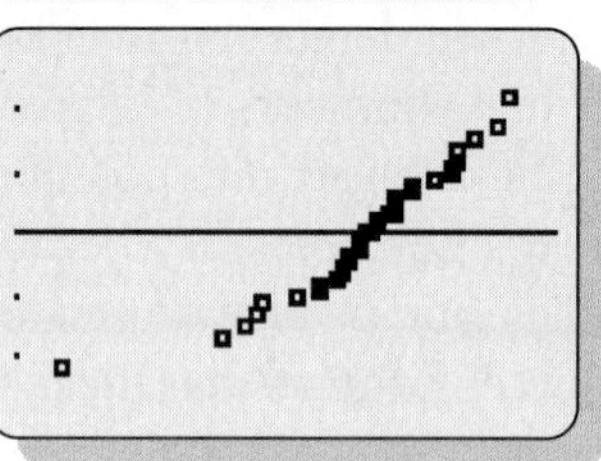

EXAMPLE Coke Versus Pepsi Data Set 12 in Appendix B includes the weights (in pounds) of samples of regular Coke and regular Pepsi. The sample statistics are summarized in the accompanying table. Use a 0.05 significance level to test the claim that the weights of regular Coke and the weights of regular Pepsi have the same standard deviation.

	Regular Coke	Regular Pepsi
n	36	36
$\bar{x}$	0.81682	0.82410
s	0.007507	0.005701

SOLUTION

REQUIREMENT ✔ First, the two populations are clearly independent of each other. The sample values are not matched or paired in any way. Second, the samples suggest that they come from populations with distributions that are approximately normal. See the paragraph preceding this example and see the normal quantile plot and histogram for the 36 weights of regular Coke. The 36 weights of regular Pepsi can be explored with a normal quantile plot and histogram, and the results would suggest that these weights are from a population having a normal distribution. The requirements are satisfied and we can proceed with the test. ✔

Instead of using the sample standard deviations to test the claim of equal population standard deviations, we use the sample variances to test the claim of equal population variances, but we can state conclusions in terms of standard deviations. Because we stipulate in this section that the larger variance is denoted by s_1^2, we let $s_1^2 = 0.007507^2$, $n_1 = 36$, $s_2^2 = 0.005701^2$, and $n_2 = 36$. We now proceed to use the traditional method of testing hypotheses as outlined in Figure 8-9.

Step 1: The claim of equal standard deviations is equivalent to a claim of equal variances, which we express symbolically as $\sigma_1^2 = \sigma_2^2$.

Step 2: If the original claim is false, then $\sigma_1^2 \neq \sigma_2^2$.

Step 3: Because the null hypothesis is the statement of equality and because the alternative hypothesis cannot contain equality, we have

$$H_0: \sigma_1^2 = \sigma_2^2 \quad \text{(original claim)} \quad H_1: \sigma_1^2 \neq \sigma_2^2$$

Step 4: The significance level is $\alpha = 0.05$.

Step 5: Because this test involves two population variances, we use the F distribution.

Step 6: The test statistic is

$$F = \frac{s_1^2}{s_2^2} = \frac{0.007507^2}{0.005701^2} = 1.7339$$

For the critical values, first note that this is a two-tailed test with 0.025 in each tail. As long as we are stipulating that the larger variance is placed in the numerator of the F test statistic, we need to find only the right-tailed critical value. From Table A-5 we see that the critical value of F is between 1.8752 and 2.0739, which we find by referring to 0.025 in the right tail, with 35 degrees of freedom for the numerator and 35 degrees of freedom for the denominator. (STATDISK and Excel provide a critical value of 1.9611.)

Step 7: Figure 9-6 shows that the test statistic $F = 1.7339$ does not fall within the critical region, so we fail to reject the null hypothesis of equal variances. It follows that there is not sufficient evidence to warrant rejection of the claim of equal standard deviations.

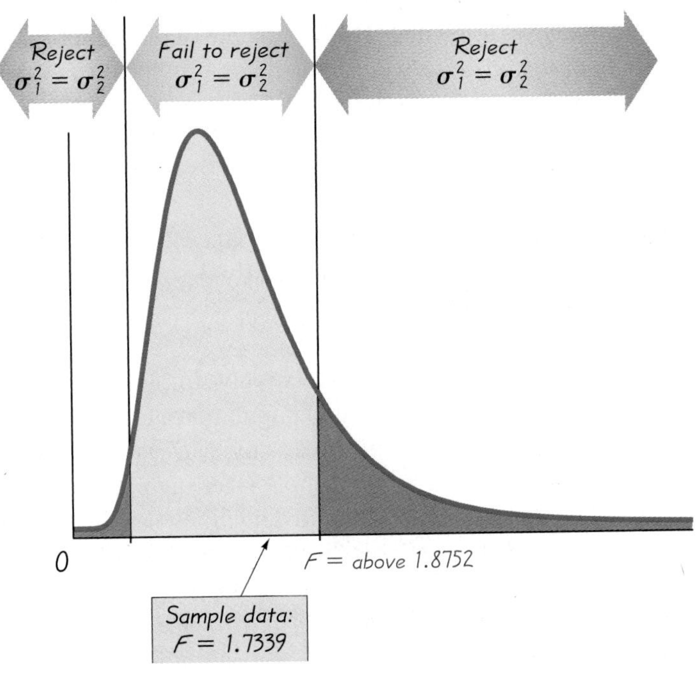

Figure 9-6 Distribution of s_1^2 / s_2^2 for Weights of Regular Coke and Regular Pepsi

INTERPRETATION There is not sufficient evidence to warrant rejection of the claim that the two standard deviations are equal. However, we should recognize that the F test is extremely sensitive to distributions that are not normally distributed, so this conclusion might make it appear that there is no significant difference between the population standard deviations when there really is a difference that was hidden by nonnormal distributions.

continued

The single weight of 0.7901 for Coke is a potential outlier, especially because its z score is -3.56. If we repeat the test without this questionable value, we again conclude that there is no significant difference between the two population standard deviations. Alternative methods discussed later in this section also lead to the conclusion that there is no significant difference between the two population standard deviations.

Now let's use some basic common sense. We know that the Coke and Pepsi cans come from two completely separate and independent manufacturing processes, so it is unlikely that the two population variances are exactly the same. However, based on our analysis, we can conclude that any difference between the two population standard deviations is not a significant difference.

In the preceding example we used a two-tailed test for the claim of equal variances. A right-tailed test would yield the same test statistic of $F = 1.7339$, but a different critical value of F.

P-Value Test and Confidence Intervals We have described the traditional method of using the F test for claims made about two population variances. The P-value method is easy to use with software capable of providing P-values. For the preceding example, the TI-83/84 Plus calculator provides a P-value of 0.1081. Exercise 24 deals with the construction of confidence intervals.

Alternative Methods

We have presented the F test for comparing variances, but that test is very sensitive to departures from normality. Here we briefly describe some alternatives that are more robust.

Count Five The *count five* method is a relatively simple alternative to the F test, and it does not require normally distributed populations. (See "A Quick, Compact, Two-Sample Dispersion Test: Count Five," by McGrath and Yeh, *American Statistician,* Vol. 59, No. 1.) If the two sample sizes are equal, and if one sample has at least five of the largest mean absolute deviations (MAD), then we conclude that its population has a larger variance. See Exercise 21 for the specific procedure.

Levene-Brown-Forsythe Test The *Levene-Brown-Forsythe test* (or modified Levene's test) is another alternative to the F test, and it is much more robust. This test begins with a transformation of each set of sample values. Within the first sample, replace each x value with $|x - \text{median}|$, and do the same for the second sample. Using the transformed values, conduct a t test of equality of means from independent samples, as described in Part 1 of Section 9-3. Because the transformed values are now deviations, the t test for equality of means is actually a test comparing variation in the two samples. See Exercise 22.

In addition to the count five test and the Levene-Brown-Forsythe test, there are other alternatives to the F test, as well as adjustments that improve the performance of the F test. See "Fixing the F Test for Equal Variances," by Shoemaker, *American Statistician,* Vol. 57, No. 2.

TI

Using a TI-83/84 Plus Calculator for Hypothesis Tests with Two Variances or Two Standard Deviations

The TI-83/84 Plus calculator is designed for hypothesis tests (but not confidence intervals) involving two variances or two standard deviations. In order to simplify the hypothesis-testing procedure, the method described in this section requires that the larger sample variance be designated as s_1^2, but this is not necessary if a TI-83/84 Plus calculator is used. In the following procedure, either sample can be designated as Sample 1. (However, the answers in Appendix F are based on the assumption that $s_1^2 > s_2^2$.)

1. Press **STAT**, then press ▷ twice to select **TESTS.**
2. Using the ▽ arrow key, scroll down to select the menu item of **2-SampFTest.**
3. For the Inpt entry, select **Data** if the two lists of sample data are already stored in the calculator, or select **Stats** if the summary statistics of s_1, n_1, s_2, n_2 are all known.
4. If **Data** was selected in Step 3, enter the list names or labels, such as L1 and L2. If **Stats** was selected, enter the values of s_1, n_1, s_2, and n_2.
5. Select the type of test (two-tailed, left-tailed, right-tailed) by selecting the format of the alternative hypothesis:

$$\sigma_1 \neq \sigma_2 \qquad \sigma_1 < \sigma_2 \qquad \sigma_1 > \sigma_2$$

6. Use the arrow keys to scroll down to the bottom line, select **Calculate,** then press **ENTER**.

If this procedure is used with the sample statistics given in the example from this section, the entry screen and the screen of results will be as shown in the accompanying displays. Note that the *P*-value of 0.1082 (rounded) is provided, so the *P*-value approach to hypothesis testing is being used. Because this *P*-value is greater than the significance level of 0.05, we fail to reject the null hypothesis and get the same conclusion found in the example.

Entry Screen

```
2-SampFTest
 Inpt:Data Stats
 Sx1:.007507
 n1:36
 Sx2:.005701
 n2:36
 σ1:≠σ2 <σ2 >σ2
 Calculate Draw
```

Results

```
2-SampFTest
 σ1≠σ2
 F=1.733926762
 p=.1081582463
 Sx1=.007507
 Sx2=.005701
↓n1=36
```

9-5 BASIC SKILLS AND CONCEPTS

Statistical Literacy and Critical Thinking

1. Weakness of *F* Test What is a major weakness of the *F* test described in this section? Name two alternatives that do a better job of overcoming that weakness.

2. Effects of Nonnormal Distributions What is a consequence of using the *F* test with samples from populations with distributions that are not normal? That is, in what way does the test perform poorly?

3. *F* Distribution In the context of this section, what is the *F* distribution?

4. *F* Distribution Identify two different properties of the *F* distribution.

Hypothesis Test of Equal Variances. In Exercises 5 and 6, test the given claim. Use a significance level of $\alpha = 0.05$ and assume that all populations are normally distributed.

TI **5.** Claim: The treatment population and the placebo population have different variances.

Treatment group: $n = 16, \bar{x} = 21.33, s = 0.80$
Placebo group: $n = 41, \bar{x} = 25.34, s = 0.40$

TI **6.** Claim: IQ scores of statistics students vary less than IQ scores of other students.

Statistics students: $n = 28, \bar{x} = 118, s = 10.0$
Other students: $n = 25, \bar{x} = 112, s = 12.0$

TI-83/84 Plus

```
2-SampFTest
 σ1≠σ2
 F=2.923274259
 p=.0020697599
 Sx1=.007507372
 Sx2=.004390896
↓x̄1=.8168222222
```

TI **7.** Interpreting Display from Weights of Regular Coke and Diet Coke This section included an example about a hypothesis test of the claim that weights of regular Coke and regular Pepsi have the same standard deviation. Use a 0.05 significance level to test the claim that regular Coke and diet Coke have weights with different standard deviations. Sample weights are found in Data Set 12 in Appendix B, but see the accompanying display results of the F test from the TI-83/84 Plus calculator. If the results were to show that the standard deviations are significantly different, what would be an important factor that might explain the difference?

8. Interpreting Display for Test of Echinacea In a randomized, double-blind, placebo-controlled trial of children, echinacea was tested as a treatment for upper respiratory infections in children. "Days of fever" was one criterion used to measure effects. Among 337 children treated with echinacea, the mean number of days with fever was 0.81, with a standard deviation of 1.50 days. Among 370 children given a placebo, the mean number of days with fever was 0.64 with a standard deviation of 1.16 days (based on data from "Efficacy and Safety of Echinacea in Treating Upper Respiratory Tract Infections in Children," by Taylor et al., *Journal of the American Medical Association,* Vol. 290, No. 21). Use a 0.05 significance level to test the claim that children treated with echinacea have a larger standard deviation than those given a placebo. See the accompanying display results from Excel.

Excel

1	F-Test Two-Sample for Variances		
2			
3		*Variable 1*	*Variable 2*
4	Mean	0.043962632	-0.066075532
5	Variance	2.250000402	1.345600694
6	Observations	337	370
7	df	336	369
8	F	1.672115965	
9	P(F<=f) one-tail	7.47504E-07	
10	F Critical one-tail	1.191519202	

TI **9.** Weights of Diet Coke and Diet Pepsi This section included an example about a hypothesis test of the claim that weights of regular Coke and regular Pepsi have the same standard deviation. Use a 0.05 significance level to test the claim that Diet Coke and Diet Pepsi have weights with different standard deviations. Sample weights are found in Data Set 12 in Appendix B, but here are the summary statistics: Diet Coke: $n = 36, \bar{x} = 0.784794$ lb, $s = 0.004391$ lb; Diet Pepsi: $n = 36, \bar{x} = 0.783858$ lb, $s = 0.004362$ lb.

TI **10.** Bipolar Depression Treatment In clinical experiments involving different groups of independent samples, it is important that the groups be similar in the important ways that affect the experiment. In an experiment designed to test the effectiveness of paroxetine for treating bipolar depression, subjects were measured using the Hamilton depression scale with the results given below (based on data from "Double-Blind, Placebo-Controlled Comparison of Imipramine and Paroxetine in the Treatment of Bipolar Depression," by Nemeroff et al., *American Journal of Psychiatry,* Vol. 158, No. 6). Using a 0.05 significance level, test the claim that both populations have the same standard deviation. Based on the results, does it appear that the two populations have different standard deviations?

Placebo group: $n = 43, \bar{x} = 21.57, s = 3.87$
Paroxetine treatment group: $n = 33, \bar{x} = 20.38, s = 3.91$

TI **11.** Hypothesis Test for Magnet Treatment of Pain Researchers conducted a study to determine whether magnets are effective in treating back pain, with results given below (based on data from "Bipolar Permanent Magnets for the Treatment of Chronic Lower Back Pain: A Pilot Study," by Collacott, Zimmerman, White, and Rindone, *Journal of the American Medical Association,* Vol. 283, No. 10). The values represent measurements of pain using the visual analog scale. Use a 0.05 significance level to test the claim that those given a sham treatment (similar to a placebo) have pain reductions that vary more than the pain reductions for those treated with magnets.

Reduction in pain level after sham treatment: $n = 20, \bar{x} = 0.44, s = 1.4$
Reduction in pain level after magnet treatment: $n = 20, \bar{x} = 0.49, s = 0.96$

TI **12.** Hypothesis Test for Effect of Marijuana Use on College Students In a study of the effects of marijuana use, light and heavy users of marijuana in college were tested for memory recall, with the results given below (based on data from "The Residual Cognitive Effects of Heavy Marijuana Use in College Students," by Pope and Yurgelun-Todd, *Journal of the American Medical Association,* Vol. 275, No. 7). Use a 0.05 significance level to test the claim that the population of heavy marijuana users has a standard deviation different from that of light users.

Items sorted correctly by light marijuana users: $n = 64, \bar{x} = 53.3, s = 3.6$
Items sorted correctly by heavy marijuana users: $n = 65, \bar{x} = 51.3, s = 4.5$

TI **13.** Effects of Alcohol An experiment was conducted to test the effects of alcohol. The errors were recorded in a test of visual and motor skills for a treatment group of people who drank ethanol and another group given a placebo. The results are shown in the accompanying table (based on data from "Effects of Alcohol Intoxication on Risk Taking, Strategy, and Error Rate in Visuomotor Performance," by Streufert et al., *Journal of Applied Psychology,* Vol. 77, No. 4). Use a 0.05 significance level to test the claim that the treatment group has scores that vary more than the scores of the placebo group.

Treatment Group	Placebo Group
$n_1 = 22$	$n_2 = 22$
$\bar{x}_1 = 4.20$	$\bar{x}_2 = 1.71$
$s_1 = 2.20$	$s_2 = 0.72$

TI **14.** Ages of Faculty and Student Cars Students at the author's college randomly selected 217 student cars and found that they had ages with a mean of 7.89 years and a standard deviation of 3.67 years. They also randomly selected 152 faculty cars and found that they had ages with a mean of 5.99 years and a standard deviation of 3.65 years. Is there sufficient evidence to support the claim that the ages of faculty cars vary less than the ages of student cars?

TI **15.** Weights of Quarters Weights of quarters are used by vending machines as one way to detect counterfeit coins. Data Set 14 in Appendix B includes weights of pre-1964 silver quarters and post-1964 quarters. Here are the summary statistics: pre-1964: $n = 40, \bar{x} = 6.19267$ g, $s = 0.08700$ g; post-1964: $n = 40, \bar{x} = 5.63930$ g, $s = 0.06194$.

continued

Use a 0.05 significance level to test the claim that the two populations of quarters have the same standard deviation. If the amounts of variation are different, vending machines might need more complicated adjustments. Does it appear that such adjustments are necessary?

TI **16.** Weights of Pennies and Quarters Data Set 14 in Appendix B includes weights of post-1983 pennies and post-1964 quarters. Here are the summary statistics: post-1983 pennies: $n = 37$, $\bar{x} = 2.49910$ g, $s = 0.01648$ g; post-1964 quarters: $n = 40$, $\bar{x} = 5.63930$ g, $s = 0.06194$. Test the claim that post-1983 pennies and post-1964 quarters have the same amount of variation. Should they have the same amount of variation?

TI **17.** Blanking out on Tests Many students have had the unpleasant experience of panicking on a test because the first question was exceptionally difficult. The arrangement of test items was studied for its effect on anxiety. Sample values consisting of measures of "debilitating test anxiety" (which most of us call panic or blanking out) are obtained for a group of subjects with test questions arranged from easy to difficult, and another group with test questions arranged from difficult to easy. Here are the summary statistics: Easy-to-difficult group: $n = 25$, $\bar{x} = 27.115$, $s^2 = 47.020$; difficult-to-easy group: $n = 16$, $\bar{x} = 31.728$, $s^2 = 18.150$ (based on data from "Item Arrangement, Cognitive Entry Characteristics, Sex and Test Anxiety as Predictors of Achievement in Examination Performance," by Klimko, *Journal of Experimental Education,* Vol. 52, No. 4). Test the claim that the two samples come from populations with the same variance.

TI **18.** Effect of Birth Weight on IQ Score When investigating a relationship between birth weight and IQ, researchers found that 258 subjects with extremely low birth weights (less than 1000 g) had Wechsler IQ scores at age 8 with a mean of 95.5 and a variance of 256.0. For 220 subjects with normal birth weights, the mean at age 8 is 104.9 and the variance is 198.8 (based on data from "Neurobehavioral Outcomes of School-age Children Born Extremely Low Birth Weight or Very Preterm in the 1990s," by Anderson et al., *Journal of the American Medical Association,* Vol. 289, No. 24). Using a 0.05 significance level, test the claim that babies with extremely low birth weights and babies with normal birth weights have different amounts of variation. (*Hint:* The conclusion is not clear from Table A-5, so use this upper critical value: $F = 1.2928$.)

TI **19.** Appendix B Data Set: Rainfall on Weekends *USA Today* and other newspapers reported on a study that supposedly showed that it rains more on weekends. The study referred to areas on the East Coast near the ocean. Data Set 10 in Appendix B lists the rainfall amounts in Boston for one year.

a. Assuming that we want to use the methods of this section to test the claim that Wednesday and Sunday rainfall amounts have the same standard deviation, identify the F test statistic, critical value, and conclusion. Use a 0.05 significance level.

b. Consider the prerequisite of normally distributed populations. Instead of constructing histograms or normal quantile plots, simply examine the numbers of days with no rainfall. Are Wednesday rainfall amounts normally distributed? Are Sunday rainfall amounts normally distributed?

c. What can be concluded from the results of parts (a) and (b)?

TI **20.** Appendix B Data Set: Tobacco and Alcohol Use in Animated Children's Movies Data Set 5 in Appendix B lists times (in seconds) that animated children's movies show tobacco use and alcohol use.

a. Assuming that we want to use the methods of this section to test the claim that the times of tobacco use and the times of alcohol use have different standard deviations, identify the F test statistic, critical value, and conclusion. Use a 0.05 significance level.

b. Do the data appear to satisfy the requirement of independent populations? Explain.
c. Do the data appear to satisfy the requirement of normally distributed populations? Instead of constructing histograms or normal quantile plots, simply examine the numbers of movies showing no tobacco or alcohol use. Are the times for tobacco use normally distributed? Are the times for alcohol use normally distributed?
d. What can be concluded from the preceding results?

9-5 BEYOND THE BASICS

21. Count Five Test for Comparing Variation in Two Populations See the example in this section and, instead of using the F test, use the following procedure for a "count five" test of equal variation. What do you conclude?
a. For the first sample, find the mean absolute deviation (MAD) of each value. (Recall from Section 3-3 that the MAD of a sample value x is $|x - \bar{x}|$.) Sort the MAD values. Do the same for the second sample.
b. Let c_1 be the count of the number of MAD values in the first sample that are greater than the largest MAD value in the other sample. Also, let c_2 be the count of the number of MAD values in the second sample that are greater than the largest MAD in the other sample. (One of these counts will always be zero.)
c. If the sample sizes are equal ($n_1 = n_2$), use a critical value of 5. If $n_1 \neq n_2$, calculate the critical value shown below.

$$\frac{\log(\alpha/2)}{\log\left(\frac{n_1}{n_1 + n_2}\right)}$$

d. If $c_1 \geq$ critical value, then conclude that $\sigma_1^2 > \sigma_2^2$. If $c_2 \geq$ critical value, then conclude that $\sigma_2^2 > \sigma_1^2$. Otherwise, fail to reject the null hypothesis of $\sigma_1^2 = \sigma_2^2$.

22. Levene-Brown-Forsythe Test for Comparing Variation in Two Populations See the example in this section and, instead of using the F test, use the Levene-Brown-Forsythe test described near the end of this section. What do you conclude?

23. Finding Lower Critical F Values In this section, for hypothesis tests that were two-tailed, we need to find only the upper critical value. Let's denote that value by F_R, where the subscript indicates the critical value for the right tail. The lower critical value F_L (for the left tail) can be found as follows: First interchange the degrees of freedom, and then take the reciprocal of the resulting F value found in Table A-5. Find the critical values F_L and F_R for two-tailed hypothesis tests based on the following values.
a. $n_1 = 10, n_2 = 10, \alpha = 0.05$
b. $n_1 = 10, n_2 = 7, \alpha = 0.05$
c. $n_1 = 7, n_2 = 10, \alpha = 0.05$

24. Constructing Confidence Intervals In addition to testing claims involving σ_1^2 and σ_2^2, we can also construct confidence interval estimates of the ratio σ_1^2/σ_2^2 using the following expression:

$$\left(\frac{s_1^2}{s_2^2} \cdot \frac{1}{F_R}\right) < \frac{\sigma_1^2}{\sigma_2^2} < \left(\frac{s_1^2}{s_2^2} \cdot \frac{1}{F_L}\right)$$

Here F_L and F_R are as described in Exercise 23. Refer to the sample data given in Exercise 5 and construct a 95% confidence interval estimate for the ratio of the treatment group variance to the placebo group variance.

Review

We use methods of inferential statistics when we use sample data to form conclusions about populations. Two major activities of inferential statistics are (1) constructing confidence interval estimates of population parameters (such as p, μ, σ), and (2) testing hypotheses or claims made about population parameters. In Chapters 7 and 8 we discussed the estimation of population parameters and methods of testing hypotheses made about population parameters, but Chapters 7 and 8 considered only cases involving a single population. In this chapter we considered two samples drawn from two populations.

- Section 9-2 considered inferences made about two population proportions. Given conditions in which the listed requirements are satisfied, we use the normal distribution for constructing confidence interval estimates of the difference $p_1 - p_2$ and for testing claims, such as the claim that $p_1 = p_2$.
- Section 9-3 considered inferences made about the means of two independent populations. Section 9-3 included three different methods, but one method is rarely used because it requires that the two population standard deviations be known. Another method involves pooling the two sample standard deviations to develop an estimate of the standard error, but this method is based on the assumption that the two population standard deviations are known to be equal, and that assumption is often risky. See Figure 9-3 for help in determining which method to apply. The procedure generally recommended is the t test that does not assume equal population variances.
- Section 9-4 considered inferences made about the mean difference for a population consisting of matched pairs.
- Section 9-5 presented the F test for testing claims about the equality of two population standard deviations or variances. It is important to know that the F test is not robust, meaning that it performs poorly with populations not having normal distributions. Alternatives to the F test were briefly described.

Statistical Literacy and Critical Thinking

1. Which Method? A candidate for political office is concerned about reports of a "gender gap" claiming that he is preferred more by male voters than by women voters. You have been hired to investigate the gender gap. What methods of this chapter would you use?

2. Simple Random Sample You have been hired to compare the mean credit debt of men in your state to the mean credit debt of women in your state. You have been given samples that were obtained by this process: First, a complete list of all creditors was obtained, then a computer was used to arrange the list in a random order, then a random sample of 200 creditors was selected. Does this selection process satisfy the requirement of being a simple random sample? Explain.

3. Comparing Incomes Using data from the Bureau of Labor Statistics, a researcher obtains the mean income of men and the mean income of women for each of the 50 states. She then conducts a t test of the null hypothesis that men and women in the United States have equal mean incomes. Is this procedure okay? Why or why not?

4. Independent Samples What is the difference between two samples that are independent and two samples that are not independent?

Review Exercises

TI **1.** Racial Profiling Racial profiling is the controversial practice of targeting someone for suspicion of criminal behavior on the basis of race, national origin, or ethnicity. The table below includes data from randomly selected drivers stopped by police in a recent year (based on data from the U.S. Department of Justice, Bureau of Justice Statistics).

a. Use a 0.05 significance level to test the claim that the proportion of blacks stopped by police is significantly greater than the proportion of whites.

b. Construct a confidence interval that could be used to test the claim in part (a). Be sure to use the correct level of significance. What do you conclude based on the confidence interval?

	Race and Ethnicity	
	Black and Non-Hispanic	**White and Non-Hispanic**
Drivers stopped by police	24	147
Total number of observed drivers	200	1400

TI **2.** Self-Reported and Measured Heights of Male Statistics Students Eleven male statistics students were given a survey that included a question asking them to report their height in inches. They weren't told that their height would be measured, but heights were accurately measured after the survey was completed. Anonymity was maintained through the use of code numbers instead of names, and the results are shown below. Is there sufficient evidence to support a claim that male statistics students exaggerate their heights?

Reported height	68	74	66.5	69	68	71	70	70	67	68	70
Measured height	66.8	73.9	66.1	67.2	67.9	69.4	69.9	68.6	67.9	67.6	68.8

TI **3.** Comparing Readability of J. K. Rowling and Leo Tolstoy Listed below are Flesch Reading Ease scores taken from randomly selected pages in J. K. Rowling's *Harry Potter and the Sorcerer's Stone* and Leo Tolstoy's *War and Peace*. (Higher Flesch Reading Ease scores indicate writing that is easier to read.) Use a 0.05 significance level to test the claim that *Harry Potter and the Sorcerer's Stone* is easier to read than *War and Peace*. Is the result as expected?

Rowling: 85.3 84.3 79.5 82.5 80.2 84.6 79.2 70.9 78.6 86.2 74.0 83.7
Tolstoy: 69.4 64.2 71.4 71.6 68.5 51.9 72.2 74.4 52.8 58.4 65.4 73.6

TI **4.** Variation in J. K. Rowling and Leo Tolstoy Refer to the same data used in Exercise 3 and use a 0.05 significance level to test the claim that pages from *Harry Potter and the Sorcerer's Stone* and *War and Peace* have Flesch Reading Ease scores with the same standard deviation.

TI **5.** Warmer Surgical Patients Recover Better? An article published in *USA Today* stated that "in a study of 200 colorectal surgery patients, 104 were kept warm with blankets and intravenous fluids; 96 were kept cool. The results show: Only 6 of those warmed developed wound infections vs. 18 who were kept cool."

a. Use a 0.05 significance level to test the claim of the article's headline: "Warmer surgical patients recover better." If these results are verified, should surgical patients be routinely warmed?

b. If a confidence interval is to be used for testing the claim in part (a), what confidence level should be used?

continued

c. Using the confidence level from part (b), construct a confidence interval estimate of the difference between the two population proportions.

d. In general, if a confidence interval estimate of the difference between two population proportions is used to test some claim about the proportions, will the conclusion based on the confidence interval always be the same as the conclusion from a standard hypothesis test?

TI **6.** Brain Volume and Psychiatric Disorders A study used x-ray computed tomography (CT) to collect data on brain volumes for a group of patients with obsessive-compulsive disorders and a control group of healthy persons. Sample results (in mL) are given below for total brain volumes (based on data from "Neuroanatomical Abnormalities in Obsessive-Compulsive Disorder Detected with Quantitative X-Ray Computed Tomography," by Luxenberg et al., *American Journal of Psychiatry,* Vol. 145, No. 9).

a. Construct a 95% confidence interval for the difference between the mean brain volume of obsessive-compulsive patients and the mean brain volume of healthy persons. Do not assume that the two populations have equal variances.

b. Use a 0.05 significance level to test the claim that there is no difference between the mean for obsessive-compulsive patients and the mean for healthy persons. Do not assume that the two populations have equal variances.

c. Based on the results from parts (a) and (b), does it appear that the total brain volume can be used as an indicator of obsessive-compulsive disorders?

Obsessive-compulsive patients: $n = 10, \bar{x} = 1390.03, s = 156.84$
Control group: $n = 10, \bar{x} = 1268.41, s = 137.97$

TI **7.** Variation of Brain Volumes Use the same sample data given in Exercise 6 with a 0.05 significance level to test the claim that the populations of total brain volumes for obsessive-compulsive patients and the control group have different amounts of variation.

TI **8.** Historical Data Set In 1908, "Student" (William Gosset) published the article "The Probable Error of a Mean" (*Biometrika,* Vol. 6, No. 1). He included the data listed below for two different types of straw seed (regular and kiln dried) that were used on adjacent plots of land. The listed values are the yields of straw in cwt per acre.

a. Using a 0.05 significance level, test the claim that there is no difference between the yields from the two types of seed.

b. Construct a 95% confidence interval estimate of the mean difference between the yields from the two types of seed.

c. Does it appear that either type of seed is better?

Regular	19.25	22.75	23	23	22.5	19.75	24.5	15.5	18	14.25	17
Kiln dried	25	24	24	28	22.5	19.5	22.25	16	17.25	15.75	17.25

Cumulative Review Exercises

TI **1.** Highway Speeds A section of Highway 405 in Los Angeles has a speed limit of 65 mi/h, and recorded speeds are listed below for randomly selected cars traveling on northbound and southbound lanes (based on data from Sigalert.com).

a. Using the speeds for the northbound lanes, find the mean, median, standard deviation, variance, and range.

b. Using all of the speeds combined, test the claim that the mean is greater than the posted speed limit of 65 mi/h.

c. Do the northbound speeds appear to come from a normally distributed population? Explain.

d. Assuming that the speeds are from normally distributed populations, test the claim that the mean speed on the northbound lanes is equal to the mean speed on the southbound lanes. Based on the result from part (c), does it appear that this hypothesis test is likely to be valid?

Highway 405 North: 68 68 72 73 65 74 73 72 68 65 65 73 66 71 68 74 66 71 65 73
Highway 405 South: 59 75 70 56 66 75 68 75 62 72 60 73 61 75 58 74 60 73 58 75

TI **2.** Tossing Coins An illusionist claims that she has the ability to toss a coin so that it turns up heads. Listed below are results from a test of her abilities.

a. Consider only the results from the tosses of a quarter. What is the probability of getting nine heads in nine tosses if the outcomes are determined only by chance? What does that result suggest about the claim that a coin can be tossed so that it turns up heads? Explain.

b. Are the results from the quarter independent of the results from the penny, or are the sample data matched pairs? Explain.

c. Using all of the results combined with a 0.01 significance level, test the claim that a coin can be tossed so that heads turn up more often than can be expected by chance.

Quarter:	H	H	H	H	H	H	H	H	H
Penny:	H	H	T	H	T	H	H	T	T

3. Cell Phones and Crashes: Analyzing Newspaper Report In an article from the Associated Press, it was reported that researchers "randomly selected 100 New York motorists who had been in an accident and 100 who had not. Of those in accidents, 13.7 percent owned a cellular phone, while just 10.6 percent of the accident-free drivers had a phone in the car." Analyze these results.

Cooperative Group Activities

1. Out-of-class activity Are estimates influenced by anchoring numbers? Refer to the related Chapter 3 Cooperative Group Activity. In Chapter 3 we noted that, according to author John Rubin, when people must estimate a value, their estimate is often "anchored" to (or influenced by) a preceding number. In that Chapter 3 activity, some subjects were asked to quickly estimate the value of $8 \times 7 \times 6 \times 5 \times 4 \times 3 \times 2 \times 1$, and others were asked to quickly estimate the value of $1 \times 2 \times 3 \times 4 \times 5 \times 6 \times 7 \times 8$. In Chapter 3, we could compare the two sets of results by using statistics (such as the mean) and graphs (such as boxplots). The methods of Chapter 9 now allow us to compare the results with a formal hypothesis test. Specifically, collect your own sample data and test the claim that when we begin with larger numbers (as in $8 \times 7 \times 6$), our estimates tend to be larger.

2. In-class activity Divide into groups according to gender, with about 10 or 12 students in each group. Each group member should record his or her pulse rate by counting the number of heartbeats in 1 minute, and the group statistics (n, $\bar{x}$, s) should be calculated. The groups should test the null hypothesis of no difference between their mean pulse rate and the mean of the pulse rates for the population from which subjects of the same gender were selected for Data Set 1 in Appendix B.

3. Out-of-class activity Randomly select a sample of male students and a sample of female students and ask each selected person whether they support a death penalty for people convicted of murder. Use a formal hypothesis test to determine whether there is a gender gap on this issue. Also, keep a record of the responses according to the gender of the person asking the question. Does the response appear to be influenced by the gender of the interviewer?

4. Out-of-class activity Use a watch to record the waiting times of a sample of McDonald's customers and the waiting times of a sample of Burger King customers. Use a hypothesis test to determine whether there is a significant difference.

5. Out-of-class activity Construct a short survey of just a few questions, including a question asking the subject to report his or her height. After the subject has completed the survey, measure the subject's height (without shoes) using an accurate measuring system. Record the gender, reported height, and measured height of each subject. (See Review Exercise 2.) Do male subjects appear to exaggerate their heights? Do female subjects appear to exaggerate their heights? Do the errors for males appear to have the same mean as the errors for females?

6. In-class activity Without using any measuring device, each student should draw a line believed to be 3 in. long. Then use rulers to measure and record the lengths of the lines drawn. Test for a difference between the mean length of lines drawn by males and the mean length of lines drawn by females.

7. In-class activity Use a ruler as a device for measuring reaction time. One person should suspend the ruler by holding it at the top while the subject holds his or her thumb and forefinger at the bottom edge, ready to catch the ruler when it is released. Record the distance that the ruler falls before it is caught. Convert that distance to the time (in seconds) that it took the subject to react and catch the ruler. (If the distance is measured in inches, use $t = \sqrt{d/192}$. If the distance is measured in centimeters, use $t = \sqrt{d/487.68}$.) Test each subject once with the dominant hand and once with the other hand, and record the paired data. Does there appear to be a difference between the mean of the reaction times using the dominant hand and the mean from the other hand? Do males and females appear to have different mean reaction times?

TI-83/84 Plus Calculator Project

The TI-83/84 Plus calculator is capable of generating normally distributed data drawn from a population with a specified mean and standard deviation. Use this procedure:

1. Clear lists L1 and L2. (Press STAT, select **ClrList,** and enter L1, L2 with the commas included.)
2. Press MATH and use ▷ to select the top menu item of **PRB.**
3. Select the menu item of **randInt.**
4. The format for this command is **randInt(μ, σ, n),** where μ is the population mean, σ is the population standard deviation, and n is the sample size.
5. Press ENTER and the random data will be generated. STO▸ can be used to store the data in a list, such as L1.

Use the above procedure to generate two sets of sample data that represent simulated IQ scores, as described below:

IQ scores of treatment group: Generate 10 sample values from a normally distributed population with mean 100 and standard deviation 15.

IQ scores of placebo group: Generate 12 sample values from a normally distributed population with mean 100 and standard deviation 15.

We can see from the way the data are generated that both data sets actually come from the same population, so there should be no difference between the two sample means.

a. After generating the two data sets, use a 0.10 significance level to test the claim that the two samples come from populations with the same mean.

b. If this experiment is repeated many times, what is the expected percentage of trials leading to the conclusion that the two population means are different? How does this relate to a type I error?

c. If your generated data should lead to the conclusion that the two population means are different, would this conclusion be correct or incorrect in reality? How do you know?

d. If part (a) is repeated 20 times, what is the probability that none of the hypothesis tests leads to rejection of the null hypothesis?

e. Repeat part (a) 20 times. How often was the null hypothesis of equal means rejected? Is this the result you expected?

From Data to Decision

Critical Thinking: The fear of flying

The lives of many people are affected by a fear that prevents them from flying. Sports announcer John Madden gained notoriety as he crossed the country by rail or motor home, traveling from one football stadium to another. The Marist Institute for Public Opinion conducted a poll of 1014 adults, 48% of whom were men. The results reported in *USA Today* showed that 12% of the men and 33% of the women fear flying.

Analyzing the Results

1. How many men were surveyed? How many women were surveyed? How many of the surveyed men fear flying? How many of the surveyed women fear flying?
2. Is there sufficient evidence to conclude that there is a significant difference between the percentage of men and the percentage of women who fear flying?
3. Construct a 95% confidence interval estimate of the difference between the percentage of men and the percentage of women who fear flying. Do the confidence interval limits contain 0, and what is the significance of whether they do or do not?
4. Construct a 95% confidence interval for the percentage of men who fear flying.
5. Based on the result from the confidence interval obtained in Exercise 4, complete the following statement, which is typical of the statement that would be reported in a newspaper or magazine: "Based on the Marist Institute for Public Opinion poll, the percentage of men who fear flying is 12% with a margin of error of _____."
6. Examine the completed statement in Exercise 5. What important piece of information should be included, but is not included?
7. In a separate Gallup poll, 1001 randomly selected adults were asked this question: "If you had to fly on an airplane tomorrow, how would you describe your feelings about flying? Would you be—very afraid, somewhat afraid, not very afraid, or not afraid at all?" Here are the responses: very afraid (18%), somewhat afraid (26%), not very afraid (17%), not afraid at all (38%), and no opinion (1%). Are these Gallup poll results consistent with those obtained by the survey conducted by the Marist Institute for Public Opinion? Explain. Can discrepancies be explained by the fact that the Gallup survey was conducted after the terrorist attacks of September 11, 2001, whereas the other survey was conducted before that date?
8. Construct a graph which would make the results understandable to typical newspaper readers.

Internet Project

Comparing Populations

The previous chapter showed you methods for testing hypotheses about a single population. This chapter expanded on those ideas, allowing you to test hypotheses about the relationships between two populations. In a similar fashion, the Internet Project for this chapter differs from that of the previous chapter in that you will need data for two populations or groups to conduct investigations. Go to the Internet Project for this chapter at

http://www.aw.com/triola.

There you will find several hypothesis-testing problems involving multiple populations. In these problems, you will analyze salary fairness, population demographics, and a traditional superstition. In each case you will formulate the problem as a hypothesis test, collect relevant data, then conduct and summarize the appropriate test.

Statistics @ Work

"It would be impossible to conduct archaeological research without at least a working knowledge of basic statistics."

Mark T. Lycett

Kathleen Morrison

Mark T. Lycett and Kathleen Morrison are both on the faculty of the Department of Anthropology at the University of Chicago.

Dr. Lycett's research deals with issues of economic, social, and political transformation associated with Spanish colonialism in the southwestern United States, and Dr. Morrison's research in southern India deals with problems of agricultural change, imperialism, and regional economic organization.

How important is the use of statistics in archaeology?

It would be impossible to conduct archaeological research without at least a working knowledge of basic statistics.

What concepts of statistics do you use?

Archaeologists make extensive use of both descriptive and inferential statistics on a daily basis. Exploratory data analysis using a variety of graphical and numerical summaries is increasingly common in modern archaeology. Archaeological problems routinely include studies of association for categorical variables, hypothesis testing for both 2-sample and *k*-sample data, correlation and regression problems, and a suite of nonparametric approaches.

Please give a specific example illustrating the use of statistics in your work.

We have explored the size distribution of ancient grass pollen grains to investigate changes in agriculture in both the New and Old Worlds during the first centuries of European colonial expansion. Although almost all important crops are grasses with morphologically similar pollen, New World staple crops (corn) have much larger pollen grains than wild grasses, and Old World crops (principally wheat, barley, and rice) are intermediate in size. By studying the size distribution of reference samples of these staple crops as well as fossil grass pollen from archaeological contexts, we have been able to specify the range of crops introduced and grown at colonial period sites in New Mexico and India.

Our data have been used to make inferences about the number and kind of archaeological sites that existed in our study areas; to reconstruct ancient patterns of vegetation, agriculture, and economy; and to study the effects of colonialism and imperialism on local social, economic, and religious practices.

Is your use of probability and statistics increasing, decreasing, or remaining stable?

Both the number and variety of statistical applications in archaeology are increasing, particularly as more sophisticated spatial databases become available through the widespread use of Geographic Information Systems technology.

In terms of statistics, what would you recommend for prospective employees?

When we were college students, we understood that statistics would be a part of our professional lives, but we never imagined the degree to which we would use it on a daily basis. Undergraduates interested in archaeology should begin with an introductory course in probability and statistics. Those with professional or academic goals should consider more advanced undergraduate or graduate level course work in quantitative data analysis.

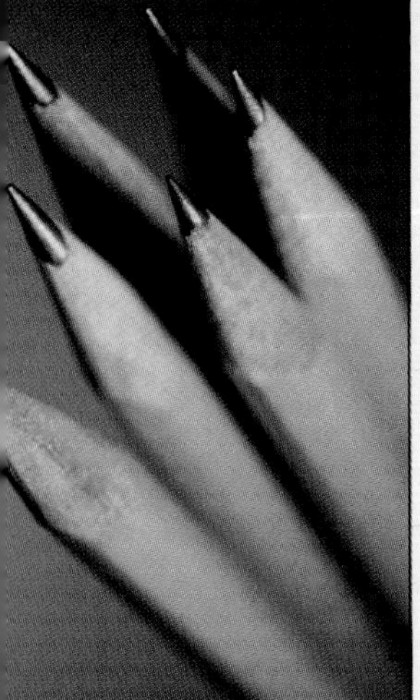

10 Correlation and Regression

10 CHAPTER PROBLEM

Can we predict the time of the next eruption of the Old Faithful geyser?

The Old Faithful geyser is the most popular attraction in Yellowstone National Park. It is located near the Old Faithful Inn, which is very possibly Yellowstone's second most popular attraction. Tourists enjoy the food, drink, lodging, and shopping facilities of the inn, but they want to be sure to see at least one eruption of the famous Old Faithful geyser. Park rangers help tourists by posting the predicted time to the next eruption. How do they make those predictions?

When Old Faithful erupts, these measurements are recorded: duration (in seconds) of the eruption, the time interval (in minutes) between the preceding eruption and the current eruption, the time interval (in minutes) between the current eruption and the following eruption, and the height (in feet) of the eruption. Table 10-1 includes measurements from eight eruptions. (The measurements in Table 10-1 are from eight of the 40 eruptions included in Data Set 11 from Appendix B. Table 10-1 includes a small sample so that calculations will be easier when the data are used in discussing the methods of the following sections.)

Once an eruption has occurred, we want to predict the time to the next eruption, which is the time interval after the eruption. To see which variables affect the "interval after" times, we might begin by constructing scatterplots such as those generated by the TI-83/84 Plus calculator and shown in Figure 10-1. (Scatterplots were first introduced in Section 2-4.) By simply examining the patterns of the points in the three scatterplots, we can make these subjective conclusions:

1. There appears to be a relationship between the time interval after an eruption and the duration of the eruption. [See Figure 10-1(a).]
2. There does not appear to be a relationship between the time interval after an eruption and the height of the eruption. [See Figure 10-1(b).]
3. There does not appear to be a relationship between the time interval after an eruption and the time interval before the eruption. [See Figure 10-1(c).]

Such conclusions based on scatterplots are largely subjective, so this chapter presents tools for addressing issues such as these:

- How can methods of statistics be used to objectively determine whether there is a relationship between two variables, such as the time intervals after eruptions and the durations of eruptions?

Table 10-1 Eruptions of the Old Faithful Geyser

Duration	240	120	178	234	235	269	255	220
Interval Before	98	90	92	98	93	105	81	108
Interval After	92	65	72	94	83	94	101	87
Height	140	110	125	120	140	120	125	150

- If there is a relationship between two variables, how can it be described? Is there an equation that can be used to predict the time to the next geyser eruption, given the duration of the current eruption?
- If we can predict the time to the next Old Faithful eruption, how accurate is that prediction likely to be?

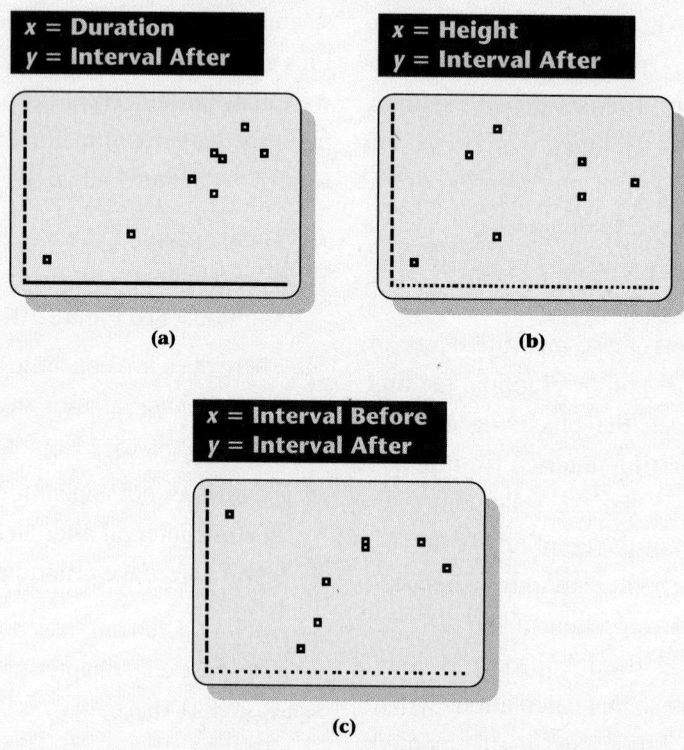

Figure 10-1 Scatterplots from Old Faithful Eruption Measurements

10-1 Overview

This chapter introduces important methods for making inferences about a *correlation* (or relationship) between two variables, and describing such a relationship with an equation that can be used for predicting the value of one variable given the value of the other variable. We consider sample data that come in *pairs*. Section 9-4 also used matched pairs, but the objective was fundamentally different. The inferences in Section 9-4 dealt with the *mean of differences* between pairs of values, but this chapter has the objective of making inferences about the *relationship* between two variables.

Section 10-2 introduces the relationship between two variables. Section 10-3 introduces regression analysis, which allows us to describe the relationship with an equation. In Section 10-4 we analyze differences between predicted values and actual observed values. Sections 10-2 through 10-4 involve relationships between *two* variables, but Section 10-5 introduces multiple regression for relationships among three or more variables. Finally, in Section 10-6 we describe some basic methods for developing a mathematical model that can be used to describe the relationship between two variables. Although Section 10-3 is limited to linear relationships, Section 10-6 includes some common nonlinear relationships.

10-2 Correlation

Key Concept This section introduces the *linear correlation coefficient* r, which is a numerical measure of the strength of the relationship between two variables representing quantitative data. Using paired sample data (sometimes called **bivariate data**), we find the value of r (usually using technology), then we use that value to conclude that there is (or is not) a relationship between the two variables. In this section we consider only *linear* relationships, which means that when graphed, the points approximate a straight-line pattern. Because software programs or calculators are usually used to find the value of r, it is important to focus on the concepts in this section, without becoming overly involved with tedious arithmetic calculations.

Part 1: Basic Concepts of Correlation

We begin with the basic definition of *correlation*, a term commonly used in the context of a relationship between two variables.

Definition

A **correlation** exists between two variables when one of them is related to the other in some way.

Table 10-1, for example, includes paired sample data consisting of durations and time intervals after eruptions of the Old Faithful geyser. (The "interval after" an eruption is the time to the next eruption.) We will determine whether there is a correlation between the variable x (duration) and the variable y (interval after).

Exploring the Data

Before working with the more formal methods of this section, we should first explore the data set to see what we can learn. We can often see a relationship between two variables by constructing a scatterplot. When we examine a scatterplot, we should study the overall pattern of the plotted points. If there is a pattern, we should note its direction. An uphill direction suggests that as one variable increases, the other also increases. A downhill direction suggests that as one variable increases, the other decreases. We should look for outliers, which are points that lie very far away from all of the other points.

Figure 10-2 shows scatterplots with different characteristics. The scatterplots in Figure 10-2(a), (b), and (c) depict a pattern of increasing values of y that correspond to increasing values of x. As you proceed from figures (a) to (c), the pattern of points becomes closer to a straight line, suggesting that the relationship between x and y becomes stronger. The scatterplots in Figure 10-2(d), (e), and (f) depict patterns in which the y-values decrease as the x-values increase. Again, as you proceed from figures (d) to (f), the relationship becomes stronger. In contrast to the first six graphs, the scatterplot of Figure 10-2(g) shows no pattern and suggests that there is no correlation (or relationship) between x and y. Finally, the scatterplot of Figure 10-2(h) shows a very distinct pattern suggesting a relationship between x and y, but it is not a linear (straight-line) pattern. (Figures 10-2(a) through (g) are from ActivStats, and Figure 10-2(h) is from Minitab.)

Linear Correlation Coefficient

Because conclusions based on visual examinations of scatterplots are largely subjective, we need more precise and objective measures. We use the linear correlation coefficient r, which is useful for detecting straight-line patterns.

Definition

The **linear correlation coefficient** r measures the strength of the linear relationship between the paired x- and y-quantitative values in a *sample*. Its value is computed by using Formula 10-1, included in the box on page 541. [The linear correlation coefficient is sometimes referred to as the **Pearson product moment correlation coefficient** in honor of Karl Pearson (1857–1936), who originally developed it.]

Because the linear correlation coefficient r is calculated using sample data, it is a sample statistic used to measure the strength of the linear correlation between x and y. If we had every pair of population values for x and y, the result of Formula 10-1 would be a population parameter, represented by ρ (Greek rho). The box on pages 540–541 includes the requirements, notation, and Formula 10-1.

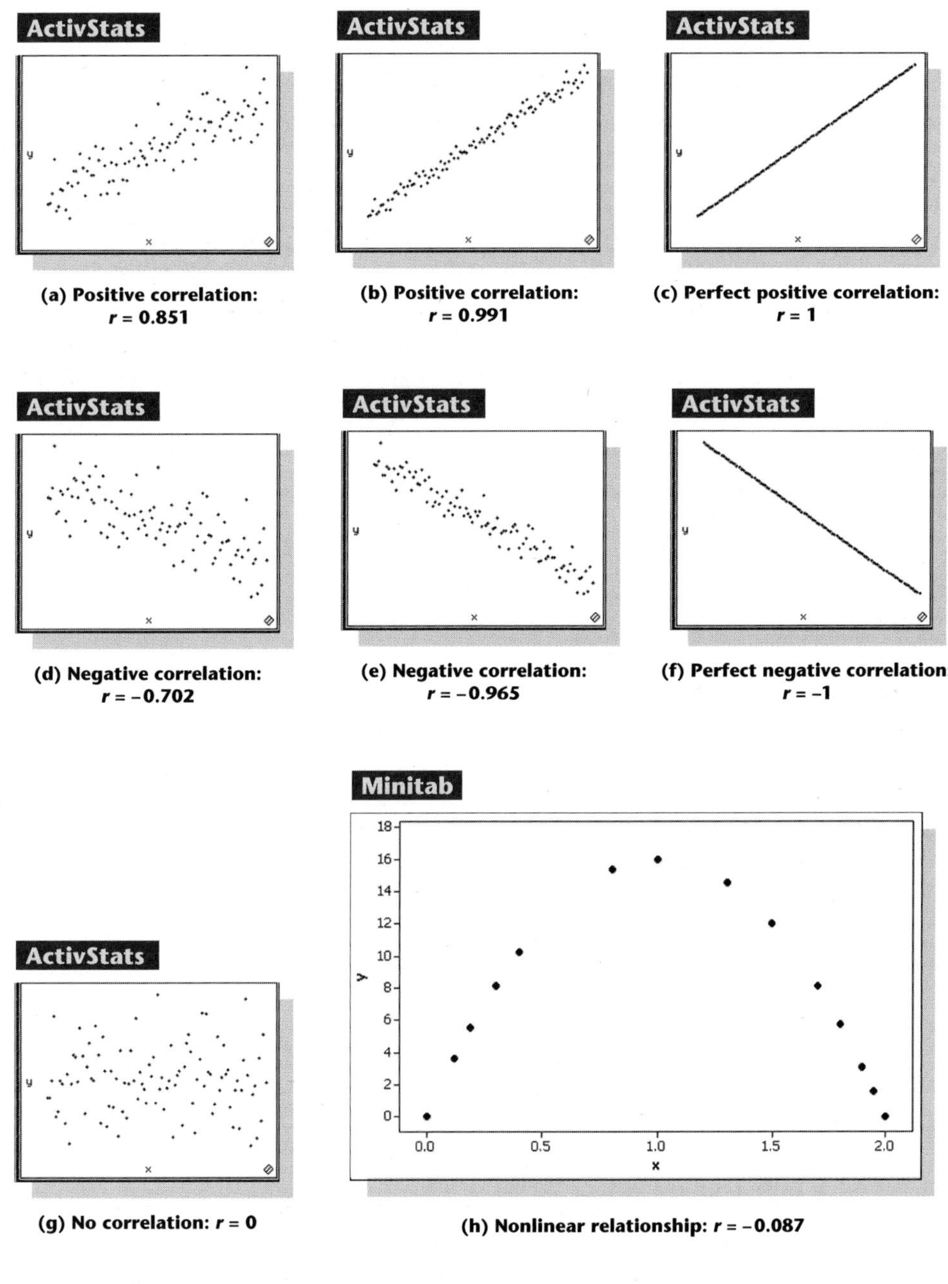

(a) Positive correlation: $r = 0.851$

(b) Positive correlation: $r = 0.991$

(c) Perfect positive correlation: $r = 1$

(d) Negative correlation: $r = -0.702$

(e) Negative correlation: $r = -0.965$

(f) Perfect negative correlation: $r = -1$

(g) No correlation: $r = 0$

(h) Nonlinear relationship: $r = -0.087$

Figure 10-2 Scatterplots

Palm Reading

Some people believe that the length of their palm's lifeline can be used to predict longevity. In a letter published in the *Journal of the American Medical Association,* authors M. E. Wilson and L. E. Mather refuted that belief with a study of cadavers. Ages at death were recorded, along with the lengths of palm lifelines. The authors concluded that there is no significant correlation between age at death and length of lifeline. Palmistry lost, hands down.

Requirements

Given any collection of sample paired data, the linear correlation coefficient r can always be computed, but the following requirements should be satisfied when testing hypotheses or making other inferences about r.

1. The sample of paired (x, y) data is a *random* sample of independent quantitative data. (It is important that the sample data have not been collected using some inappropriate method, such as using a voluntary response sample.)
2. Visual examination of the scatterplot must confirm that the points approximate a straight-line pattern.
3. Any outliers must be removed if they are known to be errors. The effects of any other outliers should be considered by calculating r with and without the outliers included.

Note: Requirements 2 and 3 above are simplified attempts at checking this formal requirement:

The pairs of (x, y) data must have a **bivariate normal distribution.** (Normal distributions are discussed in Chapter 6, but this assumption basically requires that for any fixed value of x, the corresponding values of y have a distribution that is bell-shaped, and for any fixed value of y, the values of x have a distribution that is bell-shaped.) This requirement is usually difficult to check, so for now, we will use Requirements 2 and 3 as listed above.

Notation for the Linear Correlation Coefficient

n	represents the number of pairs of data present.
Σ	denotes the addition of the items indicated.
Σx	denotes the sum of all x-values.
Σx^2	indicates that each x-value should be squared and then those squares added.
$(\Sigma x)^2$	indicates that the x-values should be added and the total then squared. It is extremely important to avoid confusing Σx^2 and $(\Sigma x)^2$
Σxy	indicates that each x-value should first be multiplied by its corresponding y-value. After obtaining all such products, find their sum.
r	represents the linear correlation coefficient for a *sample.*
ρ	Greek letter rho used to represent the linear correlation coefficient for a *population.*

Formula 10-1

$$r = \frac{n(\Sigma xy) - (\Sigma x)(\Sigma y)}{\sqrt{n(\Sigma x^2) - (\Sigma x)^2}\sqrt{n(\Sigma y^2) - (\Sigma y)^2}}$$

This shortcut formula simplifies manual calculations. The format of the formula makes it easy to use in a spreadsheet or computer program. See other equivalent formulas given later in this section, and also see a rationale for the calculation of r.

Interpreting r Using Table A-6: If the absolute value of the computed value of r exceeds the value in Table A-6, conclude that there is a linear correlation. Otherwise, there is not sufficient evidence to support the conclusion of a linear correlation.

Interpreting r Using the TI-83/84 Plus Calculator: If the computed P-value is less than or equal to the significance level, conclude that there is a linear correlation. Otherwise, there is not sufficient evidence to support the conclusion of a linear correlation.

TI-83/84 Plus Calculator: The TI-83/84 Plus calculator is designed to generate the value of the linear correlation coefficient r. See the instructions at the end of this section.

Rounding the Linear Correlation Coefficient

Round the linear correlation coefficient r to three decimal places (so that its value can be directly compared to critical values in Table A-6). If manually calculating r and other statistics in this chapter, rounding in the middle of a calculation often creates substantial errors, so try using your calculator's memory to store intermediate results and round off only at the end.

x	3	1	3	5
y	5	8	6	4

EXAMPLE **Calculating r** Using the simple random sample of data given in the margin, find the value of the linear correlation coefficient r.

SOLUTION

REQUIREMENT ✔ The data are a simple random sample. The accompanying TI-83/84 Plus calculator scatterplot shows a pattern of points that does appear to be a straight-line pattern. There are no outliers. The requirements are satisfied and it makes sense to proceed with the calculation of the linear correlation coefficient r. ✔

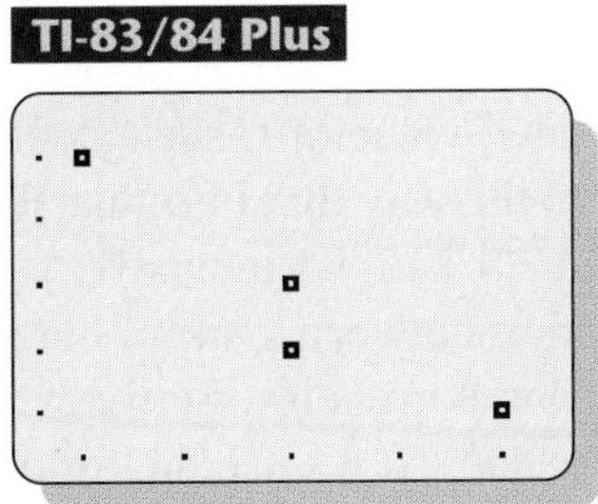

For the given sample of paired data, $n = 4$ because there are four pairs of data. The other components required in Formula 10-1 are found from the calculations in Table 10-2. Note how this vertical format makes the calculations easier.

continued

Teacher Evaluations Correlate with Grades

Student evaluations of faculty are often used to measure teaching effectiveness. Many studies reveal a correlation with higher student grades being associated with higher faculty evaluations. One study at Duke University involved student evaluations collected before and after final grades were assigned. The study showed that "grade expectations or received grades caused a change in the way students perceived their teacher and the quality of instruction." It was noted that with student evaluations, "the incentives for faculty to manipulate their grading policies in order to enhance their evaluations increase." It was concluded that "the ultimate consequence of such manipulations is the degradation of the quality of education in the United States." (See "Teacher Course Evaluations and Student Grades: An Academic Tango," by Valen Johnson, *Chance*, Vol. 15, No. 3.)

Table 10-2 Finding Statistics Used to Calculate r

	x	y	$x \cdot y$	x^2	y^2
	3	5	15	9	25
	1	8	8	1	64
	3	6	18	9	36
	5	4	20	25	16
Total	**12**	**23**	**61**	**44**	**141**
	↑	↑	↑	↑	↑
	Σx	Σy	Σxy	Σx^2	Σy^2

Using the calculated values and Formula 10-1, we can now evaluate r as follows:

$$
\begin{aligned}
r &= \frac{n(\Sigma xy) - (\Sigma x)(\Sigma y)}{\sqrt{n(\Sigma x^2) - (\Sigma x)^2}\sqrt{n(\Sigma y^2) - (\Sigma y)^2}} \\
&= \frac{4(61) - (12)(23)}{\sqrt{4(44) - (12)^2}\sqrt{4(141) - (23)^2}} \\
&= \frac{-32}{\sqrt{32}\sqrt{35}} = -0.956
\end{aligned}
$$

These calculations get quite messy with larger data sets, so it's fortunate that the linear correlation coefficient can be found automatically with many different calculators and computer programs. See the procedure at the end of this section for the TI-83/84 Plus calculator.

Interpreting the Linear Correlation Coefficient

We need to interpret a calculated value of r, such as the value of -0.956 found in the preceding example. Given the way that Formula 10-1 is constructed, the value of r must always fall between -1 and $+1$ inclusive. If r is close to 0, we conclude that there is no linear correlation between x and y, but if r is close to -1 or $+1$ we conclude that there is a linear correlation between x and y. Interpretations of "close to" 0 or 1 or -1 are vague, so we use the following very specific decision criterion:

Using Table A-6: **If the absolute value of the computed value of r exceeds the value in Table A-6, conclude that there is a linear correlation. Otherwise, there is not sufficient evidence to support the conclusion of a linear correlation.**

***Using the TI-83/84 Plus Calculator:* If the computed *P*-value is less than or equal to the significance level, conclude that there is a linear correlation. Otherwise, there is not sufficient evidence to support the conclusion of a linear correlation.**

When there really is no linear correlation between x and y, Table A-6 lists values that are "critical" in this sense: They separate *usual* values of r from those that are *unusual*. For example, Table A-6 shows us that with $n = 4$ pairs of sample data, the critical values are 0.950 (for $\alpha = 0.05$) and 0.999 (for $\alpha = 0.01$). Critical values and the role of α are described in Chapters 7 and 8. Here's how we interpret those numbers: With 4 pairs of data and no linear correlation between x and y, there is a 5% chance that the absolute value of the computed linear correlation coefficient r will exceed 0.950. With $n = 4$ and no linear correlation, there is a 1% chance that $|r|$ will exceed 0.999. Given $r = -0.956$ computed in the preceding example, if we use a 0.05 significance level we conclude that there is a linear correlation between x and y (because $|-0.956|$ exceeds the critical value of 0.950). However, if we use a 0.01 significance level, we do not conclude that there is a linear correlation (because $|-0.956|$ does not exceed the critical value of 0.999).

EXAMPLE Old Faithful Using the paired duration and time interval after eruption data from Table 10-1, find the value of the linear correlation coefficient r. Then refer to Table A-6 to determine whether there is a linear correlation between duration times and time intervals after eruptions. In Table A-6, use the critical value for $\alpha = 0.05$. (With $\alpha = 0.05$ we conclude that there is a linear correlation only if the sample is unlikely in this sense: If there is no linear correlation between the two variables, such a value of r occurs 5% of the time or less.)

SOLUTION

REQUIREMENT ✔ The data are from randomly selected eruptions. The scatterplot shown in Figure 10-1(a) shows a pattern of points that does appear to be a straight-line pattern. There do not appear to be any outliers. The requirements are satisfied and we can proceed to find r and determine whether there is a linear correlation between duration times and time intervals after eruptions. ✔

Using the same procedure illustrated in the preceding example, or using technology, we can find that the 8 pairs of duration/interval after eruption times in Table 10-1 result in $r = 0.926$. The TI-83/84 Plus calculator display is shown in the margin, and the bottom line provides the value of r.

TI-83/84 Plus

```
LinRegTTest
 y=a+bx
 β≠0 and ρ≠0
↑b=.2340614319
 s=4.973916136
 r²=.8567190634
 r=.9255911967
```

Referring to Table A-6, we locate the row for which $n = 8$ (because there are 8 pairs of data). That row contains the critical values of 0.707 (for $\alpha = 0.05$) and 0.834 (for $\alpha = 0.01$). Using the critical value for $\alpha = 0.05$, we see that there is less than a 5% chance that with no linear correlation, the absolute value of the computed r will exceed 0.707. Because $r = 0.926$, its absolute

continued

value does exceed 0.707, so we conclude that there is a linear correlation between duration and interval after eruption times.

We have already noted that Formula 10-1 requires that the calculated value of r always falls between -1 and $+1$ inclusive. We list that property along with other important properties.

Properties of the Linear Correlation Coefficient r

1. The value of r is always between -1 and $+1$ inclusive. That is,
$$-1 \leq r \leq +1$$
2. *The value of r does not change if all values of either variable are converted to a different scale.*
3. *The value of r is not affected by the choice of x or y.* Interchange all x- and y-values and the value of r will not change.
4. *r measures the strength of a linear relationship.* It is not designed to measure the strength of a relationship that is not linear.

Interpreting r: Explained Variation

If we conclude that there is a linear correlation between x and y, we can find a linear equation that expresses y in terms of x, and that equation can be used to predict values of y for given values of x. In Section 10-3 we will describe a procedure for finding such equations and show how to predict values of y when given values of x. But a predicted value of y will not necessarily be the exact result, because in addition to x, there are other factors affecting y, such as random variation and other characteristics not included in the study. In Section 10-4 we will present a rationale and more details about this important principle:

> **The value of r^2 is the proportion of the variation in y that is explained by the linear relationship between x and y.**

EXAMPLE Old Faithful Using the duration/interval after eruption times in Table 10-1, we have found that the linear correlation coefficient is $r = 0.926$. What proportion of the variation in interval after eruption times can be explained by the variation in duration times?

SOLUTION With $r = 0.926$, we get $r^2 = 0.857$.

INTERPRETATION We conclude that 0.857 (or about 86%) of the variation in interval after eruption times can be explained by the linear relationship between duration times and interval after eruption times. This implies that about 14% of the variation in interval after eruption times cannot be explained by duration times.

Common Errors Involving Correlation

We now identify three of the most common sources of errors made in interpreting results involving correlation:

1. *A common error is to conclude that correlation implies causality.* Using the sample data in Table 10-1, we can conclude that there is a correlation between duration times and interval after eruption times, but we cannot conclude that longer duration times *cause* longer interval after eruption times. The interval after eruption times may be affected by some other variable lurking in the background. (A **lurking variable** is one that affects the variables being studied, but is not included in the study.) For example, the outside air temperature might affect the duration of an eruption and the time interval after an eruption. The outside air temperature would then be a lurking variable.
2. *Another error arises with data based on averages.* Averages suppress individual variation and may inflate the correlation coefficient. One study produced a 0.4 linear correlation coefficient for paired data relating income and education among individuals, but the linear correlation coefficient became 0.7 when regional averages were used.
3. *A third error involves the property of linearity.* A relationship may exist between x and y even when there is no linear correlation. The data depicted in Figure 10-2(h) result in a value of $r = -0.087$, which is an indication of no *linear* correlation between the two variables. However, we can easily see from looking at the figure that there is a pattern reflecting a very strong *nonlinear* relationship. [Figure 10-2(h) is a scatterplot that depicts the relationship between distance above ground and time elapsed for an object thrown upward.]

Part 2: Formal Hypothesis Test (Requires Coverage of Chapter 8)

We present two methods (summarized in the accompanying box and in Figure 10-3) for using a formal hypothesis test to determine whether there is a significant linear correlation between two variables. Some instructors prefer Method 1 because it reinforces concepts introduced in earlier chapters. Others prefer Method 2 because it involves easier calculations. Method 1 uses the Student t distribution with a test statistic having the form $t = r/s_r$, where s_r denotes the sample standard deviation of r values. The test statistic given in the box (for Method 1) reflects the fact that the standard deviation of r values can be expressed as $\sqrt{(1 - r^2)/(n - 2)}$.

Figure 10-3 shows that the decision criterion is to reject the null hypothesis of $\rho = 0$ if the absolute value of the test statistic exceeds the critical values; rejection of $\rho = 0$ means that there is sufficient evidence to support a claim of a linear correlation between the two variables. If the absolute value of the test statistic does not exceed the critical values, then we fail to reject $\rho = 0$; that is, there is not sufficient evidence to conclude that there is a linear correlation between the two variables.

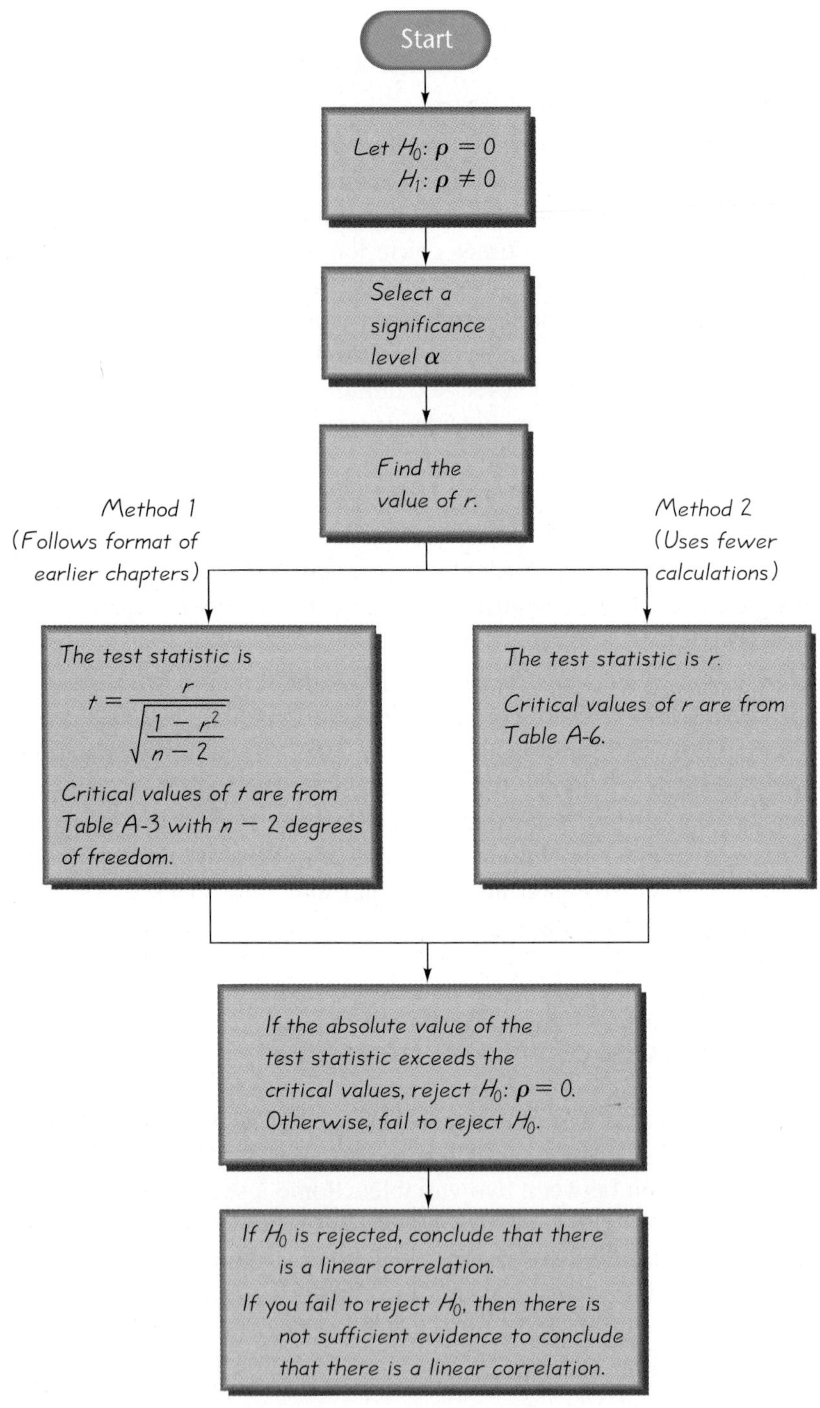

Figure 10-3 Hypothesis Test for a Linear Correlation

Hypothesis Test for Correlation (See Figure 10-3.)

H_0: $\rho = 0$ (There is no linear correlation.)

H_1: $\rho \neq 0$ (There is a linear correlation.)

Method 1: Test Statistic Is *t*

Test statistic: $$t = \frac{r}{\sqrt{\dfrac{1 - r^2}{n - 2}}}$$

Critical values: Use Table A-3 with $n - 2$ degrees of freedom.

***P*-value:** Use Table A-3 with $n - 2$ degrees of freedom.

Conclusion: If $|t| >$ critical value from Table A-3, reject H_0 and conclude that there is a linear correlation. If $|t| \leq$ critical value, fail to reject H_0; there is not sufficient evidence to conclude that there is a linear correlation.

Method 2: Test Statistic Is *r*

Test statistic: r

Critical values: Refer to Table A-6.

Conclusion: If $|r| >$ critical value from Table A-6, reject H_0 and conclude that there is a linear correlation. If $|r| \leq$ critical value, fail to reject H_0; there is not sufficient evidence to conclude that there is a linear correlation.

TI-83/84 Plus Calculator: Using the procdeure described at the end of this section, the TI-83/84 Plus calculator will provide the values of the linear correlation coefficient *r*, the *t*-test statistic, and the *P*-value. This calculator makes it easy to use the *P*-value approach to hypothesis testing.

EXAMPLE Old Faithful Using the sample data of duration and interval after eruption times in Table 10-1, test the claim that there is a linear correlation between the duration of an eruption and the time interval after that eruption (to the next eruption). For the test statistic, use both (a) Method 1 and (b) Method 2.

SOLUTION

REQUIREMENT ✔ A preceding example already includes verification that the requirements are satisfied. [The sample data have been randomly selected, the scatterplot in Figure 10-1(a) shows a pattern of points that appears to be linear, and there are no outliers.] Having completed this simplified check of requirements, we proceed with our analysis. ✔

This solution will follow the procedure summarized in Figure 10-3. To claim that there is a linear correlation is to claim that the population linear

continued

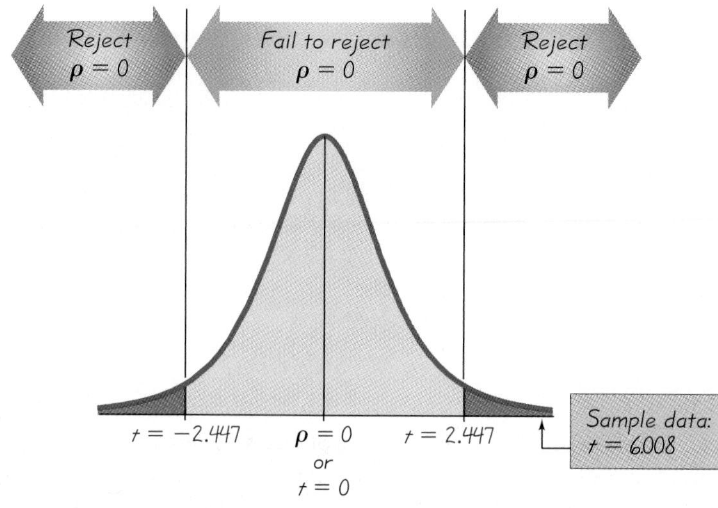

Figure 10-4 **Testing H_0: $\rho = 0$ with Method 1**

correlation coefficient ρ is different from 0. We therefore have the following hypotheses:

$$H_0: \rho = 0 \quad \text{(There is no linear correlation.)}$$

$$H_1: \rho \neq 0 \quad \text{(There is a linear correlation.)}$$

No significance level was specified, so use $\alpha = 0.05$.

In a preceding example we already found that $r = 0.926$. With that value, we now find the test statistic and critical value, using each of the two methods just described.

a. *Method 1:* The test statistic is

$$t = \frac{r}{\sqrt{\frac{1 - r^2}{n - 2}}} = \frac{0.926}{\sqrt{\frac{1 - 0.926^2}{8 - 2}}} = 6.008$$

The TI-83/84 Plus calculator yields $t = 5.990$, which is more accurate because it is calculated with the unrounded value of r. The critical values of $t = \pm 2.447$ are found in Table A-3, where 2.447 corresponds to an area of 0.05 divided between two tails and the number of degrees of freedom is $n - 2 = 6$. See Figure 10-4 for the graph that includes the test statistic and critical values. Also, the TI-83/84 Plus calculator provides a P-value of 0.000973, which is less than 0.05.

b. *Method 2:* The test statistic is $r = 0.926$. The critical values of $r = \pm 0.707$ are found in Table A-6 with $n = 8$ and $\alpha = 0.05$. See Figure 10-5 for a graph that includes this test statistic and critical values.

Using either of the two methods, we find that the absolute value of the test statistic does exceed the critical value (Method 1: $6.008 > 2.447$. Method 2: $0.926 > 0.707$); that is, the test statistic falls in the critical region. We therefore reject H_0: $\rho = 0$. There is sufficient evidence to support the claim of a linear correlation between duration times of eruptions and time intervals after eruptions.

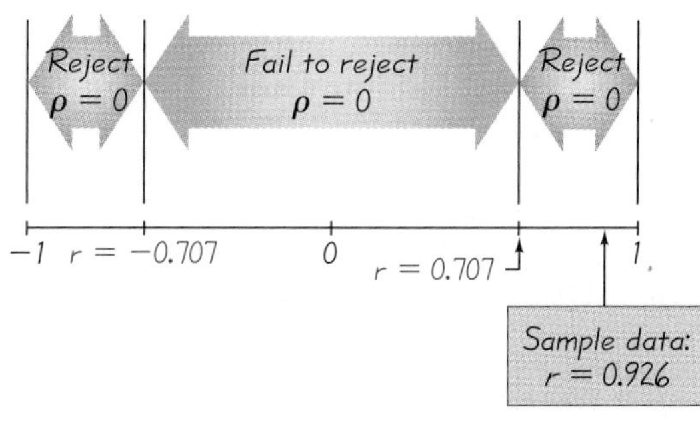

Figure 10-5 **Testing H_0: $\rho = 0$ with Method 2**

One-tailed Tests: The preceding example and Figures 10-4 and 10-5 illustrate a two-tailed hypothesis test. The examples and exercises in this section will generally involve only two-tailed tests, but one-tailed tests can occur with a claim of a positive linear correlation or a claim of a negative linear correlation. In such cases, the hypotheses will be as shown here.

Claim of *Negative* Correlation (Left-tailed test)	Claim of *Positive* Correlation (Right-tailed test)
H_0: $\rho = 0$	H_0: $\rho = 0$
H_1: $\rho < 0$	H_1: $\rho > 0$

For these one-tailed tests, Method 1 can be handled as in earlier chapters. For Method 2, either calculate the critical value as described in Exercise 38 or modify Table A-6 by replacing the column headings of $\alpha = 0.05$ and $\alpha = 0.01$ by the one-sided significance levels of $\alpha = 0.025$ and $\alpha = 0.005$, respectively.

Rationale: We have presented Formula 10-1 for calculating r and have illustrated its use. Formula 10-1 is given below along with some other formulas that are "equivalent" in the sense that they all produce the same values. Different textbook authors prefer different expressions for different reasons, and the formulas given below are the most commonly used expressions for r. These formulas are simply different versions of the same expression. (Algebra enthusiasts are welcome to have fun proving the equivalence of the formulas; others can trust the author that the various formulas for r will produce the same results.) The format of Formula 10-1 simplifies manual calculations, it simplifies calculations using spreadsheets, and it is easy to include as part of an original computer program.

$$r = \frac{n\Sigma xy - (\Sigma x)(\Sigma y)}{\sqrt{n(\Sigma x^2) - (\Sigma x)^2}\sqrt{n(\Sigma y^2) - (\Sigma y)^2}} \qquad r = \frac{\Sigma\left[\dfrac{(x - \bar{x})}{s_x}\dfrac{(y - \bar{y})}{s_y}\right]}{n - 1}$$

$$r = \frac{\Sigma(x - \bar{x})(y - \bar{y})}{(n - 1)s_x s_y} \qquad r = \frac{s_{xy}}{\sqrt{s_{xx}}\sqrt{s_{yy}}}$$

Although Formula 10-1 simplifies manual calculations, other formulas for r are better for helping us *understand* how r works. In attempting to understand the reasoning that underlies the development of the linear correlation coefficient, we will use this version of the linear correlation coefficient:

$$r = \frac{\Sigma\left[\frac{(x - \bar{x})}{s_x}\frac{(y - \bar{y})}{s_y}\right]}{n - 1}$$

We will temporarily use this latter version of Formula 10-1 because its form relates more directly to the underlying theory. We now consider the following paired data, which are depicted in the scatterplot shown in Figure 10-6.

x	1	1	2	4	7
y	4	5	8	15	23

Figure 10-6 includes the point $(\bar{x}, \bar{y}) = (3, 11)$, which is called the *centroid* of the sample points.

Definition

Given a collection of paired (x, y) data, the point $(\bar{x}, \bar{y})$ is called the **centroid.**

The statistic r, sometimes called the *Pearson product moment,* was first developed by Karl Pearson. It is based on the sum of the products $(x - \bar{x})(y - \bar{y})$. The statistic $\Sigma(x - \bar{x})(y - \bar{y})$ is the basic expression forming the foundation for r, and we now explain why that statistic is key.

In any scatterplot, vertical and horizontal lines through the centroid $(\bar{x}, \bar{y})$ divide the diagram into four quadrants, as in Figure 10-6. If the points of the scatterplot

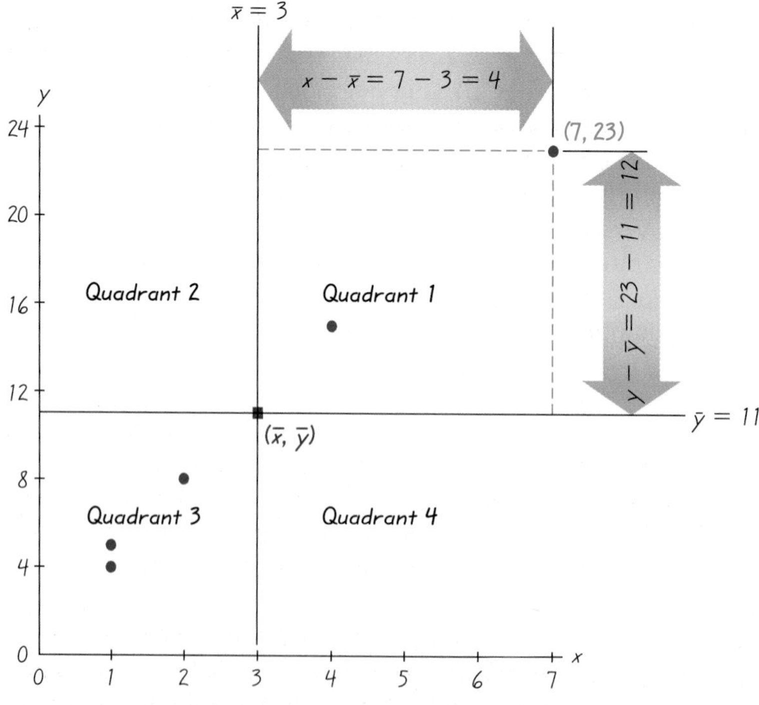

Figure 10-6 Scatterplot Partitioned into Quadrants

tend to approximate an uphill line (as in the figure), individual values of the product $(x - \bar{x})(y - \bar{y})$ tend to be positive because most of the points are found in the first and third quadrants, where the products of $(x - \bar{x})$ and $(y - \bar{y})$ are positive. If the points of the scatterplot approximate a downhill line, most of the points are in the second and fourth quadrants, where $(x - \bar{x})$ and $(y - \bar{y})$ are opposite in sign, so $\Sigma(x - \bar{x})(y - \bar{y})$ is negative. Points that follow no linear pattern tend to be scattered among the four quadrants, so the value of $\Sigma(x - \bar{x})(y - \bar{y})$ tends to be close to 0. We can therefore use $\Sigma(x - \bar{x})(y - \bar{y})$ as a measure of how the points are arranged. A large positive sum suggests that the points are predominantly in the first and third quadrants (corresponding to a positive linear correlation), a large negative sum suggests that the points are predominantly in the second and fourth quadrants (corresponding to a negative linear correlation), and a sum near zero suggests that the points are scattered among the four quadrants (with no linear correlation).

Unfortunately, the sum $\Sigma(x - \bar{x})(y - \bar{y})$ depends on the magnitude of the numbers used. For example, if you change x from inches to feet, that sum will change. To make r independent of the particular scale used, we include standardization. In Section 3-4 we saw that we could "standardize" values by converting them to z scores, as in $z = (x - \bar{x})/s$. (Here we use s_x to denote the standard deviation of the sample x values, and we use s_y to denote the standard deviation of the sample y values.) We use a similar technique as we standardize each deviation $(x - \bar{x})$ by dividing it by s_x. We also make the deviations $(y - \bar{y})$ independent of the magnitudes of the numbers by dividing by s_y. We now have the statistic

$$\Sigma\left[\frac{(x - \bar{x})}{s_x}\frac{(y - \bar{y})}{s_y}\right]$$

which we further modify by introducing the divisor of $n - 1$, which gives us a type of average instead of a sum that grows simply because we have more data. (The reasons for dividing by $n - 1$ instead of n are essentially the same reasons that relate to the standard deviation.) The end result is the expression

$$r = \frac{\Sigma\left[\frac{(x - \bar{x})}{s_x}\frac{(y - \bar{y})}{s_y}\right]}{n - 1}$$

This expression can be algebraically manipulated into the equivalent form of Formula 10-1 or any of the other expressions for r.

Confidence Intervals In preceding chapters we discussed methods of inferential statistics by addressing methods of hypothesis testing and methods for constructing confidence interval estimates. A similar procedure may be used to find confidence intervals for ρ. However, because the construction of confidence intervals involves somewhat complicated transformations, that process is presented in Exercise 39.

We can use the linear correlation coefficient to determine whether there is a linear relationship between two variables. Using the duration/interval after eruption data in Table 10-1, we have concluded that there is a linear correlation between durations of eruptions and time intervals after eruptions. Having concluded that a relationship exists, we would like to determine what that relationship is so that we can predict the time interval after an eruption for a given duration time. That is, we can use the duration time of an eruption to predict the time to the next eruption. This next stage of analysis is addressed in the following section.

Using the TI-83/84 Plus Calculator to Construct a Scatter Diagram

See Section 2-4 for the TI-83/84 Plus calculator procedure for generating a scatter diagram.

Using a TI-83/84 Plus Calculator for Correlation

Important: The following procedure applies to this section and Sections 10-3 and 10-4. This procedure requires that you have paired sample data in two different lists. The paired data can be stored in lists such as L1 and L2, or they can be stored in named lists, such as those obtained from the APP containing the data sets from Appendix B.

TI-83/84 Plus

```
LinRegTTest
 y=a+bx
 β≠0 and ρ≠0
 t=5.989638817
 p=9.7332025E-4
 df=6
↓a=34.7698041
```

Continuation of the Above Screen

```
LinRegTTest
 y=a+bx
 β≠0 and ρ≠0
↑b=.2340614319
 s=4.973916136
 r²=.8567190634
 r=.9255911967
```

1. Enter or open two lists containing paired sample data.
2. Press STAT, then press ▷ twice to select **TESTS,** then proceed to select the menu item of **LinRegTTest** (linear regression t test).
3. For Xlist, enter the label or list name for the first variable, such as L1.
4. For Ylist, enter the label or list name for the second variable, such as L2.
5. For Freq (frequency), enter 1.
6. For the row labeled "β & ρ" select the option of $\neq 0$. This choice corresponds to a test of the null hypothesis $H_0 : \rho = 0$, which means that we are testing to determine whether there is a linear correlation between the two variables.
7. Scroll down to **Calculate** and press the ENTER key.
8. There are too many results to fit on one TI-83/84 Plus calculator screen, so use the ▽ key to scroll down for more results. For example, if you use the paired sample data in Table 10-1, the results will be as shown in the two accompanying screens.

Interpretation of results: See the accompanying screen displays showing $r = 0.926$ (rounded), $t = 5.990$ (rounded), and P-value $= 0.000973$ (rounded and expressed in standard form). Because the P-value is less than 0.05, we reject the null hypothesis of $\rho = 0$ and conclude that there is a linear correlation between the two variables. The values of a, b, and s will be discussed in the following two sections.

10-2 BASIC SKILLS AND CONCEPTS

Statistical Literacy and Critical Thinking

1. SSN and Income A government researcher wants to conduct a study to determine if there is a correlation between social security numbers and income. He collects the paired data from a random sample of 100 people. Should the methods of this section be used with the linear correlation coefficient? Why or why not?

2. Adverse Drug Reaction A clinical study is conducted to investigate the effectiveness of the drug Dozenol for treating insomnia. It was found that there is a correlation between the amount of Dozenol taken and the length of sleep. Based on this statistical analysis, can we conclude that Dozenol is the cause of sleep? Why or why not?

3. Correlation and Lurking Variable What is correlation? What is a lurking variable?

4. Identify Scatterplots Given below are three scatterplots generated from STATDISK. Match the scatterplots with these values of the linear correlation coefficient: $r = 0.857$, $r = -0.658$, $r = 0.012$.

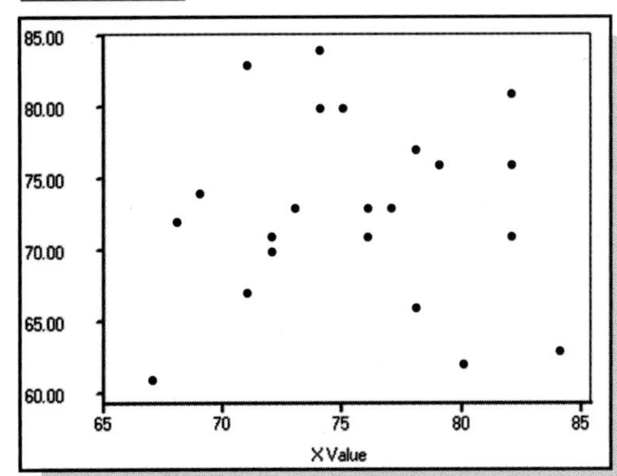

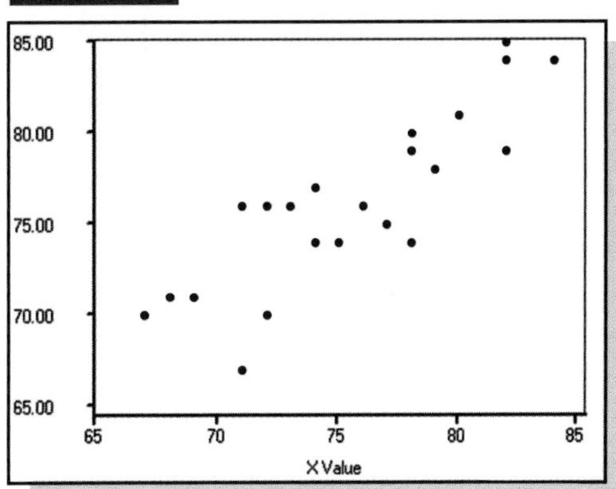

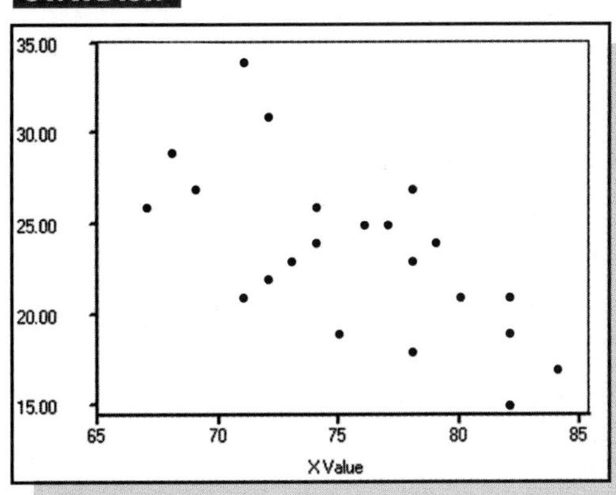

Interpreting r. In Exercises 5–8, use a significance level of $\alpha = 0.05$.

TI 5. Chest Sizes and Weights of Bears When eight bears were anesthetized, researchers measured the distances (in inches) around the bears' chests and weighed the bears (in pounds). The TI-83/84 Plus calculator was used to find that the value of the linear correlation coefficient is $r = 0.993$.

a. Is there a linear correlation between chest size and weight? Explain.

b. What proportion of the variation in weight can be explained by the linear relationship between weight and chest size?

TI 6. Guns and Murder Rate Using data collected from the FBI and the Bureau of Alcohol, Tobacco, and Firearms, the number of registered automatic weapons and the murder

continued

rate (in murders per 100,000 people) were obtained for each of eight randomly selected states. The TI-83/84 Plus calculator was used to find that the value of the linear correlation coefficient is $r = 0.885$.

a. Is there a linear correlation between the number of registered automatic weapons and the murder rate? Explain.

b. What proportion of the variation in the murder rate can be explained by the linear relationship between the murder rate and the number of registered automatic weapons?

TI **7.** Stocks and Super Bowl The Dow Jones Industrial Average (DJIA) high values and the total number of points scored in the Super Bowl were recorded for 21 different years. The TI-83/84 Plus calculator was used to find that the value of the linear correlation coefficient is $r = -0.133$.

a. Is there a linear correlation between DJIA high value and Super Bowl points? Explain.

b. What proportion of the variation in Super Bowl points can be explained by the variation in the high value of the DJIA?

TI **8.** Ages of Marathon Runners The ages and finishing times of 150 randomly selected runners who completed the New York City Marathon were recorded. A TI-83/84 Plus calculator was used to find that the value of the linear correlation coefficient is $r = 0.144$. (*Hint:* For $n = 150$, the critical values are ± 0.160.)

a. Is there a linear correlation between age and finishing time? Explain.

b. What proportion of the variation in the finishing time can be explained by the variation in the age?

Testing for a Linear Correlation. *In Exercises 9 and 10, use a scatterplot and the linear correlation coefficient r to determine whether there is a correlation between the two variables.*

TI **9.**

x	1	0	5	2	3
y	3	1	15	6	8

TI **10.**

x	0	3	3	1	4
y	1	7	2	5	5

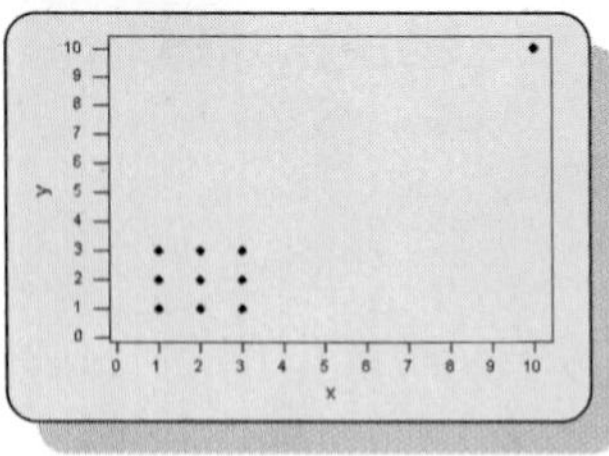

TI **11.** Effects of an Outlier Refer to the following Minitab-generated scatterplot.

a. Examine the pattern of all 10 points and subjectively determine whether there appears to be a correlation between x and y.

b. After identifying the 10 pairs of coordinates corresponding to the 10 points, find the value of the correlation coefficient r and determine whether there is a linear correlation.

c. Now remove the point with coordinates (10, 10) and repeat parts (a) and (b).

d. What do you conclude about the possible effect from a single pair of values?

TI **12.** Effects of Clusters Refer to the following Minitab-generated scatterplot. The four points in the lower left corner are measurements from women, and the four points in the upper right corner are from men.

a. Examine the pattern of the four points in the lower left corner (from women) only, and subjectively determine whether there appears to be a correlation between x and y for women.

b. Examine the pattern of the four points in the upper right corner (from men) only, and subjectively determine whether there appears to be a correlation between x and y for men.

Minitab

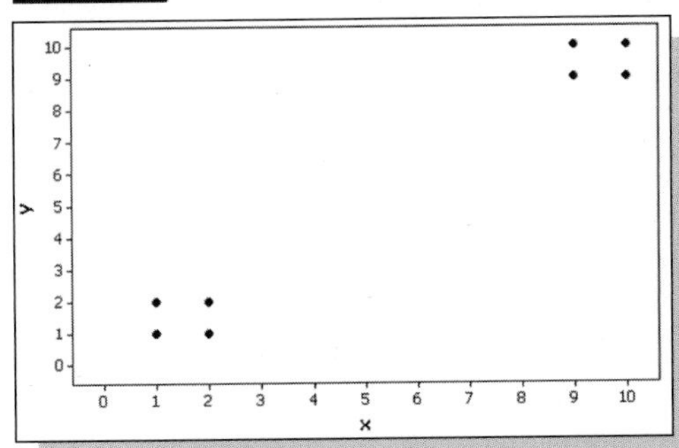

c. Find the linear correlation coefficient using only the four points in the lower left corner (for women). Will the four points in the upper right corner (for men) have the same linear correlation coefficient?

d. Find the value of the linear correlation coefficient using all eight points. What does that value suggest about the relationship between x and y?

e. Based on the preceding results, what do you conclude? Should the data from women and the data from men be considered together, or do they appear to represent two different and distinct populations that should be analyzed separately?

Testing for a Linear Correlation. *In Exercises 13–28, construct a scatterplot, find the value of the linear correlation coefficient r, find the critical value of r from Table A-6 by using $\alpha = 0.05$, and determine whether there is a linear correlation between the two variables. Save your work because the same data sets will be used in Section 10-3 exercises.*

TI 10 **13.** Old Faithful Examples in this section analyzed the correlation between duration times and time intervals after eruptions of the Old Faithful geyser. Using the data given below (from Table 10-1), is there a linear correlation between height of an eruption and the time interval after the eruption?

Height	140	110	125	120	140	120	125	150
Interval after	92	65	72	94	83	94	101	87

TI **14.** Song Audiences and Sales The table below lists the numbers of audience impressions (in hundreds of millions) listening to songs and the corresponding numbers of albums sold (in hundreds of thousands). The number of audience impressions is a count of the number of times people have heard the song. The table is based on data from *USA Today*. Does it appear that album sales are affected very strongly by the number of audience impressions?

Audience impressions	28	13	14	24	20	18	14	24	17
Albums sold	19	7	7	20	6	4	5	25	12

TI **15.** Movie Budgets and Gross Listed below are the budgets (in millions of dollars) and the gross receipts (in millions of dollars) for randomly selected movies (based on data from the Motion Picture Association of America). Does there appear to be a linear correlation between the money spent making the movie and the amount that it recovered in theaters? Apart from the budget amount, identify one other important factor that is likely to affect the amount that a movie earns.

Budget	62	90	50	35	200	100	90
Gross	65	64	48	57	601	146	47

TI **16.** Car Weight and Fuel Consumption Listed below are the weights (in pounds) and the highway fuel consumption amounts (in mi/gal) of randomly selected cars (Chrysler Sebring, Ford Mustang, BMW 3-Series, Ford Crown Victoria, Honda Civic, Mazda Protégé, Hyundai Accent). Is there a linear correlation between weight and highway fuel consumption? What does the result suggest about a national program to reduce the consumption of imported oil?

Weight	3175	3450	3225	3985	2440	2500	2290
Fuel consumption	27	29	27	24	37	34	37

TI **17.** Bear Chest Size and Weight Listed below are the chest sizes (in inches) and weights (in pounds) of randomly selected bears that were anesthetized and measured (based on data from Gary Alt and Minitab, Inc.). Because it is much more difficult to weigh a bear than to measure its chest size, the presence of a correlation could lead to a method for estimating weight based on chest size. Is there a linear correlation between chest size and weight?

Chest size	26	45	54	49	35	41	41	49	38	31
Weight	80	344	416	348	166	220	262	360	204	144

TI **18.** Supermodel Heights and Weights Listed below are heights (in inches) and weights (in pounds) for supermodels Michelle Alves, Nadia Avermann, Paris Hilton, Kelly Dyer, Christy Turlington, Bridget Hall, Naomi Campbell, Valerie Mazza, and Kristy Hume. Is there a correlation between height and weight? If there is a correlation, does it mean that there is a correlation between height and weight of all adult women?

Height (in.)	70	70.5	68	65	70	70	70	70	71
Weight (lb)	117	119	105	115	119	127	113	123	115

TI **19.** Blood Pressure Measurements Fourteen different second-year medical students took blood pressure measurements of the same patient and the results are listed below (data provided by Marc Triola, MD). Is there a correlation between systolic and diastolic values? Apart from correlation, is there some other method that might be used to address an important issue suggested by the data?

Systolic	138	130	135	140	120	125	120	130	130	144	143	140	130	150
Diastolic	82	91	100	100	80	90	80	80	80	98	105	85	70	100

TI **20.** Murders and Population Size The table below lists the numbers of murders and the population sizes (in hundreds of thousands) for large cities in America during a recent year (based on data from the *New York Times*). What do you conclude?

Murders	258	264	402	253	111	648	288	654	256	60	590
Population	4	6	9	6	3	29	15	38	20	6	81

TI **21.** Buying a TV Audience The *New York Post* published the annual salaries (in millions) and the number of viewers (in millions), with results given below for Oprah Winfrey, David Letterman, Jay Leno, Kelsey Grammer, Barbara Walters, Dan Rather, James

Gandolfini, and Susan Lucci, respectively. Is there a correlation between salary and number of viewers? Which of the listed stars has the lowest cost per viewer? Highest cost per viewer?

Salary	100	14	14	35.2	12	7	5	1
Viewers	7	4.4	5.9	1.6	10.4	9.6	8.9	4.2

TI **22.** Smoking and Cotinine When nicotine is absorbed by the body, cotinine is produced. A measurement of cotinine in the body is therefore a good indicator of how much a person smokes. Listed below are the reported numbers of cigarettes smoked per day and the measured amounts of nicotine (in ng/mL). (The values are from randomly selected subjects in the National Health Examination Survey.) Is there a linear correlation? Explain the result.

x (cigarettes per day)	60	10	4	15	10	1	20	8	7	10	10	20
y (cotinine)	179	283	75.6	174	209	9.51	350	1.85	43.4	25.1	408	344

TI **23.** Temperatures and Marathons In "The Effects of Temperature on Marathon Runner's Performance," by David Martin and John Buoncristiani (*Chance,* Vol. 12, No. 4), high temperatures and times (in minutes) were given for women who won the New York City marathon in recent years. Results are listed below. Is there a correlation between temperature and winning time? Does it appear that winning times are affected by temperature?

x (temperature)	55	61	49	62	70	73	51	57
y (time)	145.283	148.717	148.300	148.100	147.617	146.400	144.667	147.533

TI **24.** Parent–Child Heights Listed below are heights (in inches) of mothers and heights (in inches) of their daughters (based on data from the National Health Examination Survey). Does there appear to be a linear correlation between mother's heights and the heights of their daughters?

Mother's height	63	67	64	60	65	67	59	60
Daughter's height	58.6	64.7	65.3	61.0	65.4	67.4	60.9	63.1

TI **25.** Crickets and Temperature One classic application of correlation involves the association between the temperature and the number of times a cricket chirps in a minute. Listed below are the numbers of chirps in 1 min and the corresponding temperatures in degrees Fahrenheit (based on data from *The Song of Insects* by George W. Pierce, Harvard University Press). Is there sufficient evidence to conclude that there is a relationship between the number of chirps in 1 min and the temperature?

Chirps in 1 min	882	1188	1104	864	1200	1032	960	900
Temperature (°F)	69.7	93.3	84.3	76.3	88.6	82.6	71.6	79.6

TI **26.** Fires and Acres Burned Listed below are the numbers of fires (in thousands) and the acres that were burned (in millions) in 11 western states in each year of the last decade (based on data from *USA Today*). Is there a correlation? The data were listed under a headline of "Loggers seize on fires to argue for more cutting." Do the data support the argument that as loggers remove more trees, the risk of fire decreases because the forests are less dense?

Fires	73	69	58	48	84	62	57	45	70	63	48
Acres burned	6.2	4.2	1.9	2.7	5.0	1.6	3.0	1.6	1.5	2.0	3.7

TI **27.** Height and Pulse Rate A medical student hypothesizes that taller people have faster pulse rates because the blood has farther to travel. Pulse rates (beats per minute) and heights (inches) are listed below for a random sample of adult women (based on data from the National Health Examination Survey). The height and pulse entries are listed in order so that they correspond to the same people. Does the medical student's hypothesis appear to be correct?

Height

64.3 66.4 62.3 62.3 59.6 63.6
59.8 63.3 67.9 61.4 66.7 64.8

Pulse Rate

76 72 88 60 72 68
80 64 68 68 80 76

TI **28.** State Budget and Days Late New York State has become notorious for approving the state budget after the annual deadline of April 1. The amounts of the budget (in billions of dollars adjusted for inflation) and the numbers of days late are listed below with corresponding entries representing the same year. (The data are in order by row.) Does it appear that the size of the budget affects the number of days that the budget is late? Use a 0.05 significance level.

Budget

101 96 91 85 80 73 71 66 63 63
62 58 55 52 49 46 44 40 37 35

Number of Days Late

133 44 45 124 34 125 13 125 103 67
68 4 1 64 48 18 19 10 4 4

Testing for a Linear Correlation. *In Exercises 29–32, use the data from Appendix B to construct a scatterplot, find the value of the linear correlation coefficient r, and use a significance level of $\alpha = 0.05$ to determine whether there is a linear correlation between the two variables. Save your work because the same data sets will be used in the next section.*

TI **29.** Appendix B Data Set: List Price and Selling Price Refer to Data Set 18 in Appendix B and use the list prices and selling prices of homes sold.

TI **30.** Appendix B Data Set: Discarded Plastic and Household Size Refer to Data Set 16 in Appendix B and use the weights of discarded plastic and the corresponding household sizes.

TI **31.** Appendix B Data Set: Cigarette Tar and Nicotine Refer to Data Set 3 in Appendix B.

a. Use the paired data consisting of tar and nicotine. Based on the result, does there appear to be a linear correlation between cigarette tar and nicotine? If so, can researchers reduce their laboratory expenses by measuring only one of these two variables?

b. Use the paired data consisting of carbon monoxide and nicotine. Based on the result, does there appear to be a linear correlation between cigarette carbon monoxide and nicotine? If so, can researchers reduce their laboratory expenses by measuring only one of these two variables?

c. Assume that researchers want to develop a method for predicting the amount of nicotine, and they want to measure only one other variable. In choosing between tar and carbon monoxide, which is the better choice? Why?

TI **32.** Appendix B Data Set: Forecast and Actual Temperatures Refer to Data Set 8 in Appendix B.

a. Use the five-day forecast high temperatures and the actual high temperatures. Is there a correlation? Does a linear correlation imply that the five-day forecast temperatures are accurate?

b. Use the one-day forecast high temperatures and the actual high temperatures. Is there a correlation? Does a linear correlation imply that the one-day forecast temperatures are accurate?

c. Which would you expect to have a higher correlation with the actual high temperatures: the five-day forecast high temperatures or the one-day forecast high temperatures? Are the results from parts (a) and (b) what you would expect? If there is a very high correlation between forecast temperatures and actual temperatures, does it follow that the forecast temperatures are accurate?

Identifying Correlation Errors. *In Exercises 33–36, describe the error in the stated conclusion. (See the list of common errors included in this section.)*

33. *Given:* The paired sample data of the ages of subjects and their scores on a test of reasoning result in a linear correlation coefficient very close to 0.

Conclusion: Younger people tend to get higher scores.

34. *Given:* There is a linear correlation between personal income and years of education.

Conclusion: More education causes a person's income to rise.

35. *Given:* Subjects take a test of verbal skills and a test of manual dexterity, and those pairs of scores result in a linear correlation coefficient very close to 0.

Conclusion: Scores on the two tests are not related in any way.

36. *Given:* There is a linear correlation between state average tax burdens and state average incomes.

Conclusion: There is a linear correlation between individual tax burdens and individual incomes.

10-2 BEYOND THE BASICS

TI **37.** Correlations with Transformed Data In addition to testing for a linear correlation between x and y, we can often use *transformations* of data to explore for other relationships. For example, we might replace each x value by x^2 and use the methods of this section to determine whether there is a linear correlation between y and x^2. Given the paired data in the accompanying table, construct the scatterplot and then test for a linear correlation between y and each of the following. Which case results in the largest value of r?

a. x **b.** x^2 **c.** $\log x$ **d.** $\sqrt{x}$ **e.** $1/x$

x	1.3	2.4	2.6	2.8	2.4	3.0	4.1
y	0.11	0.38	0.41	0.45	0.39	0.48	0.61

38. Finding Critical r-Values The critical values of r in Table A-6 are found by solving

$$t = \frac{r}{\sqrt{\dfrac{1 - r^2}{n - 2}}}$$

continued

for r to get

$$r = \frac{t}{\sqrt{t^2 + n - 2}}$$

where the t value is found from Table A-3 by assuming a two-tailed case with $n - 2$ degrees of freedom. Table A-6 lists the results for selected values of n and α. Use the formula for r given here and Table A-3 (with $n - 2$ degrees of freedom) to find the critical values of r for the given cases.

a. H_1: $\rho \neq 0$, $n = 50$, $\alpha = 0.05$

b. H_1: $\rho \neq 0$, $n = 75$, $\alpha = 0.10$

c. H_1: $\rho < 0$, $n = 20$, $\alpha = 0.05$

d. H_1: $\rho > 0$, $n = 10$, $\alpha = 0.05$

e. H_1: $\rho > 0$, $n = 12$, $\alpha = 0.01$

39. Constructing Confidence Intervals for ρ When obtaining samples of n paired values from a population with a correlation coefficient of ρ, the distribution of linear correlation coefficients r is not a normal distribution, but values of $z = \frac{1}{2}\ln\left(\frac{1+r}{1-r}\right)$ have a distribution that is approximately normal with mean $\frac{1}{2}\ln\left(\frac{1+\rho}{1-\rho}\right)$ and standard deviation $\sqrt{\frac{1}{n-3}}$. This conversion of r values to z values is referred to as a *Fisher transformation*. This Fisher transformation can be used to construct a confidence interval estimate of the population parameter ρ. Use the following procedure to construct a 95% confidence interval for ρ, given 50 pairs of data for which $r = 0.600$.

Step a. Use Table A-2 to find $z_{\alpha/2}$ that corresponds to the desired degree of confidence.

Step b. Evaluate the interval limits w_L and w_R:

$$w_L = \frac{1}{2}\ln\left(\frac{1+r}{1-r}\right) - z_{\alpha/2} \cdot \frac{1}{\sqrt{n-3}}$$

$$w_R = \frac{1}{2}\ln\left(\frac{1+r}{1-r}\right) + z_{\alpha/2} \cdot \frac{1}{\sqrt{n-3}}$$

Step c. Now evaluate the confidence interval limits in the expression below.

$$\frac{e^{2w_L} - 1}{e^{2w_L} + 1} < \rho < \frac{e^{2w_R} - 1}{e^{2w_R} + 1}$$

40. Power of a Test A sample of size $n = 36$ is used with a 0.05 significance level to test the null hypothesis of $\rho = 0$ against the alternative hypothesis of $\rho \neq 0$. If the population actually has a correlation coefficient of $\rho = 0.55$, the value of β is 0.06. What does that value of β indicate? What is the power of the test? Write a statement interpreting the value of the power.

10-3 Regression

Key Concept Section 10-2 gave us the tools for determining whether there is a linear correlation between two variables. The key concept of this section is to describe the relationship between two variables by finding the graph and equation of the straight line that best represents the relationship. This straight line is called the *regression line,* and its equation is called the *regression equation.* Given sample paired data, we will find estimated values of the y-intercept b_0 and slope b_1 so that we can identify a straight line with the equation of $\hat{y} = b_0 + b_1x$. Under suitable conditions, that equation can be used for making predictions. Software or calculators can be used to perform the somewhat messy arithmetic, so we should focus on understanding concepts instead of number crunching. As in Section 10-2, this section is partitioned into two parts: (1) Basic Concepts of Regression; (2) Beyond the Basics of Regression. The first part includes core concepts that should be understood well before moving on to the second part.

Part 1: Basic Concepts of Regression

In some cases, two variables are related in a *deterministic* way, meaning that given a value for one variable, the value of the other variable is automatically determined without any error. For example, the total cost y of an item with a list price of x and a sales tax of 5% can be found by using the deterministic equation $y = 1.05x$. If an item is priced at $50, its total cost is $52.50. Such functions are considered extensively in algebra courses. In this chapter, we are more interested in *probabilistic* models, meaning that one variable is not determined completely by the other variable. For example, a child's height is not determined completely by the height of the father (or mother). Sir Francis Galton (1822–1911) studied the phenomenon of heredity and showed that when tall or short couples have children, the heights of those children tend to *regress,* or revert to the more typical mean height for people of the same gender. We continue to use Galton's "regression" terminology, even though our data do not involve the same height phenomena studied by Galton.

The box on page 562 includes the definition of the regression equation and regression line, as well as the notation and formulas we are using. The regression equation expresses a relationship between x (called the **explanatory variable,** or **predictor variable,** or **independent variable**) and $\hat{y}$ (called the **response variable,** or **dependent variable**). The typical equation of a straight line $y = mx + b$ is expressed in the form $\hat{y} = b_0 + b_1x$, where b_0 is the y-intercept and b_1 is the slope. The given notation shows that b_0 and b_1 are sample statistics used to estimate the population parameters β_0 and β_1. We will use paired sample data to estimate the regression equation. Using only sample data, we can't find the exact values of the population parameters β_0 and β_1, but we can use the sample data to estimate them with b_0 and b_1, which are found by using Formulas 10-2 and 10-3.

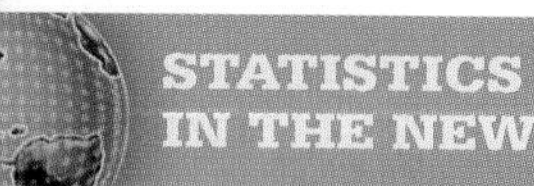

STATISTICS IN THE NEWS

1° Forecast Error = $1 Billion

Although the prediction of forecast temperatures might seem to be an inexact science, many companies are working feverishly to obtain more accurate estimates. *USA Today* reporter Del Jones wrote that "the annual cost of electricity could decrease by at least $1 billion if the accuracy of weather forecasts improved by 1 degree Fahrenheit." When referring to the Tennessee Valley Authority, he states that "forecasts over its 80,000 square miles have been wrong by an average of 2.35 degrees the last 2 years, fairly typical of forecasts nationwide. Improving that to within 1.35 degrees would save TVA as much as $100,000 a day, perhaps more." Forecast temperatures are used to determine the allocation of power from generators, nuclear plants, hydroelectric plants, coal, natural gas, and wind. Statistical forecasting techniques are being refined so that money and natural resources can be saved.

Requirements

1. The sample of paired (x, y) data is a *random* sample of quantitative data.
2. Visual examination of the scatterplot shows that the points approximate a straight-line pattern.
3. Any outliers must be removed if they are known to be errors. Consider the effects of any outliers that are not known errors.

Note: Requirements 2 and 3 above are simplified attempts at checking these formal requirements for regression analysis:

- For each fixed value of x, the corresponding values of y have a distribution that is bell-shaped.
- For the different fixed values of x, the distributions of the corresponding y-values all have the same variance. (This is violated if part of the scatterplot shows points very close to the regression line while another portion of the scatterplot shows points that are much farther away from the regression line. See the discussion of residual plots near the end of this section.)
- For the different fixed values of x, the distributions of the corresponding y-values have means that lie along the same straight line.
- The y-values are independent.

Results are not seriously affected if departures from normal distributions and equal variances are not too extreme.

Definitions

Given a collection of paired sample data, the **regression equation**

$$\hat{y} = b_0 + b_1x$$

algebraically describes the relationship between the two variables. The graph of the regression equation is called the **regression line** (or *line of best fit,* or *least-squares line*).

Notation for Regression Equation

	Population Parameter	Sample Statistic
y-intercept of regression equation	β_0	b_0
Slope of regression equation	β_1	b_1
Equation of the regression line	$y = \beta_0 + \beta_1x$	$\hat{y} = b_0 + b_1x$

Finding the slope b_1 and y-intercept b_0 in the regression equation $\hat{y} = b_0 + b_1x$

Formula 10-2 **Slope:** $$b_1 = \frac{n(\Sigma xy) - (\Sigma x)(\Sigma y)}{n(\Sigma x^2) - (\Sigma x)^2}$$

Formula 10-3 **y-intercept:** $$b_0 = \bar{y} - b_1\bar{x}$$

TI-83/84 Plus Calculator: The TI-83/84 Plus calculator is designed to generate the value of the slope and y-intercept. See the end of this section.

The y-intercept b_0 can also be found using the formula shown below, but it is much easier to use Formula 10-3 instead.

$$b_0 = \frac{(\Sigma y)(\Sigma x^2) - (\Sigma x)(\Sigma xy)}{n(\Sigma x^2) - (\Sigma x)^2}$$

Formulas 10-2 and 10-3 might look intimidating, but they are programmed into the TI-83/84 Plus calculator, so the values of b_0 and b_1 can be easily found. Once we have evaluated b_1 and b_0, we can identify the estimated regression equation, which has the following special property: *The regression line fits the sample points best.* (The specific criterion used to determine which line fits "best" is the least-squares property, which will be described later.) We will now briefly discuss rounding and then illustrate the procedure for finding and applying the regression equation.

Rounding the Slope b_1 and the y-Intercept b_0

Round b_1 and b_0 to three significant digits. It's difficult to provide a simple universal rule for rounding values of b_1 and b_0, but this rule will work for most situations in this book. Depending on how you round, this book's answers to examples and exercises may be slightly different from your answers.

EXAMPLE Finding the Regression Equation In Section 10-2 we used the simple random sample of values listed below to find the linear correlation coefficient of $r = -0.956$. (Using the methods of Section 10-2, we conclude that there is a linear correlation between x and y.) Use the given sample data to find the regression equation.

x	3	1	3	5
y	5	8	6	4

SOLUTION

REQUIREMENT ✔ The data are a simple random sample. The TI-83/84 Plus calculator scatterplot in Section 10-2 shows a pattern of points that does appear to be a straight-line pattern. There are no outliers. We can proceed to find the slope and intercept of the regression line. ✔

We will find the regression equation by using Formulas 10-2 and 10-3 and these values already found in Table 10-2 in Section 10-2:

$n = 4$ $\quad \Sigma x = 12$ $\quad \Sigma y = 23$

$\Sigma x^2 = 44$ $\quad \Sigma y^2 = 141$ $\quad \Sigma xy = 61$

First find the slope b_1 by using Formula 10-2:

$$b_1 = \frac{n(\Sigma xy) - (\Sigma x)(\Sigma y)}{n(\Sigma x^2) - (\Sigma x)^2}$$

$$= \frac{4(61) - (12)(23)}{4(44) - (12)^2} = \frac{-32}{32} = -1$$

continued

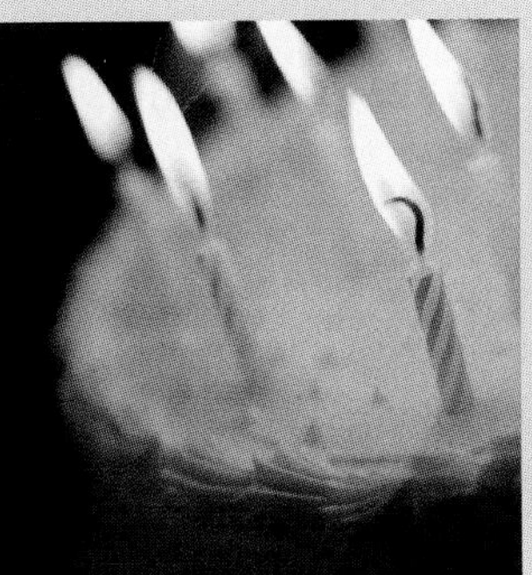

Postponing Death

Several studies addressed the ability of people to postpone their death until after an important event. For example, sociologist David Phillips analyzed death rates of Jewish men who died near Passover, and he found that the death rate dropped dramatically in the week before Passover, but rose the week after. A more recent study suggests that people have no such ability to postpone death. Based on records of 1.3 million deaths, this more recent study found no relationship between the time of death and Christmas, Thanksgiving, or the person's birthday. Dr. Donn Young, one of the researchers, said that "the fact is, death does not keep a calendar. You can't put in your Palm Pilot and say 'O.K, let's have dinner on Friday and I'll pencil in death for Sunday.' " The findings were disputed by David Phillips, who said that the study focused on cancer patients, but they are least likely to have psychosomatic effects.

Next, find the y-intercept b_0 by using Formula 10-3 (with $\bar{y} = 23/4 = 5.75$ and $\bar{x} = 12/4 = 3$):

$$\begin{aligned} b_0 &= \bar{y} - b_1\bar{x} \\ &= 5.75 - (-1)(3) = 8.75 \end{aligned}$$

Knowing the slope b_1 and y-intercept b_0, we can now express the estimated equation of the regression line as

$$\hat{y} = 8.75 - 1x$$

We should realize that this equation is an *estimate* of the true regression equation $y = \beta_0 + \beta_1 x$. This estimate is based on one particular set of sample data, but another sample drawn from the same population would probably lead to a slightly different equation.

EXAMPLE Old Faithful Using the duration/interval after eruption times from Table 10-1, we have found that the linear correlation coefficient is $r = 0.926$. Using the same sample data, find the equation of the regression line.

SOLUTION

REQUIREMENT ✔ Figure 10-1(a) is a scatterplot of these points showing a pattern that is approximately a straight-line pattern. There are no outliers, and the sample data have been randomly selected. We can now proceed to find the equation of the regression line. ✔

Using the same procedure illustrated in the preceding example, or using the TI-83/84 Plus calculator, we can find that the eight pairs of duration/interval after eruption times in Table 10-1 result in $b_0 = 34.8$ and $b_1 = 0.234$. Substituting the computed values for b_0 and b_1, we express the regression equation as $\hat{y} = 34.8 + 0.234x$. The accompanying display shows the TI-83/84 Plus scatterplot with the regression line included. We can see that the regression line fits the data reasonably well.

TI-83/84 Plus

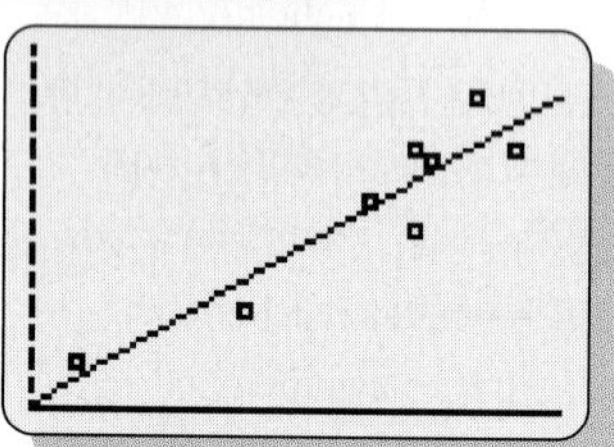

Using the Regression Equation for Predictions

Regression equations are often useful for *predicting* the value of one variable, given some particular value of the other variable. If the regression line fits the data quite well, then it makes sense to use its equation for predictions, provided that we don't go beyond the scope of the available values. Don't base predictions on values that are far beyond the boundaries of the known sample data. For example, using the sample data in Table 10-1, a scatterplot suggests that the durations and time intervals after eruptions fit a straight-line pattern reasonably well, so if we observe an eruption that is 180 sec, we can predict the time interval after the eruption (to the next eruption) by substituting $x = 180$ into the regression equation of $\hat{y} = 34.8 + 0.234x$. The following result shows that if an eruption has a duration of 180 sec, the best predicted time interval after the eruption is 76.9 min. (Remember, durations are in seconds and time intervals after eruptions are in minutes.)

$$\hat{y} = 34.8 + 0.234x$$
$$= 34.8 + 0.234(180) = 76.9 \text{ min} \quad \text{(rounded)}$$

However, *we should use the equation of the regression line only if the regression equation is a good model for the data.* The suitability of the regression equation can be judged by testing the significance of the linear correlation coefficient r. See the following procedure for using the regression equation for making predictions. See also the following example that formalizes the procedure for using a duration time of 180 sec to predict the time interval after the eruption.

Part 2: Beyond the Basics of Regression

Using the Regression Equation for Predictions (*Continued*)

We noted that we should use the equation of the regression line for predictions only if the regression equation is a good model for the data. To be more precise, *we should use the regression equation for predictions only if there is a linear correlation. In the absence of a linear correlation, we should not use the regression equation for projecting or predicting; instead, our best estimate of the second variable is simply its sample mean.*

In predicting a value of y based on some given value of x . . .

1. **If there is *not* a linear correlation, the best predicted y-value is $\bar{y}$.**
2. **If there is a linear correlation, the best predicted y-value is found by substituting the x-value into the regression equation.**

How good is the regression equation as a model for the population data? Figure 10-7 summarizes the process for making predictions, and that process is easier to understand if we think of r as a measure of how well the regression line fits the sample data. In addition to thinking about the presence or absence of a linear correlation, we might also consider this: Regression equations obtained from paired sample data with r very close to -1 or $+1$ (because the points in the scatterplot are very close to the regression line) are likely to be much better models for the population data than regression equations from data sets with values of r that are not so close to -1 or $+1$ (even if r is found to be significant). With some very large samples of paired data, we might find that r is significant, even though it is a relatively small value, such as 0.200. In this case, the regression equation might be an acceptable model, but predictions might not be very accurate because r is not very close to -1 or $+1$. If r is not significant, then the regression line fits the data poorly, and the regression equation should not be used for predictions.

Predicting Condo Prices

A massive study involved 99,491 sales of condominiums and cooperatives in Manhattan. The study used 41 different variables used to predict the value of the condo or co-op. The variables include condition of the unit, the neighborhood, age, size, and whether there are doormen. Some conclusions: With all factors equal, a condo is worth 15.5% more than a co-op; a fireplace increases the value of a condo 9.69% and it increases the value of a co-op 11.36%; an additional bedroom in a condo increases the value by 7.11% and it increases the value in a co-op by 18.53%. This use of statistical methods allows buyers and sellers to estimate value with much greater accuracy. Methods of multiple regression (Section 10-5) are used when there is more than one predictor variable, as in this study. (Based on data from "So How Much Is That . . . Worth," by Dennis Hevesi, *New York Times*.)

Figure 10-7
Procedure for Predicting

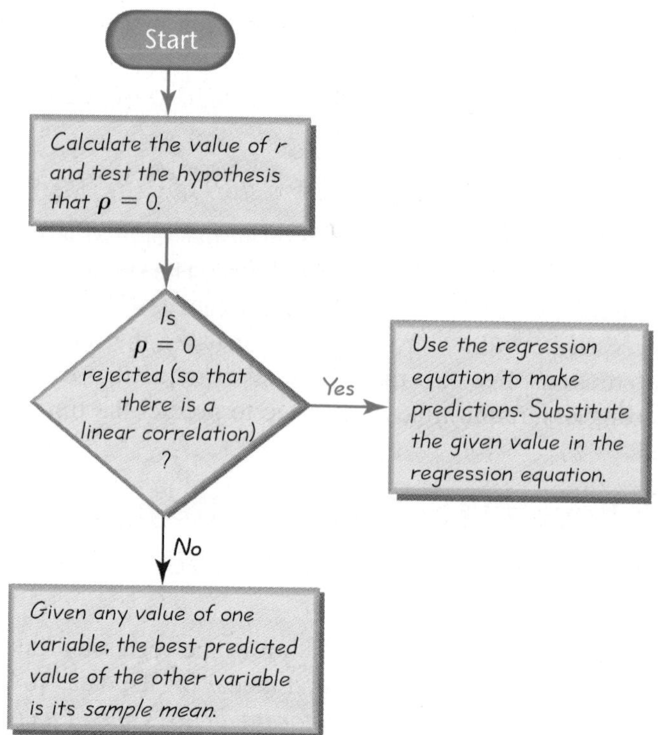

CHAPTER 10 PROBLEM

EXAMPLE Predicting Eruptions of Old Faithful Using the duration times and time intervals after eruptions in Table 10-1, we found that there is a linear correlation between the two variables, and we also found that the regression equation is $\hat{y} = 34.8 + 0.234x$. Assuming that the current eruption has a duration of $x = 180$ sec, find the best predicted value of y, the time interval after this eruption (which is the predicted time to the next eruption).

SOLUTION

REQUIREMENT ✔ There's a strong temptation to jump in and substitute 180 for x in the regression equation, but we should first consider whether there is a linear correlation that justifies the use of that equation. In this example, the scatterplot shows that the points approximate a straight-line pattern, and we do have a linear correlation (with $r = 0.926$), so our predicted value is found as shown below. ✔

Because there is a linear correlation, our predicted value can be found by substitution in the regression equation, as follows:

$$\begin{aligned}\hat{y} &= 34.8 + 0.234x \\ &= 34.8 + 0.234(180) = 76.9 \text{ min} \quad \text{(rounded)}\end{aligned}$$

The predicted time interval after the current eruption to the next eruption is predicted to be 76.9 min. (If there had not been a linear correlation, our best predicted value would have been $\bar{y} = 688/8 = 86.0$ min.)

EXAMPLE Hat Size and IQ There is obviously no linear correlation between hat sizes and IQ scores of adults. Given that an individual has a hat size of 7, find the best predicted value of this person's IQ score.

SOLUTION

REQUIREMENT ✔ Because there is no linear correlation, we do not use the equation of the regression line for making predictions. ✔

Because there is no linear correlation, we do not use a regression equation. There is no need to collect paired sample data consisting of hat size and IQ score for a sample of randomly selected adults. Instead, the best predicted IQ score is simply the mean IQ of all adults, which is 100.

Carefully compare the solutions to the preceding two examples and note that we used the regression equation when there was a linear correlation, but in the absence of such a correlation, the best predicted value of y is simply the value of the sample mean $\bar{y}$. A common error is to use the regression equation for making a prediction when there is no linear correlation. That error violates the first of the following guidelines.

Guidelines for Using the Regression Equation

1. *If there is no linear correlation, don't use the regression equation to make predictions.*
2. *When using the regression equation for predictions, stay within the scope of the available sample data.* If you find a regression equation that relates women's heights and shoe sizes, it's absurd to predict the shoe size of a woman who is 10 ft tall.
3. *A regression equation based on old data is not necessarily valid now.* The regression equation relating used-car prices and ages of cars is no longer usable if it's based on data from the 1990s.
4. *Don't make predictions about a population that is different from the population from which the sample data were drawn.* If we collect sample data from men and develop a regression equation relating age and TV remote-control usage, the results don't necessarily apply to women. If we use state *averages* to develop a regression equation relating SAT math scores and SAT verbal scores, the results don't necessarily apply to *individuals*.

Interpreting the Regression Equation: Marginal Change

We can use the regression equation to see the effect on one variable when the other variable changes by some specific amount.

Definition

In working with two variables related by a regression equation, the **marginal change** in a variable is the amount that it changes when the other variable changes by exactly one unit. The slope b_1 in the regression equation represents the marginal change in y that occurs when x changes by one unit.

For the Table 10-1 data of duration times and time intervals after eruptions of Old Faithful, the regression line has a slope of 0.234, which shows that if we increase x (the duration time) by 1 sec, the predicted time interval after the eruption will increase by 0.234 min. That is, for every additional 1 sec of duration time, we expect the time interval after the eruption (to the next eruption) to increase by an additional 0.234 min.

Outliers and Influential Points

A correlation/regression analysis of bivariate (paired) data should include an investigation of *outliers* and *influential points,* defined as follows.

Definitions

In a scatterplot, an **outlier** is a point lying far away from the other data points.

Paired sample data may include one or more **influential points,** which are points that strongly affect the graph of the regression line.

An outlier is easy to identify: Examine the scatterplot and identify a point that is far away from the others. Here's how to determine whether a point is an influential point: Graph the regression line resulting from the data with the point included, then graph the regression line resulting from the data with the point excluded. If the graph changes by a considerable amount, the point is influential. Influential points are often found by identifying those outliers that are *horizontally* far away from the other points.

For example, see the two accompanying scatterplots. The first scatterplot represents the paired duration and interval after times listed in Table 10-1. The second scatterplot includes this additional pair of data: $x = 300$, $y = 1$ (an eruption lasts for 300 sec and the time interval after the eruption is 1 min). This additional point would be an influential point because the graph of the regression line would change considerably, as shown. Compare the two regression lines and you will see clearly that the addition of that one pair of values has a very dramatic effect on the regression line.

Original Data

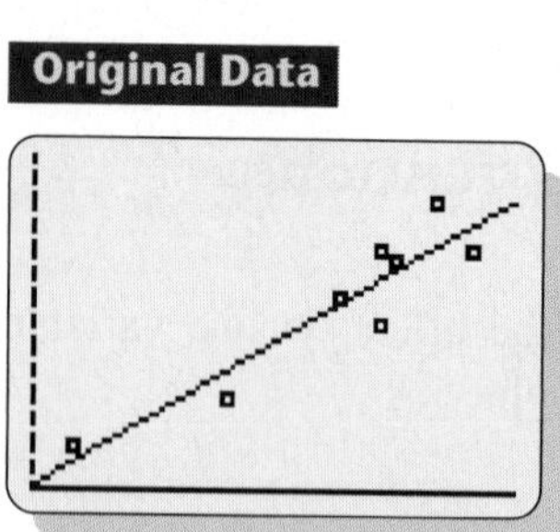

Point added: $x = 300$, $y = 1$

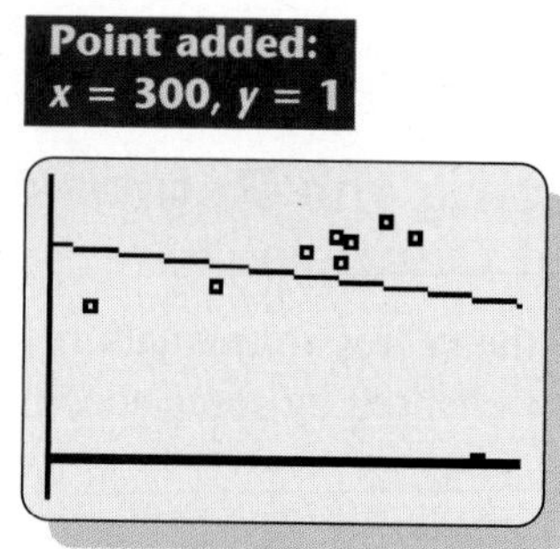

Residuals and the Least-Squares Property

We have stated that the regression equation represents the straight line that fits the data "best," and we will now describe the criterion used in determining the line that

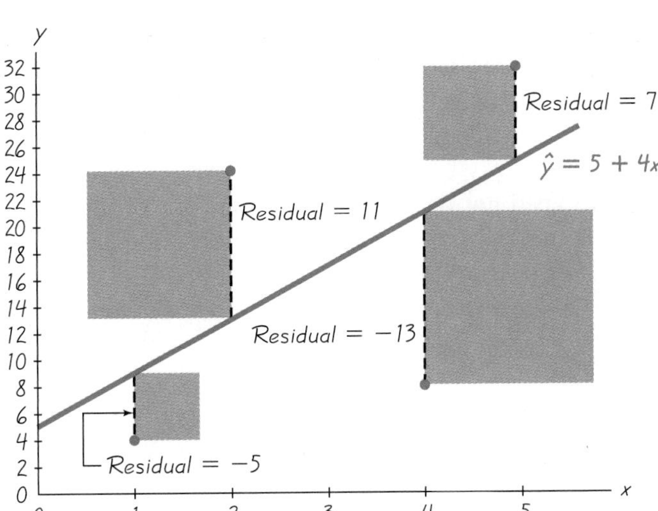

Figure 10-8

Residuals and Squares of Residuals

is better than all others. This criterion is based on the vertical distances between the original data points and the regression line. Such distances are called *residuals.*

Definition

For a sample of paired (x, y) data, a **residual** is the difference $(y - \hat{y})$ between an observed sample y-value and the value of $\hat{y}$, which is the value of y that is predicted by using the regression equation. That is,

$$\text{residual} = \text{observed } y - \text{predicted } y = y - \hat{y}$$

This definition might seem as clear as tax-form instructions, but you can easily understand residuals by referring to Figure 10-8, which corresponds to the paired sample data shown in the margin. In Figure 10-8, the residuals are represented by the dashed lines. For a specific example, see the residual indicated as 7, which is directly above $x = 5$. If we substitute $x = 5$ into the regression equation $\hat{y} = 5 + 4x$, we get a predicted value of $\hat{y} = 25$. When $x = 5$, the *predicted* value of y is $\hat{y} = 25$, but the actual *observed* sample value is $y = 32$. The difference $y - \hat{y} = 32 - 25 = 7$ is a residual.

x	1	2	4	5
y	4	24	8	32

The regression equation represents the line that fits the points "best" according to the following *least-squares property.*

Definition

A straight line satisfies the **least-squares property** if the sum of the squares of the residuals is the smallest sum possible.

From Figure 10-8, we see that the residuals are −5, 11, −13, and 7, so the sum of their squares is

$$(-5)^2 + 11^2 + (-13)^2 + 7^2 = 364$$

We can visualize the least-squares property by referring to Figure 10-8, where the squares of the residuals are represented by the red-square areas. The sum of the red-square areas is 364, which is the smallest sum possible. Use any other straight line, and the red squares will combine to produce an area larger than the combined red area of 364.

Fortunately, we need not deal directly with the least-squares property when we want to find the equation of the regression line. Calculus has been used to build the least-squares property into Formulas 10-2 and 10-3. Because the derivations of these formulas require calculus, we don't include them in this text.

Residual Plots

In this section and the preceding section we listed simplified requirements for the effective analyses of correlation and regression results. We noted that we should always begin with a scatterplot, and we should verify that the pattern of points is approximately a straight-line pattern. We should also consider outliers. A *residual plot* can be another helpful device for analyzing correlation and regression results and for checking the requirements necessary for making inferences about correlation and regression.

Definition

A **residual plot** is a scatterplot of the (x, y) values after each of the y-coordinate values has been replaced by the residual value $y - \hat{y}$ (where $\hat{y}$ denotes the predicted value of y). That is, a residual plot is a graph of the points $(x, y - \hat{y})$.

To construct a residual plot, use the same x-axis as the scatterplot, but use a vertical axis of residual values. Draw a horizontal reference line through the residual value of 0, then plot the paired values of $(x, y - \hat{y})$. Because the manual construction of residual plots can be what mathematicians refer to as "tedious," it is recommended that software be used. When analyzing a residual plot, look for a pattern in the way the points are configured, and use these criteria:

If a residual plot does not reveal any pattern, the regression equation is a good representation of the association between the two variables.

If a residual plot reveals some systematic pattern, the regression equation is not a good representation of the association between the two variables.

Consider the accompanying examples of Minitab displays. With Case 1, all is well. The points are close to the regression line, so the regression equation is a good model for describing the association between the two variables. The corresponding residual plot does not reveal any distinct pattern.

Case 2 results in a scatterplot showing an association between the two variables, but that association is not linear. The corresponding residual plot shows a distinct pattern, confirming that the linear model is not a good model in this case.

Case 3 has a scatterplot in which the points are getting farther away from the regression line, and the residual plot does reveal a pattern of increasing variation, which violates the requirement that for different values of x, the distributions of y values have the same variance. In this case, the regression equation is probably not a good model.

After acquiring the ability to obtain and analyze residual plots along with scatterplots, we are better prepared to check the requirements necessary to ensure the validity of inferences made from using the correlation and regression procedures.

Case 1

x	0	1	2	3	4	5	7	8	9	10
y	1	4	8	18	19	24	36	43	42	47

The regression line fits the data well.

Minitab

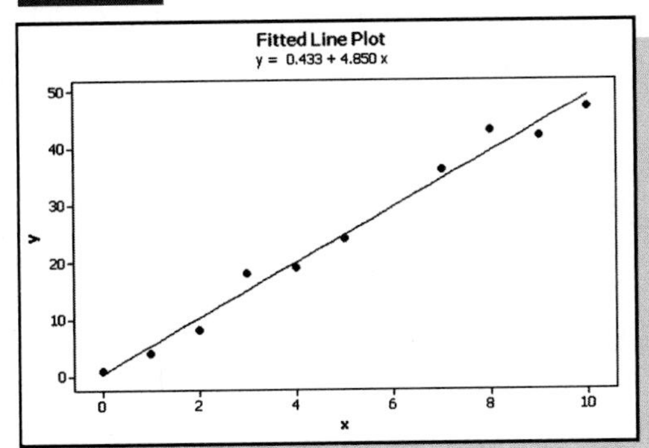

The residual plot reveals no pattern.

Minitab

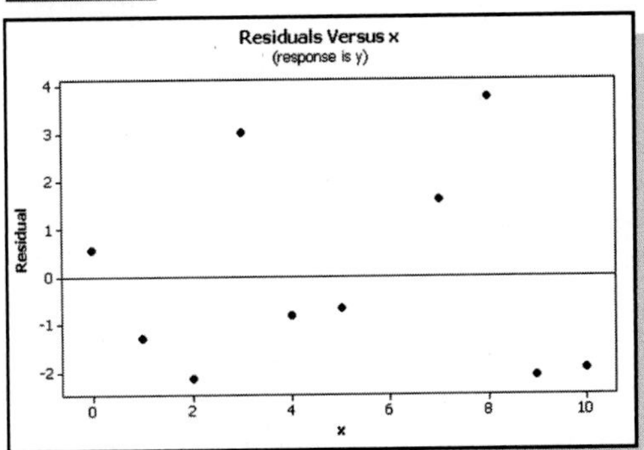

Case 2

x	0	1	2	3	4	5	7	8	9	10
y	1	0	2	5	10	20	15	10	7	3

The scatterplot shows that the association is not linear.

Minitab

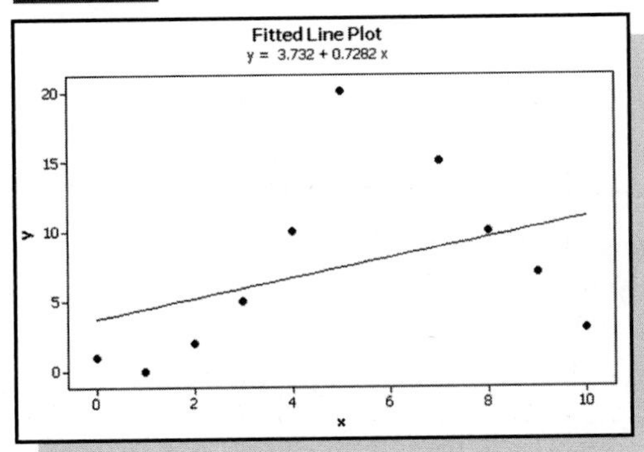

The residual plot reveals a distinct pattern.

Minitab

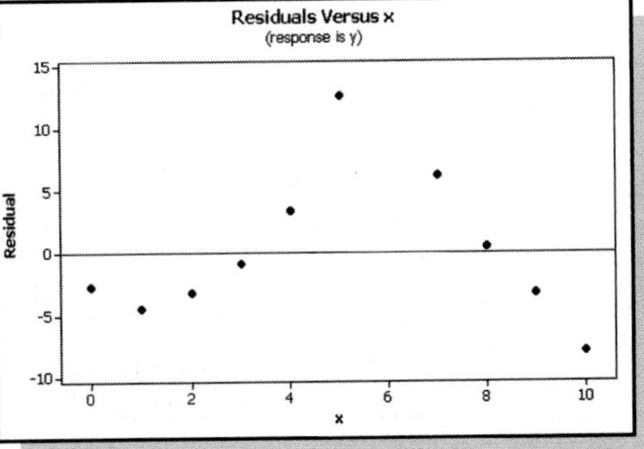

Case 3

x	0	1	2	3	4	5	7	8	9	10
y	0	6	9	15	10	35	15	60	75	20

The scatterplot shows increasing variation of points away from the regression line.

The residual plot reveals this pattern: Going from left to right, the points show more spread. (This is contrary to the requirement that for the different values of x, the distributions of y-values have the same variance.)

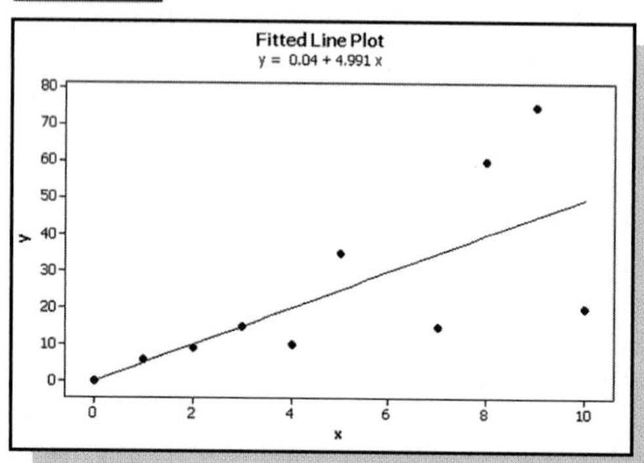

Minitab

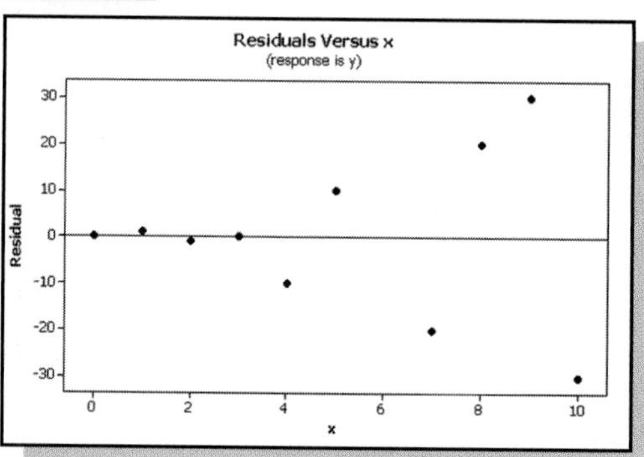

TI Using a TI-83/84 Plus Calculator for Regression

Regression Equation To find the equation of the regression line, use the same TI-83/84 Plus calculator procedure given at the end of Section 10-2. The results will include values of a and b, and the second line of the display shows that the regression equation has the format of $y = a + bx$, so a and b represent the y-intercept and slope, respectively.

TI-83/84 Plus Calculator	Equivalent Notation	Description
a	b_0	y-intercept
b	b_1	slope

Graphing Regression Line

```
Plot1 Plot2 Plot3
\Y1■34.8+0.234X
\Y2=
\Y3=
\Y4=
\Y5=
\Y6=
\Y7=
```

Graph of Scatterplot and Regression Line To obtain a TI-83/84 Plus graph of a scatterplot with the regression line included, first generate the scatterplot using the procedure described in Section 2-4. Next, press the Y= key and complete the equation for Y1. See the accompanying screen display for the entry of the regression equation corresponding to the sample data for duration/interval after eruption times from Table 10-1 ($\hat{y} = 34.8 + 0.234x$). The variable x can be entered by pressing ALPHA STO› (because the alphabet character x is located above the STO› key). Press the GRAPH key and the regression line will be printed along with the scatterplot in the same screen.

10-3 BASIC SKILLS AND CONCEPTS

Statistical Literacy and Critical Thinking

1. Predictor and Response Variables What is a predictor variable? What is a response variable? Why do you suppose that they were given those particular names?

2. Best-Fitting Line In what sense is the regression line the straight line that "best" fits the points in a scatterplot?

3. Predictions Why is it unwise to use a regression equation for predicting a value of y by using a value of x that is far beyond the scope of the available sample data?

4. Requirements Are the requirements for regression analysis identical to the requirements for correlation analysis? If not, how do the requirements differ?

Making Predictions. *In Exercises 5–8, use the given data to find the best predicted value of the dependent variable. Be sure to follow the prediction procedure described in this section.*

5. IQ Scores of Twins Separated at Birth IQ scores were obtained from randomly selected twins separated at birth. For 20 such twins, the linear correlation coefficient is 0.870 and the equation of the regression line is $\hat{y} = -3.22 + 1.02x$, where x represents the IQ score of the twin that was born first. Also, the 20 x values have a mean of 104.2 and the 20 y values have a mean of 103.1. What is the best predicted IQ of a twin who was born second, given that the first-born twin has an IQ of 110?

6. IQ Scores of Adopted Children Living Together IQ scores were obtained from randomly selected adopted children living together. For 20 such pairs of children, the linear correlation coefficient is 0.027 and the equation of the regression line is $\hat{y} = -3.22 + 1.02x$, where x represents the IQ score of the older child. Also, the 20 x values have a mean of 104.2 and the 20 y values have a mean of 103.1. What is the best predicted IQ of a younger adopted child, given that the older adopted child in the same family has an IQ of 110?

7. Sugar and Calories in Cereal The author collected data from 16 cereals consisting of the sugar contents (in grams per gram of cereal) and the calories (per gram of cereal). STATDISK was used to find that the linear correlation coefficient is $r = 0.765$ and the equation of the regression line is $\hat{y} = 3.46 + 1.01x$, where x represents the sugar content. Also, the mean calorie amount is 3.76 calories per gram of cereal. What is the best predicted calorie content for a cereal with 0.40 gram of sugar per gram of cereal?

8. Stocks and Super Bowl The Dow Jones Industrial Average (DJIA) high value and the total number of points scored in the Super Bowl were recorded for 21 different years. Excel was used to find that the value of the linear correlation coefficient is $r = -0.133$ and the regression equation is $\hat{y} = 53.3 - 0.000442x$, where x is the high value of the DJIA. Also, the mean number of Super Bowl points is 51.4. What is the best predicted value for the total number of Super Bowl points scored in a year with a DJIA high of 1200?

Finding the Equation of the Regression Line. *In Exercises 9 and 10, use the given data to find the equation of the regression line.*

TI 9.

x	1	1	2	5	3
y	0	2	2	3	5

TI **10.**

x	0	2	1	4	8
y	6	5	4	0	6

TI **11.** Effects of an Outlier Refer to the Minitab-generated scatterplot given in Exercise 11 of Section 10-2.

a. Using the pairs of values for all 10 points, find the equation of the regression line.

b. After removing the point with coordinates (10, 10), use the pairs of values for the remaining nine points and find the equation of the regression line.

c. Compare the results from parts (a) and (b).

TI **12.** Effects of Clusters Refer to the Minitab-generated scatterplot given in Exercise 12 of Section 10-2.

a. Using the pairs of values for all 8 points, find the equation of the regression line.

b. Using only the pairs of values for the four points in the lower left corner, find the equation of the regression line.

c. Using only the pairs of values for the four points in the upper right corner, find the equation of the regression line.

d. Compare the results from parts (a), (b), and (c).

Finding the Equation of the Regression Line and Making Predictions. *Exercises 13–32 use the same data sets as Exercises 13–32 in Section 10-2. In each case, find the regression equation, letting the first variable be the independent (x) variable. Find the indicated predicted values.* Caution: *When finding predicted values, be sure to follow the prediction procedure described in this section.*

TI 10 **13.** Old Faithful Find the best predicted time of the interval after an eruption (to the next eruption) given that the current eruption has a height of 100 ft.

Height	140	110	125	120	140	120	125	150
Interval after	92	65	72	94	83	94	101	87

TI **14.** Song Audiences and Sales Find the best predicted number of albums sold for a song with 20 (hundred million) audience impressions. (In the table below, audience impressions are in hundreds of millions and the numbers of albums sold are in hundreds of thousands).

Audience impressions	28	13	14	24	20	18	14	24	17
Albums sold	19	7	7	20	6	4	5	25	12

TI **15.** Movie Budgets and Gross Find the best predicted gross amount for a movie with a budget of 40 million dollars. (In the table below, all amounts are in millions of dollars.)

Budget	62	90	50	35	200	100	90
Gross	65	64	48	57	601	146	47

TI **16.** Car Weight and Fuel Consumption Find the best predicted highway fuel consumption amount (in mi/gal) for a car that weighs 3000 lb.

Weight	3175	3450	3225	3985	2440	2500	2290
Fuel consumption	27	29	27	24	37	34	37

TI **17.** Bear Chest Size and Weight Find the best predicted weight (in pounds) of a bear with a chest size of 50 in.

Chest size	26	45	54	49	35	41	41	49	38	31
Weight	80	344	416	348	166	220	262	360	204	144

TI **18.** Supermodel Heights and Weights Find the best predicted weight of a supermodel who is 72 in. tall.

Height (in.)	70	70.5	68	65	70	70	70	70	71
Weight (lb)	117	119	105	115	119	127	113	123	115

TI **19.** Blood Pressure Measurements Find the best predicted diastolic blood pressure for a person with a systolic reading of 140.

Systolic	138	130	135	140	120	125	120	130	130	144	143	140	130	150
Diastolic	82	91	100	100	80	90	80	80	80	98	105	85	70	100

TI **20.** Murders and Population Size Find the best predicted population size for a city with 120 murders. (The population sizes are in hundreds of thousands.)

Murders	258	264	402	253	111	648	288	654	256	60	590
Population	4	6	9	6	3	29	15	38	20	6	81

TI **21.** Buying a TV Audience Find the best predicted number of viewers for a television star with a salary of $2 million. (In the table below, the salaries are in millions of dollars and the numbers of viewers are in millions.)

Salary	100	14	14	35.2	12	7	5	1
Viewers	7	4.4	5.9	1.6	10.4	9.6	8.9	4.2

TI **22.** Smoking and Nicotine Find the best predicted cotinine level for a person who smokes 40 cigarettes per day.

x (cigarettes per day)	60	10	4	15	10	1	20	8	7	10	10	20
y (cotinine)	179	283	75.6	174	209	9.51	350	1.85	43.4	25.1	408	344

TI **23.** Temperatures and Marathons Find the best predicted winning time for the 1990 marathon when the temperature was 73 degrees. How does that predicted winning time compare to the actual winning time of 150.750 min?

x (temperature)	55	61	49	62	70	73	51	57
y (time)	145.283	148.717	148.300	148.100	147.617	146.400	144.667	147.533

TI **24.** Parent/Child Heights Find the best predicted height of a daughter to be born to a mother who is 66 in. tall.

Mother's height	63	67	64	60	65	67	59	60
Daughter's height	58.6	64.7	65.3	61.0	65.4	67.4	60.9	63.1

TI **25.** Crickets and Temperature Find the best predicted temperature for a time when a cricket is chirping at the rate of 1000 chirps per minute.

Chirps in 1 min	882	1188	1104	864	1200	1032	960	900
Temperature (°F)	69.7	93.3	84.3	76.3	88.6	82.6	71.6	79.6

TI **26.** Fires and Acres Burned Find the best predicted number of acres burned given that there were 50 thousand fires. (In the table below, the numbers of fires are in thousands and acres are in millions.)

Fires	73	69	58	48	84	62	57	45	70	63	48
Acres burned	6.2	4.2	1.9	2.7	5.0	1.6	3.0	1.6	1.5	2.0	3.7

TI **27.** Height and Pulse Rate Find the best predicted pulse rate for a woman who is 66 in. tall.

Height (in.)					
64.3	66.4	62.3	62.3	59.6	63.6
59.8	63.3	67.9	61.4	66.7	64.8

Pulse Rate (beats/min)					
76	72	88	60	72	68
80	64	68	68	80	76

TI **28.** State Budget and Days Late Find the best predicted number of days that a New York State budget is late, given that the size of the budget is \$104 billion. (In the data given below, the budget amounts are in billions of dollars.)

Budget									
101	96	91	85	80	73	71	66	63	63
62	58	55	52	49	46	44	40	37	35

Number of Days Late									
133	44	45	124	34	125	13	125	103	67
68	4	1	64	48	18	19	10	4	4

TI **29.** Appendix B Data Set: List Price and Selling Price Refer to Data Set 18 in Appendix B and use the list prices and selling prices of homes sold. Find the best predicted selling price of a home having a list price of \$400,000.

TI **30.** Appendix B Data Set: Discarded Plastic and Household Size Refer to Data Set 16 in Appendix B and use the weights of discarded plastic and the corresponding household sizes. Find the best predicted household size given that the household discards 5.00 lb of plastic.

TI **31.** Appendix B Data Set: Bad Stuff in Cigarettes Refer to Data Set 3 in Appendix B.

a. Use the paired data consisting of tar (x) and nicotine (y). What is the best predicted nicotine content of a cigarette with 15 mg of tar?

b. Use the paired data consisting of carbon monoxide (x) and nicotine (y). What is the best predicted nicotine level for a cigarette with 15 mg of carbon monoxide?

TI **32.** Appendix B Data Set: Forecast and Actual Temperatures Refer to Data Set 8 in Appendix B.

a. Use the five-day forecast high temperatures (x) and the actual high temperatures (y). What is the best predicted actual high temperature if the five-day forecast high temperature is 70 degrees?

b. Use the one-day forecast high temperatures (x) and the actual high temperatures (y). What is the best predicted actual high temperature if the one-day forecast high temperature is 70 degrees?

10-3 BEYOND THE BASICS

33. Equivalent Hypothesis Tests Explain why a test of the null hypothesis H_0: $\rho = 0$ is equivalent to a test of the null hypothesis H_0: $\beta_1 = 0$ where ρ is the linear correlation coefficient for a population of paired data, and β_1 is the slope of the regression line for that same population.

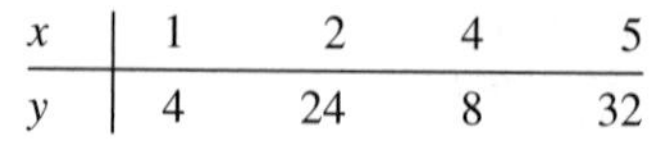

x	1	2	4	5
y	4	24	8	32

34. Testing Least-Squares Property According to the least-squares property, the regression line minimizes the sum of the squares of the residuals. We noted that with the paired data in the margin, the regression equation is $\hat{y} = 5 + 4x$ and the sum of the squares of the residuals is 364. Show that the equation $\hat{y} = 8 + 3x$ results in a sum of squares of residuals that is greater than 364.

TI **35.** Using Logarithms to Transform Data If a scatterplot reveals a nonlinear (not a straight line) pattern that you recognize as another type of curve, you may be able to apply the methods of this section. For the data given in the margin, find the linear equation ($y = b_0 + b_1x$) that best fits the sample data, and find the logarithmic equation ($y = a + b \ln x$) that best fits the sample data. (*Hint:* Begin by replacing each x-value with $\ln x$.) Which of these two equations fits the data better? Why?

x	2.0	2.5	4.2	10.0
y	12.0	18.7	53.0	225.0

TI **36.** Residual Plot Consider the data in the table below.

a. Examine the data and identify the relationship between x and y.

b. Find the linear correlation coefficient and use it to determine whether there appears to be a significant linear correlation between x and y.

c. Construct a scatterplot. What does it suggest about the relationship between x and y?

d. Construct a residual plot. Are there any noticeable patterns? What does the residual plot suggest about the relationship between x and y?

x	0	1	2	3	4	5	6	7	8	9	10
y	0	1	4	9	16	25	36	49	64	81	100

10-4 Variation and Prediction Intervals

Key Concept Using paired (x, y) data, we describe the variation that can be explained by the linear correlation between x and y and the variation that is unexplained. We then proceed to consider a method for constructing a *prediction interval,* which is an interval estimate of a predicted value of y. (Interval estimates of parameters are usually referred to as *confidence intervals,* whereas interval estimates of variables are usually called *prediction intervals.*)

Explained and Unexplained Variation

We will now examine measures of *deviation* and *variation* for a pair of values (x, y). Instead of contemplating abstractions, let's consider the specific case depicted in Figure 10-9. Imagine a sample of paired (x, y) data that includes (5, 19). Assume that we use this sample of paired data to find the following results:

- There is a linear correlation (with r significantly different from 0).
- The equation of the regression line is $\hat{y} = 3 + 2x$.
- The mean of the y-values is given by $\bar{y} = 9$.
- One of the pairs of sample data is $x = 5$ and $y = 19$.
- The point (5, 13) is one of the points on the regression line, because substituting $x = 5$ into the regression equation yields $\hat{y} = 13$.

Figure 10-9 shows that the point (5, 13) lies on the regression line, but the point (5, 19) from the original data set does not lie on the regression line. If we completely ignore correlation and regression concepts and want to predict a value of y given a value of x and a collection of paired (x, y) data, our best guess would be the mean $\bar{y}$. But in this case with a significant linear correlation, the way to predict the value of y when $x = 5$ is to use the regression equation to get $\hat{y} = 13$. We can explain the discrepancy between $\bar{y} = 9$ and $\hat{y} = 13$ by noting that there is a linear relationship best described by the regression line. Consequently, when $x = 5$, the predicted y value is 13, not the mean value of 9. For $x = 5$, the predicted y value is 13, but the observed sample value of y is actually 19. The discrepancy between $\hat{y} = 13$ and $y = 19$ cannot be explained by the regression line, and it is called an *unexplained deviation,* or a *residual.* This unexplained deviation can be expressed in symbols as $y - \hat{y}$.

Super Bowl as Stock Market Predictor

The "Super Bowl omen" states that a Super Bowl victory by a team with NFL origins is followed by a year in which the New York Stock Exchange index rises; otherwise, it falls. (In 1970, the NFL and AFL merged into the current NFL.) After the first 29 Super Bowl games, the prediction was correct 90% of the time, but it has been much less successful in recent years. As of this writing, it has been correct in 29 of 38 Super Bowl games, for a 76% success rate. Forecasting and predicting are important goals of statistics and investment advisors, but common sense suggests that no one should base investments on the outcome of one football game. Other indicators used to forecast stock market performance include rising skirt hemlines, aspirin sales, limousines on Wall Street, orders for cardboard boxes, sales of beer versus wine, and elevator traffic at the New York Stock Exchange.

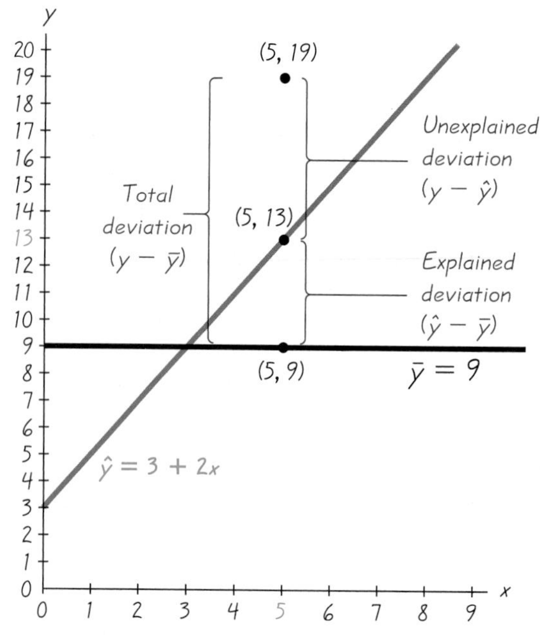

Figure 10-9 Unexplained, Explained, and Total Deviation

As in Section 3-3 where we defined the standard deviation, we again consider a *deviation* to be a difference between a value and the mean. (In this case, the mean is $\bar{y} = 9$.) Examine Figure 10-9 carefully and note these specific deviations from $\bar{y} = 9$:

Total deviation (from $\bar{y} = 9$) of the point (5, 19) $= y - \bar{y} = 19 - 9 = 10$

Explained deviation (from $\bar{y} = 9$) of the point (5, 19) $= \hat{y} - \bar{y} = 13 - 9 = 4$

Unexplained deviation (from $\bar{y} = 9$) of the point (5, 19) $= y - \hat{y} = 19 - 13 = 6$

These deviations from the mean are generalized and formally defined as follows.

Definitions

Assume that we have a collection of paired data containing the sample point (x, y), that $\hat{y}$ is the predicted value of y (obtained by using the regression equation), and that the mean of the sample y-values is $\bar{y}$.

The **total deviation** of (x, y) is the vertical distance $y - \bar{y}$, which is the distance between the point (x, y) and the horizontal line passing through the sample mean $\bar{y}$.

The **explained deviation** is the vertical distance $\hat{y} - \bar{y}$, which is the distance between the predicted y-value and the horizontal line passing through the sample mean $\bar{y}$.

The **unexplained deviation** is the vertical distance $y - \hat{y}$, which is the vertical distance between the point (x, y) and the regression line. (The distance $y - \hat{y}$ is also called a *residual,* as defined in Section 10-3.)

We can see the following relationship in Figure 10-9:

$$\begin{array}{ccccc} \text{(total deviation)} & = & \text{(explained deviation)} & + & \text{(unexplained deviation)} \\ (y - \bar{y}) & = & (\hat{y} - \bar{y}) & + & (y - \hat{y}) \end{array}$$

The above expression involves deviations away from the mean, and it applies to any one particular point (x, y). If we sum the squares of deviations using all points (x, y), we get amounts of *variation*, and the same relationship applies to the sums of squares shown in Formula 10-4, even though the above expression is not algebraically equivalent to Formula 10-4. In Formula 10-4, the **total variation** is expressed as the sum of the squares of the total deviation values, the **explained variation** is the sum of the squares of the explained deviation values, and the **unexplained variation** is the sum of the squares of the unexplained deviation values.

Formula 10-4

$$\begin{array}{cccccc} & \text{(total variation)} & = & \text{(explained variation)} & + & \text{(unexplained variation)} \\ \text{or} & \Sigma(y - \bar{y})^2 & = & \Sigma(\hat{y} - \bar{y})^2 & + & \Sigma(y - \hat{y})^2 \end{array}$$

In Section 10-2 we saw that the linear correlation coefficient r can be used to find the proportion of the total variation in y that can be explained by the linear correlation. This statement was made in Section 10-2:

The value of r^2 is the proportion of the variation in y that is explained by the linear relationship between x and y.

This statement about the explained variation is formalized with the following definition.

Definition

The **coefficient of determination** is the amount of the variation in y that is explained by the regression line. It is computed as

$$r^2 = \frac{\text{explained variation}}{\text{total variation}}$$

We can compute r^2 by using the definition just given with Formula 10-4, or we can simply square the linear correlation coefficient r.

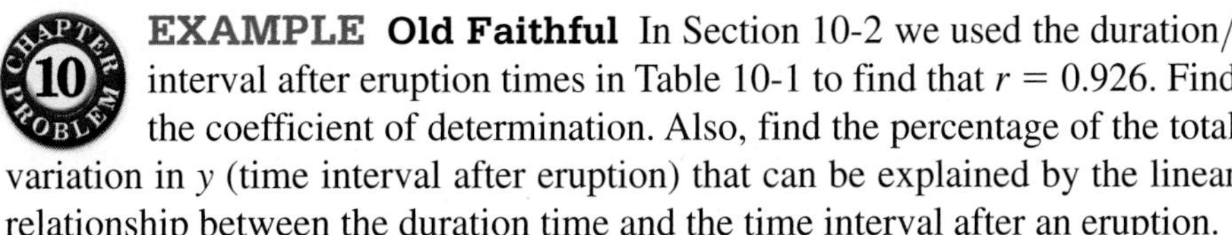

EXAMPLE Old Faithful In Section 10-2 we used the duration/interval after eruption times in Table 10-1 to find that $r = 0.926$. Find the coefficient of determination. Also, find the percentage of the total variation in y (time interval after eruption) that can be explained by the linear relationship between the duration time and the time interval after an eruption.

SOLUTION The coefficient of determination is $r^2 = 0.926^2 = 0.857$. Because r^2 is the proportion of total variation that is explained, we conclude that about 86% of the total variation in time interval after eruptions can be explained by the duration times. This means that about 14% of the total variation in time interval after eruptions can be explained by factors other than the duration times. But remember that these results are estimates based on the given sample data. Other sample data will likely result in different estimates.

Prediction Intervals

In Section 10-3 we used the Table 10-1 sample data to find the regression equation $\hat{y} = 34.8 + 0.234x$, where $\hat{y}$ represents the predicted time interval (in minutes) after an eruption to the next eruption and x represents the duration of an eruption (in seconds). We then used that equation to predict the y-value, given that an eruption has a duration of $x = 180$ sec. We found that the best predicted time interval after the eruption is 76.9 min. Because 76.9 is a single value, it is referred to as a *point estimate*. In Chapter 7 we saw that point estimates have the serious disadvantage of not giving us any information about how accurate they might be. Here, we know that 76.9 is the best predicted value, but we don't know how accurate that value is. In Chapter 7 we developed confidence interval estimates to overcome that disadvantage, and in this section we follow the same approach. We will use a **prediction interval,** which is an interval estimate of a predicted value of y. [An interval estimate of a *parameter* (such as the mean of all time intervals after eruptions) is usually referred to as a *confidence interval,* whereas an interval estimate of a *variable* (such as the estimated time interval after an eruption with a duration of 180 sec) is usually called a *prediction interval*.]

The development of a prediction interval requires a measure of the spread of sample points about the regression line. Recall that the unexplained deviation (or residual) is the vertical distance between a sample point and the regression line, as illustrated in Figure 10-9. The *standard error of estimate* is a collective measure of the spread of the sample points about the regression line, and it is formally defined as follows.

Definition

The **standard error of estimate,** denoted by s_e, is a measure of the differences (or distances) between the observed sample y-values and the predicted values $\hat{y}$ that are obtained using the regression equation. It is given as

$$s_e = \sqrt{\frac{\Sigma(y - \hat{y})^2}{n - 2}} \quad \text{(where } \hat{y} \text{ is the predicted } y\text{-value)}$$

or as the following equivalent formula:

Formula 10-5
$$s_e = \sqrt{\frac{\Sigma y^2 - b_0\Sigma y - b_1\Sigma xy}{n - 2}}$$

The TI-83/84 Plus calculator provides the value of s_e. See the end of this section.

The development of the standard error of estimate s_e closely parallels that of the ordinary standard deviation introduced in Section 3-3. Just as the standard deviation is a measure of how values deviate from their mean, the standard error of estimate s_e is a measure of how sample data points deviate from their regression line. The reasoning behind dividing by $n - 2$ is similar to the reasoning that led to division by $n - 1$ for the ordinary standard deviation. It is important to note that relatively smaller values of s_e reflect points that stay close to the regression line, and relatively larger values occur with points farther away from the regression line.

Formula 10-5 is algebraically equivalent to the other equation in the definition, but Formula 10-5 is generally easier to work with because it doesn't require that we compute each of the predicted values $\hat{y}$ by substitution in the regression equation. However, Formula 10-5 does require that we find the y-intercept b_0 and the slope b_1 of the estimated regression line.

EXAMPLE Finding s_e Use Formula 10-5 to find the standard error of estimate s_e for the Old Faithful duration/interval after eruption times listed in Table 10-1.

SOLUTION Using the sample data in Table 10-1, we find these values:

$$n = 8 \qquad \Sigma y^2 = 60{,}204 \qquad \Sigma y = 688 \qquad \Sigma xy = 154{,}378$$

In Section 10-3 we used the Table 10-1 sample data to find the y-intercept and the slope of the regression line. Those values are given here with extra decimal places for greater precision.

$$b_0 = 34.7698041 \qquad b_1 = 0.2340614319$$

We can now use these values in Formula 10-5 to find the standard error of estimate s_e.

$$s_e = \sqrt{\frac{\Sigma y^2 - b_0\Sigma y - b_1\Sigma xy}{n - 2}}$$

$$= \sqrt{\frac{60{,}204 - (34.7698041)(688) - (0.2340614319)(154{,}378)}{8 - 2}}$$

$$= 4.973916052 = 4.97 \quad \text{(rounded)}$$

We can measure the spread of the sample points about the regression line with the standard error of estimate $s_e = 4.97$. We can use the standard error of estimate s_e to construct interval estimates that will help us see how dependable our point estimates of y really are. Assume that for each fixed value of x, the corresponding sample values of y are normally distributed about the regression line, and those normal distributions have the same variance. The following interval estimate applies to an *individual* y-value. (For a confidence interval used to predict the *mean* of all y-values for some given x-value, see Exercise 26.)

Prediction Interval for an Individual y

Given the fixed value x_0, the prediction interval for an individual y is

$$\hat{y} - E < y < \hat{y} + E$$

where the margin of error E is

$$E = t_{\alpha/2}\, s_e \sqrt{1 + \frac{1}{n} + \frac{n(x_0 - \bar{x})^2}{n(\Sigma x^2) - (\Sigma x)^2}}$$

and x_0 represents the given value of x, $t_{\alpha/2}$ has $n - 2$ degrees of freedom, and s_e is found from Formula 10-5.

EXAMPLE Old Faithful For the paired duration/interval after eruption times in Table 10-1, we have found that for a duration of 180 sec, the best predicted time interval after the eruption is 76.9 min. Construct a 95% prediction interval for the time interval after an eruption, given that the duration of the eruption is 180 sec (so that $x = 180$). This will provide a sense of how accurate the predicted value of 76.9 min really is.

SOLUTION In previous sections we have shown that $r = 0.926$, so that there is a linear correlation (at the 0.05 significance level), and the regression equation is $\hat{y} = 34.8 + 0.234x$. In the preceding example we found that $s_e =$ 4.973916052, and the following statistics are obtained from the duration/interval after eruption times in Table 10-1:

$$n = 8 \qquad \bar{x} = 218.875 \qquad \Sigma x = 1751 \qquad \Sigma x^2 = 399{,}451$$

From Table A-3 we find $t_{\alpha/2} = 2.447$. (We used $8 - 2 = 6$ degrees of freedom with $\alpha = 0.05$ in two tails.) We first calculate the margin of error E by letting $x_0 = 180$, because we want the prediction interval of the interval after eruption time given that the duration time is $x = 180$.

$$E = t_{\alpha/2} s_e \sqrt{1 + \frac{1}{n} + \frac{n(x_0 - \bar{x})^2}{n(\Sigma x^2) - (\Sigma x)^2}}$$

$$= (2.447)(4.973916052)\sqrt{1 + \frac{1}{8} + \frac{8(180 - 218.875)^2}{8(399{,}451) - (1751)^2}}$$

$$= (2.447)(4.973916052)(1.103758562) = 13.4 \qquad \text{(rounded)}$$

With $\hat{y} = 76.9$ and $E = 13.4$, we get the prediction interval as follows:

$$\hat{y} - E < y < \hat{y} + E$$

$$76.9 - 13.4 < y < 76.9 + 13.4$$

$$63.5 < y < 90.3$$

That is, for an eruption with a duration of 180 sec, we have 95% confidence that the time interval after the eruption is between 63.5 min and 90.3 min. That's a relatively large range. (One major factor contributing to the large range is that the sample size is very small because we are using only 8 pairs of sample data.)

In addition to knowing that for an eruption duration of $x = 180$ sec, the predicted time interval after the eruption to the next eruption is 76.9 min, we now have a sense of how reliable that estimate really is. The 95% prediction interval found in this example shows that the actual value of y can vary substantially from the predicted value of 76.9 min.

TI

Using the TI-83/84 Plus Calculator to Construct a Prediction Interval

If using some given fixed value of x to construct a prediction interval $\hat{y} - E < y < \hat{y} + E$, you must calculate the control limits of $\hat{y} - E$ and $\hat{y} + E$.

1. Find the regression equation by using the procedure at the end of Section 10-3, then substitute the given value of x to find the value of $\hat{y}$. The result is the value of y that is predicted from the given value of x.
2. The margin of error E is calculated using the components listed below.

$t_{\alpha/2}$	found from Table A-3 by using $n - 2$ degrees of freedom	
$s_{\hat{e}}$	provided by the TI-83/84 Plus calculator by using the procedure given at the end of Section 10-2	
x_0	given value of x used for the prediction	
n	number of pairs of data	Press **STAT**, select **CALC, 1-VarStats,** enter L1, press **ENTER**.
Σx	sum of the x values	
Σx^2	sum of the squares of the x values	

Proceed to complete the calculation of the control limits of $\hat{y} - E$ and $\hat{y} + E$ as shown in this section. Although the prediction interval is not completely generated, the TI-83/84 Plus calculator is quite helpful in finding it.

10-4 BASIC SKILLS AND CONCEPTS

Statistical Literacy and Critical Thinking

1. Prediction Interval Use your own words to describe a prediction interval.
2. Coefficient of Determination What is a coefficient of determination in general? What is the coefficient of determination for $r = 0.400$? What practical information does it give us?
3. Explained Deviation and Variation What is the difference between explained *deviation* and explained *variation?*
4. Explained and Unexplained Variation What is the difference between explained variation and unexplained variation?

Interpreting the Coefficient of Determination. *In Exercises 5–8, use the value of the linear correlation coefficient r to find the coefficient of determination and the percentage of the total variation that can be explained by the linear relationship between the two variables.*

5. $r = 0.3$
6. $r = -0.2$
7. $r = -0.901$
8. $r = 0.747$

TI-83/84 Plus

```
LinRegTTest
 y=a+bx
 β≠0 and ρ≠0
 t=18.14770789
 p=1.178633E-16
 df=27
↓a=.1540298507
```

TI-83/84 Plus

```
LinRegTTest
 y=a+bx
 β≠0 and ρ≠0
↑b=.0650516648
 s=.0878540347
 r²=.9242295198
 r=.961368566
```

Interpreting a Calculator Display. *In Exercises 9–12, refer to the TI-83/84 Plus calculator displays that were obtained by using the paired data consisting of tar and nicotine for a sample of 29 cigarettes, as listed in Data Set 3 in Appendix B.*

TI **9.** Testing for Correlation Use the information provided in the display to determine the value of the linear correlation coefficient. Given that there are 29 pairs of data, is there a linear correlation between the amount of tar and the amount of nicotine in a cigarette?

TI **10.** Identifying Total Variation What percentage of the total variation in nicotine can be explained by the linear relationship between tar and nicotine?

TI **11.** Predicting Nicotine Amount If a cigarette has 17 mg of tar, what is the single value that is the best predicted amount of nicotine? (Assume that there is a linear correlation between tar and nicotine.)

TI **12.** Finding Prediction Interval For a given tar amount of 17 mg, identify the 95% prediction interval estimate of the amount of nicotine, and write a statement interpreting that interval.

Finding Measures of Variation. *In Exercises 13–16, find the (a) explained variation, (b) unexplained variation, (c) total variation, (d) coefficient of determination, and (e) standard error of estimate s_e. In each case, there is a linear correlation so that it is reasonable to use the regression equation when making predictions. (Results are used in Exercises 17–20.)*

TI **13.** Car Weight and Fuel Consumption Listed below are the weights (in pounds) and the highway fuel consumption amounts (in mi/gal) of randomly selected cars (Chrysler Sebring, Ford Mustang, BMW 3-Series, Ford Crown Victoria, Honda Civic, Mazda Protégé, Hyundai Accent).

Weight	3175	3450	3225	3985	2440	2500	2290
Fuel consumption	27	29	27	24	37	34	37

TI **14.** Movie Budgets and Gross Listed below are the budgets (in millions of dollars) and the gross receipts (in millions of dollars) for randomly selected movies (based on data from the Motion Picture Association of America).

Budget	62	90	50	35	200	100	90
Gross	65	64	48	57	601	146	47

TI **15.** Blood Pressure Measurements Fourteen different second-year medical students took blood pressure measurements of the same patient and the results are listed below (data provided by Marc Triola, MD).

Systolic	138	130	135	140	120	125	120	130	130	144	143	140	130	150
Diastolic	82	91	100	100	80	90	80	80	80	98	105	85	70	100

TI **16.** Crickets and Temperature One classic application of correlation involves the association between the temperature and the number of times a cricket chirps in a minute. Listed below are the numbers of chirps in 1 min and the corresponding temperatures in degrees Fahrenheit (based on data from *The Song of Insects* by George W. Pierce, Harvard University Press).

Chirps in 1 min	882	1188	1104	864	1200	1032	960	900
Temperature (°F)	69.7	93.3	84.3	76.3	88.6	82.6	71.6	79.6

TI **17.** Effect of Variation on Prediction Interval Refer to the data given in Exercise 13 and assume that the necessary conditions of normality and variance are met.
a. Find the predicted fuel consumption rate for a car that weighs 3700 lb.
b. Find a 95% prediction interval estimate of the fuel consumption rate for a car that weighs 3700 lb.

TI **18.** Finding Predicted Value and Prediction Interval Refer to Exercise 14 and assume that the necessary conditions of normality and variance are met.
a. Find the predicted gross amount for a movie with a budget of $100 million.
b. Find a 95% prediction interval estimate of the gross amount for a movie with a budget of $100 million.

TI **19.** Finding Predicted Value and Prediction Interval Refer to the data given in Exercise 15 and assume that the necessary conditions of normality and variance are met.
a. Find the predicted diastolic reading given that the systolic reading is 120.
b. Find a 95% prediction interval estimate of the diastolic reading given that the systolic reading is 120.

TI **20.** Finding Predicted Value and Prediction Interval Refer to the data described in Exercise 16 and assume that the necessary conditions of normality and variance are met.
a. Find the predicted temperature when a cricket chirps 1000 times in 1 min.
b. Find a 99% prediction interval estimate of the temperature when a cricket chirps 1000 times in 1 min.

Finding a Prediction Interval. *In Exercises 21–24, refer to the Table 10-1 sample data. Let x represent the duration time (in seconds) and let y represent the time interval (in minutes) after the eruption to the next eruption. Use the given duration time and the given confidence level to construct a prediction interval estimate of the time interval after the eruption to the next eruption. (See the example in this section.)*

TI **21.** $x = 180$ sec; 99% confidence

TI **22.** $x = 180$ sec; 90% confidence

TI **23.** $x = 200$ sec; 95% confidence

TI **24.** $x = 120$ sec; 99% confidence

10-4 BEYOND THE BASICS

TI **25.** Confidence Intervals for β_0 and β_1 Confidence intervals for the y-intercept β_0 and slope β_1 for a regression line ($y = \beta_0 + \beta_1 x$) can be found by evaluating the limits in the intervals below.

$$b_0 - E < \beta_0 < b_0 + E$$

where

$$E = t_{\alpha/2}\, s_e \sqrt{\frac{1}{n} + \frac{\bar{x}^2}{\Sigma x^2 - \frac{(\Sigma x)^2}{n}}}$$

continued

$$b_1 - E < \beta_1 < b_1 + E$$

where
$$E = t_{\alpha/2} \cdot \frac{s_e}{\sqrt{\Sigma x^2 - \frac{(\Sigma x)^2}{n}}}$$

In these expressions, the y-intercept b_0 and the slope b_1 are found from the sample data and $t_{\alpha/2}$ is found from Table A-3 by using $n - 2$ degrees of freedom. Using the duration/interval after eruption times in Table 10-1, find the 95% confidence interval estimates of β_0 and β_1.

26. Confidence Interval for Mean Predicted Value From the expression given in this section for the margin of error corresponding to a prediction interval for y, we get

$$s_{\hat{y}} = s_e\sqrt{1 + \frac{1}{n} + \frac{n(x_0 - \bar{x})^2}{n(\Sigma x^2) - (\Sigma x)^2}}$$

which is the *standard error of the prediction* when predicting for a *single* y, given that $x = x_0$. When predicting for the *mean* of all values of y for which $x = x_0$, the point estimate $\hat{y}$ is the same, but $s_{\hat{y}}$ is as follows:

$$s_{\hat{y}} = s_e\sqrt{\frac{1}{n} + \frac{n(x_0 - \bar{x})^2}{n(\Sigma x^2) - (\Sigma x)^2}}$$

Use the data from Table 10-1 to find a point estimate and a 95% confidence interval estimate of the mean of time intervals after eruptions that have duration times of 180 sec.

10-5 Multiple Regression

Key Concept Although preceding sections of this chapter apply to a relationship between *two* variables, this section presents a method for analyzing a linear relationship involving *more than two* variables. We focus on three key elements: (1) the multiple regression equation, (2) the value of adjusted R^2, and (3) the P-value. Because of the excruciatingly complex nature of the required calculations, manual calculations are impractical and a threat to mental health, so this section emphasizes the use and interpretation of results from statistical software or a TI-83/84 Plus calculator.

Multiple Regression Equation

As in the preceding sections of this chapter, we will consider *linear* relationships only. We use the following *multiple regression equation* to describe linear relationships involving more than two variables.

Definition

A **multiple regression equation** expresses a linear relationship between a response variable y and two or more predictor variables $(x_1, x_2, \ldots, x_k)$. The general form of a multiple regression equation is

$$\hat{y} = b_0 + b_1x_1 + b_2x_2 + \cdots + b_kx_k.$$

We will use the following notation, which follows naturally from the notation used in Section 9-3.

Notation

$\hat{y} = b_0 + b_1x_1 + b_2x_2 + \cdots + b_kx_k$ (General form of the estimated multiple regression equation)

n = sample size

k = number of *predictor* variables. (The predictor variables are also called *independent variables* or x variables.)

$\hat{y}$ = predicted value of y (computed by using the multiple regression equation)

$x_1, x_2, \ldots, x_k$ are the predictor variables

β_0 = the y-intercept, or the value of y when all of the predictor variables are 0 (This value is a population parameter.)

b_0 = estimate of β_0 based on the sample data (b_0 is a sample statistic.)

$\beta_1, \beta_2, \ldots, \beta_k$ are the coefficients of the predictor variables $x_1, x_2, \ldots, x_k$

$b_1, b_2, \ldots, b_k$ are the sample estimates of the coefficients $\beta_1, \beta_2, \ldots, \beta_k$

For any specific set of x values, the regression equation is associated with a random error often denoted by ε, and we assume that such errors are normally distributed with a mean of 0 and a standard deviation of σ, and the random errors are independent. Such assumptions are difficult to check. We assume throughout this section that the necessary requirements are satisfied.

The computations required for multiple regression are so complicated that a statistical software package *must* be used, so we will focus on *interpreting* calculator displays. Instructions for using a TI-83/84 Plus calculator are included at the end of this section.

EXAMPLE Old Faithful In Section 10-3 we discussed methods for predicting the time interval after an eruption of the Old Faithful geyser, but that section included only one predictor variable. Using the sample data in Table 10-1 included with the Chapter Problem, find the multiple regression equation in which the response (y) variable is the time interval after an eruption and the predictor (x) variables are the duration time of the eruption and the height of the eruption. The TI-83/84 Plus calculator results are shown below.

continued

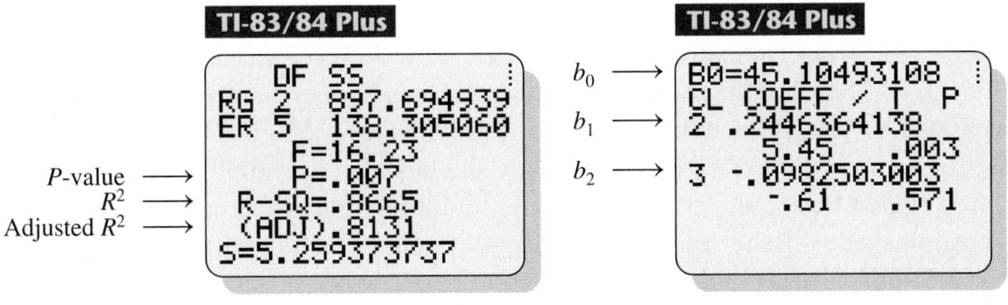

Predictors for Success

When a college accepts a new student, it would like to have some positive indication that the student will be successful in his or her studies. College admissions deans consider SAT scores, standard achievement tests, rank in class, difficulty of high school courses, high school grades, and extracurricular activities. In a study of characteristics that make good predictors of success in college, it was found that class rank and scores on standard achievement tests are better predictors than SAT scores. A multiple regression equation with college grade-point average predicted by class rank and achievement test score was not improved by including another variable for SAT score. This particular study suggests that SAT scores should not be included among the admissions criteria, but supporters argue that SAT scores are useful for comparing students from different geographic locations and high school backgrounds.

Making Music with Multiple Regression

Sony manufactures millions of compact discs in Terre Haute, Indiana. At one step in the manufacturing process, a laser exposes a photographic plate so that a musical signal is transferred into a digital signal coded with 0s and 1s. This process was statistically analyzed to identify the effects of different variables, such as the length of exposure and the thickness of the photographic emulsion. Methods of multiple regression showed that among all of the variables considered, four were most significant. The photographic process was adjusted for optimal results based on the four critical variables. As a result, the percentage of defective discs dropped and the tone quality was maintained. The use of multiple regression methods led to lower production costs and better control of the manufacturing process.

Interpreting TI-83/84 Plus calculator results: The two screens on the previous page might not win many awards for clarity of presentation, but they do contain the key elements that we need in this section. The important results from the first screen are listed below. In the second screen, see that CL (column) is the top heading for the *column number*, and each column number is followed by the coefficient used in the multiple regression equation, a t test statistic value, and a P-value for the variable corresponding to the column. Listed below are the key elements that should be extracted from the second screen.

First screen: P-value $= 0.007$
$R^2 = 0.867$ (rounded)
Adjusted $R^2 = 0.813$ (rounded)

Second screen: $b_0 = 45.1$ (rounded)
$b_1 = 0.245$ [Coefficient for column 2 (rounded)]
$b_2 = -0.098$ [Coefficient for column 3 (rounded)]

SOLUTION Using the TI-83/84 Plus calculator results, the multiple regression equation is

$$\hat{y} = 45.1 + 0.245x_1 - 0.098x_2$$

If a multiple regression equation fits the sample data well, it can be used for predictions. For example, if we determine that the equation is suitable for predictions, and if we have an eruption with a duration of 180 sec and a height of 130 ft, we can predict the time interval after the eruption by substituting those values into the regression equation to get a predicted time of 76.5 min. (Remember, duration times are in seconds, heights are in feet, and time intervals after eruptions are in minutes.) Also, the coefficients $b_1 = 0.245$ and $b_2 = -0.098$ can be used to determine marginal change, as described in Section 10-3. For example, the coefficient $b_1 = 0.245$ shows that when the height of an eruption remains constant, the predicted time interval after the eruption increases by 0.245 min for each 1-sec increase in the duration of the eruption.

Adjusted R^2

R^2 denotes the **multiple coefficient of determination,** which is a measure of how well the multiple regression equation fits the sample data. A perfect fit would result in $R^2 = 1$, and a very good fit results in a value near 1. A very poor fit results in a value of R^2 close to 0. The value of $R^2 = 0.867$ (rounded) in the TI-83/84 Plus display indicates that 86.7% of the variation in time intervals after eruptions can be explained by the duration time x_1 and the height x_2. However, the multiple coefficient of determination R^2 has a serious flaw: As more variables are included, R^2 increases. (R^2 could remain the same, but it usually increases.) The largest R^2 is obtained by simply including all of the available variables, but the best multiple regression equation does not necessarily use all of the available variables. Because of that flaw, comparison of different multiple regression equations is better accomplished with the adjusted coefficient of determination, which is R^2 adjusted for the number of variables and the sample size.

Definition

The **adjusted coefficient of determination** is the multiple coefficient of determination R^2 modified to account for the number of variables and the sample size. It is calculated by using Formula 10-6.

Formula 10-6 $$\text{adjusted } R^2 = 1 - \frac{(n - 1)}{[n - (k + 1)]}(1 - R^2)$$

where n = sample size

k = number of predictor (x) variables

The preceding TI-83/84 Plus calculator display shows the adjusted coefficient of determination as 0.813 (rounded). If we use Formula 10-6 with the R^2 value of 0.867, $n = 8$ and $k = 2$, we find that the adjusted R^2 value is 0.813. (We actually get 0.814, but we get 0.813 if we use more digits to minimize the rounding error.) The R^2 value of 0.867 indicates that 86.7% of the variation in time interval after eruptions can be explained by duration time x_1 and height x_2, but when we compare this multiple regression equation to others, it is better to use the adjusted R^2 of 0.813.

P-Value

The *P*-value is a measure of the overall significance of the multiple regression equation. The displayed TI-83/84 Plus *P*-value of 0.007 is small, indicating that the multiple regression equation has good overall significance and is usable for predictions. That is, it makes sense to predict time intervals after eruptions based on eruption durations and heights. Like the adjusted R^2, this *P*-value is a good measure of how well the equation fits the sample data. The value of 0.007 results from a test of the null hypothesis that $\beta_1 = \beta_2 = 0$. Rejection of $\beta_1 = \beta_2 = 0$ implies that at least one of β_1 and β_2 is not 0, indicating that this regression equation is effective in determining time interval after eruptions. A complete analysis of results might include other important elements, such as the significance of the individual coefficients, but we will limit our discussion to the three key components—multiple regression equation, adjusted R^2, and *P*-value.

Finding the Best Multiple Regression Equation

The preceding TI-83/84 Plus displays are based on using the predictor variables of duration and height with the sample data in Table 10-1. But if we want to predict the time interval after an eruption, is there some other combination of predictor variables that might be better than duration and height? Table 10-3 lists different combinations of variables, and we are now confronted with the important objective of finding the *best* multiple regression equation. Because determination of the best multiple regression equation requires a good dose of judgment, there is no exact and automatic procedure that can be used. *Determination of the best multiple regression equation is often quite difficult and beyond the scope of this book,* but the following guidelines should provide some help.

NBA Salaries and Performance

Researcher Matthew Weeks investigated the correlation between NBA salaries and basketball game statistics. In addition to salary (S), he considered minutes played (M), assists (A), rebounds (R), and points scored (P), and he used data from 30 players. The multiple regression equation is $S = -0.716 - 0.0756M - 0.425A + 0.0536R + 0.742P$ with $R^2 = 0.458$. Because of a high correlation between minutes played (M) and points scored (P), and because points scored had a higher correlation with salary, the variable of minutes played was removed from the multiple regression equation. Also, the variables of assists (A) and rebounds (R) were not found to be significant, so they were removed as well. The single variable of points scored appeared to be the best choice for predicting NBA salaries, but the predictions were found to be not very accurate because of other variables not considered, such as popularity of the player.

Highest adjusted R^2 → and lowest P-value

Table 10-3 Searching for the Best Multiple Regression Equation

	R^2	Adjusted R^2	Overall Significance
DURATION	0.857	0.833	0.001
INTERVAL BEFORE	0.011	0.000	0.802
HEIGHT	0.073	0.000	0.519
DURATION and INTERVAL BEFORE	0.872	0.820	0.006
DURATION and HEIGHT	0.867	0.813	0.007
INTERVAL BEFORE and HEIGHT	0.073	0.000	0.828
DURATION and INTERVAL BEFORE and HEIGHT	0.875	0.781	0.028

Guidelines for Finding the Best Multiple Regression Equation

1. *Use common sense and practical considerations to include or exclude variables.* For example, we might exclude the variable of height after learning that height is a visual estimate instead of an accurate measurement.
2. *Consider the P-value.* Select an equation having overall significance, as determined by the P-value found in the computer display. For example, see the values of overall significance in Table 10-3. The P-values of 0.802, 0.519, and 0.828 correspond to combinations of variables that do not result in overall significance, so those combinations should be excluded.
3. *Consider equations with high values of adjusted R^2, and try to include only a few variables.* Instead of including almost every available variable, try to include relatively few predictor (x) variables. Use these guidelines:

   - Select an equation having a value of adjusted R^2 with this property: If an additional predictor variable is included, the value of adjusted R^2 does not increase by a substantial amount.
   - For a given number of predictor (x) variables, select the equation with the largest value of adjusted R^2.
   - In weeding out predictor (x) variables that don't have much of an effect on the response (y) variable, it might be helpful to find the linear correlation coefficient r for each pair of variables being considered. If two predictor values have a very high linear correlation coefficient, there is no need to include them both, and we should exclude the variable with the lower value of r.

Using these guidelines in an attempt to find the best equation for predicting time interval after eruptions of Old Faithful, we find that for the data of Table 10-1, the best regression equation uses the single predictor (x) variable of duration time. The best regression equation appears to be

$$\text{INTERVAL AFTER} = 34.8 + 0.234 \text{ DURATION}$$

or

$$\hat{y} = 34.8 + 0.234x_1$$

The preceding guidelines are based on the adjusted R^2 and the P-value, but we could also conduct individual hypothesis tests based on values of the regression coefficients. Consider the regression coefficient of β_1. A test of the null hypothesis $\beta_1 = 0$ can tell us whether the corresponding predictor variable should be included in the regression equation. Rejection of $\beta_1 = 0$ suggests that β_1 has a nonzero value and is therefore helpful for predicting the value of the response variable. Procedures for such tests are described in Exercise 17.

Icing the Kicker

Just as a kicker in football is about to attempt a field goal, it is a common strategy for the opposing coach to call a time-out to "ice" the kicker. The theory is that the kicker has time to think and become nervous and less confident, but does the practice actually work? In "The Cold-Foot Effect" by Scott M. Berry in *Chance* magazine, the author wrote about his statistical analysis of results from two NFL seasons. He uses a logistic regression model with variables such as wind, clouds, precipitation, temperature, the pressure of making the kick, and whether a time-out was called prior to the kick. He writes that "the conclusion from the model is that icing the kicker works—it is likely icing the kicker reduces the probability of a successful kick."

Dummy Variables and Logistic Regression

In this section, all variables have been continuous in nature. The time interval after an eruption can be any value over a continuous range of minutes, so it is a good example of a continuous variable. However, many applications involve a **dichotomous variable,** which has only *two* possible discrete values (such as male/female or dead/alive or cured/not cured). A common procedure is to represent the two possible discrete values by 0 and 1, where 0 represents a "failure" (such as death) and 1 represents a success. A dichotomous variable with the two possible values of 0 and 1 is called a **dummy variable.**

Procedures of analysis differ dramatically, depending on whether the dummy variable is a predictor (x) variable or the response (y) variable. If we include a dummy variable as another *predictor* (x) variable, we can use the methods of this section, as illustrated in the following example.

EXAMPLE Using a Dummy Variable Use the height, weight, waist, and pulse rates of the combined data set of 80 women and men as listed in Data Set 1 in Appendix B. Let the response y variable represent height and, for the first predictor variable, use the dummy variable of *gender* (coded as 0 = female, 1 = male). Given a weight of 150 lb, a waist size of 80 cm, and a pulse rate of 75 beats per minute, find the multiple regression equation and use it to predict the height of (a) a female and (b) a male.

SOLUTION Using the methods of this section with software, we get this regression equation:

$$\text{HT} = 64.4 + 3.47(\text{GENDER}) + 0.118(\text{WT}) - 0.222(\text{WAIST}) + 0.00602(\text{PULSE})$$

continued

To find the predicted height of a female, we substitute 0 for the gender variable. Also substituting 150 for weight, 80 for waist, and 75 for pulse results in a predicted height of 64.8 in. (or 5 ft 5 in.) for a female.

To find the predicted height of a male, we substitute 1 for the gender variable. Also substituting the other values results in a predicted height of 68.3 in. (or 5 ft 8 in.) for a male. Note that when all other variables are the same, a male will have a predicted height that is 3.47 in. more than the height of a female.

In the preceding example, we could use the methods of this section because the dummy variable of gender is a *predictor* variable. If the dummy variable is the response (y) variable, we cannot use the methods of this section, and we should use a method known as **logistic regression.** Suppose, for example, that we use height, weight, waist, and pulse rates of women and men as listed in Data Set 1 in Appendix B. Let the response y variable represent gender (0 = female, 1 = male). Using the 80 values of y (with female coded as 0 and male coded as 1) and the combined list of corresponding heights, weights, waist sizes, and pulse rates, we can use logistic regression to obtain this model:

$$\ln\left(\frac{p}{1-p}\right) = -41.8193 + 0.679195(\text{HT}) - 0.0106791(\text{WT}) + 0.0375373(\text{WAIST}) - 0.0606805(\text{PULSE})$$

In the above expression, p represents a probability. A value of $p = 0$ indicates that the person is a female and $p = 1$ indicates a male. A value of $p = 0.2$ indicates that there is a 0.2 chance of the person being a male, so it follows that there is a 0.8 chance that the person is a female. If we use the above model and substitute a height of 72 in., a weight of 200 lb, a waist circumference of 90 cm, and a pulse rate of 85 beats per minute, we can solve for p to get $p = 0.960$, indicating that such a large person is very likely to be a male. In contrast, a small person with a height of 60 in., a weight of 90 lb, a waist size of 68 cm, and a pulse rate of 85 beats per minute results in a value of $p = 0.00962$, indicating that such a small person is unlikely to be a male and is very likely to be a female. This book does not include detailed procedures for using logistic regression, but several books are devoted to this topic, and several other textbooks include detailed information about this method.

When we discussed regression in Section 10-3, we listed four common errors that should be avoided when using regression equations to make predictions. These same errors should be avoided when using multiple regression equations. Be especially careful about concluding that a cause-effect association exists.

TI Using the TI-83/84 Plus Calculator for Multiple Regression

The TI-83/84 Plus program A2MULREG must be downloaded to your calculator from the CD-ROM included with this book. On the CD-ROM, select the *software* folder, then select *TI-83 TI-84 Plus Programs*. After the program A2MULREG has been downloaded to your calculator, use the following procedure:

1. *Manual entry of sample data:* Enter the sample data as matrix D, with the first column containing the values of the dependent (y) variable. (A matrix is a rectangular array of data, and it consists of multiple rows and columns.) To manually enter the sample data in matrix D, press 2ND and the x^{-1} key to get the **Matrix** menu. Select **EDIT,** scroll down for **[D],** then press ENTER. Begin by entering the dimension of the matrix in the format of rows by columns. For the number of rows, enter the number of sample values listed for each variable. For the number of columns, enter the total number of variables (including the dependent variable). For example, to enter all of the data in Table 10-1 as matrix [D], enter 8 for the number of rows (because there are 8 sample values listed for each variable) and enter 4 for the number of columns (because there are 4 different variables). See the accompanying screen showing the creation of matrix [D] for the data in Table 10-1. After entering the dimension of the matrix, enter the first value, press ENTER, then proceed to enter the other values followed by the ENTER key after each value.

   Using existing lists: If the data are already stored as individual lists, those lists can be combined and stored in matrix [D]. To combine lists into a matrix, press 2ND and the x^{-1} key to get the **Matrix** menu. Select the top menu item of **MATH,** then select the item of **List→matr.** Now enter the list names (with the first list corresponding to the dependent y variable), and end with the matrix name [D], all separated by commas. Enter the matrix name of [D] by opening the MATRIX menu, scrolling to [D], and pressing enter. This is all likely to appear confusing, so consider the objective of doing multiple regression with the dependent variable named NICOT and the independent variables of TAR and CO, all stored in the calculator as lists. (The lists of NICOT, TAR, and CO can be opened with the APP named CIGARET.) Here is the command that puts these lists into matrix [D]: **List→dmatr(NICOT, TAR, CO, [D]).** It sure beats manual entry of all of the sample data.

TI-83/84 Plus

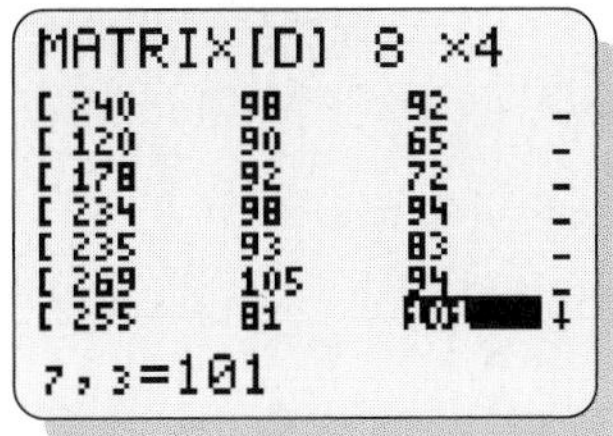

2. Now press **PRGM,** select **A2MULREG** and press **ENTER** three times, then select **MULT REGRESSION** and press **ENTER.**
3. When prompted, enter the number of *independent* variables, then enter the column numbers of the independent variables that you want to include. (Remember, the y-variable is the dependent variable, and it is not one of the independent variables.)

10-5 BASIC SKILLS AND CONCEPTS

Statistical Literacy and Critical Thinking

1. Multiple Regression What is *multiple regression,* and how does it differ from the regression discussed in Section 10-3?
2. Adjusted Coefficient of Determination When comparing different multiple regression equations, why is the adjusted R^2 a better measure than R^2?
3. Predicting Eye Color A geneticist wants to develop a method for predicting the eye color of a baby given the eye color of each parent. Can the methods of this section be used? Why or why not?
4. Variables What is the difference between a *response variable* and a *predictor variable*?

Interpreting a Calculator Display. In Exercises 5–8, refer to the TI-83/84 Plus calculator displays given here and answer the given questions or identify the indicated items. The displays are based on the sample of 54 bears listed in Data Set 6 in Appendix B. Here is the order of the variables and columns: weight, head length, length, chest size.

TI **5.** Bear Measurements Identify the multiple regression equation that expresses weight in terms of head length, length, and chest size.

TI **6.** Bear Measurements Identify the following.

a. The P-value corresponding to the overall significance of the multiple regression equation

b. The value of the multiple coefficient of determination R^2

c. The adjusted value of R^2

TI **7.** Bear Measurements Is the multiple regression equation usable for predicting a bear's weight based on its head length, length, and chest size? Why or why not?

TI **8.** Bear Measurements A bear is found to have a head length of 14.0 in., a length of 70.0 in., and a chest size of 50.0 in.

a. Find the predicted weight of the bear.

b. The bear in question actually weighed 320 lb. How accurate is the predicted weight from part (a)?

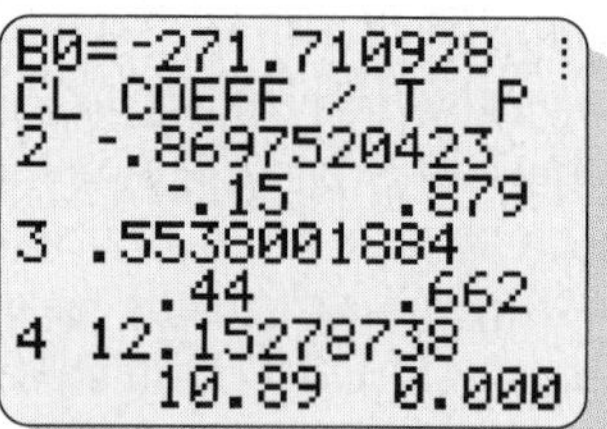

Health Data: Finding the Best Multiple Regression Equation. In Exercises 9–12, refer to the accompanying table, which was obtained by using the data for males in Data Set 1 in Appendix B. The response (y) variable is weight (in pounds), and the predictor (x) variables are HT (height in inches), WAIST (waist circumference in cm), and CHOL (cholesterol in mg).

Predictor (x) Variables	P-value	R^2	Adjusted R^2	Regression Equation
HT, WAIST, CHOL	0.000	0.880	0.870	$\hat{y} = -199 + 2.55\text{ HT} + 2.18\text{ WAIST} - 0.00534\text{ CHOL}$
HT, WAIST	0.000	0.877	0.870	$\hat{y} = -206 + 2.66\text{ HT} + 2.15\text{ WAIST}$
HT, CHOL	0.002	0.277	0.238	$\hat{y} = -148 + 4.65\text{ HT} + 0.00589\text{ CHOL}$
WAIST, CHOL	0.000	0.804	0.793	$\hat{y} = -42.8 + 2.41\text{ WAIST} - 0.0106\text{ CHOL}$
HT	0.001	0.273	0.254	$\hat{y} = -139 + 4.55\text{ HT}$
WAIST	0.000	0.790	0.785	$\hat{y} = -44.1 + 2.37\text{ WAIST}$
CHOL	0.874	0.001	0.000	$\hat{y} = 173 - 0.00233\text{ CHOL}$

9. If only one predictor (x) variable is used to predict weight, which single variable is best? Why?

10. If exactly two predictor (x) variables are to be used to predict weight, which two variables should be chosen? Why?

11. Which regression equation is best for predicting the weight? Why?

12. If a male has a height of 72 in., a waist circumference of 105 cm, and a cholesterol level of 250 mg, what is the best predicted value of his weight? Is that predicted value likely to be a good estimate? Is that predicted value likely to be very accurate?

TI 13. Appendix B Data Set: Predicting Nicotine in Cigarettes Refer to Data Set 3 in Appendix B.

a. Find the regression equation that expresses the response variable (y) of nicotine amount in terms of the predictor variable (x) of the tar amount.

b. Find the regression equation that expresses the response variable (y) of nicotine amount in terms of the predictor variable (x) of the carbon monoxide amount.

c. Find the regression equation that expresses the response variable (y) of nicotine amount in terms of predictor variables (x) of tar amount and carbon monoxide amount.

d. For the regression equations found in parts (a), (b), and (c), which is the best equation for predicting the nicotine amount?

e. Is the best regression equation identified in part (d) a *good* equation for predicting the nicotine amount? Why or why not?

TI 14. Appendix B Data Set: Using Garbage to Predict Population Size Refer to Data Set 16 in Appendix B.

a. Find the regression equation that expresses the response variable (y) of household size in terms of the predictor variable of the weight of discarded food.

b. Find the regression equation that expresses the response variable (y) of household size in terms of the predictor variable (x) of the weight of discarded plastic.

c. Find the regression equation that expresses the response variable (y) of household size in terms of predictor variables (x) of the weight of discarded food and the weight of discarded plastic.

d. For the regression equations found in parts (a), (b), and (c), which is the best equation for predicting the household size? Why?

e. Is the best regression equation identified in part (d) a *good* equation for predicting the household size? Why or why not?

TI 15. Appendix B Data Set: Home Selling Price Refer to Data Set 18 in Appendix B and find the best multiple regression equation with selling price as the response (y) variable. Is this "best" equation good for predicting the selling price of a home?

TI 16. Appendix B Data Set: Old Faithful This section used the Old Faithful data from 8 eruptions, as listed in Table 10-1. Refer to Data Set 11 in Appendix B and use the complete data set from 40 eruptions. Determine the best multiple regression equation that expresses the response variable (y) of time interval after an eruption in terms of one or more of the other variables. Explain your choice.

10-5 BEYOND THE BASICS

17. Dummy Variable Refer to Data Set 6 in Appendix B and use the sex, age, and weight of the bears. For sex, let 0 represent female and let 1 represent male. (In Data Set 6, males are already represented by 1, but for females change the sex values of 2 to 0.) Letting the response (y) variable be weight, use the variable of age and the dummy variable of sex to find the multiple regression equation, then use it to find the predicted weight of a bear with the characteristics given below. Does sex appear to have much of an effect on the weight of a bear?

a. Female bear that is 20 years of age

b. Male bear that is 20 years of age

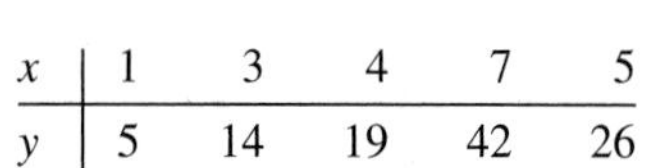

x	1	3	4	7	5
y	5	14	19	42	26

18. Using Multiple Regression for Equation of Parabola In some cases, the best-fitting multiple regression equation is of the form $\hat{y} = b_0 + b_1x + b_2x^2$. The graph of such an equation is a parabola. Using the data set listed in the margin, let $x_1 = x$, let $x_2 = x^2$, and find the multiple regression equation for the parabola that best fits the given data. Based on the value of the multiple coefficient of determination, how well does this equation fit the data?

10-6 Modeling

Key Concept This section introduces some basic concepts of developing a **mathematical model,** which is a mathematical function that "fits" or describes real-world data. For example, we might want a mathematical model consisting of an equation relating a variable for population size to another variable representing time. Unlike Section 10-3, we are not restricted to a model that must be linear. Also, instead of using randomly selected sample data, we will consider data collected periodically over time or some other basic unit of measurement. There are some powerful statistical methods that we could discuss (such as *time series*), but the main objective of this section is to describe briefly how technology can be used to find a good mathematical model.

The following are some generic models as listed in a menu from the TI-83/84 Plus calculator (press **STAT,** then select **CALC**):

Linear:	$y = a + bx$	Quadratic:	$y = ax^2 + bx + c$
Logarithmic:	$y = a + b \ln x$	Exponential:	$y = ab^x$
Power:	$y = ax^b$		

The particular model that you select depends on the nature of the sample data, and a scatterplot can be very helpful in making that determination. The illustrations that follow are graphs of some common models displayed on a TI-83/84 Plus calculator.

TI-83/84 Plus

Linear: $y = 1 + 2x$

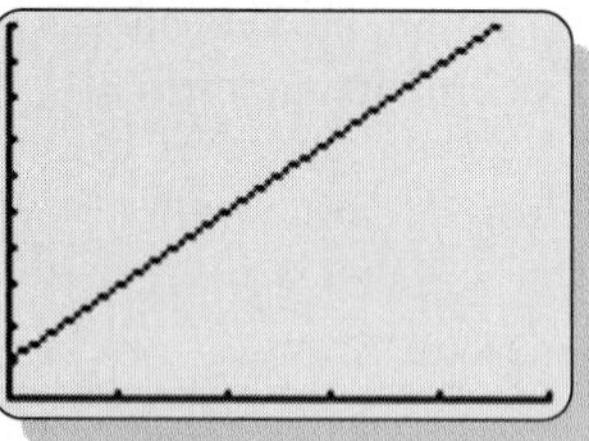

Quadratic: $y = x^2 - 8x + 18$

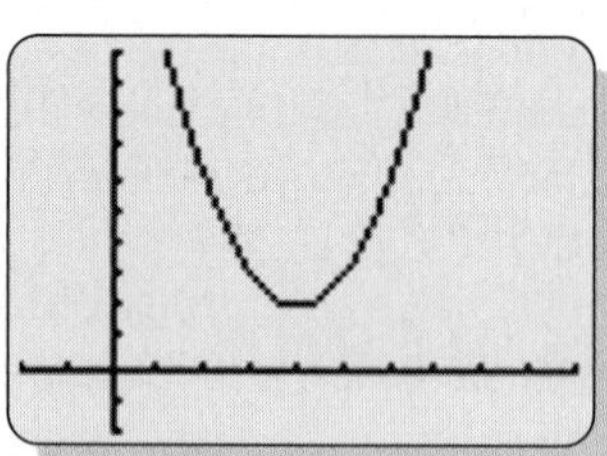

Logarithmic: $y = 1 + 2 \ln x$

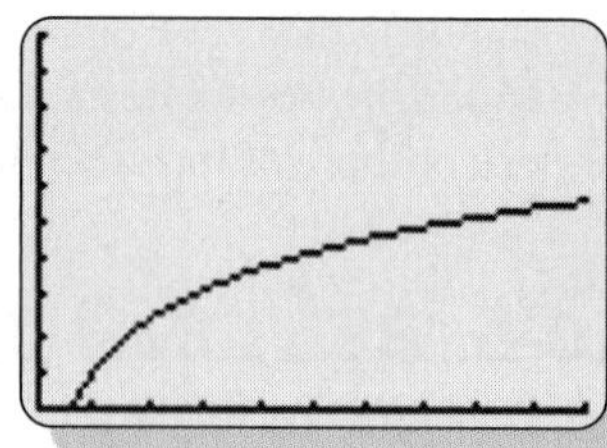

Exponential: $y = 2^x$

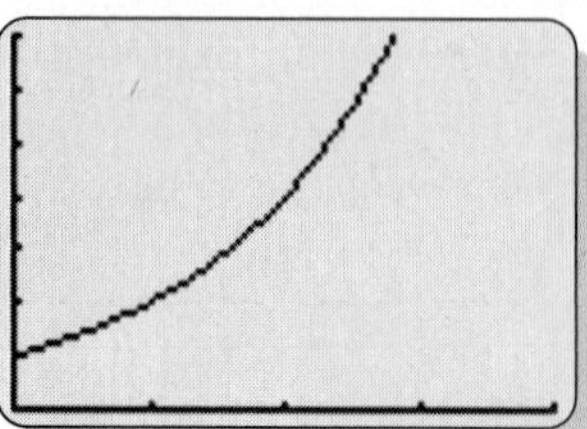

Power: $y = 3x^{2.5}$

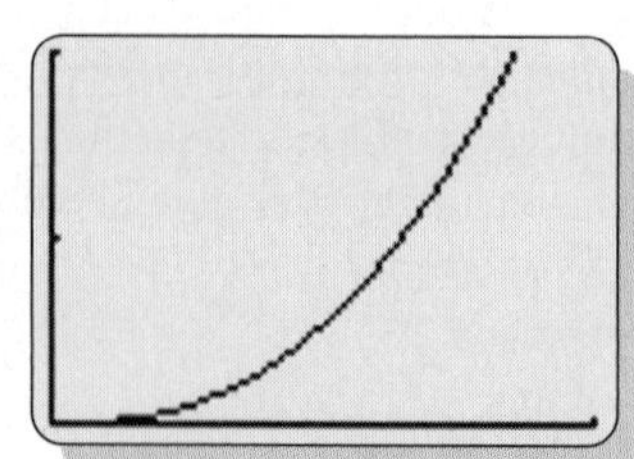

Here are three basic rules for developing a good mathematical model:

1. *Look for a pattern in the graph.* Examine the graph of the plotted points and compare the basic pattern to the known generic graphs of a linear function, quadratic function, exponential function, power function, and so on. (Refer to the accompanying graphs shown in the examples of the TI-83/84 Plus calculator displays.) When trying to select a model, consider only those functions that visually appear to fit the observed points reasonably well.
2. *Find and compare values of R^2.* For each model being considered, use computer software or a TI-83/84 Plus calculator to find the value of the coefficient of determination R^2. Values of R^2 can be interpreted here the same way that they were interpreted in Section 10-5. When narrowing your possible models, select functions that result in larger values of R^2, because such larger values correspond to functions that better fit the observed points. However, don't place much importance on small differences, such as the difference between $R^2 = 0.984$ and $R^2 = 0.989$. (Another measurement used to assess the quality of a model is the sum of squares of the residuals. See Exercise 15.)
3. *Think.* Use common sense. Don't use a model that leads to predicted values known to be totally unrealistic. Use the model to calculate future values, past values, and values for missing years, then determine whether the results are realistic.

EXAMPLE Table 10-4 lists the population of the United States for different years. Find a good mathematical model for the population size, then predict the size of the U.S. population in the year 2020.

SOLUTION First, we "code" the year values by using 1, 2, 3 . . . , instead of 1800, 1820, 1840. . . . The reason for this coding is to use values of x that are much smaller and much less likely to cause the computational difficulties that are likely to occur with really large x-values.

Look for a pattern in the graph. Examine the pattern of the data values in the TI-83/84 Plus display (shown in the margin) and compare that pattern to the generic models shown earlier in this section. The pattern of those points is clearly not a straight line, so we rule out a linear model. Good candidates for the model appear to be the quadratic, exponential, and power functions.

TI-83/84 Plus

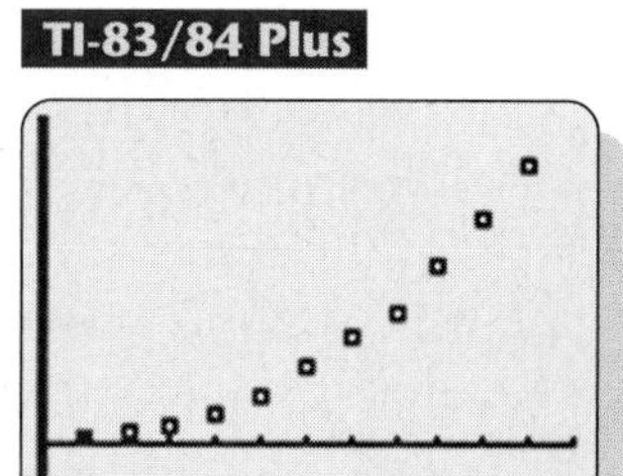

Table 10-4 Population (in millions) of the United States

Year	1800	1820	1840	1860	1880	1900	1920	1940	1960	1980	2000
Coded year	1	2	3	4	5	6	7	8	9	10	11
Population	5	10	17	31	50	76	106	132	179	227	281

continued

Find and compare values of R^2. The following displays show the TI-83/84 Plus results based on the quadratic, exponential, and power models. Comparing the values of the coefficient R^2, it appears that the quadratic model is best because it has the highest value of 0.9992, but the other displayed values are also quite high. If we select the quadratic function as the best model, we conclude that the equation $y = 2.77x^2 - 6.00x + 10.01$ best describes the relationship between the year x (coded with $x = 1$ representing 1800, $x = 2$ representing 1820, and so on) and the population y (in millions).

TI-83/84 Plus

```
QuadReg
 y=ax²+bx+c
 a=2.766899767
 b=-6.002797203
 c=10.01212121
 R²=.9991688446
```

TI-83/84 Plus

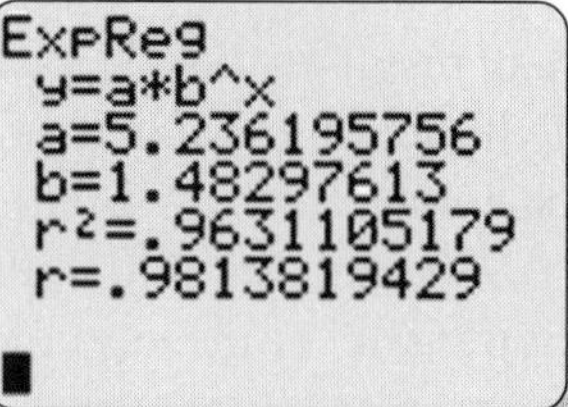

TI-83/84 Plus

```
PwrReg
 y=a*x^b
 a=3.353115397
 b=1.766059823
 r²=.976406226
 r=.9881326966
```

To predict the U.S. population for the year 2020, first note that the year 2020 is coded as $x = 12$ (see Table 10-4). Substituting $x = 12$ into the quadratic model of $y = 2.77x^2 - 6.00x + 10.01$ results in $y = 337$, which indicates that the U.S. population is estimated to be 337 million in the year 2020.

Think. The forecast result of 337 million in 2020 seems reasonable. (A U.S. Bureau of the Census projection suggests that the population in 2020 will be around 325 million.) However, there is considerable danger in making estimates for times that are beyond the scope of the available data. For example, the quadratic model suggests that in 1492, the U.S. population was 671 million—an absurd result. The quadratic model appears to be good for the available data (1800–2000), but other models might be better if it is absolutely necessary to make population estimates far beyond that time frame.

In "Modeling the U.S. Population" (*AMATYC Review,* Vol. 20, No. 2), Sheldon Gordon uses more data than Table 10-4, and he uses much more advanced techniques to find better population models. In that article, he makes this important point:

> **"The best choice (of a model) depends on the set of data being analyzed and requires an exercise in judgment, not just computation."**

TI Using the TI-83/84 Plus Calculator for Modeling

To use TI-83/84 Plus for creating mathematical models such as those discussed in this section, use the following procedure:

1. First enter or open the sample data. The data can be manually entered in lists L1 and L2, or existing named lists can be opened by using their names.

2. Turn on the diagnostics feature by pressing **2ND** **0** to get the **CATALOG** menu. Use the arrow key ⌄ to scroll down to **DiagnosticsOn** and press the **ENTER** key twice.
3. Press **STAT** and select **CALC.**
4. Select the desired model from the list of available options, then press **ENTER**.
5. Enter the two list names or labels, using a comma to separate them. Press **ENTER** when finished.

The display will be similar to those given in this section. The format of the equation will be given along with the coefficients used in the equation. The value of R^2 is also included for many of the models.

Statistics: Jobs and Employers

Here is a small sample of advertised jobs in the field of statistics: forecaster, database analyst, marketing scientist, credit-risk manager, cancer researcher and evaluator, insurance-risk analyst, educational testing researcher, biostatistician, statistician for pharmaceutical products, cryptologist, statistical programmer.

Here is a small sample of firms offering jobs in the field of statistics: Centers for Disease Control and Prevention, Cardiac Pacemakers, Inc., National Institutes of Health, National Cancer Institute, CNA Insurance Companies, Educational Testing Service, Roswell Park Cancer Institute, Cleveland Clinic Foundation, National Security Agency, Quantiles, 3M, IBM, Nielsen Media Research, AT&T Labs, Bell Labs, Hewlett Packard, Johnson & Johnson, Smith Hanley.

10-6 BASIC SKILLS AND CONCEPTS

Statistical Literacy and Critical Thinking

1. Model What is a mathematical model?
2. R^2 How are values of R^2 used to compare different models being considered?
3. Projections In this section we used the population values from the year 1800 to the year 2000, and we found that the best model is described by $y = 2.77x^2 - 6.00x + 10.01$, where the population value of y is in millions. What is wrong with using this model to project the population size for the year 3000?
4. Best Model Assume that we use a sample with the methods of this section to find that among the five different possible models, the best model is $y = 4x^{1.2}$ with $R^2 = 0.200$. Does this best model appear to be a good model? Why or why not?

Finding the Best Model. *In Exercises 5–12, construct a scatterplot and identify the mathematical model that best fits the given data. Assume that the model is to be used only for the scope of the given data, and consider only linear, quadratic, logarithmic, exponential, and power models.*

TI 5.

x	1	2	3	4	5	6
y	5	7	9	11	13	15

TI 6.

x	1	2	3	4	5	6
y	2	4	8	16	32	64

TI 7.

x	1	2	3	4	5	6
y	1	7	17	31	49	71

TI 8.

x	1	2	3	4	5	6
y	3	8.485	15.588	24	33.541	44.091

TI 9. Manatee Deaths from Boats The accompanying table lists the number of Florida manatee deaths related to encounters with watercraft (based on data from Florida Fish and Wildlife Conservation).

Year	1983	1984	1985	1986	1987	1988	1989	1990	1991	1992	1993
Deaths	15	34	33	33	39	43	50	47	53	38	35

Year	1994	1995	1996	1997	1998	1999	2000	2001	2002	2003	2004
Deaths	49	42	60	54	67	82	78	81	95	73	69

TI **10.** Manatee Deaths from Natural Causes The accompanying table lists the number of Florida manatee deaths from natural causes (based on data from Florida Fish and Wildlife Conservation). Does the best model appear to be a reasonably good model?

Year	1983	1984	1985	1986	1987	1988	1989	1990	1991	1992	1993
Deaths	6	24	19	1	10	15	18	21	13	20	22

Year	1994	1995	1996	1997	1998	1999	2000	2001	2002	2003	2004
Deaths	33	35	101	42	12	37	37	34	59	102	25

TI **11.** Physics Experiment An experiment in a physics class involves dropping a golf ball and recording the distance (in meters) it falls for different times (in seconds) after it was released. The data are given in the table below. Project the distance for a time of 12 sec, given that the golf ball is dropped from a building that is 50 m tall.

Time	0	0.5	1	1.5	2	2.5	3
Distance	0	1.2	4.9	11.0	19.5	30.5	44.0

TI **12.** Stock Market Listed below in order by row are the annual high values of the Dow Jones Industrial Average for each year beginning with 1980. What is the best predicted value for the year 2004? Given that the actual high value in 2004 was 10,855, how good was the predicted value? What does the pattern suggest about the stock market for investment purposes? (Acts of terrorism and bad economic conditions caused substantial stock market losses in 2002.)

1000 1024 1071 1287 1287 1553 1956 2722 2184 2791 3000 3169 3413
3794 3978 5216 6561 8259 9374 11,568 11,401 11,350 10,635 10,454

10-6 BEYOND THE BASICS

TI **13.** Moore's Law In 1965, Intel co-founder Gordon Moore initiated what has since become known as *Moore's law:* the number of transistors per square inch on integrated circuits will double approximately every 18 months. The table below lists the number of transistors (in thousands) for different years.

Year	1971	1974	1978	1982	1985	1989	1993	1997	1999	2000	2002	2003
Transistors	2.3	5	29	120	275	1180	3100	7500	24,000	42,000	220,000	410,000

a. Assuming that Moore's law is correct and transistors double every 18 months, which mathematical model best describes this law: linear, quadratic, logarithmic, exponential, power? What specific function describes Moore's law?

b. Which mathematical model best fits the listed sample data?

c. Compare the results from parts (a) and (b). Does Moore's law appear to be working reasonably well?

14. Population in 2050 When the exercises in this section were written, the United Nations used its own model to predict a population of 394 million for the United States in 2050. Based on the data in Table 10-4, which of the models discussed in Section 10-6 yields a projected population closest to 394 million in 2050?

15. Using the Sum of Squares Criterion In addition to the value of R^2, another measurement used to assess the quality of a model is the *sum of squares of the residuals.* A residual is the difference between an observed y value and the value of y predicted from the model, which is denoted as $\hat{y}$. Better models have smaller sums of squares. Refer to the example in this section.

a. Find $\Sigma(y - \hat{y})^2$, the sum of squares of the residuals resulting from the linear model.

b. Find the sum of squares of residuals resulting from the quadratic model.

c. Verify that according to the sum of squares criterion, the quadratic model is better than the linear model.

Review

This chapter presents basic methods for investigating relationships or correlations between two or more variables.

- Section 10-2 used scatter diagrams and the linear correlation coefficient to decide whether there is a linear correlation between two variables.
- Section 10-3 presented methods for finding the equation of the regression line that (by the least-squares criterion) best fits the paired data. When there is a significant linear correlation, the regression equation can be used to predict the value of a variable, given some value of the other variable.
- Section 10-4 introduced the concept of total variation, with components of explained and unexplained variation. The coefficient of determination r^2 gives us the proportion of the variation in the response variable (y) that can be explained by the linear correlation between x and y. We also developed methods for constructing prediction intervals, which are helpful in judging the accuracy of predicted values.
- In Section 10-5, we considered multiple regression, which allows us to investigate relationships involving more than one predictor (x) variable. We discussed procedures for obtaining a multiple regression equation, as well as the value of the multiple coefficient of determination R^2, the adjusted R^2, and a P-value for the overall significance of the equation.
- In Section 10-6 we explored basic concepts of developing a mathematical model, which is a function that can be used to describe a relationship between two variables. Unlike the preceding sections of this chapter, Section 10-6 included several nonlinear functions.

Statistical Literacy and Critical Thinking

1. Correlation and Regression In your own words, describe correlation, regression, and the difference between them.

2. Correlation Given a collection of paired data, the linear correlation coefficient is found to be $r = 0$. Does that mean that there is no relationship between the two variables?

3. Causation A medical researcher finds that there is a significant linear correlation between the amount of a drug taken and the cholesterol level of the subject. Is she justified in writing in a journal article that the drug causes lower cholesterol levels? Why or why not?

4. Predictions After finding that there is a significant linear correlation between two variables, a predicted value of y is obtained by using the regression equation. Given that there is a significant linear correlation, will the projected value be very accurate?

Review Exercises

TI **1.** Manatee Deaths The table below lists the number of Florida manatee deaths related to encounters with watercraft and natural causes for each of several different years (based on data from Florida Fish and Wildlife Conservation).

a. Find the value of the linear correlation coefficient and determine whether there is a significant linear correlation between the two variables.

b. Find the equation of the regression line. Let the number of natural deaths represent the response (y) variable. What is the best predicted number of natural deaths in a year with 50 deaths from encounters with watercraft?

Watercraft	49	42	60	54	67	82	78	81	95	73	69
Natural	33	35	101	42	12	37	37	34	59	102	25

TI  **2.** Old Faithful Use the data given below (from Table 10-1). The duration times are in seconds and the heights are in feet.

a. Is there a significant linear correlation between duration of an eruption of the Old Faithful geyser and the height of the eruption?

b. Find the equation of the regression line with height representing the response (y) variable.

c. What is the best predicted height of an eruption that has a duration of 180 sec?

Duration	240	120	178	234	235	269	255	220
Height	140	110	125	120	140	120	125	150

Predicting Cost of Electricity. *Given below are measurements from the author's home taken from Data Set 9 in Appendix B. Use these data for Exercises 3–5.*

kWh	3375	2661	2073	2579	2858	2296	2812	2433	2266	3128
Heating Degree Days	2421	1841	438	15	152	1028	1967	1627	537	26
Average Daily Temp	26	34	58	72	67	48	33	39	66	71
Cost (dollars)	321.94	221.11	205.16	251.07	279.8	183.84	244.93	218.59	213.09	333.49

TI **3. a.** Use a 0.05 significance level to test for a linear correlation between the cost of electricity and the kWh of electricity consumed.

b. What percentage of the variation in cost can be explained by the linear relationship between electricity consumption (in kWh) and cost?

c. Find the equation of the regression line that expresses cost (y) in terms of the amount of electricity consumed (in kWh).

d. What is the best predicted cost for a time when 3000 kWh of electricity is used?

TI **4. a.** Use a 0.05 significance level to test for a linear correlation between the average daily temperature and the cost.

b. What percentage of the variation in cost can be explained by the linear relationship between cost and average daily temperature?

c. Find the equation of the regression line that expresses cost (y) in terms of the average daily temperature.

d. What is the best predicted cost at a time when the average daily temperature is 40?

TI **5.** Use the TI-83/84 Plus calculator to find the multiple regression equation of the form $\hat{y} = b_0 + b_1x_1 + b_2x_2$, where the response variable y represents cost, x_1 represents electricity consumption in kWh, and x_2 represents average daily temperature. Also identify the value of the multiple coefficient of determination R^2, the adjusted R^2, and the P-value representing the overall significance of the multiple regression equation. Can the regression equation be used to predict cost? Are either of the regression equations from Exercises 3 and 4 better?

Cumulative Review Exercises

Super Bowl Points and DJIA. *Listed below are the total numbers of points scored in Super Bowl football games and the high value of the Dow Jones Industrial Average (DJIA). The data are paired according to year, and they represent recent and consecutive years. Use those sample data for Exercises 1-8.*

Super Bowl Points	56	55	53	39	41	37	69	61
DJIA	6561	8259	9374	11,568	11,401	11,350	10,635	10,454

TI **1.** Test for a correlation between Super Bowl points and the DJIA. Is the result as you expected?

TI **2.** Find the regression equation in which the DJIA high value is the response (y) variable. What is the best predicted DJIA value for a year in which there are 50 points scored in the Super Bowl?

TI **3.** Is it possible to test the claim that the mean number of points scored in the Super Bowl is equal to the mean value of the DJIA? Would such a test make sense?

TI **4.** Construct a 95% confidence interval estimate for the mean number of points scored in Super Bowl games.

5. Why would it be a bad idea to try to estimate the next consecutive DJIA high value by constructing a confidence interval estimate of the DJIA values?

TI **6.** Do the Super Bowl points appear to come from a population with a normal distribution? Why or why not?

TI **7.** Find the mean and standard deviation of the sample of Super Bowl points.

TI **8.** The mean and standard deviation from Exercise 7 are sample statistics, but treat them as population parameters for a normally distributed population, and find the probability that a random Super Bowl game will have less than 40 total points scored.

Cooperative Group Activities

1. In-class activity Divide into groups of 8 to 12 people. For each group member, measure the person's height and also measure his or her navel height, which is the height from the floor to the navel. Is there a correlation between height and navel height? If so, find the regression equation with height expressed in terms of navel height. According to an old theory, the average person's ratio of height to navel height is the golden ratio: $(1 + \sqrt{5})/2 \approx 1.6$. Does this theory appear to be reasonably accurate?

2. In-class activity Divide into groups of 8 to 12 people. For each group member, *measure* height and arm span. For the arm span, the subject should stand with arms extended, like the wings on an airplane. It's easy to mark the height and arm span on a chalkboard, then measure the distances there. Using the paired sample data, is there a correlation between height and arm span? If so, find the regression equation with height expressed in terms of arm span. Can arm span be used as a reasonably good predictor of height?

3. In-class activity Divide into groups of 8 to 12 people. For each group member, use a string and ruler to measure head circumference and forearm length. Is there a relationship between these two variables? If so, what is it?

4. In-class activity Use a ruler as a device for measuring reaction time. One person should suspend the ruler by holding it at the top while the subject holds his or her thumb and forefinger at the bottom edge ready to catch the ruler when it is released. Record the distance that the ruler falls before it is caught. Convert that distance to the time (in seconds) that it took the subject to react and catch the ruler. (If the distance is measured in inches, use $t = \sqrt{d/192}$. If the distance is measured in centimeters, use $t = \sqrt{d/487.68}$.) Test each subject once with the right hand and once with the left hand, and record the paired data. Test for a correlation. Find the equation of the regression line. Does the equation of the regression line suggest that the dominant hand has a faster reaction time?

5. In-class activity Divide into groups of 8 to 12 people. For each group member, record the pulse rate by counting the number of heartbeats in 1 min. Also record height. Is there a relationship between pulse rate and height? If so, what is it?

6. In-class activity Collect data from each student consisting of the number of credit cards and the number of keys that the student has in his or her possession. Is there a correlation? If so, what is it? Try to identify at least one reasonable explanation for the presence or absence of a correlation.

7. In-class activity Divide into groups of three or four people. Appendix B includes many data sets not yet included in examples or exercises in this chapter. Search Appendix B for a pair of variables of interest, then investigate correlation and regression. State your conclusions and try to identify practical applications.

8. Out-of-class activity Divide into groups of three or four people. Investigate the relationship between two variables by collecting your own paired sample data and using the methods of this chapter to determine whether there is a significant linear correlation. Also identify the regression equation and describe a procedure for predicting values of one of the variables when given values of the other variable. Suggested topics:
   - Is there a relationship between taste and cost of different brands of chocolate chip cookies (or colas)? Taste can be measured on some number scale, such as 1 to 10.
   - Is there a relationship between salaries of professional baseball (or basketball, or football) players and their season achievements?
   - Is there a relationship between the lengths of men's (or women's) feet and their heights?
   - Is there a relationship between student grade-point averages and the amount of television watched? If so, what is it?

TI-83/84 Plus Calculator Project

Much effort is spent studying identical twins that were separated at birth and raised apart. Identical twins occur when a single fertilized egg splits in two, so that both twins share the same genetic makeup. By obtaining IQ scores of identical twins separated at birth, researchers hope to identify the effects of heredity and environment on intelligence. In this project, we will simulate 100 sets of twin births, but we will generate their IQ scores in a way that has no common genetic or environmental influences. Using the random number generator feature of the TI-83/84 Plus calculator, generate a list of 100 simulated IQ scores randomly selected from a normally distributed population having a mean of 100 and a standard deviation of 15. This is accomplished by pressing MATH, selecting **PRB,** then selecting **randNorm.** The command of **randNorm** (100, 15, 100) produces 100 values from a normally distributed population with mean 100 and standard deviation 15. Now use the same procedure to generate a second list of 100 simulated IQ scores that are also randomly selected from a normally distributed population with a mean of 100 and a standard deviation of 15. Even though the two lists were independently generated, treat them as paired data, so that the first score from each list represents the first set of twins, the second score from each list represents the second set of twins, and so on. Before doing any calculations, first estimate a value of the linear correlation coefficient that you would expect. Now use the methods of Section 10-2 with a 0.05 significance level to test for a significant linear correlation and state your results.

Consider the preceding procedure to be one trial. Given the way that the sample data were generated, what proportion of such trials should lead to the incorrect conclusion that there is a significant linear correlation? By repeating the trials, we can verify that the proportion is approximately correct. Either repeat the trial or combine your results with others to verify that the proportion is approximately correct. Note that a type I error is the mistake of rejecting a true null hypothesis which, in this case, means that we conclude that there is a significant linear correlation when there really is no such linear correlation.

From Data to Decision

Critical Thinking: Is Duragesic effective in reducing pain?

Listed below are measures of pain intensity before and after using the proprietary drug Duragesic (based on data from Janssen Pharmaceutical Products, L.P.) The data are listed in order by row, and corresponding measures are from the same subject before and after the treatment. For example, the first subject had a measure of 1.2 before the treatment and a measure of 0.4 after the treatment. Each pair of measurements is from one subject, and the intensity of pain was measured using the standard visual analog scale.

Pain Intensity Before Duragesic Treatment

1.2 1.3 1.5 1.6 8.0 3.4 3.5 2.8 2.6 2.2
3.0 7.1 2.3 2.1 3.4 6.4 5.0 4.2 2.8 3.9
5.2 6.9 6.9 5.0 5.5 6.0 5.5 8.6 9.4 10.0
7.6

Pain Intensity After Duragesic Treatment

0.4 1.4 1.8 2.9 6.0 1.4 0.7 3.9 0.9 1.8
0.9 9.3 8.0 6.8 2.3 0.4 0.7 1.2 4.5 2.0
1.6 2.0 2.0 6.8 6.6 4.1 4.6 2.9 5.4 4.8
4.1

Analyzing the Results

1. Use the given data to construct a scatterplot, then use the methods of Section 10-2 to test for a linear correlation between the pain intensity before the treatment and after the treatment. If there is a significant linear correlation, does it follow that the drug treatment is effective?

2. Use the given data to find the equation of the regression line. Let the response (y) variable be the pain intensity after the treatment. What would be the equation of the regression line for a treatment having absolutely no effect?

3. The methods of Section 9-3 can be used to test the claim that two populations have the same mean. Identify the specific claim that the treatment is effective, then use the methods of Section 9-3 to test that claim. The methods of Section 9-3 are based on the requirement that the samples are independent. Are they independent in this case?

4. The methods of Section 9-4 can be used to test a claim about matched data. Identify the specific claim that the treatment is effective, then use the methods of Section 9-4 to test that claim.

5. Which of the preceding results is best for determining whether the drug treatment is effective in reducing pain? Which of the preceding results is least effective in determining whether the drug treatment is effective in reducing pain? Based on the preceding results, does the drug appear to be effective?

Internet Project

Linear Regression

The linear correlation coefficient is a tool that is used to measure the strength of the linear relationship between two sets of measurements. From a strictly computational point of view, the correlation coefficient may be found for any two data sets of paired values, regardless of what the data values represent. For this reason, certain questions should be asked whenever a correlation is being investigated. Is it reasonable to expect a linear correlation? Could a perceived correlation be caused by a third quantity related to each of the variables being studied? Go to the Web site for this textbook:

http://www.aw.com/triola

The Internet Project for this chapter will guide you to several sets of paired data in the fields of sports, medicine, and economics. You will then apply the methods of this chapter, computing correlation coefficients and determining regression lines, while considering the true relationships between the variables involved.

Statistics @ Work

"In a business world that is fascinated with numbers and data, statistics is a key to being able to properly analyze and summarize vast quantities of data."

Mark D. Haskell

Director, Forecasting and Analysis
Walt Disney World Resort

As Director of Forecasting and Analysis for Walt Disney World Resort, Mark leads a team of people responsible for planning and forecasting values such as attendance, hotel occupancy, and projected revenue. By analyzing various factors, Mark and his team help Disney continue to work to ensure that each guest has an enjoyable and memorable experience at Walt Disney World Resort.

What do you do in your work?

I lead a team of people responsible for planning and forecasting such metrics as theme park attendance, occupancy at each of our resort hotels, and the revenue the Walt Disney World will realize from these key business drives.

How do you use statistics and what specific statistical concepts do you use?

Statistics is central to the forecasting process. Many of our forecasting tools are based upon multiple regression techniques, with some of those models more complex than others. We also use very basic statistical concepts on a daily basis, whether reporting the "mean absolute percent error" for our forecasts, understanding measures of central tendency, distributions, and sampling techniques when reviewing marketing research, or running correlations to understand how different variables align with our key business drivers. There are many approaches that can be used in creating high quality forecasts, but statistics is a basic building block for almost all of those approaches.

Describe a specific example of how the use of statistics was useful in improving a product or service.

My team recently used correlation analysis to help us understand what sources of data would be most helpful in predicting attendance and spending at one of our retail centers. Based upon that work, we developed a regression model that helps company leaders understand revenue potential, determine staffing needs, set operating hours, identify new product opportunities, and identify capital investment needs, to name just a few.

What background in statistics is required to obtain a job like yours?

I have a Masters Degree in Economics, specializing in quantitative analysis methods. Generally some form of advanced degree with emphasis in statistical analysis would be required to succeed in my role.

Do you feel job applicants at your company are viewed more favorably if they have studied some statistics?

Some level of experience with statistics is required for roles on the Forecasting and Analysis team. There are many other roles at Walt Disney World that would look favorably on those who have studied statistics.

Do you recommend that today's college students study statistics? Why?

Absolutely. In a business world that is fascinated with numbers and data, statistics is a key to being able to properly analyze and summarize vast quantities of data. Even if you aren't responsible for conducting the analysis, you need a basic understanding to properly use the information for decision making. You need to learn how to use statistics properly, or you risk having those with a better understanding of statistics use them against you.

What other skills are important for today's college students?

Communication skills, both verbal and written. There is tremendous value in having people who can analyze complex information, then simplify it and clearly communicate it for easy consumption.

Multinomial Experiments and Contingency Tables

11

11 CHAPTER PROBLEM

Using statistics to detect fraud

In the *New York Times* article "Following Benford's Law, or Looking Out for No. 1," Malcolm Browne writes that "the income tax agencies of several nations and several states, including California, are using detection software based on Benford's Law, as are a score of large companies and accounting businesses." According to Benford's law, a variety of different data sets include numbers with leading (first) digits that follow the distribution shown in the first two rows of Table 11-1. Data sets with values having leading digits that conform to Benford's law include stock market values, population sizes, numbers appearing on the front page of a newspaper, amounts on tax returns, lengths of rivers, and check amounts.

When working for the Brooklyn District Attorney, investigator Robert Burton used Benford's law to identify fraud by analyzing the leading digits on 784 checks. If the 784 checks follow Benford's law perfectly, 30.1% of the checks should have amounts with a leading digit of 1. The expected number of checks with amounts having a leading digit of 1 is 235.984 (because 30.1% of 784 is 235.984). The other expected frequencies are listed in the third row of Table 11-1. The bottom row of Table 11-1 lists the frequencies of the leading digits from amounts on 784 checks issued by seven different companies. A quick visual comparison shows that there appear to be major discrepancies between the frequencies expected by Benford's law and the frequencies observed in the check amounts, but how do we measure that disagreement? Are those discrepancies *significant?* Is there enough evidence to justify the conclusion that fraud has been committed? Is the evidence beyond a "reasonable doubt"? We will address these questions in this chapter.

Table 11-1 Benford's Law: Distribution of Leading Digits

Leading Digit	1	2	3	4	5	6	7	8	9
Benford's law: frequency distribution of leading digits	30.1%	17.6%	12.5%	9.7%	7.9%	6.7%	5.8%	5.1%	4.6%
Expected frequencies of leading digits from 784 checks following Benford's law	235.984	137.984	98.000	76.048	61.936	52.528	45.472	39.984	36.064
Observed leading digits of 784 actual checks analyzed for fraud	0	15	0	76	479	183	8	23	0

11-1 Overview

This chapter involves categorical (or qualitative, or attribute) data that can be separated into different cells. For example, we might separate a sample of M&Ms into the color categories of red, orange, yellow, brown, blue, and green. After finding the frequency count for each category, we might proceed to test the claim that the frequencies fit (or agree with) the color distribution claimed by the manufacturer (Mars, Inc.). The main objective of this chapter is to test claims about categorical data consisting of frequency counts for different categories. In Section 11-2 we consider multinomial experiments, which consist of observed frequency counts arranged in a single row or column (called a one-way frequency table), and we will test the claim that the observed frequency counts agree with some claimed distribution. In Section 11-3 we will consider contingency tables (or two-way frequency tables), which consist of frequency counts arranged in a table with at least two rows and two columns. In Section 11-4 we consider two-way tables involving data consisting of matched pairs.

The methods of this chapter use the same χ^2 (chi-square) distribution that was first introduced in Section 7-5. As a quick review, here are important properties of the chi-square distribution:

1. The chi-square distribution is not symmetric. (See Figure 11-1.)
2. The values of the chi-square distribution can be 0 or positive, but they cannot be negative. (See Figure 11-1.)
3. The chi-square distribution is different for each number of degrees of freedom. (See Figure 11-2.)

Critical values of the chi-square distribution are found in Table A-4.

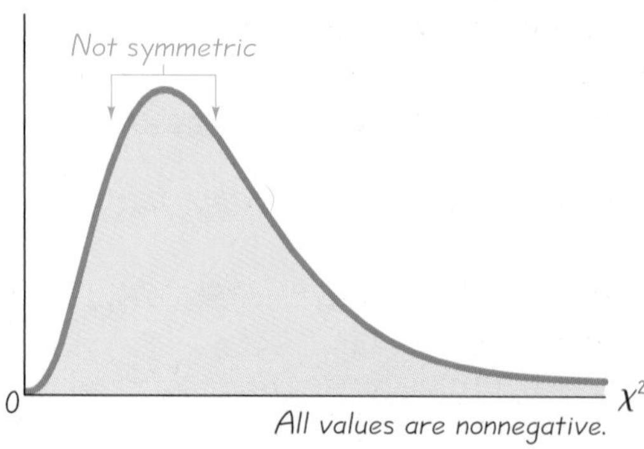

Figure 11-1 The Chi-Square Distribution

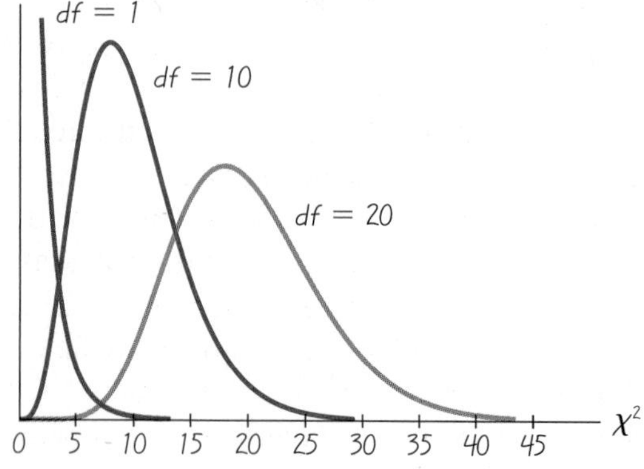

Figure 11-2 Chi-Square Distribution for 1, 10, and 20 Degrees of Freedom

11-2 Multinomial Experiments: Goodness-of-Fit

Key Concept Given data separated into different categories, we will test the hypothesis that the distribution of the data agrees with or "fits" some claimed distribution. The hypothesis test will use the chi-square distribution with the observed frequency counts and the frequency counts that we would expect with the claimed distribution. The chi-square test statistic is a measure of the discrepancy between the observed and expected frequencies.

We begin with the definition of a *multinomial experiment* that is very similar to the definition of a binomial experiment given in Section 5-3, except that a multinomial experiment has more than two categories (unlike a binomial experiment, which has exactly two categories).

Definition

A **multinomial experiment** is an experiment that meets the following conditions:

1. The number of trials is fixed.
2. The trials are independent.
3. All outcomes of each trial must be classified into exactly one of several different categories.
4. The probabilities for the different categories remain constant for each trial.

EXAMPLE Last Digits of Weights Thousands of subjects are routinely studied as part of the National Health Examination Survey. The examination procedures are quite exact. For example, when obtaining weights of subjects, it is extremely important to actually weigh the individuals instead of asking them to report their weights. When asked, people have been known to provide weights that are somewhat lower than their actual weights. So how can researchers verify that weights were obtained through actual measurements instead of asking subjects? One method is to analyze the *last digits* of the weights. When people report weights, they tend to round down—sometimes *way* down. Such reported weights tend to have last digits with disproportionately more 0s and 5s than the last digits of weights obtained through a measurement process. In contrast, if people are actually weighed, the weights tend to have last digits that are uniformly distributed, with 0, 1, 2, . . . , 9 all occurring with roughly the same frequencies. The author obtained weights from 80 randomly selected students, and those weights had last digits summarized in Table 11-2. Later, we will analyze the data, but for now, simply verify that the four conditions of a multinomial experiment are satisfied.

continued

Table 11-2
Last Digits of Weights

Last Digit	Frequency
0	35
1	0
2	2
3	1
4	4
5	24
6	1
7	4
8	7
9	2

SOLUTION Here is the verification that the four conditions of a multinomial experiment are all satisfied:

1. The number of trials (last digits) is the fixed number 80.
2. The trials are independent, because the last digit of any individual weight does not affect the last digit of any other weight.
3. Each outcome (last digit) is classified into exactly 1 of 10 different categories. The categories are identified as 0, 1, 2, . . . , 9.
4. In testing the claim that the 10 digits are equally likely, each possible digit has a probability of $1/10$, and by assumption, that probability remains constant for each subject.

In this section we are presenting a method for testing a claim that in a multinomial experiment, the frequencies observed in the different categories fit some claimed distribution. Because we test for how well an observed frequency distribution fits some specified theoretical distribution, this method is often called a *goodness-of-fit test.*

Definition

A **goodness-of-fit test** is used to test the hypothesis that an observed frequency distribution fits (or conforms to) some claimed distribution.

For example, using the data in Table 11-2, we can test the hypothesis that the data fit a uniform distribution, with all of the digits being equally likely. Our goodness-of-fit tests will incorporate the following notation.

Notation

O represents the *observed frequency* of an outcome.
E represents the *expected frequency* of an outcome.
k represents the *number of different categories* or outcomes.
n represents the total *number of trials*.

Finding Expected Frequencies

In Table 11-2 the observed frequencies O are 35, 0, 2, 1, 4, 24, 1, 4, 7, and 2. The sum of the observed frequencies is 80, so $n = 80$. If we assume that the 80 digits were obtained from a population in which all digits are equally likely, then we *expect* that each digit should occur in $1/10$ of the 80 trials, so each of the 10 expected frequencies is given by $E = 8$. If we generalize this result, we get an easy procedure for finding expected frequencies whenever we are assuming that all of the expected frequencies are equal: Simply divide the total number of observations by the number of different categories ($E = n/k$). In other cases where the expected frequencies are not all equal, we can often find the expected frequency for each category by multiplying the sum of all observed frequencies and the probability p for the category, so $E = np$. We summarize these two procedures here.

- **If all expected frequencies are equal, then each expected frequency is the sum of all observed frequencies divided by the number of categories, so that $E = n/k$.**
- **If the expected frequencies are not all equal, then each expected frequency is found by multiplying the sum of all observed frequencies by the probability for the category, so $E = np$ for each category.**

As good as these two formulas for E might be, it would be better to use an informal approach based on an understanding of the circumstances. Just ask, "How can the observed frequencies be split up among the different categories so that there is perfect agreement with the claimed distribution?" Also, recognize that the *observed* frequencies must all be whole numbers because they represent actual counts, but *expected* frequencies need not be whole numbers. For example, when rolling a single die 33 times, the expected frequency for each possible outcome is $33/6 = 5.5$. The expected frequency for the number of 3s occurring is 5.5, even though it is impossible to have the outcome of 3 occur exactly 5.5 times.

We know that sample frequencies typically deviate somewhat from the values we theoretically expect, so we now present the key question: Are the differences between the actual *observed* values O and the theoretically *expected* values E statistically significant? We need a measure of the discrepancy between the O and E values, so we use the test statistic that is given with the requirements and critical values. (Later, we will explain how this test statistic was developed, but you can see that it has differences of $O - E$ as a key component.)

Requirements

1. The data have been randomly selected.
2. The sample data consist of frequency counts for each of the different categories.
3. For each category, the *expected* frequency is at least 5. (The expected frequency for a category is the frequency that would occur if the data actually have the distribution that is being claimed. There is no requirement that the *observed* frequency for each category must be at least 5.)

Test Statistic for Goodness-of-Fit Tests in Multinomial Experiments

$$\chi^2 = \sum \frac{(O - E)^2}{E}$$

Critical values

1. Critical values are found in Table A-4 by using $k - 1$ degrees of freedom, where k = number of categories.
2. Goodness-of-fit hypothesis tests are always *right-tailed.*

TI-83/84 Plus Calculator: The TI-83/84 Plus calculator generates P-values, so the P-value method of hypothesis testing becomes easy. See the instructions at the end of this section.

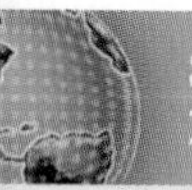

STATISTICS IN THE NEWS

Safest Airplane Seats

Many of us believe that the rear seats are safest in an airplane crash. Safety experts do not agree that any particular part of an airplane is safer than others. Some planes crash nose first when they come down, but others crash tail first on take-off. Matt McCormick, a survival expert for the National Transportation Safety Board, told *Travel* magazine that "there is no one safe place to sit." Goodness-of-fit tests can be used with a null hypothesis that all sections of an airplane are equally safe. Crashed airplanes could be divided into the front, middle, and rear sections. The observed frequencies of fatalities could then be compared to the frequencies that would be expected with a uniform distribution of fatalities. The χ^2 test statistic reflects the size of the discrepancies between observed and expected frequencies, and it would reveal whether some sections are safer than others.

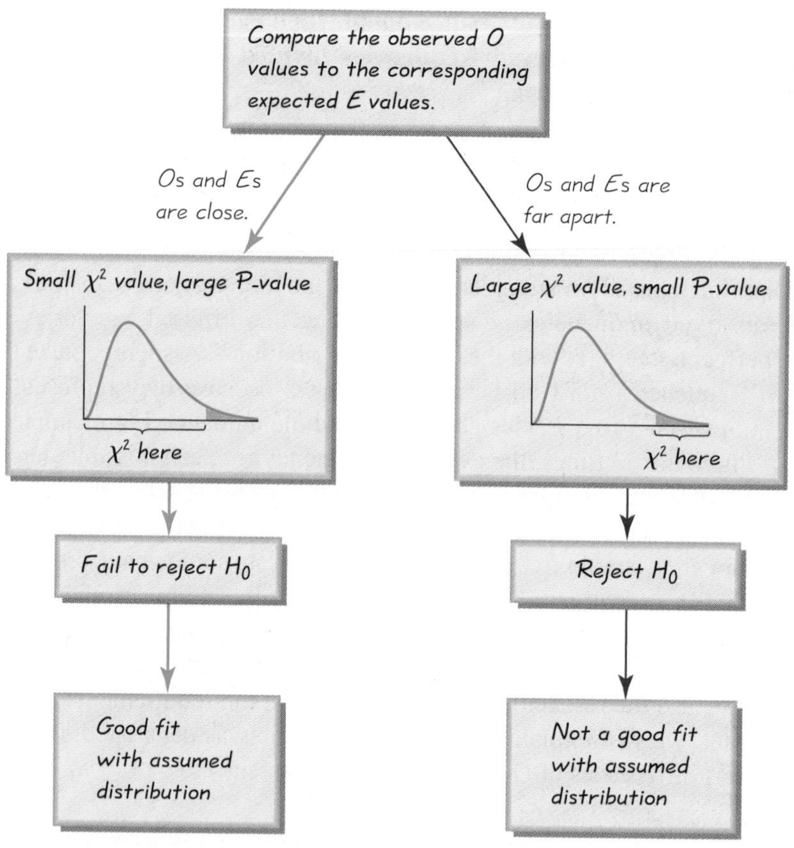

Figure 11-3 Relationships Among the χ^2 Test Statistic, *P*-Value, and Goodness-of-Fit

The χ^2 test statistic is based on differences between observed and expected values, so *close agreement* between observed and expected values will lead to a *small* value of χ^2 and a *large* *P*-value. A large discrepancy between observed and expected values will lead to a *large* value of χ^2 and a *small* *P*-value. The hypothesis tests of this section are therefore always right-tailed, because the critical value and critical region are located at the extreme right of the distribution. These relationships are summarized and illustrated in Figure 11-3.

Once we know how to find the value of the test statistic and the critical value, we can test hypotheses by using the same general procedures introduced in Chapter 8.

EXAMPLE Last Digit Analysis of Weights: Equal Expected Frequencies See Table 11-2 for the last digits of 80 weights. Test the claim that the digits do *not* occur with the same frequency. Based on the results, what can we conclude about the procedure used to obtain the weights?

SOLUTION

REQUIREMENT ✔ We require that the sample data are randomly selected, they consist of frequency counts, the data come from a multinomial experiment, and each expected frequency must be at least 5. We have noted earlier that the data come from randomly selected students. The data do consist of frequency counts. The preceding example established that the conditions for a multinomial experiment are satisfied. The preceding discussion of expected values included the result that each expected frequency is 8, so each expected frequency does satisfy the requirement of being a value of at least 5. All of the requirements are satisfied and we can proceed with the hypothesis test. ✔

The claim that the digits do not occur with the same frequency is equivalent to the claim that the relative frequencies or probabilities of the 10 cells (p_0, $p_1, \ldots, p_9$) are not all equal. We will use the traditional method for testing hypotheses (see Figure 8-9).

Step 1: The original claim is that the digits do not occur with the same frequency. That is, at least one of the probabilities $p_0, p_1, \ldots, p_9$ is different from the others.

Step 2: If the original claim is false, then all of the probabilities are the same. That is, $p_0 = p_1 = \cdots = p_9$.

Step 3: The null hypothesis must contain the condition of equality, so we have

H_0: $p_0 = p_1 = p_2 = p_3 = p_4 = p_5 = p_6 = p_7 = p_8 = p_9$

H_1: At least one of the probabilities is different from the others.

Step 4: No significance level was specified, so we select $\alpha = 0.05$, a very common choice.

Step 5: Because we are testing a claim about the distribution of the last digits being a uniform distribution, we use the goodness-of-fit test described in this section. The χ^2 distribution is used with the test statistic given earlier.

Step 6: The observed frequencies O are listed in Table 11-2. Each corresponding expected frequency E is equal to 8 (because the 80 digits would be uniformly distributed through the 10 categories). Table 11-3 shows the computation of the χ^2 test statistic. The test statistic is $\chi^2 = 156.500$. The critical value is $\chi^2 = 16.919$ (found in Table A-4 with $\alpha = 0.05$ in the right tail and degrees of freedom equal to $k - 1 = 9$). The test statistic and critical value are shown in Figure 11-4.

Step 7: Because the test statistic falls within the critical region, there is sufficient evidence to reject the null hypothesis.

Step 8: There is sufficient evidence to support the claim that the last digits do not occur with the same relative frequency. We now have very strong evidence suggesting that the weights were not actually measured. It is reasonable to speculate that they were reported values instead of actual measurements.

The preceding example dealt with the null hypothesis that the probabilities for the different categories are all equal. The methods of this section can also be used when the hypothesized probabilities (or frequencies) are different, as shown in the next example.

Table 11-3 Calculating the χ^2 Test Statistic for the Last Digits of Weights

Last Digit	Observed Frequency O	Expected Frequency E	$O - E$	$(O - E)^2$	$\frac{(O - E)^2}{E}$
0	35	8	27	729	91.1250
1	0	8	−8	64	8.0000
2	2	8	−6	36	4.500
3	1	8	−7	49	6.125
4	4	8	−4	16	2.000
5	24	8	16	256	32.000
6	1	8	−7	49	6.125
7	4	8	−4	16	2.000
8	7	8	−1	1	0.125
9	2	8	−6	36	4.500
	80	80			$\chi^2 = \sum\frac{(O - E)^2}{E} = 156.500$

(Except for rounding errors, these two totals must agree.)

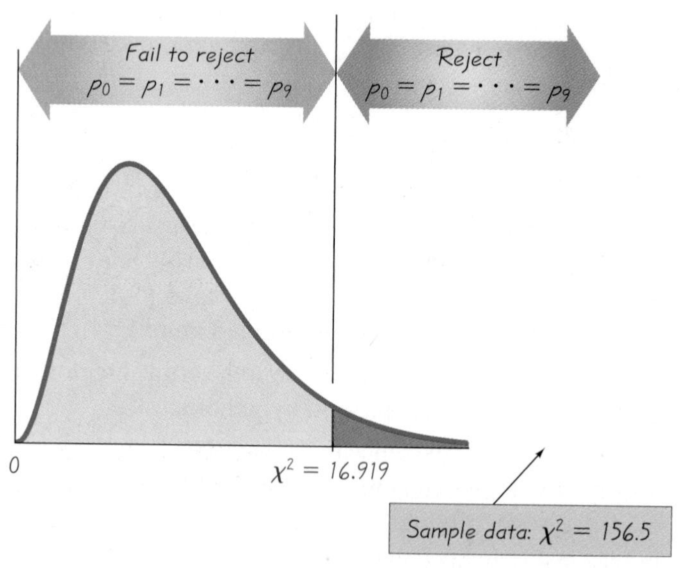

Figure 11-4 Test of $p_0 = p_1 = p_2 = p_3 = p_4 = p_5 = p_6 = p_7 = p_8 = p_9$

EXAMPLE Detecting Fraud: Unequal Expected Frequencies In the Chapter Problem, it was noted that statistics is sometimes used to detect fraud. The second row of Table 11-1 lists percentages for leading digits as expected from Benford's law, and the third row lists the frequency counts expected when the Benford's law percentages are applied to 784 leading digits. The bottom row of Table 11-1 lists the observed frequencies of the leading digits from amounts on 784 checks issued by seven different companies. Test the claim that there is a significant discrepancy between the leading digits expected from Benford's law and the leading digits observed on the 784 checks. Use a significance level of 0.01.

SOLUTION

REQUIREMENTS ✔ In checking the three requirements listed earlier, we begin by noting that the leading digits from the checks are not actually random. However, we treat them as random for the purpose of determining whether they are typical results that might be obtained from a random sample following Benford's law. The data are listed as frequency counts. They satisfy the requirements of a multinomial experiment. Each expected frequency (shown in Table 11-1) is at least 5. All of the requirements are satisfied and we can proceed with the hypothesis test. ✔

Step 1: The original claim is that the leading digits do not have the same distribution as claimed by Benford's law. That is, at least one of the following equations is wrong: $p_1 = 0.301$ and $p_2 = 0.176$ and $p_3 = 0.125$ and $p_4 = 0.097$ and $p_5 = 0.079$ and $p_6 = 0.067$ and $p_7 = 0.058$ and $p_8 = 0.051$ and $p_9 = 0.046$. (The proportions are the decimal equivalent values of the percentages listed for Benford's law in Table 11-1.)

Step 2: If the original claim is false, then the following are all true: $p_1 = 0.301$ and $p_2 = 0.176$ and $p_3 = 0.125$ and $p_4 = 0.097$ and $p_5 = 0.079$ and $p_6 = 0.067$ and $p_7 = 0.058$ and $p_8 = 0.051$ and $p_9 = 0.046$.

Step 3: The null hypothesis must contain the condition of equality, so we have

H_0: $p_1 = 0.301$ and $p_2 = 0.176$ and $p_3 = 0.125$ and $p_4 = 0.097$ and $p_5 = 0.079$ and $p_6 = 0.067$ and $p_7 = 0.058$ and $p_8 = 0.051$ and $p_9 = 0.046$

H_1: At least one of the proportions is not equal to the given claimed value.

Step 4: The significance level of $\alpha = 0.01$ was specified.

Step 5: Because we are testing a claim about the distribution of digits conforming to the distribution from Benford's law, we use the goodness-of-fit test described in this section. The χ^2 distribution is used with the test statistic given earlier.

Step 6: The observed frequencies O and the expected frequencies E are shown in Table 11-1. Adding the nine $(O - E)^2/E$ values results in the test statistic of $\chi^2 = 3650.251$. The critical value is $\chi^2 = 20.090$ (found in Table A-4 with $\alpha = 0.01$ in the right tail and degrees of freedom equal to $k - 1 = 8$). The test statistic and critical value are shown in Figure 11-5.

continued

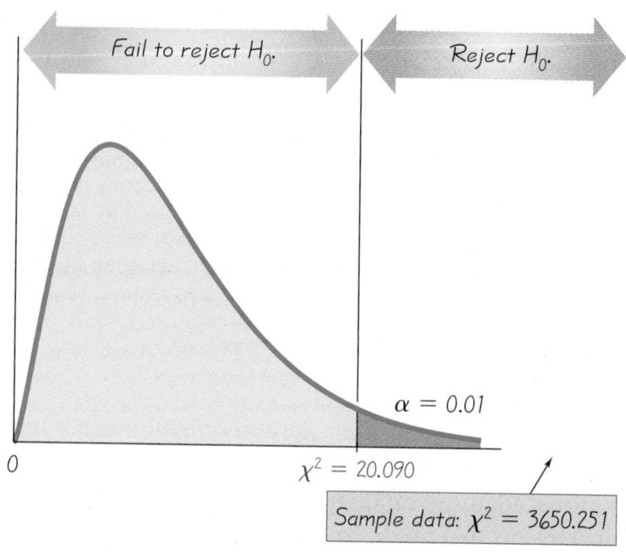

Figure 11-5
Testing for Agreement Between Observed Frequencies and Frequencies Expected with Benford's Law

Step 7: Because the test statistic falls within the critical region, there is sufficient evidence to reject the null hypothesis.

Step 8: There is sufficient evidence to support the claim that there is a discrepancy between the distribution expected from Benford's law and the observed distribution of leading digits from the checks.

In Figure 11-6(a) we graph the claimed proportions of 0.301, 0.176, 0.125, 0.097, 0.079, 0.067, 0.058, 0.051, and 0.046 along with the observed proportions of 0.000, 0.019, 0.000, 0.097, 0.611, 0.233, 0.010, 0.029, and 0.000, so that we can visualize the discrepancy between the Benford's law distribution that was claimed and the frequencies that were observed. The points along the red line represent the claimed proportions, and the points along the green line represent the observed proportions. The corresponding pairs of points are far apart, showing that the expected frequencies are very different from the corresponding observed frequencies. The great disparity between the green line for observed frequencies and the red line for expected frequencies suggests that the check amounts are not the result of typical transactions. It appears that fraud may be involved. In fact, the Brooklyn District Attorney charged fraud by using this line of reasoning. For comparison, see Figure 11-6(b), which is based on the leading digits from the amounts on the last 200 checks written by the author. Note how the observed proportions from the author's checks agree quite well with the proportions expected with Benford's law. The author's checks appear to be typical instead of showing a pattern that might suggest fraud. In general, graphs such as Figure 11-6 are helpful in visually comparing expected frequencies and observed frequencies, as well as suggesting which categories result in the major discrepancies.

P-Values

The examples in this section used the traditional approach to hypothesis testing, but the *P*-value approach can also be used. *P*-values are automatically provided by the TI-83/84 Plus calculator, or they can be obtained by using the methods

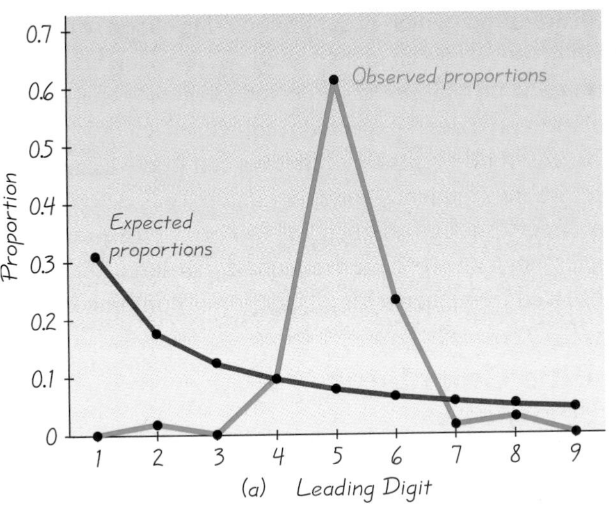

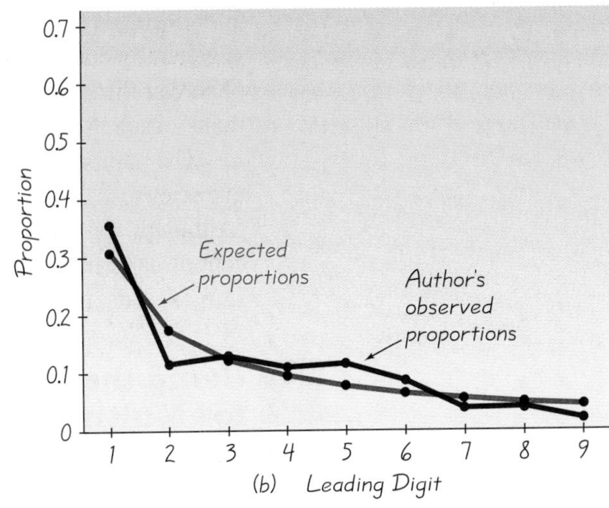

Figure 11-6 Comparison of Observed Frequencies and Frequencies Expected with Benford's Law

described in Chapter 8. For example, the preceding example resulted in a test statistic of $\chi^2 = 3650.251$. That example had $k = 9$ categories, so there were $k - 1 = 8$ degrees of freedom. Referring to Table A-4, we see that for the row with 8 degrees of freedom, the test statistic of 3650.251 is greater than the highest value in the row (21.955). Because the test statistic of $\chi^2 = 3650.251$ is farther to the right than 21.955, the P-value is less than 0.005. If the calculations for the preceding example are run on the TI-83/84 Plus calculator, the display will include a P-value of 0. The small P-value suggests that the null hypothesis should be rejected. (Remember, we reject the null hypothesis when the P-value is equal to or less than the significance level.) While the traditional method of testing hypotheses led us to reject the claim that the 784 check amounts have leading digits that conform to Benford's law, the P-value of 0 indicates that the probability of getting leading digits like those that were obtained is extremely small. This appears to be evidence "beyond a reasonable doubt" that the check amounts are not the result of typical honest transactions.

Rationale for the Test Statistic: The preceding examples should be helpful in developing a sense for the role of the χ^2 test statistic. It should be clear that we want to measure the amount of disagreement between observed and expected frequencies. Simply summing the differences between observed and expected values does not result in an effective measure because that sum is always 0. Squaring the $O - E$ values provides a better statistic. (The reasons for squaring the $O - E$ values are essentially the same as the reasons for squaring the $x - \bar{x}$ values in the formula for standard deviation.) The value of $\Sigma(O - E)^2$ measures only the magnitude of the differences, but we need to find the magnitude of the differences relative to what was expected. This relative magnitude is found through division by the expected frequencies, as in the test statistic.

The theoretical distribution of $\Sigma(O - E)^2/E$ is a discrete distribution because the number of possible values is limited to a finite number. The distribution can be approximated by a chi-square distribution, which is continuous. This approximation is generally considered acceptable, provided that all expected values E are at least 5. (There are ways of circumventing the problem of an expected frequency that is less than 5, such as combining categories so that all expected frequencies are at least 5.)

The number of degrees of freedom reflects the fact that we can freely assign frequencies to $k - 1$ categories before the frequency for every category is determined. (Although we say that we can "freely" assign frequencies to $k - 1$ categories, we cannot have negative frequencies nor can we have frequencies so large that their sum exceeds the total of the observed frequencies for all categories combined.)

TI Using the TI-83/84 Plus Calculator for Multinomial Experiments

The TI-84 Plus calculator can execute the methods of this section, but the TI-83 Plus calculator requires Michael Lloyd's program X2GOF that must be downloaded from the CD-ROM included with this book or the Web site www.aw.com/triola. Here are the procedures:

TI-84 Plus Calculator for Multinomial Experiments

1. Enter the observed frequencies in list L1 and enter the expected frequencies in list L2.
2. Press STAT, press ▷ twice to select **TESTS,** then scroll to χ^2 **GOF-Test** and press ENTER.
3. For Observed entry, enter L1, and for the Expected entry, enter L2.
4. Scroll to the df line and enter the number of degrees of freedom, which is 1 less than the number of categories.
5. Scroll down to **Calculate** and press ENTER.

Results will include the χ^2 test statistic and the P-value.

TI-83 Plus Calculator for Multinomial Experiments

1. Download the program X2GOF from the CD-ROM or the Web site www.aw.com/triola.
2. Enter the observed frequencies in list L1 and enter the expected frequencies in list L2.
3. Press PRGM and select X2GOF and press ENTER.
4. At the prompt of OBS LIST, enter L1, which contains the list of observed frequencies. Press ENTER.
5. At the prompt of EXP LIST, enter L2, which contains the list of expected frequencies. Press ENTER.

TI-83/84 Plus

Results will be displayed. The value of X2 is the χ^2 test statistic. The value of P is the P-value. The accompanying screen display results from using the program X2GOF on the TI-83/84 Plus calculator with the observed and expected frequencies listed in Table 11-1. This screen display shows the test statistic of $\chi^2 = 3650.251$ (rounded). Also, the P-value is shown as 0, indicating that the P-value is a positive number that is so small, it can be considered to be 0 for all practical purposes. This small P-value suggests that we should reject the null hypothesis that the leading digits follow the distribution described by Benford's law. There is sufficient evidence to support a claim that there is a discrepancy between the distribution expected with Benford's law and the observed distribution of leading digits from checks.

11-2 BASIC SKILLS AND CONCEPTS

Statistical Literacy and Critical Thinking

1. Goodness-of-Fit What does it mean when we say that we test for "goodness-of-fit"?

2. Right-Tailed Test Why is the hypothesis test for goodness-of-fit always a right-tailed test?

3. Observed/Expected Frequencies What is an observed frequency? What is an expected frequency?

4. Weights of Students A researcher collects weights of 20 male students randomly selected from each of four different classes, then he finds the total of those weights and summarizes them in the table below (based on data from the National Health Examination Survey). Can the methods of this section be used to test the claim that the weights come from populations with the same mean? Why or why not?

	Grade 1	Grade 2	Grade 3	Grade 4
Total weight (lb)	1034	1196	1440	1584

In Exercises 5 and 6, identify the components of the hypothesis test.

TI **5.** Testing for Equally Likely Categories Here are the observed frequencies from three categories: 5, 5, 20. Assume that we want to use a 0.05 significance level to test the claim that the three categories are all equally likely.
a. What is the null hypothesis?
b. What is the expected frequency for each of the three categories?
c. What is the value of the test statistic?
d. What is the critical value?
e. What do you conclude about the given claim?

TI **6.** Testing for Categories with Different Proportions Here are the observed frequencies from four categories: 5, 10, 10, 20. Assume that we want to use a 0.05 significance level to test the claim that the four categories have proportions of 0.20, 0.25, 0.25, and 0.30, respectively.
a. What is the null hypothesis?
b. What are the expected frequencies for the four categories?
c. What is the value of the test statistic?
d. What is the critical value?
e. What do you conclude about the given claim?

7. Testing Fairness of Roulette Wheel The author observed 500 spins of a roulette wheel at the Mirage Resort and Casino. (To the IRS: Isn't that Las Vegas trip now a tax deduction?) For each spin, the ball can land in any one of 38 different slots that are supposed to be equally likely. When the TI-83/84 Plus calculator was used to test the claim that the slots are in fact equally likely, the test statistic $\chi^2 = 38.232$ was obtained.

continued

a. Find the critical value assuming that the significance level is 0.10.
b. The TI-83/84 Plus calculator displayed a P-value of 0.41331, but what do you know about the P-value if you must use only Table A-4 along with the given test statistic of 38.232, which results from the 500 spins?
c. Write a conclusion about the claim that the 38 results are equally likely.

8. Testing a Slot Machine The author purchased a slot machine (Bally Model 809), and tested it by playing it 1197 times. When testing the claim that the observed outcomes agree with the expected frequencies, a test statistic of $\chi^2 = 8.185$ was obtained. There are 10 different categories of outcome, including no win, win jackpot, win with three bells, and so on.
a. Find the critical value assuming that the significance level is 0.05.
b. What can you conclude about the P-value from Table A-4 if you know that the test statistic is $\chi^2 = 8.185$ and there are 10 categories?
c. State a conclusion about the claim that the observed outcomes agree with the expected frequencies. Does the author's slot machine appear to be working correctly?

TI **9.** Loaded Die The author drilled a hole in a die and filled it with a lead weight, then proceeded to roll it 200 times. Here are the observed frequencies for the outcomes of 1, 2, 3, 4, 5, and 6, respectively: 27, 31, 42, 40, 28, 32. Use a 0.05 significance level to test the claim that the outcomes are not equally likely. Does it appear that the loaded die behaves differently than a fair die?

TI **10.** Flat Tire and Missed Class A classic tale involves four car-pooling students who missed a test and gave as an excuse a flat tire. On the makeup test, the instructor asked the students to identify the particular tire that went flat. If they really didn't have a flat tire, would they be able to identify the same tire? The author asked 41 other students to identify the tire they would select. The results are listed in the following table (except for one student who selected the spare). Use a 0.05 significance level to test the author's claim that the results fit a uniform distribution. What does the result suggest about the ability of the four students to select the same tire when they really didn't have a flat?

Tire	Left front	Right front	Left rear	Right rear
Number selected	11	15	8	6

TI **11.** Deaths from Car Crashes Randomly selected deaths from car crashes were obtained, and the results are included in the table below (based on data from the Insurance Institute for Highway Safety). Use a 0.05 significance level to test the claim that car crash fatalities occur with equal frequency on the different days of the week. How might the results be explained? Why does there appear to be an exceptionally large number of car crash fatalities on Saturday?

Day	Sun	Mon	Tues	Wed	Thurs	Fri	Sat
Number of fatalities	132	98	95	98	105	133	158

TI **12.** Births Randomly selected birth records were obtained and results are listed in the table below (based on data from the *National Vital Statistics Report*, Vol. 49, No. 1). Use a 0.05 significance level to test the reasonable claim that births occur with equal frequency on the different days of the week. How might the apparent lower frequencies on Saturday and Sunday be explained?

Day	Sun	Mon	Tues	Wed	Thurs	Fri	Sat
Births	36	55	62	60	60	58	48

TI **13.** Motorcycle Deaths Randomly selected deaths of motorcycle riders are summarized in the table below (based on data from the Insurance Institute for Highway Safety). Use a 0.05 significance level to test the claim that such fatalities occur with equal frequency in the different months. How might the results be explained?

Month	Jan.	Feb.	March	April	May	June	July	Aug.	Sept.	Oct.	Nov.	Dec.
Number	6	8	10	16	22	28	24	28	26	14	10	8

TI **14.** Grade and Seating Location Do "A" students tend to sit in a particular part of the classroom? The author recorded the locations of the students who received grades of A, with these results: 17 sat in the front, 9 sat in the middle, and 5 sat in the back of the classroom. Is there sufficient evidence to support the claim that the "A" students are not evenly distributed throughout the classroom? If so, does that mean you can increase your likelihood of getting an A by sitting in the front?

TI **15.** Oscar-Winning Actresses The author collected data consisting of the month of birth of actresses who won Oscars. Use a 0.05 significance level to test the claim that Oscar-winning actresses are born in the different months with the same frequency. Is there any reason why Oscar-winning actresses would be born in some months more often than others?

Month	Jan.	Feb.	March	April	May	June	July	Aug.	Sept.	Oct.	Nov.	Dec.
Number	7	3	7	7	8	7	6	6	5	6	9	5

TI **16.** Oscar-Winning Actors The author collected data consisting of the month of birth of actors who won Oscars. Use a 0.05 significance level to test the claim that Oscar-winning actors are born in the different months with the same frequency. Compare the results to those found in Exercise 15.

Month	Jan.	Feb.	March	April	May	June	July	Aug.	Sept.	Oct.	Nov.	Dec.
Number	9	5	7	14	8	1	7	6	4	5	1	9

TI **17.** June Bride A wedding caterer randomly selects clients from the past few years and records the months in which the wedding receptions were held. The results are listed below (based on data from *The Amazing Almanac*). Use a 0.05 significance level to test the claim that weddings are held in the different months with the same frequency. Do the results support or refute the belief that most marriages occur in June?

Month	Jan.	Feb.	March	April	May	June	July	Aug.	Sept.	Oct.	Nov.	Dec.
Number	5	8	6	8	11	14	10	9	10	12	8	9

TI **18.** Eye Color Experiment A researcher has developed a theoretical model for predicting eye color. After examining a random sample of parents, she predicts the eye color of the first child. The table below lists the eye colors of offspring. Based on her theory, she predicted that 87% of the offspring would have brown eyes, 8% would have blue eyes, and 5% would have green eyes. Use a 0.05 significance level to test the claim that the actual frequencies correspond to her predicted distribution.

	Brown Eyes	Blue Eyes	Green Eyes
Frequency	132	17	0

TI **19.** World Series Games The *USA Today* headline of "Seven-game series defy odds" referred to a claim that seven-game World Series contests occur more often than expected by chance. Listed below are the numbers of games of World Series contests (omitting two that lasted eight games) along with the proportions that would be expected with teams of equal abilities. Use a 0.05 significance level to test the claim that the observed frequencies agree with the theoretical proportions. Based on the results, does there appear to be evidence to support the claim that seven-game series occur more often than expected?

Games	4	5	6	7
Actual World Series contests	18	20	22	37
Expected proportion	2/16	4/16	5/16	5/16

TI **20.** Genetics Experiment Based on the genotypes of parents, offspring are expected to have genotypes distributed in such a way that 25% have genotypes denoted by AA, 50% have genotypes denoted by Aa, and 25% have genotypes denoted by aa. When 145 offspring are obtained, it is found that 20 of them have AA genotypes, 90 have Aa genotypes, and 35 have aa genotypes. Test the claim that the observed genotype offspring frequencies fit the expected distribution of 25% for AA, 50% for Aa, and 25% for aa. Use a significance level of 0.05.

TI **21.** M&M Candies Mars, Inc. claims that its M&M plain candies are distributed with the following color percentages: 16% green, 20% orange, 14% yellow, 24% blue, 13% red, and 13% brown. Refer to Data Set 13 in Appendix B and use the sample data to test the claim that the color distribution is as claimed by Mars, Inc. Use a 0.05 significance level.

TI **22.** Measuring Pulse Rates An example in this section was based on the principle that when certain quantities are measured, the last digits tend to be uniformly distributed, but if they are estimated or reported, the last digits tend to have disproportionately more 0s or 5s. Refer to Data Set 1 in Appendix B and use the last digits of the pulse rates of the 80 men and women. Those pulse rates were obtained as part of the National Health Examination Survey. Test the claim that the last digits of 0, 1, 2, . . . , 9 occur with the same frequency. Based on the observed digits, what can be inferred about the procedure used to obtain the pulse rates?

TI **23.** Participation in Clinical Trials by Race A study was conducted to investigate racial disparity in clinical trials of cancer. Among the randomly selected participants, 644 were white, 23 were Hispanic, 69 were black, 14 were Asian/Pacific Islander, and 2 were American Indian/Alaskan Native. The proportions of the U.S. population of the same groups are 0.757, 0.091, 0.108, 0.038, and 0.007, respectively. (Based on data from "Participation in Clinical Trials," by Murthy, Krumholz, and Gross, *Journal of the American Medical Association*, Vol. 291, No. 22.) Use a 0.05 significance level to test the claim that the participants fit the same distribution as the U.S. population. Why is it important to have proportionate representation in such clinical trials?

TI **24.** Do World War II Bomb Hits Fit a Poisson Distribution? In analyzing hits by V-1 buzz bombs in World War II, South London was subdivided into regions, each with an area of 0.25 km^2. In Section 5-5 we presented an example and included a table of actual frequencies of hits and the frequencies expected with the Poisson distribution. Use the values listed on the next page and test the claim that the actual frequencies fit a Poisson distribution. Use a 0.05 significance level.

Number of bomb hits	0	1	2	3	4 or more
Actual number of regions	229	211	93	35	8
Expected number of regions (from Poisson distribution)	227.5	211.4	97.9	30.5	8.7

TI **25.** Author's Check Amounts and Benford's Law Figure 11-6(b) illustrates the observed frequencies of the leading digits from the amounts of the last 200 checks that the author wrote. The observed frequencies of those leading digits are listed below. Using a 0.05 significance level, test the claim that they come from a population of leading digits that conform to Benford's law. (See the first two rows of Table 11-1 included in the Chapter Problem.)

Leading digit	1	2	3	4	5	6	7	8	9
Frequency	72	23	26	20	21	18	8	8	4

11-2 BEYOND THE BASICS

TI **26.** Testing Effects of Outliers In conducting a test for the goodness-of-fit as described in this section, does an outlier have much of an effect on the value of the χ^2 test statistic? Test for the effect of an outlier by repeating Exercise 10 after changing the frequency for the right rear tire from 6 to 60. Describe the general effect of an outlier.

27. Detecting Altered Experimental Data When Gregor Mendel conducted his famous hybridization experiments with peas, it appears that his gardening assistant knew the results that Mendel expected, and he altered the results to fit Mendel's expectations. Subsequent analysis of the results led to the conclusion that there is a probability of only 0.00004 that the expected results and reported results would agree so closely. How could the methods of this section be used to detect such results that are just too perfect to be realistic?

28. Equivalent Test In this exercise we will show that a hypothesis test involving a multinomial experiment with only two categories is equivalent to a hypothesis test for a proportion (Section 8-3). Assume that a particular multinomial experiment has only two possible outcomes, A and B, with observed frequencies of f_1 and f_2, respectively.

a. Find an expression for the χ^2 test statistic, and find the critical value for a 0.05 significance level. Assume that we are testing the claim that both categories have the same frequency, $(f_1 + f_2)/2$.

b. The test statistic $z = (\hat{p} - p)/\sqrt{pq/n}$ is used to test the claim that a population proportion is equal to some value p. With the claim that $p = 0.5$, $\alpha = 0.05$, and $\hat{p} = f_1/(f_1 + f_2)$, show that z^2 is equivalent to χ^2 [from part (a)]. Also show that the square of the critical z score is equal to the critical χ^2 value from part (a).

TI **29.** Testing Goodness-of-Fit with a Binomial Distribution An observed frequency distribution is as follows:

Number of successes	0	1	2	3
Frequency	89	133	52	26

a. Assuming a binomial distribution with $n = 3$ and $p = 1/3$, use the binomial probability formula to find the probability corresponding to each category of the table.

b. Using the probabilities found in part (a), find the expected frequency for each category.

c. Use a 0.05 significance level to test the claim that the observed frequencies fit a binomial distribution for which $n = 3$ and $p = 1/3$.

30. Testing Goodness-of-Fit with a Normal Distribution An observed frequency distribution of sample IQ scores is as follows:

IQ score	Less than 80	80–95	96–110	111–120	More than 120
Frequency	20	20	80	40	40

a. Assuming a normal distribution with $\mu = 100$ and $\sigma = 15$, use the methods given in Chapter 6 to find the probability of a randomly selected subject belonging to each class. (Use class boundaries of 79.5, 95.5, 110.5, and 120.5.)

b. Using the probabilities found in part (a), find the expected frequency for each category.

c. Use a 0.01 significance level to test the claim that the IQ scores were randomly selected from a normally distributed population with $\mu = 100$ and $\sigma = 15$.

11-3 Contingency Tables: Independence and Homogeneity

Key Concept In this section we consider *contingency tables* (or *two-way frequency tables*), which include frequency counts for categorical data arranged in a table with at least two rows and at least two columns. We present a method for testing the claim that the row and column variables are independent of each other. We will use the same method for a test of homogeneity, whereby we test the claim that different populations have the same proportion of some characteristics.

We begin with the definition of a contingency table.

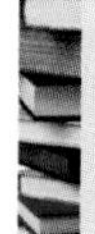

Definition

A **contingency table** (or **two-way frequency table**) is a table in which frequencies correspond to two variables. (One variable is used to categorize rows, and a second variable is used to categorize columns.)

Table 11-4 is an example of a contingency table with two rows and three columns, and the cell entries are frequency counts. The data in Table 11-4 are from a

Table 11-4 Case-Control Study of Motorcycle Drivers

	Color of Helmet		
	Black	White	Yellow/Orange
Controls (not injured)	491	377	31
Cases (injured or killed)	213	112	8

retrospective (or case-control) study. The row variable has two categories: controls and cases. Subjects in the control group were motorcycle riders randomly selected at roadside locations. Subjects in the case group were motorcycle drivers seriously injured or killed. The column variable is used for the color of the helmet they were wearing. Here is the key issue: Is the color of the motorcycle helmet somehow related to the risk of crash related injuries? (The data are based on "Motorcyle Rider Conspicuity and Crash Related Injury: Case-Control Study," by Wells et al, *BMJ USA,* Vol. 4.)

This section presents two types of hypothesis testing based on contingency tables. We first consider tests of independence, used to determine whether a contingency table's row variable is independent of its column variable. We then consider tests of homogeneity, used to determine whether different populations have the same proportions of some characteristic. Both types of tests use the *same* basic methods. We begin with tests of independence.

Test of Independence

One of the two tests included in this section is a *test of independence* between the row variable and column variable.

Definition

A **test of independence** tests the null hypothesis that there is no association between the row variable and the column variable in a contingency table. (For the null hypothesis, we will use the statement that "the row and column variables are independent.")

It is very important to recognize that in this context, the word *contingency* refers to dependence, but this is only a statistical dependence, and it cannot be used to establish a direct cause-and-effect link between the two variables in question. When testing the null hypothesis of independence between the row and column variables in a contingency table, the requirements, test statistic, and critical values are described on the following page.

The test statistic allows us to measure the amount of disagreement between the frequencies actually observed and those that we would theoretically expect when the two variables are independent. Large values of the χ^2 test statistic are in the rightmost region of the chi-square distribution, and they reflect significant differences between observed and expected frequencies. In repeated large samplings, the distribution of the test statistic χ^2 can be approximated by the chi-square distribution, provided that all expected frequencies are at least 5. The number of degrees of freedom $(r - 1)(c - 1)$ reflects the fact that because we know the total of all frequencies in a contingency table, we can freely assign frequencies to only $r - 1$ rows and $c - 1$ columns before the frequency for every cell is determined. [However, we cannot have negative frequencies or frequencies so large that any row (or column) sum exceeds the total of the observed frequencies for that row (or column).]

Requirements

1. The sample data are randomly selected, and are represented as frequency counts in a two-way table.
2. The null hypothesis H_0 is the statement that the row and column variables are *independent;* the alternative hypothesis H_1 is the statement that the row and column variables are *dependent.*
3. For every cell in the contingency table, the *expected* frequency E is at least 5. (There is no requirement that every *observed* frequency must be at least 5. Also, there is no requirement that the population must have a normal distribution or any other specific distribution.)

Test Statistic for a Test of Independence

$$\chi^2 = \sum \frac{(O - E)^2}{E}$$

Critical values

1. The critical values are found in Table A-4 by using

   degrees of freedom = $(r - 1)(c - 1)$

   where r is the number of rows and c is the number of columns.
2. In a test of independence with a contingency table, the critical region is located in the *right tail only.*

TI-83/84 Plus Calculator: The TI-83/84 Plus calculator generates P-values, so the P-value method of hypothesis testing becomes easy. See the instructions at the end of this section.

The expected frequency E can be calculated for each cell by simply multiplying the total of the row frequencies by the total of the column frequencies, then dividing by the grand total of all frequencies, as shown below.

Expected Frequency for a Cell in a Contingency Table

$$\text{expected frequency} = \frac{(\text{row total})(\text{column total})}{(\text{grand total})}$$

EXAMPLE Finding Expected Frequency Refer to Table 11-4 and find the expected frequency for the first cell, where the frequency is 491.

SOLUTION The first cell lies in the first row (with total 899) and the first column (with total 704), and the sum of all frequencies in the table is 1232. The expected frequency is

$$E = \frac{(\text{row total})(\text{column total})}{(\text{grand total})} = \frac{(899)(704)}{1232} = 513.714$$

INTERPRETATION To interpret this result for the first cell, we can say that although 491 motorcycle drivers in the control group actually wore black helmets, we would have expected 513.714 of them to wear black helmets if the group (controls or cases) is independent of the color of helmet worn. There is a discrepancy between $O = 491$ and $E = 513.714$, and such discrepancies are key components of the test statistic.

To better understand expected frequencies, pretend that we know only the row and column totals, as in Table 11-5, and that we must fill in the cell expected frequencies by assuming independence (or no relationship) between the row and column variables. In the first row, 899 of the 1232 subjects are in the control group, so $P(\text{control group}) = 899/1232$. In the first column, 704 of the 1232 drivers wore black helmets, so $P(\text{black helmet}) = 704/1232$. Because we are assuming independence between the group and helmet color, the multiplication rule for independent events $[P(A \text{ and } B) = P(A) \cdot P(B)]$ is expressed as

$$P(\text{control group and black helmet}) = P(\text{control group}) \cdot P(\text{black helmet}) = \frac{899}{1232} \cdot \frac{704}{1232}$$

Table 11-5 Case-Control Study of Motorcycle Drivers

	Color of Helmet			
	Black	White	Yellow/Orange	Row totals:
Controls				899
Cases				333
Column totals:	**704**	**489**	**39**	**Grand total: 1232**

Knowing the probability of being in the upper left cell, we can now find the *expected value* for that cell, which we get by multiplying the probability for that cell by the total number of subjects, as shown in the following equation:

$$E = n \cdot p = 1232\left[\frac{899}{1232} \cdot \frac{704}{1232}\right] = 513.714$$

The form of this product suggests a general way to obtain the expected frequency of a cell:

$$\text{Expected frequency } E = (\text{grand total}) \cdot \frac{(\text{row total})}{(\text{grand total})} \cdot \frac{(\text{column total})}{(\text{grand total})}$$

This expression can be simplified to

$$E = \frac{(\text{row total}) \cdot (\text{column total})}{(\text{grand total})}$$

Knowing how to find expected values, we can now proceed to use contingency table data for testing hypotheses, as in the following example.

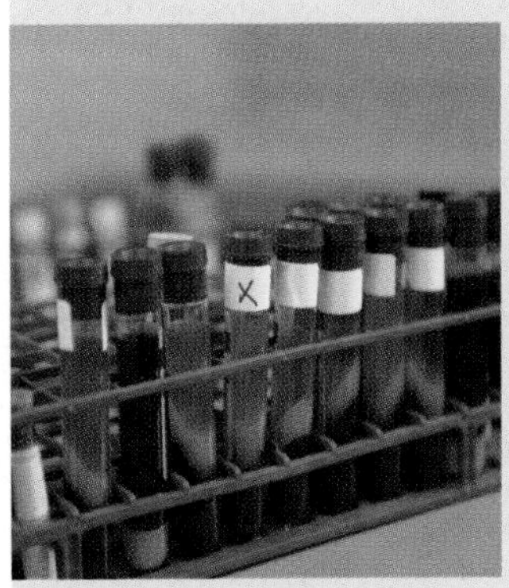

An Eight-Year False Positive

The Associated Press recently released a report about Jim Malone, who had received a positive test result for an HIV infection. For eight years, he attended group support meetings, fought depression, and lost weight while fearing a death from AIDS. Finally, he was informed that the original test was wrong. He did not have an HIV infection. A follow-up test was given after the first positive test result, and the confirmation test showed that he did not have an HIV infection, but nobody told Mr. Malone about the new result. Jim Malone agonized for eight years because of a test result that was actually a false positive.

EXAMPLE Injuries and Color of Motorcycle Helmet Refer to the data in Table 11-4. Using a 0.05 significance level, test the claim that the group (control or case) is independent of the helmet color.

SOLUTION

REQUIREMENT ✔ As required, the data have been randomly selected, they do consist of frequency counts in a two-way table, we are testing the null hypothesis that the variables are independent, and the expected frequencies are all at least 5. (The expected frequencies are 513.714, 356.827, 28.459, 190.286, 132.173, 10.541.) Because all of the requirements are satisfied, we can proceed with the hypothesis test. ✔

The null hypothesis and alternative hypothesis are as follows:

H_0: Whether a subject is in the control group or case group is independent of the helmet color. (This is equivalent to saying that injuries are independent of helmet color.)

H_1: The group and helmet color are dependent.

The significance level is $\alpha = 0.05$.

Because the data are in the form of a contingency table, we use the χ^2 distribution with this test statistic:

$$\chi^2 = \Sigma \frac{(O - E)^2}{E} = \frac{(491 - 513.714)^2}{513.714} + \cdots + \frac{(8 - 10.541)^2}{10.541}$$
$$= 8.775$$

The critical value is $\chi^2 = 5.991$ and it is found from Table A-4 by noting that $\alpha = 0.05$ in the right tail and the number of degrees of freedom is given by $(r - 1)(c - 1) = (2 - 1)(3 - 1) = 2$. The test statistic and critical value are shown in Figure 11-7. Because the test statistic falls within the critical region, we reject the null hypothesis of independence between group and helmet color.

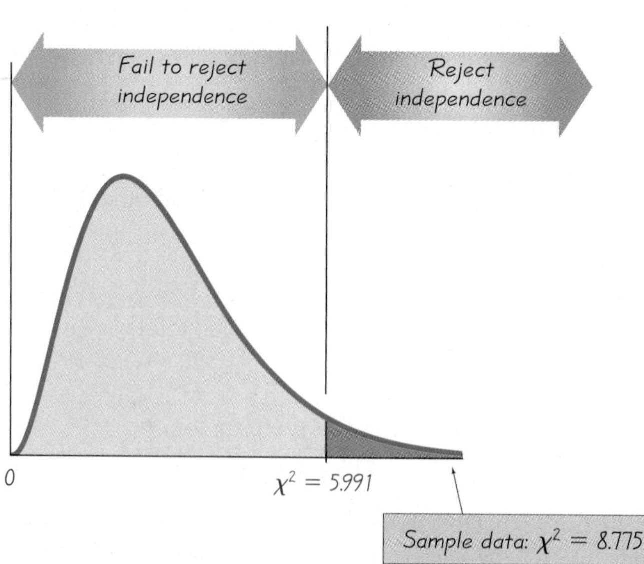

Figure 11-7
Test of Independence for the Motorcycle Data

It appears that helmet color and group (control or case) are dependent. Because the controls were uninjured and the cases were injured or killed, it appears that there is an association between helmet color and motorcycle safety. The authors of the journal article stated that the study supports the introduction of laws requiring greater visibility of motorcycle riders.

P-Values

The preceding example used the traditional approach to hypothesis testing, but we can easily use the *P*-value approach. The TI-83/84 Plus calculator provides *P*-values for tests of independence in contingency tables. If you don't have a suitable calculator or statistical software, estimate *P*-values from Table A-4 by finding where the test statistic falls in the row corresponding to the appropriate number of degrees of freedom. For the preceding example, see the row for 2 degrees of freedom and note that the test statistic of 8.775 falls between the row entries of 7.378 and 9.210. The *P*-value must therefore fall between 0.025 and 0.01, so we conclude that $0.01 < P\text{-value} < 0.025$. (The actual *P*-value is 0.0124.) Knowing that the *P*-value is less than the significance level of 0.05, we reject the null hypothesis as we did in the preceding example.

As in Section 11-2, if observed and expected frequencies are close, the χ^2 test statistic will be small and the *P*-value will be large. If observed and expected frequencies are far apart, the χ^2 test statistic will be large and the *P*-value will be small. These relationships are summarized and illustrated in Figure 11-8.

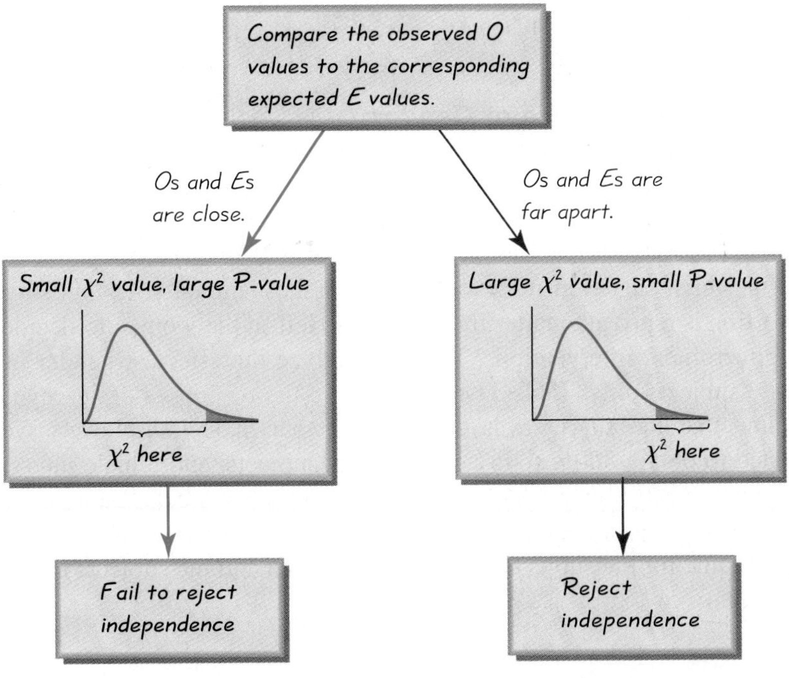

Figure 11-8 Relationships Among Key Components in Test of Independence

Home Field Advantage

In the *Chance* magazine article "Predicting Professional Sports Game Outcomes from Intermediate Game Scores," authors Harris Cooper, Kristina DeNeve, and Frederick Mosteller used statistics to analyze two common beliefs: Teams have an advantage when they play at home, and only the last quarter of professional basketball games really counts. Using a random sample of hundreds of games, they found that for the four top sports, the home team wins about 58.6% of games. Also, basketball teams ahead after 3 quarters go on to win about 4 out of 5 times, but baseball teams ahead after 7 innings go on to win about 19 out of 20 times. The statistical methods of analysis included the chi-square distribution applied to a contingency table.

Survey Medium Can Affect Results

In a survey of Catholics in Boston, the subjects were asked if contraceptives should be made available to unmarried women. In personal interviews, 44% of the respondents said yes. But among a similar group contacted by mail or telephone, 75% of the respondents answered yes to the same question.

Test of Homogeneity

In the preceding example, we illustrated a test of independence between two variables and we used a population of motorcycle riders. However, some other samples are drawn from *different* populations, and we want to determine whether those populations have the same proportions of the characteristics being considered. The *test of homogeneity* can be used in such cases. (The word *homogeneous* means "having the same quality," and in this context, we are testing to determine whether the proportions are the same.)

Definition

In a **test of homogeneity,** we test the claim that *different populations* have the same proportions of some characteristics.

In conducting a test of homogeneity, we can use the requirements, test statistic, critical value, and the same procedures already presented in this section, with one exception: Instead of testing the null hypothesis of independence between the row and column variables, we test the null hypothesis that the different populations have the same proportions of some characteristics.

Table 11-6 Gender and Survey Responses

	Gender of Interviewer	
	Man	Woman
Men who agree	560	308
Men who disagree	240	92

EXAMPLE Influence of Gender Does a pollster's gender have an effect on poll responses by men? A *U.S. News & World Report* article about polls stated: "On sensitive issues, people tend to give 'acceptable' rather than honest responses; their answers may depend on the gender or race of the interviewer." To support that claim, data were provided for an Eagleton Institute poll in which surveyed men were asked if they agreed with this statement: "Abortion is a private matter that should be left to the woman to decide without government intervention." We will analyze the effect of gender on male survey subjects only. Table 11-6 is based on the responses of surveyed men. Assume that the survey was designed so that male interviewers were instructed to obtain 800 responses from male subjects, and female interviewers were instructed to obtain 400 responses from male subjects. Using a 0.05 significance level, test the claim that the proportions of agree/disgree responses are the same for the subjects interviewed by men and the subjects interviewed by women.

SOLUTION

REQUIREMENT ✔ The data consist of independent frequency counts, each observation can be categorized according to two variables, and the expected frequencies (578.67, 289.33, 221.33, and 110.67) are all at least 5. [The two variables are (1) gender of interviewer, and (2) whether the subject agreed or disagreed.] Because this is a test of homogeneity, we test the claim that the proportions of agree/disagree responses are the same for the subjects interviewed by males and the subjects interviewed by females. All of the requirements are satisfied, so we can proceed with the hypothesis test. ✔

Because we have two separate populations (subjects interviewed by men and subjects interviewed by women), we test for homogeneity with these hypotheses:

H_0: The proportions of agree/disgree responses are the same for the subjects interviewed by men and the subjects interviewed by women.

H_1: The proportions are different.

The significance level is $\alpha = 0.05$. We use the same χ^2 test statistic described earlier, and it is calculated by using the same procedure. Instead of listing the details of that calculation, we provide the TI-83/84 Plus calculator display that results from the data in Table 11-6.

The calculator display shows the test statistic of $\chi^2 = 6.529$ and the P-value of 0.0106. Using the P-value approach to hypothesis testing, we reject the null hypothesis of equal (homogeneous) proportions (because the P-value of 0.0106 is less than 0.05). There is sufficient evidence to warrant rejection of the claim that the proportions are the same. It appears that response and the gender of the interviewer are dependent. Although this statistical analysis cannot be used to justify any statement about causality, it does appear that men are influenced by the gender of the interviewer.

TI-83/84 Plus

```
χ²-Test
 χ²=6.529343179
 P=.0106109132
 df=1
```

EXAMPLE Flipping and Spinning Pennies When flipping a penny or spinning a penny, is the probability of getting heads the same? Use the data in Table 11-7 with a 0.05 significance level to test the claim that the proportion of heads is the same with flipping as with spinning. (The data are from experimental results given in *Chance News*.)

TABLE 11-7 Coin Experiments

	Heads	Tails
Flipping	2048	1992
Spinning	953	1047

SOLUTION

REQUIREMENTS ✔ As required, the data are random and they do consist of frequency counts in a two-way table. Here we are testing the null hypothesis that the proportion of heads with flipping is the same as the proportion of heads with spinning. The expected frequencies are all at least 5. (The expected frequencies are 2007.291, 2032.709, 993.709, and 1006.291.) Because all of the requirements are satisfied, we can proceed with the hypothesis test. ✔

Because we have two separate populations (coins that were flipped in one experiment and coins that were spun in a different experiment), we want to test for homogeneity with these hypotheses:

H_0: The proportions of heads is the same for flipping and spinning.

H_1: The proportions are different.

continued

TI-83/84 Plus

```
χ²-Test
 χ²=4.955360917
 p=.026009895
 df=1
```

The significance level is $\alpha = 0.05$. We use the same χ^2 test statistic described earlier, and it is calculated by using the same procedure. Instead of listing the details of that calculation, we provide the TI-83/84 Plus calculator display that results from the data in Table 11-7.

The calculator display shows the test statistic of $\chi^2 = 4.955$ and the *P*-value of 0.026. Using the *P*-value approach to hypothesis testing, we reject the null hypothesis of equal (homogeneous) proportions (because the *P*-value of 0.026 is less than 0.05). There is sufficient evidence to warrant rejection of the claim that the proportions are the same. It appears that flipping a penny and spinning a penny result in different proportions of heads.

Fisher Exact Test

For the analysis of 2 × 2 tables, we have included the requirement that every cell must have an expected frequency of 5 or greater. This requirement is necessary for the χ^2 distribution to be a suitable approximation to the exact distribution of the test statistic $\Sigma\frac{(O - E)^2}{E}$. Consequently, if a 2 × 2 table has a cell with an expected frequency less than 5, the preceding procedures should not be used, because the distribution is not a suitable approximation. The *Fisher exact test* is often used for such a 2 × 2 table, because it provides an *exact* *P*-value and does not require an approximation technique.

Consider the data in Table 11-8, with expected frequencies shown in parentheses below the observed frequencies. The first cell has an expected frequency less than 5, so the preceding methods should not be used. With the Fisher exact test, we calculate the probability of getting the observed results by chance (assuming that wearing a helmet and receiving facial injuries are independent), and we also calculate the probability of any result that is *more extreme*. (This use of "more extreme" results can be a somewhat confusing concept, so it might be helpful to again see the Section 5-2 subsection of "Using Probabilities to Determine When Results Are Unusual.") When testing the null hypothesis of independence between wearing a helmet and receiving a facial injury, the frequencies of 2, 13, 6, 19 can be replaced by 1, 14, 7, 18, respectively, to obtain *more extreme* results with the same row and column totals. (The Fisher exact test is sometimes criticized because the use of fixed row and column totals is often unrealistic.) The

Table 11-8 Helmets and Facial Injuries in Bicycle Accidents (Expected frequencies are in parentheses)

	Helmet Worn	No Helmet
Facial injuries received	2 (3)	13 (12)
All injuries nonfacial	6 (5)	19 (20)

Fisher exact test requires that we find the probabilities for the observed frequencies and each set of more extreme frequencies. Those probabilities are then added to provide an exact *P*-value.

Because the calculations are typically quite complex, it's a good idea to use software. For the data in Table 11-8, STATDISK, SPSS, SAS, and Minitab use Fisher's exact test to obtain an exact *P*-value of 0.686. Because this exact *P*-value is not small (such as less than 0.05), we fail to reject the null hypothesis that wearing a helmet and receiving facial injuries are independent.

Matched Pairs In addition to the requirement that each cell must have an expected frequency of at least 5, the methods of this section also require that the individual observations must be *independent*. If a 2 × 2 table consists of frequency counts that result from *matched pairs*, we do not have the required independence. For such cases, we can use McNemar's test, introduced in the following section.

TI

Using the TI-83/84 Plus Calculator for Contingency Tables

The TI-83/84 Plus calculator includes a procedure designed for the methods of this section. We can find the test statistic, number of degrees of freedom, and the *P*-value. Because the *P*-value is generated, we can use the *P*-value method of hypothesis testing, and there is no need to find critical values from Table A-4. Use the following procedure:

1. Enter the observed frequencies as a *matrix,* which is a rectangular arrangement of values. The matrix will consist of rows and columns, and it should contain the same data in the same arrangement as the original contingency table. Press 2ND x⁻¹ to get the **MATRIX** menu.
2. Use the arrow key ▷ to select the top menu item of **EDIT,** and press ENTER.
3. Enter the dimensions of the matrix in the format of $r \times c$, where r is the number of rows and c is the number of columns. Press ENTER after entering each dimension.
4. Enter all of the observed frequencies in the matrix. See the accompanying TI-83/84 Plus screen that shows the entry of the matrix representing Table 11-4.
5. Press STAT, select **TESTS,** and then select the menu item of **χ^2-Test.** Be sure that the observed matrix is the one you entered, such as matrix A. Scroll down to **Calculate** and press ENTER to get the test statistic, *P*-value, and number of degrees of freedom. See the accompanying screen display showing the results based on Table 11-4. The results include the test statistic of $\chi^2 = 8.775$ (rounded) and the *P*-value of 0.0124 (rounded).

TI-83/84 Plus

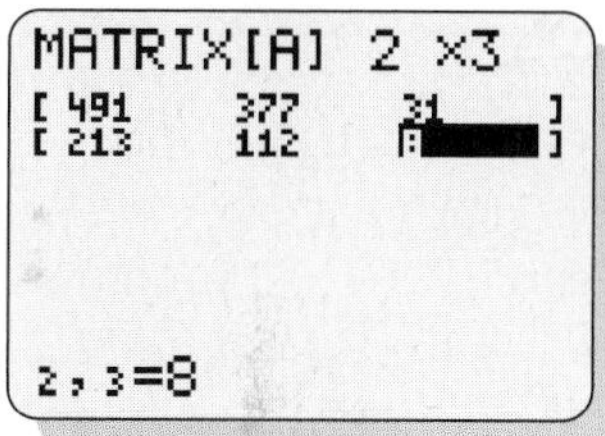

TI-83/84 Plus

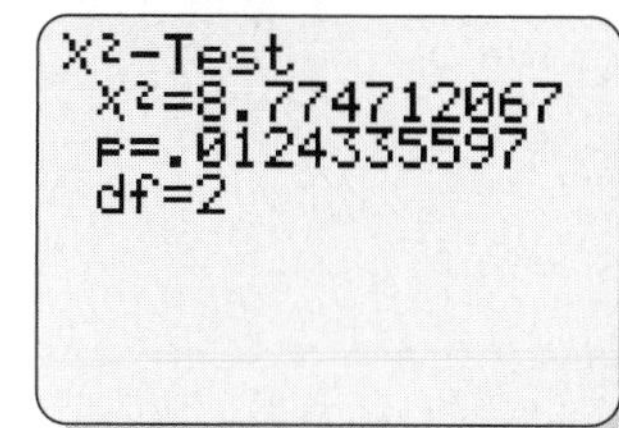

11-3 BASIC SKILLS AND CONCEPTS

Statistical Literacy and Critical Thinking

1. Chi-Square Test Statistic Use your own words to describe what the chi-square test statistic measures when used in this section.

2. Right-Tailed Test Why are the hypothesis tests described in this section always right-tailed?

3. Contingency What does the word "contingency" mean in the context of this section?

4. Causation Assume that we reject the null hypothesis of independence between the row variable of whether a subject smokes and the column variable of whether the subject can pass a standard test of physical endurance. Can we conclude that smoking causes people to fail the test? Why or why not?

In Exercises 5 and 6, test the given claim using the displayed software results.

5. Is There Racial Profiling? *Racial profiling* is the controversial practice of targeting someone for criminal behavior on the basis of the person's race, national origin, or ethnicity. The accompanying table summarizes results for randomly selected drivers stopped by police in a recent year (based on data from the U.S. Department of Justice, Bureau of Justice Statistics). Using the data in this table results in the Minitab display. Use a 0.05 significance level to test the claim that being stopped is independent of race and ethnicity. Based on the available evidence, can we conclude that racial profiling is being used?

Race and Ethnicity

	Black and Non-Hispanic	White and Non-Hispanic
Stopped by police	24	147
Not stopped by police	176	1253

Minitab

```
Chi-Sq = 0.413, DF = 1, P-Value = 0.521
```

TI **6.** No Smoking The accompanying table summarizes successes and failures when subjects used different methods in trying to stop smoking. The determination of smoking or not smoking was made five months after the treatment was begun, and the data are based on results from the Centers for Disease Control and Prevention. Use the accompanying TI-83/84 Plus results with a 0.05 significance level to test the claim that success is independent of the method used. If someone wants to stop smoking, does the choice of the method make a difference?

	Nicotine Gum	Nicotine Patch
Smoking	191	263
Not smoking	59	57

TI-83/84 Plus

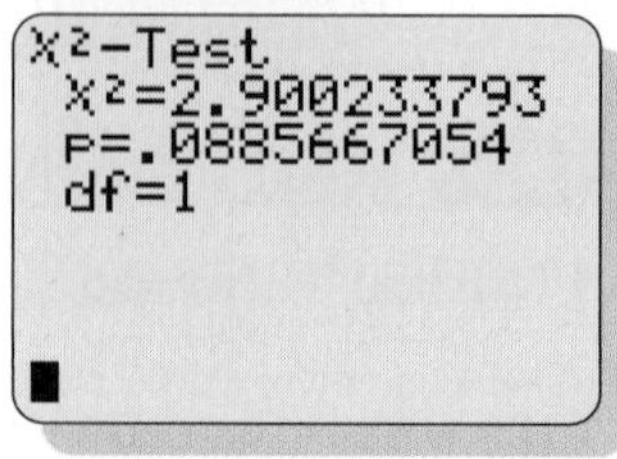

7. Is the Vaccine Effective? In a *USA Today* article about an experimental vaccine for children, the following statement was presented: "In a trial involving 1602 children, only 14 (1%) of the 1070 who received the vaccine developed the flu, compared with 95 (18%) of the 532 who got a placebo." The data are shown in the table below. Use a 0.05 significance level to test for independence between the variable of treatment (vaccine or placebo) and the variable representing flu (developed flu, did not develop flu). Does the vaccine appear to be effective?

	Developed Flu?	
	Yes	No
Vaccine treatment	14	1056
Placebo	95	437

8. Pedestrian Fatalities A study was conducted of the association between intoxication and pedestrian deaths, with the results shown in the accompanying table (based on data from the National Highway Traffic Safety Administration). Use a 0.05 significance level to test the claim that pedestrian fatalities are independent of the intoxication of the driver and the intoxication of the pedestrian.

	Pedestrian Intoxicated	Pedestrian Not Intoxicated
Driver intoxicated	59	79
Driver not intoxicated	266	581

9. Left-Handedness and Gender The table below is based on data from a Scripps Survey Research Center poll. Use a 0.05 significance level to test the claim that gender and left-handedness are independent.

	Left-handed	Not Left-handed
Male	83	17
Female	184	16

10. Birth Weight and Graduation The data in the table below are based on data from a *Time* magazine article. Use a 0.05 significance level to test the claim that whether a subject had low birth weight or normal birth weight is independent of whether the subject graduates from high school by age 19. Do the results show that low birth weight causes people to not graduate from high school by age 19?

	Graduated from high school by age 19	Did not graduate from high school by age 19
Low birth weight	8	42
Normal birth weight	86	64

11. Accuracy of Polygraph Tests The data in the accompanying table summarize results from tests of the accuracy of polygraphs (based on data from the Office of Technology Assessment). Use a 0.05 significance level to test the claim that whether the subject lies is independent of the polygraph indication. What do the results suggest about the effectiveness of polygraphs?

	Polygraph Indicated Truth	Polygraph Indicated Lie
Subject actually told the truth	65	15
Subject actually told a lie	3	17

TI **12.** Can Dogs Detect Cancer? An experiment was conducted to test the ability of dogs to detect bladder cancer. Dogs were trained with urine samples from bladder cancer patients and people in a control group who did not have bladder cancer. Results are given in the table below (based on data from the *New York Times*). Using a 0.01 significance level, test the claim that the source of the sample (healthy or with bladder cancer) is independent of the dog's selections. What do the results suggest about the ability of dogs to detect bladder cancer? If the dogs did significantly better than random guessing, did they do well enough to be used for accurate diagnoses?

	Sample from subject with bladder cancer	Sample from subject without bladder cancer
Dog identified subject as cancerous	22	32
Dog did not identify subject as cancerous	32	282

TI **13.** Is Sentence Independent of Plea? Many people believe that criminals who plead guilty tend to get lighter sentences than those who are convicted in trials. The accompanying table summarizes randomly selected sample data for San Francisco defendants in burglary cases. All of the subjects had prior prison sentences. At the 0.05 significance level, test the claim that the sentence (sent to prison or not sent to prison) is independent of the plea. If you were an attorney defending a guilty defendant, would these results suggest that you should encourage a guilty plea?

	Guilty Plea	Not Guilty Plea
Sent to prison	392	58
Not sent to prison	564	14

Based on data from "Does It Pay to Plead Guilty? Differential Sentencing and the Functioning of the Criminal Courts," by Brereton and Casper, *Law and Society Review,* Vol. 16, No. 1.

TI **14.** Which Treatment Is Better? A randomized controlled trial was designed to compare the effectiveness of splinting against surgery in the treatment of carpal tunnel syndrome. Results are given in the table below (based on data from "Splinting vs. Surgery in the Treatment of Carpal Tunnel Syndrome," by Gerritsen et al., *Journal of the American Medical Association,* Vol. 288, No. 10). The results are based on evaluations made one year after the treatment. Using a 0.01 significance level, test the claim that success is independent of the type of treatment. What do the results suggest about treating carpal tunnel syndrome?

	Successful Treatment	Unsuccessful Treatment
Splint treatment	60	23
Surgery treatment	67	6

TI **15.** Flipping and Spinning Pennies When flipping a penny or spinning a penny, is the probability of getting heads the same? Use the data in the table below with a 0.05 significance level to test the claim that the proportion of heads is the same with flipping as with spinning. (The data are from experimental results from Professor Robin Lock as given in *Chance News*.)

	Heads	Tails
Flipping	14,709	14,306
Spinning	9197	11,225

TI **16.** Testing Influence of Gender Table 11-6 summarizes data for male survey subjects, but the accompanying table summarizes data for a sample of women (based on data from an Eagleton Institute poll). Using a 0.01 significance level, and assuming that the sample sizes of 800 men and 400 women are predetermined, test the claim that the proportions of agree/disagree responses are the same for the subjects interviewed by men and the subjects interviewed by women. Does it appear that the gender of the interviewer affected the responses of women?

	Gender of Interviewer	
	Man	Woman
Women who agree	512	336
Women who disagree	288	64

TI **17.** Occupational Hazards Use the data in the table to test the claim that occupation is independent of whether the cause of death was homicide. The table is based on data from the U.S. Department of Labor, Bureau of Labor Statistics. Does any particular occupation appear to be most prone to homicides? If so, which one?

	Police	Cashiers	Taxi Drivers	Guards
Homicide	82	107	70	59
Cause of death other than homicide	92	9	29	42

TI **18.** Is Scanner Accuracy the Same for Specials? In a study of store checkout scanning systems, samples of purchases were used to compare the scanned prices to the posted prices. The accompanying table summarizes results for a sample of 819 items. When stores use scanners to check out items, are the error rates the same for regular-priced items as they are for advertised-special items? How might the behavior of consumers change if they believe that disproportionately more overcharges occur with advertised-special items?

	Regular-Priced Items	Advertised-Special Items
Undercharge	20	7
Overcharge	15	29
Correct price	384	364

Based on data from "UPC Scanner Pricing Systems: Are They Accurate?" by Ronald Goodstein, *Journal of Marketing,* Vol. 58.

TI **19.** Is Seat Belt Use Independent of Cigarette Smoking? A study of seat belt users and nonusers yielded the randomly selected sample data summarized in the given table. Test the claim that the amount of smoking is independent of seat belt use. A plausible theory is that people who smoke more are less concerned about their health and safety and are therefore less inclined to wear seat belts. Is this theory supported by the sample data?

	Number of Cigarettes Smoked per Day			
	0	1–14	15–34	35 and over
Wear seat belts	175	20	42	6
Don't wear seat belts	149	17	41	9

Based on data from "What Kinds of People Do Not Use Seat Belts?" by Helsing and Comstock, *American Journal of Public Health,* Vol. 67, No. 11.

TI **20.** Is the Home Field Advantage Independent of the Sport? Winning team data were collected for teams in different sports, with the results given in the accompanying table. Use a 0.10 significance level to test the claim that home/visitor wins are independent of the sport. Given that among the four sports included here, baseball is the only sport in which the home team can modify field dimensions to favor its own players, does it appear that baseball teams are effective in using this advantage?

	Basketball	Baseball	Hockey	Football
Home team wins	127	53	50	57
Visiting team wins	71	47	43	42

Based on data from "Predicting Professional Sports Game Outcomes from Intermediate Game Scores," by Copper, DeNeve, and Mosteller, *Chance,* Vol. 5, No. 3–4.

TI **21.** Injuries and Motorcycle Helmet Color An example in this section involved data from a case-control study involving injuries and the color of helmets of motorcycle riders. Use the additional data included in the table below and test the claim that injuries are independent of helmet color. Do these data lead to the same conclusion reached with the data in the example of this section?

	Color of Helmet				
	Black	White	Yellow/Orange	Red	Blue
Controls (not injured)	491	377	31	170	55
Cases (injured or killed)	213	112	8	70	26

TI **22.** Survey Refusals and Age Bracket A study of people who refused to answer survey questions provided the randomly selected sample data shown in the table below. At the 0.01 significance level, test the claim that the cooperation of the subject (response or refusal) is independent of the age category. Does any particular age group appear to be particularly uncooperative?

	Age					
	18–21	22–29	30–39	40–49	50–59	60 and over
Responded	73	255	245	136	138	202
Refused	11	20	33	16	27	49

Based on data from "I Hear You Knocking but You Can't Come In," by Fitzgerald and Fuller, *Sociological Methods and Research,* Vol. 11, No. 1.

11-3 BEYOND THE BASICS

23. Using Yates' Correction for Continuity The chi-square distribution is continuous, whereas the test statistic used in this section is discrete. Some statisticians use *Yates' correction for continuity* in cells with an expected frequency of less than 10 or in all cells of a contingency table with two rows and two columns. With Yates' correction, we replace

$$\sum\frac{(O-E)^2}{E} \quad \text{with} \quad \sum\frac{(|O-E|-0.5)^2}{E}$$

Given the contingency table in Exercise 5, find the value of the χ^2 test statistic with and without Yates' correction. What effect does Yates' correction have?

24. Equivalent Tests Assume that a contingency table has two rows and two columns with frequencies of a and b in the first row and frequencies of c and d in the second row.

a. Verify that the test statistic can be expressed as

$$\chi^2 = \frac{(a + b + c + d)(ad - bc)^2}{(a + b)(c + d)(b + d)(a + c)}$$

b. Let $\hat{p}_1 = a/(a + c)$ and let $\hat{p}_2 = b/(b + d)$. Show that the test statistic

$$z = \frac{(\hat{p}_1 - \hat{p}_2) - 0}{\sqrt{\frac{\bar{p}\bar{q}}{n_1} + \frac{\bar{p}\bar{q}}{n_2}}}$$

where $$\bar{p} = \frac{a + b}{a + b + c + d}$$

and $$\bar{q} = 1 - \bar{p}$$

is such that $z^2 = \chi^2$ [the same result as in part (a)]. This result shows that the chi-square test involving a 2×2 table is equivalent to the test for the difference between two proportions, as described in Section 9-2.

11-4 McNemar's Test for Matched Pairs

Key Concept The contingency table procedures in Section 11-3 are based on *independent* data. For 2×2 tables consisting of frequency counts that result from *matched pairs*, we do not have independence and, for such cases, we can use McNemar's test for matched pairs. We will test the null hypothesis that frequencies from the discordant (different) categories occur in the same proportion.

Assume that each of several test subjects is afflicted with tinea pedis (athlete's foot) on each foot, and each subject is given a treatment X on one foot and a treatment Y on the other foot. Table 11-9 is a general table summarizing the frequency counts that result from the matched pairs of feet given the two different treatments. If $a = 12$ in Table 11-9, then 12 subjects enjoyed a cure on each foot. If $b = 8$ in Table 11-9, then each of 8 subjects had one foot not cured by treatment X while their other foot was cured by treatment Y. *Important:* Note that the entries in Table 11-9 are frequency counts of *people*, not feet.

Table 11-9 2 × 2 Table with Frequency Counts from Matched Pairs

		Treatment X	
		Cured	Not Cured
Treatment Y	Cured	a	b
	Not cured	c	d

Because the frequency counts in Table 11-9 result from *matched* pairs of feet, the data are not independent and we cannot use the contingency table procedures from Section 11-3. Instead, we use McNemar's test.

Definition

McNemar's test uses frequency counts from *matched pairs* of nominal data from two categories to test the null hypothesis that for a table such as Table 11-9, the frequencies b and c occur in the same proportion.

Requirements

1. The sample data have been randomly selected.
2. The sample data consist of *matched pairs* of frequency counts.
3. The data are at the nominal level of measurement, and each observation can be classified two ways: (1) According to the category distinguishing values with each matched pair (such as left foot and right foot), and (2) according to another category with two possible values (such as cured/not cured).
4. For tables such as Table 11-9, the frequencies are such that $b + c \geq 10$.

Test Statistic (for testing the null hypothesis that for tables such as Table 11-9, the frequencies b and c occur in the same proportion):

$$\chi^2 = \frac{(|b - c| - 1)^2}{b + c}$$

where the frequencies of b and c are obtained from the 2×2 table with a format similar to Table 11-9. (The frequencies b and c must come from "discordant" pairs, as described later in this section.)

Critical values

1. The critical region is located in the *right tail only*.
2. The critical values are found in Table A-4 by using **degrees of freedom = 1.**

EXAMPLE Comparing Treatments Two different creams are used to treat tinea pedis (athlete's foot). Each subject with this fungal infection on both feet is given a treatment of Pedacream on one foot while their other foot is treated with Fungacream. The sample results are summarized in Table 11-10. Using a 0.05 significance level, apply McNemar's test to test the null hypothesis that the following two proportions are the same:

- The proportion of subjects with no cure on the Pedacream-treated foot and a cure on the Fungacream-treated foot.
- The proportion of subjects with a cure on the Pedacream-treated foot and no cure on the Fungacream-treated foot.

Based on the results, does there appear to be a difference between the two treatments? Does one of the treatments appear to be better than the other?

SOLUTION

REQUIREMENT ✔ The data consist of matched pairs of frequency counts from randomly selected subjects, and each observation can be categorized according to two variables. (One variable has values of "Pedacream" and

Table 11-10 Clinical Trials of Treatments for Athlete's Foot

		Treatment with Pedacream	
		Cured	Not Cured
Treatment with Fungacream	Cured	12	8
	Not cured	40	20

80 subjects treated on 160 feet:
12 had both feet cured.
20 had neither foot cured.
8 had cures with Fungacream, but not Pedacream.
40 had cures with Pedacream, but not Fungacream.

"Fungacream," and the other variable has values of "cured" and "not cured.") Also, for tables such as Table 11-9, the frequencies must be such that $b + c \geq 10$. For Table 11-10, $b = 8$ and $c = 40$, so that $b + c = 48$, which is at least 10. All of the requirements are therefore satisfied. Although Table 11-10 might appear to be a 2×2 contingency table, we cannot use the procedures of Section 11-3 because the data come from *matched pairs* (instead of being independent). Instead, we use McNemar's test. ✔

After comparing the frequency counts in Table 11-9 to those given in Table 11-10, we see that $b = 8$ and $c = 40$, so the test statistic can be calculated as follows:

$$\chi^2 = \frac{(|b - c| - 1)^2}{b + c} = \frac{(|8 - 40| - 1)^2}{8 + 40} = 20.021$$

With a 0.05 significance level and degrees of freedom given by df = 1, we refer to Table A-4 to find the critical value of $\chi^2 = 3.841$ for this right-tailed test. The test statistic of $\chi^2 = 20.021$ exceeds the critical value of $\chi^2 = 3.841$, so we reject the null hypothesis. It appears that the two creams produce different results. Analyzing the frequencies of 8 and 40, we see that many more feet were cured with Pedacream than Fungacream, so the Pedacream treatment appears to be more effective.

Note that in the calculation of the test statistic in the preceding example, we did not use the 12 subjects with both feet cured (one foot from each cream) and we did not use the 20 subjects with neither foot cured. Instead of including the cure/cure results and the no cure/no cure results, we used only the cure/no cure results and the no cure/cure results. That is, we are using only the results from the categories that are *different*. Such different categories are referred to as *discordant pairs*.

Definition

Discordant pairs of results come from pairs of categories in which the two categories are different (as in cure/no cure or no cure/cure).

When trying to determine whether there is a significant difference between the two cream treatments in Table 11-10, we are not helped by the subjects with both feet cured, and we are not helped by those subjects with neither foot cured. The differences are reflected in the discordant results from the subjects with one foot cured while the other foot was not cured. Consequently, the test statistic includes only the two frequencies that result from the two discordant (or different) pairs of categories.

Caution: When applying McNemar's test, be careful to use only the frequencies from the pairs of categories that are different. Do not blindly use the frequencies in the upper right and lower left corners, because they do not necessarily represent the discordant pairs. If Table 11-10 were reconfigured as shown below, it would be inconsistent in its format, but it would be technically correct in summarizing the same results as the preceding table; however, blind use of the frequencies of 20 and 12 would result in the *wrong* test statistic.

		Treatment with Pedacream	
		Cured	Not cured
Treatment with Fungacream	Not cured	40	20
	Cured	12	8

In this reconfigured table, the discordant pairs of frequencies are these:

Cured/Not cured: 40

Not cured/Cured: 8

With this reconfigured table, we should again use the frequencies of 40 and 8, not 20 and 12. In a more perfect world, all such 2×2 tables would be configured with a consistent format, and we would be much less likely to use the wrong frequencies.

In addition to comparing treatments given to matched pairs (as in the preceding example), McNemar's test is often used to test a null hypothesis of no change in before/after types of experiments. (See Exercises 5–12.)

Using the TI-83/84 Plus Calculator for McNemar's Test

The TI-83/84 Plus calculator does not include a procedure designed for McNemar's test, but it can be used to find the *P*-value for the test. Use the following procedure:

1. Use the methods of this section to calculate the χ^2 test statistic.
2. Press 2ND VARS to select the **DISTR** (distribution) menu.
3. Select the menu item of **χ^2 cdf(** and proceed to enter the χ^2 test statistic, the number 99999999, followed by 1, with the three numbers separated by commas.

(Note: The format for the χ^2 cdf command is to enter the lower boundary of χ^2, the upper boundary of χ^2, and the number of degrees of freedom. Because we want the area to the right of the χ^2 test statistic, we enter the value of 99999999 for the upper boundary, which essentially means that there is no upper boundary. For the example of this section, the screen should appear as follows:

$$\chi^2 \text{ cdf}(20.021, 99999999, 1).$$

4. Press ENTER and the *P*-value for the test will be displayed.

11-4 BASIC SKILLS AND CONCEPTS

Statistical Literacy and Critical Thinking

1. McNemar's Test When conducting hypothesis tests with 2 × 2 tables, what circumstances indicate that McNemar's test is suitable while the methods of Section 11-3 are not?

2. McNemar's Test Can McNemar's test be used on two-way tables with more than two rows or more than two columns? Why or why not?

3. Discordant Pairs What are discordant pairs of results?

4. Discordant Pairs Why does McNemar's test involve only discordant pairs of data while ignoring the other data?

In Exercises 5–12, refer to the following table. The table summarizes results from an experiment in which subjects were first classified as smokers or nonsmokers, then they were given a treatment, then later they were again classified as smokers or nonsmokers.

		Before Treatment	
		Smoke	Don't Smoke
After treatment	Smoke	50	6
	Don't smoke	8	80

5. Sample Size How many subjects are included in the experiment?

6. Treatment Effectiveness How many subjects changed their smoking status after the treatment?

7. Treatment Ineffectiveness How many subjects appear to be unaffected by the treatment one way or the other?

8. Why not *t* test? Section 9-4 presented procedures for dealing with data consisting of matched pairs. Why can't we use the procedures of Section 9-4 for the analysis of the results summarized in the table?

9. Discordant Pairs Which of the following pairs of before/after results are *discordant*?
   a. smoke/smoke
   b. smoke/don't smoke
   c. don't smoke/smoke
   d. don't smoke/don't smoke

10. Test statistic Using the appropriate frequencies, find the value of the test statistic.

11. Critical value Using a 0.01 significance level, find the critical value.

12. Conclusion Based on the preceding results, what do you conclude? How does the conclusion make sense in terms of the original sample results?

TI **13.** Treating Athlete's Foot As in the example of this section, assume that subjects are inflicted with athlete's foot on each of their feet. Also assume that for each subject, one foot is treated with a fungicide solution while the other foot is given a placebo. The results are given in the accompanying table. Using a 0.05 significance level, test the effectiveness of the treatment.

		Fungicide Treatment	
		Cure	No Cure
Placebo	Cure	5	12
	No cure	22	55

TI **14.** Treating Athlete's Foot Repeat Exercise 13 after changing the frequency of 22 to 66.

TI **15.** PET/CT Compared to MRI In the article "Whole-Body Dual-Modality PET/CT and Whole Body MRI for Tumor Staging in Oncology" (Antoch et al., *Journal of the American Medical Association,* Vol. 290, No. 24), the authors cite the importance of accurately identifying the stage of a tumor. Accurate staging is critical for determining appropriate therapy. The article discusses a study involving the accuracy of positron emission tomography (PET) and computed tomography (CT) compared to magnetic resonance imaging (MRI). Using the data in the given table for 50 tumors analyzed with both technologies, does there appear to be a difference in accuracy? Does either technology appear to be better?

		PET/CT	
		Correct	Incorrect
MRI	Correct	36	1
	Incorrect	11	2

TI **16.** Testing a Treatment In the article "Eradication of Small Intestinal Bacterial Overgrowth Reduces Symptoms of Irritable Bowel Syndrome" (Pimentel, Chow, Lin, *American Journal of Gastroenterology,* Vol. 95, No. 12), the authors include a discussion of whether antibiotic treatment of bacteria overgrowth reduces intestinal complaints. McNemar's test was used to analyze results for those subjects with eradication of bacterial overgrowth. Using the data in the given table, does the treatment appear to be effective against abdominal pain?

		Abdominal Pain Before Treatment?	
		Yes	No
Abdominal pain after treatment?	Yes	11	1
	No	14	3

11-4 BEYOND THE BASICS

TI **17.** Correction for Continuity The test statistic given in this section includes a correction for continuity. The test statistic given below does not include the correction for continuity, and it is sometimes used as the test statistic for McNemar's test. Refer to the example in this section, find the value of the test statistic using the expression given below, and compare the result to the one found in the example.

$$\chi^2 = \frac{(b - c)^2}{b + c}$$

TI **18.** Using Common Sense Consider the table given below, and use a 0.05 significance level.

a. What does McNemar's test suggest about the effectiveness of the treatment?

b. The values of a and d are not used in the calculations, but what does common sense suggest if $a = 5000$ and $d = 4000$?

		Before Treatment	
		Smoke	Don't Smoke
After treatment	Smoke	a	5
	Don't smoke	20	d

TI **19.** Small Sample Case The requirements for McNemar's test include the condition that $b + c \geq 10$ so that the distribution of the test statistic can be approximated by the chi-square distribution. Refer to the example in this section and replace the table data with the values given below. McNemar's test should not be used because the condition of $b + c \geq 10$ is not satisfied with $b = 2$ and $c = 6$. Instead, use the binomial distribution to find the probability that among 8 equally likely outcomes, the results consist of 6 items in one category and 2 in the other category, or the results are more extreme. That is, use a probability of 0.5 to find the probability that among $n = 8$ trials, the number of successes x is 6 or 7 or 8. Double that probability to find the P-value for this test. Compare the result to the P-value of 0.289 that results from using the chi-square approximation, even though the condition of $b + c \geq 10$ is violated. What do you conclude about the two treatments?

		Treatment with Pedacream	
		Cured	Not Cured
Treatment with Fungacream	Cured	12	2
	Not cured	6	20

Review

In this chapter we worked with data summarized as frequency counts for different categories. In Section 11-2 we described methods for testing goodness-of-fit in a multinomial experiment, which is similar to a binomial experiment except that there are more than two categories of outcomes. Multinomial experiments result in frequency counts arranged in a single row or column, and we tested to determine whether the observed sample frequencies agree with (or "fit") some claimed distribution.

In Section 11-3 we described methods for testing claims involving contingency tables (or two-way frequency tables), which have at least two rows and two columns. Contingency tables incorporate two variables: One variable is used for determining the row that describes a sample value, and the second variable is used for determining the column that describes a sample value. Section 11-3 included two types of hypothesis tests: (1) a test of independence between the row and column variables; (2) a test of homogeneity to determine whether different populations have the same proportions of some characteristics.

Section 11-4 introduced McNemar's test for testing the null hypothesis that a sample of matched pairs of data comes from a population in which the discordant (different) pairs occur in the same proportion.

The following are some key components of the methods discussed in this chapter:

- *Section 11-2 (Test for goodness-of-fit):*

  $$\text{Test statistic is } \chi^2 = \sum \frac{(O - E)^2}{E}$$

  Test is right-tailed with $k - 1$ degrees of freedom. All expected frequencies must be at least 5.

- *Section 11-3 (Contingency table test of independence or homogeneity):*

  $$\text{Test statistic is } \chi^2 = \sum \frac{(O - E)^2}{E}$$

  Test is right-tailed with $(r - 1)(c - 1)$ degrees of freedom. All expected frequencies must be at least 5.

- *Section 11-4 (2 × 2 table with frequencies from matched pairs of data):*

  $$\text{Test statistic is } \chi^2 = \frac{(|b - c| - 1)^2}{b + c}$$

  where the frequencies of b and c must come from "discordant" pairs. Test is right-tailed with 1 degree of freedom. The frequencies b and c must be such that $b + c \geq 10$.

Statistical Literacy and Critical Thinking

1. Categorical Data This chapter introduced a few different methods for the analysis of *categorical* data. What are categorical data?

2. Conducting a Survey A student conducts a research project by asking 200 classmates if they have had a credit card stolen. She constructs a contingency table with row categories of gender (male/female) and column categories of response (yes, no, refused to answer). She uses the methods of Section 11-3 to conclude that gender is independent of response. What is wrong with her project?

3. Chi-Square Distribution This chapter presented different methods involving application of the chi-square distribution. Which of the following properties of a chi-square distribution are true?

a. Values of a chi-square test statistic are always positive or zero, but never negative.
b. A chi-square distribution is symmetric.
c. There is a different chi-square distribution for each number of degrees of freedom.

d. When using a chi-square distribution, the number of degrees of freedom is always the sample size minus 1.

e. When using the chi-square distribution, sample data need not be random if the sample size is very large.

4. Checking Requirements The methods of testing for goodness-of-fit and the methods of testing for independence between two variables used for a contingency table require that all expected frequencies must be at least 5. Can those methods be used if there is a cell with an observed frequency count less than 5? Why or why not?

Review Exercises

TI **1.** Are DWI Fatalities the Result of Weekend Drinking? Many people believe that fatal DWI crashes occur because of casual drinkers who tend to binge on Friday and Saturday nights, whereas others believe that fatal DWI crashes are caused by people who drink every day of the week. In a study of fatal car crashes, 216 cases are randomly selected from the pool in which the driver was found to have a blood alcohol content over 0.10. These cases are broken down according to the day of the week, with the results listed in the accompanying table (based on data from the Dutchess County STOP-DWI Program). At the 0.05 significance level, test the claim that such fatal crashes occur on the different days of the week with equal frequency. Does the evidence support the theory that fatal DWI car crashes are due to casual drinkers or that they are caused by those who drink daily?

Day	Sun	Mon	Tues	Wed	Thurs	Fri	Sat
Number	40	24	25	28	29	32	38

TI **2.** E-Mail and Privacy Workers and senior-level bosses were asked if it was seriously unethical to monitor employee e-mail, and the results are summarized in the table (based on data from a Gallup poll). Use a 0.05 significance level to test the claim that the response is independent of whether the subject is a worker or a senior-level boss. Does the conclusion change if a significance level of 0.01 is used instead of 0.05? Do workers and bosses appear to agree on this issue?

	Yes	No
Workers	192	244
Bosses	40	81

TI **3.** Crime and Strangers The accompanying table lists survey results obtained from a random sample of different crime victims (based on data from the U.S. Department of Justice). At the 0.05 significance level, test the claim that the type of crime is independent of whether the criminal is a stranger. How might the results affect the strategy police officers use when they investigate crimes?

	Homicide	Robbery	Assault
Criminal was a stranger	12	379	727
Criminal was an acquaintance or relative	39	106	642

TI **4.** Comparing Treatments Two different creams are used to treat subjects with poison ivy irritation on both hands. Each subject is given a treatment of Ivy Ease on one hand while their other hand is treated with a placebo. The sample results are summarized in the table below. Use a 0.05 significance level to test the null hypothesis that the following two proportions are the same: (1) the proportion of subjects with relief on the hand treated with Ivy Ease and no relief on the hand treated with a placebo; (2) the proportion of subjects with no relief on the hand treated with Ivy Ease and relief on the hand treated with a placebo. Does the Ivy Ease treatment appear to be effective?

		Treatment with Ivy Ease	
		Relief	No Relief
Placebo	Relief	12	8
	No relief	32	19

Cumulative Review Exercises

Table 11-11

	A	B	C	D
x	85	90	80	75
y	80	84	73	70

TI **1.** Finding Statistics Assume that in Table 11-11, the row and column titles have no meaning so that the table contains test scores for eight randomly selected prisoners who were convicted of removing labels from pillows. Find the mean, median, range, variance, standard deviation, and 5-number summary.

2. Finding Probability Assume that in Table 11-11, the letters A, B, C, and D represent the choices on the first question of a multiple-choice quiz. Also assume that x represents men and y represents women and that the table entries are frequency counts, so 85 men chose answer A, 80 women chose answer A, 90 men chose answer B, and so on.

a. If one response is randomly selected, find the probability that it is response C.

b. If one response is randomly selected, find the probability that it was made by a man.

c. If one response is randomly selected, find the probability that it is response C or was made by a man.

d. If two different responses are randomly selected, find the probability that they were both made by a woman.

e. If one response is randomly selected, find the probability that it was response B, given that the response was made by a woman.

TI **3.** Testing for Equal Proportions Using the same assumptions as in Exercise 2, test the claim that men and women choose the different answers in the same proportions.

TI **4.** Testing for a Relationship Assume that Table 11-11 lists test scores for four people, where the x-score is from a test of memory and the y-score is from a test of reasoning. Test the claim that there is a linear correlation between the x- and y-scores.

TI **5.** Testing for Effectiveness of Training Assume that Table 11-11 lists test scores for four people, where the x-score is from a pretest taken before a training session on memory improvement and the y-score is from a posttest taken after the training. Test the claim that the training session has no effect.

TI **6.** Testing for Equality of Means Assume that in Table 11-11, the letters A, B, C, and D represent different versions of the same test of reasoning. The x-scores were obtained by four randomly selected men and the y-scores were obtained by four randomly selected women. Test the claim that men and women have the same mean score.

Cooperative Group Activities

1. Out-of-class activity Divide into groups of four or five students. See the first two rows of Table 11-1 in the Chapter Problem for the distribution of leading digits expected with Benford's law. Collect data and use the methods of Section 11-2 to verify that the data conform reasonably well to Benford's law. Here are some possibilities that might be considered:
   - The amounts on the checks you wrote
   - The prices of stocks
   - Populations of counties in the United States
   - Numbers on street addresses

2. Out-of-class activity Divide into groups of four or five students and collect past results from a state lottery. Such results are often available on Web sites for individual state lotteries. Use the methods of Section 11-2 to test that the numbers are selected in such a way that all possible outcomes are equally likely.

3. Out-of-class activity Divide into groups of four or five students. Each group member should survey at least 15 male students and 15 female students at the same college by asking two questions: (1) Which political party does the subject favor most? (2) If the subject were to make up an absence excuse of a flat tire, which tire would he or she say went flat if the instructor asked? (See Exercise 10 in Section 11-2.) Ask the subject to write the two responses on an index card, and also record the gender of the subject and whether the subject wrote with the right or left hand. Use the methods of this chapter to analyze the data collected. Include these tests:
   - The four possible choices for a flat tire are selected with equal proportions.
   - The tire identified as being flat is independent of the gender of the subject.
   - Political party choice is independent of the gender of the subject.
   - Political party choice is independent of whether the subject is right- or left-handed.
   - The tire identified as being flat is independent of whether the subject is right- or left-handed.
   - Gender is independent of whether the subject is right- or left-handed.
   - Political party choice is independent of the tire identified as being flat.

4. Out-of-class activity Divide into groups of four or five students. Each group member should select about 15 other students and first ask them to "randomly" select four digits each. After the four digits have been recorded, ask each subject to write the last four digits of his or her social security number. Take the "random" sample results and mix them into one big sample, then mix the social security digits into a second big sample. Using the "random" sample set, test the claim that students select digits randomly. Then use the social security digits to test the claim that they come from a population of random digits. Compare the results. Does it appear that students can randomly select digits? Are they likely to select any digits more often than others? Are they likely to select any digits less often than others? Do the last digits of social security numbers appear to be randomly selected?

5. In-class activity Divide into groups of three or four students. Each group should be given a die along with the instruction that it should be tested for "fairness." Is the die fair or is it biased? Describe the analysis and results.

6. Out-of-class activity Divide into groups of two or three students. Some examples and exercises of this chapter were based on the analysis of last digits of values. It was noted that the analysis of last digits can sometimes reveal whether values are the results of actual measurements or whether they are reported estimates. Refer to an almanac and find the lengths of rivers in the world, then analyze the last digits to determine whether those lengths appear to be actual measurements or whether they appear to be reported estimates. (Instead of lengths of rivers, you could use heights of mountains, heights of the tallest buildings, lengths of bridges, and so on.)

TI-83/84 Plus Calculator Project

A TI-83/84 Plus calculator can be used to generate equally likely random digits between 0 and 9 inclusive. This project will test the randomness of such generated values to determine whether the calculator appears to be working correctly. We will use the TI-83/84 Plus calculator to randomly generate 500 digits between 0 and 9 inclusive, then we will test those digits to determine whether they appear to be from a population with a uniform distribution. Use the procedure given below.

1. Press MATH, then use the arrow key ▷ to select the top menu item of **PRB.**
2. Select the menu item of **randInt** (for random integer).
3. The format for the randInt command is this: (lower limit, upper limit, number of values). To generate 500 random digits between 0 and 9 inclusive, enter the command of **randInt(0, 9, 500).** (If this command creates a memory error message, delete some or all of the lists that have been created. Press 2ND + to get the **MEM** (memory) menu, then select **Mem Mgmt/Del,** then select **List** and proceed to delete unneeded lists.)
4. Store the results in list L1 by pressing STO› 2ND 1.
5. Sort the results. (Press STAT, select **SortA,** enter L1, then press ENTER).
6. Use the sorted values in list L1 to enter the frequencies in the following table.

Digit	0	1	2	3	4	5	6	7	8	9
Frequency										

7. Use a 0.05 significance level with the sample data from Step 6 to test the claim that the sample digits come from a population with a uniform distribution (so that all digits are equally likely). Does the calculator's random number generator appear to be working as it should?

From Data to Decision

Critical Thinking: Is the defendant guilty of fraud?

In the trial of *State of Arizona vs. Wayne James Nelson,* the defendant was accused of issuing checks to a vendor that did not really exist. The amounts of the checks are listed below in order by row.

\$1,927.48	\$27,902.31	\$86,241.90	\$72,117.46	\$81,321.75	\$97,473.96
\$93,249.11	\$89,658.16	\$87,776.89	\$92,105.83	\$79,949.16	\$87,602.93
\$96,879.27	\$91,806.47	\$84,991.67	\$90,831.83	\$93,766.67	\$88,336.72
\$94,639.49	\$83,709.26	\$96,412.21	\$88,432.86	\$71,552.16	

Analyzing the Results

Do the leading digits conform to Benford's law described in the Chapter Problem? When testing for goodness-of-fit with the proportions expected with Benford's law, it is necessary to combine categories because not all expected values are at least 5. Use one category with leading digits of 1, a second category with leading digits of 2, 3, 4, 5, and a third category with leading digits of 6, 7, 8, 9. Are the expected values for these three categories all at least 5? Is there sufficient evidence to conclude that the leading digits on the checks do not conform to Benford's law? Apart from the leading digits, are there any other patterns suggesting that the check amounts were created by the defendant instead of being the result of typical and real transactions? Based on the evidence, if you were a juror, would you conclude that the check amounts are the result of fraud? What would be one argument that you might present if you were the attorney for the defendant?

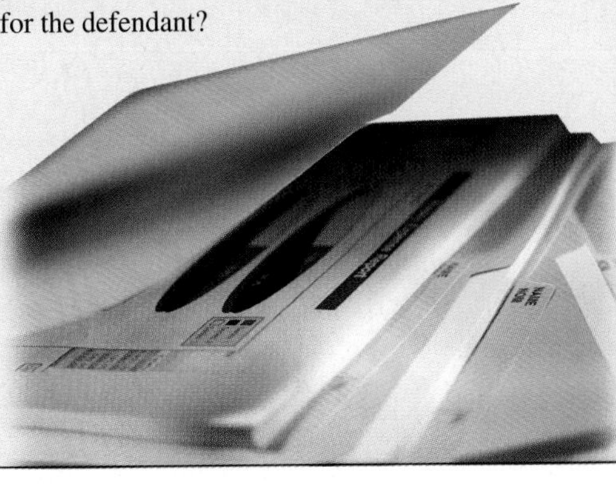

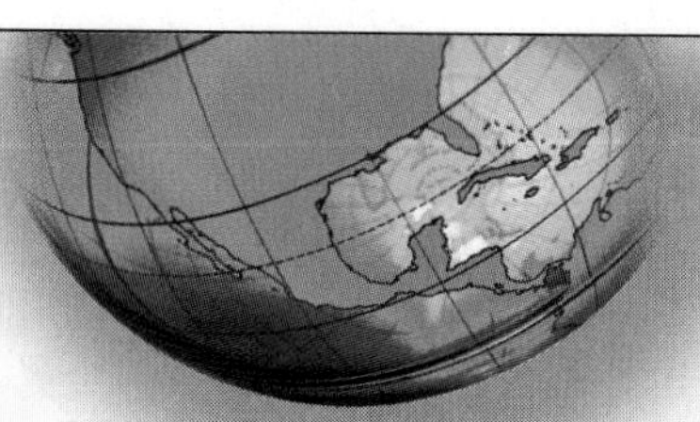

Internet Project

Contingency Tables

An important characteristic of tests of independence with contingency tables is that the data collected need not be quantitative in nature. A contingency table summarizes observations by the categories or labels of the rows and columns. As a result, characteristics such as gender, race, and political party all become fair game for formal hypothesis testing procedures.

The Internet Project for this chapter is found at the *Elementary Statistics* Web site:

http://www.aw.com/triola

You will find links to a variety of demographic data. With these data sets, you will conduct tests in areas as diverse as academics, politics, and the entertainment industry. In each test, you will draw conclusions related to the independence of interesting characteristics.

Statistics @ Work

"Even if you're not a numbers cruncher, [statistical] knowledge can be helpful in any situation that requires prediction, decision making, or evaluation."

Nabil Lebbos

Graphics illustrator, Published Image

As analyst for Standard & Poor's *Published Image,* Nabil's studies on investment performance are published in newspapers read by over one million investors.

Please describe your occupation.

I work for *Published Image* where I use statistics to generate the charts and data that we use in our financial publications—using loads of statistics and applications. We write newsletters for banks and mutual funds.

What concepts of statistics do you use?

I use standard deviation to measure risk, regression to measure an investment's relationship to its benchmark, and correlation to determine an investment's movement in relation to other investments.

How do you use statistics on the job?

I start with a given set of raw data. These are usually monthly, daily, or annual returns on an investment. I then use Excel to chart the data so I can get a picture of what I'm dealing with. From there I proceed to perform an analysis. Sometimes, the results do not back up a point that the accompanying article is trying to make strongly enough. In such situations, I look at other possibilities.

Please describe one specific example illustrating how the use of statistics was successful in improving a product or service.

One of our clients wanted to make the point that although their mutual fund did not outperform all others, it did succeed in consistently avoiding large negative returns. I ran some tests on skewness and downside risk and showed that, in fact, the fund's returns were positively skewed. We created histograms comparing this fund with an average of all funds, and that clearly made the point.

In terms of statistics, what would you recommend for prospective employees?

It's a logical tool that, when used informatively, can convince you and your audience of the point you're trying to make much more effectively than words. Even if you're not a numbers cruncher, [statistical] knowledge can be helpful in any situation that requires prediction, decision making, or evaluation.

Do you feel job applicants are viewed more favorably if they have studied some statistics?

Yes.

While a college student, did you expect to be using statistics on the job?

No. I studied architecture as an undergrad and business as a grad student.

12 Analysis of Variance

CHAPTER PROBLEM

Do different treatments affect the weights of poplar trees?

Data Set 7 in Appendix B includes weights (in kilograms) of poplar trees given different treatments at different sites. Let's consider only those weights given for year 1 at Site 1. Site 1 has rich and moist soil, and it is located near a creek. The weights we will consider are summarized in Table 12-1.

In the spirit of exploring data by investigating center, variation, distribution, outliers, and changing patterns over time (CVDOT), we begin by obtaining the sample statistics listed at the bottom of Table 12-1. Examining the sample means, we see that they appear to vary considerably, going from a low of 0.164 kg to a high of 1.334 kg. Also, the sample standard deviations appear to vary considerably, going from a low of 0.126 kg to a high of 0.859 kg. It's difficult to analyze distributions because each sample consists of only 5 values, but normal quantile plots suggest that three of the samples are from populations having distributions that are approximately normal. However, analysis of the weights of poplar trees given a fertilizer treatment suggests that the weight of 1.34 kg is an outlier when compared to the other weights of fertilized trees. With only one outlier present, we will proceed under the assumption that the samples come from populations with distributions that are approximately normal. We could do additional analyses later to determine whether the weight of 1.34 kg has much of an effect on our results. (See Exercise 5 in Section 12-2.)

It appears that the differences among the sample means indicate that the samples come from populations with different means, but instead of considering only the sample means, we should also consider the amounts of variation, the sample sizes, and the nature of the distribution of sample means. One way of taking all of these relevant factors into account is to conduct a formal hypothesis test that automatically includes them. Such a test will be introduced in this chapter, and we will use it to determine whether there is sufficient evidence to conclude that the means are not all equal. We will then know whether the different treatments have an effect.

Table 12-1 Weights (kg) of Poplar Trees

	Treatment			
	None	Fertilizer	Irrigation	Fertilizer and Irrigation
	0.15	1.34	0.23	2.03
	0.02	0.14	0.04	0.27
	0.16	0.02	0.34	0.92
	0.37	0.08	0.16	1.07
	0.22	0.08	0.05	2.38
n	5	5	5	5
$\bar{x}$	0.184	0.332	0.164	1.334
s	0.127	0.565	0.126	0.859

12-1 Overview

In Section 12-2 we introduce an important method for testing the equality of three or more population means. In Section 9-3 we already presented procedures for testing the hypothesis that *two* population means are equal, but the methods of that section do not apply when three or more means are involved. Instead of referring to the main objective of testing for equal means, the term *analysis of variance* refers to the *method* we use, which is based on an analysis of sample variances.

Definition

Analysis of variance (ANOVA) is a method of testing the equality of three or more population means by analyzing sample variances.

Why Can't We Just Test Two Samples at a Time? Why do we need a new procedure when we can test for equality of two means by using the methods presented in Chapter 9? For example, if we want to use the sample data from Table 12-1 to test the claim that the three populations have the same mean, why not simply pair them off and do two at a time by testing H_0: $\mu_1 = \mu_2$, then H_0: $\mu_2 = \mu_3$, and so on? For the data in Table 12-1, the approach of testing equality of two means at a time requires six different hypothesis tests, so the degree of confidence could be as low as 0.95^6 (or 0.735). In general, as we increase the number of individual tests of significance, we increase the risk of finding a difference by chance alone (instead of a real difference in the means). The risk of a type I error—finding a difference in one of the pairs when no such difference actually exists—is far too high. The method of analysis of variance helps us avoid that particular pitfall (rejecting a true null hypothesis) by using one test for equality of several means.

F Distribution

The ANOVA methods of this chapter require the F distribution, which was first introduced in Section 9-5. In Section 9-5 we noted that the F distribution has the following important properties (see Figure 12-1):

1. The F distribution is not symmetric; it is skewed to the right.
2. The values of F can be 0 or positive, but they cannot be negative.
3. There is a different F distribution for each pair of degrees of freedom for the numerator and denominator.

Critical values of F are given in Table A-5.

Analysis of variance (ANOVA) is based on a comparison of two different estimates of the variance common to the different populations. Those estimates (the *variance between samples* and the *variance within samples*) will be described in Section 12-2. The term *one-way* is used because the sample data are separated into groups according to one characteristic, or factor. In Section 12-3 we will introduce two-way analysis of variance, which allows us to compare populations separated into categories using two characteristics (or factors). For example, we might categorize heights of people using the following two factors: (1) gender (male or female) and (2) right- or left-handedness.

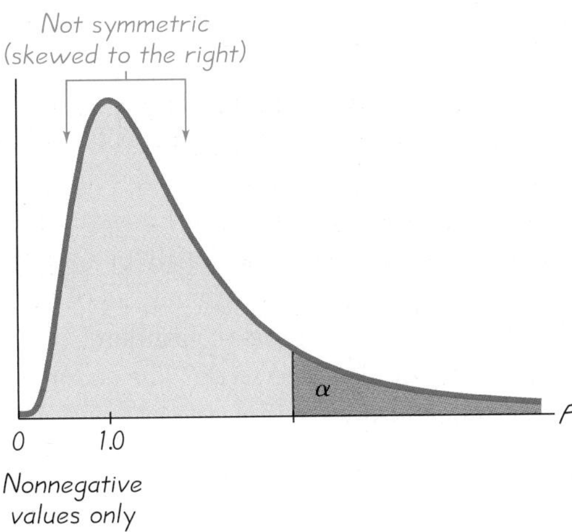

Figure 12-1

***F* Distribution**

There is a different *F* distribution for each different pair of degrees of freedom for numerator and denominator.

Suggested Study Strategy: Because the procedures used in this chapter require complicated calculations, we will emphasize the use and interpretation of computer software, such as STATDISK, Minitab, and Excel, or a TI-83/84 Plus calculator. We suggest that you begin Section 12-2 by focusing on this key concept: We are using a procedure to test a claim that three or more means are equal. Although the details of the calculations are complicated, our procedure will be easy because it is based on a *P*-value. If the *P*-value is small, such as 0.05 or lower, reject equality of means. Otherwise, fail to reject equality of means. After understanding that basic and simple procedure, proceed to understand the underlying rationale.

12-2 One-Way ANOVA

Key Concept This section introduces the method of *one-way analysis of variance,* which is used for tests of hypotheses that three or more population means are all equal, as in H_0: $\mu_1 = \mu_2 = \mu_3$. Because the calculations are very complicated, we emphasize the interpretation of results obtained by using software or a TI-83/84 Plus calculator. We suggest this study strategy:

1. Understand that a small *P*-value (such as 0.05 or less) leads to rejection of the null hypothesis of equal means. With a large *P*-value (such as greater than 0.05), fail to reject the null hypothesis of equal means.
2. Develop an understanding of the underlying rationale by studying the examples in this section.
3. Become acquainted with the nature of the SS (sum of squares) and MS (mean square) values and their role in determining the *F* test statistic, but use statistical software packages or a calculator for finding those values.

The method we use is called **one-way analysis of variance** (or **single-factor analysis of variance**) because we use a single property, or characteristic, for categorizing the populations. This characteristic is sometimes referred to as a *treatment,* or *factor.*

Definition

A **treatment** (or **factor**) is a property, or characteristic, that allows us to distinguish the different populations from one another.

For example, the weights of poplar trees in Table 12-1 are distinguished according to the treatment (none, fertilizer, irrigation, fertilizer and irrigation). The term *treatment* is used because early applications of analysis of variance involved agricultural experiments in which different plots of farmland were treated with different fertilizers, seed types, insecticides, and so on. The accompanying box includes the requirements and the procedure we will use.

Requirements

1. The populations have distributions that are approximately normal. (This is a loose requirement, because the method works well unless a population has a distribution that is very far from normal. If a population does have a distribution that is far from normal, use the Kruskal-Wallis test described in Section 13-5.)
2. The populations have the same variance σ^2 (or standard deviation σ). (This is a loose requirement, because the method works well unless the population variances differ by large amounts. University of Wisconsin statistician George E. P. Box showed that as long as the sample sizes are equal (or nearly equal), the variances can differ by amounts that make the largest up to nine times the smallest and the results of ANOVA will continue to be essentially reliable.)
3. The samples are simple random samples. (That is, samples of the same size have the same probability of being selected.)
4. The samples are independent of each other. (The samples are not matched or paired in any way.)
5. The different samples are from populations that are categorized in only one way. (This is the basis for the name of the method: *one-way* analysis of variance.)

Procedure for Testing H_0: $\mu_1 = \mu_2 = \mu_3 = \ldots$

1. **Use STATDISK, Minitab, Excel, or a TI-83/84 Plus calculator to obtain results.**
2. **Identify the *P*-value from the display.**
3. **Form a conclusion based on these criteria:**
   - **If the *P*-value $\leq \alpha$, reject the null hypothesis of equal means and conclude that at least one of the population means is different from the others.**
   - **If the *P*-value $> \alpha$, fail to reject the null hypothesis of equal means.**

Caution when interpreting results: When we conclude that there is sufficient evidence to reject the claim of equal population means, we cannot conclude from ANOVA that any particular mean is different from the others. (There are several other tests that can be used to identify the specific means that are different, and those procedures are called *multiple comparison procedures,* and they are discussed later in this section.

EXAMPLE Weights of Poplar Trees Given the weights of poplar trees listed in Table 12-1 and a significance level of $\alpha = 0.05$, use STATDISK, Minitab, Excel, or a TI-83/84 Plus calculator to test the claim that the four samples come from populations with means that are not all the same. See the following displays.

STATDISK

Source:	DF:	SS:	MS:	Test Stat, F:	Cirtical F:	P-Value:
Treatment:	3	4.682415	1.560805	5.731353	3.238868	0.0073483
Error:	16	4.35724	0.2723275			
Total:	19	9.039655	0.4757713			

Reject the Null Hypothesis
Reject equality of means

Minitab

One-way ANOVA: None, Fert, Irrig, Fert&Irrig

```
Source  DF     SS     MS     F      P
Factor   3  4.682  1.561  5.73  0.007
Error   16  4.357  0.272
Total   19  9.040

S = 0.5219   R-Sq = 51.80%   R-Sq(adj) = 42.76%
```

Excel

ANOVA						
Source of Variation	*SS*	*df*	*MS*	*F*	*P-value*	*F crit*
Between Groups	4.682415	3	1.560805	**5.731352875**	**0.007348294**	**3.238871522**
Within Groups	4.35724	16	0.2723275			
Total	9.039655	19				

TI-83/84 Plus

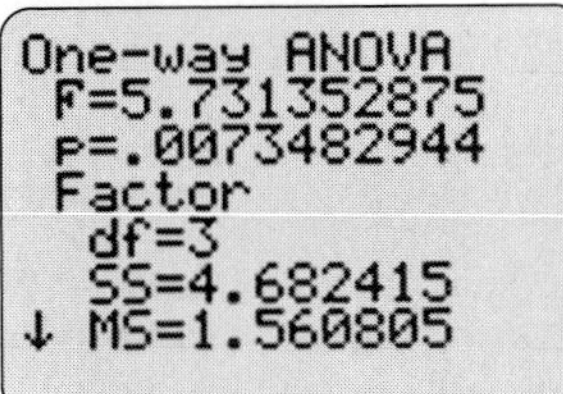


```
One-way ANOVA
↑ MS=1.560805
 Error
  df=16
  SS=4.35724
  MS=.2723275
 Sxp=.521850074
```

SOLUTION

REQUIREMENT ✔ When investigating the normality requirement for the four different populations, the only questionable sample is the second sample of the fertilizer treatment group. For that sample alone, the normal quantile plot and the histogram suggest that normality might not be satisfied, and the problem is the value of 1.34 kg, which appears to be an outlier, so we should be careful here. It would be wise to do the analysis with and without that value included.

continued

See Exercise 5 where we see that in this case, the outlier of 1.34 kg does not have a dramatic effect on the results. The variances are very different, but the largest is no more than nine times the smallest, so the loose requirement of equal variances is satisfied. We are assuming that we have simple random samples. We know that the samples are independent, and each value belongs to exactly one group. The requirements are therefore satisfied and we can proceed with the hypothesis test. ✔

The null hypothesis is H_0: $\mu_1 = \mu_2 = \mu_3 = \mu_4$ and the alternative hypothesis is the claim that at least one of the means is different from the others.

Step 1: Use technology to obtain ANOVA results, such as one of those shown in this example.

Step 2: The displays all show that the P-value is 0.007 when rounded.

Step 3: Because the P-value of 0.007 is less than the significance level of $\alpha = 0.05$, we reject the null hypothesis of equal means.

INTERPRETATION There is sufficient evidence to support the claim that the four population means are not all the same. Based on the sample of weights listed in Table 12-1, we conclude that those weights come from populations having means that are not all the same. On the basis of this ANOVA test, we cannot conclude that any particular mean is different from the others.

Rationale

Assuming that the populations have the same variance σ^2, the F test statistic is the ratio of these two estimates of σ^2: (1) variation *between* samples (based on variation among sample means); and (2) variation *within* samples (based on the sample variances). A significantly *large* F test statistic (located far to the right in the F distribution graph) is evidence against equal population means, so the test is right-tailed. Figure 12-2 shows the relationship between the F test statistic and the P-value.

Test Statistic for One-Way ANOVA

$$F = \frac{\text{variance between samples}}{\text{variance within samples}}$$

The numerator of the test statistic F measures variation between sample means. The estimate of variance in the denominator depends only on the sample variances and is not affected by differences among the sample means. Consequently, sample means that are close in value result in a small F test statistic and we conclude that there are no significant differences among the sample means. But if the value of F is excessively *large,* then we reject the claim of equal means. (The vague terms "small" and "excessively large" are made objective by the corresponding P-value, which tells us whether the F test statistic is or is not in the critical region.) Because excessively large values of F reflect unequal means, the test is right-tailed.

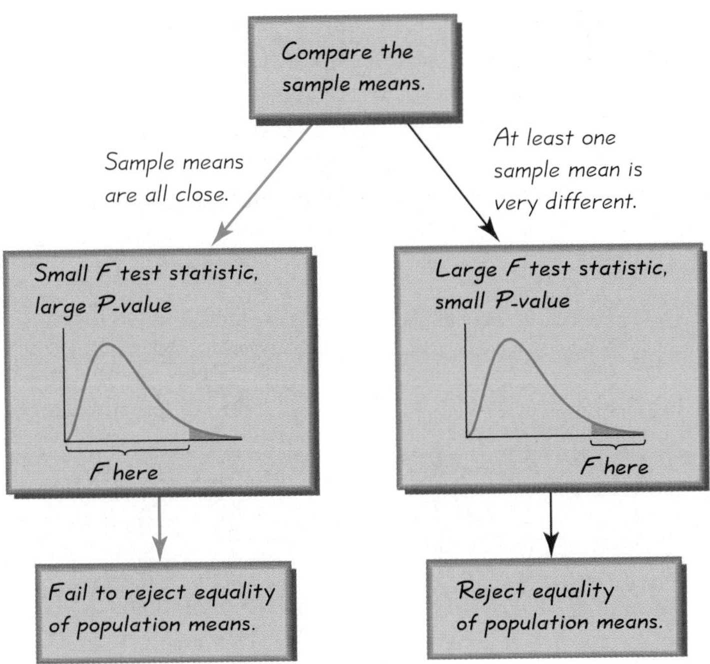

Figure 12-2

Relationship Between the *F* Test Statistic and *P*-Value

Calculations with Equal Sample Sizes *n*

To really understand the effects of one-way analysis of variance, see Table 12-2. Compare Data Set A to Data Set B and observe that the only difference is that we added 10 to each value in Sample 1 from Data Set A to get the values of Sample 1 in Data Set B. If the data sets all have the same sample size (as in $n = 4$ for Table 12-2), the calculations aren't overwhelmingly difficult. First, find the variance *between* samples by evaluating $ns_{\bar{x}}^2$, where $s_{\bar{x}}^2$ is the variance of the sample means and n is the size of each of the samples. That is, consider the sample means to be an ordinary set of values and calculate the variance. (From the central limit theorem, $\sigma_{\bar{x}} = \sigma/\sqrt{n}$ can be solved for σ to get $\sigma = \sqrt{n} \cdot \sigma_{\bar{x}}$, so that we can estimate σ^2 with $ns_{\bar{x}}^2$.) For example, the sample means for Data Set A in Table 12-2 are 5.5, 6.0, and 6.0. Those three values have a variance of $s_{\bar{x}}^2 = 0.0833$, so that

$$\text{variance between samples} = ns_{\bar{x}}^2 = 4(0.0833) = 0.3332$$

Next, estimate the variance *within* samples by calculating s_p^2, which is the pooled variance obtained by finding the mean of the sample variances. The sample variances in Table 12-2 are 3.0, 2.0, and 2.0, so that

$$\begin{aligned}\text{variance within samples} &= s_p^2 \\ &= \frac{3.0 + 2.0 + 2.0}{3} = 2.3333\end{aligned}$$

Finally, evaluate the F test statistic as follows:

$$F = \frac{\text{variance between samples}}{\text{variance within samples}} = \frac{ns_{\bar{x}}^2}{s_p^2} = \frac{0.3332}{2.3333} = 0.1428$$

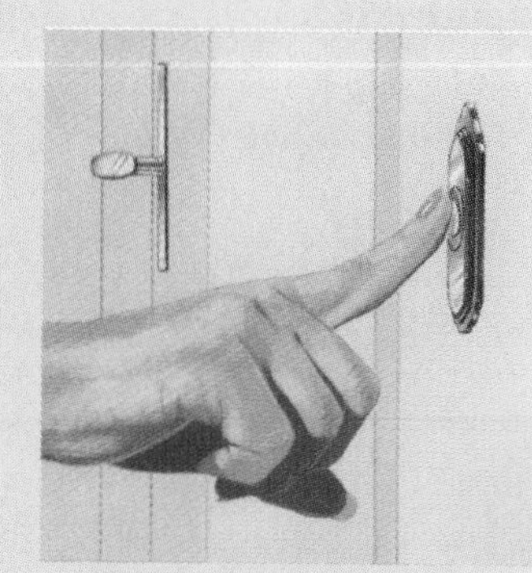

Poll Resistance

Surveys based on relatively small samples can be quite accurate, provided the sample is random or representative of the population. However, increasing survey refusal rates are now making it more difficult to obtain random samples. The Council of American Survey Research Organizations reported that in a recent year, 38% of consumers refused to respond to surveys. The head of one market research company said, "Everyone is fearful of self-selection and worried that generalizations you make are based on cooperators only." Results from the multi-billion-dollar market research industry affect the products we buy, the television shows we watch, and many other facets of our lives.

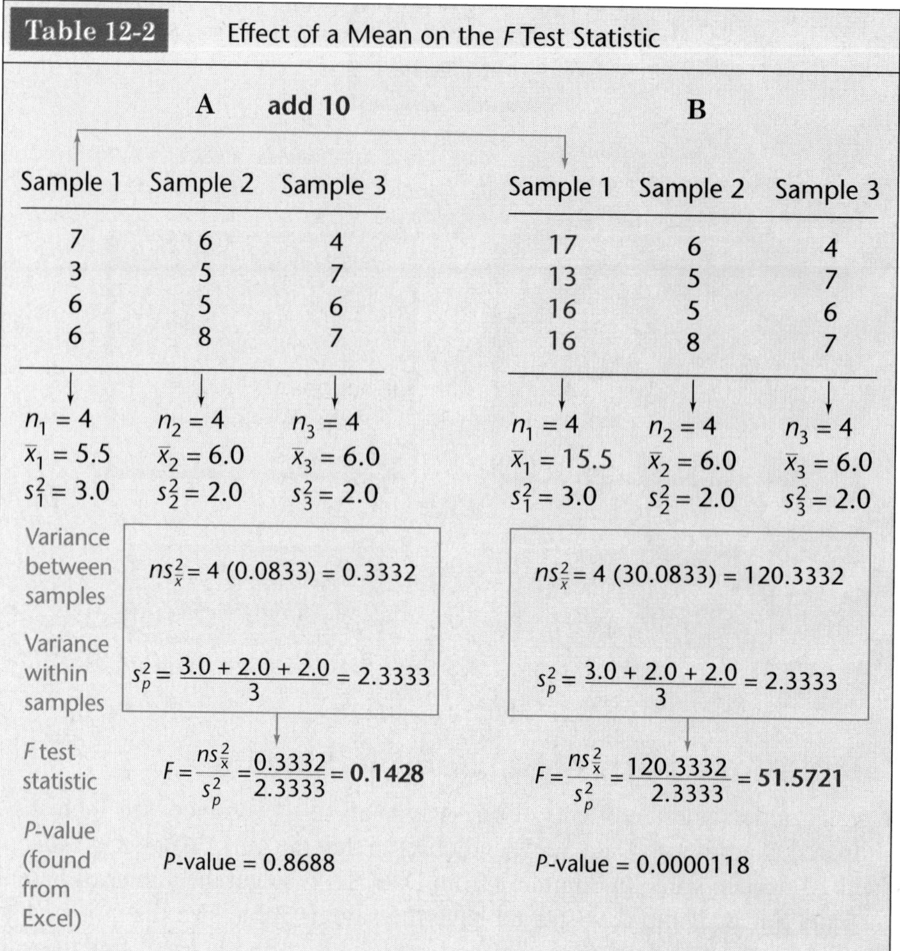

Table 12-2 Effect of a Mean on the F Test Statistic

A (add 10 to Sample 1 values) → B

	A: Sample 1	A: Sample 2	A: Sample 3	B: Sample 1	B: Sample 2	B: Sample 3
	7	6	4	17	6	4
	3	5	7	13	5	7
	6	5	6	16	5	6
	6	8	7	16	8	7
	$n_1 = 4$	$n_2 = 4$	$n_3 = 4$	$n_1 = 4$	$n_2 = 4$	$n_3 = 4$
	$\bar{x}_1 = 5.5$	$\bar{x}_2 = 6.0$	$\bar{x}_3 = 6.0$	$\bar{x}_1 = 15.5$	$\bar{x}_2 = 6.0$	$\bar{x}_3 = 6.0$
	$s_1^2 = 3.0$	$s_2^2 = 2.0$	$s_3^2 = 2.0$	$s_1^2 = 3.0$	$s_2^2 = 2.0$	$s_3^2 = 2.0$
Variance between samples	$ns_{\bar{x}}^2 = 4\,(0.0833) = 0.3332$			$ns_{\bar{x}}^2 = 4\,(30.0833) = 120.3332$		
Variance within samples	$s_p^2 = \frac{3.0 + 2.0 + 2.0}{3} = 2.3333$			$s_p^2 = \frac{3.0 + 2.0 + 2.0}{3} = 2.3333$		
F test statistic	$F = \frac{ns_{\bar{x}}^2}{s_p^2} = \frac{0.3332}{2.3333} = \mathbf{0.1428}$			$F = \frac{ns_{\bar{x}}^2}{s_p^2} = \frac{120.3332}{2.3333} = \mathbf{51.5721}$		
P-value (found from Excel)	P-value = 0.8688			P-value = 0.0000118		

The critical value of F is found by assuming a right-tailed test, because large values of F correspond to significant differences among means. With k samples each having n values, the numbers of degrees of freedom are computed as follows.

Degrees of Freedom:
(k = number of samples and n = sample size)

numerator degrees of freedom $= k - 1$

denominator degrees of freedom $= k\,(n - 1)$

For Data Set A in Table 12-2, $k = 3$ and $n = 4$, so the degrees of freedom are 2 for the numerator and $3(4 - 1) = 9$ for the denominator. With $\alpha = 0.05$, 2 degrees of freedom for the numerator, and 9 degrees of freedom for the denominator, the critical F value from Table A-5 is 4.2565. If we were to use the traditional method of hypothesis testing with Data Set A in Table 12-2, we would see that this right-tailed test has a test statistic of $F = 0.1428$ and a critical value of $F = 4.2565$, so the test statistic is not in the critical region and we therefore fail to reject the null hypothesis of equal means.

To really see how the F test statistic works, consider both collections of sample data in Table 12-2. Note that the three samples in Part A are identical to the three samples in Part B, except that in Part B we have added 10 to each value of Sample 1 from Part A. The three sample means in Part A are very close, but there are substantial differences in Part B. The three sample variances in Part A are identical to those in Part B.

Adding 10 to each data value in the first sample of Table 12-2 has a dramatic effect on the test statistic, with F changing from 0.1428 to 51.5721. Adding 10 to each data value in the first sample also has a dramatic effect on the P-value, which changes from 0.8688 (not significant) to 0.0000118 (significant). Note that the variance between samples in Part A is 0.3332, but for Part B it is 120.3332 (indicating that the sample means in Part B are farther apart). Note also that the variance within samples is 2.3333 in both parts, because the variance within a sample isn't affected when we add a constant to every sample value. *The change in the F test statistic and the P-value is attributable only to the change in* $\bar{x}_1$. This illustrates that the F test statistic is very sensitive to sample *means,* even though it is obtained through two different estimates of the common population *variance.*

Here is the key point of Table 12-2: Data Sets A and B are identical except that in Data Set B, 10 is added to each value of the first sample. Adding 10 to each value of the first sample causes the three sample means to grow farther apart, with the result that the F test statistic increases and the P-value decreases.

Ethics in Reporting

The American Association for Public Opinion Research developed a voluntary code of ethics to be used in news reports of survey results. This code requires that the following be included: (1) identification of the survey sponsor, (2) date the survey was conducted, (3) size of the sample, (4) nature of the population sampled, (5) type of survey used, (6) exact wording of survey questions. Surveys funded by the U.S. government are subject to a prescreening that assesses the risk to those surveyed, the scientific merit of the survey, and the guarantee of the subject's consent to participate.

Calculations with Unequal Sample Sizes

While the calculations required for cases with equal sample sizes are reasonable, they become really complicated when the sample sizes are not all the same. The same basic reasoning applies because we calculate an F test statistic that is the ratio of two different estimates of the common population variance σ^2, but those estimates involve *weighted* measures that take the sample sizes into account, as shown below.

$$F = \frac{\text{variance between samples}}{\text{variance within samples}} = \frac{\left[\dfrac{\Sigma n_i(\bar{x}_i - \bar{\bar{x}})^2}{k - 1}\right]}{\left[\dfrac{\Sigma(n_i - 1)s_i^2}{\Sigma(n_i - 1)}\right]}$$

where
$\bar{\bar{x}}$ = mean of all sample values combined
k = number of population means being compared
n_i = number of values in the ith sample
$\bar{x}_i$ = mean of values in the ith sample
s_i^2 = variance of values in the ith sample

The factor of n_i is included so that larger samples carry more weight. The denominator of the test statistic is simply the mean of the sample variances, but it is a weighted mean with the weights based on the sample sizes.

Because calculating this test statistic can lead to large rounding errors, the various software packages typically use a different (but equivalent) expression that involves SS (for sum of squares) and MS (for mean square) notation.

Although the following notation and components are complicated and involved, the basic idea is the same: The test statistic F is a ratio with a numerator reflecting variation *between* the means of the samples and a denominator reflecting variation *within* the samples. If the populations have equal means, the F ratio tends to be small, but if the population means are not equal, the F ratio tends to be significantly large. Key components in our ANOVA method are described as follows.

SS(total), or total sum of squares, is a measure of the total variation (around $\bar{\bar{x}}$) in all of the sample data combined.

Formula 12-1 $$\text{SS(total)} = \Sigma(x - \bar{\bar{x}})^2$$

SS(total) can be broken down into the components of SS(treatment) and SS(error), described as follows.

SS(treatment), also referred to as SS(factor) or SS(between groups) or SS(between samples), is a measure of the variation *between* the sample means.

Formula 12-2

$$\begin{aligned}\text{SS(treatment)} &= n_1(\bar{x}_1 - \bar{\bar{x}})^2 + n_2(\bar{x}_2 - \bar{\bar{x}})^2 + \cdots + n_k(\bar{x}_k - \bar{\bar{x}})^2 \\ &= \Sigma n_i(\bar{x}_i - \bar{\bar{x}})^2\end{aligned}$$

If the population means $(\mu_1, \mu_2, \ldots, \mu_k)$ are equal, then the sample means $\bar{x}_1, \bar{x}_2, \ldots, \bar{x}_k$ will all tend to be close together and also close to $\bar{\bar{x}}$. The result will be a relatively small value of SS(treatment). If the population means are not all equal, however, then at least one of $\bar{x}_1, \bar{x}_2, \ldots, \bar{x}_k$ will tend to be far apart from the others and also far apart from $\bar{\bar{x}}$. The result will be a relatively large value of SS(treatment).

SS(error), also referred to as SS(within groups) or SS(within samples), is a sum of squares representing the variation that is assumed to be common to all the populations being considered.

Formula 12-3

$$\begin{aligned}\text{SS(error)} &= (n_1 - 1)s_1^2 + (n_2 - 1)s_2^2 + \cdots + (n_k - 1)s_k^2 \\ &= \Sigma(n_i - 1)s_i^2\end{aligned}$$

Given the preceding expressions for SS(total), SS(treatment), and SS(error), the following relationship will always hold.

Formula 12-4 $$\text{SS(total)} = \text{SS(treatment)} + \text{SS(error)}$$

SS(treatment) and SS(error) are both sums of squares, and if we divide each by its corresponding number of degrees of freedom, we get *mean* squares. Some of the following expressions for mean squares include the notation N:

N = total number of values in all samples combined

MS(treatment) is a mean square for treatment, obtained as follows:

Formula 12-5 $$\text{MS(treatment)} = \frac{\text{SS(treatment)}}{k - 1}$$

MS(error) is a mean square for error, obtained as follows:

Formula 12-6 $$\text{MS(error)} = \frac{\text{SS(error)}}{N - k}$$

MS(total) is a mean square for the total variation, obtained as follows:

Formula 12-7 $$\text{MS(total)} = \frac{\text{SS(total)}}{N - 1}$$

Test Statistic for ANOVA with Unequal Sample Sizes

In testing the null hypothesis H_0: $\mu_1 = \mu_2 = \cdots = \mu_k$ against the alternative hypothesis that these means are not all equal, the test statistic

Formula 12-8 $$F = \frac{\text{MS(treatment)}}{\text{MS(error)}}$$

has an F distribution (when the null hypothesis H_0 is true) with degrees of freedom given by

$$\text{numerator degrees of freedom} = k - 1$$
$$\text{denominator degrees of freedom} = N - k$$

This test statistic is essentially the same as the one given earlier, and its interpretation is also the same as described earlier. The denominator depends only on the sample variances that measure variation within the treatments and is not affected by the differences among the sample means. In contrast, the numerator is affected by differences among the sample means. If the differences among the sample means are extreme, they will cause the numerator to be excessively large, so F will also be excessively large. Consequently, very large values of F suggest unequal means, and the ANOVA test is therefore right-tailed.

Tables are a convenient format for summarizing key results in ANOVA calculations, and Table 12-3 has a format often used in computer displays. The entries in Table 12-3 result from the data in Table 12-1.

Designing the Experiment With one-way (or single-factor) analysis of variance, we use one factor as the basis for partitioning the data into different categories. If we conclude that the differences among the means are significant, we can't be

Table 12-3 ANOVA Table for Data in Table 12-1

Source of Variation	Sum of Squares (SS)	Degrees of Freedom	Mean Square (MS)	*F* Test Statistic	*P*-Value
Treatments	4.6824	3	1.5608	5.7314	0.0073
Error	4.3572	16	0.2723		
Total	9.0397	19			

absolutely sure that the differences can be explained by the factor being used. It is possible that the variation of some other unknown factor is responsible. One way to reduce the effect of the extraneous factors is to design the experiment so that it has a **completely randomized design,** in which each element is given the same chance of belonging to the different categories, or treatments. For example, you might assign subjects to two different treatment groups and a placebo group through a process of random selection equivalent to picking slips from a bowl. Another way to reduce the effect of extraneous factors is to use a **rigorously controlled design,** in which elements are carefully chosen so that all other factors have no variability. In general, good results require that the experiment be carefully designed and executed.

Identifying Means That Are Different

After conducting an analysis of variance test, we might conclude that there is sufficient evidence to reject a claim of equal population means, but we cannot conclude from ANOVA that any *particular* mean is different from the others. There are several formal and informal procedures that can be used to identify the specific means that are different. One informal approach is to use the same scale for constructing boxplots of the data sets to see if one or more of the data sets is very different from the others. Another approach is to construct confidence interval estimates of the means from the data sets, then compare those confidence intervals to see if one or more of them does not overlap with the others.

We noted earlier that it is unwise to pair off the samples and conduct individual hypothesis tests using the procedure described in Section 9-3. With four populations, this approach (doing two at a time) requires six different hypothesis tests, so if each test is conducted with a 0.05 significance level, the overall level of confidence for the six tests could be as low as 0.95^6 (or 0.735), so the significance level could be as high as $1 - 0.735 = 0.265$. This high significance level indicates that the risk of a type I error—finding a difference in one of the pairs when no such difference actually exists—is far too high.

There are several procedures for identifying which means differ from the others. Some of the tests, called **range tests,** allow us to identify subsets of means that are not significantly different from each other. Other tests, called **multiple comparison tests,** use pairs of means, but they make adjustments to overcome the problem of having a significance level that increases as the number of individual tests increases. There is no consensus on which test is best, but some of the more common tests are the Duncan test, Student-Newman-Keuls test (or SNK test), Tukey test (or Tukey honestly significant difference test), Scheffé test, Dunnett test, least significant difference test, and the Bonferroni test.

TI

Using the TI-83/84 Plus Calculator for One-Way Analysis of Variance

To use the TI-83/84 Plus calculator for one-way analysis of variance, use the following procedure:

1. First enter the data in lists L1, L2, L3, . . . , or use previously created named lists, such as those available from the APP.
2. Press STAT, select the top menu item of **TESTS,** and proceed to select the menu item of **ANOVA.**
3. The format for the ANOVA command is to enter the list labels or names, all separated by commas. See the examples below.

TI-83/84 Plus Command	Lists of Sample Data
ANOVA(L1, L2, L3)	L1, L2, L3
ANOVA(SMKR, ETS, NOETS)	SMKR, ETS, NOETS

4. Press ENTER and the results will be displayed.
5. The results do not all fit in a single window, so press the arrow key ⌄ to scroll down for more of the results. However, the top of the first window contains the most important results, which are the *F* test statistic and the *P*-value.

12-2 BASIC SKILLS AND CONCEPTS

Statistical Literacy and Critical Thinking

1. ANOVA What is *one-way analysis of variance,* and what is it used for?
2. Treatment In the context of analysis of variance, what is a treatment? In this context, is the meaning the same as in general usage of the word "treatment"?
3. Between/Within What is variance between samples, and what is variance within samples?
4. Comparing Majors A student at the College of Newport administers a test of abstract reasoning to randomly selected English, mathematics, and science majors at her college. She then uses analysis of variance with the sample data and concludes that the means are not all equal. Can she conclude that in the United States, English, mathematics, and science majors have mean abstract reasoning scores that are not all the same? Why or why not?

Exercises 5 and 6 are based on data from Data Set 7 in Appendix B.

5. Weights of Poplar Trees We noted in the Chapter Problem that the weight of 1.34 kg appears to be an outlier. If we delete that value, the STATDISK results are as on the next page.
   a. What is the null hypothesis?
   b. What is the alternative hypothesis?

continued

c. Identify the value of the test statistic.
d. Identify the critical value for a 0.05 significance level.
e. Identify the *P*-value.
f. Based on the preceding results, what do you conclude about equality of the population means?
g. Compare these results to those obtained with the weight of 1.34 kg included. Does the apparent outlier of 1.34 kg have much of an effect on the results? Does the conclusion change?

STATDISK

Source:	DF:	SS:	MS:	Test Stat, F:	Critical F:	P-Value:
Treatment:	3	5.216	1.739	8.448	3.2874	0.0016
Error:	15	3.087	0.206			
Total:	18	8.303	0.461			

TI-83/84 Plus

```
One-way ANOVA
 F=6.141270387
 p=.0055664528
 Factor
  df=3
  SS=3.346455
↓ MS=1.115485
```

```
One-way ANOVA
↑ MS=1.115485
 Error
  df=16
  SS=2.9062
  MS=.1816375
 Sxp=.426189512
```

TI 12

6. Weights of Poplar Trees The Chapter Problem uses the weights of poplar trees for Year 1 and Site 1. (See Data Set 7 in Appendix B.) If we use the weights for Year 2 and Site 1, the analysis of variance results from the TI-83/84 Plus calculator are as shown in the accompanying display. Assume that we want to use a 0.05 significance level in testing the null hypothesis that the four treatments result in weights with the same population mean.
a. What is the null hypothesis?
b. What is the alternative hypothesis?
c. Identify the value of the test statistic.
d. Find the critical value for a 0.05 significance level.
e. Identify the *P*-value.
f. Based on the preceding results, what do you conclude about equality of the population means?

7. Weight Loss from Different Diets In a test of the Atkins, Zone, Weight Watchers, and Ornish weight loss programs, 160 subjects followed the diet programs, with 40 subjects using each diet. They were weighed one year after being on the diet, and the ANOVA results from Excel are given below (based on data from "Comparison of the Atkins, Ornish, Weight Watchers, and Zone Diets for Weight Loss and Heart Disease Risk Reduction," by Dansinger et al., *Journal of the American Medical Association,* Vol. 293, No. 1). Use a 0.05 significance level to test the claim that the mean weight loss is the same for the diets. Given that the mean amounts of weight loss after one year are −2.1 lb, −3.2 lb, −3.0 lb, and −3.3 lb for the four diets, do the diets appear to be effective?

Excel

ANOVA

Source of Variation	*SS*	*df*	*MS*	*F*	*P-value*	*F crit*
Between Groups	35.99984	3	11.99995	0.35206	0.787709	2.662569
Within Groups	5317.256	156	34.08497			
Total	5353.256	159				

8. Fabric Flammability Tests in Different Laboratories Flammability tests were conducted on children's sleepwear. The Vertical Semirestrained Test was used, in which pieces of fabric were burned under controlled conditions. After the burning stopped, the length of the charred portion was measured and recorded. The same fabric samples were tested at five different laboratories. The data are from Minitab, and the analysis of variance results from Minitab are shown below. Using a 0.05 significance level, test the claim that the five laboratories produce the same mean test scores.

Minitab

Source	DF	SS	MS	F	P
Lab	4	2.987	0.747	4.53	0.003
Error	50	8.233	0.165		
Total	54	11.219			

9. Mean Weights of M&Ms Data Set 13 in Appendix B includes weights (in grams) of M&M plain candies categorized according to color (red, orange, yellow, brown, blue, and green). When STATDISK is used for analysis of variance with those data, the result is as shown below. Use a 0.05 significance level to test the claim that the mean weight of M&Ms is the same for each of the six different color populations. If it is the intent of Mars, Inc. to make the candies so that the different color populations have the same mean weight, do these results suggest that the company has a problem requiring corrective action?

STATDISK

Source:	DF:	SS:	MS:	Test Stat, F:	Critical F:	P-Value:
Treatment:	5	0.006	0.001	0.443	2.3113	0.8173
Error:	94	0.259	0.003			
Total:	99	0.266	0.003			

TI-83/84 Plus

```
One-way ANOVA
 F=38.03789731
 p=1.3340195E-6
 Factor
  df=2
  SS=6.91444444
↓ MS=3.45722222
```

```
One-way ANOVA
↑ MS=3.45722222
 Error
  df=15
  SS=1.36333333
  MS=.090888889
 Sxp=.301477841
```

TI **10.** Solar Energy in Different Weather A student of the author lives in a home with a solar electric system. At the same time each day, she collected voltage readings from a meter connected to the system and analysis of variance was used with readings obtained on three different types of day: sunny, cloudy, and rainy. The TI-83/84 Plus calculator results are in the margin. Use a 0.05 significance level to test the claim that the mean voltage reading is the same for the three different types of day. Is there sufficient evidence to support a claim of different population means? We might expect that a solar system would provide more electrical energy on sunny days than on cloudy or rainy days. Can we conclude that sunny days result in greater amounts of electrical energy?

In Exercises 11 and 12, use the listed sample data from car crash experiments conducted by the National Transportation Safety Administration. New cars were purchased and crashed into a fixed barrier at 35 mi/h, and the listed measurements were recorded for the dummy in the driver's seat. The subcompact cars are the Ford Escort, Honda Civic, Hyundai Accent, Nissan Sentra, and Saturn SL4. The compact cars are Chevrolet Cavalier, Dodge Neon, Mazda 626 DX, Pontiac Sunfire, and Subaru Legacy. The midsize cars are Chevrolet Camaro, Dodge Intrepid, Ford Mustang, Honda Accord, and Volvo S70. The full-size cars are Audi A8, Cadillac Deville, Ford Crown Victoria, Oldsmobile Aurora, and Pontiac Bonneville.

TI **11.** Head Injury in a Car Crash The head injury data (in hic) are given below. Use a 0.05 significance level to test the null hypothesis that the different weight categories have the same mean. Do the data suggest that larger cars are safer?

Subcompact:	681	428	917	898	420
Compact:	643	655	442	514	525
Midsize:	469	727	525	454	259
Full-size:	384	656	602	687	360

TI **12.** Chest Deceleration in a Car Crash The chest deceleration data (g) are given below. Use a 0.05 significance level to test the null hypothesis that the different weight categories have the same mean. Do the data suggest that larger cars are safer?

Subcompact:	55	47	59	49	42
Compact:	57	57	46	54	51
Midsize:	45	53	49	51	46
Full-size:	44	45	39	58	44

TI **13.** Exercise and Stress A study was conducted to investigate the effects of exercise on stress. The table below lists systolic blood pressure readings (in mmHg) of subjects from the time preceding 25 minutes of aerobic bicycle exercise and preceding the introduction of stress through arithmetic and speech tests (based on data from "Sympathoadrenergic Mechanisms in Reduced Hemodynamic Stress Responses after Exercise," by Kim Brownley et al., *Medicine and Science in Sports and Exercise,* Vol. 35, No. 6). Use a 0.05 significance level to test the claim that the different groups of subjects have the same mean blood pressure. Based on the results, can those groups be considered to be samples all from the same population?

Female/Black	Male/Black	Female/White	Male/White
117.00	115.67	119.67	124.33
130.67	120.67	106.00	111.00
102.67	133.00	108.33	99.67
93.67	120.33	107.33	128.33
96.33	124.67	117.00	102.00
92.00	118.33	113.33	127.33

TI **14.** Archaeology: Skull Breadths from Different Epochs Samples of head breadths were obtained by measuring skulls of Egyptian males from three different epochs, and the measurements are listed below (based on data from *Ancient Races of the Thebaid,* by Thomson and Randall-Maciver). Changes in head shape over time suggest that interbreeding occurred with immigrant populations. Use a 0.05 significance level to test the claim that the different epochs do not all have the same mean. What do you conclude?

4000 B.C.	1850 B.C.	150 A.D.
131	129	128
138	134	138
125	136	136
129	137	139
132	137	141
135	129	142
132	136	137
134	138	145
138	134	137

In Exercises 15 and 16, use the data set from Appendix B.

TI **15.** Appendix B Data Set: Weights of Pennies Refer to Data Set 14 in Appendix B and use the weights of the Indian head pennies, wheat pennies, pre-1983 pennies, and post-1983 pennies. Use a 0.05 significance level to test the claim that the mean weight of pennies is the same for the four different categories. Based on the results, does it appear that a coin-counting machine can treat the weights of the pennies the same way?

TI **16.** Appendix B Data Set: Home Run Distances Refer to Data Set 17 in Appendix B. Use a 0.05 significance level to test the claim that the home runs hit by Barry Bonds, Mark McGwire, and Sammy Sosa have mean distances that are not all the same. Do the home run distances explain the fact that as of this writing, Barry Bonds has the most home runs in one season, while Mark McGwire has the second highest number of runs? (In recent years, there have been claims that baseball players used steroids to increase their strength, which helped them hit baseballs farther.)

12-2 BEYOND THE BASICS

TI 12 **17.** Equivalent Tests In this exercise you will verify that when you have two sets of sample data, the pooled t test for independent samples and the ANOVA method of this section are equivalent. Refer to the weights of the poplar trees in Table 12-1, but use only the data for the trees given no treatment and those given fertilizer only.

a. Use a 0.05 significance level and the method of Section 9-3 to test the claim that the two samples come from populations with the same mean. (Assume that both populations have the same variance.)

b. Use a 0.05 significance level and the ANOVA method of this section to test the claim made in part (a).

c. Verify that the squares of the t test statistic and critical value from part (a) are equal to the F test statistic and critical value from part (b).

18. Finding Components of ANOVA Exercise 9 is based on the weights of the 100 M&Ms listed in Data Set 13 in Appendix B. If the weights of all of the M&Ms from the full bag are used, SS(treatment) = 0.00644, df(treatment) = 5, SS(error) = 1.47320, and df(error) = 459.

a. How many M&Ms are in the full bag?

b. Find the values of MS(treatment) and MS(error).

c. Find the value of the test statistic F and the P-value.

d. Compare the results to those found in Exercise 9.

12-3 Two-Way ANOVA

Key Concept The analysis of variance procedure introduced in Section 12-2 is referred to as *one*-way analysis of variance (or single-factor analysis of variance) because the data are categorized into groups according to a *single* factor (or treatment). In this section we introduce the method of *two-way analysis of variance,* which is used with data partitioned into categories according to *two* factors. The

Table 12-4 Poplar Tree Weights (kg)

	No Treatment	Fertilizer	Irrigation	Fertilizer and Irrigation
Site 1	0.15	1.34	0.23	2.03
(rich, moist)	0.02	0.14	0.04	0.27
	0.16	0.02	0.34	0.92
	0.37	0.08	0.16	1.07
	0.22	0.08	0.05	2.38
Site 2	0.60	1.16	0.65	0.22
(sandy, dry)	1.11	0.93	0.08	2.13
	0.07	0.30	0.62	2.33
	0.07	0.59	0.01	1.74
	0.44	0.17	0.03	0.12

method of this section requires that we begin by testing for an *interaction* between the two factors. Then we test to determine whether the row factor has an effect and we also test to determine whether the column factor has an effect.

Table 12-4 is an example of data categorized with two factors. One factor is the row variable of site (Site 1 and Site 2), and the second factor is the column variable of treatment (none, fertilizer, irrigation, fertilizer and irrigation). The subcategories in Table 12-4 are often called *cells,* so Table 12-4 has eight cells containing five values each.

In analyzing the sample data in Table 12-4, we have already discussed one-way analysis of variance for a single factor, so it might seem reasonable to simply proceed with one-way ANOVA for the factor of site and another one-way ANOVA for the factor of treatment. Unfortunately, conducting two separate one-way ANOVA tests wastes information and totally ignores a very important feature: the effect of an interaction between the two factors.

Definition

There is an **interaction** between two factors if the effect of one of the factors changes for different categories of the other factor.

As an example of an *interaction* between two factors, consider food pairings. Peanut butter and jelly interact well, but ketchup and ice cream interact in a way that results in a bad taste. Physicians must be careful to avoid prescribing drugs with interactions that produce adverse effects. It was found that the antifungal drug Nizoral (ketoconazole) interacted with the antihistamine drug Seldane (terfenadine) in such a way that Seldane was not metabolized properly, causing abnormal heart rhythms in some patients. Seldane was subsequently removed from the market.

Exploring Data Let's explore the data in Table 12-4 by calculating the mean for each cell and by constructing a graph. The individual cell means are shown in Table 12-5. Those means vary from a low of 0.164 kg to a high of 1.334 kg, so they appear to vary considerably. Figure 12-3 shows graphs of those means, and that figure shows that the Site 2 means appear to be greater than the Site 1 means for three of the four treatment categories. Because the Site 2 line segments appear to be approximately *parallel* to the corresponding Site 1 line segments, it appears that the site weights behave the same for the different treatment categories, so there does not appear to be an interaction effect. In general, if a graph such as Figure 12-3 results in line segments that are approximately *parallel,* we have evidence that there is *not an interaction* between the row and column variables. If the Site 2 line segments were far from parallel to the Site 1 line segments, we would have evidence of an interaction between site and treatment. These observations based on Table 12-5 and Figure 12-3 are largely subjective, so we will proceed with the more objective method of two-way analysis of variance.

Polls and Psychologists

Poll results can be dramatically affected by the wording of questions. A phrase such as "over the last few years" is interpreted differently by different people. Over the last few years (actually, since 1980), survey researchers and psychologists have been working together to improve surveys by decreasing bias and increasing accuracy. In one case, psychologists studied the finding that 10 to 15 percent of those surveyed say they voted in the last election when they did not. They experimented with theories of faulty memory, a desire to be viewed as responsible, and a tendency of those who usually vote to say that they voted in the most recent election, even if they did not. Only the last theory was actually found to be part of the problem.

Table 12-5 Means (kg) of Cells from Table 12-4

	Treatment			
	None	Fertilizer	Irrigation	Fertilizer and Irrigation
Site 1 (rich, moist)	0.184	0.332	0.164	1.334
Site 2 (sandy, dry)	0.458	0.630	0.278	1.308

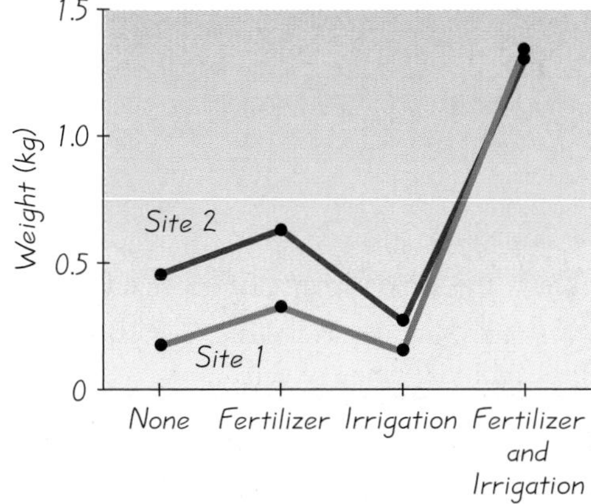

Figure 12-3 Interaction Graph of Means (kg) of Cells in Table 12-4

In using two-way ANOVA for the data of Table 12-4, we consider three possible effects on the poplar tree weights: (1) the effects of an interaction between site and treatment; (2) the effects of site; (3) the effects of treatment. The calculations are quite involved, so *we will assume that a software package or TI-83/84 Plus calculator is being used.* (Procedures for using the TI-83/84 Plus calculator are described at the end of this section.) The Minitab and TI-83/84 Plus calculator displays for the data in Table 12-4 are shown here.

Minitab

Source	DF	SS	MS	F	P
Site	1	0.2722	0.27225	**0.81**	**0.374**
Treatment	3	7.5470	2.51567	**7.50**	**0.001**
Interaction	3	0.1716	0.05721	**0.17**	**0.915**
Error	32	10.7267	0.33521		
Total	39	18.7176			

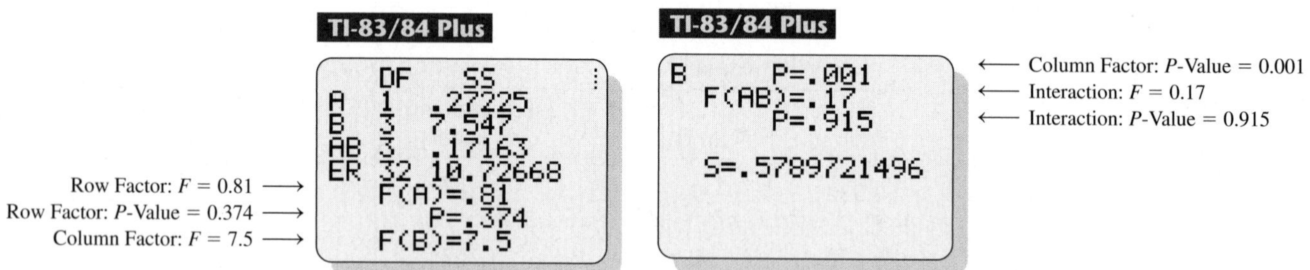

Here are the requirements and basic procedure for two-way analysis of variance (ANOVA). The procedure is also summarized in Figure 12-4.

Requirements

1. For each cell, the sample values come from a population with a distribution that is approximately normal.
2. The populations have the same variance σ^2 (or standard deviation σ).
3. The samples are simple random samples. (That is, samples of the same size have the same probability of being selected.)
4. The samples are independent of each other. (The samples are not matched or paired in any way.)
5. The sample values are categorized two ways. (This is the basis for the name of the method: *two-way* analysis of variance.)
6. All of the cells have the same number of sample values. (This is called a *balanced* design.)

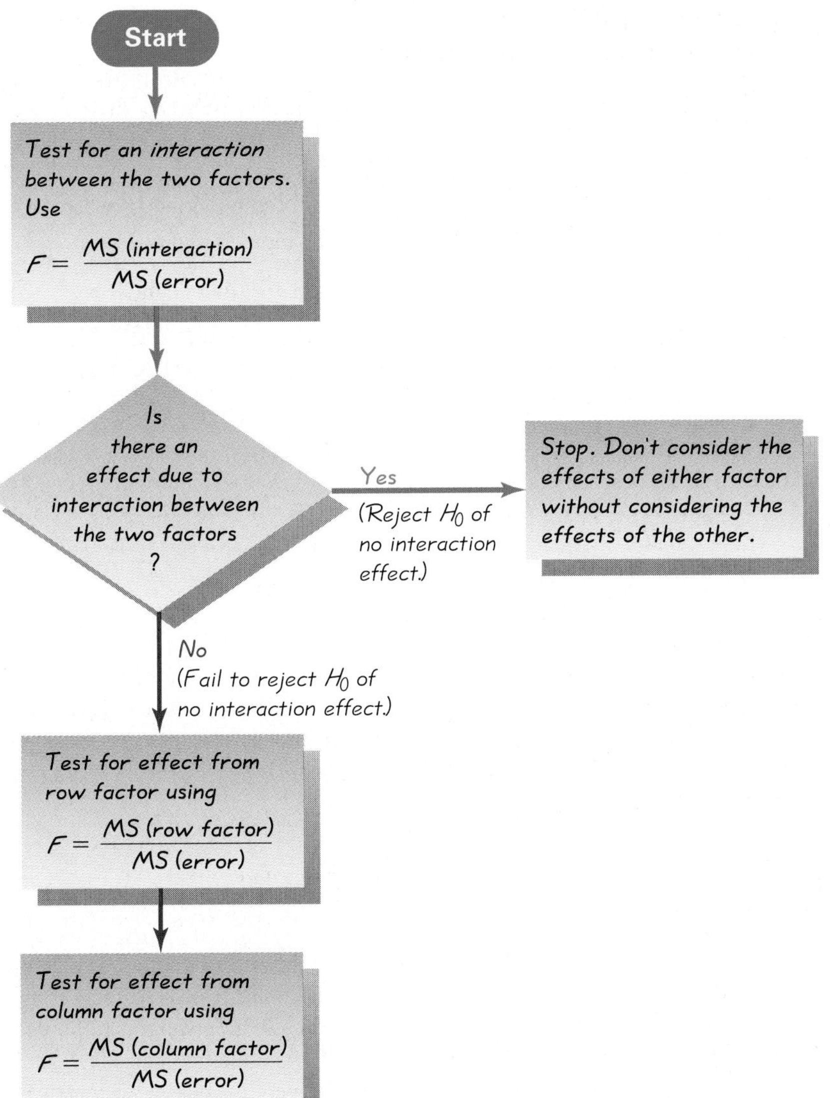

Figure 12-4

Procedure for Two-Way ANOVAS

Procedure for Two-Way ANOVAS (See Figure 12-4.)

Step 1: *Interaction Effect:* In two-way analysis of variance, begin by testing the null hypothesis that there is no interaction between the two factors. Using Minitab for the data in Table 12-4, we get the following test statistic:

$$F = \frac{\text{MS(interaction)}}{\text{MS(error)}} = \frac{0.05721}{0.33521} = 0.17$$

Interpretation: The corresponding P-value is shown in the Minitab and TI-83/84 Plus displays as 0.915, so we fail to reject the null hypothesis of no interaction between the two factors. It does not appear that the poplar tree weights are affected by an interaction between site and treatment.

continued

Step 2: *Row/Column Effects:* If we do reject the null hypothesis of no interaction between factors, then we should stop now; we should not proceed with the two additional tests. (If there is an interaction between factors, we shouldn't consider the effects of either factor without considering those of the other.)

If we fail to reject the null hypothesis of no interaction between factors, then we should proceed to test the following two hypotheses:

H_0: There are no effects from the row factor (that is, the row means are equal).

H_0: There are no effects from the column factor (that is, the column means are equal).

In Step 1, we failed to reject the null hypothesis of no interaction between factors, so we proceed with the next two hypotheses tests identified in Step 2.

For the row factor of site we get

$$F = \frac{\text{MS(site)}}{\text{MS(error)}} = \frac{0.27225}{0.33521} = 0.81$$

Interpretation: This value is not significant because the corresponding *P*-value is shown in the Minitab and TI-83/84 Plus displays as 0.374. We fail to reject the null hypothesis of no effects from site. That is, the site does not appear to have an effect on poplar tree weight.

For the column factor of treatment we get

$$F = \frac{\text{MS(treatment)}}{\text{MS(error)}} = \frac{2.51567}{0.33521} = 7.50$$

Interpretation: This value is significant because the corresponding *P*-value is shown as 0.001. We therefore reject the null hypothesis of no effects from treatment. The treatment does appear to have an effect on weights of poplar trees. Based on the sample data in Table 12-4, we conclude that the different treatments do appear to have unequal mean weights, but the weights appear to have equal means for both sites. (See Figure 12-3 and note that the line segments appear to change dramatically for the different treatment categories, but the Site 1 and Site 2 line segments are not dramatically different from each other.)

Special Case: One Observation per Cell and No Interaction Table 12-4 contains 5 observations per cell. If our sample data consist of only one observation per cell, we lose MS(interaction), SS(interaction), and df(interaction) because those values are based on sample variances computed for each individual cell. If there is only one observation per cell, there is no variation within individual cells and those sample variances cannot be calculated. Here's how we proceed when there is one observation per cell: *If it seems reasonable to assume (based on knowledge about the circumstances) that there is no interaction between the two*

Table 12-6 Weights (kg) of Poplar Trees in Year 1

	Treatment			
	None	Fertilizer	Irrigation	Fertilizer and Irrigation
Site 1 (rich, moist)	0.15	1.34	0.23	2.03
Site 2 (sandy, dry)	0.60	1.16	0.65	0.22

factors, make that assumption and then proceed as before to test the following two hypotheses separately:

H_0: There are no effects from the row factor.

H_0: There are no effects from the column factor.

As an example, if we know only the first value in each cell of Table 12-4, we have the data shown in Table 12-6. In Table 12-6, the two row means are 0.9375 and 0.6575. Is that difference significant, suggesting that there is an effect due to site? In Table 12-6, the four column means are 0.3750, 1.2500, 0.4400, and 1.1250. Are those differences significant, suggesting that there is an effect due to treatment? Assuming that weights are not affected by some interaction between site and treatment, the Minitab and TI-83/84 Plus calculator displays are as shown below. (If we believe there is an interaction, the method described here does not apply.)

Minitab

Source	DF	SS	MS	F	P
Site	1	0.15680	0.156800	**0.28**	**0.634**
Treatment	3	1.23665	0.412217	**0.73**	**0.598**
Error	3	1.68690	0.562300		
Total	7	3.08035			

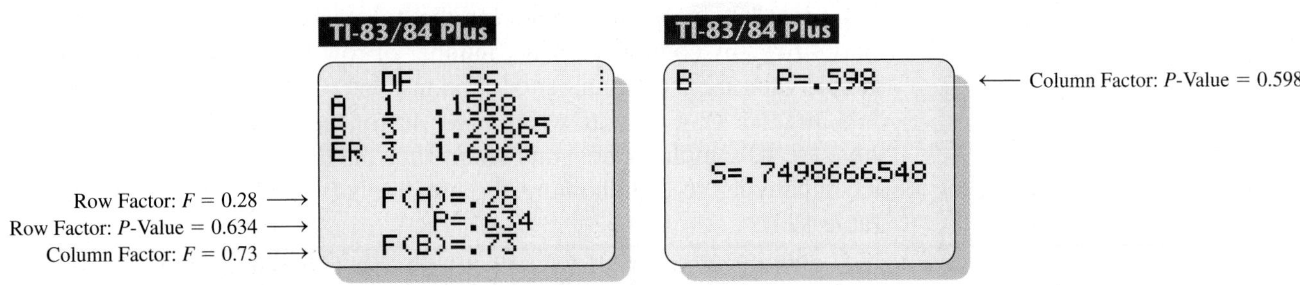

Row factor: We first use the results from the Minitab display to test the null hypothesis of no effects from the row factor of site.

$$F = \frac{\text{MS(site)}}{\text{MS(error)}} = \frac{0.156800}{0.562300} = 0.28$$

This test statistic is not significant, because the corresponding P-value in the Minitab and TI-83/84 Plus displays is 0.634. We fail to reject the null hypothesis; it appears that the poplar tree weights are not affected by the site.

Column factor: We now use the Minitab display to test the null hypothesis of no effect from the column factor of treatment category. The test statistic is

$$F = \frac{\text{MS(treatment)}}{\text{MS(error)}} = \frac{0.412217}{0.562300} = 0.73$$

This test statistic is not significant because the corresponding P-value is given in the Minitab and TI-83/84 Plus displays as 0.598. We fail to reject the null hypothesis, so it appears that the poplar tree weight is not affected by the treatment. Using the data in Table 12-6, we conclude that the poplar tree weights do not appear to be affected by either site or treatment, but when we used 5 values from each cell (Table 12-4), we concluded that the weights appear to be affected by treatment. Such is the power of larger samples.

In this section we have briefly discussed an important branch of statistics. We have emphasized the interpretation of computer displays while omitting the manual calculations and formulas, which are quite formidable.

TI Using the TI-83/84 Plus Calculator for Two-Way Analysis of Variance

This procedure requires that we first create a matrix [D] with the sample data, and we must use the program A1ANOVA that is on the CD-ROM included with this book. On the CD-ROM, select the **Software** folder, then select **TI-83 TI-84 Plus Programs.** After the program A1ANOVA has been downloaded to your calculator, use the following procedure.

TI-83/84 Plus

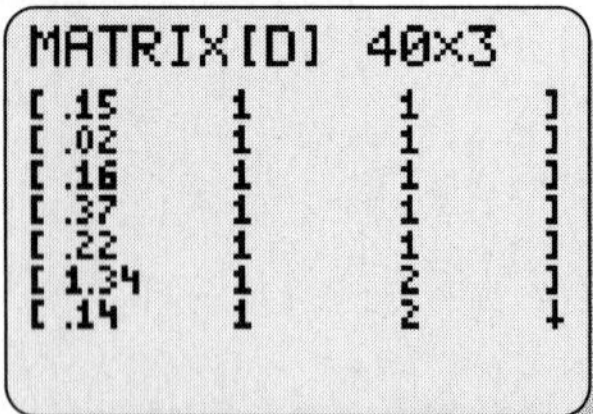

1. *Manual entry of sample data:* Enter the sample data as matrix D, with the first column containing the sample, the second column containing the corresponding row numbers, and the third column containing the corresponding column numbers. To manually enter the sample data in matrix D, press **2ND** and the **x^{-1}** key to get the Matrix menu. Select **EDIT,** scroll down for **[D],** then press **ENTER**. Begin by entering the dimension of the matrix in the format of rows by columns. For the number of rows, enter the total number of sample values. For the number of columns, enter 3. For example, to enter the data in Table 12-4 as matrix [D], enter 40 for the number of rows (because there are 40 sample values) and enter 3 for the number of columns. See the accompanying screen showing the creation of matrix [D] for the data in Table 12-4.

   Using existing lists: If the data are already stored as an individual list, it can be entered into matrix [D]. Store the list in L1, then create a list L2 consisting of the corresponding row numbers, and also create a column L3 consisting of the corresponding column numbers. To combine lists L1, L2, L3 into matrix [D], press **2ND** and the **x^{-1}** key to get the Matrix menu. Select the top menu item of **MATH,** then select the 9th item of **List→matr.** Now enter L1, L2, L3, and end with the matrix name [D], all separated by commas. Enter the matrix name of [D] by opening the MATRIX menu, scrolling to [D], and pressing enter. Here is the command that puts lists L1, L2, L3 into matrix [D]: **List→matr(L1, L2, L3, [D]).**

2. Now press **PRGM,** select **A1ANOVA** and press **ENTER** twice, then select **RAN BLOCK DESI** (randomized block design) and press **ENTER.**
3. A message will be displayed beginning with "DATA IN N*3 MAT [D]." Press ENTER.
4. Select the option of **CONTINUE** and press ENTER.
5. The program will work with the data that were previously stored in matrix [D], and the results will be displayed. The results will not all fit in a single window, so press ENTER to see the remaining results.

Interpreting a TI-83/84 Plus Calculator Display The TI-83/84 Plus calculator display is not in a format that is easy to read, so focus on the following key components.

Interaction:	Look for **F(AB)=.** The *F* test statistic is the value shown, and the *P*-value for the interaction is on the line below this test statistic.
Effect from row factor:	Look for **F(A)=.** The value shown is the *F* test statistic for the effect from the row factor, and the *P*-value is on the line below this test statistic.
Effect from column factor:	Look for **F(B)=.** The value shown is the *F* test statistic for the effect from the column factor, and the *P*-value is on the line below this test statistic.

12-3 BASIC SKILLS AND CONCEPTS

Statistical Literacy and Critical Thinking

1. Two-Way ANOVA What is *two-way analysis of variance*, and what is it used for?

2. Interaction What is an interaction, and what role does it play in two-way analysis of variance?

3. Interaction When conducting two-way analysis of variance, if there is an interaction between the two factors, why should we not continue with separate tests for effects from the row and column factors?

4. Interaction Shown on the next page is a Minitab-generated interaction plot similar in nature to Figure 12-3. [Like Figure 12-3, this interaction plot is based on weights of trees, and the factors are site (Site 1 and Site 2) and treatments (none, fertilizer, irrigation, fertilizer and irrigation).] What does this graph suggest about the interaction between the two factors?

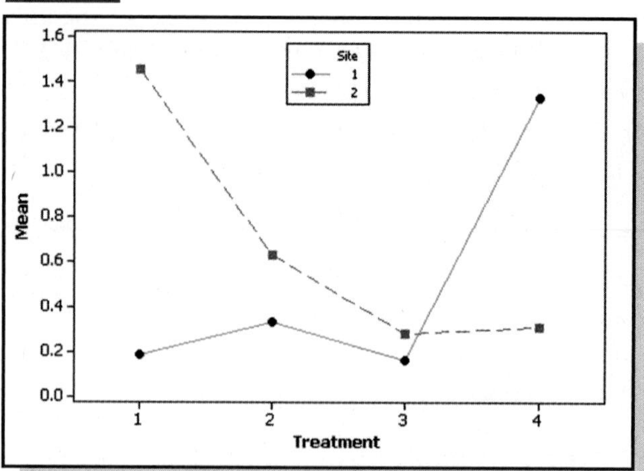

TI-83/84 Plus

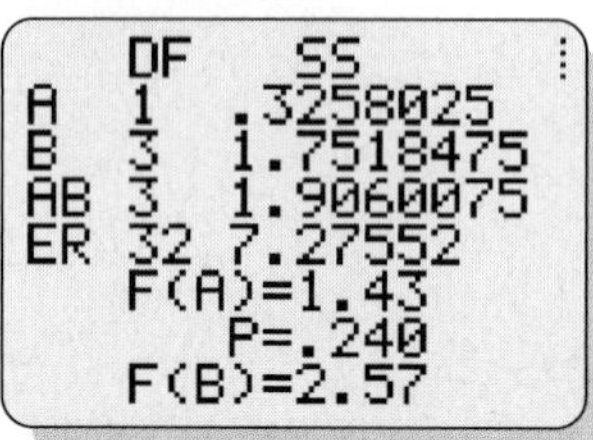

TI-83/84 Plus

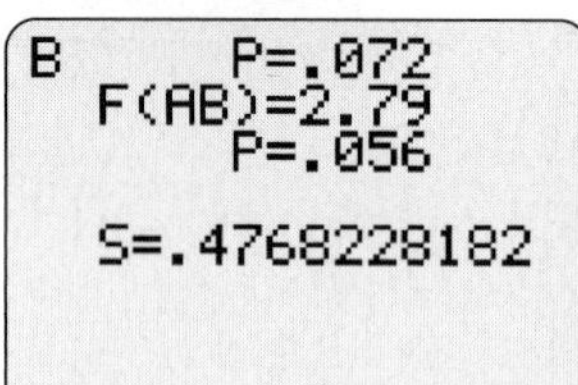

Interpreting a Computer Display: Weights of Poplar Trees. Exercises 5–7 require the given TI-83/84 Plus calculator displays, which result from the weights of poplar trees listed for the second year in Data Set 7 in Appendix B. (Table 12-4 in this section includes the weights for the first *year, but the displays given here are based on the weights from the second year. The row factor is site (Site 1 and Site 2) and the column factor is treatment (none, fertilizer, irrigation, fertilizer and irrigation).*

TI **5.** Interaction Effect Refer to the TI-83/84 Plus displays and test the null hypothesis that the weights of poplar trees are not affected by an interaction between site and treatment. What do you conclude?

TI **6.** Effect of Site Refer to the TI-83/84 Plus displays and assume that the weights of poplar trees are not affected by an interaction between site and treatment. Is there sufficient evidence to support the claim that site has an effect on weight?

TI **7.** Effect of Treatment Refer to the TI-83/84 Plus displays and assume that the weights of poplar trees are not affected by an interaction between site and treatment. Is there sufficient evidence to support the claim that treatment has an effect on weight?

TI-83/84 Plus

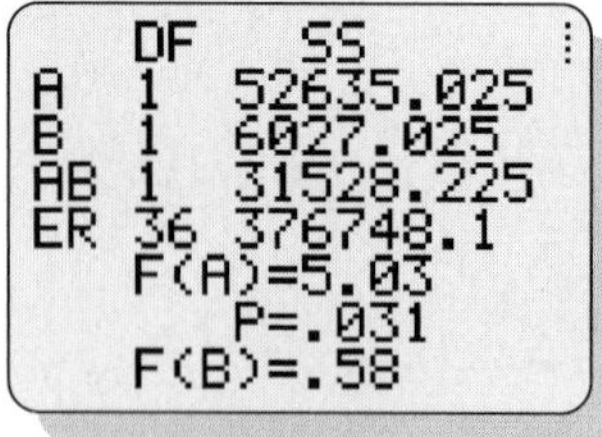

TI-83/84 Plus

B P=.453
F(AB)=3.01
P=.091
S=102.2996823

Interpreting a Computer Display. *In Exercises 8–10, use the TI-83/84 Plus displays, which result from the scores listed in the accompanying table. The sample data are SAT scores on the verbal and math portions of SAT-I and are based on reported statistics from the College Board.*

Verbal

Female	646	539	348	623	478	429	298	782	626	533
Male	562	525	512	576	570	480	571	555	519	596

Math

Female	484	489	436	396	545	504	574	352	365	350
Male	547	678	464	651	645	673	624	624	328	548

TI **8.** Interaction Effect Test the null hypothesis that SAT scores are not affected by an interaction between gender and test (verbal/math). What do you conclude?

TI **9.** Effect of Gender Assume that SAT scores are not affected by an interaction between gender and the type of test (verbal/math). Is there sufficient evidence to support the claim that gender has an effect on SAT scores?

TI **10.** Effect of Type of SAT Test Assume that SAT scores are not affected by an interaction between gender and the type of test (verbal/math). Is there sufficient evidence to support the claim that the type of test (verbal/math) has an effect on SAT scores?

Interpreting a Computer Display: One Observation Per Cell. *In Exercises 11 and 12, refer to the given TI-83/84 Plus calculator displays. These displays result from a study in which 24 subjects were given hearing tests using four different lists of words. The 24 subjects had normal hearing and the tests were conducted with no background noise. The main objective was to determine whether the four lists are equally difficult to understand. In the original table of hearing test scores, each cell has one entry. The original data are from* A Study of the Interlist Equivalency of the CID W-22 Word List Presented in Quiet and in Noise, *by Faith Loven, University of Iowa. The original data are available on the Internet through DASL (Data and Story Library).*

TI-83/84 Plus

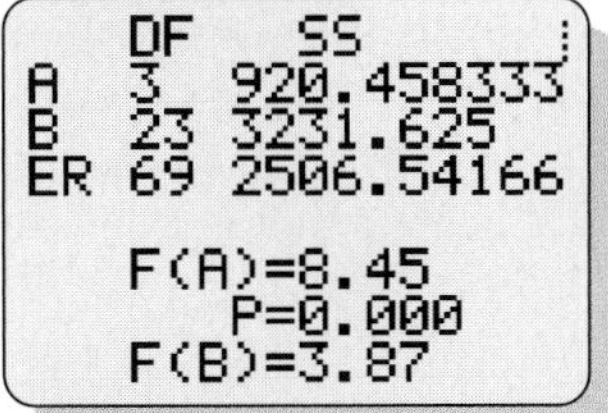

TI-83/84 Plus

TI **11.** Hearing Tests: Effect of Subject Assuming that there is no effect on hearing test scores from an interaction between subject and list, is there sufficient evidence to support the claim that the choice of subject has an effect on the hearing test score? Interpret the result by explaining why it makes practical sense.

TI **12.** Hearing Tests: Effect of Word List Assuming that there is no effect on hearing test scores from an interaction between subject and list, is there sufficient evidence to support the claim that the choice of word list has an effect on the hearing test score?

In Exercises 13 and 14, use software or a TI-83/84 Plus calculator to obtain the results from two-way analysis of variance.

TI **13.** Pulse Rates The following table lists pulse rates from Data Set 1 in Appendix B. Are pulse rates affected by an interaction between gender and age? Are pulse rates affected by gender? Are pulse rates affected by age?

	Age		
	Under 20	20–40	Over 40
Male	96 64 68 60	64 88 72 64	68 72 60 88
Female	76 64 76 68	72 88 72 68	60 68 72 64

TI **14.** Marathon Times Listed below are New York Marathon running times (in seconds) for randomly selected runners who completed the marathon. Are the running times affected by an interaction between gender and age bracket? Are running times affected by gender? Are running times affected by age bracket?

Times (in seconds) for New York Marathon Runners

	Age		
	21–29	30–39	40 and over
Male	13,615	14,677	14,528
	18,784	16,090	17,034
	14,256	14,086	14,935
	10,905	16,461	14,996
	12,077	20,808	22,146
Female	16,401	15,357	17,260
	14,216	16,771	25,399
	15,402	15,036	18,647
	15,326	16,297	15,077
	12,047	17,636	25,898

12-3 BEYOND THE BASICS

15. Transformations of Data Assume that two-way ANOVA is used to analyze sample data consisting of more than one entry per cell. How are the ANOVA results affected in each of the following cases?

a. The same constant is added to each sample value.

b. Each sample value is multiplied by the same nonzero constant.

c. The format of the table is transposed, so that the row and column factors are interchanged.

d. The first sample value in the first cell is changed so that it becomes an outlier.

Review

Section 9-3 introduced a procedure for testing equality between *two* population means, but Section 12-2 introduced one-way analysis of variance as a method for testing equality of three or more population means. This method requires (1) normally distributed populations, (2) populations with the same standard deviation (or variance), and (3) simple random samples that are independent of each other. The methods of one-way analysis of variance are used when we have three or more samples taken from populations that are characterized according to a single factor. The following are key features of one-way analysis of variance:

- The F test statistic is based on the ratio of two different estimates of the common population variance σ^2, as shown below:

$$F = \frac{\text{variance between samples}}{\text{variance within samples}} = \frac{MS(treatment)}{MS(error)}$$

- Critical values of F can be found in Table A-5, but we focused on the interpretation of P-values that are included as part of a computer display.

In Section 12-3 we considered two-way analysis of variance, with data categorized according to two different factors. One factor is used to arrange the sample data in different rows, while the other factor is used for different columns. The procedure for two-way analysis of variance is summarized in Figure 12-4, and it requires that we first test for an interaction between the two factors. If there is no significant interaction, we then proceed to conduct individual tests for effects from each of the two factors. We also considered the use of two-way analysis of variance for the special case in which there is only one observation per cell.

Because of the nature of the calculations required throughout this chapter, we emphasized the interpretation of computer displays.

Statistical Literacy and Critical Thinking

1. Analysis of Variance Why is the method used to test equality of three or more population means referred to as analysis of "variance," when means are the parameters of interest?

2. Which Test? A medical researcher conducts a clinical trial of a drug intended to reduce LDL cholesterol. Fifty subjects are given 10 mg of the drug, 50 other subjects are given 20 mg of the drug, and 50 other subjects are given a placebo. What method should be used to test for equality of the three mean amounts of cholesterol reductions?

3. Which Test? Assume that the clinical trial in Exercise 2 is modified to exclude the treatment consisting of 20 mg of the drug. What method should be used to test for equality of the mean amounts of cholesterol reductions in the placebo group and the 10-mg treatment group?

4. Two-Way ANOVA A geneticist collects data consisting of eye colors, and the colors are categorized by gender (male, female) and age bracket (under 30, 30 and older). Can the methods of this chapter be used to test for effects of gender and age bracket on eye color? Why or why not?

Review Exercises

TI **1.** Interpreting Computer Display: Drinking and Driving The Associated Insurance Institute sponsors studies of the effects of drinking on driving. In one such study, three groups of adult men were randomly selected for an experiment designed to measure their blood alcohol levels after consuming five drinks. Members of group A were tested after 1 hour, members of group B were tested after 2 hours, and members of group C were tested after 4 hours. The results are given in the accompanying table; the TI-83/84 Plus display for these data is also shown. At the 0.05 significance level, test the claim that the three groups have the same mean level.

TI-83/84 Plus

One-way ANOVA
F=46.9
p=6.2311029E-7
Factor
df=2
SS=.007657143
↓ MS=.003828571

A	B	C
0.11	0.08	0.04
0.10	0.09	0.04
0.09	0.07	0.05
0.09	0.07	0.05
0.10	0.06	0.06
		0.04
		0.05

TI **2.** Readability Scores Listed below are Flesch Reading Ease scores from each of 12 randomly selected pages in Tom Clancy's *The Bear and the Dragon,* J. K. Rowling's *Harry Potter and the Sorcerer's Stone*, and Leo Tolstoy's *War and Peace*. Use a 0.05 significance level to test the claim that the three books have the same mean Flesch Reading Ease score. Based on the results, do the three books appear to have the same reading level?

Clancy:	58.2	73.4	73.1	64.4	72.7	89.2	43.9	76.3	76.4	78.9	69.4	72.9
Rowling:	85.3	84.3	79.5	82.5	80.2	84.6	79.2	70.9	78.6	86.2	74.0	83.7
Tolstoy:	69.4	64.2	71.4	71.6	68.5	51.9	72.2	74.4	52.8	58.4	65.4	73.6

Interpreting a Computer Display. In Exercises 3–5, use the TI-83/84 Plus displays, which result from the values listed in the accompanying table. The sample data are student estimates (in feet) of the length of their classroom. The actual length of the classroom is 24 ft 7.5 in.

TI-83/84 Plus

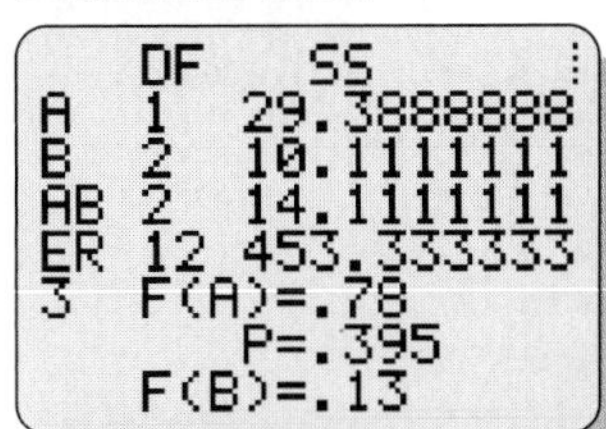

TI-83/84 Plus

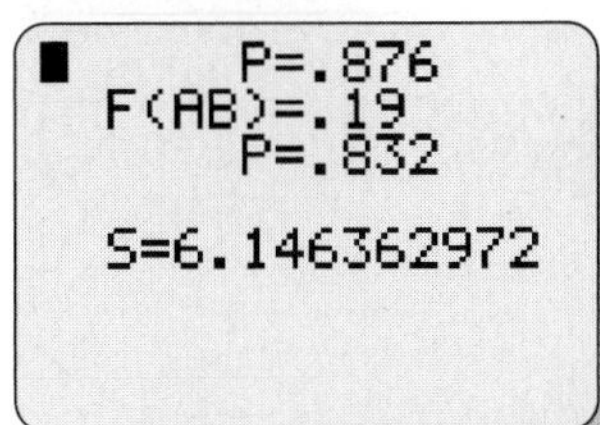

	Major		
	Math	Business	Liberal Arts
Female	28 25 30	35 25 20	40 21 30
Male	25 30 20	30 24 25	25 20 32

TI **3.** Interaction Effect Test the null hypothesis that the estimated lengths are not affected by an interaction between gender and major.

TI **4.** Effect of Gender Assume that estimated lengths are not affected by an interaction between gender and major. Is there sufficient evidence to support the claim that estimated length is affected by gender?

TI **5.** Effect of Major Assume that estimated lengths are not affected by an interaction between gender and major. Is there sufficient evidence to support the claim that estimated length is affected by major?

TI **6.** Smoking, Body Temperature, Gender The table below lists body temperatures obtained from randomly selected subjects (based on Data Set 2 in Appendix B). The temperatures are categorized according to gender and whether the subject smokes. Using a 0.05 significance level, test for an interaction between gender and smoking, test for an effect from gender, and test for an effect from smoking. What do you conclude?

	Smokes				Does Not Smoke			
Male	98.4	98.4	99.4	98.6	98.0	98.0	98.8	97.0
Female	98.8	98.0	98.7	98.4	97.7	98.0	98.2	99.1

TI **7.** Auto Pollution The accompanying table lists the amounts of greenhouse gases emitted by different cars in one year.

a. Assuming that there is no interaction effect, is there sufficient evidence to support the claim that amounts of emitted greenhouse gases are affected by the type of transmission (automatic/manual)?

b. Assuming that there is no interaction effect, is there sufficient evidence to support the claim that amounts of emitted greenhouse gases are affected by the number of cylinders?

c. Based on the results from parts (a) and (b), can we conclude that greenhouse gas emissions are not affected by the type of transmission or the number of cylinders? Why or why not?

Emission of Greenhouse Gases (tons/year)

	4 Cylinders	6 Cylinders	8 Cylinders
Automatic	10	12	14
Manual	10	12	12

TI **8.** Longevity Refer to the accompanying table that lists the numbers of years that U.S. presidents and popes and British monarchs (since 1690) lived after their inauguration, election, or coronation. Determine whether the survival times for the three groups differ. (Table based on data from *Computer-Interactive Data Analysis,* by Lunn and McNeil, John Wiley & Sons.)

Presidents		Popes		Kings and Queens	
Washington	10	Alex VIII	2	James II	17
J. Adams	29	Innoc XII	9	Mary II	6
Jefferson	26	Clem XI	21	William III	13
Madison	28	Innoc XIII	3	Anne	12
Monroe	15	Ben XIII	6	George I	13
J. Q. Adams	23	Clem XII	10	George II	33

continued

Presidents		Popes		Kings and Queens	
Jackson	17	Ben XIV	18	George III	59
Van Buren	25	Clem XIII	11	George IV	10
Harrison	0	Clem XIV	6	William IV	7
Tyler	20	Pius VI	25	Victoria	63
Polk	4	Pius VII	23	Edward VII	9
Taylor	1	Leo XII	6	George V	25
Fillmore	24	Pius VIII	2	Edward VIII	36
Pierce	16	Greg XVI	15	George VI	15
Buchanan	12	Pius IX	32		
Lincoln	4	Leo XIII	25		
A. Johnson	10	Pius X	11		
Grant	17	Ben XV	8		
Hayes	16	Pius XI	17		
Garfield	0	Pius XII	19		
Arthur	7	John XXIII	5		
Cleveland	24	Paul VI	15		
Harrison	12	John Paul I	0		
McKinley	4	John Paul II	26		
T. Roosevelt	18				
Taft	21				
Wilson	11				
Harding	2				
Coolidge	9				
Hoover	36				
F. Roosevelt	12				
Truman	28				
Kennedy	3				
Eisenhower	16				
L. Johnson	9				
Nixon	25				
Reagan	23				
Ford	32				

Cumulative Review Exercises

TI **1.** Longevity Refer to the numbers of years that U.S. presidents and popes and British monarchs lived after their inauguration, election, or coronation. The data are listed in the table used for Review Exercise 8.

a. Find the mean for each of the three groups.

b. Find the standard deviation for each of the three groups.

c. Test the claim that there is a difference between the mean for presidents and the mean for British monarchs.

continued

d. Use the longevity times for presidents and determine whether they appear to come from a population having a normal distribution. Explain why the distribution does or does not appear to be normal.

e. Use the longevity times for presidents and construct a 95% confidence interval estimate of the population mean.

TI **2.** M&M Treatment The table below lists 60 SAT scores separated into categories according to the color of the M&M candy used as a treatment. The SAT scores are based on data from the College Board, and the M&M color element is based on author whimsy.

a. Find the mean of the 20 SAT scores in each of the three categories. Do the three means appear to be approximately equal?

b. Find the median of the 20 SAT scores in each of the three categories. Do the three medians appear to be approximately equal?

c. Find the standard deviation of the 20 SAT scores in each of the three categories. Do the three standard deviations appear to be approximately equal?

d. Test the null hypothesis that there is no difference between the mean SAT score of subjects treated with red M&Ms and the mean SAT score of subjects treated with green M&Ms.

e. Construct a 95% confidence interval estimate of the mean SAT score for the population of subjects receiving the red M&M treatment.

f. Test the null hypothesis that the three populations (red, green, and blue M&M treatments) have the same mean SAT score.

Red	1130	621	813	996	1030	1257	898	743	921	1179
	1092	855	896	858	1095	1133	896	1190	908	699
Green	996	630	583	828	1121	993	1025	907	1111	1147
	780	916	793	1188	499	1180	1229	1450	1071	1153
Blue	706	1068	1013	892	1370	1590	939	1004	821	915
	866	848	1408	793	1097	1244	996	1131	1039	1159

TI **3.** Blue Genes Some couples have genetic characteristics configured so that one-quarter of all their offspring have blue eyes. A study is conducted of 100 couples believed to have those characteristics, with the result that 19 of their 100 offspring have blue eyes. Assuming that one-quarter of all offspring have blue eyes, estimate the probability that among 100 offspring, 19 or fewer have blue eyes. Based on that probability, does it seem that the one-quarter rate is wrong? Why or why not?

TI **4.** Weights of Babies: Finding Probabilities In the United States, weights of newborn babies are normally distributed with a mean of 7.54 lb and a standard deviation of 1.09 lb (based on data from "Birth Weight and Perinatal Mortality," by Wilcox, Skjaerven, Buekens, and Kiely, *Journal of the American Medical Association,* Vol. 273, No. 9).

a. If a newborn baby is randomly selected, what is the probability that he or she weighs more than 8.00 lb?

b. If 16 newborn babies are randomly selected, what is the probability that their mean weight is more than 8.00 lb?

c. What is the probability that each of the next three babies will have a birth weight greater than 7.54 lb?

Cooperative Group Activities

1. Out-of-class activity The *World Almanac and Book of Facts* includes a section called "Noted Personalities," with subsections comprised of architects, artists, business leaders, cartoonists, social scientists, military leaders, philosophers, political leaders, scientists, writers, composers, entertainers, and others. Design and conduct an observational study that begins with the selection of samples from select groups, to be followed by a comparison of life spans of people from the different categories. Do any particular groups appear to have life spans that are different from the other groups? Can you explain such differences?

2. In-class activity Begin by asking each student in the class to estimate the length of the classroom. Specify that the length is the distance between the chalkboard and the opposite wall. (See Review Exercises 3–5.) On the same sheet of paper, each student should also write his or her gender (male/female) and major. Then divide into groups of three or four, and use the data from the entire class to address these questions:
   - Is there a significant difference between the mean estimate for males and the mean estimate for females?
   - Is there sufficient evidence to reject equality of the mean estimates for different majors? Describe how the majors were categorized.
   - Does an interaction between gender and major have an effect on the estimated length?
   - Does gender appear to have an effect on estimated length?
   - Does major appear to have an effect on estimated length?

3. Out-of-class activity Divide into groups of three or four students. Each group should survey other students at the same college by asking them to identify their major and gender. You might include other factors, such as employment (none, part-time, full-time) and age (under 21, 21–30, over 30). For each surveyed subject, determine the accuracy of the time on his or her wristwatch. First set your own watch to the correct time using an accurate and reliable source ("At the tone, the time is . . . "). For watches that are ahead of the correct time, record positive times. For watches that are behind the correct time, record negative times. Use the sample data to address questions such as these:
   - Does gender appear to have an effect on the accuracy of the wristwatch?
   - Does major have an effect on wristwatch accuracy?
   - Does an interaction between gender and major have an effect on wristwatch accuracy?

TI-83/84 Plus Calculator Project

Data Set 10 in Appendix B lists rainfall amounts for different days of the week. That data set is named BOSTRAIN in the TI-83/84 Plus APP that is on the CD-ROM included with this book. Use the APP to install the seven individual files.

Consider and test the requirements that should be satisfied before using analysis of variance to test the claim that the different days of the week have the same mean amount of rainfall. Should analysis of variance be used to test that claim? Why or why not?

Is it possible to use the values in Data Set 10 to conduct an analysis of variance test? If so, conduct the test and state the conclusion suggested by the results.

Considering the requirements and the results, what do you conclude?

From Data to Decision

Critical Thinking: Should you approve this drug?

Drugs must undergo thorough testing before being approved for general use. In addition to testing for adverse reactions, they must also be tested for their effectiveness, and the analysis of such test results typically involves methods of statistics. Consider the development of Xynamine—a new drug designed to lower pulse rates. In order to obtain more consistent results that do not have a confounding variable of gender, the drug is tested using males only. Given below are pulse rates for a placebo group, a group of men treated with Xynamine in 10-mg doses, and a group of men treated with 20-mg doses of Xynamine. The project manager for the drug conducts research and finds that for adult males, pulse rates are normally distributed with a mean around 70 beats per minute and a standard deviation of approximately 11 beats per minute. His summary report states that the drug is effective, based on this evidence: The placebo group has a mean pulse rate of 68.9, which is close to the value of 70 beats per minute for adult males in general, but the group treated with 10-mg doses of Xynamine has a lower mean of 66.2, and the group treated with 20-mg doses of Xynamine has the lowest mean of 65.2.

Analyzing the Results

Analyze the data using the methods of this chapter. Based on the results, does it appear that there is sufficient evidence to support the claim that the drug lowers pulse rates? Are there any serious problems with the design of the experiment? Given that only males were involved in the experiment, do the results also apply to females? The project manager compared the posttreatment pulse rates to the mean pulse rate for adult males. Is there a better way to measure the drug's effectiveness in lowering pulse rates? How would you characterize the overall validity of the experiment? Based on the available results, should the drug be approved? Write a brief report summarizing your findings.

Placebo Group	10-mg Treatment Group	20-mg Treatment Group
77	67	72
61	48	94
66	79	57
63	67	63
81	57	69
75	71	59
66	66	64
79	85	82
66	75	34
75	77	76
48	57	59
70	45	53

Internet Project

Analysis of Variance

Go to the *Elementary Statistics* Web site at

http://www.aw.com/triola.

Follow the link to the Internet Project for this chapter. The project provides the background for experiments in areas as varied as athletic performance, consumer product labeling, and biology of the human body. In each case, the associated data will lend itself naturally to groupings ideal for application of this chapter's techniques. You will formulate the appropriate hypotheses, then conduct and summarize ANOVA tests.

Statistics @ Work

"If I didn't have any background in statistics, I would not be able to fully understand the data my company produces . . . help protect our workers and customers."

Jeffrey Foy

Jeffrey Foy is a toxicologist working for the Cabot Corporation, a chemical company.

Jeffrey Foy is also responsible for the hazard evaluation of the chemicals Cabot Corporation produces. It is his job to understand how the company's products may affect humans or the environment and help decide on the best ways to protect both.

What do you do in your job?

My responsibilities include arranging and evaluating toxicological studies, writing material safety data sheets, and helping our research and development groups produce materials that are safe for both people and the environment or to understand what potential hazards the materials might have.

What concepts of statistics do you use?

The primary concept I use is hypothesis testing (probability testing).

How do you use statistics on the job?

I use statistics daily. Statistical methods have been and are used in two ways in my work. First, statistics is used to help determine how I design my experiments. Second, statistics is used to determine if the data generated are significant or sometimes if they're even good enough to use.

Studies that I am involved in can cost from as little as $1000 to as much as $500,000 or more, and if you don't properly determine how you are going to evaluate the data, you could cost your company a great deal of time and money. If the experiment is done properly, then we move on to analyze the data. The data from the studies we perform are used to assess any potential health effects our products may have on our workers, customers, or the environment. The results are used to determine how chemicals can be sold or handled. When performing experiments at a testing laboratory or drug company you want to determine if your materials have an effect, whether desired (a drug curing a disease) or undesired (that same drug being toxic). Statistics plays an enormous role in our evaluation of the significance of the effects.

Please describe one specific example illustrating how the use of statistics was successful in improving a product or service.

A toxicology study costing about $300,000 was recently conducted. The data from the study were to be used to help determine if a particular chemical caused any effects in the subjects studied. After the study was performed, flaws were found in both the data and statistics used. It took an additional two years to properly review the data and finish the health evaluation. If the proper methods and endpoints had been chosen, then the additional time and money may not have been necessary. It was the understanding of the data and correct statistical evaluation that helped prevent the failure and potential repeat of the study.

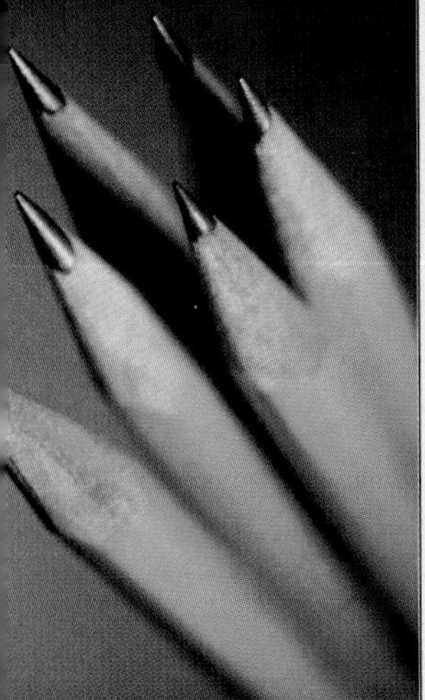

Nonparametric Statistics

13

CHAPTER 13 PROBLEM

Do students rank colleges the same as *U.S. News and World Report?*

Each year, *U.S. News and World Report* magazine publishes rankings of colleges based on statistics such as admission rates, graduation rates, class size, faculty-student ratio, faculty salaries, and peer ratings of administrators. Economists Christopher Avery, Mark Glickman, Caroline Minter Hoxby, and Andrew Metrick took an alternative approach of analyzing the college choices of 3240 high-achieving school seniors. They examined the colleges that offered admission along with the colleges that the students chose to attend. Table 13-1 lists rankings for a small sample of colleges. Table 13-1 shows some agreement between the student preference rankings and the magazine rankings, but it also shows some disagreement. For example, among the eight colleges considered, Harvard was ranked first by both the students and *U.S. News and World Report*. However, among the eight colleges considered, the University of Pennsylvania was ranked 7th by students but 3rd by *U.S. News and World Report*.

Let's consider the issue of a correlation between the student rankings and the magazine rankings. The concept of correlation was discussed in Section 10-2, where the linear correlation coefficient r was used to measure the association between two variables. The methods of Section 10-2 require paired data, and the data in Table 13-1 are paired. However, there is a very important difference: The methods of Section 10-2 have requirements that include normal distributions, and ranks such as those in Table 13-1 do not satisfy such requirements. The methods of Section 10-2 cannot be used with the sample data in Table 13-1. This chapter introduces several different methods that can be used with data that do not satisfy a requirement of a normal distribution. In particular, several methods of this section can be used with sample data that are in the form of ranks, as in Table 13-1. Section 13-6 will introduce a method for testing for a correlation with paired data that are in the form of ranks. We will then be able to analyze the agreement and disagreement between the student rankings and magazine rankings in Table 13-1. We can then test for a correlation between the student preference rankings and the magazine rankings. We can then address this key question: Do the students agree with the magazine?

Table 13-1 Colleges Ranked by Students and *U.S. News and World Report*

College	Rank by Student Preference	Rank by *U.S. News and World Report*
Harvard	1	1
Yale	2	2
Cal. Inst. of Tech.	3	5
M.I.T.	4	4
Brown	5	7
Columbia	6	6
U. of Penn.	7	3
Notre Dame	8	8

13-1 Overview

The methods of inferential statistics presented in Chapters 7, 8, 9, 10, and 12 are called *parametric methods* because they are based on sampling from a population with specific parameters, such as the mean μ, standard deviation σ, or proportion p. Those parametric methods usually must conform to some fairly strict conditions, such as a requirement that the sample data come from a normally distributed population. This chapter introduces nonparametric methods, which do not have such strict requirements.

Definitions

Parametric tests have requirements about the nature or shape of the populations involved; **nonparametric tests** do not require that samples come from populations with normal distributions or any other particular distributions. Consequently, nonparametric tests of hypotheses are often called **distribution-free tests.**

Although the term *nonparametric* suggests that the test is not based on a parameter, there are some nonparametric tests that do depend on a parameter such as the median. The nonparametric tests do not, however, require a particular distribution, so they are sometimes referred to as *distribution-free* tests. Although *distribution-free* is a more accurate description, the term *nonparametric* is more commonly used. The following are major advantages and disadvantages of nonparametric methods.

Advantages of Nonparametric Methods

1. Nonparametric methods can be applied to a wide variety of situations because they do not have the more rigid requirements of the corresponding parametric methods. In particular, nonparametric methods do not require normally distributed populations.
2. Unlike parametric methods, nonparametric methods can often be applied to categorical data, such as the genders of survey respondents.
3. Nonparametric methods usually involve simpler computations than the corresponding parametric methods and are therefore easier to understand and apply (since technology has simplified calculations, however, easier computations might not be too important).

Disadvantages of Nonparametric Methods

1. Nonparametric methods tend to waste information because exact numerical data are often reduced to a qualitative form. For example, in the nonparametric sign test (described in Section 13-2), weight losses by dieters are

recorded simply as negative signs; the actual magnitudes of the weight losses are ignored.

2. Nonparametric tests are not as efficient as parametric tests, so with a nonparametric test we generally need stronger evidence (such as a larger sample or greater differences) in order to reject a null hypothesis.

When the requirements of population distributions are satisfied, nonparametric tests are generally less efficient than their parametric counterparts, but the reduced efficiency can be compensated for by an increased sample size. For example, Section 13-6 will present a concept called *rank correlation,* which has an efficiency rating of 0.91 when compared to the linear correlation presented in Chapter 10. This means that with all other things being equal, nonparametric rank correlation requires 100 sample observations to achieve the same results as 91 sample observations analyzed through parametric linear correlation, assuming the stricter requirements for using the parametric method are met. Table 13-2 lists the nonparametric methods covered in this chapter, along with the corresponding parametric approach and **efficiency** rating. Table 13-2 shows that several nonparametric tests have efficiency ratings above 0.90, so the lower efficiency might not be a critical factor in choosing between parametric and nonparametric methods. However, because parametric tests do have higher efficiency ratings than their nonparametric counterparts, it's generally better to use the parametric tests when their required assumptions are satisfied.

Ranks

Sections 13-3 through 13-6 use methods based on ranks, which we now describe.

Definition

Data are *sorted* when they are arranged according to some criterion, such as smallest to largest or best to worst. A **rank** is a number assigned to an individual sample item according to its order in the sorted list. The first item is assigned a rank of 1, the second item is assigned a rank of 2, and so on.

Table 13-2 Efficiency: Comparison of Parametric and Nonparametric Tests

Application	Parametric Test	Nonparametric Test	Efficiency Rating of Nonparametric Test with Normal Population
Matched pairs of sample data	*t* test or *z* test	Sign test	0.63
		Wilcoxon signed-ranks test	0.95
Two independent samples	*t* test or *z* test	Wilcoxon rank-sum test	0.95
Several independent samples	Analysis of variance (*F* test)	Kruskal-Wallis test	0.95
Correlation	Linear correlation	Rank correlation test	0.91
Randomness	No parametric test	Runs test	No basis for comparison

Handling ties in ranks: If a tie in ranks occurs, the usual procedure is to find the mean of the ranks involved and then assign this mean rank to each of the tied items, as in the following example.

EXAMPLE The numbers 4, 5, 5, 5, 10, 11, 12, and 12 are given ranks of 1, 3, 3, 3, 5, 6, 7.5, and 7.5, respectively. See the table below and note the procedure for handling ties.

Sorted Data	Preliminary Ranking		Rank
4	1		**1**
5	2	Mean is 3.	**3**
5	3		**3**
5	4		**3**
10	5		**5**
11	6		**6**
12	7	Mean is 7.5.	**7.5**
12	8		**7.5**

13-2 Sign Test

Key Concept The main objective of this section is to understand the *sign test* procedure, which involves converting data values to plus and minus signs, then testing for disproportionately more of either sign.

Definition

The **sign test** is a nonparametric (distribution-free) test that uses plus and minus signs to test different claims, including:

1. Claims involving matched pairs of sample data
2. Claims involving nominal data
3. Claims about the median of a single population

Basic Concept of the Sign Test

The basic idea underlying the sign test is to analyze the frequencies of the plus and minus signs to determine whether they are significantly different. For example, suppose that we test a treatment designed to increase the likelihood that a baby is a girl. If 100 mothers are treated and 51 of them have girls, common sense suggests that there is not sufficient evidence to say that the treatment is effective, because 51 girls out of 100 is not significant. But what about 52 girls and 48 boys? Or 90 girls and 10 boys? The sign test allows us to determine when such results are significant.

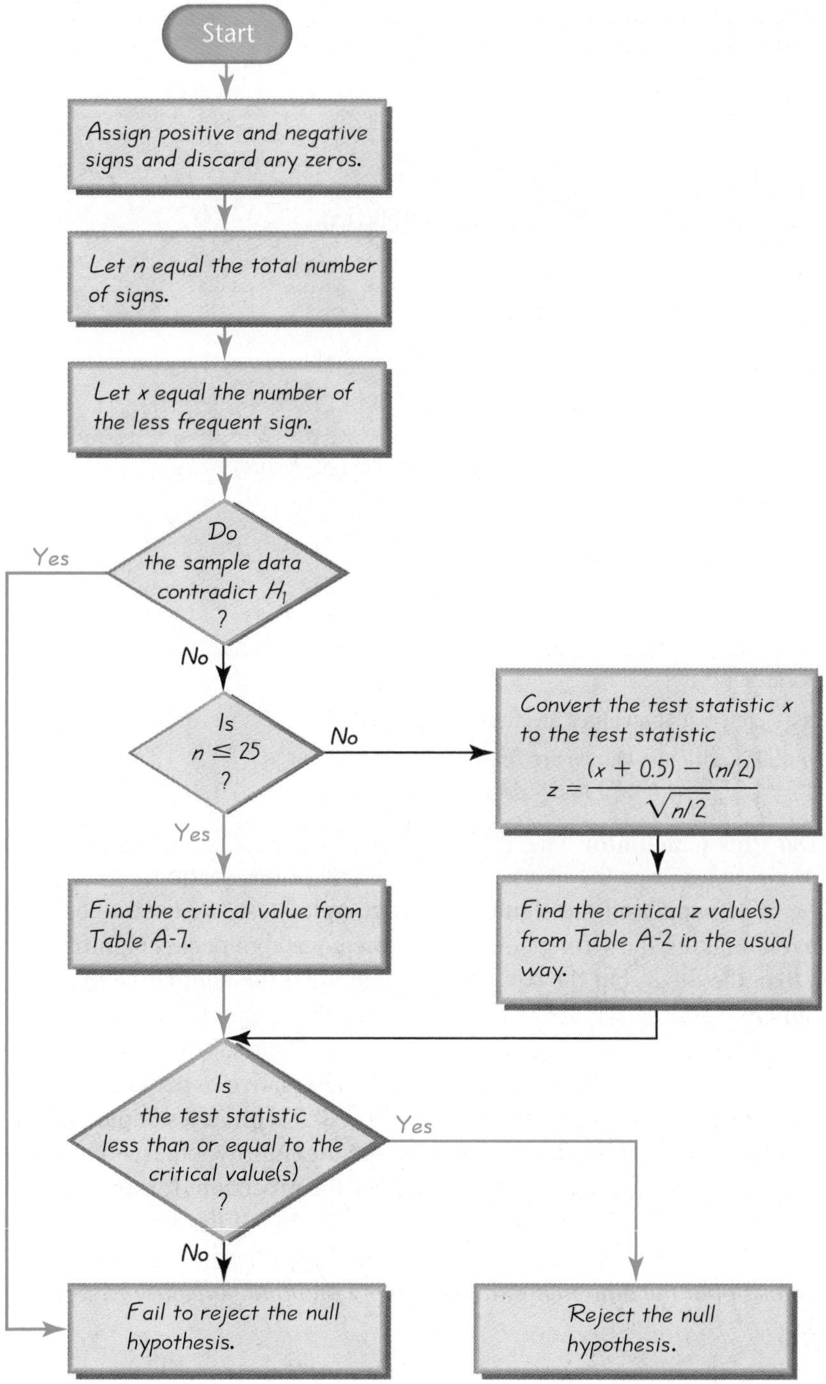

Start

Assign positive and negative signs and discard any zeros.

Let n equal the total number of signs.

Let x equal the number of the less frequent sign.

Do the sample data contradict H_1?

Yes

No

Is $n \leq 25$?

No

Convert the test statistic x to the test statistic

$$z = \frac{(x + 0.5) - (n/2)}{\sqrt{n}/2}$$

Yes

Find the critical value from Table A-7.

Find the critical z value(s) from Table A-2 in the usual way.

Is the test statistic less than or equal to the critical value(s)?

Yes

No

Fail to reject the null hypothesis.

Reject the null hypothesis.

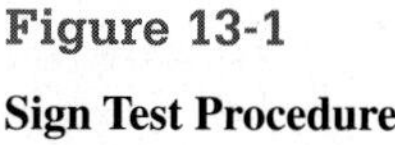

Figure 13-1

Sign Test Procedure

For consistency and simplicity, we will use a test statistic based on the number of times that the *less frequent* sign occurs. The relevant assumptions, notation, test statistic, and critical values are summarized in the accompanying box. Figure 13-1 summarizes the sign test procedure, which will be illustrated with examples that follow.

Class Attendance and Grades

In a study of 424 undergraduates at the University of Michigan, it was found that students with the worst attendance records tended to get the lowest grades. (Is anybody surprised?) Those who were absent less than 10% of the time tended to receive grades of B or above. The study also showed that students who sit in the front of the class tend to get significantly better grades.

Sign Test

Requirements

1. The sample data have been randomly selected.
2. There is *no* requirement that the sample data come from a population with a particular distribution, such as a normal distribution.

Notation

x = the number of times the *less frequent* sign occurs

n = the total number of positive and negative signs combined

Test Statistic

For $n \leq 25$: x (the number of times the less frequent sign occurs)

For $n > 25$: $z = \dfrac{(x + 0.5) - \left(\dfrac{n}{2}\right)}{\dfrac{\sqrt{n}}{2}}$

Critical values

1. For $n \leq 25$, critical x values are found in Table A-7.
2. For $n > 25$, critical z values are found in Table A-2.

TI-83/84 Plus Calculator The TI-83/84 Plus calculator can be used to find P-values for a sign test. See the instructions at the end of this section.

Caution: When applying the sign test in a one-tailed test, we need to be very careful to avoid making the wrong conclusion when one sign occurs significantly more often than the other, but the sample data contradict the alternative hypothesis. For example, suppose we are testing the claim that a gender-selection technique favors boys, but we get a sample of 10 boys and 90 girls. With a sample proportion of boys equal to 0.10, the data contradict the alternative hypothesis H_1: $p >$ 0.5. There is no way we can support a claim of $p > 0.5$ with any sample proportion less than 0.5, so we immediately fail to reject the null hypothesis and don't proceed with the sign test. Figure 13-1 summarizes the procedure for the sign test and includes this check: Do the sample data contradict H_1? If the sample data are in the opposite direction of H_1, fail to reject the null hypothesis. *It is always important to think about the data and to avoid relying on blind calculations or computer results.*

Claims Involving Matched Pairs

When using the sign test with data that are matched by pairs, we convert the raw data to plus and minus signs as follows:

1. We subtract each value of the second variable from the corresponding value of the first variable.
2. We record only the *sign* of the difference found in Step 1. We *exclude ties:* that is, we exclude any matched pairs in which both values are equal.

The key concept underlying this use of the sign test is this:

If the two sets of data have equal medians, the number of positive signs should be approximately equal to the number of negative signs.

EXAMPLE Does the Type of Seed Affect Corn Growth? In 1908, William Gosset published the article "The Probable Error of a Mean" under the pseudonym of "Student" (*Biometrika,* Vol. 6, No. 1). He included the data listed in Table 13-3 for two different types of corn seed (regular and kiln dried) that were used on *adjacent* plots of land. The listed values are the yields of head corn in pounds per acre. Use the sign test with a 0.05 significance level to test the claim that there is no difference between the yields from the regular and kiln-dried seed.

SOLUTION

REQUIREMENT ✔ The only requirement is that the sample data are randomly selected. There is no requirement about the distribution of the population, such as a common requirement that the sample data come from a normally distributed population. Based on the design of this experiment, we assume that the sample data are random. ✔

Here's the basic idea: If there is no difference between the yields from regular seeds and the yields from kiln-dried seeds, the numbers of positive and negative signs should be approximately equal. In Table 13-3 we have 7 negative and 4 positive signs. Are the numbers of positive and negative signs approximately equal, or are they significantly different? We follow the same basic steps for testing hypotheses as outlined in Figure 8-9, and we apply the sign test procedure summarized in Figure 13-1.

Steps 1, 2, 3: The null hypothesis is the claim of no difference between the yields from the regular seed and the yields from the kiln-dried seed, and the alternative hypothesis is the claim that there is a difference.

H_0: There is no difference. (The median of the differences is equal to 0.)

H_1: There is a difference. (The median of the differences is not equal to 0.)

Step 4: The significance level is $\alpha = 0.05$.

Step 5: We are using the nonparametric sign test.

continued

Table 13-3 Yields of Corn from Different Seeds

Regular	1903	1935	1910	2496	2108	1961	2060	1444	1612	1316	1511
Kiln dried	2009	1915	2011	2463	2180	1925	2122	1482	1542	1443	1535
Sign of difference	−	+	−	+	−	+	−	−	+	−	−

Step 6: The test statistic x is the number of times the less frequent sign occurs. Table 13-3 includes differences with 7 negative signs and 4 positive signs. (If there had been any differences of 0, we would have discarded them.) We let x equal the smaller of 7 and 4, so $x = 4$. Also, $n = 11$ (the total number of positive and negative signs combined). Our test is two-tailed with $\alpha = 0.05$. We refer to Table A-7 where the critical value of 1 is found for $n = 11$ and $\alpha = 0.05$ in two tails. (See Figure 13-1.)

Step 7: With a test statistic of $x = 4$ and a critical value of 1, we fail to reject the null hypothesis of no difference. [See Note 2 included with Table A-7: "Reject the null hypothesis if the number of the less frequent sign (x) is less than or equal to the value in the table." Because $x = 4$ is *not* less than or equal to the critical value of 1, we fail to reject the null hypothesis.]

We could also use the TI-83/84 Plus calculator to find that the *P*-value is 0.5488.

Step 8: There is not sufficient evidence to warrant rejection of the claim that the median of the differences is equal to 0. That is, there is not sufficient evidence to warrant rejection of the claim of no difference between the yields from the regular seed and the yields from the kiln-dried seed. This is the same conclusion that would be reached using the parametric t test with matched pairs in Section 9-4, but sign test results do not always agree with parametric test results.

Claims Involving Nominal Data

In Chapter 1 we defined nominal data to be data that consist of names, labels, or categories only. The nature of nominal data limits the calculations that are possible, but we can identify the *proportion* of the sample data that belong to a particular category, and we can test claims about the corresponding population proportion p. The following example uses nominal data consisting of genders (girls/boys). The sign test is used by representing girls with positive (+) signs and boys with negative (−) signs. (Those signs are chosen arbitrarily, honest.) Also note the procedure for handling cases in which $n > 25$.

EXAMPLE Gender Selection The Genetics and IVF Institute conducted a clinical trial of its methods for gender selection. As this book was written, results included 325 babies born to parents using the XSORT method to increase the probability of conceiving a girl, and 295 of those babies were girls. Use the sign test and a 0.05 significance level to test the claim that this method of gender selection has no effect.

SOLUTION

REQUIREMENT ✔ The only requirement is that the sample data are randomly selected. Based on the design of this experiment, we can assume that the sample data are random. We can now proceed with the sign test. ✔

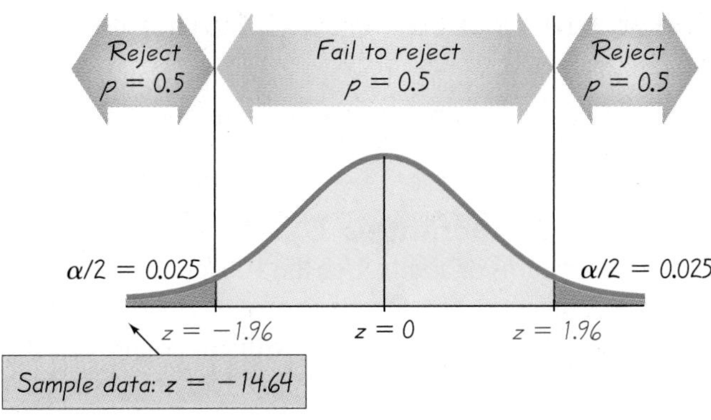

Figure 13-2 **Testing the Effect of a Gender-Selection Method**

Let p denote the population proportion of baby girls. The claim of no effect implies that the proportions of girls and boys are both equal to 0.5, so that $p = 0.5$. The null and alternative hypotheses can therefore be stated as follows:

H_0: $p = 0.5$ (the proportion of girls is equal to 0.5)

H_1: $p \neq 0.5$

Denoting girls by the positive sign (+) and boys by the negative sign (−), we have 295 positive signs and 30 negative signs. Refer now to the sign test procedure summarized in Figure 13-1. The test statistic x is the smaller of 295 and 30, so $x = 30$. This test involves two tails because a disproportionately high or low number of girls will cause us to reject the claim of $p = 0.5$. The sample data do not contradict the alternative hypothesis because 295 and 30 are not precisely equal. (That is, the sample data are consistent with the alternative hypothesis of a difference.) Continuing with the procedure in Figure 13-1, we note that the value of $n = 325$ is above 25, so the test statistic x is converted (using a correction for continuity) to the test statistic z as follows:

$$z = \frac{(x + 0.5) - \left(\frac{n}{2}\right)}{\frac{\sqrt{n}}{2}}$$

$$= \frac{(30 + 0.5) - \left(\frac{325}{2}\right)}{\frac{\sqrt{325}}{2}} = -14.64$$

With $\alpha = 0.05$ in a two-tailed test, the critical values are $z = \pm 1.96$. The test statistic $z = -14.64$ is less than -1.96 (see Figure 13-2), so we reject the null hypothesis that the proportion of girls is equal to 0.5. There is sufficient sample evidence to warrant rejection of the claim that the method of gender selection has no effect (with the proportions of girls and boys both equal to 0.5). This method does appear to affect the genders of babies.

Claims About the Median of a Single Population

The next example illustrates the procedure for using the sign test in testing a claim about the median of a single population. See how the negative and positive signs are based on the claimed value of the median.

EXAMPLE Body Temperatures Data Set 2 in Appendix B includes measured body temperatures of adults. Use the 106 temperatures listed for 12 A.M. on Day 2 with the sign test to test the claim that the median is less than 98.6°F. The data set has 106 subjects—68 subjects with temperatures below 98.6°F, 23 subjects with temperatures above 98.6°F, and 15 subjects with temperatures equal to 98.6°F.

SOLUTION

REQUIREMENT ✔ The only requirement is that the sample data are randomly selected and, based on the design of this experiment, we assume that the sample data are random. We can now proceed with the sign test. ✔

The claim that the median is less than 98.6°F is the alternative hypothesis, while the null hypothesis is the claim that the median is equal to 98.6°F.

H_0: Median is equal to 98.6°F. (median = 98.6°F)

H_1: Median is less than 98.6°F. (median < 98.6°F)

Following the procedure outlined in Figure 13-1, we discard the 15 zeros, we use the negative sign (−) to denote each temperature that is below 98.6°F, and we use the positive sign (+) to denote each temperature that is above 98.6°F. We therefore have 68 negative signs and 23 positive signs, so $n = 91$ and $x = 23$ (the number of the less frequent sign). The sample data do not contradict the alternative hypothesis, because most of the 91 temperatures are below 98.6°F. (If the sample data did conflict with the alternative hypothesis, we could immediately terminate the test by concluding that we fail to reject the null hypothesis.) The value of n exceeds 25, so we convert the test statistic x to the test statistic z:

$$z = \frac{(x + 0.5) - \left(\frac{n}{2}\right)}{\frac{\sqrt{n}}{2}}$$

$$= \frac{(23 + 0.5) - \left(\frac{91}{2}\right)}{\frac{\sqrt{91}}{2}} = -4.61$$

In this one-tailed test with $\alpha = 0.05$, we use Table A-2 to get the critical z value of -1.645. From Figure 13-3 we can see that the test statistic of $z = -4.61$ does fall within the critical region. We therefore reject the null hypothesis. On the basis of the available sample evidence, we support the claim that the median body temperature of healthy adults is less than 98.6°F.

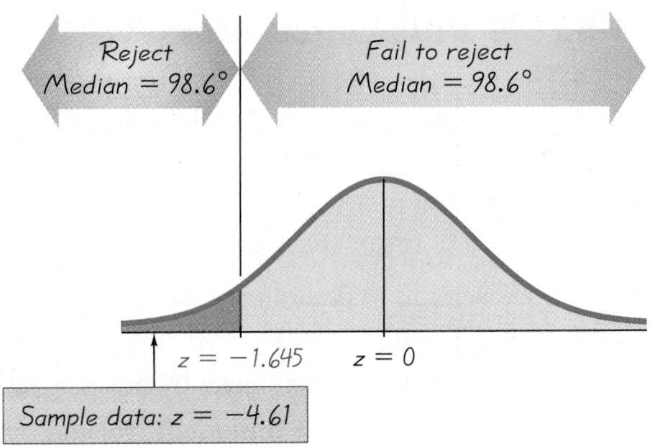

Figure 13-3

Testing the Claim That the Median Is Less Than 98.6°F

In this sign test of the claim that the median is below 98.6°F, we get a test statistic of $z = -4.61$ with a P-value of 0.00000202, but a parametric test of the claim that $\mu < 98.6$°F results in a test statistic of $t = -6.611$ with a P-value of 0.000000000813. Because the P-value from the sign test is not as low as the P-value from the parametric test, we see that the sign test isn't as sensitive as the parametric test. Both tests lead to rejection of the null hypothesis, but the sign test doesn't consider the sample data to be as extreme, partly because the sign test uses only information about the *direction* of the data, ignoring the *magnitudes* of the data values. The next section introduces the Wilcoxon signed-ranks test, which largely overcomes that disadvantage.

Rationale for the test statistic used when $n > 25$: When finding critical values for the sign test, we use Table A-7 only for n up to 25. When $n > 25$, the test statistic z is based on a normal approximation to the binomial probability distribution with $p = q = 1/2$. Recall that in Section 6-6 we saw that the normal approximation to the binomial distribution is acceptable when both $np \geq 5$ and $nq \geq 5$. Recall also that in Section 5-4 we saw that $\mu = np$ and $\sigma = \sqrt{npq}$ for binomial probability distributions. Because this sign test assumes that $p = q = 1/2$, we meet the $np \geq 5$ and $nq \geq 5$ prerequisites whenever $n \geq 10$. Also, with the assumption that $p = q = 1/2$, we get $\mu = np = n/2$ and $\sqrt{npq} = \sqrt{n/4} = \sqrt{n}/2$, so

$$z = \frac{x - \mu}{\sigma}$$

becomes

$$z = \frac{x - \left(\dfrac{n}{2}\right)}{\dfrac{\sqrt{n}}{2}}$$

Finally, we replace x by $x + 0.5$ as a correction for continuity. That is, the values of x are discrete, but because we are using a continuous probability distribution, a discrete value such as 10 is actually represented by the interval from 9.5 to 10.5. Because x represents the less frequent sign, we act conservatively by concerning ourselves only with $x + 0.5$; we thus get the test statistic z, as given in the equation and in Figure 13-1.

Using the TI-83/84 Plus Calculator for the Sign Test

The TI-83/84 Plus calculator does not have a built-in function dedicated to the sign test, but it is easy to use the calculator's **binomcdf** function to find the *P*-value for a sign test. The binomcdf function uses the binomial probability distribution for finding the probability of x or fewer successes.

1. Use the methods of this section to determine the value of x (the less frequent sign) and the value of n (the sample size).
2. Gain access to the **DISTR** (distributions) menu by pressing 2ND VARS.
3. Scroll down to select the menu item of **binomcdf** (for the binomial cumulative density function), then press ENTER.
4. The format for this function is binomcdf(n, p, x), so enter **binomcdf(n, 0.5, x),** using the values of n and x found in Step 1. (For the sign test, always use $p = 0.5$.)
5. The result of Step 4 is the probability of getting x or fewer successes among n trials. *Double this value for two-tailed tests.* The final result is the *P*-value for the sign test, so reject the null hypothesis if the *P*-value is less than or equal to the significance level. Otherwise, fail to reject the null hypothesis.

For example, the last example of this section involved a sign test with $n = 91$ and $x = 23$, and the result of binomcdf(91, 0.5, 23) is 1.261076831E^{-}6, or 0.00000126 rounded. Because that particular sign test was one-tailed, the *P*-value is 0.00000126. (The normal approximation yielded a z test statistic of -4.61, which corresponds to a *P*-value of 0.00000202, but the *P*-value of 0.00000126 is more accurate.)

13-2 BASIC SKILLS AND CONCEPTS

Statistical Literacy and Critical Thinking

1. Nonparametric Test Why is the sign test considered to be a "nonparametric" test or a "distribution-free" test?

2. Sign Test Why is the procedure introduced in this section referred to as the "sign" test?

3. Sign Test Procedure You have been given the task of testing the claim that a method of gender selection has the effect of increasing the likelihood that a baby will be a girl, and sample data consist of 20 girls among 80 newborn babies. Without applying the formal sign test procedure, what do you conclude about the claim? Why?

4. Efficiency of the Sign Test Refer to Table 13-2 and identify the efficiency of the sign test. What does that value tell us about the sign test?

In Exercises 5–8, assume that matched pairs of data result in the given number of signs when the value of the second variable is subtracted from the corresponding value of the first variable. Use the sign test with a 0.05 significance level to test the null hypothesis of no difference.

TI 5. Positive signs: 15; negative signs: 4; ties: 1

TI 6. Positive signs: 3; negative signs: 12; ties: 2

TI **7.** Positive signs: 30; negative signs: 35; ties: 3

8. Positive signs: 50; negative signs: 40; ties: 4

In Exercises 9–18, use the sign test.

TI **9.** Is Friday the 13th Unlucky? Researchers collected data on the numbers of hospital admissions resulting from motor vehicle crashes, and results are given below for Fridays on the 6th of a month and Fridays on the following 13th of the same month (based on data from "Is Friday the 13th Bad for Your Health?" by Scanlon et al., *BMJ,* Vol. 307, as listed in the *Data and Story Line* online resource of data sets). Use a 0.05 significance level to test the claim that when the 13th day of a month falls on a Friday, the numbers of hospital admissions from motor vehicle crashes are not affected.

Friday the 6th:	9	6	11	11	3	5
Friday the 13th:	13	12	14	10	4	12

TI **10.** Testing Corn Seeds In 1908, William Gosset published the article "The Probable Error of a Mean" under the pseudonym of "Student" (*Biometrika,* Vol. 6, No. 1). He included the data listed below for yields from two different types of seed (regular and kiln dried) that were used on adjacent plots of land. The listed values are the yields of straw in cwt per acre, where cwt represents 100 lb. Using a 0.05 significance level, test the claim that there is no difference between the yields from the two types of seed. Does it appear that either type of seed is better?

Regular	19.25	22.75	23	23	22.5	19.75	24.5	15.5	18	14.25	17
Kiln dried	25	24	24	28	22.5	19.5	22.25	16	17.25	15.75	17.25

TI **11.** Testing for a Difference Between Reported and Measured Male Heights As part of the National Health Examination Survey conducted by the Department of Health and Human Services, self-reported heights and measured heights were obtained for males aged 12–16. Listed below are sample results. Is there sufficient evidence to support the claim that there is a difference between self-reported heights and measured heights of males aged 12–16? Use a 0.05 significance level.

Reported height	68	71	63	70	71	60	65	64	54	63	66	72
Measured height	67.9	69.9	64.9	68.3	70.3	60.6	64.5	67.0	55.6	74.2	65.0	70.8

TI **12.** Heights of Winners and Runners-Up Listed below are the heights of candidates who won presidential elections and the heights of the candidates with the next highest number of popular votes. The data are in chronological order, so the corresponding heights from the two lists are matched. For candidates who won more than once, only the heights from the first election are included, and no elections before 1900 are included. A popular theory is that winning candidates tend to be taller than the corresponding losing candidates. Use a 0.05 significance level to test that theory. Does height appear to be an important factor in winning the presidency?

Won Presidency

71	74.5	74	73	69.5	71.5	75	72
70.5	69	74	70	71	72	70	67

Runner-Up

73	74	68	69.5	72	71	72	71.5
70	68	71	72	70	72	72	72

TI **13.** Testing for a Median Body Temperature of 98.6°F A premed student in a statistics class is required to do a class project. Intrigued by the body temperatures in Data Set 2 of

Appendix B, she plans to collect her own sample data to test the claim that the median body temperature is less than 98.6°F. Because of time constraints, she finds that she has time to collect data from only 12 people. After carefully planning a procedure for obtaining a simple random sample of 12 healthy adults, she measures their body temperatures and obtains the results listed below. Use a 0.05 significance level to test the claim that these body temperatures come from a population with a median that is less than 98.6°F.

97.6 97.5 98.6 98.2 98.0 99.0 98.5 98.1 98.4 97.9 97.9 97.7

TI **14.** Testing for Median Weight of Quarters Listed below are weights (in grams) of randomly selected quarters that were minted after 1964. The quarters are supposed to have a median weight of 5.670 g. Use a 0.05 significance level to test the claim that the median is equal to 5.670 g. Do quarters appear to be minted according to specifications?

5.7027 5.7495 5.7050 5.5941 5.7247 5.6114 5.6160 5.5999 5.7790 5.6841

TI **15.** Nominal Data: Gender Selection for Boys The Genetics and IVF Institute conducted a clinical trial of the YSORT method designed to increase the probability of conceiving a boy. As this book was being written, 51 babies were born to parents using the YSORT method, and 39 of them were boys. Use the sample data with a 0.01 significance level to test the claim that with this method, the probability of a baby being a boy is greater than 0.5. Does the method appear to work?

TI **16.** Nominal Data: Car Crashes In a study of 11,000 car crashes, it was found that 5720 of them occurred within 5 miles of home (based on data from Progressive Insurance). Use a 0.01 significance level to test the claim that more than 50% of car crashes occur within 5 miles of home. Are the results questionable because they are based on a survey sponsored by an insurance company?

TI **17.** Nominal Data: Travel Through the Internet Among 734 randomly selected Internet users, it was found that 360 of them use the Internet for making travel plans (based on data from a Gallup poll). Use a 0.01 significance level to test the claim that among Internet users, less than 50% use it for making travel plans. Are the results important for travel agents?

TI **18.** Postponing Death An interesting and popular hypothesis is that individuals can temporarily postpone their death to survive a major holiday or important event such as a birthday. In a study of this phenomenon, 6062 deaths were recorded in the week before Thanksgiving, and there were 5938 deaths the week after Thanksgiving (based on data from "Holidays, Birthdays, and Postponement of Cancer Death," by Young and Hade, *Journal of the American Medical Association,* Vol. 292, No. 24). If people can postpone their deaths until after Thanksgiving, then the proportion of deaths in the week before should be less than 0.5. Use a 0.05 significance level to test the claim that the proportion of deaths in the week before Thanksgiving is less than 0.5. Based on the result, does there appear to be any indication that people can temporarily postpone their death to survive the Thanksgiving holiday?

13-2 BEYOND THE BASICS

TI **19.** Procedures for Handling Ties In the sign test procedure described in this section, we excluded ties (represented by 0 instead of a sign of + or −). A second approach is to treat half of the 0s as positive signs and half as negative signs. (If the number of 0s is odd, exclude one so that they can be divided equally.) With a third approach, in two-tailed tests make half of the 0s positive and half negative; in one-tailed tests make all 0s either positive or negative, whichever supports the null hypothesis. Assume that in using the sign test on a claim that the median value is less than 100, we get 60 values below 100, 40 values above 100, and 21 values equal to 100. Identify the test statistic and conclusion for the three different ways of handling ties (with differences of 0). Assume a 0.05 significance level in all three cases.

20. Finding Critical Values Table A-7 lists critical values for limited choices of α. Use Table A-1 to add a new column in Table A-7 (down to $n = 15$) that represents a significance level of 0.03 in one tail or 0.06 in two tails. For any particular n, use $p = 0.5$, because the sign test requires the assumption that $P(\text{positive sign}) = P(\text{negative sign}) = 0.5$. The probability of x or fewer like signs is the sum of the probabilities for values up to and including x.

TI **21.** Normal Approximation Error The Compulife.com company has hired 18 women among the last 54 new employees. Job applicants are about half men and half women, all of whom are qualified. Using a 0.01 significance level with the sign test, is there sufficient evidence to charge bias? Does the conclusion change if the binomial distribution is used instead of the approximating normal distribution?

13-3 Wilcoxon Signed-Ranks Test for Matched Pairs

Key Concept This section introduces the *Wilcoxon signed-ranks test,* which uses ranks of sample data consisting of matched pairs. This test is used with a null hypothesis that the population of differences from the matched pairs has a median equal to zero.

The sign test (Section 13-2) can also be used with matched pairs, but the sign test uses only the signs of the differences. By using ranks, the Wilcoxon signed-ranks test takes the magnitudes of the differences into account. Because the Wilcoxon signed-ranks test incorporates and uses more information than the sign test, it tends to yield conclusions that better reflect the true nature of the data.

Definition

The **Wilcoxon signed-ranks test** is a nonparametric test that uses ranks of sample data consisting of matched pairs. It is used to test the null hypothesis that the population of differences has a median of zero, so the null and alternative hypotheses are as follows:

H_0: The matched pairs have differences that come from a population with a median equal to zero.

H_1: The matched pairs have differences that come from a population with a nonzero median.

(The Wilcoxon signed-ranks test can also be used to test the claim that a sample comes from a population with a specified median. See Exercise 13 for this application.)

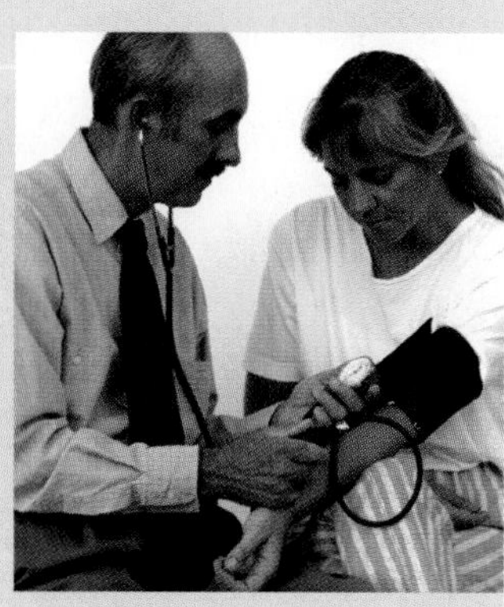

Gender Gap in Drug Testing

A study of the relationship between heart attacks and doses of aspirin involved 22,000 male physicians. This study, like many others, excluded women. The General Accounting Office recently criticized the National Institutes of Health for not including both sexes in many studies because results of medical tests on males do not necessarily apply to females. For example, women's hearts are different from men's in many important ways. When forming conclusions based on sample results, we should be wary of an inference that extends to a population larger than the one from which the sample was drawn.

Wilcoxon Signed-Ranks Test

Requirements

1. The data consist of matched pairs that have been randomly selected.
2. The population of differences (found from the pairs of data) has a distribution that is approximately *symmetric,* meaning that the left half of its histogram is roughly a mirror image of its right half. (There is *no* requirement that the data have a normal distribution.)

Notation

The procedure for finding the rank sum T follows this box.

T = the smaller of the following two sums:

1. The sum of the absolute values of the negative ranks of the nonzero differences d
2. The sum of the positive ranks of the nonzero differences d

Test Statistic

If $n \leq 30$, the test statistic is T.

If $n > 30$, the test statistic is $z = \dfrac{T - \dfrac{n(n+1)}{4}}{\sqrt{\dfrac{n(n+1)(2n+1)}{24}}}$

Critical values

1. If $n \leq 30$, the critical T value is found in Table A-8.
2. If $n > 30$, the critical z values are found in Table A-2.

Wilcoxon Signed-Ranks Procedure

Step 1: For each pair of data, find the difference d by subtracting the second value from the first value. Keep the signs, but discard any pairs for which $d = 0$.

Step 2: *Ignore the signs of the differences,* then sort the differences from lowest to highest and replace the differences by the corresponding rank value (as described in Section 13-1). When differences have the same numerical value, assign to them the mean of the ranks involved in the tie.

Step 3: Attach to each rank the sign of the difference from which it came. That is, insert those signs that were ignored in Step 2.

Step 4: Find the sum of the absolute values of the negative ranks. Also find the sum of the positive ranks.

Step 5: Let T be the *smaller* of the two sums found in Step 4. Either sum could be used, but for a simplified procedure we arbitrarily select the smaller of the two sums. (See the notation for T in the preceding box.)

Step 6: Let n be the number of pairs of data for which the difference d is not 0.

Step 7: Determine the test statistic and critical values based on the sample size, as shown in the preceding box.

Step 8: When forming the conclusion, reject the null hypothesis if the sample data lead to a test statistic that is in the critical region—that is, the test statistic is less than or equal to the critical value(s). Otherwise, fail to reject the null hypothesis.

EXAMPLE Does the Type of Seed Affect Corn Growth? In 1908, William Gosset published the article "The Probable Error of a Mean" under the pseudonym of "Student" (*Biometrika,* Vol. 6, No. 1). He included the data listed in Table 13-4 for two different types of corn seed (regular and kiln dried) that were used on *adjacent* plots of land. The listed values are the yields of head corn in pounds per acre. Use the Wilcoxon signed-ranks test with a 0.05 significance level to test the claim that there is no difference between the yields from the regular and kiln-dried seed.

SOLUTION

REQUIREMENTS ✔ We must have matched pairs of randomly selected data. The data are matched and, given the design of this experiment, it is reasonable to assume that the matched pairs have been randomly selected. Also, the accompanying Minitab-generated histogram shows that the distribution of differences is approximately symmetric, as required. (That is, the left side of the graph is roughly a mirror image of the right side. Visually, they might not appear to be symmetric, but with only 11 values, the difference between the left side and the right side is not too extreme.) ✔

continued

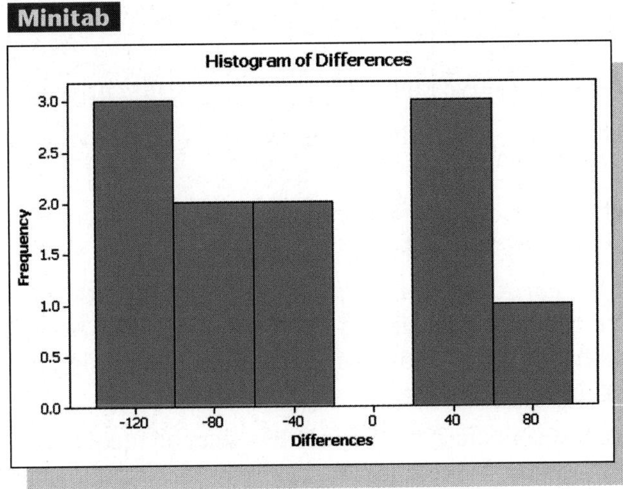

Table 13-4 Yields of Corn from Different Seeds

Regular	1903	1935	1910	2496	2108	1961	2060	1444	1612	1316	1511
Kiln dried	2009	1915	2011	2463	2180	1925	2122	1482	1542	1443	1535
Differences *d*	−106	20	−101	33	−72	36	−62	−38	70	−127	−24
Ranks of |differences|	10	1	9	3	8	4	6	5	7	11	2
Signed ranks	−10	1	−9	3	−8	4	−6	−5	7	−11	−2

The null and alternative hypotheses are as follows:

H_0: The yields from regular seed and kiln-dried seed are such that the median of the population of differences is equal to zero.

H_1: The median of the population of differences is nonzero.

The significance level is $\alpha = 0.05$. We are using the Wilcoxon signed-ranks test procedure, so the test statistic is calculated by using the eight-step procedure presented earlier in this section.

Step 1: In Table 13-4, the row of differences is obtained by computing this difference for each pair of data:

$$d = \text{yield from regular seed} - \text{yield from kiln-dried seed}$$

Step 2: Ignoring their signs, we rank the absolute differences from lowest to highest. (If there had been any ties in ranks, they would have been handled by assigning the mean of the involved ranks to each of the tied values. Also, any differences of 0 would have been discarded.)

Step 3: The bottom row of Table 13-4 is created by attaching to each rank the sign of the corresponding difference. If there really is no difference between the yields from the two types of seed (as in the null hypothesis), we expect the sum of the positive ranks to be approximately equal to the sum of the absolute values of the negative ranks.

Step 4: We now find the sum of the absolute values of the negative ranks, and we also find the sum of the positive ranks.

Sum of absolute values of negative ranks: 51 (from $10 + 9 + 8 + 6 + 5 + 11 + 2$)

Sum of positive ranks: 15 (from $1 + 3 + 4 + 7$)

Step 5: Letting T be the smaller of the two sums found in Step 4, we find that $T = 15$.

Step 6: Letting n be the number of pairs of data for which the difference d is not 0, we have $n = 11$.

Step 7: Because $n = 11$, we have $n \leq 30$, so we use a test statistic of $T = 15$ (and we do not calculate a z test statistic). Also, because $n \leq 30$, we use Table A-8 to find the critical value of 11 (using $n = 11$ and $\alpha = 0.05$ in two tails).

Step 8: The test statistic $T = 15$ is not less than or equal to the critical value of 11, so we fail to reject the null hypothesis. It appears that there is no difference between yields from regular seed and kiln-dried seed.

If we use the sign test with the preceding example, we will arrive at the same conclusion. Although the sign test and the Wilcoxon signed-ranks test agree in this particular case, there are other cases in which they do not agree.

Rationale: In this example the unsigned ranks of 1 through 11 have a total of 66, so if there are no significant differences, each of the two signed-rank totals should be around $66 \div 2$, or 33. That is, the negative ranks and positive ranks should split up as 33–33 or something close, such as 31–35. The table of critical values shows that at the 0.05 significance level with 11 pairs of data, a split of 11–55 represents a significant departure from the null hypothesis, and any split that is farther apart (such as 10–56 or 2–64) will also represent a significant departure from the null hypothesis. Conversely, splits like 12–52 do not represent significant departures from a 33–33 split, and they would not justify rejecting the null hypothesis. The Wilcoxon signed-ranks test is based on the lower rank total, so instead of analyzing both numbers constituting the split, we consider only the lower number.

The sum $1 + 2 + 3 + \cdots + n$ of all the ranks is equal to $n(n + 1)/2$, and if this is a rank sum to be divided equally between two categories (positive and negative), each of the two totals should be near $n(n + 1)/4$, which is half of $n(n + 1)/2$. Recognition of this principle helps us understand the test statistic used when $n > 30$. The denominator in that expression represents a standard deviation of T and is based on the principle that

$$1^2 + 2^2 + 3^3 + \cdots + n^2 = \frac{n(n + 1)(2n + 1)}{6}$$

The Wilcoxon signed-ranks test can be used for matched pairs of data only. The next section will describe a rank-sum test that can be applied to two sets of independent data that are not matched in pairs.

TI Using the TI-83/84 Plus Calculator for the Wilcoxon Signed-Ranks Test

The TI-83/84 Plus calculator is not programmed for the Wilcoxon signed-ranks test, but the program SRTEST (by Michael Lloyd) can be used. Use the following procedure:

1. First download and install the program SRTEST from the CD-ROM or the Web site www.aw.com/triola. (If using the CD-ROM, select the **Software** folder, select **TI-83 TI-84 Plus Programs,** select **Michael Lloyd Programs,** then select **programs.** The program SRTEST should now be available for download.)
2. Also download the program ZZRANK, which is used by the program SRTEST.
3. Create a list of differences between values in the matched pairs. (The first set of values can be entered in list L1, the second set of values can be entered in list L2, then the differences can be stored in list L3 by entering L1 − L2 → L3, where the STO› key is used for the arrow.)
4. Press the PRGM key and select **SRTEST.** Press the **ENTER** key twice.
5. When given the prompt of DATA=, enter the list containing the differences, such as L3, then press ENTER.
6. The screen will display the sample size indicated by N, the sum of the positive ranks indicated by $T+$, and the sum of the negative ranks indicated by $T-$.
7. Press ENTER again to see the mean and standard deviation.
8. Press ENTER once again to see the z score.

9. Using the calculator results:

- If $n \leq 30$, let the test statistic T equal the smaller of the two rank sums found in Step 6, then get the critical T value from Table A-8.
- If $n > 30$, let the test statistic z equal the value given in Step 8, then get the critical z values from Table A-2.

If using the program SRTEST with the 11 matched pairs of sample data in the example of this section, the above procedure yields the values of $N = 11$, $T+ = 15$, $T- = 51$, and $z = 1.60$. Because the sample size is $n = 11$ (which is less than or equal to 30), the test statistic T is the smaller of 15 and 51, which is 15. The hypothesis test is then completed as shown in Steps 7 and 8 of the solution to the example given in this section.

13-3 BASIC SKILLS AND CONCEPTS

Statistical Literacy and Critical Thinking

1. Wilcoxon Signed-Ranks Test Why would we use the Wilcoxon signed-ranks test for matched pairs instead of the methods presented in Section 9-4?

2. Wilcoxon Signed-Ranks Test and Sign Test The Wilcoxon signed-ranks test and the sign test can both be used with sample data consisting of matched pairs. Could they both lead to the same conclusion? Do they both always lead to the same conclusion?

3. Wilcoxon Signed-Ranks Test Given sample data consisting of matched pairs, we can compare the values using the sign test or the Wilcoxon signed-ranks test. What important advantage does the Wilcoxon signed-ranks test have over the sign test?

4. Wilcoxon Signed-Ranks Test A market researcher designs an experiment that involves a survey of married couples in randomly selected shopping malls. After separating the male and female, each spouse is asked how much has been spent shopping in the past hour. The results are analyzed with the Wilcoxon signed-ranks test. Do the conclusions apply to the population of married couples in the United States? Why or why not?

Using the Wilcoxon Signed-Ranks Test. *In Exercises 5 and 6, refer to the given paired sample data and use the Wilcoxon signed-ranks test to test the claim that the matched pairs have differences that come from a population with a median equal to zero. Use a 0.05 significance level.*

TI **5.**

x	60	55	89	92	78	84	93	87
y	35	27	47	44	39	48	51	54

TI **6.**

x	90	93	112	97	102	115	148	152	121
y	88	91	115	95	103	116	150	147	119

Using the Wilcoxon Signed-Ranks Test. *In Exercises 7–10, refer to the sample data for the given exercises in Section 13-2. Instead of the sign test, use the Wilcoxon signed-ranks test to test the claim that the matched pairs have differences that come from a population with a median equal to zero. Use a 0.05 significance level.*

TI **7.** Exercise 9

TI **8.** Exercise 10

TI **9.** Exercise 11

TI **10.** Exercise 12

Appendix B Data Sets. In Exercises 11 and 12, use the Wilcoxon signed-ranks test with the data in Appendix B.

TI **11.** Testing for Difference Between Forecast and Actual Temperatures Refer to Data Set 8 in Appendix B and use the actual high temperatures and the three-day forecast high temperatures. Does there appear to be a difference?

TI **12.** Testing for Difference Between Times Depicting Alcohol Use and Tobacco Use Refer to Data Set 5 in Appendix B. Use only those movies that showed some use of tobacco or alcohol. (That is, ignore those movies with times of zero for both tobacco use and alcohol use.) Does there appear to be a difference?

13-3 BEYOND THE BASICS

TI **13.** Using the Wilcoxon Signed-Ranks Test for Claims About a Median The Wilcoxon signed-ranks test can be used to test the claim that a sample comes from a population with a specified median. The procedure used is the same as the one described in this section, except that the differences (Step 1) are obtained by subtracting the value of the hypothesized median from each value. Use the sample data consisting of the 106 body temperatures listed for 12 A.M. on Day 2 in Data Set 2 in Appendix B. At the 0.05 significance level, test the claim that healthy adults have a median body temperature that is equal to 98.6°F.

13-4 Wilcoxon Rank-Sum Test for Two Independent Samples

Key Concept This section introduces the *Wilcoxon rank-sum test,* which uses ranks of values from two independent sets of sample data to test the null hypothesis that the two populations have equal medians.

The Wilcoxon signed-ranks test (Section 13-3) involves matched pairs of data, but the Wilcoxon rank-sum test of this section involves two independent samples that are not related or somehow matched or paired. (To avoid confusion between the Wilcoxon rank-sum test for two independent samples and the Wilcoxon signed-ranks test for matched pairs, consider using the Internal Revenue Service for the mnemonic of IRS to remind us of "independent: rank sum.")

Definition

The **Wilcoxon rank-sum test** is a nonparametric test that uses ranks of sample data from two independent populations. It is used to test the null hypothesis that the two independent samples come from populations with equal medians. The alternative hypothesis is the claim that the two populations have different medians.

H_0: The two samples come from populations with equal medians.

H_1: The two samples come from populations with different medians.

Basic Concept: The Wilcoxon rank-sum test is equivalent to the **Mann-Whitney *U* test** (see Exercise 13), which is included in some other textbooks and software packages. The basic idea underlying the Wilcoxon rank-sum test is this: If two samples are drawn from identical populations and the individual values are all ranked as one combined collection of values, then the high and low ranks should fall evenly between the two samples. If the low ranks are found predominantly in one sample and the high ranks are found predominantly in the other sample, we suspect that the two populations have different medians.

Wilcoxon Rank-Sum Test

Requirements

1. There are two independent samples of randomly selected data.
2. Each of the two samples has more than 10 values. (For samples with 10 or fewer values, special tables are available in reference books, such as *CRC Standard Probability and Statistics Tables and Formulae,* published by CRC Press.)
3. There is *no* requirement that the two populations have a normal distribution or any other particular distribution.

Notation

n_1 = size of Sample 1

n_2 = size of Sample 2

R_1 = sum of ranks for Sample 1, found by using the procedure that follows

R_2 = sum of ranks for Sample 2, found by using the procedure that follows

R = same as R_1 (sum of ranks for Sample 1)

μ_R = mean of the sample R values that is expected when the two populations have equal medians

σ_R = standard deviation of the sample R values that is expected with two populations having equal medians

Test Statistic

$$z = \frac{R - \mu_R}{\sigma_R}$$

where

$$\mu_R = \frac{n_1(n_1 + n_2 + 1)}{2}$$

$$\sigma_R = \sqrt{\frac{n_1 n_2(n_1 + n_2 + 1)}{12}}$$

n_1 = size of the sample from which the rank sum R is found

n_2 = size of the other sample

R = sum of ranks of the sample with size n_1

Critical values: Critical values can be found in Table A-2 (because the test statistic is based on the normal distribution).

Procedure for Finding the Value of the Test Statistic

1. Temporarily combine the two samples into one big sample, then replace each sample value with its rank. (The lowest value gets a rank of 1, the next lowest value gets a rank of 2, and so on. If values are tied, assign to them the mean of the ranks involved in the tie. See Section 13-1 for a description of ranks and the procedure for handling ties.)
2. Find the sum of the ranks for either one of the two samples.
3. Calculate the value of the z test statistic as shown in the preceding box, where either sample can be used as "Sample 1." (If both sample sizes are greater than 10, then the sampling distribution of R is approximately normal with mean μ_R and standard deviation σ_R, and the test statistic is as shown in the preceding box.)

Note that unlike the corresponding hypothesis tests in Section 9-3, the Wilcoxon rank-sum test does *not* require normally distributed populations. Also, the Wilcoxon rank-sum test can be used with data at the ordinal level of measurement, such as data consisting of ranks. In contrast, the parametric methods of Section 9-3 cannot be used with data at the ordinal level of measurement. In Table 13-2 we noted that the Wilcoxon rank-sum test has a 0.95 efficiency rating when compared with the parametric t test or z test. Because this test has such a high efficiency rating and involves easier calculations, it is often preferred over the parametric tests presented in Section 9-3, even when the requirement of normality is satisfied.

EXAMPLE BMI of Men and Women Refer to Data Set 1 in Appendix B and use only the first 13 sample values of the body mass index (BMI) of men and the first 12 sample values of the BMI of women. The sample BMI values are listed in Table 13-5. (Only parts of the available sample values are used so that the calculations in this example are easier to follow.) Use a 0.05 significance level to test the claim that the median BMI of men is equal to the median BMI of women.

Table 13-5
BMI Measurements

Men	Women
23.8 (11.5)	19.6 (2.5)
23.2 (9)	23.8 (11.5)
24.6 (14)	19.6 (2.5)
26.2 (17)	29.1 (22)
23.5 (10)	25.2 (15.5)
24.5 (13)	21.4 (5)
21.5 (6)	22.0 (7)
31.4 (24)	27.5 (19)
26.4 (18)	33.5 (25)
22.7 (8)	20.6 (4)
27.8 (20)	29.9 (23)
28.1 (21)	17.7 (1)
25.2 (15.5)	
$n_1 = 13$	$n_2 = 12$
$R_1 = 187$	$R_2 = 138$

SOLUTION

REQUIREMENTS ✔ The Wilcoxon rank-sum test requires two independent and random samples, each with more than 10 values. The sample data are independent and random, and the sample sizes are 13 and 12. The requirements are satisfied and we can proceed with the test. ✔

The null and alternative hypotheses are as follows:

H_0: Men and women have BMI values with equal medians

H_1: Men and women have BMI values with medians that are not equal.

Rank all 25 BMI measurements combined, beginning with a rank of 1 (assigned to the lowest value of 17.7). Ties in ranks are handled as described in Section 13-1: Find the mean of the ranks involved and assign this mean rank to each of the tied values. (The 2nd and 3rd values are both 19.6, so assign a rank of 2.5 to each of those values. The 11th and 12th values are both 23.8, so assign

continued

the rank of 11.5 to each of those values. The 15th and 16th values are both 25.2, so assign a rank of 15.5 to each of those values.) The ranks corresponding to the individual sample values are shown in parentheses in Table 13-5. R denotes the sum of the ranks for the sample we choose as Sample 1. If we choose the BMI values for men, we get

$$R = 11.5 + 9 + 14 + \cdots + 15.5 = 187$$

Because there are 13 values for men, we have $n_1 = 13$. Also, $n_2 = 12$ because there are 12 values for women. We can now find the values of μ_R, σ_R, and the test statistic z.

$$\mu_R = \frac{n_1(n_1 + n_2 + 1)}{2} = \frac{13(13 + 12 + 1)}{2} = 169$$

$$\sigma_R = \sqrt{\frac{n_1 n_2(n_1 + n_2 + 1)}{12}} = \sqrt{\frac{(13)(12)(13 + 12 + 1)}{12}} = 18.385$$

$$z = \frac{R - \mu_R}{\sigma_R} = \frac{187 - 169}{18.385} = 0.98$$

The test is two-tailed because a large positive value of z would indicate that the higher ranks are found disproportionately in Sample 1, and a large negative value of z would indicate that Sample 1 had a disproportionate share of lower ranks. In either case, we would have strong evidence against the claim that the two samples come from populations with equal medians.

The significance of the test statistic z can be treated in the same manner as in previous chapters. We are now testing (with $\alpha = 0.05$) the hypothesis that the two populations have equal medians, so we have a two-tailed test with critical z values of 1.96 and -1.96. The test statistic of $z = 0.98$ does *not* fall within the critical region, so we fail to reject the null hypothesis that men and women have BMI values with equal medians. It appears that BMI values of men and women are basically the same.

We can verify that if we interchange the two sets of sample values and consider the sample of BMI values of women to be first, $R = 138$, $\mu_R = 156$, $\sigma_R = 18.385$, and $z = -0.98$, so the conclusion is exactly the same.

EXAMPLE BMI of Men and Women The preceding example used only 13 of the 40 sample BMI values for men listed in Data Set 1 in Appendix B, and it used only 12 of the 40 BMI values for women. Do the results change if we use all 40 sample values for men and all 40 sample values for women? Use the Wilcoxon rank-sum test.

SOLUTION

REQUIREMENTS ✔ As in the preceding example, the sample data are independent and random. Also, both sample sizes are greater than 10. The requirements are satisfied and we can proceed with the test. ✔

The null and alternative hypotheses are the same as in the preceding example. Instead of manually calculating the rank sums, we use the TI-83/84 Plus program RSTEST to obtain the test statistic $z = 1.03$ (rounded). The critical values found in Table A-2 are $z = -1.96$ and $z = 1.96$. Because the test statistic is between the critical values, we fail to reject the null hypothesis. There is not sufficient evidence to warrant rejection of the claim that men and women have BMI values with equal medians. It is common knowledge that men are generally taller and heavier than women, but the data used here suggest that BMI values are about the same for men and women.

TI

Using the TI-83/84 Plus Calculator for the Wilcoxon Rank-Sum Test

The TI-83/84 Plus calculator is not programmed for the Wilcoxon rank-sum test, but the program RSTEST can be used. Use the following procedure:

1. First download and install the program RSTEST (rank-sum test) from the CD-ROM or the Web site www.aw.com/triola. (If using the CD-ROM, select the **Software** folder, select **TI-83 TI-84 Plus Programs,** select **Michael Lloyd Programs,** then select **programs.** The program RSTEST should now be available for download.)
2. Also download the program ZZRANK, which is used by the program RSTEST.
3. Enter the two sets of sample data as lists in L1 and L2 (or use named lists).
4. Press the **PRGM** key and select RSTEST and press ENTER twice.
5. When given the prompt of GROUP A=, enter L1 (or the name of a list) and press ENTER.
6. When given the prompt of GROUP B=, enter L2 (or the name of a list) and press ENTER.
7. Press ENTER once again to get the test statistic z.
8. Find the critical values by referring to Table A-2.
9. Using the calculator results: The program RSTEST will provide a test statistic z, rank sum R, mean, and standard deviation by using calculations based on the sample with the fewer number of values (instead of using the first sample). Consequently, the program RSTEST might provide the same test statistic that would be calculated with the methods of this section, or it might provide a test statistic with an opposite sign. *Use the test statistic provided by RSTEST and the conclusion will be the same as the conclusion reached by using the methods of this section.*

For the first example in this section, the test statistic will be displayed as $z = -0.98$ (rounded). The critical values are $z = -1.96$ and $z = 1.96$. Because the test statistic is between the critical values, we fail to reject the null hypothesis that the median BMI of men is equal to the median BMI of women.

13-4 BASIC SKILLS AND CONCEPTS

Statistical Literacy and Critical Thinking

1. Wilcoxon Rank-Sum Test What is the most fundamental difference between the Wilcoxon signed-ranks test and the Wilcoxon rank-sum test?

2. Wilcoxon Rank-Sum Test The test statistic for the Wilcoxon rank-sum test is based on the rank sum R, which has a distribution that is approximately normal. Is the Wilcoxon rank-sum test a parametric test because it requires a normal distribution?

3. Wilcoxon Rank-Sum Test The Wilcoxon rank-sum test and the methods of hypothesis testing described in Section 9-3 both apply to two independent samples. What advantage does the Wilcoxon rank-sum test have over the test described in Section 9-3?

4. Efficiency Refer to Table 13-2 and identify the efficiency of the Wilcoxon rank-sum test. What does that value tell us about the test?

Identifying Rank Sums. *In Exercises 5 and 6, use a 0.05 significance level with the methods of this section to identify the rank sums R_1 and R_2, μ_R, σ_R, the test statistic z, the critical z values, and then state the conclusion about a claim of equal medians.*

TI **5.**

Sample 1 values:	2	7	10	16	20	22	23	26	27	30	33	
Sample 2 values:	3	4	11	14	28	35	40	46	47	52	53	60

TI **6.**

Sample 1 values:	8	15	27	39	45	62	68	72	77	80	87
Sample 2 values:	3	5	9	11	14	21	33	44	61	70	85

Using the Wilcoxon Rank-Sum Test. *In Exercises 7–12, use the Wilcoxon rank-sum test.*

TI **7.** Are Severe Psychiatric Disorders Related to Biological Factors? One study used x-ray computed tomography (CT) to collect data on brain volumes for a group of patients with obsessive-compulsive disorders and a control group of healthy persons. The accompanying list shows sample results (in milliliters) for volumes of the right cordate (based on data from "Neuroanatomical Abnormalities in Obsessive-Compulsive Disorder Detected with Quantitative X-Ray Computed Tomography," by Luxenberg et al., *American Journal of Psychiatry,* Vol. 145, No. 9). Use a 0.01 significance level to test the claim that obsessive-compulsive patients and healthy persons have the same median brain volumes. Based on this result, can we conclude that obsessive-compulsive disorders have a biological basis?

Obsessive-Compulsive Patients				**Control Group**			
0.308	0.210	0.304	0.344	0.519	0.476	0.413	0.429
0.407	0.455	0.287	0.288	0.501	0.402	0.349	0.594
0.463	0.334	0.340	0.305	0.334	0.483	0.460	0.445

TI **8.** Testing the Anchoring Effect Randomly selected statistics students were given 5 sec to estimate the value of a product of numbers with the results given in the accompanying table. (See the Cooperative Group Activities at the end of Chapter 3.) Do the samples appear to be significantly different?

Estimates from Students Given $1 \times 2 \times 3 \times 4 \times 5 \times 6 \times 7 \times 8$									
1560	169	5635	25	842	40,320	5000	500	1110	10,000
200	1252	4000	2040	175	856	42,200	49,654	560	800

Estimates from Students Given 8 × 7 × 6 × 5 × 4 × 3 × 2 × 1										
100,000	2000	42,000	1500	52,836	2050	428	372	300	225	64,582
23,410	500	1200	400	49,000	4000	1876	3600	354	750	640

TI **9.** Hypothesis Test for Difference in Car/Taxi Ages When the author visited Dublin in Ireland, he recorded the ages of randomly selected passenger cars and randomly selected taxis. The ages (in years) are listed below. Use a 0.05 significance level to test the claim that there is a difference between the median age of a Dublin car and the median age of a Dublin taxi. We might expect that taxis would be newer, but what do the results suggest?

Cars	Taxis
4 0 8 11 14 3 4 4 3 5	8 8 0 3 8 4 3 3 6 11
8 3 3 7 4 6 6 1 8 2 15	7 7 6 9 5 10 8 4 3 4
11 4 1 6 1 8	

TI **10.** Pulse Rates Refer to Data Set 1 in Appendix B for the pulse rates of men and women. Use only the first 13 pulse rates of men and use only the first 12 pulse rates of women to test the claim that the two samples of pulse rates come from populations with the same median. Use a 0.05 significance level.

TI **11.** Appendix B Large Data Set: Pulse Rates Repeat Exercise 10 using all 40 pulse rates of men and all 40 pulse rates of women.

TI **12.** Appendix B Large Data Set: Weights of Pennies Refer to Data Set 14 in Appendix B and use the weights of the wheat pennies and the weights of the pre-1983 pennies. Use a 0.05 significance level to test the claim that those two samples are from populations with the same median.

13-4 BEYOND THE BASICS

TI **13.** Using the Mann-Whitney U Test The Mann-Whitney U test is equivalent to the Wilcoxon rank-sum test for independent samples in the sense that they both apply to the same situations and always lead to the same conclusions. In the Mann-Whitney U test we calculate

$$z = \frac{U - \dfrac{n_1 n_2}{2}}{\sqrt{\dfrac{n_1 n_2 (n_1 + n_2 + 1)}{12}}}$$

where

$$U = n_1 n_2 + \frac{n_1(n_1 + 1)}{2} - R$$

Using the BMI measures listed in Table 13-5 in this section, find the z test statistic for the Mann-Whitney U test and compare it to the z test statistic that was found using the Wilcoxon rank-sum test.

14. Finding Critical Values Assume that we have two treatments (A and B) that produce quantitative results, and we have only two observations for treatment A and two observations for treatment B. We cannot use the test statistic given in this section because both sample sizes do not exceed 10.

Rank				Rank sum for
1	2	3	4	treatment A
A	A	B	B	3

a. Complete the accompanying table by listing the five rows corresponding to the other five cases, and enter the corresponding rank sums for treatment A.
b. List the possible values of R, along with their corresponding probabilities. [Assume that the rows of the table from part (a) are equally likely.]
c. Is it possible, at the 0.10 significance level, to reject the null hypothesis that there is no difference between treatments A and B? Explain.

13-5 Kruskal-Wallis Test

Key Concept This section introduces the *Kruskal-Wallis test,* which uses ranks of data from three or more independent samples to test the null hypothesis that the samples come from populations with equal medians.

In Section 12-2 we used one-way analysis of variance (ANOVA) to test the null hypothesis that three or more populations have the same mean, but ANOVA requires that all of the involved populations have normal distributions. The Kruskal-Wallis test does not require normal distributions.

Definition

The **Kruskal-Wallis Test** (also called the ***H*** **test**) is a nonparametric test that uses ranks of sample data from three or more independent populations. It is used to test the null hypothesis that the independent samples come from populations with equal medians; the alternative hypothesis is the claim that the populations have medians that are not all equal.

H_0: The samples come from populations with equal medians.

H_1: The samples come from populations with medians that are not all equal.

In applying the Kruskal-Wallis test, we compute the *test statistic H, which has a distribution that can be approximated by the chi-square distribution as long as each sample has at least five observations.* When we use the chi-square distribution in this context, the number of degrees of freedom is $k - 1$, where k is the number of samples. (For a quick review of the key features of the chi-square distribution, see Section 7-5.)

Kruskal-Wallis Test

Requirements

1. We have at least three independent samples, all of which are randomly selected.
2. Each sample has at least five observations. (If samples have fewer than five observations, refer to special tables of critical values, such as *CRC Standard Probability and Statistics Tables and Formulae,* published by CRC Press.)
3. There is *no* requirement that the populations have a normal distribution or any other particular distribution.

Notation

N = total number of observations in all samples combined

k = number of samples

R_1 = sum of ranks for Sample 1 calculated with the procedure that follows

n_1 = number of observations in Sample 1

For Sample 2, the sum of ranks is R_2 and the number of observations is n_2, and similar notation is used for the other samples.

Test Statistic

$$H = \frac{12}{N(N+1)}\left(\frac{R_1^2}{n_1} + \frac{R_2^2}{n_2} + \cdots + \frac{R_k^2}{n_k}\right) - 3(N+1)$$

Critical values

1. The test is *right-tailed.*
2. df = $k - 1$. (Because the test statistic H can be approximated by a chi-square distribution, use Table A-4 with $k - 1$ degrees of freedom, where k is the number of different samples.)

Procedure for Finding the Value of the Test Statistic *H*

1. Temporarily combine all samples into one big sample and assign a rank to each sample value. (Sort the values from lowest to highest, and in cases of ties, assign to each observation the mean of the ranks involved.)
2. For each sample, find the sum of the ranks and find the sample size.
3. Calculate H by using the results of Step 2 and the notation and test statistic given in the preceding box.

The test statistic H is basically a measure of the variance of the rank sums R_1, $R_2, \ldots, R_k$. If the ranks are distributed evenly among the sample groups, then H should be a relatively small number. If the samples are very different, then the ranks will be excessively low in some groups and high in others, with the net effect that H will be large. Consequently, only large values of H lead to rejection of the null hypothesis that the samples come from identical populations. *The Kruskal-Wallis test is therefore a right-tailed test.*

Table 13-6 Weights (kg) of Poplar Trees

Treatment			
None	Fertilizer	Irrigation	Fertilizer and Irrigation
0.15 (8)	1.34 (18)	0.23 (12)	2.03 (19)
0.02 (1.5)	0.14 (7)	0.04 (3)	0.27 (13)
0.16 (9.5)	0.02 (1.5)	0.34 (14)	0.92 (16)
0.37 (15)	0.08 (5.5)	0.16 (9.5)	1.07 (17)
0.22 (11)	0.08 (5.5)	0.05 (4)	2.38 (20)
$n_1 = 5$	$n_2 = 5$	$n_3 = 5$	$n_4 = 5$
$R_1 = 45$	$R_2 = 37.5$	$R_3 = 42.5$	$R_4 = 85$

EXAMPLE Effects of Treatments on Poplar Tree Weights Table 13-6 lists weights (in kg) of poplar trees given different treatments. In Section 12-2 we used analysis of variance to test the null hypothesis that the four samples of weights come from populations with the same mean. We will now use the Kruskal-Wallis test of the null hypothesis that the four samples come from populations with equal medians.

SOLUTION

REQUIREMENT ✔ The Kruskal-Wallis test requires three or more independent and random samples, each with at least 5 values. Each of the four samples is independent and random, and each sample size is 5. Having satisfied the requirements, we can proceed with the test. ✔

The null and alternative hypotheses are as follows:

H_0: The populations of poplar tree weights from the four treatments have equal medians.

H_1: The four population medians are not all equal.

In determining the value of the test statistic H, we must first rank all of the data. We begin with the lowest values of 0.02 and 0.02. Because there is a tie between the values corresponding to ranks 1 and 2, we assign the mean rank of 1.5 to each of those tied items. In Table 13-6, ranks are shown in parentheses next to the original tree weights. Next we find the sample size, n, and sum of ranks, R, for each sample, and those values are listed at the bottom of Table 13-6. Because the total number of observations is 20, we have $N = 20$. We can now evaluate the test statistic as follows:

$$H = \frac{12}{N(N+1)}\left(\frac{R_1^2}{n_1} + \frac{R_2^2}{n_2} + \cdots + \frac{R_k^2}{n_k}\right) - 3(N+1)$$

$$= \frac{12}{20(20+1)}\left(\frac{45^2}{5} + \frac{37.5^2}{5} + \frac{42.5^2}{5} + \frac{85^2}{5}\right) - 3(20+1)$$

$$= 8.214$$

Because each sample has at least five observations, the distribution of H is approximately a chi-square distribution with $k - 1$ degrees of freedom. The number of samples is $k = 4$, so we have $4 - 1 = 3$ degrees of freedom. Refer to Table A-4 to find the critical value of 7.815, which corresponds to 3 degrees of freedom and a 0.05 significance level (with an area of 0.05 in the right tail).

The test statistic $H = 8.214$ is in the critical region bounded by 7.815, so we reject the null hypothesis of equal medians. (In Section 12-2, we rejected the null hypothesis of equal means.)

INTERPRETATION There is sufficient evidence to reject the claim that the populations of poplar tree weights from the four treatments have equal medians. At least one of the medians appears to be different from the others.

Rationale: The Kruskal-Wallis test statistic H is the rank version of the test statistic F used in the analysis of variance discussed in Chapter 12. When we deal with ranks R instead of original values x, many components are predetermined. For example, the sum of all ranks can be expressed as $N(N + 1)/2$, where N is the total number of values in all samples combined. The expression

$$H = \frac{12}{N(N+1)} \Sigma n_i(\overline{R}_i - \overline{\overline{R}})^2$$

where

$$\overline{R}_i = \frac{R_i}{n_i} \quad \text{and} \quad \overline{\overline{R}} = \frac{\Sigma R_i}{\Sigma n_i}$$

combines weighted variances of ranks to produce the test statistic H given here. This expression for H is algebraically equivalent to the expression for H given earlier as the test statistic.

TI

Using the TI-83/84 Plus Calculator for the Kruskal-Wallis Test

The TI-83/84 Plus calculator is not programmed for the Kruskal-Wallis test, but the program KWTEST can be used. Use the following procedure:

1. First download and install the program KWTEST (Kruskal-Wallis test) from the CD-ROM or the Web site www.aw.com/triola. (If using the CD-ROM, select the **Software** folder, select **TI-83 TI-84 Plus Programs,** select **Michael Lloyd Programs,** then select **programs.** The program KWTEST should now be available for download.)
2. Also download the program ZZRANK, which is used by the program KWTEST.
3. Enter the lists of sample data in separate columns of matrix [A]. The margin display shows a portion of the matrix corresponding to the sample data in Table 13-6 of this section. (Press 2ND x^{-1} to get the **MATRIX** menu, then press ▷ twice to scroll to **EDIT,** then press ENTER to select the first item of [A]. Enter the dimensions of the matrix, where the first dimension is the sample size of the largest sample and the second dimension is the number of different sample lists. *Note:* If the samples have different sizes and one of the data values is zero, add some convenient constant to all of the sample values so that no zeros are present.)

Entering Lists in Matrix [A]

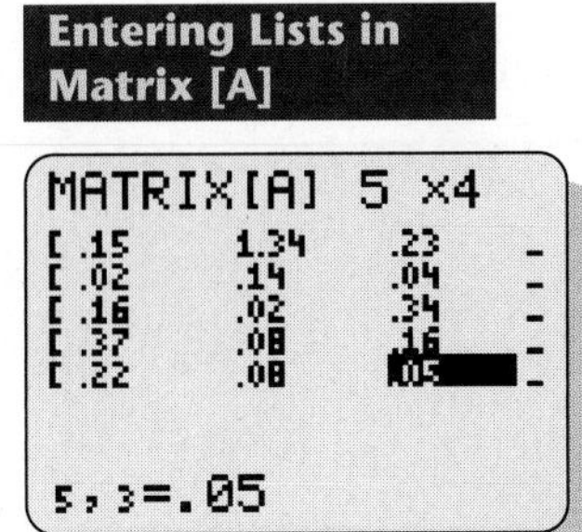

4. Press the **PRGM** key, select KWTEST, then press ENTER. The value of the H test statistic and the number of degrees of freedom will be provided. Refer to Table A-4 to find the critical value.

If the program KWTEST is used with the sample data in Table 13-6, the display will show the test statistic of $H = 8.214$ (rounded) and $DF = 3$, indicating that the number of degrees of freedom is 3. See the example and solution in this section.

13-5 BASIC SKILLS AND CONCEPTS

Statistical Literacy and Critical Thinking

1. Kruskal-Wallis Test What major advantage does the Kruskal-Wallis test have over the one-way analysis of variance test?

2. Kruskal-Wallis Test If the Kruskal-Wallis test and one-way analysis of variance are both used with three sets of independent samples, will both tests always result in the same conclusion?

3. Efficiency Refer to Table 13-2 and identify the efficiency of the Kruskal-Wallis test. What does that value tell us about the test?

4. Requirements Fifty families are randomly selected and IQ tests are given to the mother, father, and firstborn child. Can the Kruskal-Wallis test be used to test the claim that the three populations of mothers, fathers, and firstborn children have IQ scores with the same median? Why or why not?

Using the Kruskal-Wallis Test. In Exercises 5–10, use the Kruskal-Wallis test.

TI 5. Does the Weight of a Car Affect Head Injuries in a Crash? Data were obtained from car crash experiments conducted by the National Transportation Safety Administration. New cars were purchased and crashed into a fixed barrier at 35 mi/h, and measurements were recorded for the dummy in the driver's seat. Use the sample data listed below to test for differences in head injury measurements (in hic) among the four weight categories. Is there sufficient evidence to conclude that head injury measurements for the four car weight categories are not all the same? Do the data suggest that heavier cars are safer in a crash?

Subcompact:	681	428	917	898	420
Compact:	643	655	442	514	525
Midsize:	469	727	525	454	259
Full-size:	384	656	602	687	360

TI 6. Does the Weight of a Car Affect Chest Injuries in a Crash? Data were obtained from car crash experiments conducted by the National Transportation Safety Administration. New cars were purchased and crashed into a fixed barrier at 35 mi/h, and the chest deceleration data (g) are given below. Use the sample data listed below to test the null hypothesis that the different weight categories have medians that are not all the same. Do the data suggest that heavier cars are safer in a crash?

Subcompact:	55	47	59	49	42
Compact:	57	57	46	54	51
Midsize:	45	53	49	51	46
Full-size:	44	45	39	58	44

TI **7.** Is Solar Energy the Same Every Day? A student of the author lives in a home with a solar electric system. At the same time each day, she collected voltage readings from a meter connected to the system and the results are listed in the accompanying table. Use a 0.05 significance level to test the claim that voltage readings have the same median for the three different types of day. We might expect that a solar system would provide more electrical energy on sunny days than on cloudy or rainy days. Can we conclude that sunny days result in greater amounts of electrical energy?

Data for Exercise 7

Sunny	Cloudy	Rainy
13.5	12.7	12.1
13.0	12.5	12.2
13.2	12.6	12.3
13.9	12.7	11.9
13.8	13.0	11.6
14.0	13.0	12.2

TI **8.** Testing for Skull-Breadth Differences in Different Times The accompanying values are measured maximum breadths of male Egyptian skulls from different epochs (based on data from *Ancient Races of the Thebaid,* by Thomson and Randall-Maciver). Changes in head shape over time suggest that interbreeding occurred with immigrant populations. Use a 0.05 significance level to test the claim that the three samples come from identical populations. Is interbreeding of cultures suggested by the data?

Data for Exercise 8

4000 B.C.	1850 B.C.	150 A.D.
131	129	128
138	134	138
125	136	136
129	137	139
132	137	141
135	129	142
132	136	137
134	138	145
138	134	137

TI **9.** Exercise and Stress A study was conducted to investigate the effects of exercise on stress. The table below lists systolic blood pressure readings (in mmHg) of subjects from the time preceding 25 minutes of aerobic bicycle exercise and preceding the introduction of stress through arithmetic and speech tests (based on data from "Sympathoadrenergic Mechanisms in Reduced Hemodynamic Stress Responses after Exercise," by Kim Brownley et al., *Medicine and Science in Sports and Exercise*, Vol. 35, No. 6). Use a 0.05 significance level to test the claim that the different groups of subjects have the same median blood pressure. Based on the results, can those groups be considered to be samples all from the same population?

Female/Black	Male/Black	Female/White	Male/White
117.00	115.67	119.67	124.33
130.67	120.67	106.00	111.00
102.67	133.00	108.33	99.67
93.67	120.33	107.33	128.33
96.33	124.67	117.00	102.00
92.00	118.33	113.33	127.33

TI **10.** Laboratory Testing of Flammability of Children's Sleepwear Flammability tests were conducted on children's sleepwear. The Vertical Semirestrained Test was used, in which pieces of fabric were burned under controlled conditions. After the burning stopped, the length of the charred portion was measured and recorded. Results are given in the margin for the same fabric tested at different laboratories. Because the same fabric was used, the different laboratories should have obtained the same results. Did they?

Data for Exercise 10

Laboratory				
1	2	3	4	5
2.9	2.7	3.3	3.3	4.1
3.1	3.4	3.3	3.2	4.1
3.1	3.6	3.5	3.4	3.7
3.7	3.2	3.5	2.7	4.2
3.1	4.0	2.8	2.7	3.1
4.2	4.1	2.8	3.3	3.5
3.7	3.8	3.2	2.9	2.8
3.9	3.8	2.8	3.2	
3.1	4.3	3.8	2.9	
3.0	3.4	3.5		
2.9	3.3			

In Exercises 11 and 12, use the Kruskal-Wallis test with the data set from Appendix B.

TI **11.** Appendix B Data Set: Weights of Pennies Refer to Data Set 14 in Appendix B and use the weights of the Indian pennies, wheat pennies, pre-1983 pennies, and post-1983 pennies. Use a 0.05 significance level to test the claim that the median weight of pennies is the same for the four different categories. Based on the results, does it appear that a coin-counting machine can treat the weights of the pennies the same way?

TI **12.** Appendix B Data Set: Do All Colors of M&Ms Weigh the Same? Refer to Data Set 13 in Appendix B. At the 0.05 significance level, test the claim that the median weights of M&Ms are the same for each of the six different color populations. If it is the intent of Mars, Inc. to make the candies so that the different color populations are the same, do your results suggest that the company has a problem that requires corrective action?

13-5 BEYOND THE BASICS

13. Correcting the H Test Statistic for Ties In using the Kruskal-Wallis test, there is a correction factor that should be applied whenever there are many ties: Divide H by

$$1 - \frac{\Sigma T}{N^3 - N}$$

For each group of tied observations in the combined set of all sample data, calculate $T = t^3 - t$, where t is the number of observations that are tied within the individual group. Find t for each group of tied values, then compute the value of T for each group, then add the T values to get ΣT. The total number of observations in all samples combined is N. Use this procedure to find the corrected value of H for Exercise 7. Does the corrected value of H differ substantially from the value found in Exercise 7?

14. Equivalent Tests Show that for the case of two samples, the Kruskal-Wallis test is equivalent to the Wilcoxon rank-sum test. This can be done by showing that for the case of two samples, the test statistic H equals the square of the test statistic z used in the Wilcoxon rank-sum test. Also note that with 1 degree of freedom, the critical values of χ^2 correspond to the square of the critical z score.

13-6 Rank Correlation

Key Concept This section describes the nonparametric method of rank correlation, which uses paired data to test for an association between two variables. In Chapter 10 we used paired sample data to compute values for the linear correlation coefficient r, but in this section we use *ranks* as the basis for computing the rank correlation coefficient r_s.

Definition

The **rank correlation test** (or **Spearman's rank correlation test**) is a nonparametric test that uses ranks of sample data consisting of matched pairs. It is used to test for an association between two variables, so the null and alternative hypotheses are as follows (where ρ_s denotes the rank correlation coefficient for the entire population):

H_0: $\rho_s = 0$ (There is *no* correlation between the two variables.)

H_1: $\rho_s \neq 0$ (There is a correlation between the two variables.)

Advantages: Rank correlation has these advantages over the parametric methods discussed in Chapter 10:

1. The nonparametric method of rank correlation can be used in a wider variety of circumstances than the parametric method of linear correlation. With rank correlation, we can analyze paired data that are ranks or can be converted to ranks. For example, if two judges rank 30 different gymnasts, we can use rank correlation, but not linear correlation. Unlike the parametric methods of Chapter 10, the method of rank correlation does *not* require a normal distribution for any population.
2. Rank correlation can be used to detect some (not all) relationships that are not linear. (An example will be given later in this section.)

Disadvantage: A disadvantage of rank correlation is its efficiency rating of 0.91, as described in Section 13-1. This efficiency rating shows that with all other circumstances being equal, the nonparametric approach of rank correlation requires 100 pairs of sample data to achieve the same results as only 91 pairs of sample observations analyzed through the parametric approach, assuming that the stricter requirements of the parametric approach are met.

We use the notation r_s for the rank correlation coefficient so that we don't confuse it with the linear correlation coefficient r. The subscript s has nothing to do with standard deviation; it is used in honor of Charles Spearman (1863–1945), who originated the rank correlation approach. In fact, r_s is often called **Spearman's rank correlation coefficient.** The rank correlation procedure is summarized in Figure 13-4.

Rank Correlation

Requirements

1. The sample paired data have been randomly selected.
2. Unlike the parametric methods of Section 10-2, there is *no* requirement that the sample pairs of data have a bivariate normal distribution (as described in Section 10-2). There is *no* requirement of a normal distribution for any population.

Notation

r_s = rank correlation coefficient for sample paired data (r_s is a sample statistic)

ρ_s = rank correlation coefficient for all the population data (ρ_s is a population parameter)

n = number of pairs of sample data

d = difference between ranks for the two values within a pair

Test Statistic

No ties: After converting the data in each sample to ranks, if there are no ties among ranks for the first variable and there are no ties among ranks for the second variable, the exact value of the test statistic can be calculated using this formula:

$$r_s = 1 - \frac{6\Sigma d^2}{n(n^2 - 1)}$$

Ties: After converting the data in each sample to ranks, if either variable has ties among its ranks, the exact value of the test statistic r_s can be found by using Formula 10-1 with the ranks:

$$r_s = \frac{n\Sigma xy - (\Sigma x)(\Sigma y)}{\sqrt{n(\Sigma x^2) - (\Sigma x)^2}\sqrt{n(\Sigma y^2) - (\Sigma y)^2}}$$

Critical values

1. If $n \leq 30$, critical values are found in Table A-9.
2. If $n > 30$, critical values of r_s are found by using Formula 13-1.

Formula 13-1 $$r_s = \frac{\pm z}{\sqrt{n - 1}}$$ (critical values when $n > 30$)

where the value of z corresponds to the significance level. (For example, if $\alpha = 0.05$, $z = 1.96$.)

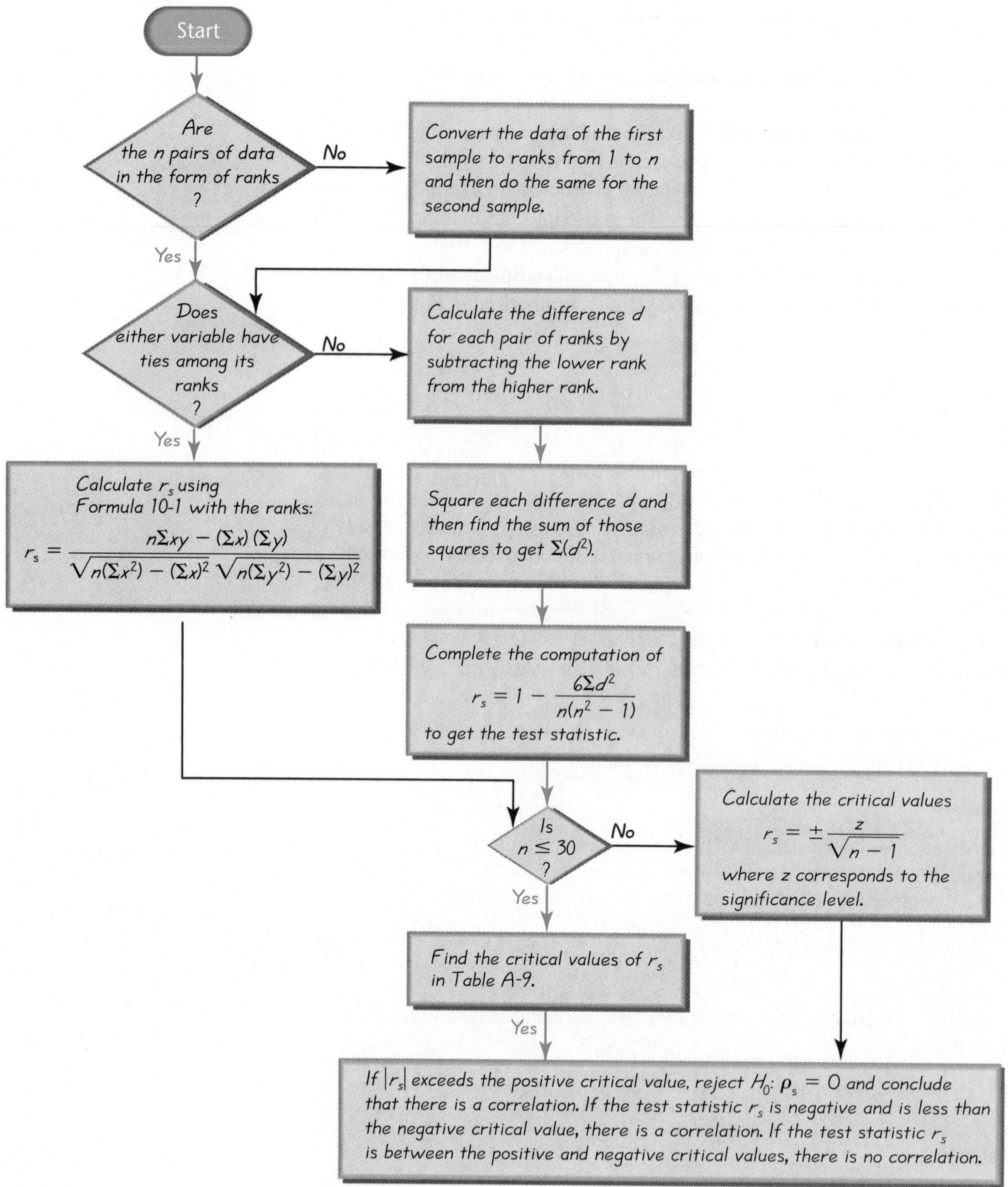

Figure 13-4 Rank Correlation Procedure for Testing H_0: $\rho_s = 0$

Table 13-7 Colleges Ranked by Students and *U.S. News and World Report*

College	Student Ranks	*U.S. News and World Report* Ranks	Difference d	d^2
Harvard	1	1	0	0
Yale	2	2	0	0
Cal. Inst. of Tech.	3	5	2	4
M.I.T.	4	4	0	0
Brown	5	7	2	4
Columbia	6	6	0	0
U. of Penn.	7	3	4	16
Notre Dame	8	8	0	0
			Total:	24

CHAPTER 13 PROBLEM

EXAMPLE Student and *U.S. News and World Report* Rankings of Colleges The Chapter Problem includes rankings of colleges by students and *U.S. News and World Report* magazine. Those rankings are included in Table 13-7, which also includes the differences d and the squares of the differences d^2. Find the value of the rank correlation coefficient and use it to determine whether there is a correlation between the student rankings and the rankings of the magazine. Use a 0.05 significance level.

SOLUTION

REQUIREMENT ✔ The only requirement is that the sample paired data have been randomly selected. The colleges included have been randomly selected from those available, so we can proceed with the test. ✔

The linear correlation coefficient r (Section 10-2) should not be used because it requires normal distributions, but the data consist of ranks, which are not normally distributed. Instead, we use the rank correlation coefficient to test for a relationship between the ranks of students and the magazine.

The null and alternative hypotheses are as follows:

$$H_0: \rho_s = 0$$
$$H_1: \rho_s \neq 0$$

Following the procedure of Figure 13-4, the data are in the form of ranks and neither of the two variables has ties among ranks, so the exact value of the test statistic can be calculated as shown below. We use $n = 8$ (for 8 pairs of data) and $\Sigma d^2 = 24$ (as shown in Table 13-7) to get

$$r_s = 1 - \frac{6\Sigma d^2}{n(n^2 - 1)} = 1 - \frac{6(24)}{8(8^2 - 1)}$$

$$= 1 - \frac{144}{504} = 0.714$$

continued

Direct Link Between Smoking and Cancer

When we find a statistical correlation between two variables, we must be extremely careful to avoid the mistake of concluding that there is a cause-effect link. The tobacco industry has consistently emphasized that correlation does not imply causality. However, Dr. David Sidransky of Johns Hopkins University now says that "we have such strong molecular proof that we can take an individual cancer and potentially, based on the patterns of genetic change, determine whether cigarette smoking was the cause of that cancer." Based on his findings, he also said that "the smoker had a much higher incidence of the mutation, but the second thing that nailed it was the very distinct pattern of mutations . . . so we had the smoking gun." Although statistical methods cannot prove that smoking *causes* cancer, such proof can be established with physical evidence of the type described by Dr. Sidransky.

Now we refer to Table A-9 to determine that the critical values are ±0.738 (based on $\alpha = 0.05$ and $n = 8$). Because the test statistic $r_s = 0.714$ does not exceed the critical value of 0.738, we fail to reject the null hypothesis. There is not sufficient evidence to support a claim of a correlation between the rankings of the students and the magazine. It appears that when it comes to ranking colleges, students and the magazine do not agree. (If they did agree, there would be a significant correlation, but there is not.)

EXAMPLE Large Sample Case Assume that the preceding example is expanded by including a total of 40 colleges and that the test statistic r_s is found to be 0.300. If the significance level is $\alpha = 0.05$, what do you conclude about the correlation?

SOLUTION Because there are 40 pairs of data, we have $n = 40$. Because n exceeds 30, we find the critical values from Formula 13-1 instead of Table A-9. With $\alpha = 0.05$ in two tails, we let $z = 1.96$ to get

$$r_s = \frac{\pm 1.96}{\sqrt{40 - 1}} = \pm 0.314$$

The test statistic of $r_s = 0.300$ does not exceed the critical value of 0.314, so we fail to reject the null hypothesis. There is not sufficient evidence to support the claim of a correlation between students and the magazine.

The next example is intended to illustrate the principle that rank correlation can sometimes be used to detect relationships that are not linear.

EXAMPLE Detecting a Nonlinear Pattern A *Raiders of the Lost Ark* pinball machine (model L-7) is used to measure learning that results from repeating manual functions. Subjects were selected so that they are similar in important characteristics of age, gender, intelligence, education, and so on. Table 13-8 lists the numbers of games played and the last scores (in millions) for subjects randomly selected from the group with similar characteristics. We expect that there should be an association between the number of games played and the pinball score. Is there sufficient evidence to support the claim that there is such an association?

Table 13-8 Pinball Scores (Ranks in parentheses)

Number of Games Played	9 (2)	13 (4)	21 (5)	6 (1)	52 (7)	78 (8)	33 (6)	11 (3)	120 (9)
Score	22 (2)	62 (4)	70 (6)	10 (1)	68 (5)	73 (8)	72 (7)	58 (3)	75 (9)
d	0	0	1	0	2	0	1	0	0
d^2	0	0	1	0	4	0	1	0	0

SOLUTION We will test the null hypothesis of no rank correlation ($\rho_s = 0$).

$$H_0: \rho_s = 0 \quad \text{(no correlation)}$$
$$H_1: \rho_s \neq 0 \quad \text{(correlation)}$$

Refer to Figure 13-4, which we follow in this solution. The original scores are not ranks, so we converted them to ranks and entered the results in parentheses in Table 13-8. (Section 13-1 describes the procedure for converting scores into ranks.) There are no ties among the ranks for the numbers of games played, nor are there ties among the ranks for the scores, so we proceed by calculating the differences, d, and squaring them, and finding the sum of the d^2 values, which is 6. We now calculate

$$r_s = 1 - \frac{6\Sigma d^2}{n(n^2 - 1)} = 1 - \frac{6(6)}{9(9^2 - 1)}$$
$$= 1 - \frac{36}{720} = 0.950$$

Proceeding with Figure 13-4, we have $n = 9$, so we answer yes when asked if $n \leq 30$. We use Table A-9 to get the critical values of ± 0.700. Finally, the sample statistic of 0.950 exceeds 0.700, so we conclude that there is a significant correlation. Higher numbers of games played appear to be associated with higher scores. Subjects appeared to better learn the game by playing more.

In the preceding example, if we compute the linear correlation coefficient r (using Formula 9-1) for the original data, we get $r = 0.586$, which leads to the conclusion that there is not enough evidence to support the claim of a significant linear correlation at the 0.05 significance level. If we examine the TI-83/84 Plus scatter diagram, we can see that the pattern of points is not a straight-line pattern. This last example illustrates an advantage of the nonparametric approach over the parametric approach: With rank correlation, we can sometimes detect relationships that are not linear.

TI-83/84 Plus

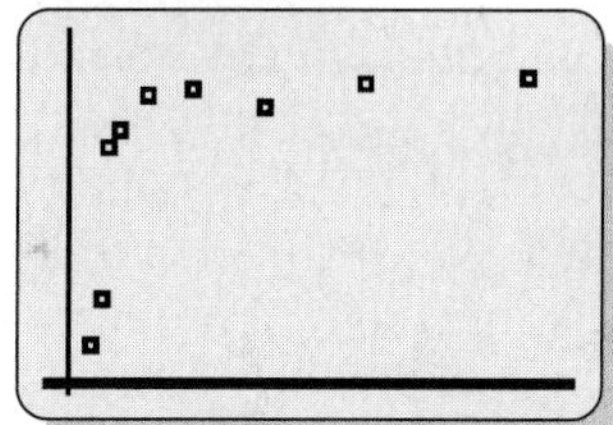

TI

Using the TI-83/84 Plus Calculator for Rank Correlation

The TI-83/84 Plus calculator is not designed to calculate the rank correlation coefficient from original sample values, but we can use it for that purpose. The basic idea is to replace the original sample data with ranks, then proceed as if we were calculating the linear correlation coefficient as in Section 10-2.

1. If the original sample values are not already ranks, replace each of the original sample values by its corresponding rank. [Enter the values for one of the variables in list L1, then use the command **SortA(L1)** to sort those values. The ranks can be easily identified from the sorted list of sample values. Then do the same for the second sample. Be sure that the original values are replaced by their ranks in a way that matched pairs stay together.]
2. Press STAT, select the top menu item of **TESTS,** then select the menu item of **LinRegTTest.** For the entry labeled Xlist, enter the list containing the ranks for the first variable. For the entry labeled Ylist, enter the list containing the ranks for the second variable.

3. For the entry labeled Freq (frequency), enter 1.
4. For the row labeled β & ρ, choose **≠0.**
5. Scroll down to **Calculate** and press ENTER.

The value of the rank correlation coefficient will be displayed as the value labeled *r*. Also, the *P*-value will be displayed, so it could be used to determine whether there is a correlation.

13-6 BASIC SKILLS AND CONCEPTS

Statistical Literacy and Critical Thinking

1. Rank Correlation What major advantage does the rank correlation method of this section have over the linear correlation method of Section 10-2?

2. Rank Correlation If two judges each rank 25 different gymnasts from 1 through 25 with no ties, and their rankings agree perfectly, what is the value of the rank correlation coefficient?

3. Notation We represent the rank correlation coefficient test statistic with the notation r_s, and the corresponding population parameter is represented as ρ_s. Why is the subscript *s* used? Does the subscript *s* represent the same standard deviation *s* introduced in Section 3-3?

4. Efficiency Refer to Table 13-2 and identify the efficiency of the rank correlation test. What does that value tell us about the test?

In Exercises 5 and 6, sketch a scatter diagram, find the value of r_s, and determine whether there appears to be a correlation between x and y.

TI **5.**

x	2	4	1	5	3
y	2	4	1	5	3

TI **6.**

x	1	2	3	4	5
y	5	4	3	2	1

Finding Critical Values. *In Exercises 7 and 8, find the critical value(s) r_s by using either Table A-9 or Formula 13-1, as appropriate. Assume that the null hypothesis is $\rho_s = 0$, so the test is two-tailed. Also, n denotes the number of pairs of data.*

7. **a.** $n = 12, \alpha = 0.05$

b. $n = 20, \alpha = 0.01$

c. $n = 60, \alpha = 0.05$

d. $n = 80, \alpha = 0.01$

8. **a.** $n = 14, \alpha = 0.05$

b. $n = 23, \alpha = 0.01$

c. $n = 120, \alpha = 0.01$

d. $n = 75, \alpha = 0.05$

Testing for Rank Correlation. *In Exercises 9–16, use the rank correlation coefficient to test for a correlation between the two variables. Use a significance level of $\alpha = 0.05$.*

TI **9.** Stock Market and Car Sales For a recent series of 10 years, the yearly high values of the Dow Jones Industrial Average (DJIA) and the corresponding numbers of cars (in thousands) sold in the United States were obtained. The table below lists the ranks of each set of values. Test for a correlation between the DJIA and the numbers of cars sold.

DJIA high	1	2	3	4	5	6	7	8	10	9
Car sales	2	3	5	10	7	6	4	1	8	9

TI **10.** Sunspots and Super Bowl Points For a recent series of 10 years, the sunspot numbers and the total points scored in the Super Bowl were obtained. The table below lists the ranks of each set of values. Test for a correlation between sunspot number and Super Bowl points. Does the result agree with what might be expected?

Sunspot number	10	8	5	4	2	1	3	6	7	9
Super Bowl points	8	9	3	10	4	7	6	5	1	2

TI **11.** Correlation Between Salary and Stress The accompanying table lists salary rankings and stress rankings for randomly selected jobs (based on data from *The Jobs Rated Almanac*). Does it appear that salary increases as stress increases?

Job	Salary Rank	Stress Rank
Stockbroker	2	2
Zoologist	6	7
Electrical engineer	3	6
School principal	5	4
Hotel manager	7	5
Bank officer	10	8
Occupational safety inspector	9	9
Home economist	8	10
Psychologist	4	3
Airline pilot	1	1

TI **12.** Correlation Between Salary and Physical Demand Exercise 11 includes paired salary and stress level ranks for 10 randomly selected jobs. The physical demands of the jobs were also ranked; the salary and physical demand ranks are given below (based on data from *The Jobs Rated Almanac*). Does there appear to be a relationship between the salary of a job and its physical demands?

Salary	2	6	3	5	7	10	9	8	4	1
Physical demand	5	2	3	8	10	9	1	7	6	4

TI **13.** Crickets and Temperature The association between the temperature and the number of times a cricket chirps in 1 min was studied. Listed below are the numbers of chirps in 1 min and the corresponding temperatures in degrees Fahrenheit (based on data from *The Song of Insects* by George W. Pierce, Harvard University Press). Is there sufficient evidence to conclude that there is a relationship between the number of chirps in 1 min and the temperature?

Chirps in 1 min	882	1188	1104	864	1200	1032	960	900
Temperature (°F)	69.7	93.3	84.3	76.3	88.6	82.6	71.6	79.6

TI **14.** Deaths from Motor Vehicles and Murders Listed below are the number of motor vehicle deaths (in hundreds) and the number of murders (in hundreds) in the United States for each of several different years. Test for a correlation between those two variables.

Motor vehicle deaths	435	410	418	425	434	436	434	435	413	430
Murders	247	238	245	233	216	197	182	170	155	156

TI **15.** Song Audiences and Sales The table below lists the numbers of audience impressions (in hundreds of millions) listening to songs and the corresponding numbers of albums sold (in hundreds of thousands). The number of audience impressions is a count of the number of times people have heard the song. The table is based on data from *USA Today*. Does it appear that album sales are affected very strongly by the number of audience impressions?

Audience impressions	28	13	14	24	20	18	14	24	17
Albums sold	19	7	7	20	6	4	5	25	12

TI **16.** Movie Budgets and Gross Listed below are the budgets (in millions of dollars) and the gross receipts (in millions of dollars) for randomly selected movies (based on data from the Motion Picture Association of America). Does there appear to be a significant correlation between the money spent making the movie and the amount that it recovered in theaters? Apart from the budget amount, identify one other important factor that is likely to affect the amount that a movie earns.

Budget	62	90	50	35	200	100	90
Gross	65	64	48	57	601	146	47

Appendix B Data Sets: *In Exercises 17 and 18, use the data sets from Appendix B to test for rank correlation with a 0.05 significance level.*

TI **17.** Appendix B Data Set: Bad Stuff in Cigarettes Refer to Data Set 3 in Appendix B.

a. Use the paired data consisting of tar and nicotine. Based on the result, does there appear to be a significant correlation between cigarette tar and nicotine? If so, can researchers reduce their laboratory expenses by measuring only one of these two variables?

b. Use the paired data consisting of carbon monoxide and nicotine. Based on the result, does there appear to be a significant correlation between cigarette carbon monoxide and nicotine? If so, can researchers reduce their laboratory expenses by measuring only one of these two variables?

c. Assume that researchers want to develop a method for predicting the amount of nicotine, and they want to measure only one other item. In choosing between tar and carbon monoxide, which is the better choice? Why?

TI **18.** Appendix B Data Set: Forecasting Weather Refer to Data Set 8 in Appendix B.

a. Use the five-day forecast high temperatures and the actual high temperatures. Is there a correlation? Does a significant correlation imply that the five-day forecast temperatures are accurate?

b. Use the one-day forecast high temperatures and the actual high temperatures. Is there a correlation? Does a significant correlation imply that the one-day forecast temperatures are accurate?

c. Which would you expect to have a higher correlation with the actual high temperatures: the five-day forecast high temperatures or the one-day forecast high temperatures? Are the results from parts (a) and (b) what you would expect? If there is a very high correlation between forecast temperatures and actual temperatures, does it follow that the forecast temperatures are accurate?

13-6 BEYOND THE BASICS

TI **19.** Learning Curve A psychologist designs a test of learning. Subjects are given letters to memorize and their study times (in seconds) are matched with the numbers of letters they can recall. Results are listed in the table below. Is there a correlation between study time and the number of letters that can be recalled? Also, sketch a scatter diagram and find the results obtained by using the linear correlation coefficient described in Section 10-2. Compare the results.

Time	1	25	31	33	36	38	48	55	95	165	300
Words	1	2	20	36	51	72	74	75	77	78	79

20. Effect of Ties on r_s Refer to Data Set 5 in Appendix B for the times (in seconds) of tobacco use and alcohol use depicted in animated children's movies. Calculate the value of the test statistic r_s by using each of the two formulas (Formulas 13-1 and 10-1) given in this section. Is there a substantial difference between the two results? Which result is better? Is the conclusion affected by the formula used?

13-7 Runs Test for Randomness

Key Concept This section introduces the runs test for randomness, which can be used to determine whether the sample data in a sequence are in a random order. This test is based on sample data that have two characteristics, and it analyzes runs of those characteristics to determine whether the runs appear to result from some random process, or whether the runs suggest that the order of the data is not random.

Definitions

A **run** is a sequence of data having the same characteristic; the sequence is preceded and followed by data with a different characteristic or by no data at all.

The **runs test** uses the number of runs in a sequence of sample data to test for randomness in the order of the data.

Fundamental Principle of the Runs Test

The fundamental principle of the runs test can be briefly stated as follows:

Reject randomness if the number of runs is very low or very high.

- Example: The sequence of genders FFFFFMMMMM is not random because it has only 2 runs, so the number of runs is very *low*.
- Example: The sequence of genders FMFMFMFMFM is not random because there are 10 runs, which is very *high*.

Sports Hot Streaks

It is a common belief that athletes often have "hot streaks"—that is, brief periods of extraordinary success. Stanford University psychologist Amos Tversky and other researchers used statistics to analyze the thousands of shots taken by the Philadelphia 76ers for one full season and half of another. They found that the number of "hot streaks" was no different than you would expect from random trials with the outcome of each trial independent of any preceding results. That is, the probability of a hit doesn't depend on the preceding hit or miss.

The exact criteria for determining whether a number of runs is very high or low are found in the box below, which summarizes the key elements of the runs test for randomness. The procedure for the runs test for randomness is also summarized in Figure 13-5.

It is important to note that the runs test for randomness is based on the *order* in which the data occur; it is *not* based on the *frequency* of the data. For example, a sequence of 3 men and 20 women might appear to be random, but the issue of whether 3 men and 20 women constitute a *biased* sample (with disproportionately more women) is not addressed by the runs test.

Runs Test for Randomness

Requirements

1. The sample data are arranged according to some ordering scheme, such as the order in which the sample values were obtained.
2. Each data value can be categorized into one of *two* separate categories (such as male/female).

Notation

n_1 = number of elements in the sequence that have one particular characteristic. (The characteristic chosen for n_1 is arbitrary.)

n_2 = number of elements in the sequence that have the other characteristic

G = number of runs

Test Statistic

For Small Samples and $\alpha = 0.05$: If $n_1 \leq 20$ and $n_2 \leq 20$ and the significance level is $\alpha = 0.05$, the test statistic is the number of runs G. Critical values are found in Table A-10. Here is the decision criterion:

Reject randomness if the number of runs G is

- less than or equal to the smaller critical value found in Table A-10.
- or greater than or equal to the larger critical value found in Table A-10.

For Large Samples or $\alpha \neq 0.05$: If $n_1 > 20$ or $n_2 > 20$ or $\alpha \neq 0.05$, use the following test statistic and critical values.

Test statistic: $$z = \frac{G - \mu_G}{\sigma_G}$$

where $$\mu_G = \frac{2n_1n_2}{n_1 + n_2} + 1$$

and $$\sigma_G = \sqrt{\frac{(2n_1n_2)(2n_1n_2 - n_1 - n_2)}{(n_1 + n_2)^2(n_1 + n_2 - 1)}}$$

Critical values of z: Use Table A-2.

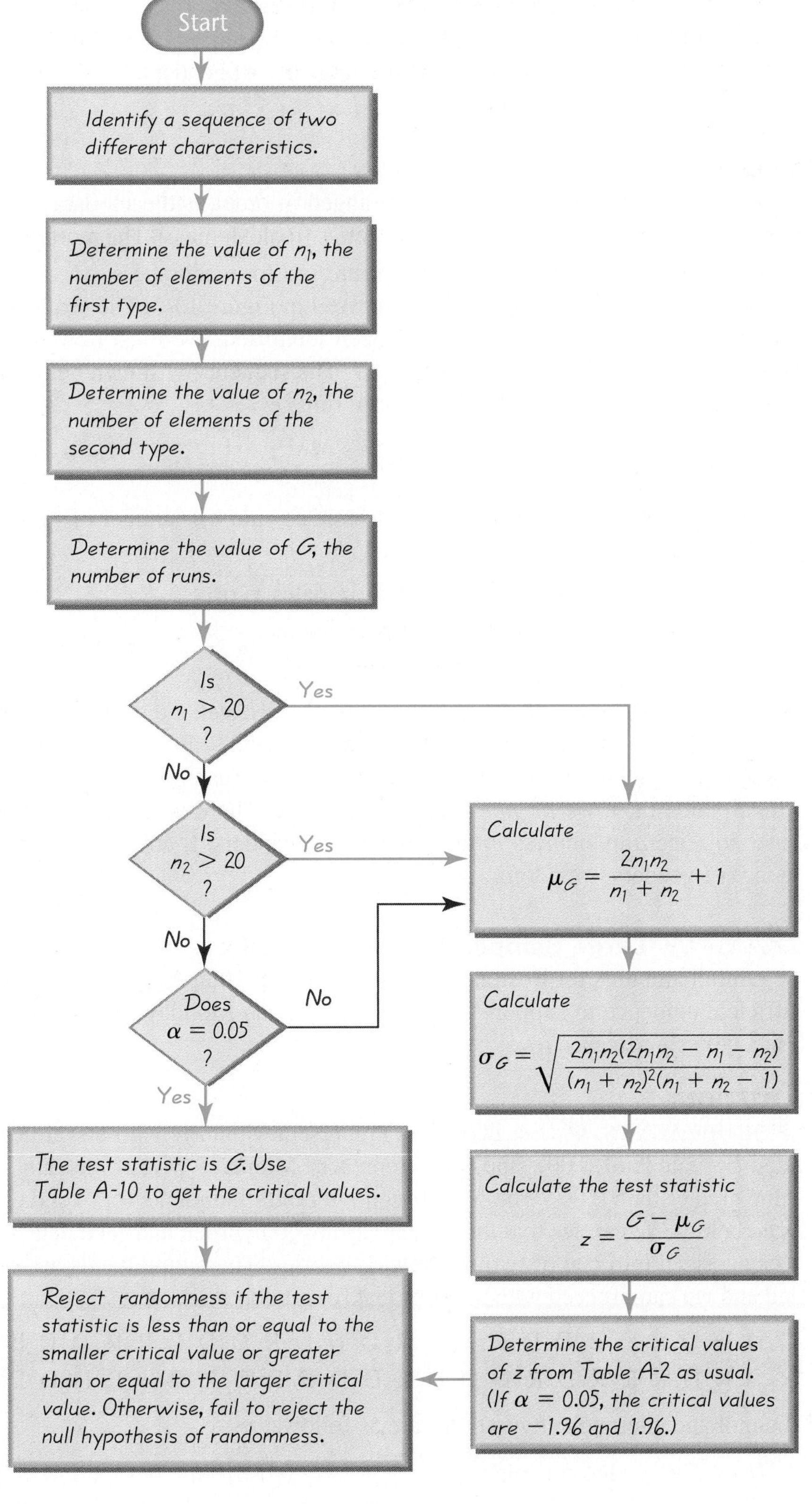

Figure 13-5

Procedure for Runs Test for Randomness

EXAMPLE Small Sample: Genders of Bears Listed below are the genders of the first 10 bears from Data Set 6 in Appendix B. Use a 0.05 significance level to test for randomness in the sequence of genders.

M M M M F F M M F F

SOLUTION

REQUIREMENTS ✔ The data are arranged in order, and each data value is categorized into one of two separate categories (male/female). The requirements are satisfied and we can proceed with the runs test for randomness. ✔

We will follow the procedure summarized in Figure 13-5. The sequence of two characteristics (male/female) has been identified. We must now find the values of n_1, n_2, and the number of runs G. The sequence is shown below with spacing used to better identify the separate runs.

$$\underbrace{\text{MMMM}}_{\text{1st run}} \quad \underbrace{\text{FF}}_{\text{2nd run}} \quad \underbrace{\text{MM}}_{\text{3rd run}} \quad \underbrace{\text{FF}}_{\text{4th run}}$$

We can see that there are 6 males and 4 females, and the number of runs is 4. We therefore have

$$n_1 = \text{total number of males} = 6$$
$$n_2 = \text{total number of females} = 4$$
$$G = \text{number of runs} = 4$$

Because $n_1 \leq 20$ and $n_2 \leq 20$ and $\alpha = 0.05$, the test statistic is $G = 4$, and we refer to Table A-10 to find the critical values of 2 and 9. Because $G = 4$ is not less than or equal to the critical value of 2, nor is it greater than or equal to the critical value of 9, *we do not reject randomness.* There is not sufficient evidence to reject randomness in the sequence of genders. It appears that the sequence of genders is random.

EXAMPLE Large Sample: Boston Rainfall on Mondays Refer to the rainfall amounts for Boston as listed in Data Set 10 in Appendix B. Is there sufficient evidence to support the claim that rain on Mondays is not random? Use a 0.05 significance level.

SOLUTION

REQUIREMENTS ✔ Let D (for dry) represent Mondays with no rain (indicated by values of 0.00), and let R represent Mondays with some rain (any value greater than 0.00). The 52 consecutive rainfall amounts for Monday are listed below. We can see that the data are arranged in order, and each data value is categorized into one of two separate categories. The requirements are satisfied and we can proceed with the runs test for randomness. ✔

D D D D R D R D D R D D R D D D R D D R R R D D D D
R D R D R R R D R D D D R D D D R D R D D R D D D R

The null and alternative hypotheses are as follows:

H_0: The sequence is random.

H_1: The sequence is not random.

The test statistic is obtained by first finding the number of Ds, the number of Rs, and the number of runs. It's easy to examine the sequence to find that

$$n_1 = \text{number of Ds} = 33$$
$$n_2 = \text{number of Rs} = 19$$
$$G = \text{number of runs} = 30$$

As we follow Figure 13-5, we answer yes to the question "Is $n_1 > 20$?" We therefore need to evaluate the test statistic z given in the box summarizing the key elements of the runs test for randomness. We must first evaluate μ_G and σ_G. We get

$$\mu_G = \frac{2n_1n_2}{n_1 + n_2} + 1 = \frac{2(33)(19)}{33 + 19} + 1 = 25.115$$

$$\sigma_G = \sqrt{\frac{(2n_1n_2)(2n_1n_2 - n_1 - n_2)}{(n_1 + n_2)^2(n_1 + n_2 - 1)}}$$

$$= \sqrt{\frac{(2)(33)(19)[2(33)(19) - 33 - 19]}{(33 + 19)^2(33 + 19 - 1)}} = 3.306$$

We can now find the test statistic:

$$z = \frac{G - \mu_G}{\sigma_G} = \frac{30 - 25.115}{3.306} = 1.48$$

Because the significance level is $\alpha = 0.05$ and we have a two-tailed test, the critical values are $z = -1.96$ and $z = 1.96$. The test statistic of $z = 1.48$ does not fall within the critical region, so we fail to reject the null hypothesis of randomness. The given sequence does appear to be random.

Numerical Data: Randomness Above and Below the Mean or Median

In each of the preceding examples, the data clearly fit into two categories, but we can also test for randomness in the way numerical data fluctuate above or below a mean or median. To test for randomness above and below the median, for example, use the sample data to find the value of the median, then replace each individual value with the letter A if it is *above* the median, and replace it with B if it is *below* the median. Delete any values that are equal to the median. It is helpful to write the As and Bs directly above the numbers they represent because this makes checking easier and also reduces the chance of having the wrong number of letters. After finding the sequence of A and B letters, we can proceed to apply the runs test as described. Economists use the runs test for randomness above and below the median in an attempt to identify trends or cycles. An upward economic trend would contain a predominance of Bs at the beginning and As at the end, so the number of runs would be small. A downward trend would have As dominating at the beginning and Bs at the end, with a low number of runs. A cyclical pattern would yield a sequence that systematically changes, so the number of runs would tend to be large. (See Exercise 14.)

Using Technology for the Runs Test for Randomness

The TI-83/84 Plus calculator is not yet programmed for the runs test for randomness, but STATDISK can be used instead. STATDISK is a program included with the CD-ROM that is packaged with this book. Load STATDISK, then use the following procedure:

1. Determine the values of n_1, n_2, and the number of runs G as described in this section.
2. Select **Analysis** from the main menu bar, then select **Runs Test.**
3. Proceed to enter the required data in the dialog box.
4. Click **Evaluate** and the results will be displayed. The results will include the test statistic (G or z as appropriate), critical values, and conclusion.

13-7 BASIC SKILLS AND CONCEPTS

Statistical Literacy and Critical Thinking

1. Runs Test A pollster conducts a survey by filling out a page of responses from each subject he interviews. He then shuffles the pages before submitting them for analysis. Can the runs test for randomness be used to determine whether his survey subjects were selected in a random sequence?

2. Runs Test What is a run?

3. Runs Test If the runs test leads to the conclusion that a particular sample is random, can we conclude that the data have been selected in a way that is appropriate for statistical purposes?

4. Biased Sample After using the runs test for randomness, it is found that a sequence of 20 males and 80 females appears to be random. Does this imply that the sample is representative of the population of adult Americans?

Identifying Runs and Finding Critical Values. *In Exercises 5–8, use the given sequence to determine the values of n_1, n_2, the number of runs G, and the critical values from Table A-10, and use those results to determine whether the sequence appears to be random.*

5. Y N N Y N N Y N N Y N N

6. T T T T T F F F F T T T T

7. M M M F F M F F M F F M

8. X Y X X Y Y X X X Y Y Y X X X X Y Y Y Y

Using the Runs Test for Randomness. *In Exercises 9–18, use the runs test of this section to determine whether the given sequence is random. Use a significance level of $\alpha = 0.05$. (All data are listed in order by row.)*

9. Genders of Bears The first example of this section used the genders of the first 10 bears listed in Data Set 6 in Appendix B. Conduct a runs test for randomness using the genders of the first 20 bears from Data Set 6. Those genders are listed below.

M M M M F F M M F F M M F M F M M F M M

10. Olympic Gold Medal Winners Listed below are gold medal winners in the men's 400-meter dash, where U represents the United States and O represents any other country. Does it appear that the countries of the winners are in a random sequence?

U U U U O U O O U U U O O
U U U U U O O U U U U U U

11. Testing for Randomness of Baseball World Series Victories Test the claim that the sequence of World Series wins by American League and National League teams is random. Given below are recent results, with American and National League teams represented by A and N, respectively. What does the result suggest about the abilities of the two leagues?

N A A A N N A A N N N N A A A N A
N A N A A A N A N A A A N A N A

12. Testing for Randomness of Presidential Election Winners For a recent sequence of presidential elections, the political party of the winner is indicated by D for Democrat and R for Republican. Does it appear that we elect Democrat and Republican candidates in a sequence that is random?

R R D R D R R R R D D R R R D D D
D D R R D D R R D R R R D D R R

13. Indoor Movie Theaters: Testing for Randomness Above and Below Median Trends in business and economics applications are often analyzed with the runs test. Given below are the numbers of indoor movie theaters, listed in order by row for each year beginning with 1987 (based on data from the National Association of Theater Owners). First find the median, then replace each value by A if it is above the median and B if is below the median. Then apply the runs test to the resulting sequence of As and Bs. What does the result suggest about the trend in the number of indoor movie theaters?

20,595 21,632 21,907 22,904 23,740 24,344 24,789 25,830 26,995
28,905 31,050 33,418 36,448 35,567 34,490 35,170 35,361

14. Stock Market: Testing for Randomness Above and Below the Median Listed below are the annual high points of the Dow Jones Industrial Average for a recent sequence of years. First find the median of the values, then replace each value by A if it is above the median and B if it is below the median. Then apply the runs test to the resulting sequence of As and Bs. What does the result suggest about the stock market as an investment consideration?

969	995	943	985	969	842	951	1036	1052	892
882	1015	1000	908	898	1000	1024	1071	1287	1287
1553	1956	2722	2184	2791	3000	3169	3413	3794	3978
5216	6561	8259	9374	11,497	11,723	11,338	10,635	10,454	10,895

15. Large Sample: Genders of Bears The first example of this section used the genders of the first 10 bears listed in Data Set 6 in Appendix B. Conduct a runs test for randomness using the genders of all bears from Data Set 6.

16. Large Sample: Saturday Rainfall The second example of this section used the Monday rainfall amounts from Data Set 10 in Appendix B. Use the Saturday rainfall amounts to test for randomness of days with precipitation. Consider a day to be dry if the precipitation amount is 0.00. Consider a day to be rainy if the precipitation amount is any value greater than 0.00.

17. Large Sample: Testing for Randomness of Odd and Even Digits in Pi A *New York Times* article about the calculation of decimal places of π noted that "mathematicians are pretty sure that the digits of π are indistinguishable from any random sequence." Given below are the first 100 decimal places of π. Test for randomness of odd (O) and even (E) digits.

1 4 1 5 9 2 6 5 3 5 8 9 7 9 3 2 3 8 4 6 2 6 4 3 3 8 3 2 7 9 5 0 2 8 8 4 1 9 7 1
6 9 3 9 9 3 7 5 1 0 5 8 2 0 9 7 4 9 4 4 5 9 2 3 0 7 8 1 6 4 0 6 2 8 6 2 0 8 9 9
8 6 2 8 0 3 4 8 2 5 3 4 2 1 1 7 0 6 7 9

18. Large Sample: Testing for Randomness of Baseball World Series Victories Test the claim that the sequence of World Series wins by American League and National League teams is random. Given below are recent results, with American and National League teams represented by A and N, respectively.

A N A N N N A A A A N A A A A N A N N A A N N A A A A N A N
N A A A A A N A N A N A N A A A A A A A A N N A N A N N A A N
N N A N A N A N A A A N N A A N N N N A A A N A N A N A A A
N A N A A A N A N A

13-7 BEYOND THE BASICS

19. Finding Critical Numbers of Runs Using the elements A, A, B, B, what is the minimum number of possible runs that can be arranged? What is the maximum number of runs? Now refer to Table A-10 to find the critical G values for $n_1 = n_2 = 2$. What do you conclude about this case?

20. Finding Critical Values

a. Using all of the elements A, A, A, B, B, B, B, B, B, list the 84 different possible sequences.

b. Find the number of runs for each of the 84 sequences.

c. Use the results from parts (a) and (b) to find your own critical values for G.

d. Compare your results to those given in Table A-10.

Review

In this chapter we examined six different nonparametric tests for analyzing sample data. Nonparametric tests are also called distribution-free tests because they do not require that the populations have a particular distribution, such as a normal distribution. However, nonparametric tests are not as efficient as parametric tests, so we generally need stronger evidence before we reject a null hypothesis.

Table 13-9 lists the nonparametric tests presented in this chapter, along with their functions. The table also lists the corresponding parametric tests.

Table 13-9 Summary of Nonparametric Tests

Nonparametric Test	Function	Parametric Test
Sign test (Section 13-2)	Test for claimed value of average with one sample	z test or t test (Sections 8-4, 8-5)
	Test for differences between matched pairs	t test (Section 9-4)
	Test for claimed value of a proportion	z test (Section 8-3)
Wilcoxon signed-ranks test (Section 13-3)	Test for differences between matched pairs	t test (Section 9-4)
Wilcoxon rank-sum test (Section 13-4)	Test for difference between two independent samples	t test or z test (Section 9-3)
Kruskal-Wallis test (Section 13-5)	Test that more than two independent populations have the same median	Analysis of variance (Section 12-2)
Rank correlation (Section 13-6)	Test for relationship between two variables	Linear correlation (Section 10-2)
Runs test (Section 13-7)	Test for randomness of sample data	(No parametric test)

Statistical Literacy and Critical Thinking

1. Nonparametric Test What is a *nonparametric* test?

2. Distribution-Free Test What is the difference between a nonparametric test and a distribution-free test?

3. Rank Many nonparametric tests are based on ranks. What is a rank?

4. Efficiency Nonparametric tests are typically not as *efficient* as a corresponding parametric test, assuming that the necessary requirements are satisfied. What does the efficiency measure?

Review Exercises

Using Nonparametric Tests. In Exercises 1–8, use a 0.05 significance level with the indicated test. If no particular test is specified, use the appropriate nonparametric test from this chapter.

TI **1.** Measuring Intelligence in Children The table below lists matched pairs of times (in seconds) obtained from a random sample of children who were given blocks and instructed to build a tower as tall as possible (based on data from "Tower Building," by Johnson and Courtney, *Child Development,* Vol. 3). This procedure is used to measure intelligence in children. Use the sign test and a 0.05 significance level to test the claim that there is no difference between the times of the first and second trials.

Child	A	B	C	D	E	F	G	H	I	J	K	L	M	N	O
First trial	30	19	19	23	29	178	42	20	12	39	14	81	17	31	52
Second trial	30	6	14	8	14	52	14	22	17	8	11	30	14	17	15

TI **2.** Measuring Intelligence in Children Repeat Exercise 1 using the Wilcoxon signed-ranks test.

TI **3.** Business and Law School Rankings *U.S. News and World Report* magazine ranked business schools and law schools, and the ranks in the table below are based on those results. Is there a correlation between the business school rankings and law school rankings?

College	Hvd	Yale	Sfd	NYU	Chi	UPenn	BU	Ohio	Gtwn	USC
Business	1	6	2	5	4	3	10	8	9	7
Law	2	1	3	4	5	6	9	10	7	8

TI **4.** Medical School Rankings *U.S. News and World Report* magazine ranked research medical schools and listed the corresponding "average" MCAT scores. The table below is based on those results. Is there a correlation between the ranking and average MCAT score?

College	Hvd	NYU	Yale	Chi	USC	Ohio	BU	Utah
Rank	1	4	2	3	5	6	7	8
MCAT	11.3	11.0	11.4	10.3	10.8	10.5	9.5	9.6

TI **5.** Testing for Gender Discrimination The Axipon accounting firm claims that hiring is done without any gender bias. Among the last 40 new employees hired, 15 are women. Job applicants are about half men and half women, who are all qualified. Is there sufficient evidence to charge bias in favor of men?

TI **6.** Do Beer Drinkers and Liquor Drinkers Have Different BAC Levels? The sample data in the following list show BAC (blood alcohol concentration) levels at arrest of randomly selected jail inmates who were convicted of DWI or DUI offenses. The data are categorized by the type of drink consumed (based on data from the U.S. Department of Justice). Test the claim that beer drinkers and liquor drinkers have the same BAC levels. Based on these results, do both groups seem equally dangerous, or is one group more dangerous than the other?

Beer

0.129	0.146	0.148	0.152
0.154	0.155	0.187	0.212
0.203	0.190	0.164	0.165

Liquor

0.220	0.225	0.185	0.182
0.253	0.241	0.227	0.205
0.247	0.224	0.226	0.234
0.190	0.257		

7. Is the Lottery Random? Listed below are the first digits selected in 40 consecutive drawings of the New York State Win 4 lottery game. Do odd and even digits appear to be drawn in a sequence that is random?

9 7 0 7 5 5 1 9 0 0 8 7 6 0 1 6 7 2 4 7
5 5 5 2 0 4 4 9 9 0 5 3 3 1 9 2 5 6 8 2

TI **8.** Does the Weight of a Car Affect Leg Injuries in a Crash? Data were obtained from car crash experiments conducted by the National Transportation Safety Administration. New cars were purchased and crashed into a fixed barrier at 35 mi/h, and measurements were recorded for the dummy in the driver's seat. Use the sample data on the top of the next page to test for differences in left femur load measurements (in lb) among the four weight categories. Is there sufficient evidence to conclude that leg injury measurements for the four car weight categories are not all the same? Do the data suggest that heavier cars are safer in a crash?

Subcompact:	595	1063	885	519	422
Compact:	1051	1193	946	984	584
Midsize:	629	1686	880	181	645
Full-size:	1085	971	996	804	1376

Cumulative Review Exercises

In Exercises 1–8, use the data in the table below. The values are based on a scatterplot in *"An SAT Coaching Program That Works,"* by Jack Kaplan, *Chance*, Vol. 15, No. 1. The values are math SAT scores of students before and after taking an SAT preparation course.

Before	460	470	490	490	510	510	600	620	610
After	480	510	500	610	590	630	630	660	690

TI **1.** Finding Statistics Find the mean, median, range, standard deviation, and variance of the before scores.

TI **2.** Graph Construct a scatterplot. Does there appear to be a correlation between the before scores and the after scores?

TI **3.** Linear Correlation Test for a linear correlation between the before scores and the after scores. If there is a linear correlation, does that mean that the preparation course is effective? Why or why not?

TI **4.** Rank Correlation Use rank correlation to test for a correlation between the before scores and the after scores.

TI **5.** Matched Data Use a t test to test the claim that the mean difference between the before and after scores is equal to zero.

TI **6.** Sign Test Use the sign test to test the claim that there is no difference between the before and after scores.

TI **7.** Wilcoxon Signed-Ranks Test Use the Wilcoxon signed-ranks test to test the claim that there is no difference between the before and after scores.

8. Test Effectiveness Which of the preceding exercises leads to the results that are most helpful in determining whether the preparation course appears to be effective? Does the preparation course appear to be effective? Why or why not?

Cooperative Group Activities

1. In-class activity Use the existing seating arrangement in your class and apply the runs test to determine whether the students are arranged randomly according to gender. After recording the seating arrangement, analysis can be done in subgroups of three or four students.

2. In-class activity Divide into groups of 8 to 12 people. For each group member, *measure* his or her height and *measure* his or her arm span. For the arm span, the subject should stand with arms extended, like the wings on an airplane. It's easy to mark the height and arm span on a chalkboard, then measure the distances there. Divide the following tasks among subgroups of three or four people.

a. Use rank correlation with the paired sample data to determine whether there is a correlation between height and arm span.

b. Use the sign test to test for a difference between the two variables.

c. Use the Wilcoxon signed-ranks test to test for a difference between the two variables.

3. In-class activity Do Activity 2 using pulse rate instead of arm span. Measure pulse rates by counting the number of heartbeats in 1 min.

4. Out-of-class activity Divide into groups of three or four students. Investigate the relationship between two variables by collecting your own paired sample data and using the methods of Section 13-6 to determine whether there is a correlation. Suggested topics:
   - Is there a relationship between taste and cost of different brands of chocolate chip cookies (or colas)? (Taste can be measured on some number scale, such as 1 to 10.)
   - Is there a relationship between salaries of professional baseball (or basketball or football) players and their season achievements?
   - Rates versus weights: Is there a relationship between car fuel-consumption rates and car weights?
   - Is there a relationship between the lengths of men's (or women's) feet and their heights?
   - Is there a relationship between student grade-point averages and the amount of television watched?
   - Is there a relationship between heights of fathers (or mothers) and heights of their first sons (or daughters)?

5. Out-of-class activity See this chapter's "From Data to Decision" project, which involves analysis of the 1970 lottery used for drafting men into the U.S. Army. Because the 1970 results raised concerns about the randomness of selecting draft priority numbers, design a new procedure for generating the 366 priority numbers. Use your procedure to generate the 366 numbers and test your results by using the techniques suggested in parts (a), (b), and (c) of the "From Data to Decision" project. How do your results compare to those obtained in 1970? Does your random selection process appear to be better than the one used in 1970? Write a report that clearly describes the process you designed. Also include your analyses and conclusions.

6. Out-of-class activity Divide into groups of three or four. Survey students by asking them to identify their major and gender. For each surveyed subject, determine the accuracy of the time on his or her wristwatch. First set your own watch to the correct time using an accurate and reliable source ("At the tone, the time is . . . "). For watches that are ahead of the correct time, record positive times. For watches that are behind the correct time, record negative times. Use the sample data to address these questions:
   - Do the errors appear to be the same for both genders?
   - Do the errors appear to be the same for the different majors?

7. In-class activity Divide into groups of 8 to 12 people. For each group member, measure the person's height and also measure his or her navel height, which is the height from the floor to the navel. Use the rank correlation coefficient to determine whether there is a correlation between height and navel height.

8. In-class activity Divide into groups of three or four people. Appendix B includes many data sets not yet addressed by the methods of this chapter. Search Appendix B for variables of interest, then investigate using appropriate methods of nonparametric statistics. State your conclusions and try to identify practical applications.

TI-83/84 Plus Calculator Project

Randomly generated digits are commonly used for a variety of different applications. Pollsters often select telephone numbers by using random digits, so that unlisted numbers can be included. Simulations, such as those discussed in Section 4-6, often require the random generation of digits. This project involves the random generation of digits 0, 1, 2, . . . , 9 in such a way that they are all equally likely.

1. Listed at the right are data sets that were generated by using the random features of different technologies. (Excel, STATDISK, and Minitab are statistical software packages.) Use the appropriate nonparametric method from this chapter to determine whether the listed randomly generated digits come from identical populations.

TI-83/84 Plus:	5 8 4 9 5 4 0 1 4 2 4 3 1 4 9 7 5 3 7 5
Excel:	3 1 5 9 8 8 4 8 1 2 0 0 1 2 0 2 3 5 3 9
STATDISK:	1 7 8 3 0 0 4 9 7 5 6 5 9 6 1 1 5 5 8 5
Minitab:	1 2 5 1 1 3 1 8 8 5 3 5 5 3 5 3 5 2 3 3

2. Use a TI-83/84 Plus calculator to generate your own list of 100 randomly selected digits. Generate the data by pressing MATH, selecting the top menu item of **PROB,** then selecting **randInt.** The format for the randInt command is this: randInt(minimum integer, maximum integer, number of generated values). Enter the command **randInt(0, 9, 100)** to generate 100 randomly selected digits between 0 and 9 inclusive. (*Hint:* Use the STO key to store the generated digits in list L1.) Test for randomness. Describe the method used and the conclusion reached.

From Data to Decision

Critical Thinking: Was the draft lottery random?

In 1970, a lottery was used to determine who would be drafted into the U.S. Army. The 366 dates in the year were placed in individual capsules. First, the 31 January capsules were placed in a box; then the 29 February capsules were added and the two months were mixed. Then the 31 March capsules were added and the three months were mixed. This process continued until all months were included. The first capsule selected was September 14, so men born on that date were drafted first. The accompanying list shows the 366 dates in the order of selection.

Analyzing the Results

a. Use the runs test to test the sequence for randomness above and below the median of 183.5.

b. Use the Kruskal-Wallis test to test the claim that the 12 months had priority numbers drawn from the same population.

c. Calculate the 12 monthly means. Then plot those 12 means on a graph. (The horizontal scale lists the 12 months, and the vertical scale ranges from 100 to 260.) Note any pattern suggesting that the original priority numbers were not randomly selected.

d. Based on the results from parts (a), (b), and (c), decide whether this particular draft lottery was fair. Write a statement either supporting your position that it was fair or explaining why you believe that it was not fair. If you decided that this lottery was unfair, describe a process for selecting lottery numbers that would have been fair.

Jan:	305	159	251	215	101	224	306	199	194	325	329	221	318	238	017	121
	235	140	058	280	186	337	118	059	052	092	355	077	349	164	211	
Feb:	086	144	297	210	214	347	091	181	338	216	150	068	152	004	089	212
	189	292	025	302	363	290	057	236	179	365	205	299	285			
Mar:	108	029	267	275	293	139	122	213	317	323	136	300	259	354	169	166
	033	332	200	239	334	265	256	258	343	170	268	223	362	217	030	
Apr:	032	271	083	081	269	253	147	312	219	218	014	346	124	231	273	148
	260	090	336	345	062	316	252	002	351	340	074	262	191	208		
May:	330	298	040	276	364	155	035	321	197	065	037	133	295	178	130	055
	112	278	075	183	250	326	319	031	361	357	296	308	226	103	313	
Jun:	249	228	301	020	028	110	085	366	335	206	134	272	069	356	180	274
	073	341	104	360	060	247	109	358	137	022	064	222	353	209		
Jul:	093	350	115	279	188	327	050	013	277	284	248	015	042	331	322	120
	098	190	227	187	027	153	172	023	067	303	289	088	270	287	193	
Aug:	111	045	261	145	054	114	168	048	106	021	324	142	307	198	102	044
	154	141	311	344	291	339	116	036	286	245	352	167	061	333	011	
Sep:	225	161	049	232	082	006	008	184	263	071	158	242	175	001	113	207
	255	246	177	063	204	160	119	195	149	018	233	257	151	315		
Oct:	359	125	244	202	024	087	234	283	342	220	237	072	138	294	171	254
	288	005	241	192	243	117	201	196	176	007	264	094	229	038	079	
Nov:	019	034	348	266	310	076	051	097	080	282	046	066	126	127	131	107
	143	146	203	185	156	009	182	230	132	309	047	281	099	174		
Dec:	129	328	157	165	056	010	012	105	043	041	039	314	163	026	320	096
	304	128	240	135	070	053	162	095	084	173	078	123	016	003	100	

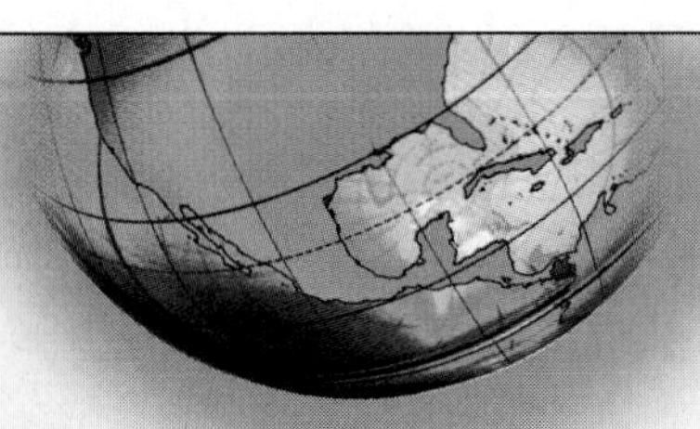

Internet Project

Nonparametric Tests

This chapter introduced hypothesis-testing methods of the nonparametric or distribution-free variety. Nonparametric methods allow you to test hypotheses without making assumptions regarding the underlying distribution of the population being sampled. To continue your work with this important class of statistical testing methods, go to the *Elementary Statistics* Web site:

http://www.aw.com/Triola

Statistics @ Work

"Prospective employees should have a fundamental grasp of statistics and its implications in the business world."

Angela Gillespie

Traffic Analyst, Lycos.com

As a Traffic Analyst for Lycos, Inc., Angela reports on major and minor traffic metrics. She monitors changes in trends and behavior patterns for use, enhancing the site to increase reach and stickiness (the amount of time people spend online at any particular Web site).

What is your job at Lycos?

I produce traffic reports of our site's activities each week. These are reviewed by our product group teams and senior management. They see what is increasing, what is decreasing, and make decisions about where our resources are spent.

My reports basically analyze trends on the sites and give projections for where we will be in a year or in any given time frame.

What concepts of statistics do you use?

Regression analysis and *R*-squared values.

How do you use statistics on the job?

To determine what is working and what is not working for our users. To determine the effectiveness of advertising campaigns, and to project future growth.

Please describe one specific example illustrating how the use of statistics was successful in improving a product or service.

At the end of our last fiscal year our CEO, Bob Davis, presented the company with an average daily pageview goal to be reached by the end of the next fiscal year. Using two years' worth of pageview data, I put together a projection showing where we would be at the end of the next fiscal year if things remained static. Using an *R*-squared value gave these charts the oomph I needed to be effective. I updated the charts each week and presented them to the Product Management team. The data helped them understand what adjustments to make to their products and each week they got closer and closer to their goals. When Bob first presented the pageview goal, we all thought he had gone mad, but I am happy to say that at the end of the next fiscal year, we will have either reached our goal or be within 98% of it. Without the representation I supplied, Product Management would not have known where to focus their energy and resources. Because they were an efficient team, we have reached our unreachable goal.

Is your use of probability and statistics increasing, decreasing, or remaining stable?

As Lycos gets more sophisticated, they (management) expect more and more sophisticated reporting. It is increasing.

Do you feel job applicants are viewed more favorably if they have studied some statistics?

Absolutely, and not just within Lycos Reporting, but also in Product Marketing and Finance.

14 Statistical Process Control

14 CHAPTER PROBLEM

Is the production of aircraft altimeters dangerous for those who fly?

The Altigauge Manufacturing Company produces aircraft altimeters, which provide pilots with readings of their heights above sea level. The accuracy of altimeters is important because pilots rely on them to maintain altitudes with safe vertical clearance above mountains, towers, and tall buildings, as well as vertical separation from other aircraft. The accuracy of altimeters is especially important when pilots fly approaches to landing while not being able to see the ground. In the past, pilots and passengers have been killed in crashes caused by wrong altimeter readings that led pilots to believe that they were safely above the ground when they were actually flying dangerously low.

Because aircraft altimeters are so critically important to aviation safety, their accuracy is carefully controlled by government regulations. According to Federal Aviation Administration Regulation Part 43, Appendix E, an altimeter must give a reading with an error of no more than 20 ft when tested for an altitude of 1000 ft.

At the Altigauge Manufacturing Company, four altimeters are randomly selected from production on each of 20 consecutive business days, and Table 14-1 lists the errors (in feet) when they are tested in a pressure chamber that simulates an altitude of 1000 ft. On Day 1, for example, the actual altitude readings for the four selected altimeters are 1002 ft, 992 ft, 1005 ft, and 1011 ft, so the corresponding errors (in feet) are 2, −8, 5, and 11.

In this chapter we will evaluate this altimeter manufacturing process by analyzing the behavior of the errors over time. We will see how methods of statistics can be used to monitor a manufacturing process with the goal of identifying and correcting any serious problems. In addition to helping companies stay in business, methods of statistics can positively affect our safety in very significant ways.

Table 14-1 Aircraft Altimeter Errors (in feet)

Day	Error				Mean	Median	Range	Standard Deviation
1	2	−8	5	11	2.50	3.5	19	7.94
2	−5	2	6	8	2.75	4.0	13	5.74
3	6	7	−1	−8	1.00	2.5	15	6.98
4	−5	5	−5	6	0.25	0.0	11	6.08
5	9	3	−2	−2	2.00	0.5	11	5.23
6	16	−10	−1	−8	−0.75	−4.5	26	11.81
7	13	−8	−7	2	0.00	−2.5	21	9.76
8	−5	−4	2	8	0.25	−1.0	13	6.02
9	7	13	−2	−13	1.25	2.5	26	11.32
10	15	7	19	1	10.50	11.0	18	8.06
11	12	12	10	9	10.75	11.0	3	1.50
12	11	9	11	20	12.75	11.0	11	4.92
13	18	15	23	28	21.00	20.5	13	5.72
14	6	32	4	10	13.00	8.0	28	12.91
15	16	−13	−9	19	3.25	3.5	32	16.58
16	8	17	0	13	9.50	10.5	17	7.33
17	13	3	6	13	8.75	9.5	10	5.06
18	38	−5	−5	5	8.25	0.0	43	20.39
19	18	12	25	−6	12.25	15.0	31	13.28
20	−27	23	7	36	9.75	15.0	63	27.22

14-1 Overview

In Chapter 2 we noted that when describing, exploring, or comparing data sets, it is usually important to consider center, variation, distribution, outliers, and changing characteristics over time. The main objective of this chapter is to address the last item: changing characteristics of data over time. When investigating characteristics such as center and variation, it is important to know whether we are dealing with a stable population or one that is changing with the passage of time.

There is currently a strong trend toward trying to improve the quality of American goods and services, and the methods presented in this chapter are being used by growing numbers of businesses. Evidence of the increasing importance of quality is found in its greater role in advertising and the growing number of books and articles that focus on the issue of quality. In many cases, job applicants (you?) have a definite advantage when they can tell employers that they have studied statistics and methods of quality control. This chapter will present some of the basic tools commonly used to monitor quality.

Minitab, Excel, and other software packages include programs for automatically generating charts of the type discussed in this chapter, and we will include several examples of such displays. Control charts are good examples of wonderful graphical devices that allow us to *see* and *understand* some property of data that would be more difficult or impossible to understand without graphs. The world needs more people who can construct and interpret important graphs, such as the control charts described in this chapter.

14-2 Control Charts for Variation and Mean

Key Concept The main objective of this section is to construct run charts, R charts, and $\bar{x}$ charts so that we can monitor important features of data over time. We will use such charts to determine whether some process is statistically stable (or within statistical control).

The following definition formally describes the type of data that will be considered in this chapter.

Definition

Process data are data arranged according to some time sequence. They are measurements of a characteristic of goods or services that result from some combination of equipment, people, materials, methods, and conditions.

For example, Table 14-1 includes process data consisting of the measured error (in feet) in altimeter readings over 20 consecutive days of production. Each day, four altimeters were randomly selected and tested. Because the data in Table 14-1 are arranged according to the time at which they were selected, they are process data. It is very important to recognize this point:

Important characteristics of process data can change over time.

In making altimeters, a manufacturer might use competent and well-trained personnel along with good machines that are correctly calibrated, but if the personnel are replaced or the machines wear with use, the altimeters might begin to become defective. Companies have gone bankrupt because they unknowingly allowed manufacturing processes to deteriorate without constant monitoring.

Run Charts

There are various methods that can be used to monitor a process to ensure that the important desired characteristics don't change—analysis of a *run chart* is one such method.

Definition

A **run chart** is a sequential plot of *individual* data values over time. One axis (usually the vertical axis) is used for the data values, and the other axis (usually the horizontal axis) is used for the time sequence.

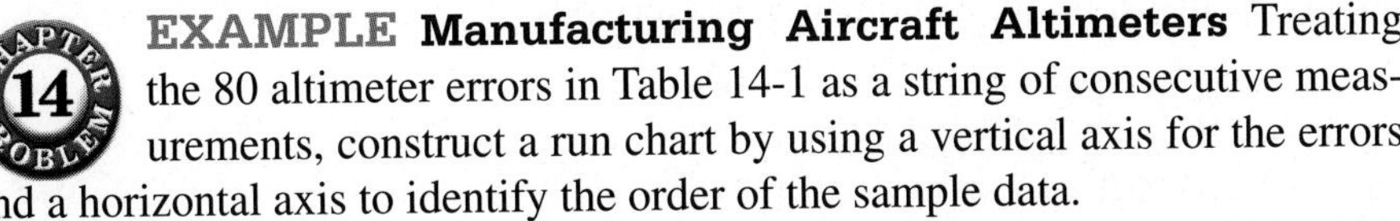

EXAMPLE Manufacturing Aircraft Altimeters Treating the 80 altimeter errors in Table 14-1 as a string of consecutive measurements, construct a run chart by using a vertical axis for the errors and a horizontal axis to identify the order of the sample data.

SOLUTION Figure 14-1 includes a Minitab-generated run chart and a TI-83/84 Plus run chart for the data in Table 14-1. The vertical scale is designed to be

continued

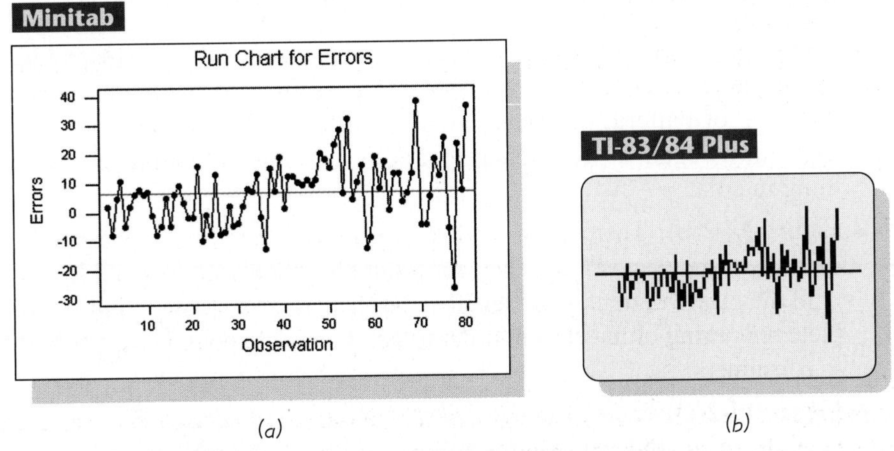

(a) (b)

Figure 14-1 Run Charts of Individual Altimeter Errors in Table 14-1

The Flynn Effect: Upward Trend in IQ Scores

A run chart or control chart of IQ scores would reveal that they exhibit an upward trend, because IQ scores have been steadily increasing since they began to be used about 70 years ago. The trend is worldwide, and it is the same for different types of IQ tests, even those that rely heavily on abstract and nonverbal reasoning with minimal cultural influence. This upward trend has been named the *Flynn effect,* because political scientist James R. Flynn discovered the trend in his studies of U.S. military recruits. The amount of the increase is quite substantial: Based on a current mean IQ score of 100, it is estimated that the mean IQ in 1920 would be about 77. The typical student of today is therefore brilliant when compared to his or her great-grandparents. So far, there is no generally accepted explanation for the Flynn effect.

suitable for altimeter errors ranging from -27 ft to 38 ft, which are the minimum and maximum values in Table 14-1. The horizontal scale is designed to include the 80 values arranged in sequence. The first point represents the first value of 2 ft, the second point represents the second value of -8 ft, and so on.

In Figure 14-1, the horizontal scale identifies the sample number, so the number 20 indicates the 20th sample item. The vertical scale represents the altimeter error (in feet). Now examine Figure 14-1 and try to identify any *patterns* that jump out begging for attention. Figure 14-1 does reveal this problem: As time progresses from left to right, the heights of the points appear to show a pattern of increasing variation. See how the points at the left fluctuate considerably less than the points farther to the right. The Federal Aviation Administration regulations require errors less than 20 ft (or between 20 ft and -20 ft), so the altimeters represented by the points at the left are okay, whereas several of the points farther to the right correspond to altimeters not meeting the required specifications. It appears that the manufacturing process started out well, but deteriorated as time passed. If left alone, this manufacturing process will cause the company to go out of business.

Interpreting Run Charts Only when a process is *statistically stable* can its data be treated as if they came from a population with a constant mean, standard deviation, distribution, and other characteristics.

Definition

A process is **statistically stable** (or **within statistical control**) if it has only natural variation, with no patterns, cycles, or unusual points.

Figure 14-2 consists of run charts illustrating typical patterns showing ways in which the process of filling 12-oz cola cans may not be statistically stable.

- **Figure 14-2(a):** There is an obvious *upward trend* that corresponds to values that are increasing over time. If the filling process were to follow this type of pattern, the cans would be filled with more and more cola until they began to overflow, eventually leaving the employees swimming in cola.
- **Figure 14-2(b):** There is an obvious *downward trend* that corresponds to steadily decreasing values. The cans would be filled with less and less cola until they were extremely underfilled. Such a process would require a complete reworking of the cans in order to get them full enough for distribution to consumers.
- **Figure 14-2(c):** There is an *upward shift.* A run chart such as this one might result from an adjustment to the filling process, making all subsequent values higher.

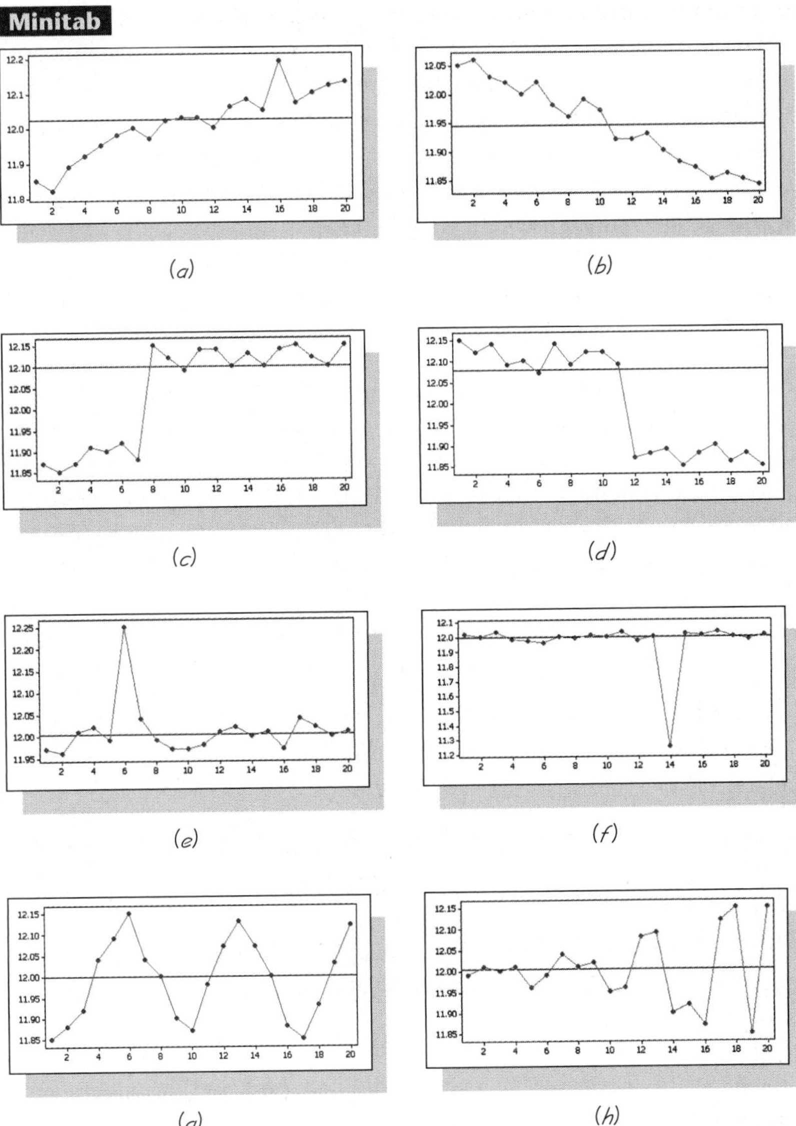

Figure 14-2

Processes That Are Not Statistically Stable

- **Figure 14-2(d):** There is a *downward shift*—the first few values are relatively stable, and then something happened so that the last several values are relatively stable, but at a much lower level.
- **Figure 14-2(e):** The process is stable except for one *exceptionally high value.* The cause of that unusual value should be investigated. Perhaps the cans became temporarily stuck and one particular can was filled twice instead of once.
- **Figure 14-2(f):** There is an *exceptionally low value.*

- **Figure 14-2(g):** There is a *cyclical pattern* (or repeating cycle). This pattern is clearly nonrandom and therefore reveals a statistically unstable process. Perhaps periodic overadjustments are being made to the machinery, with the effect that some desired value is continually being chased but never quite captured.
- **Figure 14-2(h):** The *variation is increasing over time.* This is a common problem in quality control. The net effect is that products vary more and more until almost all of them are worthless. For example, some cola cans will be overflowing with wasted cola, and some will be underfilled and unsuitable for distribution to consumers.

A common goal of many different methods of quality control is this: *reduce variation* in the product or service. For example, Ford became concerned with variation when it found that its transmissions required significantly more warranty repairs than the same type of transmissions made by Mazda in Japan. A study showed that the Mazda transmissions had substantially less variation in the gearboxes; that is, crucial gearbox measurements varied much less in the Mazda transmissions. Although the Ford transmissions were built within the allowable limits, the Mazda transmissions were more reliable because of their lower variation. Variation in a process can result from two types of causes.

Definitions

Random variation is due to chance; it is the type of variation inherent in any process that is not capable of producing every good or service exactly the same way every time.

Assignable variation results from causes that can be identified (such factors as defective machinery, untrained employees, and so on).

Later in the chapter we will consider ways to distinguish between assignable variation and random variation.

The run chart is one tool for monitoring the stability of a process. We will now consider *control charts,* which are also extremely useful for that same purpose.

Control Chart for Monitoring Variation: The *R* Chart

In the article "The State of Statistical Process Control as We Proceed into the 21st Century" (Stoumbos, Reynolds, Ryan, and Woodall, *Journal of the American Statistical Association,* Vol. 95, No. 451), the authors state that "control charts are among the most important and widely used tools in statistics. Their applications have now moved far beyond manufacturing into engineering, environmental science, biology, genetics, epidemiology, medicine, finance, and even law enforcement and athletics." We begin with the definition of a control chart.

Definition

A **control chart** of a process characteristic (such as mean or variation) consists of values plotted sequentially over time, and it includes a **centerline** as well as a **lower control limit** (LCL) and an **upper control limit** (UCL). The centerline represents a central value of the characteristic measurements, whereas the control limits are boundaries used to separate and identify any points considered to be *unusual.*

We will assume that the population standard deviation σ is not known as we consider only two of several different types of *control charts:* (1) R charts (or range charts) used to monitor variation and (2) $\bar{x}$ charts used to monitor means. When using control charts to monitor a process, it is common to consider R charts and $\bar{x}$ charts together, because a statistically unstable process may be the result of increasing variation or changing means or both.

An ***R* chart** (or **range chart**) is a plot of the sample ranges instead of individual sample values, and it is used to monitor the *variation* in a process. (It might make more sense to use standard deviations, but range charts are used more often in practice. This is a carryover from times when calculators and computers were not available. See Exercise 17 for a control chart based on standard deviations.) In addition to plotting the range values, we include a centerline located at $\bar{R}$, which denotes the mean of all sample ranges, as well as another line for the lower control limit and a third line for the upper control limit. Following is a summary of notation for the components of the R chart.

Notation

Given: Process data consisting of a sequence of samples all of the same size n, and the distribution of the process data is essentially normal.

n = size of each sample, or *subgroup*

$\bar{R}$ = mean of the sample ranges (that is, the sum of the sample ranges divided by the number of samples)

Monitoring Process Variation: Control Chart for *R*

Points plotted: Sample ranges

Centerline: $\bar{R}$

Upper control limit (UCL): $D_4\bar{R}$ (where D_4 is found in Table 14-2)

Lower control limit (LCL): $D_3\bar{R}$ (where D_3 is found in Table 14-2)

The values of D_4 and D_3 were computed by quality-control experts, and they are intended to simplify calculations. The upper and lower control limits of $D_4\bar{R}$ and $D_3\bar{R}$ are values that are roughly equivalent to 99.7% confidence interval limits. It is therefore highly unlikely that values from a statistically stable process would fall beyond those limits. If a value does fall beyond the control limits, it's very likely that the process is not statistically stable.

Costly Assignable Variation

The Mars Climate Orbiter was launched by NASA and sent to Mars, but it was destroyed when it flew too close to its destination planet. The loss was estimated at $125 million. The cause of the crash was found to be confusion between the use of units used for calculations. Acceleration data were provided in the English units of pounds of force, but the Jet Propulsion Laboratory assumed that those units were in metric "newtons" instead of pounds. The thrusters of the spacecraft subsequently provided wrong amounts of force in adjusting the position of the spacecraft. The errors caused by the discrepancy were fairly small at first, but the cumulative error over months of the spacecraft's journey proved to be fatal to its success.

In 1962, the rocket carrying the Mariner 1 satellite was destroyed by ground controllers when it went off course due to a missing minus sign in a computer program.

Don't Tamper!

Nashua Corp. had trouble with its paper-coating machine and considered spending a million dollars to replace it. The machine was working well with a stable process, but samples were taken every so often and, based on the results, adjustments were made. These over-adjustments, called *tampering*, caused shifts away from the distribution that had been good. The effect was an increase in defects. When statistician and quality expert W. Edwards Deming studied the process, he recommended that no adjustments be made unless warranted by a signal that the process had shifted or had become unstable. The company was better off with no adjustments than with the tampering that took place.

Table 14-2 Control Chart Constants

n: Number of Observations in Subgroup	$\overline{x}$		*s*		*R*	
	A_2	A_3	B_3	B_4	D_3	D_4
2	1.880	2.659	0.000	3.267	0.000	3.267
3	1.023	1.954	0.000	2.568	0.000	2.574
4	0.729	1.628	0.000	2.266	0.000	2.282
5	0.577	1.427	0.000	2.089	0.000	2.114
6	0.483	1.287	0.030	1.970	0.000	2.004
7	0.419	1.182	0.118	1.882	0.076	1.924
8	0.373	1.099	0.185	1.815	0.136	1.864
9	0.337	1.032	0.239	1.761	0.184	1.816
10	0.308	0.975	0.284	1.716	0.223	1.777
11	0.285	0.927	0.321	1.679	0.256	1.744
12	0.266	0.886	0.354	1.646	0.283	1.717
13	0.249	0.850	0.382	1.618	0.307	1.693
14	0.235	0.817	0.406	1.594	0.328	1.672
15	0.223	0.789	0.428	1.572	0.347	1.653
16	0.212	0.763	0.448	1.552	0.363	1.637
17	0.203	0.739	0.466	1.534	0.378	1.622
18	0.194	0.718	0.482	1.518	0.391	1.608
19	0.187	0.698	0.497	1.503	0.403	1.597
20	0.180	0.680	0.510	1.490	0.415	1.585
21	0.173	0.663	0.523	1.477	0.425	1.575
22	0.167	0.647	0.534	1.466	0.434	1.566
23	0.162	0.633	0.545	1.455	0.443	1.557
24	0.157	0.619	0.555	1.445	0.451	1.548
25	0.153	0.606	0.565	1.435	0.459	1.541

Source: Adapted from *ASTM Manual on the Presentation of Data and Control Chart Analysis*, © 1976 ASTM, pp. 134–136. Reprinted with permission of American Society for Testing and Materials.

EXAMPLE Manufacturing Aircraft Altimeters Refer to the altimeter errors in Table 14-1. Using the samples of size $n = 4$ collected each day of manufacturing, construct a control chart for *R*.

SOLUTION We begin by finding the value of $\overline{R}$, the mean of the sample ranges.

$$\overline{R} = \frac{19 + 13 + \cdots + 63}{20} = 21.2$$

The centerline for our *R* chart is therefore located at $\overline{R} = 21.2$. To find the upper and lower control limits, we must first find the values of D_3 and D_4. Referring to

Table 14-2 for $n = 4$, we get $D_3 = 0.000$ and $D_4 = 2.282$, so the control limits are as follows:

$$\text{Upper control limit: } D_4\overline{R} = (2.282)(21.2) = 48.4$$
$$\text{Lower control limit: } D_3\overline{R} = (0.000)(21.2) = 0.0$$

Using a centerline value of $\overline{R} = 21.2$ and control limits of 48.4 and 0.0, we now proceed to plot the sample ranges. The result is shown in the Minitab R chart and the TI-83/84 Plus R chart.

Minitab

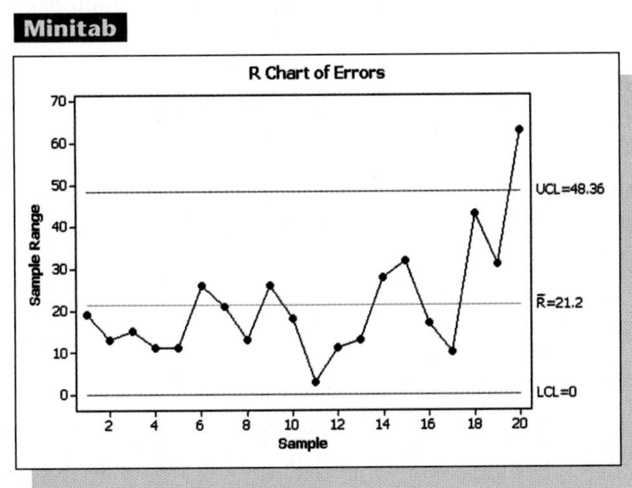

TI-83/84 Plus *R* Chart

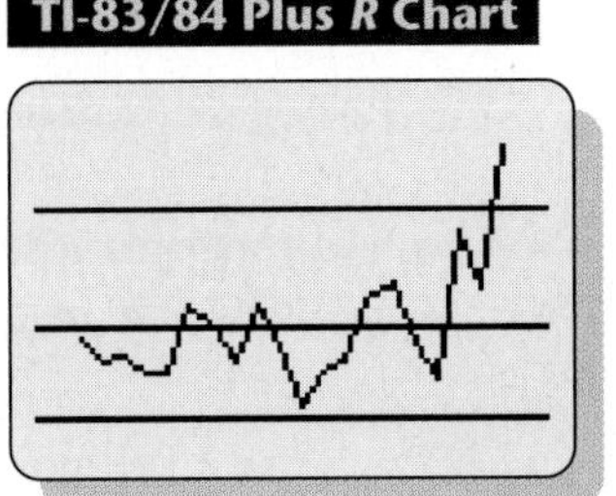

Interpreting Control Charts

When interpreting control charts, the following point is extremely important:

> **Upper and lower control limits of a control chart are based on the *actual* behavior of the process, not the *desired* behavior. Upper and lower control limits are totally unrelated to any process *specifications* that may have been decreed by the manufacturer.**

When investigating the quality of some process, there are typically two key questions that need to be addressed:

1. Based on the current behavior of the process, can we conclude that the process is within statistical control?
2. Do the process goods or services meet design specifications?

The methods of this chapter are intended to address the first question, but not the second. That is, we are focusing on the behavior of the process with the objective of determining whether the process is within statistical control. Whether the process results in goods or services that meet some stated specifications is another issue not addressed by the methods of this chapter. For example, the preceding Minitab R chart includes upper and lower control limits of 48.36 and 0, which result from the sample values listed in Table 14-1. Government regulations require that altimeters have errors between -20 ft and 20 ft, but those desired (or required) specifications are not included in the control chart for R.

Also, we should clearly understand the specific criteria for determining whether a process is in statistical control (that is, whether it is statistically stable). So far, we have noted that a process is not statistically stable if its pattern resembles any of the patterns shown in Figure 14-2. This criterion is included with some others in the following list.

Criteria for Determining When a Process Is Not Statistically Stable (Out of Statistical Control)

1. There is a pattern, trend, or cycle that is obviously not random (such as those depicted in Figure 14-2).
2. There is a point lying outside of the region between the upper and lower control limits. (That is, there is a point above the upper control limit or below the lower control limit.)
3. *Run of 8 Rule:* There are eight consecutive points all above or all below the centerline. (With a statistically stable process, there is a 0.5 probability that a point will be above or below the centerline, so it is very unlikely that eight consecutive points will all be above the centerline or all below it.)

We will use only the three out-of-control criteria listed above, but some businesses use additional criteria such as these:

- There are six consecutive points all increasing or all decreasing.
- There are 14 consecutive points all alternating between up and down (such as up, down, up, down, and so on).
- Two out of three consecutive points are beyond control limits that are 2 standard deviations away from the centerline.
- Four out of five consecutive points are beyond control limits that are 1 standard deviation away from the centerline.

EXAMPLE Statistical Process Control Examine the R charts shown in the preceding example and determine whether the process variation is within statistical control.

SOLUTION We can interpret control charts for R by applying the three out-of-control criteria just listed. Applying the three criteria to the R charts from the preceding example, we conclude that variation in this process is out of statistical control. There are not eight consecutive points all above or all below the centerline, so the third condition is not violated, but the first two conditions are violated.

1. There is a pattern, trend, or cycle that is obviously not random: Going from left to right, there is a pattern of upward trend, as in Figure 14-2(a).
2. There is a point (the rightmost point) that lies above the upper control limit.

INTERPRETATION We conclude that the variation (not necessarily the mean) of the process is out of statistical control. Because the variation appears to be increasing with time, immediate corrective action must be taken to fix the *variation* among the altimeter errors.

Control Chart for Monitoring Means: The $\bar{x}$ Chart

An $\bar{x}$ **chart** is a plot of the sample means, and it is used to monitor the *center* in a process. In addition to plotting the sample means, we include a centerline located at $\bar{\bar{x}}$, which denotes the mean of all sample means (equal to the mean of all sample values combined), as well as another line for the lower control limit and a third line for the upper control limit. Using the approach common in business and industry, the centerline and control limits are based on ranges instead of standard deviations. See Exercise 18 for an $\bar{x}$ chart based on standard deviations.

Monitoring Process Mean: Control Chart for $\bar{x}$

Points plotted: Sample means

Centerline: $\bar{\bar{x}}$ = mean of all sample means

Upper control limit (UCL): $\bar{\bar{x}} + A_2\bar{R}$ (where A_2 is found in Table 14-2)

Lower control limit (LCL): $\bar{\bar{x}} - A_2\bar{R}$ (where A_2 is found in Table 14-2)

EXAMPLE Manufacturing Aircraft Altimeters Refer to the altimeter errors in Table 14-1. Using samples of size $n = 4$ collected each working day, construct a control chart for $\bar{x}$. Based on the control chart for $\bar{x}$ only, determine whether the process mean is within statistical control.

SOLUTION Before plotting the 20 points corresponding to the 20 values of $\bar{x}$, we must first find the value for the centerline and the values for the control limits. We get

$$\bar{\bar{x}} = \frac{2.50 + 2.75 + \cdots + 9.75}{20} = 6.45$$

$$\bar{R} = \frac{19 + 13 + \cdots + 63}{20} = 21.2$$

Referring to Table 14-2, we find that for $n = 4$, $A_2 = 0.729$. Knowing the values of $\bar{\bar{x}}$, A_2, and $\bar{R}$, we can now evaluate the control limits.

Upper control limit: $\bar{\bar{x}} + A_2\bar{R} = 6.45 + (0.729)(21.2) = 21.9$

Lower control limit: $\bar{\bar{x}} - A_2\bar{R} = 6.45 - (0.729)(21.2) = -9.0$

INTERPRETATION The resulting control chart for $\bar{x}$ will be as shown in the following displays. Examination of the control charts shows that the process mean is out of statistical control because at least one of the three out-of-control criteria is not satisfied. Specifically, the third criterion is not satisfied because there are eight (or more) consecutive points all below the centerline. Also, there does appear to be a pattern of an upward trend. Again, immediate corrective action is required to fix the production process.

Bribery Detected with Control Charts

Control charts were used to help convict a person who bribed Florida jai alai players to lose. (See "Using Control Charts to Corroborate Bribery in Jai Alai," by Charnes and Gitlow, *The American Statistician,* Vol. 49, No. 4.) An auditor for one jai alai facility noticed that abnormally large sums of money were wagered for certain types of bets, and some contestants didn't win as much as expected when those bets were made. R charts and $\bar{x}$ charts were used in court as evidence of highly unusual patterns of betting. Examination of the control charts clearly shows points well beyond the upper control limit, indicating that the process of betting was out of statistical control. The statistician was able to identify a date at which assignable variation appeared to stop, and prosecutors knew that it was the date of the suspect's arrest.

Excel

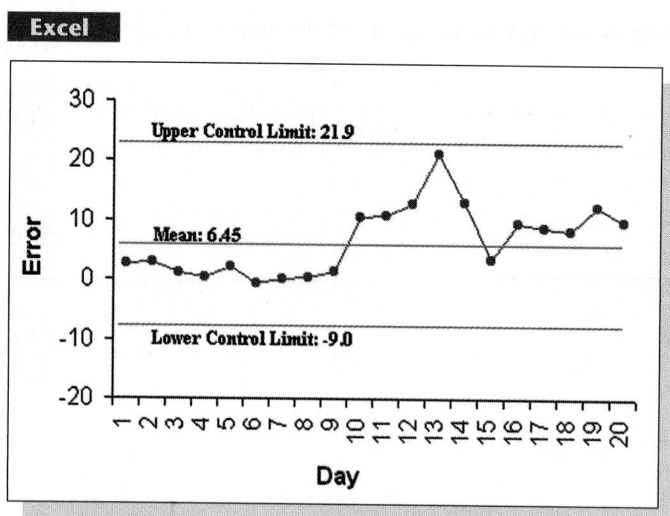

TI-83/84 Plus

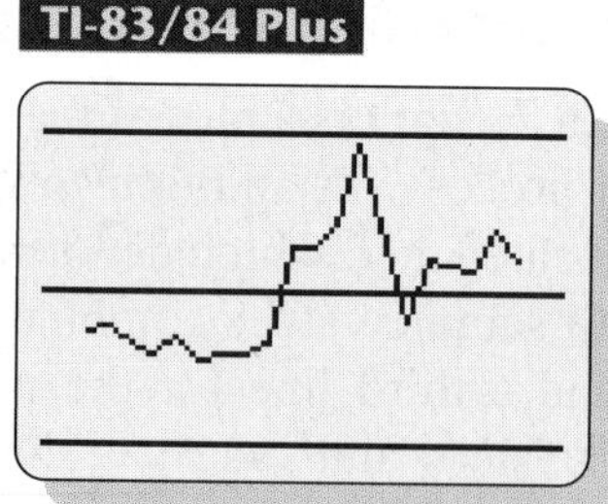

Using a TI-83/84 Plus Calculator for Run Charts, *R* Charts, and $\bar{x}$ Charts

This section includes examples of a TI-83/84 Plus-generated run chart, *R* chart, and $\bar{x}$ chart. This section also includes examples of Minitab and Excel charts. Comparisons should lead to the obvious conclusion that the Minitab and Excel versions are better in the sense that they include more information and are easier to read. The Excel charts generally require substantially more effort than Minitab charts. Consequently, if choosing among a TI-83/84 Plus calculator, Minitab, or Excel for the purpose of creating a run chart, *R* chart, or $\bar{x}$ chart, Minitab is the best choice. (Along with some other advantages, Minitab automatically produces the upper and lower control limits as well as the centerline, so there is no need to manually determine them.) The following instructions describe the use of the TI-83/84 Plus calculator.

TI-83/84 Plus

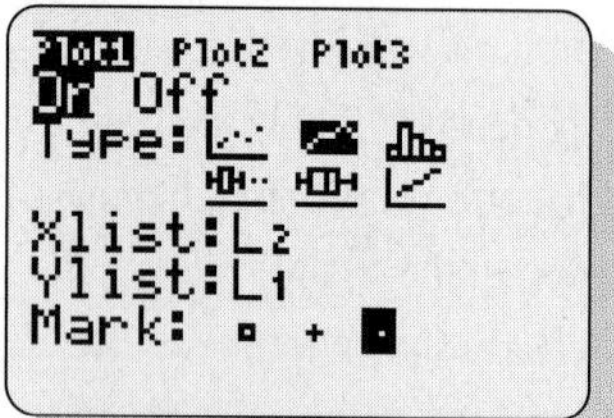

TI-83/84 Plus

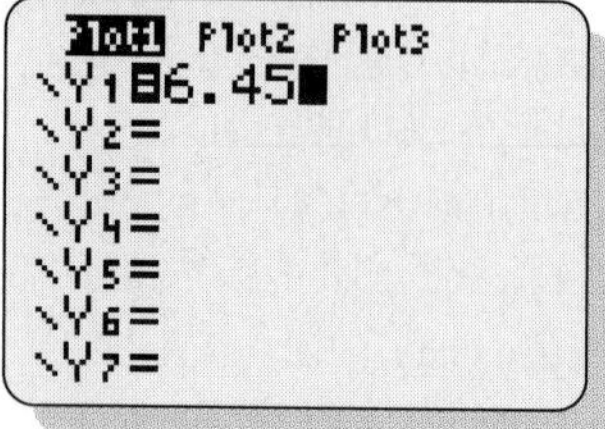

Run Chart

1. To construct a run chart using a TI-83/84 Plus calculator, first enter all of the sample data in list L1.
2. In list L2 enter the consecutive integers 1, 2, 3, . . . (stopping when L2 has the same length as L1).
3. Find the mean of all of the sample data. (Press STAT, select **CALC,** press ENTER, enter L1, and press ENTER again.)
4. Set up Plot1 as shown in the accompanying screen display. (Press 2ND Y= to get to the **STAT PLOTS** menu.) Note that list L2 is used for the Xlist entry, and list L1 is used for the Ylist entry.
5. Press Y= and enter the mean found in Step 3. For the run chart example in this section, the mean is 6.45, so the screen should appear as shown here.
6. Press 2ND ZOOM and select **AxesOff.**

7. Press ZOOM 9 to select the **Zoomstat** menu item, which automatically configures the display window so that all of the data points are shown.
8. After the run chart is displayed, change back to AxesOn. Press 2ND ZOOM, use the arrow keys to highlight **AxesON,** then press ENTER.

R Chart

1. In list L1, enter the sample *ranges* (instead of the original list of sample data).
2. In list L2 enter the consecutive integers 1, 2, 3, . . . (stopping when L2 has the same length as L1).
3. Find the mean of all of the sample ranges, and find the values used for the upper and lower control limits (as described in this section).
4. Set up **Plot1** as shown in Step 4 of the above procedure for generating a run chart.
5. Press Y= and enter the mean of all of the sample ranges (as shown in Step 5 for the above procedure for generating a run chart.) For **Y2=** enter the value used for the upper control limit, and for **Y3=** enter the value used for the lower control limit.
6. Press 2ND ZOOM and select **AxesOff.**
7. Press ZOOM 9 to select the **Zoomstat** menu item, which automatically configures the display window so that all of the data points are shown.
8. After the run chart is displayed, change back to AxesOn. Press 2ND ZOOM, use the arrow keys to highlight **AxesON,** then press ENTER.

$\bar{x}$ Chart

1. In list L1, enter the sample *means* (instead of the original list of sample data).
2. In list L2 enter the consecutive integers 1, 2, 3, . . . (stopping when L2 has the same length as L1).
3. Find the mean of all of the sample means, and find the values used for the upper and lower control limits (as described in this section).
4. Set up Plot1 as shown in Step 4 of the above procedure for generating a run chart.
5. Press Y= and enter the mean of all of the sample means (as shown in Step 5 for the above procedure for generating a run chart.) For **Y2=** enter the value used for the upper control limit, and for **Y3=** enter the value used for the lower control limit.
6. Press 2ND ZOOM and select **AxesOff.**
7. Press ZOOM 9 to select the **Zoomstat** menu item, which automatically configures the display window so that all of the data points are shown.
8. After the run chart is displayed, change back to AxesOn. Press 2ND ZOOM, use the arrow keys to highlight **AxesON,** then press ENTER.

14-2 BASIC SKILLS AND CONCEPTS

Statistical Literacy and Critical Thinking

1. Process Data What are *process data*?

2. Statistical Control What does it mean for a process to be out of statistical control?

3. Control Charts What is a control chart? What is an R chart? What is an $\bar{x}$ chart? What is the difference between an R chart and an $\bar{x}$ chart?

4. Variation What is the difference between random variation and assignable variation?

Interpreting Run Charts. *In Exercises 5–8, examine the run chart from a process of filling 12-oz cans of cola and do the following: (a) Determine whether the process is within statistical control; (b) if the process is not within statistical control, identify reasons why it is not; (c) apart from being within statistical control, does the process appear to be behaving as it should?*

5.

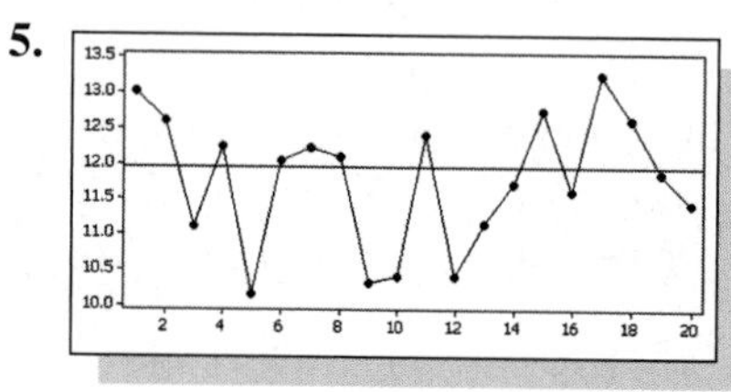

6.

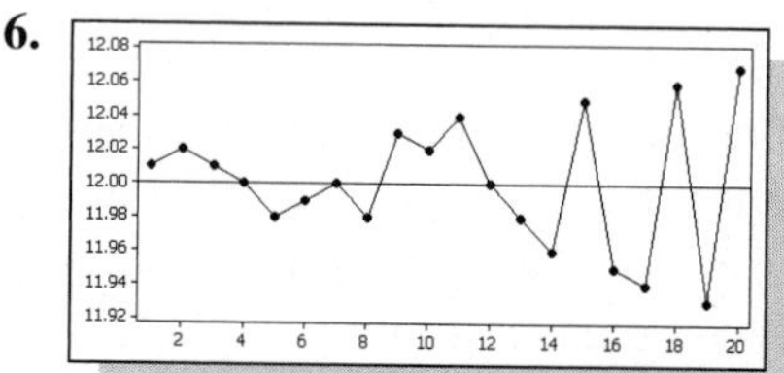

7.

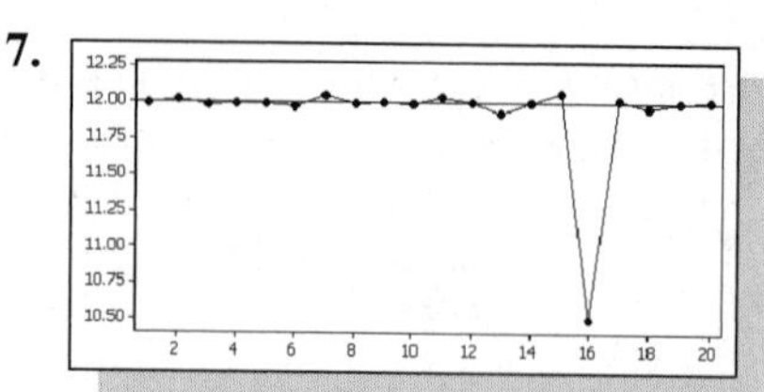

8.

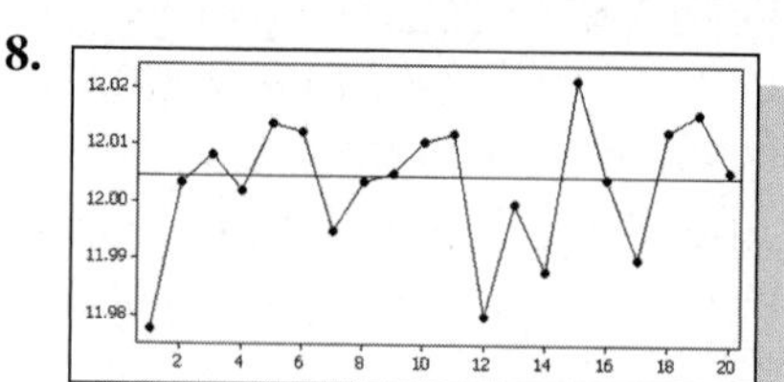

Constructing Control Charts for Aluminum Cans. *Exercises 9 and 10 are based on the axial loads (in pounds) of aluminum cans that are 0.0109 in. thick, as listed in Data Set 15 in Appendix B. An axial load of a can is the maximum weight supported by its side, and it is important to have an axial load high enough so that the can isn't crushed when the top lid is pressed into place. The data are from a real manufacturing process, and they were provided by a student who used an earlier edition of this book.*

TI **9.** R Chart In each day of production, seven aluminum cans with thickness 0.0109 in. were randomly selected and the axial loads were measured. The ranges for the different days are listed below, but they can also be found from the values given in Data Set 15 in Appendix B. Construct an R chart and determine whether the process variation is within statistical control. If it is not, identify which of the three out-of-control criteria lead to rejection of statistically stable variation.

78	77	31	50	33	38	84	21	38	77	26	78	78
17	83	66	72	79	61	74	64	51	26	41	31	

TI **10.** $\bar{x}$ Chart In each day of production, seven aluminum cans with thickness 0.0109 in. were randomly selected and the axial loads were measured. The means for the different days are listed on the next page, but they can also be found from the values given in Data Set 15 in Appendix B. Construct an $\bar{x}$ chart and determine whether the process mean is within statistical control. If it is not, identify which of the three out-of-control criteria lead to rejection of statistically stable variation.

252.7 247.9 270.3 267.0 281.6 269.9 257.7 272.9 273.7 259.1 275.6 262.4 256.0
277.6 264.3 260.1 254.7 278.1 259.7 269.4 266.6 270.9 281.0 271.4 277.3

Monitoring the Minting of Quarters. *In Exercises 11–13, use the following information: The U.S. Mint has a goal of making quarters with a weight of 5.670 g, but any weight between 5.443 g and 5.897 g is considered acceptable. A new minting machine is placed into service and the weights are recorded for a quarter randomly selected every 12 min for 20 consecutive hours. The results are listed in the accompanying table.*

TI **11.** Minting Quarters: Run Chart Construct a run chart for the 100 values. Does there appear to be a pattern suggesting that the process is not within statistical control? What are the practical implications of the run chart?

TI **12.** Minting Quarters: *R* Chart Construct an *R* chart and determine whether the process variation is within statistical control. If it is not, identify which of the three out-of-control criteria lead to rejection of statistically stable variation.

TI **13.** Minting Quarters: $\bar{x}$ Chart Construct an $\bar{x}$ chart and determine whether the process mean is within statistical control. If it is not, identify which of the three out-of-control criteria lead to rejection of a statistically stable mean. Does this process need corrective action?

Weights (in grams) of Minted Quarters

Hour	Weight (g)					$\bar{x}$	s	Range
1	5.639	5.636	5.679	5.637	5.691	5.6564	0.0265	0.055
2	5.655	5.641	5.626	5.668	5.679	5.6538	0.0211	0.053
3	5.682	5.704	5.725	5.661	5.721	5.6986	0.0270	0.064
4	5.675	5.648	5.622	5.669	5.585	5.6398	0.0370	0.090
5	5.690	5.636	5.715	5.694	5.709	5.6888	0.0313	0.079
6	5.641	5.571	5.600	5.665	5.676	5.6306	0.0443	0.105
7	5.503	5.601	5.706	5.624	5.620	5.6108	0.0725	0.203
8	5.669	5.589	5.606	5.685	5.556	5.6210	0.0545	0.129
9	5.668	5.749	5.762	5.778	5.672	5.7258	0.0520	0.110
10	5.693	5.690	5.666	5.563	5.668	5.6560	0.0534	0.130
11	5.449	5.464	5.732	5.619	5.673	5.5874	0.1261	0.283
12	5.763	5.704	5.656	5.778	5.703	5.7208	0.0496	0.122
13	5.679	5.810	5.608	5.635	5.577	5.6618	0.0909	0.233
14	5.389	5.916	5.985	5.580	5.935	5.7610	0.2625	0.596
15	5.747	6.188	5.615	5.622	5.510	5.7364	0.2661	0.678
16	5.768	5.153	5.528	5.700	6.131	5.6560	0.3569	0.978
17	5.688	5.481	6.058	5.940	5.059	5.6452	0.3968	0.999
18	6.065	6.282	6.097	5.948	5.624	6.0032	0.2435	0.658
19	5.463	5.876	5.905	5.801	5.847	5.7784	0.1804	0.442
20	5.682	5.475	6.144	6.260	6.760	6.0642	0.5055	1.285

Appendix B Data Set: Constructing Control Charts for Boston Rainfall. *In Exercises 14–16, refer to the daily amounts of rainfall in Boston for one year, as listed in Data Set 10 in Appendix B. Omit the last entry for Wednesday so that each day of the week has exactly 52 values.*

TI **14.** Boston Rainfall: Constructing a Run Chart Using only the 52 rainfall amounts for Monday, construct a run chart. Does the process appear to be within statistical control?

TI **15.** Boston Rainfall: Constructing an R Chart Using the 52 samples of seven values each, construct an R chart and determine whether the process variation is within statistical control. If it is not, identify which of the three out-of-control criteria lead to rejection of statistically stable variation.

TI **16.** Boston Rainfall: Constructing an $\bar{x}$ Chart Using the 52 samples of seven values each, construct an $\bar{x}$ chart and determine whether the process mean is within statistical control. If it is not, identify which of the three out-of-control criteria lead to rejection of a statistically stable mean. If not, what can be done to bring the process within statistical control?

14-2 BEYOND THE BASICS

TI **17.** Constructing an s Chart In this section we described control charts for R and $\bar{x}$ based on ranges. Control charts for monitoring variation and center (mean) can also be based on standard deviations. An *s chart* for monitoring variation is made by plotting sample standard deviations with a centerline at $\bar{s}$ (the mean of the sample standard deviations) and control limits at $B_4\bar{s}$ and $B_3\bar{s}$, where B_4 and B_3 are found in Table 14-2. Construct an s chart for the data of Table 14-1. Compare the result to the R chart given in this section.

TI **18.** Constructing an $\bar{x}$ Chart Based on Standard Deviations An $\bar{x}$ chart based on standard deviations (instead of ranges) is made by plotting sample means with a centerline at $\bar{\bar{x}}$ and control limits at $\bar{\bar{x}} + A_3\bar{s}$ and $\bar{\bar{x}} - A_3\bar{s}$, where A_3 is found in Table 14-2 and $\bar{s}$ is the mean of the sample standard deviations. Use the data in Table 14-1 to construct an $\bar{x}$ chart based on standard deviations. Compare the result to the $\bar{x}$ chart based on sample ranges (shown in this section).

Quality Control at Perstorp

Perstorp Components, Inc. uses a computer that automatically generates control charts to monitor the thicknesses of the floor insulation the company makes for Ford Rangers and Jeep Grand Cherokees. The \$20,000 cost of the computer was offset by a first-year savings of \$40,000 in labor, which had been used to manually generate control charts to ensure that insulation thicknesses were between the specifications of 2.912 mm and 2.988 mm. Through the use of control charts and other quality-control methods, Perstorp reduced its waste by more than two-thirds.

14-3 Control Charts for Attributes

Key Concept This section presents a method for constructing a control chart to monitor the proportion p for some *attribute,* such as whether a service or manufactured item is defective or nonconforming. (A good or a service is nonconforming if it doesn't meet specifications or requirements; nonconforming goods are sometimes discarded, repaired, or called "seconds" and sold at reduced prices.) The control chart is interpreted by using the same three criteria from Section 14-2 to determine whether the process is statistically stable.

Section 14-2 discussed control charts for *quantitative* data, but this section describes the construction of control charts for *qualitative* data. As in Section 14-2, we select samples of size n at regular time intervals and plot points in a sequential graph with a centerline and control limits. (There are ways to deal with samples of different sizes, but we don't consider them here.)

Definition

A **control chart for p** (or **p chart**) is a graph of proportions plotted sequentially over time, and it includes a centerline, a lower control limit (LCL), and an upper control limit (UCL).

The notation and control chart values are as follows (where the attribute of "defective" can be replaced by any other relevant attribute).

Notation

$\bar{p}$ = pooled estimate of the proportion of defective items in the process

$$= \frac{\text{total number of defects found among all items sampled}}{\text{total number of items sampled}}$$

$\bar{q}$ = pooled estimate of the proportion of process items that are *not* defective

$= 1 - \bar{p}$

n = size of each sample (not the number of samples)

Control Chart for p

Centerline: $\bar{p}$

Upper control limit: $\bar{p} + 3\sqrt{\dfrac{\bar{p}\,\bar{q}}{n}}$

Lower control limit: $\bar{p} - 3\sqrt{\dfrac{\bar{p}\,\bar{q}}{n}}$

(If the calculation for the lower control limit results in a negative value, use 0 instead. If the calculation for the upper control limit exceeds 1, use 1 instead.)

We use $\bar{p}$ for the centerline because it is the best estimate of the proportion of defects from the process. The expressions for the control limits correspond to 99.7% confidence interval limits as described in Section 7-2.

EXAMPLE Defective Aircraft Altimeters The Chapter Problem describes the process of manufacturing aircraft altimeters. Section 14-2 includes examples of control charts for monitoring the errors in altimeter readings. An altimeter is considered to be defective if it cannot be calibrated or corrected to give accurate readings (within 20 ft of the true altitude). The Altigauge Manufacturing Company produces altimeters in batches of 100, and each altimeter is tested and determined to be acceptable or defective. Listed below are the numbers of defective altimeters in successive batches of 100. Construct a control chart for the proportion p of defective altimeters and determine whether the process is within statistical control. If not, identify which of the three out-of-control criteria apply.

Defects: 2 0 1 3 1 2 2 4 3 5 3 7

continued

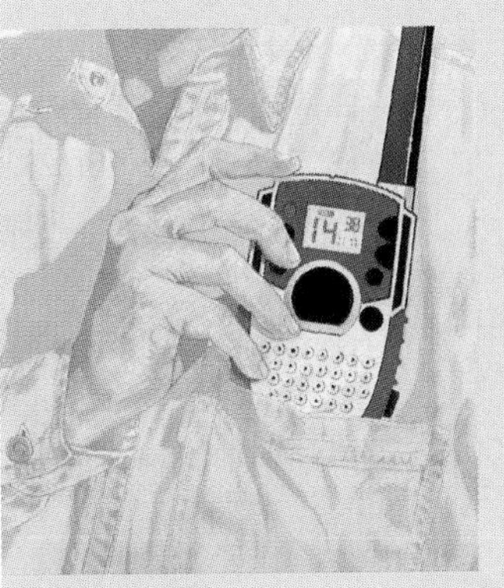

Six Sigma in Industry

Six Sigma is the term used in industry to describe a process that results in a rate of no more than 3.4 defects out of a million. The reference to Six Sigma suggests six standard deviations away from the center of a normal distribution, but the assumption of a perfectly stable process is replaced with the assumption of a process that drifts slightly, so the defect rate is no more than 3 or 4 defects per million.

Started around 1985 at Motorola, Six Sigma programs now attempt to improve quality and increase profits by reducing variation in processes. Motorola saved more than $940 million in three years. Allied Signal reported a savings of $1.5 billion. GE, Polaroid, Ford, Honeywell, Sony, and Texas Instruments are other major companies that have adopted the Six Sigma goal.

STATISTICS IN THE NEWS

High Cost of Low Quality

The Federal Drug Administration recently reached an agreement whereby a pharmaceutical company, the Schering-Plough Corporation, would pay a record $500 million for failure to correct problems in manufacturing drugs. According to a *New York Times* article by Melody Petersen, "Some of the problems relate to the lack of controls that would identify faulty medicines, while others stem from outdated equipment. They involve some 200 medicines, including Claritin, the allergy medicine that is Schering's top-selling product."

SOLUTION The centerline for the control chart is located by the value of $\bar{p}$:

$$\bar{p} = \frac{\text{total number of defects from all samples combined}}{\text{total number of altimeters sampled}}$$

$$= \frac{2 + 0 + 1 + \cdots + 7}{12 \cdot 100} = \frac{33}{1200} = 0.0275$$

Because $\bar{p} = 0.0275$, it follows that $\bar{q} = 1 - \bar{p} = 0.9725$. Using $\bar{p} = 0.0275$, $\bar{q} = 0.9725$, and $n = 100$, we find the control limits as follows:

Upper control limit:

$$\bar{p} + 3\sqrt{\frac{\bar{p}\,\bar{q}}{n}} = 0.0275 + 3\sqrt{\frac{(0.0275)(0.9725)}{100}} = 0.0766$$

Lower control limit:

$$\bar{p} - 3\sqrt{\frac{\bar{p}\,\bar{q}}{n}} = 0.0275 - 3\sqrt{\frac{(0.0275)(0.9725)}{100}} = -0.0216$$

Because the lower control limit is less than 0, we use 0 instead. Having found the values for the centerline and control limits, we can proceed to plot the proportions of defective altimeters. The Minitab control chart for p and the TI-83/84 Plus control chart for p are shown here.

Minitab

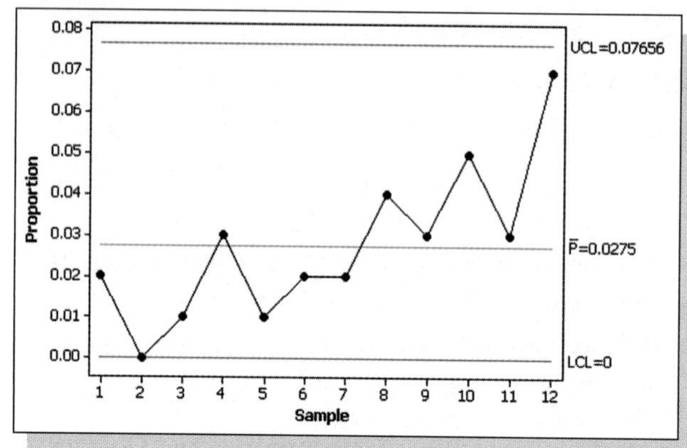

TI-83/84 Plus

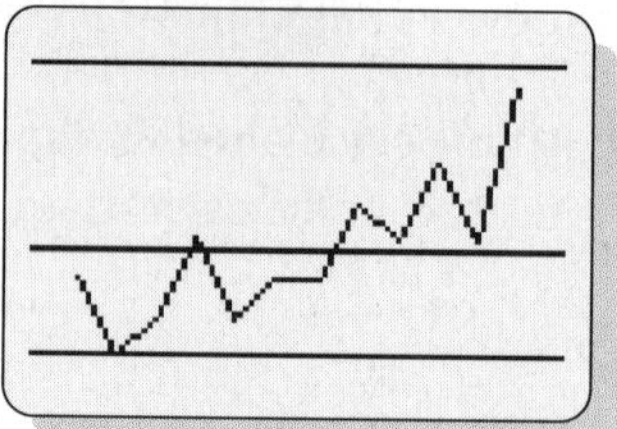

INTERPRETATION We can interpret the control chart for p by considering the three out-of-control criteria listed in Section 14-2. Using those criteria, we conclude that this process is out of statistical control for this reason: There appears to be an upward trend. The company should take immediate action to correct the increasing proportion of defects.

TI Using a TI-83/84 Plus Calculator for Control Charts for *p*

If choosing among a TI-83/84 Plus calculator, Excel, and Minitab, it is recommended that Minitab be used to generate control charts for p. Minitab automatically determines the upper and lower control limits as well as the centerline, and it provides a display that is easy to read. The following instructions describe the use of the TI-83/84 Plus calculator:

1. To construct a control chart for p using a TI-83/84 Plus calculator, first enter all of the sample proportions in list L1.
2. In list L2 enter the consecutive integers 1, 2, 3, . . . (stopping when L2 has the same length as L1).
3. Find the three values to be used for the centerline, upper control limit, and lower control limit (as described in this section).
4. Set up Plot1 as shown in the accompanying screen display. (Press 2ND Y= to get to the **STAT PLOTS** menu.) Note that list L2 is used for the Xlist entry, and list L1 is used for the Ylist entry.
5. Press Y= and use three separate lines to enter the three values used for the centerline, the upper control limit, and the lower control limit (from Step 3).
6. Press 2ND ZOOM and select **AxesOff.**
7. Press ZOOM 9 to select the **Zoomstat** menu item, which automatically configures the display window so that all of the data points are shown. If the lines for the upper and lower control limits are not both shown, press WINDOW and change the values used for Ymin and Ymax.
8. After the p chart is displayed, change back to AxesOn. (Press 2ND ZOOM, use the arrow keys to highlight **AxesON,** then press ENTER.)

TI-83/84 Plus

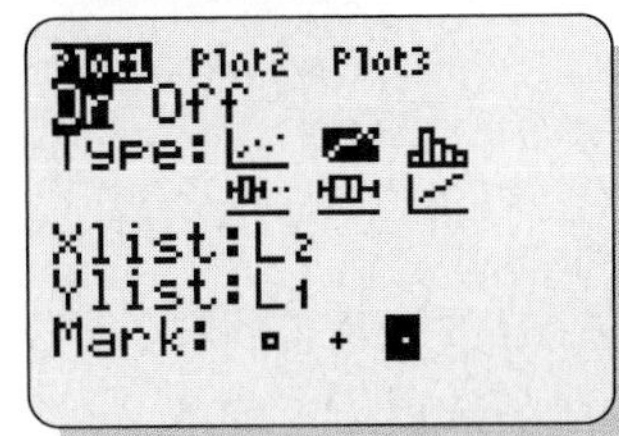

14-3 BASIC SKILLS AND CONCEPTS

Statistical Literacy and Critical Thinking

1. *p* Chart What is a p chart, and what is its purpose?

2. Contract Specification The Paper Chase office supply company requires a supplier of pens to maintain a production process with a defect rate less than 2%. If the supplier uses a p chart to determine that the manufacturing process is within statistical control, does that indicate that the defect rate is less than 2%?

3. Control Limits Using the methods of this section, a process is analyzed and the upper and lower control limits are found to be 0.250 and -0.050 respectively. What upper and lower control limits are used to construct the control chart?

4. Interpreting a Control Chart When monitoring the process of producing altimeters, a company finds that the process is out of statistical control because there is a downward pattern of defects that is not random. Should the downward pattern be corrected? What should the company do?

Determining Whether a Process Is in Control. *In Exercises 5–8, examine the given control chart for p and determine whether the process is within statistical control. If it is not, identify which of the three out-of-control criteria apply.*

5.

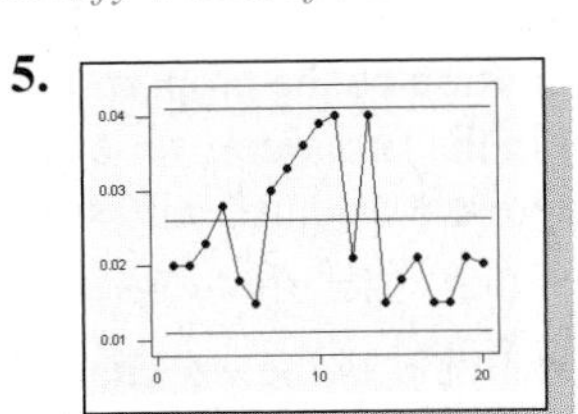

6.

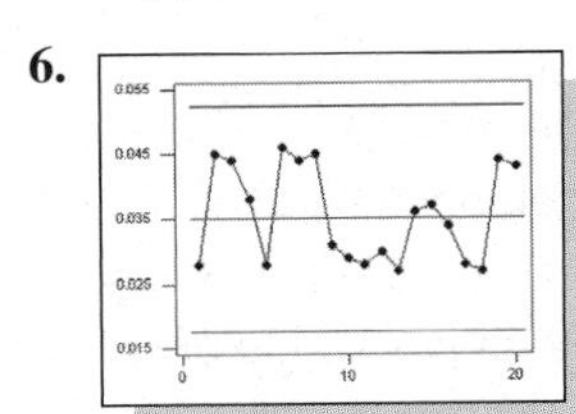

7. 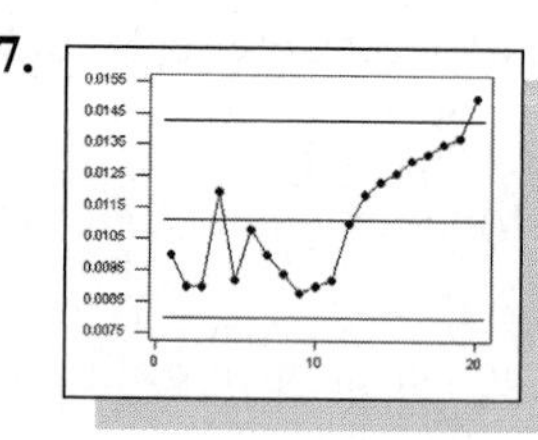

8. 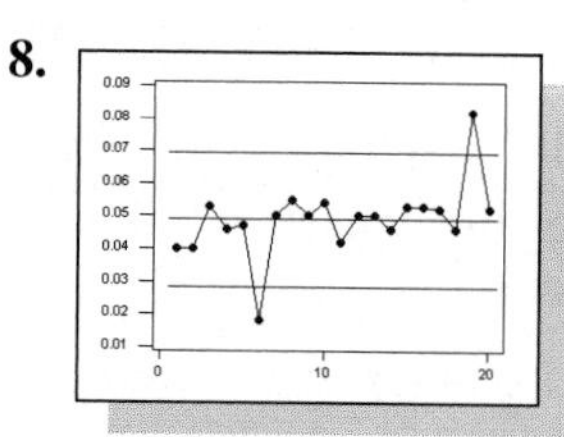

Constructing Control Charts for p. In Exercises 9–12, use the given process data to construct a control chart for p. In each case, use the three out-of-control criteria listed in Section 14-2 and determine whether the process is within statistical control. If it is not, identify which of the three out-of-control criteria apply.

TI **9.** *p* Chart for Birth Rate In each of 10 consecutive and recent years, 10,000 people were randomly selected and the numbers of births they generated were found, with the results given below (based on data from the National Center for Health Statistics). How might the results be explained?

Births: 155 152 148 147 145 146 145 144 141 139

TI **10.** *p* Chart for Divorce Rate In each of 10 consecutive and recent years, 10,000 people were randomly selected and the numbers of divorces were found, with the results given below (based on data from the National Center for Health Statistics). How might the results be explained?

Divorces: 48 46 46 44 43 43 42 41 42 40

TI **11.** *p* Chart for Defective Car Batteries Defective car batteries are a nuisance because they can strand and inconvenience drivers. A car battery is considered to be defective if it fails before its warranty expires. Defects are identified when the batteries are returned under the warranty program. The Powerco Battery corporation manufactures car batteries in batches of 1000 and the numbers of defects are listed below for each of 12 consecutive batches. Does the manufacturing process require correction?

Defects: 8 6 5 9 10 7 7 4 6 11 5 8

TI **12.** Polling When the Infopop polling organization conducts a telephone survey, a call is considered to be a defect if the respondent is unavailable or refuses to answer questions. For one particular poll about consumer preferences, 200 people are called each day, and the numbers of defects are listed below. Does the calling process require corrective action?

Defects: 92 83 85 87 98 108 96 115 121 125 112 127 109 131 130

14-3 BEYOND THE BASICS

TI **13.** *p* Chart for Boston Rainfall Refer to the Boston rainfall amounts in Data Set 10 in Appendix B. For each of the 52 weeks, let the sample proportion be the proportion of days that it rained. (Delete the 53rd value for Wednesday.) In the first week, for example, the sample proportion is $3/7 = 0.429$. Do the data represent a statistically stable process?

TI **14.** Constructing an *np* Chart A variation of the control chart for p is the ***np* chart** in which the *actual numbers* of defects are plotted instead of the *proportions* of defects. The *np* chart will have a centerline value of $n\bar{p}$, and the control limits will have values of $n\bar{p} + 3\sqrt{n\bar{p}\,\bar{q}}$ and $n\bar{p} - 3\sqrt{n\bar{p}\,\bar{q}}$. The p chart and the np chart differ only in the scale of values used for the vertical axis. Construct the np chart for the example given in this section. Compare the result with the control chart for p given in this section.

Review

In Chapter 2 we identified important characteristics of data: center, variation, distribution, outliers, and changing pattern of data over time. The focus of this chapter is the changing pattern of data over time. Process data were defined to be data arranged according to some time sequence, and such data can be analyzed with run charts and control charts. Control charts have a centerline, an upper control limit, and a lower control limit. A process is statistically stable (or within statistical control) if it has only natural variation with no patterns, cycles, or unusual points. Decisions about statistical stability are based on how a process is actually behaving, not how we might like it to behave because of such factors as manufacturer specifications. The following graphs were described:

- *Run chart:* a sequential plot of *individual* data values over time
- *R chart:* a control chart that uses ranges in an attempt to monitor the *variation* in a process
- *$\bar{x}$ chart:* a control chart used to determine whether the process *mean* is within statistical control
- *p chart:* a control chart used to monitor the proportion of some process *attribute,* such as whether items are defective

Statistical Literacy and Critical Thinking

1. Pattern over Time Why is it important to monitor a changing pattern of data over time?

2. Statistical Process Control The title of this chapter is "Statistical Process Control." What does that mean?

3. Manufacturing Run Amok What would be a possible adverse consequence of a Pepsi bottling plant running a process that is not monitored?

4. Control Charts When monitoring the times it takes technicians to repair computers, why is it important to use an $\bar{x}$ chart and an R chart together?

Review Exercises

Constructing Control Charts for Consumption of Electricity. The following table lists amounts of electrical consumption (in kWh) for the author's home, as given in Data Set 9 in Appendix B. Use the data for Exercises 1–3.

Electrical Consumption (kWh)

Year 1: 1st Half	3375	2661	2073
Year 1: 2nd Half	2579	2858	2296
Year 2: 1st Half	2812	2433	2266
Year 2: 2nd Half	3128	3286	2749
Year 3: 1st Half	3427	578	3792

TI **1.** Run Chart Construct a run chart for the 15 values. Does there appear to be a pattern suggesting that the process is not within statistical control?

TI **2.** R Chart Using subgroups of size $n = 3$ corresponding to the rows of the table, construct an R chart and determine whether the process variation is within statistical control. If it is not, identify which of the three out-of-control criteria lead to rejection of statistically stable variation.

TI **3.** $\bar{x}$ Chart Using subgroups of size $n = 3$ corresponding to the rows of the table, construct an $\bar{x}$ chart and determine whether the process mean is within statistical control. Does the process appear to be statistically stable? If it is not, identify which of the three out-of-control criteria lead to rejection of statistically stable variation.

TI **4.** Constructing a Control Chart for Infectious Diseases In each of 13 consecutive and recent years, 100,000 adults 65 years of age or older were randomly selected and the number who died from infectious diseases is recorded, with the results given below (based on data from "Trends in Infectious Diseases Mortality in the United States," by Pinner et al., *Journal of the American Medical Association,* Vol. 275, No. 3). Construct an appropriate control chart and determine whether the process is within statistical control. If not, identify which criteria lead to rejection of statistical stability.

Number who died: 270 264 250 278 302 334 348 347 377 357 362 351 343

TI **5.** Control Chart for Defects The Medassist Pharmaceutical Company manufactures aspirin tablets. Each day, 100 tablets are randomly selected and tested. A tablet is considered defective if it has obvious physical deformities or the aspirin content is less than 490 mg or greater than 510 mg. The numbers of defects are listed below for consecutive days. Construct an appropriate control chart and determine whether the process is within statistical control. If not, identify which criteria lead to rejection of statistical stability.

Defects: 4 2 2 3 5 2 9 12 1 11 3 2 12 14

Cumulative Review Exercises

TI **1.** Control Chart for Defective Seat Belts The Flint Accessory Corporation manufactures seat belts for cars. Federal specifications require that the webbing must have a breaking strength of at least 5000 lb. During each week of production, 200 belts are randomly selected and tested for breaking strength, and a belt is considered defective if it breaks before reaching the force of 5000 lb. The numbers of defects are listed below for a sequence of 10 weeks. Use a control chart for p to verify that the process is within statistical control. If it is not in control, explain why it is not.

6 4 12 3 7 2 3 5 4 2

TI **2.** Confidence Interval for Defective Seat Belts Refer to the data in Exercise 1 and, using all of the data from the 2000 seat belts that were tested, construct a 95% confidence interval for the proportion of defects.

TI **3.** Hypothesis Test for Defective Seat Belts Refer to the data in Exercise 1 and, using all of the data from the 2000 seat belts that were tested, use a 0.05 significance level to test the claim that the rate of defects is greater than 1%.

4. Using Probability in Control Charts When interpreting control charts, one of the three out-of-control criteria is that there are eight consecutive points all above or all below the centerline. For a statistically stable process, there is a 0.5 probability that a point will be above the centerline and there is a 0.5 probability that a point will be below the centerline. In each of the following, assume that sample values are independent and the process is statistically stable.
   a. Find the probability that when eight consecutive points are randomly selected, they are all above the centerline.
   b. Find the probability that when eight consecutive points are randomly selected, they are all below the centerline.
   c. Find the probability that when eight consecutive points are randomly selected, they are all above or all below the centerline.

Cooperative Group Activities

1. Out-of-class activity Collect your own process data and analyze them using the methods of this section. It would be ideal to collect data from a real manufacturing process, but that may be difficult to accomplish. If so, consider using a simulation or referring to published data, such as those found in an almanac. Here are some suggestions:
   - Shoot five basketball foul shots (or shoot five crumpled sheets of paper into a wastebasket) and record the number of shots made; then repeat this procedure 20 times, and use a p chart to test for statistical stability in the proportion of shots made.
   - Your pulse rate can be measured by counting the number of times your heart beats in 1 min. Measure your pulse rate four times each hour for several hours, then construct appropriate control charts. What factors contribute to random variation? Assignable variation?
   - Go through newspapers for the past 12 weeks and record the closing of the Dow Jones Industrial Average (DJIA) for each business day. Use run and control charts to explore the statistical stability of the DJIA. Identify at least one practical consequence of having this process statistically stable, and identify at least one practical consequence of having this process out of statistical control.
   - Find the marriage rate per 10,000 population for several years. (See the *Information Please Almanac* or the *Statistical Abstract of the United States.*) Assume that in each year 10,000 people were randomly selected and surveyed to determine whether they were married. Use a p chart to test for statistical stability of the marriage rate. (Other possible rates: death, accident fatality, crime.)

   Obtain a printed copy of computer results, and write a report summarizing your conclusions.

2. In-class activity If the instructor can distribute the numbers of absences for each class meeting, groups of three or four students can analyze them for statistical stability and make recommendations based on the conclusions.

3. Out-of-class activity Conduct research to identify *Deming's funnel experiment,* then use a funnel and marbles to collect data for the different rules for adjusting the funnel location. Construct appropriate control charts for the different rules of funnel adjustment. What does the funnel experiment illustrate? What do you conclude?

TI-83/84 Plus Calculator Project

The Biotyme company manufactures heart pacemakers with an overall defect rate of 1%. For each of 20 consecutive workdays, 200 heart pacemakers are manufactured, and the proportion of defects is recorded for each of the 20 days.

This project involves the simulation of the manufacturing process. The pacemakers for each day are simulated by randomly generating 200 numbers between 1 and 100. If we consider an outcome of 1 to be a defective pacemaker, while outcomes of 2, 3, 4, . . . , 100 are acceptable pacemakers, we have the same 1% defect rate as the actual production process. Here are the TI-83/84 Plus calculator instructions for conducting the simulation:

1. Press MATH, select **PRB,** then select the menu item of **randInt.**
2. The format for the randInt command is randInt(minimum, maximum, number of generated values), so use the command **randInt(1, 100, 200)** to generate 200 integers between 1 and 100 inclusive.
3. Store the results in list L1 (by pressing STO› 2ND 1).
4. Examine list L1 and find the number of "defective pacemakers" (or the number of times that the value of 1 was generated). This becomes easy if the values in list L1 are first sorted using SortA. [Press STAT and scroll down to select **Sort A,** then enter L1 to get the command **Sort(L1).**]
5. Repeat the preceding steps until results from 20 days are simulated.

Use the proportions of defects from the 20 simulated days of production to construct a p chart, then determine whether the process is within statistical control. Since we know that the process is actually stable with $p = 0.01$, the conclusion that it is not stable would be a type I error; that is, we would have a false positive signal, causing us to believe that the process needed to be adjusted when in fact it should be left alone.

The preceding simulation represents 20 days. Now simulate another 10 days of manufacturing pacemakers, but modify these last 10 days so that the defect rate is 3% instead of 1%. Combine all of the generated data to represent a total of 30 days of sample results. Construct a p chart for this combined data set. Is the process out of control? If we concluded that the process was not out of control, we would be making a type II error; that is, we would believe that the process was okay when in fact it should be repaired or adjusted to correct the shift to the 3% nonconforming rate.

From Data to Decision

Critical Thinking: Are the axial loads within statistical control?

Is the process of manufacturing cans proceeding as it should?

Exercises 9 and 10 in Section 14-2 used process data from a New York company that manufactures 0.0109-in. thick aluminum cans for a major beverage supplier. Refer to Data Set 15 in Appendix B and conduct an analysis of the process data for the cans that are 0.0111 in. thick. The values in the data set are the measured axial loads of cans, and the top lids are pressed into place with pressures that vary between 158 lb and 165 lb.

Analyzing the Results

Based on the given process data, should the company take any corrective action? Write a report summarizing your conclusions. Address not only the issue of statistical stability, but also the ability of the cans to withstand the pressures applied when the top lids are pressed into place. Also compare the behavior of the 0.0111-in. cans to the behavior of the 0.0109-in. cans and recommend which thickness should be used.

Internet Project

Control Charts

This chapter introduces different charting techniques used to summarize and study data associated with a process along with methods for analyzing the stability of that process. With the exception of the run chart, individual data points are not needed to construct a chart. For example, the *R* chart is constructed from sample ranges while the *p* chart is based on sample proportions. This is an important point, as data collected from third-party sources are often given in terms of summarizing statistics. Go to the *Elementary Statistics* Web site:

http://www.aw.com/Triola.

Locate the Internet Project dealing with control charts. There you will be directed to data sets and sources of data for use in constructing control charts. From the resulting charts you will be asked to interpret and discuss trends in the underlying processes.

Statistics @ Work

"There is a certain amount of respect that is given to someone who knows statistics and can explain it to someone who doesn't know it."

Dan O'Toole

Account Executive: A. C. Nielsen

In his work in the Advanced Analytics Group at A. C. Nielsen, Dan develops statistical solutions to help clients like Polaroid, Ocean Spray, and Gillette understand which of their marketing vehicles drives sales most profitably. Dan has a Masters Degree in Business Economics from Bentley College.

What concepts of statistics do you use?

I have worked with analyses as simple as correlation and general significance tests, all the way to multiple regression, factor analysis, correspondence analysis, and cluster analysis.

How do you use statistics on the job?

My job is to discover or uncover client issues, and then find out if we can apply one of our statistical techniques to their specific issue. If a technique won't help a client, then you need to know that. An example of how I use stats: A client may say, I sell product "X," whether it is juice, bread, or a camera. Right now, they may control 20% of the market. They may come to us to see if they can increase market share by lowering their price. My job would be to design a study to analyze this question. To do this I have to design a study that will take into account everything that affects the sales of a product. Using techniques like regression, if I am able to create a model with good significance, I will be able to isolate specific influences on the sales and offer recommendations. Things like seasonality distribution as well as any marketing efforts that may have taken place, must be included. In addition, I will have to take into account complementary products' price (butter is complementary to bread; while film is for a camera) and also competitive products. For instance, bread may compete with English muffins (I know it does for me).

Do you feel job applicants are viewed more favorably if they have studied some statistics?

By far. There is a certain amount of respect that is given to someone who knows statistics and can explain it to someone who doesn't know it (because it means you really know it and aren't reciting from a textbook). Almost every job uses statistics (particularly correlations and regressions). People will say things like, "Oh, check if they're correlated."

Is your use of probability and statistics increasing, decreasing, or remaining stable?

It definitely is increasing. In this business (consulting), you are constantly challenged to learn a new technique or look at an old technique in order to improve it. In addition, since we are constantly coming out with new products, our understanding of statistics has to increase to use these techniques effectively.

How beneficial do you find your knowledge of statistics for performing your responsibilities?

It is not a question of beneficial, but rather it is a necessity. In fact, we find that we have to know it so well, so that we can explain it in "layman's" terms to our clients.

Projects, Procedures, Perspectives

15

15-1 Projects

Key Concept This section includes suggestions for a final project in the introductory statistics course. One fantastic advantage of this course is that it deals with skills and concepts that can be applied immediately to the real world. After only one fun semester, students are able to conduct their own studies. Some of the suggested topics can be addressed by actually conducting experiments, whereas others might be observational studies that require research of results already available. For example, testing the effectiveness of air bags by actually crashing cars is strongly discouraged, but destructive taste tests of chocolate chip cookies can be an easy and somewhat enjoyable experiment. Here is a suggested format, followed by a list of suggested topics.

Group Project vs. Individual Project Topics can be assigned to individuals, but group projects are particularly effective because they help develop the interpersonal skills that are so necessary in today's working environment. One study showed that the "inability to get along with others" is the main reason for firing employees, so a group project can be very helpful in preparing students for their future work environments.

Oral Report A 10- to 15-minute-long class presentation should involve all group members in a coordinated effort to clearly describe the important components of the study. Students typically have some reluctance to speak in public, so a brief oral report can be very helpful in building the confidence that they so well deserve. The oral report is an activity that helps students to be better prepared for future professional activities.

Written Report The main objective of the project is not to produce a written document equivalent to a term paper, but a brief written report should be submitted, and it should include the following components:

1. List of data collected along with a description of how the data were obtained.
2. Description of the method of analysis
3. Relevant graphs and/or statistics, including STATDISK, Minitab, Excel, or TI-83/84 Plus displays
4. Statement of conclusions
5. Reasons why the results might not be correct, along with a description of ways in which the study could be improved, given sufficient time and money

Large Classes or Online Classes: Posters or PowerPoint Some classes are too large for individual projects or group projects with three or four or five students per group. Some online classes are not able to meet as a group. For such cases, reports of individual or small group projects can be presented through posters similar to those found at conference poster sessions. Posters summarizing important elements of a project can be submitted to professors for evaluation. PowerPoint presentations can also be used.

Survey A survey can be an excellent source of data. See the accompanying sample survey that provides opportunities for many interesting projects that address questions such as these:

1. When people "randomly" select digits (as in Question 2), are the results actually random?
2. Do the last four digits of social security numbers appear to be random?
3. Do males and females carry different amounts of change?
4. Do males and females have different numbers of credit cards?
5. Is there a difference in pulse rates between those who exercise and those who do not?
6. Is there a difference in pulse rates between those who smoke and those who do not?
7. Is there a relationship between exercise and smoking?
8. Is there a relationship between eye color and exercise?
9. Is there a relationship between exercise and the number of hours worked each week?
10. Is there a correlation between height and pulse rate?

Survey

1. _____ Female _____ Male
2. Randomly select four digits and enter them here: __ __ __ __
3. Eye color: ______________
4. Enter your height in inches: ______________
5. What is the total value of all coins now in your possession? ______________
6. How many keys are in your possession at this time? ______________
7. How many credit cards are in your possession at this time? __
8. Enter the last four digits of your social security number: __ __ __ __
9. Record your pulse rate by counting the number of heartbeats for 1 minute: ______________
10. Do you exercise vigorously (such as running, swimming, cycling, tennis, basketball, etc.) for at least 20 minutes at least twice a week? _____ Yes _____ No
11. How many credit hours of courses are you taking this semester? ________
12. Are you currently employed? _____ Yes _____ No
    If yes, how many hours do you work each week? ________
13. During the past 12 months, have you been the driver of a car that was involved in a crash? _____ Yes _____ No
14. Do you smoke? _____ Yes _____ No
15. _____ Left-handed _____ Right-handed _____ Ambidextrous

Project Topics The preceding survey questions are a source of good project ideas. Also see the "Cooperative Group Activities" listed near the end of each chapter. The following list gives additional project suggestions.

1. Graph from a newspaper or magazine redrawn to better describe the data
2. Newspaper article about a survey rewritten to better inform the reader
3. Using coin toss to get better survey results from sensitive question
4. Ages of student cars compared to faculty/staff cars
5. Proportion of foreign cars driven by students compared to the proportion of foreign cars driven by faculty
6. Car ages in the parking lot of a discount store compared to car ages in the parking lot of an upscale department store
7. Are husbands older than their wives?
8. Are husband/wife age differences the same for young married couples as for older married couples?
9. Analysis of the ages of books in the college library

10. How do the ages of books in the college library compare with those in the library of a nearby college?
11. Comparison of the ages of science books and English books in the college library
12. Estimate the hours that students study each week
13. Is there a relationship between hours studied and grades earned?
14. Is there a relationship between hours worked and grades earned?
15. A study of *reported* heights compared to *measured* heights
16. A study of the accuracy of wristwatches
17. Is there a relationship between taste and cost of different brands of chocolate chip cookies?
18. Is there a relationship between taste and cost of different brands of peanut butter?
19. Is there a relationship between taste and cost of different brands of cola?
20. Is there a relationship between salaries of professional baseball (or basketball or football) players and their season achievements?
21. Rates versus weights: Is there a relationship between car fuel-consumption rates and car weights? If so, what is it?
22. Is there a relationship between the lengths of men's (or women's) feet and their heights?
23. Are there differences in taste between ordinary tap water and different brands of bottled water?
24. Were auto fatality rates affected by laws requiring the use of seat belts?
25. Were auto fatality rates affected when the national speed limit of 55 mi/h was eliminated?
26. Were auto fatality rates affected by the presence of air bags?
27. Is there a difference in taste between Coke and Pepsi?
28. Is there a relationship between student grade-point averages and the amount of television watched? If so, what is it?
29. Is there a relationship between the selling price of a home and its living area (in square feet), lot size (in acres), number of rooms, number of baths, and the annual tax bill?
30. Is there a relationship between the height of a person and the height of his or her navel?
31. Is there support for the theory that the ratio of a person's height to his or her navel height is the Golden Ratio of about 1.6:1?
32. A comparison of the numbers of keys carried by males and females
33. A comparison of the numbers of credit cards carried by males and females
34. Are murderers now younger than they were in the past?
35. Do people who exercise vigorously tend to have lower pulse rates than those who do not?
36. Do people who exercise vigorously tend to have reaction times that are different from those of people who do not?

37. Do people who smoke tend to have higher pulse rates than those who do not?
38. For people who don't exercise, how is pulse rate affected by climbing a flight of stairs?
39. Do statistics students tend to have pulse rates that are different from those of people not studying statistics?
40. A comparison of GPAs of statistics students with those of students not taking statistics
41. Do left-handed people tend to be involved in more car crashes?
42. Do men have more car crashes than women?
43. Do young drivers have more car crashes than older drivers?
44. Are drivers who get tickets more likely to be involved in crashes?
45. Do smokers tend to be involved in more car crashes?
46. Do people with higher pulse rates tend to be involved in more/fewer car crashes?
47. A comparison of reaction times measured with right and left hands
48. Are the proportions of male and female smokers equal?
49. Do statistics students tend to smoke more (or less) than the general population?
50. Are people more likely to smoke if their parents smoked?
51. Evidence to support/refute the belief that smoking tends to stunt growth
52. Does a sports team have an advantage by playing at home instead of away?
53. Analysis of service times (in seconds) for a car drive-up window at a bank
54. A comparison of service times for car drive-up windows at two different banks
55. Analysis of times that McDonald's' patrons are seated at a table
56. Analysis of times that McDonald's' patrons wait in line
57. Analysis of times cars require for refueling
58. Is the state lottery a wise investment?
59. Comparison of casino games: craps versus roulette
60. Starting with $1, is it easier to win a million dollars by playing casino craps or by playing a state lottery?
61. Bold versus cautious strategies of gambling: When gambling with $100, does it make any difference if you bet $1 at a time or if you bet the whole $100 at once?
62. Designing and analyzing results from a test for extrasensory perception
63. Analyzing paired data consisting of heights of fathers (or mothers) and heights of their first sons (or daughters)
64. Gender differences in preferences of dinner partners among the options of Brad Pitt, Tiger Woods, the President, Nicole Kidman, Cameron Diaz, Julia Roberts, and the Pope
65. Gender differences in preferences of activities among the options of dinner, movie, watching television, reading a book, golf, tennis, swimming, attending a baseball game, attending a football game

66. Is there support for the theory that cereals with high sugar content are placed on shelves at eye level with children?
67. Is there support for the claim that the mean body temperature is less than 98.6°F?
68. Is there a relationship between smoking and drinking coffee?
69. Is there a relationship between course grades and time spent playing video games?
70. Is there support for the theory that a Friday is unlucky if it falls on the 13th day of a month?

15-2 Procedures

Key Concept This section describes a general procedure for conducting a statistical analysis of data. The data can be collected through experiments or observational studies. It is absolutely essential to critique the method used to collect the data, because a poor method of data collection destroys the usefulness of the data. Look carefully for bias in the way data are collected, as well as bias on the part of the person or group collecting the data. Many of the procedures in this book are based on the assumption that we are working with a simple random sample, meaning that every possible sample of the same size has the same chance of being selected. If a sample is self-selected (voluntary response), it is worthless for making inferences about a population.

Exploring, Comparing, Describing After collecting data, first consider exploring, describing, and comparing data sets using the basic tools included in Chapters 2 and 3. Be sure to address the following:

1. *Center:* Find the mean and median, which are measures of center that are representative or average values giving us an indication of where the middle of the data set is located.
2. *Variation:* Find the range and standard deviation, which are measures of the amount that the sample values vary among themselves.
3. *Distribution:* Construct a histogram to see the nature or shape of the distribution of the data. Also construct a normal quantile plot and determine if the data are from a population having a normal distribution.
4. *Outliers:* Identify any sample values that lie very far away from the vast majority of the other sample values. If there are outliers, try to determine whether they are errors that should be corrected. If the outliers are correct values, study their effects by repeating the analysis with the outliers excluded.
5. *Time:* Determine if the population is stable or if its characteristics are changing over time.

Inferences: Estimating Parameters and Hypothesis Testing When trying to use sample data for making inferences about a population, it is often difficult to choose the particular procedure that should be applied. This text includes a wide variety of procedures that apply to many different circumstances. Here are some key questions that should be answered:

- What is the level of measurement (nominal, ordinal, interval, ratio) of the data?
- Does the study involve one, two, or more populations?
- Is there a claim to be tested or a parameter to be estimated?
- What is the relevant parameter (mean, standard deviation, proportion)?
- Is the population standard deviation known? (The answer is almost always "no.")

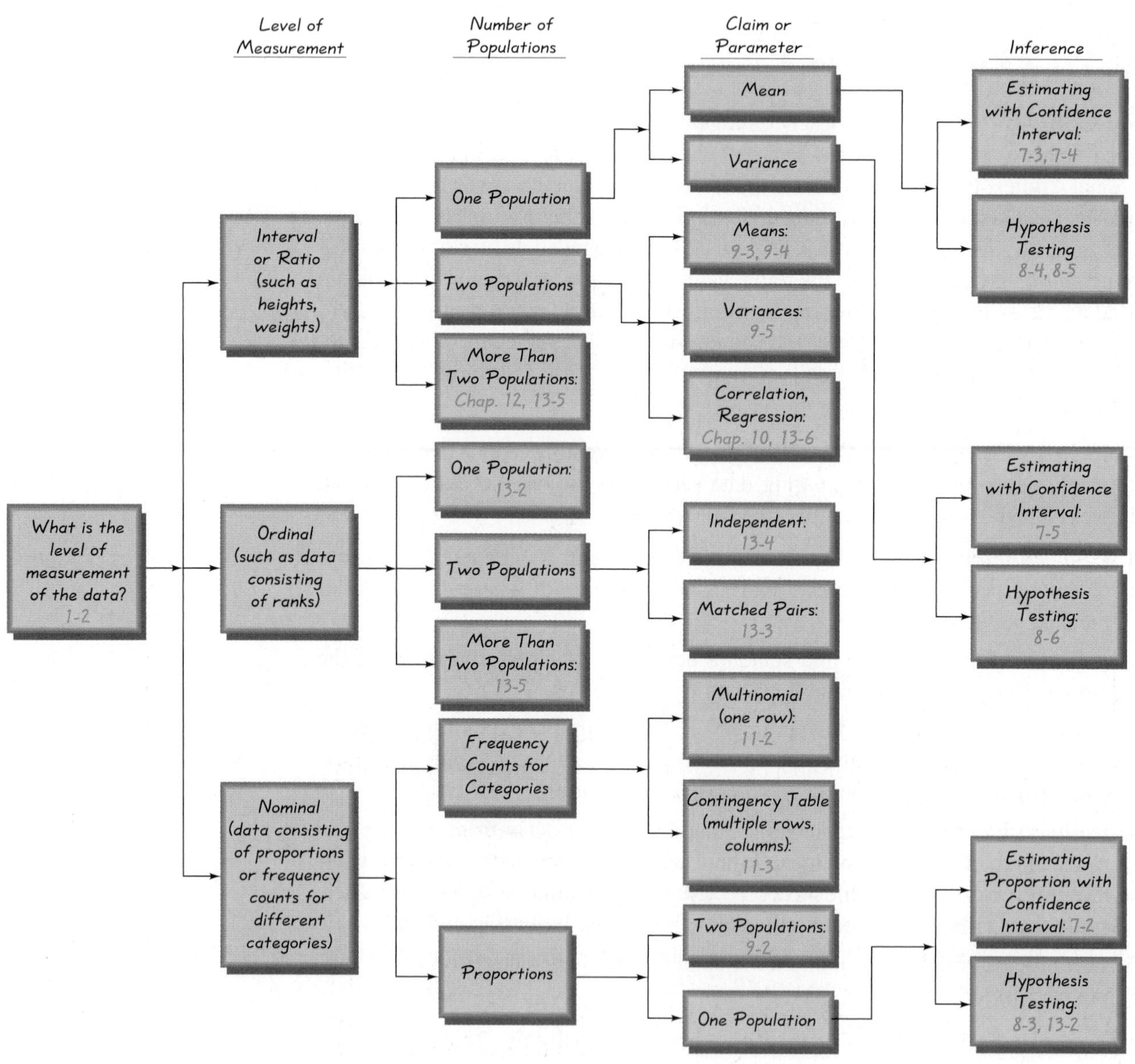

Figure 15-1 Selecting the Appropriate Procedure

- Is there reason to believe that the population is normally distributed?
- What is the basic question or issue that you want to address?

In Figure 15-1 we list the major methods included in this book, along with a scheme for determining which of those methods should be used. To use Figure 15-1, start at the extreme left side of the figure and begin by identifying the level of measurement of the data. Proceed to follow the path suggested by the level of measurement, the number of populations, and the claim or parameter being considered.

Note: This figure applies to a fixed population. If the data are from a process that may change over time, construct a control chart (see Chapter 14) to determine whether the process is statistically stable. This figure applies to process data only if the process is statistically stable.

Figure 15-1 can be used for statistical methods presented in this book, but there may be other methods that might be more suitable for a particular statistical analysis. Consult your friendly professional statistician for help with other methods.

15-3 Perspectives

Key Concept No single introductory statistics course can transform anyone into an expert statistician. The introductory course has a limited scope and does not include many important topics. Know that professional help is available from expert statisticians, and this introductory statistics course will help you in discussions with one of these experts.

Successful completion of an introductory statistics course results in benefits that extend far beyond the attainment of credit toward a college degree. You will have improved job marketability. You will be better prepared to critically analyze reports in the media and professional journals. You will understand the basic concepts of probability and chance. You will know that in attempting to gain insight into a set of data, it is important to investigate measures of center (such as mean and median), measures of variation (such as range and standard deviation), the nature of the distribution (via a frequency distribution or graph), the presence of outliers, and whether the population is stable or is changing over time. You will know and understand the importance of estimating population parameters (such as a mean, standard deviation, and proportion), as well as testing claims made about population parameters.

Throughout this text we have emphasized the importance of good sampling. You should recognize that a bad sample may be beyond repair by even the most expert statisticians using the most sophisticated techniques. There are many mail, magazine, and telephone call-in surveys that allow respondents to be "self-selected." The results of such surveys are generally worthless when judged according to the criteria of sound statistical methodology. Keep this in mind when you encounter voluntary response (self-selected) surveys, so that you don't let them affect your beliefs and decisions. You should also recognize, however, that many surveys and polls obtain very good results, even though the sample sizes might seem to be relatively small. Although many people refuse to believe it, a nationwide survey of only 1200 voters can provide good results if the sampling is carefully planned and executed.

Throughout this text we have emphasized the *interpretation* of results. A final conclusion to "reject the null hypothesis" is basically worthless to all of those other people who lacked the vision and wisdom to take a statistics course. Computers and calculators are quite good at yielding results, but such results typically require the careful interpretation that breathes life into an otherwise meaningless result. We should recognize that a result is not automatically valid simply because it was computer-generated. Computers don't think, and they are quite capable of providing results that are quite ridiculous when considered in the context of the real world. We should always apply the most important and indispensable tool in all of statistics: *common sense!*

The Role of Statistics in Education There was once a time that a person was considered to be educated if he or she could simply read. But we now live in a time that demands so much more. Today, an educated person is capable of critical thinking. An educated person is capable of learning, instead of just doing. An educated person has intellectual curiosity. An educated person can communicate effectively both orally and in writing. An educated person can relate to all other people, including those from different cultures, as well as those who might not be so educated. The introductory statistics course can provide so much more than the mere attainment of technical skills. Successful completion of the introductory statistics course can enable students to grow as individuals and professionals so that they can make substantial progress toward becoming productive professionals, responsible citizens, and people who are truly educated.

Appendices

Appendix A: Tables

TABLE A-1 Binomial Probabilities

		p													
n	*x*	.01	.05	.10	.20	.30	.40	.50	.60	.70	.80	.90	.95	.99	*x*
2	0	.980	.902	.810	.640	.490	.360	.250	.160	.090	.040	.010	.002	0+	0
	1	.020	.095	.180	.320	.420	.480	.500	.480	.420	.320	.180	.095	.020	1
	2	0+	.002	.010	.040	.090	.160	.250	.360	.490	.640	.810	.902	.980	2
3	0	.970	.857	.729	.512	.343	.216	.125	.064	.027	.008	.001	0+	0+	0
	1	.029	.135	.243	.384	.441	.432	.375	.288	.189	.096	.027	.007	0+	1
	2	0+	.007	.027	.096	.189	.288	.375	.432	.441	.384	.243	.135	.029	2
	3	0+	0+	.001	.008	.027	.064	.125	.216	.343	.512	.729	.857	.970	3
4	0	.961	.815	.656	.410	.240	.130	.062	.026	.008	.002	0+	0+	0+	0
	1	.039	.171	.292	.410	.412	.346	.250	.154	.076	.026	.004	0+	0+	1
	2	.001	.014	.049	.154	.265	.346	.375	.346	.265	.154	.049	.014	.001	2
	3	0+	0+	.004	.026	.076	.154	.250	.346	.412	.410	.292	.171	.039	3
	4	0+	0+	0+	.002	.008	.026	.062	.130	.240	.410	.656	.815	.961	4
5	0	.951	.774	.590	.328	.168	.078	.031	.010	.002	0+	0+	0+	0+	0
	1	.048	.204	.328	.410	.360	.259	.156	.077	.028	.006	0+	0+	0+	1
	2	.001	.021	.073	.205	.309	.346	.312	.230	.132	.051	.008	.001	0+	2
	3	0+	.001	.008	.051	.132	.230	.312	.346	.309	.205	.073	.021	.001	3
	4	0+	0+	0+	.006	.028	.077	.156	.259	.360	.410	.328	.204	.048	4
	5	0+	0+	0+	0+	.002	.010	.031	.078	.168	.328	.590	.774	.951	5
6	0	.941	.735	.531	.262	.118	.047	.016	.004	.001	0+	0+	0+	0+	0
	1	.057	.232	.354	.393	.303	.187	.094	.037	.010	.002	0+	0+	0+	1
	2	.001	.031	.098	.246	.324	.311	.234	.138	.060	.015	.001	0+	0+	2
	3	0+	.002	.015	.082	.185	.276	.312	.276	.185	.082	.015	.002	0+	3
	4	0+	0+	.001	.015	.060	.138	.234	.311	.324	.246	.098	.031	.001	4
	5	0+	0+	0+	.002	.010	.037	.094	.187	.303	.393	.354	.232	.057	5
	6	0+	0+	0+	0+	.001	.004	.016	.047	.118	.262	.531	.735	.941	6
7	0	.932	.698	.478	.210	.082	.028	.008	.002	0+	0+	0+	0+	0+	0
	1	.066	.257	.372	.367	.247	.131	.055	.017	.004	0+	0+	0+	0+	1
	2	.002	.041	.124	.275	.318	.261	.164	.077	.025	.004	0+	0+	0+	2
	3	0+	.004	.023	.115	.227	.290	.273	.194	.097	.029	.003	0+	0+	3
	4	0+	0+	.003	.029	.097	.194	.273	.290	.227	.115	.023	.004	0+	4
	5	0+	0+	0+	.004	.025	.077	.164	.261	.318	.275	.124	.041	.002	5
	6	0+	0+	0+	0+	.004	.017	.055	.131	.247	.367	.372	.257	.066	6
	7	0+	0+	0+	0+	0+	.002	.008	.028	.082	.210	.478	.698	.932	7
8	0	.923	.663	.430	.168	.058	.017	.004	.001	0+	0+	0+	0+	0+	0
	1	.075	.279	.383	.336	.198	.090	.031	.008	.001	0+	0+	0+	0+	1
	2	.003	.051	.149	.294	.296	.209	.109	.041	.010	.001	0+	0+	0+	2
	3	0+	.005	.033	.147	.254	.279	.219	.124	.047	.009	0+	0+	0+	3
	4	0+	0+	.005	.046	.136	.232	.273	.232	.136	.046	.005	0+	0+	4
	5	0+	0+	0+	.009	.047	.124	.219	.279	.254	.147	.033	.005	0+	5
	6	0+	0+	0+	.001	.010	.041	.109	.209	.296	.294	.149	.051	.003	6
	7	0+	0+	0+	0+	.001	.008	.031	.090	.198	.336	.383	.279	.075	7
	8	0+	0+	0+	0+	0+	.001	.004	.017	.058	.168	.430	.663	.923	8

NOTE: 0+ represents a positive probability less than 0.0005.

(*continued*)

TABLE A-1 Binomial Probabilities (*continued*)

n	x	.01	.05	.10	.20	.30	.40	.50	.60	.70	.80	.90	.95	.99	x
								p							
9	0	.914	.630	.387	.134	.040	.010	.002	0+	0+	0+	0+	0+	0+	0
	1	.083	.299	.387	.302	.156	.060	.018	.004	0+	0+	0+	0+	0+	1
	2	.003	.063	.172	.302	.267	.161	.070	.021	.004	0+	0+	0+	0+	2
	3	0+	.008	.045	.176	.267	.251	.164	.074	.021	.003	0+	0+	0+	3
	4	0+	.001	.007	.066	.172	.251	.246	.167	.074	.017	.001	0+	0+	4
	5	0+	0+	.001	.017	.074	.167	.246	.251	.172	.066	.007	.001	0+	5
	6	0+	0+	0+	.003	.021	.074	.164	.251	.267	.176	.045	.008	0+	6
	7	0+	0+	0+	0+	.004	.021	.070	.161	.267	.302	.172	.063	.003	7
	8	0+	0+	0+	0+	0+	.004	.018	.060	.156	.302	.387	.299	.083	8
	9	0+	0+	0+	0+	0+	0+	.002	.010	.040	.134	.387	.630	.914	9
10	0	.904	.599	.349	.107	.028	.006	.001	0+	0+	0+	0+	0+	0+	0
	1	.091	.315	.387	.268	.121	.040	.010	.002	0+	0+	0+	0+	0+	1
	2	.004	.075	.194	.302	.233	.121	.044	.011	.001	0+	0+	0+	0+	2
	3	0+	.010	.057	.201	.267	.215	.117	.042	.009	.001	0+	0+	0+	3
	4	0+	.001	.011	.088	.200	.251	.205	.111	.037	.006	0+	0+	0+	4
	5	0+	0+	.001	.026	.103	.201	.246	.201	.103	.026	.001	0+	0+	5
	6	0+	0+	0+	.006	.037	.111	.205	.251	.200	.088	.011	.001	0+	6
	7	0+	0+	0+	.001	.009	.042	.117	.215	.267	.201	.057	.010	0+	7
	8	0+	0+	0+	0+	.001	.011	.044	.121	.233	.302	.194	.075	.004	8
	9	0+	0+	0+	0+	0+	.002	.010	.040	.121	.268	.387	.315	.091	9
	10	0+	0+	0+	0+	0+	0+	.001	.006	.028	.107	.349	.599	.904	10
11	0	.895	.569	.314	.086	.020	.004	0+	0+	0+	0+	0+	0+	0+	0
	1	.099	.329	.384	.236	.093	.027	.005	.001	0+	0+	0+	0+	0+	1
	2	.005	.087	.213	.295	.200	.089	.027	.005	.001	0+	0+	0+	0+	2
	3	0+	.014	.071	.221	.257	.177	.081	.023	.004	0+	0+	0+	0+	3
	4	0+	.001	.016	.111	.220	.236	.161	.070	.017	.002	0+	0+	0+	4
	5	0+	0+	.002	.039	.132	.221	.226	.147	.057	.010	0+	0+	0+	5
	6	0+	0+	0+	.010	.057	.147	.226	.221	.132	.039	.002	0+	0+	6
	7	0+	0+	0+	.002	.017	.070	.161	.236	.220	.111	.016	.001	0+	7
	8	0+	0+	0+	0+	.004	.023	.081	.177	.257	.221	.071	.014	0+	8
	9	0+	0+	0+	0+	.001	.005	.027	.089	.200	.295	.213	.087	.005	9
	10	0+	0+	0+	0+	0+	.001	.005	.027	.093	.236	.384	.329	.099	10
	11	0+	0+	0+	0+	0+	0+	0+	.004	.020	.086	.314	.569	.895	11
12	0	.886	.540	.282	.069	.014	.002	0+	0+	0+	0+	0+	0+	0+	0
	1	.107	.341	.377	.206	.071	.017	.003	0+	0+	0+	0+	0+	0+	1
	2	.006	.099	.230	.283	.168	.064	.016	.002	0+	0+	0+	0+	0+	2
	3	0+	.017	.085	.236	.240	.142	.054	.012	.001	0+	0+	0+	0+	3
	4	0+	.002	.021	.133	.231	.213	.121	.042	.008	.001	0+	0+	0+	4
	5	0+	0+	.004	.053	.158	.227	.193	.101	.029	.003	0+	0+	0+	5
	6	0+	0+	0+	.016	.079	.177	.226	.177	.079	.016	0+	0+	0+	6
	7	0+	0+	0+	.003	.029	.101	.193	.227	.158	.053	.004	0+	0+	7
	8	0+	0+	0+	.001	.008	.042	.121	.213	.231	.133	.021	.002	0+	8
	9	0+	0+	0+	0+	.001	.012	.054	.142	.240	.236	.085	.017	0+	9
	10	0+	0+	0+	0+	0+	.002	.016	.064	.168	.283	.230	.099	.006	10
	11	0+	0+	0+	0+	0+	0+	.003	.017	.071	.206	.377	.341	.107	11
	12	0+	0+	0+	0+	0+	0+	0+	.002	.014	.069	.282	.540	.886	12

NOTE: 0+ represents a positive probability less than 0.0005.

(*continued*)

TABLE A-1 Binomial Probabilities (*continued*)

		p													
n	*x*	.01	.05	.10	.20	.30	.40	.50	.60	.70	.80	.90	.95	.99	*x*
13	0	.878	.513	.254	.055	.010	.001	0+	0+	0+	0+	0+	0+	0+	0
	1	.115	.351	.367	.179	.054	.011	.002	0+	0+	0+	0+	0+	0+	1
	2	.007	.111	.245	.268	.139	.045	.010	.001	0+	0+	0+	0+	0+	2
	3	0+	.021	.100	.246	.218	.111	.035	.006	.001	0+	0+	0+	0+	3
	4	0+	.003	.028	.154	.234	.184	.087	.024	.003	0+	0+	0+	0+	4
	5	0+	0+	.006	.069	.180	.221	.157	.066	.014	.001	0+	0+	0+	5
	6	0+	0+	.001	.023	.103	.197	.209	.131	.044	.006	0+	0+	0+	6
	7	0+	0+	0+	.006	.044	.131	.209	.197	.103	.023	.001	0+	0+	7
	8	0+	0+	0+	.001	.014	.066	.157	.221	.180	.069	.006	0+	0+	8
	9	0+	0+	0+	0+	.003	.024	.087	.184	.234	.154	.028	.003	0+	9
	10	0+	0+	0+	0+	.001	.006	.035	.111	.218	.246	.100	.021	0+	10
	11	0+	0+	0+	0+	0+	.001	.010	.045	.139	.268	.245	.111	.007	11
	12	0+	0+	0+	0+	0+	0+	.002	.011	.054	.179	.367	.351	.115	12
	13	0+	0+	0+	0+	0+	0+	0+	.001	.010	.055	.254	.513	.878	13
14	0	.869	.488	.229	.044	.007	.001	0+	0+	0+	0+	0+	0+	0+	0
	1	.123	.359	.356	.154	.041	.007	.001	0+	0+	0+	0+	0+	0+	1
	2	.008	.123	.257	.250	.113	.032	.006	.001	0+	0+	0+	0+	0+	2
	3	0+	.026	.114	.250	.194	.085	.022	.003	0+	0+	0+	0+	0+	3
	4	0+	.004	.035	.172	.229	.155	.061	.014	.001	0+	0+	0+	0+	4
	5	0+	0+	.008	.086	.196	.207	.122	.041	.007	0+	0+	0+	0+	5
	6	0+	0+	.001	.032	.126	.207	.183	.092	.023	.002	0+	0+	0+	6
	7	0+	0+	0+	.009	.062	.157	.209	.157	.062	.009	0+	0+	0+	7
	8	0+	0+	0+	.002	.023	.092	.183	.207	.126	.032	.001	0+	0+	8
	9	0+	0+	0+	0+	.007	.041	.122	.207	.196	.086	.008	0+	0+	9
	10	0+	0+	0+	0+	.001	.014	.061	.155	.229	.172	.035	.004	0+	10
	11	0+	0+	0+	0+	0+	.003	.022	.085	.194	.250	.114	.026	0+	11
	12	0+	0+	0+	0+	0+	.001	.006	.032	.113	.250	.257	.123	.008	12
	13	0+	0+	0+	0+	0+	0+	.001	.007	.041	.154	.356	.359	.123	13
	14	0+	0+	0+	0+	0+	0+	0+	.001	.007	.044	.229	.488	.869	14
15	0	.860	.463	.206	.035	.005	0+	0+	0+	0+	0+	0+	0+	0+	0
	1	.130	.366	.343	.132	.031	.005	0+	0+	0+	0+	0+	0+	0+	1
	2	.009	.135	.267	.231	.092	.022	.003	0+	0+	0+	0+	0+	0+	2
	3	0+	.031	.129	.250	.170	.063	.014	.002	0+	0+	0+	0+	0+	3
	4	0+	.005	.043	.188	.219	.127	.042	.007	.001	0+	0+	0+	0+	4
	5	0+	.001	.010	.103	.206	.186	.092	.024	.003	0+	0+	0+	0+	5
	6	0+	0+	.002	.043	.147	.207	.153	.061	.012	.001	0+	0+	0+	6
	7	0+	0+	0+	.014	.081	.177	.196	.118	.035	.003	0+	0+	0+	7
	8	0+	0+	0+	.003	.035	.118	.196	.177	.081	.014	0+	0+	0+	8
	9	0+	0+	0+	.001	.012	.061	.153	.207	.147	.043	.002	0+	0+	9
	10	0+	0+	0+	0+	.003	.024	.092	.186	.206	.103	.010	.001	0+	10
	11	0+	0+	0+	0+	.001	.007	.042	.127	.219	.188	.043	.005	0+	11
	12	0+	0+	0+	0+	0+	.002	.014	.063	.170	.250	.129	.031	0+	12
	13	0+	0+	0+	0+	0+	0+	.003	.022	.092	.231	.267	.135	.009	13
	14	0+	0+	0+	0+	0+	0+	0+	.005	.031	.132	.343	.366	.130	14
	15	0+	0+	0+	0+	0+	0+	0+	0+	.005	.035	.206	.463	.860	15

NOTE: 0+ represents a positive probability less than 0.0005.

From Frederick C. Mosteller, Robert E. K. Rourke, and George B. Thomas, Jr., *Probability with Statistical Applications*, 2nd ed., © 1970 Addison-Wesley Publishing Co., Reading, MA. Reprinted with permission.

NEGATIVE z Scores

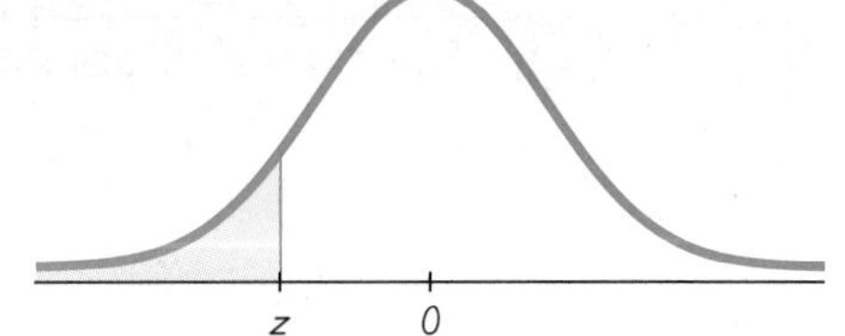

TABLE A-2 Standard Normal (z) Distribution: Cumulative Area from the LEFT

z	.00	.01	.02	.03	.04	.05	.06	.07	.08	.09
−3.50 and lower	.0001									
−3.4	.0003	.0003	.0003	.0003	.0003	.0003	.0003	.0003	.0003	.0002
−3.3	.0005	.0005	.0005	.0004	.0004	.0004	.0004	.0004	.0004	.0003
−3.2	.0007	.0007	.0006	.0006	.0006	.0006	.0006	.0005	.0005	.0005
−3.1	.0010	.0009	.0009	.0009	.0008	.0008	.0008	.0008	.0007	.0007
−3.0	.0013	.0013	.0013	.0012	.0012	.0011	.0011	.0011	.0010	.0010
−2.9	.0019	.0018	.0018	.0017	.0016	.0016	.0015	.0015	.0014	.0014
−2.8	.0026	.0025	.0024	.0023	.0023	.0022	.0021	.0021	.0020	.0019
−2.7	.0035	.0034	.0033	.0032	.0031	.0030	.0029	.0028	.0027	.0026
−2.6	.0047	.0045	.0044	.0043	.0041	.0040	.0039	.0038	.0037	.0036
−2.5	.0062	.0060	.0059	.0057	.0055	.0054	.0052	.0051	* .0049	.0048
−2.4	.0082	.0080	.0078	.0075	.0073	.0071	.0069	.0068	.0066	.0064
−2.3	.0107	.0104	.0102	.0099	.0096	.0094	.0091	.0089	.0087	.0084
−2.2	.0139	.0136	.0132	.0129	.0125	.0122	.0119	.0116	.0113	.0110
−2.1	.0179	.0174	.0170	.0166	.0162	.0158	.0154	.0150	.0146	.0143
−2.0	.0228	.0222	.0217	.0212	.0207	.0202	.0197	.0192	.0188	.0183
−1.9	.0287	.0281	.0274	.0268	.0262	.0256	.0250	.0244	.0239	.0233
−1.8	.0359	.0351	.0344	.0336	.0329	.0322	.0314	.0307	.0301	.0294
−1.7	.0446	.0436	.0427	.0418	.0409	.0401	.0392	.0384	.0375	.0367
−1.6	.0548	.0537	.0526	.0516	.0505	* .0495	.0485	.0475	.0465	.0455
−1.5	.0668	.0655	.0643	.0630	.0618	.0606	.0594	.0582	.0571	.0559
−1.4	.0808	.0793	.0778	.0764	.0749	.0735	.0721	.0708	.0694	.0681
−1.3	.0968	.0951	.0934	.0918	.0901	.0885	.0869	.0853	.0838	.0823
−1.2	.1151	.1131	.1112	.1093	.1075	.1056	.1038	.1020	.1003	.0985
−1.1	.1357	.1335	.1314	.1292	.1271	.1251	.1230	.1210	.1190	.1170
−1.0	.1587	.1562	.1539	.1515	.1492	.1469	.1446	.1423	.1401	.1379
−0.9	.1841	.1814	.1788	.1762	.1736	.1711	.1685	.1660	.1635	.1611
−0.8	.2119	.2090	.2061	.2033	.2005	.1977	.1949	.1922	.1894	.1867
−0.7	.2420	.2389	.2358	.2327	.2296	.2266	.2236	.2206	.2177	.2148
−0.6	.2743	.2709	.2676	.2643	.2611	.2578	.2546	.2514	.2483	.2451
−0.5	.3085	.3050	.3015	.2981	.2946	.2912	.2877	.2843	.2810	.2776
−0.4	.3446	.3409	.3372	.3336	.3300	.3264	.3228	.3192	.3156	.3121
−0.3	.3821	.3783	.3745	.3707	.3669	.3632	.3594	.3557	.3520	.3483
−0.2	.4207	.4168	.4129	.4090	.4052	.4013	.3974	.3936	.3897	.3859
−0.1	.4602	.4562	.4522	.4483	.4443	.4404	.4364	.4325	.4286	.4247
−0.0	.5000	.4960	.4920	.4880	.4840	.4801	.4761	.4721	.4681	.4641

NOTE: For values of z below −3.49, use 0.0001 for the area.
*Use these common values that result from interpolation:

z score	Area
−1.645	0.0500
−2.575	0.0050

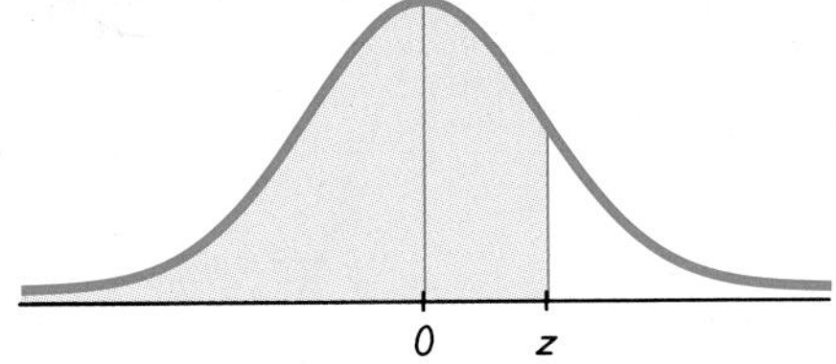

POSITIVE *z* Scores

TABLE A-2 (*continued*) Cumulative Area from the LEFT

z	.00	.01	.02	.03	.04	.05	.06	.07	.08	.09
0.0	.5000	.5040	.5080	.5120	.5160	.5199	.5239	.5279	.5319	.5359
0.1	.5398	.5438	.5478	.5517	.5557	.5596	.5636	.5675	.5714	.5753
0.2	.5793	.5832	.5871	.5910	.5948	.5987	.6026	.6064	.6103	.6141
0.3	.6179	.6217	.6255	.6293	.6331	.6368	.6406	.6443	.6480	.6517
0.4	.6554	.6591	.6628	.6664	.6700	.6736	.6772	.6808	.6844	.6879
0.5	.6915	.6950	.6985	.7019	.7054	.7088	.7123	.7157	.7190	.7224
0.6	.7257	.7291	.7324	.7357	.7389	.7422	.7454	.7486	.7517	.7549
0.7	.7580	.7611	.7642	.7673	.7704	.7734	.7764	.7794	.7823	.7852
0.8	.7881	.7910	.7939	.7967	.7995	.8023	.8051	.8078	.8106	.8133
0.9	.8159	.8186	.8212	.8238	.8264	.8289	.8315	.8340	.8365	.8389
1.0	.8413	.8438	.8461	.8485	.8508	.8531	.8554	.8577	.8599	.8621
1.1	.8643	.8665	.8686	.8708	.8729	.8749	.8770	.8790	.8810	.8830
1.2	.8849	.8869	.8888	.8907	.8925	.8944	.8962	.8980	.8997	.9015
1.3	.9032	.9049	.9066	.9082	.9099	.9115	.9131	.9147	.9162	.9177
1.4	.9192	.9207	.9222	.9236	.9251	.9265	.9279	.9292	.9306	.9319
1.5	.9332	.9345	.9357	.9370	.9382	.9394	.9406	.9418	.9429	.9441
1.6	.9452	.9463	.9474	.9484	.9495 *	.9505	.9515	.9525	.9535	.9545
1.7	.9554	.9564	.9573	.9582	.9591	.9599	.9608	.9616	.9625	.9633
1.8	.9641	.9649	.9656	.9664	.9671	.9678	.9686	.9693	.9699	.9706
1.9	.9713	.9719	.9726	.9732	.9738	.9744	.9750	.9756	.9761	.9767
2.0	.9772	.9778	.9783	.9788	.9793	.9798	.9803	.9808	.9812	.9817
2.1	.9821	.9826	.9830	.9834	.9838	.9842	.9846	.9850	.9854	.9857
2.2	.9861	.9864	.9868	.9871	.9875	.9878	.9881	.9884	.9887	.9890
2.3	.9893	.9896	.9898	.9901	.9904	.9906	.9909	.9911	.9913	.9916
2.4	.9918	.9920	.9922	.9925	.9927	.9929	.9931	.9932	.9934	.9936
2.5	.9938	.9940	.9941	.9943	.9945	.9946	.9948	.9949 *	.9951	.9952
2.6	.9953	.9955	.9956	.9957	.9959	.9960	.9961	.9962	.9963	.9964
2.7	.9965	.9966	.9967	.9968	.9969	.9970	.9971	.9972	.9973	.9974
2.8	.9974	.9975	.9976	.9977	.9977	.9978	.9979	.9979	.9980	.9981
2.9	.9981	.9982	.9982	.9983	.9984	.9984	.9985	.9985	.9986	.9986
3.0	.9987	.9987	.9987	.9988	.9988	.9989	.9989	.9989	.9990	.9990
3.1	.9990	.9991	.9991	.9991	.9992	.9992	.9992	.9992	.9993	.9993
3.2	.9993	.9993	.9994	.9994	.9994	.9994	.9994	.9995	.9995	.9995
3.3	.9995	.9995	.9995	.9996	.9996	.9996	.9996	.9996	.9996	.9997
3.4	.9997	.9997	.9997	.9997	.9997	.9997	.9997	.9997	.9997	.9998
3.50 and up	.9999									

NOTE: For values of *z* above 3.49, use 0.9999 for the area.

*Use these common values that result from interpolation:

z score	Area
1.645	0.9500
2.575	0.9950

Common Critical Values

Confidence Level	Critical Value
0.90	1.645
0.95	1.96
0.99	2.575

TABLE A-3	*t* Distribution: Critical *t* Values				
	Area in One Tail				
	0.005	0.01	0.025	0.05	0.10
Degrees of Freedom	Area in Two Tails				
	0.01	0.02	0.05	0.10	0.20
1	**63.657**	31.821	**12.706**	6.314	3.078
2	**9.925**	6.965	**4.303**	2.920	1.886
3	**5.841**	4.541	**3.182**	2.353	1.638
4	**4.604**	3.747	**2.776**	2.132	1.533
5	**4.032**	3.365	**2.571**	2.015	1.476
6	**3.707**	3.143	**2.447**	1.943	1.440
7	**3.499**	2.998	**2.365**	1.895	1.415
8	**3.355**	2.896	**2.306**	1.860	1.397
9	**3.250**	2.821	**2.262**	1.833	1.383
10	**3.169**	2.764	**2.228**	1.812	1.372
11	**3.106**	2.718	**2.201**	1.796	1.363
12	**3.055**	2.681	**2.179**	1.782	1.356
13	**3.012**	2.650	**2.160**	1.771	1.350
14	**2.977**	2.624	**2.145**	1.761	1.345
15	**2.947**	2.602	**2.131**	1.753	1.341
16	**2.921**	2.583	**2.120**	1.746	1.337
17	**2.898**	2.567	**2.110**	1.740	1.333
18	**2.878**	2.552	**2.101**	1.734	1.330
19	**2.861**	2.539	**2.093**	1.729	1.328
20	**2.845**	2.528	**2.086**	1.725	1.325
21	**2.831**	2.518	**2.080**	1.721	1.323
22	**2.819**	2.508	**2.074**	1.717	1.321
23	**2.807**	2.500	**2.069**	1.714	1.319
24	**2.797**	2.492	**2.064**	1.711	1.318
25	**2.787**	2.485	**2.060**	1.708	1.316
26	**2.779**	2.479	**2.056**	1.706	1.315
27	**2.771**	2.473	**2.052**	1.703	1.314
28	**2.763**	2.467	**2.048**	1.701	1.313
29	**2.756**	2.462	**2.045**	1.699	1.311
30	**2.750**	2.457	**2.042**	1.697	1.310
31	**2.744**	2.453	**2.040**	1.696	1.309
32	**2.738**	2.449	**2.037**	1.694	1.309
34	**2.728**	2.441	**2.032**	1.691	1.307
36	**2.719**	2.434	**2.028**	1.688	1.306
38	**2.712**	2.429	**2.024**	1.686	1.304
40	**2.704**	2.423	**2.021**	1.684	1.303
45	**2.690**	2.412	**2.014**	1.679	1.301
50	**2.678**	2.403	**2.009**	1.676	1.299
55	**2.668**	2.396	**2.004**	1.673	1.297
60	**2.660**	2.390	**2.000**	1.671	1.296
65	**2.654**	2.385	**1.997**	1.669	1.295
70	**2.648**	2.381	**1.994**	1.667	1.294
75	**2.643**	2.377	**1.992**	1.665	1.293
80	**2.639**	2.374	**1.990**	1.664	1.292
90	**2.632**	2.368	**1.987**	1.662	1.291
100	**2.626**	2.364	**1.984**	1.660	1.290
200	**2.601**	2.345	**1.972**	1.653	1.286
300	**2.592**	2.339	**1.968**	1.650	1.284
400	**2.588**	2.336	**1.966**	1.649	1.284
500	**2.586**	2.334	**1.965**	1.648	1.283
750	**2.582**	2.331	**1.963**	1.647	1.283
1000	**2.581**	2.330	**1.962**	1.646	1.282
2000	**2.578**	2.328	**1.961**	1.646	1.282
Large	**2.576**	2.326	**1.960**	1.645	1.282

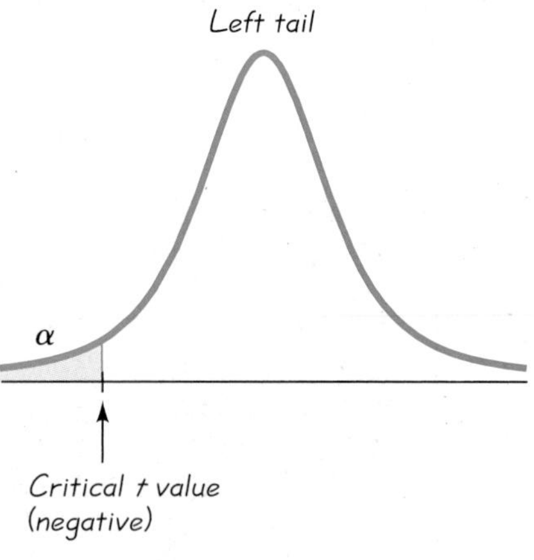

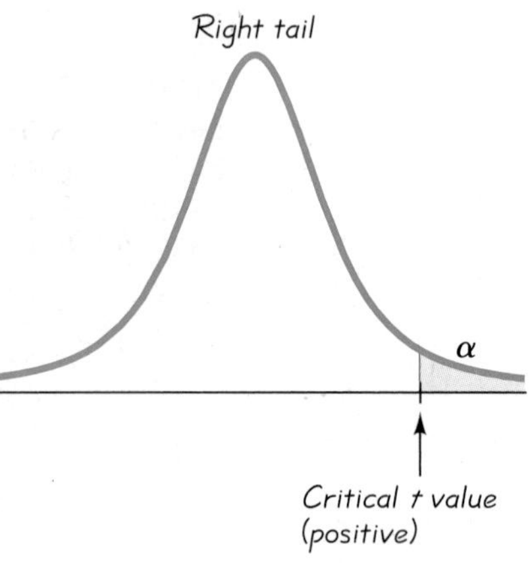

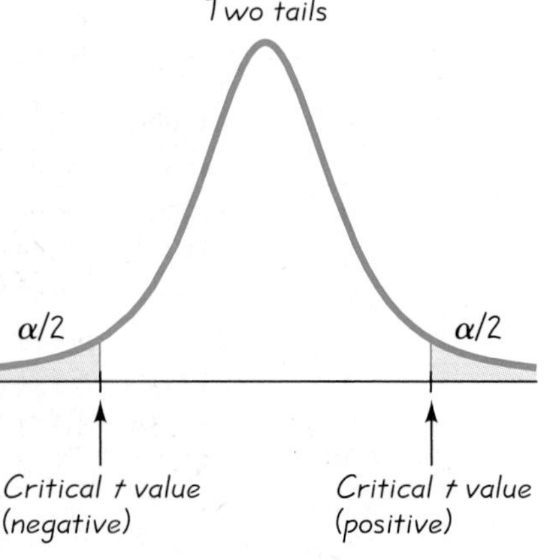

TABLE A-4 Chi-Square (χ^2) Distribution

Degrees of Freedom	Area to the Right of the Critical Value									
	0.995	0.99	0.975	0.95	0.90	0.10	0.05	0.025	0.01	0.005
1	—	—	0.001	0.004	0.016	2.706	3.841	5.024	6.635	7.879
2	0.010	0.020	0.051	0.103	0.211	4.605	5.991	7.378	9.210	10.597
3	0.072	0.115	0.216	0.352	0.584	6.251	7.815	9.348	11.345	12.838
4	0.207	0.297	0.484	0.711	1.064	7.779	9.488	11.143	13.277	14.860
5	0.412	0.554	0.831	1.145	1.610	9.236	11.071	12.833	15.086	16.750
6	0.676	0.872	1.237	1.635	2.204	10.645	12.592	14.449	16.812	18.548
7	0.989	1.239	1.690	2.167	2.833	12.017	14.067	16.013	18.475	20.278
8	1.344	1.646	2.180	2.733	3.490	13.362	15.507	17.535	20.090	21.955
9	1.735	2.088	2.700	3.325	4.168	14.684	16.919	19.023	21.666	23.589
10	2.156	2.558	3.247	3.940	4.865	15.987	18.307	20.483	23.209	25.188
11	2.603	3.053	3.816	4.575	5.578	17.275	19.675	21.920	24.725	26.757
12	3.074	3.571	4.404	5.226	6.304	18.549	21.026	23.337	26.217	28.299
13	3.565	4.107	5.009	5.892	7.042	19.812	22.362	24.736	27.688	29.819
14	4.075	4.660	5.629	6.571	7.790	21.064	23.685	26.119	29.141	31.319
15	4.601	5.229	6.262	7.261	8.547	22.307	24.996	27.488	30.578	32.801
16	5.142	5.812	6.908	7.962	9.312	23.542	26.296	28.845	32.000	34.267
17	5.697	6.408	7.564	8.672	10.085	24.769	27.587	30.191	33.409	35.718
18	6.265	7.015	8.231	9.390	10.865	25.989	28.869	31.526	34.805	37.156
19	6.844	7.633	8.907	10.117	11.651	27.204	30.144	32.852	36.191	38.582
20	7.434	8.260	9.591	10.851	12.443	28.412	31.410	34.170	37.566	39.997
21	8.034	8.897	10.283	11.591	13.240	29.615	32.671	35.479	38.932	41.401
22	8.643	9.542	10.982	12.338	14.042	30.813	33.924	36.781	40.289	42.796
23	9.260	10.196	11.689	13.091	14.848	32.007	35.172	38.076	41.638	44.181
24	9.886	10.856	12.401	13.848	15.659	33.196	36.415	39.364	42.980	45.559
25	10.520	11.524	13.120	14.611	16.473	34.382	37.652	40.646	44.314	46.928
26	11.160	12.198	13.844	15.379	17.292	35.563	38.885	41.923	45.642	48.290
27	11.808	12.879	14.573	16.151	18.114	36.741	40.113	43.194	46.963	49.645
28	12.461	13.565	15.308	16.928	18.939	37.916	41.337	44.461	48.278	50.993
29	13.121	14.257	16.047	17.708	19.768	39.087	42.557	45.722	49.588	52.336
30	13.787	14.954	16.791	18.493	20.599	40.256	43.773	46.979	50.892	53.672
40	20.707	22.164	24.433	26.509	29.051	51.805	55.758	59.342	63.691	66.766
50	27.991	29.707	32.357	34.764	37.689	63.167	67.505	71.420	76.154	79.490
60	35.534	37.485	40.482	43.188	46.459	74.397	79.082	83.298	88.379	91.952
70	43.275	45.442	48.758	51.739	55.329	85.527	90.531	95.023	100.425	104.215
80	51.172	53.540	57.153	60.391	64.278	96.578	101.879	106.629	112.329	116.321
90	59.196	61.754	65.647	69.126	73.291	107.565	113.145	118.136	124.116	128.299
100	67.328	70.065	74.222	77.929	82.358	118.498	124.342	129.561	135.807	140.169

From Donald B. Owen, *Handbook of Statistical Tables*, ©1962 Addison-Wesley Publishing Co., Reading, MA. Reprinted with permission of the publisher.

Degrees of Freedom

$n - 1$	for confidence intervals or hypothesis tests with a standard deviation or variance
$k - 1$	for multinomial experiments or goodness-of-fit with k categories
$(r - 1)(c - 1)$	for contingency tables with r rows and c columns
$k - 1$	for Kruskal-Wallis test with k samples

TABLE A-5 *F* Distribution ($\alpha = 0.025$ in the right tail)

Numerator degrees of freedom (df_1); rows: Denominator degrees of freedom (df_2)

df_2	1	2	3	4	5	6	7	8	9
1	647.79	799.50	864.16	899.58	921.85	937.11	948.22	956.66	963.28
2	38.506	39.000	39.165	39.248	39.298	39.331	39.335	39.373	39.387
3	17.443	16.044	15.439	15.101	14.885	14.735	14.624	14.540	14.473
4	12.218	10.649	9.9792	9.6045	9.3645	9.1973	9.0741	8.9796	8.9047
5	10.007	8.4336	7.7636	7.3879	7.1464	6.9777	6.8531	6.7572	6.6811
6	8.8131	7.2599	6.5988	6.2272	5.9876	5.8198	5.6955	5.5996	5.5234
7	8.0727	6.5415	5.8898	5.5226	5.2852	5.1186	4.9949	4.8993	4.8232
8	7.5709	6.0595	5.4160	5.0526	4.8173	4.6517	4.5286	4.4333	4.3572
9	7.2093	5.7147	5.0781	4.7181	4.4844	4.3197	4.1970	4.1020	4.0260
10	6.9367	5.4564	4.8256	4.4683	4.2361	4.0721	3.9498	3.8549	3.7790
11	6.7241	5.2559	4.6300	4.2751	4.0440	3.8807	3.7586	3.6638	3.5879
12	6.5538	5.0959	4.4742	4.1212	3.8911	3.7283	3.6065	3.5118	3.4358
13	6.4143	4.9653	4.3472	3.9959	3.7667	3.6043	3.4827	3.3880	3.3120
14	6.2979	4.8567	4.2417	3.8919	3.6634	3.5014	3.3799	3.2853	3.2093
15	6.1995	4.7650	4.1528	3.8043	3.5764	3.4147	3.2934	3.1987	3.1227
16	6.1151	4.6867	4.0768	3.7294	3.5021	3.3406	3.2194	3.1248	3.0488
17	6.0420	4.6189	4.0112	3.6648	3.4379	3.2767	3.1556	3.0610	2.9849
18	5.9781	4.5597	3.9539	3.6083	3.3820	3.2209	3.0999	3.0053	2.9291
19	5.9216	4.5075	3.9034	3.5587	3.3327	3.1718	3.0509	2.9563	2.8801
20	5.8715	4.4613	3.8587	3.5147	3.2891	3.1283	3.0074	2.9128	2.8365
21	5.8266	4.4199	3.8188	3.4754	3.2501	3.0895	2.9686	2.8740	2.7977
22	5.7863	4.3828	3.7829	3.4401	3.2151	3.0546	2.9338	2.8392	2.7628
23	5.7498	4.3492	3.7505	3.4083	3.1835	3.0232	2.9023	2.8077	2.7313
24	5.7166	4.3187	3.7211	3.3794	3.1548	2.9946	2.8738	2.7791	2.7027
25	5.6864	4.2909	3.6943	3.3530	3.1287	2.9685	2.8478	2.7531	2.6766
26	5.6586	4.2655	3.6697	3.3289	3.1048	2.9447	2.8240	2.7293	2.6528
27	5.6331	4.2421	3.6472	3.3067	3.0828	2.9228	2.8021	2.7074	2.6309
28	5.6096	4.2205	3.6264	3.2863	3.0626	2.9027	2.7820	2.6872	2.6106
29	5.5878	4.2006	3.6072	3.2674	3.0438	2.8840	2.7633	2.6686	2.5919
30	5.5675	4.1821	3.5894	3.2499	3.0265	2.8667	2.7460	2.6513	2.5746
40	5.4239	4.0510	3.4633	3.1261	2.9037	2.7444	2.6238	2.5289	2.4519
60	5.2856	3.9253	3.3425	3.0077	2.7863	2.6274	2.5068	2.4117	2.3344
120	5.1523	3.8046	3.2269	2.8943	2.6740	2.5154	2.3948	2.2994	2.2217
∞	5.0239	3.6889	3.1161	2.7858	2.5665	2.4082	2.2875	2.1918	2.1136

TABLE A-5 *F* Distribution ($\alpha = 0.025$ in the right tail) (*continued*)

Numerator degrees of freedom (df_1); Denominator degrees of freedom (df_2)

df_2 \ df_1	10	12	15	20	24	30	40	60	120	∞
1	968.63	976.71	984.87	993.10	997.25	1001.4	1005.6	1009.8	1014.0	1018.3
2	39.398	39.415	39.431	39.448	39.456	39.465	39.473	39.481	39.490	39.498
3	14.419	14.337	14.253	14.167	14.124	14.081	14.037	13.992	13.947	13.902
4	8.8439	8.7512	8.6565	8.5599	8.5109	8.4613	8.4111	8.3604	8.3092	8.2573
5	6.6192	6.5245	6.4277	6.3286	6.2780	6.2269	6.1750	6.1225	6.0693	6.0153
6	5.4613	5.3662	5.2687	5.1684	5.1172	5.0652	5.0125	4.9589	4.9044	4.8491
7	4.7611	4.6658	4.5678	4.4667	4.4150	4.3624	4.3089	4.2544	4.1989	4.1423
8	4.2951	4.1997	4.1012	3.9995	3.9472	3.8940	3.8398	3.7844	3.7279	3.6702
9	3.9639	3.8682	3.7694	3.6669	3.6142	3.5604	3.5055	3.4493	3.3918	3.3329
10	3.7168	3.6209	3.5217	3.4185	3.3654	3.3110	3.2554	3.1984	3.1399	3.0798
11	3.5257	3.4296	3.3299	3.2261	3.1725	3.1176	3.0613	3.0035	2.9441	2.8828
12	3.3736	3.2773	3.1772	3.0728	3.0187	2.9633	2.9063	2.8478	2.7874	2.7249
13	3.2497	3.1532	3.0527	2.9477	2.8932	2.8372	2.7797	2.7204	2.6590	2.5955
14	3.1469	3.0502	2.9493	2.8437	2.7888	2.7324	2.6742	2.6142	2.5519	2.4872
15	3.0602	2.9633	2.8621	2.7559	2.7006	2.6437	2.5850	2.5242	2.4611	2.3953
16	2.9862	2.8890	2.7875	2.6808	2.6252	2.5678	2.5085	2.4471	2.3831	2.3163
17	2.9222	2.8249	2.7230	2.6158	2.5598	2.5020	2.4422	2.3801	2.3153	2.2474
18	2.8664	2.7689	2.6667	2.5590	2.5027	2.4445	2.3842	2.3214	2.2558	2.1869
19	2.8172	2.7196	2.6171	2.5089	2.4523	2.3937	2.3329	2.2696	2.2032	2.1333
20	2.7737	2.6758	2.5731	2.4645	2.4076	2.3486	2.2873	2.2234	2.1562	2.0853
21	2.7348	2.6368	2.5338	2.4247	2.3675	2.3082	2.2465	2.1819	2.1141	2.0422
22	2.6998	2.6017	2.4984	2.3890	2.3315	2.2718	2.2097	2.1446	2.0760	2.0032
23	2.6682	2.5699	2.4665	2.3567	2.2989	2.2389	2.1763	2.1107	2.0415	1.9677
24	2.6396	2.5411	2.4374	2.3273	2.2693	2.2090	2.1460	2.0799	2.0099	1.9353
25	2.6135	2.5149	2.4110	2.3005	2.2422	2.1816	2.1183	2.0516	1.9811	1.9055
26	2.5896	2.4908	2.3867	2.2759	2.2174	2.1565	2.0928	2.0257	1.9545	1.8781
27	2.5676	2.4688	2.3644	2.2533	2.1946	2.1334	2.0693	2.0018	1.9299	1.8527
28	2.5473	2.4484	2.3438	2.2324	2.1735	2.1121	2.0477	1.9797	1.9072	1.8291
29	2.5286	2.4295	2.3248	2.2131	2.1540	2.0923	2.0276	1.9591	1.8861	1.8072
30	2.5112	2.4120	2.3072	2.1952	2.1359	2.0739	2.0089	1.9400	1.8664	1.7867
40	2.3882	2.2882	2.1819	2.0677	2.0069	1.9429	1.8752	1.8028	1.7242	1.6371
60	2.2702	2.1692	2.0613	1.9445	1.8817	1.8152	1.7440	1.6668	1.5810	1.4821
120	2.1570	2.0548	1.9450	1.8249	1.7597	1.6899	1.6141	1.5299	1.4327	1.3104
∞	2.0483	1.9447	1.8326	1.7085	1.6402	1.5660	1.4835	1.3883	1.2684	1.0000

From Maxine Merrington and Catherine M. Thompson, "Tables of Percentage Points of the Inverted Beta (*F*) Distribution," *Biometrika 33* (1943): 80–84. Reproduced with permission of the Biometrika Trustees.

(*continued*)

TABLE A-5 *F* Distribution ($\alpha = 0.05$ in the right tail)

Denominator degrees of freedom (df_2)	Numerator degrees of freedom (df_1): 1	2	3	4	5	6	7	8	9
1	161.45	199.50	215.71	224.58	230.16	233.99	236.77	238.88	240.54
2	18.513	19.000	19.164	19.247	19.296	19.330	19.353	19.371	19.385
3	10.128	9.5521	9.2766	9.1172	9.0135	8.9406	8.8867	8.8452	8.8123
4	7.7086	6.9443	6.5914	6.3882	6.2561	6.1631	6.0942	6.0410	6.9988
5	6.6079	5.7861	5.4095	5.1922	5.0503	4.9503	4.8759	4.8183	4.7725
6	5.9874	5.1433	4.7571	4.5337	4.3874	4.2839	4.2067	4.1468	4.0990
7	5.5914	4.7374	4.3468	4.1203	3.9715	3.8660	3.7870	3.7257	3.6767
8	5.3177	4.4590	4.0662	3.8379	3.6875	3.5806	3.5005	3.4381	3.3881
9	5.1174	4.2565	3.8625	3.6331	3.4817	3.3738	3.2927	3.2296	3.1789
10	4.9646	4.1028	3.7083	3.4780	3.3258	3.2172	3.1355	3.0717	3.0204
11	4.8443	3.9823	3.5874	3.3567	3.2039	3.0946	3.0123	2.9480	2.8962
12	4.7472	3.8853	3.4903	3.2592	3.1059	2.9961	2.9134	2.8486	2.7964
13	4.6672	3.8056	3.4105	3.1791	3.0254	2.9153	2.8321	2.7669	2.7144
14	4.6001	3.7389	3.3439	3.1122	2.9582	2.8477	2.7642	2.6987	2.6458
15	4.5431	3.6823	3.2874	3.0556	2.9013	2.7905	2.7066	2.6408	2.5876
16	4.4940	3.6337	3.2389	3.0069	2.8524	2.7413	2.6572	2.5911	2.5377
17	4.4513	3.5915	3.1968	2.9647	2.8100	2.6987	2.6143	2.5480	2.4943
18	4.4139	3.5546	3.1599	2.9277	2.7729	2.6613	2.5767	2.5102	2.4563
19	4.3807	3.5219	3.1274	2.8951	2.7401	2.6283	2.5435	2.4768	2.4227
20	4.3512	3.4928	3.0984	2.8661	2.7109	2.5990	2.5140	2.4471	2.3928
21	4.3248	3.4668	3.0725	2.8401	2.6848	2.5727	2.4876	2.4205	2.3660
22	4.3009	3.4434	3.0491	2.8167	2.6613	2.5491	2.4638	2.3965	2.3419
23	4.2793	3.4221	3.0280	2.7955	2.6400	2.5277	2.4422	2.3748	2.3201
24	4.2597	3.4028	3.0088	2.7763	2.6207	2.5082	2.4226	2.3551	2.3002
25	4.2417	3.3852	2.9912	2.7587	2.6030	2.4904	2.4047	2.3371	2.2821
26	4.2252	3.3690	2.9752	2.7426	2.5868	2.4741	2.3883	2.3205	2.2655
27	4.2100	3.3541	2.9604	2.7278	2.5719	2.4591	2.3732	2.3053	2.2501
28	4.1960	3.3404	2.9467	2.7141	2.5581	2.4453	2.3593	2.2913	2.2360
29	4.1830	3.3277	2.9340	2.7014	2.5454	2.4324	2.3463	2.2783	2.2229
30	4.1709	3.3158	2.9223	2.6896	2.5336	2.4205	2.3343	2.2662	2.2107
40	4.0847	3.2317	2.8387	2.6060	2.4495	2.3359	2.2490	2.1802	2.1240
60	4.0012	3.1504	2.7581	2.5252	2.3683	2.2541	2.1665	2.0970	2.0401
120	3.9201	3.0718	2.6802	2.4472	2.2899	2.1750	2.0868	2.0164	1.9588
∞	3.8415	2.9957	2.6049	2.3719	2.2141	2.0986	2.0096	1.9384	1.8799

(continued)

Table A-5 *F* Distribution ($\alpha = 0.05$ in the right tail) *(continued)*

Denominator degrees of freedom (df_2)	Numerator degrees of freedom (df_1)									
	10	12	15	20	24	30	40	60	120	∞
1	241.88	243.91	245.95	248.01	249.05	250.10	251.14	252.20	253.25	254.31
2	19.396	19.413	19.429	19.446	19.454	19.462	19.471	19.479	19.487	19.496
3	8.7855	8.7446	8.7029	8.6602	8.6385	8.6166	8.5944	8.5720	8.5494	8.5264
4	5.9644	5.9117	5.8578	5.8025	5.7744	5.7459	5.7170	5.6877	5.6581	5.6281
5	4.7351	4.6777	4.6188	4.5581	4.5272	4.4957	4.4638	4.4314	4.3985	4.3650
6	4.0600	3.9999	3.9381	3.8742	3.8415	3.8082	3.7743	3.7398	3.7047	3.6689
7	3.6365	3.5747	3.5107	3.4445	3.4105	3.3758	3.3404	3.3043	3.2674	3.2298
8	3.3472	3.2839	3.2184	3.1503	3.1152	3.0794	3.0428	3.0053	2.9669	2.9276
9	3.1373	3.0729	3.0061	2.9365	2.9005	2.8637	2.8259	2.7872	2.7475	2.7067
10	2.9782	2.9130	2.8450	2.7740	2.7372	2.6996	2.6609	2.6211	2.5801	2.5379
11	2.8536	2.7876	2.7186	2.6464	2.6090	2.5705	2.5309	2.4901	2.4480	2.4045
12	2.7534	2.6866	2.6169	2.5436	2.5055	2.4663	2.4259	2.3842	2.3410	2.2962
13	2.6710	2.6037	2.5331	2.4589	2.4202	2.3803	2.3392	2.2966	2.2524	2.2064
14	2.6022	2.5342	2.4630	2.3879	2.3487	2.3082	2.2664	2.2229	2.1778	2.1307
15	2.5437	2.4753	2.4034	2.3275	2.2878	2.2468	2.2043	2.1601	2.1141	2.0658
16	2.4935	2.4247	2.3522	2.2756	2.2354	2.1938	2.1507	2.1058	2.0589	2.0096
17	2.4499	2.3807	2.3077	2.2304	2.1898	2.1477	2.1040	2.0584	2.0107	1.9604
18	2.4117	2.3421	2.2686	2.1906	2.1497	2.1071	2.0629	2.0166	1.9681	1.9168
19	2.3779	2.3080	2.2341	2.1555	2.1141	2.0712	2.0264	1.9795	1.9302	1.8780
20	2.3479	2.2776	2.2033	2.1242	2.0825	2.0391	1.9938	1.9464	1.8963	1.8432
21	2.3210	2.2504	2.1757	2.0960	2.0540	2.0102	1.9645	1.9165	1.8657	1.8117
22	2.2967	2.2258	2.1508	2.0707	2.0283	1.9842	1.9380	1.8894	1.8380	1.7831
23	2.2747	2.2036	2.1282	2.0476	2.0050	1.9605	1.9139	1.8648	1.8128	1.7570
24	2.2547	2.1834	2.1077	2.0267	1.9838	1.9390	1.8920	1.8424	1.7896	1.7330
25	2.2365	2.1649	2.0889	2.0075	1.9643	1.9192	1.8718	1.8217	1.7684	1.7110
26	2.2197	2.1479	2.0716	1.9898	1.9464	1.9010	1.8533	1.8027	1.7488	1.6906
27	2.2043	2.1323	2.0558	1.9736	1.9299	1.8842	1.8361	1.7851	1.7306	1.6717
28	2.1900	2.1179	2.0411	1.9586	1.9147	1.8687	1.8203	1.7689	1.7138	1.6541
29	2.1768	2.1045	2.0275	1.9446	1.9005	1.8543	1.8055	1.7537	1.6981	1.6376
30	2.1646	2.0921	2.0148	1.9317	1.8874	1.8409	1.7918	1.7396	1.6835	1.6223
40	2.0772	2.0035	1.9245	1.8389	1.7929	1.7444	1.6928	1.6373	1.5766	1.5089
60	1.9926	1.9174	1.8364	1.7480	1.7001	1.6491	1.5943	1.5343	1.4673	1.3893
120	1.9105	1.8337	1.7505	1.6587	1.6084	1.5543	1.4952	1.4290	1.3519	1.2539
∞	1.8307	1.7522	1.6664	1.5705	1.5173	1.4591	1.3940	1.3180	1.2214	1.0000

From Maxine Merrington and Catherine M. Thompson, "Tables of Percentage Points of the Inverted Beta (*F*) Distribution," *Biometrika 33* (1943): 80–84. Reproduced with permission of the Biometrika Trustees.

TABLE A-6 Critical Values of the Pearson Correlation Coefficient r

n	$\alpha = .05$	$\alpha = .01$
4	.950	.999
5	.878	.959
6	.811	.917
7	.754	.875
8	.707	.834
9	.666	.798
10	.632	.765
11	.602	.735
12	.576	.708
13	.553	.684
14	.532	.661
15	.514	.641
16	.497	.623
17	.482	.606
18	.468	.590
19	.456	.575
20	.444	.561
25	.396	.505
30	.361	.463
35	.335	.430
40	.312	.402
45	.294	.378
50	.279	.361
60	.254	.330
70	.236	.305
80	.220	.286
90	.207	.269
100	.196	.256

NOTE: To test H_0: $\rho = 0$ against H_1: $\rho \neq 0$, reject H_0 if the absolute value of r is greater than the critical value in the table.

TABLE A-7 Critical Values for the Sign Test

n	α .005 (one tail) .01 (two tails)	.01 (one tail) .02 (two tails)	.025 (one tail) .05 (two tails)	.05 (one tail) .10 (two tails)
1	*	*	*	*
2	*	*	*	*
3	*	*	*	*
4	*	*	*	*
5	*	*	*	0
6	*	*	0	0
7	*	0	0	0
8	0	0	0	1
9	0	0	1	1
10	0	0	1	1
11	0	1	1	2
12	1	1	2	2
13	1	1	2	3
14	1	2	2	3
15	2	2	3	3
16	2	2	3	4
17	2	3	4	4
18	3	3	4	5
19	3	4	4	5
20	3	4	5	5
21	4	4	5	6
22	4	5	5	6
23	4	5	6	7
24	5	5	6	7
25	5	6	7	7

NOTES:

1. * indicates that it is not possible to get a value in the critical region.
2. Reject the null hypothesis if the number of the less frequent sign (x) is less than or equal to the value in the table.
3. For values of n greater than 25, a normal approximation is used with

$$z = \frac{(x + 0.5) - \left(\frac{n}{2}\right)}{\frac{\sqrt{n}}{2}}$$

TABLE A-8 Critical Values of *T* for the Wilcoxon Signed-Ranks Test

	α			
n	.005 (one tail) .01 (two tails)	.01 (one tail) .02 (two tails)	.025 (one tail) .05 (two tails)	.05 (one tail) .10 (two tails)
5	*	*	*	1
6	*	*	1	2
7	*	0	2	4
8	0	2	4	6
9	2	3	6	8
10	3	5	8	11
11	5	7	11	14
12	7	10	14	17
13	10	13	17	21
14	13	16	21	26
15	16	20	25	30
16	19	24	30	36
17	23	28	35	41
18	28	33	40	47
19	32	38	46	54
20	37	43	52	60
21	43	49	59	68
22	49	56	66	75
23	55	62	73	83
24	61	69	81	92
25	68	77	90	101
26	76	85	98	110
27	84	93	107	120
28	92	102	117	130
29	100	111	127	141
30	109	120	137	152

NOTES:

1. * indicates that it is not possible to get a value in the critical region.
2. Reject the null hypothesis if the test statistic *T* is less than or equal to the critical value found in this table. Fail to reject the null hypothesis if the test statistic *T* is greater than the critical value found in the table.

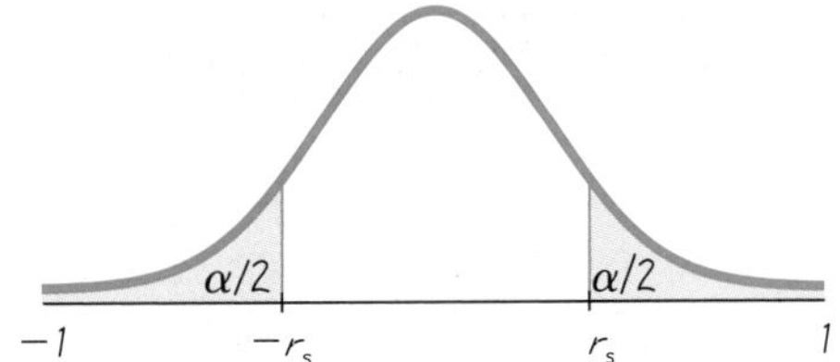

TABLE A-9 Critical Values of Spearman's Rank Correlation Coefficient r_s

n	$\alpha = 0.10$	$\alpha = 0.05$	$\alpha = 0.02$	$\alpha = 0.01$
5	.900	—	—	—
6	.829	.886	.943	—
7	.714	.786	.893	.929
8	.643	.738	.833	.881
9	.600	.700	.783	.833
10	.564	.648	.745	.794
11	.536	.618	.709	.755
12	.503	.587	.678	.727
13	.484	.560	.648	.703
14	.464	.538	.626	.679
15	.446	.521	.604	.654
16	.429	.503	.582	.635
17	.414	.485	.566	.615
18	.401	.472	.550	.600
19	.391	.460	.535	.584
20	.380	.447	.520	.570
21	.370	.435	.508	.556
22	.361	.425	.496	.544
23	.353	.415	.486	.532
24	.344	.406	.476	.521
25	.337	.398	.466	.511
26	.331	.390	.457	.501
27	.324	.382	.448	.491
28	.317	.375	.440	.483
29	.312	.368	.433	.475
30	.306	.362	.425	.467

NOTES:

1. For $n > 30$, use $r_s = \pm z/\sqrt{n-1}$ where z corresponds to the level of significance. For example, if $\alpha = 0.05$, then $z = 1.96$.
2. If the absolute value of the test statistic r_s exceeds the positive critical value, then reject H_0: $\rho_s = 0$ and conclude that there is a correlation.

Based on data from "*Biostatistical Analysis*, 4th edition," © 1999, by Jerrold Zar, Prentice Hall, Inc., Upper Saddle River, New Jersey, and "Distribution of Sums of Squares of Rank Differences to Small Numbers with Individuals," *The Annals of Mathematical Statistics,* Vol. 9, No. 2, with permission of the Institute of Mathematical Statistics.

TABLE A-10 Critical Values for Number of Runs G

Value of n_1	Value of n_2: 2	3	4	5	6	7	8	9	10	11	12	13	14	15	16	17	18	19	20
2	1	1	1	1	1	1	1	1	1	1	2	2	2	2	2	2	2	2	2
	6	6	6	6	6	6	6	6	6	6	6	6	6	6	6	6	6	6	6
3	1	1	1	1	2	2	2	2	2	2	2	2	2	3	3	3	3	3	3
	6	8	8	8	8	8	8	8	8	8	8	8	8	8	8	8	8	8	8
4	1	1	1	2	2	2	3	3	3	3	3	3	3	3	4	4	4	4	4
	6	8	9	9	9	10	10	10	10	10	10	10	10	10	10	10	10	10	10
5	1	1	2	2	3	3	3	3	3	4	4	4	4	4	4	4	5	5	5
	6	8	9	10	10	11	11	12	12	12	12	12	12	12	12	12	12	12	12
6	1	2	2	3	3	3	3	4	4	4	4	5	5	5	5	5	5	6	6
	6	8	9	10	11	12	12	13	13	13	13	14	14	14	14	14	14	14	14
7	1	2	2	3	3	3	4	4	5	5	5	5	5	6	6	6	6	6	6
	6	8	10	11	12	13	13	14	14	14	14	15	15	15	16	16	16	16	16
8	1	2	3	3	3	4	4	5	5	5	6	6	6	6	6	7	7	7	7
	6	8	10	11	12	13	14	14	15	15	16	16	16	16	17	17	17	17	17
9	1	2	3	3	4	4	5	5	5	6	6	6	7	7	7	7	8	8	8
	6	8	10	12	13	14	14	15	16	16	16	17	17	18	18	18	18	18	18
10	1	2	3	3	4	5	5	5	6	6	7	7	7	7	8	8	8	8	9
	6	8	10	12	13	14	15	16	16	17	17	18	18	18	19	19	19	20	20
11	1	2	3	4	4	5	5	6	6	7	7	7	8	8	8	9	9	9	9
	6	8	10	12	13	14	15	16	17	17	18	19	19	19	20	20	20	21	21
12	2	2	3	4	4	5	6	6	7	7	7	8	8	8	9	9	9	10	10
	6	8	10	12	13	14	16	16	17	18	19	19	20	20	21	21	21	22	22
13	2	2	3	4	5	5	6	6	7	7	8	8	9	9	9	10	10	10	10
	6	8	10	12	14	15	16	17	18	19	19	20	20	21	21	22	22	23	23
14	2	2	3	4	5	5	6	7	7	8	8	9	9	9	10	10	10	11	11
	6	8	10	12	14	15	16	17	18	19	20	20	21	22	22	23	23	23	24
15	2	3	3	4	5	6	6	7	7	8	8	9	9	10	10	11	11	11	12
	6	8	10	12	14	15	16	18	18	19	20	21	22	22	23	23	24	24	25
16	2	3	4	4	5	6	6	7	8	8	9	9	10	10	11	11	11	12	12
	6	8	10	12	14	16	17	18	19	20	21	21	22	23	23	24	25	25	25
17	2	3	4	4	5	6	7	7	8	9	9	10	10	11	11	11	12	12	13
	6	8	10	12	14	16	17	18	19	20	21	22	23	23	24	25	25	26	26
18	2	3	4	5	5	6	7	8	8	9	9	10	10	11	11	12	12	13	13
	6	8	10	12	14	16	17	18	19	20	21	22	23	24	25	25	26	26	27
19	2	3	4	5	6	6	7	8	8	9	10	10	11	11	12	12	13	13	13
	6	8	10	12	14	16	17	18	20	21	22	23	23	24	25	26	26	27	27
20	2	3	4	5	6	6	7	8	9	9	10	10	11	12	12	13	13	13	14
	6	8	10	12	14	16	17	18	20	21	22	23	24	25	25	26	27	27	28

NOTES:

1. The entries in this table are the critical G values, assuming a two-tailed test with a significance level of $\alpha = 0.05$.
2. The null hypothesis of randomness is rejected if the total number of runs G is less than or equal to the smaller entry or greater than or equal to the larger entry.

From "Tables for Testing Randomness of Groupings in a Sequence of Alternatives," *The Annals of Mathematical Statistics,* Vol. 14, No. 1. Reprinted with permission of the Institute of Mathematical Statistics.

Appendix B: Data Sets

Additional data sets are available at the Web site aw.com/Triola.

Data Set 1: Health Exam Results

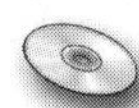

AGE is in years, HT is height (inches), WT is weight (pounds), WAIST is circumference (cm), PULSE is pulse rate (beats per minute), SYS is systolic blood pressure (mmHg), DIAS is diastolic blood pressure (mmHg), CHOL is cholesterol (mg), BMI is body mass index, LEG is upper leg length (cm), ELBOW is elbow breadth (cm), WRIST is wrist breadth (cm), and ARM is arm circumference (cm). Data are from the U.S. Department of Health and Human Services, National Center for Health Statistics, Third National Health and Nutrition Examination Survey.

STATDISK: Data set name for males is Mhealth.
Minitab: Worksheet name for males is MHEALTH.MTW.
Excel: Workbook name for males is MHEALTH.XLS.
TI-83/84 Plus: APP name for male data is MHEALTH and the file names are the same as for text files.
Text file names for males: MAGE, MHT, MWT, MWAST, MPULS, MSYS, MDIAS, MCHOL, MBMI, MLEG, MELBW, MWRST, MARM.

Male	Age	HT	WT	Waist	Pulse	SYS	DIAS	CHOL	BMI	Leg	Elbow	Wrist	Arm
	58	70.8	169.1	90.6	68	125	78	522	23.8	42.5	7.7	6.4	31.9
	22	66.2	144.2	78.1	64	107	54	127	23.2	40.2	7.6	6.2	31.0
	32	71.7	179.3	96.5	88	126	81	740	24.6	44.4	7.3	5.8	32.7
	31	68.7	175.8	87.7	72	110	68	49	26.2	42.8	7.5	5.9	33.4
	28	67.6	152.6	87.1	64	110	66	230	23.5	40.0	7.1	6.0	30.1
	46	69.2	166.8	92.4	72	107	83	316	24.5	47.3	7.1	5.8	30.5
	41	66.5	135.0	78.8	60	113	71	590	21.5	43.4	6.5	5.2	27.6
	56	67.2	201.5	103.3	88	126	72	466	31.4	40.1	7.5	5.6	38.0
	20	68.3	175.2	89.1	76	137	85	121	26.4	42.1	7.5	5.5	32.0
	54	65.6	139.0	82.5	60	110	71	578	22.7	36.0	6.9	5.5	29.3
	17	63.0	156.3	86.7	96	109	65	78	27.8	44.2	7.1	5.3	31.7
	73	68.3	186.6	103.3	72	153	87	265	28.1	36.7	8.1	6.7	30.7
	52	73.1	191.1	91.8	56	112	77	250	25.2	48.4	8.0	5.2	34.7
	25	67.6	151.3	75.6	64	119	81	265	23.3	41.0	7.0	5.7	30.6
	29	68.0	209.4	105.5	60	113	82	273	31.9	39.8	6.9	6.0	34.2
	17	71.0	237.1	108.7	64	125	76	272	33.1	45.2	8.3	6.6	41.1
	41	61.3	176.7	104.0	84	131	80	972	33.2	40.2	6.7	5.7	33.1
	52	76.2	220.6	103.0	76	121	75	75	26.7	46.2	7.9	6.0	32.2
	32	66.3	166.1	91.3	84	132	81	138	26.6	39.0	7.5	5.7	31.2
	20	69.7	137.4	75.2	88	112	44	139	19.9	44.8	6.9	5.6	25.9
	20	65.4	164.2	87.7	72	121	65	638	27.1	40.9	7.0	5.6	33.7
	29	70.0	162.4	77.0	56	116	64	613	23.4	43.1	7.5	5.2	30.3
	18	62.9	151.8	85.0	68	95	58	762	27.0	38.0	7.4	5.8	32.8
	26	68.5	144.1	79.6	64	110	70	303	21.6	41.0	6.8	5.7	31.0
	33	68.3	204.6	103.8	60	110	66	690	30.9	46.0	7.4	6.1	36.2
	55	69.4	193.8	103.0	68	125	82	31	28.3	41.4	7.2	6.0	33.6
	53	69.2	172.9	97.1	60	124	79	189	25.5	42.7	6.6	5.9	31.9
	28	68.0	161.9	86.9	60	131	69	957	24.6	40.5	7.3	5.7	32.9
	28	71.9	174.8	88.0	56	109	64	339	23.8	44.2	7.8	6.0	30.9
	37	66.1	169.8	91.5	84	112	79	416	27.4	41.8	7.0	6.1	34.0
	40	72.4	213.3	102.9	72	127	72	120	28.7	47.2	7.5	5.9	34.8
	33	73.0	198.0	93.1	84	132	74	702	26.2	48.2	7.8	6.0	33.6
	26	68.0	173.3	98.9	88	116	81	1252	26.4	42.9	6.7	5.8	31.3
	53	68.7	214.5	107.5	56	125	84	288	32.1	42.8	8.2	5.9	37.6
	36	70.3	137.1	81.6	64	112	77	176	19.6	40.8	7.1	5.3	27.9
	34	63.7	119.5	75.7	56	125	77	277	20.7	42.6	6.6	5.3	26.9
	42	71.1	189.1	95.0	56	120	83	649	26.3	44.9	7.4	6.0	36.9
	18	65.6	164.7	91.1	60	118	68	113	26.9	41.1	7.0	6.1	34.5
	44	68.3	170.1	94.9	64	115	75	656	25.6	44.5	7.3	5.8	32.1
	20	66.3	151.0	79.9	72	115	65	172	24.2	44.0	7.1	5.4	30.7

(continued)

Data Set 1: Health Exam Results (*continued*)

STATDISK:	**Data set name for females is Fhealth.**
Minitab:	**Worksheet name for females is FHEALTH.MTW.**
Excel:	**Workbook name for females is FHEALTH.XLS.**
TI-83/84 Plus:	**APP name for female data is FHEALTH and the file names are the same as for text files.**
Text file names for females:	**FAGE, FHT, FWT, FWAST, FPULS, FSYS, FDIAS, FCHOL, FBMI, FLEG, FELBW, FWRST, FARM.**

Female	Age	HT	WT	Waist	Pulse	SYS	DIAS	CHOL	BMI	Leg	Elbow	Wrist	Arm
	17	64.3	114.8	67.2	76	104	61	264	19.6	41.6	6.0	4.6	23.6
	32	66.4	149.3	82.5	72	99	64	181	23.8	42.8	6.7	5.5	26.3
	25	62.3	107.8	66.7	88	102	65	267	19.6	39.0	5.7	4.6	26.3
	55	62.3	160.1	93.0	60	114	76	384	29.1	40.2	6.2	5.0	32.6
	27	59.6	127.1	82.6	72	94	58	98	25.2	36.2	5.5	4.8	29.2
	29	63.6	123.1	75.4	68	101	66	62	21.4	43.2	6.0	4.9	26.4
	25	59.8	111.7	73.6	80	108	61	126	22.0	38.7	5.7	5.1	27.9
	12	63.3	156.3	81.4	64	104	41	89	27.5	41.0	6.8	5.5	33.0
	41	67.9	218.8	99.4	68	123	72	531	33.5	43.8	7.8	5.8	38.6
	32	61.4	110.2	67.7	68	93	61	130	20.6	37.3	6.3	5.0	26.5
	31	66.7	188.3	100.7	80	89	56	175	29.9	42.3	6.6	5.2	34.4
	19	64.8	105.4	72.9	76	112	62	44	17.7	39.1	5.7	4.8	23.7
	19	63.1	136.1	85.0	68	107	48	8	24.0	40.3	6.6	5.1	28.4
	23	66.7	182.4	85.7	72	116	62	112	28.9	48.6	7.2	5.6	34.0
	40	66.8	238.4	126.0	96	181	102	462	37.7	33.2	7.0	5.4	35.2
	23	64.7	108.8	74.5	72	98	61	62	18.3	43.4	6.2	5.2	24.7
	27	65.1	119.0	74.5	68	100	53	98	19.8	41.5	6.3	5.3	27.0
	45	61.9	161.9	94.0	72	127	74	447	29.8	40.0	6.8	5.0	35.0
	41	64.3	174.1	92.8	64	107	67	125	29.7	38.2	6.8	4.7	33.1
	56	63.4	181.2	105.5	80	116	71	318	31.7	38.2	6.9	5.4	39.6
	22	60.7	124.3	75.5	64	97	64	325	23.8	38.2	5.9	5.0	27.0
	57	63.4	255.9	126.5	80	155	85	600	44.9	41.0	8.0	5.6	43.8
	24	62.6	106.7	70.0	76	106	59	237	19.2	38.1	6.1	5.0	23.6
	37	60.6	149.9	98.0	76	110	70	173	28.7	38.0	7.0	5.1	34.3
	59	63.5	163.1	104.7	76	105	69	309	28.5	36.0	6.7	5.1	34.4
	40	58.6	94.3	67.8	80	118	82	94	19.3	32.1	5.4	4.2	23.3
	45	60.2	159.7	99.3	104	133	83	280	31.0	31.1	6.4	5.2	35.6
	52	67.6	162.8	91.1	88	113	75	254	25.1	39.4	7.1	5.3	31.8
	31	63.4	130.0	74.5	60	113	66	123	22.8	40.2	5.9	5.1	27.0
	32	64.1	179.9	95.5	76	107	67	596	30.9	39.2	6.2	5.0	32.8
	23	62.7	147.8	79.5	72	95	59	301	26.5	39.0	6.3	4.9	31.0
	23	61.3	112.9	69.1	72	108	72	223	21.2	36.6	5.9	4.7	27.0
	47	58.2	195.6	105.5	88	114	79	293	40.6	27.0	7.5	5.5	41.2
	36	63.2	124.2	78.8	80	104	73	146	21.9	38.5	5.6	4.7	25.5
	34	60.5	135.0	85.7	60	125	73	149	26.0	39.9	6.4	5.2	30.9
	37	65.0	141.4	92.8	72	124	85	149	23.5	37.5	6.1	4.8	27.9
	18	61.8	123.9	72.7	88	92	46	920	22.8	39.7	5.8	5.0	26.5
	29	68.0	135.5	75.9	88	119	81	271	20.7	39.0	6.3	4.9	27.8
	48	67.0	130.4	68.6	124	93	64	207	20.5	41.6	6.0	5.3	23.0
	16	57.0	100.7	68.7	64	106	64	2	21.9	33.8	5.6	4.6	26.4

Data Set 2: Body Temperatures (in degrees Fahrenheit) of Healthy Adults

Data provided by Dr. Steven Wasserman, Dr. Philip Mackowiak, and Dr. Myron Levine of the University of Maryland.

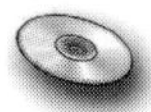

STATDISK:	**Data set name for the 12 A.M. temperatures on Day 2 is Bodytemp.**
Minitab:	**Worksheet name for the 12 A.M. temperatures on Day 2 is BODYTEMP.MTW.**
Excel:	**Workbook name for the 12 A.M. temperatures on Day 2 is BODYTEMP.XLS.**
TI-83/84 Plus:	**APP name for 12 A.M. temperatures on Day 2 is BTEMP and the file name is BTEMP.**
Text files:	**Text file name is BTEMP.**

Subject	Age	Sex	Smoke	Temperature Day 1 8 AM	Temperature Day 1 12 AM	Temperature Day 2 8 AM	Temperature Day 2 12 AM
1	22	M	Y	98.0	98.0	98.0	98.6
2	23	M	Y	97.0	97.6	97.4	—
3	22	M	Y	98.6	98.8	97.8	98.6
4	19	M	N	97.4	98.0	97.0	98.0
5	18	M	N	98.2	98.8	97.0	98.0
6	20	M	Y	98.2	98.8	96.6	99.0
7	27	M	Y	98.2	97.6	97.0	98.4
8	19	M	Y	96.6	98.6	96.8	98.4
9	19	M	Y	97.4	98.6	96.6	98.4
10	24	M	N	97.4	98.8	96.6	98.4
11	35	M	Y	98.2	98.0	96.2	98.6
12	25	M	Y	97.4	98.2	97.6	98.6
13	25	M	N	97.8	98.0	98.6	98.8
14	35	M	Y	98.4	98.0	97.0	98.6
15	21	M	N	97.6	97.0	97.4	97.0
16	33	M	N	96.2	97.2	98.0	97.0
17	19	M	Y	98.0	98.2	97.6	98.8
18	24	M	Y	—	—	97.2	97.6
19	18	F	N	—	—	97.0	97.7
20	22	F	Y	—	—	98.0	98.8
21	20	M	Y	—	—	97.0	98.0
22	30	F	Y	—	—	96.4	98.0
23	29	M	N	—	—	96.1	98.3
24	18	M	Y	—	—	98.0	98.5
25	31	M	Y	—	98.1	96.8	97.3
26	28	F	Y	—	98.2	98.2	98.7
27	27	M	Y	—	98.5	97.8	97.4
28	21	M	Y	—	98.5	98.2	98.9
29	30	M	Y	—	99.0	97.8	98.6
30	27	M	N	—	98.0	99.0	99.5
31	32	M	Y	—	97.0	97.4	97.5
32	33	M	Y	—	97.3	97.4	97.3
33	23	M	Y	—	97.3	97.5	97.6

(continued)

Data Set 2: Body Temperatures (*continued*)

Subject	Age	Sex	Smoke	Temperature Day 1		Temperature Day 2	
				8 AM	12 AM	8 AM	12 AM
34	29	M	Y	—	98.1	97.8	98.2
35	25	M	Y	—	—	97.9	99.6
36	31	M	N	—	97.8	97.8	98.7
37	25	M	Y	—	99.0	98.3	99.4
38	28	M	N	—	97.6	98.0	98.2
39	30	M	Y	—	97.4	—	98.0
40	33	M	Y	—	98.0	—	98.6
41	28	M	Y	98.0	97.4	—	98.6
42	22	M	Y	98.8	98.0	—	97.2
43	21	F	Y	99.0	—	—	98.4
44	30	M	N	—	98.6	—	98.6
45	22	M	Y	—	98.6	—	98.2
46	22	F	N	98.0	98.4	—	98.0
47	20	M	Y	—	97.0	—	97.8
48	19	M	Y	—	—	—	98.0
49	33	M	N	—	98.4	—	98.4
50	31	M	Y	99.0	99.0	—	98.6
51	26	M	N	—	98.0	—	98.6
52	18	M	N	—	—	—	97.8
53	23	M	N	—	99.4	—	99.0
54	28	M	Y	—	—	—	96.5
55	19	M	Y	—	97.8	—	97.6
56	21	M	N	—	—	—	98.0
57	27	M	Y	—	98.2	—	96.9
58	29	M	Y	—	99.2	—	97.6
59	38	M	N	—	99.0	—	97.1
60	29	F	Y	—	97.7	—	97.9
61	22	M	Y	—	98.2	—	98.4
62	22	M	Y	—	98.2	—	97.3
63	26	M	Y	—	98.8	—	98.0
64	32	M	N	—	98.1	—	97.5
65	25	M	Y	—	98.5	—	97.6
66	21	F	N	—	97.2	—	98.2
67	25	M	Y	—	98.5	—	98.5
68	24	M	Y	—	99.2	97.0	98.8
69	25	M	Y	—	98.3	97.6	98.7
70	35	M	Y	—	98.7	97.5	97.8
71	23	F	Y	—	98.8	98.8	98.0
72	31	M	Y	—	98.6	98.4	97.1
73	28	M	Y	—	98.0	98.2	97.4
74	29	M	Y	—	99.1	97.7	99.4
75	26	M	Y	—	97.2	97.3	98.4
76	32	M	N	—	97.6	97.5	98.6
77	32	M	Y	—	97.9	97.1	98.4
78	21	F	Y	—	98.8	98.6	98.5
79	20	M	Y	—	98.6	98.6	98.6

(continued)

Data Set 2: Body Temperatures (*continued*)

Subject	Age	Sex	Smoke	Temperature Day 1 8 AM	Temperature Day 1 12 AM	Temperature Day 2 8 AM	Temperature Day 2 12 AM
80	24	F	Y	—	98.6	97.8	98.3
81	21	F	Y	—	99.3	98.7	98.7
82	28	M	Y	—	97.8	97.9	98.8
83	27	F	N	98.8	98.7	97.8	99.1
84	28	M	N	99.4	99.3	97.8	98.6
85	29	M	Y	98.8	97.8	97.6	97.9
86	19	M	N	97.7	98.4	96.8	98.8
87	24	M	Y	99.0	97.7	96.0	98.0
88	29	M	N	98.1	98.3	98.0	98.7
89	25	M	Y	98.7	97.7	97.0	98.5
90	27	M	N	97.5	97.1	97.4	98.9
91	25	M	Y	98.9	98.4	97.6	98.4
92	21	M	Y	98.4	98.6	97.6	98.6
93	19	M	Y	97.2	97.4	96.2	97.1
94	27	M	Y	—	—	96.2	97.9
95	32	M	N	98.8	96.7	98.1	98.8
96	24	M	Y	97.3	96.9	97.1	98.7
97	32	M	Y	98.7	98.4	98.2	97.6
98	19	F	Y	98.9	98.2	96.4	98.2
99	18	F	Y	99.2	98.6	96.9	99.2
100	27	M	N	—	97.0	—	97.8
101	34	M	Y	—	97.4	—	98.0
102	25	M	N	—	98.4	—	98.4
103	18	M	N	—	97.4	—	97.8
104	32	M	Y	—	96.8	—	98.4
105	31	M	Y	—	98.2	—	97.4
106	26	M	N	—	97.4	—	98.0
107	23	M	N	—	98.0	—	97.0

Data Set 3: Cigarette Tar, Nicotine, and Carbon Monoxide

All measurements are in milligrams per cigarette, and all cigarettes are 100 mm long, filtered, and not menthol or light types. Data are from the Federal Trade Commission.

STATDISK: Data set name is Cigaret.
Minitab: Worksheet name is CIGARET.MTW.
Excel: Workbook name is CIGARET.XLS.
TI-83/84 Plus: APP name is CIGARET and the file names are the same as for text files.
Text file names: TAR, NICOT, CO.

Brand	Tar	Nicotine	CO
American Filter	16	1.2	15
Benson & Hedges	16	1.2	15
Camel	16	1.0	17
Capri	9	0.8	6
Carlton	1	0.1	1
Cartier Vendome	8	0.8	8
Chelsea	10	0.8	10
GPC Approved	16	1.0	17
Hi-Lite	14	1.0	13
Kent	13	1.0	13
Lucky Strike	13	1.1	13
Malibu	15	1.2	15
Marlboro	16	1.2	15
Merit	9	0.7	11
Newport Stripe	11	0.9	15
Now	2	0.2	3
Old Gold	18	1.4	18
Pall Mall	15	1.2	15
Players	13	1.1	12
Raleigh	15	1.0	16
Richland	17	1.3	16
Rite	9	0.8	10
Silva Thins	12	1.0	10
Tareyton	14	1.0	17
Triumph	5	0.5	7
True	6	0.6	7
Vantage	8	0.7	11
Viceroy	18	1.4	15
Winston	16	1.1	18

Data Set 4: Passive and Active Smoke

All values are measured levels of serum cotinine (in ng/mL), a metabolite of nicotine. (When nicotine is absorbed by the body, cotinine is produced.) Data are from the U.S. Department of Health and Human Services, National Center for Health Statistics, Third National Health and Nutrition Examination Survey.

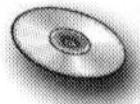

STATDISK:	Data set name is Cotinine.
Minitab:	Worksheet name is COTININE.MTW.
Excel:	Workbook name is COTININE.XLS.
TI-83/84 Plus:	APP name is COTININE and the file names are the same as for text files.
Text file names:	NOETS, ETS, SMKR.

Smokers (subjects reported tobacco use)

1	0	131	173	265	210	44	277	32	3
35	112	477	289	227	103	222	149	313	491
130	234	164	198	17	253	87	121	266	290
123	167	250	245	48	86	284	1	208	173

ETS (nonsmokers exposed to environmental tobacco smoke at home or work)

384	0	69	19	1	0	178	2	13	1
4	0	543	17	1	0	51	0	197	3
0	3	1	45	13	3	1	1	1	0
0	551	2	1	1	1	0	74	1	241

NOETS (nonsmokers with no exposure to environmental tobacco smoke at home or work)

0	0	0	0	0	0	0	0	0	0
0	9	0	0	0	0	0	0	244	0
1	0	0	0	90	1	0	309	0	0
0	0	0	0	0	0	0	0	0	0

Data Set 5: Alcohol and Tobacco Use in Animated Children's Movies

STATDISK: Data set name is Chmovie.
Minitab: Worksheet name is CHMOVIE.MTW.
Excel: Workbook name is CHMOVIE.XLS.
TI-83/84 Plus: APP name is CHMOVIE and the file names are the same as for text files.
Text file names: CHLEN, CHTOB, CHALC.

Movie lengths are in minutes, tobacco use times are in seconds, and alcohol use times are in seconds. The data are based on "Tobacco and Alcohol Use in G-Rated Children's Animated Films," by Goldstein, Sobel, and Newman, *Journal of the American Medical Association,* Vol. 281, No. 12.

Movie	Company	Length (min)	Tobacco Use (sec)	Alcohol Use (sec)
Snow White	Disney	83	0	0
Pinocchio	Disney	88	223	80
Fantasia	Disney	120	0	0
Dumbo	Disney	64	176	88
Bambi	Disney	69	0	0
Three Caballeros	Disney	71	548	8
Fun and Fancy Free	Disney	76	0	4
Cinderella	Disney	74	37	0
Alice in Wonderland	Disney	75	158	0
Peter Pan	Disney	76	51	33
Lady and the Tramp	Disney	75	0	0
Sleeping Beauty	Disney	75	0	113
101 Dalmatians	Disney	79	299	51
Sword and the Stone	Disney	80	37	20
Jungle Book	Disney	78	0	0
Aristocats	Disney	78	11	142
Robin Hood	Disney	83	0	39
Rescuers	Disney	77	0	0
Winnie the Pooh	Disney	71	0	0
Fox and the Hound	Disney	83	0	0
Black Cauldron	Disney	80	0	34
Great Mouse Detective	Disney	73	165	414
Oliver and Company	Disney	72	74	0
Little Mermaid	Disney	82	9	0
Rescuers Down Under	Disney	74	0	76
Beauty and the Beast	Disney	84	0	123
Aladdin	Disney	90	2	3
Lion King	Disney	89	0	0
Pocahontas	Disney	81	6	7
Toy Story	Disney	81	0	0
Hunchback of Notre Dame	Disney	90	23	46
James and the Giant Peach	Disney	79	206	38
Hercules	Disney	92	9	13
Secret of NIMH	MGM	82	0	0
All Dogs Go to Heaven	MGM	89	205	73
All Dogs Go to Heaven 2	MGM	82	162	72
Babes in Toyland	MGM	74	0	0
Thumbelina	Warner Bros	86	6	5
Troll in Central Park	Warner Bros	76	1	0
Space Jam	Warner Bros	81	117	0
Pippi Longstocking	Warner Bros	75	5	0
Cats Don't Dance	Warner Bros	75	91	0
An American Tail	Universal	77	155	74
Land Before Time	Universal	70	0	0
Fievel Goes West	Universal	75	24	28
We're Back: Dinosaur Story	Universal	64	55	0
Land Before Time 2	Universal	73	0	0
Balto	Universal	74	0	0
Once Upon a Forest	20th Century Fox	71	0	0
Anastasia	20th Century Fox	94	17	39

Data Set 6: Bears (wild bears anesthetized)

AGE is in months, MONTH is the month of measurement (1 = January), SEX is coded with 1 = male and 2 = female, HEADLEN is head length (inches), HEADWTH is width of head (inches), NECK is distance around neck (in inches), LENGTH is length of body (inches), CHEST is distance around chest (inches), and WEIGHT is measured in pounds. Data are from Gary Alt and Minitab, Inc.

STATDISK: Data set name is Bears.
Minitab: Worksheet name is BEARS.MTW.
Excel: Workbook name is BEARS.XLS.
TI-83/84 Plus: APP name is BEARS and the file names are the same as for text files.
Text file names: BAGE, BMNTH, BSEX, BHDLN, BHDWD, BNECK, BLEN, BCHST, BWGHT.

Age	Month	Sex	Headlen	Headwth	Neck	Length	Chest	Weight
19	7	1	11.0	5.5	16.0	53.0	26.0	80
55	7	1	16.5	9.0	28.0	67.5	45.0	344
81	9	1	15.5	8.0	31.0	72.0	54.0	416
115	7	1	17.0	10.0	31.5	72.0	49.0	348
104	8	2	15.5	6.5	22.0	62.0	35.0	166
100	4	2	13.0	7.0	21.0	70.0	41.0	220
56	7	1	15.0	7.5	26.5	73.5	41.0	262
51	4	1	13.5	8.0	27.0	68.5	49.0	360
57	9	2	13.5	7.0	20.0	64.0	38.0	204
53	5	2	12.5	6.0	18.0	58.0	31.0	144
68	8	1	16.0	9.0	29.0	73.0	44.0	332
8	8	1	9.0	4.5	13.0	37.0	19.0	34
44	8	2	12.5	4.5	10.5	63.0	32.0	140
32	8	1	14.0	5.0	21.5	67.0	37.0	180
20	8	2	11.5	5.0	17.5	52.0	29.0	105
32	8	1	13.0	8.0	21.5	59.0	33.0	166
45	9	1	13.5	7.0	24.0	64.0	39.0	204
9	9	2	9.0	4.5	12.0	36.0	19.0	26
21	9	1	13.0	6.0	19.0	59.0	30.0	120
177	9	1	16.0	9.5	30.0	72.0	48.0	436
57	9	2	12.5	5.0	19.0	57.5	32.0	125
81	9	2	13.0	5.0	20.0	61.0	33.0	132
21	9	1	13.0	5.0	17.0	54.0	28.0	90
9	9	1	10.0	4.0	13.0	40.0	23.0	40
45	9	1	16.0	6.0	24.0	63.0	42.0	220
9	9	1	10.0	4.0	13.5	43.0	23.0	46
33	9	1	13.5	6.0	22.0	66.5	34.0	154
57	9	2	13.0	5.5	17.5	60.5	31.0	116
45	9	2	13.0	6.5	21.0	60.0	34.5	182
21	9	1	14.5	5.5	20.0	61.0	34.0	150
10	10	1	9.5	4.5	16.0	40.0	26.0	65
82	10	2	13.5	6.5	28.0	64.0	48.0	356
70	10	2	14.5	6.5	26.0	65.0	48.0	316
10	10	1	11.0	5.0	17.0	49.0	29.0	94
10	10	1	11.5	5.0	17.0	47.0	29.5	86
34	10	1	13.0	7.0	21.0	59.0	35.0	150
34	10	1	16.5	6.5	27.0	72.0	44.5	270
34	10	1	14.0	5.5	24.0	65.0	39.0	202
58	10	2	13.5	6.5	21.5	63.0	40.0	202
58	10	1	15.5	7.0	28.0	70.5	50.0	365
11	11	1	11.5	6.0	16.5	48.0	31.0	79
23	11	1	12.0	6.5	19.0	50.0	38.0	148
70	10	1	15.5	7.0	28.0	76.5	55.0	446
11	11	2	9.0	5.0	15.0	46.0	27.0	62
83	11	2	14.5	7.0	23.0	61.5	44.0	236
35	11	1	13.5	8.5	23.0	63.5	44.0	212
16	4	1	10.0	4.0	15.5	48.0	26.0	60
16	4	1	10.0	5.0	15.0	41.0	26.0	64
17	5	1	11.5	5.0	17.0	53.0	30.5	114
17	5	2	11.5	5.0	15.0	52.5	28.0	76
17	5	2	11.0	4.5	13.0	46.0	23.0	48
8	8	2	10.0	4.5	10.0	43.5	24.0	29
83	11	1	15.5	8.0	30.5	75.0	54.0	514
18	6	1	12.5	8.5	18.0	57.3	32.8	140

Data Set 7: Poplar Tree Weights (kg)

The data are from a study conducted by researchers at Pennsylvania State University; the data were obtained from Minitab, Inc.

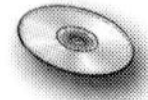

STATDISK:	Data set name is Poplar.
Minitab:	Worksheet name is POPLAR.MTW.
Excel:	Workbook name is POPLAR.XLS.
TI-83/84 Plus:	APP name is POPLAR and the file names are the same as for text files.
Text file names:	File names are POPNO, POPFR, POPIR, POPFI.

Year 1

	No Treatment	Fertilizer	Irrigation	Fertilizer and Irrigation
Site 1 (rich, moist)	0.15	1.34	0.23	2.03
	0.02	0.14	0.04	0.27
	0.16	0.02	0.34	0.92
	0.37	0.08	0.16	1.07
	0.22	0.08	0.05	2.38
Site 2 (sandy, dry)	0.60	1.16	0.65	0.22
	1.11	0.93	0.08	2.13
	0.07	0.30	0.62	2.33
	0.07	0.59	0.01	1.74
	0.44	0.17	0.03	0.12

Year 2

	No Treatment	Fertilizer	Irrigation	Fertilizer and Irrigation
Site 1 (rich, moist)	1.21	0.94	0.07	0.85
	0.57	0.87	0.66	1.78
	0.56	0.46	0.10	1.47
	0.13	0.58	0.82	2.25
	1.30	1.03	0.94	1.64
Site 2 (sandy, dry)	0.24	0.92	0.96	1.07
	1.69	0.07	1.43	1.63
	1.23	0.56	1.26	1.39
	0.99	1.74	1.57	0.49
	1.80	1.13	0.72	0.95

Data Set 8: Forecast and Actual Temperatures

Temperatures are in degrees Fahrenheit and the precipitation amounts are in inches. All measurements were recorded near the author's home.

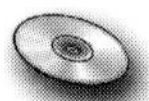

STATDISK:	**Data set name is Weather.**
Minitab:	**Worksheet name is WEATHER.MTW.**
Excel:	**Workbook name is WEATHER.XLS.**
TI-83/84 Plus:	**APP name is WEATHER and the file names are the same as for text files.**
Text file names:	**ACTHI, ACTLO, PHI1, PLO1, PHI3, PLO3, PHI5, PLO5, PREC.**

Date	Actual High	Actual Low	1 Day Predicted High	1 Day Predicted Low	3 Day Predicted High	3 Day Predicted Low	5 Day Predicted High	5 Day Predicted Low	Precip. (in.)
Sept. 1	80	54	78	52	79	52	80	56	0.00
Sept. 2	77	54	75	53	86	63	80	57	0.00
Sept. 3	81	55	81	55	79	59	79	59	0.00
Sept. 4	85	60	85	62	83	59	80	56	0.00
Sept. 5	73	64	76	53	80	63	79	64	0.00
Sept. 6	73	51	75	58	76	61	82	57	0.00
Sept. 7	80	59	79	66	80	63	76	61	0.00
Sept. 8	72	61	74	66	79	67	73	63	0.47
Sept. 9	83	68	75	66	76	66	77	59	1.59
Sept. 10	81	62	80	53	79	58	83	61	0.07
Sept. 11	75	53	75	51	78	58	77	58	0.01
Sept. 12	78	52	79	55	75	54	79	56	0.00
Sept. 13	80	56	80	53	74	48	74	50	0.01
Sept. 14	71	56	70	53	73	55	75	52	0.00
Sept. 15	73	54	72	60	73	59	76	54	0.00
Sept. 16	78	64	79	63	76	60	78	62	0.06
Sept. 17	75	62	75	60	76	56	76	58	0.01
Sept. 18	63	55	67	43	73	52	75	60	2.85
Sept. 19	63	48	64	43	75	53	77	55	0.00
Sept. 20	70	40	69	46	68	53	71	50	0.00
Sept. 21	77	47	77	50	77	51	74	54	0.00
Sept. 22	82	49	81	55	83	54	73	56	0.00
Sept. 23	81	53	81	51	78	57	75	53	0.01
Sept. 24	76	51	80	53	75	54	79	56	0.00
Sept. 25	77	54	78	54	77	51	74	53	0.00
Sept. 26	76	58	76	50	72	46	71	44	0.00
Sept. 27	74	48	76	60	74	56	70	45	0.01
Sept. 28	66	61	70	49	74	49	73	52	1.99
Sept. 29	66	57	69	41	68	41	72	48	0.67
Sept. 30	62	53	68	45	72	50	69	47	0.21
Oct. 1	71	51	75	49	72	49	70	47	0.02
Oct. 2	68	46	71	47	73	52	68	46	0.01
Oct. 3	66	44	68	42	66	43	67	46	0.05
Oct. 4	71	39	69	44	68	44	61	40	0.00
Oct. 5	58	43	56	29	62	38	64	42	0.00

Data Set 9: Electricity Consumption of a Home

All measurements are from the author's home.

STATDISK: Data set name is Electric.
Minitab: Worksheet name is ELECTRIC.MTW.
Excel: Workbook name is ELECTRIC.XLS.
TI-83/84 Plus: APP name is ELECTRIC and the file names are the same as for text files.
Text file names: Text file names are KWH, COST, HDD, ADT.

Time Period	Electricity Consumed (kWh)	Cost (dollars)	Heating Degree Days	Average Daily Temp (°F)
Year 1: Jan/Feb	3375	321.94	2421	26
Year 1: March/Apr	2661	221.11	1841	34
Year 1: May/June	2073	205.16	438	58
Year 1: July/Aug	2579	251.07	15	72
Year 1: Sept/Oct	2858	279.80	152	67
Year 1: Nov/Dec	2296	183.84	1028	48
Year 2: Jan/Feb	2812	244.93	1967	33
Year 2: March/Apr	2433	218.59	1627	39
Year 2: May/June	2266	213.09	537	66
Year 2: July/Aug	3128	333.49	26	71
Year 2: Sept/Oct	3286	370.35	116	
Year 2: Nov/Dec	2749	222.79	1457	
Year 3: Jan/Feb	3427	316.18	253	
Year 3: March/Apr	578	77.39	1811	
Year 3: May/June	3792	385.44	632	

Data Set 10: Rainfall (in inches) in Boston for One Year

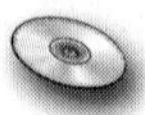

STATDISK: Data set name is Bostrain.
Minitab: Worksheet name is BOSTRAIN.MTW.
Excel: Workbook name is BOSTRAIN.XLS.
TI-83/84 Plus: APP name is BOSTRAIN and the file names are the same as for text files.
Text file names: RNMON, RNTUE, RNWED, RNTHU, RNFRI, RNSAT, RNSUN.

Mon	Tues	Wed	Thurs	Fri	Sat	Sun
0	0	0	0.04	0.04	0	0.05
0	0	0	0.06	0.03	0.1	0
0	0	0	0.71	0	0	0
0	0.44	0.14	0.04	0.04	0.64	0
0.05	0	0	0	0.01	0.05	0
0	0	0.64	0	0	0	0
0.01	0	0	0	0.3	0.05	0
0	0	0.01	0	0	0	0
0	0.01	0.01	0.16	0	0	0.09
0.12	0.06	0.18	0.39	0	0.1	0
0	0	0	0	0.78	0.49	0
0	0.02	0	0	0.01	0.17	0
1.41	0.65	0.31	0	0	0.54	0
0	0	0	0	0	0	0
0	0	0	0	0	0.4	0.28
0	0	0	0.3	0.87	0.49	0
0.47	0	0	0	0	0	0
0	0.09	0	0.24	0	0.05	0
0	0.14	0	0	0.04	0.07	0
0.92	0.36	0.02	0.09	0.27	0	0
0.01	0	0.06	0	0	0	0.27
0.01	0	0	0	0	0	0.01
0	0	0	0	0	0	0
0	0	0	0	0.71	0	0
0	0	0.27	0.08	0	0	0.33
0	0	0	0	0	0	0
0.03	0	0.08	0.14	0	0	0
0	0.11	0.06	0.02	0	0	0
0.01	0.05	0	0.01	0	0	0
0	0	0	0	0.12	0	0
0.11	0.03	0	0	0	0	0.44
0.01	0.01	0	0	0.11	0.18	0
0.49	0	0.64	0.01	0	0	0.01
0	0	0.08	0.85	0.01	0	0
0.01	0.02	0	0	0.03	0	0
0	0	0.12	0	0	0	0
0	0	0.01	0.04	0.26	0.04	0
0	0	0	0	0	0.4	0
0.12	0	0	0	0	0	0
0	0	0	0	0.24	0	0.23
0	0	0	0.02	0	0	0
0	0	0	0.02	0	0	0
0.59	0	0	0	0	0.68	0
0	0.01	0	0	0	1.48	0.21
0.01	0	0	0	0.05	0.69	1.28
0	0	0	0	0.96	0	0.01
0	0	0	0	0	0.79	0.02
0.41	0	0.06	0.01	0	0	0.28
0	0	0	0.08	0.04	0	0
0	0	0	0	0	0	0
0	0.74	0	0	0	0	0
0.43	0.3	0	0.26	0	0.02	0.01
		0				

Data Set 11: Old Faithful Geyser

Measurements are from eruptions of the Old Faithful Geyser in Yellowstone National Park. Prediction errors are based on predicted times to the eruption, where a negative value corresponds to an eruption that occurred before the predicted time.

STATDISK: Data set name is OldFaith.
Minitab: Worksheet name is OLDFAITH.MTW.
Excel: Workbook name is OLDFAITH.XLS.
TI-83/84 Plus: APP name is OLDFAITH and the file names are the same as for text files.
Text file names: Text file names are OFDUR, OFBEF, OFAFT, OFHT, OFERR.

Duration (sec)	Interval Before (min)	Interval After (min)	Height (ft)	Prediction Error (min)
240	98	92	140	4
237	92	95	140	−2
250	95	92	148	1
243	87	100	130	−7
255	96	90	125	2
120	90	65	110	−4
260	65	92	136	0
178	92	72	125	−2
259	95	93	115	1
245	93	98	120	−1
234	98	94	120	4
213	94	80	120	0
255	93	93	150	−1
235	93	83	140	−1
250	96	89	136	2
110	89	66	120	−5
245	93	89	148	−1
269	89	86	130	−5
251	86	97	130	−8
234	69	105	136	4
252	105	92	130	13
254	92	89	115	0
273	89	93	136	−3
266	93	112	130	1
284	112	88	138	20
252	95	105	120	3
269	105	94	120	13
250	94	90	120	2
261	90	98	95	−2
253	98	81	140	6
255	81	101	125	−11
280	69	94	130	4
270	94	92	130	2
241	92	106	110	0
272	106	93	110	14
294	93	96	125	1
220	108	87	150	21
253	87	97	130	−5
245	97	86	120	5
274	102	92	95	10

Data Set 12: Weights and Volumes of Cola

Weights are in pounds and volumes are in ounces.

STATDISK:	**Data set name is Cola.**
Minitab:	**Worksheet name is COLA.MTW.**
Excel:	**Workbook name is COLA.XLS.**
TI-83/84 Plus:	**APP name is COLA, and the file names are the same as for text files.**
Text file names:	**CRGWT, CRGVL, CDTWT, CDTVL, PRGWT, PRGVL, PDTWT, PDTVL.**

Weight Regular Coke	Volume Regular Coke	Weight Diet Coke	Volume Diet Coke	Weight Regular Pepsi	Volume Regular Pepsi	Weight Diet Pepsi	Volume Diet Pepsi
0.8192	12.3	0.7773	12.1	0.8258	12.4	0.7925	12.3
0.8150	12.1	0.7758	12.1	0.8156	12.2	0.7868	12.2
0.8163	12.2	0.7896	12.3	0.8211	12.2	0.7846	12.2
0.8211	12.3	0.7868	12.3	0.8170	12.2	0.7938	12.3
0.8181	12.2	0.7844	12.2	0.8216	12.2	0.7861	12.2
0.8247	12.3	0.7861	12.3	0.8302	12.4	0.7844	12.2
0.8062	12.0	0.7806	12.2	0.8192	12.2	0.7795	12.2
0.8128	12.1	0.7830	12.2	0.8192	12.2	0.7883	12.3
0.8172	12.2	0.7852	12.2	0.8271	12.3	0.7879	12.2
0.8110	12.1	0.7879	12.3	0.8251	12.3	0.7850	12.3
0.8251	12.3	0.7881	12.3	0.8227	12.2	0.7899	12.3
0.8264	12.3	0.7826	12.3	0.8256	12.3	0.7877	12.2
0.7901	11.8	0.7923	12.3	0.8139	12.2	0.7852	12.2
0.8244	12.3	0.7852	12.3	0.8260	12.3	0.7756	12.1
0.8073	12.1	0.7872	12.3	0.8227	12.2	0.7837	12.2
0.8079	12.1	0.7813	12.2	0.8388	12.5	0.7879	12.2
0.8044	12.0	0.7885	12.3	0.8260	12.3	0.7839	12.2
0.8170	12.2	0.7760	12.1	0.8317	12.4	0.7817	12.2
0.8161	12.2	0.7822	12.2	0.8247	12.3	0.7822	12.2
0.8194	12.2	0.7874	12.3	0.8200	12.2	0.7742	12.1
0.8189	12.2	0.7822	12.2	0.8172	12.2	0.7833	12.2
0.8194	12.2	0.7839	12.2	0.8227	12.3	0.7835	12.2
0.8176	12.2	0.7802	12.1	0.8244	12.3	0.7855	12.2
0.8284	12.4	0.7892	12.3	0.8244	12.2	0.7859	12.2
0.8165	12.2	0.7874	12.2	0.8319	12.4	0.7775	12.1
0.8143	12.2	0.7907	12.3	0.8247	12.3	0.7833	12.2
0.8229	12.3	0.7771	12.1	0.8214	12.2	0.7835	12.2
0.8150	12.2	0.7870	12.2	0.8291	12.4	0.7826	12.2
0.8152	12.2	0.7833	12.3	0.8227	12.3	0.7815	12.2
0.8244	12.3	0.7822	12.2	0.8211	12.3	0.7791	12.1
0.8207	12.2	0.7837	12.3	0.8401	12.5	0.7866	12.3
0.8152	12.2	0.7910	12.4	0.8233	12.3	0.7855	12.2
0.8126	12.1	0.7879	12.3	0.8291	12.4	0.7848	12.2
0.8295	12.4	0.7923	12.4	0.8172	12.2	0.7806	12.2
0.8161	12.2	0.7859	12.3	0.8233	12.4	0.7773	12.1
0.8192	12.2	0.7811	12.2	0.8211	12.3	0.7775	12.1

Data Set 13: Weights (grams) of a Sample of M&M Plain Candies

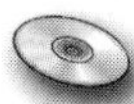

STATDISK: Data set name is M&M.
Minitab: Worksheet name is M&M.MTW.
Excel: Workbook name is M&M.XLS.
TI-83/84 Plus: APP name is MM, and the file names are the same as for text files.
Text file names: Text file names are RED, ORNG, YLLW, BROWN, BLUE, GREEN.

Red	Orange	Yellow	Brown	Blue	Green
0.751	0.735	0.883	0.696	0.881	0.925
0.841	0.895	0.769	0.876	0.863	0.914
0.856	0.865	0.859	0.855	0.775	0.881
0.799	0.864	0.784	0.806	0.854	0.865
0.966	0.852	0.824	0.840	0.810	0.865
0.859	0.866	0.858	0.868	0.858	1.015
0.857	0.859	0.848	0.859	0.818	0.876
0.942	0.838	0.851	0.982	0.868	0.809
0.873	0.863			0.803	0.865
0.809	0.888			0.932	0.848
0.890	0.925			0.842	0.940
0.878	0.793			0.832	0.833
0.905	0.977			0.807	0.845
	0.850			0.841	0.852
	0.830			0.932	0.778
	0.856			0.833	0.814
	0.842			0.881	0.791
	0.778			0.818	0.810
	0.786			0.864	0.881
	0.853			0.825	
	0.864			0.855	
	0.873			0.942	
	0.880			0.825	
	0.882			0.869	
	0.931			0.912	
				0.887	
				0.886	

Data Set 14: Coin Weights (grams)

The "pre-1983 pennies" were made after the Indian and wheat pennies, and they are 97% copper and 3% zinc. The "post-1983 pennies" are 3% copper and 97% zinc. The "pre-1964 silver quarters" are 90% silver and 10% copper. The "post-1964 quarters" are made with a copper-nickel alloy.

STATDISK:	**Data set name is Coins.**
Minitab:	**Worksheet name is COINS.MTW.**
Excel:	**Workbook name is COINS.XLS.**
TI-83/84 Plus:	**APP name is COINS, and the file names are the same as for text files.**
Text file names:	**Text file names are CPIND, CPWHT, CPPRE, CPPST, CPCAN, CQPRE, CQPST, CDOL.**

Indian Pennies	Wheat Pennies	Pre-1983 Pennies	Post-1983 Pennies	Canadian Pennies	Pre-1964 Quarters	Post-1964 Quarters	Dollar Coins
3.0630	3.1366	3.1582	2.5113	3.2214	6.2771	5.7027	8.1008
3.0487	3.0755	3.0406	2.4907	3.2326	6.2371	5.7495	8.1072
2.9149	3.1692	3.0762	2.5024	2.4662	6.1501	5.7050	8.0271
3.1358	3.0476	3.0398	2.5298	2.8357	6.0002	5.5941	8.0813
2.9753	3.1029	3.1043	2.4950	3.3189	6.1275	5.7247	8.0241
	3.0377	3.1274	2.5127	3.2612	6.2151	5.6114	8.0510
	3.1083	3.0775	2.4998	3.2441	6.2866	5.6160	7.9817
	3.1141	3.1038	2.4848	2.4679	6.0760	5.5999	8.0954
	3.0976	3.1086	2.4823	2.7202	6.1426	5.7790	8.0658
	3.0862	3.0586	2.5163	2.5120	6.3415	5.6841	8.1238
	3.0570	3.0603	2.5222		6.1309	5.6234	8.1281
	3.0765	3.0502	2.5004		6.2412	5.5928	8.0307
	3.1114	3.1028	2.5248		6.1442	5.6486	8.0719
	3.0965	3.0522	2.5058		6.1073	5.6661	8.0345
	3.0816	3.0546	2.4900		6.1181	5.5361	8.0775
	3.0054	3.0185	2.5068		6.1352	5.5491	8.1384
	3.1934	3.0712	2.5016		6.2821	5.7239	8.1041
	3.1461	3.0717	2.4797		6.2647	5.6555	8.0894
	3.0185	3.0546	2.5067		6.2908	5.6063	8.0538
	3.1267	3.0817	2.5139		6.1661	5.5709	8.0342
	3.1524	3.0704	2.4762		6.2674	5.5591	
	3.0786	3.0797	2.5004		6.2718	5.5864	
	3.0131	3.0713	2.5170		6.1949	5.6872	
	3.1535	3.0631	2.4925		6.2465	5.6274	
	3.0480	3.0866	2.4876		6.3172	5.6157	
	3.0050	3.0763	2.4933		6.1487	5.6668	
	3.0290	3.1299	2.4806		6.0829	5.7198	
	3.1038	3.0846	2.4907		6.1423	5.6694	
	3.0357	3.0917	2.5017		6.1970	5.5454	
	3.0064	3.0877	2.4950		6.2441	5.6646	
	3.0936	2.9593	2.4973		6.3669	5.5636	
	3.1031	3.0966	2.5252		6.0775	5.6485	
	3.0408	2.9800	2.4978		6.1095	5.6703	
	3.0561	3.0934	2.5073		6.1787	5.6848	
	3.0994	3.1340	2.4658		6.2130	5.5609	
			2.4529		6.1947	5.7344	
			2.5085		6.1940	5.6449	
					6.0257	5.5804	
					6.1719	5.6010	
					6.3278	5.6022	

Data Set 15: Axial Loads of Aluminum Cans

Axial loads are measured in pounds.

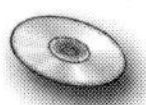

STATDISK: Data set name is Cans.
Minitab : Worksheet name is CANS.MTW.
Excel: Workbook name is CANS.XLS.
TI-83/84 Plus: APP name is CANS, and the file names are the same as for text files.
Text file names: CN109, CN111.

Sample	Aluminum cans 0.0109 in. Load (pounds)							Sample	Aluminum cans 0.0111 in. Load (pounds)						
1	270	273	258	204	254	228	282	1	287	216	260	291	210	272	260
2	278	201	264	265	223	274	230	2	294	253	292	280	262	295	230
3	250	275	281	271	263	277	275	3	283	255	295	271	268	225	246
4	278	260	262	273	274	286	236	4	297	302	282	310	305	306	262
5	290	286	278	283	262	277	295	5	222	276	270	280	288	296	281
6	274	272	265	275	263	251	289	6	300	290	284	304	291	277	317
7	242	284	241	276	200	278	283	7	292	215	287	280	311	283	293
8	269	282	267	282	272	277	261	8	285	276	301	285	277	270	275
9	257	278	295	270	268	286	262	9	290	288	287	282	275	279	300
10	272	268	283	256	206	277	252	10	293	290	313	299	300	265	285
11	265	263	281	268	280	289	283	11	294	262	297	272	284	291	306
12	263	273	209	259	287	269	277	12	263	304	288	256	290	284	307
13	234	282	276	272	257	267	204	13	273	283	250	244	231	266	504
14	270	285	273	269	284	276	286	14	284	227	269	282	292	286	281
15	273	289	263	270	279	206	270	15	296	287	285	281	298	289	283
16	270	268	218	251	252	284	278	16	247	279	276	288	284	301	309
17	277	208	271	208	280	269	270	17	284	284	286	303	308	288	303
18	294	292	289	290	215	284	283	18	306	285	289	292	295	283	315
19	279	275	223	220	281	268	272	19	290	247	268	283	305	279	287
20	268	279	217	259	291	291	281	20	285	298	279	274	205	302	296
21	230	276	225	282	276	289	288	21	282	300	284	281	279	255	210
22	268	242	283	277	285	293	248	22	279	286	293	285	288	289	281
23	278	285	292	282	287	277	266	23	297	314	295	257	298	211	275
24	268	273	270	256	297	280	256	24	247	279	303	286	287	287	275
25	262	268	262	293	290	274	292	25	243	274	299	291	281	303	269

Data Set 16: Weights of Discarded Garbage for One Week

Weights are in pounds. HHSIZE is the household size. Data provided by Masakuza Tani, the Garbage Project, University of Arizona.

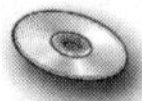

STATDISK: Data set name is Garbage.
Minitab: Worksheet name is GARBAGE.MTW.
Excel: Workbook name is GARBAGE.XLS.
TI-83/84 Plus: APP name is GARBAGE, and the file names are the same as for text files.
Text file names: HHSIZ, METAL, PAPER, PLAS, GLASS, FOOD, YARD, TEXT, OTHER, TOTAL.

Household	HHSize	Metal	Paper	Plas	Glass	Food	Yard	Text	Other	Total
1	2	1.09	2.41	0.27	0.86	1.04	0.38	0.05	4.66	10.76
2	3	1.04	7.57	1.41	3.46	3.68	0.00	0.46	2.34	19.96
3	3	2.57	9.55	2.19	4.52	4.43	0.24	0.50	3.60	27.60
4	6	3.02	8.82	2.83	4.92	2.98	0.63	2.26	12.65	38.11
5	4	1.50	8.72	2.19	6.31	6.30	0.15	0.55	2.18	27.90
6	2	2.10	6.96	1.81	2.49	1.46	4.58	0.36	2.14	21.90
7	1	1.93	6.83	0.85	0.51	8.82	0.07	0.60	2.22	21.83
8	5	3.57	11.42	3.05	5.81	9.62	4.76	0.21	10.83	49.27
9	6	2.32	16.08	3.42	1.96	4.41	0.13	0.81	4.14	33.27
10	4	1.89	6.38	2.10	17.67	2.73	3.86	0.66	0.25	35.54
11	4	3.26	13.05	2.93	3.21	9.31	0.70	0.37	11.61	44.44
12	7	3.99	11.36	2.44	4.94	3.59	13.45	4.25	1.15	45.17
13	3	2.04	15.09	2.17	3.10	5.36	0.74	0.42	4.15	33.07
14	5	0.99	2.80	1.41	1.39	1.47	0.82	0.44	1.03	10.35
15	6	2.96	6.44	2.00	5.21	7.06	6.14	0.20	14.43	44.44
16	2	1.50	5.86	0.93	2.03	2.52	1.37	0.27	9.65	24.13
17	4	2.43	11.08	2.97	1.74	1.75	14.70	0.39	2.54	37.60
18	4	2.97	12.43	2.04	3.99	5.64	0.22	2.47	9.20	38.96
19	3	1.42	6.05	0.65	6.26	1.93	0.00	0.86	0.00	17.17
20	3	3.60	13.61	2.13	3.52	6.46	0.00	0.96	1.32	31.60
21	2	4.48	6.98	0.63	2.01	6.72	2.00	0.11	0.18	23.11
22	2	1.36	14.33	1.53	2.21	5.76	0.58	0.17	1.62	27.56
23	4	2.11	13.31	4.69	0.25	9.72	0.02	0.46	0.40	30.96
24	1	0.41	3.27	0.15	0.09	0.16	0.00	0.00	0.00	4.08
25	4	2.02	6.67	1.45	6.85	5.52	0.00	0.68	0.03	23.22
26	6	3.27	17.65	2.68	2.33	11.92	0.83	0.28	4.03	42.99
27	11	4.95	12.73	3.53	5.45	4.68	0.00	0.67	19.89	51.90
28	3	1.00	9.83	1.49	2.04	4.76	0.42	0.54	0.12	20.20
29	4	1.55	16.39	2.31	4.98	7.85	2.04	0.20	1.48	36.80
30	3	1.41	6.33	0.92	3.54	2.90	3.85	0.03	0.04	19.02
31	2	1.05	9.19	0.89	1.06	2.87	0.33	0.01	0.03	15.43
32	2	1.31	9.41	0.80	2.70	5.09	0.64	0.05	0.71	20.71
33	2	2.50	9.45	0.72	1.14	3.17	0.00	0.02	0.01	17.01
34	4	2.35	12.32	2.66	12.24	2.40	7.87	4.73	0.78	45.35
35	6	3.69	20.12	4.37	5.67	13.20	0.00	1.15	1.17	49.37
36	2	3.61	7.72	0.92	2.43	2.07	0.68	0.63	0.00	18.06

(continued)

Data Set 16: Weights of Discarded Garbage for One Week (*continued*)

Household	HHSize	Metal	Paper	Plas	Glass	Food	Yard	Text	Other	Total
37	2	1.49	6.16	1.40	4.02	4.00	0.30	0.04	0.00	17.41
38	2	1.36	7.98	1.45	6.45	4.27	0.02	0.12	2.02	23.67
39	2	1.73	9.64	1.68	1.89	1.87	0.01	1.73	0.58	19.13
40	2	0.94	8.08	1.53	1.78	8.13	0.36	0.12	0.05	20.99
41	3	1.33	10.99	1.44	2.93	3.51	0.00	0.39	0.59	21.18
42	3	2.62	13.11	1.44	1.82	4.21	4.73	0.64	0.49	29.06
43	2	1.25	3.26	1.36	2.89	3.34	2.69	0.00	0.16	14.95
44	2	0.26	1.65	0.38	0.99	0.77	0.34	0.04	0.00	4.43
45	3	4.41	10.00	1.74	1.93	1.14	0.92	0.08	4.60	24.82
46	6	3.22	8.96	2.35	3.61	1.45	0.00	0.09	1.12	20.80
47	4	1.86	9.46	2.30	2.53	6.54	0.00	0.65	2.45	25.79
48	4	1.76	5.88	1.14	3.76	0.92	1.12	0.00	0.04	14.62
49	3	2.83	8.26	2.88	1.32	5.14	5.60	0.35	2.03	28.41
50	3	2.74	12.45	2.13	2.64	4.59	1.07	0.41	1.14	27.17
51	10	4.63	10.58	5.28	12.33	2.94	0.12	2.94	15.65	54.47
52	3	1.70	5.87	1.48	1.79	1.42	0.00	0.27	0.59	13.12
53	6	3.29	8.78	3.36	3.99	10.44	0.90	1.71	13.30	45.77
54	5	1.22	11.03	2.83	4.44	3.00	4.30	1.95	6.02	34.79
55	4	3.20	12.29	2.87	9.25	5.91	1.32	1.87	0.55	37.26
56	7	3.09	20.58	2.96	4.02	16.81	0.47	1.52	2.13	51.58
57	5	2.58	12.56	1.61	1.38	5.01	0.00	0.21	1.46	24.81
58	4	1.67	9.92	1.58	1.59	9.96	0.13	0.20	1.13	26.18
59	2	0.85	3.45	1.15	0.85	3.89	0.00	0.02	1.04	11.25
60	4	1.52	9.09	1.28	8.87	4.83	0.00	0.95	1.61	28.15
61	2	1.37	3.69	0.58	3.64	1.78	0.08	0.00	0.00	11.14
62	2	1.32	2.61	0.74	3.03	3.37	0.17	0.00	0.46	11.70

Data Set 17: Homerun Distances

Homerun distances are in feet for Mark McGwire (1998), Sammy Sosa (1998), and Barry Bonds (2001).

STATDISK:	**Data set name is Homeruns.**
Minitab:	**Worksheet name is HOMERUNS.MTW.**
Excel:	**Workbook name is HOMERUNS.XLS.**
TI-83/84 Plus:	**APP name is HOMERUNS, and the file names are the same as for text files.**
Text file names:	**MCGWR, SOSA, BONDS.**

McGwire

360	370	370	430	420	340	460	410	440	410
380	360	350	527	380	550	478	420	390	420
425	370	480	390	430	388	423	410	360	410
450	350	450	430	461	430	470	440	400	390
510	430	450	452	420	380	470	398	409	385
369	460	390	510	500	450	470	430	458	380
430	341	385	410	420	380	400	440	377	370

Sosa

371	350	430	420	430	434	370	420	440	410
420	460	400	430	410	370	370	410	380	340
350	420	410	415	430	380	380	366	500	380
390	400	364	430	450	440	365	420	350	420
400	380	380	400	370	420	360	368	430	433
388	440	414	482	364	370	400	405	433	390
480	480	434	344	410	420				

Bonds

420	417	440	410	390	417	420	410	380	430
370	420	400	360	410	420	391	416	440	410
415	436	430	410	400	390	420	410	420	410
410	450	320	430	380	375	375	347	380	429
320	360	375	370	440	400	405	430	350	396
410	380	430	415	380	375	400	435	420	420
488	361	394	410	411	365	360	440	435	454
442	404	385							

Data Set 18: Homes Sold in Dutchess County, New York

STATDISK: Data set name is Homes.
Minitab: Worksheet name is HOMES.MTW.
Excel: Workbook name is HOMES.XLS.
TI-83/84 Plus: APP name is HOMES, and the file names are the same as for text files.
Text file names: Text file names are HMSP, HMLST, HMLA, HMACR, HMAGE, HMTAX, HMRMS, HMBRS, HMBTH.

Selling Price (dollars)	List Price (dollars)	Living Area (sq. ft.)	Acres	Age (years)	Taxes (dollars)	Rooms	Bedrooms	Baths (full)
400000	414000	2704	2.27	27	4920	9	3	3
370000	379000	2096	0.75	21	4113	8	4	2
382500	389900	2737	1.00	36	6072	9	4	2
300000	299900	1800	0.43	34	4024	8	4	2
305000	319900	1066	3.60	69	3562	6	3	2
320000	319900	1820	1.70	34	4672	7	3	2
321000	328900	2700	0.81	35	3645	8	3	1
445000	450000	2316	2.00	19	6256	9	4	2
377500	385000	2448	1.50	40	5469	9	4	3
460000	479000	3040	1.09	20	6740	10	4	2
265000	275000	1500	1.60	39	4046	6	2	2
299000	299000	1448	0.42	44	3481	7	3	1
385000	379000	2400	0.89	33	4411	9	4	3
430000	435000	2200	4.79	6	5714	8	4	2
214900	219900	1635	0.25	49	2560	5	3	1
475000	485000	2224	11.58	21	7885	7	3	2
280000	289000	1738	0.46	49	3011	8	3	2
457000	499900	3432	1.84	14	9809	11	4	3
210000	224900	1175	0.94	64	1367	7	3	1
272500	274900	1393	1.39	44	2317	6	3	1
268000	275000	1196	0.83	44	3360	4	2	1
300000	319900	1860	0.57	32	4294	9	3	2
477000	479000	3867	1.10	19	9135	10	4	4
292000	294900	1800	0.52	47	3690	8	2	1
379000	383900	2722	1.00	29	6283	10	4	3
295000	299900	2240	0.90	144	3286	6	3	1
499000	499000	2174	5.98	62	3894	6	3	2
292000	299000	1650	2.93	52	3476	7	3	1
305000	299900	2000	0.33	36	4146	8	3	3
520000	529700	3350	1.53	6	8350	11	4	2
308000	320000	1776	0.63	42	4584	8	4	2
316000	310000	1850	2.00	25	4380	7	3	2
355500	362500	2600	0.44	46	4009	10	5	2
225000	229000	1300	0.62	49	3047	6	3	1
270000	290000	1352	0.68	24	2801	6	3	1
253000	259900	1312	0.68	44	4048	6	2	1
310000	314900	1664	1.69	53	2940	6	3	2
300000	309900	1700	0.83	33	4281	8	4	2
295000	295000	1650	2.90	34	4299	6	2	2
478000	479000	2400	2.14	6	6688	8	4	2

Appendix C: Glossary

Absolute deviation The measure of variation equal to the sum of the deviations of each value from the mean, divided by the number of values

Acceptance sampling Sampling items without replacement and rejecting the whole batch based on the number of defects obtained

Actual odds against The ratio $P(\overline{A})/P(A)$, usually expressed in the form of *a*:*b* (or "*a* to *b*")

Actual odds in favor The reciprocal of the actual odds against an event

Addition rule Rule for determining the probability that, on a single trial, either event *A* occurs, or event *B* occurs, or they both occur

Adjusted coefficient of determination Multiple coefficient of determination R^2 modified to account for the number of variables and sample size

Alpha (α) Symbol used to represent the probability of a type I error. *See also* Significance level.

Alternative hypothesis Statement that is equivalent to the negation of the null hypothesis; denoted by H_1

Analysis of variance Method of analyzing population variances in order to test hypotheses about means of populations

ANOVA *See* Analysis of variance.

Arithmetic mean Sum of a set of values divided by the number of values; usually referred to as the mean

Assignable variation Type of variation in a process that results from causes that can be identified

Attribute data Data that can be separated into different categories distinguished by some nonnumeric characteristic

Average Any one of several measures designed to reveal the center of a collection of data

Beta (β) Symbol used to represent the probability of a type II error

Bimodal Having two modes

Binomial experiment Experiment with a fixed number of independent trials, where each outcome falls into exactly one of two categories

Binomial probability formula Expression used to calculate probabilities in a binomial experiment (see Formula 5-5 in Section 5-3)

Bivariate data Data arranged as matched pairs

Bivariate normal distribution Distribution of paired data in which, for any fixed value of one variable, the values of the other variable are normally distributed

Blinding Procedure used in experiments whereby the subject doesn't know whether he or she is receiving a treatment or a placebo

Block A group of subjects that are similar in the ways that might affect the outcome of an experiment

Box-and-whisker diagram *See* Boxplot.

Boxplot Graphical representation of the spread of a set of data

Case-control study Study in which data are collected from the past by going back in time (through examination of records, interviews, and so on).

Categorical data Data that can be separated into different categories that are distinguished by some nonnumeric characteristic

Cell Category used to separate qualitative (or attribute) data

Census Collection of data from every element in a population

Centerline Line used in a control chart to represent a central value of the characteristic measurements

Central limit theorem Theorem stating that sample means tend to be normally distributed with mean μ and standard deviation $\sigma/\sqrt{n}$

Centroid The point $(\bar{x}, \bar{y})$ determined from a collection of bivariate data

Chebyshev's theorem Theorem that uses the standard deviation to provide information about the distribution of data

Chi-square distribution A continuous probability distribution (first introduced in Section 7-5)

Class boundaries Values obtained from a frequency distribution by increasing the upper class limits and decreasing the lower class limits by the same amount so that there are no gaps between consecutive classes

Classical approach to probability Approach in which the probability of an event is determined by dividing the number of ways the event can occur by the total number of possible outcomes

Classical method of testing hypotheses Method of testing hypotheses based on a comparison of the test statistic and critical values

Class midpoint In a class of a frequency distribution, the value midway between the lower class limit and the upper class limit

Class width The difference between two consecutive lower class limits in a frequency distribution

Cluster sampling Dividing the population area into sections (or clusters), then randomly selecting a few of those sections, and then choosing *all* the members from those selected sections

Coefficient of determination Amount of the variation in *y* that is explained by the regression line

Coefficient of variation (or CV) The ratio of the standard deviation to the mean, expressed as a percent

Cohort study Study of subjects in identified groups sharing common factors (called *cohorts*), with data collected in the future

Combinations rule Rule for determining the number of different combinations of selected items

Complement of an event All outcomes in which the original event does not occur

Completely randomized design Procedure in an experiment whereby each element is given the same chance of belonging to the different categories or treatments

Compound event Combination of simple events

Conditional probability The probability of an event, given that some other event has already occurred

Confidence coefficient Probability that a population parameter is contained within a particular confidence interval; also called confidence level or degree of confidence

Confidence interval Range of values used to estimate some population parameter with a specific confidence level; also called an interval estimate

Confidence interval limits Two numbers that are used as the high and low boundaries of a confidence interval

Confidence level Probability that a population parameter is contained within a particular confidence interval

Confounding A situation that occurs when the effects from two or more variables cannot be distinguished from each other

Contingency table Table of observed frequencies where the rows correspond to one variable of classification and the columns correspond to another variable of classification; also called a two-way table

Continuity correction Adjustment made when a discrete random variable is being approximated by a continuous random variable (Section 6-6)

Continuous data Data resulting from infinitely many possible values that correspond to some continuous scale that covers a range of values without gaps, interruptions, or jumps

Continuous random variable A random variable with infinite values that can be associated with points on a continuous line interval

Control chart Any one of several types of charts (Chapter 14) depicting some characteristic of a process in order to determine whether there is statistical stability

Control group A group of subjects in an experiment who are not given a particular treatment

Control limit Boundary used in a control chart for identifying unusual points

Convenience sampling Sampling in which data are selected because they are readily available

Correlation Statistical association between two variables

Correlation coefficient Measurement of the strength of the relationship between two variables

Critical region The set of all values of the test statistic that would cause rejection of the null hypothesis

Critical value Value separating the critical region from the values of the test statistic that would not lead to rejection of the null hypothesis

Cross-sectional study Study in which data are observed, measured, and collected at one point in time

Cumulative frequency Sum of the frequencies for a class and all preceding classes

Cumulative frequency distribution Frequency distribution in which each class and frequency represents cumulative data up to and including that class

Data Numbers or information describing some characteristic

Degree of confidence Probability that a population parameter is contained within a particular confidence interval; also called level of confidence

Degrees of freedom Number of values that are free to vary after certain restrictions have been imposed on all values

Denominator degrees of freedom Degrees of freedom corresponding to the denominator of the F test statistic

Density curve Graph of a continuous probability distribution

Dependent events Events for which the occurrence of any one event affects the probabilities of the occurrences of the other events

Dependent sample Sample whose values are related to the values in another sample

Dependent variable y variable in a regression or multiple regression equation

Descriptive statistics Methods used to summarize the key characteristics of known data

Deviation Amount of difference between a value and the mean; expressed as $x - \bar{x}$

Dichotomous variable Variable which has two possible discrete values

Discordant pairs Pairs of categories in which the two categories are different; used in McNemar's test

Discrete data Data with the property that the number of possible values is either a finite number or a "countable" number, which results in 0 possibilities, or 1 possibility, or 2 possibilities, and so on

Discrete random variable Random variable with either a finite number of values or a countable number of values

Disjoint events Events that cannot occur simultaneously

Distribution-free tests Tests not requiring a particular distribution, such as the normal distribution. *See also* Nonparametric tests.

Dotplot Graph in which each data value is plotted as a point (or dot) along a scale of values

Double-blind Procedure used in an experiment whereby the subject doesn't know whether he or she is receiving a treatment or placebo, and the person administering the treatment also does not know

Dummy variable A dichotomous variable with the two possible values of 0 and 1. Used in multiple regression

Efficiency Measure of the sensitivity of a nonparametric test in comparison to a corresponding parametric test

Empirical rule Rule that uses standard deviation to provide information about data with a bell-shaped distribution (Section 3-3)

Estimate Specific value or range of values used to approximate some population parameter

Estimator Sample statistic (such as the sample mean $\bar{x}$) used to approximate a population parameter

Event Result or outcome of an experiment

Expected frequency Theoretical frequency for a cell of a contingency table or multinomial table

Expected value For a discrete random variable, the mean value of the outcomes

Experiment Application of some treatment followed by observation of its effects on the subjects

Experimental units Subjects in an experiment

Explained deviation For one pair of values in a collection of bivariate data, the difference between the predicted y value and the mean of the y values

Explained variation Sum of the squares of the explained deviations for all pairs of bivariate data in a sample

Exploratory data analysis (EDA) Branch of statistics emphasizing the investigation of data

Factor In analysis of variance, a property or characteristic that allows us to distinguish the different populations from one another

Factorial rule Rule stating that n different items can be arranged $n!$ different ways

***F* distribution** Continuous probability distribution first introduced in Section 9-5

Finite population correction factor Factor for correcting the standard error of the mean when a sample size exceeds 5% of the size of a finite population

Five-number summary Minimum value, maximum value, median, and the first and third quartiles of a set of data

Fractiles Numbers that partition data into parts that are approximately equal in size

Frequency distribution Listing of data values (either individually or by groups of intervals), along with their corresponding frequencies (or counts)

Frequency polygon Graphical representation of the distribution of data using connected straight-line segments

Frequency table List of categories of values along with their corresponding frequencies

Fundamental counting rule Rule stating that, for a sequence of two events in which the first event can occur m ways and the second can occur n ways, the events together can occur a total of $m \cdot n$ ways

Goodness-of-fit test Test for how well some observed frequency distribution fits some theoretical distribution

Histogram Graph of vertical bars representing the frequency distribution of a set of data

***H* test** The nonparametric Kruskal-Wallis test

Hypothesis Statement or claim about some property of a population

Hypothesis test Method for testing claims made about populations; also called test of significance

Independent events Events for which the occurrence of any one of the events does not affect the probabilities of the occurrences of the other events

Independent sample Sample whose values are not related to the values in another sample

Independent variable The x variable in a regression equation, or one of the x variables in a multiple regression equation

Inferential statistics Methods involving the use of sample data to make generalizations or inferences about a population

Influential point Point that strongly affects the graph of a regression line

Interaction In two-way analysis of variance, the effect when one of the factors changes for different categories of the other factor

Interquartile range The difference between the first and third quartiles

Interval Level of measurement of data; characterizes data that can be arranged in order and for which differences between data values are meaningful

Interval estimate Range of values used to estimate some population parameter with a specific level of confidence; also called a confidence interval

Kruskal-Wallis test Nonparametric hypothesis test used to compare three or more independent samples; also called an H test

Least-squares property Property stating that, for a regression line, the sum of the squares of the vertical deviations of the sample points from the regression line is the smallest sum possible

Left-tailed test Hypothesis test in which the critical region is located in the extreme left area of the probability distribution

Level of confidence Probability that a population parameter is contained within a particular confidence interval; also called degree of confidence

Linear correlation coefficient Measure of the strength of the relationship between two variables

Logistic regression Method used in multiple regression when the dummy variable is the response (y) variable

Longitudinal study Study of subjects in identified groups sharing common factors (called *cohorts*), with data collected in the future

Lower class limits Smallest numbers that can actually belong to the different classes in a frequency distribution

Lower control limit Boundary used in a control chart to separate points that are unusually low

Lurking variable Variable that affects the variables being studied, but is not itself included in the study

Mann-Whitney *U* test Hypothesis test equivalent to the Wilcoxon rank-sum test for two independent samples

Marginal change For variables related by a regression equation, the amount of change in the dependent variable when one of the independent variables changes by one unit and the other independent variables remain constant

Margin of error Maximum likely (with probability $1 - \alpha$) difference between the observed sample statistic and the true value of the population parameter

Matched pairs With two samples, there is some relationship so that each value in one sample is paired with a corresponding value in the other sample

Mathematical model Mathematical function that "fits" or describes real-world data

Maximum error of estimate *See* Margin of error.

McNemar's test Uses frequency counts from matched pairs of nominal data from two categories to test the null hypothesis that the frequencies from discordant pairs occur in the same proportion

Mean The sum of a set of values divided by the number of values

Mean absolute deviation Measure of variation equal to the sum of the deviations of each value from the mean, divided by the number of values

Measure of center Value intended to indicate the center of the values in a collection of data

Measure of variation Any of several measures designed to reflect the amount of variation or spread for a set of values

Median Middle value of a set of values arranged in order of magnitude

Midquartile One-half of the sum of the first and third quartiles

Midrange One-half the sum of the highest and lowest values

Mode Value that occurs most frequently

MS(error) Mean square for error; used in analysis of variance

MS(total) Mean square for total variation; used in analysis of variance

MS(treatment) Mean square for treatments; used in analysis of variance

Multimodal Having more than two modes

Multinomial experiment Experiment with a fixed number of independent trials, where each outcome falls into exactly one of several categories

Multiple coefficient of determination Measure of how well a multiple regression equation fits the sample data

Multiple comparison procedures Procedures for identifying which particular means are different, after concluding that three or more means are not all equal

Multiple regression Study of linear relationships among three or more variables

Multiple regression equation Equation that expresses a linear relationship between a dependent variable y and two or more independent variables ($x_1, x_2, \ldots, x_k$)

Multiplication rule Rule for determining the probability that event A will occur on one trial and event B will occur on a second trial

Mutually exclusive events Events that cannot occur simultaneously

Negatively skewed Skewed to the left

Nominal Level of measurement of data; characterizes data that consist of names, labels, or categories only

Nonparametric tests Statistical procedures for testing hypotheses or estimating parameters, where there are no required assumptions about the nature or shape of population distributions; also called distribution-free tests

Nonsampling errors Errors from external factors not related to sampling

Normal distribution Bell-shaped probability distribution described algebraically by Formula 6-1 in Section 6-1

Normal quantile plot Graph of points (x, y), where each x value is from the original set of sample data, and each y value is a z score corresponding to a quantile value of the standard normal distribution

***np* chart** Control chart in which numbers of defects are plotted so that a process can be monitored

Null hypothesis Claim made about some population characteristic, usually involving the case of no difference; denoted by H_0

Numerator degrees of freedom Degrees of freedom corresponding to the numerator of the F test statistic

Numerical data Data consisting of numbers representing counts or measurements

Observational study Study in which we observe and measure specific characteristics, but don't attempt to manipulate or modify the subjects being studied

Observed frequency Actual frequency count recorded in one cell of a contingency table or multinomial table

Odds against Ratio of the probability of an event not occurring to the event occurring, usually expressed in the form of *a*:*b* where *a* and *b* are integers having no common factors

Odds in favor Ratio of the probability of an event occurring to the event not occurring, usually expressed as the ratio of two integers with no common factors

Ogive Graphical representation of a cumulative frequency distribution

One-way analysis of variance Analysis of variance involving data classified into groups according to a single criterion only

Ordinal Level of measurement of data; characterizes data that may be arranged in order, but differences between data values either cannot be determined or are meaningless

Outliers Values that are very unusual in the sense that they are very far away from most of the data

Paired samples Two samples that are dependent in the sense that the data values are matched by pairs

Parameter Measured characteristic of a population

Parametric tests Statistical procedures, based on population parameters, for testing hypotheses or estimating parameters

Pareto chart Bar graph for qualitative data, with the bars arranged in order according to frequencies

Payoff odds Ratio of net profit (if you win) to the amount bet

***p* chart** Control chart used to monitor the proportion *p* for some attribute in a process

Pearson's product moment correlation coefficient *See* Linear correlation coefficient.

Percentile The 99 values that divide ranked data into 100 groups with approximately 1% of the values in each group

Permutations rule Rule for determining the number of different arrangements of selected items

Pie chart Graphical representation of data in the form of a circle containing wedges

Placebo effect Effect that occurs when an untreated subject incorrectly believes that he or she is receiving a real treatment and reports an improvement in symptoms

Point estimate Single value that serves as an estimate of a population parameter

Poisson distribution Discrete probability distribution that applies to occurrences of some event over a specified interval of time, distance, area, volume, or some similar unit

Pooled estimate of p_1 and p_2 Probability obtained by combining the data from two sample proportions and dividing the total number of successes by the total number of observations

Pooled estimate of σ^2 Estimate of the variance σ^2 that is common to two populations, found by computing a weighted average of the two sample variances

Population Complete and entire collection of elements to be studied

Positively skewed Skewed to the right

Power of a test Probability $(1 - \beta)$ of rejecting a false null hypothesis

Predicted values Values of a dependent variable found by using values of independent variables in a regression equation

Prediction interval Confidence interval estimate of a predicted value of *y*

Predictor variables Independent variables in a regression equation

Probability Measure of the likelihood that a given event will occur; expressed as a number between 0 and 1

Probability distribution Collection of values of a random variable along with their corresponding probabilities

Probability histogram Histogram with outcomes listed along the horizontal axis and probabilities listed along the vertical axis

Probability value *See P*-value.

Process data Data, arranged according to some time sequence, that measure a characteristic of goods or services resulting from some combination of equipment, people, materials, methods, and conditions

Prospective study Study of subjects in identified groups sharing common factors (called *cohorts*), with data collected in the future

***P*-value** Probability that a test statistic in a hypothesis test is at least as extreme as the one actually obtained

Qualitative data Data that can be separated into different categories distinguished by some nonnumeric characteristic

Quantitative data Data consisting of numbers representing counts or measurements

Quartiles The three values that divide ranked data into four groups with approximately 25% of the values in each group

Randomized block design Design in which a measurement is obtained for each treatment on each of several individuals matched according to similar characteristics

Random sample Sample selected in a way that allows every member of the population to have the same chance of being chosen

Random selection Selection of sample elements in such a way that all elements available for selection have the same chance of being selected

Random variable Variable (typically represented by x) that has a single numerical value (determined by chance) for each outcome of an experiment

Random variation Type of variation in a process that is due to chance; the type of variation inherent in any process not capable of producing every good or service exactly the same way every time

Range The measure of variation that is the difference between the highest and lowest values

Range chart Control chart based on sample ranges; used to monitor variation in a process

Range rule of thumb Rule based on the principle that for typical data sets, the difference between the lowest typical value and the highest typical value is approximately 4 standard deviations ($4s$)

Rank Numerical position of an item in a sample set arranged in order

Rank correlation coefficient Measure of the strength of the relationship between two variables; based on the ranks of the values

Rare event rule If, under a given assumption, the probability of a particular observed result is extremely small, we conclude that the assumption is probably not correct

Ratio Level of measurement of data; characterizes data that can be arranged in order, for which differences between data values are meaningful, and there is an inherent zero starting point

***R* chart** Control chart based on sample ranges; used to monitor variation in a process

Regression equation Algebraic equation describing the relationship among variables

Regression line Straight line that best fits a collection of points representing paired sample data

Relative frequency Frequency for a class, divided by the total of all frequencies

Relative frequency approximation of probability Estimated value of probability based on actual observations

Relative frequency distribution Variation of the basic frequency distribution in which the frequency for each class is divided by the total of all frequencies

Relative frequency histogram Variation of the basic histogram in which frequencies are replaced by relative frequencies

Replication Repetition of an experiment

Residual Difference between an observed sample y value and the value of y that is predicted from a regression equation

Response variable y variable in a regression or multiple regression equation

Retrospective study Study in which data are collected from the past by going back in time (through examination of records, interviews, and so on)

Right-tailed test Hypothesis test in which the critical region is located in the extreme right area of the probability distribution

Rigorously controlled design Design of experiment in which all factors are forced to be constant so that effects of extraneous factors are eliminated

Run Sequence of data exhibiting the same characteristic; used in runs test for randomness

Run chart Sequential plot of individual data values over time, where one axis (usually the vertical axis) is used for the data values and the other axis (usually the horizontal axis) is used for the time sequence

Runs test Nonparametric method used to test for randomness

Sample Subset of a population

Sample size Number of items in a sample

Sample space Set of all possible outcomes or events in an experiment that cannot be further broken down

Sampling distribution of proportion The probability distribution of sample proportions, with all samples having the same sample size n

Sampling distribution of sample means Distribution of the sample means that is obtained when we repeatedly draw samples of the same size from the same population

Sampling error Difference between a sample result and the true population result; results from chance sample fluctuations

Sampling variability Variation of a statistic in different samples

Scatter diagram Graphical display of paired (x, y) data

Scatterplot Graphical display of paired (x, y) data

***s* chart** Control chart, based on sample standard deviations, that is used to monitor variation in a process

Self-selected sample Sample in which the respondents themselves decide whether to be included; also called voluntary response sample

Semi-interquartile range One-half of the difference between the first and third quartiles

Significance level Probability of making a type I error when conducting a hypothesis test

Sign test Nonparametric hypothesis test used to compare samples from two populations

Simple event Experimental outcome that cannot be further broken down

Simple random sample Sample of a particular size selected so that every possible sample of the same size has the same chance of being chosen

Simulation Process that behaves in a way that is similar to some experiment so that similar results are produced

Single factor analysis of variance *See* One-way analysis of variance.

Skewed Not symmetric and extending more to one side than the other

Slope Measure of steepness of a straight line

Sorted data Data arranged in order

Spearman's rank correlation coefficient *See* Rank correlation coefficient.

SS(error) Sum of squares representing the variability that is assumed to be common to all the populations being considered; used in analysis of variance

SS(total) Measure of the total variation (around $\bar{\bar{x}}$) in all of the sample data combined; used in analysis of variance

SS(treatment) Measure of the variation between the sample means; used in analysis of variance

Standard deviation Measure of variation equal to the square root of the variance

Standard error of estimate Measure of spread of sample points about the regression line

Standard error of the mean Standard deviation of all possible sample means $\bar{x}$

Standard normal distribution Normal distribution with a mean of 0 and a standard deviation equal to 1

Standard score Number of standard deviations that a given value is above or below the mean; also called z score

Statistic Measured characteristic of a sample

Statistically stable process Process with only natural variation and no patterns, cycles, or unusual points

Statistical process control (SPC) Use of statistical techniques such as control charts to analyze a process or its outputs so as to take appropriate actions to achieve and maintain a state of statistical control and to improve the process capability

Statistics Collection of methods for planning experiments, obtaining data, organizing, summarizing, presenting, analyzing, interpreting, and drawing conclusions based on data

Stem-and-leaf plot Method of sorting and arranging data to reveal the distribution

Stepwise regression Process of using different combinations of variables until the best model is obtained; used in multiple regression

Stratified sampling Sampling in which samples are drawn from each stratum (class)

Student *t* distribution *See t* distribution.

Subjective probability Guess or estimate of a probability based on knowledge of relevant circumstances

Symmetric Property of data for which the distribution can be divided into two halves that are approximately mirror images by drawing a vertical line through the middle

Systematic sampling Sampling in which every kth element is selected

***t* distribution** Bell-shaped distribution usually associated with sample data from a population with an unknown standard deviation.

10–90 percentile range Difference between the 10th and 90th percentiles

Test of homogeneity Test of the claim that different populations have the same proportion of some characteristic

Test of independence Test of the null hypothesis that for a contingency table, the row variable and column variable are not related

Test of significance *See* Hypothesis test.

Test statistic Sample statistic based on the sample data; used in making the decision about rejection of the null hypothesis

Time-series data Data that have been collected at different points in time

Total deviation Sum of the explained deviation and unexplained deviation for a given pair of values in a collection of bivariate data

Total variation Sum of the squares of the total deviation for all pairs of bivariate data in a sample

Traditional method of testing hypotheses Method of testing hypotheses based on a comparison of the test statistic and critical values

Treatment Property or characteristic that allows us to distinguish the different populations from one another; used in analysis of variance

Treatment group Group of subjects given some treatment in an experiment

Tree diagram Graphical depiction of the different possible outcomes in a compound event

Two-tailed test Hypothesis test in which the critical region is divided between the left and right extreme areas of the probability distribution

Two-way analysis of variance Analysis of variance involving data classified according to two different factors

Two-way table *See* Contingency table.

Type I error Mistake of rejecting the null hypothesis when it is true

Type II error Mistake of failing to reject the null hypothesis when it is false

Unbiased estimator Sample statistic that tends to target the population parameter that it is used to estimate

Unexplained deviation For one pair of values in a collection of bivariate data, the difference between the y coordinate and the predicted value

Unexplained variation Sum of the squares of the unexplained deviations for all pairs of bivariate data in a sample

Uniform distribution Probability distribution in which every value of the random variable is equally likely

Upper class limits Largest numbers that can belong to the different classes in a frequency distribution

Upper control limit Boundary used in a control chart to separate points that are unusually high

Variance Measure of variation equal to the square of the standard deviation

Variance between samples In analysis of variance, the variation among the different samples

Variation due to error *See* Variation within samples.

Variation due to treatment *See* Variance between samples.

Variation within samples In analysis of variance, the variation that is due to chance

Voluntary response sample Sample in which the respondents themselves decide whether to be included

Weighted mean Mean of a collection of values that have been assigned different degrees of importance

Wilcoxon rank-sum test Nonparametric hypothesis test used to compare two independent samples

Wilcoxon signed-ranks test Nonparametric hypothesis test used to compare two dependent samples

Within statistical control *See* Statistically stable process.

$\bar{x}$ chart Control chart used to monitor the mean of a process

y-intercept Point at which a straight line crosses the y-axis

z score Number of standard deviations that a given value is above or below the mean

Appendix D: Bibliography

***An asterisk denotes a book recommended for reading. Other books are recommended as reference texts.**

Bennett, D. 1998. *Randomness.* Cambridge: Harvard University Press.

*Best, J. 2001. *Damned Lies and Statistics.* Berkeley: University of California Press.

*Campbell, S. 2004. *Flaws and Fallacies in Statistical Thinking.* Mineola, N.Y.: Dover Publications.

*Crossen, C. 1994. *Tainted Truth: The Manipulation of Fact in America.* New York: Simon & Schuster.

*Freedman, D., R. Pisani, R. Purves, and A. Adhikari. 1997. *Statistics.* 3rd ed. New York: Norton.

*Gonick, L., and W. Smith. 1993. *The Cartoon Guide to Statistics.* New York: HarperCollins.

Halsey, J., and E. Reda. 2006. *Excel Student Laboratory Manual and Workbook.* Boston: Addison-Wesley.

*Heyde, C., and E. Seneta, eds. 2001. *Statisticians of the Centuries.* New York: Springer-Verlag.

*Hollander, M., and F. Proschan. 1984. *The Statistical Exorcist: Dispelling Statistics Anxiety.* New York: Marcel Dekker.

*Holmes, C. 1990. *The Honest Truth About Lying with Statistics.* Springfield, Ill.: Charles C. Thomas.

*Hooke, R. 1983. *How to Tell the Liars from the Statisticians.* New York: Marcel Dekker.

*Huff, D. 1993. *How to Lie with Statistics.* New York: Norton.

Humphrey, P. 2006. *Graphing Calculator Manual for the TI-83 Plus, TI-84 Plus, and the TI-89.* Boston: Addison-Wesley.

*Jaffe, A., and H. Spirer. 1987. *Misused Statistics.* New York: Marcel Dekker.

*Kimble, G. 1978. *How to Use (and Misuse) Statistics.* Englewood Cliffs, N.J.: Prentice-Hall.

Kotz, S., and D. Stroup. 1983. *Educated Guessing—How to Cope in an Uncertain World.* New York: Marcel Dekker.

*Loyer, M. 2006. *Student Solutions Manual to Accompany Elementary Statistics.* 10th ed. Boston: Addison-Wesley.

*Moore, D. 2001. *Statistics: Concepts and Controversies.* 5th ed. San Francisco: Freeman.

Morgan, J. 2006. *SAS Student Laboratory Manual and Workbook.* 3rd ed. Boston: Addison-Wesley.

*Paulos, J. 2001. *Innumeracy: Mathematical Illiteracy and Its Consequences.* New York: Hill and Wang.

Peck, R. 2006. *SPSS Student Laboratory Manual and Workbook.* Boston: Addison-Wesley.

*Reichard, R. 1974. *The Figure Finaglers.* New York: McGraw-Hill.

*Reichmann, W. 1962. *Use and Abuse of Statistics.* New York: Oxford University Press.

*Rossman, A. 1996. *Workshop Statistics: Discovery with Data.* New York: Springer.

*Salsburg, D. 2000. *The Lady Tasting Tea: How Statistics Revolutionized the Twentieth Century.* New York: W. H. Freeman.

Sheskin, D. 1997. *Handbook of Parametric and Nonparametric Statistical Procedures.* Boca Raton, Fla.: CRC Press.

Simon, J. 1992. *Resampling: The New Statistics.* Belmont, Calif.: Duxbury Press.

*Stigler, S. 1986. *The History of Statistics.* Cambridge, Mass.: Harvard University Press.

*Tanur, J., ed. 1989. *Statistics: A Guide to the Unknown.* 3rd ed. Belmont, Calif.: Wadsworth.

Triola, M. 2006. *Elementary Statistics.* 10th ed. Boston: Addison-Wesley.

Triola, M. 2006. *Minitab Student Laboratory Manual and Workbook.* 10th ed. Boston: Addison-Wesley.

Triola, M. 2006. *STATDISK 10.0 Student Laboratory Manual and Workbook.* 10th ed. Boston: Addison-Wesley.

Triola, M., and L. Franklin. 1994. *Business Statistics.* Boston: Addison-Wesley.

Triola, M., and M. Triola. 2006. *Biostatistics for the Biological and Health Sciences.* Boston: Addison-Wesley.

*Tufte, E. 2001. *The Visual Display of Quantitative Information.* 2nd ed. Cheshire, Conn.: Graphics Press.

Tukey, J. 1977. *Exploratory Data Analysis.* Boston: Addison-Wesley.

Zwillinger, D., and S. Kokoska. 2000. *CRC Standard Probability and Statistics Tables and Formulae.* Boca Raton, Fla.: CRC Press.

Appendix E: Answers to Odd-Numbered Exercises (and ALL of these exercises: Statistical Literacy and Critical Thinking Exercises located at the end of the chapter, Review Exercises, and Cumulative Review Exercises)

Chapter 1 Answers

Section 1-2

1. A parameter is a numerical measurement describing some characteristic of a population, whereas a statistic is a numerical measurement describing some characteristic of a sample.
3. Discrete data result when the number of possible values is either a finite number or a "countable" number (where the number of possible values is 0 or 1 or 2, and so on), but continuous data result from infinitely many possible values that correspond to some continuous scale that covers a range of values without gaps, interruptions, or jumps.
5. Statistic
7. Parameter
9. Discrete
11. Continuous
13. Nominal
15. Nominal
17. Interval
19. Ordinal
21. Sample: The 25 Senators selected. Population: The 100 Senators currently serving. The sample is likely to be representative.
23. Sample: The 1059 selected adults. Population: All adults. The sample is likely to be representative.
25. With no natural starting point, temperatures are at the interval level of measurement, so ratios such as "twice" are meaningless.
27. Either ordinal or interval are reasonable answers, but ordinal makes more sense because differences between values are not likely to be meaningful. For example, the difference between a food rated 1 and a food rated 2 is not likely to be the same as the difference between a food rated 9 and a food rated 10.

Section 1-3

1. A voluntary response sample (or self-selected sample) is one in which the respondents themselves decide whether to be included. It is unsuitable because people with special interests are more likely to respond, so that the sample is likely to be biased.
3. No. Because the response rate is so low, and the sample is a voluntary response sample, it is very possibly biased.
5. People who are taller have an advantage in being closer to the basket, so taller people are more likely to perform better and are more likely to play.
7. Perhaps police officers are more inclined to stop minorities than whites, so minorities are given more tickets.
9. Because the study was funded by a candy company and the Chocolate Manufacturers Association, there is a real possibility that researchers were somehow encouraged to obtain results favorable to the consumption of chocolate.
11. No, she used a voluntary response sample.
13. People with unlisted numbers are not included, and people without telephones are not included, so the sample might be biased.
15. Motorcyclists who were killed
17. Not necessarily, because population sizes should be taken into account. It would be much better to obtain a nationwide random sample of wage earners and calculate the average of their incomes.
19. Households with no children are missing. The results will not be representative.
21. a. 15%
    b. 0.567
    c. 170
    d. 78.9%
23. a. 540
    b. 5%
25. 62% of 8% of 1875 is only 93.
27. All percentages of success should be multiples of 5. The given percentages cannot be correct.

Section 1-4

1. With a random sample, each individual has the same chance of being selected. With a simple random sample, all samples of the same size have the same chance of being selected.
3. Blinding is a method whereby a subject (or researcher) in an experiment doesn't know whether the subject is given a treatment or a placebo. It is important to use blinding so that results are not distorted because of a placebo effect, whereby subjects think that they experience improvements just because they were treated.
5. Observational study
7. Observational study
9. Prospective
11. Cross-sectional
13. Systematic
15. Stratified
17. Cluster
19. Convenience
21. Random
23. Cluster
25. Yes, no. Individual returns have the same chance of being selected. Some samples are not possible, such as those that include returns from different days.
27. No, no. The 2nd M&M has no chance of being selected. Samples including the 2nd M&M have no chance of being included.
29. Yes, yes. Each student has the same chance, and every sample of size six has the same chance of being chosen.
31. Answers vary.

33. Prospective: The experiment was begun and results were followed forward in time. Randomized: Subjects were assigned to the different groups through a process of random selection, whereby they had the same chance of belonging to each group. Double-blind: The subjects did not know which of the three groups they were in, and the people who evaluated results did not know either. Placebo-controlled: There was a group of subjects who were given a placebo; by comparing the placebo group to the two treatment groups, the effects of the treatments might be better understood.

Section 1-5

1. An APP is a software application that can be downloaded to the calculator.
3. The key (−) is used for the operation of subtraction, but the key ((−)) is used for the negative sign.
5. −20 (The parentheses force the calculator to do the subtraction before the multiplication.)
7. 7 (The value of −2 is subtracted from 5. Use ((−)) for the negative sign and use (−) for subtraction.)
9. −4.5 (Use ((−)) for the negative sign.)
11. 15.29705854 (Use 2ND x^2 for the square root. Enter 2ND x^2 234.)
13. 3 (The result is the square root of 9.)
15. 25 (The result is the square of 5.)
17. {0 223 0 176 0 ... (The result is a display of the original list of values.)
19. {0 446 0 352 0 ... (The result is a display in which each of the original values is added to itself, so each entry is twice the value in the original list.)
21. 35,184,372,090,000
23. 0.0009765625
25. a. A list of the applications, including an entry of TriolaXe.
    b. The top line consists of the menu items of Elem Stats and Quit, and the rows list the file names of MHEALTH, FHEALTH, and so on.
    c. The top line consists of the menu items of Sel and All and Load, and the rows list the names of TAR, NICOT, and CO.
    d. The top line consists of the menu items of NAMES and OPS and MATH. The rows consist of L1 through L6 followed by list names that include TAR, NICOT, and CO.
    e. {16 16 16 9 1 8 ... (The list of TAR values in Data Set 5)

Chapter 1 Statistical Literacy and Critical Thinking

1. No. If a sample is collected in an inappropriate way, such as using a voluntary response sample, it has a good chance of being biased and is a bad sample, even if it is quite large.
2. a. Quantitative
   b. Continuous
   c. The runners who finished that particular marathon.
3. No, because it is a voluntary response sample. It is likely to be biased.
4. No. People are asked the values of their cars and they are likely to exaggerate. Also, the results should be weighted to account for the different numbers of car owners in the different states.

Chapter 1 Review Exercises

1. No, because it is a voluntary response sample, it might not be representative of the population.
2. Answer varies.
3. a. Ratio
   b. Ordinal
   c. Nominal
   d. Interval
4. a. Discrete
   b. Ratio
   c. Stratified
   d. Statistic
   e. The largest values because they represent stockholders that could potentially gain control of the company.
   f. The voluntary response sample is likely to be biased.
5. a. Systematic; representative
   b. Convenience; not representative
   c. Cluster; not representative
   d. Random; representative
   e. Stratified; not representative
6. a. Design the experiment so that the subjects don't know whether they are using Sleepeze or a placebo, and also design it so that those who observe and evaluate the subjects do not know which subjects are using Sleepeze and which are using a placebo.
   b. Blinding will help to distinguish between the effectiveness of Sleepeze and the placebo effect, whereby subjects and evaluators tend to believe that improvements are occurring just because some treatment is given.
   c. Subjects are put into different groups through a process of *random selection.*
   d. Subjects are very *carefully chosen* for the different groups so that those groups are made to be similar in the ways that are important.
   e. Replication is used when the experiment is repeated. It is important to have a sample of subjects that is large enough so that we can see the true nature of any effects. It is important so that we are not misled by erratic behavior of samples that are too small.
7. a. Parameter
   b. Discrete
   c. 743,005
8. a. If they have no fat at all, they have 100% less than any other amount with fat, so the 125% figure cannot be correct.
   b. 45
   c. 16.1%

Chapter 1 Cumulative Review Exercises

1. 3.02754 g
2. −0.64516129
3. −6.6423420
4. 266.77778
5. 0.55555556
6. 7
7. 2.6457513
8. 0.89735239
9. 0.0009765625
10. 1,099,511,627,776 (Most calculators won't show the last few digits, so a result such as 1,099, 511, 600,000 is acceptable.)
11. 13,841,287,201 (Most calculators won't show the last few digits, so a result such as 13,841,287,000 is acceptable.)
12. 0.000014272477

Chapter 2 Answers

Section 2-2

1. A frequency distribution lists data values (either individually or by groups of values) along with their corresponding frequency counts. It is useful for organizing and summarizing data.
3. For a value such as 10, it can belong to either of two classes, but every value should belong to one class only. Overlapping of class limits should be avoided.
5. Class width: 5. Class midpoints: 37, 42, 47, 52, 57, 62, 67. Class boundaries: 34.5, 39.5, 44.5, 49.5, 54.5, 59.5, 64.5, 69.5.
7. Class width: 5.0. Class midpoints: 62.45, 67.45, 72.45, 77.45, 82.45, 87.45, 92.45, 97.45, 102.45, 107.45. Class boundaries: 59.95, 64.95, 69.95, 74.95, 79.95, 84.95, 89.95, 94.95, 99.95, 104.95, 109.95.
9. Yes
11. The tallest man is between 105.0 in. and 109.9 in., which is more than 8 ft tall. That value is probably wrong. After deleting that error, the distribution appears to be approximately normal.

13.

Daily Low Temperature (°F)	Relative Frequency
35–39	3%
40–44	9%
45–49	14%
50–54	31%
55–59	20%
60–64	20%
65–69	3%

15.

Daily Low Temperature (°F)	Cumulative Frequency
Less than 40	1
Less than 45	4
Less than 50	9
Less than 55	20
Less than 60	27
Less than 65	34
Less than 70	35

17. Because there are disproportionately more 0s and 5s, it appears that the heights were reported instead of measured. Consequently, it is likely that the results are not very accurate.

x	Frequency
0	9
1	2
2	1
3	3
4	1
5	15
6	2
7	0
8	3
9	1

19. The distribution does not appear to be normal. Most of the days have no rainfall. The distribution is not symmetric and there are very few days with high rainfall amounts.

Rainfall	Frequency
0.00–0.19	44
0.20–0.39	6
0.40–0.59	1
0.60–0.79	0
0.80–0.99	0
1.00–1.19	0
1.20–1.39	1

21. The distribution appears to be roughly normal.

BMI	Frequency
15.0–20.9	10
21.0–26.9	15
27.0–32.9	11
33.0–38.9	2
39.0–44.9	2

23. The distribution appears to be roughly normal.

Weight	Frequency
2.9500–2.9999	2
3.0000–3.0499	3
3.0500–3.0999	22
3.1000–3.1499	7
3.1500–3.1999	1

25. Answers vary depending on the choices for class width and starting point. The relative frequency distributions are not dramatically different, except for the outlier of 504 lb which is in the list of axial loads for the cans 0.0111 in. thick.
27. 46–90, 91–181, 182–362, 363–724, 725–1448, 1449–2896

Section 2-3

1. Distribution of the data
3. The true nature of the distribution cannot be seen.
5. 18

7. In comparison to the rowers, the coxswains tend to be very light. The two crew members with the low weights are most likely coxswains.
9. The digits of 0 and 5 occur disproportionately more often than the others, so it appears that the heights were reported, not measured.

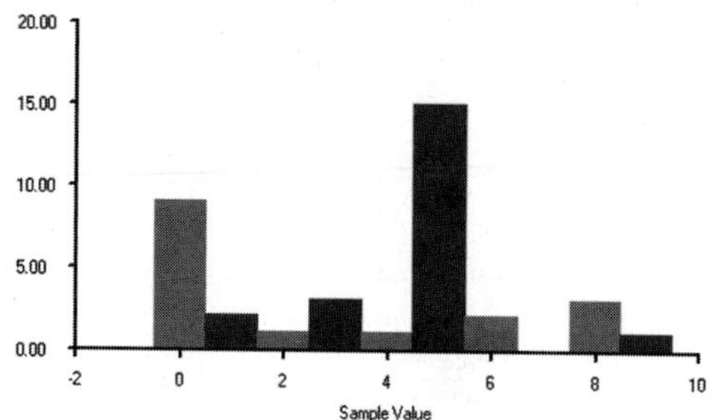

11. The data do not appear to be normally distributed.

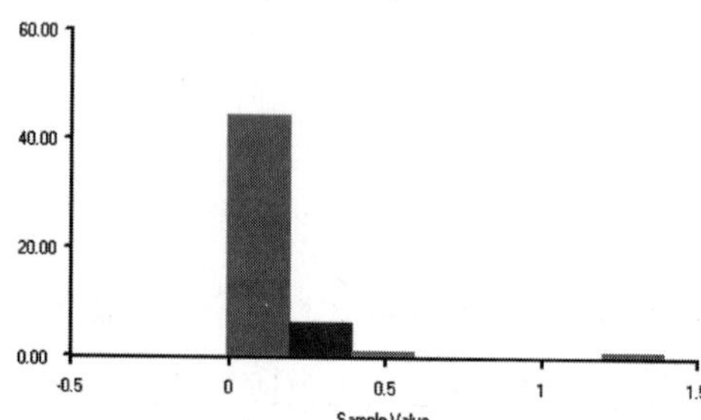

13. The data appear to have a distribution that is roughly normal.

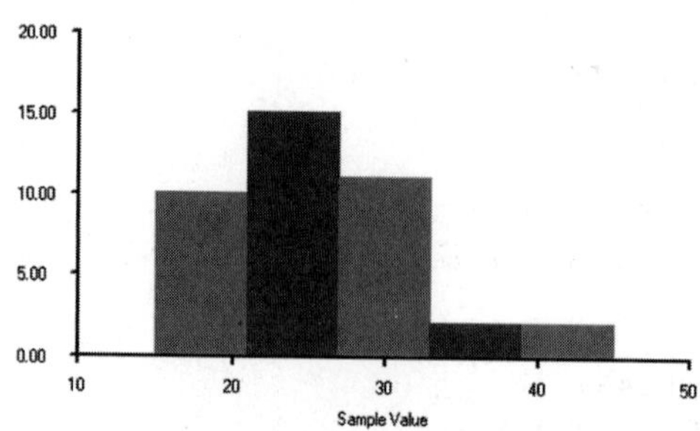

15. The data appear to have a distribution that is approximately normal.

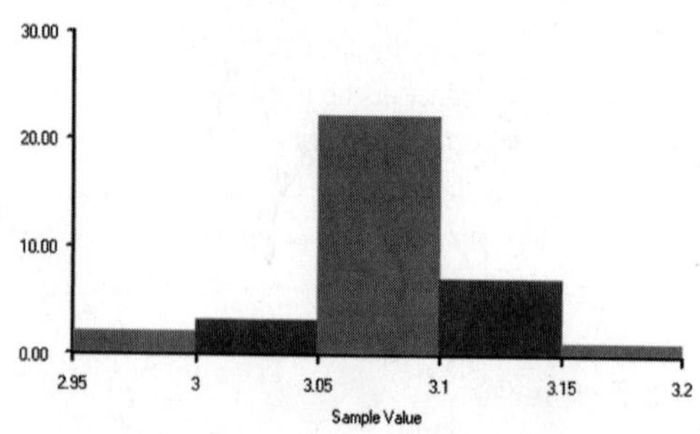

17. The distribution shapes are not dramatically different, but the ages of actresses appear to be centered around 36 years, whereas the ages of actors appear to be centered around 43 years. Actresses appear to win Oscars when they are younger than actors.

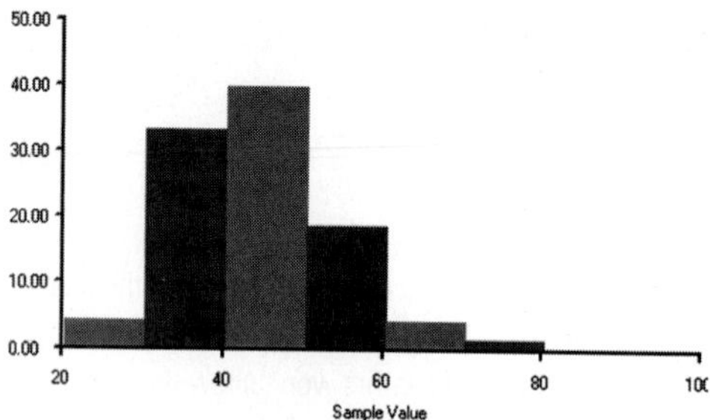

19. The back-to-back relative frequency histograms can vary somewhat, depending on the choices made for class boundaries. The graphs should show that the cans with a thickness of 0.0109 in. have axial loads that are centered around 270 lb, whereas the cans with a thickness of 0.0111 in. have axial loads that are centered around 280 lb. There is a notable difference.

Section 2-4

1. Data should be graphed so that we can see characteristics that cannot be seen by simply examining the list of data values. Graphs reveal such features as the nature of the distribution and the presence of outliers.
3. The data must be ordered in some time sequence, and the time-series graph helps to reveal trends or patterns over time.
5. The actual high temperatures range from 58 degrees to 85 degrees with most readings in the 70–80 degree range.

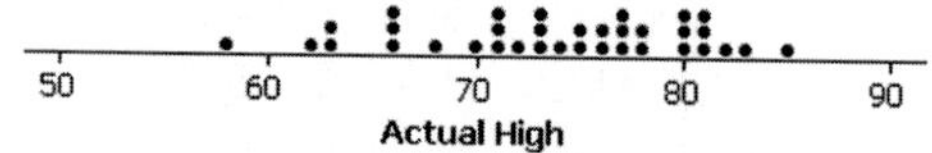

7.

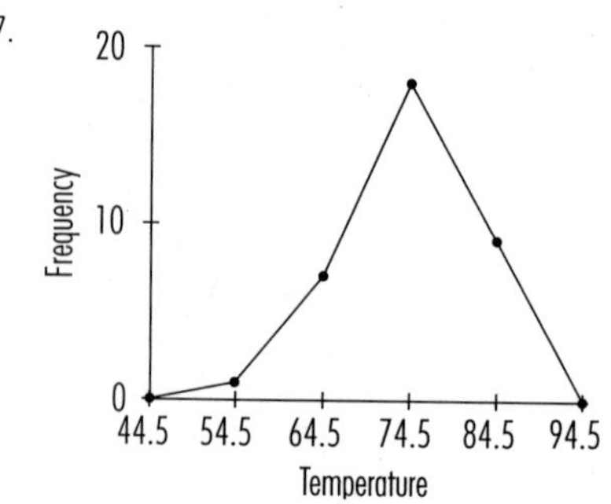

9. The distribution appears to be roughly bell-shaped.

Stem	Leaves
9	55
10	
11	00055
12	000000005555
13	0000000066668
14	000088
15	00

11. Eruptions below 120 ft: 7.

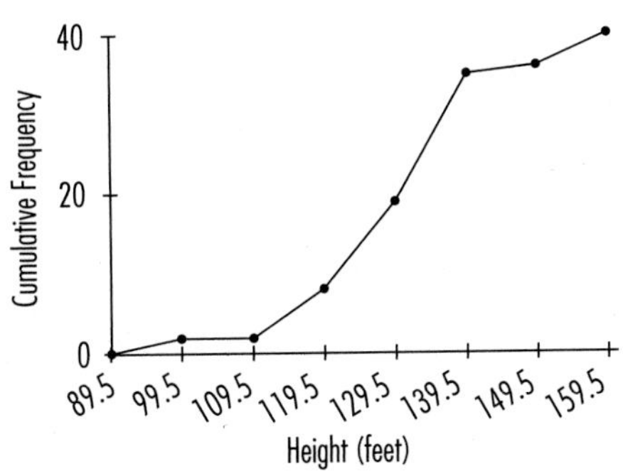

13. Networking appears to be the most effective approach in getting a job.

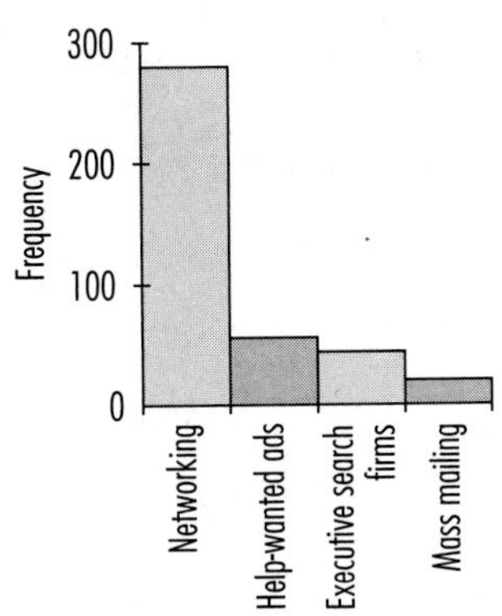

15.

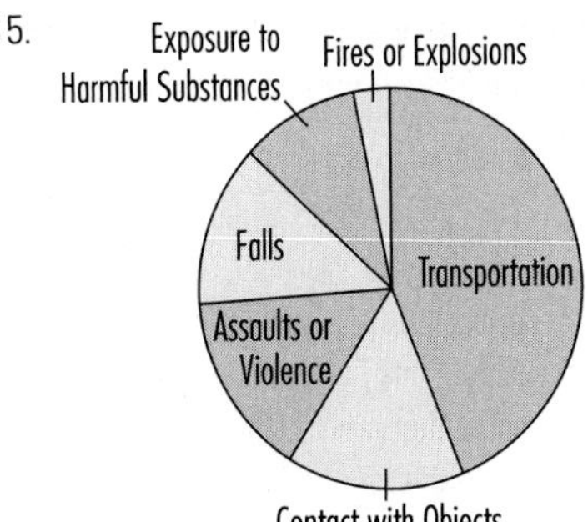

17. There appears to be a relationship whereby larger amounts of tar are associated with larger amounts of carbon monoxide.

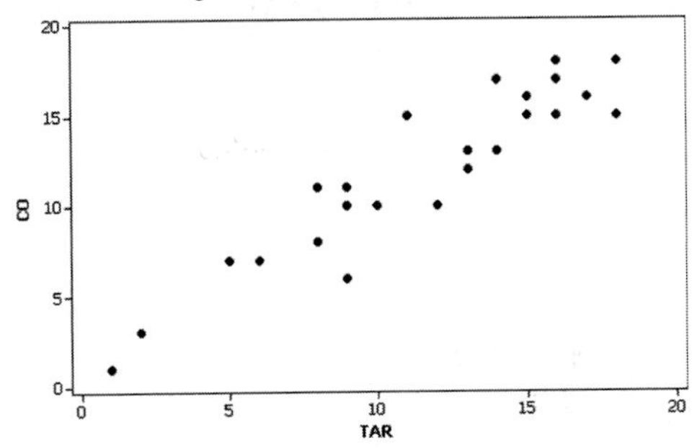

19. There appears to be a recent trend of increasing numbers of runway near-hits.

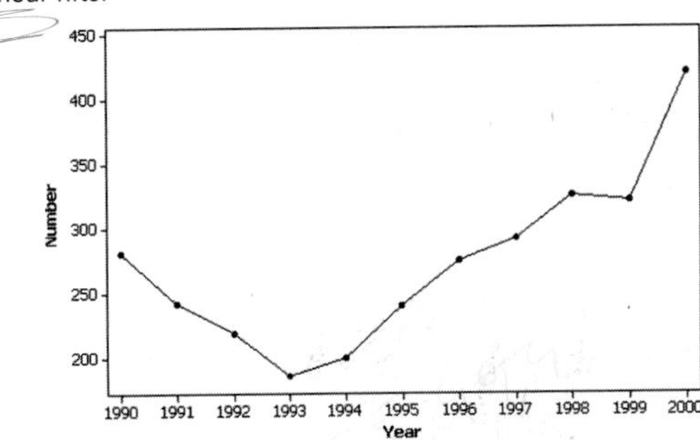

21. 10,000; 2.4%
23. 100,000; 500 miles
25. The ages of actresses are notably lower than those of actors.

Actresses' Ages (units)	Stem (tens)	Actors' Ages (units)
99999988887777666655554421	2	9
9988887765555555444333332211100	3	0012224445566677778888999
965322111110	4	00011111122222233334445556777 8999
40	5	112222346677
310	6	00222
4	7	6
0	8	

Chapter 2 Statistical Literacy and Critical Thinking

1. The histogram is generally more effective because it provides a picture, and the visual display is much easier to understand than the table of numbers.
2. If the two data sets have substantially different numbers of values, comparison of the frequency distributions is made more difficult because we must compare numbers that are very different in magnitude. Because the relative frequency distribution uses percents, the numbers are comparable. Therefore, relative frequency distributions are better.
3. A time-series graph would be better, because the selling prices have been changing dramatically over time. The histogram would hide the time factor.
4. The bars of the histogram start relatively low, then they increase to some maximum height, then they decrease. The histogram is symmetric with the left half being roughly a mirror image of the right half.

Chapter 2 Review Exercises

1. The ages of actors are greater than the ages of actresses.

Age of Actors	Frequency
21–30	3
31–40	25
41–50	30
51–60	14
61–70	3
71–80	1

2. The histogram shows that actors are older than actresses.

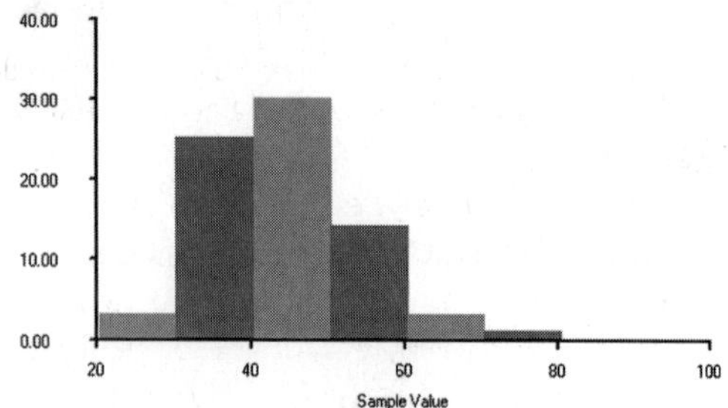

3. The dotplot shows that actors are older than actresses when they win Oscars.

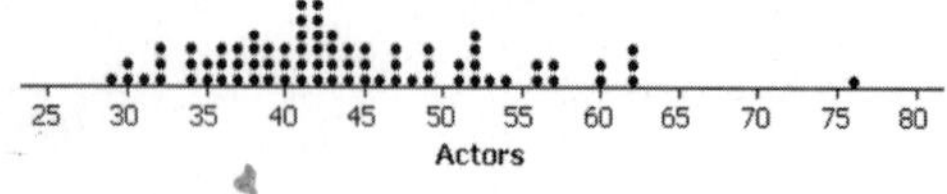

4. The stemplot shows that actors are older than actresses when they win Oscars.

```
2 | 9
3 | 001222444556667778888999
4 | 000111111222222333344455567778999
5 | 112222346677
6 | 00222
7 | 6
```

5. There does not appear to be an association between the ages of actresses and actors when they won Oscars.

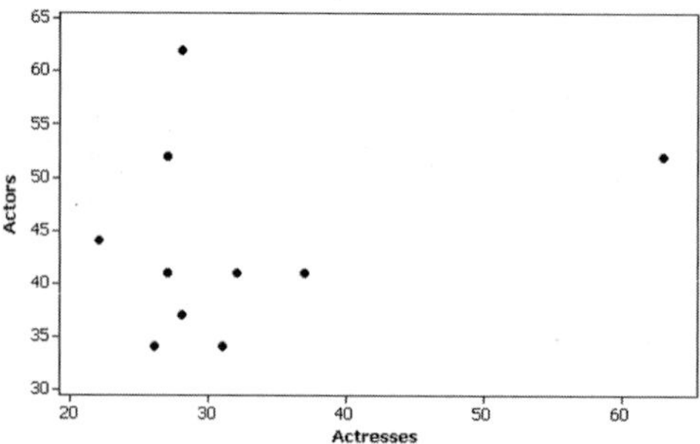

6. The graph does not reveal a trend. The ages are not systematically changing over time.

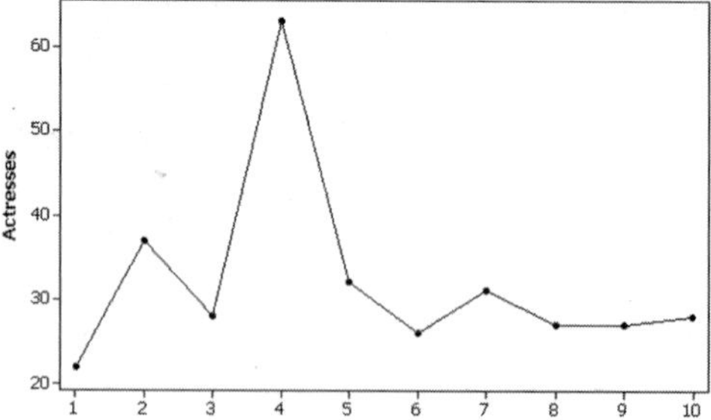

Chapter 2 Cumulative Review Exercises

1. No, the numbers are used to identify the slot in which the ball lands. Letters or some other symbols could have been used instead.
2. The results are at the nominal level of measurement. (The frequency counts are at the ratio level of measurement.)
3. The first seven classes represent five slots each, whereas the last class represents three slots. The expected frequency for the last class should therefore be about 3/5 of the typical frequency for each of the first seven classes. The frequency of 25 is not dramatically different from the frequency expected by chance.
4. Yes. The average should be around 18, so an average of five suggests that the wheel is not behaving in a random way. By betting on lower numbers, the gambler could increase the likelihood of winning.
5. No. The voluntary response survey is likely to be biased, so the results will probably not reflect the population of all car owners.

Chapter 3 Answers

Section 3-2

1. They use different approaches for providing values of the center or middle of a data set.
3. No. Because the numbers don't measure or count anything, the mean would be a meaningless statistic.
5. $\bar{x}$ = 58.3 sec; median = 55.5 sec; mode = 49 sec; midrange = 62.0 sec. The sample of students is not a representative sample from the population.
7. $\bar{x}$ = 1.9; median = 2.0; mode = 1; midrange = 2.5. The mode of 1 correctly indicates that the smooth-yellow peas occur more than any other phenotype, but the other measures of center don't make sense with these data at the nominal level of measurement.
9. $\bar{x}$ = 133.9; median = 132.5; mode = 130; midrange = 135. Given that the same person was measured, the values appear to vary considerably.
11. $\bar{x}$ = 0.807 mm; median = 0.84 mm; mode = 0.84 mm; midrange = 0.780 mm. The sample from Pocatello would not be representative of the population in the United States.
13. *Cat on a Hot Tin Roof:* $\bar{x}$ = 3.9; median = 3.0. *The Cat in the Hat:* $\bar{x}$ = 3.1; median = 3.0. Based on the means, words in the first book appear to be longer than those in the second book.
15. One day: $\bar{x}$ = 0.5 degree; median = 0.0 degree. Five days: $\bar{x}$ = −0.4 degree; median = −1.0 degree. There does not appear to be a substantial difference in accuracy.
17. Jefferson Valley: $\bar{x}$ = 7.15 min; median = 7.20 min. Providence: same results as Jefferson Valley. Although the measures of center are the same, the Providence times are much more varied than the Jefferson Valley times.
19. Before 1983: $\bar{x}$ = 3.08950 g; median = 3.09065 g. After 1983: $\bar{x}$ = 2.50251 g; median = 2.50110 g. The weights after 1983 appear to be substantially lower than the weights before 1983.
21. One day: $\bar{x}$ = −0.6 degree; median = 0 degrees. Five days: $\bar{x}$ = −0.7 degree; median = −1.0 degree. The one-day forecast appears to be slightly more accurate. There is more confidence in results from a larger data set than results from a smaller data set.
23. Before 1983: $\bar{x}$ = 3.07478 g; median = 3.07630 g. After 1983: $\bar{x}$ = 2.49910 g; median = 2.50040 g. Same conclusion.
25. 53.4 degrees, which is close to the mean of 53.8 degrees found from the original list of values.
27. 46.8 mi/h, which is very close to the value found from the original list of values.
29. a. 182.9 lb
    b. 171.0 lb
    c. 159.2 lb
    The results differ by substantial amounts, suggesting that the mean of the original set of weights is strongly affected by extreme values.
31. The mean is at least 3.42 years.
33. 48.0 mi/h
35. 109.35 volts

Section 3-3

1. It is a measure of how much values deviate (or vary) from the mean.
3. 85 is unusually high because it is above the mean by more than two standard deviations.
5. range = 26.0 sec; s^2 = 89.6 sec^2; s = 9.5 sec. The sample is very small.
7. range = 3.0; s^2 = 0.9; s = 0.9. Because the data are at the nominal level of measurement, these results make no sense.
9. range = 30.0; s^2 = 81.8; s = 9.0. Ideally, the standard deviation would be zero.
11. range = 0.280 mm; s^2 = 0.009 mm^2; s = 0.094 mm. The estimate is not reasonable because the sample is not representative.
13. *Cat on a Hot Tin Roof:* range = 10.0; s^2 = 6.8; s = 2.6. *The Cat in the Hat:* range = 3.0; s^2 = 0.8; s = 0.9. There is much less variation among the word lengths in *The Cat in the Hat.*
15. One day: range = 11.0 degrees; s^2 = 6.9 degrees^2; s = 2.6 degrees. Five days: range = 15.0 degrees; s^2 = 20.6 degrees^2; s = 4.5 degrees. There appears to be greater variation in errors from the five-day forecast.
17. Jefferson Valley: range = 1.20 min; s^2 = 0.23 min^2; s = 0.48 min Providence: range = 5.80 min; s^2 = 3.32 min^2; s = 1.82 min. The variation with the single line is much less than that for multiple lines.
19. Before 1983: range = 0.11840 g; s^2 = 0.00145 g^2; s = 0.03812 g. After 1983: range = 0.04750 g; s^2 = 0.00023 g^2; s = 0.01506 g. The pre-1983 pennies appear to have weights that vary more than the pennies made after 1983.
21. One day: range = 14.0 degrees; s^2 = 6.5 degrees^2; s = 2.5 degrees. Five days: range = 24.0 degrees; s^2 = 31.2 degrees^2; s = 5.6 degrees. There appears to be greater variation in errors from the five-day forecast. The conclusion is the same, but there is more confidence because the sample sizes are larger.
23. Before 1983: range = 0.19890 g; s^2 = 0.00153 g^2; s = 0.03910 g. After 1983: range = 0.07690 g; s^2 = 0.00027 g^2; s = 0.01648 g. The pre-1983 pennies appear to have weights that vary more than the pennies made after 1983.
25. 7.0 is quite close to 6.9 found from the original list of values.
27. 4.1 is quite close to 4.0 found from the original list of values.
29. Answer varies, but using a minimum of 23 years and a maximum of 70 years, the standard deviation is estimated to be about 12 years.
31. 1830 kWh, 3846 kWh. Yes, 578 kWh is less than the minimum usual value.
33. a. 68%
    b. 99.7%
35. Men: 3.1%. Eggs: 5.1%. The relative amounts of variation are not substantially different.
37. The effect of the outlier will be large.

Section 3-4

1. The value is below the mean by an amount equal to two standard deviations.

3. About 25% of the values are below 15 and about 75% of the values are above 15.
5. a. 6 cm
   b. 6/7 or 0.86
   c. 0.86
   d. Usual
7. a. 4.5 in.
   b. 2.14
   c. -2.14
   d. Unusual
9. a. -1.13
   b. 0.65
   c. 0
11. 2.67; unusual
13. Psychology test, because $z = -0.50$ is greater than $z = -2.00$.
15. 5
17. 75
19. 25
21. 28
23. 29
25. 21
27. a. 11.5
   b. 33.75
   c. 24
   d. Yes; yes
   e. No; no
29. a. P_{10}, P_{50}, P_{80}
   b. 25, 27, 29, 31, 33.5, 35, 38, 41, 49
   c. 27, 31, 35, 41

Section 3-5

1. 2 is the minimum value, 5 is the first quartile, 10 is the median, 12 is the third quartile, and 20 is the maximum value.
3. Sigma has less variation. Because Sigma has less variation, estimates of repair costs will tend to be more accurate, so the costs will tend to be more predictable.

5., 7.

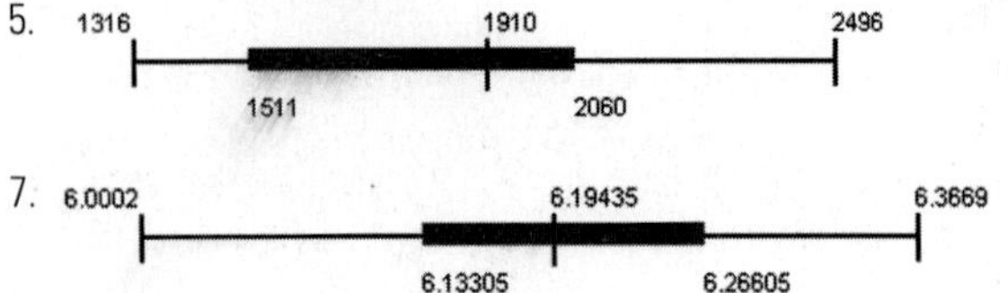

9. Roughly symmetric

11.

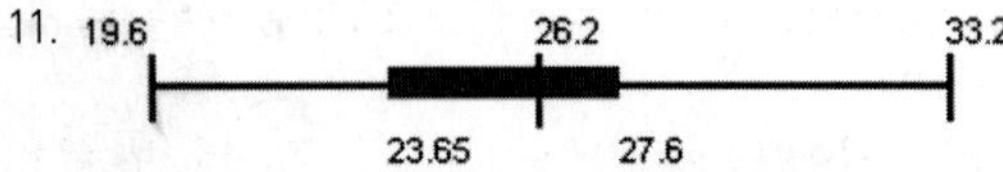

13. Select the batteries represented by the top boxplot. They have the best combination of highest mean and lowest variation.
15. Mild: 0.03; extreme: 0.05, 0.11, 0.12, 0.41, 0.43, 0.47, 0.49, 0.59, 0.92, 1.41

Chapter 3 Statistical Literacy and Critical Thinking

1. No. The reduction in variation implies that the repair times are closer together, but the mean does not necessarily change.
2. The standard deviation should not be ignored. It is possible to have a mean greater than 10 years with a standard deviation so high that some batteries will fail shortly after they are installed. That situation would be unacceptable.
3. The mean and standard deviation will change dramatically, but the median will probably change a small amount.
4. No. Because the data are from a voluntary response sample, it is very likely that the values are not representative of the population of all cars owned by Americans. Also, respondents are very likely to exaggerate the true values of their cars.

Chapter 3 Review Exercises

1. a. 4.54 ft
   b. 3.95 ft
   c. 1.8 ft, 3.7 ft, 5.1 ft
   d. 7.75 ft
   e. 11.90 ft
   f. 2.65 ft
   g. 7.03 ft^2
   h. 3.25 ft
   i. 5.15 ft
   j. 1.85 ft
2. a. 3.46
   b. Yes, because it is more than two standard deviations above the mean.
   c. No other values are unusual.
3.

x	Frequency
1.0–2.9	4
3.0–4.9	9
5.0–6.9	5
7.0–8.9	1
9.0–10.9	0
11.0–12.9	0
13.0–14.9	1

4. Skewed

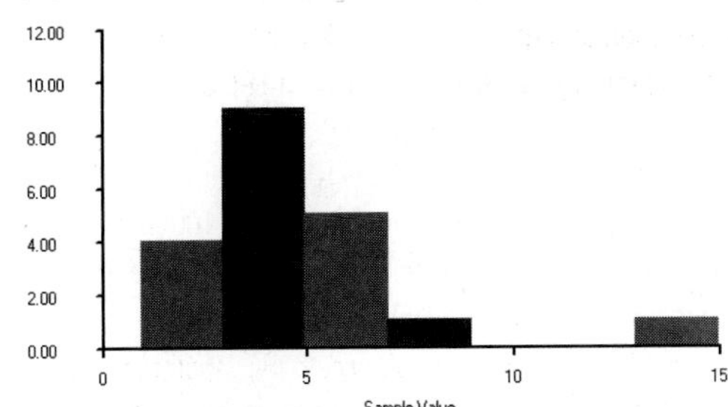

5. 1.8 3.95 13.7

3.25 5.15

6. The score of 19 is better because $z = -0.20$ is greater than $z = -0.67$.
7. a. Answer varies, but an estimate around 4 years is reasonable.
   b. Answer varies, but using a minimum of 0 years and a maximum of 20 years results in $s = 5.0$ years.
8. Minimum: 64 in.; maximum: 74 in. A height of 72 in. is not unusual because it falls within the range of usual values.

Chapter 3 Cumulative Review Exercises

1. a. Continuous
   b. Ratio
2. a. The mode is most appropriate because it identifies the most common choice. The other measures of center do not apply to data at the nominal level of measurement.
   b. Convenience
   c. Cluster
   d. The standard deviation should be lowered.

Chapter 4 Answers

Section 4-2

1. There is one chance out of 20,358,520. Such a win would be unusual because its probability is so small.
3. The statement is correct. Because the probability is so small, the event will occur on average only once in 1000 trials, so the event is unusual.
5. a. 1
   b. 1/10 or 0.1
   c. 0
7. $-1, 2, 5/3, \sqrt{2}$
9. a. 3/8
   b. 3/8
   c. 1/8
11. 0.738; yes
13. 0.908; yes
15. a. 0.27
    b. The result agrees roughly with the claim that 24% are blue.
17. a. 1/365
    b. Yes
    c. He already knew.
    d. 0
19. 0.34; no; yes
21. 0.159; no; yes
23. a. 0.135
    b. No
25. a. bb, bg, gb, gg
    b. 1/4
    c. 1/2
27. a. brown/brown, brown/blue, brown/brown, brown/blue
    b. 0
    c. 1
29. 423:77 or roughly 5.5:1 or 11:2
31. a. 18/38 or 0.474
    b. 10:9
    c. \$18
    d. \$20
33. 1/10 or 0.1
35. 1

Section 4-3

1. The two events can't happen at the same time.
3. The convenience sample might not be representative of the population. It might be biased and might therefore provide wrong results.
5. a. No
   b. No
   c. Yes
7. a. 0.95
   b. 0.9975
9. 0.410
11. 0.920
13. 0.6
15. 0.49
17. 0.14
19. 0.87
21. Top row: 5, 3. Bottom row: 4, 2.
23. 0.290
25. No. Here is one example: A = event of selecting a male under 30 years of age, B = selecting a female, C = selecting a male over 18 years of age.
27. $P(A \text{ or } B \text{ or } C) = P(A) + P(B) + P(C) - P(A \text{ and } B) - P(A \text{ and } C) - P(B \text{ and } C) + P(A \text{ and } B \text{ and } C)$

Section 4-4

1. The occurrence of one of the events does not affect the probability of the other event.
3. Sampling without replacement. The second outcome is not independent of the first.

5. a. Independent
   b. Independent
   c. Dependent
7. 1/8; no
9. a. 0.00238
   b. 0.00200
   c. If cases are to be selected for a follow-up study, it doesn't make much sense to select the same item twice, so select without replacement.
11. 0.922
13. 1/4096; yes. The likelihood of getting 12 girls in 12 births by chance is so small, that the outcome of 12 girls suggests that the method is effective.
15. a. 0.025
    b. 0.000625
    c. 0.999375
    d. Yes
17. 0.0195
19. a. 0.590
    b. 0.348
    c. 0.348
    d. The results are the same when rounded to three significant digits.
21. a. 0.992
    b. 0.973
    c. 0.431
23. 0.0192

Section 4-5

1. The number of defects is 1 or 2 or 3 or 4 or 5 or 6 or 7 or 8 or 9 or 10.
3. No. The proportion of male credit card users was ignored.
5. All six applicants test negative.
7. At least one has the gene.
9. 0.999
11. 15/16; yes
13. 0.5; no
15. 0.865
17. 0.999875; she does gain much reliability, because the probability of getting to class on time goes up from 0.95 (with one alarm clock) to 0.999875 (with three alarm clocks).
19. 0.271
21. 0.535
23. 0
25. a. 0.431
    b. 0.569
27. 1/12; 35

Section 4-6

1. A simulation is a process that behaves the same way as some procedure. If a simulation is used, the answer is usually not the exact correct answer.
3. No, because people generally favor some numbers over others so that they don't select numbers with a process that is truly random.
5. Repeat this procedure 20 times: Randomly generate an integer between 1 and 100, and consider an outcome of 1 through 95 to be a male, while an outcome of 96 through 100 is a female.
7. Randomly generate 500 integers between 1 and 100, and consider an outcome of 1 or 2 to be a defective cell phone, while any outcome of 3 through 100 is a good cell phone.
9. a. Answer varies.
   b. The results should show that it would be unusual to obtain a result consisting of 10 females.
11. a. Answer varies.
    b. It would be very unusual to find no defects among 500 cell phones.
13. Answer varies, but the results will probably show that it is easy to get 12 successes when the success rate stays at 20%, so there is not strong evidence that the drug is effective.
15. With switching, $P(\text{win}) = 2/3$. With sticking, $P(\text{win}) = 1/3$.
17. No; no

Section 4-7

1. With permutations, different orderings of the same items are counted separately, but they are not counted separately with combinations.
3. No. The methods of this section are not suitable for finding the relative frequency of a word in typical English text.
5. 120
7. 10,626
9. 2652
11. 4060
13. 1/324,632
15. 1/2,760,681
17. 1/69,090,840
19. 1/3003; yes
21. 1,048,576
23. a. 120
    b. 1/120
25. 1/7776 or 0.000129; yes
27. a. 11,880
    b. 495
29. 14,348,907
31. 144
33. a. 63
    b. $0.5^{63} \approx 1.08 \times 10^{-19}$
    c. 5,738,831,575 or about 5.7 billion
35. 1/41,416,353
37. 2,095,681,645,538 (about 2 trillion)
39. a. Calculator: 3.0414093×10^{64}; approximation: 3.0363452×10^{64}
    b. 615
41. 293

Chapter 4 Statistical Literacy and Critical Thinking

1. A probability value is a number between 0 and 1. When tossing a fair coin, the probability of heads is 0.5.
2. No, 0.27 is high enough so that chance is a reasonable explanation.
3. No. The reasoning assumes that the two outcomes (life, no life) are equally likely, but they are not equally likely.
4. Two events are disjoint if they cannot occur simultaneously. Two events are independent if the outcome of one does not affect the probability of the other.

Chapter 4 Review Exercises

1. 0.8
2. 0.32
3. 0.97
4. 0.85
5. 0.460
6. 0.638
7. 0.469
8. 0.188
9. a. 1/365
   b. 31/365
   c. Answer varies, but it is probably small, such as 0.02
   d. Yes
10. 0.130
11. 1/4096 or 0.000244; because the probability of getting 12 girls by chance is so small, it appears that chance is not a reasonable explanation, so the results support the claim that the method is effective.
12. 0.979
13. 1/10 or 0.1
14. 0.0777
15. a. 0.0027832
    b. 0.00000775
    c. 0.9944413
16. 10,000,000,000,000

Chapter 4 Cumulative Review Exercises

1. a. 4.0
   b. 4.0
   c. 2.2
   d. 4.7
   e. Yes
   f. 6/7
   g. 0.729
   h. 1/262,144; yes
2. a. 76 degrees
   b. 0.25
   c. 0.75
   d. 1/16
   e. No. The temperatures for both days could be determined by a weather pattern that affects several consecutive days at a time, so a high temperature on one day increases the likelihood of a high temperature on the following day.

Chapter 5 Answers

Section 5-2

1.

x	$P(x)$
1	1/6
2	1/6
3	1/6
4	1/6
5	1/6
6	1/6

3. No. Because the sum of the probabilities is 2.1, it is impossible to have the outcomes occur with the given probabilities. A probability distribution is not described by the list of outcomes with the corresponding probabilities.
5. a. Continuous
   b. Discrete
   c. Continuous
   d. Discrete
   e. Discrete
7. Probability distribution with $\mu = 0.7$ and $\sigma = 0.7$.
9. Not a probability distribution because $\sum P(x) = 1.2 \neq 1$.
11. Probability distribution with $\mu = 5.8$ and $\sigma = 1.1$. It is not unusual for a team to win in four games, because the probability is high (0.1818).
13. a. 0.003
    b. 0.004
    c. The result from part (b) is relevant.
    d. Yes. Because the probability of randomly selecting five or fewer Mexican-Americans is so small (0.004), that occurrence is not likely to happen by chance.
15. a. 0.206 (which is the probability of 8 or fewer Mexican-Americans)
    b. No. Because the probability of 8 or fewer Mexican-Americans is high (0.206), that event could easily occur by chance.
17. a. −26¢
    b. −26¢
    c. Don't bet, because the expected value with no bet is 0, which is better than −26¢.
19. a. Lives: −250 dollars (a loss); dies: $99,750 (a gain)
    b. −100 dollars
    c. $150
    d. The negative expected value is a relatively small price to pay for insuring for the financial security of his heirs.
21. $\mu = 1.5$; $\sigma = 0.9$ It is not unusual to get three girls, because the probability of three girls is high (1/8), indicating that three girls could easily occur by chance.

23. $\mu = 4.5$; $\sigma = 2.9$. The probability histogram is flat.
25. A frequency distribution summarizes actual observed outcomes, whereas a probability distribution describes how outcomes are likely to occur in the long run.
27. $\mu = 0.6$; $\sigma = 0.6$.

Section 5-3

1. If p represents the probability of a correct answer, then x is a count of the number of correct answers. p and x should refer to the same event.
3. Table A-1 includes only select values of p, and the table stops at $n = 15$.
5. Not binomial; the outcomes belong to more than two categories.
7. Not binomial; the outcomes belong to more than two categories.
9. Binomial
11. Binomial
13. a. 0.128
    b. WWC, WCW, CWW; 0.128 for each
    c. 0.384
15. 0.857
17. 0+
19. 0.113
21. 0.264
23. 0.234
25. 0.000672; yes
27. 0.26402 or 0.264032; no
29. a. 0.107
    b. 0.893
    c. 0.375 (or 0.376)
    d. No, because with a 20% rate, the probability of at most one is high (it's greater than 0.05).
31. 0.751
33. 0.0874; no
35. 0.000201; yes
37. 0.0524
39. 0.000535

Section 5-4

1. Yes, because the maximum usual value is 60.
3. 1.44 females^2
5. $\mu = 80.0$, $\sigma = 6.9$, minimum $= 66.1$, maximum $= 93.9$
7. $\mu = 373.0$, $\sigma = 16.7$, minimum $= 339.6$, maximum $= 406.4$
9. a. $\mu = 8.0$, $\sigma = 2.0$
   b. No. With guessing, the usual number of correct answers is between 4 and 12, so it would not be unusual to get 10 correct answers.
11. a. $\mu = 20.0$, $\sigma = 4.0$
    b. No, because 25 orange M&Ms is within the range of usual values (12 to 28). The claimed rate of 20% does not necessarily appear to be wrong, because that rate will usually result in 12 to 28 orange M&Ms (among 100), and the observed number of orange M&Ms is within that range.
13. a. $\mu = 162.5$, $\sigma = 9.0$
    b. Yes, 295 girls is well outside the range of usual values (144.5 to 180.5), and the method appears to be effective.
15. a. $\mu = 200.2$, $\sigma = 13.6$
    b. No, it is within the range of usual values (173.0 to 227.4).
17. a. $\mu = 611.2$, $\sigma = 15.4$
    b. No, because the range of usual values is 580.4 to 642.0. The result of 701 is unlikely to occur because it is outside of the range of usual values.
    c. No, it appears that substantially more people say that they voted than the proportion of people who actually did vote.
19. a. $\mu = 16.4$, $\sigma = 4.0$
    b. No, because 19 is within the range of usual values (8.4 to 24.4).
    c. No
21. a. Minimum $= 40.0$, maximum $= 60.0$
    b. Yes. The probability histogram is bell-shaped.
    c. The probability is 0.95 (because 40 and 60 are within two standard deviations of the mean).

Section 5-5

1. The random variable x is the number of occurrences of an event over some interval, the occurrences are random, the occurrences are independent of each other, and the occurrences are uniformly distributed over the interval.
3. Only the mean
5. 0.175
7. 0.0126
9. a. 0.000912
   b. 0.999
   c. 0.0296
11. a. 62.2
    b. 0.0155 (0.0156 using rounded mean)
13. a. 0.728
    b. 0.231
    c. 0.0368
    d. 0.00389
    e. 0.000309
    Using the computed probabilities, the expected frequencies are 266, 84, 13, 1.4, and 0.1, and they agree quite well with the actual frequencies.
15. 4.82×10^{-64} is so small that, for all practical purposes, we can consider it to be zero.

Chapter 5 Statistical Literacy and Critical Thinking

1. A probability distribution is a graph, table, or formula that gives the probability for each value of the random variable.
2. $\Sigma P(x) = 1$ and $0 \le P(x) \le 1$ for every individual value of x.

3. The probability distribution involves a random variable having either a finite number of values or a countable number of values, where "countable" refers to the fact that there might be infinitely many values, but they can be associated with a counting process. Another type of probability distribution consists of those that are continuous.
4. No. There are many discrete probability distributions that don't satisfy the requirements of a binomial distribution or a Poisson distribution.

Chapter 5 Review Exercises

1. a. $\Sigma P(x) = 0.999 = 1$ (with a small rounding error), and each x value is between 0 and 1.
   b. 2.0
   c. 1.3
   d. 0.033
   e. 2.0
   f. 0.892 (or 0.893)
   g. No, because it is easy (with probability 0.892) to get at least one correct by guessing the answer to each question.
2. a. 1.8
   b. 1.8
   c. 1.2
   d. 0.172
   e. Because the probability of no sets (among 12) tuned to *Cold Case* is 0.142, it is not unusual.
3. a. 0.00361
   b. This company appears to be very different because the event of at least four firings is so unlikely with a probability of only 0.00361.
4. a. 7/365
   b. 0.981
   c. 0.0188
   d. 0.0002
   e. No, because the event is so rare.

Chapter 5 Cumulative Review Exercises

1. a. $\bar{x} = 4.4, s = 3.0$
   b. Replace the frequency counts with these relative frequencies: 9%, 18%, 6%, 14%, 10%, 5%, 6%, 8%, 15%, 10%.
   c. $\mu = 4.5, \sigma = 2.9$
   d. The expected frequency for each cell is 8. Comparing the observed frequencies in the table to the expected frequencies (all 8s), we see that there are some notable differences, but the overall disagreement is not extreme.
2. a. $\mu = 15.0, \sigma = 3.7$
   b. Twelve positive cases falls within the range of usual values (7.6 to 22.4), so 12 is not unusually low. Because 12 positive cases could easily occur with an ineffective program, we do not have enough justification to say that the program is effective.
3. a. Yes, because it satisfies the two requirements that the the probabilities must be 1 and each individual probabili value between 0 and 1.
   b. The table corresponds to the population consisting of only the 25 surveyed subjects. The population is definitely not the population of all credit card holders in the United States.
   c. Yes, she used a convenience sample which describes only the 25 friends that she surveyed. If she had used a random sample, the results would have been much more representative of a larger population.
   d. 1.7
   e. 1.0

Chapter 6 Answers

Section 6-2

1. The word "normal" has a special meaning in statistics. It refers to a specific bell-shaped distribution that can be described by Formula 5-1.
3. The mean and standard deviation have the values of $\mu = 0$ and $\sigma = 1$.
5. 0.75
7. 0.5
9. 0.1587
11. 0.8413
13. 0.1056
15. 0.9599
17. 0.1359
19. 0.0157 (TI-83/84 Plus: 0.0156)
21. 0.9220
23. 0.8412 (TI-83/84 Plus: 0.8413)
25. 0.0001 (TI-83/84 Plus: 0.0002)
27. 0.5000
29. 0.6826 (TI-83/84 Plus: 0.6827)
31. 0.9974 (TI-83/84 Plus: 0.9973)
33. 0.9500
35. 0.0100
37. 1.28
39. −1.96, 1.96
41. a. 68.26% (TI-83/84 Plus: 68.27%)
    b. 95.00%
    c. 99.74% (TI-83/84 Plus: 99.73%)
    d. 81.85% (TI-83/84 Plus: 81.86%)
    e. 4.56% (TI-83/84 Plus: 4.55%)
43. a. 1.23
    b. 1.50
    c. 1.52
    d. −2.42
    e. −0.13

45. a.

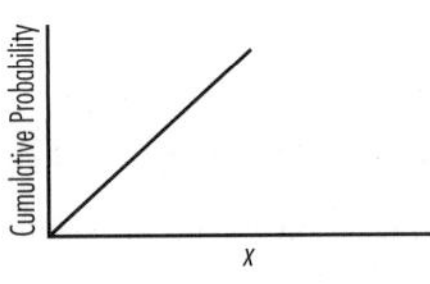

b.

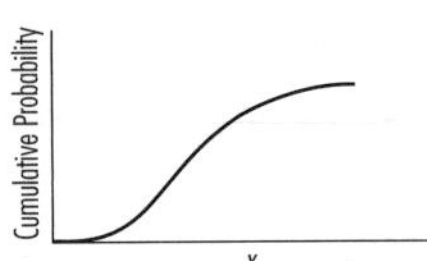

Section 6-3

1. The nonstandard normal distribution has a mean of 0 and a standard deviation of 1, but a nonstandard distribution has a different value for one or both of those parameters.
3. No, because the generated digits do not have a normal distribution. The probability of a digit less than 5 is 0.5.
5. 0.9772
7. 0.4972 (TI-83/84 Plus: 0.4950)
9. 80.8
11. 105.9 (TI-83/84 Plus: 105.8)
13. 0.52%
15. a. Men: 0.01% (TI-83/84 Plus: 0.004%); women: 0.01% (TI-83/84 Plus: 0.00%)
    b. 73.6 in.
17. 2405 g (TI-83/84 Plus: 2403 g)
19. 0.2005 (TI-83/84 Plus: 0.2015)
21. a. 0.0038; either a very rare event has occurred or the husband is not the father.
    b. 242 days
23. 4.0 in.; 8.0 in.
25. a. $\bar{x} = 118.9$, $s = 10.5$, histogram is roughly bell-shaped.
    b. 101.6, 136.2
27. a. The z scores are real numbers that have no units of measurement.
    b. $\mu = 0$; $\sigma = 1$; distribution is normal.
29. a. 75; 5
    b. No, the conversion should also account for variation.
    c. 31.4, 27.6, 22.4, 18.6
    d. Part (c), because variation is included in the conversion.

Section 6-4

1. A sampling distribution of a statistic is the distribution of all values of that statistic when all possible samples of the same size are taken from the same population.
3. Mean, proportion, variance
5. No, because of sampling variability, sample proportions will naturally vary from the true population proportion, even if the sampling is done with a perfectly valid procedure.
7. a. 10–10; 10–6; 10–5; 6–10; 6–6; 6–5; 5–10; 5–6; 5–5; the means are listed in part (b).
   b. The nine samples have means of 10.0, 8.0, 7.5, 8.0, 6.0, 5.5, 7.5, 5.5, 5.0, and they each have a probability of 1/9. (The sampling distribution can be described in a condensed form by listing the distinct means of 10.0, 8.0, 7.5, 6.0, 5.5, 5.0 with their corresponding probabilities of 1/9, 2/9, 2/9, 1/9, 2/9, 1/9.)
   c. 7.0
   d. Yes; yes
9. a. The means of 20.0, 20.5, 21.0, 31.5, 32.0, 33.5, 34.0, 43.0, 45.0, 47.0 have the corresponding probabilities of 4/25, 4/25, 1/25, 4/25, 2/25, 4/25, 2/25, 1/25, 2/25, 1/25.
   b. 30.2 (billion dollars)
   c. Yes; yes
11. a. The proportions of 0, 0.5, 1 have the corresponding probabilities of 1/16, 6/16, 9/16.
    b. 0.75
    c. Yes; yes
13. a. The means of 3, 6, 9, 13, 16, 23 have the corresponding probabilities of 1/9, 2/9, 1/9, 2/9, 2/9, 1/9.
    b. No
15. The formula yields $P(0) = 0.25$, $P(0.5) = 0.5$, and $P(1) = 0.25$, which does describe the sampling distribution of the proportions. The formula is just a different way of presenting the same information in the table that describes the sampling distribution.

Section 6-5

1. It is the standard deviation of the sample means, which is denoted as $\sigma_{\bar{x}}$ or $\sigma/\sqrt{n}$.
3. $\mu_{\bar{x}}$, $\sigma_{\bar{x}}$
5. a. 0.5636
   b. 0.8315
7. a. 0.1566 (TI-83/84 Plus: 0.1565)
   b. 0.4077
   c. If the original population has a normal distribution, the central limit theorem provides good results for any sample size.
9. a. 0.5675 (TI-83/84 Plus: 0.5684)
   b. 0.7257 (TI-83/84 Plus: 0.7248)
   c. The gondola is probably designed to safely carry a load somewhat greater than 2004 lb, but the operators would be wise to avoid a load of 12 men, especially if they appear to have high weights.
11. a. 0.0001 (TI-83/84 Plus: 0.0000 when rounded)
    b. No, but consumers are not being cheated because the cans are being overfilled, not underfilled.
13. a. 0.0062
    b. 0.0001 (TI-83/84 Plus: 0.0000 when rounded)
    c. Part (a). The behavior of the individual strobe lights is more important than the behavior of batches of 60 strobe lights.

15. a. 0.0274 (TI-83/84 Plus: 0.0272)
    b. 0.0001
    c. Because the original population is normally distributed, the sampling distribution of sample means will be normally distributed for any sample size.
    d. No, the mean can be less than 140 while individual values are above 140.
17. a. 0.5302 (TI-83/84 Plus: 0.5317)
    b. 0.7323 (TI-83/84 Plus: 0.7326)
    c. Part (a), because the seats will be occupied by individual women, not groups of women.
19. a. 15
    b. 0.9998 (TI-83/84 Plus: 1)
    c. Part (a), because the individual rejected quarters could result in lost sales and lower profits.
21. a. 267.5 in.
    b. Because college football players tend to be larger than randomly selected men, the given mean and standard deviation don't apply. The bench should be longer than the 267.5 in. length found in part (a).
23. a. $\mu = 8.0, \sigma = 5.4$
    b. 2,3 2,6 2,8 2,11 2,18 3,6 3,8 3,11 3,18 6,8 6,11 6,18 8,11 8,18 11,18
    c. 2.5, 4.0, 5.0, 6.5, 10.0, 4.5, 5.5, 7.0, 10.5, 7.0, 8.5, 12.0, 9.5, 13.0, 14.5
    d. $\mu_{\bar{x}} = 8.0, \sigma_{\bar{x}} = 3.4$

Section 6-6

1. Approximately bell-shaped (normal).
3. No. With $n = 6$ and $p = 0.001$, the requirements of $np \geq 5$ and $nq \geq 5$ are not satisfied.
5. The area to the right of 15.5
7. The area to the left of 11.5
9. The area to the left of 4.5
11. The area between 7.5 and 10.5
13. 0.227; normal approximation is not suitable.
15. 0.887; normal approximation: 0.8869 (TI-83/84 Plus: 0.8861)
17. 0.1292 (normal approximation using TI-83/84 Plus: 0.1303; exact value using TI-83/84 Plus: 0.1302). Not unusual.
19. 0.0001 (normal approximation using TI-83/84 Plus: 0.0000 when rounded; exact value using TI-83/84 Plus: 0.0000 when rounded). The results suggest that the surveyed people did not respond accurately.
21. 0.2676 (normal approximation using TI-83/84 Plus: 0.2665; exact value using TI-83/84 Plus: 0.2650); no
23. 0.2709 (normal approximation using TI-83/84 Plus: 0.2679; exact value using TI-83/84 Plus: 0.2726); media reports appear to be wrong.
25. 0.0080 (normal approximation using TI-83/84 Plus: 0.0079; exact value using TI-83/84 Plus: 0.0076); yes
27. 0.6368 (normal approximation using TI-83/84 Plus: 0.6375; exact value using TI-83/84 Plus: 0.6369); the pool is likely to be sufficient, but the probability should be much higher. It would be better to increase the pool of volunteers.
29. 0.0001 (normal approximation using TI-83/84 Plus: 0.0000; exact value using TI-83/84 Plus: 0.0000); yes
31. The probability of getting at least 18 totals above $100 is 0.0051 (normal approximation using TI-83/84 Plus: 0.0050; exact value using TI-83/84 Plus: 0.0059), which is very small. This suggests that the spending is unusual and it should be verified.
33. 6; 0.4602
35. a. 0.821
    b. 0.9993
    c. 0.0000165
    d. 0.552
37. Answer varies.

Section 6-7

1. A normal quantile plot can be used to determine whether sample data are from a population having a normal distribution.
3. There should be 100 points configured so that they are reasonably close to a straight-line pattern, and there is no systematic pattern that is not a straight-line pattern.
5. Not normal. There is a systematic pattern that is not a straight-line pattern.
7. Normal
9. Normal
11. Not normal
13. Normal

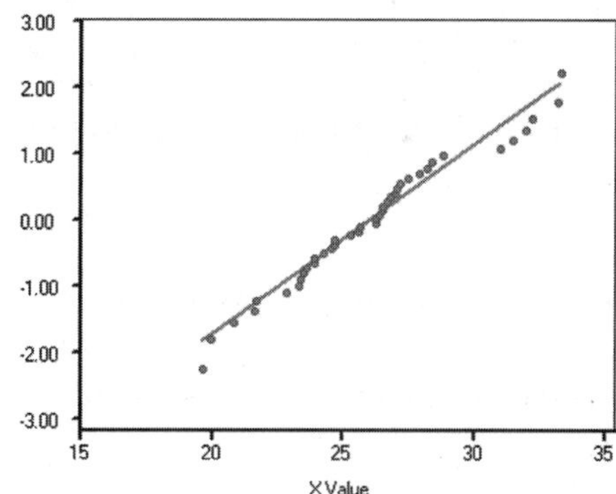

15. Not normal

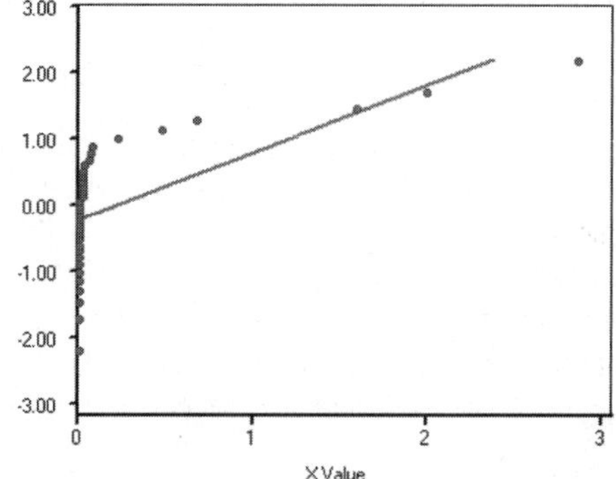

17. Heights appear to be normal, but cholesterol levels do not appear to be normal. Cholesterol levels are strongly affected by diet, and diets might vary in dramatically different ways that do not yield normally distributed results.
19. −1.28, −0.52, 0, 0.52, 1.28; normal

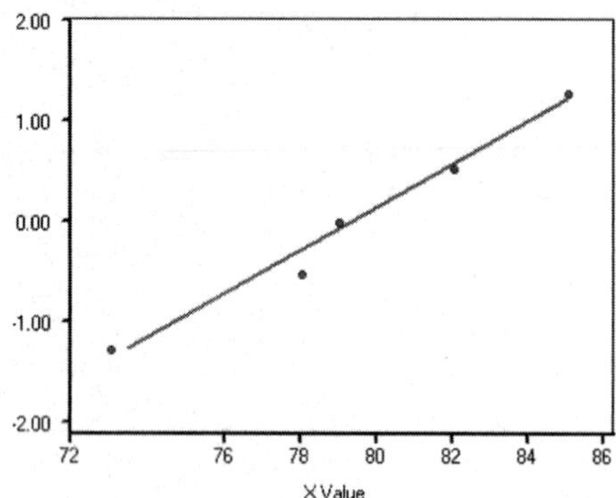

21. No, the transformation to *z* scores involves subtracting a constant and dividing by a constant, so the plot of the (x, z) points will always be a straight line, regardless of the nature of the distribution.

Chapter 6 Statistical Literacy and Critical Thinking

1. A normal distribution is the probability distribution of a continuous random variable with the properties that a graph of the distribution is symmetric and bell-shaped, and it can be described by Formula 6-1.
2. The sample means will tend to be normally distributed.
3. No, the sample might well be biased. It might have the correct distribution shape, but the mean and standard deviation might be very different from the population parameters.
4. When obtaining simple random samples of size *n* from a population with mean μ and standard deviation σ, the distribution of sample means will, as the sample size increases, approach a normal distribution, the mean of all sample means is μ, and the standard deviation of all sample means is $\sigma/\sqrt{n}$.

Chapter 6 Review Exercises

1. a. 0.3830 (TI-83/84 Plus: 0.3829)
   b. 0.9544 (TI-83/84 Plus: 0.9545)
   c. 1.28
2. a. 3.67% (TI-83/84 Plus: 3.71%)
   b. 0.0001 (TI-83/84 Plus: 0.000178)
   c. 72.6 in.
3. a. 0.0222 (TI-83/84 Plus: 0.0221)
   b. 0.2847 (TI-83/84 Plus: 0.2836)
   c. 0.6720 (TI-83/84 Plus: 0.6715)
   d. 254.6
4. 0.0004 (normal approximation using TI-83/84 Plus: 0.0004; exact value using TI-83/84 Plus: 0.0002). The very low probability suggests that random chance should be ruled out as an explanation for the fact that 18 of the incorrect answers were by women. There does appear to be strong evidence for the argument that the question is biased in favor of men.
5. 0.1112 (normal approximation using TI-83/84 Plus: 0.1113; exact value using TI-83/84 Plus: 0.1071). The probability is not very small (such as 0.05 or less), so there does not appear to be sufficient evidence to claim that the company is discriminating based on gender.
6. A histogram shows a graph that is substantially different from being bell-shaped, so the times do not appear to come from a population having a normal distribution.

Chapter 6 Cumulative Review Exercises

1. a. 10.5 g
   b. 8.5 g
   c. 10.2 g
   d. 103.7 g^2
   e. −0.74
   f. 62.5%
   g. 0.7704 (TI-83/84 Plus: 0.7689)
   h. ratio
   i. continuous
   j. No. The sample is not the typical American diet, and Americans don't consume the sample items in equal amounts.
2. a. 0.001
   b. 0.271
   c. The requirement that $np \geq 5$ is not satisfied, indicating that the normal approximation would result in errors that are too large.
   d. 5.0
   e. 2.1
   f. No, 8 is within two standard deviations of the mean and is within the range of values that could easily occur by chance.

Chapter 7 Answers

Section 7-2

1. It is a standard *z* score that can be used to distinguish between sample statistics that are likely to occur and those that are unlikely. The number $z_{\alpha/2}$ is a *z* score that separates an area of $\alpha/2$ in the right tail of the standard normal distribution.
3. The point estimate does not reveal any information about the accuracy of the estimate. By providing a range of values associated with a probability, a confidence interval does reveal information about accuracy.
5. 2.575 (TI-83/84 Plus: 2.5758293)
7. 2.33 (TI-83/84 Plus: 2.3263479)
9. 0.333 ± 0.111
11. 0.246 ± 0.040
13. 0.879; 0.011

15. 0.660; 0.053
17. 0.0429
19. 0.0309 (TI-83/84 Plus: 0.0308)
21. $0.357 < p < 0.443$
23. $0.219 < p < 0.281$
25. 2401
27. 842
29. $0.866 < p < 0.949$; the method appears effective because the proportion of girls is substantially greater than 0.5.
31. $0.496 < p < 0.514$; no, because the proportion could easily equal 0.5. The proportion is not substantially less than 0.5 the week before Thanksgiving.
33. a. $0.226 < p < 0.298$
    b. No, the confidence interval includes 0.25, so the true percentage could easily equal 25%.
35. a. $0.0267\% < p < 0.0376\%$
    b. No, because 0.0340% is included in the confidence interval.
37. $0.347 < p < 0.432$ (using $x = 339$); yes, because the proportion selected appears to be substantially less than 0.791 (or 79.1%).
39. $0.931 < p < 0.949$ (using $x = 4019$); yes, based on the confidence interval, we are quite confident that the proportion is more than 0.35 or 35%.
41. 4145 (TI-83/84 Plus: 4147)
43. 1509 (TI-83/84 Plus: 1504)
45. $0.183 < p < 0.357$; yes, because the confidence interval includes 0.24.
47. Wednesday: $0.178 < p < 0.425$; Sunday: $0.165 < p < 0.412$. The confidence intervals are not substantially different. Precipitation does not appear to occur more on either day.
49. 406
51. $0.0395 < p < 0.710$; no
53. a. The requirement of at least 5 successes and at least 5 failures is not satisfied, so the normal distribution cannot be used.
    b. 0.15

Section 7-3

1. There is 95% confidence that the limits of 2.5 and 6.0 contain the true population mean μ.
3. No, because the sample is unsuitable. It is very possible that the convenience sample is not representative of the population of all potential riders.
5. 1.96
7. 1.75
9. 2.94
11. 12.875 (TI-83/84 Plus: 12.879)
13. $\$61,605 < \mu < \$72,795$ (TI-83/84 Plus: $\$61,606 < \mu < \$72,794$)
15. $667 < \mu < 709$
17. 97
19. 390
21. 67.3849
23. 67.3849 ± 0.7295
25. 4.3 years $< \mu <$ 5.3 years; no, because the confidence interval is an estimate of the mean, and it does not imply that 95% of the times are between the confidence interval limits.
27. 55.4 sec $< \mu <$ 61.2 sec; yes, because the confidence interval limits do contain 60 sec.
29. $128.7 < \mu < 139.2$; ideally, all of the measurements would be the same, so there would not be an interval estimate.
31. 5.61161 g $< \mu <$ 5.66698 g; the confidence interval suggests that minted quarters meet the specifications and the production process is okay.
33. 217
35. 6907
37. 80,770 (TI-83/84 Plus: 80,767); the sample size is too large to be practical, so it should be reduced by increasing the acceptable margin of error.
39. $105 < \mu < 115$

Section 7-4

1. The amount refers to an average, which is likely to be the mean or median, but the margin of error is appropriate for a proportion, not a mean or median. The margin of error should be an amount in dollars, not percentage points.
3. We are 99% confident that the limits of 114.4 and 123.4 contain the true mean systolic blood pressure.
5. $t_{\alpha/2} = 2.201$
7. Neither normal nor *t* distribution applies.
9. $t_{\alpha/2} = 1.653$
11. $z_{\alpha/2} = 2.33$
13. 1.6 kg; 1.4 kg $< \mu <$ 4.6 kg
15. $113.583 < \mu < 122.417$; we have 95% confidence that the limits of 113.583 and 122.417 contain the true population mean IQ score of statistics students.
17. 3002 g $< \mu <$ 3204 g; the mean weight of babies born to mothers who used cocaine appears to be substantially less than the mean weight of babies born to mothers who did not use cocaine. Cocaine appears to affect birth weight.
19. a. $-3.5° < \mu < 0.9°$
    b. Yes, 0 is included. No, because the confidence interval indicates that the mean difference could be 0.
21. $-0.471 < \mu < 3.547$; the confidence interval is likely to be a poor estimate because the value of 5.40 appears to be an outlier, suggesting that the assumption of a normally distributed population is not correct.
23. $589.7 < \mu < 731.0$; no, there will be a substantial number of applicants who do not qualify.
25. $0.075 < \mu < 0.168$; no, it is possible that the requirement is being met, but it is also very possible that the mean is not less than 0.165 g/mi.

27. a. $65.8 < \mu < 73.0$
    b. $72.3 < \mu < 80.3$
    c. Because the two CIs overlap, we cannot conclude that the two population means are different.
29. $26.2 < \mu < 96.3$; the confidence interval changes by a very substantial amount. Confidence intervals are very sensitive to outliers. Outliers should be carefully examined and discarded if they are found to be errors. If an outlier is a correct value, it might be very helpful to construct the confidence interval with the outlier and without the outlier, so that its effects can be seen.
31. $0.8462 \text{ g} < \mu < 0.8668 \text{ g}$; $0.8474 < \mu < 0.8656$; the second confidence interval is narrower, indicating that we have a more accurate estimate when the relatively large sample is from a relatively small finite population.
33. a. No, a single value has no variation.
    b. The critical t value cannot be obtained because df = 0 is not available in the table of critical t values.
    c. $-104.16 \text{ ft} < \mu < 128.16 \text{ ft}$; no

Section 7-5

1. We have 95% confidence that the limits of 2.25 in. and 3.52 in. contain the true value of the standard deviation of the heights of all women.
3. No, the confidence interval would be an estimate of variation among sample means, not the variation among individuals. Individuals would have much greater variation.
5. 13.844, 41.923
7. 20.707, 66.766
9. $\$15{,}006 < \sigma < \$23{,}385$
11. $55 < \sigma < 86$ (TI-83/84 Plus: $56 < \sigma < 87$)
13. 767
15. 1401; the sample size is common with polls and surveys, but there are many situations in which the sample size will be impractical because it is too large.
17. $586 \text{ g} < \sigma < 717 \text{ g}$; no, because the confidence interval limits contain 696 g.
19. $0.54°\text{F} < \sigma < 0.77°\text{F}$ (TI-83/84 Plus: $0.53°\text{F} < \sigma < 0.75°\text{F}$); yes
21. $64.9 < \sigma < 173.2$
23. $1.195 < \sigma < 4.695$; yes, the confidence interval is likely to be a poor estimate because the value of 5.40 appears to be an outlier, suggesting that the assumption of a normally distributed population is not correct.
25. a. $2.62 < \sigma < 4.71$ (TI-83/84 Plus: $2.65 < \sigma < 4.79$)
    b. $4.71 < \sigma < 8.46$ (TI-83/84 Plus: $4.76 < \sigma < 8.61$)
    c. The two CIs don't overlap, so the two populations appear to have different amounts of variation. (TI-83/84 Plus: The CIs overlap slightly, so that at a 99% confidence level, the two populations appear to have amounts of variation that are not substantially different.)
27. 152.3644 and 228.4771 are close to the STATDISK values.

Chapter 7 Statistical Literacy and Critical Thinking

1. The two critical values of $z = -1.96$ and $z = 1.96$ separate tail areas of 0.025 each, so the area between $z = -1.96$ and $z = 1.96$ is 0.95 or 95%.
2. The confidence interval gives us a much better indication of the accuracy of the estimate of 27.44 lb. It gives us an indication of how good the estimate is.
3. $0.62 < p < 0.68$
4. We are 95% confident that the limits of 0.62 and 0.68 contain the true value of the population proportion.

Chapter 7 Review Exercises

1. $91.5\% < p < 94.6\%$ (using $x = 934$); we are 95% confident that the limits of 91.5% and 94.6% contain the true value of the population percentage.
2. 157 (using $\hat{p} = 0.93$ and $\hat{q} = 0.07$)
3. $10.06 \text{ oz} < \mu < 11.72 \text{ oz}$; confidence interval does not contain 12, and sample mean is 10.89, indicating that the mean is less than 12. Consumers are not given enough coffee. Also, the variation is too high, with some amounts that are unacceptably low.
4. $1.15 \text{ oz} < \sigma < 2.41 \text{ oz}$; no, 0.25 oz is not a feasible value of σ. The machine definitely requires modifications to reduce variation.
5. a. 2401
   b. 1164
   c. 2401
6. a. $5.47 \text{ years} < \mu < 8.55 \text{ years}$
   b. $2.92 \text{ years} < \sigma < 5.20 \text{ years}$
   c. 1484
   d. No, the sample would not be representative of the population of all car owners.
7. a. $15.1\% < p < 21.6\%$ (using $x = 144$)
   b. Yes, it appears that the smoking rate for the college graduates is significantly less than the rate for the general population.
8. a. $\$5403 < \mu < \$12{,}605$
   b. \$12,605

Chapter 7 Cumulative Review Exercises

1. a. 121.0 lb
   b. 123.0 lb
   c. 119 lb, 128 lb
   d. 116.5 lb
   e. 23.0 lb
   f. 56.8 lb^2
   g. 7.5 lb
   h. 119.0 lb
   i. 123.0 lb
   j. 127.0 lb
   k. ratio

l.

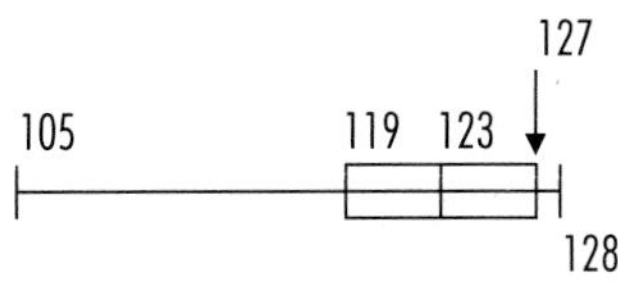

m. 112.6 lb $< \mu <$ 129.4 lb
n. 4.5 lb $< \sigma <$ 18.4 lb
o. 95
p. The individual supermodel weights do not appear to be considerably different from weights of randomly selected women, because they are all within 1.31 standard deviations of the mean of 143 lb. However, when considered as a group, their mean is significantly less than the mean of 143 lb [see part (m)].

2. a. 0.0089 (TI-83/84 Plus using binomial: 0.0103)
   b. $0.260 < p < 0.390$
   c. Because the confidence interval limits do not contain 0.25, it is unlikely that the expert is correct.
3. a. 9.00%
   b. $7.40\% < p < 10.6\%$
   c. 2653 (TI-83/84 Plus: 2654)

Chapter 8 Answers

Section 8-2

1. Hypothesis testing cannot be used with survey data to "prove" that a population percentage is equal to some specific value. The correct statement would be that there is not sufficient evidence to conclude that the 50% rate is wrong, or to say that based on the survey, the percentage appears to be about 50%.
3. A test statistic is based on the sample data, but the critical value is based on the distribution that is being used and the significance level that has been chosen.
5. There is not sufficient evidence to support the claim (because it is very easy to get 11 heads in 20 tosses by chance).
7. There does appear to be sufficient evidence to support the claim.
9. H_0: $p = 0.25$. H_1: $p > 0.25$.
11. H_0: $\mu = 121$ lb. H_1: $\mu \neq 121$ lb.
13. H_0: $\sigma = 15$. H_1: $\sigma < 15$.
15. H_0: $\mu = 0.8535$ g. H_1: $\mu < 0.8535$ g.
17. $z = \pm 1.96$
19. $z = 2.33$
21. $z = \pm 1.645$
23. $z = -2.05$
25. -2.21
27. 7.77
29. 0.1587
31. 0.0500
33. 0.0668
35. 0.0802 (TI-83/84 Plus: 0.0801)
37. There is sufficient evidence to support the claim that the proportion of male golfers is less than 0.5.
39. There is not sufficient evidence to support the claim that the proportion of red M&Ms is different from 0.13.
41. Type I error: Reject the claim that the proportion of settled malpractice suits is 0.25 when that proportion actually is 0.25.
    Type II error: Fail to reject the claim that the proportion of settled malpractice suits is 0.25 when that proportion is actually different from 0.25.
43. Type I error: Reject the claim that the proportion of murders cleared by arrests is 0.62 when that proportion actually is 0.62.
    Type II error: Fail to reject the claim that the proportion of murders cleared by arrests is 0.62 when that proportion is actually different from 0.62.
45. P-value = 0.9999. With an alternative hypothesis of $p > 0.5$, it is impossible for a sample statistic of 0.27 to fall in the critical region. No sample proportion less than 0.5 can ever support a claim that $p > 0.5$.
47. a. 0.7852 (TI-83/84 Plus: 0.7857)
    b. 0.2148 (TI-83/84 Plus: 0.2143)

Section 8-3

1. Normal, because the normal distribution is a suitable approximation to a binomial distribution, provided that the requirements are satisfied.
3. No. Because the sample is a voluntary-response sample, it is not suitable for making inferences about the general population.
5. a. $z = -0.12$
   b. $z = \pm 1.96$
   c. 0.9044 (TI-83/84 Plus: 0.9025)
   d. There is not sufficient evidence to warrant rejection of the claim that green-flowered peas occur at a rate of 25%.
   e. No, a hypothesis test cannot be used to prove that a proportion is equal to some claimed value.
7. H_0: $p = 0.29$. H_1: $p > 0.29$. Test statistic: $z = 0.73$. Critical value: $z = 1.645$ (assuming a 0.05 significance level). P- value: 0.2313. Fail to reject H_0. There is not sufficient evidence to support the claim that more than 29% of federal crimes were for drug offenses.
9. H_0: $p = 0.5$. H_1: $p > 0.5$. Test statistic: $z = 14.70$. Critical value: $z = 2.33$. P-value: 0.0001 (TI-83/84 Plus: 0.0000). Reject H_0. There is sufficient evidence to support the claim that the probability of a baby being a girl is greater than 0.5. The method does appear to be effective.
11. H_0: $p = 0.5$. H_1: $p > 0.5$. Test statistic: $z = 4.20$. Critical value: $z = 2.33$. P-value: 0.0001 (TI-83/84 Plus: 0.00001). Reject H_0. There is sufficient evidence to support the claim that more than 50% of car crashes occur within 5 miles of home. The results might be questionable because the sponsor might have a self-serving interest in the results.

13. H_0: $p = 0.15$. H_1: $p > 0.15$. Test statistic: $z = 1.60$. Critical value: $z = 1.645$. *P*-value: 0.0548 (TI-83/84 Plus: 0.0543). Fail to reject H_0. There is not sufficient evidence to support the claim that more than 15% of U.S. households use e-mail. The conclusion is not valid today because the population characteristics (use of e-mail) are changing quickly over time.
15. H_0: $p = 0.000340$. H_1: $p \neq 0.000340$. Test statistic: $z = -0.66$. Critical values: $z = \pm 2.81$. *P*-value: 0.5092 (TI-83/84 Plus: 0.5122). Fail to reject H_0. There is not sufficient evidence to support the claim that the rate is different from 0.0340%. Cell phone users should not be concerned about cancer of the brain or nervous system.
17. H_0: $p = 0.01$. H_1: $p \neq 0.01$. Test statistic: $z = 2.19$. Critical values: $z = \pm 1.96$. *P*-value: 0.0286 (TI-83/84 Plus: 0.0284). Reject H_0: $p = 0.01$. There is sufficient evidence to warrant rejection of the claim that 1% of sales are overcharges. Because 1.62% of the sampled items are overcharges, it appears that the error rate is worse with scanners, not better.
19. H_0: $p = 1/3$. H_1: $p < 1/3$. Test statistic: $z = -0.91$. Critical value: $z = -1.645$. *P*-value: 0.1814 (TI-83/84 Plus: 0.1826). Fail to reject H_0. There is not sufficient evidence to support the claim that less than 1/3 of all adults never drink. The wording of the question appears to elicit honest responses, although the nature of the subject might cause people to be less than honest.
21. H_0: $p = 0.5$. H_1: $p \neq 0.5$. Test statistic: $z = -2.00$. Critical values: $z = \pm 1.96$. *P*-value: 0.0456 (TI-83/84 Plus: 0.0458). Reject H_0. There is sufficient evidence to warrant rejection of the claim that the percentage of adults who fly never or rarely is equal to 50%. Yes, it would have been better to somehow get sample data on actual observed flying habits instead of relying on the honesty of the subjects. It is possible that they might be embarrassed to say that they fly never or rarely.
23. H_0: $p = 0.012$. H_1: $p > 0.012$. Test statistic: $z = 3.42$ (using $x = 35$). Critical value: $z = 2.33$. *P*-value: 0.0003. Reject H_0. There is sufficient evidence to support the claim that more than 1.2% of the Clarinex users experience fatigue. Fatigue appears to be an adverse reaction of Clarinex.
25. H_0: $p = 0.24$. H_1: $p \neq 0.24$. Test statistic: $z = 0.70$. Critical values: $z = \pm 1.96$ (assuming a 0.05 significance level). *P*-value: 0.4840 (TI-83/84 Plus: 0.4824). Fail to reject H_0. There is not sufficient evidence to warrant rejection of the claim that 24% of the M&Ms are blue.
27. H_0: $p = 0.5$. H_1: $p < 0.5$. Test statistic: $z = -2.87$. Critical value: $z = -1.645$. *P*-value: 0.0021 (TI-83/84 Plus: 0.0020). Reject H_0. There is sufficient evidence to support the claim that $p < 0.5$. The result suggests that the forecast high is off by 2° less than 50% of the time.
29. a. H_0: $p = 0.10$. H_1: $p \neq 0.10$. Test statistic: $z = 2.00$. Critical values: $z = \pm 1.96$. Reject H_0. There is sufficient evidence to warrant rejection of the claim that the proportion of zeros is 0.1.
    b. H_0: $p = 0.10$. H_1: $p \neq 0.10$. Test statistic: $z = 2.00$. *P*-value: 0.0456 (TI-83/84 Plus: 0.0452). There is sufficient evidence to warrant rejection of the claim that the proportion of zeros is 0.1.
    c. $0.0989 < p < 0.139$; because 0.1 is contained within the confidence interval, fail to reject H_0: $p = 0.10$. There is not sufficient evidence to warrant rejection of the claim that the proportion of zeros is 0.1.
    d. The traditional and *P*-value methods both lead to rejection of the claim, but the confidence interval method does not lead to rejection.
31. a. With $n = 80$ and $p = 0.0025$, the conditions $np \geq 5$ and $nq \geq 5$ are not both satisfied.
    b. 0.00000000165 is found by using the binomial probability distribution.
    c. If the probability of a man being color-blind is 0.0025, the probability of getting at least 7 color-blind men among 80 is extremely small, so it appears that the color-blindness rate is actually greater than 0.0025 or 0.25%. There is sufficient evidence to support the claim that the color-blindness rate for men is greater than 0.25%.
33. a. 0.7224 (TI-83/84 Plus: 0.7219)
    b. 0.2776 (TI-83/84 Plus: 0.2781)
    c. The power of 0.7224 shows that there is a reasonably good chance of making the correct decision of rejecting the false null hypothesis. It would be better if the power were even higher, such as greater than 0.8 or 0.9.

Section 8-4

1. No, the requirements can be satisfied when $n \leq 30$ provided that the population has a normal distribution. (Also, σ must be known and the sample must be a simple random sample.)
3. 0.98 or 98%
5. The requirements are satisfied.
7. No, the standard deviation must be known.
9. a. $z = 0.23$
   b. $z = \pm 1.96$
   c. 0.8180 (TI-83/84 Plus: 0.8182)
   d. Fail to reject H_0.
   e. There is not sufficient evidence to warrant rejection of the claim that the mean weight of M&Ms is equal to 0.8535 g.
11. H_0: $\mu = 30.0°C$. H_1: $\mu > 30.0°C$. Test statistic: $z = 1.732$. Critical value: $z = 1.645$. *P*-value: 0.0416. Reject H_0. There is sufficient evidence to support the claim that the mean is greater than 30.0°C.
13. H_0: $\mu = 60$ sec. H_1: $\mu \neq 60$ sec. Test statistic: $z = -1.13$. Critical value: $z = \pm 1.96$. *P*-value: 0.2584 (TI-83/84 Plus: 0.2577). Fail to reject H_0. There is not sufficient evidence to warrant rejection of the claim that the mean is equal to 60 sec. The mean overall perception is reasonably accurate, but individual values might be in error by fairly large amounts.
15. H_0: $\mu = 0$. H_1: $\mu < 0$. Test statistic: $z = -2.77$. Critical value: $z = -1.645$. *P*-value: 0.0028. Reject H_0. There is sufficient evidence to support the claim that the mean weight change is less than 0. The diet appears to be effective, but the mean weight loss of only 2.1 lb suggests that the diet isn't practical.

17. H_0: $\mu = 140$ mm Hg. H_1: $\mu < 140.0$ mmHg. Test statistic: $z = -2.27$. Critical value: $z = -1.645$. P-value: 0.0116. Reject H_0. There is sufficient evidence to support the claim that the mean is less than 140.0 mmHg. Yes.
19. H_0: $\mu = 5.670$ g. H_1: $\mu \neq 5.670$ g. Test statistic: $z = -2.86$. Critical values: $z = \pm 2.575$. P-value: 0.0042 (TI-83/84 Plus: 0.0043). Reject H_0. There is sufficient evidence to warrant rejection of the claim that the mean equals 5.670 g. These quarters do not appear to have weights meeting the specifications, but that might be due to wear from use.
21. Power = 0.9582 (TI-83/84 Plus: 0.9580); $\beta = 0.0418$ (TI-83/84 Plus: 0.0420). The high power value shows that if $\mu = 108$, the test is very effective in recognizing that $\mu > 100$.

Section 8-5

1. 4; 4
3. If $\bar{x} > 12$ oz, the hypothesis test can never support a claim that $\mu <$ 12 oz. If $\bar{x} > 12$ oz, it can't possibly be significantly less than 12 oz.
5. Student t
7. Neither normal nor Student t
9. Table: $0.005 < P\text{-value} < 0.01$; TI-83/84 Plus: 0.00641.
11. Table: $P\text{-value} < 0.01$; TI-83/84 Plus: 0.0000.
13. H_0: $\mu = 120$. H_1: $\mu > 120$. Test statistic: $t = 3.464$. Critical value: $t = 1.796$. $P\text{-value} < 0.005$ (TI-83/84 Plus: 0.0026). There is sufficient evidence to support the claim that the mean IQ is greater than 120.
15. H_0: $\mu = 110$. H_1: $\mu > 110$. Test statistic: $t = 3.74$. Critical value: $t = 1.711$. P-value: 0.001. Reject H_0. There is sufficient evidence to support the claim that the mean is greater than 110.
17. H_0: $\mu = 98.6$. H_1: $\mu < 98.6$. Test statistic: $t = -6.642$. Critical value: $t = -1.660$ (approximately). P-value: Less than 0.005 (TI-83/84 Plus: 0.0000). Reject H_0. There is sufficient evidence to support the claim that the mean is less than 98.6.
19. H_0: $\mu = 3103$ g. H_1: $\mu < 3103$ g. Test statistic: $t = -8.612$. Critical value: $t = -2.345$. P-value is less than 0.005 (TI-83/84 Plus: 0.0000). Reject H_0. There is sufficient evidence to support the claim that the mean weight is less than 3103 g. It appears that birth weights are affected by cocaine use by the mother.
21. H_0: $\mu = 0$. H_1: $\mu > 0$. Test statistic: $t = 8.447$. Critical value: $t = 2.528$. P-value is less than 0.005 (TI-83/84 Plus: 0.0000). Reject H_0. There is sufficient evidence to support a claim that the mean is positive. The treatment appears to be effective.
23. H_0: $\mu = 63.6$ in. H_1: $\mu > 63.6$ in. Test statistic: $t = 13.200$. Critical value: $t = 2.896$. P-value is less than 0.005 (TI-83/84 Plus: 0.0000). Reject H_0. There is sufficient evidence to support the claim that supermodels have a mean height greater than 63.6 in. Yes, the evidence is strong.
25. H_0: $\mu = 1.5$. H_1: $\mu > 1.5$. Test statistic: $t = 0.049$. Critical value: $t = 2.015$. P-value is greater than 0.10 (TI-83/84 Plus: 0.4814). Fail to reject H_0. There is not sufficient evidence to support a claim that the mean is greater than 1.5 $\mu g/m^3$. The assumption of a normal distribution is questionable because 5.40 appears to be an outlier.
27. H_0: $\mu = 1.8$ g. H_1: $\mu \neq 1.8$ g. Test statistic: $t = -1.297$. Critical values: $t = \pm 2.145$ (assuming $\alpha = 0.05$). P-value is greater than 0.20 (TI-83/84 Plus: 0.2155). Fail to reject H_0. There is not sufficient evidence to warrant rejection of the claim that the bats have a mean weight equal to 1.8 g.
29. H_0: $\mu = 5.670$ g. H_1: $\mu \neq 5.670$ g. Test statistic: $t = -3.135$. Critical values: $t = \pm 2.024$ (approximately, assuming $\alpha = 0.05$). P-value is less than 0.01 (TI-83/84 Plus: 0.0033). Reject H_0. There is sufficient evidence to warrant rejection of the claim that the mean is equal to 5.670 g.
31. H_0: $\mu = 60$. H_1: $\mu > 60$. Test statistic: $t = 5.262$. Critical value: $t = 1.686$ (approximately, assuming $\alpha = 0.05$). P-value is less than 0.005 (TI-83/84 Plus: 0.0000). Reject H_0. There is sufficient evidence to support the claim that the population has a mean pulse rate greater than 60.
33. H_0: $\mu = 100$ sec. H_1: $\mu \neq 100$ sec. Test statistic: $t = 1.999$. Critical values: $t = \pm 2.040$. $0.05 < P\text{-value} < 0.10$ (TI-83/84 Plus: 0.0545). Fail to reject H_0. There is not sufficient evidence to warrant rejection of the claim that the mean equals 100. Using the alternative method with 15.0 used for σ, we get test statistic $z = 2.00$, critical values $z = \pm 1.96$, P-value $= 0.0456$, so we reject H_0 and conclude that there is sufficient evidence to warrant rejection of the claim that the mean equals 100. The conclusions are different.
35. The test statistic changes to $t = 0.992$ and the P-value changes to 0.182. An outlier can change the test statistic and P-value substantially. Although the conclusion does not change here, it could change in other cases.
37. Power = 0.4127, P(type II error) $= \beta = .5873$. No, the power of 0.4127 indicates that the likelihood of rejecting $\mu = 1.8$ g (when $\mu = 1.7$ g) is not very high. The power should be 0.8 or higher.

Section 8-6

1. The chi-square test is not resistant to departures from normality, so that it is very sensitive to departures from normality and performs poorly with distributions that are far from normal. When testing the requirement of a normal distribution, we use much stricter criteria for determining that a distribution is normal.
3. No, the distribution of the outcomes is likely to be a uniform distribution, not a normal distribution as required.
5. Test statistic: $\chi^2 = 20.250$. Critical values: $\chi^2 = 2.700, 19.023$. P-value: Between 0.02 and 0.05 (TI-83/84 Plus: 0.0329). Reject H_0. There is sufficient evidence to support the claim that $\sigma \neq 2.00$.
7. Test statistic: $\chi^2 = 8.889$. Critical value: $\chi^2 = 8.260$. P-value: Between 0.01 and 0.025 (TI-83/84 Plus: 0.0158). Fail to reject H_0. There is not sufficient evidence to support the claim that $\sigma < 15$.

9. H_0: $\sigma = 696$ g. H_1: $\sigma \neq 696$ g. Test statistic: $\chi^2 = 162.317$. Critical values are given. *P*-value: Greater than 0.05 (TI-83/84 Plus: 0.1591). Fail to reject H_0. There is not sufficient evidence to support the claim that the standard deviation of birth weights of infants born to mothers who use cocaine is different from 696 g.
11. H_0: $\sigma = 0.056$ g. H_1: $\sigma > 0.056$ g. Test statistic: $\chi^2 = 1225.765$. Critical value: $\chi^2 = 63.691$. *P*-value: Less than 0.005 (TI-83/84 Plus: 0.0000). Reject H_0. There is sufficient evidence to support the claim that peanut M&Ms have weights that vary more than the weights of plain M&Ms. The weights of the peanut M&Ms vary more because the weights of the peanuts have large variation.
13. H_0: $\sigma = 83$. H_1: $\sigma \neq 83$. Test statistic: $\chi^2 = 20.024$. Critical values: $\chi^2 = 6.262, 27.488$. *P*-value: Greater than 0.20 (TI-83/84 Plus: 0.3420). Fail to reject H_0. There is not sufficient evidence to support the claim that the standard deviation is different from 83. Applicants from the branch do not appear to have credit ratings that vary more than those at the main bank.
15. H_0: $\sigma = 29$. H_1: $\sigma < 29$. Test statistic: $\chi^2 = 0.540$. Critical value: $\chi^2 = 1.646$. *P*-value: Less than 0.005 (TI-83/84 Plus: 0.0002). Reject H_0. There is sufficient evidence to support the claim that the weights of supermodels vary less than the weights of women in general.
17. H_0: $\sigma = 0.4$. H_1: $\sigma > 0.4$. Test statistic: $\chi^2 = 114.506$. Critical value: $\chi^2 = 11.071$. *P*-value: Less than 0.005 (TI-83/84 Plus: 0.0000). Reject H_0. There is sufficient evidence to support the claim that the standard deviation is greater than 0.4. The sample value of 5.40 appears to be an outlier, and the requirement of a normal distribution is questionable.
19. H_0: $\sigma = 0.068$ g. H_1: $\sigma \neq 0.068$ g. Test statistic: $\chi^2 = 32.355$. Critical values: $\chi^2 = 24.433, 59.342$ (approximately). *P*-value: Greater than 0.20 (TI-83/84 Plus: 0.4695). Fail to reject H_0. There is not sufficient evidence to warrant rejection of the claim that the standard deviation is equal to 0.068 g. The variation of the weights appears to be as desired.
21. The approximate value of 152.3645 is close to the STATDISK value of 152.8222, and the approximate value of 228.4771 is close to the STATDISK value of 228.9638. The approximation is quite good.
23. Yes, because an outlier dramatically affects the value of the sample standard deviation.

Chapter 8 Statistical Literacy and Critical Thinking

1. The mean weight loss is statistically significant, but it does not appear to have practical significance. It should not be recommended because it results in such a small amount of lost weight.
2. The population is limited to the respondents. Because the sample is a voluntary response sample, results of a hypothesis test do not apply to the population of adult Americans.
3. 0.001, because it corresponds to the strongest evidence of an effective treatment. Because the *P*-value of 0.001 is so low, it indicates that it is highly unlikely that the results could be obtained by chance, assuming that the treatment has no effect.
4. There is not sufficient evidence to support the claim that the mean amount of cola is greater than 12 oz. Assuming that the mean amount of cola in cans is equal to 12 oz, there is a 0.25 probability of getting a sample mean with a value equal to or greater than the sample mean that was obtained.

Chapter 8 Review Exercises

1. a. H_1: $\mu < \$10{,}000$; *t* distribution
   b. H_1: $p > 0.5$; normal distribution
   c. H_1: $\mu \neq 100$; normal distribution
   d. H_1: $\sigma > 1.8$ sec; the methods of this chapter should not be used.
   e. H_1: $\sigma < 15$; chi-square distribution
2. H_0: $p = 0.5$. H_1: $p < 0.5$. Test statistic: $z = -1.47$. Critical value: $z = -1.645$. *P*-value: 0.0708. Fail to reject H_0. There is not sufficient evidence to support the claim that less than half of all executives identify "little or no knowledge of the company" as the most common interview mistake.
3. H_0: $p = 0.62$. H_1: $p < 0.62$. Test statistic: $z = -2.06$. Critical value: $z = -2.33$. *P*-value: 0.0197. Fail to reject H_0. There is not sufficient evidence to support the claim that less than 62% of brides spend less than \$750 on their wedding gown. The results would not be valid if they were obtained from a voluntary response sample.
4. H_0: $\mu = 3.5$ g. H_1: $\mu \neq 3.5$ g. Test statistic: $t = 9.723$. Critical values: $t = \pm 1.994$ (approximately, assuming $\alpha = 0.05$). *P*-value: Less than 0.01 (TI-83/84 Plus: 0.0000). Reject H_0. There is sufficient evidence to warrant rejection of the claim that the mean equals 3.5 g. The difference does not seem large enough to create health problems for consumers.
5. H_0: $\mu = 12$ oz. H_1: $\mu < 12$ oz. Test statistic: $t = -4.741$. Critical value: $t = -1.714$ (assuming $\alpha = 0.05$). *P*-value is less than 0.005 (TI-83/84 Plus: 0.00004). Reject H_0. There is sufficient evidence to support the claim that the mean is less than 12 oz. Windsor's argument is not valid.
6. H_0: $p = 0.019$. H_1: $p > 0.019$. Test statistic: $z = 0.65$. Critical value: $z = 2.33$. *P*-value: 0.2578 (TI-83/84 Plus: 0.2582). Fail to reject H_0. There is not sufficient evidence to support the claim that the percentage of treated patients with flu symptoms is greater than 1.9%. Flu symptoms do not appear to be an adverse reaction to the treatment.
7. H_0: $\sigma = 15$. H_1: $\sigma \neq 15$. Test statistic: $\chi^2 = 2.765$. Critical values: $\chi^2 = 4.404, 23.337$. *P*-value: Less than 0.01 (TI-83/84 Plus: 0.0060). Reject H_0. There is sufficient evidence to warrant rejection of the claim that $\sigma = 15$. The standard deviation of statistics professors appears to be less than 15.
8. H_0: $\mu = 100$. H_1: $\mu \neq 100$. Test statistic: $z = -0.75$. Critical value: $z = \pm 1.645$. *P*-value: 0.4532 (TI-83/84 Plus: 0.4507). Fail to reject H_0. There is not sufficient evidence to warrant rejection of the claim that the mean is equal to 100. Based on these results, the calculator is working okay.
9. H_0: $\mu = 98.6°$F. H_1: $\mu < 98.6°$F. Test statistic: $t = -1.349$. Critical value: $t = -1.796$. *P*-value: Greater than 0.10 (TI-83/84 Plus:

0.1023). Fail to reject H_0. There is not sufficient evidence to support the claim that the mean body temperature is less than 98.6°F.

10. H_0: $\sigma = 2.52$ in. H_1: $\sigma \neq 2.52$ in. Test statistic: $\chi^2 = 46.107$. Critical values: $\chi^2 = 24.433, 59.342$ (approximately). *P*- value: Greater than 0.20 (TI-83/84 Plus: 0.4038). Fail to reject H_0. There is not sufficient evidence to warrant rejection of the claim that $\sigma = 2.52$ in. Designs based on a standard deviation that is too small will result in too many women not being able to use products.
11. H_0: $p = 0.610$. H_1: $p > 0.610$. Test statistic: $z = 1.60$. Critical value: $z = 1.645$. *P*-value: 0.0548 (TI-83/84 Plus: 0.0552). Fail to reject H_0. There is not sufficient evidence to support the claim that the strike rate is greater than 61.0%.
12. H_0: $\mu = 40$ mg. H_1: $\mu > 40$ mg. Test statistic: $t = 2.746$. Critical value: $t = 2.821$. *P*-value: Greater than 0.01 (TI-83/84 Plus: 0.0113). Fail to reject H_0. There is not sufficient evidence to support the claim that the mean body nicotine content is greater than 40 mg.

Chapter 8 Cumulative Review Exercises

1. a. 71.5 in.
   b. 71.5 in.
   c. 2.1 in.
   d. 4.6 in.^{2}
   e. 8.0 in.
   f. 70.4 in. $< \mu <$ 72.6 in.
   g. H_0: $\mu = 69.0$ in. H_1: $\mu > 69.0$ in. Test statistic: $t = 4.908$ (or 4.875 if using greater precision with the original data). Critical value: $t = 1.746$. *P*-value: Less than 0.005 (TI-83/84 Plus: 0.0001). Reject H_0. There is sufficient evidence to support the claim that the mean height is greater than 69.0 in. Presidents appear to be taller than the typical man.
2. Yes, the points are quite close to a straight-line pattern, and there are no other patterns that are not straight-line patterns.
3. a. 0.4840 (TI-83/84 Plus: 0.4852)
   b. 0.0266 (TI-83/84 Plus: 0.0269)
   c. 0.4681 (TI-83/84 Plus: 0.4670)
   d. 634

Chapter 9 Answers

Section 9-2

1. It is represented by $\bar{p}$ and it is the proportion that consists of the total number of successes from both samples divided by the total number of trials from both samples. It is used as an estimate for the common population proportion in hypothesis tests, but it is not used for constructing confidence intervals for the difference between two proportions.
3. We have 95% confidence that the limits 0.200 and 0.300 contain the true value of the difference $p_1 - p_2$. This means that if we were to repeat the sampling process many times, the confidence interval limits will contain the true difference between the population proportions in 95% of the cases.
5. 748
7. 68
9. a. $150/900 = 0.167$
   b. 3.00
   c. ± 1.96
   d. 0.0026 (TI-83/84 Plus: 0.0027)
11. H_0: $p_1 = p_2$. H_1: $p_1 \neq p_2$. Test statistic: $z = 1.10$. *P*-value: 0.2719. Critical values for $\alpha = 0.05$: $z = \pm 1.96$. Fail to reject H_0. There is not sufficient evidence to support a claim that there is a significant difference between the proportions of home wins. Based on these results, the home field advantage appears to be about the same for basketball and football.
13. a. $0.0224 < p_1 - p_2 < 0.264$
    b. Yes. The confidence interval limits do not include 0, suggesting that there is a difference between the two proportions. Both the XSORT and YSORT methods appear to be effective because they both produce results that are not likely to occur by chance.
15. H_0: $p_1 = p_2$. H_1: $p_1 > p_2$. Test statistic: $z = 9.59$. *P*-value: 0.0000. Critical value: $z = 2.33$. Reject H_0. There is sufficient evidence to support the claim that the refusal rate is lower with the rigorous survey. The rigorous survey is more likely to produce accurate results.
17. H_0: $p_1 = p_2$. H_1: $p_1 > p_2$. Test statistic: $z = 2.17$. *P*-value: 0.0150 (TI-83/84 Plus: 0.0151). Critical value: $z = 1.645$. Reject H_0. There is sufficient evidence to support the claim that for those saying that monitoring e-mail is seriously unethical, the proportion of employees is greater than the proportion of bosses.
19. H_0: $p_1 = p_2$. H_1: $p_1 \neq p_2$. Test statistic: $z = 2.43$. *P*-value: 0.0150 (TI-83/84 Plus: 0.0149). Critical values for $\alpha = 0.05$: $z = \pm 1.96$. Critical values for $\alpha = 0.01$: $z = \pm 2.575$. Difference is significant at the 0.05 level, but not the 0.01 level.
21. Use sample proportions of 35/1655 and 20/1652. H_0: $p_1 = p_2$. H_1: $p_1 > p_2$. Test statistic: $z = 2.03$. *P*-value: 0.0212 (TI-83/84 Plus: 0.0210). Critical value: $z = 1.645$. Reject H_0. There is sufficient evidence to support the claim that the incidence of fatigue is greater among those who use Clarinex. Although the incidence of fatigue does appear to be significantly greater for Clarinex users, the rate of 2.1% is not so much greater than the rate of 1.2%, so fatigue does not appear to be a major concern for Clarinex users.
23. Use sample proportions of 880/1068 and 725/1064 to get $0.106 < p_1 - p_2 < 0.179$. There does appear to be a significant difference. Major factor: Answer varies, but one reason might be that homeowners tend to live in suburbs (where they are more likely to drive), whereas renters tend to live in urban areas (where they are more likely to use public transportation). Income levels could be another factor.
25. H_0: $p_1 = p_2$. H_1: $p_1 \neq p_2$. Test statistic: $z = -9.83$. *P*-value: 0.0002 (TI-83/84 Plus: 0.0000). Critical values: $z = \pm 1.96$. Reject H_0. There is sufficient evidence to warrant rejection of the claim that both genders have the same rate of seat belt use. There does appear to be a gender gap.
27. $-0.135 < p_1 - p_2 < 0.0742$ (using $x_1 = 49$ and $x_2 = 70$); there does not appear to be a gender gap.

29. Weekdays: 92/261; weekends: 35/104. H_0: $p_1 = p_2$. H_1: $p_1 < p_2$. Test statistic: $z = 0.29$. P-value: 0.6141 (TI-83/84 Plus: 0.6136). Critical value using a 0.05 significance level: $z = -1.645$. Fail to reject H_0. There is not sufficient evidence to support the claim that the proportion of weekend days with rain is greater than the proportion of weekdays with rain.
31. H_0: $p_1 = p_2$. H_1: $p_1 < p_2$. Test statistic: $z = -0.60$. P-value: 0.2743 (TI-83/84 Plus: 0.2739). Critical value using a 0.05 significance level: $z = -1.645$. Fail to reject H_0. There is not sufficient evidence to support the claim that the proportion of children's movies showing alcohol use is less than the proportion showing tobacco use. The results do not apply to Data Set 5 because the samples are not independent.
33. a. $0.0227 < p_1 - p_2 < 0.217$; because the confidence interval limits do not contain 0, it appears that $p_1 = p_2$ can be rejected.
    b. $0.491 < p_1 < 0.629$; $0.371 < p_2 < 0.509$; because the confidence intervals do overlap, it appears that $p_1 = p_2$ cannot be rejected.
    c. H_0: $p_1 = p_2$. H_1: $p_1 \neq p_2$. Test statistic: $z = 2.40$. P-value: 0.0164. Critical values: $z = \pm 1.96$. Reject H_0. There is sufficient evidence to reject $p_1 = p_2$.
    d. Reject $p_1 = p_2$. Least effective: Using the overlap between the individual confidence intervals.
35. H_0: $p_1 - p_2 = 0.50$. H_1: $p_1 - p_2 \neq 0.50$. Test statistic: $z = 1.65$ (using $x_1 = 151$ and $x_2 = 17$). P-value: 0.0990 (TI-83/84 Plus: 0.0985). Critical values: $z = \pm 1.96$. Fail to reject H_0. There is not sufficient evidence to warrant rejection of the claim that the virus leak rate for vinyl gloves is 50 percentage points greater than the virus leak rate for latex gloves.
37. a. For the proportions 0, 0.5, 0.5, 1, $\mu = 0.5$ and $\sigma^2 = 1/8$.
    b. For the differences 0, −0.5, −0.5, −1, 0.5, 0, 0, −0.5, 0.5, 0, 0, −0.5, 1, 0.5, 0.5, 0, $\mu = 0$ and $\sigma^2 = 1/4$.
    c. $\sigma^2_{\hat{p}_b} + \sigma^2_{\hat{p}_a} = 1/8 + 1/8 = 1/4$, and $\sigma^2_{(\hat{p}_b - \hat{p}_a)} = 1/4$.

Section 9-3

1. No, 80 would be an outlier suggesting that the requirement of normal distributions is not satisfied.
3. The critical value of 1.717 is more conservative than 1.682 in the sense that rejection of the null hypothesis of equal means requires a greater difference between the sample means. The sample evidence must be stronger with the critical value of 1.717.
5. Independent
7. Matched pairs
9. H_0: $\mu_1 = \mu_2$. H_1: $\mu_1 \neq \mu_2$. Test statistic: $t = 1.674$. Critical values: $t = \pm 1.968$ (approximately). P-value is between 0.05 and 0.10 (TI-83/84 Plus: 0.0946). Fail to reject H_0. Echinacea does not appear to have an effect.
11. $6.7 < \mu_1 - \mu_2 < 12.1$. IQ score does appear to be affected by birth weight.
13. H_0: $\mu_1 = \mu_2$. H_1: $\mu_1 > \mu_2$. Test statistic: $t = 2.790$. Critical value: $t = 2.385$ (approximately). P-value is less than 0.005 (TI-83/84 Plus: 0.0031). Reject H_0. There is sufficient evidence to support the claim that heavy users have a lower mean than light users. Because marijuana appears to adversely affect mental abilities, it should be a serious concern.
15. $-0.65 < \mu_1 - \mu_2 < 3.03$ (TI-83/84 Plus: $-0.61 < \mu_1 - \mu_2 < 2.99$). Because the confidence interval does contain zero, we should not conclude that the two population means are different. The treatment does not appear to be effective, so paroxetine should not be prescribed.
17. H_0: $\mu_1 = \mu_2$. H_1: $\mu_1 > \mu_2$. Test statistic: $t = 0.132$. Critical value: $t = 1.729$. P-value > 0.10 (TI-83/84 Plus: 0.4480). Fail to reject H_0. There is not sufficient evidence to support the claim that the magnets are effective in reducing pain. It is valid to argue that the magnets might appear to be effective if the sample sizes are larger.
19. $-0.01 < \mu_1 - \mu_2 < 0.23$; because this CI contains zero, there does not appear to be a significant difference between the two population means, so it does not appear that obsessive-compulsive disorders have a biological basis. (With TI-83/84 Plus: df = 18 and $0.01 < \mu_1 - \mu_2 < 0.21$, which does not contain zero, suggesting that there is a significant difference so that obsessive-compulsive disorders appear to have a biological basis. This is a rare case where the simple and conservative estimate of df leads to a different conclusion than use of the more accurate Formula 9-1.)
21. $1.46 < \mu_1 - \mu_2 < 3.52$ (TI-83/84 Plus: $1.47 < \mu_1 - \mu_2 < 3.51$). Because the confidence interval does not contain zero, there appears to be a significant difference between the two population means. It does appear that there are significantly more errors made by those treated with alcohol.
23. H_0: $\mu_1 = \mu_2$. H_1: $\mu_1 > \mu_2$. Test statistic: $t = 3.992$. Critical value: $t = 2.132$. P-value is between 0.005 and 0.01 (TI-83/84 Plus: 0.0035). Reject H_0. There is sufficient evidence to support the claim that oceanfront homes have greater value than oceanside homes. The large difference between the means more than compensates for the fact that the samples are small.
25. Filtered: $n_1 = 21$, $\bar{x}_1 = 13.3$, $s_1 = 3.7$. Nonfiltered: $n_2 = 8$, $\bar{x}_2 = 24.0$, $s_2 = 1.7$. H_0: $\mu_1 = \mu_2$. H_1: $\mu_1 < \mu_2$. Test statistic: $t = -10.585$. Critical value: $t = -1.895$. P-value < 0.005 (TI-83/84 Plus: 0.0000). Reject H_0. There is sufficient evidence to support the claim that the mean amount of tar in filtered king-size cigarettes is less than the mean amount of tar in nonfiltered king-size cigarettes.
27. $n_1 = 40$, $\bar{x}_1 = 6.192672$, $s_1 = 0.086995$, $n_2 = 40$, $\bar{x}_2 = 5.639298$, $s_2 = 0.0619368$. H_0: $\mu_1 = \mu_2$. H_1: $\mu_1 \neq \mu_2$. Test statistic: $t = 32.773$. Critical values: $t = \pm 2.024$ (assuming a 0.05 significance level). P-value < 0.01 (TI-83/84 Plus: 0.000). Reject H_0. There is sufficient evidence to warrant rejection of the claim that the two populations have equal means. The difference is highly significant, even though the sample is relatively small.
29. $-1.3 < \mu_1 - \mu_2 < 3.1$. The results are the same as in Exercise 12, so in this case they are not affected by the assumption of equal standard deviations.

31. The test statistic is the same: $t = 0.132$. Critical value changes from $t = 1.729$ to $t = 1.686$. P-value > 0.10 (TI-83/84 Plus: 0.4479). In this case, the conclusions are the same.
33. H_0: $\mu_1 = \mu_2$. H_1: $\mu_1 \neq \mu_2$. Test statistic: $t = 0.697$. Critical values: $t = \pm 2.093$. P-value > 0.20 (TI-83/84 Plus: 0.4904). The test statistic changes by a considerable amount. Although the conclusion is the same in this case, the results are dramatically affected.
35. a. 50/3
    b. 2/3
    c. 52/3
    d. $50/3 + 2/3 = 52/3$
    e. The range of the x-y values equals the range of the x values plus the range of the y values.
37. df = 18 (instead of 9), the critical values become $t = \pm 2.878$ (instead of ± 3.250), the confidence interval limits become 0.007 and 0.213, and the P-value is less than 0.01 (instead of between 0.01 and 0.02). Using Formula 9-1, the confidence interval is a little narrower, the critical value is a little smaller, and the P-value is a little smaller. With df = 9 it does not appear that obsessive-compulsive disorders have a biological basis; with df = 18 from Formula 9-1, it does appear that obsessive-compulsive disorders have a biological basis. Using the smaller of $n_1 - 1$ and $n_2 - 1$ for df is more conservative (than the use of Formula 9-1) in the sense that the sample data need to be more extreme to be considered significant, as can be seen by the different conclusions.

Section 9-4

1. No. Even though the data consist of matched pairs, the two variables measure different quantities, so it makes no sense to test a claim about the mean of the differences.
3. $\bar{d}$ represents the mean of the differences between the matched sample values, and s_d denotes the standard deviation of those differences.
5. a. -0.4
   b. 2.1
   c. -0.431
   d. ± 2.776
7. $-3.0 < \mu_d < 2.2$
9. a. H_0: $\mu_d = 0$. H_1: $\mu_d \neq 0$. Test statistic: $t = -0.41$. P-value: 0.691. Fail to reject H_0. There is not sufficient evidence to support the claim that astemizole has an effect. Don't take astemizole for motion sickness.
   b. 0.3455; there is not sufficient evidence to support the claim that astemizole prevents motion sickness.
11. a. H_0: $\mu_d = 0$. H_1: $\mu_d \neq 0$. Test statistic: $t = 0.218$. Critical values: $t = \pm 2.776$. P-value > 0.20 (TI-83/84 Plus: 0.8379). Fail to reject H_0. There is not sufficient evidence to warrant rejection of the claim that the differences have a mean of zero. It appears that the forecast values are reasonably accurate.
    b. $-2.3 < \mu_d < 2.7$. The confidence interval contains zero, so we cannot reject the claim that the differences have a mean of zero.
13. $1.7 < \mu_d < 13.9$. Because the confidence interval does not contain 0, it appears the differences do not have a mean equal to 0. The results suggest that there is not a stable blood pressure that is measured consistently and with accuracy.
15. H_0: $\mu_d = 0$. H_1: $\mu_d \neq 0$. Test statistic: $t = -0.132$. Critical values: $t = \pm 2.306$ (using a 0.05 significance level). P-value > 0.20 (TI-83/84 Plus: 0.8981). Fail to reject H_0. There is not sufficient evidence to warrant rejection of the claim that the differences have a mean of 0. The results suggest that the scores did not change much for those in the control group.
17. a. $0.69 < \mu_d < 5.56$.
    b. H_0: $\mu_d = 0$. H_1: $\mu_d > 0$. Test statistic: $t = 3.036$. Critical value: $t = 1.895$. P-value is between 0.005 and 0.01 (TI-83/84 Plus: 0.0095). Reject H_0. There is sufficient evidence to support the claim that the sensory measurements are lower after hypnosis.
    c. Yes
19. a. H_0: $\mu_d = 0$. H_1: $\mu_d \neq 0$. Test statistic: $t = -1.690$. Critical values: $t = \pm 2.228$. P-value is between 0.10 and 0.20 (TI-83/84 Plus: 0.1218). Fail to reject H_0. There is not sufficient evidence to warrant rejection of the claim that there is no difference between the yields from the two types of seed.
    b. $-78.2 < \mu_d < 10.7$
    c. No
21. a. H_0: $\mu_d = 0$. H_1: $\mu_d \neq 0$. Test statistic: $t = 0.155$. Critical values: $t = \pm 2.032$. P-value > 0.20 (TI-83/84 Plus: 0.8775). Fail to reject H_0. There is not sufficient evidence to warrant rejection of the claim of no difference between the actual and predicted low temperatures.
    b. $-1.7 < \mu_d < 2.0$
    c. Same conclusion as example. It appears that the forecast temperatures are accurate.
23. a. H_0: $\mu_d = 0$. H_1: $\mu_d \neq 0$. Test statistic: $t = -5.354$. Critical values: $t = \pm 2.024$. P-value < 0.01 (TI-83/84 Plus: 0.0000). Reject H_0. There is sufficient evidence to warrant rejection of the claim of no difference between selling prices and list prices. There appears to be a significant difference.
    b. $-\$10{,}011.50 < \mu_d < -\$4{,}518.50$ (TI-83/84 Plus: $-\$10{,}010 < \mu_d < -\$4{,}520$)
25. Because the values must be estimated from the graph, answers will vary somewhat. H_0: $\mu_d = 0$. H_1: $\mu_d < 0$. Test statistic: $t = -4.334$. Critical value: $t = -1.860$. P-value < 0.005 (TI-83/84 Plus: 0.0012). Reject H_0. There is sufficient evidence to support the claim that the preparatory course is effective in raising scores.
27. a. Test statistic: $t = 1.861$. Critical values: $t = 1.833$. P-value is between 0.025 and 0.05 (TI-83/84 Plus: 0.0479). Reject H_0. There is sufficient evidence to support $\mu_d > 0$.
    b. Test statistic: $t = 1.627$. Critical value: $t = 1.833$. P-value is between 0.05 and 0.10 (TI-83/84 Plus: 0.0620). Fail to reject H_0. There is not sufficient evidence to support $\mu_1 > \mu_2$.
    c. Yes, the conclusion is affected by the test that is used.

Section 9-5

1. The F test is too sensitive to departures from normal distributions. Alternatives: Count five test, and Levene-Brown-Forsythe test.
3. It is the sampling distribution of values of s_1^2/s_2^2.
5. H_0: $\sigma_1^2 = \sigma_2^2$. H_1: $\sigma_1^2 \neq \sigma_2^2$. Test statistic: $F = 4.0000$. Upper critical value: $F = 2.1819$. P-value $= 0.0005$. Reject H_0. There is sufficient evidence to support the claim that the treatment and placebo populations have different variances.
7. H_0: $\sigma_1 = \sigma_2$. H_1: $\sigma_1 \neq \sigma_2$. Test statistic: $F = 2.9233$. P-value $=$ 0.0021. Reject H_0. There is sufficient evidence to support the claim that regular Coke and Diet Coke have weights with different standard deviations. The difference is probably due to the sugar in regular Coke.
9. H_0: $\sigma_1 = \sigma_2$. H_1: $\sigma_1 \neq \sigma_2$. Test statistic: $F = 1.0133$. Critical value is between 1.8752 and 2.0739. (TI-83/84 Plus: Upper critical value is $F = 1.9611$ and P-value $= 0.9690$). Fail to reject H_0. There is not sufficient evidence to support the claim that Diet Coke and Diet Pepsi have weights with different standard deviations.
11. H_0: $\sigma_1^2 = \sigma_2^2$. H_1: $\sigma_1^2 > \sigma_2^2$. Test statistic: $F = 2.1267$. The critical F value is between 2.1555 and 2.2341. (TI-83/84 Plus: P-value $=$ 0.0543.) Fail to reject H_0. There is not sufficient evidence to support the claim that the pain reductions for the sham treatment group vary more than the pain reductions for the magnet treatment group.
13. H_0: $\sigma_1^2 = \sigma_2^2$. H_1: $\sigma_1^2 > \sigma_2^2$. Test statistic: $F = 9.3364$. Critical value of F is between 2.0540 and 2.0960. (TI-83/84 Plus: P-value $= 0.0000$.) Reject H_0. There is sufficient evidence to support the claim that the treatment group has scores that vary more than the scores of the placebo group.
15. H_0: $\sigma_1 = \sigma_2$. H_1: $\sigma_1 \neq \sigma_2$. Test statistic: $F = 1.9729$. The upper critical value of F is between 1.8752 and 2.0739, but it should be much closer to 1.8752. (TI-83/84 Plus: P-value $= 0.0368$.) Reject H_0. There is sufficient evidence to warrant rejection of the claim that the two populations of quarters have the same standard deviation.
17. H_0: $\sigma_1^2 = \sigma_2^2$. H_1: $\sigma_1^2 \neq \sigma_2^2$. Test statistic: $F = 2.5906$. The upper critical value is 2.7006 (assuming a 0.05 significance level). (TI-83/84 Plus: P-value: 0.0599.) Fail to reject H_0. There is not sufficient evidence to warrant rejection of the claim that the two samples come from populations with the same variance.
19. a. Test statistic: $F = 2.2080$. Upper critical value of F is between 1.6668 and 1.8752. (TI-83/84 Plus: P-value $= 0.0052$.) Reject H_0. There is sufficient evidence to warrant rejection of the claim that Wednesday and Sunday rainfall amounts have the same standard deviation.
    b. Because they have so many zeros as the lowest values, neither the Wednesday rainfall amounts nor the Sunday rainfall amounts are normally distributed.
    c. Because the populations do not appear to be normally distributed, the conclusion given in part (a) is not necessarily valid. The methods of Section 9-5 do not apply.
21. $c_1 = 1$, $c_2 = 0$, critical value is 5. Fail to reject $\sigma_1^2 = \sigma_2^2$.
23. a. $F_L = 0.2484$, $F_R = 4.0260$
    b. $F_L = 0.2315$, $F_R = 5.5234$
    c. $F_L = 0.1810$, $F_R = 4.3197$

Chapter 9 Statistical Literacy and Critical Thinking

1. Use the methods of Section 9-2 to make inferences about the difference between the proportion of male voters who favor the candidate and the proportion of female voters who favor the candidate.
2. Yes. Every sample of size 200 has the same chance of being selected.
3. It is not okay. The different states have different populations, so weighted means should be used. A better procedure is to use simple random samples selected from the U.S. population.
4. If samples are not independent, they are dependent. Dependent samples are often referred to as matched pairs. With matched pairs, each value in one sample is somehow paired with a particular value from the other sample.

Chapter 9 Review Exercises

1. a. H_0: $p_1 = p_2$. H_1: $p_1 > p_2$. Test statistic: $z = 0.64$. Critical value: $z = 1.645$. P-value: 0.2611 (TI-83/84 Plus: 0.2603). Fail to reject H_0. There is not sufficient evidence to support the claim that the proportion of blacks stopped by police is significantly greater than the proportion of whites.
   b. 90%: $-0.0251 < p_1 - p_2 < 0.0551$; because the confidence interval contains 0, there does not appear to be a significant difference between the two proportions.
2. H_0: $\mu_d = 0$. H_1: $\mu_d > 0$. Test statistic: $t = 2.701$. Critical value: $t = 1.812$ (assuming a 0.05 significance level). P-value is between 0.01 and 0.025 (TI-83/84 Plus: 0.0111). Reject H_0. There is sufficient evidence to support the claim that male statistics students exaggerate their heights.
3. H_0: $\mu_1 = \mu_2$. H_1: $\mu_1 > \mu_2$. Test statistic: $t = 5.529$. Critical value: $t = 1.796$. P-value < 0.005 (TI-83/84 Plus: 0.0000). Reject H_0. There is sufficient evidence to support the claim that *Harry Potter* is easier to read than *War and Peace*.
4. H_0: $\sigma_1 = \sigma_2$. H_1: $\sigma_1 \neq \sigma_2$. Test statistic: $F = 2.8176$. The upper critical value of F is between 3.4296 and 3.5257. (TI-83/84 Plus: P-value $= 0.1000$.) Fail to reject H_0. There is not sufficient evidence to warrant rejection of the claim that pages from the two books have Flesch Reading Ease scores with the same standard deviation.
5. a. H_0: $p_1 = p_2$. H_1: $p_1 < p_2$. Test statistic: $z = -2.82$. Critical value: $z = -1.645$. P-value: 0.0024. Reject H_0. There is sufficient evidence to support the stated claim. It appears that surgical patients should be routinely warmed.
   b. 90%
   c. $-0.205 < p_1 - p_2 < -0.0543$
   d. No, the conclusions may be different.

6. a. $-27.80 < \mu_1 - \mu_2 < 271.04$
   (TI-83/84 Plus: $-17.32 < \mu_1 - \mu_2 < 260.56$)
   b. H_0: $\mu_1 = \mu_2$. H_1: $\mu_1 \neq \mu_2$. Test statistic: $t = 1.841$. Critical values: $t = \pm 2.262$. *P*-value: between 0.05 and 0.10 (TI-83/84 Plus: 0.0824). Fail to reject H_0. There is not sufficient evidence to warrant rejection of the claim that there is no difference between the two population means.
   c. No
7. H_0: $\sigma_1 = \sigma_2$. H_1: $\sigma_1 \neq \sigma_2$. Test statistic: $F = 1.2922$. Upper critical value: $F = 4.0260$. (TI-83/84 Plus: *P*-value = 0.7087). Fail to reject H_0. There is not sufficient evidence to support the claim that the two populations have different amounts of variation.
8. a. H_0: $\mu_d = 0$. H_1: $\mu_d \neq 0$. Test statistic: $t = -1.532$. Critical values: $t = \pm 2.228$. *P*-value is between 0.10 and 0.20 (TI-83/84 Plus: 0.1565). Fail to reject H_0. There is not sufficient evidence to warrant rejection of the claim that there is no difference. There does not appear to be a difference.
   b. $-2.7 < \mu_d < 0.5$
   c. No, there is not a significant difference.

Chapter 9 Cumulative Review Exercises

1. a. Mean: 69.5 mi/h; median: 69.5 mi/h; $s = 3.4$ mi/h; $s^2 = 11.6$ mi^2/h^2; range: 9.0 mi/h
   b. H_0: $\mu = 65$. H_1: $\mu > 65$. Test statistic: $t = 3.765$. Critical value: $t = 1.686$ (approximately). *P*-value < 0.005 (TI-83/84 Plus: *P*-value = 0.0003). Reject H_0. There is sufficient evidence to support the claim that the mean is greater than 65 mi/h.
   c. Using a histogram or normal quantile plot, the distribution is not very normal, but it is not a dramatic departure from a normal distribution.
   d. H_0: $\mu_1 = \mu_2$. H_1: $\mu_1 \neq \mu_2$. Test statistic: $t = 1.265$. Critical values: $t = \pm 2.093$ (assuming a 0.05 significance level). *P*-value > 0.20 (TI-83/84 Plus: 0.2167). Fail to reject H_0. There is not sufficient evidence to warrant rejection of the claim that the mean speed on the northbound lanes is equal to the mean speed on the southbound lanes. The distribution is not normal, but there are no outliers and the distribution is not heavily skewed, so the robustness of the *t* test probably makes it okay in this situation.
2. a. 1/512 or 0.00195. The probability indicates that it is very unlikely that she would get nine heads by chance, so it appears that she has the ability to toss the quarter so that it comes up heads.
   b. Independent. The outcomes are not matched or paired in any way.
   c. H_0: $p = 0.5$. H_1: $p > 0.5$. Test statistic: $z = 2.36$. Critical value: $z = 2.33$. *P*-value = 0.0091 (TI-83/84 Plus: 0.0092). Reject H_0. There is sufficient evidence to support the claim that a coin can be tossed so that heads turns up more often than can be expected by chance.
3. There must be an error, because the rates of 13.7% and 10.6% are not possible with sample sizes of 100.

Chapter 10 Answers

Section 10-2

1. No, because social security numbers do not satisfy the necessary requirements. Because they serve as identification labels that do not measure or count anything, they are not quantitative data.
3. A correlation is a relationship between two variables. A lurking variable is one that affects the variables being studied, but is not included in the study.
5. a. Yes, because the absolute value of the test statistic exceeds the critical values $r = \pm 0.707$.
   b. 0.986
7. a. No, because the absolute value of the test statistic does not exceed the critical values $r = \pm 0.444$ (approximately).
   b. 0.0177
9. The scatterplot suggests that there is a linear correlation. With $r = 0.994$ and critical values of $r = \pm 0.878$ (for a 0.05 significance level), there is a significant linear correlation.
11. a. There appears to be a linear correlation.
   b. $r = 0.906$. Critical values: $r = \pm 0.632$ (for a 0.05 significance level). There is a linear correlation.
   c. $r = 0$. Critical values: $r = \pm 0.666$ (for a 0.05 significance level). There does not appear to be a linear correlation.
   d. The effect from a single pair of values can be very substantial, and it can change the conclusion.
13. $r = 0.269$. Critical values: $r = \pm 0.707$. No linear correlation.
15. $r = 0.926$. Critical values: $r = \pm 0.754$. There is a linear correlation. Other factors include the quality of the movie, the type of movie, and the stars that are featured.
17. $r = 0.983$. Critical values: $r = \pm 0.632$. There is a linear correlation.
19. $r = 0.658$. Critical values: $r = \pm 0.532$. There is a linear correlation. Another issue is the accuracy of the measurements, which appear to vary widely. A study might be conducted to determine whether the subject's blood pressure really does vary considerably, or whether the measurements are in error because of other factors.
21. $r = -0.118$. Critical values: $r = \pm 0.707$. No linear correlation. Lowest: Susan Lucci. Highest: Kelsey Grammer.
23. $r = 0.183$. Critical values: $r = \pm 0.707$. No linear correlation. It appears that winning time is not affected by temperature.
25. $r = 0.874$. Critical values: $r = \pm 0.707$. There is a linear correlation. Yes.
27. $r = -0.038$. Critical values: $r = \pm 0.576$. No linear correlation. The hypothesis does not appear to be correct.
29. $r = 0.995$. Critical values: $r = \pm 0.312$. There is a linear correlation.
31. a. $r = 0.961$. Critical values: $r = \pm 0.361$ approximately. There is a linear correlation. Yes.
   b. $r = 0.863$. Critical values: $r = \pm 0.361$ approximately. There is a linear correlation. Yes.
   c. Tar, because it has a higher correlation with nicotine.

33. With a linear correlation coefficient very close to 0, we should conclude that there is not a linear correlation, so there does not appear to be an association between age and score, as the conclusion incorrectly suggests.
35. Although there is no linear correlation, the variables might be related in some other nonlinear way.
37. a. 0.972
    b. 0.905
    c. 0.999 (largest)
    d. 0.992
    e. −0.984
39. $0.386 < \rho < 0.753$

Section 10-3

1. In a regression equation $\hat{y} = b_0 + b_1x$, the variable x is the predictor variable and the variable y is the response variable. They were given those names because the value of x is commonly used to predict the value of y, so that the value of y responds to the value of x.
3. Different conditions might apply so that the predicted value might be off by a substantial amount.
5. 109
7. 3.86 calories per gram of cereal
9. $\hat{y} = 0.857 + 0.643x$
11. a. $\hat{y} = 0.264 + 0.906x$
    b. $\hat{y} = 2 + 0x$ (or $\hat{y} = 2$)
    c. The results are very different, indicating that one point can dramatically affect the regression equation.
13. $\hat{y} = 54.3 + 0.246x$; 86 min
15. $\hat{y} = -164 + 3.47x$; −$25,200,000 (which is a loss)
17. $\hat{y} = -252 + 12.4x$; 368 lb
19. $\hat{y} = -14.4 + 0.769x$; 93
21. $\hat{y} = 6.76 - 0.0111x$; 6.5 million.
23. $\hat{y} = 145 + 0.0316x$; 147.077
25. $\hat{y} = 27.6 + 0.0523x$; 79.9°F
27. $\hat{y} = 79.9 - 0.113x$; 73 beats per minute
29. $\hat{y} = 99.2 + 0.979x$; $391,699
31. a. $\hat{y} = 0.154 + 0.0651x$; 1.1
    b. $\hat{y} = 0.192 + 0.0606x$; 1.1
33. With $\beta_1 = 0$, the regression line is horizontal so that different values of x result in the same y value, and there is no correlation between x and y.
35. The equation $\hat{y} = -49.9 + 27.2x$ is better because it has $r = 0.997$, which is higher than $r = 0.963$ for $\hat{y} = -103.2 + 134.9x \ln x$.

Section 10-4

1. A prediction interval is a confidence interval (or interval estimate) of a predicted value of y.
3. Given paired data, $\bar{y}$ is the mean of the observed y values and, for an observed value of x, we can find the predicted y value (denoted as $\hat{y}$). The explained deviation is $\hat{y} - \bar{y}$, and the explained variation is the sum of the squares of the explained deviation values.
5. 0.09; 9%
7. 0.812; 81.2%
9. 0.961; yes
11. 1.3
13. a. 147.3965
    b. 18.03204
    c. 165.4286
    d. 0.891
    e. 1.899055
15. a. 628.9603
    b. 824.254
    c. 1453.214
    d. 0.433
    e. 8.287812
17. a. 25 mi/gal
    b. 19.5 mi/gal $< y <$ 30.9 mi/gal
19. a. 78
    b. $57.7 < y < 98.2$
21. 56.6 min $< y <$ 97.3 min
23. 68.6 min $< y <$ 94.6 min
25. $13.4 < \beta_0 < 56.1$; $0.138 < \beta_1 < 0.330$

Section 10-5

1. A multiple regression equation expresses a linear relationship between a response (y) variable and two or more predictor (x) variables. It differs from the regression equations discussed in Section 10-3, because that section included cases with only one predictor (x) variable, whereas a multiple regression equation has two or more predictor (x) variables.
3. No, because the data are categorical (or qualitative). The methods of this section require quantitative data. (Exceptions can occur with logistic regression.)
5. WEIGHT = −272 − 0.870 HEADLEN + 0.554 LENGTH + 12.2 CHEST or $\hat{y} = -272 - 0.870x_1 + 0.554x_2 + 12.2x_3$, where x_1 represents head length, x_2 represents length, and x_3 represents chest size.
7. Yes, because the overall significance of 0.000 is small and the adjusted R^2 of 0.924 is high.
9. The waist size, because it has the lowest P-value of 0.000 and the highest adjusted R^2 value of 0.785.
11. $\hat{y} = -206 + 2.66$ HT + 2.15 WAIST or $\hat{y} = -206 + 2.66x_1 + 2.15x_2$ (where x_1 represents height and x_2 represents waist size). Using predictor variables of height and waist size yields the lowest P-value of 0.000 and the highest adjusted R^2 of 0.870. Using predictor variables of height, waist size, and cholesterol level also yields a P-value of 0.000 and an adjusted R^2 of 0.870, but it is better to use two predictor variables instead of three.
13. a. $\hat{y} = 0.154 + 0.0651x$
    b. $\hat{y} = 0.192 + 0.0606x$

c. $\hat{y} = 0.182 + 0.0818x_1 - 0.0186x_2$, where x_1 represents the amount of tar and x_2 represents the amount of carbon monoxide.
d. Part (a). All three yield a *P*-value of 0.000, but the adjusted R^2 values are 0.924, 0.736, and 0.928, respectively. Choose part (a) because it has only one predictor variable, whereas part (c) has a slightly higher adjusted R^2, but part (c) has two predictor variables. If the values of adjusted R^2 are not too far apart, it is better to use the regression equation with fewer predictor variables.
e. Yes, because the *P*-value is low (0.000) and the adjusted R^2 is high (0.924).

15. $\hat{y} = 99.2 + 0.979x$, where *x* represents the list price. With a *P*-value of 0.000 and an adjusted R^2 of 0.990, this regression equation is likely to be very good for predicting selling prices based on list prices.
17. $\hat{y} = 3.06 + 82.4x_1 + 2.91x_2$, where x_1 represents sex and x_2 represents age. Female: 61 lb; male: 144 lb. The sex of the bear does appear to have an effect on its weight. The regression equation indicates that the predicted weight of a male bear is about 82 lb more than the predicted weight of a female bear with other characteristics being the same.

Section 10-6

1. A mathematical model is a mathematical function that fits or describes real-world data.
3. The year 3000 is too far beyond the scope of the available data. Conditions might change so that the model should not be used for the year 3000.
5. Linear: $y = 2x + 3$
7. Quadratic: $y = 2x^2 - 1$
9. Power: $y = 18.1x^{0.455}$ (where 1983 is coded as 1). With $R^2 = 0.789$, the model appears to be okay.
11. Quadratic: $y = 4.90x^2 - 0.0286x + 0.00476$. The model yields a distance of 705 m, but the building is only 50 m tall, so the distance cannot exceed 50 m.
13. a. Exponential: $y = 2^{\frac{2}{3}(x-1)}$ [or $y = (0.629961)(1.587401)^x$ for an initial value of 1 that doubles every 1.5 years]
    b. Exponential: $y = (1.38)(1.42)^x$
    c. Moore's law does appear to be working reasonably well. With $R^2 = 0.984$, the model appears to be very good.
15. a. 6641.8
    b. 73.2
    c. The quadratic sum of squares of residuals (73.2) is less than the sum of squares of residuals from the linear model (6641.8).

Chapter 10 Statistical Literacy and Critical Thinking

1. With correlation we investigate whether there is a relationship between variables, but with regression we attempt to identify the relationship with an equation.
2. No. There might be a relationship that is not linear.
3. No. The linear correlation coefficient can be used to establish that there is an association between taking the drug and the cholesterol level, but it cannot be used to establish that the drug *causes* lower cholesterol levels.
4. Not necessarily. It is very possible that there is a significant linear correlation, but the predicted values are not very accurate.

Chapter 10 Review Exercises

1. a. $r = 0.090$. Critical values: $r = \pm 0.602$ (assuming a 0.05 significance level). There is no linear correlation.
   b. $\hat{y} = 35.7 + 0.166x$. The best predicted number of natural deaths is 47 (the mean).
2. a. $r = 0.389$. Critical values: $r = \pm 0.707$ (assuming a 0.05 significance level). There is no linear correlation.
   b. $\hat{y} = 105 + 0.108x$.
   c. 128.75 ft (the mean)
3. a. $r = 0.915$. Critical values: $r = \pm 0.632$. There is a linear correlation.
   b. 84%
   c. $\hat{y} = -51.7 + 0.113x$
   d. $287.30
4. a. $r = 0.085$. Critical values: $r = \pm 0.632$. No linear correlation.
   b. 0.7%
   c. $\hat{y} = 235 + 0.243x$
   d. $247.30 (the mean)
5. $\hat{y} = -128 + 0.123x_1 + 0.955x_2$. $R^2 = 0.942$; adjusted $R^2 = 0.925$; *P*-value = 0.000. The equation can be used to predict cost, and it is better than the regression equations from Exercises 3 and 4 because it has the highest adjusted R^2 and the lowest *P*-value.

Chapter 10 Cumulative Review Exercises

1. $r = -0.431$. Critical values: $r = \pm 0.707$ (assuming a 0.05 significance level). There is no linear correlation.
2. $\hat{y} = 13{,}403 + 67.2x$; 9950 (the mean)
3. Yes, but the test would make no sense because the two variables measure different quantities.
4. $41.9 < \mu < 60.9$
5. Values of the DJIA are changing over time. The confidence interval would be an estimate of the mean based on the years included in the sample, but it would not necessarily provide a good estimate of the DJIA value for the next year.
6. Yes. A normal quantile plot shows that the points are reasonably close to a straight-line pattern, and there is no pattern that is not a straight-line pattern.
7. $\bar{x} = 51.4$; $s = 11.4$
8. 0.1587

Chapter 11 Answers

Section 11-2

1. To test for "goodness-of-fit" is to test the hypothesis that sample data agree with or fit a particular distribution that has been identified.
3. An observed frequency is a count of the number of sample values that correspond to one of the categories being considered. An expected frequency is the number of sample values expected for a category, assuming that the frequencies follow the claimed distribution.
5. a. H_0: $p_1 = p_2 = p_3$
   b. 10, 10, 10
   c. $\chi^2 = 15.000$
   d. $\chi^2 = 5.991$
   e. There is sufficient evidence to warrant rejection of the claim that the three categories are equally likely.
7. a. df = 37, so $\chi^2 = 51.805$ (approximately).
   b. $0.10 < P\text{-value} < 0.90$
   c. There is not sufficient evidence to warrant rejection of the claim that the roulette slots are equally likely.
9. Test statistic: $\chi^2 = 5.860$. Critical value: $\chi^2 = 11.071$. (TI-83/84 Plus: *P*-value = 0.3201.) There is not sufficient evidence to support the claim that the outcomes are not equally likely. The outcomes appear to be equally likely, so the loaded die does not appear to behave differently from a fair die.
11. Test statistic: $\chi^2 = 30.017$. Critical value: $\chi^2 = 12.592$. (TI-83/84 Plus: *P*-value = 0.0000.) There is sufficient evidence to warrant rejection of the claim that the car crash fatalities occur with equal frequency on the different days of the week. Drinking on Friday night might be a cause of an exceptionally large number of deaths on early Saturday morning.
13. Test statistic: $\chi^2 = 47.200$. Critical value: $\chi^2 = 19.675$. (TI-83/84 Plus: *P*-value = 0.0000.) There is sufficient evidence to warrant rejection of the claim that motorcycle fatalities occur with equal frequencies in the different months. Fatalities might be lower in winter months when colder weather is associated with substantially less use of motorcycles.
15. Test statistic: $\chi^2 = 4.211$. Critical value: $\chi^2 = 19.675$. (TI-83/84 Plus: *P*-value = 0.9633.) There is not sufficient evidence to warrant rejection of the claim that Oscar-winning actresses are born in the different months with the same frequency.
17. Test statistic: $\chi^2 = 7.382$. Critical value: $\chi^2 = 19.675$. (TI-83/84 Plus: *P*-value = 0.7674.) There is not sufficient evidence to warrant rejection of the claim that weddings are held in the different months with the same frequency. The results refute the belief that most marriages occur in June.
19. Test statistic: $\chi^2 = 7.346$. Critical value: $\chi^2 = 7.815$. (TI-83/84 Plus: *P*-value = 0.0616.) There is not sufficient evidence to warrant rejection of the claim that the observed frequencies agree with the theoretical proportions. Based on these results, there is not enough evidence to support a claim that seven-game series occur more often than expected.
21. Test statistic: $\chi^2 = 6.682$. Critical value: $\chi^2 = 11.071$. (TI-83/84 Plus: *P*-value = 0.2454.) There is not sufficient evidence to warrant rejection of the claim that the color distribution is as claimed by Mars, Inc.
23. Test statistic: $\chi^2 = 51.270$. Critical value: $\chi^2 = 9.488$. (TI-83/84 Plus: *P*-value: 0.0000.) There is sufficient evidence to warrant rejection of the claim that the participants fit the same distribution as the U.S. population. If study participants are not representative of the population, the results might be misleading because some groups might have cancer rates different from others, and they might skew the outcomes.
25. Test statistic: $\chi^2 = 14.421$. Critical value: $\chi^2 = 15.507$. (TI-83/84 Plus: *P*-value = 0.0714.) There is not sufficient evidence to warrant rejection of the claim that the digits come from a population of leading digits that conform to Benford's law. They appear to conform.
27. Instead of a right-tailed test, use a left-tailed test. If the sample data fit the claimed distribution almost perfectly, the test statistic will be located near the extreme left position of zero.
29. a. 0.296, 0.444, 0.222, 0.037
    b. 88.9, 133.3, 66.7, 11.1
    c. Test statistic: $\chi^2 = 23.241$. Critical value: $\chi^2 = 7.815$. Reject the claim that the observed frequencies fit a binomial distribution with $n = 3$ and $p = 1/3$.

Section 11-3

1. The chi-square test statistic measures the amount of disagreement between the observed frequencies and the frequencies that would be expected if the null hypothesis of independence (or equal proportions) is true.
3. Contingency means *dependence*, and it refers to the test of dependence (or independence) between the row and column variables.
5. Test statistic: $\chi^2 = 0.413$. *P*-value: 0.521. There is not sufficient evidence to warrant rejection of the claim that race and ethnicity are independent of whether someone is stopped by police. There is not sufficient evidence to support a claim of racial profiling.
7. Test statistic: $\chi^2 = 153.462$. Critical value: $\chi^2 = 3.841$. (TI-83/84 Plus: *P*-value = 0.0000.) There is sufficient evidence to warrant rejection of independence between the treatment and whether flu developed. The vaccine does appear to be effective.
9. Test statistic: $\chi^2 = 5.516$. Critical value: $\chi^2 = 3.841$. (TI-83/84 Plus: *P*-value = 0.0188.) There is sufficient evidence to warrant rejection of the claim of independence between gender and left-handedness.
11. Test statistic: $\chi^2 = 32.273$. Critical value: $\chi^2 = 3.841$. (TI-83/84 Plus: *P*-value = 0.0000.) There is sufficient evidence to warrant rejection of the claim that whether a subject lies is independent of the polygraph indication. The results suggest that the polygraph is effective, but it is not always correct.
13. Test statistic: $\chi^2 = 42.557$. Critical value: $\chi^2 = 3.841$. (TI-83/84 Plus: *P*-value = 0.0000.) There is sufficient evidence to warrant rejection of the claim that the sentence is independent of the plea. The results encourage pleas for guilty defendants.

15. Test statistic: $\chi^2 = 153.739$. Critical value: $\chi^2 = 3.841$. (TI-83/84 Plus: *P*-value = 0.0000.) There is sufficient evidence to warrant rejection of the claim that the proportion of heads is the same with flipping as with spinning.
17. Test statistic: $\chi^2 = 65.524$. Critical value: $\chi^2 = 7.815$ (assuming a 0.05 significance level). (TI-83/84 Plus: *P*-value = 0.0000.) There is sufficient evidence to warrant rejection of the claim that occupation is independent of whether the cause of death was homicide. Cashiers appear to be most vulnerable to homicide.
19. Test statistic: $\chi^2 = 1.358$. Critical value: $\chi^2 = 7.815$ (assuming a 0.05 significance level). (TI-83/84 Plus: *P*-value = 0.7154.) There is not sufficient evidence to warrant rejection of the claim that the amount of smoking is independent of seat belt use. The theory is not supported by the given data.
21. Test statistic: $\chi^2 = 9.971$. Critical value: $\chi^2 = 9.488$ (assuming a 0.05 significance level). (TI-83/84 Plus: *P*-value = 0.0409.) There is sufficient evidence to warrant rejection of the claim that injuries are independent of helmet color. The conclusions are the same as those from the example.
23. Without Yates' correction: $\chi^2 = 0.413$. With Yates' correction: $\chi^2 = 0.270$. Yates' correction decreases the test statistic so that sample data must be more extreme to be considered significant.

Section 11-4

1. McNemar's test is used only with data that result from matched pairs, whereas the methods of Section 11-3 do not apply to such data.
3. Discordant pairs of results come from pairs of categories in which the two categories are different.
5. 144
7. 130
9. b, c
11. 6.635
13. Test statistic: $\chi^2 = 2.382$. Critical value: $\chi^2 = 3.841$. (TI-83/84 Plus: *P*-value = 0.123.) There is not sufficient evidence to warrant rejection of the null hypothesis that the following two proportions are the same: (1) the proportion of subjects with no cure on the Fungicide-treated foot and a cure on the foot treated with a placebo; (2) the proportion of subjects with a cure on the Fungicide-treated foot and no cure on the foot treated with a placebo. The Fungicide treatment does not appear to be effective.
15. Test statistic: $\chi^2 = 6.750$. Critical value: $\chi^2 = 3.841$ (assuming a 0.05 significance level). (TI-83/84 Plus: *P*-value = 0.009.) Reject the null hypothesis that the following two proportions are the same: (1) the proportion of tumors with incorrect staging from MRI and correct staging from PET/CT; (2) the proportion of tumors with correct staging from MRI and incorrect staging from PET/CT. The PET/CT technology appears to be more accurate.
17. The uncorrected test statistic is 21.333. The uncorrected value is slightly larger than the corrected value of 20.021. The conclusion is the same in this case. There could be cases in which the uncorrected test statistic leads to rejection of the null hypothesis while the corrected test statistic does not.
19. When rounded to three decimal places, the same *P*-value of 0.289 is obtained. With a *P*-value of 0.289, fail to reject the null hypothesis that the following two proportions are the same: (1) the proportion of subjects with one foot cured with the Pedacream treatment while the other foot is not cured with the Fungacream treatment; (2) the proportion of subjects with one foot not cured with the Pedacream treatment while the other foot is cured with the Fungacream treatment. There does not appear to be a significant difference between the two treatments.

Chapter 11 Statistical Literacy and Critical Thinking

1. Categorical data (or qualitative data) are data that can be separated into different categories that are distinguished by some nonnumeric characteristic.
2. She is using a convenience sample that may be biased and not representative of a population other than the selected classmates.
3. a, c
4. There is a requirement that all *expected* frequencies must be at least 5, but there is no requirement that the observed frequencies for all cells must be at least 5. The methods could be used if all requirements are satisfied, even if there are cells with observed frequencies less than 5.

Chapter 11 Review Exercises

1. Test statistic: $\chi^2 = 7.417$. Critical value: $\chi^2 = 12.592$. (TI-83/84 Plus: *P*-value = 0.284.) There is not sufficient evidence to warrant rejection of the claim that the fatal crashes occur on the days of the week with equal frequency. The crashes appear to be caused by those who drink daily.
2. Test statistic: $\chi^2 = 4.698$. Critical value: $\chi^2 = 3.841$. (TI-83/84 Plus: *P*-value = 0.030.) There is sufficient evidence to warrant rejection of the claim that the response is independent of whether the subject is a worker or boss. The conclusion changes if a 0.01 significance level is used. Workers and bosses do not appear to agree on this issue.
3. Test statistic: $\chi^2 = 119.330$. Critical value: $\chi^2 = 5.991$. (TI-83/84 Plus: *P*-value = 0.000.) There is sufficient evidence to warrant rejection of the claim that the type of crime is independent of whether the criminal is a stranger. Police should place a higher priority on acquaintances and relatives when investigating homicides.
4. Test statistic: $\chi^2 = 13.225$. Critical value: $\chi^2 = 3.841$. (TI-83/84 Plus: *P*-value = 0.000.) There is sufficient evidence to warrant rejection of the claim that the two given proportions are the same. The treatment appears to be effective.

Chapter 11 Cumulative Review Exercises

1. $\bar{x} = 79.6$; median: 80.0; range: 20.0; $s^2 = 44.8$; $s = 6.7$; 5-number summary: 70, 74, 80, 84.5, 90.

2. a. 0.240
   b. 0.518
   c. 0.633
   d. 0.232
   e. 0.274
3. Contingency table; see Section 11-3. Test statistic: $\chi^2 = 0.021$. Critical value: $\chi^2 = 7.815$ (assuming a 0.05 significance level). (TI-83/84 Plus: *P*-value = 0.999.) There is not sufficient evidence to warrant rejection of the claim that men and women choose the different answers in the same proportions.
4. Use correlation; see Section 10-2. Test statistic: $r = 0.989$. Critical values: $r = \pm 0.950$ (assuming a 0.05 significance level). There is sufficient evidence to support the claim that there is a relationship between the memory and reasoning scores.
5. Use the test for matched pairs; see Section 9-4. $\bar{d} = 5.75$; $s_d = 0.957$. Test statistic: $t = 12.011$. Critical values: $t = \pm 3.182$ (assuming a 0.05 significance level). (TI-83/84 Plus: *P*-value = 0.001.) Reject H_0. There is sufficient evidence to warrant rejection of the claim that the training has no effect. The training does appear to have an effect.
6. Test for the difference between two independent samples; see Section 9-3. Test statistic: $t = 1.265$. Critical values: $t = \pm 3.182$ (assuming a 0.05 significance level). (TI-83/84 Plus: Critical values are $t = \pm 2.447$ and *P*-value = 0.253.) Fail to reject H_0: $\mu_1 = \mu_2$. There is not sufficient evidence to warrant rejection of the claim that men and women have the same mean score.

Chapter 12 Answers

Section 12-2

1. One-way analysis of variance is a method of testing the equality of three or more population means by analyzing sample variances. The method is called *one-way* analysis of variance because a single characteristic is used for categorizing the populations.
3. Variance between samples is a measure of variation among sample means. Variance within samples is a measure of variation based on the sample variances.
5. a. $\mu_1 = \mu_2 = \mu_3 = \mu_4$
   b. At least one of the four means is different from the others.
   c. $F = 8.448$
   d. $F = 3.2874$
   e. 0.0016
   f. There is sufficient evidence to warrant rejection of the claim that the four populations have equal means.
   g. Removing the outlier changes the test statistic from $F = 5.7314$ to $F = 8.448$, and the *P*-value changes from 0.0073 to 0.0016. The conclusion does not change.
7. Test statistic: $F = 0.3521$. Critical value: $F = 2.6626$. *P*-value: 0.7877. Fail to reject H_0: $\mu_1 = \mu_2 = \mu_3 = \mu_4$. There is not sufficient evidence to warrant rejection of the claim that the mean weight loss is the same for the diets. The diets do not appear to be very effective.
9. Test statistic: $F = 0.443$. Critical value: $F = 2.3113$. *P*-value: 0.8173. Fail to reject H_0: $\mu_1 = \mu_2 = \mu_3 = \mu_4 = \mu_5 = \mu_6$. There is not sufficient evidence to warrant rejection of the claim that the six colors of M&Ms have the same mean. No corrective action is required.
11. Test statistic: $F = 0.9922$. Critical value: $F = 3.2389$. *P*-value: 0.4216. Fail to reject H_0: $\mu_1 = \mu_2 = \mu_3 = \mu_4$. There is not sufficient evidence to support the claim that larger cars are safer.
13. Test statistic: $F = 2.4749$. Critical value: $F = 3.0984$. *P*-value: 0.0911. Fail to reject H_0: $\mu_1 = \mu_2 = \mu_3 = \mu_4$. There is not sufficient evidence to warrant rejection of the claim of equal means. The groups can be considered to be samples from the same population.
15. Test statistic: $F = 1713.725$. Critical value: $F = 2.6802$ (approximately). *P*-value: 0.0000. Reject H_0: $\mu_1 = \mu_2 = \mu_3 = \mu_4$. There is sufficient evidence to warrant rejection of the claim that the four categories of pennies have the same mean weight. Coin counting machines cannot treat the weights the same way.
17. a. Test statistic: $t = -0.57139$. Critical values: $t = \pm 2.306$. (TI-83/84 Plus: *P*-value = 0.583.) Fail to reject H_0: $\mu_1 = \mu_2$.
    b. Test statistic: $F = 0.3265$. Critical value: $F = 5.3177$. *P*-value = 0.583. Fail to reject H_0: $\mu_1 = \mu_2$.
    c. Test statistics: $t^2 = F = 0.3265$. Critical values: $t^2 = F = 5.3177$.

Section 12-3

1. "Analysis of variance" refers to the method used, which is based on two different estimates of the assumed common population variance. "Two-way" refers to the inclusion of two different factors, which are properties or characteristics used to distinguish different populations from one another. Two-way analysis of variance is used to test for an effect from each of two different factors, as well as testing for an effect from an interaction between the two factors.
3. If there is an interaction between factors, we should not consider the effects of either factor without considering those of the other factor.
5. Test statistic: $F = 2.79$. *P*-value: 0.056. Assuming a 0.05 significance level, fail to reject the null hypothesis of no interaction. There does not appear to be a significant effect from the interaction between site and treatment.
7. Test statistic: $F = 2.57$. *P*-value: 0.072. Fail to reject the null hypothesis that treatment has no effect on weight. There is not sufficient evidence to support the claim that treatment has an effect on weight.
9. Test statistic: $F = 5.03$. *P*-value: 0.031. Assuming a 0.05 significance level, reject the null hypothesis that gender has no effect on SAT scores. There is sufficient evidence to support the claim that gender has an effect on SAT scores.
11. Test statistic: $F = 3.87$. *P*-value: 0.000. Reject the null hypothesis that the choice of subject has no effect on the hearing test score. There is sufficient evidence to support the claim that the choice of subject has an effect on the hearing test score.

13. For interaction, the test statistic is $F = 0.36$ and the P-value is 0.701, so there is no significant interaction effect. For gender, the test statistic is $F = 0.09$ and the P-value is 0.762, so there is no significant effect from gender. For age, the test statistic is $F = 0.36$ and the P-value is 0.701, so there is no significant effect from age.
15. a. The test statistics, critical values, P-values, and conclusions do not change.
    b. The test statistics, critical values, P-values, and conclusions do not change.
    c. The test statistics, critical values, P-values, and conclusions do not change.
    d. An outlier can dramatically affect and change all of the results and conclusions.

Chapter 12 Statistical Literacy and Critical Thinking

1. The method for testing equality of means is based on two different estimates of a common population variance.
2. One-way analysis of variance
3. t test for two independent means (as in Section 9-3)
4. No, because the data are qualitative (or categorical) instead of being quantitative.

Chapter 12 Review Exercises

1. Test statistic: $F = 46.90$. P-value: 0.000. Reject H_0: $\mu_1 = \mu_2 = \mu_3$. There is sufficient evidence to warrant rejection of the claim of equal population means.
2. Test statistic: $F = 9.469$. Critical value: $F = 3.3158$ (approximately). (TI-83/84 Plus: P-value $= 0.000562$.) Reject H_0: $\mu_1 = \mu_2 = \mu_3$. There is sufficient evidence to warrant rejection of the claim that the three books have the same mean score. The three books do not appear to have the same reading level.
3. Test statistic: $F = 0.19$. P-value: 0.832. Fail to reject the null hypothesis of no interaction. There does not appear to be a significant effect from the interaction between gender and major.
4. Test statistic: $F = 0.78$. P-value: 0.395. Fail to reject the null hypothesis that gender has no effect on estimated length. There is not sufficient evidence to support the claim that estimated length is affected by gender.
5. Test statistic: $F = 0.13$. P-value: 0.876. Fail to reject the null hypothesis that major has no effect on estimated length. There is not sufficient evidence to support the claim that estimated length is affected by major.
6. For interaction, the test statistic is $F = 0.8733$ and the P-value is 0.3685, so there is no significant interaction effect. For gender, the test statistic is $F = 0.0178$ and the P-value is 0.8960, so there is no significant effect from gender. For smoking, the test statistic is $F = 3.0119$ and the P-value is 0.1082, so there is no significant effect from smoking.
7. a. Test statistic: $F = 1.0000$. P-value: 0.4226. There is not sufficient evidence to support the claim that amounts of emitted greenhouse gases are affected by the type of transmission.
   b. Test statistic: $F = 7.0000$. P-value: 0.1250. There is not sufficient evidence to support the claim that amounts of greenhouse gases are affected by the number of cylinders.
   c. Perhaps greenhouse gases *are* affected by the type of transmission and/or the number of cylinders; however, the given sample data do not provide sufficient evidence to support such claims.
8. Test statistic: $F = 3.110$. Critical F value is between 3.0718 and 3.1504. (TI-83/84 Plus: P-value $= 0.0506$.) Using a 0.05 significance level, fail to reject H_0: $\mu_1 = \mu_2 = \mu_3$. There is not sufficient evidence to warrant rejection of the claim that the three groups have the same mean longevity. The survival times appear to have means that are the same.

Chapter 12 Cumulative Review Exercises

1. a. 15.5, 13.1, 22.7
   b. 9.7, 9.0, 18.6
   c. Test statistic: $t = -1.383$. Critical values: $t = \pm 2.160$ (assuming a 0.05 significance level). (TI-83/84 Plus: P-value $= 0.1860$.) Fail to reject H_0: $\mu_1 = \mu_2$. There is not sufficient evidence to support the claim that there is a difference between the mean for the two groups.
   d. Normal, because the histogram is approximately bell-shaped.
   e. 12.3 years $< \mu <$ 18.7 years
2. a. 960.5, 980.0, 1045.0; no
   b. 914.5, 1010.5, 1008.5; no
   c. 174.6, 239.6, 224.1; no
   d. Test statistic: $t = -0.294$. Critical values: $t = \pm 2.093$ (assuming a 0.05 significance level). (TI-83/84 Plus: P-value $= 0.7704$.) Fail to reject H_0: $\mu_1 = \mu_2$.
   e. $878.8 < \mu < 1042.2$
   f. Test statistic: $F = 0.8495$. P-value: 0.4330. Fail to reject H_0: $\mu_1 = \mu_2 = \mu_3$. There is not sufficient evidence to warrant rejection of the claim that the three populations have the same mean SAT score.
3. Using normal as approximation to binomial: 0.1020. (Exact result using TI-83/84 Plus: 0.0995.) Because the probability of getting 19 or fewer offspring with blue eyes is so high, there is not sufficient evidence to conclude that the one-quarter rate is wrong.
4. a. 0.3372 (TI-83/84 Plus: 0.3365)
   b. 0.0455 (TI-83/84 Plus: 0.0457)
   c. 1/8 or 0.125

Chapter 13 Answers

Section 13-2

1. The sign test does not have a requirement that the sample data come from a population with a particular distribution.

3. Because the proportion of girls is less than 0.5, there is not sufficient evidence to support the claim that the method increases the likelihood that a baby will be a girl. With a proportion of girls less than 0.5, there is no way that we could support a claim that the population proportion is significantly greater than 0.5.
5. The test statistic of $x = 4$ is less than or equal to the critical value of 4. (TI-83/84 Plus P-value $= 0.0192$.) There is sufficient evidence to warrant rejection of the claim of no difference.
7. The test statistic of $z = -0.50$ is not less than or equal to the critical value of -1.96. (TI-83/84 Plus P-value $= 0.620$.) There is not sufficient evidence to reject the claim of no difference.
9. The test statistic of $x = 1$ is not less than or equal to the critical value of 0. (TI-83/84 Plus P-value $= 0.219$.) There is not sufficient evidence to reject the claim of no effect. Based on the sample data, it appears that when Friday falls on the 13th, the numbers of hospital admissions are not affected.
11. The test statistic of $x = 5$ is not less than or equal to the critical value of 2. (TI-83/84 Plus P-value $= 0.774$.) There is not sufficient evidence to support the claim that there is a difference between self-reported heights and measured heights.
13. The test statistic of $x = 1$ is less than or equal to the critical value of 2. (TI-83/84 Plus P-value $= 0.00586$.) There is sufficient evidence to support the claim that the population has a median less than 98.6°F.
15. The test statistic of $z = -3.64$ is less than or equal to the critical value of $z = -2.33$. (TI-83/84 Plus P-value $= 0.0000990$.) There is sufficient evidence to support the claim that with the YSORT method, the probability of a boy is greater than 0.5. The method appears to work.
17. The test statistic of $z = -0.48$ is not less than or equal to the critical value of $z = -2.33$. (TI-83/84 Plus P-value $= 0.316$.) There is not sufficient evidence to support the claim that among Internet users, less than 50% use it for making travel plans. The results are important for travel agents because their business could be affected.
19. First approach: $z = -1.90$; reject H_0.
    Second approach: $z = -1.73$; reject H_0.
    Third approach: $z = 0$; fail to reject H_0.
21. Convert $x = 18$ to the test statistic $z = -2.31$. Critical value: $z = -2.33$. (TI-83/84 Plus: P-value $= 0.0104$.) There is not sufficient evidence to support a charge of gender bias. If the binomial distribution is used instead of the normal approximation, the P-value is 0.0099, which is less than 0.01, so there is sufficient evidence to support a charge of gender bias. Using the normal approximation, the test statistic is just barely outside of the critical region; using the binomial distribution, the test statistic is just barely inside the critical region.

Section 13-3

1. The Wilcoxon signed-ranks test should be used when the population of differences has a distribution that is approximately symmetric, and the number of matched pairs of sample data is small ($n \leq 30$), and the pairs of values have differences that are not from a population having a distribution that is approximately normal.
3. By using ranks instead of signs of differences, the Wilcoxon signed-ranks test includes information about the magnitudes of the differences instead of simply the sign of the differences. The Wilcoxon signed-ranks test uses more information and is more likely to reflect the true nature of the data.
5. Test statistic: $T = 0$. Critical value: $T = 4$. Reject the null hypothesis that the population of differences has a median of 0.
7. Test statistic: $T = 1.5$. Critical value: $T = 1$. Fail to reject the null hypothesis that the population of differences has a median of 0. Based on the sample data, it appears that when Friday falls on the 13th, the numbers of hospital admissions are not affected.
9. Test statistic: $T = 34$. Critical value: $T = 14$. Fail to reject the null hypothesis that the population of differences has a median of 0. There is not sufficient evidence to support the claim that there is a difference between self-reported heights and measured heights of males aged 12–16.
11. Test statistic: $T = 158.5$. Critical value: $T = 127$ (assuming a 0.05 significance level). Fail to reject the null hypothesis that the population of differences has a median of 0. The differences between the actual and forecast high temperatures do not appear to be significant.
13. Convert $T = 661$ to test statistic $z = -5.67$. (TI-83/84 Plus: P-value $=$ 0.0000.) Critical values: $z = \pm 1.96$. There is sufficient evidence to warrant rejection of the claim that healthy adults have a median body temperature that is equal to 98.6°F.

Section 13-4

1. The Wilcoxon signed-ranks test applies to sample data consisting of matched pairs, but the Wilcoxon rank-sum test applies to two independent samples that are not matched or paired in any way.
3. Unlike the parametric test in Section 9-3, the nonparametric Wilcoxon rank-sum test does not require normally distributed populations, so it can be used in more situations.
5. $R_1 = 104$, $R_2 = 172$, $\mu_R = 132$, $\sigma_R = 16.248$, test statistic: $z = -1.72$. Critical values: $z = \pm 1.96$. (TI-83/84 Plus: P-value $=$ 0.0848.) Fail to reject the null hypothesis that the populations have the same median.
7. $\mu_R = 150$, $\sigma_R = 17.321$, $R = 96.5$, $z = -3.09$. Test statistic: $z = -3.09$. Critical values: $z = \pm 2.575$. (TI-83/84 Plus: P-value $=$ 0.0020.) There is sufficient evidence to warrant rejection of the claim that the two samples come from populations with the same median. Based on these results, it appears that obsessive-compulsive disorders do have a biological basis.
9. $\mu_R = 648$, $\sigma_R = 46.4758$, $R = 616.5$, $z = -0.68$. Test statistic: $z = -0.68$. Critical values: $z = \pm 1.96$. (TI-83/84 Plus: P-value $=$ 0.4979.) Fail to reject the null hypothesis that the two samples come from populations with the same median. The results do not suggest that taxis are newer.

11. $\mu_R = 1620$, $\sigma_R = 103.923$, $R = 1339.5$, $z = -2.70$. Test statistic: $z = -2.70$. Critical values: $z = \pm 1.96$. (TI-83/84 Plus: P-value = 0.0070.) Reject the null hypothesis that the two samples come from populations with the same median.
13. $z = -0.98$; the test statistic is the same value with opposite sign.

Section 13-5

1. The Kruskal-Wallis test does not require that the populations have normal distributions or any other specific distribution.
3. The efficiency rating of 0.95 indicates that with all other factors being the same, the Kruskal-Wallis test requires 100 sample observations to achieve the same results as 95 observations with the parametric one-way analysis of variance test, assuming that the stricter requirements for using the parametric test are met.
5. Test statistic: $H = 1.1914$. Critical value: $\chi^2 = 7.815$. (TI-83/84 Plus: P-value = 0.755.) There is not sufficient evidence to support the claim that the head injuries for the four weight categories have measures with medians that are not all the same. (The medians appear to be equal.) The given data do not provide sufficient evidence to conclude that heavier cars are safer in a crash.
7. Test statistic: $H = 14.7485$. Critical value: $\chi^2 = 5.991$. (TI-83/84 Plus: P-value = 0.0006.) There is sufficient evidence to warrant rejection of the claim that the medians of the voltage readings are the same for the three different types of day. Sunny days appear to result in greater energy amounts.
9. Test statistic: $H = 6.0317$. Critical value: $\chi^2 = 7.815$. (TI-83/84 Plus: P-value = 0.1101.) There is not sufficient evidence to warrant rejection of the claim that the different groups have the same median blood pressure. The groups can be considered to be samples from the same population.
11. Test statistic: $H = 74.8519$. Critical value: $\chi^2 = 7.815$. (TI-83/84 Plus: P-value = 0.000.) There is sufficient evidence to warrant rejection of the claim that the four categories of pennies have the same median weight. Coin counting machines cannot treat the weights the same way.
13. 14.840 (using $T = 6, 6, 24$); no

Section 13-6

1. Rank correlation does not require normal distributions or any other specific distributions.
3. The subscript s is used so that the rank correlation coefficient can be distinguished from the linear correlation coefficient r. The subscript does not represent the standard deviation s. It is used to honor Charles Spearman, who introduced the rank correlation method.
5. $r_s = 1$ and there appears to be a correlation between x and y. (TI-83/84 Plus P-value = 0.)
7. a. ± 0.587
   b. ± 0.570
   c. ± 0.255
   d. ± 0.290
9. $r_s = 0.345$. Critical values: $r_s = \pm 0.648$. (TI-83/84 Plus P-value = 0.328.) No correlation. There does not appear to be a correlation between the DJIA and the numbers of cars sold.
11. $r_s = 0.855$. Critical values: $r_s = \pm 0.648$. (TI-83/84 Plus P-value = 0.00164.) There is a correlation. There appears to be a correlation between salary and stress.
13. $r_s = 0.857$. Critical values: $r_s = \pm 0.738$. (TI-83/84 Plus P-value = 0.00653.) There is a correlation. There appears to be a correlation between the number of chirps in 1 min and the temperature.
15. $r_s = 0.561$. Critical values: $r_s = \pm 0.700$. (TI-83/84 Plus P-value = 0.116.) No correlation. There does not appear to be a correlation between the number of audience impressions and the numbers of albums sold. Sales do not appear to be affected very strongly by the number of audience impressions.
17. a. $r_s = 0.918$. Critical values: $r_s = \pm 0.368$. (TI-83/84 Plus P-value = 0.000.) There is a correlation between tar and nicotine.
    b. $r_s = 0.739$. Critical values: $r_s = \pm 0.368$. (TI-83/84 Plus P-value = 0.000.) There is a correlation between carbon monoxide and nicotine.
    c. Tar is the better choice, because it has a higher correlation with nicotine.
19. $r_s = 1$. Critical values: $r_s = \pm 0.618$ (assuming a 0.05 significance level). There is a linear correlation. Using linear correlation (Section 10-2), $r = 0.572$, the critical values are $r = \pm 0.602$ (assuming a 0.05 significance level), and there is no *linear* correlation. The scatterplot shows a pattern, but it is not a straight-line pattern. The results from the two methods are different.

Section 13-7

1. No, because shuffling changed the order in which the subjects were interviewed.
3. No. There might be problems with the data selection process. For example, a voluntary response sample might appear to be random, but it would not be suitable for most statistical purposes.
5. $n_1 = 4$, $n_2 = 8$, $G = 8$, critical values: 3, 10; fail to reject randomness.
7. $n_1 = 6$, $n_2 = 6$, $G = 7$, critical values: 3, 11; fail to reject randomness.
9. $n_1 = 13$, $n_2 = 7$, $G = 11$, critical values: 5, 15; fail to reject randomness.
11. $n_1 = 14$, $n_2 = 19$, $G = 20$, critical values: 11, 23; fail to reject randomness. The leagues appear to win in a random sequence.
13. $n_1 = 8$, $n_2 = 8$, $G = 2$, critical values: 4, 14; reject randomness. The result suggests that there is an upward trend.
15. $n_1 = 35$, $n_2 = 19$, $G = 23$, $\mu_G = 25.6296$, $\sigma_G = 3.3137$. Test statistic: $z = -0.79$. Critical values: $z = \pm 1.96$. (TI-83/84 Plus: P-value = 0.4274.) Fail to reject randomness. The sequence of genders appears to be random.

17. $n_1 = 49$, $n_2 = 51$, $G = 43$, $\mu_G = 50.98$, $\sigma_G = 4.9727$. Test statistic: $z = -1.60$. Critical values: $z = \pm 1.96$. (TI-83/84 Plus: P-value = 0.1085.) Fail to reject randomness. The digits appear to be in a random sequence.
19. Minimum is 2, maximum is 4. Critical values of 1 and 6 can never be realized so that the null hypothesis of randomness can never be rejected.

Chapter 13 Statistical Literacy and Critical Thinking

1. It is a hypothesis test that does not require that sample data have a normal distribution or any other particular distribution.
2. There is no difference. They are different names for the same category of hypothesis tests that do not require that populations have normal distributions or any other particular distribution.
3. A rank is a number assigned to an individual sample item according to its order in the sorted list. The first item is assigned a rank of 1, the second item is assigned a rank of 2, and so on.
4. Efficiency is a measure of how much stronger sample evidence must be for the nonparametric test to produce the same results as a corresponding parametric test. For example, the sign test has an efficiency of 0.63, meaning that under the same conditions, the sign test requires 100 sample observations to achieve the same results as 63 sample observations analyzed with a corresponding parametric test.

Chapter 13 Review Exercises

1. The test statistic $x = 2$ is less than or equal to the critical value of 2. (TI-83/84 Plus P-value = 0.0129.) There is sufficient evidence to warrant rejection of the claim that there is no difference between the times of the first and second trials.
2. Test statistic: $T = 5.5$. Critical value: $T = 21$. There is sufficient evidence to warrant rejection of the claim that there is no difference between the times of the first and second trials.
3. Rank correlation: $r_s = 0.709$. Critical values: $r_s = \pm 0.648$. (TI-83/84 Plus P-value = 0.0217.) There appears to be a correlation between the business school rankings and the law school rankings.
4. Rank correlation: $r_s = -0.810$. Critical values: $r_s = \pm 0.738$. There does appear to be a correlation between the ranking and average MCAT score.
5. Sign test: Convert $x = 15$ to the test statistic $z = -1.42$. Critical value: $z = -1.645$. (TI-83/84 Plus: P-value = 0.0769.) There is not sufficient evidence to support the claim of bias in favor of men.
6. Wilcoxon rank-sum test: $\mu_R = 162$, $\sigma_R = 19.442$, $R = 89.5$, $z = -3.73$. Test statistic: $z = -3.73$. Critical values: $z = \pm 1.96$. (TI-83/84 Plus: P-value = 0.0002.) Reject the null hypothesis that the two samples come from populations with the same median. There is sufficient evidence to warrant rejection of the claim that beer drinkers and liquor drinkers have the same BAC levels.
7. Runs test: $n_1 = 22$, $n_2 = 18$, $G = 18$, $\mu_G = 20.8$, $\sigma_G = 3.0894$. Test statistic: $z = -0.91$. Critical values: $z = \pm 1.96$. (TI-83/84 Plus: P-value = 0.3648.) Fail to reject randomness. Odd and even digits appear to occur randomly.
8. Kruskal-Wallis test: Test statistic: $H = 4.234$. Critical value: $\chi^2 = 7.815$. (TI-83/84 Plus: P-value = 0.237.) There is not sufficient evidence to support the claim that the injury measurements are not the same for the four categories. There is not sufficient evidence to support the claim that heavier cars are safer.

Chapter 13 Cumulative Review Exercises

1. $\bar{x} = 528.9$; median = 510.0; range = 160.0; $s = 63.1$; $s^2 = 3986.1$.
2. There does appear to be a correlation.

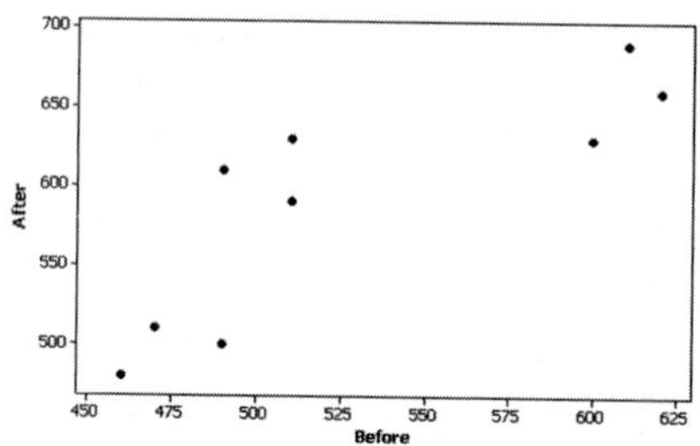

3. $r = 0.833$. Critical values: $r = \pm 0.666$ (assuming a 0.05 significance level). There is a linear correlation, but that does not necessarily mean that the course is effective.
4. $r_s = 0.907$. Critical values: $r_s = \pm 0.700$ (assuming a 0.05 significance level). (TI-83/84 Plus P-value = 0.000732.) There is a correlation.
5. H_0: $\mu_d = 0$. H_1: $\mu_d \neq 0$. Test statistic: $t = -4.334$. Critical values: $t = \pm 2.306$ (assuming a 0.05 significance level). (TI-83/84 Plus: P-value = 0.0025.) Reject H_0. There is sufficient evidence to warrant rejection of the claim that the mean difference between before and after scores is equal to zero.
6. The test statistic $x = 0$ is less than or equal to the critical value of 1 (assuming a 0.05 significance level). (TI-83/84 Plus P-value = 0.00391.) There is sufficient evidence to reject the claim of no difference between before and after scores.
7. $T = 0$. Critical value: $T = 6$ (assuming a 0.05 significance level). There is sufficient evidence to reject the claim of no difference between before and after scores.
8. Because the differences appear to be from a normally distributed population, the t test in Exercise 5 is most helpful. The preparation course does appear to be effective, because the after scores are significantly higher than the corresponding before scores.

Chapter 14 Answers

Section 14-2

1. *Process data* are data arranged according to some time sequence. They are measurements of a characteristic of goods or services that result from some combination of equipment, people, materials, methods, and conditions.

3. A control chart of a process characteristic (such as mean or variation) consists of values plotted sequentially over time, and it includes a centerline as well as a lower control limit and an upper control limit. An *R* chart is a plot of the sample ranges. An $\overline{x}$ chart is a plot of the sample means. An *R* chart is used to monitor the variation in a process, but an $\overline{x}$ chart is used to monitor the center in a process.
5. a. Within statistical control
   b. Not applicable
   c. The variation is too large, so that some cans are overfilled too much while others are underfilled too much.
7. a. Out of statistical control.
   b. There is an exceptionally low value.
   c. The low amount of 10.5 oz is unacceptable, so the process is not behaving as it should.
9. The process variation appears to be within statistical control.

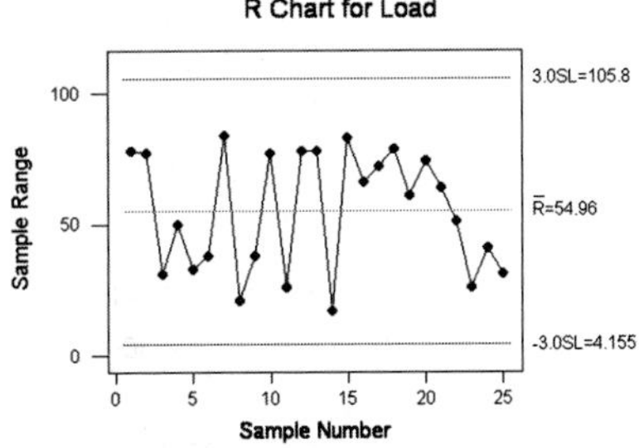

11. There is a pattern of increasing variation, so the process is out of statistical control. The increasing variation will result in more and more defects.

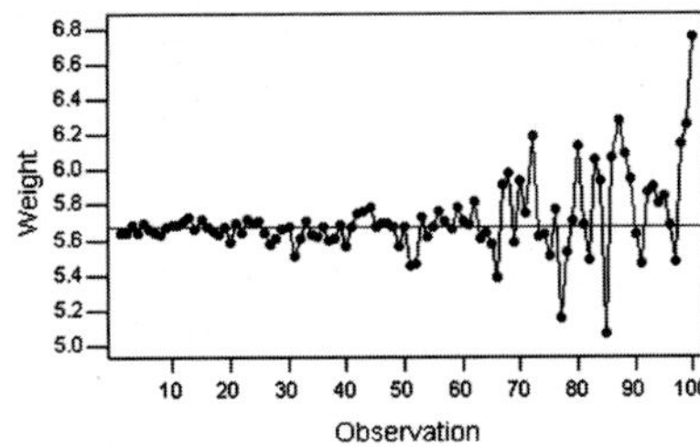

13. There is a pattern of increasing variation, there are points lying beyond the upper control limit, and there are eight consecutive points all below the centerline, so the process mean is out of statistical control. This process does need corrective action.

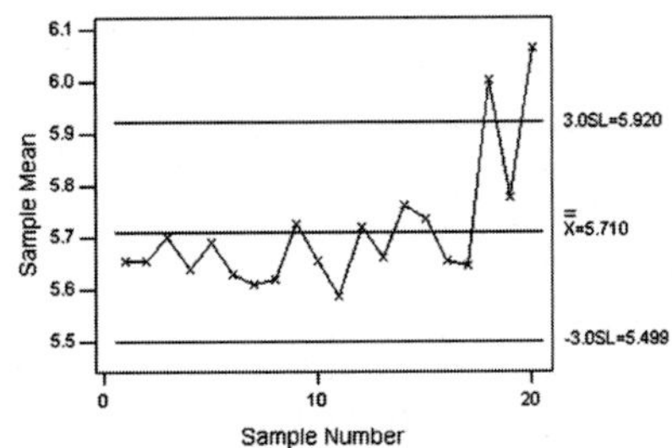

15. The process variation appears to be out of statistical control. There are points that lie beyond the control limits.

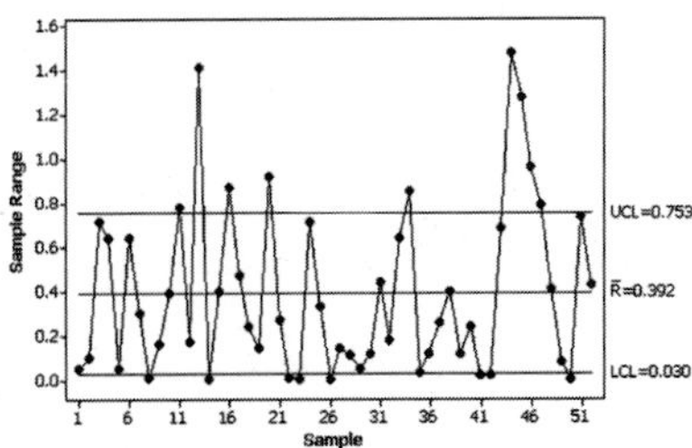

17. The process variation appears to be out of statistical control. There is a point beyond the upper control limit and there appears to be an upward trend.

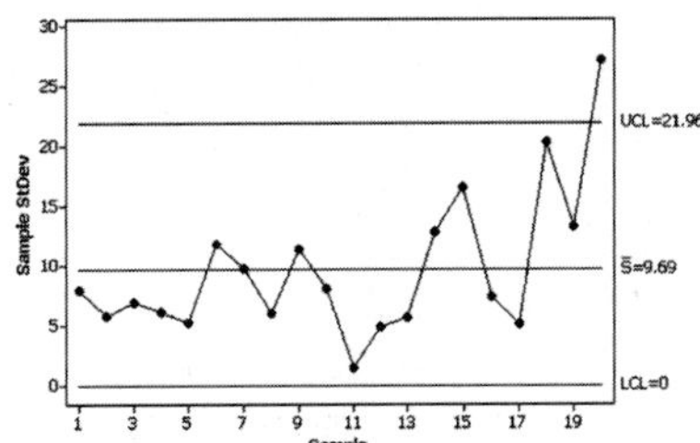

Section 14-3

1. It is a control chart of the proportion of some attribute, and it is used to monitor the proportion in order to determine whether the process is within statistical control.
3. 0.250 and 0
5. Within statistical control
7. Process appears to be out of statistical control because there is a pattern of an upward trend and there is a point that lies beyond the upper control limit.
9. The process is out of statistical control because of a downward pattern. Some possible explanations: There has been a substantial decrease in the birth rate among teenagers; the increasing life span results in a smaller proportion of women who are of child-bearing age.

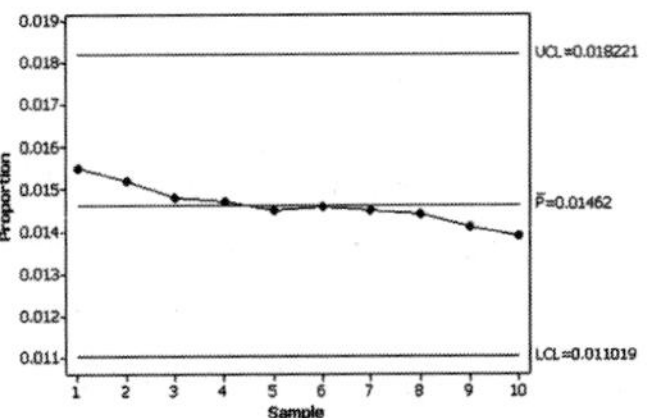

11. The process is within statistical control. Although no corrective action appears to be crucial, the company should work toward the goal of zero defects.

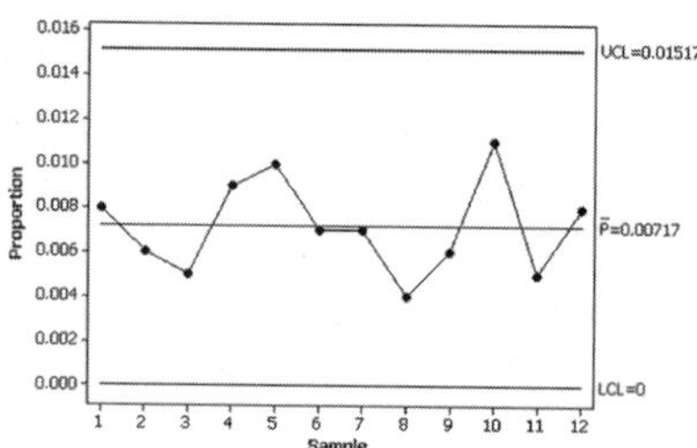

13. The process is statistically stable.

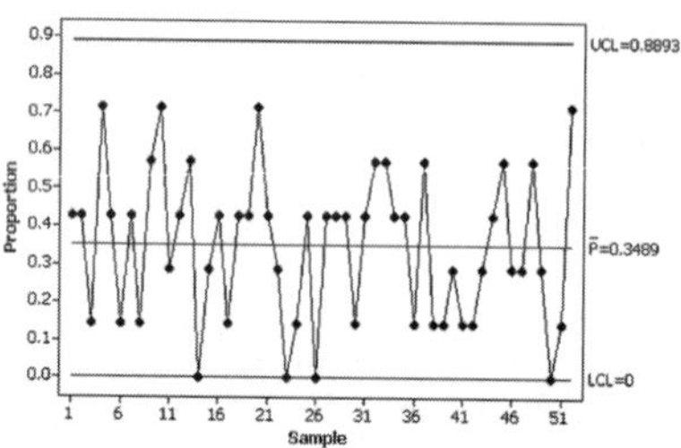

Chapter 14 Statistical Literacy and Critical Thinking

1. If data change over time, important characteristics such as the mean, standard deviation, or proportion of defects are not stable and their values change. If it is important to control one or more characteristics, then the data must be monitored.
2. Statistical process control consists of methods used to monitor data over time and to determine whether the data are within statistical control, meaning that the data do not meet any of the specific out-of-control criteria being used.
3. If the bottling process goes out of statistical control, bottling costs could rise due to increasing numbers of defects, and the rising costs could force the plant to close because of lost profitability.
4. The process could go out of statistical control because of a changing mean, or increasing variation, or both.

Chapter 14 Review Exercises

1. The process is not within statistical control because there is one point that is exceptionally low.

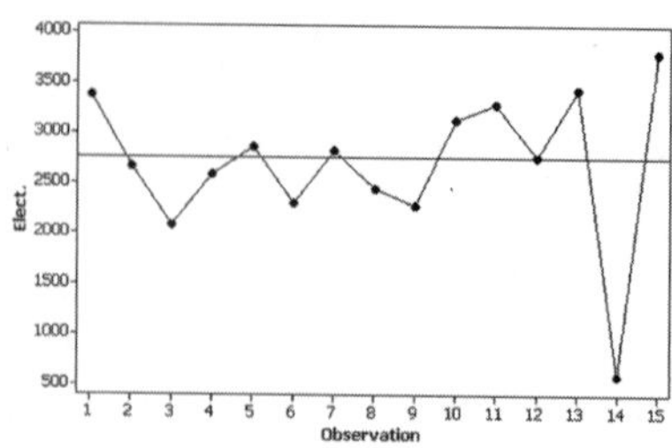

2. Because there is a point beyond the upper control limit, the process variation is out of statistical control.

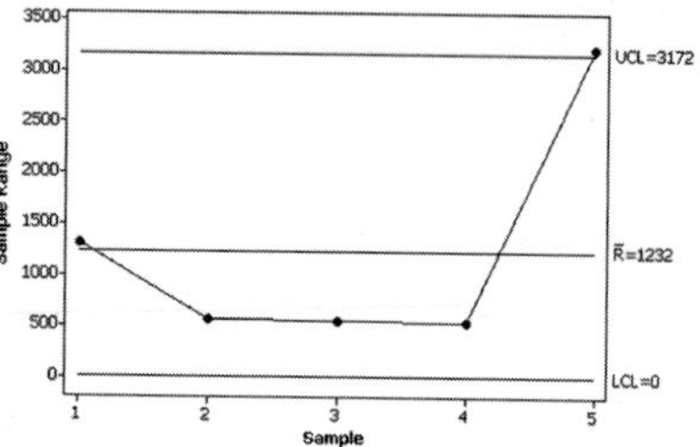

3. The process mean is within statistical control.

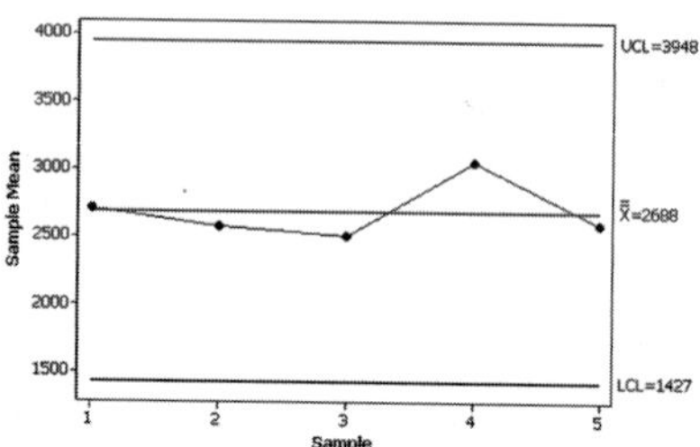

4. The process is out of control because there is a shift up and there are points beyond the control limits.

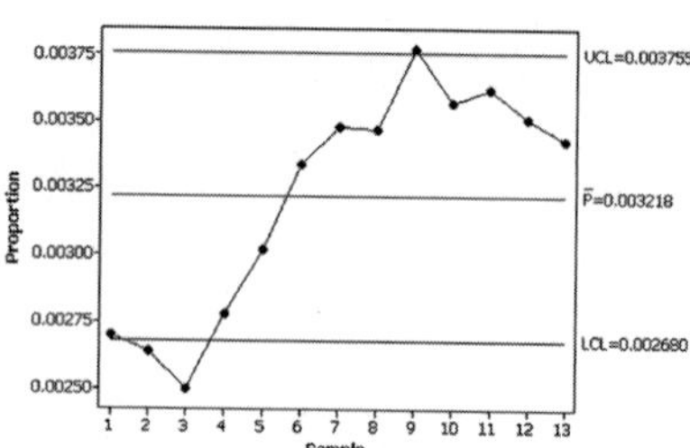

5. The process is out of statistical control because there is a point lying beyond the control limits.

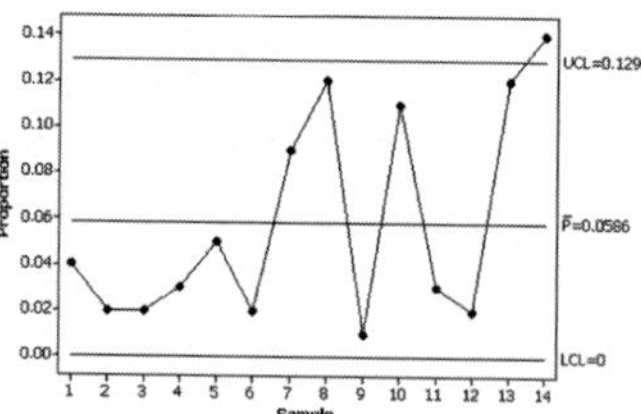

Chapter 14 Cumulative Review Exercises

1. The process is out of statistical control because there is a point beyond the upper control limit.

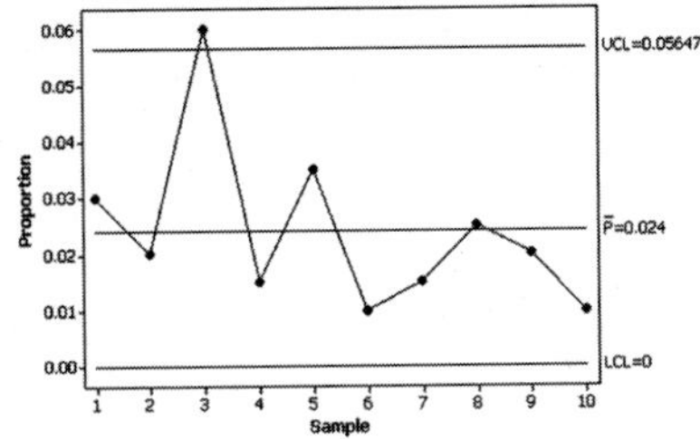

2. $0.0173 < p < 0.0307$
3. H_0: $p = 0.01$. H_1: $p > 0.01$. Test statistic: $z = 6.29$. Critical values: $z = 1.645$. P-value: 0.0000. Reject H_0. There is sufficient evidence to support the claim that the rate of defects is greater than 1%.
4. a. 1/256
   b. 1/256
   c. 1/128

Credits

Photographs

Chapter Openers

Chapter 1: Six Degrees of Kevin Bacon © AFP/Getty Images
Chapter 2: Academy Awards: Best Actress Age © AP Photo
Chapter 3: Academy Awards: Best Actor © Getty Images
Chapter 4: Job Interview Drug Testing © Corbis
Chapter 5: Jury Selection: Benjamin Spock © Digital Vision
Chapter 6: Passenger Safety © Thinkstock
Chapter 7: Touch Therapy © Beth Anderson
Chapter 8: Career Networking © PhotoDisc Red
Chapter 9: Carpal Tunnel Syndrome Treatment © Stockbyte Platinum
Chapter 10: Old Faithful Geyser © Digital Vision
Chapter 11: Detecting Check Fraud © Corbis
Chapter 12: Poplar Trees © Corbis
Chapter 13: College Ranking © Comstock
Chapter 14: Aircraft Altimeters © Digital Vision

The following margin photos are all copyright PhotoDisc, Inc.

Chapter 1
p. 5, The State of Statistics
p. 13, Literary Digest Poll
p. 17, Publication Bias

Chapter 3
p. 87, Class Size Paradox
p. 103, Reliability and Validity
p. 108, More Stocks, Less Risk

Chapter 4
p. 148, You Bet
p. 149, How Probable
p. 159, Boys and Girls Are Not Equally Likely
p. 179, Composite Sampling
p. 182, Probability of an Event That Has Never Occurred
p. 188, The Phone Number Crunch
p. 189, Making Cents of the Lottery
p. 191, How Many Shuffles?
p. 192, The Random Secretary
p. 193, Too Few Bar Codes

Chapter 5
p. 212, Picking Lottery Numbers
p. 226, Not at Home

Chapter 6
p. 282, Prophets for Profits
p. 305, Is Parachuting Safe?
p. 315, States Rig Lottery Selections

Chapter 7
p. 350, Estimating Wildlife Population Sizes

Chapter 8
p. 402, Lie Detectors
p. 406, Large Sample Size Isn't Good Enough
p. 433, Commercials

Chapter 9
p. 476, The Lead Margin of Error
p. 478, Does Aspirin Help Prevent Heart Attacks?
p. 492, Using Statistics to Identify Thieves
p. 506, Do Air Bags Save Lives?
p. 516, Lower Variation; Higher Quality
p. 518, Research in Twins

Chapter 10
p. 542, Student Ratings of Teachers
p. 563, Postponing Death
p. 587, Predictors for Success
p. 588, Making Music with Multiple Regression

Chapter 11
p. 629, An Eight-Year False Positive
p. 631, Home Field Advantage

Chapter 12
p. 675, Polls and Psychologists

Chapter 13
p. 708, Gender Gap in Drug Testing
p. 729, Direct Link Between Smoking and Cancer

Misc. Photos

Chapter 1
p. 26, Prospective National Children's Study; Copyright PhotoDisc Red
p. 29, Should You Believe A Statistical Study?; Copyright Jeff Greenberg/PhotoEdit

Chapter 2
p. 53, Growth Charts Updated; Copyright Kelly/Mooney Photgraphy/Corbis
p. 59, Missing Data; Copyright Brand X Pictures
p. 66, The Power of a Graph; Copyright Stockbyte Gold

Chapter 3
p. 105, Where are the 0.400 Hitters?; Copyright AP/Wide World Photos
p. 131, An Outlier Tip; Copyright Reuters NewMedia, Inc./Corbis
p. 134, Human Lie Detectors; Copyright Getty/Image Bank

Chapter 4
p. 153, Most Common Birthday: October 5; Copyright Dora Kingsley
p. 172, Perfect SAT Scores; Copyright Taxi/Getty Images

Chapter 6
p. 272, Clinical Trial Cut Short; Copyright Seits/Photo Researchers, Inc.
p. 281, Do Boys or Girls Run in the Family?; Copyright Anthony Redpath/Corbis

Chapter 7
p. 336, Curbstoning; Copyright PhotoDisc Red
p. 368, Estimates to Improve Census; Copyright Corbis
p. 369, Estimating Crowd Size; Copyright Beth Anderson

Chapter 8
p. 407, Win $1,000,000 for ESP; Copyright PhotoDisc Blue
p. 422, Process of Drug Approval; Copyright Digital Vision
p. 444, Lefties Die Sooner?; Copyright PhotoDisc Blue
p. 446, Meta-analysis; Copyright Beth Anderson

Chapter 9
p. 477, Author as a Witness; Copyright Bill Wisser/Getty Images
p. 493, Expensive Diet Pill; Copyright Lance Manning/Corbis
p. 495, Super Bowls; Copyright Comstock

Chapter 10
p. 565, Predicting Condo Prices; Copyright Comstock
p. 578, Super Bowl as Stock Market Predictor; Copyright PhotoDisc Red
p. 589, NBA Salaries and Performance; Copyright Marc Asnin/Corbis Saba
p. 592, Icing the Kicker; Copyright PhotoDisc Red

Chapter 12
p. 665, Ethics in Reporting; Copyright AP/Wide World

Chapter 13
p. 736, Sports Hot Streaks; Copyright 2002 NBAE Photo by Noren Trotman/NBAE/Getty Images

Chapter 14
p. 757, Costly Assignable Variation; Copyright NASA
p. 758, Don't Tamper! Copyright Davis Lees/Corbis
p. 761, Bribery Detected with Control Charts; Copyright Owen Franken/Corbis

Illustrations

The following margin illustrations were rendered by Bob Giuliani.

Chapter 1
p. 16, Detecting Phony Data
p. 24, Hawthorne and Experimenter Effects

Chapter 2
p. 50, Authors Identified
p. 70, Florence Nightingale

Chapter 3
p. 119, Cost of Laughing Index

Chapter 4
p. 147, Probabilities That Change Intuition
p. 160, Shakespeare's Vocabulary
p. 185, Monkey Typists

Chapter 6
p. 273, Sensitive Surveys

Chapter 7
p. 351, Captured Tank Serial Numbers Reveal Population Size
p. 363, Excerpts from a Department of Transportation Circular
p. 367, Estimating Sugar in Oranges

Chapter 8
p. 442, Ethics in Experiments

Chapter 9
p. 480, Death Penalty as Deterrent
p. 505, Crest and Dependent Samples

Chapter 12
p. 664, Poll Resistance

Chapter 14
p. 754, The Flynn Effect
p. 767, Six Sigma in Industry

The following margin illustrations were rendered by James Bryant.

Chapter 1
p. 7, Measuring Disobedience

Chapter 3
p. 129, Data Mining

Chapter 4
p. 168, Independent Jet Engines
p. 171, Convicted by Probability
p. 178, Coincidences?
p. 190, Choosing Personal Security Codes

Chapter 6
p. 298, The Fuzzy Central Limit Theorem

Chapter 8
p. 455, Push Polling

Chapter 9
p. 479, Polio Experiment

Chapter 10
p. 539, Palm Reading

Chapter 11
p. 632, Survey Medium Can Affect Results

Chapter 13
p. 698, Class Attendance and Grades

Chapter 14
p. 766, Quality Control at Perstop

The following margin illustrations were rendered by Darwin Hennings.

Chapter 5
p. 241, Queues

Chapter 6
p. 297, The Placebo Effect

Chapter 9
p. 496, Better Results with More Class Size

Chapter 10
p. 599, Statistics: Jobs and Employers

Figure 2.9
p. 70, Copyright 2003 by Consumers Union of U.S., Inc. Yonkers, NY 10703-1057, a nonprofit organization. Reprinted with permission from the April 2003 issue of Consumer Reports® for educational purposes only. No commercial use or reproduction permitted. <http://www.consumerreports.org/> www.ConsumerReports.org®

Tufte Illustration:

p. 71, Losses of Soldiers in Napoleon's Army During the Russian Campaign (1812–1813): Edward R. Tufte, *The Visual Display of Quantitative Information* (Cheshire, CT: Graphics Press, 1983). Reprinted with permission

Index

TABLE A-3 *t* Distribution: Critical *t* Values

	Area in One Tail				
	0.005	0.01	0.025	0.05	0.10
Degrees of Freedom	**Area in Two Tails**				
	0.01	0.02	0.05	0.10	0.20
1	63.657	31.821	12.706	6.314	3.078
2	9.925	6.965	4.303	2.920	1.886
3	5.841	4.541	3.182	2.353	1.638
4	4.604	3.747	2.776	2.132	1.533
5	4.032	3.365	2.571	2.015	1.476
6	3.707	3.143	2.447	1.943	1.440
7	3.499	2.998	2.365	1.895	1.415
8	3.355	2.896	2.306	1.860	1.397
9	3.250	2.821	2.262	1.833	1.383
10	3.169	2.764	2.228	1.812	1.372
11	3.106	2.718	2.201	1.796	1.363
12	3.055	2.681	2.179	1.782	1.356
13	3.012	2.650	2.160	1.771	1.350
14	2.977	2.624	2.145	1.761	1.345
15	2.947	2.602	2.131	1.753	1.341
16	2.921	2.583	2.120	1.746	1.337
17	2.898	2.567	2.110	1.740	1.333
18	2.878	2.552	2.101	1.734	1.330
19	2.861	2.539	2.093	1.729	1.328
20	2.845	2.528	2.086	1.725	1.325
21	2.831	2.518	2.080	1.721	1.323
22	2.819	2.508	2.074	1.717	1.321
23	2.807	2.500	2.069	1.714	1.319
24	2.797	2.492	2.064	1.711	1.318
25	2.787	2.485	2.060	1.708	1.316
26	2.779	2.479	2.056	1.706	1.315
27	2.771	2.473	2.052	1.703	1.314
28	2.763	2.467	2.048	1.701	1.313
29	2.756	2.462	2.045	1.699	1.311
30	2.750	2.457	2.042	1.697	1.310
31	2.744	2.453	2.040	1.696	1.309
32	2.738	2.449	2.037	1.694	1.309
34	2.728	2.441	2.032	1.691	1.307
36	2.719	2.434	2.028	1.688	1.306
38	2.712	2.429	2.024	1.686	1.304
40	2.704	2.423	2.021	1.684	1.303
45	2.690	2.412	2.014	1.679	1.301
50	2.678	2.403	2.009	1.676	1.299
55	2.668	2.396	2.004	1.673	1.297
60	2.660	2.390	2.000	1.671	1.296
65	2.654	2.385	1.997	1.669	1.295
70	2.648	2.381	1.994	1.667	1.294
75	2.643	2.377	1.992	1.665	1.293
80	2.639	2.374	1.990	1.664	1.292
90	2.632	2.368	1.987	1.662	1.291
100	2.626	2.364	1.984	1.660	1.290
200	2.601	2.345	1.972	1.653	1.286
300	2.592	2.339	1.968	1.650	1.284
400	2.588	2.336	1.966	1.649	1.284
500	2.586	2.334	1.965	1.648	1.283
750	2.582	2.331	1.963	1.647	1.283
1000	2.581	2.330	1.962	1.646	1.282
2000	2.578	2.328	1.961	1.646	1.282
Large	2.576	2.326	1.960	1.645	1.282

NEGATIVE z Scores

TABLE A-2 Standard Normal (z) Distribution: Cumulative Area from the LEFT

z	.00	.01	.02	.03	.04	.05	.06	.07	.08	.09
−3.50 and lower	.0001									
−3.4	.0003	.0003	.0003	.0003	.0003	.0003	.0003	.0003	.0003	.0002
−3.3	.0005	.0005	.0005	.0004	.0004	.0004	.0004	.0004	.0004	.0003
−3.2	.0007	.0007	.0006	.0006	.0006	.0006	.0006	.0005	.0005	.0005
−3.1	.0010	.0009	.0009	.0009	.0008	.0008	.0008	.0008	.0007	.0007
−3.0	.0013	.0013	.0013	.0012	.0012	.0011	.0011	.0011	.0010	.0010
−2.9	.0019	.0018	.0018	.0017	.0016	.0016	.0015	.0015	.0014	.0014
−2.8	.0026	.0025	.0024	.0023	.0023	.0022	.0021	.0021	.0020	.0019
−2.7	.0035	.0034	.0033	.0032	.0031	.0030	.0029	.0028	.0027	.0026
−2.6	.0047	.0045	.0044	.0043	.0041	.0040	.0039	.0038	.0037	.0036
−2.5	.0062	.0060	.0059	.0057	.0055	.0054	.0052	.0051 *	.0049	.0048
−2.4	.0082	.0080	.0078	.0075	.0073	.0071	.0069	.0068	.0066	.0064
−2.3	.0107	.0104	.0102	.0099	.0096	.0094	.0091	.0089	.0087	.0084
−2.2	.0139	.0136	.0132	.0129	.0125	.0122	.0119	.0116	.0113	.0110
−2.1	.0179	.0174	.0170	.0166	.0162	.0158	.0154	.0150	.0146	.0143
−2.0	.0228	.0222	.0217	.0212	.0207	.0202	.0197	.0192	.0188	.0183
−1.9	.0287	.0281	.0274	.0268	.0262	.0256	.0250	.0244	.0239	.0233
−1.8	.0359	.0351	.0344	.0336	.0329	.0322	.0314	.0307	.0301	.0294
−1.7	.0446	.0436	.0427	.0418	.0409	.0401	.0392	.0384	.0375	.0367
−1.6	.0548	.0537	.0526	.0516	.0505 *	.0495	.0485	.0475	.0465	.0455
−1.5	.0668	.0655	.0643	.0630	.0618	.0606	.0594	.0582	.0571	.0559
−1.4	.0808	.0793	.0778	.0764	.0749	.0735	.0721	.0708	.0694	.0681
−1.3	.0968	.0951	.0934	.0918	.0901	.0885	.0869	.0853	.0838	.0823
−1.2	.1151	.1131	.1112	.1093	.1075	.1056	.1038	.1020	.1003	.0985
−1.1	.1357	.1335	.1314	.1292	.1271	.1251	.1230	.1210	.1190	.1170
−1.0	.1587	.1562	.1539	.1515	.1492	.1469	.1446	.1423	.1401	.1379
−0.9	.1841	.1814	.1788	.1762	.1736	.1711	.1685	.1660	.1635	.1611
−0.8	.2119	.2090	.2061	.2033	.2005	.1977	.1949	.1922	.1894	.1867
−0.7	.2420	.2389	.2358	.2327	.2296	.2266	.2236	.2206	.2177	.2148
−0.6	.2743	.2709	.2676	.2643	.2611	.2578	.2546	.2514	.2483	.2451
−0.5	.3085	.3050	.3015	.2981	.2946	.2912	.2877	.2843	.2810	.2776
−0.4	.3446	.3409	.3372	.3336	.3300	.3264	.3228	.3192	.3156	.3121
−0.3	.3821	.3783	.3745	.3707	.3669	.3632	.3594	.3557	.3520	.3483
−0.2	.4207	.4168	.4129	.4090	.4052	.4013	.3974	.3936	.3897	.3859
−0.1	.4602	.4562	.4522	.4483	.4443	.4404	.4364	.4325	.4286	.4247
−0.0	.5000	.4960	.4920	.4880	.4840	.4801	.4761	.4721	.4681	.4641

NOTE: For values of z below −3.49, use 0.0001 for the area.
*Use these common values that result from interpolation:

z score	Area
−1.645	0.0500
−2.575	0.0050